Save time... learn faster... see genetic processes come to life!

When you see this icon you can access one of twelve interactive animations.

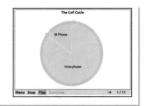

Chapter 2. Cell Cycle and Mitosis: Discover the process that allowed you to multiply from one cell to trillions.

Chapter 13. Overview of Eukaryotic Gene Expression: Tying it all together—gene expression, from DNA to protein, in one handy place.

Chapter 2. Meiosis: You say mitosis; I say meiosis. What's the big difference anyway? Find out by viewing this animation.

Chapter 14. Overview of mRNA Processing: Cap, tail, splice—new snowboarding tricks? No, just a little genetic processing. Check it out.

Chapter 2. Genetic Variation in Meiosis: Ever wonder why you don't look exactly like your siblings? The answer is in this animation.

Chapter 15. Bacterial Translation: Watch the body's very own United Nations interpreter, the ribosome, hard at work translating the language of DNA into proteins. Then give it some text of your own to translate.

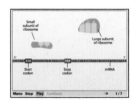

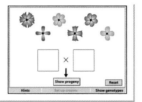

Chapter 3. Genetic Crosses Including Multiple Loci: This one is a real brain teaser, compliments of Mendel.

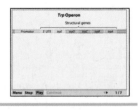

Chapter 16. The *lac* Operon: So many genotypes, so little time. This animation can help you sort it out.

Chapter 12. Overview of Replication: Does DNA polymerase I act before DNA polymerase III? And when do you need gyrase vs. primase? Test your knowledge of the steps of DNA replication with this animation.

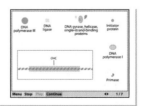

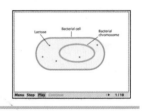

Chapter 16. Attenuation: Do you even remember what attenuation means? No? This animation will tell you.

Chapter 13. Bacterial Transcription: Build a transcription unit and then watch it at work in this animation.

Chapter 23. The Hardy-Weinberg Law and the Effects of Inbreeding and Natural Selection: Deconstructing Hardy... Weinberg. Plus, what would happen if we all mated with our cousins?

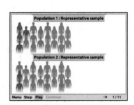

Animations produced by Sumanas, Inc. www.sumanasinc.com

Genetics in Motion at www.whfreeman.com/pierce

Genetics

A Conceptual Approach

Benjamin A. Pierce

Baylor University

W. H. Freeman and Company • New York

ABOUT THE COVER:

DNA (deoxyribonucleic acid) is a complex nucleic acid molecule composed of two strands of nucleotides joined together forming a double helix. It is the primary genetic and hereditary material of nearly all living organisms and forms the basis of chromosomes.

ACQUISITIONS EDITOR: Jason Noe
DEVELOPMENT EDITOR: Rachel Warren
EDITORIAL ASSISTANT: Jeff Ciprioni
SENIOR PROJECT EDITOR: Georgia Lee Hadler
LINE EDITOR: Patricia Zimmerman
PRODUCTION EDITOR: Bradley Umbaugh
COVER AND TEXT DESIGNER: Diana Blume
COVER ILLUSTRATION: Paul Morrell/Getty Images
MARKETING DIRECTOR: John Britch
MEDIA/SUPPLEMENTS EDITOR: Joy Ohm
ILLUSTRATION COORDINATOR: Cecilia Varas
ILLUSTRATIONS: J/B Woolsey and Associates
PHOTO EDITOR: Vikii Wong
PRODUCTION COORDINATOR: Paul W. Rohloff
COMPOSITION: Progressive Information Technologies
LAYOUT: Jerry Wilke Design
MANUFACTURING: RR Donnelley & Sons Company

LIBRARY OF CONGRESS CATALOGING-IN-PUBLICATION DATA
Pierce, Benjamin A.
    Genetics : a conceptual approach / by Benjamin A. Pierce
        p. cm.
    ISBN 1-57259-160-9
1. Genetics. I. Title.

QH430 .P54 2002
576.5—dc21

2002029473

Printed in the United States of America

First printing 2002

W. H. Freeman and Company
41 Madison Avenue
New York, NY 10010
Houndmills, Basingstoke RG21 6XS, England
www.whfreeman.com

To my parents, Rush and Amanda Pierce;

my children, Sarah and Michael Pierce;

and my partner, friend, and soul mate of 22 years, Marlene Tyrrell

Contents in Brief

Contents

15 The Genetic Code and Translation   404

16 Control of Gene Expression   434

17 Gene Mutations and DNA Repair   472

Preface

When I was 14 years old, I went on a two-week canoe trip through the Quetico, a vast wilderness in southern Canada with thousands of interconnected lakes. Our guide on this trip was an experienced outdoorsman who had traveled the region many times before; he taught us how to find trails that connected one lake to the next, how to use new tools and equipment necessary for the journey, and how to survive in the wilderness. He kept us motivated by telling stories of fur trappers and traders who traveled the region in the past. Perhaps most important, this excellent teacher and motivator never carried our gear for us or paddled our canoes, recognizing that each of us had to make the journey for ourselves.

My goal as a teacher is to become a trusted guide on students' journeys through introductory genetics. I have taught genetics for more than 22 years, and one of my strengths is helping students create a mental map of genetics—one that shows where we've been, where we'll go, and how we'll get there. I provide advice and encouragement at places that are often rough spots for students, and I tell stories of the people, places, and experiments of genetics—past and present—to keep the subject interesting and alive. I help students learn the necessary details, concepts, and problem-solving skills, and I also encourage them see the beauty of the larger landscape.

Features of This Book

In writing *Genetics: A Conceptual Approach,* I have tried to keep in mind what my students have taught me about the art of teaching genetics. As a result, this text focuses on several key elements that are critical for success in an introductory genetics textbook.

● Connecting Key Concepts   *Genetics: A Conceptual Approach* focuses on important concepts, giving students the big picture without overwhelming them with detail. I have emphasized the topics and experiments that are most important to students of introductory genetics. These concepts are driven home by brief Concepts boxes, which appear immediately after a topic is discussed in the text and summarize the key points. The important concepts are reinforced again at the end of each chapter in the Concepts Summary, a bulleted list of essential points covered in the chapter.

Over the years, I've learned that students understand and remember new material best when they can relate it to something they already know; for this reason, in *Genetics*: *A Conceptual Approach* I provide a clear picture of where we've been and where we're going as we move through the topics. At the beginning of each chapter, I briefly describe the chapter organization; within chapters, Connecting Concepts boxes allow students to pause and to synthesize and integrate new concepts. Finally, a feature called Connecting Concepts Across Chapters, which highlights connections between material covered in the current chapter and topics covered in other chapters, closes the text presentation.

Conceptual Approach

• **Concepts** Boxes containing key concepts appear throughout the text, succinctly summarizing important concepts and stating the take-home message.

• **Connecting Concepts** These features, sprinkled throughout each chapter, consider the "big picture," synthesizing newly learned concepts and integrating them with previously covered topics.

• **Connecting Concepts Across Chapters** Appearing at the end of each chapter, these sections reinforce major themes of the chapter and highlight connections to related material in other chapters.

• **Concepts Summary** A bulleted list of important concepts appears at the end of each chapter.

Problem Solving

• **In-Text and End-of-Chapter Worked Problems** Worked Problems carefully guide students through each step in finding the solution to the types of problems they need the most practice solving.

• **End-of-Chapter Questions and Problems** Extensive class-tested questions and fresh, unique problems are presented at the end of each chapter. They are presented as Comprehension Questions, Application Questions and Problems, and Challenge Questions.

• **The New Genetics MINING GENOMES** Ten interactive tutorials guide students through live analyses and searches of online databases, giving them experience using the latest tools of bioinformatics. Each step simply and clearly explains what is occurring on the screen and directs students to the next step.

• **Accessibility** *Genetics: A Conceptual Approach* is written in a lively, conversational style to engage, interest, and motivate the student. Each chapter opens with a short vignette that draws the reader into the topic. One of my goals is for students to view this textbook as an old friend, one who serves as their tutor and engages them in a dialog, speaking with understanding and clarity. During my years in the classroom, I've become familiar with the topics in genetics that often give students difficulty; at these points, I slow the pace and provide careful explanation.

Today's students are interested in the implications and applications of genetics; these issues have been given emphasis in The New Genetics: Ethics · Science · Technology features, written by leading bioethicists, which examine ethical, social, and economic issues associated with recent research in genetics. Throughout the text, a Web icon appears to alert, the reader to sources of additional information on the Internet that can be accessed through the textbook Web site.

• **Emphasis on Problem Solving** If there is one thing I have learned in my many years of teaching, it is that practice with problems is essential to mastery of genetics, for it is through the application of knowledge that students become self-sufficient. *Genetics: A Conceptual Approach* places strong emphasis on problem solving. Fresh, original problem sets, many drawing on original genetics research, were developed specifically for this book and have undergone thorough class testing by more than 100 professors across the country. One question centers around the fictional use of bonobos as future organ donors in the book by Robin Cooke titled *Chromosome 6*. Another features a pig farmer who is feeding his pigs food from the local university cafeteria and needs to calculate the variance in weight gain ascribable to genetics. In-chapter Worked Problems provide students with step-by-step solutions when appropriate topics arise in the text, and additional Worked Problems are found at the conclusion of every chapter. The New Genetics: Mining Genomes tutorials and problems give students experience using the latest tools of bioinformatics and genetic databases.

• **Clear, Simple Art Program** Because many students are visual learners, and because simple, clear illustrations are essential for developing an understanding of genetics, special care has been taken to create a superior art program. The artist and I worked side by side to develop every figure in the book, integrating text and art and attempting to create figures that are clear and instructive. Extensive balloon text walks the student through the figure step by step and makes each piece of art tell its own story. Many figures include a conclusion box that emphasizes the take-home message; critical experiments are illustrated in special figures that begin by posing a question and end with the answer.

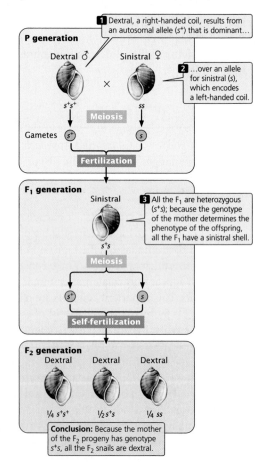

1 Dextral, a right-handed coil, results from an autosomal allele (s^+) that is dominant…

P generation

Dextral ♂   Sinistral ♀

s^+s^+ × ss

2 …over an allele for sinistral (s), which encodes a left-handed coil.

Meiosis

Gametes s^+ s

Fertilization

F₁ generation

Sinistral

s^+s

3 All the F_1 are heterozygous (s^+s); because the genotype of the mother determines the phenotype of the offspring, all the F_1 have a sinistral shell.

Meiosis

s^+ s

Self-fertilization

F₂ generation

Dextral   Dextral   Dextral

¼ s^+s^+ ½ s^+s ¼ ss

Conclusion: Because the mother of the F_2 progeny has genotype s^+s, all the F_2 snails are dextral.

• People, Places, and Experiments   It is not only important that students know the facts and concepts of genetics, but they should also have understanding of the process of science and how we know the information. At appropriate places in the text, I tell stories of the people who have contributed to our knowledge of genetics and the circumstances of their discoveries. Key experiments are outlined in special figures, which begin by posing a question, lead students through the experiment, and end with an answer to the question that the experiment addresses.

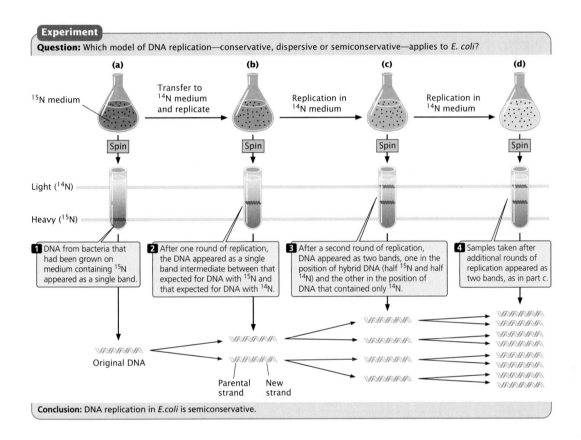

Experiment

Question: Which model of DNA replication—conservative, dispersive or semiconservative—applies to *E. coli*?

Conclusion: DNA replication in *E.coli* is semiconservative.

In *Genetics: A Conceptual Approach*, I not only help students develop an understanding of the fundamentals of classic genetic analysis, but I also present students with the latest developments in molecular genetics. I have enlisted the assistance of leading experts in biotechnology and bioethics to contribute essays and to carefully review every chapter to make sure the most essential current techniques and topics are covered appropriately. Of special note are The New Genetics: Mining Genomes, interactive bioinformatics tutorials, which introduce students to the basics of how to access and use on-line resources, tools that future biologists will have to master.

Art

• Balloon text Much of the explanatory information for the figures, which traditionally appears in figure legends, is placed in balloon text that leads students through complex processes step by step.

• Conclusion boxes At the bottom of many of the figures is a conclusion box that emphasizes the important concept illustrated by the figure.

Media and Supplements

A complete package of media resources and supplements is available and provides teachers and students with an array of tools that can enhance and extend the material presented in the textbook. All resources have been integrated with the text's style and goals.

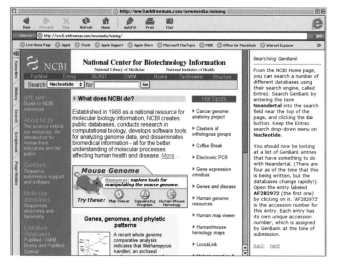

For Students

INTERACTIVE MEDIA

• Genetics On-line Web Site at www.whfreeman.com/pierce

▶ **The New Genetics: Mining Genomes** are ten interactive tutorials that guide students through live analyses and searches of online databases, giving them experience using the latest tools of bioinformatics. Each step simply and clearly explains what is occurring on the screen and directs students to the next step.

▶ **The Time Saver Series: Genetics in Motion** animated tutorials help students get right to the heart of important processes where visualization and interactivity are most important for understanding. Students choose their own paths through these animations and directly engage the dynamic processes they are studying. For example, they can arrange chromosomes on the metaphase plate in meiosis, choose crossing-over points, and see the resulting gametes. Watch your students master these concepts more quickly as a result of working with the Genetics in Motion animations.

▶ **Web links** are called out in the text of the book and provide additional timely information.

PROBLEM SOLVING

• The Solutions and Problem-Solving MegaManual contains everything a student needs to become adept at solving problems and thinking like a geneticist.

▶ **Solutions Manual** by Jung H. Choi of Georgia Institute of Technology and Mark McCallum of Pfeiffer University, contains complete worked-out solutions to all questions and problems that appear in the textbook. It has been reviewed extensively by instructors around the country. The solutions manual also contains the following CD-ROM.

▶ **Interactive Genetics CD-ROM** by Lianna Johnson and John Merriam of UCLA. This program leads students through many typical genetics problems one step at a time, with specific feedback to provide guidance along the way. Common mistakes and misunderstandings are addressed, and useful problem-solving strategies are suggested.

▶ **The New Genetics: Mining Genomes** by Mark S. Wilson of Humboldt State University. The text for these interactive bioinformatics Web tutorials makes it easier for students to review in a print format the information they've learned while doing the on-line exercises.

For Instructors

IMAGES

• All text images are available already placed in PowerPoint® and as high-resolution JPEG files, with large type and a vibrant color palette that projects clearly even in large lecture presentations. Images are available on the Instructor's Resource CD-ROM and on the Instructor's Resource Web Site.

• The overhead transparencies set includes 130 images from the text, with large type for easy reading. Complicated figures are broken down into parts so they project more clearly.

ASSESSMENT

• The test bank by Ben Pierce, Chris Tachibana of University of Washington, and Daniel Bergey of Montana State University is carefully designed to match the pedagogical intent of the text. It contains 50 problems per chapter, including multiple choice, true/false, short answer, and application, and it has been reviewed extensively by other instructors. It is provided as chapter-by-chapter Microsoft Word files that are easy to download, edit, and print from the Instructor's Resource CD-ROM or the Instructor's Resource Web Site. It is also available as a printed booklet.

ACCESS

- The Instructor's Resource CD-ROM contains all text images in PowerPoint® and as high-resolution JPEG files, all animations, and the Solutions Manual and Test Bank in chapter-by-chapter Microsoft Word files that are easy to download, edit, and print.
- The Password-Protected Instructor's Resource Web Site includes all the resources listed here (www.whfreeman.com/pierce). Contact your W. H. Freeman sales representative to learn how to log on as an instructor.

Acknowledgments

I am indebted to more than 3000 genetics students who have filled my classes over the past 22 years, first at Connecticut College and later at Baylor University. Their intelligence, enthusiasm, curiosity, and humor have been a constant source of motivation and pleasure throughout my professional life. I thank my teachers, Dr. Raymond Canham and Dr. Jeffrey Mitton, for introducing me to genetics and teaching me to be a lifelong learner of the subject.

Baylor University provided continual support throughout the development of this textbook. My department chair, W. Keith Hartberg, has been a friend and colleague; his advice, support, and encouragement are greatly appreciated. I am especially grateful to Wallace Daniel, Dean of the College of Arts and Sciences, who made it possible for me to complete the textbook while simultaneously serving as Associate Dean for Sciences. Without his leadership and example, this project would never have been finished. I appreciate the collegiality and expert technical advice from all my colleagues in the Department of Biology.

Modern science textbooks are a team effort, and I have been blessed with an outstanding team. I am grateful to Kerry Baruth for initiating this project and to Robert Worth for believing that I could successfully write a new genetics textbook. Jason Noe and Sara Tenney skillfully piloted this project at W. H. Freeman and Company; I appreciate their commitment to the project and creative contributions. Valerie Neal served as developmental editor during the first half of the project; her devotion to the task and detailed editing shaped much of the text in the book. Rachel Warren served as developmental editor for the latter part of the project. Her skilled editing, keen sense of organization, and tremendous energy are largely responsible for bringing the book to completion, and her good humor and encouragement kept me motivated. Randi Rossignol contributed many good ideas and helped with development of the art.

Special credit goes to John Woolsey and the staff at J/B Woolsey Associates for the outstanding art program. John was true collaborator; we worked together to develop the art in the book. I am indebted to Joy Ohm for organizing and managing the media and supplements, and to Georgia Lee Hadler for effective management of the book's production, Patricia Zimmerman for copy editing, Diana Blume for the book's design, Vikii Wong and Patricia Marx for photo research, Paul Rohloff for production coordination, and John Britch for directing marketing of the book.

Brigitte Raumann, Tanya Awabdy, and Tracy Washburn's creative skills and exceptional technical expertise produced the wonderful animations that accompany the book. Chris Tachibana and Daniel Bergey created the Test Bank, Jung Choi and Mark McCallum wrote the Solutions Manual, Lianna Johnson and John Merriam created the interactive problem-solving tutorials, and Mark Wilson wrote The New Genetics: Mining Genomes tutorials and exercises. Ronald Green, Art Caplan, and Kelly Carroll wrote The New Genetics: Ethics · Science · Technology essays.

A large number of colleagues served as reviewers and class testers for the text, kindly lending me their technical expertise and teaching experience. Their assistance is gratefully acknowledged; any remaining errors are entirely my own responsibility.

I would especially like to thank the panel of expert contributors whose currency and accuracy checking during the final stages of manuscript development proved invaluable: Craig Coates, Texas A&M University; Karen Hicks, Kenyon College; David Hyde, University of Notre Dame; Wendy Raymond, Williams College; and Jeff Sekelsky, University of North Carolina.

As we all know, problem solving lies at the heart of learning and practicing genetics. I have done my best to provide students and instructors with interesting and innovative problem sets. I would like to thank the more than 100 instructors who agreed to class test my problems. Their attention in this regard was an enormous help in ensuring the accuracy of the problems, as well as the careful explanation of the solutions that accompany them.

Last, but certainly not least, I am indebted to my family—Sarah, Michael, and Marlene—who provided encouragement, support, and advice through every stage of the project and who are the inspiration for everything I do. ▪

Reviewers and Class Testers

Joan Abramowitz	University of Houston, Downtown	Patricia Conklin	State University of New York College at Cortland
James Allan	Edinburgh University		
Fred Allendorf	University of Montana	Victor Corces	Johns Hopkins University
James O. Allen	Florida International University	Victor Cox	Parkland College
Ross Anderson	Master's College	Drew Cressman	Sarah Lawrence College
Pablo Arenaz	University of Texas at El Paso	David Crowley	Mercer Community College
Alan Atherly	Iowa State University	Laszlo Csonka	Purdue University
Charles Atkins	Ohio University	Michael R. Culbertson	University of Wisconsin at Madison
Vance Baird	Clemson University	Mike Dalbey	University of California at Santa Cruz
Stephanie Baker	Erskine College		
Phillip T. Barnes	Connecticut College	Alix Darden	The Citadel
George Bates	Florida State University	Terry Davin	Penn Valley Community College
Amy Bejsovec	Duke University	Sandra Davis	University of Louisiana at Monroe
John Belote	Syracuse University	Thomas Davis	University of New Hampshire
David Benner	East Tennessee State University	Ann Marie Davison	Kwantlen University College
Spencer Benson	University of Maryland at College Park	Steven Denison	Eckerd College
		Carter Denniston	University of Wisconsin at Madison
Dan Bergey	Montana State University	Myra Derbyshire	Mount Saint Mary's College
Andrew Bohonak	San Diego State University	Andrew A. Dewees	Sam Houston State University
J. Hoyt Bowers	Wayland Baptist University	AnnMarie DiLorenzo	Montclair State University
Jane Bradley	Des Moines Area Community College	Judith A. Dilts	William Jewell College
		Stephen DiNardo	University of Pennsylvania
Elizabeth Bryda	Marshall University	Linda Dixon	University of Colorado at Denver
Michael Buratovich	Spring Arbor University	Frank Doe	University of Dallas
Peter Burgers	Washington University in St. Louis	Diana Downs	University of Wisconsin at Madison
Jeffrey Byrd	St. Mary's College of Maryland	Richard Duhrkopf	Baylor University
Patrick Calie	Eastern Kentucky University at Richmond	Maureen Dunbar	Penn State, Berks College
		Lynn Ebersole	Northern Kentucky University
Arthur Champlin	Colby College	Larry Eckroat	Penn State, Erie
Lee Anne Chaney	Whitworth College	Johnny El-Rady	University of South Florida
Bruce Chase	University of Nebraska at Omaha	Ted English	North Carolina State University
Christian Chavret	Indiana University, Kokomo	Scott Erdman	Syracuse University
Carol Chihara	University of San Francisco	Asim Esen	Virginia Polytechnic Institute and State University
Joseph Chinnici	VA Commonwealth University		
Jung H. Choi	Georgia Institute of Technology	Paul Evans	Brigham Young University
Craig Coates	Texas A & M University	Elsa Q. Falls	Randolph-Macon College
John Condie	Fort Lewis College	Robert Farrell	Penn State, York

Kathleen M. Fisher	*San Diego State University*
Valerie Flechtner	*John Carroll University*
Mary Flynn	*MCP Hahnemann University*
Thomas Fogle	*Saint Mary's College*
Rosemary Ford	*Washington College*
Laurie Freeman	*Fulton-Montgomery Community College*
Julia Frugoli	*Clemson University*
Maria Galb-Meagher	*University of Florida*
Arupa Ganguly	*University of Pennsylvania*
Dan Garza	*Novartis Pharmaceutical Company*
Ivan Gepner	*Monmouth University*
Elliot S. Goldstein	*Arizona State University*
Paul Goldstein	*University of Texas El Paso*
Javier Gonzalez	*University of Texas at Brownsville*
Deborah Good	*University of Massachusetts*
Harvey F. Good	*University of Le Verne*
Myron Goodman	*University of Southern California*
Richard L. Gourse	*University of Wisconsin at Madison*
Michael W. Gray	*Dalhousie University*
Michael Grotewiel	*Michigan State University*
James Haber	*Brandeis University*
Randall Harris	*William Carey College*
Stan Hattman	*University of Rochester*
Martha Haviland	*Rutgers University*
John Hays	*Oregon State University*
James L. Hayward	*Andrews University*
Donna Hazelwood	*Dakota State University*
Kaius Helenurm	*University of South Dakota*
Curtis Henderson	*MacMurray College*
Jerald Hendrix	*Kennesaw State Universtiy*
Richard P. Hershberger	*Carlow College*
Karen Hicks	*Kenyon College*
Jerry Higginbotham	*Transylvania University*
David Hillis	*University of Texas*
Alan Hinnebusch	*National Institutes of Health*
Deborah Hinson	*Dallas Baptist University*
Margaret Hollingsworth	*State University of New York at Buffalo*
George Hudock	*Indiana University*
Kim Hunter	*Salisbury State University*
David Hyde	*University of Notre Dame*
Colleen Jacks	*Gustavus Adolphus College*
William Jackson	*University of South Carolina, Aiken*
Tony Jilek	*University of Wisconsin at River Falls*
Rick Johns	*Northern Illinois University*
Casonya M. Johnson	*Morgan State University*
Hugh Johnson	*Southern Arkansas University*
J. Spencer Johnston	*Texas A & M University*
Greg Jones	*Santa Fe Community College*
Michael Jones	*Indiana University*
Gregg Jongeward	*University of the Pacific*
Chris Kapicka	*Northwest Nazarene University*
Clifford Kiel	*University of Delaware*
Hai Kinal	*Springfield College*
Olga Ruiz Kopp	*University of Tennessee*
Gae Kovalik	*University of Texas of the Permian Basin*
Rhonda J. Kuykindoll	*Rust College*
Trip Lamb	*East Carolina University*
Franz Lang	*Universite de Montreal*
Allan Larsen	*Washington University, St. Louis*
Chris Lawrence	*University of Rochester*
Alicia Lesnikowska	*Georgia Southwestern State University*
Ricki Lewis	*University of Albany*
Alice Lindgren	*Bemidji State University*
Malcolm Lippert	*Saint Michael's College*
John Locke	*University of Alberta*
Pat Lord	*Wake Forest University*
Charles Louis	*Georgia State University*
Carl Luciano	*Indiana University of Pennsylvania*
Paul Lurquin	*Washington State University*
Aldons Lusis	*University of California, Los Angeles*
Peter Luykx	*University of Miami*
J. David MacDonald	*Wichita State University*
Paul Mangum	*Midland College*
Michael Markovitz	*Midland Lutheran College*
Marcie Marston	*Roger Williams University*
Alfred Martin	*Benedictine University*
Robert Martinez	*Quinnipiac College*
David Matthes	*San Jose State University*
Joyce Maxwell	*California Institute of Technology*
Mark McCallum	*Pfeiffer University*
Jim McGivern	*Gannon University*
Denise McGuire	*Saint Cloud State University*
Lauren McIntyre	*Purdue University*
Sandra Michael	*Binghamton University*
Gail Miller	*York College*
John Mishler	*Delaware Valley College*
Jeffrey Mitton	*University of Colorado at Boulder*
Aaron Moe	*Concordia University*
Mary Montgomery	*Macalaster College*
Ammini Moorthy	*Wagner College*
Nancy Morvillo	*Florida Southern College*
Mary Murnik	*Ferris State College*
Jennifer Myka	*Brescia University*
Elbert Myles	*Tennessee State University*
William Nelson	*Morgan State University*
Bryan Ness	*Pacific Union College*
John Newell	*Shaw University*
Jeffrey Newman	*Lycoming College*
Harry Nickla	*Creighton University*
Timothy Nilsen	*Center for RNA Molecular Biology*
Donna Nofziger-Plank	*Pepperdine University*
Marcia O'Connell	*State University of New York, Stony Brook*
Ronald Ostrowski	*University of North Carolina at Charlotte*

Tony Palombella	*Longwood College*	Martha Stauderman	*University of San Diego*
Louise Paquin	*Western Maryland College*	Todd Steck	*University of North Carolina at Charlotte*
Ann Paterson	*Williams Baptist College*		
Jackie Peltier	*Houston Baptist University*	Anne Stone	*University of New Mexico*
Dorene Petrosky	*Delaware State University*	Susan Strome	*Indiana University*
Lynn Petrullo	*College of New Rochelle*	Heidi Super	*Minot State University*
Bernie Possidente	*Skidmore College*	Chris Tachibana	*University of Washington*
Daphne Preuss	*University of Chicago*	Jennifer Thomson	*University of Cape Town*
Jim Price	*Utah Valley State College*	Grant Thorsett	*Willamette University*
Frank Pugh	*Pennsylvania State University*	Tammy Tobin-Janzen	*Susquehanna University*
Mary Puterbaugh	*University of Pittsburgh at Bradford*	John Tonzetich	*Bucknell University*
		Andrew Travers	*Medical Research Council, Cambridge England*
Todd Rainey	*Bluffton College*		
Wendy Raymond	*Williams College*	Carol Trent	*Western Washington University*
Peggy Redshaw	*Austin College*		
William S. Reznikoff	*University of Wisconsin at Madison*	Callie Vanderbilt	*San Juan College*
R.H. Richardson	*University of Texas at Austin*	Albrecht Von Arnim	*University of Tennessee*
Todd Rimkus	*Marymount College*	Alan Waldman	*University of South Carolina*
John Ringo	*University of Maine*	Melinda Wales	*Texas A & M University*
Tara Robinson	*Oregon State Univesity*	Nancy Walker	*Clemson University*
Torbert R. Rocheford	*University of Illinois*	Karen Weiler	*Idaho State University*
Stephen Roof	*Fairmont State College*	William Wellnitz	*Augusta State University*
Jim Sanders	*Texas A&M Univ.*	David Weber	*Illinois State University*
Mark Sanders	*University of California, Davis*	Michelle Whaley	*University of Notre Dame*
Andrew Scala	*Dutchess Community College*	Matt White	*Ohio University*
Joanne Scalzitti	*University of Nebraska at Kearney*	Mark Wilson	*Humboldt State University*
Marilyn Schendel	*Carroll College*	Betsy Wilson	*University of North Carolina at Asheville*
Jennifer Schisa	*Central Michigan University*		
Kathy Schmeidler	*Irvine Valley College*	Tom Wiltshire	*Culver-Stockton College*
Malcolm Schug	*University of North Carolina at Greensboro*	Mark Winey	*University of Colorado*
		David Wing	*Shepherd College*
Terry Schwaner	*Southern Utah University*	Darla Wise	*Concord College*
Ralph Seelke	*University of Wisconsin at Superior*	D. S. Wofford	*University of Florida*
Jeanine Seguin	*Keuka College*	Kathleen Wood	*University of Mary Hardin-Baylor*
Jeff Sekelsky	*University of North Carolina*		
David Sherratt	*University of Oxford*	Bonnie Wood	*University of Maine at Presque Isle*
DeWayne Shoemaker	*University of Wisconsin, Madison*		
Nancy N. Shontz	*Grand Valley State University*	David Woods	*Rhodes University*
J. Kenneth Shull	*Appalachian State University*	Michael Wooten	*Auburn Univesity*
Laura Sigismondi	*University of Rio Grande*	Debbie Wygal	*College of Saint Catherine*
Pat Singer	*Simpson College*	Krassimir (Joseph)Yankulov	*University of Guelph*
Laurie Smith	*University of California at San Diego*	Jeff Young	*Western Washington University*
		Malcolm Zellars	*Georgia State University*
Ken Spitze	*University of Illinois at Urbana-Champaign*	Hong Zhang	*Texas Tech University*
		Suzanne Ziesmann	*Sweet Briar College*

1 Introduction to Genetics

Alexis, heir to the Russian throne, and his father, Tsar Nicholas Romanov II. (Hulton/Archive by Getty Images.)

Royal Hemophilia and Romanov DNA

On August 12, 1904, Tsar Nicholas Romanov II of Russia wrote in his diary: "A great never-to-be forgotten day when the mercy of God has visited us so clearly." That day Alexis, Nicholas's first son and heir to the Russian throne, had been born.

At birth, Alexis was a large and vigorous baby with yellow curls and blue eyes, but at 6 weeks of age he began spontaneously hemorrhaging from the navel. The bleeding persisted for several days and caused great alarm. As he grew and began to walk, Alexis often stumbled and fell, as all children do. Even his small scrapes bled profusely, and minor bruises led to significant internal bleeding. It soon became clear that Alexis had hemophilia.

Hemophilia results from a genetic deficiency of blood clotting. When a blood vessel is severed, a complex cascade of reactions swings into action, eventually producing a protein called fibrin. Fibrin molecules stick together to form a clot, which stems the flow of blood. Hemophilia, marked by slow clotting and excessive bleeding, is the result if any one of the factors in the clotting cascade is missing or faulty. In those with hemophilia, life-threatening blood loss can occur with minor injuries, and spontaneous bleeding into joints erodes the bone with crippling consequences.

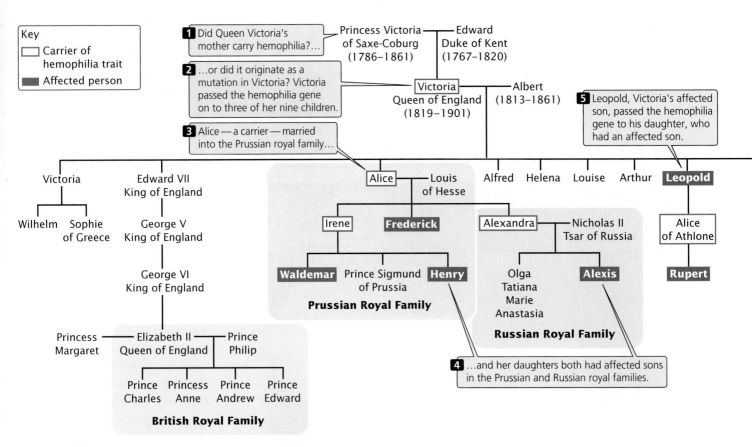

Key

☐ Carrier of hemophilia trait
■ Affected person

1 Did Queen Victoria's mother carry hemophilia?...

2 ...or did it originate as a mutation in Victoria? Victoria passed the hemophilia gene on to three of her nine children.

3 Alice—a carrier—married into the Prussian royal family...

5 Leopold, Victoria's affected son, passed the hemophilia gene to his daughter, who had an affected son.

4 ...and her daughters both had affected sons in the Prussian and Russian royal families.

1.1 Hemophilia was passed down through the royal families of Europe.

Alexis suffered from classic hemophilia, which is caused by a defective copy of a gene on the X chromosome. Females possess two X chromosomes per cell and may be unaffected carriers of the gene for hemophilia. A carrier has one normal version and one defective version of the gene; the normal version produces enough of the clotting factor to prevent hemophilia. A female exhibits hemophilia only if she inherits two defective copies of the gene, which is rare. Because males have a single X chromosome per cell, if they inherit a defective copy of the gene, they develop hemophilia. Consequently, hemophilia is more common in males than in females.

Alexis inherited the hemophilia gene from his mother, Alexandra, who was a carrier. The gene appears to have originated with Queen Victoria of England (1819–1901), (◄ FIGURE 1.1). One of her sons, Leopold, had hemophilia and died at the age of 31 from brain hemorrhage following a minor fall. At least two of Victoria's daughters were carriers; through marriage, they spread the hemophilia gene to the royal families of Prussia, Spain, and Russia. In all, 10 of Queen Victoria's male descendants suffered from hemophilia. Six female descendants, including her granddaughter Alexandra (Alexis's mother), were carriers.

Nicholas and Alexandra constantly worried about Alexis's health. Although they prohibited his participation in sports and other physical activities, cuts and scrapes were inevitable, and Alexis experienced a number of severe bleeding episodes. The royal physicians were helpless during these crises—they had no treatment that would stop the bleeding. Gregory Rasputin, a monk and self-proclaimed "miracle worker," prayed over Alexis during one bleeding crisis, after which Alexis made a remarkable recovery. Rasputin then gained considerable influence over the royal family.

At this moment in history, the Russian Revolution broke out. Bolsheviks captured the tsar and his family and held them captive in the city of Ekaterinburg. On the night of July 16, 1918, a firing squad executed the royal family and their attendants, including Alexis and his four sisters. Eight days later, a pro-tsarist army fought its way into Ekaterinburg. Although army investigators searched vigorously for the bodies of Nicholas and his family, they found only a few personal effects and a single finger. The Bolsheviks eventually won the revolution and instituted the world's first communist state.

Historians have debated the role that Alexis's illness may have played in the Russian Revolution. Some have argued that the revolution was successful because the tsar and Alexandra were distracted by their son's illness and under the influence of Rasputin. Others point out that many factors contributed to the overthrow of the tsar. It is probably naive to attribute the revolution entirely to one sick boy, but it is

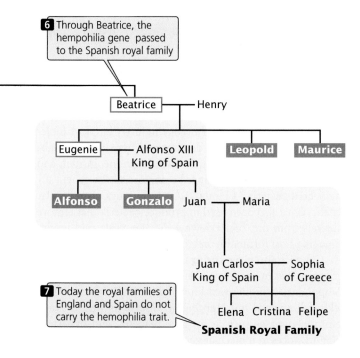

6 Through Beatrice, the hempohilia gene passed to the Spanish royal family

Beatrice — Henry

Eugenie — Alfonso XIII King of Spain   Leopold   Maurice

Alfonso   Gonzalo   Juan — Maria

Juan Carlos — Sophia
King of Spain   of Greece

Elena Cristina Felipe

7 Today the royal families of England and Spain do not carry the hemophilia trait.

Spanish Royal Family

Romanov line. The samples matched at all but one nucleotide position: the living relatives possessed a cytosine (C) residue at this position, whereas some of the skeletal DNA possessed a thymine (T) residue and some possessed a C. This difference could be due to normal variation in the DNA; so experts concluded that the skeleton was almost certainly that of Tsar Nicholas. The finding remained controversial, however, until July 1994, when the body of Nicholas's younger brother Georgij, who died in 1899, was exhumed. Mitochondrial DNA from Georgij also contained both C and T at the controversial position, proving that the skeleton was indeed that of Tsar Nicholas.

This chapter introduces you to genetics and reviews some concepts that you may have encountered briefly in a preceding biology course. We begin by considering the importance of genetics to each of us, to society at large, and to students of biology. We then turn to the history of genetics, how the field as a whole developed. The final part of the chapter reviews some fundamental terms and principles of genetics that are used throughout the book.

There has never been a more exciting time to undertake the study of genetics than now. Genetics is one of the frontiers of science. Pick up almost any major newspaper or news magazine and chances are that you will see something related to genetics: the discovery of cancer-causing genes; the use of gene therapy to treat diseases; or reports of possible hereditary influences on intelligence, personality, and sexual orientation. These findings often have significant economic and ethical implications, making the study of genetics relevant, timely, and interesting.

www.whfreeman.com/pierce   More information about the history of Nicholas II and other tsars of Russia and about hemophilia

The Importance of Genetics

Alexis's hemophilia illustrates the important role that genetics plays in the life of an individual. A difference in one gene, of the 35,000 or so genes that each human possesses, changed Alexis's life, affected his family, and perhaps even altered history. We all possess genes that influence our lives. They affect our height and weight, our hair color and skin pigmentation. They influence our susceptibility to many diseases and disorders (◀ FIGURE 1.2) and even contribute to our intelligence and personality. Genes are fundamental to who and what we are.

Although the science of genetics is relatively new, people have understood the hereditary nature of traits and have practiced genetics for thousands of years. The rise of agriculture began when humans started to apply genetic principles to the domestication of plants and animals. Today, the major crops and animals used in agriculture have undergone extensive genetic alterations to greatly increase their yields and provide many desirable traits, such as disease and pest

clear that a genetic defect, passed down through the royal family, contributed to the success of the Russian Revolution.

More than 80 years after the tsar and his family were executed, an article in the *Moscow News* reported the discovery of their skeletons outside Ekaterinburg. The remains had first been located in 1979; however, because of secrecy surrounding the tsar's execution, the location of the graves was not made public until the breakup of the Soviet government in 1989. The skeletons were eventually recovered and examined by a team of forensic anthropologists, who concluded that they were indeed the remains of the tsar and his wife, three of their five children, and the family doctor, cook, maid, and footman. The bodies of Alexis and his sister Anastasia are still missing.

To prove that the skeletons were those of the royal family, mitochondrial DNA (which is inherited only from the mother) was extracted from the bones and amplified with a molecular technique called the polymerase chain reaction (PCR). DNA samples from the skeletons thought to belong to Alexandra and the children were compared with DNA taken from Prince Philip of England, also a direct descendant of Queen Victoria. Analysis showed that mitochondrial DNA from Prince Philip was identical with that from these four skeletons.

DNA from the skeleton presumed to be Tsar Nicholas was compared with that of two living descendants of the

(a)

(b)

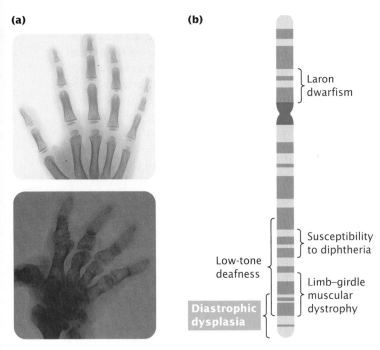

Chromosome 5

Laron dwarfism

Susceptibility to diphtheria

Low-tone deafness

Limb–girdle muscular dystrophy

Diastrophic dysplasia

◀1.2 **Genes influence susceptibility to many diseases and disorders.** (a) X-ray of the hand of a person suffering from diastrophic dysplasia (bottom), a hereditary growth disorder that results in curved bones, short limbs, and hand deformities, compared with an X-ray of a normal hand (top). (b) This disorder is due to a defect in a gene on chromosome 5. Braces indicate regions on chromosome 5 where genes giving rise to other disorders are located. (Part a: top, Biophoto Associates/Science Source/Photo Researchers; bottom, courtesy of Eric Lander, Whitehead Institute, MIT.)

resistance, special nutritional qualities, and characteristics that facilitate harvest. The Green Revolution, which expanded global food production in the 1950s and 1960s, relied heavily on the application of genetics (◀FIGURE 1.3). Today, genetically engineered corn, soybeans, and other crops constitute a significant proportion of all the food produced worldwide.

The pharmaceutical industry is another area where genetics plays an important role. Numerous drugs and food additives are synthesized by fungi and bacteria that have been genetically manipulated to make them efficient producers of these substances. The biotechnology industry employs molecular genetic techniques to develop and mass-produce substances of commercial value. Growth hormone, insulin, and clotting factor are now produced commercially by genetically engineered bacteria (◀FIGURE 1.4). Techniques of molecular genetics have also been used to produce bacteria that remove minerals from ore, break down toxic chemicals, and inhibit damaging frost formation on crop plants.

Genetics also plays a critical role in medicine. Physicians recognize that many diseases and disorders have a hereditary component, including well-known genetic disorders such as sickle-cell anemia and Huntington disease as well as many common diseases such as asthma, diabetes, and hypertension. Advances in molecular genetics have allowed important insights into the nature of cancer and permitted the development of many diagnostic tests. Gene therapy—the direct alteration of genes to treat human diseases—has become a reality.

www.whfreeman.com/pierce   Information about biotechnology, including its history and applications

(a)

(b)

◀1.3 **The Green Revolution used genetic techniques to develop new strains of crops that greatly increased world food production during the 1950s and 1960s.** (a) Norman Borlaug, a leader in the development of new strains of wheat that led to the Green Revolution. Borlaug received the Nobel Peace Prize in 1970. (b) Traditional rice plant (left) and modern, high-yielding rice plant (right). (Part a, UPI/Corbis-Bettman; part b, IRRI.)

1.4 **The biotechnology industry uses molecular genetic methods to produce substances of economic value.** In the flask, mammalian cells are being cultured for the commercial production of recombinant proteins. (James Holmes/Celltech Ltd./Science Photo Library/ Photo Researchers.)

The Role of Genetics in Biology

Although an understanding of genetics is important to all people, it is critical to the student of biology. Genetics provides one of biology's unifying principles: all organisms use nucleic acids for their genetic material and all encode their genetic information in the same way. Genetics undergirds the study of many other biological disciplines. Evolution, for example, is genetic change taking place through time; so

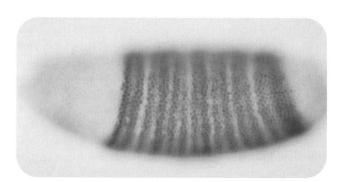

1.5 **The key to development lies in the regulation of gene expression.** This early fruit-fly embryo illustrates the localized production of proteins from two genes, *ftz* (stained gray) and *eve* (stained brown), which determine the development of body segments in the adult fly. (Peter Lawrence, 1992. *The Making of a Fly,* Blackwell Scientific Publications.)

the study of evolution requires an understanding of basic genetics. Developmental biology relies heavily on genetics: tissues and organs form through the regulated expression of genes (◀FIGURE 1.5). Even such fields as taxonomy, ecology, and animal behavior are making increasing use of genetic methods. The study of almost any field of biology or medicine is incomplete without a thorough understanding of genes and genetic methods.

Genetic Variation Is the Foundation of Evolution

Life on Earth exists in a tremendous array of forms and features that occupy almost every conceivable environment. All life has a common origin (see Chapter 2); so this diversity has developed during Earth's 4-billion-year history. Life is also characterized by adaptation: many organisms are exquisitely suited to the environment in which they are found. The history of life is a chronicle of new forms of life emerging, old forms disappearing, and existing forms changing.

Life's diversity and adaptation are a product of evolution, which is simply genetic change through time. Evolution is a two-step process: first, genetic variants arise randomly and, then, the proportion of particular variants increases or decreases. Genetic variation is therefore the foundation of all evolutionary change and is ultimately the basis of all life as we know it. Genetics, the study of genetic variation, is critical to understanding the past, present, and future of life.

Concepts

Heredity affects many of our physical features as well as our susceptibility to many diseases and disorders. Genetics contributes to advances in agriculture, pharmaceuticals, and medicine and is fundamental to modern biology. Genetic variation is the foundation of the diversity of all life.

Divisions of Genetics

Traditionally, the study of genetics has been divided into three major subdisciplines: transmission genetics, molecular genetics, and population genetics (◀FIGURE 1.6). Also known as classical genetics, **transmission genetics** encompasses the basic principles of genetics and how traits are passed from one generation to the next. This area addresses the relation between chromosomes and heredity, the arrangement of genes on chromosomes, and gene mapping. Here the focus is on the individual organism—how an individual organism inherits its genetic makeup and how it passes its genes to the next generation.

Molecular genetics concerns the chemical nature of the gene itself: how genetic information is encoded, replicated, and expressed. It includes the cellular processes of replication, transcription, and translation—by which genetic information is transferred from one molecule to another—and gene

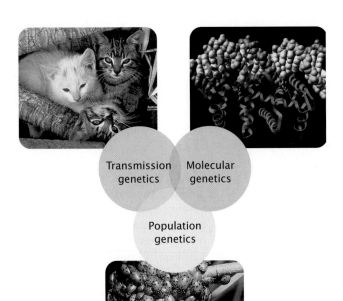

◀1.6 Genetics can be subdivided into three interrelated fields. (Top left, Alan Carey/Photo Researchers; top right, MONA file M0214602tif; bottom, J. Alcock/Visuals Unlimited.)

regulation—the processes that control the expression of genetic information. The focus in molecular genetics is the gene—its structure, organization, and function.

Population genetics explores the genetic composition of groups of individual members of the same species (populations) and how that composition changes over time and space. Because evolution is genetic change, population genetics is fundamentally the study of evolution. The focus of population genetics is the group of genes found in a population.

It is convenient and traditional to divide the study of genetics into these three groups, but we should recognize that the fields overlap and that each major subdivision can be further divided into a number of more specialized fields, such as chromosomal genetics, biochemical genetics, quantitative genetics, and so forth. Genetics can alternatively be subdivided by organism (fruit fly, corn, or bacterial genetics), and each of these organisms can be studied at the level of transmission, molecular, and population genetics. Modern genetics is an extremely broad field, encompassing many interrelated subdisciplines and specializations.

Concepts

The three major divisions of genetics are transmission genetics, molecular genetics, and population genetics. Transmission genetics

examines the principles of heredity; molecular genetics deals with the gene and the cellular processes by which genetic information is transferred and expressed; population genetics concerns the genetic composition of groups of organisms and how that composition changes over time and space.

www.whfreeman.com/pierce   Information about careers in genetics

A Brief History of Genetics

Although the science of genetics is young—almost entirely a product of the past 100 years—people have been using genetic principles for thousands of years.

Prehistory

The first evidence that humans understood and applied the principles of heredity is found in the domestication of plants and animals, which began between approximately 10,000 and 12,000 years ago. Early nomadic people depended on hunting and gathering for subsistence but, as human populations grew, the availability of wild food resources declined. This decline created pressure to develop new sources of food; so people began to manipulate wild plants and animals, giving rise to early agriculture and the first fixed settlements.

Initially, people simply selected and cultivated wild plants and animals that had desirable traits. Archeological evidence of the speed and direction of the domestication process demonstrates that people quickly learned a simple but crucial rule of heredity: like breeds like. By selecting and breeding individual plants or animals with desirable traits, they could produce these same traits in future generations.

The world's first agriculture is thought to have developed in the Middle East, in what is now Turkey, Iraq, Iran, Syria, Jordan, and Israel, where domesticated plants and animals were major dietary components of many populations by 10,000 years ago. The first domesticated organisms included wheat, peas, lentils, barley, dogs, goats, and sheep. Selective breeding produced woollier and more manageable goats and sheep and seeds of cereal plants that were larger and easier to harvest. By 4000 years ago, sophisticated genetic techniques were already in use in the Middle East. Assyrians and Babylonians developed several hundred varieties of date palms that differed in fruit size, color, taste, and time of ripening. An Assyrian bas-relief from 2880 years ago depicts the use of artificial fertilization to control crosses between date palms (◀FIGURE 1.7). Other crops and domesticated animals were developed by cultures in Asia, Africa, and the Americas in the same period.

◀ 1.7 **Ancient peoples practiced genetic techniques in agriculture.** (Top) Modern wheat, with larger and more numerous seeds that do not scatter before harvest, was produced by interbreeding at least three different wild species. (Bottom) Assyrian bas-relief sculpture showing artificial pollination of date palms at the time of King Assurnasirpalli II, who reigned 883–859 B.C. (Top, Scott Bauer/ARS/USDA; bottom, The Metropolitan Museum of Art, gift of John D. Rockefeller, Jr., 1932.)

Early Written Records

Ancient writings demonstrate that early humans were aware of their own heredity. Hindu sacred writings dating to 2000 years ago attribute many traits to the father and suggest that differences between siblings can be accounted for by effects from the mother. These same writings advise that one should avoid potential spouses having undesirable traits that might be passed on to one's children. The Talmud, the Jewish book of religious laws based on oral traditions dating back thousands of years, presents an uncannily accurate understanding of the inheritance of hemophilia. It directs that, if a woman bears two sons who die of bleeding after circumcision, any additional sons that she bears should not be circumcised; nor should the sons of her sisters be circumcised, although the sons of her brothers should. This advice accurately reflects the X-linked pattern of inheritance of hemophilia (discussed further in Chapter 6).

The ancient Greeks gave careful consideration to human reproduction and heredity. The Greek physician Alcmaeon (circa 520 B.C.) conducted dissections of animals and proposed that the brain was not only the principle site of perception, but also the origin of semen. This proposal sparked a long philosophical debate about where semen was produced and its role in heredity. The debate culminated in the concept of **pangenesis,** which proposed that specific particles, later called gemmules, carry information from various parts of the body to the reproductive organs, from where they are passed to the embryo at the moment of conception (◀ FIGURE 1.8a). Although incorrect, the concept of pangenesis was highly influential and persisted until the late 1800s.

Pangenesis led the ancient Greeks to propose the notion of the **inheritance of acquired characteristics,** in which traits acquired during one's lifetime become incorporated into one's hereditary information and are passed on to

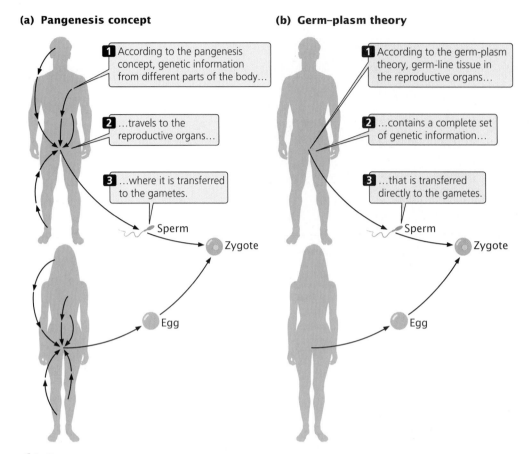

(a) Pangenesis concept

1 According to the pangenesis concept, genetic information from different parts of the body...

2 ...travels to the reproductive organs...

3 ...where it is transferred to the gametes.

Sperm

Zygote

Egg

(b) Germ–plasm theory

1 According to the germ-plasm theory, germ-line tissue in the reproductive organs...

2 ...contains a complete set of genetic information...

3 ...that is transferred directly to the gametes.

Sperm

Zygote

Egg

◀1.8 **Pangenesis, an early concept of inheritance, compared with the modern germ-plasm theory.**

offspring; for example, people who developed musical ability through diligent study would produce children who are innately endowed with musical ability. The notion of the inheritance of acquired characteristics also is no longer accepted, but it remained popular through the twentieth century.

The Greek philosopher Aristotle (384–322 B.C.) was keenly interested in heredity. He rejected the concepts of both pangenesis and the inheritance of acquired characteristics, pointing out that people sometimes resemble past ancestors more than their parents and that acquired characteristics such as mutilated body parts are not passed on. Aristotle believed that both males and females made contributions to the offspring and that there was a struggle of sorts between male and female contributions.

Although the ancient Romans contributed little to the understanding of human heredity, they successfully developed a number of techniques for animal and plant breeding; the techniques were based on trial and error rather than any general concept of heredity. Little new was added to the understanding of genetics in the next 1000 years. The ancient ideas of pangenesis and the inheritance of acquired characteristics, along with techniques of plant and animal breeding, persisted until the rise of modern science in the seventeenth and eighteenth centuries.

The Rise of Modern Genetics

Dutch spectacle makers began to put together simple microscopes in the late 1500s, enabling Robert Hooke (1653–1703) to discover cells in 1665. Microscopes provided naturalists with new and exciting vistas on life, and perhaps it was excessive enthusiasm for this new world of the very small that gave rise to the idea of **preformationism.** According to preformationism, inside the egg or sperm existed a tiny miniature adult, a *homunculus*, which simply enlarged during development. Ovists argued that the homunculus resided in the egg, whereas spermists insisted that it was in the sperm (◀FIGURE 1.9). Preformationism meant that all traits would be inherited from only one parent—from the father if the homunculus was in the sperm or from the mother if it was in the egg. Although many observations suggested that offspring possess a mixture of traits from both parents, preformationism remained a popular concept throughout much of the seventeenth and eighteenth centuries.

Another early notion of heredity was **blending inheritance,** which proposed that offspring are a blend, or mixture,

The New Genetics
ETHICS · SCIENCE · TECHNOLOGY

Sex Selection

Ron Green

A couple visits a genetic counselor requesting prenatal diagnosis for the purposes of selecting the sex of their unborn child. They already have four girls and are desperate for a boy; they tell the counselor that, if the fetus is a girl, they will abort it and will keep trying until they conceive a boy. They also say that, if the counselor refuses to do prenatal diagnosis for sex selection, they will abort the fetus rather than risk having another girl. The clinic for which the counselor works has no regulations prohibiting the use of prenatal diagnosis for sex selection.

The issue of prenatal sex selection for nondisease conditions (where there is no X-linked disorder) has been the center of controversy both in the United States and elsewhere. The practice is currently widespread both in India and in China, where critics contend that it threatens major social dislocations by leading to a significantly unequal proportion of male and female births. But even where it is less prevalent and used only for family balancing (as in the case described herein), some maintain that the practice reinforces discriminatory attitudes toward women.

Genetic professionals in the United States have been reluctant to reject requests such as this one. One poll of U.S. geneticists indicated that about one-third of those questioned would assent to the parents' request. Another one-third would not provide the service themselves but would feel obligated to refer the couple to a counselor who would. These poll results illustrate once again how important the principle of respect for patient autonomy has been in the thinking of genetic professionals.

In the future, this conflict between justice and autonomy in genetic medicine will undoubtedly increase. More sensitive, earlier, and less-invasive means of prenatal testing, including maternal blood tests for fetal sex, are in the works. Genetic research also promises to enhance our understanding of other genetic conditions whose discovery will raise justice issues. One example is sexual orientation. Within the past few years, research has suggested a possible genetic component in male homosexuality. Although initially applauded by many in the gay community, who were pleased with this support for their view that sexual orientation is not chosen but shaped by biology, this research has recently drawn criticism from some gay activists and others. They argue that it may open the way to widespread decisions by parents to prevent the birth of a gay child, reinforcing prejudicial attitudes and, possibly, threatening the very existence of the gay community.

Issues of justice also arise in the conduct of genetic research. Many genetic diseases are more common in minority ethnic groups. For example, sickle-cell anemia is associated with persons of African descent; Tay-Sachs with persons of Ashkenazic Jewish heritage, and Mediterranean fever with Armenians. Identifying the genetic basis of a disorder common to an ethnic community can place all members of that community at risk for stigmatization or discrimination.

This issue has already emerged in research on cancer genetics in the Jewish community. Although this research was initially greeted with applause because of its life-saving potential, the identification of gene mutation after gene mutation in this relatively small community has recently raised questions about heightened discrimination. Will this research lead to a situation where, merely by being a member of this group, one is exposed to negative judgments by employers, insurers, or prospective marriage partners?

Behavioral genetics research on alcoholism in Native American communities raises similar questions. Although individual persons or families can benefit from such research by learning of their immediate risks, the finding that many members of a particular group have a heightened genetic propensity to develop a behavioral disorder can serve to reinforce existing stereotypes and prejudices. Individual benefit may thus be won at the price of group harm. Faced with these challenges and criticisms, genetic professionals will have to either temper their commitment to autonomy and nondirective counseling or make a much better case, to themselves and others, explaining why the single principle of respect for autonomy deserves the pride of place that it has been given. Some believe that we must also expand our concept of informed consent in research beyond the individual person under study to include representatives of the ethnic groups or communities to which that person belongs.

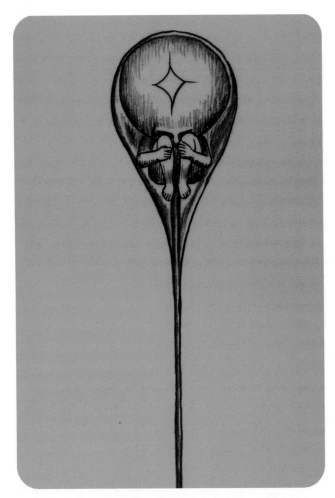

<1.9 **Preformationism was a popular idea of inheritance in the seventeenth and eighteenth centuries.** Shown here is a drawing of a homunculus inside a sperm. (Science VU/Visuals Unlimited.)

work set the foundation for the modern study of genetics. Subsequent to his work, a number of other botanists began to experiment with hybridization, including Gregor Mendel (1822–1884) (<Figure 1.10), who went on to discover the basic principles of heredity. Mendel's conclusions, which were unappreciated for 35 years, laid the foundation for our modern understanding of heredity, and he is generally recognized today as the father of genetics.

Developments in cytology (the study of cells) in the 1800s had a strong influence on genetics. Robert Brown (1773–1858) described the cell nucleus in 1833. Building on the work of others, Matthis Jacob Schleiden (1804–1881) and Theodor Schwann (1810–1882) proposed the concept of the **cell theory** in 1839. According to this theory, all life is composed of cells, cells arise only from preexisting cells, and the cell is the fundamental unit of structure and function in living organisms. Biologists began to examine cells to see how traits were transmitted in the course of cell division.

Charles Darwin (1809–1882), one of the most influential biologists of the nineteenth century, put forth the theory of evolution through natural selection and published his ideas in *On the Origin of Species* in 1856. Darwin recognized that heredity was fundamental to evolution, and he

of parental traits. This idea suggested that the genetic material itself blends, much as blue and yellow pigments blend to make green paint. Once blended, genetic differences could not be separated out in future generations, just as green paint cannot be separated out into blue and yellow pigments. Some traits do *appear* to exhibit blending inheritance; however, we realize today that individual genes do not blend.

Nehemiah Grew (1641–1712) reported that plants reproduce sexually by using pollen from the male sex cells. With this information, a number of botanists began to experiment with crossing plants and creating hybrids. Foremost among these early plant breeders was Joseph Gottlieb Kölreuter (1733–1806), who carried out numerous crosses and studied pollen under the microscope. He observed that many hybrids were intermediate between the parental varieties. Because he crossed plants that differed in many traits, Kölreuter was unable to discern any general pattern of inheritance. In spite of this limitation, Kölreuter's

<1.10 **Gregor Mendel was the founder of modern genetics.** Mendel first discovered the principles of heredity by crossing different varieties of pea plants and analyzing the pattern of transmission of traits in subsequent generations. (Hulton/Archive by Getty Images.)

conducted extensive genetic crosses with pigeons and other organisms. However, he never understood the nature of inheritance, and this lack of understanding was a major omission in his theory of evolution.

In the last half of the nineteenth century, the invention of the microtome (for cutting thin sections of tissue for microscopic examination) and the development of improved histological stains stimulated a flurry of cytological research. Several cytologists demonstrated that the nucleus had a role in fertilization. Walter Flemming (1843–1905) observed the division of chromosomes in 1879 and published a superb description of mitosis. By 1885, it was generally recognized that the nucleus contained the hereditary information.

Near the close of the nineteenth century, August Weismann (1834–1914) finally laid to rest the notion of the inheritance of acquired characteristics. He cut off the tails of mice for 22 consecutive generations and showed that the tail length in descendants remained stubbornly long. Weismann proposed the **germ-plasm theory,** which holds that the cells in the reproductive organs carry a complete set of genetic information that is passed to the gametes (see Figure 1.8b).

Twentieth-Century Genetics

The year 1900 was a watershed in the history of genetics. Gregor Mendel's pivotal 1866 publication on experiments with pea plants, which revealed the principles of heredity, was rediscovered, as discussed in more detail in Chapter 3. The significance of his conclusions was recognized, and other biologists immediately began to conduct similar genetic studies on mice, chickens, and other organisms. The results of these investigations showed that many traits indeed follow Mendel's rules.

Walter Sutton (1877–1916) proposed in 1902 that genes are located on chromosomes. Thomas Hunt Morgan (1866–1945) discovered the first genetic mutant of fruit flies in 1910 and used fruit flies to unravel many details of transmission genetics. Ronald A. Fisher (1890–1962), John B. S. Haldane (1892–1964), and Sewall Wright (1889–1988) laid the foundation for population genetics in the 1930s.

Geneticists began to use bacteria and viruses in the 1940s; the rapid reproduction and simple genetic systems of these organisms allowed detailed study of the organization and structure of genes. At about this same time, evidence accumulated that DNA was the repository of genetic information. James Watson (b. 1928) and Francis Crick (b. 1916) described the three-dimensional structure of DNA in 1953, ushering in the era of molecular genetics.

By 1966, the chemical structure of DNA and the system by which it determines the amino acid sequence of proteins had been worked out. Advances in molecular genetics led to the first recombinant DNA experiments in 1973, which touched off another revolution in genetic research. Walter Gilbert (b. 1932) and Frederick Sanger (b. 1918) developed methods for sequencing DNA in 1977. The polymerase chain reaction, a technique for quickly amplifying tiny amounts of DNA, was developed by Kary Mullis (b. 1944) and others in 1983. In 1990, gene therapy was used for the first time to treat human genetic disease in the United States (◀FIGURE 1.11), and the Human Genome Project was launched. By 1995, the first complete DNA sequence of a free-living organism—the bacterium *Haemophilus influenzae*—was determined, and the first complete sequence of a eukaryotic organism (yeast) was reported a year later. At the beginning of the twenty-first century, the human genome sequence was determined, ushering in a new era in genetics.

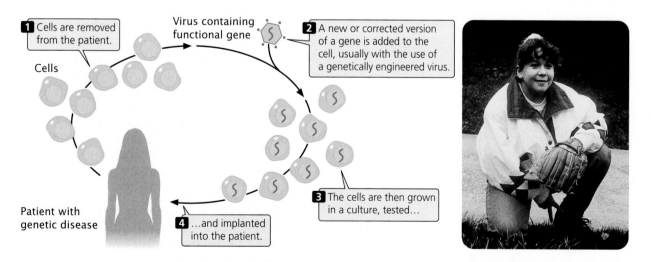

◀1.11 **Gene therapy applies genetic engineering to the treatment of human diseases.** Method (left) used to treat the first gene therapy patient (right) in 1990. Today, this patient is healthy and benefiting from the treatment. (J. Coate, MDBD/Science VU/Visuals Unlimited.)

The Future of Genetics

The information content of genetics now doubles every few years. The genome sequences of many organisms are added to DNA databases every year, and new details about gene structure and function are continually expanding our knowledge of heredity. All of this information provides us with a better understanding of numerous biological processes and evolutionary relationships. The flood of new genetic information requires the continuous development of sophisticated computer programs to store, retrieve, compare, and analyze genetic data and has given rise to the field of bioinformatics, a merging of molecular biology and computer science.

In the future, the focus of DNA-sequencing efforts will shift from the genomes of different species to individual differences within species. It is reasonable to assume that each person may some day possess a copy of his or her entire genome sequence. New genetic microchips that simultaneously analyze thousands of RNA molecules will provide information about the activity of thousands of genes in a given cell, allowing a detailed picture of how cells respond to external signals, environmental stresses, and disease states. The use of genetics in the agricultural, chemical, and health-care fields will continue to expand; some predict that biotechnology will be to the twenty-first century what the electronics industry was to the twentieth century. This ever-widening scope of genetics will raise significant ethical, social, and economic issues.

This brief overview of the history of genetics is not intended to be comprehensive; rather it is designed to provide a sense of the accelerating pace of advances in genetics. In the chapters to come, we will learn more about the experiments and the scientists who helped shape the discipline of genetics.

www.whfreeman.com/pierce   More information about the history of genetics

Concepts

Developments in plant hybridization and cytology in the eighteenth and nineteenth centuries laid the foundation for the field of genetics today. After Mendel's work was rediscovered in 1900, the science of genetics developed rapidly and today is one of the most active areas of science.

Basic Concepts in Genetics

Undoubtedly, you learned some genetic principles in other biology classes. Let's take a few moments to review some of these fundamental genetic concepts.

Cells are of two basic types: eukaryotic and prokaryotic- Structurally, cells consist of two basic types, although, evolutionarily, the story is more complex (see Chapter 2). Prokaryotic cells lack a nuclear membrane and possess no membrane-bounded cell organelles, whereas eukaryotic cells are more complex, possessing a nucleus and membrane-bounded organelles such as chloroplasts and mitochondria.

The gene is the fundamental unit of heredity- The precise way in which a gene is defined often varies. At the simplest level, we can think of a gene as a unit of information that encodes a genetic characteristic. We will enlarge this definition as we learn more about what genes are and how they function.

Genes come in multiple forms called alleles- A gene that specifies a characteristic may exist in several forms, called alleles. For example, a gene for coat color in cats may exist in alleles that encode either black or orange fur.

Genes encode phenotypes- One of the most important concepts in genetics is the distinction between traits and genes. Traits are not inherited directly. Rather, genes are inherited and, along with environmental factors, determine the expression of traits. The genetic information that an individual organism possesses is its genotype; the trait is its phenotype. For example, the A blood type is a phenotype; the genetic information that encodes the blood type A antigen is the genotype.

Genetic information is carried in DNA and RNA- Genetic information is encoded in the molecular structure of nucleic acids, which come in two types: deoxyribonucleic acid (DNA) and ribonucleic acid (RNA). Nucleic acids are polymers consisting of repeating units called nucleotides; each nucleotide consists of a sugar, a phosphate, and a nitrogenous base. The nitrogenous bases in DNA are of four types (abbreviated A, C, G, and T), and the sequence of these bases encodes genetic information. Most organisms carry their genetic information in DNA, but a few viruses carry it in RNA. The four nitrogenous bases of RNA are abbreviated A, C, G, and U.

Genes are located on chromosomes- The vehicles of genetic information within the cell are chromosomes (◀ FIGURE 1.12), which consist of DNA and associated proteins. The cells of each species have a characteristic number of chromosomes; for example, bacterial cells normally possess a single chromosome; human cells possess 46; pigeon cells possess 80. Each chromosome carries a large number of genes.

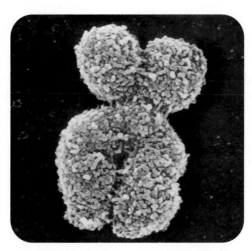

◀1.12 **Genes are carried on chromosomes.**
(Biophoto Associates/Science Source/Photo Researchers.)

Chromosomes separate through the processes of mitosis and meiosis- The processes of mitosis and meiosis ensure that each daughter cell receives a complete set of an organism's chromosomes. Mitosis is the separation of replicated chromosomes during the division of somatic (nonsex) cells. Meiosis is the pairing and separation of replicated chromosomes during the division of sex cells to produce gametes (reproductive cells).

Genetic information is transferred from DNA to RNA to protein- Many genes encode traits by specifying the structure of proteins. Genetic information is first transcribed from DNA into RNA, and then RNA is translated into the amino acid sequence of a protein.

Mutations are permanent, heritable changes in genetic information- Gene mutations affect only the genetic information of a single gene; chromosome mutations alter the number or the structure of chromosomes and therefore usually affect many genes.

Some traits are affected by multiple factors- Some traits are influenced by multiple genes that interact in complex ways with environmental factors. Human height, for example, is affected by hundreds of genes as well as environmental factors such as nutrition.

Evolution is genetic change- Evolution can be viewed as a two-step process: first, genetic variation arises and, second, some genetic variants increase in frequency, whereas other variants decrease in frequency.

 www.whfreeman.com/pierce   A glossary of genetics terms

Connecting Concepts Across Chapters

This chapter introduces the study of genetics, outlining its history, relevance, and some fundamental concepts. One of the themes that emerges from our review of the history of genetics is that humans have been interested in, and using, genetics for thousands of years, yet our understanding of the mechanisms of inheritance is relatively new. A number of ideas about how inheritance works have been proposed throughout history, but many of them have turned out to be incorrect. This is to be expected, because science progresses by constantly evaluating and challenging explanations. Genetics, like all science, is a self-correcting process, and thus many ideas that are proposed will be discarded or modified through time.

CONCEPTS SUMMARY

- Genetics is central to the life of every individual: it influences our physical features, susceptibility to numerous diseases, personality, and intelligence.

- Genetics plays important roles in agriculture, the pharmaceutical industry, and medicine. It is central to the study of biology.

- Genetic variation is the foundation of evolution and is critical to understanding all life.

- The study of genetics can be divided into transmission genetics, molecular genetics, and population genetics.

- The use of genetics by humans began with the domestication of plants and animals.

- The ancient Greeks developed the concept of pangenesis and the concept of the inheritance of acquired characteristics.

- Ancient Romans developed practical measures for the breeding of plants and animals.

- In the seventeenth century, biologists proposed the idea of preformationism, which suggested that a miniature adult is present inside the egg or the sperm and that a person inherits all of his or her traits from one parent.

- Another early idea, blending inheritance, proposed that genetic information blends during reproduction and offspring are a mixture of the parental traits.

- By studying the offspring of crosses between varieties of peas, Gregor Mendel discovered the principles of heredity.

- Darwin developed the concept of evolution by natural selection in the 1800s, but he was unaware of Mendel's work and was not able to incorporate genetics into his theory.

- Developments in cytology in the nineteenth century led to the understanding that the cell nucleus is the site of heredity.

- In 1900, Mendel's principles of heredity were rediscovered. Population genetics was established in the early 1930s, followed closely by biochemical genetics and bacterial and viral genetics. Watson and Crick discovered the structure of DNA in 1953, stimulating the rise of molecular genetics.

- Advances in molecular genetics have led to gene therapy and the Human Genome Project.

- Cells are of two basic types: prokaryotic and eukaryotic.

- Genetics is the study of genes, which are the fundamental units of heredity.

- The genes that determine a trait are termed the genotype; the trait that they produce is the phenotype.

- Genes are located on chromosomes, which are made up of nucleic acids and proteins and are partitioned into daughter cells through the process of mitosis or meiosis.

- Genetic information is expressed through the transfer of information from DNA to RNA to proteins.

- Evolution requires genetic change in populations.

IMPORTANT TERMS

transmission genetics (p. 5)
molecular genetics (p. 5)
population genetics (p. 6)

pangenesis (p. 7)
inheritance of acquired
   characteristics (p. 7)

preformationism (p. 8)
blending inheritance (p. 8)
cell theory (p. 10)

germ-plasm theory (p. 11)

COMPREHENSION QUESTIONS

Answers to questions and problems preceded by an asterisk will be found at the end of the book.

1. Outline some of the ways in which genetics is important to each of us.

* 2. Give at least three examples of the role of genetics in society today.

3. Briefly explain why genetics is crucial to modern biology.

* 4. List the three traditional subdisciplines of genetics and summarize what each covers.

5. When and where did agriculture first arise? What role did genetics play in the development of the first domesticated plants and animals?

* 6. Outline the notion of pangenesis and explain how it differs from the germ-plasm theory.

* 7. What does the concept of the inheritance of acquired characteristics propose and how is it related to the notion of pangenesis?

* 8. What is preformationism? What did it have to say about how traits are inherited?

9. Define blending inheritance and contrast it with preformationism.

10. How did developments in botany in the seventeenth and eighteenth centuries contribute to the rise of modern genetics?

11. How did developments in cytology in the nineteenth century contribute to the rise of modern genetics?

*12. Who first discovered the basic principles that laid the foundation for our modern understanding of heredity?

13. List some advances in genetics that have occurred in the twentieth century.

*14. Briefly define the following terms: (a) gene; (b) allele; (c) chromosome; (d) DNA; (e) RNA; (f) genetics; (g) genotype; (h) phenotype; (i) mutation; (j) evolution.

15. What are the two basic cell types (from a structural perspective) and how do they differ?

16. Outline the relations between genes, DNA, and chromosomes.

APPLICATION QUESTIONS AND PROBLEMS

*17. Genetics is said to be both a very old science and a very young science. Explain what is meant by this statement.

18. Find at least one newspaper article that covers some aspect of genetics. Briefly summarize the article. Does this article focus on transmission, molecular, or population genetics?

19. The following concepts were widely believed at one time but are no longer accepted as valid genetic theories. What experimental evidence suggests that these concepts are incorrect and what theories have taken their place? (a) pangenesis; (b) the inheritance of acquired characteristics; (c) preformationism; (d) blending inheritance.

CHALLENGE QUESTIONS

20. Describe some of the ways in which your own genetic makeup affects you as a person. Be as specific as you can.

21. Pick one of the following ethical or social issues and give your opinion on this issue. For background information, you might read one of the articles on ethics listed and marked with an asterisk in Suggested Readings below.

    (a) Should a person's genetic makeup be used in determining his or her eligibility for life insurance?

    (b) Should biotechnology companies be able to patent newly sequenced genes?

    (c) Should gene therapy be used on people?

    (d) Should genetic testing be made available for inherited conditions for which there is no treatment or cure?

    (e) Should governments outlaw the cloning of people?

SUGGESTED READINGS

Articles on ethical issues in genetics are preceded by an asterisk.

*American Society of Human Genetics Board of Directors and the American College of Medical Genetics Board of Directors. 1995. Points to consider: ethical, legal, pyschosocial implications of genetic testing in children. *American Journal of Human Genetics* 57:1233–1241.

An official statement on some of the ethical, legal, and psychological considerations in conducting genetic tests on children by two groups of professional geneticists.

Dunn, L. C. 1965. *A Short History of Genetics.* New York: McGraw-Hill.

An excellent history of major developments in the field of genetics.

*Friedmann, T. 2000. Principles for human gene therapy studies. *Science* 287:2163–2165.

An editorial that outlines principles that serve as the foundation for clinical gene therapy.

Kottak, C. P. 1994. *Anthropology: The Exploration of Human Diversity,* 6th ed. New York: McGraw-Hill.

Contains a summary of the rise of agriculture and initial domestication of plants and animals.

Lander, E. S., and R. A. Weinberg. 2000. Genomics: journey to the center of biology. *Science* 287:1777–1782.

A succinct history of genetics and, more specifically, genomics written by two of the leaders of modern genetics.

McKusick, V. A. 1965. The royal hemophilia. *Scientific American* 213(2):88–95.

Contains a history of hemophilia in Queen Victoria's descendants.

Massie, R. K. 1967. *Nicholas and Alexandra.* New York: Atheneum.

One of the classic histories of Tsar Nicholas and his family.

Massie, R. K. 1995. *The Romanovs: The Final Chapter.* New York: Random House.

Contains information about the finding of the Romanov remains and the DNA testing that verified the identity of the skeletons.

*Rosenberg, K., B. Fuller, M. Rothstein, T. Duster, et al. 1997. Genetic information and workplace: legislative approaches and policy challenges. *Science* 275:1755–1757.

Deals with the use of genetic information in employment.

*Shapiro, H. T. 1997. Ethical and policy issues of human cloning. *Science* 277:195–196.

Discussion of the ethics of human cloning.

Stubbe, H. 1972. *History of Genetics: From Prehistoric Times to the Rediscovery of Mendel's Laws.* Translated by T. R. W. Waters. Cambridge, MA: MIT Press.

A good history of genetics, especially for pre-Mendelian genetics.

Sturtevant, A. H. 1965. *A History of Genetics.* New York: Harper and Row.

An excellent history of genetics.

*Verma, I. M., and N. Somia. 1997. Gene therapy: promises, problems, and prospects. *Nature* 389:239–242.

An update on the status of gene therapy.

2 Chromosomes and Cellular Reproduction

Life exists in a tremendous diversity of forms, all of which utilize the same genetic system. (Art Wolfe/Photo Researchers.)

The Diversity of Life

More than by any other feature, life is characterized by diversity: 1.4 million species of plants, animals, and microorganisms have already been described, but this number vastly underestimates the total number of species on Earth. Consider the arthropods—insects, spiders, crustaceans, and related animals with hard exoskeletons. About 875,000 arthropods have been described by scientists worldwide. The results of recent studies, however, suggest that as many as 5 million to 30 million species of arthropods may be living in tropical rain forests *alone.* Furthermore, many species contain numerous genetically distinct populations, and each population contains genetically unique individuals.

Despite their tremendous diversity, living organisms have an important feature in common: all use the same genetic system. A complete set of genetic instructions for any organism is its **genome,** and all genomes are encoded in nucleic acids, either DNA or RNA. The coding system for genomic information also is common to all life—genetic instructions are in the same format and, with rare exceptions, the code words are identical. Likewise, the process by which genetic information is copied and decoded is remarkably similar for all forms of life. This universal genetic system is a consequence of the common origin of living organisms; all life on Earth evolved from the same primordial ancestor that arose between 3.5 billion and 4 billion years ago. The biologist Richard Dawkins describes life as a river of DNA that runs through time, connecting all organisms past and present.

That all organisms have a common genetic system means that the study of one organism's genes reveals principles that apply to other organisms. Investigations of how bacterial DNA is copied (replicated), for example, provides information that applies to the replication of human DNA. It also means that genes will function in foreign cells, which makes genetic engineering possible. Unfortunately, this common genetic system is also the basis for diseases such as AIDS (acquired immune deficiency syndrome), in which viral genes are able to function—sometimes with alarming efficiency—in human cells.

This chapter explores cell reproduction and how genetic information is transmitted to new cells. In prokaryotic cells, cell division is relatively simple because a prokaryotic cell usually possesses only a single chromosome. In eukaryotic cells, multiple chromosomes must be copied and distributed to each of the new cells. Cell division in eukaryotes takes place through mitosis and meiosis, processes that serve as the foundation for much of genetics; so it is essential to understand them well.

Grasping mitosis and meiosis requires more than simply memorizing the sequences of events that take place in each stage, although these events are important. The key is to understand how genetic information is apportioned during cell reproduction through a dynamic interplay of DNA synthesis, chromosome movement, and cell division. These processes bring about the transmission of genetic information and are the bases of similarities and differences between parents and progeny.

Basic Cell Types: Structure and Evolutionary Relationships

Biologists traditionally classify all living organisms into two major groups, the *prokaryotes* and the *eukaryotes*. A **prokaryote** is a unicellular organism with a relatively simple cell structure (◀ FIGURE 2.1). A **eukaryote** has a compartmentalized cell structure divided by intracellular membranes; eukaryotes may be unicellular or multicellular.

Prokaryote

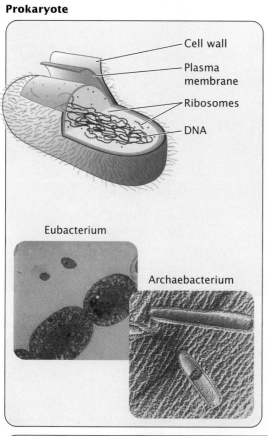

Eukaryote

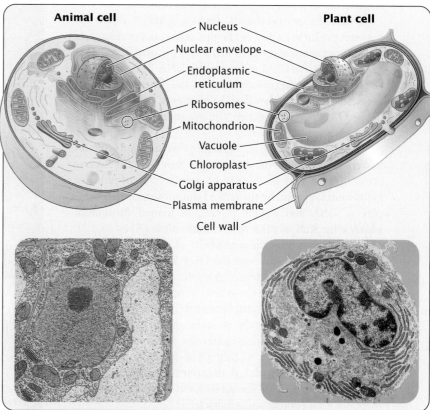

	Prokaryotic cells	Eukaryotic cells
Nucleus	Absent	Present
Cell diameter	Relatively small, from 1 to 10 μm	Relatively large, from 10 to 100 μm
Genome	Usually one circular DNA molecule	Multiple linear DNA molecules
DNA	Not complexed with histones in eubacteria; some histones in archaea	Complexed with histones
Amount of DNA	Relatively small	Relatively large
Membrane-bounded organelles	Absent	Present
Cytoskeleton	Absent	Present

◀ 2.1 **Prokaryotic and eukaryotic cells differ in structure.** (Left to right: T. J. Beveridge/Visuals Unlimited; W. Baumeister/Science Photo Library/Photo Researchers; Biophoto Associates/Photo Researchers; G. Murti/Phototake.)

Research indicates that a division of life into two major groups, the prokaryotes and eukaryotes, is not useful. Although similar in cell structure, prokaryotes include at least two fundamentally distinct types of bacteria. These distantly related groups are termed **eubacteria** (the true bacteria) and **archaea** (ancient bacteria). An examination of equivalent DNA sequences reveals that eubacteria and archaea are as distantly related to one another as they are to the eukaryotes. Although eubacteria and archaea are similar in cell structure, some genetic processes in archaea (such as transcription) are more similar to those in eukaryotes, and the archaea may actually be evolutionarily closer to eukaryotes than to eubacteria. Thus, from an evolutionary perspective, there are three major groups of organisms: eubacteria, archaea, and eukaryotes. In this book, the prokaryotic–eukaryotic distinction will be used frequently, but important eubacterial–archaeal differences also will be noted.

From the perspective of genetics, a major difference between prokaryotic and eukaryotic cells is that a eukaryote has a *nuclear envelope,* which surrounds the genetic material to form a **nucleus** and separates the DNA from the other cellular contents. In prokaryotic cells, the genetic material is in close contact with other components of the cell—a property that has important consequences for the way in which genes are controlled.

Another fundamental difference between prokaryotes and eukaryotes lies in the packaging of their DNA. In eukaryotes, DNA is closely associated with a special class of proteins, the **histones,** to form tightly packed chromosomes. This complex of DNA and histone proteins is termed **chromatin,** which is the stuff of eukaryotic chromosomes (◀ FIGURE 2.2). Histone proteins limit the accessibility of enzymes and other proteins that copy and read the DNA but they enable the DNA to fit into the nucleus. Eukaryotic DNA must separate from the histones before the genetic information in the DNA can be accessed. Archaea also have some histone proteins that complex with DNA, but the structure of their chromatin is different from that found in eukaryotes. However, eubacteria do not possess histones, so their DNA does not exist in the highly ordered, tightly packed arrangement found in eukaryotic cells (◀ FIGURE 2.3). The copying and reading of DNA are therefore simpler processes in eubacteria.

Genes of prokaryotic cells are generally on a single, circular molecule of DNA, the chromosome of the prokaryotic cell. In eukaryotic cells, genes are located on multiple, usually linear DNA molecules (multiple chromosomes). Eukaryotic cells therefore require mechanisms that ensure that a copy of each chromosome is faithfully transmitted to each new cell. This generalization—a single, circular chromosome in prokaryotes and multiple, linear chromosomes in eukaryotes—is not always true. A few bacteria have more than one chromosome, and important bacterial genes are frequently found on other DNA molecules called plasmids. Furthermore, in some eukaryotes, a few genes are located on circular DNA molecules found outside the nucleus (see Chapter 20).

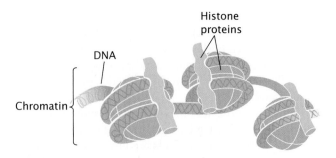

◀ 2.2 **In eukaryotic cells, DNA is complexed with histone proteins to form chromatin.**

(a)

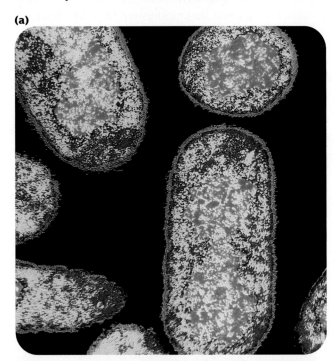

(b)

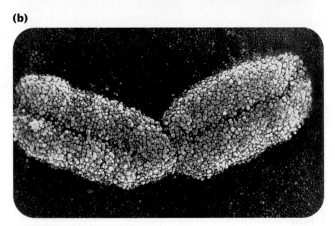

◀ 2.3 **Prokaryotic DNA (a) is not surrounded by a nuclear membrane nor is the DNA complexed with histone proteins.** Bacterial DNA is shown in red. Eukaryotic DNA (b) is complexed to histone proteins to form chromosomes that are located in the nucleus. (Part a, A. B. Dowsett/Science Photo Library/Photo Researchers; part b, Biophoto Associates/Photo Researchers.)

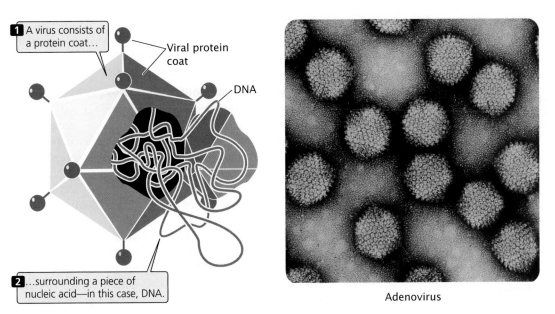

1 A virus consists of a protein coat...

Viral protein coat

DNA

2 ...surrounding a piece of nucleic acid—in this case, DNA.

Adenovirus

◀2.4 A virus is a simple replicative structure consisting of protein and nucleic acid. (Micrograph, Hans Gelderblom/Visuals Unlimited.)

Concepts

Organisms are classified as prokaryotes or eukaryotes, and prokaryotes comprise archaea and eubacteria. A prokaryote is a unicellular organism that lacks a nucleus, its DNA is not complexed to histone proteins, and its genome is usually a single chromosome. Eukaryotes are either unicellular or multicellular, their cells possess a nucleus, their DNA is complexed to histone proteins, and their genomes consist of multiple chromosomes.

Viruses are relatively simple structures composed of an outer protein coat surrounding nucleic acid (either DNA or RNA; ◀FIGURE 2.4). Viruses are neither cells nor primitive forms of life: they can reproduce only within host cells, which means that they must have evolved after, rather than before, cells. In addition, viruses are not an evolutionarily distinct group but are most closely related to their hosts— the genes of a plant virus are more similar to those in a plant cell than to those in animal viruses, which suggests that viruses evolved from their hosts, rather than from other viruses. The close relationship between the genes of virus and host makes viruses useful for studying the genetics of host organisms.

www.whfreeman.com/pierce More information on the diversity of life and the evolutionary relationships among organisms

Cell Reproduction

For any cell to reproduce successfully, three fundamental events must take place: (1) its genetic information must be copied, (2) the copies of genetic information must be separated from one another, and (3) the cell must divide. All cellular reproduction includes these three events, but the processes that lead to these events differ in prokaryotic and eukaryotic cells.

Prokaryotic Cell Reproduction

When prokaryotic cells reproduce, the circular chromosome of the bacterium is replicated (◀FIGURE 2.5). The two resulting identical copies are attached to the plasma membrane, which grows and gradually separates the two chromosomes. Finally, a new cell wall forms between the two chromosomes, producing two cells, each with an identical copy of the chromosome. Under optimal conditions, some bacterial cells divide every 20 minutes. At this rate, a single bacterial cell could produce a billion descendants in a mere 10 hours.

Eukaryotic Cell Reproduction

Like prokaryotic cell reproduction, eukaryotic cell reproduction requires the processes of DNA replication, copy separation, and division of the cytoplasm. However, the presence of multiple DNA molecules requires a more complex mechanism to ensure that one copy of each molecule ends up in each of the new cells.

Eukaryotic chromosomes are separated from the cytoplasm by the nuclear envelope. The nucleus was once thought to be a fluid-filled bag in which the chromosomes

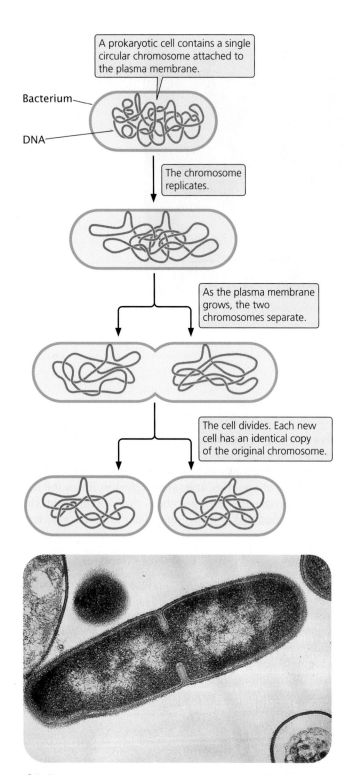

A prokaryotic cell contains a single circular chromosome attached to the plasma membrane.

Bacterium

DNA

The chromosome replicates.

As the plasma membrane grows, the two chromosomes separate.

The cell divides. Each new cell has an identical copy of the original chromosome.

◀ **2.5 Prokaryotic cells reproduce by simple division.** (Micrograph, Lee D. Simon/Photo Researchers.)

floated, but we now know that the nucleus has a highly organized internal scaffolding called the *nuclear matrix*. This matrix consists of a network of protein fibers that maintains precise spatial relations among the nuclear components and takes part in DNA replication, the expression

of genes, and the modification of gene products before they leave the nucleus. We will now take a closer look at the structure of eukaryotic chromosomes.

Eukaryotic chromosomes Each eukaryotic species has a characteristic number of chromosomes per cell: potatoes have 48 chromosomes, fruit flies have 8, and humans have 46. There appears to be no special significance between the complexity of an organism and its number of chromosomes per cell.

In most eukaryotic cells, there are two *sets* of chromosomes. The presence of two sets is a consequence of sexual reproduction; one set is inherited from the male parent and the other from the female parent. Each chromosome in one set has a corresponding chromosome in the other set, together constituting a **homologous pair** (◀ FIGURE 2.6). Human cells, for example, have 46 chromosomes, comprising 23 homologous pairs.

The two chromosomes of a homologous pair are usually alike in structure and size, and each carries genetic information for the same set of hereditary characteristics. (An exception is the sex chromosomes, which will be discussed in Chapter 4.) For example, if a gene on a particular chromosome encodes a characteristic such as hair color, another gene (called an allele) at the same position on that chromosome's homolog *also* encodes hair color. However, these two alleles need not be identical: one might produce red hair and the other might produce blond hair. Thus, most cells carry two sets of genetic information; these cells are **diploid.** But not all eukaryotic cells are diploid: reproductive cells (such as eggs, sperm, and spores) and even nonreproductive cells in some organisms may contain a single set of chromosomes. Cells with a single set of chromosomes are **haploid.** Haploid cells have only one copy of each gene.

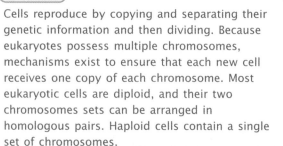

Concepts

Cells reproduce by copying and separating their genetic information and then dividing. Because eukaryotes possess multiple chromosomes, mechanisms exist to ensure that each new cell receives one copy of each chromosome. Most eukaryotic cells are diploid, and their two chromosomes sets can be arranged in homologous pairs. Haploid cells contain a single set of chromosomes.

Chromosome structure The chromosomes of eukaryotic cells are larger and more complex than those found in prokaryotes, but each unreplicated chromosome nevertheless consists of a single molecule of DNA. Although linear, the DNA molecules in eukaryotic chromosomes are highly folded and condensed; if stretched out, some human chromosomes

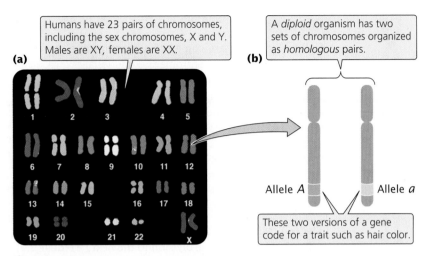

2.6 Diploid eukaryotic cells have two sets of chromosomes.
(a) A set of chromosomes from a female human cell. Each pair of
chromosomes is hybridized to a uniquely colored probe, giving it a distinct
color. (b) The chromosomes are present in homologous pairs, which consist of
chromosomes that are alike in size and structure and carry information for
the same characteristics. (Part a, courtesy of Dr. Thomas Ried and Dr. Evelin Schrock.)

would be several centimeters long—thousands of times
longer than the span of a typical nucleus. To package such a
tremendous length of DNA into this small volume, each DNA
molecule is coiled again and again and tightly packed around
histone proteins, forming the rod-shaped chromosomes. Most
of the time the chromosomes are thin and difficult to observe
but, before cell division, they condense further into thick,
readily observed structures; it is at this stage that chromosomes are usually studied (◀FIGURE 2.7).

A functional chromosome has three essential elements:
a centromere, a pair of telomeres, and origins of replication.
The *centromere* is the attachment point for *spindle microtubules*, which are the filaments responsible for moving
chromosomes during cell division. The centromere appears
as a constricted region that often stains less strongly than
does the rest of the chromosome. Before cell division, a
protein complex called the *kinetochore* assembles on the centromere, to which spindle microtubules later attach. Chromosomes without a centromere cannot be drawn into the
newly formed nuclei; these chromosomes are lost, often with
catastrophic consequences to the cell. On the basis of the location of the centromere, chromosomes are classified into
four types: metacentric, submetacentric, acrocentric, and telocentric (◀FIGURE 2.8). One of the two arms of a chromosome (the short arm of a submetacentric or acrocentric
chromosome) is designated by the letter p and the other arm
is designated by q.

Telomeres are the natural ends, the tips, of a linear
chromosome (see Figure 2.7); they serve to stabilize the
chromosome ends. If a chromosome breaks, producing new
ends, these ends have a tendency to stick together, and the
chromosome is degraded at the newly broken ends.
Telomeres provide chromosome stability. The results of

research (discussed in Chapter 12) suggest that telomeres
also participate in limiting cell division and may play
important roles in aging and cancer.

Origins of replication are the sites where DNA synthesis begins; they are not easily observed by microscopy. Their
structure and function will be discussed in more detail in
Chapters 11 and 12. In preparation for cell division, each

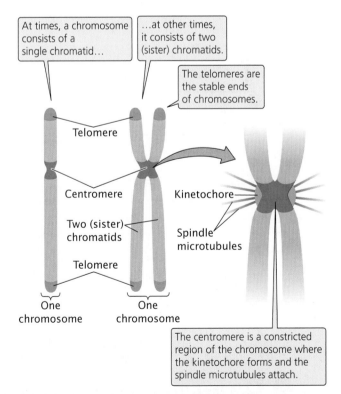

2.7 Structure of a eukaryotic chromosome.

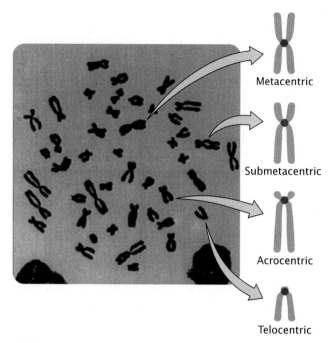

Metacentric

Submetacentric

Acrocentric

Telocentric

◀ 2.8 **Eukaryotic chromosomes exist in four major types based on the position of the centromere.**
(L. Lisco, D. W. Fawcett/Visuals Unlimited.)

chromosome replicates, making a copy of itself. These two initially identical copies, called **sister chromatids,** are held together at the centromere (see Figure 2.7). Each sister chromatid consists of a single molecule of DNA.

Concepts

Sister chromatids are copies of a chromosome held together at the centromere. Functional chromosomes contain centromeres, telomeres, and origins of replication. The kinetochore is the point of attachment for the spindle microtubules; telomeres are the stabilizing ends of a chromosome; origins of replication are sites where DNA synthesis begins.

The Cell Cycle and Mitosis

The **cell cycle** is the life story of a cell, the stages through which it passes from one division to the next (◀ FIGURE 2.9). This process is critical to genetics because, through the cell cycle, the genetic instructions for all characteristics are passed from parent to daughter cells. A new cycle begins after a cell has divided and produced two new cells. A new cell metabolizes, grows, and develops. At the end of its cycle, the cell divides to produce two cells, which can then undergo additional cell cycles.

The cell cycle consists of two major phases. The first is **interphase,** the period between cell divisions, in which the cell grows, develops, and prepares for cell division. The second is **M phase** (mitotic phase), the period of active cell division. M phase includes **mitosis,** the process of nuclear division, and **cytokinesis,** or cytoplasmic division. Let's take a closer look at the details of interphase and M phase.

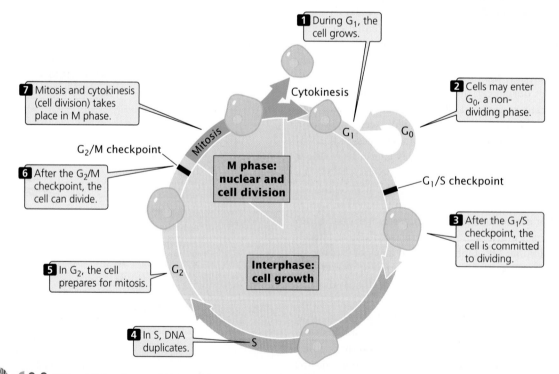

1 During G_1, the cell grows.

Cytokinesis

2 Cells may enter G_0, a non-dividing phase.

7 Mitosis and cytokinesis (cell division) takes place in M phase.

G_2/M checkpoint

6 After the G_2/M checkpoint, the cell can divide.

M phase: nuclear and cell division

G_1/S checkpoint

3 After the G_1/S checkpoint, the cell is committed to dividing.

5 In G_2, the cell prepares for mitosis.

Interphase: cell growth

4 In S, DNA duplicates.

◀ 2.9 **The cell cycle consists of interphase (a period of cell growth) and M phase (the period of nuclear and cell division).**

Interphase Interphase is the extended period of growth and development between cell divisions. Although little activity can be observed with a light microscope, the cell is quite busy: DNA is being synthesized, RNA and proteins are being produced, and hundreds of biochemical reactions are taking place.

By convention, interphase is divided into three phases: G_1, S, and G_2 (see Figure 2.9). Interphase begins with G_1 (for gap 1). In G_1, the cell grows, and proteins necessary for cell division are synthesized; this phase typically lasts several hours. There is a critical point in the cell cycle, termed the G_1/S *checkpoint*, in G_1; after this checkpoint has been passed, the cell is committed to divide.

Before reaching the G_1/S checkpoint, cells may exit from the active cell cycle in response to regulatory signals and pass into a nondividing phase called G_0 (see Figure 2.9), which is a stable state during which cells usually maintain a constant size. They can remain in G_0 for an extended period of time, even indefinitely, or they can reenter G_1 and the active cell cycle. Many cells never enter G_0; rather, they cycle continuously.

After G_1, the cell enters the *S* phase (for DNA synthesis), in which each chromosome duplicates. Although the cell is committed to divide after the G_1/S checkpoint has been passed, DNA synthesis must take place before the cell can proceed to mitosis. If DNA synthesis is blocked (with drugs or by a mutation), the cell will not be able to undergo mitosis. Before S phase, each chromosome is composed of one chromatid; following S phase, each chromosome is composed of two chromatids.

After the S phase, the cell enters G_2 (gap 2). In this phase, several additional biochemical events necessary for cell division take place. The important G_2/M *checkpoint* is reached in G_2; after this checkpoint has been passed, the cell is ready to divide and enters M phase. Although the length of interphase varies from cell type to cell type, a typical dividing mammalian cell spends about 10 hours in G_1, 9 hours in S, and 4 hours in G_2 (see Figure 2.9).

Throughout interphase, the chromosomes are in a relatively relaxed, but by no means uncoiled, state, and individual chromosomes cannot be seen with the use of a microscope. This condition changes dramatically when interphase draws to a close and the cell enters M phase.

M phase M phase is the part of the cell cycle in which the copies of the cell's chromosomes (sister chromatids) are separated and the cell undergoes division. A critical process in M phase is the separation of sister chromatids to provide a complete set of genetic information for each of the resulting cells. Biologists usually divide M phase into six stages: the five stages of mitosis (prophase, prometaphase, metaphase, anaphase, and telophase) and cytokinesis (◀ FIGURE 2.10). It's important to keep in mind that M phase is a continuous process, and its separation into these six stages is somewhat artificial.

During interphase, the chromosomes are relaxed and are visible only as diffuse chromatin, but they condense during **prophase**, becoming visible under a light microscope. Each chromosome possesses two chromatids because the chromosome was duplicated in the preceding S phase. The *mitotic spindle*, an organized array of microtubules that move the chromosomes in mitosis, forms. In animal cells, the spindle grows out from a pair of *centrosomes* that migrate to opposite sides of the cell. Within each centrosome is a special organelle, the *centriole*, which is also composed of microtubules. (Higher plant cells do not have centrosomes or centrioles, but they do have mitotic spindles).

Disintegration of the nuclear membrane marks the start of **prometaphase**. Spindle microtubules, which until now have been outside the nucleus, enter the nuclear region. The ends of certain microtubules make contact with the chromosome and anchor to the kinetochore of *one* of the sister chromatids; a microtubule from the opposite centrosome then attaches to the *other* sister chromatid, and so each chromosome is anchored to both of the centrosomes. The microtubules lengthen and shorten, pushing and pulling the chromosomes about. Some microtubules extend from each centrosome toward the center of the spindle but do not attach to a chromosome.

During **metaphase**, the chromosomes arrange themselves in a single plane, the *metaphase plate*, between the two centrosomes. The centrosomes, now at opposite ends of the cell with microtubules radiating outward and meeting in the middle of the cell, center at the spindle pole. **Anaphase** begins when the sister chromatids separate and move toward opposite spindle poles. After the chromatids have separated, each is considered a separate chromosome. **Telophase** is marked by the arrival of the chromosomes at the spindle poles. The nuclear membrane re-forms around each set of chromosomes, producing two separate nuclei within the cell. The chromosomes relax and lengthen, once again disappearing from view. In many cells, division of the cytoplasm (cytokinesis) is simultaneous with telophase. The major features of the cell cycle are summarized in Table 2.1.

Concepts

The active cell-cycle phases are interphase and M phase. Interphase consists of G_1, S, and G_2. In G_1, the cell grows and prepares for cell division; in the S phase, DNA synthesis takes place; in G_2, other biochemical events necessary for cell division take place. Some cells enter a quiescent phase called G_0. M phase includes mitosis and cytokinesis and is divided into prophase, prometaphase, metaphase, anaphase, and telophase.

---

www.whfreeman.com/pierce   Mitosis animations, tutorials, and pictures of dividing cells

Movement of Chromosomes in Mitosis

Each microtubule of the spindle is composed of subunits of a protein called tubulin, and each microtubule has direction

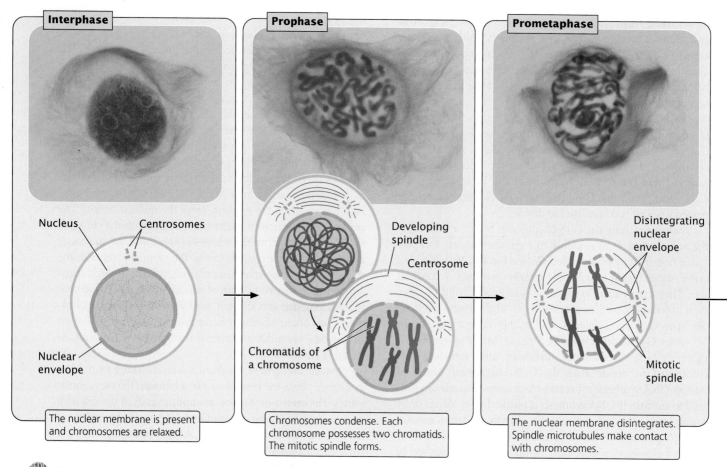

Interphase

Nucleus   Centrosomes

Nuclear
envelope

The nuclear membrane is present
and chromosomes are relaxed.

Prophase

Developing
spindle

Centrosome

Chromatids of
a chromosome

Chromosomes condense. Each
chromosome possesses two chromatids.
The mitotic spindle forms.

Prometaphase

Disintegrating
nuclear
envelope

Mitotic
spindle

The nuclear membrane disintegrates.
Spindle microtubules make contact
with chromosomes.

2.10 **The cell cycle is divided into stages.** (Photos © Andrew S. Bajer, University of Oregon.)

Table 2.1	Features of the cell cycle
Stage	**Major Features**
G_0 phase	Stable, nondividing period of variable length
Interphase	
$\quad$ G_1 phase	Growth and development of the cell; G_1/S checkpoint
$\quad$ S phase	Synthesis of DNA
$\quad$ G_2 phase	Preparation for division; G_2/S checkpoint
M phase	
$\quad$ Prophase	Chromosomes condense and mitotic spindle forms
$\quad$ Prometaphase	Nuclear envelope disintegrates, spindle microtubules anchor to kinetochores
$\quad$ Metaphase	Chromosomes align on the metaphase plate
$\quad$ Anaphase	Sister chromatids separate, becoming individual chromosomes that migrate toward spindle poles
$\quad$ Telophase	Chromosomes arrive at spindle poles, the nuclear envelope re-forms, and the condensed chromosomes relax
$\quad$ Cytokinesis	Cytoplasm divides; cell wall forms in plant cells

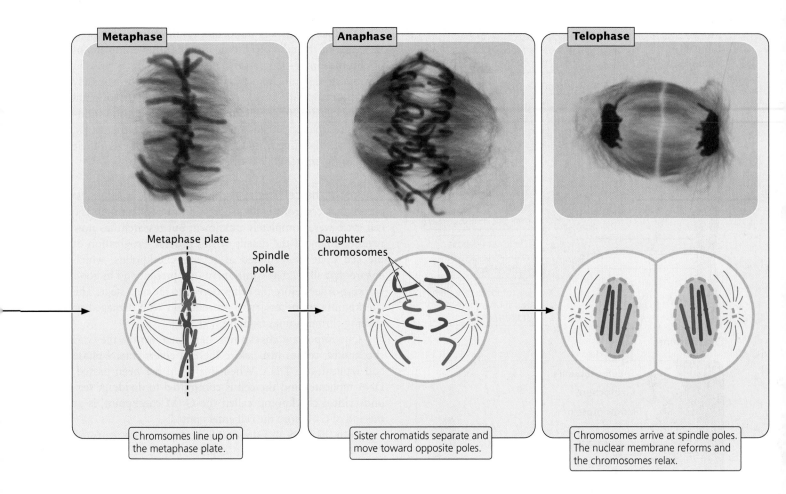

Metaphase

Metaphase plate

Spindle pole

Chromsomes line up on the metaphase plate.

Anaphase

Daughter chromosomes

Sister chromatids separate and move toward opposite poles.

Telophase

Chromosomes arrive at spindle poles. The nuclear membrane reforms and the chromosomes relax.

or polarity. Like a flashlight battery, one end is referred to as plus (+) and the other end as minus (−). The "−" end is always oriented toward the centrosome, and the "+" end is always oriented away from the centrosome; microtubules lengthen and shorten by the addition and removal of subunits primarily at the "+" end.

At one time, chromosomes were viewed as passive carriers of genetic information that were pushed about by the active spindle microtubules. Research findings now indicate that chromosomes actively control and generate the forces responsible for their movement in the course of mitosis and meiosis. Chromosome movement is accomplished through complex interactions between the kinetochore of the chromosome and the microtubules of the spindle apparatus.

The forces responsible for the poleward movement of chromosomes during anaphase are known to be generated at the kinetochore itself but are not completely understood. Located within each kinetochore are specialized proteins called *molecular motors,* which may help pull a chromosome toward the spindle pole (◀ FIGURE 2.11). The poleward force is created primarily by the removal of the tubulin at the "+" end of the microtubule.

In mitosis, depolymerization of tubulin and perhaps also molecular motors pull the chromosome toward the pole, but this force is initially counterbalanced by the attachment of the two chromatids. Throughout prophase, prometaphase, and metaphase, the sister chromatids are held together by a gluelike material. This cohesion material breaks down at the onset of anaphase, allowing the two chromatids to separate and the resulting newly formed chromosomes to move toward the spindle pole. While the chromosomal microtubules shorten, other microtubules elongate, pushing the two spindle poles farther apart. As the chromosomes near the spindle poles, they contract to form a compact mass. In spite of much study, the precise role of the poles, kinetochores, and microtubules in the formation and function of the spindle apparatus is still incompletely understood.

Genetic consequences of the cell cycle What are the genetically important results of the cell cycle? From a single cell, the cell cycle produces two cells that contain the same genetic instructions. These two cells are identical with each other and with the cell that gave rise to them. They are identical because DNA synthesis in S phase creates an exact copy of each DNA molecule, giving rise to two genetically

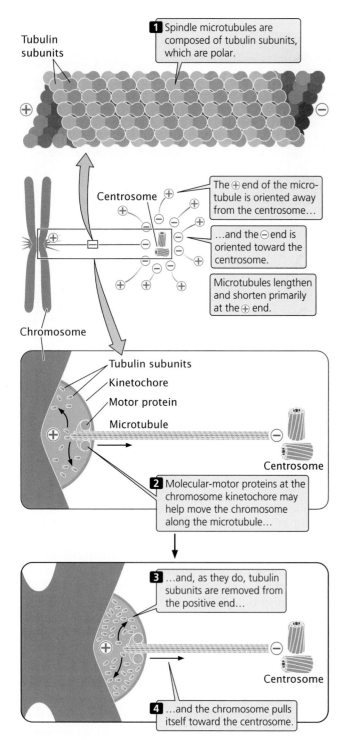

Tubulin subunits

1 Spindle microtubules are composed of tubulin subunits, which are polar.

Chromosome

Centrosome

The ⊕ end of the microtubule is oriented away from the centrosome…

…and the ⊖ end is oriented toward the centrosome.

Microtubules lengthen and shorten primarily at the ⊕ end.

Tubulin subunits
Kinetochore
Motor protein
Microtubule

Centrosome

2 Molecular-motor proteins at the chromosome kinetochore may help move the chromosome along the microtubule…

3 …and, as they do, tubulin subunits are removed from the positive end…

Centrosome

4 …and the chromosome pulls itself toward the centrosome.

◀ 2.11 **Removal of the tubulin subunits from microtubules at the kinetochore, and perhaps molecular motors, are responsible for the poleward movement of chromosomes during anaphase.**

identical sister chromatids. Mitosis then ensures that one chromatid from each replicated chromosome passes into each new cell.

Another genetically important result of the cell cycle is that each of the cells produced contains a full complement of chromosomes—there is no net reduction or increase in chromosome number. Each cell also contains approximately half the cytoplasm and organelle content of the original parental cell, but no precise mechanism analogous to mitosis ensures that organelles are evenly divided. Consequently, not all cells resulting from the cell cycle are identical in their cytoplasmic content.

Control of the cell cycle For many years, the biochemical events that controlled the progression of cells through the cell cycle were completely unknown, but research has now revealed many of the details of this process. Progression of the cell cycle is regulated at several checkpoints, which ensure that all cellular components are present and in good working order before the cell proceeds to the next stage. The checkpoints are necessary to prevent cells with damaged or missing chromosomes from proliferating.

One important checkpoint mentioned earlier, the G_1/S checkpoint, comes just before the cell enters into S phase and replicates its DNA. When this point has been passed, DNA replicates and the cell is committed to divide. A second critical checkpoint, called the G_2/M checkpoint, is at the end of G_2, before the cell enters mitosis.

Both the G_1/S and the G_2/M checkpoints are regulated by a mechanism in which two proteins interact. The concentration of the first protein, *cyclin,* oscillates during the cell cycle (◀ FIGURE 2.12a). The second protein, *cyclin-dependent kinase* (CDK), cannot function unless it is bound to cyclin. Cyclins and CDKs are called by different names in different organisms, but here we will use the terms applied to these molecules in yeast.

Let's begin by looking at the G_2/M checkpoint. This checkpoint is regulated by cyclin B, which combines with CDK to form *M-phase promoting factor* (MPF). After MPF is formed, it must be activated by the addition of a phosphate group to one of the amino acids of CDK (◀ FIGURE 2.12b).

Whereas the amount of cyclin B changes throughout the cell cycle, the amount of CDK remains constant. During G_1, cyclin B levels are low; so the amount of MPF also is low (see Figure 2.12a). As more cyclin B is produced, it combines with CDK to form increasing amounts of MPF. Near the end of G_2, the amount of active MPF reaches a critical level, which commits the cell to divide. The MPF concentration continues to increase, reaching a peak in mitosis (see Figure 2.12a).

The active form of MPF is a protein kinase, an enzyme that adds phosphate groups to certain other proteins. Active MPF brings about many of the events associated with mitosis, such as nuclear-membrane breakdown, spindle formation, and chromosome condensation. At the end of metaphase, cyclin is abruptly degraded, which lowers the amount of MPF and, initiating anaphase, sets in motion a chain of events that ultimately brings mitosis to a close

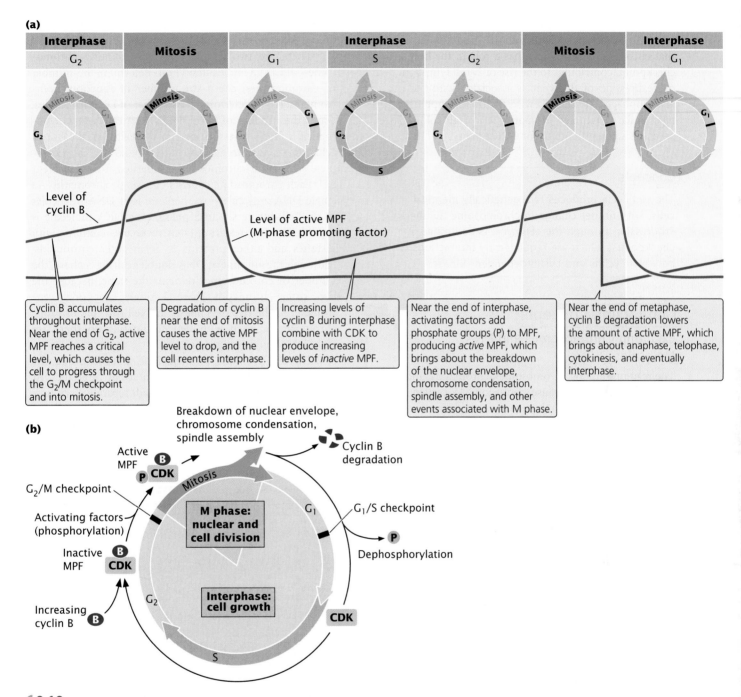

(a)

Interphase	Mitosis	Interphase			Mitosis	Interphase
G_2		G_1	S	G_2		G_1

Level of cyclin B

Level of active MPF (M-phase promoting factor)

Cyclin B accumulates throughout interphase. Near the end of G_2, active MPF reaches a critical level, which causes the cell to progress through the G_2/M checkpoint and into mitosis.

Degradation of cyclin B near the end of mitosis causes the active MPF level to drop, and the cell reenters interphase.

Increasing levels of cyclin B during interphase combine with CDK to produce increasing levels of *inactive* MPF.

Near the end of interphase, activating factors add phosphate groups (P) to MPF, producing *active* MPF, which brings about the breakdown of the nuclear envelope, chromosome condensation, spindle assembly, and other events associated with M phase.

Near the end of metaphase, cyclin B degradation lowers the amount of active MPF, which brings about anaphase, telophase, cytokinesis, and eventually interphase.

(b)

Breakdown of nuclear envelope, chromosome condensation, spindle assembly

Active MPF **P** **CDK** **B**

Cyclin B degradation

G_2/M checkpoint

Activating factors (phosphorylation)

Inactive MPF **B** **CDK**

Increasing cyclin B **B**

Mitosis

M phase: nuclear and cell division

Interphase: cell growth

G_1

G_1/S checkpoint

P

Dephosphorylation

CDK

G_2

S

2.12 Progression through the cell cycle is regulated by cyclins and CDKs. The regulation of the G_2/M checkpoint in yeast is shown here.

(see Figure 2.12b). Ironically, active MPF brings about its own demise by destroying cyclin. In brief, high levels of active MPF stimulate mitosis, and low levels of MPF bring a return to interphase conditions.

A number of factors stimulate the synthesis of cyclin B and the activation of MPF, whereas other factors inhibit MPF. Together these factors determine whether the cell passes through the G_2/M checkpoint and ensure that mitosis is not initiated until conditions are appropriate for cell division. For example, DNA damage inhibits the activation of MPF; the cell is arrested in G_2 and does not undergo division.

The G_1/S checkpoint is regulated in a similar manner. In fission yeast (*Shizosaccharomyces pombe*), the same CDK is used, but it combines with G_1 cyclins. Again, the level of CDK remains relatively constant, whereas the level of G_1 cyclins increases throughout G_1. When the activated CDK–G_1–cyclin complex reaches a critical concentration, proteins necessary for replication are activated and the cell enters S phase.

Many cancers are caused by defects in the cell cycle's regulatory machinery. For example, mutation in the gene that encodes cyclin D, which has a role in the human G_1/S checkpoint, contributes to the rise of B-cell lymphoma. The overexpression of this gene is associated with both breast and esophageal cancer. Likewise, the tumor-suppressor gene *p53*, which is mutated in about 75% of all colon cancers, regulates a potent inhibitor of CDK activity.

Concepts

The cell cycle produces two genetically identical cells, with no net change in chromosome number. Progression through the cell cycle is controlled at checkpoints, which are regulated by interactions between cyclins and cyclin-dependent kinases.

Connecting Concepts

Counting Chromosomes and DNA Molecules

The relations among chromosomes, chromatids, and DNA molecules frequently cause confusion. At certain times, chromosomes are unreplicated; at other times, each possesses two chromatids (see Figure 2.7). Chromosomes sometimes consist of a single DNA molecule; at other times, they consist of two DNA molecules. How can we keep track of the number of these structures in the cell cycle?

There are two simple rules for counting chromosomes and DNA molecules: (1) to determine the number of chromosomes, count the number of functional centromeres; (2) to determine the number of DNA molecules, count the number of chromatids. Let's examine a hypothetical cell as it passes through the cell cycle (◀FIGURE 2.13). At the beginning of G_1, this diploid cell has a complete set of four chromosomes, inherited from its parent cell. Each chromosome consists of a single chromatid—a single DNA molecule—so there are four DNA molecules in the cell during G_1. In S phase, each DNA molecule is copied. The two resulting DNA molecules combine with histones and other proteins to form sister chromatids. Although the amount of DNA doubles during S phase, the number of chromosomes remains the same, because the two sister chromatids share a single functional centromere. At the end of S phase, this cell still contains four chromosomes, each with two chromatids; so there are eight DNA molecules present.

Through prophase, prometaphase, and metaphase, the cell has four chromosomes and eight DNA molecules. At anaphase, however, the sister chromatids separate. Each now has its own functional centromere, and so each is considered a separate chromosome. Until cytokinesis, each cell contains eight chromosomes, each consisting of a single chromatid;

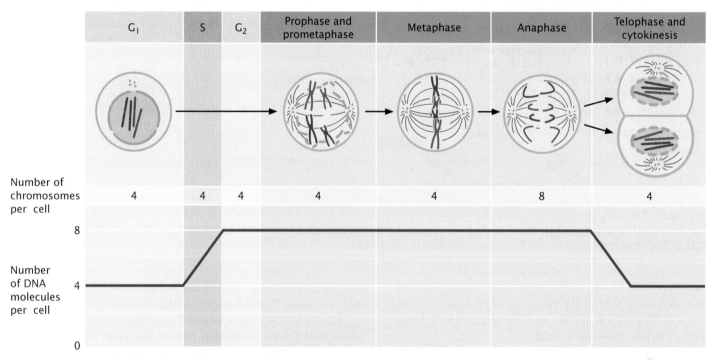

◀2.13 **The number of chromosomes and DNA molecules changes in the course of the cell cycle.** The number of chromosomes per cell equals the number of functional centromeres, and the number of DNA molecules per cell equals the number of chromatids.

thus, there are still eight DNA molecules present. After cytokinesis, the eight chromosomes (eight DNA molecules) are distributed equally between two cells; so each new cell contains four chromosomes and four DNA molecules, the number present at the beginning of the cell cycle.

Sexual Reproduction and Genetic Variation

If all reproduction were accomplished through the cell cycle, life would be quite dull, because mitosis produces only genetically identical progeny. With only mitosis, you, your children, your parents, your brothers and sisters, your cousins, and many people you didn't even know would be clones—copies of one another. Only the occasional mutation would introduce any genetic variability. This is how all organisms reproduced for the first 2 billion years of Earth's existence (and the way in which some organisms still reproduce today). Then, some 1.5 billion to 2 billion years ago, something remarkable evolved: cells that produce genetically variable offspring through sexual reproduction.

The evolution of sexual reproduction is one of the most significant events in the history of life. As will be discussed in Chapters 22 and 23, the pace of evolution depends on the amount of genetic variation present. By shuffling the genetic information from two parents, sexual reproduction greatly increases the amount of genetic variation and allows for accelerated evolution. Most of the tremendous diversity of life on Earth is a direct result of sexual reproduction.

Sexual reproduction consists of two processes. The first is **meiosis,** which leads to gametes in which chromosome number is reduced by half. The second process is **fertilization,** in which two haploid gametes fuse and restore chromosome number to its original diploid value.

Meiosis

The words *mitosis* and *meiosis* are sometimes confused. They sound a bit alike, and both refer to chromosome division and cytokinesis. Don't let this deceive you. The outcomes of mitosis and meiosis are radically different, and several unique events that have important genetic consequences take place only in meiosis.

How is meiosis different from mitosis? Mitosis consists of a single nuclear division and is usually accompanied by a single cell division. Meiosis, on the other hand, consists of two divisions. After mitosis, chromosome number in newly formed cells is the same as that in the original cell, whereas meiosis causes chromosome number in the newly formed cells to be reduced by half. Finally, mitosis produces genetically identical cells, whereas meiosis produces genetically variable cells. Let's see how these differences arise.

Like mitosis, meiosis is preceded by an interphase stage that includes G_1, S, and G_2 phases. Meiosis consists of two distinct phases: *meiosis I* and *meiosis II,* each of which

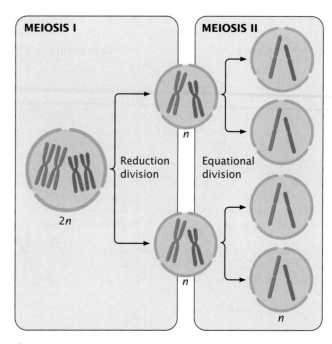

◀ **2.14 Meiosis includes two cell divisions. In this figure, the original cell is 2n=4. After two meiotic divisions, each resulting cell is 1n=2.**

includes a cell division. The first division is termed the reduction division because the number of chromosomes per cell is reduced by half (◀ FIGURE 2.14). The second division is sometimes termed the equational division; the events in this phase are similar to those of mitosis. However, meiosis II differs from mitosis in that chromosome number has already been halved in meiosis I, and the cell does not begin with the same number of chromosomes as it does in mitosis (see Figure 2.14).

The stages of meiosis are outlined in ◀ FIGURE 2.15. During interphase, the chromosomes are relaxed and visible as diffuse chromatin. **Prophase I** is a lengthy stage, divided into five substages (◀ FIGURE 2.16). In *leptotene,* the chromosomes contract and become visible. In *zygotene,* the chromosomes continue to condense; homologous chromosomes begin to pair up and begin **synapsis,** a very close pairing association. Each homologous pair of synapsed chromosomes consists of four chromatids called a **bivalent** or **tetrad.** In *pachytene,* the chromosomes become shorter and thicker, and a three-part *synaptonemal complex* develops between homologous chromosomes. **Crossing over** takes place, in which homologous chromosomes exchange genetic information. The centromeres of the paired chromosomes move apart during *diplotene;* the two homologs remain attached at each *chiasma* (plural, *chiasmata*), which is the result of crossing over. In *diakinesis,* chromosome condensation continues, and the chiasmata move toward the ends of the chromosomes as the strands slip apart; so the homologs remained paired only at the tips. Near the end of prophase I, the nuclear membrane breaks down and the spindle forms.

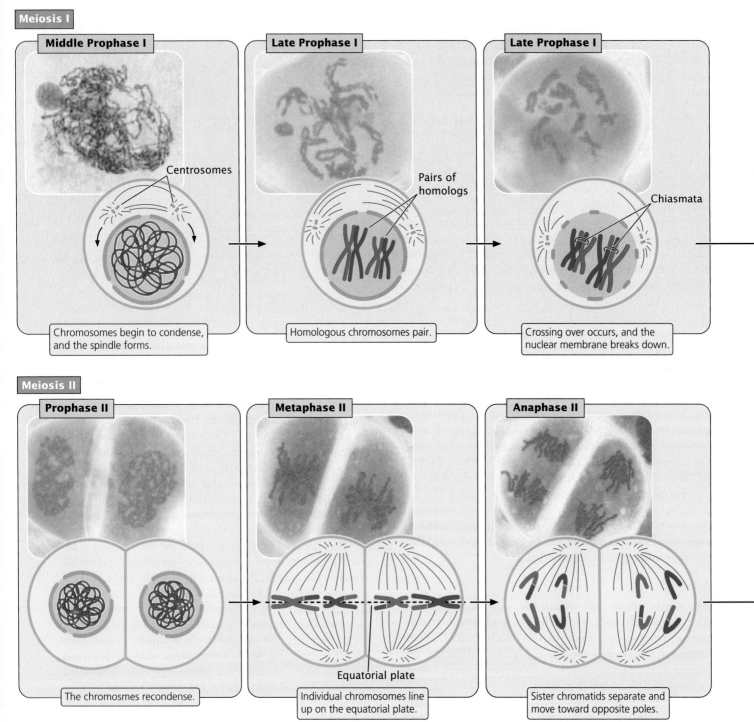

Meiosis I

Middle Prophase I

Centrosomes

Chromosomes begin to condense, and the spindle forms.

Late Prophase I

Pairs of homologs

Homologous chromosomes pair.

Late Prophase I

Chiasmata

Crossing over occurs, and the nuclear membrane breaks down.

Meiosis II

Prophase II

The chromosmes recondense.

Metaphase II

Equatorial plate

Individual chromosomes line up on the equatorial plate.

Anaphase II

Sister chromatids separate and move toward opposite poles.

Metaphase I is initiated when homologous pairs of chromosomes align along the metaphase plate (see Figure 2.15). A microtubule from one pole attaches to one chromosome of a homologous pair, and a microtubule from the other pole attaches to the other member of the pair. **Anaphase I** is marked by the separation of homologous chromosomes. The two chromosomes of a homologous pair are pulled toward opposite poles. Although the homol-ogous chromosomes separate, the sister chromatids remain attached and travel together. In **telophase I,** the chromosomes arrive at the spindle poles and the cytoplasm divides.

The period between meiosis I and meiosis II is **interkinesis,** in which the nuclear membrane re-forms around the chromosomes clustered at each pole, the spindle breaks down, and the chromosomes relax. These cells then pass through **Prophase II,** in which these events are reversed: the

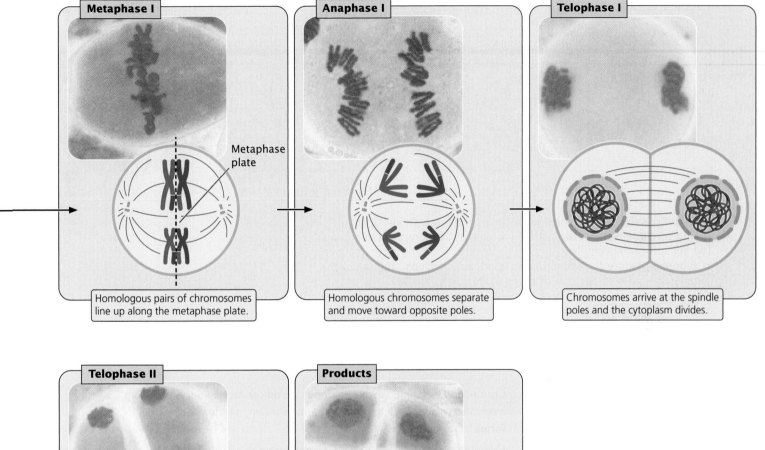

Metaphase I

Homologous pairs of chromosomes line up along the metaphase plate.

Metaphase plate

Anaphase I

Homologous chromosomes separate and move toward opposite poles.

Telophase I

Chromosomes arrive at the spindle poles and the cytoplasm divides.

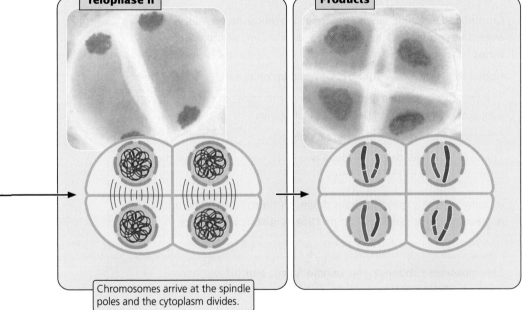

Telophase II

Chromosomes arrive at the spindle poles and the cytoplasm divides.

Products

◀ **2.15 Meiosis is divided into stages.** (Photos © C. A. Hasenkampf/BPS.)

chromosomes recondense, the spindle re-forms, and the nuclear envelope once again breaks down. In interkinesis in some types of cells, the chromosomes remain condensed, and the spindle does not break down. These cells move directly from cytokinesis into **metaphase II**, which is similar to metaphase of mitosis: the individual chromosomes line up on the metaphase plate, with the sister chromatids facing opposite poles.

In **anaphase II,** the kinetochores of the sister chromatids separate and the chromatids are pulled to opposite poles. Each chromatid is now a distinct chromosome. In **telophase II,** the chromosomes arrive at the spindle poles, a nuclear envelope re-forms around the chromosomes, and the cytoplasm divides. The chromosomes relax and are no longer visible. The major events of meiosis are summarized in Table 2.2.

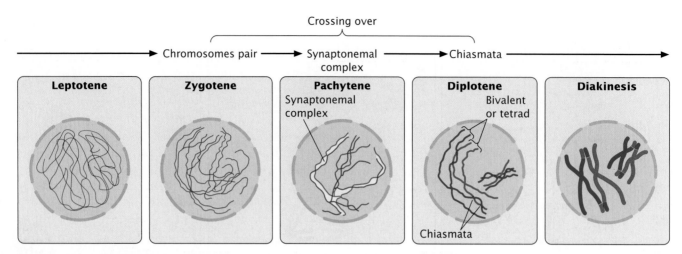

Crossing over

Chromosomes pair → Synaptonemal → Chiasmata
                     complex

| Leptotene | Zygotene | Pachytene | Diplotene | Diakinesis |

Pachytene: Synaptonemal complex

Diplotene: Bivalent or tetrad; Chiasmata

◀2.16 **Crossing over takes place in prophase I.** In yeast, rough pairing of chromosomes begins in leptotene and continues in zygotene. The synaptonemal complex forms in pachytene. Crossing over is initiated in zygotene, before the synaptonemal complex develops, and is not completed until near the end of prophase I.

Table 2.2	Major events in each stage of meiosis
Stage	**Major Events**
Meiosis I	
Prophase I	Chromosomes condense, homologous chromosomes synapse, crossing over takes place, nuclear envelope breaks down, and mitotic spindle forms
Metaphase I	Homologous pairs of chromosomes line up on the metaphase plate
Anaphase I	The two chromosomes (each with two chromatids) of each homologous pair separate and move toward opposite poles
Telophase I	Chromosomes arrive at the spindle poles
Cytokinesis	The cytoplasm divides to produce two cells, each having half the original number of chromosomes
Interkinesis	In some cells the spindle breaks down, chromosomes relax, and a nuclear envelope re-forms, but no DNA synthesis takes place
Meiosis II	
Prophase II*	Chromosomes condense, the spindle forms, and the nuclear envelope disintegrates
Metaphase II	Individual chromosomes line up on the metaphase plate
Anaphase II	Sister chromatids separate and migrate as individual chromosomes toward the spindle poles
Telophase II	Chromosomes arrive at the spindle poles; the spindle breaks down and a nuclear envelope re-forms
Cytokinesis	The cytoplasm divides

*Only in cells in which the spindle has broken down, chromosomes have relaxed, and the nuclear envelope has re-formed in telophase I. Other types of cells skip directly to metaphase II after cytokinesis.

Consequences of Meiosis

What are the overall consequences of meiosis? First, meiosis comprises two divisions; so each original cell produces four cells (there are exceptions to this generalization, as, for example, in many female animals; see Figure 2.22b). Second, chromosome number is reduced by half; so cells produced by meiosis are haploid. Third, cells produced by meiosis are genetically different from one another and from the parental cell.

Genetic differences among cells result from two processes that are unique to meiosis. The first is crossing over, which takes place in prophase I. Crossing over refers to the exchange of genes between nonsister chromatids (chromatids from different homologous chromosomes). At one time, this process was thought to take place in pachytene (Figure 2.15), and the synaptonemal complex was believed to be a requirement for crossing over. However, recent evidence from yeast suggests that the situation is more complex, as shown in Figure 2.16. Crossing over is initiated in zygotene, before the synaptonemal complex develops, and is not completed until near the end of prophase I.

After crossing over has taken place, the sister chromatids may no longer be identical. Crossing over is the basis for intrachromosomal **recombination,** creating new combinations of alleles on a chromatid. To see how crossing over produces genetic variation, consider two pairs of alleles, which we will abbreviate Aa and Bb. Assume that one chromosome possesses the A and B alleles and its homolog possesses the a and b alleles (◀ FIGURE 2.17a). When DNA is replicated in the S stage, each chromosome duplicates, and so the resulting sister chromatids are identical (◀ FIGURE 2.17b).

In the process of crossing over, breaks occur in the DNA strands and the breaks are repaired in such a way that segments of nonsister chromatids are exchanged (◀ FIGURE 2.17c). The molecular basis of this process will be described in more detail in Chapter 12; the important thing here is that, after crossing over has taken place, the two sister chromatids are no longer identical—one chromatid has alleles A and B, whereas its sister chromatid (the chromatid that underwent crossing over) has alleles a and B. Likewise, one chromatid of the other chromosome has alleles a and b, and the other has alleles A and b. Each of the four chromatids now carries a unique combination of alleles: $\underline{A\ B}$, $\underline{a\ B}$, $\underline{A\ b}$, and $\underline{a\ b}$. Eventually, the two homologous chromosomes separate, each going into a different cell. In meiosis II, the two chromatids of each chromosome separate, and thus each of the four cells resulting from meiosis carries a different combination of alleles (◀ FIGURE 2.17d).

The second process of meiosis that contributes to genetic variation is the random distribution of chromosomes in anaphase I of meiosis following their random alignment during metaphase I. To illustrate this process, consider a cell with three pairs of chromosomes, I, II, and III (◀ FIGURE 2.18a). One chromosome of each pair is maternal in origin (I_m, II_m, and III_m); the other is paternal in origin (I_p, II_p, and III_p). The chromosome pairs line up in the center of the cell in metaphase I and, in anaphase I, the chromosomes of each homologous pair separate.

How each pair of homologs aligns and separates is random and independent of how other pairs of chromosomes align and separate (◀ FIGURE 2.18b). By chance, all the maternal chromosomes might migrate to one side, with all the paternal chromosomes migrating to the other. After division, one cell would contain chromosomes I_m, II_m, and III_m, and the other, I_p, II_p, and III_p. Alternatively, the I_m, II_m, and III_p chromosomes might move to one side, and the I_p, II_p, and III_m chromosomes to the other. The different migrations would produce different combinations of chromosomes in the resulting cells (◀ FIGURE 2.18c). There are four ways in which a diploid cell with three pairs of chromosomes can divide, producing a total of eight different

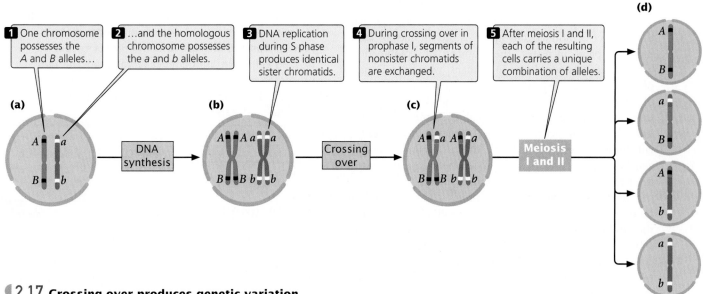

(a)

1 One chromosome possesses the A and B alleles...

2 ...and the homologous chromosome possesses the a and b alleles.

(b)

3 DNA replication during S phase produces identical sister chromatids.

(c)

4 During crossing over in prophase I, segments of nonsister chromatids are exchanged.

5 After meiosis I and II, each of the resulting cells carries a unique combination of alleles.

(d)

DNA synthesis

Crossing over

Meiosis I and II

◀ 2.17 **Crossing over produces genetic variation.**

(a)

1 This cell has three homologous pairs of chromosomes.

2 One of each pair is maternal in origin (I_m, II_m, III_m)...

3 ...and the other is paternal (I_p, II_p, III_p).

4 There are four possible ways for the three pairs to align in metaphase I.

DNA replication

(b)

(c) Gametes

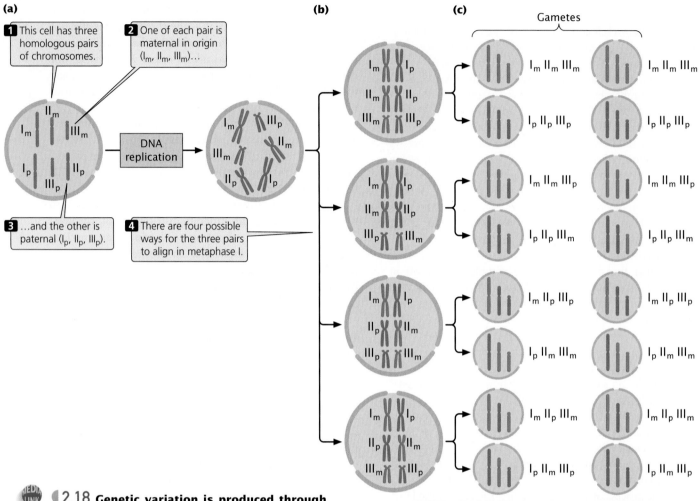

Conclusion: Eight different combinations of chromosomes in the gametes are possible, depending on how the chromosomes align and separate in meiosis I and II.

◀2.18 **Genetic variation is produced through the random distribution of chromosomes in meiosis.** In this example, the cell possesses three homologous pairs of chromosomes.

combinations of chromosomes in the gametes. In general, the number of possible combinations is 2^n, where n equals the number of homologous pairs. As the number of chromosome pairs increases, the number of combinations quickly becomes very large. In humans, who have 23 pairs of chromosomes, there are 8,388,608 different combinations of chromosomes possible from the random separation of homologous chromosomes. Through the random distribution of chromosomes in anaphase I, alleles located on different chromosomes are sorted into different combinations. The genetic consequences of this process, termed independent assortment, will be explored in more detail in Chapter 3.

In summary, crossing over shuffles alleles on the *same* homologous chromosomes into new combinations, whereas the random distribution of maternal and paternal chromosomes shuffles alleles on *different* chromosomes into new combinations. Together, these two processes are capable of producing tremendous amounts of genetic variation among the cells resulting from meiosis.

Concepts

Meiosis consists of two distinct divisions: meiosis I and meiosis II. Meiosis (usually) produces four haploid cells that are genetically variable. The two processes responsible for genetic variation are crossing over and the random distribution of maternal and paternal chromosomes.

www.whfreeman.com/pierce   A tutorial and animations of meiosis

Connecting Concepts

Comparison of Mitosis and Meiosis

Now that we have examined the details of mitosis and meiosis, let's compare the two processes (◀FIGURE 2.19). In both mitosis and meiosis, the chromosomes contract and

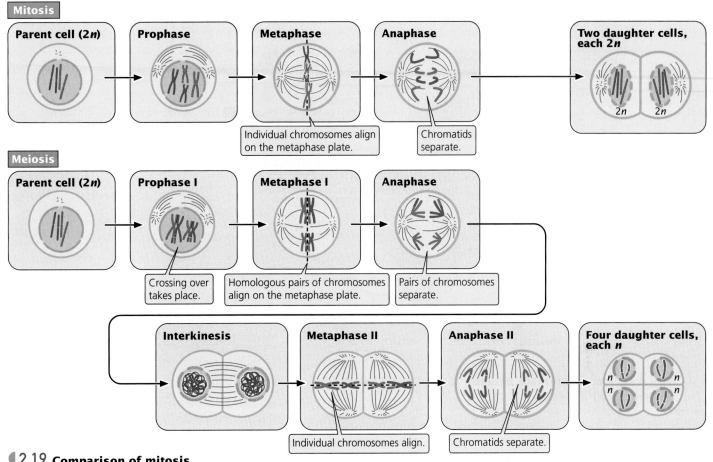

◀2.19 Comparison of mitosis and meiosis.

become visible; both processes include the movement of chromosomes toward the spindle poles, and both are accompanied by cell division. Beyond these similarities, the processes are quite different.

Mitosis entails a single cell division and usually produces two daughter cells. Meiosis, in contrast, comprises two cell divisions and usually produces four cells. In diploid cells, homologous chromosomes are present before both meiosis and mitosis, but the pairing of homologs takes place only in meiosis.

Another difference is that, in meiosis, chromosome number is reduced by half in anaphase I, but no chromosome reduction takes place in mitosis. Furthermore, meiosis is characterized by two processes that produce genetic variation: crossing over (in prophase I) and the random distribution of maternal and paternal chromosomes (in anaphase I). There are normally no equivalent processes in mitosis.

Mitosis and meiosis also differ in the behavior of chromosomes in metaphase and anaphase. In metaphase I of meiosis, *homologous pairs* of chromosomes line up on the metaphase plate, whereas *individual chromosomes* line up on the metaphase plate in metaphase of mitosis (and

metaphase II of meiosis). In anaphase I of meiosis, *paired chromosomes* separate, and each of the chromosomes that migrate toward a pole possesses two chromatids attached at the centromere. In contrast, in anaphase of mitosis (and anaphase II of meiosis), *chromatids* separate, and each chromosome that moves toward a spindle pole consists of a single chromatid.

Meiosis in the Life Cycles of Plants and Animals

The overall result of meiosis is four haploid cells that are genetically variable. Let's now see where meiosis fits into the life cycles of a multicellular plant and a multicellular animal.

Sexual reproduction in plants Most plants have a complex life cycle that includes two distinct generations (stages): the diploid *sporophyte* and the haploid *gametophyte*. These two stages alternate; the sporophyte produces haploid spores through meiosis, and the gametophyte produces haploid gametes through mitosis (◀FIGURE 2.20). This type of life cycle is sometimes called *alternation of generations*. In this cycle, the immediate products of meiosis

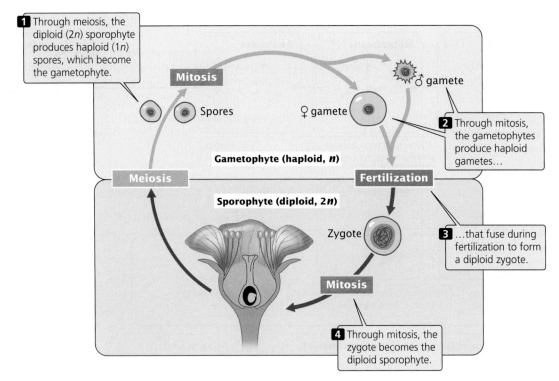

1 Through meiosis, the diploid (2*n*) sporophyte produces haploid (1*n*) spores, which become the gametophyte.

Mitosis

Spores

♀ gamete

♂ gamete

2 Through mitosis, the gametophytes produce haploid gametes...

Gametophyte (haploid, *n*)

Meiosis

Fertilization

Sporophyte (diploid, 2*n*)

Zygote

3 ...that fuse during fertilization to form a diploid zygote.

Mitosis

4 Through mitosis, the zygote becomes the diploid sporophyte.

◀ **2.20 Plants alternate between diploid and haploid life stages (female, ♀; male, ♂).**

are called spores, not gametes; the spores undergo one or more mitotic divisions to produce gametes. Although the terms used for this process are somewhat different from those commonly used in regard to animals (and from some of those employed so far in this chapter), the processes in plants and animals are basically the same: in both, meiosis leads to a reduction in chromosome number, producing haploid cells.

In flowering plants, the sporophyte is the obvious, vegetative part of the plant; the gametophyte consists of only a few haploid cells within the sporophyte. The flower, which is part of the sporophyte, contains the reproductive structures. In some plants, both male and female reproductive structures are found in the same flower; in other plants, they exist in different flowers. In either case, the male part of the flower, the stamen, contains diploid reproductive cells called **microsporocytes,** each of which undergoes meiosis to produce four haploid **microspores** (◀ FIGURE 2.21a). Each microspore divides mitotically, producing an immature pollen grain consisting of two haploid nuclei. One of these nuclei, called the tube nucleus, directs the growth of a pollen tube. The other, termed the generative nucleus, divides mitotically to produce two sperm cells. The pollen grain, with its two haploid nuclei, is the male gametophyte.

The female part of the flower, the ovary, contains diploid cells called **megasporocytes,** each of which undergoes meiosis to produce four haploid **megaspores** (◀ FIGURE 2.21b), only one of which survives. The nucleus of the surviving

megaspore divides mitotically three times, producing a total of eight haploid nuclei that make up the female gametophyte, the embryo sac. Division of the cytoplasm then produces separate cells, one of which becomes the *egg*.

When the plant flowers, the stamens open and release pollen grains. Pollen lands on a flower's stigma—a sticky platform that sits on top of a long stalk called the style. At the base of the style is the ovary. If a pollen grain germinates, it grows a tube down the style into the ovary. The two sperm cells pass down this tube and enter the embryo sac (◀ FIGURE 2.21c). One of the sperm cells fertilizes the egg cell, producing a diploid zygote, which develops into an embryo. The other sperm cell fuses with two nuclei enclosed in a single cell, giving rise to a 3*n* (triploid) endosperm, which stores food that will be used later by the embryonic plant. These two fertilization events are termed *double fertilization*.

Concepts

In the stamen of a flowering plant, meiosis produces haploid microspores that divide mitotically to produce haploid sperm in a pollen grain. Within the ovary, meiosis produces four haploid megaspores, only one of which divides mitotically three times to produce eight haploid nuclei. During pollination, one sperm fertilizes the egg cell, producing a diploid zygote; the other fuses with two nuclei to form the endosperm.

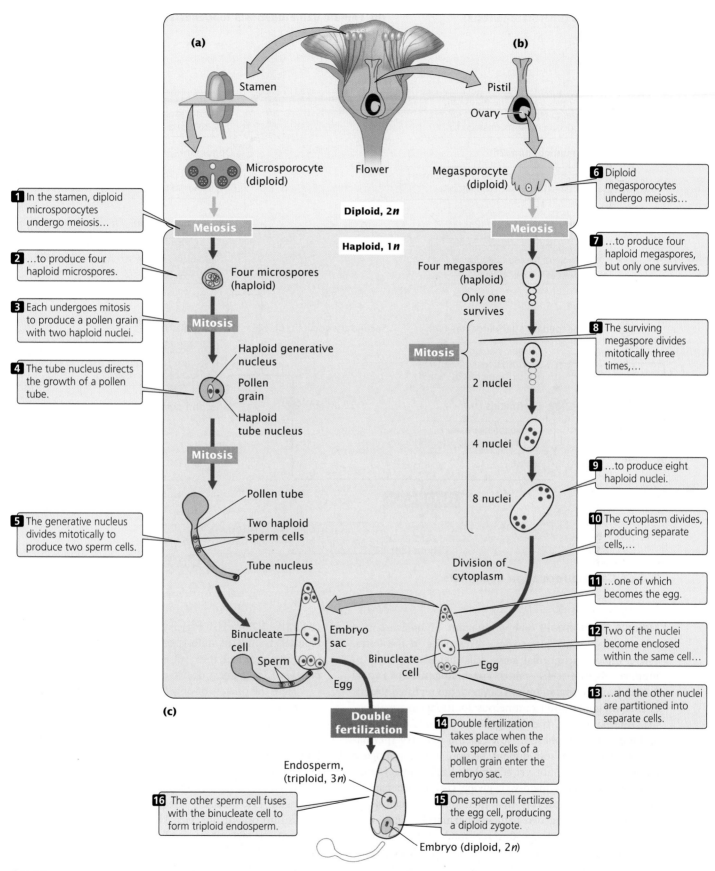

(a) Stamen

Flower

(b) Pistil

Ovary

Microsporocyte (diploid)

Megasporocyte (diploid)

1 In the stamen, diploid microsporocytes undergo meiosis…

Diploid, 2n

Meiosis

Meiosis

6 Diploid megasporocytes undergo meiosis…

Haploid, 1n

2 …to produce four haploid microspores.

Four microspores (haploid)

Four megaspores (haploid)

Only one survives

7 …to produce four haploid megaspores, but only one survives.

3 Each undergoes mitosis to produce a pollen grain with two haploid nuclei.

Mitosis

Mitosis

8 The surviving megaspore divides mitotically three times,…

4 The tube nucleus directs the growth of a pollen tube.

Haploid generative nucleus

Pollen grain

Haploid tube nucleus

2 nuclei

4 nuclei

Mitosis

8 nuclei

9 …to produce eight haploid nuclei.

Pollen tube

Two haploid sperm cells

5 The generative nucleus divides mitotically to produce two sperm cells.

Tube nucleus

Division of cytoplasm

10 The cytoplasm divides, producing separate cells,…

11 …one of which becomes the egg.

Binucleate cell

Embryo sac

Sperm

Egg

Binucleate cell

Egg

12 Two of the nuclei become enclosed within the same cell…

13 …and the other nuclei are partitioned into separate cells.

(c)

Double fertilization

14 Double fertilization takes place when the two sperm cells of a pollen grain enter the embryo sac.

Endosperm, (triploid, 3n)

16 The other sperm cell fuses with the binucleate cell to form triploid endosperm.

15 One sperm cell fertilizes the egg cell, producing a diploid zygote.

Embryo (diploid, 2n)

2.21 Sexual reproduction in flowering plants.

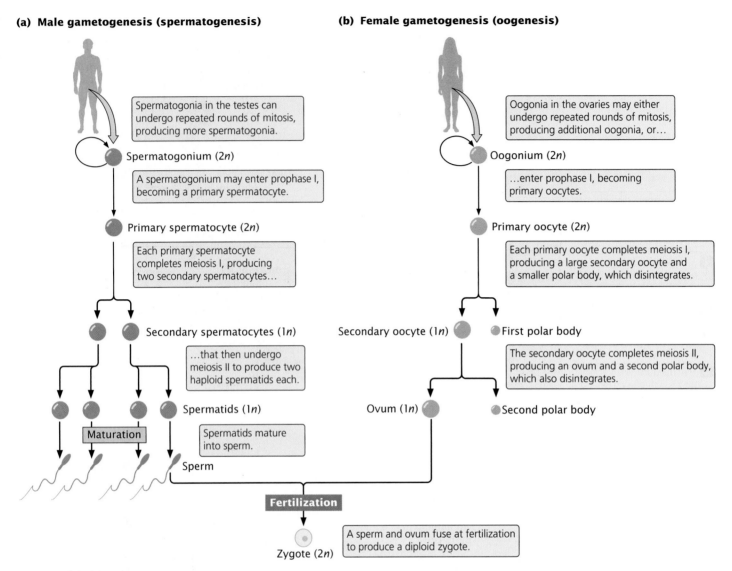

(a) Male gametogenesis (spermatogenesis)

Spermatogonia in the testes can undergo repeated rounds of mitosis, producing more spermatogonia.

Spermatogonium (2*n*)

A spermatogonium may enter prophase I, becoming a primary spermatocyte.

Primary spermatocyte (2*n*)

Each primary spermatocyte completes meiosis I, producing two secondary spermatocytes…

Secondary spermatocytes (1*n*)

…that then undergo meiosis II to produce two haploid spermatids each.

Spermatids (1*n*)

Maturation

Spermatids mature into sperm.

Sperm

(b) Female gametogenesis (oogenesis)

Oogonia in the ovaries may either undergo repeated rounds of mitosis, producing additional oogonia, or…

Oogonium (2*n*)

…enter prophase I, becoming primary oocytes.

Primary oocyte (2*n*)

Each primary oocyte completes meiosis I, producing a large secondary oocyte and a smaller polar body, which disintegrates.

Secondary oocyte (1*n*)   First polar body

The secondary oocyte completes meiosis II, producing an ovum and a second polar body, which also disintegrates.

Ovum (1*n*)   Second polar body

Fertilization

Zygote (2*n*)

A sperm and ovum fuse at fertilization to produce a diploid zygote.

◀ 2.22 **Gamete formation in animals.**

Meiosis in animals The production of gametes in a male animal (**spermatogenesis**) takes place in the testes. There, diploid primordial germ cells divide mitotically to produce diploid cells called **spermatogonia** (◀ FIGURE 2.22a). Each spermatogonium can undergo repeated rounds of mitosis, giving rise to numerous additional spermatogonia. Alternatively, a spermatogonium can initiate meiosis and enter into prophase I. Now called a **primary spermatocyte,** the cell is still diploid because the homologous chromosomes have not yet separated. Each primary spermatocyte completes meiosis I, giving rise to two haploid **secondary spermatocytes** that then undergo meiosis II, with each producing two haploid **spermatids.** Thus, each primary spermatocyte produces a total of four haploid spermatids, which mature and develop into sperm.

The production of gametes in the female (**oogenesis**) begins much like spermatogenesis. Diploid primordial germ cells within the ovaries divide mitotically to produce **oogonia**

(◀ FIGURE 2.22b). Like spermatogonia, oogonia can undergo repeated rounds of mitosis or they can enter into meiosis. Once in prophase I, these still-diploid cells are called **primary oocytes.** Each primary oocyte completes meiosis I and divides.

Here the process of oogenesis begins to differ from that of spermatogenesis. In oogenesis, cytokinesis is unequal: most of the cytoplasm is allocated to one of the two haploid cells, the **secondary oocyte.** The smaller cell, which contains half of the chromosomes but only a small part of the cytoplasm, is called the **first polar body;** it may or may not divide further. The secondary oocyte completes meiosis II, and again cytokinesis is unequal—most of the cytoplasm passes into one of the cells. The larger cell, which acquires most of the cytoplasm, is the **ovum,** the mature female gamete. The smaller cell is the **second polar body.** Only the ovum is capable of being fertilized, and the polar bodies usually disintegrate. Oogenesis, then, produces a single mature gamete from each primary oocyte.

We have now examined the place of meiosis in the sexual cycle of two organisms, a flowering plant and a typical multicellular animal. These cycles are just two of the many variations found among eukaryotic organisms. Although the cellular events that produce reproductive cells in plants and animals differ in the number of cell divisions, the number of haploid gametes produced, and the relative size of the final products, the overall result is the same: meiosis gives rise to haploid, genetically variable cells that then fuse during fertilization to produce diploid progeny.

Concepts

In the testes, a diploid spermatogonium undergoes meiosis, producing a total of four haploid sperm cells. In the ovary, a diploid oogonium undergoes meiosis to produce a single large ovum and smaller polar bodies that normally disintegrate.

Connecting Concepts Across Chapters

This chapter focused on the processes that bring about cell reproduction, the starting point of all genetics. We have examined four major concepts: (1) the differences that exist in the organization and packaging of genetic material in prokaryotic and eukaryotic cells; (2) the cell cycle and its genetic results; (3) meiosis, its genetic results, and how it differs from mitosis of the cell cycle; and (4) how meiosis fits into the reproductive cycles of plants and animals.

Several of the concepts presented in this chapter serve as an important foundation for topics in other chapters of this book. The fundamental differences in the organization of genetic material of prokaryotes and eukaryotes are important to keep in mind as we explore the molecular functioning of DNA. The presence of histone proteins in eukaryotes affects the way that DNA is copied (Chapter 12) and read (Chapter 13). The direct contact between DNA and cytoplasmic organelles in prokaryotes and the separation of DNA by the nuclear envelope in eukaryotes have important implications for gene regulation (Chapter 16) and the way that gene products are modified before they are translated into proteins (Chapter 14). The smaller amount of DNA per cell in prokaryotes also affects the organization of genes on chromosomes (Chapter 11).

A critical concept in this chapter is meiosis, which serves as the cellular basis of genetic crosses in most eukaryotic organisms. It is the basis for the rules of inheritance presented in Chapters 3 through 6 and provides a foundation for almost all of the remaining chapters of this book.

CONCEPTS SUMMARY

- A prokaryotic cell possesses a simple structure, with no nuclear envelope and usually a single, circular chromosome. A eukaryotic cell possesses a more complex structure, with a nucleus and multiple linear chromosomes consisting of DNA complexed to histone proteins.

- Cell reproduction requires the copying of the genetic material, separation of the copies, and cell division.

- In a prokaryotic cell, the single chromosome replicates, and each copy attaches to the plasma membrane; growth of the plasma membrane separates the two copies, which is followed by cell division.

- In eukaryotic cells, reproduction is more complex than in prokaryotic cells, requiring mitosis and meiosis to ensure that a complete set of genetic information is transferred to each new cell.

- In eukaryotic cells, chromosomes are typically found in homologous pairs.

- Each functional chromosome consists of a centromere, a telomere, and multiple origins of replication. Centromeres are the points at which kinetochores assemble and to which microtubules attach. Telomeres are the stable ends of chromosomes. After a chromosome is copied, the two copies remain attached at the centromere, forming sister chromatids.

- The cell cycle consists of the stages through which a eukaryotic cell passes between cell divisions. It consists of: (1) interphase, in which the cell grows and prepares for division and (2) M phase, in which nuclear and cell division take place. M phase consists of mitosis, the process of nuclear division, and cytokinesis, the division of the cytoplasm.

- Interphase begins with G_1, in which the cell grows and synthesizes proteins necessary for cell division, followed by S phase, during which the cell's DNA is replicated. The cell then enters G_2, in which additional biochemical events necessary for cell division take place. Some cells exit G_1 and enter a nondividing state called G_0.

- M phase consists of prophase, prometaphase, metaphase, anaphase, telophase, and cytokinesis. In these stages, the chromosomes contract, the nuclear membrane breaks down, and the spindle forms. The chromosomes line up in the center of the cell. Sister chromatids separate and become independent chromosomes, which then migrate to opposite ends of the cell. The nuclear membrane reforms around chromosomes at each end of the cell, and the cytoplasm divides.

- The usual result of mitosis is the production of two genetically identical cells.

- Progression through the cell cycle is controlled by interactions between cyclins and cyclin-dependent kinases.

- Sexual reproduction produces genetically variable progeny and allows for accelerated evolution. It includes meiosis, in which haploid sex cells are produced, and fertilization, the fusion of sex cells. Meiosis includes two cell divisions. In meiosis I, crossing over occurs and homologous chromosomes separate. In meiosis II, chromatids separate.

- The usual result of meiosis is the production of four haploid cells that are genetically variable.

- Genetic variation in meiosis is produced by crossing over and by the random distribution of maternal and paternal chromosomes.

- In plants, diploid microsporocytes in the stamens undergo meiosis, each microsporocyte producing four haploid microspores. Each microspore divides mitotically to produce a haploid tube nucleus and two haploid sperm cells. In the ovary, diploid megasporocytes undergo meiosis, each megasporocyte producing four haploid macrospores, only one of which survives. The surviving megaspore divides mitotically three times to produce eight haploid nuclei, one of which forms the egg. During pollination, one sperm fertilizes the egg cell and the other fuses with two haploid nuclei to form a $3n$ endosperm.

- In animals, diploid spermatogonia initiate meiosis and become diploid primary spermatocytes, which then complete meiosis I, producing two haploid secondary spermatocytes. Each secondary spermatocyte undergoes meiosis II, producing a total of four haploid sperm cells from each primary spermatocyte. Diploid oogonia in the ovary enter meiosis and become diploid primary oocytes, each of which then completes meiosis I, producing one large haploid secondary oocyte and one small haploid polar body. The secondary oocyte completes meiosis II to produce a large haploid ovum and a smaller second polar body.

IMPORTANT TERMS

genome (p. 16)
prokaryote (p. 17)
eukaryote (p. 17)
eubacteria (p. 18)
archaea (p. 18)
nucleus (p. 18)
histone (p. 18)
chromatin (p. 18)
homologous pair (p. 20)
diploid (p. 20)
haploid (p. 20)
telomere (p. 21)
origin of replication (p. 21)
sister chromatid (p. 22)

cell cycle (p. 22)
interphase (p. 22)
M phase (p. 22)
mitosis (p. 22)
cytokinesis (p. 22)
prophase (p. 23)
prometaphase (p. 23)
metaphase (p. 23)
anaphase (p. 23)
telophase (p. 23)
meiosis (p. 29)
fertilization (p. 29)
prophase I (p. 29)
synapsis (p. 29)

bivalent (p. 29)
tetrad (p. 29)
crossing over (p. 29)
metaphase I (p. 30)
anaphase I (p. 30)
telophase I (p. 30)
interkinesis (p. 30)
prophase II (p. 30)
metaphase II (p. 31)
anaphase II (p. 31)
telophase II (p. 31)
recombination (p. 33)
microsporocyte (p. 36)
microspore (p. 36)

megasporocyte (p. 36)
megaspore (p. 36)
spermatogenesis (p. 38)
spermatogonium (p. 38)
primary spermatocyte (p. 38)
secondary spermatocyte (p. 38)
spermatid (p. 38)
oogenesis (p. 38)
oogonium (p. 38)
primary oocyte (p. 38)
secondary oocyte (p. 38)
first polar body (p. 38)
ovum (p. 38)
second polar body (p. 38)

Worked Problems

1. A student examines a thin section of an onion root tip and records the number of cells that are in each stage of the cell cycle. She observes 94 cells in interphase, 14 cells in prophase, 3 cells in prometaphase, 3 cells in metaphase, 5 cells in anaphase, and 1 cell in telophase. If the complete cell cycle in an onion root tip requires 22 hours, what is the average duration of each stage in the cycle? Assume that all cells are in active cell cycle (not G_0).

• Solution

This problem is solved in two steps. First, we calculate the proportions of cells in each stage of the cell cycle, which correspond to the amount of time that an average cell spends in each stage. For example, if cells spend 90% of their time in interphase, then, at any given moment, 90% of the cells will be in interphase. The second step is to convert the proportions into lengths of time, which is done by multiplying the proportions by the total time of the cell cycle (22 hours).

Step 1. Calculate the proportion of cells at each stage. The proportion of cells at each stage is equal to the number of cells found in that stage divided by the total number of cells examined:

Interphase	$^{94}/_{120}$	$= 0.783$
Prophase	$^{14}/_{120}$	$= 0.117$
Prometaphase	$^{3}/_{120}$	$= 0.025$
Metaphase	$^{3}/_{120}$	$= 0.025$
Anaphase	$^{5}/_{120}$	$= 0.042$
Telophase	$^{1}/_{120}$	$= 0.008$

We can check our calculations by making sure that the proportions sum to 1.0, which they do.

Step 2. Determine the average duration of each stage. To determine the average duration of each stage, multiply the proportion of cells in each stage by the time required for the entire cell cycle:

Interphase	0.783 × 22 hours = 17.23 hours
Prophase	0.117 × 22 hours = 2.57 hours
Prometaphase	0.025 × 22 hours = 0.55 hour
Metaphase	0.025 × 22 hours = 0.55 hour
Anaphase	0.042 × 22 hours = 0.92 hour
Telophase	0.008 × 22 hours = 0.18 hour

2. A cell in G_1 of interphase has 8 chromosomes. How many chromosomes and how many DNA molecules will be found per cell as this cell progresses through the following stages: G_2, metaphase of mitosis, anaphase of mitosis, after cytokinesis in mitosis, metaphase I of meiosis, metaphase II of meiosis, and after cytokinesis of meiosis II?

• **Solution**

Remember the rules about counting chromosomes and DNA molecules: (1) to determine the number of chromosomes, count the functional centromeres; and (2) to determine the number of DNA molecules, count the chromatids. Think carefully about when and how the numbers of chromosomes and DNA molecules change in the course of mitosis and meiosis.

The number of DNA molecules increases only in S phase, when DNA replicates; the number of DNA molecules decreases only when the cell divides. Chromosome number increases only when sister chromatids separate in anaphase of mitosis and anaphase II of meiosis (homologous chromosomes, not chromatids, separate in anaphase I of meiosis). Chromosome number, like the number of DNA molecules, is reduced only by cell division.

Let us now apply these principles to the problem. A cell in G_1 has 8 chromosomes, each consisting of a single chromatid; so 8 DNA molecules are present in G_1. DNA replicates in S stage; so, in G_2, 16 DNA molecules are present per cell. However, the two copies of each DNA molecule remain attached at the centromere; so there are still only 8 chromosomes present. As the cell passes through prophase and metaphase of the cell cycle, the number of chromosomes and DNA molecules remains the same; so, at metaphase, there are 16 DNA molecules and 8 chromosomes. In anaphase, the chromatids separate and each becomes an independent chromosome; at this point, the number of chromosomes increases from 8 to 16. This increase is temporary, lasting only until the cell divides in telophase or subsequent to it. The number of DNA molecules remains at 16 in anaphase. The number of DNA molecules and chromosomes per cell is reduced by cytokinesis after telophase, because the 16 chromosomes and DNA molecules are now distributed between two cells. Therefore, after cytokinesis, each cell has 8 DNA molecules and 8 chromosomes, the same numbers that were present at the beginning of the cell cycle.

Now, let's trace the numbers of DNA molecules and chromosomes through meiosis. At G_1, there are 8 chromosomes and 8 DNA molecules. The number of DNA molecules increases to 16 in S stage, but the number of chromosomes remains at 8 (each chromosome has two chromatids). The cell therefore enters metaphase I with 16 DNA molecules and 8 chromosomes. In anaphase I of meiosis, homologous chromosomes separate, but the number of chromosomes remains at 8. After cytokinesis, the original 8 chromosomes are distributed between two cells; so the number of chromosomes per cell falls to 4 (each with two chromatids). The original 16 DNA molecules also are distributed between two cells; so the number of DNA molecules per cell is 8. There is no DNA synthesis during interkinesis, and each cell still maintains 4 chromosomes and 8 DNA molecules through metaphase II. In anaphase II, the two chromatids of each chromosome separate, temporarily raising the number of chromosomes per cell to 8, whereas the number of DNA molecules per cell remains at 8. After cytokinesis, the chromosomes and DNA molecules are again distributed between two cells, providing 4 chromosomes and 4 DNA molecules per cell. These results are summarized in the following table:

Stage	Number of chromosomes per cell	Number of DNA molecules per cell
G_1	8	8
G_2	8	16
Metaphase of mitosis	8	16
Anaphase of mitosis	16	16
After cytokinesis of mitosis	8	8
Metaphase I of meiosis	8	16
Metaphase II of meiosis	4	8
After cytokinesis of meiosis II	4	4

(COMPREHENSION QUESTIONS)

1. All organisms have the same universal genetic system. What are the implications of this universal genetic system?

2. Why are the viruses that infect mammalian cells useful for studying the genetics of mammals?

* 3. List three fundamental events that must take place in cell reproduction.

4. Outline the process by which prokaryotic cells reproduce.

5. Name three essential structural elements of a functional eukaryotic chromosome and describe their functions.

* 6. Sketch and label four different types of chromosomes based on the position of the centromere.

7. List the stages of interphase and the major events that take place in each stage.

* 8. List the stages of mitosis and the major events that take place in each stage.

* 9. What are the genetically important results of the cell cycle?

10. Why are the two cells produced by the cell cycle genetically identical?

11. What are checkpoints? What two general classes of compounds regulate progression through the cell cycle?

12. What are the stages of meiosis and what major events take place in each stage?

*13. What are the major results of meiosis?

14. What two processes unique to meiosis are responsible for genetic variation? At what point in meiosis do these processes take place?

*15. List similarities and differences between mitosis and meiosis. Which differences do you think are most important and why?

16. Outline the process by which male gametes are produced in plants. Outline the process of female gamete formation in plants.

17. Outline the process of spermatogenesis in animals. Outline the process of oogenesis in animals.

APPLICATION QUESTIONS AND PROBLEMS

18. A certain species has three pairs of chromosomes: an acrocentric pair, a metacentric pair, and a submetacentric pair. Draw a cell of this species as it would appear in metaphase of mitosis.

19. A biologist examines a series of cells and counts 160 cells in interphase, 20 cells in prophase, 6 cells in prometaphase, 2 cells in metaphase, 7 cells in anaphase, and 5 cells in telophase. If the complete cell cycle requires 24 hours, what is the average duration of M phase in these cells? Of metaphase?

*20. A cell in G_1 of interphase has 12 chromosomes. How many chromosomes and DNA molecules will be found per cell when this original cell progresses to the following stages?

  (a) G_2 of interphase

  (b) Metaphase I of meiosis

  (c) Prophase of mitosis

  (d) Anaphase I of meiosis

  (e) Anaphase II of meiosis

  (f) Prophase II of meiosis

  (g) After cytokinesis following mitosis

  (h) After cytokinesis following meiosis II

*21. All of the following cells, shown in various stages of mitosis and meiosis, come from the same rare species of plant. What is the diploid number of chromosomes in this plant? Give the names of each stage of mitosis or meiosis shown.

22. A cell has x amount of DNA in G_1 of interphase. How much DNA (in multiples or fractions of x) will be present per cell at the following stages?

  (a) G_2

  (b) Anaphase of mitosis

  (c) Prophase II of meiosis

  (d) After cytokinesis associated with meiosis II

23. A cell in prophase II of meiosis contains 12 chromosomes. How many chromosomes would be present in a cell from the same organism if it were in prophase of mitosis? Prophase I of meiosis?

*24. The fruit fly *Drosophila melanogaster* has four pairs of chromosomes, whereas the house fly *Musca domestica* has six pairs of chromosomes. Other things being equal, in which species would you expect to see more genetic variation among the progeny of a cross? Explain your answer.

*25. A cell has two pairs of submetacentric chromosomes, which we will call chromosomes I_a, I_b, II_a, and II_b (chromosomes I_a and I_b are homologs, and chromosomes II_a and II_b are homologs). Allele *M* is located on the long arm of chromosome I_a, and allele *m* is located at the same position on chromosome I_b. Allele *P* is located on the short arm of chromosome I_a, and allele *p* is located at the same position on chromosome I_b. Allele *R* is located on chromosome II_a and allele *r* is located at the same position on chromosome II_b.

  (a) Draw these chromosomes, labeling genes *M*, *m*, *P*, *p*, *R*, and *r*, as they might appear in metaphase I of meiosis. Assume that there is no crossing over.

  (b) Considering the random separation of chromosomes in anaphase I, draw the chromosomes (with labeled genes) present in all possible types of gametes that might result from this cell going through meiosis. Assume that there is no crossing over.

26. A horse has 64 chromosomes and a donkey has 62 chromosomes. A cross between a female horse and a male donkey produces a mule, which is usually sterile. How many chromosomes does a mule have? Can you think of any reasons for the fact that most mules are sterile?

CHALLENGE QUESTIONS

27. Suppose that life exists elsewhere in the universe. All life must contain some type of genetic information, but alien genomes might not consist of nucleic acids and have the same features as those found in the genomes of life on Earth. What do you think might be the common features of all genomes, no matter where they exist?

28. On average, what proportion of the genome in the following pairs of humans would be exactly the same if no crossing over occurred? (For the purposes of this question only, we will ignore the special case of the X and Y sex chromosomes and assume that all genes are located on nonsex chromosomes.)

    (a) Father and child

    (b) Mother and child

    (c) Two full siblings (offspring that have the same two biological parents)

    (d) Half siblings (offspring that have only one biological parent in common)

    (e) Uncle and niece

    (f) Grandparent and grandchild

29. Females bees are diploid and male bees are haploid. The haploid males produce sperm and can successfully mate with diploid females. Fertilized eggs develop into females and unfertilized eggs develop into males. How do you think the process of sperm production in male bees differs from sperm production in other animals?

30. Rec8 is a protein that is found in yeast chromosome arms and centromeres. Rec8 persists throughout meiosis I but breaks down at anaphase II. When the gene that encodes Rec8 is deleted, sister chromatids separate in anaphase I.

    (a) From these observations, propose a mechanism for the role of Rec8 in meiosis that helps to explain why sister chromatids normally separate in anaphase II but not anaphase I.

    (b) Make a prediction about the presence or absence of Rec8 during the various stages of mitosis.

SUGGESTED READINGS

Hawley, R. S., and T. Arbel. 1993. Yeast genetics and the fall of the classical view of meiosis. *Cell* 72:301–303.

Contains information about where in meiosis crossing over takes place and the role of the synaptonemal complex in recombination.

Jarrell, K. F., D. P. Bayley, J. D. Correia, and N. A. Thomas. 1999. Recent excitement about Archaea. *Bioscience* 49:530–541.

An excellent review of differences between eubacteria, archaea, and eukaryotes.

King, R. W., P. K. Jackson, and M. W. Kirschner. 1994. Mitosis in transition. *Cell* 79:563–571.

A good review of how the cell cycle is controlled.

Kirschner, M. 1992. The cell cycle then and now. *Trends in Biochemical Sciences.* 17:281–285.

A good review of the history of research into control of the cell cycle.

Koshland, D. 1994. Mitosis: back to basics. *Cell* 77:951–954.

Reviews research on mitosis and chromosome movement.

McIntosh, J. R., and M. P. Koonce. 1989. Mitosis. *Science* 246:622–628.

A review of the process of mitosis.

McIntosh, J. R., and K. L. McDonald. 1989. The mitotic spindle. *Scientific American* 261(4):48–56.

A review of the mitotic spindle.

McIntosh, J. R., and C. M. Pfarr. 1991. Mini-review: mitotic motors. *Journal of Cell Biology* 115:577–583.

Considers some of the experimental evidence concerning the role of molecular motors in the organization of the spindle and in chromosome movement.

McKim, K. S., and R. S. Hawley. 1995. Chromosomal control of meiotic cell division. *Science* 270:1595–1601.

Reviews evidence that chromosomes actively take part in their own movement and in controlling the cell cycle.

Morgan, D. O. 1995. Principles of CDK regulation. *Nature* 34:131–134.

An excellent short review of cell-cycle control.

Nasmyth, K. 1999. Separating sister chromatids. *Trends in Biochemical Sciences* 24:98–103.

Considers the role of cohesion in the separation of sister chromatids.

Pennisi, E. 1998. Cell division gatekeepers identified. *Science* 279:477–478.

Short review of work on the role of kinetochores in chromosome separation.

Pluta, A. F., A. M. MacKay, A. M. Ainsztein, I. G. Goldberg, and W. C. Earnshaw. 1995. Centromere: the hub of chromosome activities. *Science* 270:1591–1594.

An excellent review of centromere structure and function.

Rothfield, L., S. Justice, and J. Garcia-Lara. 1999. Bacterial cell division. *Annual Review of Genetics* 33:423–428.

Comprehensive review of how bacterial cells divide.

Uhlmann, F., F. Lottespeich, and K. Nasmyth. 1999. Sister-chromatid separation at anaphase onset is promoted by cleavage of the cohesion subunit Scc1. *Nature* 400:37–42.

Report that cleavage of cohesion protein has a role in chromatid separation.

Zickler, D., and N. Kleckner. 1999. Meiotic chromosomes: integrating structure and function. *Annual Review of Genetics* 33:603–754.

A review of chromosomes in meiosis, their structure and function.

3 Basic Principles of Heredity

Archibald Garrod was an early twentieth-century English scientist whose discoveries in genetics, though unnoticed for many years, contributed significantly to our understanding of the nature of genes. (CSHL Archives)

Black Urine and First Cousins

Voiding black urine is a rare and peculiar trait. In 1902, Archibald Garrod discovered the hereditary basis of black urine and, in the process, contributed to our understanding of the nature of genes.

Garrod was an English physician who was more interested in chemical explanations of disease than in the practice of medicine. He became intrigued by several of his patients who produced black urine, a condition known as alkaptonuria. The urine of alkaptonurics contains homogentisic acid, a compound that, on exposure to air, oxidizes and turns the urine black. Garrod observed that alkaptonuria appears at birth and remains for life. He noted that often several children in the same family were affected: of the 32 cases that he knew about, 19 appeared in only seven families. Furthermore, the parents of these alkaptonurics were frequently first cousins. With the assistance of geneticist William Bateson, Garrod recognized that this pattern of inheritance is precisely the pattern produced by the transmission of a rare, recessive gene.

Garrod later proposed that several other human disorders, including albinism and cystinuria, are inherited in the same way as alkaptonuria. He concluded that each gene

45

encodes an enzyme that controls a biochemical reaction. When there is a flaw in a gene, its enzyme is deficient, resulting in a biochemical disorder. He called these flaws "inborn errors of metabolism." Garrod was the first to apply the basic principles of genetics, which we will learn about in this chapter, to the inheritance of a human disease. His idea—that genes code for enzymes—was revolutionary and correct. Unfortunately, Garrod's ideas were not recognized as being important at the time and were appreciated only after they had been rediscovered 30 years later.

This chapter is about the principles of heredity: how genes are passed from generation to generation. These principles were first put forth by Gregor Mendel, so we begin by examining his scientific achievements. We then turn to simple genetic crosses, those in which a single characteristic is examined. We learn some techniques for predicting the outcome of genetic crosses and then turn to crosses in which two or more characteristics are examined. We will see how the principles applied to simple genetic crosses and the ratios of offspring that they produce serve as the key for understanding more complicated crosses. We end the chapter by considering statistical tests for analyzing crosses and factors that vary their outcome.

Throughout this chapter, a number of concepts are interwoven: Mendel's principles of segregation and independent assortment, probability, and the behavior of chromosomes. These might at first appear to be unrelated, but they are actually different views of the same phenomenon, because the genes that undergo segregation and independent assortment are located on chromosomes. The principle aim of this chapter is to examine these different views and to clarify their relations.

www.whfreeman.com/pierce   Archibald Garrod's original paper on the genetics of alkaptonuria

Mendel: The Father of Genetics

In 1902, the basic principles of genetics, which Archibald Garrod successfully applied to the inheritance of alkaptonuria, had just become widely known among biologists. Surprisingly, these principles had been discovered some 35 years earlier by Johann Gregor Mendel (1822–1884).

Mendel was born in what is now part of the Czech Republic. Although his parents were simple farmers with little money, he was able to achieve a sound education and was admitted to the Augustinian monastery in Brno in September 1843. After graduating from seminary, Mendel was ordained a priest and appointed to a teaching position in a local school. He excelled at teaching, and the abbot of the monastery recommended him for further study at the University of Vienna, which he attended from 1851 to 1853. There, Mendel enrolled in the newly opened Physics Institute and took courses in mathematics, chemistry, entomology, paleontology, botany, and plant physiology. It was

probably here that Mendel acquired the scientific method, which he later applied so successfully to his genetics experiments. After 2 years of study in Vienna, Mendel returned to Brno, where he taught school and began his experimental work with pea plants. He conducted breeding experiments from 1856 to 1863 and presented his results publicly at meetings of the Brno Natural Science Society in 1865. Mendel's paper from these lectures was published in 1866. In spite of widespread interest in heredity, the effect of his research on the scientific community was minimal. At the time, no one seems to have noticed that Mendel had discovered the basic principles of inheritance.

In 1868, Mendel was elected abbot of his monastery, and increasing administrative duties brought an end to his teaching and eventually to his genetics experiments. He died at the age of 61 on January 6, 1884, unrecognized for his contribution to genetics.

The significance of Mendel's discovery was unappreciated until 1900, when three botanists—Hugo de Vries, Erich von Tschermak, and Carl Correns—began independently conducting similar experiments with plants and arrived at conclusions similar to those of Mendel. Coming across Mendel's paper, they interpreted their results in terms of his principles and drew attention to his pioneering work.

Concepts

Gregor Mendel put forth the basic principles of inheritance, publishing his findings in 1866. The significance of his work did not become widely appreciated until 1900.

Mendel's Success

Mendel's approach to the study of heredity was effective for several reasons. Foremost was his choice of experimental subject, the pea plant *Pisum sativum* (◀ FIGURE 3.1), which offered clear advantages for genetic investigation. It is easy to cultivate, and Mendel had the monastery garden and greenhouse at his disposal. Peas grow relatively rapidly, completing an entire generation in a single growing season. By today's standards, one generation per year seems frightfully slow—fruit flies complete a generation in 2 weeks and bacteria in 20 minutes—but Mendel was under no pressure to publish quickly and was able to follow the inheritance of individual characteristics for several generations. Had he chosen to work on an organism with a longer generation time—horses, for example—he might never have discovered the basis of inheritance. Pea plants also produce many offspring—their seeds—which allowed Mendel to detect meaningful mathematical ratios in the traits that he observed in the progeny.

The large number of varieties of peas that were available to Mendel was also crucial, because these varieties differed in various traits and were genetically pure. Mendel was therefore able to begin with plants of variable, known genetic makeup.

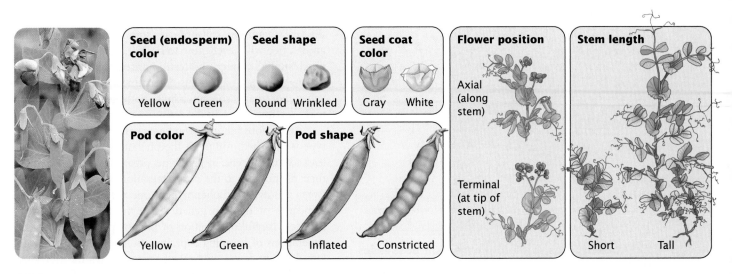

3.1 Mendel used the pea plant *Pisum sativum* in his studies of heredity. He examined seven characteristics that appeared in the seeds and in plants grown from the seeds. (Photo, Wally Eberhart/Visuals Unlimited.)

Much of Mendel's success can be attributed to the seven characteristics that he chose for study (see Figure 3.1). He avoided characteristics that display a range of variation; instead, he focused his attention on those that exist in two easily differentiated forms, such as white versus gray seed coats, round versus wrinkled seeds, and inflated versus constricted pods.

Finally, Mendel was successful because he adopted an experimental approach. Unlike many earlier investigators who just described the *results* of crosses, Mendel formulated *hypotheses* based on his initial observations and then conducted additional crosses to test his hypotheses. He kept careful records of the numbers of progeny possessing each type of trait and computed ratios of the different types. He paid close attention to detail, was adept at seeing patterns in detail, and was patient and thorough, conducting his experiments for 10 years before attempting to write up his results.

www.whfreeman.com/pierce Mendel's original paper (in German, with an English translation), as well as references, essays, and commentaries on Mendel's work

Genetic Terminology

Before we examine Mendel's crosses and the conclusions that he made from them, it will be helpful to review some terms commonly used in genetics (Table 3.1). The term *gene* was a word that Mendel never knew. It was not coined until 1909, when the Danish geneticist Wilhelm Johannsen first used it. The definition of a gene varies with the context of its use, and so its definition will change as we explore different aspects of heredity. For our present use in the context of genetic crosses, we will define a **gene** as an inherited factor that determines a characteristic.

Table 3.1	Summary of important genetic terms
Term	**Definition**
Gene	A genetic factor (region of DNA) that helps determine a characteristic
Allele	One of two or more alternate forms of a gene
Locus	Specific place on a chromosome occupied by an allele
Genotype	Set of alleles that an individual possesses
Heterozygote	An individual possessing two different alleles at a locus
Homozygote	An individual possessing two of the same alleles at a locus
Phenotype or trait	The appearance or manifestation of a character
Character or characteristic	An attribute or feature

Genes frequently come in different versions called **alleles** (◄FIGURE 3.2). In Mendel's crosses, seed shape was determined by a gene that exists as two different alleles: one allele codes for round seeds and the other codes for wrinkled seeds. All alleles for any particular gene will be found at a specific place on a chromosome called the **locus** for that gene. (The plural of locus is loci; it's bad form in genetics—and incorrect—to speak of locuses.) Thus, there is a specific place—a locus—on a chromosome in pea plants

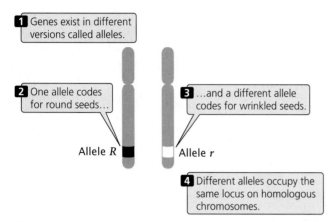

1 Genes exist in different versions called alleles.

2 One allele codes for round seeds...

3 ...and a different allele codes for wrinkled seeds.

Allele *R*   Allele *r*

4 Different alleles occupy the same locus on homologous chromosomes.

◀ **3.2 At each locus, a diploid organism possesses two alleles located on different homologous chromosomes.**

where the shape of seeds is determined. This locus might be occupied by an allele for round seeds or one for wrinkled seeds. We will use the term *allele* when referring to a specific version of a gene; we will use the term *gene* to refer more generally to any allele at a locus.

The **genotype** is the set of alleles that an individual organism possesses. A diploid organism that possesses two identical alleles is **homozygous** for that locus. One that possesses two different alleles is **heterozygous** for the locus.

Another important term is **phenotype,** which is the manifestation or appearance of a characteristic. A phenotype can refer to any type of characteristic: physical, physiological, biochemical, or behavioral. Thus, the condition of having round seeds is a phenotype, a body weight of 50 kg is a phenotype, and having sickle-cell anemia is a phenotype. In this book, the term *characteristic* or *character* refers to a general feature such as eye color; the term *trait* or *phenotype* refers to specific manifestations of that feature, such as blue or brown eyes.

A given phenotype arises from a genotype that develops within a particular environment. The genotype determines the potential for development; it sets certain limits, or boundaries, on that development. How the phenotype develops within those limits is determined by the effects of other genes and environmental factors, and the balance between these influences varies from character to character. For some characters, the differences between phenotypes are determined largely by differences in genotype; in other words, the genetic limits for that phenotype are narrow. Seed shape in Mendel's peas is a good example of a characteristic for which the genetic limits are narrow and the phenotypic differences are largely genetic. For other characters, environmental differences are more important; in this case, the limits imposed by the genotype are broad. The height that an oak tree reaches at maturity is a phenotype that is strongly influenced by environmental factors, such as the availability of water, sunlight, and nutrients. Nevertheless,

the tree's genotype still imposes some limits on its height: an oak tree will never grow to be 300 m tall no matter how much sunlight, water, and fertilizer are provided. Thus, even the height of an oak tree is determined to some degree by genes. For many characteristics, both genes and environment are important in determining phenotypic differences.

An obvious but important concept is that only the genotype is inherited. Although the phenotype is determined, at least to some extent, by genotype, organisms do not transmit their phenotypes to the next generation. The distinction between genotype and phenotype is one of the most important principles of modern genetics. The next section describes Mendel's careful observation of phenotypes through several generations of breeding experiments. These experiments allowed him to deduce not only the genotypes of the individual plants, but also the rules governing their inheritance.

Concepts

Each phenotype results from a genotype developing within a specific environment. The genotype, not the phenotype, is inherited.

Monohybrid Crosses

Mendel started with 34 varieties of peas and spent 2 years selecting those varieties that he would use in his experiments. He verified that each variety was genetically pure (homozygous for each of the traits that he chose to study) by growing the plants for two generations and confirming that all offspring were the same as their parents. He then carried out a number of crosses between the different varieties. Although peas are normally self-fertilizing (each plant crosses with itself), Mendel conducted crosses between different plants by opening the buds before the anthers were fully developed, removing the anthers, and then dusting the stigma with pollen from a different plant.

Mendel began by studying **monohybrid crosses**—those between parents that differed in a single characteristic. In one experiment, Mendel crossed a pea plant homozygous for round seeds with one that was homozygous for wrinkled seeds (◀ FIGURE 3.3). This first generation of a cross is the **P** (parental) **generation.**

After crossing the two varieties in the P generation, Mendel observed the offspring that resulted from the cross. In regard to seed characteristics, such as seed shape, the phenotype develops as soon as the seed matures, because the seed traits are determined by the newly formed embryo within the seed. For characters associated with the plant itself, such as stem length, the phenotype doesn't develop until the plant grows from the seed; for these characters, Mendel had to wait until the following spring, plant the seeds, and then observe the phenotypes on the plants that germinated.

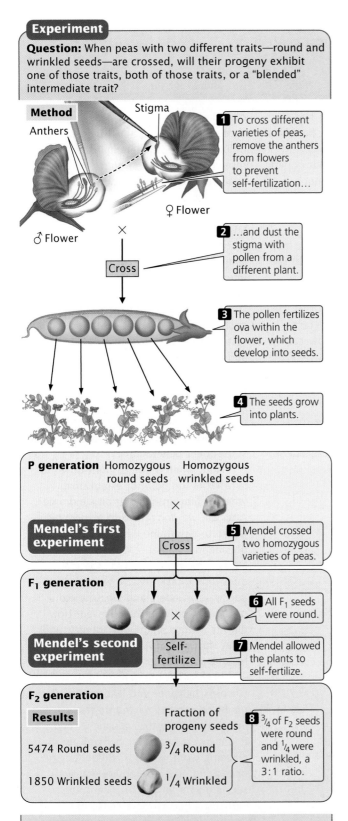

Experiment

Question: When peas with two different traits—round and wrinkled seeds—are crossed, will their progeny exhibit one of those traits, both of those traits, or a "blended" intermediate trait?

Method

Stigma

Anthers

♀ Flower

♂ Flower

× Cross

1 To cross different varieties of peas, remove the anthers from flowers to prevent self-fertilization…

2 …and dust the stigma with pollen from a different plant.

3 The pollen fertilizes ova within the flower, which develop into seeds.

4 The seeds grow into plants.

P generation Homozygous round seeds   Homozygous wrinkled seeds

Mendel's first experiment   Cross

5 Mendel crossed two homozygous varieties of peas.

F₁ generation

Mendel's second experiment   Self-fertilize

6 All F₁ seeds were round.

7 Mendel allowed the plants to self-fertilize.

F₂ generation

Results   Fraction of progeny seeds

5474 Round seeds   ³/₄ Round

1850 Wrinkled seeds   ¹/₄ Wrinkled

8 ³/₄ of F₂ seeds were round and ¹/₄ were wrinkled, a 3:1 ratio.

Conclusion: The traits of the parent plants do not blend. Although F₁ plants display the phenotype of one parent, both traits are passed to F₂ progeny in a 3:1 ratio.

◀ 3.3 **Mendel conducted monohybrid crosses.**

The offspring from the parents in the P generation are the F_1 (first filial) **generation.** When Mendel examined the F_1 of this cross, he found that they expressed only one of the phenotypes present in the parental generation: all the F_1 seeds were round. Mendel carried out 60 such crosses and always obtained this result. He also conducted **reciprocal crosses:** in one cross, pollen (the male gamete) was taken from a plant with round seeds and, in its reciprocal cross, pollen was taken from a plant with wrinkled seeds. Reciprocal crosses gave the same result: all the F_1 were round.

Mendel wasn't content with examining only the seeds arising from these monohybrid crosses. The following spring, he planted the F_1 seeds, cultivated the plants that germinated from them, and allowed the plants to self-fertilize, producing a second generation (the F_2 **generation**). Both of the traits from the P generation emerged in the F_2; Mendel counted 5474 round seeds and 1850 wrinkled seeds in the F_2 (see Figure 3.3). He noticed that the number of the round and wrinkled seeds constituted approximately a 3 to 1 ratio; that is, about $^3/_4$ of the F_2 seeds were round and $^1/_4$ were wrinkled. Mendel conducted monohybrid crosses for all seven of the characteristics that he studied in pea plants, and in all of the crosses he obtained the same result: all of the F_1 resembled only one of the two parents, but both parental traits emerged in the F_2 in approximately a 3:1 ratio.

What Monohybrid Crosses Reveal

Mendel drew several important conclusions from the results of his monohybrid crosses. First, he reasoned that, although the F_1 plants display the phenotype of only one parent, they must inherit genetic factors from both parents because they transmit both phenotypes to the F_2 generation. The presence of both round and wrinkled seeds in the F_2 could be explained only if the F_1 plants possessed both round and wrinkled genetic factors that they had inherited from the P generation. He concluded that each plant must therefore possess two genetic factors coding for a character.

The genetic factors that Mendel discovered (alleles) are, by convention, designated with letters; the allele for round seeds is usually represented by R, and the allele for wrinkled seeds by r. The plants in the P generation of Mendel's cross possessed two identical alleles: RR in the round-seeded parent and rr in the wrinkled-seeded parent (◀ FIGURE 3.4a).

A second conclusion that Mendel drew from his monohybrid crosses was that the two alleles in each plant separate when gametes are formed, and one allele goes into each gamete. When two gametes (one from each parent) fuse to produce a zygote, the allele from the male parent unites with the allele from the female parent to produce the genotype of the offspring. Thus, Mendel's F_1 plants inherited an R allele from the round-seeded plant and an r allele from the wrinkled-seeded plant (◀ FIGURE 3.4b). However, only the trait encoded by round allele (R) was *observed* in the F_1—all the F_1 progeny had round seeds. Those traits that appeared unchanged in the F_1 heterozygous offspring

The New Genetics
ETHICS · SCIENCE · TECHNOLOGY

Should Genetics Researchers Probe Abraham Lincoln's Genes?

Arthur L. Caplan

Many people agree that no one should be forced to have a genetic test without his or her consent, yet for various reasons this ethical principle is difficult to follow when dealing with those who have died. There are all sorts of reasons why genetic testing on certain deceased persons might prove important, but one of the primary reasons is for purposes of identification. In anthropology, genetic analysis might help tell us whether we have found the body of a Romanov, Hitler, or Mengele. In cases of war or terrorist attacks, such as those on September 11, 2001, there might be no other way to determine the identity of a deceased person except by matching tissue samples with previously stored biological tissue or with samples from close relatives.

One historically interesting case, which highlights the ethical issues faced when determining genetic facts about the dead, is that which centers on Abraham Lincoln. Medical geneticists and advocates for patients with Marfan syndrome have long wondered whether President Lincoln had this particular genetic disease. After all, Lincoln had the tall gangly build often associated with Marfan syndrome, which affects the connective tissues and cartilage of the body. Biographers and students of this man, whom many consider to be our greatest president, would like to know whether the depression that Lincoln suffered throughout his life might have been linked to the painful, arthritis-like symptoms of Marfan syndrome.

Lincoln was assassinated on April 14, 1865, and died early the next morning. An autopsy was performed, and samples of his hair, bone, and blood were preserved and stored at the National Museum of Health and Medicine; they are still there. The presence of a recently found genetic marker indicates whether someone has Marfan syndrome. With this

advancement, it would be possible to use some of the stored remains of Abraham Lincoln to see if he had this condition. However, would it be ethical to perform this test?

We must be careful about genetic testing, because often too much weight is assigned to the results of such tests. There is a temptation to see DNA as the essence, the blueprint, of a person— the factor that forms who we are and what we do. Given this tendency, should society be cautious about letting people explore the genes of the dead? And, if we should not test without permission, then how can we obtain permission in cases where the person in question is dead? In Lincoln's case, the "patient" is deceased and has no direct descendants; there is no one to consent. But allowing testing without consent sets a dangerous precedent.

Abraham Lincoln had the tall, gangly build often associated with Marfan syndrome. (Cartoon by Frank Billew, 1864. Bettmann/Corbis.)

Should we apply the notions of privacy and consent to the dead? But, considering that most people today agree that consent should be obtained before these tests are administered, do researchers have the right to pry into Lincoln's DNA simply because neither he nor his descendants are around to say that they can't? Are we to say that anyone's body is open to examination whenever a genetic test becomes available that might tell us an interesting fact about that person's biological makeup?

Many prominent people from the past have taken special precautions to restrict access to their diaries, papers and letters; for instance, Sigmund Freud locked away his personal papers for 100 years. Will future Lincolns and Freuds need to embargo their mortal remains for eternity to prevent unwanted genetic snooping by subsequent generations?

And, when it comes right down to it, what is the point of establishing whether Lincoln had Marfan syndrome? After all, we don't need to inspect his genes to determine whether he was presidential timber—Marfan or no Marfan, he obviously was. The real questions to ask are, Do we adequately understand what he did as president and what he believed? How did his actions shape our country, and what can we learn from them that will benefit us today?

In the end, the genetic basis for Lincoln's behavior and leadership might be seen as having no relevance. Some would say that genetic testing might divert our attention from Lincoln's work, writings, thoughts, and deeds and, instead, require that we see him as a jumble of DNA output. Perhaps it makes more sense to encourage efforts to understand and appreciate Lincoln's legacy through his actions rather than through reconstituting and analyzing his DNA.

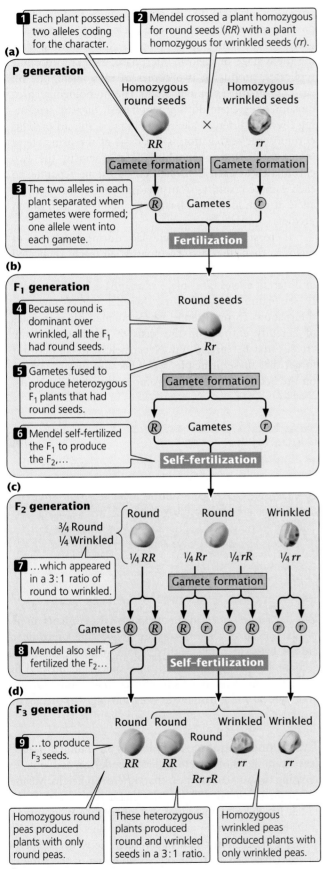

(a)

P generation

1 Each plant possessed two alleles coding for the character.

2 Mendel crossed a plant homozygous for round seeds (*RR*) with a plant homozygous for wrinkled seeds (*rr*).

Homozygous round seeds — *RR* × Homozygous wrinkled seeds — *rr*

Gamete formation | Gamete formation

3 The two alleles in each plant separated when gametes were formed; one allele went into each gamete.

Ⓡ Gametes ⓡ

Fertilization

(b)

F₁ generation

Round seeds — *Rr*

4 Because round is dominant over wrinkled, all the F₁ had round seeds.

5 Gametes fused to produce heterozygous F₁ plants that had round seeds.

Gamete formation

Ⓡ Gametes ⓡ

6 Mendel self-fertilized the F₁ to produce the F₂,…

Self-fertilization

(c)

F₂ generation

3/4 Round 1/4 Wrinkled

Round | Round | Wrinkled

1/4 *RR* | 1/4 *Rr* | 1/4 *rR* | 1/4 *rr*

7 …which appeared in a 3:1 ratio of round to wrinkled.

Gamete formation

Gametes Ⓡ Ⓡ | Ⓡ ⓡ | ⓡ Ⓡ | ⓡ ⓡ

8 Mendel also self-fertilized the F₂…

Self-fertilization

(d)

F₃ generation

9 …to produce F₃ seeds.

Round | Round | Round | Wrinkled | Wrinkled

RR | *RR* | *Rr rR* | *rr* | *rr*

Homozygous round peas produced plants with only round peas.

These heterozygous plants produced round and wrinkled seeds in a 3:1 ratio.

Homozygous wrinkled peas produced plants with only wrinkled peas.

◀3.4 **Mendel's monohybrid crosses revealed the principle of segregation and the concept of dominance.**

Mendel called **dominant,** and those traits that disappeared in the F₁ heterozygous offspring he called **recessive.** When dominant and recessive alleles are present together, the recessive allele is masked, or suppressed. The concept of dominance was a third important conclusion that Mendel derived from his monohybrid crosses.

Mendel's fourth conclusion was that the two alleles of an individual plant separate with equal probability into the gametes. When plants of the F₁ (with genotype *Rr*) produced gametes, half of the gametes received the *R* allele for round seeds and half received the *r* allele for wrinkled seeds. The gametes then paired randomly to produce the following genotypes in equal proportions among the F₂: *RR, Rr, rR, rr* (◀FIGURE 3.4c). Because round (*R*) is dominant over wrinkled (*r*), there were three round progeny in the F₂ (*RR, Rr, rR*) for every one wrinkled progeny (*rr*) in the F₂. This 3:1 ratio of round to wrinkled progeny that Mendel observed in the F₂ could occur only if the two alleles of a genotype separated into the gametes with equal probability.

The conclusions that Mendel developed about inheritance from his monohybrid crosses have been further developed and formalized into the principle of segregation and the concept of dominance. The **principle of segregation** (Mendel's first law) states that each individual diploid organism possesses two alleles for any particular characteristic. These two alleles segregate (separate) when gametes are formed, and one allele goes into each gamete. Furthermore, the two alleles segregate into gametes in equal proportions. The **concept of dominance** states that, when two different alleles are present in a genotype, only the trait of one of them—the "dominant" allele—is observed in the phenotype.

Mendel confirmed these principles by allowing his F₂ plants to self-fertilize and produce an F₃ generation. He found that the F₂ plants grown from the wrinkled seeds—those displaying the recessive trait (*rr*)—produced an F₃ in which all plants produced wrinkled seeds. Because his wrinkled-seeded plants were homozygous for wrinkled alleles (*rr*) they could pass on only wrinkled alleles to their progeny (◀FIGURE 3.4d).

The F₂ plants grown from round seeds—the dominant trait—fell into two types (see Figure 3.4c). On self-fertilization, about 2/3 of the F₂ plants produced both round and wrinkled seeds in the F₃ generation. These F₂ plants were heterozygous (*Rr*); so they produced 1/4 *RR* (round), 1/2 *Rr* (round), and 1/4 *rr* (wrinkled) seeds, giving a 3:1 ratio of round to wrinkled in the F₃. About 1/3 of the F₂ plants were of the second type; they produced only the dominant round-seeded trait in the F₃. These F₂ plants were homozygous for the round allele (*RR*) and thus could produce only round offspring in the F₃ generation. Mendel planted the seeds obtained in the F₃ and carried these plants through three more rounds of self-fertilization. In each generation, 2/3 of the round-seeded plants produced round and wrinkled offspring, whereas 1/3 produced only round offspring. These results are entirely consistent with the principle of segregation.

Concepts

The principle of segregation states that each individual organism possesses two alleles that can code for a characteristic. These alleles segregate when gametes are formed, and one allele goes into each gamete. The concept of dominance states that, when the two alleles are different, only the trait of one of them—the "dominant" allele—is observed.

Connecting Concepts

Relating Genetic Crosses to Meiosis

We have now seen how the results of monohybrid crosses are explained by Mendel's principle of segregation. Many students find that they enjoy working genetic crosses but are frustrated by the abstract nature of the symbols. Perhaps you feel the same at this point. You may be asking "What do these symbols really represent? What does the genotype *RR* mean in regard to the biology of the organism?" The answers to these questions lie in relating the abstract symbols of crosses to the structure and behavior of chromosomes, the repositories of genetic information (Chapter 2).

In 1900, when Mendel's work was rediscovered and biologists began to apply his principles of heredity, the relation between genes and chromosomes was still unclear. The theory that genes are located on chromosomes (the **chromosome theory of heredity**) was developed in the early 1900s by Walter Sutton, then a graduate student at Columbia University. Through the careful study of meiosis in insects, Sutton documented the fact that each homologous pair of chromosomes consists of one maternal chromosome and one paternal chromosome. Showing that these pairs segregate independently into gametes in meiosis, he concluded that this process is the biological basis for Mendel's principles of heredity. The German cytologist and embryologist Theodor Boveri came to similar conclusions at about the same time.

Sutton knew that diploid cells have two sets of chromosomes. Each chromosome has a pairing partner, its homologous chromosome. One chromosome of each homologous pair is inherited from the mother and the other is inherited from the father. Similarly, diploid cells possess two alleles at each locus, and these alleles constitute the genotype for that locus. The principle of segregation indicates that one allele of the genotype is inherited from each parent.

This similarity between the number of chromosomes and the number of alleles is not accidental—the two alleles of a genotype are located on homologous chromosomes. The symbols used in genetic crosses, such as *R* and *r,* are just shorthand notations for particular sequences of DNA in the chromosomes that code for particular phenotypes. The two alleles of a genotype are found on different but homologous chromosomes. During the S stage of meiotic interphase, each chromosome replicates, producing two copies of each allele, one on each chromatid (◀FIGURE 3.5a). The homologous chromosomes segregate during anaphase I, thereby separating the two different alleles (◀FIGURE 3.5b and c). This chromosome segregation is the basis of the principle of segregation. During anaphase II of meiosis, the two chromatids of each replicated chromosome separate; so each gamete resulting from meiosis carries only a single allele at each locus, as Mendel's principle of segregation predicts.

If crossing over has taken place during prophase I of meiosis, then the two chromatids of each replicated chromosome are no longer identical, and the segregation of different alleles takes place at anaphase I and anaphase II (see Figure 3.5c). Of course, Mendel didn't know anything about chromosomes; he formulated his principles of heredity entirely on the basis of the results of the crosses that he carried out. Nevertheless, we should not forget that these principles work because they are based on the behavior of actual chromosomes during meiosis.

Concepts

The chromosome theory of inheritance states that genes are located on chromosomes. The two alleles of a genotype segregate during anaphase I of meiosis, when homologous chromosomes separate. The alleles may also segregate during anaphase II of meiosis if crossing over has taken place.

Predicting the Outcomes of Genetic Crosses

One of Mendel's goals in conducting his experiments on pea plants was to develop a way to predict the outcome of crosses between plants with different phenotypes. In this section, we will first learn a simple, shorthand method for predicting outcomes of genetic crosses (the Punnett square), and then we will learn how to use probability to predict the results of crosses.

The Punnett square   To illustrate the Punnett square, let's examine another cross that Mendel carried out. By crossing two varieties of peas that differed in height, Mendel established that tall (*T*) was dominant over short (*t*). He tested his theory concerning the inheritance of dominant traits by crossing an F₁ tall plant that was heterozygous (*Tt*) with the short homozygous parental variety (*tt*). This type of cross, between an F₁ genotype and either of the parental genotypes, is called a **backcross.**

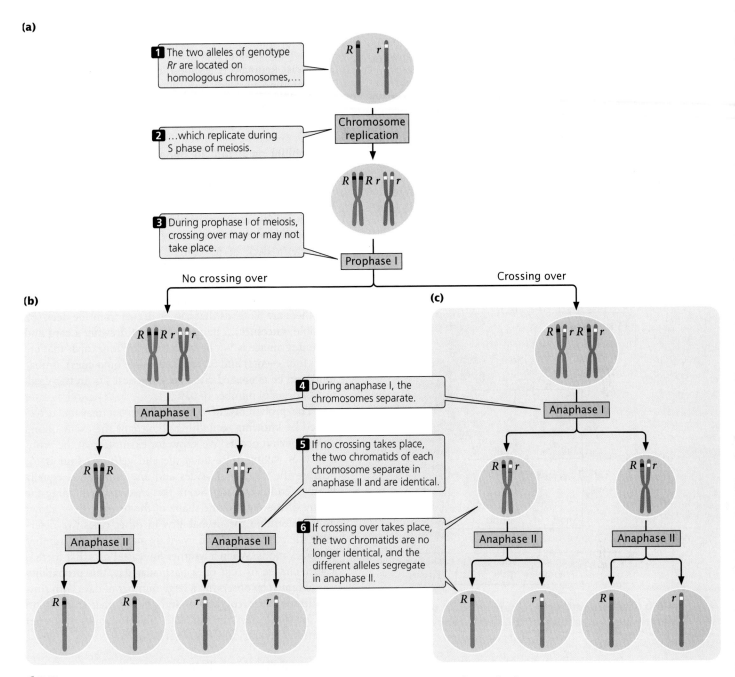

(a)

1 The two alleles of genotype *Rr* are located on homologous chromosomes,...

R   *r*

2 ...which replicate during S phase of meiosis.

Chromosome replication

R *R* *r* *r*

3 During prophase I of meiosis, crossing over may or may not take place.

Prophase I

No crossing over   Crossing over

(b)   **(c)**

R *R* *r* *r*   *R* *r* *R* *r*

4 During anaphase I, the chromosomes separate.

Anaphase I   Anaphase I

5 If no crossing takes place, the two chromatids of each chromosome separate in anaphase II and are identical.

R *R*   *r* *r*   *R* *r*   *R* *r*

6 If crossing over takes place, the two chromatids are no longer identical, and the different alleles segregate in anaphase II.

Anaphase II   Anaphase II   Anaphase II   Anaphase II

R   *R*   *r*   *r*   *R*   *r*   *R*   *r*

◀ **3.5 Segregation occurs because homologous chromosomes separate in meiosis.**

To predict the types of offspring that result from this cross, we first determine which gametes will be produced by each parent (◀ FIGURE 3.6a). The principle of segregation tells us that the two alleles in each parent separate, and one allele passes to each gamete. All gametes from the homozygous *tt* short plant will receive a single short (*t*) allele. The tall plant in this cross is heterozygous (*Tt*); so 50% of its gametes will receive a tall allele (*T*) and the other 50% will receive a short allele (*t*).

A **Punnett square** is constructed by drawing a grid, putting the gametes produced by one parent along the

upper edge and the gametes produced by the other parent down the left side (◀ FIGURE 3.6b). Each cell (a block within the Punnett square) contains an allele from each of the corresponding gametes, generating the genotype of the progeny produced by fusion of those gametes. In the upper left-hand cell of the Punnett square in Figure 3.6b, a gamete containing *T* from the tall plant unites with a gamete containing *t* from the short plant, giving the genotype of the progeny (*Tt*). It is useful to write the phenotype expressed by each genotype; here the progeny will be tall, because the tall allele is dominant over the short allele.

(a)

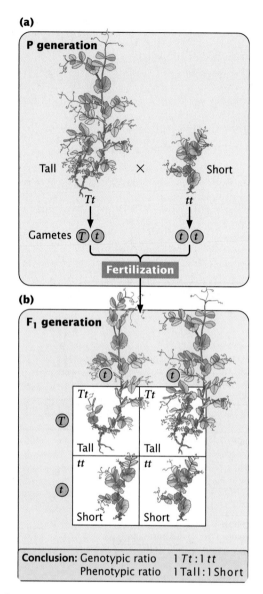

P generation

Tall   ×   Short

Tt   *tt*

Gametes T t   t t

Fertilization

(b)

F₁ generation

	T	t
T	*Tt* Tall	*Tt* Tall
t	*tt* Short	*tt* Short

Conclusion: Genotypic ratio   1 *Tt* : 1 *tt*
Phenotypic ratio   1 Tall : 1 Short

◀ **3.6 The Punnett square can be used to determine the results of a genetic cross.**

This process is repeated for all the cells in the Punnett square.

By simply counting, we can determine the types of progeny produced and their ratios. In Figure 3.6b, two cells contain tall (*Tt*) progeny and two cells contain short (*tt*) progeny; so the genotypic ratio expected for this cross is 2 *Tt* to 2 *tt* (a 1:1 ratio). Another way to express this result is to say that we expect ½ of the progeny to have genotype *Tt* (and phenotype tall) and ½ of the progeny to have genotype *tt* (and phenotype short). In this cross, the genotypic ratio and the phenotypic ratio are the same, but this outcome need not be the case. Try completing a Punnett square for the cross in which the F₁ round-seeded plants in Figure 3.4 undergo self-fertilization (you should obtain a phenotypic ratio of 3 round to 1 wrinkled and a genotypic ratio of 1 *RR* to 2 *Rr* to 1 *rr*).

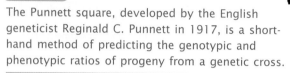

Concepts

The Punnett square, developed by the English geneticist Reginald C. Punnett in 1917, is a shorthand method of predicting the genotypic and phenotypic ratios of progeny from a genetic cross.

Probability as a tool in genetics Another method for determining the outcome of a genetic cross is to use the rules of probability, as Mendel did with his crosses. **Probability** expresses the likelihood of a particular event occurring. It is the number of times that a particular event occurs, divided by the number of all possible outcomes. For example, a deck of 52 cards contains only one king of hearts. The probability of drawing one card from the deck at random and obtaining the king of hearts is $\frac{1}{52}$, because there is only one card that is the king of hearts (one event) and there are 52 cards that can be drawn from the deck (52 possible outcomes). The probability of drawing a card and obtaining an ace is $\frac{4}{52}$, because there are four cards that are aces (four events) and 52 cards (possible outcomes). Probability can be expressed either as a fraction ($\frac{1}{52}$ in this case) or as a decimal number (0.019).

The probability of a particular event may be determined by knowing something about *how* the event occurs or *how often* it occurs. We know, for example, that the probability of rolling a six-sided die and getting a four is $\frac{1}{6}$, because the die has six sides and any one side is equally likely to end up on top. So, in this case, understanding the nature of the event—the shape of the thrown die—allows us to determine the probability. In other cases, we determine the probability of an event by making a large number of observations. When a weather forecaster says that there is a 40% chance of rain on a particular day, this probability was obtained by observing a large number of days with similar atmospheric conditions and finding that it rains on 40% of those days. In this case, the probability has been determined empirically (by observation).

The multiplication rule Two rules of probability are useful for predicting the ratios of offspring produced in genetic crosses. The first is the **multiplication rule,** which states that the probability of two or more independent events occurring together is calculated by multiplying their independent probabilities.

To illustrate the use of the multiplication rule, let's again consider the roll of dice. The probability of rolling one die and obtaining a four is $\frac{1}{6}$. To calculate the probability of rolling a die twice and obtaining 2 fours, we can apply the multiplication rule. The probability of obtaining a four on the first roll is $\frac{1}{6}$ and the probability of obtaining a four on the second roll is $\frac{1}{6}$; so the probability of rolling a four on both is $\frac{1}{6} \times \frac{1}{6} = \frac{1}{36}$ (◀ FIGURE 3.7a). The key indicator for applying the multiplication rule is the word *and;*

(a) The multiplication rule

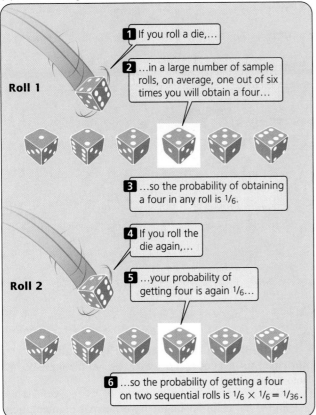

1 If you roll a die,...

Roll 1

2 ...in a large number of sample rolls, on average, one out of six times you will obtain a four...

3 ...so the probability of obtaining a four in any roll is $\frac{1}{6}$.

4 If you roll the die again,...

Roll 2

5 ...your probability of getting four is again $\frac{1}{6}$...

6 ...so the probability of getting a four on two sequential rolls is $\frac{1}{6} \times \frac{1}{6} = \frac{1}{36}$.

(b) The addition rule

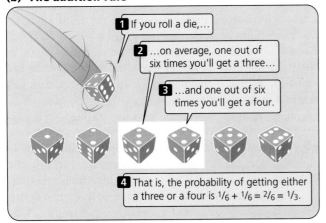

1 If you roll a die,...

2 ...on average, one out of six times you'll get a three...

3 ...and one out of six times you'll get a four.

4 That is, the probability of getting either a three or a four is $\frac{1}{6} + \frac{1}{6} = \frac{2}{6} = \frac{1}{3}$.

◀ **3.7 The multiplication and addition rules can be used to determine the probability of combinations of events.**

in the example just considered, we wanted to know the probability of obtaining a four on the first roll *and* a four on the second roll.

For the multiplication rule to be valid, the events whose joint probability is being calculated must be independent—the outcome of one event must not influence the outcome of the other. For example, the number that comes up on one roll of the die has no influence on the number that

comes up on the other roll, so these events are independent. However, if we wanted to know the probability of being hit on the head with a hammer and going to the hospital on the same day, we could not simply multiply the probability of being hit on the head with a hammer by the probability of going to the hospital. The multiplication rule cannot be applied here, because the two events are not independent—being hit on the head with a hammer certainly influences the probability of going to the hospital.

The addition rule The second rule of probability frequently used in genetics is the **addition rule,** which states that the probability of any one of two or more mutually exclusive events is calculated by adding the probabilities of these events. Let's look at this rule in concrete terms. To obtain the probability of throwing a die once and rolling *either* a three *or* a four, we would use the addition rule, adding the probability of obtaining a three ($\frac{1}{6}$) to the probability of obtaining a four (again, $\frac{1}{6}$), or $\frac{1}{6} + \frac{1}{6} = \frac{2}{6} = \frac{1}{3}$ (◀ FIGURE 3.7b). The key indicator for applying the addition rule are the words *either* and *or*.

For the addition rule to be valid, the events whose probability is being calculated must be mutually exclusive, meaning that one event excludes the possibility of the other occurring. For example, you cannot throw a single die just once and obtain both a three and a four, because only one side of the die can be on top. These events are mutually exclusive.

Concepts

The multiplication rule states that the probability of two or more independent events occurring together is calculated by multiplying their independent probabilities. The addition rule states that the probability that any one of two or more mutually exclusive events occurring is calculated by adding their probabilities.

The application of probability to genetic crosses The multiplication and addition rules of probability can be used in place of the Punnett square to predict the ratios of progeny expected from a genetic cross. Let's first consider a cross between two pea plants heterozygous for the locus that determines height, $Tt \times Tt$. Half of the gametes produced by each plant have a T allele, and the other half have a t allele; so the probability for each type of gamete is $\frac{1}{2}$.

The gametes from the two parents can combine in four different ways to produce offspring. Using the multiplication rule, we can determine the probability of each possible type. To calculate the probability of obtaining TT progeny, for example, we multiply the probability of receiving a T allele from the first parent ($\frac{1}{2}$) times the probability of receiving a T allele from the second parent ($^1/_2$). The multiplication rule should be used here because we need the probability of receiving a T allele from the first parent *and* a T allele from the second parent—two independent events. The four types of progeny from this cross and their associated probabilities are:

TT	(T gamete and T gamete)	$^1/_2 \times \, ^1/_2 = \, ^1/_4$	tall
Tt	(T gamete and t gamete)	$^1/_2 \times \, ^1/_2 = \, ^1/_4$	tall
tT	(t gamete and T gamete)	$^1/_2 \times \, ^1/_2 = \, ^1/_4$	tall
tt	(t gamete and t gamete)	$^1/_2 \times \, ^1/_2 = \, ^1/_4$	short

Notice that there are two ways for heterozygous progeny to be produced: a heterozygote can either receive a T allele from the first parent and a t allele from the second or receive a t allele from the first parent and a T allele from the second.

After determining the probabilities of obtaining each type of progeny, we can use the addition rule to determine the overall phenotypic ratios. Because of dominance, a tall plant can have genotype TT, Tt, or tT; so, using the addition rule, we find the probability of tall progeny to be $^1/_4 + \, ^1/_4 + \, ^1/_4 = \, ^3/_4$. Because only one genotype codes for short (tt), the probability of short progeny is simply $^1/_4$.

Two methods have now been introduced to solve genetic crosses: the Punnett square and the probability method. At this point, you may be saying "Why bother with probability rules and calculations? The Punnett square is easier to understand and just as quick." For simple monohybrid crosses, the Punnett square is simpler and just as easy to use. However, when tackling more complex crosses concerning genes at two or more loci, the probability method is both clearer and quicker than the Punnett square.

The binomial expansion and probability When probability is used, it is important to recognize that there may be several different ways in which a set of events can occur. Consider two parents who are both heterozygous for albinism, a recessive condition in humans that causes reduced pigmentation in the skin, hair, and eyes (◀FIGURE 3.8). When two parents heterozygous for albinism mate ($Aa \times Aa$), the probability of their having a child with albinism (aa) is $^1/_4$ and the probability of having a child with normal pigmentation (AA or Aa) is $^3/_4$. Suppose we want to know the probability of this couple having three children, all with albinism. In this case, there is only one way in which they can have three children with albinism—their first child has albinism *and* their second child has albinism *and* their third child has albinism. Here we simply apply the multiplication rule: $^1/_4 \times \, ^1/_4 \times \, ^1/_4 = \, ^1/_{64}$.

Suppose we now ask, What is the probability of this couple having three children, one with albinism and two with normal pigmentation. This situation is more complicated. The first child might have albinism, whereas the second and third are unaffected; the probability of this sequence of events is $^1/_4 \times \, ^3/_4 \times \, ^3/_4 = \, ^9/_{64}$. Alternatively, the

◀ **3.8 Albinism in human beings is usually inherited as a recessive trait.** (Richard Dranitzke/SS/Photo Researchers.)`

first and third children might have normal pigmentation, whereas the second has albinism; the probability of this sequence is $^3/_4 \times \, ^1/_4 \times \, ^3/_4 = \, ^9/_{64}$. Finally, the first two children might have normal pigmentation and the third albinism; the probability of this sequence is $^3/_4 \times \, ^3/_4 \times \, ^1/_4 = \, ^9/_{64}$. Because *either* the first sequence *or* the second sequence *or* the third sequence produces one child with albinism and two with normal pigmentation, we apply the addition rule and add the probabilities: $^9/_{64} + \, ^9/_{64} + \, ^9/_{64} = \, ^{27}/_{64}$.

If we want to know the probability of this couple having five children, two with albinism and three with normal pigmentation, figuring out the different combinations of children and their probabilities becomes more difficult. This task is made easier if we apply the binomial expansion.

The binomial takes the form $(a + b)^n$, where a equals the probability of one event, b equals the probability of the alternative event, and n equals the number of times the event occurs. For figuring the probability of two out of five children with albinism:

a = the probability of a child having albinism = $^1/_4$

b = the probability of a child having normal
pigmentation = $^3/_4$

The binomial for this situation is $(a + b)^5$ because there are five children in the family ($n = 5$). The expansion is:

$$(a + b)^5 = a^5 + 5a^4b + 10a^3b^2 + 10a^2b^3 + 5ab^4 + b^5$$

The first term in the expansion (a^5) equals the probability of having five children all with albinism, because a is the probability of albinism. The second term ($5a^4b$) equals the probability of having four children with albinism and one with normal pigmentation, the third term ($10a^3b^2$) equals the probability of having three children with albinism and two with normal pigmentation, and so forth.

To obtain the probability of any combination of events, we insert the values of a and b; so the probability of having two out of five children with albinism is:

$$10a^2b^3 = 10(^1/_4)^2(^3/_4)^3 = {}^{270}/_{1024} = .26$$

We could easily figure out the probability of any desired combination of albinism and pigmentation among five children by using the other terms in the expansion.

How did we expand the binomial in this example? In general, the expansion of any binomial $(a + b)^n$ consists of a series of $n + 1$ terms. In the preceding example, $n = 5$; so there are $5 + 1 = 6$ terms: a^5, $5a^4b$, $10a^3b^2$, $10a^2b^3$, $5ab^4$, and b^5. To write out the terms, first figure out their exponents. The exponent of a in the first term always begins with the power to which the binomial is raised, or n. In our example, n equals 5, so our first term is a^5. The exponent of a decreases by one in each successive term; so the exponent of a is 4 in the second term (a^4), 3 in the third term (a^3), and so forth. The exponent of b is 0 (no b) in the first term and increases by 1 in each successive term, increasing from 0 to 5 in our example.

Next, determine the coefficient of each term. The coefficient of the first term is always 1; so in our example the first term is $1a^5$, or just a^5. The coefficient of the second term is always the same as the power to which the binomial is raised; in our example this coefficient is 5 and the term is $5a^4b$. For the coefficient of the third term, look back at the preceding term; multiply the coefficient of the preceding term (5 in our example) by the exponent of a in that term (4) and then divide by the number of that term (second term, or 2). So the coefficient of the third term in our example is $(5 \times 4)/2 = {}^{20}/_2 = 10$ and the term is $10a^3b^2$. Follow this same procedure for each successive term.

Another way to determine the probability of any particular combination of events is to use the following formula:

$$P = \frac{n!}{s!t!}a^sb^t$$

where P equals the overall probability of event X with probability a occurring s times and event Y with probability b occurring t times. For our albinism example, event X would be the occurrence of a child with albinism and event Y the occurrence of a child with normal pigmentation; s would equal the number of children with albinism (2) and t, the number of children with normal pigmentation (3). The ! symbol is termed factorial, and it means the product of all the integers from n to 1. In this example,

$n = 5$; so $n! = 5 \times 4 \times 3 \times 2 \times 1$. Applying this formula to obtain the probability of two out of five children having albinism, we obtain:

$$P = \frac{5!}{2!3!}(^1/_4)^2(^3/_4)^3$$

$$P = \frac{5 \times 4 \times 3 \times 2 \times 1}{2 \times 1 \times 3 \times 2 \times 1}(^1/_4)^2(^3/_4)^3 = .26$$

This value is the same as that obtained with the binomial expansion.

The Testcross

A useful tool for analyzing genetic crosses is the **testcross,** in which one individual of unknown genotype is crossed with another individual with a homozygous recessive genotype for the trait in question. Figure 3.6 illustrates a testcross (as well as a backcross). A testcross tests, or reveals, the genotype of the first individual.

Suppose you were given a tall pea plant with no information about its parents. Because tallness is a dominant trait in peas, your plant could be either homozygous (TT) or heterozygous (Tt), but you would not know which. You could determine its genotype by performing a testcross. If the plant were homozygous (TT), a testcross would produce all tall progeny ($TT \times tt \rightarrow$ all Tt); if the plant were heterozygous (Tt), the testcross would produce half tall progeny and half short progeny ($Tt \times tt \rightarrow {}^1/_2\ Tt$ and $^1/_2\ tt$). When a testcross is performed, any recessive allele in the unknown genotype is expressed in the progeny, because it will be paired with a recessive allele from the homozygous recessive parent.

Concepts

The bionomial expansion may be used to determine the probability of a particular set of of events. A testcross is a cross between an individual with an unknown genotype and one with a homozygous recessive genotype. The outcome of the testcross can reveal the unknown genotype.

Incomplete Dominance

The seven characters in pea plants that Mendel chose to study extensively all exhibited dominance, but Mendel did realize that not all characters have traits that exhibit dominance. He conducted some crosses concerning the length of time that pea plants take to flower. When he crossed two homozygous varieties that differed in their flowering time by an average of 20 days, the length of time taken by the F_1 plants to flower was intermediate between those of the two parents. When the heterozygote has a phenotype intermediate between the phenotypes of the two homozygotes, the trait is said to display **incomplete dominance.**

Incomplete dominance is also exhibited in the fruit color of eggplants. When a homozygous plant that produces purple fruit (*PP*) is crossed with a homozygous plant that produces white fruit (*pp*), all the heterozygous F₁ (*Pp*) produce violet fruit (◀FIGURE 3.9a). When the F₁ are crossed with each other, ¼ of the F₂ are purple (*PP*), ½ are violet (*Pp*), and ¼ are white (*pp*), as shown in ◀FIGURE 3.9b. This 1:2:1 ratio is different from the 3:1 ratio that we would observe if eggplant fruit color exhibited dominance. When a

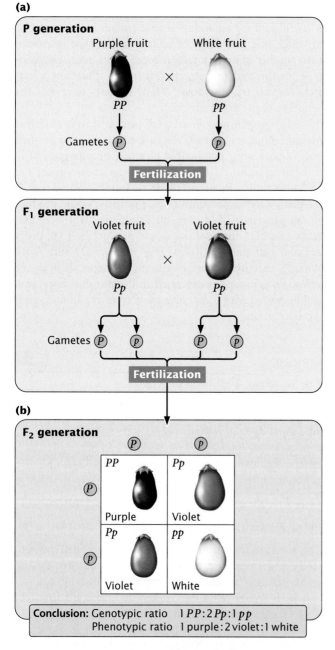

(a)

P generation

Purple fruit   White fruit

×

PP   *pp*

Gametes ⓅP   Ⓟp

Fertilization

F₁ generation

Violet fruit   Violet fruit

×

Pp   *Pp*

Gametes ⓅP Ⓟp   ⓅP Ⓟp

Fertilization

(b)

F₂ generation

          ⓅP   Ⓟp

ⓅP   *PP*   *Pp*
      Purple   Violet

Ⓟp   *Pp*   *pp*
      Violet   White

Conclusion: Genotypic ratio   1 *PP* : 2 *Pp* : 1 *pp*
            Phenotypic ratio   1 purple : 2 violet : 1 white

◀3.9 **Fruit color in eggplant is inherited as an incompletely dominant trait.**

◀3.10 **Leopard spotting in horses exhibits incomplete dominance.** (Frank Oberle/Bruce Coleman.)

trait displays incomplete dominance, the genotypic ratios and phenotypic ratios of the offspring are the *same,* because each genotype has its own phenotype. It is impossible to obtain eggplants that are pure breeding for violet fruit, because all plants with violet fruit are heterozygous.

Another example of incomplete dominance is feather color in chickens. A cross between a homozygous black chicken and a homozygous white chicken produces F₁ chickens that are gray. If these gray F₁ are intercrossed, they produce F₂ birds in a ratio of 1 black : 2 gray : 1 white. Leopard white spotting in horses is incompletely dominant over unspotted horses: *LL* horses are white with numerous dark spots, heterozygous *Ll* horses have fewer spots, and *ll* horses have no spots (◀FIGURE 3.10). The concept of dominance and some of its variations are discussed further in Chapter 5.

Concepts

Incomplete dominance is exhibited when the heterozygote has a phenotype intermediate between the phenotypes of the two homozygotes. When a trait exhibits incomplete dominance, a cross between two heterozygotes produces a 1:2:1 phenotypic ratio in the progeny.

Genetic Symbols

As we have seen, genetic crosses are usually depicted with the use of symbols to designate the different alleles. Lowercase letters are traditionally used to designate recessive alleles, and uppercase letters are for dominant alleles. Two or three letters may be used for a single allele: the recessive allele for heart-shaped leaves in cucumbers is designated *hl*, and the recessive allele for abnormal sperm head shape in mice is designated *azh*.

The common allele for a character—called the **wild type** because it is the allele most often found in the wild—

Table 3.2	Phenotypic ratios for simple genetic crosses (crosses for a single locus)		
Ratio	**Genotypes of Parents**	**Genotypes of Progeny**	**Type of Dominance**
3:1	$Aa \times Aa$	$\frac{3}{4}$ $A\_\_$: $\frac{1}{4}$ aa	Dominance
1:2:1	$Aa \times Aa$	$\frac{1}{4}$ AA: $\frac{1}{2}$ Aa: $\frac{1}{4}$ aa	Incomplete dominance
1:1	$Aa \times aa$	$\frac{1}{2}$ Aa: $\frac{1}{2}$ aa	Dominance or incomplete dominance
	$Aa \times AA$	$\frac{1}{2}$ Aa: $\frac{1}{2}$ AA	Incomplete dominance
Uniform progeny	$AA \times AA$	All AA	Dominance or incomplete dominance
	$aa \times aa$	All aa	Dominance or incomplete dominance
	$AA \times aa$	All Aa	Dominance or incomplete dominance
	$AA \times Aa$	All $A\_\_$	Dominance

Note: A line in a genotype, such as $A\_\_$, indicates that any allele is possible.

is often symbolized by one or more letters and a plus sign (+). The letter(s) chosen are usually based on the mutant (unusual) phenotype. The first letter is lowercase if the mutant phenotype is recessive, uppercase if the mutant phenotype is dominant. For example, the recessive allele for yellow eyes in the Oriental fruit fly is represented by ye, whereas the allele for wild-type eye color is represented by ye^+. At times, the letters for the wild-type allele are dropped and the allele is represented simply by a plus sign. Superscripts and subscripts are sometimes added to distinguish between genes: Lfr_1 and Lfr_2 represent dominant alleles at different loci that produce lacerate leaf margins in opium poppies; El^R represents an allele in goats that restricts the length of the ears.

A slash may be used to distinguish alleles present in an individual genotype. The genotype of a goat that is heterozygous for restricted ears might be written El^+/El^R or simply $+/El^R$. If genotypes at more than one locus are presented together, a space may separate them. A goat heterozygous for a pair of alleles that produce restricted ears and heterozygous for another pair of alleles that produce goiter can be designated by El^+/El^R G/g.

Connecting Concepts

Ratios in Simple Crosses

Now that we have had some experience with genetic crosses, let's review the ratios that appear in the progeny of simple crosses, in which a single locus is under consideration. Understanding these ratios and the parental genotypes that produce them will allow you to work simple genetic crosses quickly, without resorting to the Punnett square. Later, we will use these ratios to work more complicated crosses entailing several loci.

There are only four phenotypic ratios to understand (Table 3.2). The 3:1 ratio arises in a simple genetic cross when both of the parents are heterozygous for a dominant trait ($Aa \times Aa$). The second phenotypic ratio is the 1:2:1 ratio, which arises in the progeny of crosses between two parents heterozygous for a character that exhibits incom-

plete dominance ($Aa \times Aa$). The third phenotypic ratio is the 1:1 ratio, which results from the mating of a homozygous parent and a heterozygous parent. If the character exhibits dominance, the homozygous parent in this cross must carry two recessive alleles ($Aa \times aa$) to obtain a 1:1 ratio, because a cross between a homozygous dominant parent and a heterozygous parent ($AA \times Aa$) produces only offspring displaying the dominant trait. For a character with incomplete dominance, a 1:1 ratio results from a cross between the heterozygote and either homozygote ($Aa \times aa$ or $Aa \times AA$).

The fourth phenotypic ratio is not really a ratio—all the offspring have the same phenotype. Several combinations of parents can produce this outcome (Table 3.2). A cross between any two homozygous parents—either between two of the same homozygotes ($AA \times AA$ and $aa \times aa$) or between two different homozygotes ($AA \times aa$)—produces progeny all having the same phenotype. Progeny of a single phenotype can also result from a cross between a homozygous dominant parent and a heterozygote ($AA \times Aa$).

If we are interested in the ratios of genotypes instead of phenotypes, there are only three outcomes to remember (Table 3.3): the 1:2:1 ratio, produced by a cross between

Table 3.3	Genotypic ratios for simple genetic crosses (crosses for a single locus)	
Ratio	**Genotypes of Parents**	**Genotypes of Progeny**
1:2:1	$Aa \times Aa$	$\frac{1}{4}$ AA: $\frac{1}{2}$ Aa: $\frac{1}{4}$ aa
1:1	$Aa \times aa$	$\frac{1}{2}$ Aa: $\frac{1}{2}$ aa
	$Aa \times AA$	$\frac{1}{2}$ Aa: $\frac{1}{2}$ AA
Uniform progeny	$AA \times AA$	All AA
	$aa \times aa$	All aa
	$AA \times aa$	All Aa

two heterozygotes; the 1:1 ratio, produced by a cross between a heterozygote and a homozygote; and the uniform progeny produced by a cross between two homozygotes. These simple phenotypic and genotypic ratios and the parental genotypes that produce them provide the key to understanding crosses for a single locus and, as you will see in the next section, for multiple loci.

Multiple-Loci Crosses

We will now extend Mendel's principle of segregation to more complex crosses for alleles at multiple loci. Understanding the nature of these crosses will require an additional principle, the principle of independent assortment.

Dihybrid Crosses

In addition to his work on monohybrid crosses, Mendel also crossed varieties of peas that differed in *two* characteristics (**dihybrid crosses**). For example, he had one homozygous variety of pea that produced round seeds and yellow endosperm; another homozygous variety produced wrinkled seeds and green endosperm. When he crossed the two, all the F₁ progeny had round seeds and yellow endosperm. He then self-fertilized the F₁ and obtained the following progeny in the F₂: 315 round, yellow seeds; 101 wrinkled, yellow seeds; 108 round, green seeds; and 32 wrinkled, green seeds. Mendel recognized that these traits appeared approximately in a 9:3:3:1 ratio; that is, $9/16$ of the progeny were round and yellow, $3/16$ were wrinkled and yellow, $3/16$ were round and green, and $1/16$ were wrinkled and green.

The Principle of Independent Assortment

Mendel carried out a number of dihybrid crosses for pairs of characteristics and always obtained a 9:3:3:1 ratio in the F₂. This ratio makes perfect sense in regard to segregation and dominance if we add a third principle, which Mendel recognized in his dihybrid crosses: the **principle of independent assortment** (Mendel's second law). This principle states that alleles at different loci separate independently of one another.

A common mistake is to think that the principle of segregation and the principle of independent assortment refer to two different processes. The principle of independent assortment is really an extension of the principle of segregation. The principle of segregation states that the two alleles of a locus separate when gametes are formed; the principle of independent assortment states that, when these two alleles separate, their separation is independent of the separation of alleles at *other* loci.

Let's see how the principle of independent assortment explains the results that Mendel obtained in his dihybrid cross. Each plant possesses two alleles coding for each characteristic, so the parental plants must have had genotypes *RRYY* and *rryy* (◀ FIGURE 3.11a). The principle of segrega-

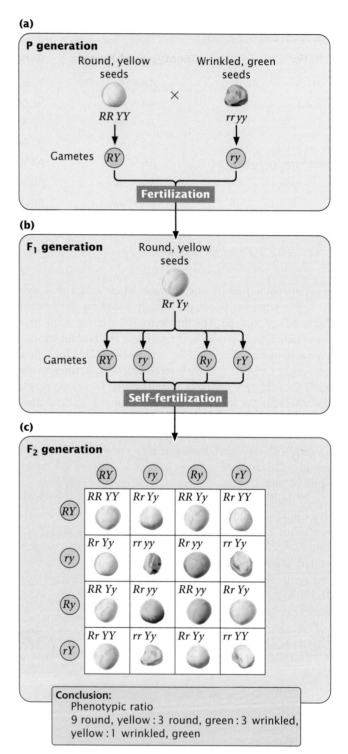

(a)

P generation

Round, yellow seeds × Wrinkled, green seeds

RR YY   *rr yy*

Gametes   *RY*   *ry*

Fertilization

(b)

F₁ generation   Round, yellow seeds

Rr Yy

Gametes   *RY*   *ry*   *Ry*   *rY*

Self-fertilization

(c)

F₂ generation

	RY	*ry*	*Ry*	*rY*
RY	*RR YY*	*Rr Yy*	*RR Yy*	*Rr YY*
ry	*Rr Yy*	*rr yy*	*Rr yy*	*rr Yy*
Ry	*RR Yy*	*Rr yy*	*RR yy*	*Rr Yy*
rY	*Rr YY*	*rr Yy*	*Rr Yy*	*rr YY*

Conclusion:
Phenotypic ratio
9 round, yellow : 3 round, green : 3 wrinkled, yellow : 1 wrinkled, green

◀ 3.11 **Mendel's dihybrid crosses revealed the principle of independent assortment.**

tion indicates that the alleles for each locus separate, and one allele for each locus passes to each gamete. The gametes produced by the round, yellow parent therefore contain alleles *RY*, whereas the gametes produced by the wrinkled, green parent contain alleles *ry*. These two types of gametes unite to produce the F₁, all with genotype *RrYy*. Because round is dominant over wrinkled and yellow is dominant over green, the phenotype of the F_1 will be round and yellow.

When Mendel self-fertilized the F_1 plants to produce the F_2, the alleles for each locus separated, with one allele going into each gamete. This is where the principle of independent assortment becomes important. Each pair of alleles can separate in two ways: (1) R separates with Y and r separates with y to produce gametes RY and ry or (2) R separates with y and r separates with Y to produce gametes Ry and rY. The principle of independent assortment tells us that the alleles at each locus separate independently; thus, both kinds of separation occur equally and all four type of gametes (RY, ry, Ry, and rY) are produced in equal proportions (◀FIGURE 3.11b). When these four types of gametes are combined to produce the F_2 generation, the progeny consist of $9/16$ round and yellow, $3/16$ wrinkled and yellow, $3/16$ round and green, and $1/16$ wrinkled and green, resulting in a 9:3:3:1 phenotypic ratio (◀FIGURE 3.11c).

The Relation of the Principle of Independent Assortment to Meiosis

An important qualification of the principle of independent assortment is that it applies to characters encoded by loci located on different chromosomes because, like the principle of segregation, it is based wholly on the behavior of chromosomes during meiosis. Each pair of homologous chromosomes separates independently of all other pairs in anaphase I of meiosis (see Figure 2.18); so genes located on different pairs of homologs will assort independently. Genes that happen to be located on the same chromosome will travel together during anaphase I of meiosis and will arrive at the same destination—within the same gamete (unless crossing over takes place). Genes located on the same chromosome therefore do not assort independently (unless they are located sufficiently far apart that crossing over takes place every meiotic division, as will be discussed fully in Chapter 7).

Concepts

The principle of independent assortment states that genes coding for different characteristics separate independently of one another when gametes are formed, owing to independent separation of homologous pairs of chromosomes during meiosis. Genes located close together on the same chromosome do not, however, assort independently.

Applying Probability and the Branch Diagram to Dihybrid Crosses

When the genes at two loci separate independently, a dihybrid cross can be understood as two monohybrid crosses. Let's examine Mendel's dihybrid cross ($RrYy \times RrYy$) by considering each characteristic separately (◀FIGURE 3.12a). If we consider only the shape of the seeds, the cross was

$Rr \times Rr$, which yields a 3:1 phenotypic ratio ($3/4$ round and $1/4$ wrinkled progeny, see Table 3.2). Next consider the other characteristic, the color of the endosperm. The cross was $Yy \times Yy$, which produces a 3:1 phenotypic ratio ($3/4$ yellow and $1/4$ green progeny).

We can now combine these monohybrid ratios by using the multiplication rule to obtain the proportion of progeny with different combinations of seed shape and color. The proportion of progeny with round and yellow seeds is $3/4$ (the probability of round) $\times$ $3/4$ (the probability of yellow) $= 9/16$. The proportion of progeny with round and green seeds is $3/4 \times 1/4 = 3/16$; the proportion of progeny with wrinkled and yellow seeds is $1/4 \times 3/4 = 3/16$; and the

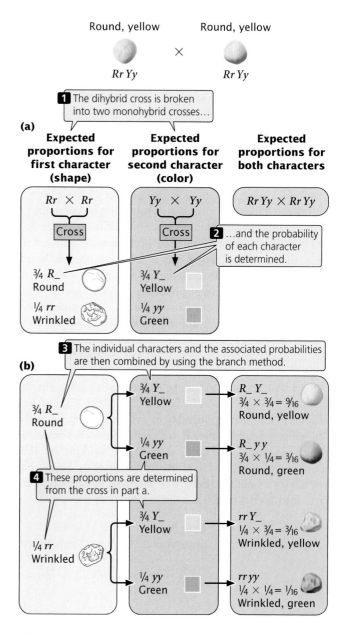

◀ 3.12 **A branch diagram can be used to determine the phenotypes and expected proportions of offspring from a dihybrid cross ($RrYy \times RrYy$).**

proportion of progeny with wrinkled and green seeds is $\frac{1}{4} \times \frac{1}{4} = \frac{1}{16}$.

Branch diagrams are a convenient way of organizing all the combinations of characteristics (◀FIGURE 3.12b). In the first column, list the proportions of the phenotypes for one character (here, $\frac{3}{4}$ round and $\frac{1}{4}$ wrinkled). In the second column, list the proportions of the phenotypes for the second character ($\frac{3}{4}$ yellow and $\frac{1}{4}$ green) next to each of the phenotypes in the first column: put $\frac{3}{4}$ yellow and $\frac{1}{4}$ green next to the round phenotype and again next to the wrinkled phenotype. Draw lines between the phenotypes in the first column and each of the phenotypes in the second column. Now follow each branch of the diagram, multiplying the probabilities for each trait along that branch. One branch leads from round to yellow, yielding round and yellow progeny. Another branch leads from round to green, yielding round and green progeny, and so on. The probability of progeny with a particular combination of traits is calculated by using the multiplicative rule: the probability of round ($\frac{3}{4}$) and yellow ($\frac{3}{4}$) seeds is $\frac{3}{4} \times \frac{3}{4} = \frac{9}{16}$. The advantage of the branch diagram is that it helps keep track of all the potential combinations of traits that may appear in the progeny. It can be used to determine phenotypic or genotypic ratios for any number of characteristics.

Using probability is much faster than using the Punnett square for crosses that include multiple loci. Genotypic and phenotypic ratios can quickly be worked out by combining, with the multiplication rule, the simple ratios in Tables 3.2 and 3.3. The probability method is particularly efficient if we need the probability of only a *particular* phenotype or genotype among the progeny of a cross. Suppose we needed to know the probability of obtaining the genotype *Rryy* in the F_2 of the dihybrid cross in Figure 3.11. The probability of obtaining the *Rr* genotype in a cross of *Rr* × *Rr* is $\frac{1}{2}$ and that of obtaining *yy* progeny in a cross of *Yy* × *Yy* is $\frac{1}{4}$ (see Table 3.3). Using the multiplication rule, we find the probability of *Rryy* to be $\frac{1}{2} \times \frac{1}{4} = \frac{1}{8}$.

To illustrate the advantage of the probability method, consider the cross *AaBbccDdEe* × *AaBbCcddEe*. Suppose we wanted to know the probability of obtaining offspring with the genotype *aabbccddee*. If we used a Punnett square to determine this probability, we might be working on the solution for months. However, we can quickly figure the probability of obtaining this one genotype by breaking this cross into a series of single-locus crosses:

Cross	Progeny genotype	Probability
Aa × *Aa*	*aa*	$\frac{1}{4}$
Bb × *Bb*	*bb*	$\frac{1}{4}$
cc × *Cc*	*cc*	$\frac{1}{2}$
Dd × *dd*	*dd*	$\frac{1}{2}$
Ee × *Ee*	*ee*	$\frac{1}{4}$

The probability of an offspring from this cross having genotype *aabbccddee* is now easily obtained by using the multiplication

rule: $\frac{1}{4} \times \frac{1}{4} \times \frac{1}{2} \times \frac{1}{2} \times \frac{1}{4} = \frac{1}{256}$. This calculation assumes that genes at these five loci all assort independently.

Concepts

A cross including several characteristics can be worked by breaking the cross down into single-locus crosses and using the multiplication rule to determine the proportions of combinations of characteristics (provided the genes assort independently).

The Dihybrid Testcross

Let's practice using the branch diagram by determining the types and proportions of phenotypes in a dihybrid testcross between the round and yellow F_1 plants (*Rr Yy*) that Mendel obtained in his dihybrid cross and the wrinkled and green plants (*rryy*) (◀FIGURE 3.13). Break the cross down into a series of single-locus crosses. The cross *Rr* × *rr* yields $\frac{1}{2}$ round (*Rr*) progeny and $\frac{1}{2}$ wrinkled (*rr*) progeny. The cross *Yy* × *yy* yields $\frac{1}{2}$ yellow (*Yy*) progeny and $\frac{1}{2}$ green (*yy*)

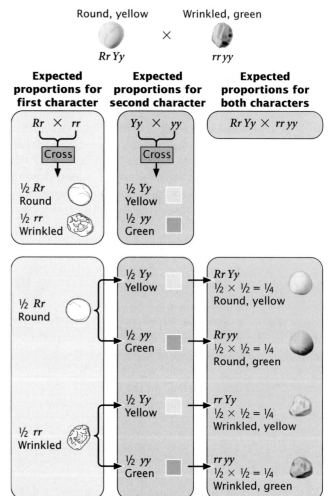

◀3.13 **A branch diagram can be used to determine the phenotypes and expected proportions of offspring from a dihybrid testcross (*RrYy* × *rryy*).**

progeny. Using the multiplication rule, we find the proportion of round and yellow progeny to be $\frac{1}{2}$ (the probability of round) $\times$ $\frac{1}{2}$ (the probability of yellow) $=\frac{1}{4}$. Four combinations of traits with the following proportions appear in the offspring: $\frac{1}{4}$ *RrYy*, round yellow; $\frac{1}{4}$ *Rryy*, round green; $\frac{1}{4}$ *rrYy*, wrinkled yellow; and $\frac{1}{4}$ *rryy*, wrinkled green.

Trihybrid Crosses

The branch diagram can also be applied to crosses including three characters (called **trihybrid crosses**). In one trihybrid cross, Mendel crossed a pure-breeding variety that

possessed round seeds, yellow endosperm, and gray seed coats with another pure-breeding variety that possessed wrinkled seeds, green endosperm, and white seed coats (◀FIGURE 3.14). The branch diagram shows that the expected phenotypic ratio in the F_2 is 27:9:9:9:3:3:3:1, and the numbers that Mendel obtained from this cross closely fit these expected ones.

In monohybrid crosses, we have seen that three genotypes (*RR*, *Rr*, and *rr*) are produced in the F_2. In dihybrid crosses, nine genotypes (3 genotypes for the first locus $\times$ 3 genotypes for the second locus = 9) are produced in the F_2:

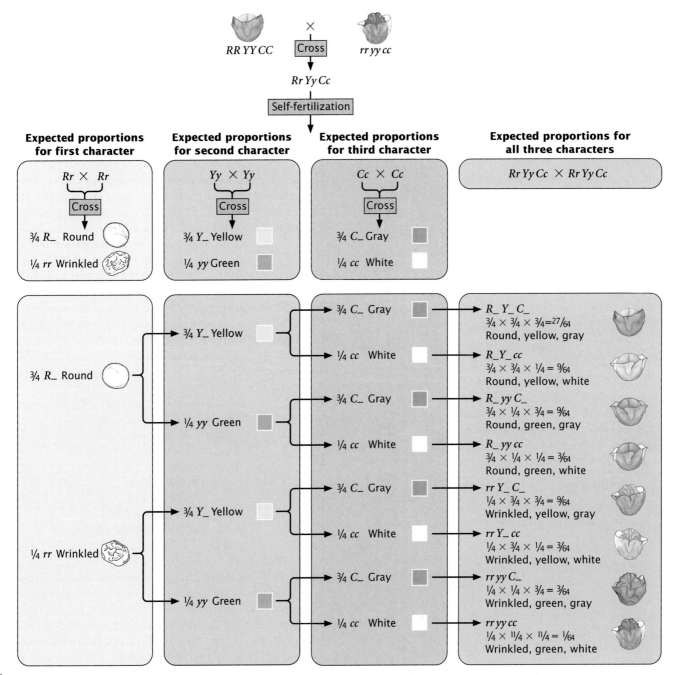

◀3.14 **A branch diagram can be used to determine the phenotypes and expected proportions of offspring from a trihybrid cross (*RrYyCc* × *RrYyCc*).**

RRYY, RRYy, RRyy, RrYY, RrYy, Rryy, rrYY, rrYy, and *rryy.* There are three possible genotypes at each locus (when there are two alternative alleles); so the number of *genotypes* produced in the F_2 of a cross between individuals heterozygous for *n* loci will be 3^n. If there is incomplete dominance, the number of *phenotypes* also will be 3^n because, with incomplete dominance, each genotype produces a different phenotype. If the traits exhibit dominance, the number of phenotypes will be 2^n.

Worked Problem

Not only are the principles of segregation and independent assortment important because they explain how heredity works, but they also provide the means for predicting the outcome of genetic crosses. This predictive power has made genetics a powerful tool in agriculture and other fields, and the ability to apply the principles of heredity is an important skill for all students of genetics. Practice with genetic problems is essential for mastering the basic principles of heredity—no amount of reading and memorization can substitute for the experience gained by deriving solutions to specific problems in genetics.

Students may have difficulty with genetics problems when they are unsure where to begin or how to organize the problem and plan a solution. In genetics, every problem is different, so there is no common series of steps that can be applied to all genetics problems. One must use logic and common sense to analyze a problem and arrive at a solution. Nevertheless, certain steps can facilitate the process, and solving the following problem will serve to illustrate these steps.

In mice, black coat color (*B*) is dominant over brown (*b*), and a solid pattern (*S*) is dominant over white spotted (*s*). Color and spotting are controlled by genes that assort independently. A homozygous black, spotted mouse is crossed with a homozygous brown, solid mouse. All the F_1 mice are black and solid. A testcross is then carried out by mating the F_1 mice with brown, spotted mice.

(**a**) Give the genotypes of the parents and the F_1 mice.

(**b**) Give the genotypes and phenotypes, along with their expected ratios, of the progeny expected from the testcross.

• Solution

Step 1: Determine the questions to be answered. What question or questions is the problem asking? Is it asking for genotypes, genotypic ratios, or phenotypic ratios? This problem asks you to provide the *genotypes* of the parents and the F_1, the *expected genotypes* and *phenotypes* of the progeny of the testcross, and their *expected proportions.*

Step 2: Write down the basic information given in the problem. This problem provides important information

about the dominance relations of the characters and about the mice being crossed. Black is dominant over brown and solid is dominant over white spotted. Furthermore, the genes for the two characters assort independently. In this problem, symbols are provided for the different alleles (*B* for black, *b* for brown, *S* for solid, and *s* for spotted); had these symbols not been provided, you would need to choose symbols to represent these alleles. It is useful to record these symbols at the beginning of the solution:

$$B\text{—black} \qquad S\text{—solid}$$
$$b\text{—brown} \qquad s\text{—white-spotted}$$

Next, write out the crosses given in the problem.

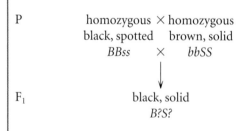

Step 3: Write down any genetic information that can be determined from the phenotypes alone. From the phenotypes and the statement that they are homozygous, you know that the P-generation mice must be *BBss* and *bbSS.* The F_1 mice are black and solid, both dominant traits, so the F_1 mice must possess at least one black allele (*B*) and one spotted allele (*S*). At this point, you cannot be certain about the other alleles, so represent the genotype of the F_1 as *B?S?*. The brown, spotted mice in the testcross must be *bbss,* because both brown and spotted are recessive traits that will be expressed only if two recessive alleles are present. Record these genotypes on the crosses that you wrote out in step 2:

P   homozygous × homozygous
              black, spotted   brown, solid
                    BBss   ×   *bbSS*

F_1 black, solid
                        B?S?

Testcross   black, solid × brown, spotted
                    B?S?   ×   *bbss*

Step 4: Break down the problem into smaller parts. First, determine the genotype of the F_1. After this genotype has been determined, you can predict the results of the testcross and determine the genotypes and phenotypes of the progeny from the testcross. Second, because this cross includes two independently assorting loci, it can be conveniently broken down into two single-locus crosses: one for coat color and another for spotting.

Third, use a branch diagram to determine the proportion of progeny of the testcross with different combinations of the two traits.

Step 5: Work the different parts of problem. Start by determining the genotype of the F$_1$ progeny. Mendel's first law indicates that the two alleles at a locus separate, one going into each gamete. Thus, the gametes produced by the black, spotted parent contain *Bs* and the gametes produced by the brown, spotted parent contain *bS*, which combine to produce F$_1$ progeny with the genotype *BbSs:*

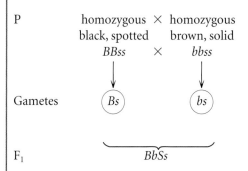

P   homozygous × homozygous
               black, spotted   brown, solid
                  BBss   ×   *bbss*

Gametes   (Bs)   (bs)

F$_1$ *BbSs*

Use the F$_1$ genotype to work the testcross (*BbSs* × *bbss*), breaking it into two single-locus crosses. First, consider the cross for coat color: *Bb* × *bb*. Any cross between a heterozygote and a homozygous recessive genotype produces a 1:1 phenotypic ratio of progeny (see Table 3.2):

BB × *bb*

$^1/_2$ *Bb* black

$^1/_2$ *bb* brown

Next do the cross for spotting: *Ss* × *ss*. This cross also is between a heterozygote and a homozygous recessive genotype and will produce $^1/_2$ solid (*Ss*) and $^1/_2$ spotted (*ss*) progeny (see Table 3.2).

Ss × *ss*

$^1/_2$ *Ss* solid

$^1/_2$ *ss* spotted

Finally, determine the proportions of progeny with combinations of these characters by using the branch diagram.

$^1/_2$ *Ss* solid ⟶ *BbSs* black, solid
$^1/_2$ × $^1/_2$ = $^1/_4$

$^1/_2$ *Bb* black

$^1/_2$ *ss* spotted ⟶ *Bbss* black, spotted
$^1/_2$ × $^1/_2$ = $^1/_4$

$^1/_2$ *Ss* solid ⟶ *bbss* brown, solid
$^1/_2$ × $^1/_2$ = $^1/_4$

$^1/_2$ *bb* brown

$^1/_2$ *ss* spotted ⟶ *bbss* brown, spotted
$^1/_2$ × $^1/_2$ = $^1/_4$

Step 6: Check all work. As a last step, reread the problem, checking to see if your answers are consistent with the information provided. You have used the genotypes *BBss* and *bbSS* in the P generation. Do these genotypes code for the phenotypes given in the problem? Are the F$_1$ progeny phenotypes consistent with the genotypes that you assigned? The answers are consistent with the information.

Observed and Expected Ratios

When two individuals of known genotype are crossed, we expect certain ratios of genotypes and phenotypes in the progeny; these expected ratios are based on the Mendelian principles of segregation, independent assortment, and dominance. The ratios of genotypes and phenotypes *actually* observed among the progeny, however, may deviate from these expectations.

For example, in German cockroaches, brown body color (*Y*) is dominant over yellow body color (*y*). If we cross a brown, heterozygous cockroach (*Yy*) with a yellow cockroach (*yy*), we expect a 1:1 ratio of brown (*Yy*) and yellow (*yy*) progeny. Among 40 progeny, we would therefore expect to see 20 brown and 20 yellow offspring. However, the observed numbers might deviate from these expected values; we might in fact see 22 brown and 18 yellow progeny.

Chance plays a critical role in genetic crosses, just as it does in flipping a coin. When you flip a coin, you expect a 1:1 ratio—$^1/_2$ heads and $^1/_2$ tails. If you flip a coin 1000 times, the proportion of heads and tails obtained would probably be very close to that expected 1:1 ratio. However, if you flip the coin 10 times, the ratio of heads to tails might be quite different from 1:1. You could easily get 6 heads and 4 tails, or 3 and 7 tails, just by chance. It is possible that you might even get 10 heads and 0 tails. The same thing happens in genetic crosses. We may expect 20 brown and 20 yellow cockroaches, but 22 brown and 18 yellow progeny *could* arise as a result of chance.

The Goodness-of-Fit Chi-Square Test

If you expected a 1:1 ratio of brown and yellow cockroaches but the cross produced 22 brown and 18 yellow, you probably wouldn't be too surprised even though it wasn't a perfect 1:1 ratio. In this case, it seems reasonable to assume that chance produced the deviation between the expected and the observed results. But, if you observed 25 brown and 15 yellow, would the ratio still be 1:1? Something other than chance might have caused the deviation. Perhaps the inheritance of this character is more complicated than was assumed or perhaps some of the yellow progeny died before they were counted. Clearly, we need some means of evaluating how likely it is that chance is responsible for the deviation between the observed and the expected numbers.

To evaluate the role of chance in producing deviations between observed and expected values, a statistical test called the **goodness-of-fit chi-square test** is used. This test provides information about how well observed values fit expected values. Before we learn how to calculate the chi square, it is important to understand what this test does and does not indicate about a genetic cross.

The chi-square test cannot tell us whether a genetic cross has been correctly carried out, whether the results are correct, or whether we have chosen the correct genetic explanation for the results. What it does indicate is the *probability* that the difference between the observed and the expected values is due to chance. In other words, it indicates the likelihood that chance alone could produce the deviation between the expected and the observed values.

If we expected 20 brown and 20 yellow progeny from a genetic cross, the chi-square test gives the probability that we might observe 25 brown and 15 yellow progeny simply owing to chance deviations from the expected 20:20 ratio. When the probability calculated from the chi-square test is high, we assume that chance alone produced the difference. When the probability is low, we assume that some factor other than chance—some significant factor—produced the deviation.

To use the goodness-of-fit chi-square test, we first determine the expected results. The chi-square test must always be applied to numbers of progeny, not to proportions or percentages. Let's consider a locus for coat color in domestic cats, for which black color (B) is dominant over gray (b). If we crossed two heterozygous black cats ($Bb \times Bb$), we would expect a 3:1 ratio of black and gray kittens. A series of such crosses yields a total of 50 kittens—30 black and 20 gray. These numbers are our *observed* values. We can obtain the *expected* numbers by multiplying the expected proportions by the total number of observed progeny. In this case, the expected number of black kittens is $3/4 \times 50 = 37.5$ and the expected number of gray kittens is $1/4 \times 50 = 12.5$. The chi-square (χ^2) value is calculated by using the following formula:

$$\chi^2 = \Sigma \frac{(\text{observed} - \text{expected})^2}{\text{expected}}$$

where Σ means the sum of all the squared differences between observed and expected divided by the expected values. To calculate the chi-square value for our black and gray kittens, we would first subtract the number of *expected* black kittens from the number of *observed* black kittens ($30 - 37.5 = -7.5$) and square this value: $-7.5^2 = 56.25$. We then divide this result by the expected number of black kittens, $56.25/37.5, = 1.5$. We repeat the calculations on the number of expected gray kittens: $(20 - 12.5)^2/12.5 = 4.5$. To obtain the overall chi-square value, we sum the (observed $-$ expected)^2/expected values: $1.5 + 4.5 = 6.0$.

Table 3.4	Critical values of the χ^2 distribution							
					P				
df	.995	.975	.9	.5	.1	.05	.025	.01	.005
1	.000	.000	0.016	0.455	2.706	3.841	5.024	6.635	7.879
2	0.010	0.051	0.211	1.386	4.605	5.991	7.378	9.210	10.597
3	0.072	0.216	0.584	2.366	6.251	7.815	9.348	11.345	12.838
4	0.207	0.484	1.064	3.357	7.779	9.488	11.143	13.277	14.860
5	0.412	0.831	1.610	4.351	9.236	11.070	12.832	15.086	16.750
6	0.676	1.237	2.204	5.348	10.645	12.592	14.449	16.812	18.548
7	0.989	1.690	2.833	6.346	12.017	14.067	16.013	18.475	20.278
8	1.344	2.180	3.490	7.344	13.362	15.507	17.535	20.090	21.955
9	1.735	2.700	4.168	8.343	14.684	16.919	19.023	21.666	23.589
10	2.156	3.247	4.865	9.342	15.987	18.307	20.483	23.209	25.188
11	2.603	3.816	5.578	10.341	17.275	19.675	21.920	24.725	26.757
12	3.074	4.404	6.304	11.340	18.549	21.026	23.337	26.217	28.300
13	3.565	5.009	7.042	12.340	19.812	22.362	24.736	27.688	29.819
14	4.075	5.629	7.790	13.339	21.064	23.685	26.119	29.141	31.319
15	4.601	6.262	8.547	14.339	22.307	24.996	27.488	30.578	32.801

P, probability; df, degrees of freedom.

The next step is to determine the probability associated with this calculated chi-square value, which is the probability that the deviation between the observed and the expected results could be due to chance. This step requires us to compare the calculated chi-square value (6.0) with theoretical values that have the same degrees of freedom in a chi-square table. The degrees of freedom represent the number of ways in which the observed classes are free to vary. For a goodness-of-fit chi-square test, the degrees of freedom are equal to $n - 1$, where n is the number of different expected phenotypes. In our example, there are two expected phenotypes (black and gray); so $n = 2$ and the degree of freedom equals $2 - 1 = 1$.

Now that we have our calculated chi-square value and have figured out the associated degrees of freedom, we are ready to obtain the probability from a chi-square table (Table 3.4). The degrees of freedom are given in the left-hand column of the table and the probabilities are given at the top; within the body of the table are chi-square values associated with these probabilities. First, find the row for the appropriate degrees of freedom; for our example with 1 degree of freedom, it is the first row of the table. Find where our calculated chi-square value (6.0) lies among the theoretical values in this row. The theoretical chi-square values increase from left to right and the probabilities decrease from left to right. Our chi-square value of 6.0 falls between the value of 5.024, associated with a probability of .025, and the value of 6.635, associated with a probability of .01.

Thus, the probability associated with our chi-square value is less than .025 and greater than .01. So, there is less than a 2.5% probability that the deviation that we observed between the expected and the observed numbers of black and gray kittens could be due to chance.

Most scientists use the .05 probability level as their cutoff value: if the probability of chance being responsible for the deviation is greater than or equal to .05, they accept that chance may be responsible for the deviation between the observed and the expected values. When the probability is less than .05, scientists assume that chance is not responsible and a significant difference exists. The expression *significant difference* means that some factor other than chance is responsible for the observed values being different from the expected values. In regard to the kittens, perhaps one of the genotypes experienced increased mortality before the progeny were counted or perhaps other genetic factors skewed the observed ratios.

In choosing .05 as the cutoff value, scientists have agreed to assume that chance is responsible for the deviations between observed and expected values unless there is strong evidence to the contrary. It is important to bear in mind that even if we obtain a probability of, say, .01, there is still a 1% probability that the deviation between the observed and expected numbers is due to nothing more than chance. Calculation of the chi-square value is illustrated in (◀ FIGURE 3.15).

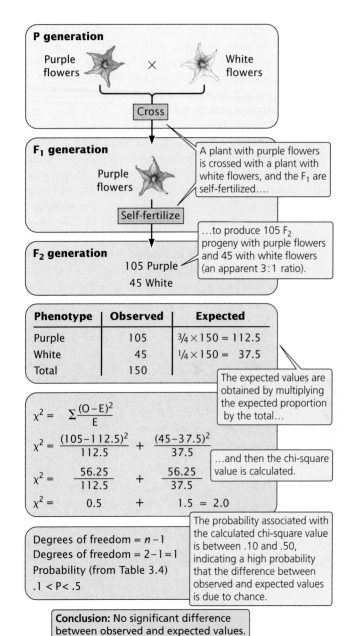

Phenotype	Observed	Expected
Purple	105	$\tfrac{3}{4} \times 150 = 112.5$
White	45	$\tfrac{1}{4} \times 150 = 37.5$
Total	150

$$\chi^2 = \sum \frac{(O - E)^2}{E}$$

$$\chi^2 = \frac{(105 - 112.5)^2}{112.5} + \frac{(45 - 37.5)^2}{37.5}$$

$$\chi^2 = \frac{56.25}{112.5} + \frac{56.25}{37.5}$$

$$\chi^2 = 0.5 + 1.5 = 2.0$$

Degrees of freedom = $n - 1$
Degrees of freedom = $2 - 1 = 1$
Probability (from Table 3.4)
$.1 < P < .5$

Conclusion: No significant difference between observed and expected values.

◀ 3.15 **A chi-square test is used to determine the probability that the difference between observed and expected values is due to chance.**

Concepts

Differences between observed and expected ratios can arise by chance. The goodness-of-fit chi-square test can be used to evaluate whether deviations between observed and expected numbers are likely to be due to chance or to some other significant factor.

Penetrance and Expressivity

In the genetic crosses considered thus far, we have assumed that every individual with a particular genotype expresses the expected phenotype. We assumed, for example, that the genotype *Rr* always produces round seeds and that the genotype *rr* always produces wrinkled seeds. For some characters, such an assumption is incorrect: the genotype does not always produce the expected phenotype, a phenomenon termed **incomplete penetrance.**

Incomplete penetrance is seen in human polydactyly, the condition of having extra fingers and toes (◀FIGURE 3.16). There are several different forms of human polydactyly, but the trait is usually caused by a dominant allele. Occasionally, people possess the allele for polydactyly (as evidenced by the fact that their children inherit the polydactyly) but nevertheless have a normal number of fingers and toes. In these cases, the gene for polydactyly is not fully penetrant. **Penetrance** is defined as the percentage of individuals having a particular genotype that express the expected phenotype. For example, if we examined 42 people having an allele for polydactyly and found that only 38 of them were polydactylous, the penetrance would be $38/42 = 0.90$ (90%).

A related concept is that of **expressivity,** the degree to which a character is expressed. In addition to incomplete penetrance, polydactyly exhibits variable expressivity. Some polydactylous persons possess extra fingers and toes that are fully functional, whereas others possess only a small tag of extra skin.

Incomplete penetrance and variable expressivity are due to the effects of other genes and to environmental factors that can alter or completely suppress the effect of a particular gene. A gene might encode an enzyme that produces a particular phenotype only within a limited temperature range. At higher or lower temperatures, the enzyme would not function and the phenotype would not be expressed; the allele encoding such an enzyme is therefore penetrant only within a particular temperature range. Many characters exhibit incomplete penetrance and variable expressivity,

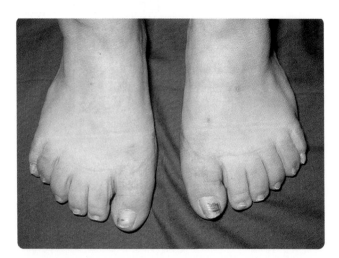

◀3.16 Human polydactyly (extra digits) exhibits incomplete penetrance and variable expressivity.
(Biophoto Associates/Science Source/Photo Researchers.)

emphasizing the fact that the mere presence of a gene does not guarantee its expression.

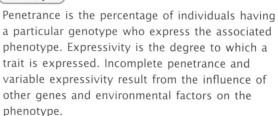

Concepts

Penetrance is the percentage of individuals having a particular genotype who express the associated phenotype. Expressivity is the degree to which a trait is expressed. Incomplete penetrance and variable expressivity result from the influence of other genes and environmental factors on the phenotype.

Connecting Concepts Across Chapters

This chapter has introduced several important concepts of heredity and presented techniques for making predictions about the types of offspring that parents will produce. Two key principles of inheritance were introduced: the principles of segregation and independent assortment. These principles serve as the foundation for understanding much of heredity. In this chapter, we also discussed some essential terminology and techniques for discussing and analyzing genetic crosses. A critical concept is the connection between the behavior of chromosomes during meiosis (Chapter 2) and the seemingly abstract symbols used in genetic crosses.

The principles taught in this chapter provide important links to much of what follows in this book. In Chapters 4 through 7, we will learn about additional factors that affect the outcome of genetic crosses: sex, interactions between genes, linkage between genes, and environment. These factors build on the principles of segregation and independent assortment. In Chapters 10 through 21, where we focus on molecular aspects of heredity, the importance of these basic principles is not so obvious, but most nuclear processes are based on the inheritance of chromosomal genes. In Chapters 22 and 23, we turn to quantitative and population genetics. These chapters build directly on the principles of heredity and can only be understood with a firm grasp of how genes are inherited. The material covered in the present chapter therefore serves as a foundation for almost all of heredity.

Finally, this chapter introduces problem solving, which is at the heart of genetics. Developing hypotheses to explain genetic phenomenon (such as the types and proportions of progeny produced in a genetic cross) and testing these hypotheses by doing genetic crosses and collecting additional data are common to all of genetics. The ability to think analytically and draw logical conclusions from observations is emphasized throughout this book.

CONCEPTS SUMMARY

- Gregor Mendel, an Austrian monk living in what is now the Czech Republic, first discovered the principles of heredity by conducting experiments on pea plants.

- Mendel's success can be attributed to his choice of the pea plant as an experimental organism, the use of characters with a few, easily distinguishable phenotypes, his experimental approach, and careful attention to detail.

- Genes are inherited factors that determine a character. Alternate forms of a gene are called alleles. The alleles are located at a specific place, a locus, on a chromosome, and the set of genes that an individual possesses is its genotype. Phenotype is the manifestation or appearance of a characteristic and may refer to physical, biochemical, or behavioral characteristics.

- Phenotypes are produced by the combined effects of genes and environmental factors. Only the genotype—not the phenotype—is inherited.

- The principle of segregation states that an individual possesses two alleles coding for a trait and that these two alleles separate in equal proportions when gametes are formed.

- The concept of dominance indicates that, when two different alleles are present in a heterozygote, only the trait of one of them, the "dominant" allele, is observed in the phenotype. The other allele is said to be "recessive".

- The two alleles of a genotype are located on homologous chromosomes, which separate during anaphase I of meiosis. The separation of homologous chromosomes brings about the segregation of alleles.

- The types of progeny produced from a genetic cross can be predicted by applying the Punnett square or probability.

- Probability is the likelihood of a particular event occurring. The multiplication rule of probability states that the probability of two or more independent events occurring together is calculated by multiplying the probabilities of the independent events. The addition rule of probability states that the probability of any of two or more mutually exclusive events occurring is calculated by adding the probabilities of the events.

- The binomial expansion may be used to determine the probability of a particular combination of events.

- A testcross reveals the genotype (homozygote or heterozygote) of an individual having a dominant trait and consists of crossing that individual with one having the homozygous recessive genotype.

- Incomplete dominance occurs when a heterozygote has a phenotype that is intermediate between the phenotypes of the two homozygotes.

- The principle of independent assortment states that genes coding for different characters assort independently when gametes are formed.

- Independent assortment is based on the random separation of homologous pairs of chromosomes during anaphase I of meiosis; it occurs when genes coding for two characters are located on different pairs of chromosomes.

- When genes assort independently, the multiplication rule of probability can be used to obtain the probability of inheriting more than one trait: a cross including more than one trait can be broken down into simple crosses, and the probability of any combination of traits can be obtained by multiplying the probabilities for each trait.

- Observed ratios of progeny from a genetic cross may deviate from the expected ratios owing to chance. The goodness-of-fit chi-square test can be used to determine the probability that a difference between observed and expected numbers is due to chance.

- Penetrance is the percentage of individuals with a particular genotype that exhibit the expected phenotype. Expressivity is the degree to which a character is expressed. Incomplete penetrance and variable expressivity result from the influence of other genes and environmental effects on the phenotype.

IMPORTANT TERMS

gene (p. 47)
allele (p. 47)
locus (p. 47)
genotype (p. 48)
homozygous (p. 48)
heterozygous (p. 48)
phenotype (p. 48)
monohybrid cross (p. 48)
P (parental) generation (p. 48)
F_1 (filial 1) generation (p. 49)
reciprocal crosses (p. 49)

F_2 (filial 2) generation (p. 49)
dominant (p. 51)
recessive (p. 51)
principle of segregation (Mendel's first law) (p. 51)
concept of dominance (p. 51)
chromosome theory of heredity (p. 52)
backcross (p. 52)
Punnett square (p. 53)
probability (p. 54)

multiplication rule (p. 54)
addition rule (p. 55)
testcross (p. 57)
incomplete dominance (p. 58)
wild type (p. 58)
dihybrid cross (p. 60)
principle of independent assortment (Mendel's second law) (p. 60)
trihybrid cross (p. 63)

goodness-of-fit chi-square test (p. 66)
incomplete penetrance (p. 68)
penetrance (p. 68)
expressivity (p. 68)

Worked Problems

1. Short hair in rabbits (S) is dominant over long hair (s). The following crosses are carried out, producing the progeny shown. Give all possible genotypes of the parents in each cross.

Parents	Progeny
(a) short × short	4 short and 2 long
(b) short × short	8 short
(c) short × long	12 short
(d) short × long	3 short and 1 long
(e) long × long	2 long

• Solution

For this problem, it is useful to first gather as much information about the genotypes of the parents as possible on the basis of their phenotypes. We can then look at the types of progeny produced to provide the missing information. Notice that the problem asks for *all* possible genotypes of the parents.

(a) short × short   4 short and 2 long
Because short hair is dominant over long hair, a rabbit having short hair could be either *SS* or *Ss*. The two long-haired offspring must be homozygous (*ss*) because long hair is recessive and will appear in the phenotype only when both alleles for long hair are present. Because each parent contributes one of the two alleles found in the progeny, each parent must be carrying the *s* allele and must therefore be *Ss*.

(b) short × short   8 short
The short-haired parents could be *SS* or *Ss*. All 8 of the offspring are short (*S_*), and so at least one of the parents is likely to be homozygous (*SS*); if both parents were heterozygous, ¼ long-haired (*ss*) progeny would be expected, but we do not observe any long-haired progeny. The other parent could be homozygous (*SS*) or heterozygous (*Ss*); as long as one parent is homozygous, all the offspring will be short haired. It is theoretically possible, although unlikely, that both parents are heterozygous (*Ss* × *Ss*). If this were the case, we would expect 2 of the 8 progeny to be long haired. Although no long-haired progeny are observed, it is possible that just by chance no long-haired rabbits would be produced among the 8 progeny of the cross.

(c) short × long   12 short
The short-haired parent could be *SS* or *Ss*. The long-haired parent must be *ss*. If the short-haired parent were heterozygous (*Ss*), half of the offspring would be expected to be long haired, but we don't see any long-haired progeny. Therefore this parent is most likely homozygous (*SS*). It is theoretically possible, although unlikely, that the parent is heterozygous and just by chance no long-haired progeny were produced.

(d) short × long   3 short and 1 long
On the basis of its phenotype, the short-haired parent could be homozygous (*SS*) or heterozygous (*Ss*), but the presence of one long-haired offspring tells us that the short-haired parent must be heterozygous (*Ss*). The long-haired parent must be homozygous (*ss*).

(e) long × long   2 long
Because long hair is recessive, both parents must be homozygous for a long-hair allele (*ss*).

2. In cats, black coat color is dominant over gray. A female black cat whose mother is gray mates with a gray male. If this female has a litter of six kittens, what is the probability that three will be black and three will be gray?

• Solution

Because black (*G*) is dominant over gray (*g*), a black cat may be homozygous (*GG*) or heterozygous (*Gg*). The black female in this problem must be heterozygous (*Bb*) because her mother is gray (*gg*) and she must inherit one of her mother's alleles. The gray male is homozygous (*gg*) because gray is recessive. Thus the cross is:

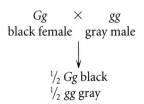

We can use the binomial expansion to determine the probability of obtaining three black and three gray kittens in a litter of six. Let *a* equal the probability of a kitten being black and *b* equal the probability of a kitten being gray. The binomial is $(a + b)^6$, the expansion of which is:

$$(a + b)^6 = a^6 + 6a^5b + 15a^4b^2 + 20a^3b^3 + 15a^2b^4 + 6a^1b^5 + b^6$$

(See text for an explanation of how to expand the binomial.) The probability of obtaining three black and three gray kittens in a litter of six is provided by the term $20a^3b^3$. The probabilities of *a* and *b* are both ½, so the overall probability is $20(\frac{1}{2})^3(\frac{1}{2})^3 = \frac{20}{64} = \frac{5}{16}$.

3. The following genotypes are crossed: *AaBbCdDd* × *AaBbCcDd*. Give the proportion of the progeny of this cross having the following genotypes: **(a)** *AaBbCcDd*, **(b)** *aabbccdd*, **(c)** *AaBbccDd*.

• Solution

This problem is easily worked if the cross is broken down into simple crosses and the multiplication rule is used to find the different combinations of genotypes:

Locus 1	*Aa* × *Aa* = ¼ *AA*, ½ *Aa*, ¼ *aa*
Locus 2	*Bb* × *Bb* = ¼ *BB*, ½ *Bb*, ¼ *bb*

Locus 3 $Cc \times Cc = \frac{1}{4} CC, \frac{1}{2} Cc, \frac{1}{4} cc$
Locus 4 $Dd \times Dd = \frac{1}{4} DD, \frac{1}{2} Dd, \frac{1}{4} dd$

To find the probability of any combination of genotypes, simply multiply the probabilities of the different genotypes:

(a) *AaBbCcDd* $\frac{1}{2} (Aa) \times \frac{1}{2} (Bb) \times \frac{1}{2} (Cc) \times \frac{1}{2} (Dd) = \frac{1}{16}$

(b) *aabbccdd* $\frac{1}{4} (aa) \times \frac{1}{4} (bb) \times \frac{1}{4} (cc) \times \frac{1}{4} (dd) = \frac{1}{256}$

(c) *AaBbccDd* $\frac{1}{2} (Aa) \times \frac{1}{2} (Bb) \times \frac{1}{4} (cc) \times \frac{1}{2} (Dd) = \frac{1}{32}$

4. In corn, purple kernels are dominant over yellow kernels, and full kernels are dominant over shrunken kernels. A corn plant having purple and full kernels is crossed with a plant having yellow and shrunken kernels, and the following progeny are obtained:

purple, full	112
purple, shrunken	103
yellow, full	91
yellow, shrunken	94

What are the most likely genotypes of the parents and progeny? Test your genetic hypothesis with a chi-square test.

• **Solution**

The best way to begin this problem is by breaking the cross down into simple crosses for a single characteristic (seed color or seed shape):

P	purple × yellow	full × shrunken
F_1	112 + 103 = 215 purple	112 + 91 = 203 full
	91 + 94 = 185 yellow	103 + 94 = 197 shrunken

Purple × yellow produces approximately $\frac{1}{2}$ purple and $\frac{1}{2}$ yellow. A 1:1 ratio is usually caused by a cross between a heterozygote and a homozygote. Because purple is dominant, the purple parent must be heterozygous (*Pp*) and the yellow parent must be homozygous (*pp*). The purple progeny produced by this cross will be heterozygous (*Pp*) and the yellow progeny must be homozygous (*pp*).

Now let's examine the other character. Full × shrunken produces $\frac{1}{2}$ full and $\frac{1}{2}$ shrunken, or a 1:1 ratio, and so these progeny phenotypes also are produced by a cross between a heterozygote (*Ff*) and a homozygote (*ff*); the full-kernel progeny will be heterozygous (*Ff*) and the shrunken-kernel progeny will be homozygous (*ff*).

Now combine the two crosses and use the multiplication rule to obtain the overall genotypes and the proportions of each genotype:

P	purple, full × yellow, shrunken
	PpFf × *ppyy*
F_1	*PpFf* = $\frac{1}{2}$ purple × $\frac{1}{2}$ full = $\frac{1}{4}$ purple, full
	Ppff = $\frac{1}{2}$ purple × $\frac{1}{2}$ shrunken = $\frac{1}{4}$ purple, shrunken
	ppFf = $\frac{1}{2}$ yellow × $\frac{1}{2}$ full = $\frac{1}{4}$ yellow, full
	ppff = $\frac{1}{2}$ yellow × $\frac{1}{2}$ shrunken = $\frac{1}{4}$ yellow shrunken

Our genetic explanation predicts that, from this cross, we should see $\frac{1}{4}$ purple, full-kernel progeny; $\frac{1}{4}$ purple, shrunken-kernel progeny; $\frac{1}{4}$ yellow, full-kernel progeny; and $\frac{1}{4}$ yellow, shrunken-kernel progeny. A total of 400 progeny were produced; so $\frac{1}{4} \times 400 = 100$ of each phenotype are expected. These observed numbers do not fit the expected numbers exactly. Could the difference between what we observe and what we expect be due to chance? If the probability is high that chance alone is responsible for the difference between observed and expected, we will assume that the progeny have been produced in the 1:1:1:1 ratio predicted by the cross. If the probability that the difference between observed and expected is due to chance is low, the progeny are not really in the predicted ratio and some other, *significant* factor must be responsible for the deviation.

The observed and expected numbers are:

Phenotype	Observed	Expected
purple full	112	$\frac{1}{4} \times 400 = 100$
purple shrunken	103	$\frac{1}{4} \times 400 = 100$
yellow full	91	$\frac{1}{4} \times 400 = 100$
yellow shrunken	94	$\frac{1}{4} \times 400 = 100$

To determine the probability that the difference between observed and expected is due to chance, we calculate a chi-square value with the formula $\chi^2 = \Sigma[(\text{observed} - \text{expected})^2/\text{expected}]$:

$$\chi^2 = \frac{(112 - 100)^2}{100} + \frac{(103 - 100)^2}{100} + \frac{(91 - 100)^2}{100}$$
$$+ \frac{(94 - 100)^2}{100}$$
$$= \frac{12^2}{100} + \frac{3^2}{100} + \frac{9^2}{100} + \frac{6^2}{100}$$
$$= \frac{144}{100} + \frac{9}{100} + \frac{81}{100} + \frac{36}{100}$$
$$= 1.44 + 0.09 + 0.81 + 0.36 = 2.70$$

Now that we have the chi-square value, we must determine the probability that this chi-square value is due to chance. To obtain this probability, we first calculate the degrees of freedom, which for a goodness-of-fit chi-square test are $n - 1$, where n equals the number of expected phenotypic classes. In this case, there are four expected phenotypic classes; so the degrees of freedom equal $4 - 1 = 3$. We must now look up the chi-square value in a chi-square table (see Table 3.4). We select the row corresponding to 3 degrees of freedom and look along this row to find our calculated chi-square value. The calculated chi-square value of 2.7 lies between 2.366 (a probability of .5) and 6.251 (a probability of .1). The probability (*P*) associated with the calculated chi-square value is therefore $.5 < P < .1$. This is the probability that the difference between what we observed and what we expect is due to chance, which in this case is relatively high, and so chance is likely responsible for the deviation. We can conclude that the progeny *do* appear in the 1:1:1:1 ratio predicted by our genetic explanation.

COMPREHENSION QUESTIONS

* 1. Why was Mendel's approach to the study of heredity so successful?

2. What is the relation between the terms *allele, locus, gene,* and *genotype?*

* 3. What is the principle of segregation? Why is it important?

4. What is the concept of dominance? How does dominance differ from incomplete dominance?

5. Give the phenotypic ratios that may appear among the progeny of simple crosses and the genotypes of the parents that may give rise to each ratio.

6. Give the genotypic ratios that may appear among the progeny of simple crosses and the genotypes of the parents that may give rise to each ratio.

* 7. What is the chromosome theory of inheritance? Why was it important?

8. What is the principle of independent assortment? How is it related to the principle of segregation?

9. How is the principle of independent assortment related to meiosis?

10. How is the goodness-of-fit chi-square test used to analyze genetic crosses? What does the probability associated with a chi-square value indicate about the results of a cross?

11. What is incomplete penetrance and what causes it?

APPLICATION QUESTIONS AND PROBLEMS

*12. In cucumbers, orange fruit color (R) is dominant over cream fruit color (r). A cucumber plant homozygous for orange fruits is crossed with a plant homozygous for cream fruits. The F_1 are intercrossed to produce the F_2.

(a) Give the genotypes and phenotypes of the parents, the F_1, and the F_2.

(b) Give the genotypes and phenotypes of the offspring of a backcross between the F_1 and the orange parent.

(c) Give the genotypes and phenotypes of a backcross between the F_1 and the cream parent.

*13. In rabbits, coat color is a genetically determined characteristic. Some black females always produce black progeny, whereas other black females produce black progeny and white progeny. Explain how these outcomes occur.

*14. In cats, blood type A results from an allele (I^A) that is dominant over an allele (i^B) that produces blood type B. There is no O blood type. The blood types of male and female cats that were mated and the blood types of their kittens follow. Give the most likely genotypes for the parents of each litter.

Male parent	Female parent	Kittens
(a) blood type A	blood type B	4 kittens with blood type A, 3 with blood type B
(b) blood type B	blood type B	6 kittens with blood type B
(c) blood type B	blood type A	8 kittens with blood type A
(d) blood type A	blood type A	7 kittens with blood type A, 2 kittens with blood type B
(e) blood type A	blood type A	10 kittens with blood type A
(f) blood type A	blood type B	4 kittens with blood type A, 1 kitten with blood type B

15. In sheep, lustrous fleece (L) results from an allele that is dominant over an allele for normal fleece (l). A ewe (adult female) with lustrous fleece is mated with a ram (adult male) with normal fleece. The ewe then gives birth to a single lamb with normal fleece. From this single offspring, is it possible to determine the genotypes of the two parents? If so, what are their genotypes? If not, why not?

*16. In humans, alkaptonuria is a metabolic disorder in which affected persons produce black urine (see the introduction to this chapter). Alkaptonuria results from an allele (a) that is recessive to the allele for normal metabolism (A). Sally has normal metabolism, but her brother has alkaptonuria. Sally's father has alkaptonuria, and her mother has normal metabolism.

(a) Give the genotypes of Sally, her mother, her father, and her brother.

(b) If Sally's parents have another child, what is the probability that this child will have alkaptonuria?

(c) If Sally marries a man with alkaptonuria, what is the probability that their first child will have alkaptonuria?

17. Suppose that you are raising Mongolian gerbils. You notice that some of your gerbils have white spots, whereas others have solid coats. What type of crosses could you carry out to determine whether white spots are due to a recessive or a dominant allele?

*18. Hairlessness in American rat terriers is recessive to the presence of hair. Suppose that you have a rat terrier with hair. How can you determine whether this dog is homozygous or heterozygous for the hairy trait?

19. In snapdragons, red flower color (R) is incompletely dominant over white flower color (r); the heterozygotes produce pink flowers. A red snapdragon is crossed with a white snapdragon, and the F_1 are intercrossed to produce the F_2.
(a) Give the genotypes and phenotypes of the F_1 and F_2, along with their expected proportions.
(b) If the F_1 are backcrossed to the white parent, what will the genotypes and phenotypes of the offspring be?
(c) If the F_1 are backcrossed to the red parent, what are the genotypes and phenotypes of the offspring?

20. What is the probability of rolling one six-sided die and obtaining the following numbers?
(a) 2
(c) An even number
(b) 1 or 2
(d) Any number but a 6

*21. What is the probability of rolling two six-sided dice and obtaining the following numbers?
(a) 2 and 3
(b) 6 and 6
(c) At least one 6
(d) Two of the same number (two 1s, or two 2s, or two 3s, etc.)
(e) An even number on both dice
(f) An even number on at least one die

*22. In a family of seven children, what is the probability of obtaining the following numbers of boys and girls?
(a) All boys
(b) All children of the same sex
(c) Six girls and one boy
(d) Four boys and three girls
(e) Four girls and three boys

23. Phenylketonuria (PKU) is a disease that results from a recessive gene. Two normal parents produce a child with PKU.
(a) What is the probability that a sperm from the father will contain the PKU allele?
(b) What is the probability that an egg from the mother will contain the PKU allele?
(c) What is the probability that their next child will have PKU?
(d) What is the probability that their next child will be heterozygous for the PKU gene?

*24. In German cockroaches, curved wing (cv) is recessive to normal wing (cv^+). A homozygous cockroach having normal wings is crossed with a homozygous cockroach having curved wings. The F_1 are intercrossed to produce the F_2. Assume that the pair of chromosomes containing the locus for wing shape is metacentric. Draw this pair of chromosomes as it would appear in the parents, the F_1, and each class of F_2 progeny at metaphase I of meiosis. Assume that no crossing over takes place. At each stage, label a location for the alleles for wing shape (cv and cv^+) on the chromosomes.

*25. In guinea pigs, the allele for black fur (B) is dominant over the allele for brown (b) fur. A black guinea pig is crossed with a brown guinea pig, producing five F_1 black guinea pigs and six F_1 brown guinea pigs.
(a) How many copies of the black allele (B) will be present in each cell from an F_1 black guinea pig at the following stages: G_1, G_2, metaphase of mitosis, metaphase I of meiosis, metaphase II of meiosis, and after the second cytokinesis following meiosis? Assume that no crossing over takes place.
(b) How may copies of the brown allele (b) will be present in each cell from an F_1 brown guinea pig at the same stages? Assume that no crossing over takes place.

26. In watermelons, bitter fruit (B) is dominant over sweet fruit (b), and yellow spots (S) are dominant over no spots (s). The genes for these two characteristics assort independently. A homozygous plant that has bitter fruit and yellow spots is crossed with a homozygous plant that has sweet fruit and no spots. The F_1 are intercrossed to produce the F_2.
(a) What will be the phenotypic ratios in the F_2?
(b) If an F_1 plant is backcrossed with the bitter, yellow spotted parent, what phenotypes and proportions are expected in the offspring?
(c) If an F_1 plant is backcrossed with the sweet, nonspotted parent, what phenotypes and proportions are expected in the offspring?

27. In cats, curled ears (Cu) result from an allele that is dominant over an allele for normal ears (cu). Black color results from an independently assorting allele (G) that is dominant over an allele for gray (g). A gray cat homozygous for curled ears is mated with a homozygous black cat with normal ears. All the F_1 cats are black and have curled ears.
(a) If two of the F_1 cats mate, what phenotypes and proportions are expected in the F_2?
(b) An F_1 cat mates with a stray cat that is gray and possesses normal ears. What phenotypes and proportions of progeny are expected from this cross?

*28. The following two genotypes are crossed: $AaBbCcddEe \times AabbCcDdEe$. What will the proportion of the following genotypes be among the progeny of this cross?
(a) $AaBbCcDdEe$
(b) $AabbCcddee$
(c) $aabbccddee$
(d) $AABBCCDDEE$

29. In mice, an allele for apricot eyes (a) is recessive to an allele for brown eyes (a^+). At an independently assorting locus, an allele for tan (t) coat color is recessive to an allele for black (t^+) coat color. A mouse that is homozygous for brown eyes and black coat color is crossed with a mouse having apricot eyes and a tan coat. The resulting F_1 are intercrossed to produce the F_2. In a litter of eight F_2 mice, what is the probability that two will have apricot eyes and tan coats?

30. In cucumbers, dull fruit (D) is dominant over glossy fruit (d), orange fruit (R) is dominant over cream fruit (r), and bitter cotyledons (B) are dominant over nonbitter cotyledons (b). The three characters are encoded by genes located on different pairs of chromosomes. A plant homozygous for dull, orange fruit and bitter cotyledons is crossed with a plant that has glossy, cream fruit and nonbitter cotyledons. The F_1 are intercrossed to produce the F_2.

    (a) Give the phenotypes and their expected proportions in the F_2.

    (b) An F_1 plant is crossed with a plant that has glossy, cream fruit and nonbitter cotyledons. Give the phenotypes and expected proportions among the progeny of this cross.

* 31. A and a are alleles located on a pair of metacentric chromosomes. B and b are alleles located on a pair of acrocentric chromosomes. A cross is made between individuals having the following genotypes: $AaBb \times aabb$.

    (a) Draw the chromosomes as they would appear in each type of gamete produced by the individuals of this cross.

    (b) For each type of progeny resulting from this cross, draw the chromosomes as they would appear in a cell at G_1, G_2, and metaphase of mitosis.

32. Ptosis (droopy eyelid) may be inherited as a dominant human trait. Among 40 people who are heterozygous for the ptosis allele, 13 have ptosis and 27 have normal eyelids.

    (a) What is the penetrance for ptosis?

    (b) If ptosis exhibited variable expressivity, what would that mean?

33. In sailfin mollies (fish), gold color is due to an allele (g) that is recessive to the allele for normal color (G). A gold fish is crossed with a normal fish. Among the offspring, 88 are normal and 82 are gold.

    (a) What are the most likely genotypes of the parents in this cross?

    (b) Assess the plausibility of your hypothesis by performing a chi-square test.

34. In guinea pigs, the allele for black coat color (B) is dominant over the allele for white coat color (b). At an independently assorting locus, an allele for rough coat (R) is dominant over an allele for smooth coat (r). A guinea pig that is homozygous for black color and rough coat is crossed with a guinea pig that has a white and smooth coat. In a series of matings, the F_1 are crossed with guinea pigs having white, smooth coats. From these matings, the following phenotypes appear in the offspring: 24 black, rough guinea pigs; 26 black, smooth guinea pigs; 23 white, rough guinea pigs; and 5 white, smooth guinea pigs.

    (a) Using a chi-square test, compare the observed numbers of progeny with those expected from the cross.

    (b) What conclusions can you draw from the results of the chi-square test?

    (c) Suggest an explanation for these results.

CHALLENGE QUESTIONS

35. Dwarfism is a recessive trait in Hereford cattle. A rancher in western Texas discovers that several of the calves in his herd are dwarfs, and he wants to eliminate this undesirable trait from the herd as rapidly as possible. Suppose that the rancher hires you as a genetic consultant to advise him on how to breed the dwarfism trait out of the herd. What crosses would you advise the rancher to conduct to ensure that the allele causing dwarfism is eliminated from the herd?

36. A geneticist discovers an obese mouse in his laboratory colony. He breeds this obese mouse with a normal mouse. All the F_1 mice from this cross are normal in size. When he interbreeds two F_1 mice, eight of the F_2 mice are normal in size and two are obese. The geneticist then intercrosses two of his obese mice, and he finds that all of the progeny from this cross are obese. These results lead the geneticist to conclude that obesity in mice results from a recessive allele.

    A second geneticist at a different university also discovers an obese mouse in her laboratory colony. She carries out the same crosses as the first geneticist did and obtains the same results. She also concludes that obesity in mice results from a recessive allele. One day the two geneticists meet at a genetics conference, learn of each other's experiments, and decide to exchange mice. They both find that, when they cross two obese mice from the different laboratories, all the offspring are normal; however, when they cross two obese mice from the same laboratory, all the offspring are obese. Explain their results.

37. Albinism is a recessive trait in humans. A geneticist studies a series of families in which both parents are normal and at least one child has albinism. The geneticist reasons that both parents in these families must be heterozygotes and that albinism should appear in $1/4$ of the children of these families. To his surprise, the geneticist finds that the

frequency of albinism among the children of these families is considerably greater than $\frac{1}{4}$. There is no evidence that normal pigmentation exhibits incomplete penetrance. Can you think of an explanation for the higher-than-expected frequency of albinism among these families?

38. Two distinct phenotypes are found in the salamander *Plethodon cinereus*: a red form and a black form. Some biologists have speculated that the red phenotype is due to an autosomal allele that is dominant over an allele for black. Unfortunately, these salamanders will not mate in captivity; so the hypothesis that red is dominant over black has never been tested.

    One day a genetics student is hiking through the forest and finds 30 female salamanders, some red and some black, laying eggs. The student places each female and her eggs (about 20–30 eggs per female) in separate plastic bags and takes them back to the lab. There, the student successfully raises the eggs until they hatch. After the eggs have hatched, the student records the phenotypes of the juvenile salamanders, along with the phenotypes of their mothers. Thus, the student has the phenotypes for 30 females and their progeny, but no information is available about the phenotypes of the fathers.

    Explain how the student can determine whether red is dominant over black with this information on the phenotypes of the females and their offspring.

SUGGESTED READINGS

Corcos, A., and F. Monaghan. 1985. Some myths about Mendel's experiments. *The American Biology Teacher* 47:233–236.

An excellent discussion of some misconceptions surrounding Mendel's life and discoveries.

Dronamraju, K. 1992. Profiles in genetics: Archibald E. Garrod. *American Journal of Human Genetics* 51:216–219.

A brief biography of Archibald Garrod and his contributions to genetics.

Dunn, L. C. 1965. *A Short History of Genetics*. New York: McGraw-Hill.

An older but very good history of genetics.

Garrod, A. E. 1902. The incidence of alkaptonuria: a study in chemical individuality. *Lancet* 2:1616–1620.

Garrod's original paper on the genetics of alkaptonuria.

Henig, R. M. 2001. *The Monk in the Garden: The Lost and Found Genius of Gregor Mendel, the Father of Genetics*. Boston: Houghton Mifflin.

A biography of Gregor Mendel, in which the author has used historical research to create a vivid portrait of Mendel's life and work.

Monaghan, F. V., and A. F. Corcos. 1987. Reexamination of the fate of Mendel's paper. *Journal of Heredity* 78:116–118.

A good discussion of why Mendel's paper was unappreciated by his peers.

Orel, V. 1984. *Mendel*. Oxford: Oxford University Press.

An excellent and authoritative biography of Mendel.

Weiling, F. 1991. Historical study: Johann Gregor Mendel 1822–1884. *American Journal of Medical Genetics* 40:1–25.

A fascinating account that contains much recent research on Mendel's life as a scientist.

Sex Determination and Sex-Linked Characteristics

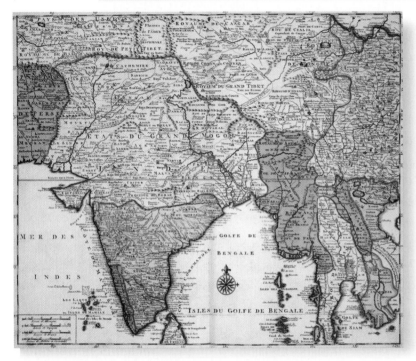

In 1875, Charles Darwin described the X-linked inheritance of a genetic condition (anhiorotic ectodermal dysplasia) that appeared in a family from northwest India. (Historical Picture Archive/Corbis.)

The Toothless, Hairless Men of Sind

In 1875, Charles Darwin, author of *On the Origin of Species,* wrote of a peculiar family of Sind, a province in northwest India,

> in which ten men, in the course of four generations, were furnished in both jaws taken together, with only four small and weak incisor teeth and with eight posterior molars. The men thus affected have little hair on the body, and become bald early in life. They also suffer much during hot weather from excessive dryness of the skin. It is remarkable that no instance has occurred of a daughter being thus affected. . . . Though daughters in the above family are never affected, they transmit the tendency to their sons; and no case has occurred of a son transmitting it to his sons.

These men possessed a genetic condition now known as anhidrotic ectodermal dysplasia, which (as noted by Darwin) is characterized by small teeth, no sweat glands, and sparse body hair. Darwin also noted several key features of the inheritance of this disorder: although it occurs primarily in men, fathers never transmit the trait to their sons; unaffected daughters, however, may pass the trait to their sons (the grandsons of affected men). These features of inheritance are the hallmarks of a sex-linked trait, a major focus of this chapter. Although Darwin didn't understand the mechanism of heredity, his attention to detail and remarkable ability to focus on crucial observations allowed him to identify the essential features of this genetic disease 25 years before Mendel's principles of heredity became widely known.

Darwin claimed that the daughters of this Hindu family were never affected, but it's now known that some women do have mild cases of anhidrotic ectodermal dysplasia. In these women, the symptoms of the disorder appear on only some parts of the body. For example, some regions of the jaw are missing teeth, whereas other regions have normal teeth. There are irregular patches of skin having few or no sweat

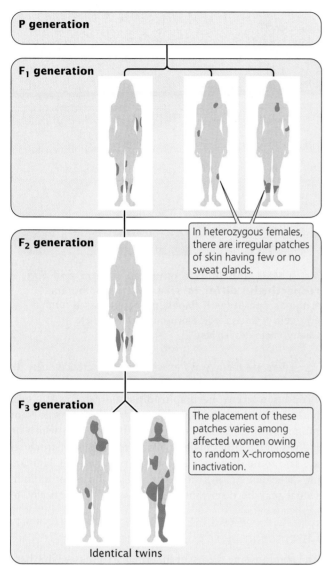

P generation

F₁ generation

In heterozygous females, there are irregular patches of skin having few or no sweat glands.

F₂ generation

F₃ generation

The placement of these patches varies among affected women owing to random X-chromosome inactivation.

Identical twins

◀ 4.1 **Three generations of women heterozygous for the X-linked recessive disorder anhidrotic ectodermal dysplasia, which is inherited as an X-linked recessive trait.** (After A. P. Mance and J. Mance, *Genetics: Human Aspects*, Sinauer, 1990, p. 133.)

glands; the placement of these patches varies among affected women (◀ FIGURE 4.1). The patchy occurrence of these features is explained by the fact that the gene for anhidrotic ectodermal dysplasia is located on a sex chromosome.

www.whfreeman.com/pierce   Additional information about anhidrotic ectodermal dysplasia, including symptoms, history, and genetics

In Chapter 3, we studied Mendel's principles of segregation and independent assortment and saw how these principles explain much about the nature of inheritance. After Mendel's principles were rediscovered in 1900, biologists began to conduct genetic studies on a wide array of different organisms. As they applied Mendel's principles more widely, exceptions were observed, and it became necessary to devise extensions to his basic principles of heredity.

In this chapter, we explore one of the major extensions to Mendel's principles: the inheritance of characteristics encoded by genes located on the sex chromosomes, which differ in males and females (◀ FIGURE 4.2). These characteristics and the genes that produce them are referred to as sex linked. To understand the inheritance of sex-linked characteristics, we must first know how sex is determined—why some members of a species are male and others are female. Sex determination is the focus of the first part of the chapter. The second part examines how characteristics encoded by genes on the sex chromosomes are inherited. In Chapter 5, we will explore some additional ways in which sex and inheritance interact.

As we consider sex determination and sex-linked characteristics, it will be helpful to think about two important principles. First, there are several different mechanisms of sex determination and, ultimately, the mechanism of sex determination controls the inheritance of sex-linked characteristics. Second, like other pairs of chromosomes, the X and Y sex chromosomes may pair in the course of meiosis and segregate, but throughout most of their length they are not homologous (their gene sequences don't code for the same characteristics): most genes on the X chromosome are different from genes on the Y chromosome. Consequently, males and females do not possess the same number of alleles at sex-linked loci. This difference in the number of sex-linked alleles produces the distinct patterns of inheritance in males and females.

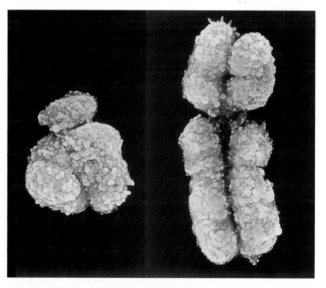

◀ 4.2 **The sex chromosomes of males (Y) and females (X) differ in size and shape.** (Biophoto Associates/Photo Researchers.)

Sex Determination

Sexual reproduction is the formation of offspring that are genetically distinct from their parents; most often, two parents contribute genes to their offspring. Among most eukaryotes, sexual reproduction consists of two processes that lead to an alternation of haploid and diploid cells: meiosis produces haploid gametes, and fertilization produces diploid zygotes (◀FIGURE 4.3).

The term **sex** refers to sexual phenotype. Most organisms have only two sexual phenotypes: male and female. The fundamental difference between males and females is gamete size: males produce small gametes; females produce relatively large gametes (◀FIGURE 4.4).

The mechanism by which sex is established is termed **sex determination.** We define the sex of an individual in terms of the individual's phenotype—ultimately, the type of gametes that it produces. Sometimes an individual has chromosomes or genes that are normally associated with one sex but a morphology corresponding to the opposite sex. For instance, the cells of female humans normally have two X chromosomes, and the cells of males have one X chromosome and one Y chromosome. A few rare persons have male anatomy, although their cells each contain two X chromosomes. Even though these people are genetically female, we refer to them as male because their sexual phenotype is male.

> **Concepts**
>
> In sexual reproduction, parents contribute genes to produce an offspring that is genetically distinct from both parents. In eukaryotes, sexual reproduction consists of meiosis, which produces haploid gametes, and fertilization, which produces a diploid zygote.

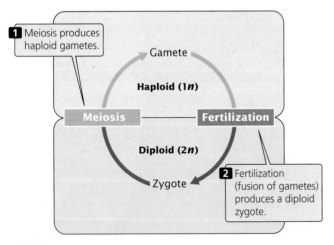

◀4.3 In most eukaryotic organisms, sexual reproduction consists of an alternation of haploid (1*n*) and diploid (2*n*) cells.

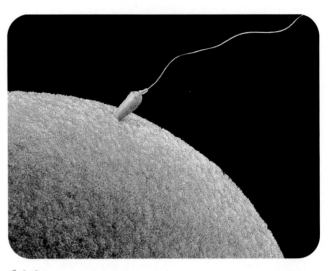

◀4.4 Male and female gametes (sperm and egg, respectively) differ in size. In this photograph, a human sperm (with flagellum) penetrates a human egg cell. (Francis Leroy, Biocosmos/Science Photo Library/Photo Researchers.)

There are many ways in which sex differences arise. In some species, both sexes are present in the same individual, a condition termed **hermaphroditism;** organisms that bear both male and female reproductive structures are said to be **monoecious** (meaning "one house"). Species in which an individual has either male or female reproductive structures are said to be **dioecious** (meaning "two houses"). Humans are dioecious. Among dioecious species, the sex of an individual may be determined chromosomally, genetically, or environmentally.

Chromosomal Sex-Determining Systems

The chromosome theory of inheritance (discussed in Chapter 3) states that genes are located on chromosomes, which serve as the vehicles for gene segregation in meiosis. Definitive proof of this theory was provided by the discovery that the sex of certain insects is determined by the presence or absence of particular chromosomes.

In 1891, Hermann Henking noticed a peculiar structure in the nuclei of cells from male insects. Understanding neither its function nor its relation to sex, he called this structure the X body. Later, Clarence E. McClung studied Henking's X body in grasshoppers and recognized that it was a chromosome. McClung called it the accessory chromosome, but eventually it became known as the X chromosome, from Henking's original designation. McClung observed that the cells of female grasshoppers had one more chromosome than the cells of male grasshoppers, and he concluded that accessory chromosomes played a role in sex determination. In 1905, Nettie Stevens and Edmund Wilson demonstrated that, in grasshoppers and other insects, the cells of females have two X chromosomes, whereas the cells of males have a single X. In some insects, they counted the same number of chromosomes in

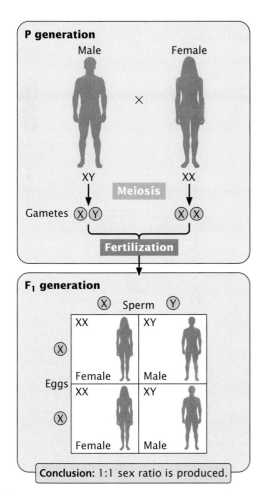

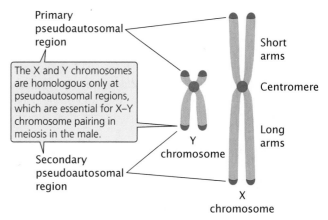

4.5 Inheritance of sex in organisms with X and Y chromosomes results in equal numbers of male and female offspring.

cells of males and females but saw that one chromosome pair was different: two X chromosomes were found in female cells, whereas a single X chromosome plus a smaller chromosome, which they called Y, was found in male cells.

Stevens and Wilson also showed that the X and Y chromosomes separate into different cells in sperm formation; half of the sperm receive an X chromosome and half receive a Y. All egg cells produced by the female in meiosis receive one X chromosome. A sperm containing a Y chromosome unites with an X-bearing egg to produce an XY male, whereas a sperm containing an X chromosome unites with an X-bearing egg to produce an XX female (FIGURE 4.5). This accounts for the 50:50 sex ratio observed in most dioecious organisms. Because sex is inherited like other genetically determined characteristics, Stevens and Wilson's discovery that sex was associated with the inheritance of a particular chromosome also demonstrated that genes are on chromosomes.

As Stevens and Wilson found for insects, sex is frequently determined by a pair of chromosomes, the **sex chromosomes,** which differ between males and females. The nonsex chromosomes, which are the same for males and females, are called

autosomes. We think of sex in these organisms as being determined by the presence of the sex chromosomes, but in fact the individual genes located on the sex chromosomes are usually responsible for the sexual phenotypes.

XX-XO sex determination   The mechanism of sex determination in the grasshoppers studied by McClung is one of the simplest mechanisms of chromosomal sex determination and is called the XX-XO system. In this system, females have two X chromosomes (XX), and males possess a single X chromosome (XO). There is no O chromosome; the letter O signifies the absence of a sex chromosome.

In meiosis in females, the two X chromosomes pair and then separate, with one X chromosome entering each haploid egg. In males, the single X chromosome segregates in meiosis to half the sperm cells—the other half receive no sex chromosome. Because males produce two different types of gametes with respect to the sex chromosomes, they are said to be the **heterogametic sex.** Females, which produce gametes that are all the same with respect to the sex chromosomes, are the **homogametic sex.** In the XX-XO system, the sex of an individual is therefore determined by which type of male gamete fertilizes the egg. X-bearing sperm unite with X-bearing eggs to produce XX zygotes, which eventually develop as females. Sperm lacking an X chromosome unite with X-bearing eggs to produce XO zygotes, which develop into males.

XX-XY sex determination   In many species, the cells of males and females have the same number of chromosomes, but the cells of females have two X chromosomes (XX) and the cells of males have a single X chromosome and a smaller sex chromosome called the Y chromosome (XY). In humans and many other organisms, the Y chromosome is acrocentric (FIGURE 4.6), not Y shaped as is commonly assumed. In this type of sex-determining system, the male is the heterogametic sex—half of his gametes have an X chromosome and half have a Y chromosome. The female is the

4.6 The X and Y chromosomes in humans differ in size and genetic content. They are homologous only at the pseudoautosomal regions.

homogametic sex—all her egg cells contain a single X chromosome. Many organisms, including some plants, insects, and reptiles, and all mammals (including humans), have the XX-XY sex-determining system.

Although the X and Y chromosomes are not generally homologous, they do pair and segregate into different cells in meiosis. They can pair because these chromosomes are homologous at small regions called the **pseudoautosomal regions** (see Figure 4.6), in which they carry the same genes. Genes found in these regions will display the same pattern of inheritance as that of genes located on autosomal chromosomes. In humans, there are pseudoautosomal regions at both tips of the X and Y chromosomes.

ZZ-ZW sex determination In this system, the female is heterogametic and the male is homogametic. To prevent confusion with the XX-XY system, the sex chromosomes in this system are labeled Z and W, but the chromosomes do not resemble Zs and Ws. Females in this system are ZW; after meiosis, half of the eggs have a Z chromosome and the other half have a W. Males are ZZ; all sperm contain a single Z chromosome. The ZZ-ZW system is found in birds, moths, some amphibians, and some fishes.

> **Concepts**
>
> In XX-XO sex determination, the male is XO and heterogametic, and the female is XX and homogametic. In XX-XY sex determination, the male is XY and the female is XX; in this system the male is heterogametic. In ZZ-ZW sex determination, the female is ZW and the male is ZZ; in this system the female is the heterogametic sex.

Haplodiploidy Some insects in the order Hymenoptera (bees, wasps, and ants) have no sex chromosomes; instead, sex is based on the number of chromosome sets found in the nucleus of each cell. Males develop from unfertilized eggs, and females develop from fertilized eggs. The cells of male hymenopterans possess only a single set of chromosomes (they are haploid) inherited from the mother. In contrast, the cells of females possess two sets of chromosomes (they are diploid), one set inherited from the mother and the other set from the father (◀FIGURE 4.7).

The haplodiploid method of sex determination produces some odd genetic relationships. When both parents are diploid, siblings on average have half their genes in common because they have a 50% chance of receiving the same allele from each parent. In these insects, males produce sperm by mitosis (they are already haploid); so all offspring receive the same set of paternal genes. The diploid females produce eggs by normal meiosis. Therefore, sisters have a 50% chance of

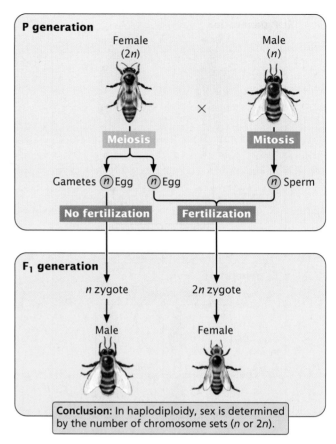

◀4.7 In insects with haplodiploidy, males develop from unfertilized eggs and are haploid; females develop from fertilized eggs and are diploid.

receiving the same allele from their mother and a 100% chance of receiving the same allele from their father; the average relatedness between sisters is therefore 75%. Brothers have a 50% chance of receiving the same copy of each of their mother's two alleles at any particular locus; so their average relatedness is only 50%. The greater genetic relatedness among female siblings in insects with haplodiploid sex determination may contribute to the high degree of social cooperation that exists among females (the workers) of these insects.

> **Concepts**
>
> Some insects possess haplodiploid sex determination, in which males develop from unfertilized eggs and are haploid; females develop from fertilized eggs and are diploid.

Genetic Sex-Determining Systems

In some plants and protozoans, sex is genetically determined, but there are no obvious differences in the chromosomes of males and females—there are no sex chromosomes. These organisms have **genic sex determination;** genotypes at one or more loci determine the sex of an individual.

It is important to understand that, even in chromosomal sex-determining systems, sex is actually determined by individual genes. For example, in mammals, a gene (*SRY,* discussed later in this chapter) located on the Y chromosome determines the male phenotype. In both genic sex determination and chromosomal sex determination, sex is controlled by individual genes; the difference is that, with chromosomal sex determination, the chromosomes that carry those genes *appear* different in males and females.

Environmental Sex Determination

Genes have had a role in all of the examples of sex determination discussed thus far, but sex is determined fully or in part by environmental factors in a number of organisms.

One fascinating example of environmental sex determination is seen in the marine mollusk *Crepidula fornicata,* also known as the common slipper limpet (◀ FIGURE 4.8). Slipper limpets live in stacks, one on top of another. Each limpet begins life as a swimming larva. The first larva to settle on a solid, unoccupied substrate develops into a female limpet. It then produces chemicals that attract other larvae, which settle on top of it. These larvae develop into males, which then serve as mates for the limpet below. After a period of time, the males on top develop into females and, in turn, attract additional larvae that settle on top of the stack, develop into males, and serve as mates for the limpets under them. Limpets can form stacks of a dozen or more animals; the uppermost animals are always male. This type of sexual development is called **sequential hermaphroditism;** each individual animal can be both male and female, although not at the same time. In *Crepidula fornicata,* sex is determined environmentally by the limpet's position in the stack.

Environmental factors are also important in determining sex in many reptiles. Although most snakes and lizards have sex chromosomes, in many turtles, crocodiles, and alligators, temperature during embryonic development determines sexual phenotype. In turtles, for example, warm temperatures produce females during certain times of the year, whereas cool temperatures produce males. In alligators, the reverse is true.

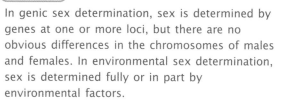

Concepts

In genic sex determination, sex is determined by genes at one or more loci, but there are no obvious differences in the chromosomes of males and females. In environmental sex determination, sex is determined fully or in part by environmental factors.

Sex Determination in *Drosophila*

The fruit fly *Drosophila melanogaster,* has eight chromosomes: three pairs of autosomes and one pair of sex chromosomes (◀ FIGURE 4.9). Normally, females have two X chromosomes and males have an X chromosome and a Y chromosome. However, the presence of the Y chromosome does not determine maleness in *Drosophila;* instead, each fly's sex is determined by a balance between genes on the autosomes and genes on the X chromosome. This type of sex determination is called the **genic balance system.** In this system, a number of genes seem to influence sexual development. The X chromosome contains genes with female-producing effects, whereas the autosomes contain genes with male-producing effects. Consequently, a fly's sex is determined by the **X:A ratio,** the number of X chromosomes divided by the number of haploid sets of autosomal chromosomes.

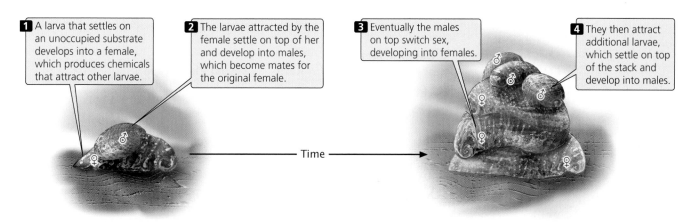

1 A larva that settles on an unoccupied substrate develops into a female, which produces chemicals that attract other larvae.

2 The larvae attracted by the female settle on top of her and develop into males, which become mates for the original female.

3 Eventually the males on top switch sex, developing into females.

4 They then attract additional larvae, which settle on top of the stack and develop into males.

Time

◀ 4.8 In *Crepidula fornicata,* the common slipper limpet, sex is determined by an environmental factor, the limpet's position in a stack of limpets.

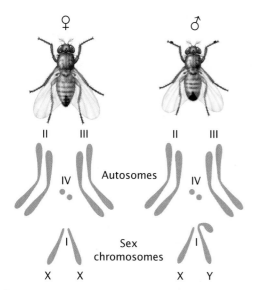

4.9 The chromosomes of *Drosophila melanogaster* (2*n* = 8) consist of three pairs of autosomes (labeled II, III, and IV) and one pair of sex chromosomes (labeled I, X and Y).

An X:A ratio of 1.0 produces a female fly; an X:A ratio of 0.5 produces a male. If the X:A ratio is less than 0.5, a male phenotype is produced, but the fly is weak and sterile—such flies are sometimes called metamales. An X:A ratio between 1.0 and 0.50 produces an intersex fly, with a mixture of male and female characteristics. If the X:A ratio is greater than 1.0, a female phenotype is produced, but these flies (called metafemales) have serious developmental problems and many never emerge from the pupal case. Table 4.1 presents some different chromosome complements in *Drosophila* and their associated sexual phenotypes. Flies with two sets of autosomes and XXY sex chromosomes (an X:A ratio of 1.0) develop as fully fertile

females, in spite of the presence of a Y chromosome. Flies with only a single X (an X:A ratio of 0.5), develop as males, although they are sterile. These observations confirm that the Y chromosome does not determine sex in *Drosophila*.

Mutations in genes that affect sexual phenotype in *Drosophila* have been isolated. For example, the *transformer* mutation converts a female with an X:A ratio of 1.0 into a phenotypic male, whereas the *doublesex* mutation transforms normal males and females into flies with intersex phenotypes. Environmental factors, such as the temperature of the rearing conditions, also can affect the development of sexual characteristics.

Concepts

The sexual phenotype of a fruit fly is determined by the ratio of the number of X chromosomes to the number of haploid sets of autosomal chromosomes (the X:A ratio).

www.whfreeman.com/pierce   Links to many Internet resources on the genetics of *Drosophila melanogaster*

Sex Determination in Humans

Humans, like *Drosophila*, have XX-XY sex determination, but in humans the presence of a gene on the Y chromosome determines maleness. The phenotypes that result from abnormal numbers of sex chromosomes, which arise when the sex chromosomes do not segregate properly in meiosis or mitosis, illustrate the importance of the Y chromosome in human sex determination.

Turner syndrome Persons who have **Turner syndrome** are female; they do not undergo puberty and their female

Table 4.1	Chromosome complements and sexual phenotypes in *Drosophila*		
Sex-Chromosome Complement	**Haploid Sets of Autosomes**	**X:A Ratio**	**Sexual Phenotype**
XX	AA	1.0	Female
XY	AA	0.5	Male
XO	AA	0.5	Male
XXY	AA	1.0	Female
XXX	AA	1.5	Metafemale
XXXY	AA	1.5	Metafemale
XX	AAA	0.67	Intersex
XO	AAA	0.33	Metamale
XXXX	AAA	1.3	Metafemale

(a) 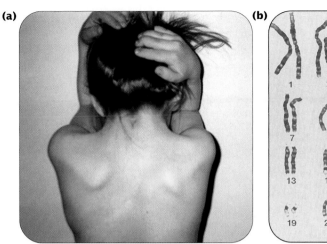   **(b)**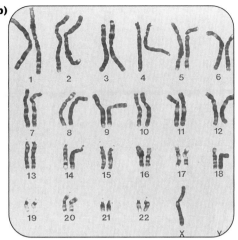

◀ 4.10 **Persons with Turner syndrome have a single X chromosome in their cells.** (a) Characteristic physical features include a low hairline and folds of skin on the neck. (b) Chromosomes from a person with Turner syndrome. (Part a, courtesy of Dr. Daniel C. Postellon, DeVos Children's Hospital; Part b, Dept. of Clinical Cytogenics, Addenbrookes Hospital/Science Photo Library/Photo Reseachers.)

secondary sex characteristics remain immature: menstruation is usually absent, breast development is slight, and pubic hair is sparse. This syndrome is seen in 1 of 3000 female births. Affected women are frequently short and have a low hairline, a relatively broad chest, and folds of skin on the neck (◀ FIGURE 4.10). Their intelligence is usually normal. Most women who have Turner syndrome are sterile. In 1959, C. E. Ford used new techniques to study human chromosomes and discovered that cells from a 14-year-old girl with Turner syndrome had only a single X chromosome; this chromosome complement is usually referred to as XO.

There are no known cases in which a person is missing both X chromosomes, an indication that at least one X chromosome is necessary for human development. Presumably, embryos missing both Xs are spontaneously aborted in the early stages of development.

Klinefelter syndrome

Persons who have **Klinefelter syndrome,** which occurs with a frequency of about 1 in 1000 male births, have cells with one or more Y chromosomes and multiple X chromosomes. The cells of most males having this condition are XXY, but cells of a few Klinefelter males are XXXY, XXXXY, or XXYY. Persons with this condition, though male, frequently have small testes, some breast enlargement, and reduced facial and pubic hair (◀ FIGURE 4.11). They are often taller than normal and sterile; most have normal intelligence.

Poly-X females

In about 1 in 1000 female births, the child's cells possess three X chromosomes, a condition often referred to as **triplo-X syndrome.** These persons have no distinctive features other than a tendency to be tall and thin. Although a few are sterile, many menstruate regularly and are fertile. The incidence of mental retardation among triple-X females is slightly greater than in the general population, but most XXX females have normal intelligence. Much rarer are women whose cells contain four or five X chromosomes. These women usually have normal female anatomy but are mentally retarded and have a number of physical problems. The severity of mental retardation increases as the number of X chromosomes increases beyond three.

www.whfreeman.com/pierce   Further information about sex-chromosomal abnormalities in humans

The role of sex chromosomes The phenotypes associated with sex-chromosome anomalies allow us to make several inferences about the role of sex chromosomes in human sex determination.

1. The X chromosome contains genetic information essential for both sexes; at least one copy of an X chromosome is required for human development.

2. The male-determining gene is located on the Y chromosome. A single copy of this chromosome, even in the presence of several X chromosomes, produces a male phenotype.

3. The absence of the Y chromosome results in a female phenotype.

4. Genes affecting fertility are located on the X and Y chromosomes. A female usually needs at least two copies of the X chromosome to be fertile.

5. Additional copies of the X chromosome may upset normal development in both males and females, producing physical and mental problems that increase as the number of extra X chromosomes increases.

(a)
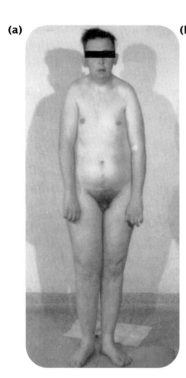

(b)
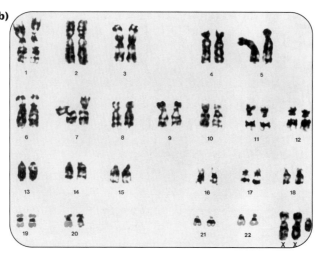

◀4.11 **Persons with Klinefelter syndrome have a Y chromosome and two or more X chromosomes in their cells.** (a) Characteristic physical features include small testes, some breast enlargement, and reduced facial and pubic hair. (b) Chromosomes of a person with Klinefelter syndrome. (Part a, from Plomin et al., *Behavioral Genetics*, 3rd edition, W. H. Freeman, 1997; part b, Biophoto Associates/Science Source/Photo Researchers.)

The male-determining gene in humans The Y chromosome in humans and all other mammals is of paramount importance in producing a male phenotype. However, scientists discovered a few rare XX males whose cells apparently lack a Y chromosome. For many years, these males presented a real enigma: How could a male phenotype exist without a Y chromosome? Close examination eventually revealed a small part of the Y chromosome attached to another chromosome. This finding indicates that it is not the entire Y chromosome that determines maleness in humans; rather, it is a gene on the Y chromosome.

Early in development, all humans possess undifferentiated gonads and both male and female reproductive ducts. Then, about 6 weeks after fertilization, a gene on the Y chromosome becomes active. By an unknown mechanism, this gene causes the neutral gonads to develop into testes, which begin to secrete two hormones: testosterone and Mullerian-inhibiting substance. Testosterone induces the development of male characteristics, and Mullerian-inhibiting substance causes the degeneration of the female reproductive ducts. In the absence of this male-determining gene, the neutral gonads become ovaries, and female features develop.

In 1987, David Page and his colleagues at the Massachusetts Institute of Technology located what appeared to be the male-determining gene near the tip of the short arm of the Y chromosome. They had examined the DNA of several XX males and XY females. The cells of one XX male that they studied possessed a very small piece of a Y chromosome attached to one of the Xs. This piece came from a section, called 1A, of the Y chromosome. Because this person had a male phenotype, they reasoned that the male-determining gene must reside within the 1A section of the Y chromosome.

Examination of the Y chromosome of a 12 year-old XY girl seemed to verify this conclusion. In spite of the fact that she possessed more than 99.8% of a Y chromosome, this XY person had a female phenotype. Page and his colleagues assumed that the male-determining gene must reside within the 0.2% of the Y chromosome that she was missing. Further examination showed that this Y chromosome was indeed missing part of section 1A. They then sequenced the DNA within section 1A of normal males and found a gene called *ZFY,* which appeared to be the testis-determining factor.

Within a few months, however, results from other laboratories suggested that *ZFY* might not in fact be the male-determining gene. Marsupials (pouched mammals), which also have XX-XY sex determination, were found to possess a *ZFY* gene on an autosomal chromosome, not on the Y chromosome. Furthermore, several human XX males were found who did not possess a copy of the *ZFY* gene.

A new candidate for the male-determining gene, called the **sex-determining region Y** (*SRY*) **gene,** was discovered in 1990 (◀ FIGURE 4.12). This gene is found in XX males and is missing from all XY females; it is also found on the Y chromosome of all mammals examined to date. Definitive proof that *SRY* is the male-determining gene came when scientists placed a copy of this gene into XX mice by means of genetic engineering. The XX mice that received this gene, although sterile, developed into anatomical males.

The *SRY* gene encodes a protein that binds to DNA and causes a sharp bend in the molecule. This alteration of DNA structure may affect the expression of other genes that

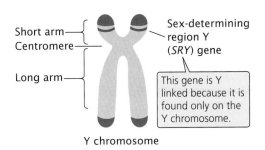

Short arm
Centromere
Long arm
Y chromosome

Sex-determining region Y (*SRY*) gene

This gene is Y linked because it is found only on the Y chromosome.

◀ 4.12 **The *SRY* gene is on the Y chromosome and causes the development of male characteristics.**

encode testis formation. Although *SRY* is the primary determinant of maleness in humans, other genes (some X linked, others Y linked, and still others autosomal) also play a role in fertility and the development of sex differences.

> **Concepts**
>
> The presence of the *SRY* gene on the Y chromosome causes a human embryo to develop as a male. In the absence of this gene, a human embryo develops as a female.

www.whfreeman.com/pierce   Additional information on the *SRY* gene

Androgen-insensitivity syndrome Several genes besides *SRY* influence sexual development in humans, as illustrated by women with androgen-insensitivity syndrome. These persons have female external sexual characteristics and psychological orientation. Indeed, most are unaware of their condition until they reach puberty and fail to menstruate. Examination by a gynecologist reveals that the vagina ends blindly and that the uterus, oviducts, and ovaries are absent. Inside the abdominal cavity lies a pair of testes, which produce levels of testosterone normally seen in males. The cells of a woman with androgen-insensitivity syndrome contain an X and a Y chromosome.

How can a person be female in appearance when her cells contain a Y chromosome and she has testes that produce testosterone? The answer lies in the complex relation between genes and sex in humans. In a human embryo with a Y chromosome, the *SRY* gene causes the gonads to develop into testes, which produce testosterone. Testosterone stimulates embryonic tissues to develop male characteristics. But, for testosterone to have its effects, it must bind to an androgen receptor. This receptor is defective in females with androgen-insensitivity syndrome; consequently, their cells are insensitive to testosterone, and female characteristics develop. The gene for the androgen receptor is located on the X chromosome; so persons with this condition always inherit it from their mothers. (All XY persons inherit the X chromosome from their mothers.)

Androgen-insensitivity syndrome illustrates several important points about the influence of genes on a person's sex. First, this condition demonstrates that human sexual development is a complex process, influenced not only by the *SRY* gene on the Y chromosome, but also by other genes found elsewhere. Second, it shows that most people carry genes for both male and female characteristics, as illustrated by the fact that those with androgen-insensitivity syndrome have the capacity to produce female characteristics, even though they have male chromosomes. Indeed, the genes for most male and female secondary sex characteristics are present not on the sex chromosomes but on autosomes. The key to maleness and femaleness lies not in the genes but in the control of their expression.

www.whfreeman.com/pierce   Additional information on androgen-insensitivity syndrome

Sex-Linked Characteristics

Sex-linked characteristics are determined by genes located on the sex chromosomes. Genes on the X chromosome determine **X-linked characteristics;** those on the Y chromosome determine **Y-linked characteristics.** Because little genetic information exists on the Y chromosome in many organisms, most sex-linked characteristics are X linked. Males and females differ in their sex chromosomes; so the pattern of inheritance for sex-linked characteristics differs from that exhibited by genes located on autosomal chromosomes.

X-Linked White Eyes in *Drosophila*

The first person to explain sex-linked inheritance was the American biologist Thomas Hunt Morgan (◀ FIGURE 4.13a). Morgan began his career as an embryologist, but the discovery of Mendel's principles inspired him to begin conducting genetic experiments, initially on mice and rats. In 1909, Morgan switched to *Drosophila melanogaster;* a year later, he discovered among the flies of his laboratory colony a single male that possessed white eyes, in stark contrast with the red eyes of normal fruit flies. This fly had a tremendous effect on the future of genetics and on Morgan's career as a biologist. With his white-eyed male, Morgan unraveled the mechanism of X-linked inheritance, ushering in the "golden age" of *Drosophila* genetics that lasted from 1910 until 1930.

Morgan's laboratory, located on the top floor of Schermerhorn Hall at Columbia University, became known as the Fly Room (◀ FIGURE 4.13b). To say that the Fly Room was unimpressive is an understatement. The cramped room, only about 16 X 23 feet, was filled with eight desks, each occupied by a student and his experiments. The primitive laboratory equipment consisted of little more than milk bottles for rearing the flies and hand-held lenses for observing their traits. Later, microscopes replaced the hand-held lenses, and crude incubators were added to maintain the fly

(a)

(b)

◄4.13 **Thomas Hunt Morgan's work with *Drosophila* helped unravel many basic principles in genetics, including X-linked inheritance.** (a) Morgan. (b) The Fly Room, where Morgan and his students conducted genetic research. (Part a, Wide World Photos; part b, American Philosophical Society.)

cultures, but even these additions did little to increase the physical sophistication of the laboratory. Morgan and his students were not tidy: cockroaches were abundant (living off spilled *Drosophila* food), dirty milk bottles filled the sink, ripe bananas—food for the flies—hung from the ceiling, and escaped fruit flies hovered everywhere.

In spite of its physical limitations, the Fly Room was the source of some of the most important research in the history of biology. There was daily excitement among the students, some of whom initially came to the laboratory as undergraduates. The close quarters facilitated informality and the free flow of ideas. Morgan and the Fly Room illustrate the tremendous importance of "atmosphere" in producing good science.

To explain the inheritance of the white-eyed characteristic in fruit flies, Morgan systematically carried out a series of genetic crosses. First, he crossed pure-breeding, red-eyed females with his white-eyed male, producing F_1 progeny that all had red eyes (◄Figure 4.14a). (In fact, Morgan found three white-eyed males among the 1237 progeny, but he assumed that the white eyes were due to new mutations.) Morgan's results from this initial cross were consistent with Mendel's principles: a cross between a homozygous dominant individual and a homozygous recessive individual produces heterozygous offspring exhibiting the dominant trait. His results suggested that white eyes were a simple recessive trait. However, when Morgan crossed the F_1 flies with one another, he found that all the female F_2 flies possessed red eyes but that half the male F_2 flies had red eyes and the other half had white eyes. This finding was clearly not the expected result for a simple recessive

trait, which should appear in $\frac{1}{4}$ of both male and female F_2 offspring.

To explain this unexpected result, Morgan proposed that the locus affecting eye color was on the X chromosome (that eye color was X linked). He recognized that the eye-color alleles were present only on the X chromosome—no homologous allele was present on the Y chromosome. Because the cells of females possess two X chromosomes, females could be homozygous or heterozygous for the eye-color alleles. The cells of males, on the other hand, possess only a single X chromosome and can carry only a single eye-color allele. Males therefore cannot be either homozygous or heterozygous but are said to be **hemizygous** for X-linked loci.

To verify his hypothesis that the white-eye trait is X linked, Morgan conducted additional crosses. He predicted that a cross between a white-eyed female and a red-eyed male would produce all red-eyed females and all white-eyed males (◄Figure 4.14b). When Morgan performed this cross, the results were exactly as predicted. Note that this cross is the reciprocal of the original cross and that the two reciprocal crosses produced different results in the F_1 and F_2 generations. Morgan also crossed the F_1 heterozygous females with their white-eyed father, the red-eyed F_2 females with white-eyed males, and white-eyed females with white-eyed males. In all of these crosses, the results were consistent with Morgan's conclusion that white eyes is an X-linked characteristic.

www.whfreeman.com/pierce   More information on the life of Thomas Hunt Morgan

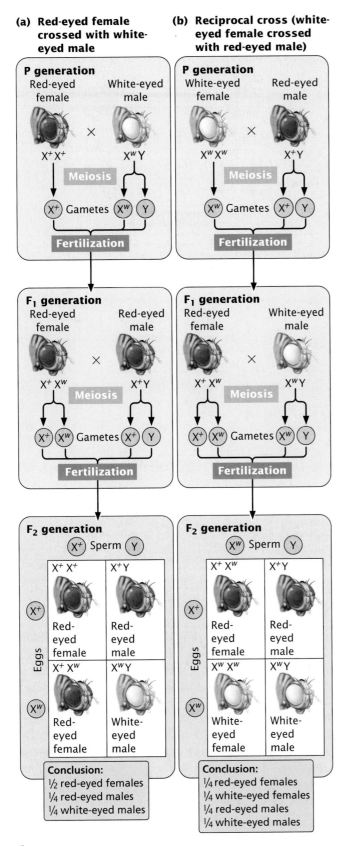

(a) Red-eyed female crossed with white-eyed male

(b) Reciprocal cross (white-eyed female crossed with red-eyed male)

◀ **4.14 Morgan's X-linked crosses for white eyes in fruit flies.** (a) Original and F₁ crosses. (b) Reciprocal crosses.

Nondisjunction and the Chromosome Theory of Inheritance

When Morgan crossed his original white-eyed male with homozygous red-eyed females, all 1237 of the progeny had red eyes, except for three white-eyed males. As already mentioned, Morgan attributed these white-eyed F₁ males to the occurrence of further mutations. However, flies with these unexpected phenotypes continued to appear in his crosses. Although uncommon, they appeared far too often to be due to mutation. Calvin Bridges, who was one of Morgan's students, set out to investigate the genetic basis of these exceptions.

Bridges found that, when he crossed a white-eyed female ($X^w X^w$) with a red-eyed male ($X^+ Y$), about 2.5% of the male offspring had red eyes and about 2.5% of the female offspring had white eyes. In this cross, every male fly should inherit its mother's X chromosome and should be $X^w Y$ with white eyes (◀ FIGURE 4.15a). Every female fly should inherit a dominant red-eye allele on its father's X chromosome, along with a white-eyed allele on its mother's X chromosome; thus, all the female progeny should be $X^+ X^w$ and have red eyes. The appearance of red-eyed males and white-eyed females in this cross was therefore unexpected.

To explain this result, Bridges hypothesized that, occasionally, the two X chromosomes in females fail to separate during anaphase I of meiosis. Bridges termed this failure of chromosomes to separate **nondisjunction.** When nondisjunction occurs, some of the eggs receive two copies of the X chromosome and others do not receive an X chromosome (◀ FIGURE 4.15b). If these eggs are fertilized by sperm from a red-eyed male, four combinations of sex chromosomes are produced. When an egg carrying two X chromosomes is fertilized by a Y-bearing sperm, the resulting zygote is $X^w X^w Y$. Sex in *Drosophila* is determined by the X:A ratio (see Table 4.1); in this case the X:A ratio is 1.0, so the $X^w X^w Y$ zygote develops into a white-eyed female. An egg with two X chromosomes that is fertilized by an X-bearing sperm produces $X^w X^w X^+$, which usually dies. An egg with no X chromosome that is fertilized by an X-bearing sperm produces $X^+ O$, which develops into a red-eyed male. If the egg with no X chromosome is fertilized by a Y-bearing sperm, the resulting zygote with only a Y chromosome and no X chromosome dies. Rare nondisjunction of the X chromosomes among white-eyed females therefore produces a few red-eyed males and white-eyed females, which is exactly what Bridges found in his crosses.

Bridges's hypothesis predicted that the white-eyed females would possess two X chromosomes and one Y and that red-eyed males would possess a single X chromosome. To verify his hypothesis, Bridges examined the chromosomes of his flies and found precisely what he predicted. The significance of Bridges's study was not that it explained

(a) White-eyed female and red-eyed male

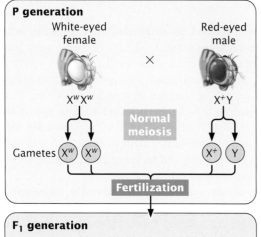

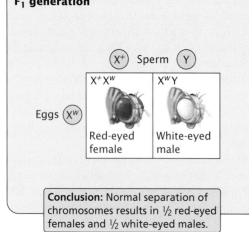

Conclusion: Normal separation of chromosomes results in ½ red-eyed females and ½ white-eyed males.

(b) White-eyed female and red-eyed male with nondisjunction

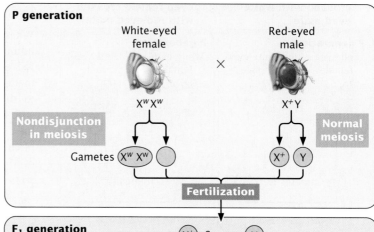

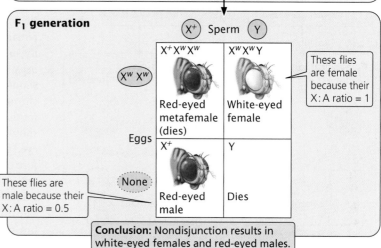

These flies are female because their X : A ratio = 1

These flies are male because their X : A ratio = 0.5

Conclusion: Nondisjunction results in white-eyed females and red-eyed males.

◀ 4.15 **Bridges conducted experiments that proved that the gene for white eyes is located on the X chromosome.** (a) A white-eyed female was crossed with a red-eyed male. (b) Rare nondisjunction produced a few eggs with two copies of the X^W chromosome and other eggs with no X chromosome.

the appearance of an occasional odd fly in his culture but that he was able to predict a fly's chromosomal makeup on the basis of its eye-color genotype. This association between genotype and chromosomes gave unequivocal evidence that sex-linked genes were located on the X chromosome and confirmed the chromosome theory of inheritance.

Concepts

By showing that the appearance of rare phenotypes was associated with the inheritance of particular chromosomes, Bridges proved that sex-linked genes are located on the X chromosome and that the chromosome theory of inheritance is correct.

X-Linked Color Blindness in Humans

To further examine X-linked inheritance, let's consider another X-linked characteristic: red–green color blindness in humans. Within the human eye, color is perceived in light-sensing cone cells that line the retina. Each cone cell contains one of three pigments capable of absorbing light of a particular wavelength; one absorbs blue light, a second absorbs red light, and a third absorbs green light. The human eye actually detects only three colors—red, green, and blue—but the brain mixes the signals from different cone cells to create the wide spectrum of colors that we perceive. Each of the three pigments is encoded by a separate locus; the locus for the blue pigment is found on chromosome 7, and those for green and red pigments lie close together on the X chromosome.

The most common types of human color blindness are caused by defects of the red and green pigments; we will refer to these conditions as red–green color blindness. Mutations that produce defective color vision are generally recessive and, because the genes coding for the red and green pigments are located on the X chromosome, red–green color blindness is inherited as an X-linked recessive characteristic.

We will use the symbol X^c to represent an allele for red–green color blindness and the symbol X^+ to represent an allele for normal color vision. Females possess two X chromosomes; so there are three possible genotypes among females: X^+X^+ and X^+X^c, which produce normal vision, and X^cX^c, which produces color blindness. Males have only a single X chromosome and two possible genotypes: X^+Y, which produces normal vision, and X^cY which produces color blindness.

If a color-blind man mates with a woman homozygous for normal color vision (◀FIGURE 4.16a), all of the gametes produced by the woman will contain an allele for normal color vision. Half of the man's gametes will receive the X chromosome with the color-blind allele, and the other half will receive the Y chromosome, which carries no alleles affecting color vision. When an X^c-bearing sperm unites with the X^+-bearing egg, a heterozygous female with normal vision (X^+X^c) is produced. When a Y-bearing sperm unites with the X^+-bearing egg, a hemizygous male with normal vision (X^+Y) is produced (see Figure 4.16a).

In the reciprocal cross between a color-blind woman and a man with normal color vision (◀FIGURE 4.16b), the woman produces only X^c-bearing gametes. The man produces some gametes that contain the X^+ chromosome and others that contain the Y chromosome. Males inherit the X chromosome

from their mothers; because both of the mother's X chromosomes bear the X^c allele in this case, all the male offspring will be color blind. In contrast, females inherit an X chromosome from both parents; thus the female offspring of this reciprocal cross will all be heterozygous with normal vision. Females are color blind only when color-blind alleles have been inherited from both parents, whereas a color-blind male need inherit a color-blind allele from his mother only; for this reason, color blindness and most other rare X-linked recessive characteristics are more common in males.

In these crosses for color blindness, notice that an affected woman passes the X-linked recessive trait to her sons but not to her daughters, whereas an affected man passes the trait to his grandsons through his daughters but never to his sons. X-linked recessive characteristics seem to alternate between the sexes, appearing in females one generation and in males the next generation; thus, this pattern of inheritance exhibited by X-linked recessive characteristics is sometimes called *crisscross inheritance.*

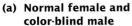

Characteristics determined by genes on the sex chromosomes are called sex-linked characteristics. Diploid females have two alleles at each X-linked locus, whereas diploid males possess a single allele at each X-linked locus. Females inherit X-linked alleles from both parents, but males inherit a single X-linked allele from their mothers.

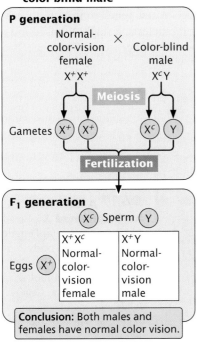

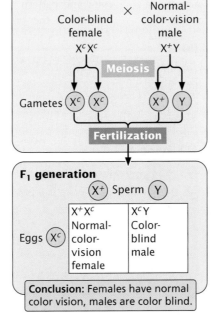

◀4.16 Red–green color blindness is inherited as an X-linked recessive trait in humans.

Symbols for X-Linked Genes

There are several different ways to record genotypes for X-linked traits. Sometimes the genotypes are recorded in the same fashion as for autosomal characteristics — the hemizygous males are simply given a single allele: the genotype of a female *Drosophila* with white eyes would be *ww*, and the genotype of a white-eyed hemizygous male would be *w*. Another method is to include the Y chromosome, designating it with a diagonal slash (/). With this method, the white-eyed female's genotype would still be *ww* and the white-eyed male's genotype would be *w/*. Perhaps the most useful method is to write the X and Y chromosomes in the genotype, designating the X-linked alleles with superscripts, as we have done in this chapter. With this method, a white-eyed female would be X^wX^w and a white-eyed male X^wY. Using Xs and Ys in the genotype has the advantage of reminding us that the genes are X linked and that the male must always have a single allele, inherited from the mother.

Dosage Compensation

The presence of different numbers of X chromosomes in males and females presents a special problem in development. Because females have two copies of every X-linked gene and males possess one copy, the amount of gene product (protein) from X-linked genes would normally differ in the two sexes — females would produce twice as much gene product as males. This difference could be highly detrimental because protein concentration plays a critical role in development. Animals overcome this potential problem through **dosage compensation,** which equalizes the amount of protein produced by X-linked genes in the two sexes. In fruit flies, dosage compensation is achieved by a doubling of the activity of the genes on the X chromosome of the male. In the worm *Caenorhabditis elegans*, it is achieved by a halving of the activity of genes on both of the X chromosomes in the female. Pla-

cental mammals use yet another mechanism of dosage compensation; genes on one of the X chromosomes in the female are completely inactivated.

In 1949, Murray Barr observed condensed, darkly staining bodies in the nuclei of cells from female cats (◀FIGURE 4.17); this darkly staining structure became known as a **Barr body.** Mary Lyon proposed in 1961 that the Barr body was an inactive X chromosome; her hypothesis (now proved) has become known as the **Lyon hypothesis.** She suggested that, within each female cell, one of the two X chromosomes becomes inactive; which X chromosome is inactivated is random. If a cell contains more than two X chromosomes, all but one of them is inactivated. The number of Barr bodies present in human cells with different complements of sex chromosomes is shown in Table 4.2.

As a result of X inactivation, females are functionally hemizygous at the cellular level for X-linked genes. In females that are heterozygous at an X-linked locus, approximately 50% of the cells will express one allele and 50% will express the other allele; thus, in heterozygous females, proteins encoded by both alleles are produced, although not within the same cell. This functional hemizygosity means that cells in females are not identical with respect to the expression of the genes on the X chromosome; females are mosaics for the expression of X-linked genes.

X inactivation takes place relatively early in development — in humans, within the first few weeks of development. Once an X chromosome becomes inactive in a cell, it remains inactivated and is inactive in all somatic cells that descend from the cell. Thus, neighboring cells tend to have the same X chromosome inactivated, producing a patchy pattern (mosaic) for the expression of an X-linked characteristic in heterozygous females.

This patchy distribution can be seen in tortoiseshell cats (◀FIGURE 4.18). Although many genes contribute to coat color and pattern in domestic cats, a single X-linked locus determines the presence of orange color. There are two

(a)

(b)

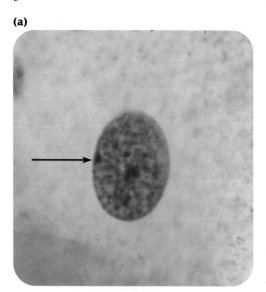

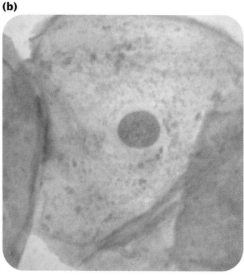

◀4.17 **A Barr body is an inactivated X chromosome.** (a) Female cell with a Barr body (indicated by arrow). (b) Male cell without a Barr body. (Part a, George Wilder/Visuals Unlimited; part b, M. Abbey/Photo Researchers.)

Table 4.2	Number of Barr bodies in human cells with different complements of sex chromosomes	
Sex Chromosomes	**Syndrome**	**Number of Barr Bodies**
XX	None	1
XY	None	0
XO	Turner	0
XXY	Klinefelter	1
XXYY	Klinefelter	1
XXXY	Klinefelter	2
XXXXY	Klinefelter	3
XXX	Triplo-X	2
XXXX	Poly-X female	3
XXXXX	Poly-X female	4

◀ **4.18 The patchy distribution of color on tortoiseshell cats results from the random inactivation of one X chromosome in females.** (David Falconer/Words & Pictures/Picture Quest.)

possible alleles at this locus: X^+, which produces nonorange (usually black) fur, and X^o, which produces orange fur. Males are hemizygous and thus may be black (X^+Y) or orange (X^oY) but not black *and* orange. (Rare tortoiseshell males can arise from the presence of two X chromosomes, X^+X^oY.) Females may be black (X^+X^+), orange (X^oX^o), or tortoiseshell (X^+X^o), the tortoiseshell pattern arising from a patchy mixture of black and orange fur. Each orange patch is a clone of cells derived from an original cell with the black allele inactivated, and each black patch is a clone of cells derived from an original cell with the orange allele inactivated. The mosaic pattern of gene expression associated with dosage compensation also produces the patchy distribution of sweat glands in women heterozygous for anhidrotic ectodermal dysplasia (see introduction to this chapter).

Lyon's hypothesis suggests that the presence of variable numbers of X chromosomes should not be detrimental in mammals, because any X chromosomes beyond one should be inactivated. However, persons with Turner syndrome (XO) differ from normal females, and those with Klinefelter syndrome (XXY) differ from normal males. How do these conditions arise in the face of dosage compensation? The reason may lie partly in the fact that there is a short period of time, very early in development, when all X chromosomes are active. If the number of X chromosomes is abnormal, any X-linked genes expressed during this early period will produce abnormal levels of gene product. Furthermore, the phenotypic abnormalities may arise because some X-linked genes escape inactivation, although how they do so isn't known.

Exactly how an X chromosome becomes inactivated is not completely understood either, but it appears to entail the addition of methyl groups ($-CH_3$) to the DNA. The *XIST*

(for X inactive-specific transcript) gene, located on the X chromosome, is required for inactivation. Only the copy of *XIST* on the inactivated X chromosome is expressed, and it continues to be expressed during inactivation (unlike most other genes on the inactivated X chromosome). Interestingly, *XIST* does not encode a protein; it produces an RNA molecule that binds to the inactivated X chromosome. This binding is thought to prevent the attachment of other proteins that participate in transcription and, in this way, it brings about X inactivation.

Concepts

In mammals, dosage compensation ensures that the same amount of X-linked gene product will be produced in the cells of both males and females. All but one X chromosome is randomly inactivated in each cell; which X chromosome is inactivated is random and varies from cell to cell.

www.whfreeman.com/pierce   Current information on *XIST* and X-chromosome inactivation in humans

Z-Linked Characteristics

In organisms with ZZ-ZW sex determination, the males are the homogametic sex (ZZ) and carry two sex-linked (usually referred to as Z-linked) alleles; thus males may be homozygous or heterozygous. Females are the heterogametic sex (ZW) and possess only a single Z-linked allele. Inheritance of Z-linked characteristics is the same as that of X-linked characteristics, except that the pattern of inheritance in males and females is reversed.

An example of a Z-linked characteristic is the cameo phenotype in Indian blue peafowl (*Pavo cristatus*). In these birds, the wild-type plumage is a glossy, metallic blue. The female peafowl is ZW and the male is ZZ. Cameo plumage, which produces brown feathers, results from a Z-linked allele (Z^{ca}) that is recessive to the wild-type blue allele (Z^{Ca+}). If a blue-colored female ($Z^{Ca+}W$) is crossed with a cameo male ($Z^{ca}Z^{ca}$), all the F_1 females are cameo ($Z^{ca}W$) and all the F_1 males are blue ($Z^{Ca+}Z^{ca}$) (◀ FIGURE 4.19). When the F_1 are interbred, $1/4$ of the F_2 are blue males ($Z^{Ca+}Z^{ca}$), $1/4$ are blue females ($Z^{Ca+}W$), $1/4$ are cameo males ($Z^{ca}Z^{ca}$), and $1/4$ are cameo females ($Z^{ca}W$). The reciprocal cross of a cameo female with a homozygous blue male produces an F_1 generation in which all offspring are blue and an F_2 consisting of $1/2$ blue males ($Z^{Ca+}Z^{ca}$ and $Z^{Ca+}Z^{Ca+}$), $1/4$ blue females ($Z^{Ca+}W$), and $1/4$ cameo females ($Z^{ca}W$).

In organisms with ZZ-ZW sex determination, the female always inherits her W chromosome from her mother, and she inherits her Z chromosome, along with any Z-linked alleles, from her father. In this system, the male inherits Z chromosomes, along with any Z-linked alleles, from both the mother and the father. This pattern of inheritance is the reverse of X-linked alleles in organisms with XX-XY sex determination.

Y-Linked Characteristics

Y-linked traits exhibit a distinct pattern of inheritance and are present only in males, because only males possess a Y chromosome. All male offspring of a male with a Y-linked trait will display the trait (provided that the penetrance—see Chapter 3—is 100%), because every male inherits the Y chromosome from his father.

In humans and many other organisms, there is relatively little genetic information on the Y chromosome, and few characteristics exhibit Y-linked inheritance. Only a few more than 20 genes have been identified outside the pseudoautosomal region on the human Y chromosome, including the *SRY* gene and the *ZFY* gene. A possible Y-linked human trait is hairy ears, a trait that is common among men in some parts of the Middle East and India, affecting as many as 70% of adult men in some regions. This trait displays variable expressivity—some men have only a few hairs on the outer ear, whereas others have ears that are covered with hair. The age at which this trait appears also is quite variable.

Only men have hairy ears and, in many families, the occurrence of the trait is entirely consistent with Y-linked inheritance. In a few families, however, not all sons of an affected man display the trait, which implies that the trait has incomplete penetrance. Some investigators have concluded that the hairy-ears trait is not Y-linked, but instead is an autosomal dominant trait expressed only in men (sex-limited expression, discussed more fully in Chapter 5). Distinguishing between a Y-linked characteristic with incomplete penetrance and an autosomal dominant characteristic expressed only in males is difficult, and the pattern of inheritance of hairy ears is consistent with both modes of inheritance.

The function of most Y-linked genes is poorly understood, but some appear to influence male sexual development

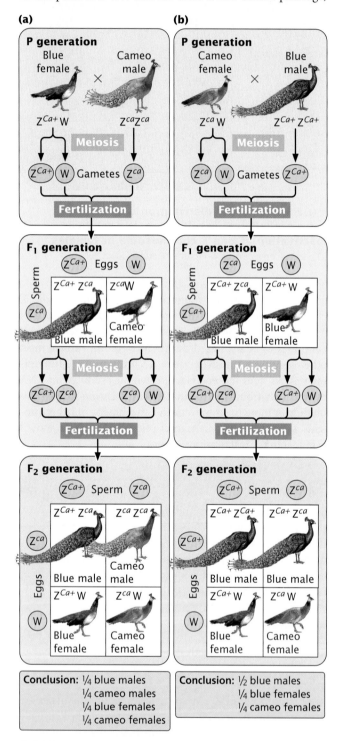

◀ **4.19 Inheritance of the cameo phenotype in Indian blue peafowl is inherited as a Z-linked recessive trait.** (a) Blue female crossed with cameo male. (b) Reciprocal cross of cameo female crossed with homozygous blue male.

and fertility. Some Y-linked genes have counterparts on the X chromosome that encode similar proteins in females.

DNA sequences in the Y chromosome undergo mutation over time and vary among individuals. Like Y-linked traits, these variants—called genetic markers—are passed from father to son and can be used to study male ancestry. Although the markers themselves do not code for any physical traits, they can be detected with molecular methods. Much of the Y chromosome is nonfunctional; so mutations readily accumulate. Many of these mutations are unique; they arise only once and are passed down through the generations without recombination. Individuals possessing the same set of mutations are therefore related, and the distribution of these genetic markers on Y chromosomes provides clues about genetic relationships of present-day people.

Y-linked markers have been used to study the offspring of Thomas Jefferson, principal author of the Declaration of Independence and third president of the United States. In 1802, Jefferson was accused by a political enemy of fathering a child by his slave Sally Hemings, but the evidence was circumstantial. Hemings, who worked in the Jefferson household and accompanied Jefferson on a trip to Paris, had five children. Jefferson was accused of fathering the first child, Tom, but rumors about the paternity of the other children circulated as well. Hemings's last child, Eston, bore a striking resemblance to Jefferson, and her fourth child, Madison, testified late in life that Jefferson was the father of all Hemings's children. Ancestors of Hemings's children maintained that they were descendants of the Jefferson line, but some Jefferson descendants refused to recognize their claim.

To resolve this long-standing controversy, geneticists examined markers from the Y chromosomes of male-line descendants of Hemings's first son (Thomas Woodson), her last son (Eston Hemings), and a paternal uncle of Thomas Jefferson with whom Jefferson had Y chromosomes in common. (Descendants of Jefferson's uncle were used because Jefferson himself had no verified male descendants.) Geneticists determined that Jefferson possessed a rare and distinctive set of genetic markers on his Y chromosome. The same markers were also found on the Y chromosomes of the male-line descendants of Eston Hemings. The probability of such a match arising by chance is less than 1%. (The markers were not found on the Y chromosomes of the descendants of Thomas Woodson.) Together with the circumstantial historical evidence, these matching markers strongly suggest that Jefferson fathered Eston Hemings but not Thomas Woodson.

Another study utilizing Y-linked genetic markers focused on the origins of the Lemba, an African tribe comprising 50,000 people who reside in South Africa and parts of Zimbabwe. Members of the Lemba tribe are commonly referred to as the black Jews of South Africa. This name derives from cultural practices of the tribe, including circumcision and food taboos, which superficially resemble those of Jewish people. Lemba oral tradition suggests that the tribe came

from "Sena in the north by boat," Sena being variously identified as Sanaa in Yemen, Judea, Egypt, or Ethiopia. Legend says that the original group was entirely male, that half of their number was lost at sea, and that the survivors made their way to the coast of Africa, where they settled.

Today, most Lemba belong to Christian churches, are Muslims, or claim to be Lemba in religion. Their religious practices have little in common with Judaism and, with the exception of their oral tradition and a few cultural practices, there is little to suggest a Jewish origin.

To reveal the genetic origin of the Lemba, scientists examined genetic markers on their Y chromosomes. Swabs of cheek cells were collected from 399 males in several populations: the Lemba in Africa, Bantu (another South African tribe), two groups from Yemen, and several groups of Jews. DNA was extracted and analyzed for alleles at 12 loci. This analysis of genetic markers revealed that Y chromosomes in the Lemba were of two types: those of Bantu origin and those similar to chromosomes found in Jewish and Yemen populations. Most importantly, members of one Lemba clan carried a large number of Y chromosomes that had a rare combination of alleles also found on the Y chromosomes of members of the Jewish priesthood. This set of alleles is thought to be an important indicator of Judaic origin. These findings are consistent with the Lemba oral tradition and strongly suggest a genetic contribution from Jewish populations.

> **Concepts**
>
> Y-linked characteristics exhibit a distinct pattern of inheritance: they are present only in males, and all male offspring of a male with a Y-linked trait inherit the trait.

 www.whfreeman.com/pierce   An overview of the use of Y-linked markers in studies of ancestry

Connecting Concepts

Recognizing Sex-linked Inheritance

What features should we look for to identify a trait as sex linked? A common misconception is that any genetic characteristic in which the phenotypes of males and females differ must be sex linked. In fact, the expression of many *autosomal* characteristics differs between males and females. The genes that code for these characteristics are the same in both sexes, but their expression is influenced by sex hormones. The different sex hormones of males and females cause the same genes to generate different phenotypes in males and females.

Another misconception is that any characteristic that is found more frequently in one sex is sex linked. A number of

3. Chickens, like all birds, have ZZ-ZW sex determination. The bar-feathered phenotype in chickens results from a Z-linked allele that is dominant over the allele for nonbar feathers. A barred female is crossed with a nonbarred male. The F_1 from this cross are intercrossed to produce the F_2. What will the phenotypes and their proportions be in the F_1 and F_2 progeny?

• Solution

With the ZZ-ZW system of sex determination, females are the heterogametic sex, possessing a Z chromosome and a W chromosome; males are the homogametic sex, with two Z chromosomes. In this problem, the barred female is hemizygous for the bar phenotype (Z^BW). Because bar is dominant over nonbar, the nonbarred male must be homozygous for nonbar (Z^bZ^b). Crossing these two chickens, we obtain:

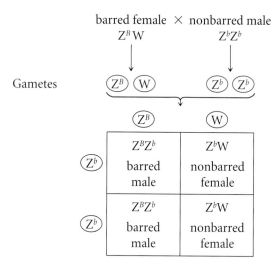

Thus, all the males in the F_1 will be barred (Z^BZ^b), and all the females will be nonbarred (Z^bW).

The F_1 are now crossed to produce the F_2:

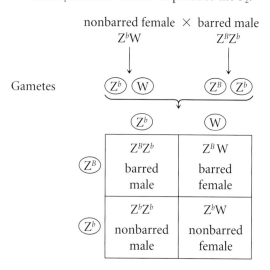

So, ¼ of the F_2 are barred males, ¼ are nonbarred males, ¼ are barred females, and ¼ are nonbarred females.

4. In *Drosophila melanogaster*, forked bristles are caused by an allele (X^f) that is X linked and recessive to an allele for normal bristles (X^+). Brown eyes are caused by an allele (b) that is autosomal and recessive to an allele for red eyes (b^+). A female fly that is homozygous for normal bristles and red eyes mates with a male fly that has forked bristles and brown eyes. The F_1 are intercrossed to produce the F_2. What will the phenotypes and proportions of the F_2 flies be from this cross?

• Solution

This problem is best worked by breaking the cross down into two separate crosses, one for the X-linked genes that determine the type of bristles and one for the autosomal genes that determine eye color.

Let's begin with the autosomal characteristics. A female fly that is homozygous for red eyes (b^+b^+) is crossed with a male with brown eyes. Because brown eyes are recessive, the male fly must be homozygous for the brown-eyed allele (bb). All of the offspring of this cross will be heterozygous (b^+b) and will have brown eyes:

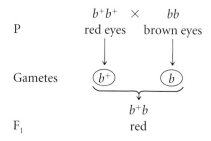

The F_1 are then intercrossed to produce the F_2. Whenever two individuals heterozygous for an autosomal recessive characteristic are crossed, ¾ of the offspring will have the dominant trait and ¼ will have the recessive trait; thus, ¾ of the F_2 flies will have red eyes and ¼ will have brown eyes:

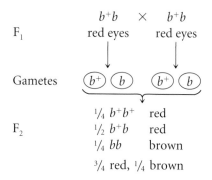

Next, we work out the results for the X-linked characteristic. A female that is homozygous for normal bristles (X^+X^+) is crossed with a male that has forked bristles (X^fY). The female F_1 from this cross are heterozygous (X^+X^f), receiving an X chromosome with a normal-bristle allele from their mother (X^+) and an X chromosome with a forked-bristle allele (X^f) from their father. The male F_1 are hemizygous (X^+Y), receiving an X chromosome with a normal-bristle allele from their mother (X^+) and a Y chromosome from their father:

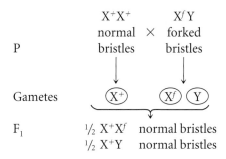

P $\quad$ X^+X^+ normal bristles $\times$ $X^f Y$ forked bristles

Gametes $\quad$ X^+ $\quad$ X^f Y

F_1 $\quad$ $\frac{1}{2}$ X^+X^f normal bristles
$\quad$ $\frac{1}{2}$ X^+Y normal bristles

When these F_1 are intercrossed, $\frac{1}{2}$ of the F_2 will be normal-bristle females, $\frac{1}{4}$ will be normal-bristle males, and $\frac{1}{4}$ will be forked-bristle males:

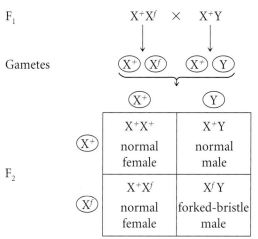

F_1 $\quad$ X^+X^f $\times$ X^+Y

Gametes $\quad$ X^+ X^f $\quad$ X^+ Y

F_2

	X^+	Y
X^+	X^+X^+ normal female	X^+Y normal male
X^f	X^+X^f normal female	$X^f Y$ forked-bristle male

$\frac{1}{2}$ normal female, $\frac{1}{4}$ normal male, $\frac{1}{4}$ forked-bristle male

To obtain the phenotypic ratio in the F_2, we now combine these two crosses by using the multiplicative rule of probability and the branch diagram:

Eye color	Bristle and sex	F_2 phenotype	Probability
red ($\frac{3}{4}$)	normal female ($\frac{1}{2}$)	red normal female	$\frac{3}{4} \times \frac{1}{2} = \frac{3}{8}$ $= \frac{6}{16}$
	normal male ($\frac{1}{4}$)	red normal male	$\frac{3}{4} \times \frac{1}{4} = \frac{3}{16}$
	forked-bristle male ($\frac{1}{4}$)	red forked-bristle male	$\frac{3}{4} \times \frac{1}{4} = \frac{3}{16}$
brown ($\frac{1}{4}$)	normal female ($\frac{1}{2}$)	brown normal female	$\frac{1}{4} \times \frac{1}{2} = \frac{1}{8}$ $= \frac{2}{16}$
	normal male ($\frac{1}{4}$)	brown normal male	$\frac{1}{4} \times \frac{1}{4} = \frac{1}{16}$
	forked-bristle male ($\frac{1}{4}$)	brown forked-bristle male	$\frac{1}{4} \times \frac{1}{4} = \frac{1}{16}$

COMPREHENSION QUESTIONS

* 1. What is the most defining difference between males and females?

2. How do monoecious organisms differ from dioecious organisms?

3. Describe the XX-XO system of sex determination. In this system, which is the heterogametic sex and which is the homogametic sex?

4. How does sex determination in the XX-XY system differ from sex determination in the ZZ-ZW system?

* 5. What is the pseudoautosomal region? How does the inheritance of genes in this region differ from the inheritance of other Y-linked characteristics?

* 6. How is sex determined in insects with haplodiploid sex determination?

7. What is meant by genic sex determination?

8. How does sex determination in *Drosophila* differ from sex determination in humans?

9. Give the typical sex chromosomes found in the cells of people with Turner syndrome, Klinefelter syndrome, and androgen insensitivity syndrome, as well as in poly-X females.

* 10. What characteristics are exhibited by an X-linked trait?

11. Explain how Bridges's study of nondisjunction in *Drosophila* helped prove the chromosome theory of inheritance.

12. Explain why tortoiseshell cats are almost always female and why they have a patchy distribution of orange and black fur.

13. What is a Barr body? How is it related to the Lyon hypothesis?

* 14. What characteristics are exhibited by a Y-linked trait?

*15. What is the sexual phenotype of fruit flies having the following chromosomes?

	Sex chromosomes	Autosomal chromosomes
(a)	XX	all normal
(b)	XY	all normal
(c)	XO	all normal
(d)	XXY	all normal
(e)	XYY	all normal
(f)	XXYY	all normal
(g)	XXX	all normal
(h)	XX	four haploid sets
(i)	XXX	four haploid sets
(j)	XXX	three haploid sets
(k)	X	three haploid sets
(l)	XY	three haploid sets
(m)	XX	three haploid sets

16. For parts a through g in problem 15 what would the human sexual phenotype (male or female) be?

*17. Joe has classic hemophilia, which is an X-linked recessive disease. Could Joe have inherited the gene for this disease from the following persons?

	Yes	No
(a) His mother's mother	____	____
(b) His mother's father	____	____
(c) His father's mother	____	____
(d) His father's father	____	____

*18. In *Drosophila*, yellow body is due to an X-linked gene that is recessive to the gene for gray body.

(a) A homozygous gray female is crossed with a yellow male. The F_1 are intercrossed to produce F_2. Give the genotypes and phenotypes, along with the expected proportions, of the F_1 and F_2 progeny.

(b) A yellow female is crossed with a gray male. The F_1 are intercrossed to produce the F_2. Give the genotypes and phenotypes, along with the expected proportions, of the F_1 and F_2 progeny.

(c) A yellow female is crossed with a gray male. The F_1 females are backcrossed with gray males. Give the genotypes and phenotypes, along with the expected proportions, of the F_2 progeny.

(d) If the F_2 flies in part b mate randomly, what are the expected phenotypic proportions of flies in the F_3?

*19. Both John and Cathy have normal color vision. After 10 years of marriage to John, Cathy gave birth to a color-blind daughter. John filed for divorce, claiming he is not the father of the child. Is John justified in his claim of nonpaternity? Explain why. If Cathy had given birth to a color-blind son, would John be justified in claiming nonpaternity?

20. Red–green color blindness in humans is due to an X-linked recessive gene. A woman whose father is color blind possesses one eye with normal color vision and one eye with color blindness.

(a) Propose an explanation for this woman's vision pattern.

(b) Would it be possible for a man to have one eye with normal color vision and one eye with color blindness?

*21. Bob has XXY chromosomes (Klinefelter syndrome) and is color blind. His mother and father have normal color vision, but his maternal grandfather is color blind. Assume that Bob's chromosome abnormality arose from nondisjunction in meiosis. In which parent and in which meiotic division did nondisjunction occur? Explain your answer.

22. In certain salamanders, it is possible to alter the sex of a genetic female, making her into a functional male; these salamanders are called sex-reversed males. When a sex-reversed male is mated with a normal female, approximately $2/3$ of the offspring are female and $1/3$ are male. How is sex determined in these salamanders? Explain the results of this cross.

23. In some mites, males pass genes to their grandsons, but they never pass genes to male offspring. Explain.

24. The Talmud, an ancient book of Jewish civil and religious laws, states that if a woman bears two sons who die of bleeding after circumcision (removal of the foreskin from the penis), any additional sons that she has should not be circumcised. (The bleeding is most likely due to the X-linked disorder hemophilia.) Furthermore, the Talmud states that the sons of her sisters must not be circumcised, whereas the sons of her brothers should. Is this religious law consistent with sound genetic principles? Explain your answer.

*25. Miniature wings (X^m) in *Drosophila* result from an X-linked allele that is recessive to the allele for long wings (X^+). Give the genotypes of the parents in the following crosses.

Male parent	Female parent	Male offspring	Female offspring
(a) long	long	231 long, 250 miniature	560 long
(b) miniature	long	610 long	632 long
(c) miniature	long	410 long, 417 miniature	412 long, 415 miniature
(d) long	miniature	753 miniature	761 long
(e) long	long	625 long	630 long

*26. In chickens, congenital baldness results from a Z-linked recessive gene. A bald rooster is mated with a normal hen.

The F_1 from this cross are interbred to produce the F_2. Give the genotypes and phenotypes, along with their expected proportions, among the F_1 and F_2 progeny.

27. In the eastern mosquito fish *(Gambusia affinis holbrooki)*, which has XX-XY sex determination, spotting is inherited as a Y-linked trait. The trait exhibits 100% penetrance when the fish are raised at 22°C, but the penetrance drops to 42% when the fish are raised at 26°C. A male with spots is crossed with a female without spots, and the F_1 are intercrossed to produce the F_2. If all the offspring are raised at 22°C, what proportion of the F_1 and F_2 will have spots? If all the offspring are raised at 26°C, what proportion of the F_1 and F_2 will have spots?

*28. How many Barr bodies would you expect to see in human cells containing the following chromosomes?

(a) XX	**(d)** XXY	**(g)** XYY
(b) XY	**(e)** XXYY	**(h)** XXX
(c) XO	**(f)** XXXY	**(i)** XXXX

29. Red–green color blindness is an X-linked recessive trait in humans. Polydactyly (extra fingers and toes) is an autosomal dominant trait. Martha has normal fingers and toes and normal color vision. Her mother is normal in all respects, but her father is color blind and polydactylous. Bill is color blind and polydactylous. His mother has normal color vision and normal fingers and toes. If Bill and Martha marry, what types and proportions of children can they produce?

*30. Miniature wings in *Drosophila melanogaster* result from an X-linked gene (X^m) that is recessive to an allele for long wings (X^+). Sepia eyes are produced by an autosomal gene (*s*) that is recessive to an allele for red eyes (s^+).

(a) A female fly that has miniature wings and sepia eyes is crossed with a male that has normal wings and is homozygous for red eyes. The F_1 are intercrossed to produce the F_2. Give the phenotypes and their proportions expected in the F_1 and F_2 flies from this cross.

(b) A female fly that is homozygous for normal wings and has sepia eyes is crossed with a male that has miniature wings and is homozygous for red eyes. The F_1 are intercrossed to produce the F_2. Give the phenotypes and proportions expected in the F_1 and F_2 flies from this cross.

31. Suppose that a recessive gene that produces a short tail in mice is located in the pseudoautosomal region. A short-tailed male is mated with a female mouse that is homozygous for a normal tail. The F_1 from this cross are intercrossed to produce the F_2. What will the phenotypes and proportions of the F_1 and F_2 mice be from this cross?

*32. A color-blind female and a male with normal vision have three sons and six daughters. All the sons are color blind. Five of the daughters have normal vision, but one of them is color blind. The color-blind daughter is 16 years old, is short for her age, and has never undergone puberty. Propose an explanation for how this girl inherited her color blindness.

CHALLENGE QUESTIONS

33. On average, what proportion of the X-linked genes in the first individual is the same as that in the second individual?

(a) A male and his mother
(b) A female and her mother
(c) A male and his father
(d) A female and her father
(e) A male and his brother
(f) A female and her sister
(g) A male and his sister
(h) A female and her brother

34. A geneticist discovers a male mouse in his laboratory colony with greatly enlarged testes. He suspects that this trait results from a new mutation that is either Y linked or autosomal dominant. How could he determine whether the trait is autosomal dominant or Y linked?

35. Amanda is a genetics student at a small college in Connecticut. While counting her fruit flies in the laboratory one afternoon, she observed a strange species of fly in the room. Amanda captured several of the flies and began to raise them. After having raised the flies for several generations, she discovered a mutation in her colony that produces yellow eyes, in contrast with normal red eyes, and Amanda determined that this trait is definitely X-linked recessive. Because yellow eyes are X linked, she assumed that either this species has the XX-XY system of sex determination with genic balance similar to *Drosophila* or it has the XX-XO system of sex determination.

How can Amanda determine whether sex determination in this species is XX-XY or XX-XO? The chromosomes of this species are very small and hard for Amanda to see with her student microscope, so she can only conduct crosses with flies having the yellow-eye mutation. Outline the crosses that Amanda should conduct and explain how they will prove XX-XY or XX-XO sex determination in this species.

36. Occasionally, a mouse X chromosome is broken into two pieces and each piece becomes attached to a different autosomal chromosome. In this event, only the genes on one of the two pieces undergo X inactivation. What does this observation indicate about the mechanism of X-chromosome inactivation?

SUGGESTED READINGS

Allen, G. E. 1978. *Thomas Hunt Morgan: The Man and His Science.* Princeton, NJ: Princeton University Press.
An excellent history of one of the most important biologists of the early twentieth century.

Bogan, J. S., and D. C. Page. 1994. Ovary? Testis? A mammalian dilemma. *Cell* 76:603–607.
A concise review of the molecular nature of sex determination in mammals.

Bridges, C. B. 1916. Nondisjunction as proof of the chromosome theory of heredity. *Genetics* 1:1–52.
Bridges's original paper describing his use of nondisjunction of X chromosomes to prove the chromosome theory of heredity.

Foster, E. A., M. A. Jobling, P. G. Taylor, P. Donnelly, P. de Knijff, R. Mieremet, T. Zerjal, and C. Tyler-Smith. 1998. Jefferson fathered slave's last child. *Nature* 396:27–28.
Report on the use of Y-linked markers to establish the paternity of children of Thomas Jefferson's slave.

Kohler, R. E. 1994. *Lords of the Fly:* Drosophila *Genetics and the Experimental Life.* Chicago: University of Chicago Press.
A comprehensive history of *Drosophila* genetics from 1910 to the early 1940s.

Marx, J. 1995. Tracing how the sexes develop. *Science* 269: 1822–1824.
A short, easy-to-read review of research on sex determination in fruit flies and the worm *Caenorhabditis elegans.*

McClung, C. E. 1902. The accessory chromosome: sex determinant. *Biological Bulletin* 3:43–84.
McClung's original description of the X chromosome.

Morgan, T. H. 1910. Sex-limited inheritance in *Drosophila. Science* 32:120–122.
First description of an X-linked trait.

Penny, G. D., G. F. Kay, S. A. Sheardown, S. Rastan, and N. Brockdorff. 1996. Requirement for Xist in X chromosome inactivation. *Nature* 379:131–137.
This article provides evidence that the *XIST* gene has a role in X-chromosome inactivation.

Ryner, L. C., and A. Swain. 1995. Sex in the 90s. *Cell* 81:483–493.
A review of research findings about sex determination and dosage compensation.

Thomas, M. G., T. Parfitt, D. A. Weiss, K. Skorecki, J. F. Wilson, M. le Roux, N. Bradman, and D. B. Goldstein. 2000. Y chromosomes traveling south: the Cohen modal haplotype and the origins of the Lemba—the "Black Jews of Southern Africa." *American Journal of Human Genetics* 66:674–686.
A fascinating report of the use of Y-linked genetic markers to trace the male ancestry of the Lemba tribe of South Africa.

Williams, N. 1995. How males and females achieve X equality. *Science* 269:1826–1827.
A brief, readable review of recent research on dosage compensation.

Extensions and Modifications of Basic Principles

A geneticist examines a pedigree from the largest known family with Huntington disease. The Juvenile form of this autosomal dominant disease is almost always inherited from the father, an exception to Mendel's principles of inheritance. (Nancy Wexler, HDF/Neurology, Columbia University.)

Was Mendel Wrong?

In 1872, a physician from Long Island, New York named George Huntington described a medical condition characterized by jerky, involuntary movements. Now known as Huntington disease, the condition typically appears in middle age. The initial symptoms are subtle, consisting of mild behavioral and neurological changes; but, as the disease progresses, speech is impaired, walking becomes difficult, and psychiatric problems develop that frequently lead to insanity. Most people who have Huntington disease live for 10 to 30 years after the disease begins; there is currently no cure or effective treatment.

Huntington disease appears with equal frequency in males and females, rarely skips generations and, when one parent has the disorder, approximately half of the children will be similarly affected. These are the hallmarks of an autosomal dominant trait—with one exception. The disorder occasionally arises before the age of 15 and, in these cases, progresses much more rapidly than it does when it arises in middle age. Among younger patients, the trait is almost always inherited from the father. According to Mendel's principles of heredity (Chapter 3), males and females transmit autosomal traits with equal frequency, and reciprocal crosses should yield identical results; yet, for juvenile cases of Huntington

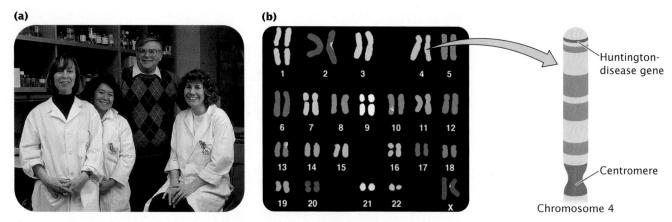

(a)   **(b)**

Huntington-disease gene

Centromere

Chromosome 4

◀5.1 **The gene for Huntington disease.** (a) James Gusella and colleagues, whose research located the Huntington gene. (b) The gene has been mapped to the tip of chromosome 4. (Part a, Sam Ogden; part b, left, courtesy of Dr. Thomas Ried and Dr. Evelin Schrock.)

disease, Mendel's principles do not apply. Was Mendel wrong?

In 1983, a molecular geneticist at Massachusetts General Hospital named James Gusella determined that the gene causing Huntington disease is located near the tip of the short arm of chromosome 4. Gusella determined its location by analyzing DNA from members of the largest known family with Huntington disease, about 7000 people who live near Lake Maracaibo in Venezuela, more than 100 of whom have Huntington disease. Many experts predicted that, with the general location of the Huntington gene pinned down, the actual DNA sequence would be isolated within a few years. Despite intensive efforts, finding the gene took 10 years. When it was finally isolated in the spring of 1993 (◀ FIGURE 5.1), the gene turned out to be quite different from any of those that code for the traits studied by Mendel.

The mutation that causes Huntington disease consists of an unstable region of DNA capable of expanding and contracting as it is passed from generation to generation. When the region expands, Huntington disease results. The degree of expansion affects the severity and age of onset of symptoms; the juvenile form of Huntington disease results from rapid expansion of the region, which occurs primarily when the gene is transmitted from father to offspring.

This genetic phenomenon—the earlier appearance of a trait as it is passed from generation to generation—is called anticipation. Like a number of other genetic phenomena, anticipation does not adhere to Mendel's principles of heredity. This lack of adherence doesn't mean that Mendel was wrong; rather, it means that Mendel's principles are not, by themselves, sufficient to explain the inheritance of all genetic characteristics. Our modern understanding of genetics has been greatly enriched by the discovery of a number of modifications and extensions of Mendel's basic principles, which are the focus of this chapter.

An important extension of Mendel's principles of heredity—the inheritance of sex-linked characteristics—

was introduced in Chapter 4. In this chapter, we will examine a number of additional refinements of Mendel's basic tenets. We begin by reviewing the concept of dominance, emphasizing that dominance entails interactions between genes at one locus (allelic genes) and affects the way in which genes are expressed in the phenotype. Next, we consider lethal alleles and their effect on phenotypic ratios, followed by a discussion of multiple alleles. We then turn to interaction among genes at different loci (nonallelic genes). The phenotypic ratios produced by gene interaction are related to the ratios encountered in Chapter 3. In the latter part of the chapter, we will consider ways in which sex interacts with heredity. Our last stop will be a discussion of environmental influences on gene expression.

The modifications and extensions of hereditary principles discussed in this chapter do not invalidate Mendel's important contributions; rather, they enlarge our understanding of heredity by building on the framework provided by his principles of segregation and independent assortment. These modifications rarely alter the way in which the genes are inherited; rather, they affect the ways in which the genes determine the phenotype.

www.whfreeman.com/pierce   Additional information about Huntington disease

Dominance Revisited

One of Mendel's important contributions to the study of heredity is the concept of *dominance*—the idea that an individual possesses two different alleles for a characteristic, but the trait encoded by only one of the alleles is observed in the phenotype. With dominance, the heterozygote possesses the same phenotype as one of the homozygotes. When biologists began to apply Mendel's principles to organisms other then peas, it quickly became apparent that many characteristics do not exhibit this type of dominance. Indeed, Mendel himself was aware that dominance is not universal, because he observed that a pea plant heterozygous for long and short flowering times had a flowering time that was intermediate between those of its homozygous parents. This situation, in which the heterozygote is intermediate in phenotype between the two homozygotes, is termed *incomplete dominance.*

Dominance can be understood in regard to how the phenotype of the heterozygote relates to the phenotypes of the homozygotes. In the example presented in ◀ FIGURE 5.2, flower color potentially ranges from red to white. One homozygous genotype, A^1A^1, codes for red flowers, and another, A^2A^2, codes for white flowers. Where the heterozygote falls on the range of phenotypes determines the type of dominance. If the heterozygote (A^1A^2) has flowers that are the same color as those of the A^1A^1 homozygote (red), then the A^1 allele is *completely dominant* over the A^2 allele; that is, red is dominant over white. If, on the other hand, the heterozygote has flowers that are the same color as the A^2A^2 homozygote (white), then the A^2 allele is completely dominant, and white is dominant over red. When the heterozygote falls in between the phenotypes of the two homozygotes, dominance is incomplete. With incomplete dominance, the heterozygote need not be exactly intermediate (pink in our example) between the two homozygotes; it might be a slightly lighter shade of red or a slightly pink shade of white. As long as the heterozygote's phenotype can be differentiated and falls within the range of the two homozygotes, dominance is

Table 5.1	Differences among dominance, incomplete dominance, and codominance
Type of Dominance	**Definition**
Dominance	Phenotype of the heterozygote is the same as the phenotype of one of the homozygotes
Incomplete dominance	Phenotype of the heterozygote is intermediate (falls within the range) between the phenotypes of the two homozygotes
Codominance	Phenotype of the heterozygote includes the phenotypes of both homozygotes

incomplete. The important thing to remember about dominance is that it affects the phenotype that genes produce, but not the way in which genes are *inherited.*

Another type of interaction between alleles is **codominance,** in which the phenotype of the heterozygote is not intermediate between the phenotypes of the homozygotes; rather, the heterozygote simultaneously expresses the phenotypes of both homozygotes. An example of codominance is seen in the MN blood types.

The MN locus codes for one of the types of antigens on red blood cells. Unlike antigens foreign to the ABO and Rh blood groups (which also code for red-blood-cell antigens), foreign MN antigens do not elicit a strong immunological reaction, and therefore the MN blood types are not routinely considered in blood transfusions. At the MN locus, there are two alleles: the L^M allele, which codes for the M antigen; and the L^N allele, which codes for the N antigen. Homozygotes with genotype L^ML^M express the M antigen on their red blood cells and have the M blood type. Homozygotes with genotype L^NL^N express the N antigen and have the N blood type. Heterozygotes with genotype L^ML^N exhibit codominance and express both the M and the N antigens; they have blood type MN. The differences between dominance, incomplete dominance, and codominance are summarized in Table 5.1.

The type of dominance that a character exhibits frequently depends on the level of the phenotype examined. An example is cystic fibrosis, one of the more common genetic disorders found in Caucasians and usually considered to be a recessive disease. People who have cystic fibrosis produce large quantities of thick, sticky mucus, which plugs up the airways of the lungs and clogs the ducts leading from the pancreas to the intestine, causing frequent respiratory infections and digestive problems. Even with medical treatment, patients with cystic fibrosis suffer chronic, life-threatening medical problems.

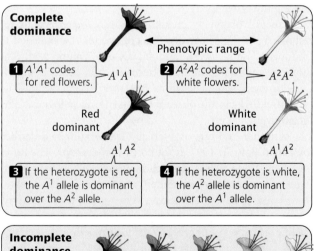
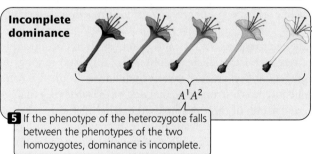

◀ 5.2 **The type of dominance exhibited by a trait depends on how the phenotype of the heterozygote relates to the phenotypes of the homozygotes.**

The gene responsible for cystic fibrosis resides on the long arm of chromosome 7. It encodes a protein termed *cystic fibrosis transmembrane conductance regulator,* mercifully abbreviated CFTR, which acts as a gate in the cell membrane and regulates the movement of chloride ions into and out of the cell. Patients with cystic fibrosis have a mutated, dysfunctional form of CFTR that causes the channel to stay closed, and so chloride ions build up in the cell. This buildup causes the formation of thick mucus and produces the symptoms of the disease.

Most people have two copies of the normal allele for CFTR, and produce only functional CFTR protein. Those with cystic fibrosis possess two copies of the mutated CFTR allele, and produce only the defective CFTR protein. Heterozygotes, with one normal and one defective CFTR allele, produce both functional and defective CFTR protein. Thus, at the molecular level, the alleles for normal and defective CFTR are codominant, because both alleles are expressed in the heterozygote. However, because one normal allele produces enough functional CFTR protein to allow normal chloride transport, the heterozygote exhibits no adverse effects, and the mutated CFTR allele appears to be recessive at the physiological level.

In summary, several important characteristics of dominance should be emphasized. First, dominance is a result of interactions between genes at the same locus; in other words, dominance is *allelic* interaction. Second, dominance does not alter the way in which the genes are inherited; it only influences the way in which they are expressed as a phenotype. The allelic interaction that characterizes dominance is therefore interaction between the *products* of the genes. Finally, dominance is frequently "in the eye of the beholder," meaning that the classification of dominance depends on the level at which the phenotype is examined. As we saw with cystic fibrosis, an allele may exhibit codominance at one level and be recessive at another level.

> **Concepts**
>
> Dominance entails interactions between genes at the same locus (allelic genes) and is an aspect of the phenotype; dominance does not affect the way in which genes are inherited. The type of dominance exhibited by a characteristic frequently depends on the level of the phenotype examined.

Lethal Alleles

In 1905, Lucien Cuenot reported a peculiar pattern of inheritance in mice. When he mated two yellow mice, approximately $2/3$ of their offspring were yellow and $1/3$ were nonyellow. When he test-crossed the yellow mice, he found that all were heterozygous; he was never able to obtain a yellow mouse that bred true. There was a great deal of

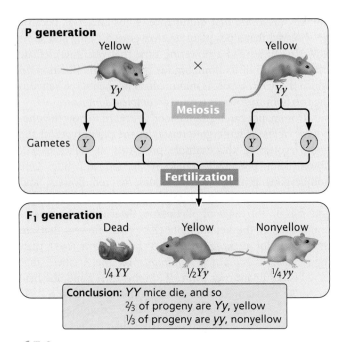

P generation

Yellow $\times$ Yellow

Yy Yy

Meiosis

Gametes Y y Y y

Fertilization

F₁ generation

Dead   Yellow   Nonyellow

$1/4\ YY$ $1/2 Yy$ $1/4 yy$

Conclusion: *YY* mice die, and so
$2/3$ of progeny are *Yy*, yellow
$1/3$ of progeny are *yy*, nonyellow

◀ **5.3 A 2 : 1 ratio among the progeny of a cross results from the segregation of a lethal allele.**

discussion about Cuenot's results among his colleagues, but it was eventually realized that the yellow allele must be lethal when homozygous (◀ FIGURE 5.3). A **lethal allele** is one that causes death at an early stage of development—often before birth—and so some genotypes may not appear among the progeny.

Cuenot originally crossed two mice heterozygous for yellow: $Yy \times Yy$. Normally, this cross would be expected to produce $1/4$ YY, $1/2$ Yy, and $1/4$ yy (see Figure 5.3). The homozygous YY mice are conceived but never complete development, which leaves a 2:1 ratio of Yy (yellow) to yy (nonyellow) in the observed offspring; all yellow mice are heterozygous *(Yy)*.

Another example of a lethal allele, originally described by Erwin Baur in 1907, is found in snapdragons. The *aurea* strain in these plants has yellow leaves. When two plants with yellow leaves are crossed, $2/3$ of the progeny have yellow leaves and $1/3$ have green leaves. When green is crossed with green, all the progeny have green leaves; however, when yellow is crossed with green, $1/2$ of the progeny are green and $1/2$ are yellow, confirming that all yellow-leaved snapdragons are heterozygous. A 2:1 ratio is almost always produced by a recessive lethal allele; so observing this ratio among the progeny of a cross between individuals with the same phenotype is a strong clue that one of the alleles is lethal.

In both of these examples, the lethal alleles are recessive because they cause death only in homozygotes. Unlike its effect on *survival,* the effect of the allele on *color* is dominant; in both mice and snapdragons, a single copy of the allele in the heterozygote produces a yellow color. Lethal alleles also can be dominant; in this case, homozygotes and heterozygotes for the allele die. Truly dominant lethal alleles cannot be transmitted unless they are expressed after the onset of reproduction, as in Huntington disease.

> **Concepts**
>
> A lethal allele causes death, frequently at an early developmental stage, and so one or more genotypes are missing from the progeny of a cross. Lethal alleles may modify the ratio of progeny resulting from a cross.

Multiple Alleles

Most of the genetic systems that we have examined so far consist of two alleles. In Mendel's peas, for instance, one allele coded for round seeds and another for wrinkled seeds; in cats, one allele produced a black coat and another produced a gray coat. For some loci, more than two alleles are present within a group of individuals—the locus has **multiple alleles.** (Multiple alleles may also be referred to as an *allelic series.*) Although there may be more than two alleles present within a *group*, the genotype of each diploid *individual* still consists of only two alleles. The inheritance of characteristics encoded by multiple alleles is no different from the inheritance of characteristics encoded by two alleles, except that a greater variety of genotypes and phenotypes are possible.

Duck-Feather Patterns

An example of multiple alleles is seen at a locus that determines the feather pattern of mallard ducks. One allele, M, produces the wild-type *mallard* pattern. A second allele, M^R, produces a different pattern called *restricted,* and a third allele, m^d, produces a pattern termed *dusky.* In this allelic series, restricted is dominant over mallard and dusky, and mallard is dominant over dusky: $M^R > M > m^d$. The six genotypes possible with these three alleles and their resulting phenotypes are:

Genotype	Phenotype
$M^R M^R$	restricted
$M^R M$	restricted
$M^R m^d$	restricted
MM	mallard
Mm^d	mallard
$m^d m^d$	dusky

In general, the number of genotypes possible will be $[n(n+1)]/2$, where n equals the number of different alleles at a locus. Working crosses with multiple alleles is no different from working crosses with two alleles; Mendel's principle of segregation still holds, as shown in the cross between a restricted duck and a mallard duck (◀ **FIGURE 5.4**).

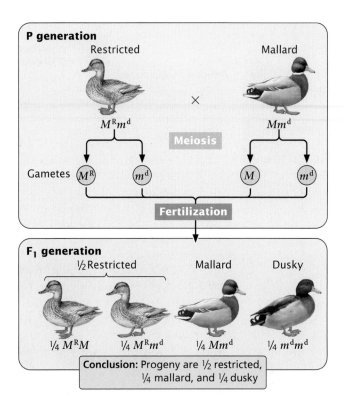

◀ **5.4 Mendel's principle of segregation applies to crosses with multiple alleles.** In this example, three alleles determine the type of plumage in mallard ducks: M^R (restricted) > M (mallard) > m^d (dusky).

The ABO Blood Group

Another multiple-allele system is at the locus for the ABO blood group. This locus determines your ABO blood type and, like the MN locus, codes for antigens on red blood cells. The three common alleles for the ABO blood group locus are: I^A, which codes for the A antigen; I^B, which codes for the B antigen; and i, which codes for no antigen (O). We can represent the dominance relations among the ABO alleles as follows: $I^A > i$, $I^B > i$, $I^A = I^B$. The I^A and I^B alleles are both dominant over i and are codominant with each other; the AB phenotype is due to the presence of an I^A allele and an I^B allele, which results in the production of A and B antigens on red blood cells. An individual with genotype ii produces neither antigen and has blood type O. The six common genotypes at this locus and their phenotypes are shown in ◀ FIGURE 5.5a.

Antibodies are produced against any foreign antigens (see Figure 5.5a). For instance, a person having blood type A produces B antibodies, because the B antigen is foreign. A person having blood type B produces A antibodies, and someone having blood type AB produces neither A nor B antibodies, because neither A nor B antigen is foreign. A person having blood type O possesses no A or B antigens; consequently that person produces both A antibodies and B antibodies. The presence of antibodies against foreign ABO antigens means

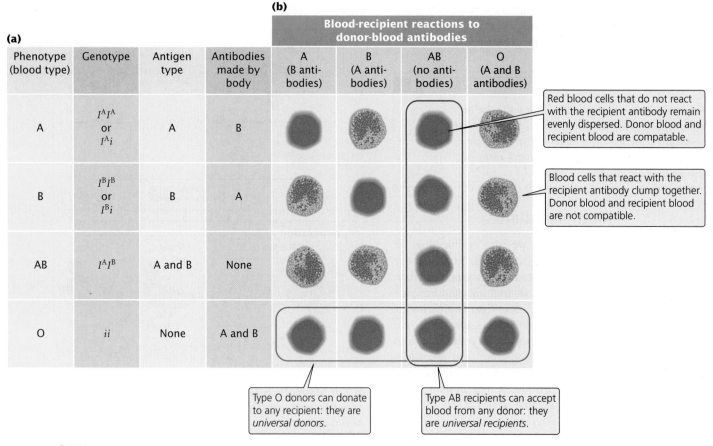

(a)

(b)

Phenotype (blood type)	Genotype	Antigen type	Antibodies made by body	Blood-recipient reactions to donor-blood antibodies			
				A (B antibodies)	B (A antibodies)	AB (no antibodies)	O (A and B antibodies)
A	$I^A I^A$ or $I^A i$	A	B				
B	$I^B I^B$ or $I^B i$	B	A				
AB	$I^A I^B$	A and B	None				
O	ii	None	A and B				

Red blood cells that do not react with the recipient antibody remain evenly dispersed. Donor blood and recipient blood are compatable.

Blood cells that react with the recipient antibody clump together. Donor blood and recipient blood are not compatible.

Type O donors can donate to any recipient: they are *universal donors.*

Type AB recipients can accept blood from any donor: they are *universal recipients.*

◀5.5 **ABO blood types and possible blood transfusions.**

that successful blood transfusions are possible only between persons with certain compatible blood types (◀FIGURE 5.5b).

The inheritance of alleles at the ABO locus can be illustrated by a paternity suit involving the famous movie actor Charlie Chaplin. In 1941, Chaplin met a young actress named Joan Barry, with whom he had an affair. The affair ended in February 1942 but, 20 months later, Barry gave birth to a baby girl and claimed that Chaplin was the father. Barry then sued for child support. At this time, blood typing had just come into widespread use, and Chaplin's attorneys had Chaplin, Barry, and the child blood typed. Barry had blood type A, her child had blood type B, and Chaplin had blood type O. Could Chaplin have been the father of Barry's child?

Your answer should be no. Joan Barry had blood type A, which can be produced by either genotype $I^A I^A$ or $I^A i$. Her baby possessed blood type B, which can be produced by either genotype $I^B I^B$ or $I^B i$. The baby could not have inherited the I^B allele from Barry (Barry could not carry an I^B allele if she were blood type A); therefore the baby must have inherited the i allele from her. Barry must have had genotype $I^A i$, and the baby must have had genotype $I^B i$. Because the baby girl inherited her i allele from Barry, she must have inherited the I^B allele from her father. With blood type O, produced only by genotype ii, Chaplin could not have been the father of Barry's child. In the course of

the trial to settle the paternity suit, three pathologists came to the witness stand and declared that it was genetically impossible for Chaplin to have fathered the child. Nevertheless, the jury ruled that Chaplin was the father and ordered him to pay child support and Barry's legal expenses.

Concepts

More than two alleles (multiple alleles) may be present within a group of individuals, although each diploid individual still has only two alleles at that locus.

Gene Interaction

In the dihybrid crosses that we examined in Chapter 3, each locus had an independent effect on the phenotype. When Mendel crossed a homozygous round and yellow plant *(RRYY)* with a homozygous wrinkled and green plant *(rryy)* and then self-fertilized the F_1, he obtained F_2 progeny in the following proportions:

$9/16$	$R\_Y\_$	round, yellow
$3/16$	R_yy	round, green
$3/16$	$rrY\_$	wrinkled, yellow
$1/16$	$rryy$	wrinkled, green

In this example, the genes showed two kinds of independence. First, the genes at each locus are independent in their *assortment* in meiosis, which is what produces the $9:3:3:1$ ratio of phenotypes in the progeny, in accord with Mendel's principle of independent assortment. Second, the genes are independent in their *phenotypic expression;* the R and r alleles affect only the shape of the seed and have no influence on the color of the endosperm; the Y and y alleles affect only color and have no influence on the shape of the seed.

Frequently, genes exhibit independent assortment but do not act independently in their phenotypic expression; instead, the effects of genes at one locus depend on the presence of genes at other loci. This type of interaction between the effects of genes at different loci (genes that are not allelic) is termed **gene interaction.** With gene interaction, the products of genes at different loci combine to produce new phenotypes that are not predictable from the single-locus effects alone. In our consideration of gene interaction, we'll focus primarily on interaction between the effects of genes at two loci, although interactions among genes at three, four, or more loci are common.

(Concepts)

In gene interaction, genes at different loci contribute to the determination of a single phenotypic characteristic.

Gene Interaction That Produces Novel Phenotypes

Let's first examine gene interaction in which genes at two loci interact to produce a single characteristic. Fruit color in the pepper *Capsicum annuum* is determined in this way. This plant produces peppers in one of four colors: red, brown, yellow, or green. If a homozygous plant with red peppers is crossed with a homozygous plant with green peppers, all the F_1 plants have red peppers (◀FIGURE 5.6a). When the F_1 are crossed with one another, the F_2 are in a ratio of 9 red : 3 brown : 3 yellow : 1 green (◀FIGURE 5.6b). This dihybrid ratio (Chapter 3) is produced by a cross between two plants that are both heterozygous for two loci ($RrCc \times RrCc$). In peppers, a dominant allele R at the first locus produces a red pigment; the recessive allele r at this locus produces no red pigment. A dominant allele C at the second locus causes decomposition of the green pigment chlorophyll; the recessive allele c allows chlorophyll to persist. The genes at the two loci then interact to produce the colors seen in F_2 peppers:

Genotype	Phenotype
R_C_	red
R_cc	brown
rrC_	yellow
rrcc	green

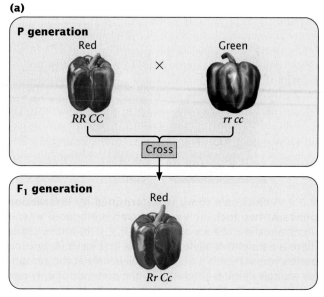

(a)

P generation

Red × Green

RR CC   rr cc

Cross

F₁ generation

Red

Rr Cc

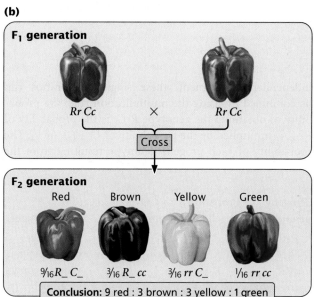

(b)

F₁ generation

Rr Cc × Rr Cc

Cross

F₂ generation

Red   Brown   Yellow   Green

$\frac{9}{16}$ R_ C_ $\frac{3}{16}$ R_ cc $\frac{3}{16}$ rr C_ $\frac{1}{16}$ rr cc

Conclusion: 9 red : 3 brown : 3 yellow : 1 green

◀5.6 **Gene interaction in which two loci determine a single characteristic, fruit color, in the pepper *Capsicum annuum.***

To illustrate how Mendel's rules of heredity can be used to understand the inheritance of characteristics determined by gene interaction, let's consider a testcross between an F_1 plant from the cross in Figure 5.6 ($RrCc$) and a plant with green peppers ($rrcc$). As outlined in Chapter 3 (pp. 61–62) for independent loci, we can work this cross by breaking it down into two simple crosses. At the first locus, the heterozygote Rr is crossed with the homozygote rr; this cross produces $\frac{1}{2}$ Rr and $\frac{1}{2}$ rr progeny. Similarly, at the second locus, the heterozygous genotype Cc is crossed with the homozygous genotype cc, producing $\frac{1}{2}$ Cc and $\frac{1}{2}$ cc progeny. In accord with Mendel's principle of

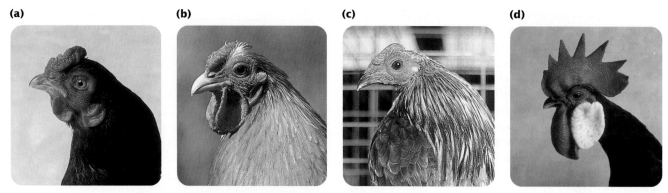

(a)   **(b)**   **(c)**   **(d)**

◀5.7 **A chicken's comb is determined by interaction between genes at two loci.** (a) A walnut comb is produced when there is a dominant allele at each of two loci (R_P_). (b) A rose comb occurs when there is a dominant allele only at the first locus (R_pp). (c) A pea comb occurs when there is a dominant allele only at the second locus (ppR_). (d) A single comb is produced by the presence of only recessive alleles at both loci (rrpp). (Parts a and d, R. OSF Dowling/Animals Animals; part b, Robert Maier/Animals Animals; part c, George Godfrey/Animals Animals.)

independent assortment, these single-locus ratios can be combined by using the multiplication rule: the probability of obtaining the genotype $RrCc$ is the probability of Rr ($\frac{1}{2}$) multiplied by the probability of Cc ($\frac{1}{2}$), or $\frac{1}{4}$. The probability of each progeny genotype resulting from the testcross is:

Progeny genotype	Probability at each locus	Overall probability	Phenotype
$RrCc$	$\frac{1}{2} \times \frac{1}{2} =$	$\frac{1}{4}$	red peppers
$Rrcc$	$\frac{1}{2} \times \frac{1}{2} =$	$\frac{1}{4}$	brown peppers
$rrCc$	$\frac{1}{2} \times \frac{1}{2} =$	$\frac{1}{4}$	yellow peppers
$rrcc$	$\frac{1}{2} \times \frac{1}{2} =$	$\frac{1}{4}$	green peppers

When you work problems with gene interaction, it is especially important to determine the probabilities of single-locus genotypes and to multiply the probabilities of *genotypes*, not phenotypes, because the phenotypes cannot be determined without considering the effects of the genotypes at all the contributing loci.

Another example of gene interaction that produces novel phenotypes is seen in the genes that determine comb shape in chickens. The comb is the fleshy structure found on the head of a chicken. Genes at two loci (R, r and P, p) interact to determine the four types of combs shown in ◀FIGURE 5.7. A walnut comb is produced when at least one dominant allele R is present at the first locus and at least one dominant allele P is present at the second locus (genotype $R\_P\_$). A chicken with at least one dominant allele at the first locus and two recessive alleles at the second locus (genotype R_pp) possesses a rose comb. If

two recessive alleles are present at the first locus and at least one dominant allele is present at the second (genotype $rrP\_$), the chicken has a pea comb. Finally, if two recessive alleles are present at both loci ($rrpp$), the bird has a single comb.

Gene Interaction with Epistasis

Sometimes the effect of gene interaction is that one gene masks (hides) the effect of another gene at a different locus, a phenomenon known as **epistasis.** This phenomenon is similar to dominance, except that dominance entails the masking of genes at the *same* locus (allelic genes). In epistasis, the gene that does the masking is called the **epistatic gene;** the gene whose effect is masked is a **hypostatic gene.** Epistatic genes may be recessive or dominant in their effects.

Recessive epistasis Recessive epistasis is seen in the genes that determine coat color in Labrador retrievers. These dogs may be black, brown, or yellow; their different coat colors are determined by interactions between genes at two loci (although a number of other loci also help to determine coat color; see pp. 112–114). One locus determines the type of pigment produced by the skin cells: a dominant allele B codes for black pigment, whereas a recessive allele b codes for brown pigment. Alleles at a second locus affect the *deposition* of the pigment in the shaft of the hair; allele E allows dark pigment (black or brown) to be deposited, whereas a recessive allele e prevents the deposition of dark pigment, causing the hair to be yellow. The presence of genotype ee at the second locus therefore masks the expression of the black and brown alleles at the first locus. The genotypes that determine coat color and their phenotypes are:

Genotype	Phenotype
B_ E_	black
bbE_	brown (frequently called chocolate)
B_ee	yellow
bbee	yellow

If we cross a black Labrador homozygous for the dominant alleles with a yellow Labrador homozygous for the recessive alleles and then intercross the F_1, we obtain progeny in the F_2 in a 9:3:4 ratio:

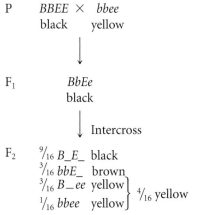

P $BBEE$ × $bbee$
     black   yellow

F_1 $BbEe$
         black

Intercross

F_2 $^9/_{16}$ $B\_E\_$ black
         $^3/_{16}$ $bbE\_$ brown
         $^3/_{16}$ B_ee yellow ⎫
         $^1/_{16}$ $bbee$ yellow ⎭ $^4/_{16}$ yellow

Notice that yellow dogs can carry alleles for either black or brown pigment, but these alleles are not expressed in their coat color.

In this example of gene interaction, allele e is epistatic to B and b, because e masks the expression of the alleles for black and brown pigments, and alleles B and b are hypostatic to e. In this case, e is a recessive epistatic allele, because two copies of e must be present to mask of the black and brown pigments.

Dominant epistasis

Dominant epistasis is seen in the interaction of two loci that determine fruit color in summer squash, which is commonly found in one of three colors: yellow, white, or green. When a homozygous plant that produces white squash is crossed with a homozygous plant that produces green squash and the F_1 plants are crossed with each other, the following results are obtained:

P   plants with   plants with
     white squash × green squash

F_1 plants with
         white squash

Intercross

F_2 $^{12}/_{16}$ plants with white squash
         $^3/_{16}$ plants with yellow squash
         $^1/_{16}$ plants with green squash

How can gene interaction explain these results?

In the F_2, $^{12}/_{16}$ or $^3/_4$ of the plants produce white squash and $^3/_{16} + ^1/_{16} = ^4/_{16} = ^1/_4$ of the plants produce squash having color. This outcome is the familiar 3:1 ratio produced by a cross between two heterozygous individuals, which suggests that a dominant allele at one locus inhibits the production of pigment, resulting in white progeny. If we use the symbol W to represent the dominant allele that inhibits pigment production, then genotype $W\_$ inhibits pigment production and produces white squash, whereas ww allows pigment and results in colored squash.

Among those ww F_2 plants with pigmented fruit, we observe $^3/_{16}$ yellow and $^1/_{16}$ green (a 3:1 ratio). This outcome is because a second locus determines the type of pigment produced in the squash, with yellow ($Y\_$) dominant over green (yy). This locus is expressed only in ww plants, which lack the dominant inhibitory allele W. We can assign the genotype $wwY\_$ to plants that produce yellow squash and the genotype $wwyy$ to plants that produce green squash. The genotypes and their associated phenotypes are:

W_Y_	white squash
W_yy	white squash
wwY_	yellow squash
wwyy	green squash

Allele W is epistatic to Y and y—it suppresses the expression of these pigment-producing genes. W is a dominant epistatic allele because, in contrast with e in Labrador retriever coat color, a single copy of the allele is sufficient to inhibit pigment production.

Summer squash provides us with a good opportunity for considering how epistasis often arises when genes affect a series of steps in a biochemical pathway. Yellow pigment in the squash is most likely produced in a two-step biochemical pathway (◀FIGURE 5.8). A colorless (white) compound (designated A in Figure 5.8) is converted by enzyme I into green compound B, which is then converted into compound C by enzyme II. Compound C is the yellow pigment in the fruit.

Plants with the genotype ww produce enzyme I and may be green or yellow, depending on whether enzyme II is present. When allele Y is present at a second locus, enzyme II is produced and compound B is converted into compound C, producing a yellow fruit. When two copies of y, which does not encode a functional form of enzyme II, are present, squash remain green. The presence of W at the first locus inhibits the conversion of compound A into compound B; plants with genotype $W\_$ do not make compound B and their fruit remains white, regardless of which alleles are present at the second locus.

Many cases of epistasis arise in this way. A gene (such as W) that has an effect on an early step in a biochemical pathway will be epistatic to genes (such as Y and y) that affect subsequent steps, because the effect of the enzyme in the later step depends on the product of the earlier reaction.

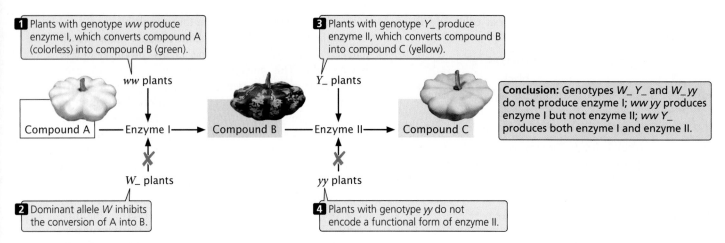

1 Plants with genotype *ww* produce enzyme I, which converts compound A (colorless) into compound B (green).

3 Plants with genotype *Y_* produce enzyme II, which converts compound B into compound C (yellow).

ww plants

*Y_* plants

Compound A ——— Enzyme I ——→ Compound B ——— Enzyme II ——→ Compound C

Conclusion: Genotypes *W_ Y_* and *W_ yy* do not produce enzyme I; *ww yy* produces enzyme I but not enzyme II; *ww Y_* produces both enzyme I and enzyme II.

*W_* plants

yy plants

2 Dominant allele *W* inhibits the conversion of A into B.

4 Plants with genotype *yy* do not encode a functional form of enzyme II.

◀ **5.8 Yellow pigment in summer squash is produced in a two-step pathway.**

Duplicate recessive epistasis Let's consider one more detailed example of epistasis. Albinism is the absence of pigment and is a common genetic trait in many plants and animals. Pigment is almost always produced through a multistep biochemical pathway; thus, albinism may entail gene interaction. Robert T. Dillon and Amy R. Wethington found that albinism in the common freshwater snail *Physa heterostroha* can result from the presence of either of two recessive alleles at two different loci. Inseminated snails were collected from a natural population and placed in cups of water, where they laid eggs. Some of the eggs hatched into albino snails. When two albino snails were crossed, all of the F_1 were pigmented. On intercrossing the F_1, the F_2 consisted of $9/16$ pigmented snails and $7/16$ albino snails. How did this 9 : 7 ratio arise?

The 9 : 7 ratio seen in the F_2 snails can be understood as a modification of the 9 : 3 : 3 : 1 ratio obtained when two individuals heterozygous for two loci are crossed. The 9 : 7 ratio arises when dominant alleles at both loci (*A_B_*) produce pigmented snails; any other genotype produces albino snails:

P   *aaBB*   *AAbb*
       albino × albino

                ↓

F_1 *AaBb*
              pigmented

                ↓ Intercross

F_2 $9/16$ *A_B_* pigmented
        $3/16$ *aaB_* albino ⎫
        $3/16$ *A_bb* albino ⎬ $7/16$ albino
        $1/16$ *aabb* albino ⎭

The 9 : 7 ratio in these snails is probably produced by a two-step pathway of pigment production (◀ FIGURE 5.9). Pigment (compound C) is produced only after compound A has been converted into compound B by enzyme I and after compound B has been converted into compound C by enzyme II. At least one dominant allele *A* at the first locus is required to produce enzyme I; similarly, at least one dominant allele *B* at the second locus is required to produce enzyme II. Albinism arises from the absence of compound C, which may happen in three ways. First, two recessive alleles at the first locus (genotype *aaB_*) may prevent the production of enzyme I, and so compound B is never produced. Second, two recessive alleles at the second locus (genotype *A_bb*) may prevent the production of enzyme II. In this case, compound B is never converted into compound C. Third, two recessive alleles may be present at both loci (*aabb*), causing the absence of both enzyme I and enzyme II. In this example of gene interaction, *a* is epistatic to *B*, and *b* is epistatic to *A; both* are recessive epistatic alleles because the presence of two copies of either allele *a* or *b* is necessary to suppress pigment production. This example differs from the suppression of coat color in Labrador retrievers in that recessive alleles at either of two loci are capable of suppressing pigment production in the snails, whereas recessive alleles at a single locus suppress pigment expression in Labs.

┌──┐
│ **Concepts**

Epistasis is the masking of the expression of one gene by another gene at a different locus. The epistatic gene does the masking; the hypostatic gene is masked. Epistatic genes can be dominant or recessive.
└──┘

┌──┐
│ **Connecting Concepts**

Interpreting Ratios Produced by Gene Interaction

A number of modified ratios that result from gene interaction are shown in Table 5.2. Each of these examples represents a modification of the basic 9 : 3 : 3 : 1 dihybrid ratio.

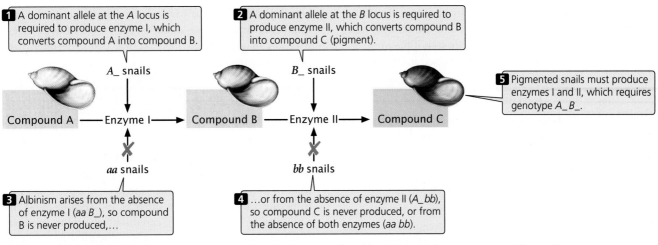

1 A dominant allele at the *A* locus is required to produce enzyme I, which converts compound A into compound B.

2 A dominant allele at the *B* locus is required to produce enzyme II, which converts compound B into compound C (pigment).

*A_* snails

*B_* snails

5 Pigmented snails must produce enzymes I and II, which requires genotype *A_B_*.

Compound A —— Enzyme I ⟶ Compound B —— Enzyme II ⟶ Compound C

aa snails

bb snails

3 Albinism arises from the absence of enzyme I (*aa B_*), so compound B is never produced,…

4 …or from the absence of enzyme II (*A_bb*), so compound C is never produced, or from the absence of both enzymes (*aa bb*).

◀ **5.9 Pigment is produced in a two-step pathway in snails.**

In interpreting the genetic basis of modified ratios, we should keep several points in mind. First, the inheritance of the genes producing these characteristics is no different from the inheritance of genes coding for simple genetic characters. Mendel's principles of segregation and independent assortment still apply; each individual possesses two alleles at each locus, which separate in meiosis, and genes at the different loci assort independently. The only difference is in how the *products* of the genotypes interact to produce the phenotype. Thus, we cannot consider the expression of genes at each locus separately, but must take into consideration how the genes at different loci interact.

A second point is that in the examples that we have considered, the phenotypic proportions were always in sixteenths because, in all the crosses, pairs of alleles segregated at two independently assorting loci. The probability of inheriting one of the two alleles at a locus is $1/2$. Because there are two loci, each with two alleles, the probability of inheriting any particular combination of genes is $(1/2)^4 = 1/16$. For a trihybrid cross, the progeny proportions should be in sixty-fourths, because $(1/2)^6 = 1/64$. In general, the progeny proportions should be in fractions of $(1/2)^{2n}$, where *n* equals the number of loci with two alleles segregating in the cross.

Table 5.2	Modified dihybrid—phenotypic ratios due to gene interaction					
	Genotype				**Type of**	
Ratio	**A_B_**	**A_bb**	**aaB_**	**aabb**	**Interaction**	**Example**
9:3:3:1	9	3	3	1	None	Seed shape and endosperm color in peas
9:3:4	9	3	4		Recessive epistasis	Coat color in Labrador retrievers
12:3:1	12		3	1	Dominant epistasis	Color in squash
9:7	9	7			Duplicate recessive epistasis	Albinism in snails
9:6:1	9	6		1	Duplicate interaction	—
15:1	15			1	Duplicate dominant epistasis	—
13:3	13		3		Dominant and recessive epistasis	—

*Each ratio is produced by a dihybrid cross (*AaBb* × *AaBb*). Shaded bars represent combinations of genotypes that give the same phenotype.

Crosses rarely produce exactly 16 progeny; therefore, modifications of a dihybrid ratio are not always obvious. Modified dihybrid ratios are more easily seen if the number of individuals of each phenotype is expressed in sixteenths:

$$\frac{x}{16} = \frac{\text{number of progeny with a phenotype}}{\text{total number of progeny}}$$

where $x/16$ equals the proportion of progeny with a particular phenotype. If we solve for x (the proportion of the particular phenotype in sixteenths), we have:

$$x = \frac{\text{number of progeny with a phenotype} \times 16}{\text{total number of progeny}}$$

For example, suppose we cross two homozygous individuals, interbreed the F_1 and obtain 63 red, 21 brown, and 28 white F_2 individuals. Using the preceding formula, the phenotypic ratio in the F_2 is: red = (63 × 16)/112 = 9; brown = (21 × 16)/112 = 3; and white = (28 × 16)/112 = 4. The phenotypic ratio is 9 : 3 : 4.

A final point to consider is how to assign genotypes to the phenotypes in modified ratios that result from gene interaction. Don't try to *memorize* the genotypes associated with all the modified ratios in Table 5.2. Instead, practice relating modified ratios to known ratios, such as the 9 : 3 : 3 : 1 dihybrid ratio. Suppose we obtain $^{15}/_{16}$ green progeny and $^{1}/_{16}$ white progeny in a cross between two plants. If we compare this 15 : 1 ratio with the standard 9 : 3 : 3 : 1 dihybrid ratio, we see that $^{9}/_{16} + ^{3}/_{16} + ^{3}/_{16}$ equals $^{15}/_{16}$. All the genotypes associated with these proportions in the dihybrid cross (*A_B_*, *A_bb*, and *aaB_*) must give the same phenotype, the green progeny. Genotype *aabb* makes up $^{1}/_{16}$ of the progeny in a dihybrid cross, the white progeny in this cross.

In assigning genotypes to phenotypes in modified ratios, students sometimes become confused about which letters to assign to which phenotype. Suppose we obtain the following phenotypic ratio: $^{9}/_{16}$ black : $^{3}/_{16}$ brown : $^{4}/_{16}$ white. Which genotype do we assign to the brown progeny, *A_bb* or *aaB_*? Either answer is correct, because the letters are just arbitrary symbols for the genetic information. The important thing to realize about this ratio is that the brown phenotype arises when two recessive alleles are present at one locus.

Concepts

Gene interaction frequently produces modified phenotypic ratios. These modified ratios can be understood by relating them to other known ratios.

The Complex Genetics of Coat Color in Dogs

Coat color in dogs is an excellent example of how complex interactions between genes may take part in the determination of a phenotype. Domestic dogs come in an amazing variety of shapes, sizes, and colors. For thousands of years, humans have been breeding dogs for particular traits, producing the large number of types that we see today. Each breed of dog carries a selection of genes from the ancestral dog gene pool; these genes define the features of a particular breed.

One of the most obvious differences between dogs is coat color. The genetics of coat color in dogs is quite complex; many genes participate, and there are numerous interactions between genes at different loci. We will consider seven loci (in the list that follows) that are important in producing many of the noticeable differences in color and pattern among breeds of dogs. In interpreting the genetic basis of differences in coat color of dogs, consider how the expression of a particular gene is modified by the effects of other genes. Keep in mind that additional loci not listed here can modify the colors produced by these seven loci and that not all geneticists agree on the genetics of color variation in some breeds.

1. Agouti (A) locus — This locus has five common alleles that determine the depth and distribution of color in a dog's coat:

A^s	Solid black pigment.
a^w	Agouti, or wolflike gray. Hairs encoded by this allele have a "salt and pepper" appearance, produced by a band of yellow pigment on a black hair.
a^y	Yellow. The black pigment is markedly reduced; so the entire hair is yellow.
a^s	Saddle markings (dark color on the back, with extensive tan markings on the head and legs).
a^t	Bicolor (dark color over most of the body, with tan markings on the feet and eyebrows).

   A^s and a^y are generally dominant over the other alleles, but the dominance relations are complex and not yet completely understood.

2. Black (B) locus — This locus determines whether black pigment can be formed. The actual color of a dog's fur depends on the effects of genes at other loci (such as the A, C, D, and E loci). Two alleles are common:

B	Allows black pigment to be produced; the dog will be black if it also possesses certain alleles at the A, C, D, and E loci.
b	Black pigment cannot be produced; pigmented dogs can be chocolate, liver, tan, or red.

   B is dominant over b.

3. Albino (C) locus — This locus determines whether full color will be expressed. There are five alleles at this locus:

C	Color fully expressed.
c^{ch}	Chinchilla. Less color is expressed, and pigment is completely absent from the base of the long hairs, producing a pale coat.
c^d	All white coat with dark nose and dark eyes.
c^b	All white coat with blue eyes.
c	Fully albino. The dogs have an all-white coat with pink eyes and nose.

(a)   **(b)**   **(c)**

◀ **5.10 Coat color in dogs is determined by interactions between genes at a number of loci.** (a) Most Labrador retrievers are genotype $A^sA^sCCDDSStt$, varying only at the B and E loci. (b) Most beagles are genotype $a^sa^sBBCCDDs^ps^ptt$. (c) Dalmations are genotype $A^sA^sCCDDEEs^ws^wTT$, varying at the B locus so that the dogs are black ($B\_$) or brown (bb). (Part a, Robert Maier/Animals Animals; part b, Ralph Reinhold/Animals Animals; part c, Robert Percy/ Animals Animals.)

The dominance relations among these alleles is presumed to be $C > c^{ch} > c^d > c^b > c$, but the c^{ch} and c alleles are rare, and crosses including all possible genotypes have not been completed.

4. Dilution (D) locus—This locus, with two alleles, determines whether the color will be diluted. For example, diluted black pigment appears bluish, and diluted yellow appears cream. The diluted effect is produced by an uneven distribution of pigment in the hair shaft:

D Intense pigmentation.
d Dilution of pigment.

D is dominant over d.

5. Extension (E) locus—Four alleles at this locus determine where the genotype at the A locus is expressed. For example, if a dog has the A^s allele (solid black) at the A locus, then black pigment will either be extended throughout the coat or be restricted to some areas, depending on the alleles present at the E locus. Areas where the A locus is not expressed may appear as yellow, red, or tan, depending on the presence of particular genes at other loci. When A^s is present at the A locus, the four alleles at the E locus have the following effects:

E^m Black mask with a tan coat.
E The A locus expressed throughout (solid black).
e^{br} Brindle, in which black and yellow are in layers to give a tiger-striped appearance.
e No black in the coat, but the nose and eyes may be black.

The dominance relations among these alleles are poorly known.

6. Spotting (S) locus—Alleles at this locus determine whether white spots will be present. There are four common alleles:

S No spots.
s^i Irish spotting; numerous white spots.
s^p Piebald spotting; various amounts of white.
s^w Extreme white piebald; almost all white.

S is completely dominant over s^i, s^p, and s^w; s^i and s^p are dominant over s^w ($S > s^i, s^p > s^w$). The relation between of s^i and s^p is poorly defined; indeed, they may not be separate alleles. Genes at other poorly known loci also modify spotting patterns.

7. Ticking (T) locus—This locus determines the presence of small colored spots on the white areas, which is called ticking:

T Ticking; small colored spots on the areas of white.
t No ticking.

T is dominant over t. Ticking cannot be expressed if a dog has a solid coat ($S\_$).

To illustrate how genes at these loci interact in determining a dog's coat color, let's consider a few examples:

Labrador retriever- Labrador retrievers (◀ FIGURE 5.10a) may be black, brown, or yellow. Most are homozygous $A^sA^sCCDDSStt$; thus, they vary only at the B and E loci. The A^s, C, and D alleles allow dark pigment to be expressed; whether a dog is black depends on which genes are present at the B and E loci. As discussed earlier in the chapter, all black Labradors must carry at least one B allele and one E allele ($B\_E\_$). Brown dogs are homozygous bb and have at least one E allele ($bbE\_$). Yellow dogs are a result of the presence of ee (B_ee or $bbee$). Labrador retrievers are homozygous for the S allele, which produces a solid color; the few white spots that appear in some dogs of this breed are due to other modifying genes. The allele for ticking, T, is presumed not to exist in Labradors; however, Labrador retrievers have solid coats and ticking is expressed only in spotted dogs; so its absence is uncertain.

Beagle- Most beagles are homozygous $a^sa^s BBCCDDs^ps^ptt$, although other alleles at these loci are occasionally present. The a^s allele produces the saddle markings—dark back and sides, with tan head and legs—that are characteristic of the breed (◀ FIGURE 5.10b). Alleles *B*, *C*, and *D* allow black to be produced, but its distribution is limited by the a^s allele. Genotype *ee* does occasionally arise, leading to a few all-tan beagles. White spotting in beagles is due to the s^p allele. Ticking can appear, but most beagles are *tt*.

Dalmatian- Dalmatians (◀ FIGURE 5.10c) have an interesting genetic makeup. Most are homozygous $A^sA^s\ CCDDEEs^ws^wTT$; so they vary only at the B locus. Notice that these dogs possess genotype $A^sA^sCCDDEE$, which allows for a solid coat that would be black, if genotype *B_* is present, or brown (called liver), if genotype *bb* is present. However, the presence of the s^w allele produces a white coat, masking the expression of the solid color. The dog's color appears only in the pigmented spots, which are due to the presence of the ticking allele *T*. Table 5.3 gives the common genotypes of other breeds of dogs.

www.whfreeman.com/pierce   Information on dog genetics, including the Dog Genome Project

Complementation: Determining Whether Mutations Are at the Same or Different Loci

How do we know whether different mutations that affect a characteristic occur at the same locus (are allelic) or at different loci? In fruit flies, for example, *white* is an X-linked mutation that produces white eyes instead of the red eyes found in wild-type flies. *Apricot* is an X-linked recessive mutation that produces light orange-colored eyes. Do the white and apricot mutations occur at the same locus or at different loci? We can use the complementation test to answer this question.

To carry out a **complementation test,** parents that are homozygous for different mutations are crossed, producing offspring that are heterozygous. If the mutations are allelic (occur at the same locus), then the heterozygous offspring have only mutant alleles (*ab*) and exhibit a mutant phenotype:

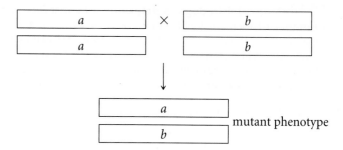

If, on the other hand, the mutations occur at different loci, each of the homozygous parents possesses wild-type genes at the other locus (*aa b^+b^+* and *a^+a^+ bb*); so the heterozygous offspring inherit a mutant and a wild-type allele at each locus. In this case, the mutations complement each other and the heterozygous offspring have the wild-type phenotype:

Table 5.3	Common genotypes in different breeds of dogs							
Breed	**Usual Homozygous Genes***	**Other Genes Present Within the Breed**						
Basset hound	$BBCCDDEEtt$	a^y, a^t	S, s^p, s^i					
Beagle	$a^sa^sBBCCDDs^ps^ptt$	E, e						
English bulldog	$BBCCDDtt$	A^s, a^y, a^t	E^m, E, e^{br}	S, s^i, s^p, s^w				
Chihuahua	tt	A^s, a^y, a^s, a^t	B, b	C, c^{ch}	D, d	E^m, E,	e^{br}, e	S, s^i s^p, s^w
Collie	$BBCCEEtt$	a^y, a^t	D, d	s^i, s^w				
Dalmatian	$A^sA^sCCDDEEs^ws^wTT$	B, b						
Doberman	$a^ta^tCCEESStt$	B, b	D, d					
German shepherd	$BBDDSStt$	a^y, a, a^s, a^t	C, c^{ch}	E^m, E, e				
Golden retriever	$A^sA^sBBDDSStt$	C, c^{ch}	E, e					
Greyhound	$BBtt$	A^s, a^y	C, c^{ch}	D, d	E, e^{br}, e	S, s^p, s^w, s^i		
Irish setter	$BBCCDDeeSStt$	A, a^t						
Labrador retriever	$A^sA^sCCDDSStt$	B, b	E, e					
Poodle	$SStt$	A^s, a^t	B, b	C, c^{ch}	D, d	E, e		
Rottweiler	$a^ta^tBBCCDDEESStt$							
St. Bernard	$a^ya^yBBCCDDtt$	E^m, E	s^i, s^p, s^w					

*Most dogs in the breed are homozygous for these genes; a few individual dogs may possess other alleles at these loci.
Source: Data from M. B. Willis, *Genetics of the Dog* (London: Witherby, 1989).

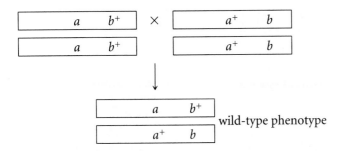

$$\begin{array}{|c c|}\hline a & b^+ \\ a & b^+ \\\hline\end{array} \times \begin{array}{|c c|}\hline a^+ & b \\ a^+ & b \\\hline\end{array}$$

$$\begin{array}{|c c|}\hline a & b^+ \\ a^+ & b \\\hline\end{array} \text{ wild-type phenotype}$$

Complementation occurs when an individual possessing two mutant genes has a wild-type phenotype and is an indicator that the mutations are nonallelic genes.

When the complementation test is applied to white and apricot mutations, all of the heterozygous offspring have light-colored eyes, demonstrating that white and apricot are produced by mutations that occur at the same locus and are allelic.

Interaction Between Sex and Heredity

In Chapter 4, we considered characteristics encoded by genes located on the sex chromosomes and how their inheritance differs from the inheritance of traits encoded by autosomal genes. Now we will examine additional influences of sex, including the effect of the sex of an individual on the expression of genes on autosomal chromosomes, characteristics determined by genes located in the cytoplasm, and characteristics for which the genotype of only the maternal parent determines the phenotype of the offspring. Finally, we'll look at situations in which the expression of genes on autosomal chromosomes is affected by the sex of the parent from whom they are inherited.

Sex-Influenced and Sex-Limited Characteristics

Sex-influenced characteristics are determined by autosomal genes and are inherited according to Mendel's principles, but they are expressed differently in males and females. In this case, a particular trait is more readily expressed in one sex; in other words, the trait has higher penetrance (see pp. 67–68 in Chapter 3) in one of the sexes.

For example, the presence of a beard on some goats is determined by an autosomal gene (B^b) that is dominant in males and recessive in females. In males, a single allele is required for the expression of this trait: both the homozygote (B^bB^b) and the heterozygote (B^bB^+) have beards, whereas the B^+B^+ male is beardless. In contrast, females require two alleles in order for this trait to be expressed: the homozygote B^bB^b has a beard, whereas the heterozygote (B^bB^+) and the other homozygote (B^+B^+) are beardless. The key to understanding the expression of the bearded gene is to look at the heterozygote. In males (for which the presence of a beard is dominant), the heterozygous genotype produces a beard but, in females (for which the presence of a beard is recessive and its absence is dominant), the heterozygous genotype produces a goat without a beard.

◀ FIGURE 5.11a illustrates a cross between a beardless male (B^+B^+) and a bearded female (B^bB^b). The alleles

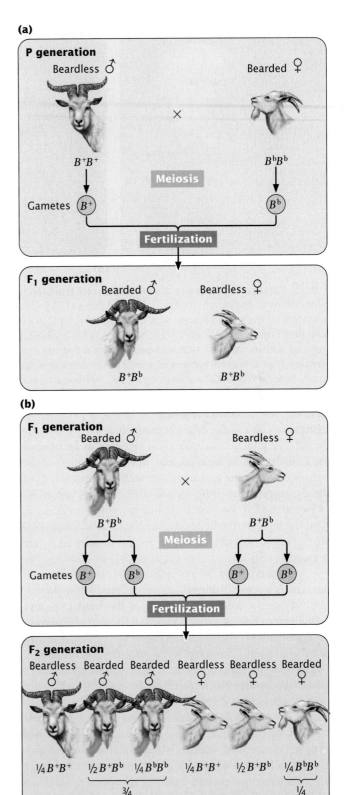

◀ 5.11 **Genes that encode sex-influenced traits are inherited according to Mendel's principles but are expressed differently in males and females.**

(a)   **(b)**   **(c)**

◀ 5.12 **Pattern baldness is a sex-influenced trait.** This trait is seen in three generations of the Adams family: (a) John Adams (1735–1826), the second president of the United States, was father to (b) John Quincy Adams (1767–1848), who was father to (c) Charles Francis Adams (1807–1886). Pattern baldness results from an autosomal allele that is thought to be dominant in males and recessive in females. (Part a, National Museum of American Art, Washington, D.C./Art Resource, NY; part b, National Portrait Gallery, Washington, D.C./Art Resource, N.Y.; part c, Bettmann/Corbis.)

separate into gametes according to Mendel's principle of segregation, and all the F_1 are heterozygous (B^+B^b). Because the trait is dominant in males and recessive in females, all the F_1 males will be bearded, and all the F_1 females will be beardless. When the F_1 are crossed with one another, $\frac{1}{4}$ of the F_2 progeny are B^bB^b, $\frac{1}{2}$ are B^bB^+, and $\frac{1}{4}$ are B^+B^+ (◀ FIGURE 5.11b). Because male heterozygotes are bearded, $\frac{3}{4}$ of the males in the F_2 possess beards; because female heterozygotes are beardless, only $\frac{1}{4}$ of the females in F_2 are bearded.

An example of a sex-influenced characteristic in humans is pattern baldness, in which hair is lost prematurely from the front and the top of the head (◀ FIGURE 5.12). Pattern baldness is an autosomal character believed to be dominant in males and recessive in females, just like beards in goats. Contrary to a popular misconception, a man does not inherit pattern baldness from his mother's side of the family (which would be the case if the character were X linked, but it isn't). Pattern baldness is autosomal; men and women can inherit baldness from either their mothers or their fathers. Men require only a single bald allele to become bald, whereas women require two bald alleles, and so pattern baldness is much more common among men. Furthermore, pattern baldness is expressed weakly in women; those with the trait usually have only a mild thinning of the hair, whereas men frequently lose all the hair on the top of the head. The expression of the allele for pattern baldness is clearly enhanced by the presence of male sex hormones; males who are castrated at an early age rarely become bald (but castration is not a recommended method for preventing baldness).

An extreme form of sex-influenced inheritance, a **sex-limited characteristic** is encoded by autosomal genes that are expressed in only one sex—the trait has zero penetrance in the other sex. In domestic chickens, some males display a plumage pattern called cock feathering (◀ FIGURE 5.13a). Other males and all females display a pattern called hen feathering (◀ FIGURE 5.13b and c). Cock feathering is an autosomal recessive trait that is sex limited to males. Because the trait is autosomal, the genotypes of males and females are the same, but the phenotypes produced by these genotypes differ in males and females:

Genotype	Male phenotype	Female phenotype
HH	hen feathering	hen feathering
Hh	hen feathering	hen feathering
hh	cock feathering	hen feathering

An example of a sex-limited characteristic in humans is male-limited precocious puberty. There are several types of precocious puberty in humans, most of which are not genetic. Male-limited precocious puberty, however, results from an autosomal dominant allele (P) that is expressed only in males; females with the gene are normal in phenotype. Males with precocious puberty undergo puberty at an early age, usually before the age of 4. At this time, the penis enlarges, the voice deepens, and pubic hair develops. There is no impairment of sexual function; affected males are fully fertile. Most are short as adults, because the long bones stop growing after puberty.

Because the trait is rare, affected males are usually heterozygous (Pp). A male with precocious puberty who mates

(a)   (b)   (c)

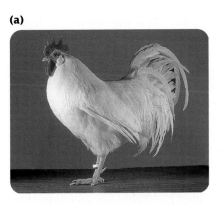

◀5.13 **A sex-limited characteristic is encoded by autosomal genes that are expressed in only one sex.** An example is cock feathering in chickens, an autosomal recessive trait that is limited to males. (a) Cock-feathered male. (b) Hen-feathered female. (c) Hen-feathered male. (Part a, b, c, Larry Lefever/Grant Heilman Photography.)

(a)

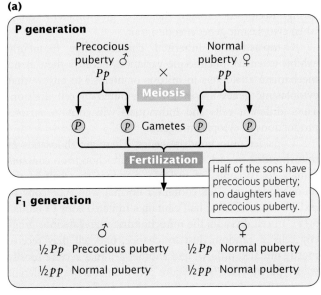

(b)

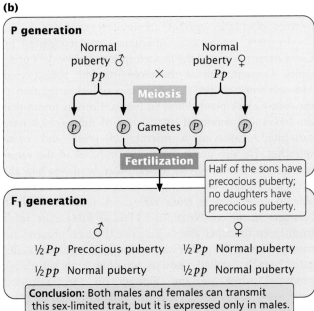

with a woman who has no family history of this condition will transmit the allele for precocious puberty to $\frac{1}{2}$ of the children (◀FIGURE 5.14a), but it will be expressed only in the sons. If one of the heterozygous daughters (Pp) mates with a male who has normal puberty (pp), $\frac{1}{2}$ of the sons will exhibit precocious puberty (◀FIGURE 5.14b). Thus a sex-limited characteristic can be inherited from either parent, although the trait appears in only one sex.

The results of molecular studies reveal that the underlying genetic defect in male-limited precocious puberty affects the receptor for luteinizing hormone (LH). This hormone normally attaches to receptors found on certain cells of the testes and stimulates these cells to produce testosterone. During normal puberty in males, high levels of LH stimulate the increased production of testosterone, which, in turn, stimulates the anatomical and physiological changes associated with puberty. The P allele for precocious puberty codes for a defective LH receptor, which stimulates testosterone production even in the absence of LH. Boys with this allele produce high levels of testosterone at an early age, when levels of LH are low. Defective LH receptors are also found in females who carry the precocious-puberty gene, but their presence does not result in precocious puberty, because additional hormones are required along with LH to induce puberty in girls.

Concepts

Sex-influenced characteristics are traits encoded by autosomal genes that are more readily expressed in one sex. Sex-limited characteristics are encoded by autosomal genes whose expression is limited to one sex.

◀5.14 **Sex-limited characteristics are inherited according to Mendel's principles.** Precocious puberty is an autosomal dominant trait that is limited to males.

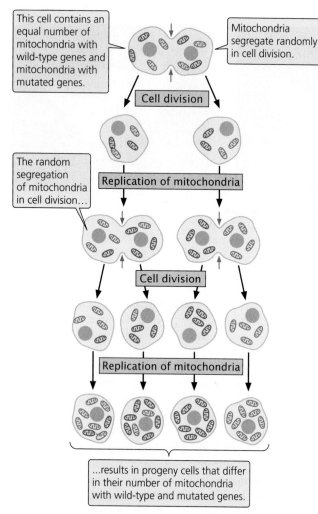

This cell contains an equal number of mitochondria with wild-type genes and mitochondria with mutated genes.

Mitochondria segregate randomly in cell division.

Cell division

Replication of mitochondria

The random segregation of mitochondria in cell division...

Cell division

Replication of mitochondria

...results in progeny cells that differ in their number of mitochondria with wild-type and mutated genes.

◄5.15 Cytoplasmically inherited characteristics frequently exhibit extensive phenotypic variation because cells and individual offspring contain various proportions of cytoplasmic genes. Mitochondria that have wild-type mtDNA are shown in red; those having mutant mtDNA are shown in blue.

Cytoplasmic Inheritance

Mendel's principles of segregation and independent assortment are based on the assumption that genes are located on chromosomes in the nucleus of the cell. For the majority of genetic characteristics, this assumption is valid, and Mendel's principles allow us to predict the types of offspring that will be produced in a genetic cross. However, not all the genetic material of a cell is found in the nucleus; some characteristics are encoded by genes located in the cytoplasm. These characteristics exhibit **cytoplasmic inheritance**.

A few organelles, notably chloroplasts and mitochondria, contain DNA. Each human mitochondrion contains about 15,000 nucleotides of DNA, encoding 37 genes. Compared with that of nuclear DNA, which contains some 3 billion nucleotides encoding perhaps 35,000 genes, the amount

of mitochondrial DNA (mtDNA) is very small; nevertheless, mitochondrial and chloroplast genes encode some important characteristics. The molecular details of this extranuclear DNA are discussed in Chapter 20; here, we will focus on *patterns* of cytoplasmic inheritance.

Cytoplasmic inheritance differs from the inheritance of characteristics encoded by nuclear genes in several important respects. A zygote inherits nuclear genes from both parents, but typically all its cytoplasmic organelles, and thus all its cytoplasmic genes, come from only one of the gametes, usually the egg. Sperm generally contributes only a set of nuclear genes from the male parent. In a few organisms, cytoplasmic genes are inherited from the male parent or from both parents; however, for most organisms, all the cytoplasm is inherited from the egg. In this case, cytoplasmically inherited traits are present in both males and females and are passed from mother to offspring, never from father to offspring. Reciprocal crosses, therefore, give different results when cytoplasmic genes encode a trait.

Cytoplasmically inherited characteristics frequently exhibit extensive phenotypic variation, because there is no mechanism analogous to mitosis or meiosis to ensure that cytoplasmic genes are evenly distributed in cell division. Thus, different cells and individuals will contain various proportions of cytoplasmic genes.

Consider mitochondrial genes. There are thousands of mitochondria in each cell, and each mitochondrion contains from 2 to 10 copies of mtDNA. Suppose that half of the mitochondria in a cell contain a normal wild-type copy of mtDNA and the other half contain a mutated copy (◄FIGURE 5.15). In cell division, the mitochondria segregate into progeny cells at random. Just by chance, one cell may receive mostly mutated mtDNA and another cell may receive mostly wild-type mtDNA (see Figure 5.15). In this way, different progeny from the same mother and even cells within an individual offspring may vary in their phenotype. Traits encoded by chloroplast DNA (cpDNA) are similarly variable.

In 1909, cytoplasmic inheritance was recognized by Carl Correns as one of the first exceptions to Mendel's principles. Correns, one of the biologists who rediscovered Mendel's work, studied the inheritance of leaf variegation in the four-o'clock plant, *Mirabilis jalapa*. Correns found that the leaves and shoots of one variety of four-o'clock were variegated, displaying a mixture of green and white splotches. He also noted that some branches of the variegated strain had all-green leaves; other branches had all-white leaves. Each branch produced flowers; so Correns was able to cross flowers from variegated, green, and white branches in all combinations (◄FIGURE 5.16). The seeds from green branches always gave rise to green progeny, no matter whether the pollen was from a green, white, or variegated branch. Similarly, flowers on white branches always produced white progeny. Flowers on the variegated branches gave rise to green, white, and variegated progeny, in no particular ratio.

The phenotype of the branch from which the pollen originated has no effect on the phenotype of the progeny.

Pollen plant (♂)

Seed plant (♀)

Pollen ← White | Pollen ← Green | Pollen ← Variegated

White | White | White | White

Green | Green | Green | Green

Variegated | White | White | White

| Green | Green | Green

| Variegated | Variegated | Variegated

Conclusion: The phenotype of the progeny is determined by the phenotype of the branch from which the seed originated.

5.16 Crosses for leaf type in four o'clocks illustrate cytoplasmic inheritance.

Corren's crosses demonstrated cytoplasmic inheritance of variegation in the four-o'clocks. The phenotypes of the offspring were determined entirely by the maternal parent, never by the paternal parent (the source of the pollen). Furthermore, the production of all three phenotypes by flowers on variegated branches is consistent with the occurrence of cytoplasmic inheritance. Variegation in these plants is caused by a defective gene in the cpDNA, which results in a failure to produce the green pigment chlorophyll. Cells from green branches contain normal chloroplasts only, cells from white branches contain abnormal chloroplasts only, and cells from variegated branches contain a mixture of normal and abnormal chloroplasts. In the flowers from variegated branches,

the random segregation of chloroplasts in the course of oogenesis produces some egg cells with normal cpDNA, which develop into green progeny; other egg cells with only abnormal cpDNA develop into white progeny; and, finally, still other egg cells with a mixture of normal and abnormal cpDNA develop into variegated progeny.

In recent years, a number of human diseases (mostly rare) that exhibit cytoplasmic inheritance have been identified. These disorders arise from mutations in mtDNA, most of which occur in genes coding for components of the electron-transport chain, which generates most of the ATP (adenosine triphosphate) in aerobic cellular respiration. One such disease is Leber hereditary optic neuropathy (LHON). Patients who have this disorder experience rapid loss of vision in both eyes, resulting from the death of cells in the optic nerve. Loss of vision typically occurs in early adulthood (usually between the ages of 20 and 24), but it can occur any time after adolescence. There is much clinical variability in the severity of the disease, even within the same family. Leber hereditary optic neuropathy exhibits maternal inheritance: the trait is always passed from mother to child.

Genetic Maternal Effect

A genetic phenomenon that is sometimes confused with cytoplasmic inheritance is **genetic maternal effect,** in which the phenotype of the offspring is determined by the genotype of the mother. In cytoplasmic inheritance, the genes for a characteristic are inherited from only one parent, usually the mother. In genetic maternal effect, the genes are inherited from both parents, but the offspring's phenotype is determined not by its own genotype but by the genotype of its mother.

Genetic maternal effect frequently arises when substances present in the cytoplasm of an egg (encoded by the mother's genes) are pivotal in early development. An excellent example is shell coiling of the snail *Limnaea peregra*. In most snails of this species, the shell coils to the right, which is termed dextral coiling. However, some snails possess a left-coiling shell, exhibiting sinistral coiling. The direction of coiling is determined by a pair of alleles; the allele for dextral (s^+) is dominant over the allele for sinistral (s). However, the direction of coiling is determined not by that snail's own genotype, but by the genotype of its *mother*. The direction of coiling is affected by the way in which the cytoplasm divides soon after fertilization, which in turn is determined by a substance produced by the mother and passed to the offspring in the cytoplasm of the egg.

If a male homozygous for dextral alleles ($s^+ s^+$) is crossed with a female homozygous for sinistral alleles (ss), all of the F_1 are heterozygous ($s^+ s$) and have a sinistral shell, because the genotype of the mother (ss) codes for sinistral (◄FIGURE 5.17). If these F_1 snails are self-fertilized, the genotypic ratio of the F_2 is $1\ s^+ s^+ : 2\ s^+ s : 1\ ss$. The phenotype of all F_2 snails will be dextral regardless of their genotypes, because the genotype of their mother ($s^+ s$) encodes a right-coiling shell and determines their phenotype.

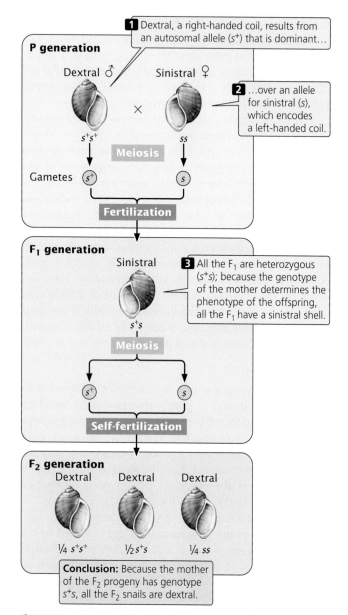

P generation

1 Dextral, a right-handed coil, results from an autosomal allele (s^+) that is dominant...

Dextral ♂   Sinistral ♀

2 ...over an allele for sinistral (s), which encodes a left-handed coil.

s^+s^+ ss

Meiosis

Gametes s^+ s

Fertilization

F₁ generation

Sinistral

3 All the F₁ are heterozygous (s^+s); because the genotype of the mother determines the phenotype of the offspring, all the F₁ have a sinistral shell.

s^+s

Meiosis

s^+ s

Self-fertilization

F₂ generation

Dextral   Dextral   Dextral

¼ s^+s^+ ½ s^+s ¼ ss

Conclusion: Because the mother of the F₂ progeny has genotype s^+s, all the F₂ snails are dextral.

◄ **5.17 In genetic maternal effect, the genotype of the maternal parent determines the phenotype of the offspring.** Shell coiling in snails is a trait that exhibits genetic maternal effect.

Notice that the phenotype of the progeny is not necessarily the same as the phenotype of the mother, because the progeny's phenotype is determined by the mother's *genotype*, not her phenotype. Neither the male parent's nor the offspring's own genotype has any role in the offspring's phenotype. A male does influence the phenotype of the F₂ generation; by contributing to the genotypes of his daughters, he affects the phenotypes of their offspring. Genes that exhibit genetic maternal effect are therefore transmitted through males to future generations. In contrast, the genes that exhibit cytoplasmic inheritance are always transmitted through only one of the sexes (usually the female).

Concepts

Characteristics exhibiting cytoplasmic inheritance are encoded by genes in the cytoplasm and are usually inherited from one parent, most commonly the mother. In genetic maternal effect, the genotype of the mother determines the phenotype of the offspring.

Genomic Imprinting

One of the basic tenets of Mendelian genetics is that the parental origin of a gene does not affect its expression—reciprocal crosses give identical results. We have seen that there are some genetic characteristics—those encoded by X-linked genes and cytoplasmic genes—for which reciprocal crosses do not give the same results. In these cases, males and females do not contribute the same genetic material to the offspring. With regard to autosomal genes, males and females contribute the same number of genes, and paternal and maternal genes have long been assumed to have equal effects. The results of recent studies, however, have identified several mammalian genes whose expression is significantly affected by their parental origin. This phenomenon, the differential expression of genetic material depending on whether it is inherited from the male or female parent, is called **genomic imprinting.**

Genomic imprinting has been observed in mice in which a particular gene has been artificially inserted into a mouse's DNA (to create a transgenic mouse). In these mice, the inserted gene is faithfully passed from generation to generation, but its expression may depend on which parent transmitted the gene. For example, when a transgenic male passes an imprinted gene to his offspring, they express the gene; but, when his daughter transmits the same gene to her offspring, they don't express it. In turn, her son's offspring express it, but her daughter's offspring don't. Both male and female offspring possess the gene for the trait; the key to whether the gene is expressed is the sex of the parent transmitting the gene. In the present example, the gene is expressed only when it is transmitted by a male parent. The reverse situation, expression of a trait when the gene is transmitted by the female parent, also occurs.

Genomic imprinting has been implicated in several human disorders, including Prader-Willi and Angelman syndromes. Children with Prader-Willi syndrome have small hands and feet, short stature, poor sexual development, and mental retardation; they develop voracious appetites and frequently become obese. Many persons with Prader-Willi syndrome are missing a small region of chromosome 15 called q11–13. The deletion of this region is always inherited from the father in persons with Prader-Willi syndrome.

The deletions of q11–13 on chromosome 15 can also be inherited from the *mother*, but this inheritance results in a completely different set of symptoms, producing Angelman

Table 5.4	Sex influences on heredity
Genetic Phenomenon	**Phenotype Determined by**
Sex-linked characteristic	genes located on the sex chromosome
Sex-influenced characteristic	genes on autosomal chromosomes that are more readily expressed in one sex
Sex-limited characteristic	autosomal genes whose expression is limited to one sex
Genetic maternal effect	nuclear genotype of the maternal parent
Cytoplasmic inheritance	cytoplasmic genes, which are usually inherited entirely from only one parent
Genomic imprinting	genes whose expression is affected by the sex of the transmitting parent

syndrome. Children with Angelman syndrome exhibit frequent laughter, uncontrolled muscle movement, a large mouth, and unusual seizures. The deletion of segment q11–13 from chromosome 15 has severe effects on the human phenotype, but the specific effects depend on which parent contributes the deletion. For normal development to take place, copies of segment q11–13 of chromosome 15 from both male and female parents are apparently required.

Several other human diseases also appear to exhibit genomic imprinting. Although the precise mechanism of this phenomenon is unknown, methylation of DNA—the addition of methyl (CH_3) groups to DNA nucleotides (see Chapters 10 and 16)—is essential to the process of genomic imprinting, as demonstrated by the observation that mice deficient in DNA methylation do not exhibit imprinting. Some of the ways in which sex interacts with heredity are summarized in Table 5.4.

Concepts

In genomic imprinting, the expression of a gene is influenced by the sex of the parent who transmits the gene to the offspring.

www.whfreeman.com/pierce   Additional information about genomic imprinting, Prader-Willi syndrome, and Angelman syndrome

Anticipation

Another genetic phenomenon that is not explained by Mendel's principles is **anticipation,** in which a genetic trait becomes more strongly expressed or is expressed at an earlier age as it is passed from generation to generation. In the early 1900s, several physicians observed that patients with moderate to severe myotonic dystrophy—an autosomal dominant muscle disorder—frequently had ancestors who were only mildly affected by the disease. These observations led to the concept of anticipation. However, the concept quickly fell out of favor with geneticists because there was no obvious mechanism to explain it; traditional genetics held that genes are passed unaltered from parents to offspring. Geneticists tended to attribute anticipation to observational bias.

The results of recent research have reestablished anticipation as a legitimate genetic phenomenon. The mutation causing myotonic dystrophy consists of an unstable region of DNA that can increase or decrease in size as the gene is passed from generation to generation, much like the gene that causes Huntington disease. The age of onset and the severity of the disease are correlated with the size of the unstable region; an increase in the size of the region through generations produces anticipation. The phenomenon has now been implicated in several genetic diseases. We will examine these interesting types of mutations in more detail in Chapter 17.

Concepts

Anticipation is the stronger or earlier expression of a genetic trait through succeeding generations. It is caused by an unstable region of DNA that increases or decreases in size.

Interaction Between Genes and Environment

In Chapter 3, we learned that each phenotype is the result of a genotype developing within a specific environment; the genotype sets the potential for development, but how the phenotype actually develops within the limits imposed by the genotype depends on environmental effects. Stated another way, each genotype may produce several different phenotypes, depending on the environmental conditions in which development occurs. For example, genotype *GG* may produce a plant that is 10 cm high when raised at 20°C, but the same genotype may produce a plant that is 18 cm high when raised at 25°C. The range of phenotypes produced by a genotype in different environments (in this case, plant height) is called the **norm of reaction** (◀ FIGURE 5.18).

For most of the characteristics discussed so far, the effect of the environment on the phenotype has been slight.

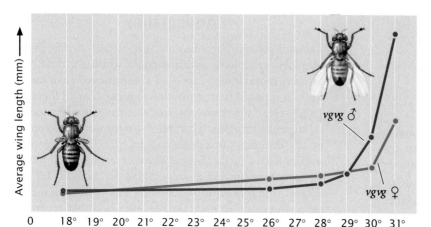

5.18 Norm of reaction is the range of phenotypes produced by a genotype in different environments. This norm of reaction is for vestigial wings in *Drosophila melanogaster.* (Data from M. H. Harnly, *Journal of Experimental Zoology* 56:363–379, 1936.)

Mendel's peas with genotype *yy,* for example, developed yellow endosperm regardless of the environment in which they were raised. Similarly, persons with genotype $I^A I^A$ have the A antigen on their red blood cells regardless of their diet, socioeconomic status, or family environment. For other phenotypes, however, environmental effects play a more important role.

Environmental Effects on Gene Expression

The expression of some genotypes is critically dependent on the presence of a specific environment. For example, the *himalayan* allele in rabbits produces dark fur at the extremities of the body—on the nose, ears, and feet (◀FIGURE 5.19). The dark pigment develops, however, only when the rabbit is reared at 25°C or less; if a Himalayan rabbit is reared at 30°C, no dark patches develop. The expression of the *himalayan* allele is thus temperature dependent—an enzyme necessary for the production of dark pigment is inactivated at higher temperatures. The pigment is restricted to the nose, feet, and ears of Himalayan rabbits because the animal's core body temperature is normally above 25°C and the enzyme is functional only in the cells of the relatively cool extremities. The *himalayan* allele is an example of a **temperature-sensitive allele,** an allele whose product is functional only at certain temperatures.

Some types of albinism in plants are temperature dependent. In barley, an autosomal recessive allele inhibits chlorophyll production, producing albinism when the plant is grown below 7°C. At temperatures above 18°C, a plant homozygous for the albino allele develops normal chlorophyll and is green. Similarly, among *Drosophila melanogaster* homozygous for the autosomal mutation *vestigial,* greatly reduced wings develop at 25°C, but wings near normal size develop at higher temperatures (see Figure 5.18).

Environmental factors also play an important role in the expression of a number of human genetic diseases. Glucose-6-phosphate dehydrogenase is an enzyme taking part in supplying energy to the cell. In humans, there are a number of genetic variants of glucose-6-phosphate dehydrogenase, some of which destroy red blood cells when the body is stressed by infection or by the ingestion of certain drugs or foods. The symptoms of the genetic disease appear only in the presence of these specific environmental factors.

Another genetic disease, phenylketonuria (PKU), is due to an autosomal recessive allele that causes mental retardation. The disorder arises from a defect in an enzyme that normally metabolizes the amino acid phenylalanine. When this enzyme is defective, phenylalanine is not metabolized, and its buildup causes brain damage in children. A simple

5.19 The expression of some genotypes depends on specific environments. The expression of a temperature-sensitive allele, *himalayan,* is shown in rabbits reared at different temperatures.

Reared at 20°C or less

Reared at temperatures above 30°C

environmental change, putting an affected child on a low-phenylalanine diet, prevents retardation.

These examples illustrate the point that genes and their products do not act in isolation; rather, they frequently interact with environmental factors. Occasionally, environmental factors alone can produce a phenotype that is the same as the phenotype produced by a genotype; this phenotype is called a **phenocopy.** In fruit flies, for example, the autosomal recessive mutation *eyeless* produces greatly reduced eyes. The eyeless phenotype can also be produced by exposing the larvae of normal flies to sodium metaborate.

Concepts

The expression of many genes is modified by the environment. The range of phenotypes produced by a genotype in different environments is called the norm of reaction. A phenocopy is a trait produced by environmental effects that mimics the phenotype produced by a genotype.

The Inheritance of Continuous Characteristics

So far, we've dealt primarily with characteristics that have only a few distinct phenotypes. In Mendel's peas, for example, the seeds were either smooth or wrinkled, yellow or green; the coats of dogs were black, brown, or yellow; blood types were of four distinct types, A, B, AB, or O. Characteristics such as these, which have a few easily distinguished phenotypes, are called **discontinuous characteristics.**

Not all characteristics exhibit discontinuous phenotypes. Human height is an example of such a character; people do not come in just a few distinct heights but, rather, display a continuum of heights. Indeed, there are so many possible phenotypes of human height that we must use a measurement to describe a person's height. Characteristics that exhibit a continuous distribution of phenotypes are termed **continuous characteristics.** Because such characteristics have many possible phenotypes and must be described in quantitative terms, continuous characteristics are also called **quantitative characteristics.**

Continuous characteristics frequently arise because genes at many loci interact to produce the phenotypes. When a single locus with two alleles codes for a characteristic, there are three genotypes possible: *AA, Aa,* and *aa.* With two loci, each with two alleles, there are $3^2 = 9$ genotypes possible. The number of genotypes coding for a characteristic is 3^n, where n equals the number of loci with two alleles that influence the characteristic. For example, when a characteristic is determined by eight loci, each with two alleles, there are $3^8 = 6561$ different genotypes possible for this character. If each genotype produces a different phenotype, many phenotypes will be possible. The slight differences between the phenotypes will be indistinguishable, and the characteristic will appear continuous. Characteristics encoded by genes at many loci are called **polygenic characteristics.**

The converse of polygeny is **pleiotropy,** in which one gene affects multiple characteristics. Many genes exhibit pleiotropy. PKU, mentioned earlier, results from a recessive allele; persons homozygous for this allele, if untreated, exhibit mental retardation, blue eyes, and light skin color.

Frequently the phenotypes of continuous characteristics are also influenced by environmental factors. Each genotype is capable of producing a range of phenotypes—it has a relatively broad norm of reaction. In this situation, the particular phenotype that results depends on both the genotype and the environmental conditions in which the genotype develops. For example, there may be only three genotypes coding for a characteristic, but, because each genotype has a broad norm of reaction, the phenotype of the character exhibits a continuous distribution. Many continuous characteristics are both polygenic and influenced by environmental factors; such characteristics are called **multifactorial** because many factors help determine the phenotype.

The inheritance of continuous characteristics may appear to be complex, but the alleles at each locus follow Mendel's principles and are inherited in the same way as alleles coding for simple, discontinuous characteristics. However, because many genes participate, environmental factors influence the phenotype, and the phenotypes do not sort out into a few distinct types, we cannot observe the distinct ratios that have allowed us to interpret the genetic basis of discontinuous characteristics. To analyze continuous characteristics, we must employ special statistical tools, as will be discussed in Chapter 22.

Concepts

Discontinuous characteristics exhibit a few distinct phenotypes; continuous characteristics exhibit a range of phenotypes. A continuous characteristic is frequently produced when genes at many loci and environmental factors combine to determine a phenotype.

Connecting Concepts Across Chapters

This chapter introduced a number of modifications and extensions of the basic concepts of heredity that we learned in Chapter 3. A major theme has been gene expression: how interactions between genes, interactions between genes and sex, and interactions between genes and the environment affect the phenotypic expression of genes. The modifications and extensions discussed in this chapter do not alter the way that genes are inherited, but they do modify the way in which the genes determine the phenotype.

A number of topics introduced in this chapter will be explored further in other chapters of the book. Here we have purposefully ignored many aspects of the nature of gene expression because our focus has been on the "big picture" of how these interactions affect phenotypic ratios in genetic crosses. In subsequent chapters, we will explore the molecular details of gene expression, including transcription (Chapter 13), translation (Chapter 15), and the control of gene expression (Chapter 16). The molecular nature of anticipation will be examined in more detail in Chapter 17, and DNA methylation, the basis of genomic imprinting, will be discussed in Chapter 10. Complementation testing will be revisited in Chapter 8, and the role of multiple genes and environmental factors in the inheritance of continuous characteristics will be studied more thoroughly in Chapter 22.

CONCEPTS SUMMARY

- Dominance always refers to genes at the same locus (allelic genes) and can be understood in regard to how the phenotype of the heterozygote relates to the phenotypes of the homozygotes.

- Dominance is complete when a heterozygote has the same phenotype as a homozygote. Dominance is incomplete when the heterozygote has a phenotype intermediate between those of two parental homozygotes. Codominance is the result when the heterozygote exhibits traits of both parental homozygotes.

- The type of dominance does not affect the inheritance of an allele; it does affect the phenotypic expression of the allele. The classification of dominance may depend on the level of the phenotype examined.

- Lethal alleles cause the death of an individual possessing them, usually at an early stage of development, and may alter phenotypic ratios.

- Multiple alleles refers to the presence of more than two alleles at a locus within a group. Their presence increases the number of genotypes and phenotypes possible.

- Gene interaction refers to interaction between genes at different loci to produce a single phenotype. An epistatic gene at one locus suppresses or masks the expression of hypostatic genes at different loci. Gene interaction frequently produces phenotypic ratios that are modifications of dihybrid ratios.

- A complementation test, in which individuals homozygous for different mutations are crossed, can be used to determine if the mutations occur at the same locus or at different loci.

- Sex-influenced characteristics are encoded by autosomal genes that are expressed more readily in one sex.

- Sex-limited characteristics are encoded by autosomal genes expressed in only one sex. Both males and females possess sex-limited genes and transmit them to their offspring.

- In cytoplasmic inheritance, the genes for the characteristic are found in the cytoplasm and are usually inherited from a single (usually maternal) parent.

- Genetic maternal effect is present when an offspring inherits genes from both parents, but the nuclear genes of the mother determine the offspring's phenotype.

- Genomic imprinting refers to characteristics encoded by autosomal genes whose expression is affected by the sex of the parent transmitting the genes.

- Anticipation refers to a genetic trait that is more strongly expressed or is expressed at an earlier age in succeeding generations.

- Phenotypes are often modified by environmental effects. The range of phenotypes that a genotype is capable of producing in different environments is the norm of reaction. A phenocopy is a phenotype produced by an environmental effect that mimics a phenotype produced by a genotype.

- Discontinuous characteristics are characteristics with a few distinct phenotypes; continuous characteristics are those that exhibit a wide range of phenotypes. Continuous characteristics are frequently produced by the combined effects of many genes and environmental effects.

IMPORTANT TERMS

codominance (p. 103)	complementation (p. 115)	genomic imprinting (p. 120)	continuous characteristic
lethal allele (p. 104)	sex-influenced characteristic	anticipation (p. 121)	(p. 123)
multiple alleles (p. 105)	(p. 115)	norm of reaction (p. 121)	quantitative characteristic
gene interaction (p. 107)	sex-limited characteristic	temperature-sensitive allele	(p. 123)
epistasis (p. 108)	(p. 116)	(p. 122)	polygenic characteristic (p. 123)
epistatic gene (p. 108)	cytoplasmic inheritance	phenocopy (p. 123)	pleiotropy (p. 123)
hypostatic gene (p. 108)	(p. 118)	discontinuous characteristic	multifactorial characteristic
complementation test (p. 114)	genetic maternal effect (p. 119)	(p. 123)	(p. 123)

Worked Problems

1. The type of plumage found in mallard ducks is determined by three alleles at a single locus: M^R, which codes for restricted plumage; M, which codes for mallard plumage; and m^d, which codes for dusky plumage. The restricted phenotype is dominant over mallard and dusky; mallard is dominant over dusky ($M^R > M > m^d$). Give the expected phenotypes and proportions of offspring produced by the following crosses.

(a) $M^R M \times m^d m^d$

(b) $M^R m^d \times M m^d$

(c) $M^R m^d \times M^R M$

(d) $M^R M \times M m^d$

• Solution

We can determine the phenotypes and proportions of offspring by (1) determining the types of gametes produced by each parent and (2) combining the gametes of the two parents with the use of a Punnett square

(a) Parents $\quad M^R M \times m^d m^d$

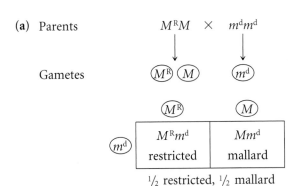

$\frac{1}{2}$ restricted, $\frac{1}{2}$ mallard

(b) Parents $\quad M^R m^d \times M m^d$

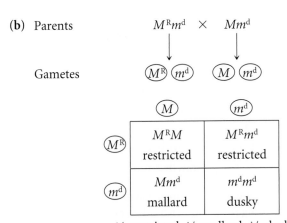

$\frac{1}{2}$ restricted, $\frac{1}{4}$ mallard, $\frac{1}{4}$ dusky

(c) Parents $\quad M^R m^d \times M^R M$

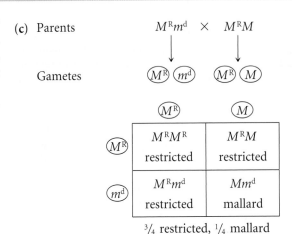

$\frac{3}{4}$ restricted, $\frac{1}{4}$ mallard

(d) Parents $\quad M^R M \times M m^d$

Gametes

	M	m^d
M^R	$M^R M$ restricted	$M^R m^d$ restricted
M	MM mallard	$M m^d$ mallard

$\frac{1}{2}$ restricted, $\frac{1}{2}$ mallard

2. A homozygous strain of yellow corn is crossed with a homozygous strain of purple corn. The F_1 are intercrossed, producing an ear of corn with 119 purple kernels and 89 yellow kernels (the progeny).

(a) What is the genotype of the yellow kernels?

(b) Give a genetic explanation for the differences in kernel color in this cross.

• Solution

(a) We should first consider whether the cross between yellow and purple strains might be a monohybrid cross for a simple dominant trait, which would produce a 3:1 ratio in the F_2 ($Aa \times Aa \rightarrow \frac{3}{4} A\_$ and $\frac{1}{4} aa$). Under this hypothesis, we would expect 156 purple progeny and 52 yellow progeny:

Phenotype	Genotype	Observed number	Expected number
purple	$A\_$	119	$\frac{3}{4} \times 208 = 156$
yellow	aa	89	$\frac{1}{4} \times 208 = 52$
total		208	

We see that the expected numbers do not closely fit the observed numbers. If we performed a chi-square test (see Chapter 3), we would obtain a calculated chi-square value of 35.08, which has a probability much less than 0.05, indicating that it is extremely unlikely that, when we expect a 3:1 ratio, we would obtain 119 purple progeny and 89 yellow progeny. Therefore we can reject the hypothesis that these results were produced by a monohybrid cross.

Another possible hypothesis is that the observed F_2 progeny are in a 1:1 ratio. However, we learned in Chapter 3 that a 1:1 ratio is produced by a cross between a heterozygote and a homozygote ($Aa \times aa$) and, from the information given, the cross was not between a heterozygote and a homozygote, because the original parental strains were both homozygous. Furthermore, a chi-square test comparing the observed numbers with an expected 1:1 ratio yields a calculated chi-square value of 4.32, which has a probability of less than .05.

Next, we should look to see if the results can be explained by a dihybrid cross ($AaBb \times AaBb$). A dihybrid cross results in phenotypic proportions that are in sixteenths. We can apply the formula given earlier in the chapter to determine the number of sixteenths for each phenotype:

$$x = \frac{\text{number of progeny with a phenotype} \times 16}{\text{total number of progeny}}$$

$$x_{(purple)} = \frac{119 \times 16}{208} = 9.15$$

$$x_{(yellow)} = \frac{89 \times 16}{208} = 6.85$$

Thus, purple and yellow appear approximately in a 9:7 ratio. We can test this hypothesis with a chi-square test:

Phenotype	Genotype	Observed number	Expected number
purple	?	119	$9/16 \times 208 = 117$
yellow	?	89	$7/16 \times 208 = 91$
total		208	

$$\chi^2 = \sum \frac{(\text{observed} - \text{expected})^2}{\text{expected}} = \frac{(119 - 117)^2}{117} + \frac{(89 - 91)^2}{91}$$
$$= 0.034 + 0.44 = 0.078$$

Degree of freedom = $n - 1 = 2 - 1 = 1$

$P > .05$

The probability associated with the chi-square value is greater than .05, indicating that there is a relatively good fit between the observed results and a 9:7 ratio.

We now need to determine how a dihybrid cross can produce a 9:7 ratio and what genotypes correspond to the two phenotypes. A dihybrid cross without epistasis produces a 9:3:3:1 ratio:

$$AaBb \times AaBb$$
$$\downarrow$$

A_B_	$9/16$
A_bb	$3/16$
aaB_	$3/16$
aabb	$1/16$

Because $9/16$ of the progeny from the corn cross are purple, purple must be produced by genotypes $A\_B\_$; in other words, individual kernels that have at least one dominant allele at the first locus and at least one dominant allele at the second locus are purple. The proportions of all the other genotypes (A_bb, $aaB\_$, and $aabb$) sum to $7/16$, which is the proportion of the progeny in the corn cross that are yellow, so any individual kernel that does not have a dominant allele at both the first and the second locus is yellow.

(b) Kernel color is an example of duplicate recessive epistasis, where the presence of two recessive alleles at either the first locus or the second locus or both suppresses the production of purple pigment.

3. A geneticist crosses two yellow mice with straight hair and obtains the following progeny:

$1/2$ yellow, straight

$1/6$ yellow, fuzzy

$1/4$ gray, straight

$1/12$ gray, fuzzy

(a) Provide a genetic explanation for the results and assign genotypes to the parents and progeny of this cross.

(b) What additional crosses might be carried out to determine if your explanation is correct?

• **Solution**

(a) This cross concerns two separate characteristics—color and type of hair; so we should begin by examining the results for each characteristic separately. First, let's look at the inheritance of color. Two yellow mice are crossed producing $1/2 + 1/6 = 3/6 + 1/6 = 4/6 = 2/3$ yellow mice and $1/4 + 1/12 = 3/12 + 1/12 = 4/12 = 1/3$ gray mice. We learned in this chapter that a 2:1 ratio is often produced when a recessive lethal gene is present:

$$Yy \times Yy$$
$$\downarrow$$

YY	$1/4$ die
Yy	$1/2$ yellow, becomes $2/3$
yy	$1/4$ gray, becomes $1/3$

Now, let's examine the inheritance of the hair type. Two mice with straight hair are crossed, producing $1/2 + 1/4 = 2/4 + 1/4 = 3/4$ mice with straight hair and $1/6 + 1/12 = 2/12 + 1/12 = 3/12 = 1/4$ mice with fuzzy hair. We learned in Chapter 3 that a

3:1 ratio is usually produced by a cross between two individuals heterozygous for a simple dominant allele:

$$Ss \times Ss$$

$$\downarrow$$

SS $\frac{1}{4}$ straight $\left. \begin{array}{l} \\ \\ \end{array} \right\}$ $\frac{3}{4}$ straight

Ss $\frac{1}{2}$ straight

ss $\frac{1}{4}$ fuzzy

We can now combine both loci and assign genotypes to all the individuals in the cross:

P   yellow, straight $\times$ yellow, straight
        $YySs$ $YySs$

$$\downarrow$$

Phenotype	Genotype	Probability at each locus	Combined probability
yellow, straight	$YyS\_$	$\frac{2}{3} \times \frac{3}{4}$	$= \frac{6}{12} = \frac{1}{2}$
yellow, fuzzy	$Yyss$	$\frac{2}{3} \times \frac{1}{4}$	$= \frac{2}{12} = \frac{1}{6}$
gray, straight	$yyS\_$	$\frac{1}{3} \times \frac{3}{4}$	$= \frac{3}{12} = \frac{1}{4}$
gray, fuzzy	$yyss$	$\frac{1}{3} \times \frac{1}{4}$	$= \frac{1}{12}$

(b) We could carry out a number of different crosses to test our hypothesis that yellow is a recessive lethal and straight is dominant over fuzzy. For example, a cross between any two yellow individuals should always produce $\frac{2}{3}$ yellow and $\frac{1}{3}$ gray, and a cross between two gray individuals should produce all gray offspring. A cross between two fuzzy individuals should always produce all fuzzy offspring.

4. In some sheep, the presence of horns is produced by an autosomal allele that is dominant in males and recessive in females. A horned female is crossed with a hornless male. One of the resulting F$_1$ females is crossed with a hornless male. What proportion of the male and female progeny from this cross will have horns?

• **Solution**

The presence of horns in these sheep is an example of a sex-influenced characteristic. Because the phenotypes associated with the genotypes differ for the two sexes, let's begin this problem by writing out the genotypes and phenotypes for each sex. We will

let H represent the allele that codes for horns and H^+ represent the allele for hornless. In males, the allele for horns is dominant over the allele for hornless, which means that males homozygous (HH) and heterozygous (H^+H) for this gene are horned. Only males homozygous for the recessive hornless allele (H^+H^+) will be hornless. In females, the allele for horns is recessive, which means that only females homozygous for this allele (HH) will be horned; females heterozygous (H^+H) and homozygous (H^+H^+) for the hornless allele will be hornless. The following table summarizes genotypes and associated phenotypes:

Genotype	Male phenotype	Female phenotype
HH	horned	horned
HH^+	horned	hornless
H^+H^+	hornless	hornless

In the problem, a horned female is crossed with a hornless male. From the preceding table, we see that a horned female must be homozygous for the allele for horns (HH) and a hornless male must be homozygous for the allele for hornless (H^+H^+); so all the F$_1$ will be heterozygous; the F$_1$ males will be horned and the F$_1$ females will be hornless, as shown below:

P $H^+H^+ \times HH$

$$\downarrow$$

F$_1$ H^+H

horned males and hornless females

A heterozygous hornless F$_1$ female (H^+H) is then crossed with a hornless male (H^+H^+):

$$H^+H \times H^+H^+$$
horned female   hornless male

$$\downarrow$$

	Males	Females
$\frac{1}{2}$ H^+H^+	hornless	hornless
$\frac{1}{2}$ H^+H	horned	hornless

Therefore, $\frac{1}{2}$ of the male progeny will be horned but none of the female progeny will be horned.

COMPREHENSION QUESTIONS

* 1. How do incomplete dominance and codominance differ?

* 2. Explain how dominance and epistasis differ.

3. What is a recessive epistatic gene?

4. What is a complementation test and what is it used for?

* 5. What is genomic imprinting?

6. What characteristics do you expect to see in a trait that exhibits anticipation?

* 7. What characteristics are exhibited by a cytoplasmically inherited trait?

8. What is the difference between genetic maternal effect and genomic imprinting?

9. What is the difference between a sex-influenced gene and a gene that exhibits genomic imprinting?

* 10. What are continuous characteristics and how do they arise?

APPLICATION QUESTIONS AND PROBLEMS

* 11. Palomino horses have a golden yellow coat, chestnut horses have a brown coat, and cremello horses have a coat that is almost white. A series of crosses between the three different types of horses produced the following offspring:

Cross	Offspring
palomino × palomino	13 palomino, 6 chestnut, 5 cremello
chestnut × chestnut	16 chestnut
cremello × cremello	13 cremello
palomino × chestnut	8 palomino, 9 chestnut
palomino × cremello	11 palomino, 11 cremello
chestnut × cremello	23 palomino

(a) Explain the inheritance of the palomino, chestnut, and cremello phenotypes in horses.

(b) Assign symbols for the alleles that determine these phenotypes, and list the genotypes of all parents and offspring given in the preceding table.

* 12. The L^M and L^N alleles at the MN blood group locus exhibit codominance. Give the expected genotypes and phenotypes and their ratios in progeny resulting from the following crosses.

(a) $L^M L^M \times L^M L^N$

(b) $L^N L^N \times L^N L^N$

(c) $L^M L^N \times L^M L^N$

(d) $L^M L^N \times L^N L^N$

(e) $L^M L^M \times L^N L^N$

13. In the pearl millet plant, color is determined by three alleles at a single locus: Rp^1 (red), Rp^2 (purple), and rp (green). Red is dominant over purple and green, and purple is dominant over green ($Rp^1 > Rp^2 > rp$). Give the expected phenotypes and ratios of offspring produced by the following crosses.

(a) $Rp^1/Rp^2 \times Rp^1/rp$

(b) $Rp^1/rp \times Rp^2/rp$

(c) $Rp^1/Rp^2 \times Rp^1/Rp^2$

(d) $Rp^2/rp \times rp/rp$

(e) $rp/rp \times Rp^1/Rp^2$

* 14. Give the expected genotypic and phenotypic ratios for the following crosses for ABO blood types.

(a) $I^A i \times I^B i$

(b) $I^A I^B \times I^A i$

(c) $I^A I^B \times I^A I^B$

(d) $ii \times I^A i$

(e) $I^A I^B \times ii$

15. If there are five alleles at a locus, how many genotypes may there be at this locus? How many different kinds of homozygotes will there be? How many genotypes and homozygotes would there be with eight alleles?

16. Turkeys have black, bronze, or black-bronze plumage. Examine the results of the following crosses:

Parents	Offspring
Cross 1: black and bronze	all black
Cross 2: black and black	$^3/_4$ black, $^1/_4$ bronze
Cross 3: black-bronze and black-bronze	all black-bronze
Cross 4: black and bronze	$^1/_2$ black, $^1/_4$ bronze, $^1/_4$ black-bronze
Cross 5: bronze and black-bronze	$^1/_2$ bronze, $^1/_2$ black-bronze
Cross 6: bronze and bronze	$^3/_4$ bronze, $^1/_4$ black-bronze

Do you think these differences in plumage arise from incomplete dominance between two alleles at a single locus? If yes, support your conclusion by assigning symbols to each allele and providing genotypes for all turkeys in the crosses. If your answer is no, provide an alternative explanation and assign genotypes to all turkeys in the crosses.

17. In rabbits, an allelic series helps to determine coat color: C (full color), c^{ch} (chinchilla, gray color), c^h (himalayan, white with black extremities), and c (albino, all white). The C allele is dominant over all others, c^{ch} is dominant over c^h and c, c^h is dominant over c, and c is recessive to all the other alleles. This dominance hierarchy can be summarized as $C > c^{ch} > c^h > c$. The rabbits in the following list are crossed and produce the progeny shown. Give the genotypes of the parents for each cross:

Phenotypes of parents	Phenotypes of offspring
(a) full color × albino	$^1/_2$ full color, $^1/_2$ albino
(b) himalayan × albino	$^1/_2$ himalayan, $^1/_2$ albino
(c) full color × albino	$^1/_2$ full color, $^1/_2$ chinchilla
(d) full color × himalayan	$^1/_2$ full color, $^1/_4$ himalayan $^1/_4$ albino
(e) full color × full color	$^3/_4$ full color, $^1/_4$ albino

18. In this chapter we considered Joan Barry's paternity suit against Charlie Chaplin and how, on the basis of blood types, Chaplin could not have been the father of her child.

(a) What blood types are possible for the father of Barry's child?

(b) If Chaplin had possessed one of these blood types, would that prove that he fathered Barry's child?

* 19. A woman has blood type A MM. She has a child with blood type AB MN. Which of the following blood types could *not* be that of the child's father? Explain your reasoning.

George	O	NN
Tom	AB	MN
Bill	B	MN
Claude	A	NN
Henry	AB	MM

20. Allele *A* is epistatic to allele *B*. Indicate whether each of the following statements is true or false. Explain why.

(a) Alleles *A* and *B* are at the same locus.

(b) Alleles *A* and *B* are at different loci.

(c) Alleles *A* and *B* are always located on the same chromosome.

(d) Alleles *A* and *B* may be located on different, homologous chromosomes.

(e) Alleles *A* and *B* may be located on different, nonhomologous chromosomes.

* 21. In chickens, comb shape is determined by alleles at two loci (*R, r* and *P, p*). A walnut comb is produced when at least one dominant allele *R* is present at one locus and at least one dominant allele *P* is present at a second locus (genotype *R_P_*). A rose comb is produced when at least one dominant allele is present at the first locus and two recessive alleles are present at the second locus (genotype *R_pp*). A pea comb is produced when two recessive alleles are present at the first locus and at least one dominant allele is present at the second (genotype *rrP_*). If two recessive alleles are present at the first and at the second locus (*rrpp*), a single comb is produced. Progeny with what types of combs and in what proportions will result from the following crosses?

(a) *RRPP* × *rrpp*

(b) *RrPp* × *rrpp*

(c) *RrPp* × *RrPp*

(d) *Rrpp* × *Rrpp*

(e) *Rrpp* × *rrPp*

(f) *Rrpp* × *rrpp*

* 22. Eye color of the Oriental fruit fly (*Bactrocera dorsalis*) is determined by a number of genes. A fly having wild-type eyes is crossed with a fly having yellow eyes. All the F_1 flies from this cross have wild-type eyes. When the F_1 are interbred, $9/16$ of the F_2 progeny have wild-type eyes, $3/16$ have amethyst eyes (a bright, sparkling blue color), and $4/16$ have yellow eyes.

(a) Give genotypes for all the flies in the P, F_1, and F_2 generations.

(b) Does epistasis account for eye color in Oriental fruit flies? If so, which gene is epistatic and which gene is hypostatic?

23. A variety of opium poppy (*Papaver somniferum* L.) having lacerate leaves was crossed with a variety that has normal leaves. All the F_1 had lacerate leaves. Two F_1 plants were interbred to produce the F_2. Of the F_2, 249 had lacerate leaves and 16 had normal leaves. Give genotypes for all the plants in the P, F_1, and F_2 generations. Explain how lacerate leaves are determined in the opium poppy.

* 24. A dog breeder liked yellow and brown Labrador retrievers. In an attempt to produce yellow and brown puppies, he bought a yellow Labrador male and a brown Labrador female and mated them. Unfortunately, all the puppies produced in this cross were black. (See pp. 108–109 for a discussion of the genetic basis of coat color in Labrador retrievers.)

(a) Explain this result.

(b) How might the breeder go about producing yellow and brown Labradors?

25. When a yellow female Labrador retriever was mated with a brown male, half of the puppies were brown and half were yellow. The same female, when mated with a different brown male, produced all brown males. Explain these results.

* 26. In summer squash, a plant that produces disc-shaped fruit is crossed with a plant that produces long fruit. All the F_1 have disc-shaped fruit. When the F_1 are intercrossed, F_2 progeny are produced in the following ratio: $9/16$ disc-shaped fruit: $6/16$ spherical fruit: $1/16$ long fruit. Give the genotypes of the F_2 progeny.

27. In sweet peas, some plants have purple flowers and other plants have white flowers. A homozygous variety of pea that has purple flowers is crossed with a homozygous variety that has white flowers. All the F_1 have purple flowers. When these F_1 are self-fertilized, the F_2 appear in a ratio of $9/16$ purple to $7/16$ white.

(a) Give genotypes for the purple and white flowers in these crosses.

(b) Draw a hypothetical biochemical pathway to explain the production of purple and white flowers in sweet peas.

28. For the following questions, refer to pp. 112–114 for a discussion of how coat color and pattern are determined in dogs.

(a) Explain why Irish setters are reddish in color.

(b) Will a cross between a beagle and a Dalmatian produce puppies with ticking? Why or why not?

(c) Can a poodle crossed with any other breed produce spotted puppies? Why or why not?

(d) If a St. Bernard is crossed with a Doberman, will the offspring have solid, yellow, saddle, or bicolor coats?

(e) If a Rottweiler is crossed with a Labrador retriever, will the offspring have solid, yellow, saddle, or bicolor coats?

*29. When a Chinese hamster with white spots is crossed with another hamster that has no spots, approximately $\frac{1}{2}$ of the offspring have white spots and $\frac{1}{2}$ have no spots. When two hamsters with white spots are crossed, $\frac{2}{3}$ of the offspring possess white spots and $\frac{1}{3}$ have no spots.

(a) What is the genetic basis of white spotting in Chinese hamsters?

(b) How might you go about producing Chinese hamsters that breed true for white spotting?

30. Male-limited precocious puberty results from a rare, sex-limited autosomal allele (P) that is dominant over the allele for normal puberty (p) and is expressed only in males. Bill undergoes precocious puberty, but his brother Jack and his sister Beth underwent puberty at the usual time, between the ages of 10 and 14. Although Bill's mother and father underwent normal puberty, two of his maternal uncles (his mother's brothers) underwent precocious puberty. All of Bill's grandparents underwent normal puberty. Give the most likely genotypes for all the relatives mentioned in this family.

*31. Pattern baldness in humans is a sex-influenced trait that is autosomal dominant in males and recessive in females. Jack has a full head of hair. JoAnn also has a full head of hair, but her mother is bald. (In women, pattern baldness is usually expressed as a thinning of the hair.) If Jack and JoAnn marry, what proportion of their children are expected to be bald?

32. In goats, a beard is produced by an autosomal allele that is dominant in males and recessive in females. We'll use the symbol B^b for the beard allele and B^+ for the beardless allele. Another independently assorting autosomal allele that produces a black coat (W) is dominant over the allele for white coat (w). Give the phenotypes and their expected proportions for the following crosses.

(a) B^+B^b Ww male $\times$ B^+B^b Ww female

(b) B^+B^b Ww male $\times$ B^+B^b ww female

(c) B^+B^+ Ww male $\times$ B^bB^b Ww female

(d) B^+B^b Ww male $\times$ B^bB^b ww female

33. In the snail *Limnaea peregra*, shell coiling results from a genetic maternal effect. An autosomal allele for a right-handed shell (s^+), called dextral, is dominant over the allele for a left-handed shell (s), called sinistral. A pet snail called Martha is sinistral and reproduces only as a female (the snails are hermaphroditic). Indicate which of the following statements are true and which are false. Explain your reasoning in each case.

(a) Martha's genotype *must* be *ss*.

(b) Martha's genotype *cannot* be s^+s^+.

(c) All the offspring produced by Martha *must* be sinistral.

(d) At least some of the offspring produced by Martha *must* be sinistral.

(e) Martha's mother *must* have been sinistral.

(f) All Martha's brothers *must* be sinistral.

34. In unicorns, two autosomal loci interact to determine the type of tail. One locus controls whether a tail is present at all; the allele for a tail (T) is dominant over the allele for tailless (t). If a unicorn has a tail, then alleles at a second locus determine whether the tail is curly or straight. Farmer Baldridge has two unicorns with curly tails. When he crosses these two unicorns, $\frac{1}{2}$ of the progeny have curly tails, $\frac{1}{4}$ have straight tails, and $\frac{1}{4}$ do not have a tail. Give the genotypes of the parents and progeny in Farmer Baldridge's cross. Explain how he obtained the 2:1:1 phenotypic ratio in his cross.

*35. Phenylketonuria (PKU) is an autosomal recessive disease that results from a defect in an enzyme that normally metabolizes the amino acid phenylalanine. When this enzyme is defective, high levels of phenylalanine cause brain damage. In the past, most children with PKU became mentally retarded. Fortunately, mental retardation can be prevented in these children today by carefully controlling the amount of phenylalanine in the diet. As a result of this treatment, many people with PKU are now reaching reproductive age with no mental retardation. By the end of the teen years, when brain development is complete, many people with PKU go off the restrictive diet. Children born to women with PKU (who are no longer on a phenylalanine-restricted diet) frequently have low birth weight, developmental abnormalities, and mental retardation, even though they are heterozygous for the recessive PKU allele. However, children of men with PKU do not have these problems. Provide an explanation for these observations.

36. In 1983, a sheep farmer in Oklahoma noticed a ram in his flock that possessed increased muscle mass in his hindquarters. Many of the offspring of this ram possessed the same trait, which became known as the callipyge mutant (*callipyge* is Greek for "beautiful buttocks"). The mutation that caused the callipyge phenotype was eventually mapped to a position on the sheep chromosome 18.

When the male callipyge offspring of the original mutant ram were crossed with normal females, they produced the following progeny: $\frac{1}{4}$ male callipyge, $\frac{1}{4}$ female callipyge, $\frac{1}{4}$ male normal, and $\frac{1}{4}$ female normal. When female callipyge offspring of the original mutant ram were crossed with normal males, all of the offspring were normal. Analysis of the chromosomes of these offspring of callipyge females showed that half of them received a chromosome 18 with the *callipyge* allele from their mother. Propose an explanation for the inheritance of the *callipyge* allele. How might you test your explanation?

CHALLENGE QUESTION

37. Suppose that you are tending a mouse colony at a genetics research institute and one day you discover a mouse with twisted ears. You breed this mouse with twisted ears and find that the trait is inherited. Both male and female mice have twisted ears, but when you cross a twisted-eared male with a normal-eared female, you obtain different results from those obtained when you cross a twisted-eared female with normal-eared male—the reciprocal crosses give different results. Describe how you would go about determining whether this trait results from a sex-linked gene, a sex-influenced gene, a genetic maternal effect, a cytoplasmically inherited gene, or genomic imprinting. What crosses would you conduct and what results would be expected with these different types of inheritance?

SUGGESTED READINGS

Barlow, D. P. 1995. Gametic imprinting in mammals. *Science* 270:1610–1613.
Discusses the phenomenon of genomic imprinting.

Harper, P. S., H. G. Harley, W. Reardon, and D. J. Shaw. 1992. Anticipation in myotonic dystrophy: new light on an old problem [Review]. *American Journal of Human Genetics* 51:10–16.
A nice review of the history of anticipation.

Li, E., C. Beard, and R. Jaenisch. 1993. Role for DNA methylation in genomic imprinting. *Nature* 366:362–365.
Reviews some of the evidence that DNA methylation is implicated in genomic imprinting.

Morell, V. 1993. Huntington's gene finally found. *Science* 260:28–30.
Report on the discovery of the gene that causes Huntington disease.

Ostrander, E. A., F. Galibert, and D. F. Patterson. 2000. Canine genetics comes of age. *Trends in Genetics* 16:117–123.
Review of the use of dog genetics for understanding human genetic diseases.

Pagel, M. 1999. Mother and father in surprise agreement. *Nature* 397:19–20.
Discusses some of the possible evolutionary reasons for genomic imprinting.

Sapienza, C. 1990. Parental imprinting of genes. *Scientific American* 263 (October):52–60.
Another review of genomic imprinting.

Shoffner, J. M., and D. C. Wallace. 1992. Mitochondrial genetics: principles and practice [Invited editorial]. *American Journal of Human Genetics* 51:1179–1186.
Discusses the characteristics of cytoplasmically inherited mitochondrial mutations.

Skuse, D. H., R. S. James, D. V. M. Bishop, B. Coppin, P. Dalton, G. Aamodt-Leeper, M. Bacarese-Hamilton, C. Creswell, R. McGurk, and P. A. Jacobs. 1997. Evidence from Turner's syndrome of an imprinted X-linked locus affecting cognitive function. *Nature* 387:705–708.
Report of imprinting in Turner syndrome.

Thomson, G., and M. S. Esposito. 1999. The genetics of complex diseases. *Trends in Genetics* 15:M17–M20.
Discussion of human multifactorial diseases and the effect of the Human Genome Project on the identification of genes influencing these diseases.

Wallace, D. C. 1989. Mitochondrial DNA mutations and neuromuscular disease. *Trends in Genetics* 5:9–13.
More discussion of cytoplasmically inherited mitochondrial mutations.

Willis, M. B. 1989. *Genetics of the Dog.* London: Witherby.
A comprehensive review of canine genetics.

6 Pedigree Analysis and Applications

Lou Gehrig at bat. Gehrig, who played baseball for the New York Yankees from 1923 to 1939, has amyotrophic lateral sclerosis, a disease that is sometimes inherited as an autosomal dominant trait. (AP/Wide World Photos)

Lou Gehrig and Superoxide Free Radicals

Lou Gehrig was the finest first baseman ever to play major league baseball. A left-handed power hitter who grew up in New York City, Gehrig played for the New York Yankees from 1923 to 1939. Throughout his career, he lived in the shadow of his teammates Babe Ruth and Joe Di Maggio, but Gehrig was a great hitter in his own right: he compiled a lifetime batting average of .340 and drove in more than 100 runs every season for 13 years. During his career, he batted in 1991 runs and hit a total of 23 grand slams (home runs with bases loaded). But Gehrig's greatest baseball record, which stood for more than 50 years and has been broken only once—by Cal Ripkin, Jr., in 1995—is his record of playing 2130 consecutive games.

In the 1938 baseball season, Gehrig fell into a strange slump. For the first time since his rookie year, his batting average dropped below .300 and, in the World Series that year, he managed only four hits—all singles. Nevertheless, he finished the season convinced that he was undergoing a temporary slump that he would overcome in the next season. He returned to training camp in 1939 with high spirits. When the season began, however, it was clear to everyone that something was terribly wrong. Gehrig had no power in his swing; he was awkward and clumsy at first base. His condition worsened and, on May 2, he voluntarily removed himself from the lineup. The Yankees sent Gehrig to the Mayo Clinic for diagnosis and, on June 20, his medical report was made public: Lou Gehrig was suffering from a rare, progressive disease known as amyotrophic lateral sclerosis (ALS). Within two years, he was dead. Since then, ALS has commonly been known as Lou Gehrig disease.

Gehrig experienced symptoms typical of ALS: progressive weakness and wasting of skeletal muscles due to

◀6.1 Some cases of amyotrophic lateral sclerosis are inherited and result from mutations in the gene that encodes the enzyme superoxide dismutase 1. A molecular model of the enzyme.

degeneration of the motor neurons. Most cases of ALS are sporadic, appearing in people with no family history of the disease. However, about 10% of cases run in families, and in these cases the disease is inherited as an autosomal dominant trait. In 1993, geneticists discovered that some familial cases of ALS are caused by a defect in a gene that encodes an enzyme called superoxide dismutase 1 (SOD1). This enzyme helps the cell to break down superoxide free radicals, which are highly reactive and extremely toxic. In families studied by the researchers, people with ALS had a defective allele for SOD1 (◀FIGURE 6.1) that produced an altered form of the enzyme. How the altered enzyme causes the symptoms of the disease has not been firmly established.

Amyotrophic lateral sclerosis is just one of a large number of human diseases that are currently the focus of intensive genetic research. This chapter will discuss human genetic characteristics and some of the techniques used to study human inheritance. A number of human characteristics have already been mentioned in discussions of general hereditary principles (Chapters 3 through 5), so by now you know that they follow the same rules of inheritance as those of characteristics in other organisms. So why do we have a separate chapter on human heredity? The answer is that the study of human inheritance requires special techniques—primarily because human biology and culture

impose certain constraints on the geneticist. In this chapter, we'll consider these constraints and examine three important techniques that human geneticists use to overcome them: pedigrees, twin studies, and adoption studies. At the end of the chapter, we will see how the information garnered with these techniques can be used in genetic counseling and prenatal diagnosis.

Keep in mind as you go through this chapter that many important characteristics are influenced by both genes and environment, and separating these factors is always difficult in humans. Studies of twins and adopted persons are designed to distinguish the effects of genes and environment, but such studies are based on assumptions that may be difficult to meet for some human characteristics, particularly behavioral ones. Therefore, it's always prudent to interpret the results of such studies with caution.

www.whfreeman.com/pierce   Information on amyotrophic lateral sclerosis, and more about Lou Gehrig, his outstanding career in baseball, and his fight with amyotrophic lateral sclerosis

The Study of Human Genetic Characteristics

Humans are the best and the worst of all organisms for genetic study. On the one hand, we know more about human anatomy, physiology, and biochemistry than we know about most other organisms; for many families, we have detailed records extending back many generations; and the medical implications of genetic knowledge of humans provide tremendous incentive for genetic studies. On the other hand, the study of human genetic characteristics presents some major obstacles.

First, controlled matings are not possible. With other organisms, geneticists carry out specific crosses to test their hypotheses about inheritance. We have seen, for example, how the testcross provides a convenient way to determine if an individual with a dominant trait is homozygous or heterozygous. Unfortunately (for the geneticist at least), matings between humans are more frequently determined by romance, family expectations, and—occasionally—accident than they are by the requirements of the geneticist.

Another obstacle is that humans have a long generation time. Human reproductive age is not normally reached until 10 to 14 years after birth, and most humans do not reproduce until they are 18 years of age or older; thus, generation time in humans is usually about 20 years. This long generation time means that, even if geneticists could control human crosses, they would have to wait on average 40 years just to observe the F_2 progeny. In contrast, generation time in *Drosophila* is 2 weeks; in bacteria, it's a mere 20 minutes.

Finally, human family size is generally small. Observation of even the simple genetic ratios that we learned in

Chapter 3 would require a substantial number of progeny in each family. When parents produce only 2 children, it's impossible to detect a 3:1 ratio. Even an extremely large family with 10 to 15 children would not permit the recognition of a dihybrid 9:3:3:1 ratio.

Although these special constraints make genetic studies of humans more complex, understanding human heredity is tremendously important. So geneticists have been forced to develop techniques that are uniquely suited to human biology and culture.

> **Concepts**
>
> Although the principles of heredity are the same in humans and other organisms, the study of human inheritance is constrained by the inability to control genetic crosses, the long generation time, and the small number of offspring.

Analyzing Pedigrees

An important technique used by geneticists to study human inheritance is the pedigree. A **pedigree** is a pictorial representation of a family history, essentially a family tree that outlines the inheritance of one or more characteristics. The symbols commonly used in pedigrees are summarized in ◀FIGURE 6.2. The pedigree shown in ◀FIGURE 6.3a illustrates a family with Waardenburg syndrome, an autosomal dominant type of deafness that may be accompanied by fair skin, a white forelock, and visual problems (◀FIGURE 6.3b). Males in a pedigree are represented by squares, females by circles. A horizontal line drawn between two symbols representing a man and a woman indicates a mating; children are connected to their parents by vertical lines extending below the parents. Persons who exhibit the trait of interest are represented by filled circles and squares; in the pedigree of Figure 6.3a, the filled symbols represent members of the family who have Waardenburg syndrome. Unaffected persons are represented by open circles and squares.

Let's look closely at Figure 6.3 and consider some additional features of a pedigree. Each generation in a pedigree is identified by a Roman numeral; within each generation, family members are assigned Arabic numerals, and children in each family are listed in birth order from left to right. Person II-4, a man with Waardenburg syndrome, mated with II-5, an unaffected woman, and they produced five children. The oldest of their children is III-8, a male with Waardenburg syndrome, and the youngest is III-14, an unaffected female. Deceased family members are indicated by a slash through the circle or square, as shown for I-1 and II-1 in Figure 6.3a. Twins are represented by diagonal lines

◀6.2 **Standard symbols are used in pedigrees.**

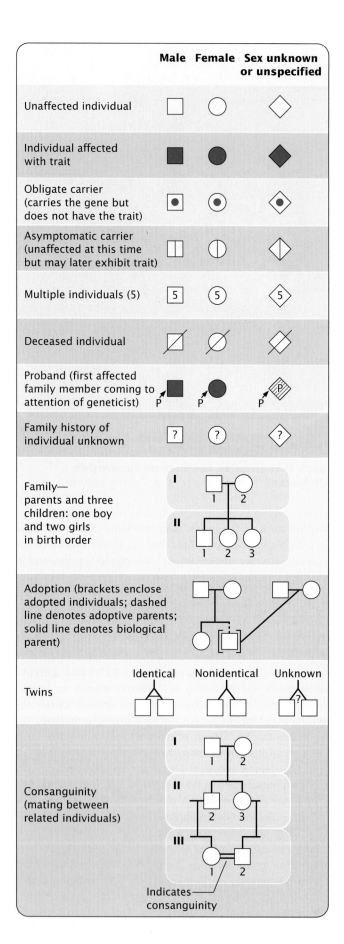

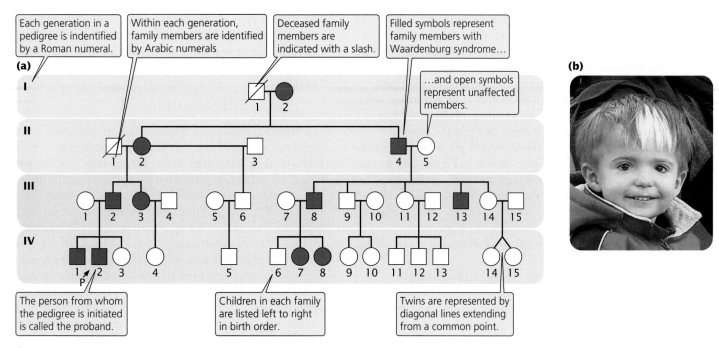

Each generation in a pedigree is indentified by a Roman numeral.

Within each generation, family members are identified by Arabic numerals

Deceased family members are indicated with a slash.

Filled symbols represent family members with Waardenburg syndrome…

…and open symbols represent unaffected members.

The person from whom the pedigree is initiated is called the proband.

Children in each family are listed left to right in birth order.

Twins are represented by diagonal lines extending from a common point.

◄6.3 Waardenburg syndrome is (a) inherited as an autosomal dominant trait and (b) is characterized by deafness, fair skin, visual problems, and a white forelock. (Photograph courtesy of Guy Rowland.)

extending from a common point (IV-14 and IV-15; nonidentical twins).

When a particular characteristic or disease is observed in a person, a geneticist studies the family of this affected person and draws a pedigree. The person from whom the pedigree is initiated is called the **proband** and is usually designated by an arrow (IV-2 in Figure 6.3a).

The limited number of offspring in most human families means that it is usually impossible to discern clear Mendelian ratios in a single pedigree. Pedigree analysis requires a certain amount of genetic sleuthing, based on recognizing patterns associated with different modes of inheritance. For example, autosomal dominant traits should appear with equal frequency in both sexes and should not skip generations, provided that the trait is fully penetrant (see pp. 67–68 in Chapter 3) and not sex influenced (see p. 115 in Chapter 5).

Certain patterns may exclude the possibility of a particular mode of inheritance. For instance, a son inherits his X chromosome from his mother. If we observe that a trait is passed from father to son, we can exclude the possibility of X-linked inheritance. In the following sections, the traits discussed are assumed to be fully penetrant and rare.

Autosomal Recessive Traits

Autosomal recessive traits normally appear with equal frequency in both sexes (unless penetrance differs in males and females), and appear only when a person inherits two alleles for the trait, one from each parent. If the trait is uncommon, most parents carrying the allele are heterozygous and

unaffected; consequently, the trait appears to skip generations (◄FIGURE 6.4). Frequently, a recessive allele may be passed for a number of generations without the trait appearing in a pedigree. Whenever both parents are heterozygous, approximately $\frac{1}{4}$ of the offspring are expected to express the trait, but this ratio will not be obvious unless the family is large. In the rare event that both parents are affected by an autosomal recessive trait, all the offspring will be affected.

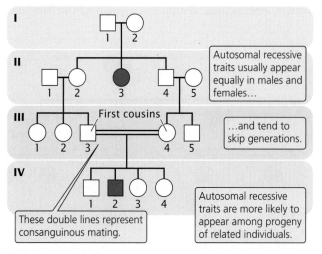

Autosomal recessive traits usually appear equally in males and females…

First cousins

…and tend to skip generations.

These double lines represent consanguinous mating.

Autosomal recessive traits are more likely to appear among progeny of related individuals.

◄6.4 Autosomal recessive traits normally appear with equal frequency in both sexes and seem to skip generations.

When a recessive trait is rare, persons from outside the family are usually homozygous for the normal allele. Thus, when an affected person mates with someone outside the family (*aa* × *AA*), usually none of the children will display the trait, although all will be carriers (i.e., heterozygous). A recessive trait is more likely to appear in a pedigree when two people within the same family mate, because there is a greater chance of both parents carrying the same recessive allele. Mating between closely related people is called **consanguinity.** In the pedigree shown in Figure 6.4, persons III-3 and III-4 are first cousins, and both are heterozygous for the recessive allele; when they mate, $1/4$ of their children are expected to have the recessive trait.

> **Concepts**
>
> Autosomal recessive traits appear with equal frequency in males and females. Affected children are commonly born to unaffected parents, and the trait tends to skip generations. Recessive traits appear more frequently among the offspring of consanguine matings.

A number of human metabolic diseases are inherited as autosomal recessive traits. One of them is Tay-Sachs disease. Children with Tay-Sachs disease appear healthy at birth but become listless and weak at about 6 months of age. Gradually, their physical and neurological conditions worsen, leading to blindness, deafness, and eventually death at 2 to 3 years of age. The disease results from the accumulation of a lipid called G_{M2} ganglioside in the brain. A normal component of brain cells, G_{M2} ganglioside is usually broken down by an enzyme called hexosaminidase A, but children with Tay-Sachs disease lack this enzyme. Excessive G_{M2} ganglioside accumulates in the brain, causing swelling and, ultimately, neurological symptoms. Heterozygotes have only one normal copy of the hexosaminidase A allele and produce only about half the normal amount of the enzyme, but this amount is enough to ensure that G_{M2} ganglioside is broken down normally, and heterozygotes are usually healthy.

Autosomal Dominant Traits

Autosomal dominant traits appear in both sexes with equal frequency, and both sexes are capable of transmitting these traits to their offspring. Every person with a dominant trait must inherit the allele from at least one parent; autosomal dominant traits therefore do not skip generations (◀FIGURE 6.5). Exceptions to this rule arise when people acquire the trait as a result of a new mutation or when the trait has reduced penetrance.

If an autosomal dominant allele is rare, most people displaying the trait are heterozygous. When one parent is affected and heterozygous and the other parent is unaffected, approximately $1/2$ of the offspring will be affected. If both parents have the trait and are heterozygous, approximately $3/4$ of the children will be affected. Provided the trait is fully penetrant, unaffected people do not transmit the trait to their descendants. In Figure 6.5, we see that none of the descendants of II-4 (who is unaffected) have the trait.

> **Concepts**
>
> Autosomal dominant traits appear in both sexes with equal frequency. Affected persons have an affected parent (unless they carry new mutations), and the trait does not skip generations. Unaffected persons do not transmit the trait.

One trait usually considered to be autosomal dominant is familial hypercholesterolemia, an inherited disease in which blood cholesterol is greatly elevated owing to a defect

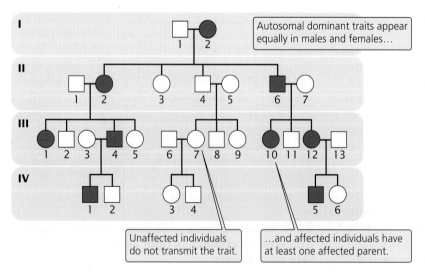

Autosomal dominant traits appear equally in males and females…

Unaffected individuals do not transmit the trait.

…and affected individuals have at least one affected parent.

◀ **6.5 Autosomal dominant traits normally appear with equal frequency in both sexes and do not skip generations.**

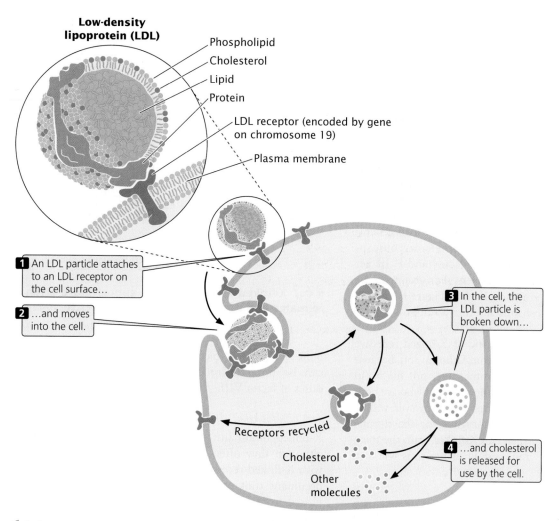

Low-density
lipoprotein (LDL)

Phospholipid

Cholesterol

Lipid

Protein

LDL receptor (encoded by gene
on chromosome 19)

Plasma membrane

1 An LDL particle attaches
to an LDL receptor on
the cell surface...

2 ...and moves
into the cell.

3 In the cell, the
LDL particle is
broken down...

Receptors recycled

Cholesterol

Other
molecules

4 ...and cholesterol
is released for
use by the cell.

◀ **6.6 Low-density lipoprotein (LDL) particles transport cholesterol.**
The LDL receptor moves LDL through the cell membrane into the cytoplasm.

in cholesterol transport. Cholesterol is an essential component of cell membranes and is used in the synthesis of bile salts and several hormones. Most of our cholesterol is obtained through foods, primarily those high in saturated fats. Because cholesterol is a lipid (a nonpolar, or uncharged, compound), it is not readily soluble in the blood (a polar, or charged, solution). Cholesterol must therefore be transported throughout the body in small soluble particles called lipoproteins (◀ FIGURE 6.6); a lipoprotein consists of a core of lipid surrounded by a shell of charged phospholipids and proteins that dissolve easily in blood. One of the principle lipoproteins in the transport of cholesterol is low-density lipoprotein (LDL). When an LDL molecule reaches a cell, it attaches to an LDL receptor, which then moves the LDL through the cell membrane into the cytoplasm, where it is broken down and its cholesterol is released for use by the cell.

Familial hypercholesterolemia is due to a defect in the gene (located on human chromosome 19) that normally codes for the LDL receptor. The disease is usually considered an autosomal dominant disorder because heterozygotes are deficient in LDL receptors. In these people, too little cholesterol is removed from the blood, leading to elevated blood levels of cholesterol and increased risk of coronary artery disease. Persons heterozygous for familial hypercholesterolemia have blood LDL levels that are twice normal and usually have heart attacks by the age of 35. About 1 in 500 people is heterozygous for familial hypercholesterolemia and is predisposed to early coronary artery disease.

Very rarely, a person inherits two defective LDL receptor alleles. Such persons don't make *any* functional LDL receptors; their blood cholesterol levels are more than six times normal and they may suffer a heart attack as early as age 2 and almost inevitably by age 20. Because homozygotes are more severely affected than heterozygotes, familial hypercholesterolemia is said to be incompletely dominant. However, homozygotes are rarely seen (occurring with a frequency of only about 1 in 1 million people), and the common heterozygous form of the disease appears as a simple dominant trait in most pedigrees.

X-Linked Recessive Traits

X-linked recessive traits have a distinctive pattern of inheritance (◀ FIGURE 6.7). First, these traits appear more frequently in males, because males need inherit only a single copy of the allele to display the trait, whereas females must inherit two copies of the allele, one from each parent, to be affected. Second, because a male inherits his X chromosome from his mother, affected males are usually born to unaffected mothers who carry an allele for the trait. Because the trait is passed from unaffected female to affected male to unaffected female, it tends to skip generations (see Figure 6.7). When a woman is heterozygous, approximately $\frac{1}{2}$ of her sons will be affected and $\frac{1}{2}$ of her daughters will be unaffected carriers. For example, we know that females I-2, II-2, and III-7 in Figure 6.7 are all carriers because they transmit the trait to approximately half of their sons.

A third important characteristic of X-linked recessive traits is that they are not passed from father to son, because a son inherits his father's Y chromosome, not his X. In Figure 6.7, there is no case of a father and son who are both affected. All daughters of an affected man, however, will be carriers (if their mother is homozygous for the normal allele). When a woman displays an X-linked trait, she must be homozygous for the trait, and all of her sons will also display the trait.

> **Concepts**
>
> Rare X-linked recessive traits appear more often in males than in females and are not passed from father to son. Affected sons are usually born to unaffected mothers; thus X-linked recessive traits tend to skip generations.

An example of an X-linked recessive trait in humans is hemophilia A, also called classical hemophilia (◀ FIGURE 6.8). This disease results from the absence of a protein necessary for blood to clot. The complex process of blood clotting consists of a cascade of reactions that includes more than 13 different factors. For this reason, there are several types of clotting disorders, each due to a glitch in a different step of the clotting pathway.

Hemophilia A results from abnormal or missing factor VIII, one of the proteins in the clotting cascade. The gene for factor VIII is located on the tip of the long arm of the X chromosome; so hemophilia A is an X-linked recessive disorder. People with hemophilia A bleed excessively; even small cuts and bruises can be life threatening. Spontaneous bleeding occurs in joints such as elbows, knees, and ankles, producing pain, swelling, and erosion of the

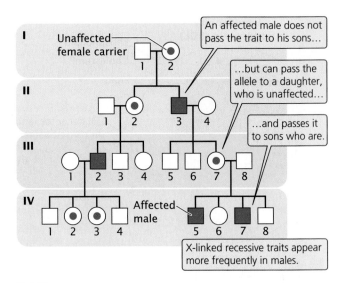

◀ 6.7 **X-linked recessive traits appear more often in males and are not passed from father to son.**

bone. Fortunately, bleeding in people with hemophilia A can be now controlled by administering concentrated doses of factor VIII.

X-Linked Dominant Traits

X-linked dominant traits appear in males and females, although they often affect more females than males. As with X-linked recessive traits, a male inherits an X-linked dominant trait only from his mother—the trait is not passed from father to son. A female, on the other hand, inherits an X chromosome from both her mother and father; so females can receive an X-linked trait from either parent. Each child with an X-linked dominant trait must have an affected parent (unless the child possesses a new mutation or the trait has reduced penetrance). X-linked dominant traits do not skip generations (◀ FIGURE 6.9); affected men pass the trait on to all their daughters and none of their sons, as is seen in the children of I-1 in Figure 6.9. In contrast, affected women (if heterozygous) pass the trait on to $\frac{1}{2}$ of their sons and $\frac{1}{2}$ of their daughters, as seen in the children of II-5 in the pedigree.

> **Concepts**
>
> X-linked dominant traits affect both males and females. Affected males must have affected mothers (unless they possess a new mutation), and they pass the trait on to all their daughters.

An example of an X-linked dominant trait in humans is hypophosphatemia, or familial vitamin D-resistant rickets. People with this trait have features that superficially resemble those produced by rickets: bone deformities, stiff

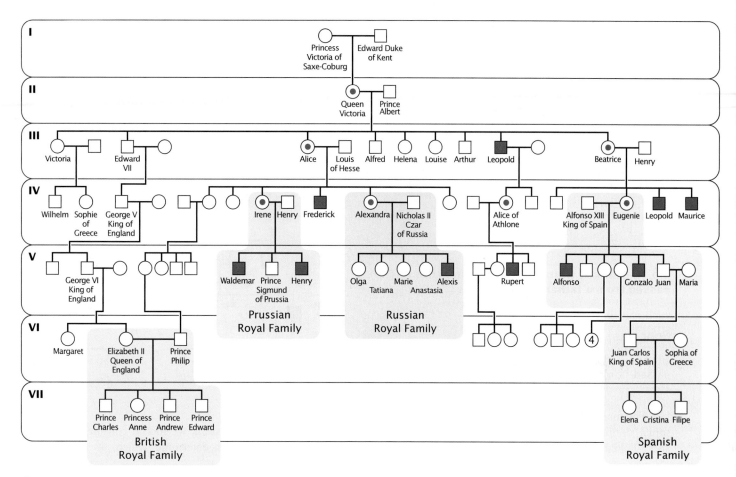

6.8 Classic hemophilia is inherited as an X-linked recessive trait. This pedigree is of hemophilia in the royal families of Europe.

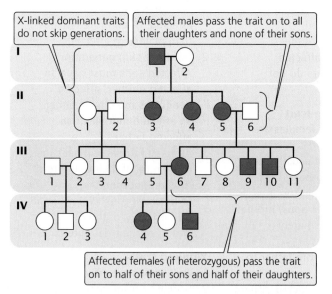

6.9 X-linked dominant traits affect both males and females. An affected male must have an affected mother.

spines and joints, bowed legs, and mild growth deficiencies. This disorder, however, is resistant to treatment with vitamin D, which normally cures rickets. X-linked hypophosphatemia results from the defective transport of phosphate, especially in cells of the kidneys. People with this disorder excrete large amounts of phosphate in their urine, resulting in low levels of phosphate in the blood and reduced deposition of minerals in the bone. As is common with X-linked dominant traits, males with hypophosphatemia are often more severely affected than females.

Y-Linked Traits

Y-linked traits exhibit a specific, easily recognized pattern of inheritance. Only males are affected, and the trait is passed from father to son. If a man is affected, all his male offspring should also be affected, as is the case for I-1, II-4, II-6, III-6, and III-10 of the pedigree in ◀ FIGURE 6.10. Y-linked traits do not skip generations. As discussed in Chapter 4, comparatively few genes reside on the human Y chromosome, and so few human traits are Y linked.

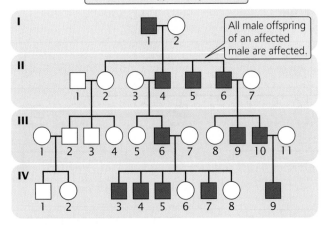

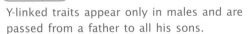

◀6.10 Y-linked traits appear only in males and are passed from a father to all his sons.

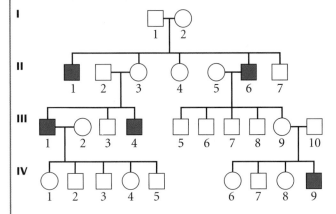

Concepts

Y-linked traits appear only in males and are passed from a father to all his sons.

The major characteristics of autosomal recessive, autosomal dominant, X-linked recessive, X-linked dominant, and Y-linked traits are summarized in Table 6.1.

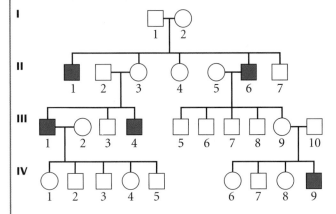

www.whfreeman.com/pierce   The *Online Mendelian Inheritance in Man,* a comprehensive database of human genes and genetic disorders

Worked Problem

The following pedigree represents the inheritance of a rare disorder in an extended family. What is the most likely mode of inheritance for this disease? (Assume that the trait is fully penetrant.)

• Solution

To answer this question, we should consider each mode of inheritance and determine which, if any, we can

Table 6.1	Pedigree characteristics of autosomal recessive, autosomal dominant, X-linked recessive, X-linked dominant, and Y-linked traits

Autosomal recessive trait
1. Appears in both sexes with equal frequency.
2. Trait tends to skip generations.
3. Affected offspring are usually born to unaffected parents.
4. When both parents are heterozygous, approximately $1/4$ of the offspring will be affected.
5. Appears more frequently among the children of consanguine marriages.

Autosomal dominant trait
1. Appears in both sexes with equal frequency.
2. Both sexes transmit the trait to their offspring.
3. Does not skip generations.
4. Affected offspring must have an affected parent, unless they possess a new mutation.

5. When one parent is affected (heterozygous) and the other parent is unaffected, approximately $1/2$ of the offspring will be affected.
6. Unaffected parents do not transmit the trait.

X-linked recessive trait
1. More males than females are affected.
2. Affected sons are usually born to unaffected mothers; thus, the trait skips generations.
3. A carrier (heterozygous) mother produces approximately $1/2$ affected sons.
4. Is never passed from father to son.
5. All daughters of affected fathers are carriers.

X-linked dominant trait
1. Both males and females are affected; often more females than males are affected.
2. Does not skip generations. Affected sons must have an affected mother; affected daughters must have either an affected mother or an affected father.
3. Affected fathers will pass the trait on to all their daughters.
4. Affected mothers (if heterozygous) will pass the trait on to $1/2$ of their sons and $1/2$ of their daughters.

Y-linked trait
1. Only males are affected.
2. Is passed from father to all sons.
3. Does not skip generations.

eliminate. Because the trait appears only in males, autosomal dominant and autosomal recessive modes of inheritance are unlikely, because these occur equally in males and females. Additionally, autosomal dominance can be eliminated because some affected persons do not have an affected parent.

The trait is observed only among males in this pedigree, which might suggest Y-linked inheritance. However, with a Y-linked trait, affected men should pass the trait to all their sons, but here this is not the case; II-6 is an affected man who has four unaffected male offspring. We can eliminate Y-linked inheritance.

X-linked dominance can be eliminated because affected men should pass an X-linked dominant trait to all of their female offspring, and II-6 has an unaffected daughter (III-9).

X-linked recessive traits often appear more commonly in males, and affected males are usually born to unaffected female carriers; the pedigree shows this pattern of inheritance. With an X-linked trait, about half the sons of a heterozygous carrier mother should be affected. II-3 and III-9 are suspected carriers, and about $1/2$ of their male children (three of five) are affected. Another important characteristic of an X-linked recessive trait is that it is not passed from father-to-son. We observe no father-to-son transmission in this pedigree.

X-linked recessive is therefore the most likely mode of inheritance.

Twin Studies

Another method that geneticists use to analyze the genetics of human characteristics is twin studies. Twins are of two types: **dizygotic** (nonidentical) **twins** arise when two separate eggs are fertilized by two different sperm, producing genetically distinct zygotes; **monozygotic** (identical) **twins** result when a single egg, fertilized by a single sperm, splits early in development into two separate embryos.

Because monozygotic twins arise from a single egg and sperm (a single, "mono," zygote), except for rare somatic mutations, they're genetically identical, having 100% of their genes in common (◀FIGURE 6.11a). Dizygotic twins (◀FIGURE 6.11b), on the other hand, have on average only 50% of their genes in common (the same percentage that any pair of siblings has in common). Like other siblings, dizygotic twins may be of the same or different sexes. The only difference between dizygotic twins and other siblings is that dizygotic twins are the same age and shared a common uterine environment.

The frequency with which dizygotic twins are born varies among populations. Among North American Caucasians, about 7 dizygotic twin pairs are born per 1000 births but among Japanese the rate is only about 3 pairs per 1000 births; among Nigerians, about 40 dizygotic

(a)

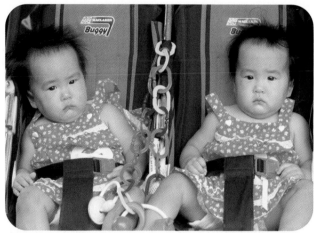

(b)

◀6.11 **Monozygotic twins (a) are identical; dizygotic twins (b) are nonidentical.** (Part a, Joe Carini/Index Stock Imagery/PictureQuest; part b, Bruce Roberts/Photo Researchers.)

twin pairs are born per 1000 births. The rate of dizygotic twinning also varies with maternal age (◀FIGURE 6.12), and dizygotic twinning tends to run in families. In contrast, monozygotic twinning is relatively constant. The frequency of monozygotic twinning in most ethnic groups is about 4 twin pairs per 1000 births, and there is relatively little tendency for monozygotic twins to run in families.

Concepts

Dizygotic twins develop from two eggs fertilized by two separate sperm; they have, on average, 50% of their genes in common. Monozygotic twins develop from a single egg, fertilized by a single sperm, that splits into two embryos; they have 100% percent of their genes in common.

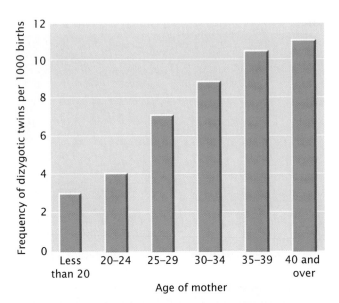

◀ 6.12 Older women tend to have more dizygotic twins than do younger women. Relation between the rate of dizygotic twinning and maternal age. [Data from J. Yerushalmy and S. E. Sheeras, *Human Biology* 12:95–113, 1940.]

Concordance

Comparisons of dizygotic and monozygotic twins can be used to estimate the importance of genetic and environmental factors in producing differences in a characteristic. This is often done by calculating the concordance for a trait. If both members of a twin pair have a trait, the twins are said to be *concordant;* if only one member of the pair has the trait, the twins are said to be *discordant.* **Concordance** is the percentage of twin pairs that are concordant for a trait. Because identical twins have 100% of their genes in com-

mon and dizygotic twins have on average only 50% in common, genetically influenced traits should exhibit higher concordance in monozygotic twins. For instance, when one member of a monozygotic twin pair has asthma, the other twin of the pair has asthma about 48% of the time, so the monozygotic concordance for asthma is 48%. However, when a dizygotic twin has asthma, the other twin has asthma only 19% of the time (19% dizygotic concordance). The higher concordance in the monozygotic twins suggests that genes influence asthma, a finding supported by other family studies of this disease. Concordance values for several human traits and diseases are listed in Table 6.2.

The hallmark of a genetic influence on a particular characteristic is higher concordance in monozygotic twins compared with concordance in dizygotic twins. High concordance in monozygotic twins by itself does *not* signal a genetic influence. Twins normally share the same environment—they are raised in the same home, have the same friends, attend the same school—so high concordance may be due to common genes *or* to common environment. If the high concordance is due to environmental factors, then dizygotic twins, who also share the same environment, should have just as high a concordance as that of monozygotic twins. When genes influence the characteristic, however, monozygotic twin pairs should exhibit higher concordance than dizygotic twin pairs, because monozygotic twins have a greater percentage of genes in common. It is important to note that any discordance among monozygotic twins must be due to environmental factors, because monozygotic twins are genetically identical.

The use of twins in genetic research rests on the important assumption that, when there is greater concordance in monozygotic twins than in dizygotic twins, it is because monozygotic twins are more similar in their genes and not because they have experienced a more similar environment.

Table 6.2	Concordance of monozygotic and dizygotic twins for several traits		
Trait	**Monozygotic Concordance (%)**	**Dizygotic Concordance (%)**	**Reference**
Heart attack (males)	39	26	1
Heart attack (females)	44	14	1
Bronchial asthma	47	24	2
Cancer (all sites)	12	15	2
Epilepsy	59	19	2
Rheumatoid arthritis	32	6	3
Multiple sclerosis	28	5	4

References: (1) B. Havald and M. Hauge, U.S. Public Health Service Publication 1103 (pp. 61–67), 1963. (2) B. Havald and M. Hauge, *Genetics and the Epidemiology of Chronic Diseases,* U.S. Departement of Health, Education, and Welfare, 1965. (3) J. S. Lawrence, *Annals of Rheumatic Diseases* 26(1970):357–379. (4) G. C. Ebers et al., *American Journal of Human Genetics* 36(1984):495.

The degree of environmental similarity between monozygotic twins and dizygotic twins is assumed to be the same. This assumption may not always be correct, particularly for human behaviors. Because they look alike, identical twins may be treated more similarly by parents, teachers, and peers than are nonidentical twins. Evidence of this similar treatment is seen in the past tendency of parents to dress identical twins alike. In spite of this potential complication, twin studies have played a pivotal role in the study of human genetics.

Twin Studies and Obesity

To illustrate the use of twins in genetic research, let's consider a genetic study of obesity. Obesity is a serious public-health problem. About 50% of adults in affluent societies are overweight and from 15% to 25% are obese. Obesity increases the risk of a number of medical conditions, including diabetes, gallbladder disease, high blood pressure, some cancers, and heart disease. Obesity is clearly familial: when both parents are obese, 80% of their children will also become obese; when both parents are not overweight, only 15% of their children will eventually become obese. The familial nature of obesity could result from genes that influence body weight; alternatively, it could be entirely environmental, resulting from the fact that family members usually have similar diets and exercise habits.

A number of genetic studies have examined twins in an effort to untangle the genetic and environmental contributions to obesity. The largest twin study of obesity was conducted on more than 4000 pairs of twins taken from the National Academy of Sciences National Research Council twin registry. This registry is a database of almost 16,000 male twin pairs, born between 1917 and 1927, who served in the U.S. armed forces during World War II or the Korean War. Albert Stunkard and his colleagues obtained weight and height for each of the twins from medical records compiled at the time of their induction into the armed forces. Equivalent data were again collected in 1967, when the men were 40 to 50 years old. The researchers then computed how overweight each man was at induction and at middle age in 1967. Concordance values for monozygotic and dizygotic twins were then computed for several weight categories (Table 6.3).

In each weight category, monozygotic twins had significantly higher concordance than did dizygotic twins at induction and in middle age 25 years later. The researchers concluded that, among the group being studied, body weight appeared to be strongly influenced by genetic factors. Using statistics that are beyond the scope of this discussion, the researchers further concluded that genetics accounted for 77% of variation in body weight at induction and 84% at middle age in 1967. (Because a characteristic such as body weight changes in a lifetime, the effects of genes on the characteristic may vary with age.)

| Table 6.3 | Concordance values for body weight among monozygotic twins (MZ) and dizygotic twins (DZ) at induction in the armed services and at follow-up |

	Concordance (%)			
	At Induction		At Follow-up in 1967	
Percent Overweight*	MZ	DZ	MZ	DZ
15	61	31	68	49
20	57	27	60	40
25	46	24	54	26
30	51	19	47	16
35	44	12	43	9
40	44	0	36	6

*Percent overweight was determined by comparing each man's actual weight with a standard recommended weight for his height.

Source: After A. J. Standard, T. T. Foch, and Z. Hrubec, A twin study of human obesity, *Journal of the American Medical Association* 256(1986):52.

This study shows that genes influence variation in body weight, yet genes *alone* do not cause obesity. In less affluent societies, obesity is rare, and no one can become overweight unless caloric intake exceeds energy expenditure. One does not inherit obesity; rather, one inherits a predisposition toward a particular body weight; geneticists say that some people are genetically more *at risk* for obesity than others.

How genes affect the risk of obesity is not yet completely understood. In 1994, scientists at Rockefeller University isolated a gene that causes an inherited form of obesity in mice (◀ FIGURE 6.13). This gene encodes a protein called leptin, named after the Greek word for "thin." Leptin is produced by fat tissue and decreases appetite by affecting

◀ 6.13 **Obesity in some mice is due to a defect in the gene that encodes the protein leptin.** Obese mouse on the left compared with normal-sized mouse on the right. (Remi Banali/Liaison.)

the hypothalamus, a part of the brain. A decrease in body fat leads to decreased leptin, which stimulates appetite; an increase in body fat leads to increased levels of leptin, which reduces appetite. Obese mice possess two mutated copies of the leptin gene and produce no functional leptin; giving leptin to these mice promotes weight loss.

The discovery of the leptin gene raised hopes that obesity in humans might be influenced by defects in the same gene and that the administration of leptin might be an effective treatment for obesity. Unfortunately, most overweight people are not deficient in leptin. Most, in fact, have elevated levels of leptin and appear to be somewhat resistant to its effects. Only a few rare cases of human obesity have been linked to genetic defects in leptin. The results of further studies have revealed that the genetic and hormonal control of body weight is quite complex; several other genes have been identified that also cause obesity in mice, and the molecular underpinnings of weight control are still being elucidated.

Concepts

Higher concordance in monozygotic twins compared with that in dizygotic twins indicates that genetic factors play a role in determining individual differences of a characteristic. Low concordance in monozygotic twins indicates that environmental factors play a significant role in the characteristic.

www.whfreeman.com/pierce   More advanced information on twin research in genetics

Adoption Studies

A third technique that geneticists use to analyze human inheritance is the study of adopted people. This approach is one of the most powerful for distinguishing the effects of genes and environment on characteristics.

For a variety of reasons, many children each year are separated from their biological parents soon after birth and adopted by adults with whom they have no genetic relationship. These adopted persons have no more genes in common with their adoptive parents than do two randomly chosen persons; however, they do share a common environment with their adoptive parents. In contrast, the adopted persons have 50% of the genes possessed by each of their biological parents but do not share the same environment with them. If adopted persons and their adoptive parents show similarities in a characteristic, these similarities can be attributed to environmental factors. If, on the other hand, adopted persons and their biological parents show similarities, these similarities are likely to be due to genetic factors.

Comparisons of adopted persons with their adoptive parents and with their biological parents can therefore help to define the roles of genetic and environmental factors in the determination of human variation.

Adoption studies assume that the environments of biological and adoptive families are independent (i.e., not more alike than would be expected by chance). This assumption may not always be correct, because adoption agencies carefully choose adoptive parents and may select a family that resembles the biological family. Offspring and their biological mother also share a common environment during prenatal development. Some of the similarity between adopted persons and their biological parents may be due to these similar environments and not due to common genetic factors.

Concepts

Similarities between adopted persons and their genetically unrelated adoptive parents indicate that environmental factors affect the characteristic; similarities between adopted persons and their biological parents indicate that genetic factors influence the characteristic.

Adoption Studies and Obesity

Like twin studies, adoption studies have played an important role in demonstrating that obesity has a genetic influence. In 1986, geneticists published the results of a study of 540 people who had been adopted in Denmark between 1924 and 1947. The geneticists obtained information concerning the adult body weight and height of the adopted persons, along with the adult weight and height of their biological parents and their unrelated adoptive parents.

Geneticists used a measurement called the body-mass index to analyze the relation between the weight of the adopted persons and that of their parents. (The body-mass index, which is a measure of weight divided by height, provides a measure of weight that is independent of height.) On the basis of body-mass index, sex, and age, the adopted persons were divided into four weight classes: thin, median weight, overweight, and obese. A strong relation was found between the weight classification of the adopted persons and the body-mass index of their biological parents: obese adoptees tended to have heavier biological parents, whereas thin adoptees tended to have lighter biological parents (FIGURE 6.14). Because the only connection between the adoptees and their biological parents was the genes that they have in common, the investigators concluded that genetic factors influence adult body weight. There was no clear relation between the weight classification of adoptees and the body-mass index of their adoptive parents (see Figure 6.14), suggesting that the rearing environment has little effect on adult body weight.

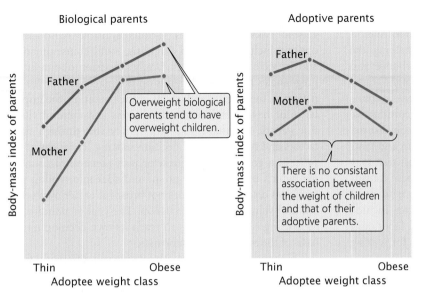

6.14 Adoption studies demonstrate that obesity has a genetic influence. (Redrawn with permission of the *New England Journal of Medicine* 314:195.)

Adoption Studies and Alcoholism

Adoption studies have also been successfully used to assess the importance of genetic factors on alcoholism. Although frequently considered a moral weakness in the past, today alcoholism is more often treated as a disease or as a psychiatric condition. An estimated 10 million people in the United States are problem drinkers, and as many as 6 million are severely addicted to alcohol. Of the U.S. population, 11% are heavy drinkers and consume as much as 50% of all alcohol sold.

A large study of alcoholism was carried out on 1775 Swedish adoptees who had been separated from their mothers at an early age and raised by biologically unrelated adoptive parents. The results of this study, along with those of others, suggest that there are at least two distinct groups of alcoholics. Type I alcoholics include men and women who typically develop problems with alcohol after the age of 25 (usually in middle age). These alcoholics lose control of the ability to drink in moderation—they drink in binges—and tend to be nonaggressive during drinking bouts. Type II alcoholics consist largely of men who begin drinking before the age of 25 (often in adolescence); they actively seek out alcohol, but do not binge, and tend to be impulsive, thrill-seeking, and aggressive while drinking.

The Swedish adoption study also found that alcohol abuse among biological parents was associated with increased alcoholism in adopted persons. Type I alcoholism usually required both a genetic predisposition and exposure to a rearing environment in which alcohol was consumed. Type II alcoholism appeared to be highly hereditary; it developed primarily among males whose biological fathers also were Type II alcoholics, regardless of whether the adoptive parents

drank. A male adoptee whose biological father was a Type II alcoholic was nine times as likely to become an alcoholic as was an adoptee whose biological father was not an alcoholic.

The results of the Swedish adoption study have been corroborated by other investigations, suggesting that some people are genetically predisposed to alcoholism. However, alcoholism is a complex behavioral characteristic that is undoubtedly influenced by many factors. It would be wrong to conclude that alcoholism is strictly a genetic characteristic. Although some people may be genetically predisposed to alcohol abuse, no gene forces a person to drink, and no one becomes alcoholic without the presence of a specific environmental factor—namely, alcohol.

Genetic Counseling and Genetic Testing

Our knowledge of human genetic diseases and disorders has expanded rapidly in the past 20 years. Victor McKusick's *Mendelian Inheritance in Man* now lists more than 13,000 human genetic diseases, disorders, and traits that have a simple genetic basis. Research has provided a great deal of information about the inheritance, chromosomal location, biochemical basis, and symptoms of many of these genetic traits. This information is often useful to people who have a genetic condition.

Genetic Counseling

Genetic counseling is a new field that provides information to patients and others who are concerned about hereditary conditions. It is also an educational process that helps patients and family members deal with many aspects of a genetic condition. Genetic counseling often includes interpreting a diagnosis of the condition; providing information about symptoms, treatment, and prognosis; helping the patient and family understand the mode of inheritance; and calculating probabilities that family members might transmit the condition to future generations. Good genetic counseling also provides information about the reproductive options that are available to those at risk for the disease. Finally, genetic counseling tries to help the patient and family cope with the psychological and physical stress that may be associated with their disorder. Clearly, all of these considerations cannot be handled by a single person; so most genetic counseling is done by a team that can include counselors, physicians, medical geneticists, and laboratory personnel. Table 6.4 lists some common reasons for seeking genetic counseling.

Genetic counseling usually begins with a diagnosis of the condition. On the bases of a physical examination, biochemical tests, chromosome analysis, family history, and other information, a physician determines the cause of the condition. An accurate diagnosis is critical, because treatment and the probability of passing on the condition may vary, depending on the diagnosis. For example, there are a number of different types of dwarfism, which may be caused by chromosome abnormalities, single-gene mutations, hormonal imbalances, or environmental factors. People who have dwarfism resulting from an autosomal dominant gene have a 50% chance of passing the condition to their children, whereas people with dwarfism caused by a rare recessive gene have a low likelihood of passing the trait to their children.

When the nature of the condition is known, a genetic counselor sits down with the patient and members of the patient's family and explains the diagnosis. A family pedigree may be constructed, and the probability of transmitting the condition to future generations can be calculated for different family members. The counselor helps the family interpret the genetic risks and explains various available reproductive options, including prenatal diagnosis, artificial insemination, and in vitro fertilization. A family's decision about future pregnancies frequently depends on the magnitude of the genetic risk, the severity and effects of the condition, the importance of having children, and religious and cultural views. The genetic counselor helps the family sort through these factors and facilitates their decision making. Throughout the process, a good genetic counselor uses *nondirected* counseling, which means that he or she provides information and facilitates discussion but does not bring his or her own opinion and values into the discussion. The goal of nondirected counseling is for the family to reach its own decision on the basis of the best available information.

Genetic conditions are often perceived differently from other diseases and medical problems, because genetic conditions are intrinsic to the individual person and can be passed on to children. Such perceptions may produce feelings of guilt about past reproductive choices and intense personal dilemmas about future choices. Genetic counselors are trained to help patients and their families recognize and cope with these feelings.

Concepts

Genetic counseling is an educational process that provides patients and their families with information about a genetic condition, its medical implications, the probabilities that others in the family may have the disease, and reproductive options. It also helps patients and their families cope with the psychological and physical stress associated with a genetic condition.

Table 6.4	Common reasons for seeking genetic counseling

1. A person knows of a genetic disease in the family.

2. A couple has given birth to a child with a genetic disease, birth defect, or chromosomal abnormality.

3. A couple has a child who is mentally retarded or has a close relative who is mentally retarded.

4. An older woman becomes pregnant or wants to become pregnant. There is disagreement about the age at which a prospective mother who has no other risk factor should seek genetic counseling; many experts suggest that any prospective mother age 35 or older should seek genetic counseling.

5. Husband and wife are closely related (e.g., first cousins).

6. A couple experiences difficulties achieving a successful pregnancy.

7. A pregnant woman is concerned about exposure to an environmental substance (drug, chemical, or virus) that causes birth defects.

8. A couple needs assistance in interpreting the results of a prenatal or other test.

9. Both parents are known carriers for a recessive genetic disease.

www.whfreeman.com/pierce Information on genetic counseling and human genetic diseases, as well as a list of genetic counseling training programs accredited by the American Board of Genetic Counseling

Genetic Testing

Improvements in our understanding of human heredity and the identification of numerous disease-causing genes have led to the development of hundreds of tests for genetic conditions. The ultimate goal of genetic testing is to recognize the potential for a genetic condition at an early stage. In some cases, genetic testing allows early intervention that may lessen or even prevent the development of the condition. In other cases, genetic testing allows people to make informed choices about reproduction. For those who know that they are at risk for a genetic condition, genetic testing may help alleviate anxiety associated with the uncertainty of their situation. Genetic testing includes newborn screening, heterozygote screening, presymptomatic diagnosis, and prenatal testing.

Newborn screening Testing for genetic disorders in newborn infants is called **newborn screening.** Most states in the United States and many other countries require that newborn infants be tested for phenylketonuria and galactosemia. These metabolic diseases are caused by autosomal recessive alleles; if not treated at an early age, they can result in mental retardation, but early intervention—through the administration of a modified diet—prevents retardation (see pp. 122–123 in Chapter 5). Testing is done by analyzing a drop of the infant's blood collected soon after birth. Because of widespread screening, the frequency of mental retardation due to these genetic conditions has dropped tremendously. Screening newborns for additional genetic diseases that benefit from treatment, such as sickle-cell anemia and hypothyroidism, also is common.

Heterozygote screening Testing members of a population to identify heterozygous carriers of recessive disease-causing alleles, who are healthy but have the potential to produce children with the particular disease, is termed **heterozygote screening.**

Testing for Tay-Sachs disease is a successful example of heterozygote screening. In the general population of North America, the frequency of Tay-Sachs disease is only about 1 person in 360,000. Among Ashkenazi Jews (descendants of Jewish people who settled in eastern and central Europe), the frequency is 100 times as great. A simple blood test is used to detect Ashkenazi Jews who carry the Tay-Sachs allele. If a man and woman are both heterozygotes, approximately one in four of their children is expected to have Tay-Sachs disease. A prenatal test for the Tay Sachs allele also is available. Screening programs have led to a significant decline in the number of children of Ashkenazi ancestry born with Tay-Sachs disease (now fewer than 10 children per year in the United States).

Presymptomatic testing Evaluating healthy people to determine whether they have inherited a disease-causing allele is known as **presymptomatic genetic testing.** For example, presymptomatic testing is available for members of families that have an autosomal dominant form of breast cancer. In this case, early identification of the disease-causing allele allows for closer surveillance and the early detection of tumors. Presymptomatic testing is also available for some genetic diseases for which no treatment is available, such as Huntington disease, an autosomal dominant disease that leads to slow physical and mental deterioration in middle age (see introduction to Chapter 5). Presymptomatic testing for untreatable conditions raises a number of social and ethical questions (Chapter 18).

Several hundred genetic diseases and disorders can now be diagnosed prenatally. The major purpose of prenatal tests is to provide families with the information that they need to make choices during pregnancies and, in some cases, to prepare for the birth of a child with a genetic condition. Several approaches to prenatal diagnosis are described in the following sections.

Ultrasonography Some genetic conditions can be detected through direct visualization of the fetus. Such visualization is most commonly done with **ultrasonography**—usually referred to as ultrasound. In this technique, high-frequency sound is beamed into the uterus; when the sound waves encounter dense tissue, they bounce back and are transformed into a picture (◀FIGURE 6.15). The size of the fetus can be determined, as can genetic conditions such as neural tube defects (defects in the development of the spinal column and the skull) and skeletal abnormalities.

Amniocentesis Most prenatal testing requires fetal tissue, which can be obtained in several ways. The most widely used method is **amniocentesis,** a procedure for obtaining a

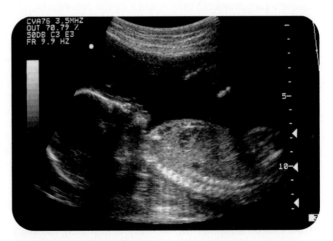

◀ **6.15 Ultrasonography can be used to detect some genetic disorders in a fetus and locate the fetus during amniocentesis and chorionic villus sampling.** (SIU School of Medicine/Photo Researchers.)

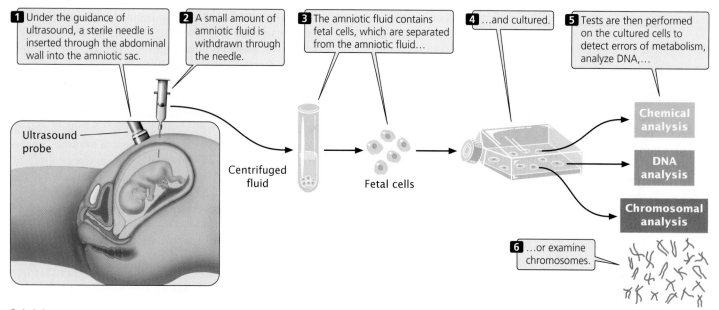

1 Under the guidance of ultrasound, a sterile needle is inserted through the abdominal wall into the amniotic sac.

2 A small amount of amniotic fluid is withdrawn through the needle.

3 The amniotic fluid contains fetal cells, which are separated from the amniotic fluid...

4 ...and cultured.

5 Tests are then performed on the cultured cells to detect errors of metabolism, analyze DNA,...

Ultrasound probe

Centrifuged fluid

Fetal cells

Chemical analysis

DNA analysis

Chromosomal analysis

6 ...or examine chromosomes.

6.16 Amniocentesis is a procedure for obtaining fetal cells for genetic testing.

sample of amniotic fluid from a pregnant woman (◀ FIGURE 6.16). Amniotic fluid—the substance that fills the amniotic sac and surrounds the developing fetus—contains fetal cells that can be used for genetic testing.

Amniocentesis is routinely performed as an outpatient procedure with the use of a local or no anesthetic. First, ultrasonography is used to locate the position of the fetus in the uterus. Next, a long, sterile needle is inserted through the abdominal wall into the amniotic sac (see Figure 6.16), and a small amount of amniotic fluid is withdrawn through the needle. Fetal cells are separated from the amniotic fluid and placed in a culture medium that stimulates them to grow and divide. Genetic tests are then performed on the cultured cells. Complications with amniocentesis (mostly miscarriage) are rare, arising in only about 1 in 400 procedures.

Chorionic villus sampling A major disadvantage with amniocentesis is that it is routinely performed in about the 16th week of a pregnancy, (although many obstetricians now successfully perform amniocentesis several weeks earlier). The cells obtained with amniocentesis must then be cultured before genetic tests can be performed, requiring yet more time. For these reasons, genetic information about the fetus may not be available until the 17th or 18th week of pregnancy. By this stage, abortion carries a risk of complications and may be stressful for the parents. **Chorionic villus sampling** (CVS) can be performed earlier (between the 10th and 11th weeks of pregnancy) and collects more fetal tissue, which eliminates the necessity of culturing the cells.

In CVS, a catheter—a soft plastic tube—is inserted into the vagina (◀ FIGURE 6.17) and, with the use of ultrasound for guidance, is pushed through the cervix into the uterus. The tip of the tube is placed into contact with the chorion, the outer layer of the placenta. Suction is then applied, and a small piece of the chorion is removed. Although the chorion is composed of fetal cells, it is a part of the placenta that is expelled from the uterus after birth; so the removal of a small sample does not endanger the fetus. The tissue that is removed contains millions of actively dividing cells that can be used directly in many genetic tests. Chorionic villus sampling has a somewhat higher risk of complication than that of amniocentesis; the results of several studies suggest that this procedure may increase the incidence of limb defects in the fetus when performed earlier than 10 weeks of gestation.

Fetal cells obtained by amniocentesis or by CVS can used to prepare a **karyotype,** which is a picture of a complete set of metaphase chromosomes. Karyotypes can be studied for chromosome abnormalities (Chapter 9). Biochemical analyses can be conducted on fetal cells to determine the presence of particular metabolic products of genes. For genetic diseases in which the DNA sequence of the causative gene has been determined, the DNA sequence (DNA testing; Chapter 18) can be examined for defective alleles.

Maternal blood tests Some genetic conditions can be detected by performing a blood test on the mother (**maternal blood testing**). For instance, α-fetoprotein is normally produced by the fetus during development and is present in the fetal blood, the amniotic fluid, and the mother's blood during pregnancy. The level of α-fetoprotein is significantly higher than normal when the fetus has a neural-tube defect or one of several other disorders. Some chromosome

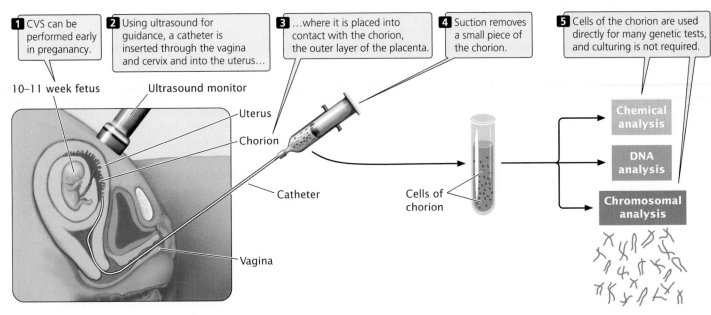

1 CVS can be performed early in preganancy.

2 Using ultrasound for guidance, a catheter is inserted through the vagina and cervix and into the uterus...

3 ...where it is placed into contact with the chorion, the outer layer of the placenta.

4 Suction removes a small piece of the chorion.

5 Cells of the chorion are used directly for many genetic tests, and culturing is not required.

10–11 week fetus   Ultrasound monitor

Uterus

Chorion

Catheter

Vagina

Cells of chorion

Chemical analysis

DNA analysis

Chromosomal analysis

◀ **6.17 Chorionic villus sampling (CVS) is another procedure for obtaining fetal cells for genetic testing.**

abnormalities produce *lower*-than-normal levels of α-feto-protein. Measuring the amount of α-fetoprotein in the mother's blood gives an indication of these conditions. However, because other factors affect the amount of α-fetoprotein in maternal blood, a high or low level by itself does not necessarily indicate a problem. Thus, when a blood test indicates that the amount of α-fetoprotein is abnormal, follow-up tests (additional α-fetoprotein determinations, ultrasound, amniocentesis, or all three) are usually performed.

Fetal cell sorting Prenatal tests that utilize only maternal blood are highly desirable because they are noninvasive and pose no risk to the fetus. During pregnancy, a few fetal cells are released into the mother's circulatory system, where they mix and circulate with her blood. Recent advances have made it possible to separate fetal cells from a maternal blood sample (a procedure called **fetal cell sorting**). With the use of lasers and automated cell-sorting machines, fetal cells can be detected and separated from maternal blood cells. The fetal cells obtained can be cultured for chromosome analysis or used as a source of fetal DNA for molecular testing (see p. 536 in Chapter 18).

A large number of genetic diseases can now be detected prenatally (Table 6.5), and the number is growing rapidly as new disease-causing genes are isolated. The Human Genome Project (Chapter 19) has accelerated the rate at which new genes are being isolated and new genetic tests are being developed. In spite of these advances, prenatal tests are still not available for many common genetic

diseases, and no test can guarantee that a "perfect" child will be born.

Preimplantation genetic diagnosis Prenatal genetic tests provide today's couples with increasing amounts of information about the health of their future children. New reproductive technologies also provide couples with options for using this information. One of these technologies is in vitro fertilization. In this procedure, hormones are used to induce ovulation. The ovulated eggs are surgically removed from the surface of the ovary, placed in a laboratory dish, and fertilized with sperm. The resulting embryo is then implanted into the uterus. Thousands of babies resulting from in vitro fertilization have now been born.

Genetic testing can be combined with in vitro fertilization to allow implantation of embryos that are free of a specific genetic defect. Called **preimplantation genetic diagnosis** (PGD), this technique allows people who carry a genetic defect to avoid producing a child with the disorder. For example, when a woman is a carrier of an X-linked recessive disease, approximately half of her sons are expected to have the disease. Through in vitro fertilization and preimplantation testing, it is possible to select an embryo without the disorder for implantation in her uterus.

The procedure begins with the production of several single-celled embryos through in vitro fertilization. The embryos are allowed to divide several times until they reach the 8- or 16-cell stage. At this point, one cell is removed from each embryo and tested for the genetic abnormality. Removing a single cell at this early stage does not harm the embryo. After determination of which embryos are free of

A couple are seeking help at a clinic that offers preimplantation genetic diagnosis (PGD), which combines in vitro fertilization with molecular analysis of the DNA from a single cell of the developing embryo and permits the selection and transfer to the uterus of embryos free of a genetic disease. Before PGD, the only alternative for those wishing to prevent the birth of a child with a serious genetic disorder was early chorionic villus sampling or amniocentesis, followed by abortion if the fetus had a disorder.

Consider a couple at risk of having a second child with severe combined immune deficiency (SCID). A child born with this condition has a seriously impaired immune system. As recently as 20 years ago, those affected died early in life, but the use of bone-marrow transplantation, which can provide the child with a supply of healthy blood stem cells, has greatly extended survival. In general, the earlier the transplantation and the closer the tissue match of the marrow donor, the better a recipient child's chances.

The couple tell the medical geneticist that they are seeking his help in identifying and transferring only embryos free of the SCID mutation so that they can begin their pregnancy knowing that it will be healthy in this regard. Some weeks later they reveal another reason for their interest in this technology: the health of their six-year-old daughter, who is affected with SCID, is on a downward course despite one partly matched bone-marrow transplant earlier in her life. Their child's best hope of survival is another bone-marrow transplant, using tissue from a compatible donor, preferably a sibling. Is it possible, they ask, to test the healthy embryos for tissue compatibility and transfer only those that match their daughter's type?

The geneticist responds that it is indeed technically possible to do so, but he wonders whether helping the couple in this way is ethically appropriate. Is it right to conceive a child for this purpose? In addition, because tissue compatibility is not a disease, would responding to this request constitute an unwise step into a world of positive, or "enhancement," genetics, where parents' desires, not medical judgment, dictate the use of genetic knowledge?

PGD offers significant new reproductive opportunities for families or persons affected by genetic disease. However, the very power of this technology raises new ethical issues that will grow in importance as PGD and related embryo manipulation procedures become more widely available. PGD offers a technology that is medically, psychologically, and, in the view of many, morally superior to the existing use of abortion for genetic selection.

The case described here is not entirely novel. Even before the advent of PGD, couples who had sought to ensure the birth of a child whose HLA (human leukocyte antigen) status would be compatible with that of an existing sibling would establish a pregnancy, undergo testing, and then abort all fetuses that did not have the appropriate HLA type. Because a woman in the United States has a right to abortion for any reason through the second trimester, this option is legal. However, it is certainly not a desirable one from a medical or psychological point of view.

Because pregnancy is never begun, PGD avoids the emotional trauma of abortion. Although some would object to the discarding of human embryos even at this early stage, the selection of viable embryos and the discarding of others is a routine part of in vitro fertilization procedures today and raises few moral questions in the minds of most people. So, from the narrow perspective of parental decision making, the alternative described in this case is a significant medical and moral step forward. We should not lose sight of this fact as we consider other ethically troubling aspects of the case.

This case of parental selection raises at least two distinct questions. First, even if the means of selection is relatively innocuous from a moral standpoint, is it appropriate for parents to bring a child into being at least partly for the purpose of saving the life of a sibling? A second question is whether genetic professionals should cooperate with a selection process that entails a nondisease trait.

With regard to the first question, some people believe that the parents' wishes in this situation violate the ethical principle not to use a person merely as a means to an end, as well as the modern principle of responsible parenthood, which judges each child to be of inestimable value. They also worry about future psychological harm to a child conceived in this way. Others argue that children are usually born for a specific purpose, whether it's to gratify the parents' need for a family, to cement the relationship of a couple, or whatever else. As a result, they argue that the important question is not whether the purpose for which the child was conceived is eithical but whether the parents will be able to accept the child in its own unique identity once it is born.

In response to the second question, some ethicists see any involvement in nondisease testing as a dangerous diversion of genetic testing down paths long since rejected for good reason. They see that a consensus has emerged in popular opinion among geneticists and ethics advisory boards that nondisease characteristics should not be subject to prenatal testing. HLA testing, however well intentioned, runs counter to this consensus. Departures from these views about genetic tests could raise very difficult questions about eugenics in the future.

Table 6.5	Examples of genetic diseases and disorders that can be detected prenatally and the techniques used in their detection
Disorder	**Method of Detection**
Chromosome abnormalities	Examination of a karyotype from cells obtained by amniocentesis or CVS
Cleft lip and palate	Ultrasound
Cystic fibrosis	DNA analysis of cells obtained by amniocentesis or CVS
Dwarfism	Ultrasound or X-ray; some forms can be detected by DNA analysis of cells obtained by amniocentesis or CVS
Hemophilia	Fetal blood sampling* or DNA analysis of cells obtained by amniocentesis or CVS
Lesch-Nyhan syndrome (deficiency of purine metabolism leading to spasms, seizures, and compulsory self-mutilation)	Biochemical tests on cells obtained by amniocentesis or CVS
Neural-tube defects	Initial screening with maternal blood test, followed by biochemical tests on amniotic fluid obtained by amniocentesis and ultrasound
Osteogenesis imperfecta (brittle bones)	Ultrasound or X-ray
Phenylketonuria	DNA analysis of cells obtained by amniocentesis or CVS
Sickle-cell anemia	Fetal blood sampling or DNA analysis of cells obtained by amniocentesis or CVS
Tay-Sachs disease	Biochemical tests on cells obtained by amniocentesis or CVS

*A sample of fetal blood is otained by inserting needle into the umblical cord.

the disorder, a healthy embryo is selected and implanted in the woman's uterus.

Preimplantation genetic diagnosis requires the ability to conduct a genetic test on a single cell. Such testing is possible with the use of the polymerase chain reaction through which minute quantities of DNA can be amplified (replicated) quickly (Chapter 18). After amplification of the cell's DNA, the DNA sequence is examined. Preimplantation diagnosis is still experimental and is available at only a few research centers. Its use raises a number of ethical concerns, because it provides a means of actively selecting for or against certain genetic traits (see The New Genetics p. 150).

Concepts

Genetic testing is used to screen newborns for genetic diseases, detect persons who are heterozygous for recessive diseases, detect disease-causing alleles in those who have not yet developed symptoms of the disease, and detect defective alleles in unborn babies. Preimplantation genetic diagnosis combined with in vitro fertilization allows for selection of embryos that are free from specific genetic diseases.

www.whfreeman.com/pierce   Additional information about genetic testing

Connecting Concepts Across Chapters

This chapter builds on the basic principles of heredity that were introduced in Chapters 1 through 5, extending them to human genetic characteristics. A dominant theme of the chapter is that human inheritance is not fundamentally different from inheritance in other organisms, but the unique biological and cultural characteristics of humans require special techniques for the study of human characteristics.

Several topics introduced in this chapter are explored further in later chapters. Molecular techniques used in genetic testing and some of the ethical implications of modern genetic testing are presented in Chapter 18. Chromosome mutations and karyotypes are studied in Chapter 9. In Chapter 22, we examine additional techniques for separating genetic and environmental contributions to characteristics in humans and other organisms.

CONCEPTS SUMMARY

- There are several difficuties in applying traditional genetic techniques to the study of human traits, including the inability to conduct controlled crosses, long generation time, small family size, and the difficulty of separating genetic and environmental influences.

- A pedigree is a pictorial representation of a family history that displays the inheritance of one or more traits through several generations.

- Autosomal recessive traits typically appear with equal frequency in both sexes. If a trait is uncommon, the parents of a child with an autosomal recessive trait are usually heterozygous and unaffected; so the trait tends to skip generations. When both parents are heterozygous, approximately $1/4$ of their offspring will have the trait. Recessive traits are more likely to appear in families with consanguinity (mating between closely related persons).

- Autosomal dominant traits usually appear equally in both sexes and do not skip generations. When one parent is affected and heterozygous, approximately $1/2$ of the offspring will have the trait. When both parents are affected and heterozygous, approximately $3/4$ of the offspring will be affected. Unaffected people do not normally transmit an autosomal dominant trait to their offspring.

- X-linked recessive traits appear more frequently in males than in females. Affected males are usually born to females who are unaffected carriers. When a woman is a heterozygous carrier and a man is unaffected, approximately $1/2$ of their sons will have the trait and $1/2$ of their daughters will be unaffected carriers. X-linked traits are not passed from father to son.

- X-linked dominant traits appear in males and females, but more frequently in females. They do not skip generations. Affected men pass an X-linked dominant trait to all of their daughters but none of their sons. Heterozygous women pass the trait to $1/2$ of their sons and $1/2$ of their daughters.

- Y-linked traits appear only in males and are passed from father to all sons.

- Analysis of twins is an important technique for the study of human genetic characteristics. Dizygotic twins arise from two separate eggs fertilized by two separate sperm; monozygotic twins arise from a single egg, fertilized by a single sperm, that splits into two separate embryos early in development.

- Concordance is the percentage of twin pairs in which both members of the pair express a trait. Higher concordance in monozygotic than in dizygotic twins indicates a genetic influence on the trait; less than 100% concordance in monozygotic twins indicates environmental influences on the trait.

- Adoption studies are used to analyze the inheritance of human characteristics. Similarities between adopted children and their biological parents indicate the importance of genetic factors in the expression of a trait; similarities between adopted children and their genetically unrelated adoptive parents indicate the influence of environmental factors.

- Genetic counseling provides information and support to people concerned about hereditary conditions in their families.

- Genetic testing includes screening for disease-causing alleles in newborns, the detection of people heterozygous for recessive alleles, presymptomatic testing for the presence of a disease-causing allele in at-risk people, and prenatal diagnosis.

- Common techniques used for prenatal diagnosis include ultrasound, amniocentesis, chorionic villus sampling, and maternal blood sampling. Preimplantation genetic diagnosis can be used to select for embryos that are free of a genetic disease.

IMPORTANT TERMS

pedigree (p. 134)
proband (p. 135)
consanguinity (p. 136)
dizygotic twins (p. 141)
monozygotic twins (p. 141)
concordance (p. 142)

genetic counseling (p. 145)
newborn screening (p. 147)
heterozygote screening (p. 147)
presymptomatic genetic testing (p. 147)

ultrasonography (p. 147)
amniocentesis (p. 147)
chorionic villus sampling (p. 148)
karyotype (p. 148)

maternal blood testing (p. 148)
fetal cell sorting (p. 149)
preimplantation genetic diagnosis (p. 149)

Worked Problems

1. Joanna has "short fingers" (brachydactyly). She has two older brothers who are identical twins; they both have short fingers. Joanna's two younger sisters have normal fingers. Joanna's mother has normal fingers, and her father has short fingers. Joanna's paternal grandmother (her father's mother) has short fingers; her paternal grandfather (her father's father), who is now deceased, had normal fingers. Both of Joanna's maternal grandparents (her mother's parents) have normal fingers. Joanna marries Tom, who has normal fingers; they adopt a son named Bill who has normal fingers. Bill's biological parents both have normal fingers. After adopting Bill, Joanna and Tom produce two children: an older daughter with short fingers and a younger son with normal fingers.

(a) Using standard symbols and labels, draw a pedigree illustrating the inheritance of short fingers in Joanna's family.

(b) What is the most likely mode of inheritance for short fingers in this family?

(c) If Joanna and Tom have another biological child, what is the probability (based on you answer to part b) that this child will have short fingers?

• Solution

(a) In the pedigree for the family, note that persons with the trait (short fingers) are indicated by filled circles (females) and filled squares (males). Joanna's identical twin brothers are connected to the line above with diagonal lines that have a horizontal line between them. The adopted child of Joanna and Tom is enclosed in brackets and is connected to the biological parents by a dashed diagonal line.

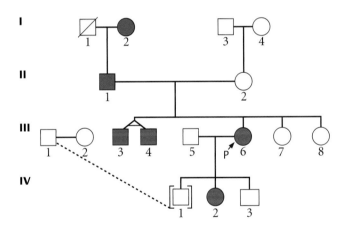

(b) The most likely mode of inheritance for short fingers in this family is autosomal dominant. The trait appears equally in males and females and does not skip generations. When one parent has the trait, it appears in approximately half of that parent's sons and daughters, although the number of children in the families is small. We can eliminate Y-linked inheritance because the trait is found in females. If short fingers were X-linked recessive, females with the trait would be expected to pass the trait to all their sons, but Joanna (III-6), who has short fingers, produced a son with normal fingers. For X-linked dominant traits, affected men should pass the trait to all their daughters; because male II-1 has short fingers and produced two daughters without short fingers (III-7 and III-8), we know that the trait cannot be X-linked dominant. It is unlikely that the trait is autosomal recessive because it does not skip generations and approximately half of the children of affected parents have the trait.

(c) If having short fingers is autosomal dominant, Tom must be homozygous (bb) because he has normal fingers. Joanna must be heterozygous (Bb) because she and Tom have produced both short- and normal-fingered offspring. In a cross between a heterozygote and homozygote, half of the progeny are expected to be heterozygous and half homozygous ($Bb \times bb \rightarrow \frac{1}{2} Bb$, $\frac{1}{2} bb$); so the probability that Joanna's and Tom's next biological child will have short fingers is $\frac{1}{2}$.

2. Concordance values for a series of traits were measured in monozygotic twins and dizygotic twins; the results are shown in the following table. For each trait, indicate whether the rates of concordance suggest genetic influences, environmental influences, or both. Explain your reasoning.

Characteristic	Monozygotic concordance (%)	Dizygotic concordance (%)
(a) ABO blood type	100	65
(b) Diabetes	85	36
(c) Coffee drinking	80	80
(d) Smoking	75	42
(e) Schizophrenia	53	16

• Solution

(a) The concordance of ABO blood type in the monozygotic twins is 100%. This high concordance in monozygotic twins does not, by itself, indicate a genetic basis for the trait. An important indicator of a genetic influence on the trait is lower concordance in dizygotic twins. Because concordance for ABO blood type is substantially lower in the dizygotic twins, we would be safe in concluding that genes play a role in determining differences in ABO blood types.

(b) The concordance for diabetes is substantially higher in monozygotic twins than in dizygotic twins; therefore, we can conclude that genetic factors play some role in susceptibility to diabetes. The fact that monozygotic twins show a concordance less than 100% suggests that environmental factors also play a role.

(c) Both monozygotic and dizygotic twins exhibit the same high concordance for coffee drinking; so we can conclude that there is little genetic influence on coffee drinking. The fact that monozygotic twins show a concordance less than 100% suggests that environmental factors play a role.

(d) There is lower concordance of smoking in dizygotic twins than in monozygotic twins, so genetic factors appear to influence the tendency to smoke. The fact that monozygotic twins show a concordance less than 100% suggests that environmental factors also play a role.

(e) Monozygotic twins exhibit substantially higher concordance for schizophrenia than do dizygotic twins; so we can conclude that genetic factors influence this psychiatric disorder. Because the concordance of monozygotic twins is substantially less than 100%, we can also conclude that environmental factors play a role in the disorder as well.

The New Genetics
MINING GENOMES

INTRODUCTION TO BIOINFORMATICS AND THE NATIONAL CENTER FOR BIOTECHNOLOGY INFORMATION (NCBI)

Biology and computer science merge in the field of bioinformatics, allowing biologists to ask questions and develop perspectives that would never be possible with traditional techniques. This project will introduce you to the diverse suite of powerful interactive bioinformatics tools at the National Center for Biotechnology Information (NCBI).

COMPREHENSION QUESTIONS

* 1. What three factors complicate the task of studying the inheritance of human characteristics?

* 2. Describe the features that will be exhibited in a pedigree in which a trait is segregating with each of the following modes of inheritance: autosomal recessive, autosomal dominant, X-linked recessive, X-linked dominant, and Y-linked inheritance.

* 3. What are the two types of twins and how do they arise?

4. Explain how a comparison of concordance in monozygotic and dizygotic twins can be used to determine the extent to which the expression of a trait is influenced by genes or by environmental factors.

5. How are adoption studies used to separate the effects of genes and environment in the study of human characteristics?

* 6. What is genetic counseling?

7. Briefly define newborn screening, heterozygote screening, presymptomatic testing, and prenatal diagnosis.

* 8. What are the differences between amniocentesis and chorionic villus sampling? What is the purpose of these two techniques?

9. What is preimplantation genetic diagnosis?

APPLICATION QUESTIONS AND PROBLEMS

* 10. Joe is color-blind. His mother and father both have normal vision, but his mother's father (Joe's maternal grandfather) is color-blind. All Joe's other grandparents have normal color vision. Joe has three sisters—Patty, Betsy, and Lora—all with normal color vision. Joe's oldest sister, Patty, is married to a man with normal color vision; they have two children, a 9-year old color-blind boy and a 4-year-old girl with normal color vision.

(a) Using standard symbols and labels, draw a pedigree of Joe's family.

(b) What is the most likely mode of inheritance for color blindness in Joe's family?

(c) If Joe marries a woman who has no family history of color blindness, what is the probability that their first child will be a color-blind boy?

(d) If Joe marries a woman who is a carrier of the color-blind allele, what is the probability that their first child will be a color-blind boy?

(e) If Patty and her husband have another child, what is the probability that the child will be a color-blind boy?

11. A man with a specific unusual genetic trait marries an unaffected woman and they have four children. Pedigrees of this family are shown in parts a through e, but the presence or absence of the trait in the children is not indicated. For each type of inheritance, indicate how many children of each sex are expected to express the trait by filling in the appropriate circles and squares. Assume that the trait is rare and fully penetrant.

(a) Autosomal recessive trait

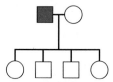

(b) Autosomal dominant trait

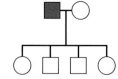

(c) X-linked recessive trait

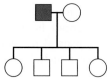

(d) X-linked dominant trait

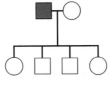

(e) Y-linked trait

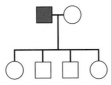

(c)

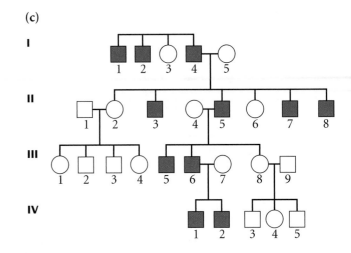

* 12. For each of the following pedigrees, give the most likely mode of inheritance, assuming that the trait is rare. Carefully explain your reasoning.

(a)

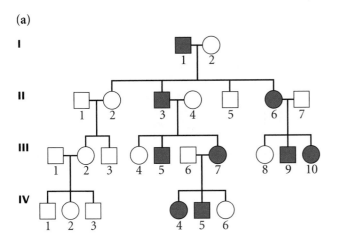

(d)

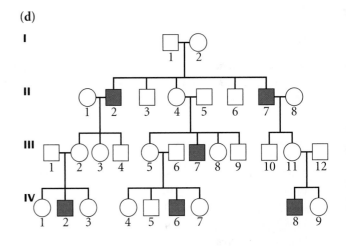

(b)

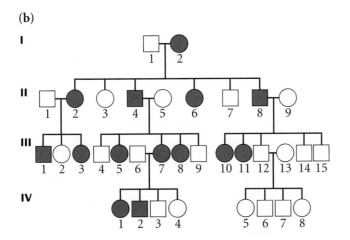

(e)

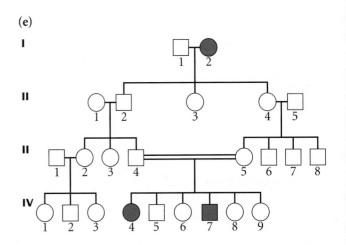

13. The trait represented in the following pedigree is expressed only in the males of the family. Is the trait Y linked? Why or why not? If you believe the trait is not Y linked, propose an alternate explanation for its inheritance.

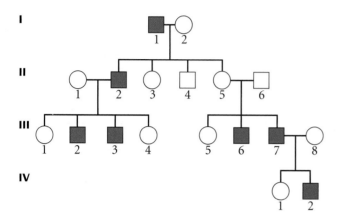

*14. A geneticist studies a series of characteristics in monozygotic twins and dizygotic twins, obtaining the following concordances. For each characteristic, indicate whether the rates of concordance suggest genetic influences, environmental influences, or both. Explain your reasoning.

Characteristic	Monozygotic concordance (%)	Dizygotic concordance (%)
Migraine headaches	60	30
Eye color	100	40
Measles	90	90
Clubfoot	30	10
High blood pressure	70	40
Handedness	70	70
Tuberculosis	5	5

15. In a study of schizophrenia (a mental disorder including disorganization of thought and withdrawal from reality), researchers looked at the prevalence of the disorder in the biological and adoptive parents of people who were adopted as children; they found the following results:

	Prevalence of schizophrenia (%)	
Adopted persons	Biological parents	Adoptive parents
With schizophrenia	12	2
Without schizophrenia	6	4

(Source: S. S. Kety, et al., The biological and adoptive families of adopted individuals who become schizophrenic: prevalence of mental illness and other characteristics, In *The Nature of Schizophrenia: New Approaches to Research and Treatment*, L. C. Wynne, R. L. Cromwell, and S. Matthysse, Eds. (New York: Wiley, 1978), pp. 25–37.)

What can you conclude from these results concerning the role of genetics in schizophrenia? Explain your reasoning.

*16. The following pedigree illustrates the inheritance of Nance-Horan syndrome, a rare genetic condition in which affected persons have cataracts and abnormally shaped teeth.

(a) On the basis of this pedigree, what do you think is the most likely mode of inheritance for Nance-Horan syndrome?

(b) If couple III-7 and III-8 have another child, what is the probability that the child will have Nance-Horan syndrome?

(c) If III-2 and III-7 mated, what is the probability that one of their children would have Nance-Horan syndrome?

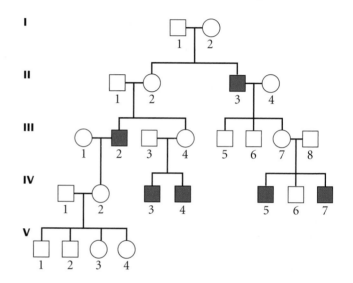

(Pedigree adapted from D. Stambolian, R. A. Lewis, K. Buetow, A. Bond, and R. Nussbaum. *American Journal of Human Genetics* 47(1990):15.)

17. The following pedigree illustrates the inheritance of ringed hair, a condition in which each hair is differentiated into light and dark zones. What mode or modes of inheritance are possible for the ringed-hair trait in this family?

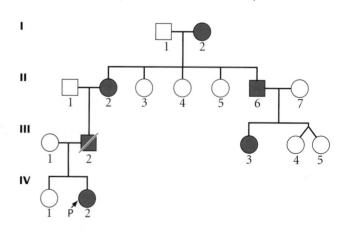

(Pedigree adapted from L. M. Ashley, and R. S. Jacques, *Journal of Heredity* 41(1950):83.)

* 18. Ectodactyly is a rare condition in which the fingers are absent and the hand is split. This condition is usually inherited as an autosomal dominant trait. Ademar Freire-Maia reported the appearance of ectodactyly in a family in São Paulo, Brazil, whose pedigree is shown here. Is this pedigree consistent with autosomal dominant inheritance? If not, what mode of inheritance is most likely? Explain your reasoning.

(Pedigree adapted from A. Freire-Maia, *Journal of Heredity* 62(1971):53.)

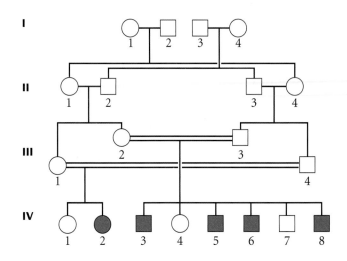

CHALLENGE QUESTIONS

19. Draw a pedigree that represent an autosomal dominant trait, sex-limited to males, and that excludes the possibility that the trait is Y linked.
20. Androgen insensitivity syndrome is a rare disorder of sexual development, in which people with an XY karyotype, genetically male, develop external female features. All persons with androgen insensitivity syndrome are infertile. In the past, some researchers proposed that androgen insensitivity syndrome is inherited as a sex-limited, autosomal dominant trait. (It is sex-limited because females cannot express the trait.) Other investigators suggested that this disorder is inherited as a X-linked recessive trait.

    Draw a pedigree that would show conclusively that androgen insensitivity syndrome is inherited as an X-linked recessive trait and that excludes the possibility that it is sex-limited, autosomal dominant. If you believe that no pedigree can conclusively differentiate between the two choices (sex-limited, X-linked recessive and sex-limited, autosomal dominant), explain why. Remember that all affected persons are infertile.

SUGGESTED READINGS

Barsh, G. S., I. S. Farooqi, and S. O'Rahilly. 2000. Genetics of body-weight regulation. *Nature* 404:644–651.

An excellent review of the genetics of body weight in humans. This issue of *Nature* has a section on obesity, with additional review articles on obesity as a medical problem, on the molecular basis of thermogenesis, on nervous-system control of food intake, and medical strategies for treatment of obesity.

Bennett, R. L., K. A. Steinhaus, S. B. Uhrich, C. K. O'Sullivan, R. G. Resta, D. Lochner-Doyle, D. S. Markel, V. Vincent, and J. Hamanishi. 1995. Recommendations for standardized human pedigree nomenclature. *American Journal of Human Genetics* 56:745–752.

Contains recommendations for standardized symbols used in pedigree construction.

Brown, M. S., and J. L. Goldstein. 1984. How LDL receptors influence cholesterol and atherosclerosis. *Scientific American* 251 November: 58–66.

Excellent review of the genetics of atherosclerosis by two scientists who received the Nobel Prize for their research on atherosclerosis.

Devor, E. J., and C. R. Cloninger. 1990. Genetics of alcoholism. *Annual Review of Genetics* 23:19–36.

A good review of how genes influence alcoholism in humans.

Gurney, M. E., A. G. Tomasselli, and R. L. Heinrikson. 2000. Stay the executioner's hand. *Science* 288:283–284.

Reports new evidence that mutated SOD1 may be implicated in apoptosis (programmed cell death) in people with amyotrophic lateral sclerosis.

Harper, P. S. 1998. *Practical Genetic Counseling*, 5th ed. Oxford: Butterworth Heineman.

A classic textbook on genetic counseling.

Jorde, L. B., J. C. Carey, M. J. Bamshad, and R. L. White. 1998. *Medical Genetics*, 2d ed. St. Louis: Mosby.

A textbook on medical aspects of human genetics.

Lewis, R. 1994. The evolution of a classical genetic tool. *Bioscience* 44:722–726.

A well-written review of the history of pedigree analysis and recent changes in symbols that have been necessitated by changing life styles and new reproductive technologies.

Mange, E. J., and A. P. Mange. 1998. *Basic Human Genetics*, 2d ed. Sunderland, MA: Sinauer.
A well-written textbook on human genetics.

MacGregor, A. J., H. Snieder, N. J. Schork, and T. D. Spector. 2000. Twins: novel uses to study complex traits and genetic diseases. *Trends in Genetics* 16:131–134.
A discussion of new methods for using twins in the study of genes.

Mahowald, M. B., M. S. Verp, and R. R. Anderson. 1998. Genetic counseling: clinical and ethical challenges. *Annual Review of Genetics* 32:547–559.
A review of genetic counseling in light of the Human Genome Project, with special consideration of the role of nondirected counseling.

McKusick, V. A. 1998. *Mendelian Inheritance in Man: A Catalog of Human Genes and Genetic Disorders*, 12th ed. Baltimore: Johns Hopkins University Press.
A comprehensive catalog of all known simple human genetic disorders and the genes responsible for them.

Pierce, B. A. 1990. *The Family Genetic Source Book*. New York: Wiley.
A book on human genetics written for the layperson; contains a catalog of more than 100 human genetic traits.

Stunkard, A. J., T. I. Sorensen, C. Hanis, T. W. Teasdale, R. Chakraborty, W. J. Schull, and F. Schulsinger. 1986. An adoption study of human obesity. *The New England Journal of Medicine* 314:193–198.
Describes the Danish adoption study of obesity.

Linkage, Recombination, and Eukaryotic Gene Mapping

Alfred Henry Sturtevant, an early geneticist, developed the first genetic map. (Institute Archives, California Institute of Technology.)

Alfred Sturtevant and the First Genetic Map

In 1909, Thomas Hunt Morgan taught the introduction to zoology class at Columbia University. Seated in the lecture hall were sophomore Alfred Henry Sturtevant and freshman Calvin Bridges. Sturtevant and Bridges were excited by Morgan's teaching style and intrigued by his interest in biological problems. They asked Morgan if they could work in his laboratory and, the following year, both young men were given desks in the "fly room," Morgan's research laboratory where the study of *Drosophila* genetics was in its infancy (see pp. 85–86 in Chapter 4). Sturtevant, Bridges,

and Morgan's other research students virtually lived in the laboratory, raising fruit flies, designing experiments, and discussing their results.

In the course of their research, Morgan and his students observed that some pairs of genes did not segregate randomly according to Mendel's principle of independent assortment but instead tended to be inherited together. Morgan suggested that possibly the genes were located on the same chromosome and thus traveled together during meiosis. He further proposed that closely linked genes—

7.1 Sturtevant's map included five genes on the X chromosome of *Drosophila*. The genes are yellow body (*y*), white eyes (*w*), vermilion eyes (*v*), miniature wings (*m*), and rudimentary wings (*r*). Sturtevant's original symbols for the genes are shown above the line; modern symbols are shown below with their current locations on the X chromosome.

those that are rarely shuffled by recombination—lie close together on the same chromosome, whereas loosely linked genes—those more frequently shuffled by recombination—lie farther apart.

One day in 1911, Sturtevant and Morgan were discussing independent assortment when, suddenly, Sturtevant had a flash of inspiration: variation in the strength of linkage indicated how genes were positioned along a chromosome, providing a way of mapping genes. Sturtevant went home and, neglecting his undergraduate homework, spent most of the night working out the first genetic map (FIGURE 7.1). Sturtevant's first chromosome map was remarkably accurate, and it established the basic methodology used today for mapping genes.

Alfred Sturtevant went on to become a leading geneticist. His research included gene mapping and basic mechanisms of inheritance in *Drosophila*, cytology, embryology, and evolution. Sturtevant's career was deeply influenced by his early years in the fly room, where Morgan's unique personality and the close quarters combined to stimulate intellectual excitement and the free exchange of ideas.

www.whfreeman.com/pierce   More details about Alfred Sturtevant's life

This chapter explores the inheritance of genes located on the same chromosome. These linked genes do not strictly obey Mendel's principle of independent assortment; rather, they tend to be inherited together. This tendency requires a new approach to understanding their inheritance and predicting the types of offspring produced. A critical piece of information necessary for predicting the results of these crosses is the arrangement of the genes on the chromosomes; thus, it will be necessary to think about the relation between genes and chromosomes. A key to understanding the inheritance of linked genes is to make the conceptual connection between the genotypes in a cross and the behavior of chromosomes during meiosis.

We will begin our exploration of linkage by first comparing the inheritance of two linked genes with the inheritance of two genes that assort independently. We will then examine how crossing over breaks up linked genes. This knowledge of linkage and recombination will be used for predicting the results of genetic crosses in which genes are linked and for mapping genes. The last section of the chapter focuses on physical methods of determining the chromosomal locations of genes.

Genes That Assort Independently and Those That Don't

Chapter 3 introduced Mendel's principles of segregation and independent assortment. Let's take a moment to review these two important concepts. The principle of segregation states that each diploid individual possesses two alleles that separate in meiosis, with one allele going into each gamete. The principle of independent assortment provides additional information about the process of segregation: it tells us that the two alleles separate independently of *alleles* at other loci.

The independent separation of alleles produces *recombination*, the sorting of alleles into new combinations. Consider a cross between individuals homozygous for two different pairs of alleles: $AABB \times aabb$. The first parent, $AABB$, produces gametes with alleles AB, and the second parent, $aabb$, produces gametes with the alleles ab, resulting in F_1 progeny with genotype $AaBb$ (FIGURE 7.2). Recombination means that, when one of the F_1 progeny reproduces, the combination of alleles in its gametes may differ from the combinations in the gametes of its parents. In other words, the F_1 may produce gametes with alleles Ab or aB in addition to gametes with AB or ab.

Mendel derived his principles of segregation and independent assortment by observing progeny of genetic crosses, but he had no idea of what biological processes produced these phenomena. In 1903, Walter Sutton proposed a biological basis for Mendel's principles, called the chromosome theory of heredity (Chapter 3). This theory holds that genes are found on chromosomes. Let's restate Mendel's two principles in terms of the chromosome theory of heredity. The principle of segregation states that each diploid individual possesses two alleles for a trait, each of which is located at the same position, or locus, on each of the two homologous chromosomes. These chromosomes segregate in meiosis, with each gamete receiving one homolog. The principle of independent assortment states that, in meiosis,

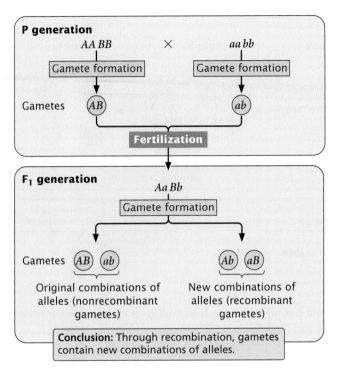

P generation

$AA\,BB$ $\times$ $aa\,bb$

Gamete formation   Gamete formation

Gametes AB ab

Fertilization

F₁ generation

$Aa\,Bb$

Gamete formation

Gametes AB ab Ab aB

Original combinations of alleles (nonrecombinant gametes)   New combinations of alleles (recombinant gametes)

Conclusion: Through recombination, gametes contain new combinations of alleles.

◄ **7.2 Recombination is the sorting of alleles into new combinations.**

each pair of homologous chromosomes assorts independently of other homologous pairs. With this new perspective, it is easy to see that the number of chromosomes in most organisms is limited and that there are certain to be more genes than chromosomes; so some genes must be present on the same chromosome and should not assort independently.

Genes located close together on the same chromosome are called **linked genes** and belong to the same **linkage group.** As we've said, linked genes travel together during meiosis, eventually arriving at the same destination (the same gamete), and are not expected to assort independently. However, all of the characteristics examined by Mendel in peas did display independent assortment and, after the rediscovery of Mendel's work, the first genetic characteristics studied in other organisms also seemed to assort independently. How could genes be carried on a limited number of chromosomes and yet assort independently?

The apparent inconsistency between the principle of independent assortment and the chromosome theory of heredity soon disappeared, as biologists began finding genetic characteristics that did *not* assort independently. One of the first cases was reported in sweet peas by William Bateson, Edith Rebecca Saunders, and Reginald C. Punnett in 1905. They crossed a homozygous strain of peas having purple flowers and long pollen grains with a homozygous strain having red flowers and round pollen grains. All the F₁ had purple flowers and long pollen grains, indicating that

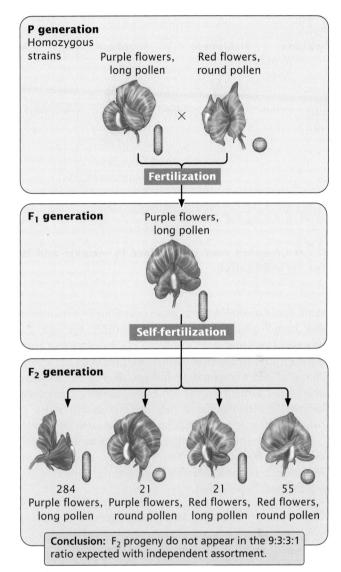

P generation
Homozygous strains

Purple flowers, long pollen   Red flowers, round pollen

$\times$

Fertilization

F₁ generation

Purple flowers, long pollen

Self-fertilization

F₂ generation

284   21   21   55
Purple flowers, long pollen   Purple flowers, round pollen   Red flowers, long pollen   Red flowers, round pollen

Conclusion: F₂ progeny do not appear in the 9:3:3:1 ratio expected with independent assortment.

◄ **7.3 Nonindependent assortment of flower color and pollen shape in sweet peas.**

purple was dominant over red and long was dominant over round. When they intercrossed the F₁, the resulting F₂ progeny did not appear in the 9:3:3:1 ratio expected with independent assortment (◄ FIGURE 7.3). An excess of F₂ plants had purple flowers and long pollen or red flowers and round pollen (the parental phenotypes). Although Bateson, Saunders, and Punnett were unable to explain these results, we now know that the two loci that they examined lie close together on the same chromosome and therefore do not assort independently.

Linkage and Recombination Between Two Genes

Genes on the same chromosome are like passengers on a charter bus: they travel together and ultimately arrive at the

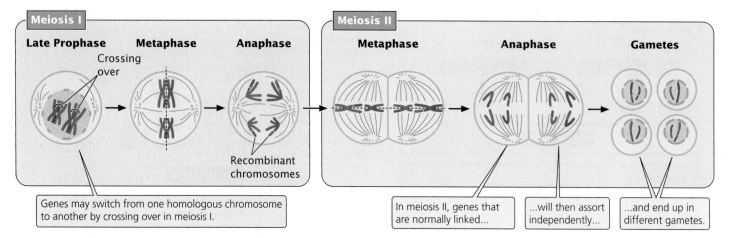

7.4 Crossing over takes place in meiosis and is responsible for recombination.

same destination. However, genes occasionally switch from one homologous chromosome to another through the process of crossing over (Chapter 2) (◀ FIGURE 7.4). Crossing over produces recombination—it breaks up the associations of genes imposed by linkage. As will be discussed later, genes located on the same chromosome can exhibit independent assortment if they are far enough apart. In summary, linkage adds a further complication to interpretations of the results of genetic crosses. With an understanding of how linkage affects heredity, we can analyze crosses for linked genes and successfully predict the types of progeny that will be produced.

Notation for Crosses with Linkage

In analyzing crosses with linked genes, we must know not only the genotypes of the individuals crossed, but also the arrangement of the genes on the chromosomes. To keep track of this arrangement, we will introduce a new system of notation for presenting crosses with linked genes. Consider a cross between an individual homozygous for dominant alleles at two linked loci and another individual homozygous for recessive alleles at those loci. Previously, we would have written these genotypes as:

$$AABB \times aabb$$

For linked genes, however, it's necessary to write out the specific alleles as they are arranged on each of the homologous chromosomes:

$$\frac{A \qquad B}{A \qquad B} \times \frac{a \qquad b}{a \qquad b}$$

In this notation, each line represents one of the two homologous chromosomes. In the first parent of the cross, each homologous chromosome contains A and B alleles; in

the second parent, each homologous chromosome contains a and b alleles. Inheriting one chromosome from each parent, the F_1 progeny will have the following genotype:

$$\frac{A \qquad B}{a \qquad b}$$

Here, the importance of designating the alleles on each chromosome is clear. One chromosome has the two dominant alleles A and B, whereas the homologous chromosome has the two recessive alleles a and b. The notation can be simplified by drawing only a single line, with the understanding that genes located on the same side of the line lie on the same chromosome:

$$\frac{A \qquad B}{a \qquad b}$$

This notation can be simplified further by separating the alleles on each chromosome with a slash: AB/ab.

Remember that the two alleles at a locus are always located on different homologous chromosomes and therefore must lie on opposite sides of the line. Consequently, we would *never* write the genotypes as:

$$\frac{A \qquad a}{B \qquad b}$$

because the alleles A and a can *never* be on the same chromosome.

It is also important to always keep the same order of the genes on both sides of the line; thus, we should *never* write:

$$\frac{A \qquad B}{b \qquad a}$$

because this would imply that alleles A and b are allelic (at the same locus).

Complete Linkage Compared with Independent Assortment

We will first consider what happens to genes that exhibit complete linkage, meaning that they are located on the same chromosome and do not exhibit crossing over. Genes rarely exhibit complete linkage but, without the complication of crossing over, the effect of linkage can be seen more clearly. We will then consider what happens when genes assort independently. Finally, we will consider the results obtained if the genes are linked but exhibit some crossing over.

A testcross reveals the effects of linkage. For example, if a heterozygous individual is test-crossed with a homozygous recessive individual ($AaBb \times aabb$), whatever alleles are present in the gametes contributed by the heterozygous parent will be expressed in the phenotype of the offspring, because the homozygous parent could not contribute dominant alleles that might mask them. Consequently, traits that appear in the progeny reveal which alleles were transmitted by the heterozygous parent.

Consider a pair of linked genes in tomato plants. One pair affects the type of leaf: an allele for mottled leaves (m) is recessive to an allele that produces normal leaves (M). Nearby on the same chromosome is another locus that determines the height of the plant: an allele for dwarf (d) is recessive to an allele for tall (D).

Testing for linkage can be done with a testcross, which requires a plant heterozygous for both traits. A geneticist might produce this heterozygous plant by crossing a variety of tomato that is homozygous for normal leaves and tall height with a variety that is homozygous for mottled leaves and dwarf height:

$$\text{P} \quad \frac{M \qquad D}{M \qquad D} \times \frac{m \qquad d}{m \qquad d}$$

$$\downarrow$$

$$\text{F}_1 \quad \frac{M \qquad D}{m \qquad d}$$

The geneticist would then use these F_1 heterozygotes in a testcross, crossing them with plants homozygous for mottled leaves and dwarf height:

$$\frac{M \qquad D}{m \qquad d} \times \frac{m \qquad d}{m \qquad d}$$

The results of this testcross are diagrammed in ◀ FIGURE 7.5a. During gamete formation, the heterozygote produces two types of gametes: some with the $\underline{M \qquad D}$ chromosome and others with the $\underline{m \qquad d}$ chromosome. Because no crossing over occurs, these gametes are the only types produced by the heterozygote. Notice that these gametes contain only combinations of alleles that were present in the original parents: either the allele for normal leaves together with the allele for tall height (M and D) or the allele for mottled leaves together with the allele for dwarf height (m and d). Gametes that contain only original combinations of alleles present in the parents are **nonrecombinant gametes,** or parental gametes.

The homozygous parent in the testcross produces only one type of gamete; it contains chromosome $\underline{m \qquad d}$ and pairs with one of the two gametes generated by the heterozygous parent (see Figure 7.5a). Two types of progeny result: half have normal leaves and are tall:

$$\frac{M \qquad D}{m \qquad d}$$

and half have mottled leaves and are dwarf:

$$\frac{m \qquad d}{m \qquad d}$$

These progeny display the original combinations of traits present in the P generation and are **nonrecombinant progeny,** or parental progeny. No new combinations of the two traits, such as normal leaves with dwarf or mottled leaves with tall, appear in the offspring, because the genes affecting the two characteristics are completely linked and are inherited together. New combinations of traits could arise only if the linkage between M and D or between m and d were broken.

These results are distinctly different from the results that are expected when genes assort independently (◀ FIGURE 7.5b). With independent assortment, the heterozygous plant ($MmDd$) would produce four types of gametes: two nonrecombinant gametes containing the original combinations of alleles (MD and md) and two gametes containing new combinations of alleles (Md and mD). Gametes with new combinations of alleles are called **recombinant gametes.** With independent assortment, nonrecombinant and recombinant gametes are produced in equal proportions. These four types of gametes join with the single type of gamete produced by the homozygous parent of the testcross to produce four kinds of progeny in equal proportions (see Figure 7.5b). The progeny with new combinations of traits formed from recombinant gametes are termed **recombinant progeny.**

In summary, a testcross in which one of the plants is heterozygous for two completely linked genes yields two types of progeny, each type displaying one of the original combinations of traits present in the P generation. Independent assortment, in contrast, produces two types of

(a) If genes are completely linked (no crossing over)

(b) If genes assort independently

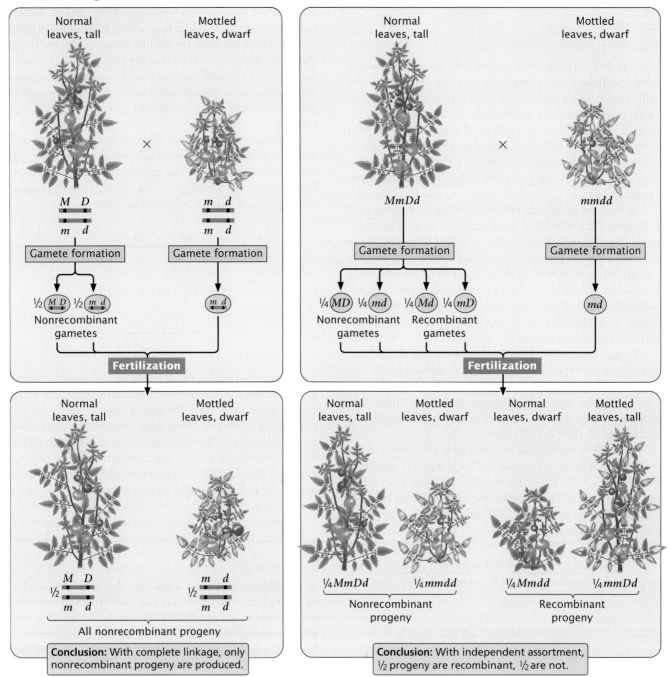

◀ 7.5 **A testcross reveals the effects of linkage.** Results of a testcross for two loci in tomatoes that determine leaf type and plant height.

recombinant progeny and two types of nonrecombinant progeny in equal proportions.

Crossing Over with Linked Genes

Linkage is rarely complete—usually, there is some crossing over between linked genes (incomple linkage), producing new combinations of traits. Let's see how this occurs.

Theory The effect of crossing over on the inheritance of two linked genes is shown in ◀ FIGURE 7.6. Crossing over, which takes place in prophase I of meiosis, is the exchange of genetic material between nonsister chromatids (see Figures 2.15 and 2.17). After a single crossover has taken place, the two chromatids that did not participate in crossing over are unchanged; gametes that receive these chromatids are

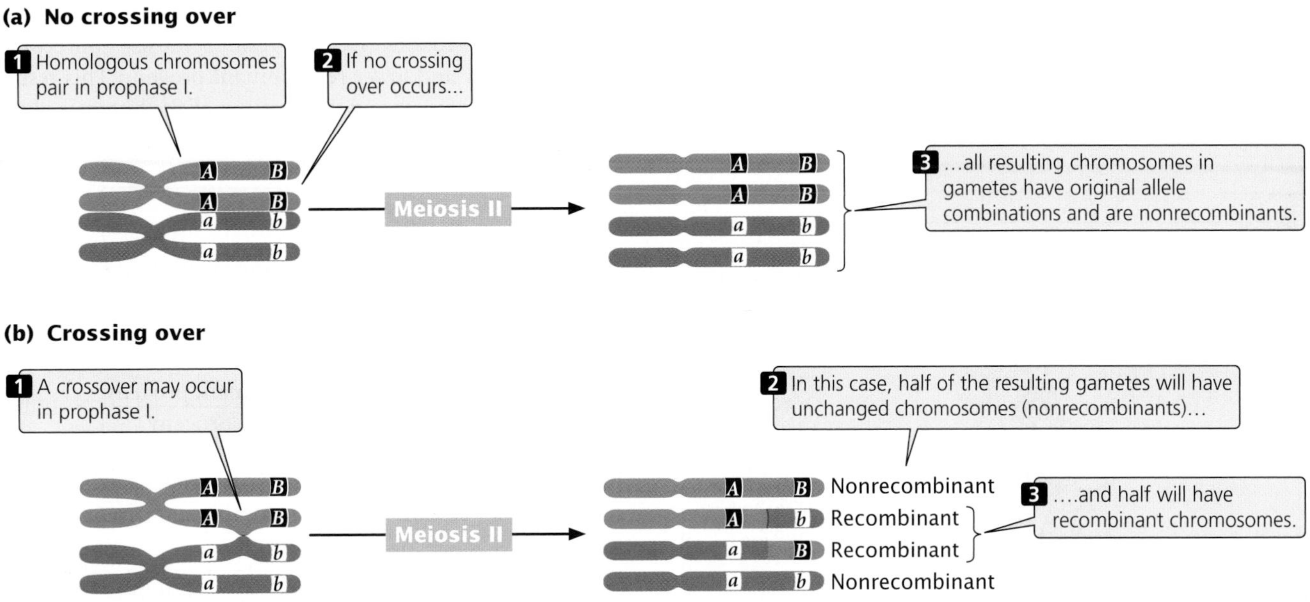

(a) No crossing over

1 Homologous chromosomes pair in prophase I.

2 If no crossing over occurs...

3 ...all resulting chromosomes in gametes have original allele combinations and are nonrecombinants.

(b) Crossing over

1 A crossover may occur in prophase I.

2 In this case, half of the resulting gametes will have unchanged chromosomes (nonrecombinants)...

3 ....and half will have recombinant chromosomes.

Nonrecombinant
Recombinant
Recombinant
Nonrecombinant

◀ 7.6 **A single crossover produces half nonrecombinant gametes and half recombinant gametes.**

nonrecombinants. The other two chromatids, which did participate in crossing over, now contain new combinations of alleles; gametes that receive these chromatids are recombinants. For each meiosis in which a single crossover takes place, then, two nonrecombinant gametes and two recombinant gametes will be produced. This result is the same as that produced by independent assortment (see Figure 7.5b); so, when crossing over between two loci takes place in every meiosis, it is impossible to determine whether the genes are linked and crossing over took place or whether the genes are on different chromosomes.

For closely linked genes, crossing over does not take place in every meiosis. In meioses in which there is no crossing over, only nonrecombinant gametes are produced. In meioses in which there is a single crossover, half the gametes are recombinants and half are nonrecombinants (because a single crossover only affects two of the four chromatids); so the total percentage of recombinant gametes is always half the percentage of meioses in which crossing over takes place. Even if crossing over between two genes takes place in every meiosis, only 50% of the resulting gametes will be recombinants. Thus, the frequency of recombinant gametes is always half the frequency of crossing over, and the maximum proportion of recombinant gametes is 50%.

(Concepts)

Linkage between genes causes them to be inherited together and reduces recombination; crossing over breaks up the associations of such genes. In a testcross for two linked genes, each crossover produces two recombinant gametes

and two nonrecombinants. The frequency of recombinant gametes is half the frequency of crossing over, and the maximum frequency of recombinant gametes is 50%.

Application Let us apply what we have learned about linkage and recombination to a cross between tomato plants that differ in the genes that code for leaf type and plant height. Assume now that these genes are linked and that some crossing over takes place between them. Suppose a geneticist carried out the testcross outlined earlier:

$$\frac{M \qquad D}{m \qquad d} \times \frac{m \qquad d}{m \qquad d}$$

When crossing over takes place between the genes for leave type and height, two of the four gametes produced will be recombinants. When there is no crossing over, all four resulting gametes will be nonrecombinants. Thus, over all, the majority of gametes will be nonrecombinants. These gametes then unite with gametes produced by the homozygous recessive parent, which contain only the recessive alleles, resulting in mostly nonrecombinant progeny and a few recombinant progeny (◀ FIGURE 7.7). In this cross, we see that 55 of the testcross progeny have normal leaves and are tall and 53 have mottled leaves and are dwarf. These plants are the nonrecombinant progeny, containing the original combinations of traits that were present in the parents. Of the 123 progeny, 15 have new combinations of traits that were not seen in the parents: 8 are normal leaved and dwarf, and 7 are mottle leaved and tall. These plants are the recombinant progeny.

The results of a cross such as the one illustrated in Figure 7.7 reveal several things. A testcross for two independently assorting genes is expected to produce a 1:1:1:1 phenotypic ratio in the progeny. The progeny of this cross clearly do not exhibit such a ratio; so we might suspect that

the genes are not assorting independently. When linked genes undergo crossing over, the result is mostly nonrecombinant progeny and fewer recombinant progeny. This result is what we observe among the progeny of the testcross illustrated in Figure 7.7; so we conclude that two genes show evidence of linkage with some crossing over.

Calculation of Recombination Frequency

The percentage of recombinant progeny produced in a cross is called the **recombination frequency,** which is calculated as follows:

$$\text{recombinant frequency} = \frac{\text{number of recombinant progeny}}{\text{total number of progeny}} \times 100\%$$

In the testcross shown in Figure 7.7, 15 progeny exhibit new combinations of traits; so the recombination frequency is:

$$\frac{8 + 7}{55 + 53 + 8 + 7} \times 100\% = \frac{15}{123} \times 100\% = 12\%$$

Thus, 12% of the progeny exhibit new combinations of traits resulting from crossing over.

Coupling and Repulsion

In crosses for linked genes, the arrangement of alleles on the homologous chromosomes is critically important in determining the outcome of the cross. For example, consider the inheritance of two linked genes in the Australian blowfly, *Lucilia cuprina*. In this species, one locus determines the color of the thorax: purple thorax (p) is recessive to the normal green thorax (p^+). A second locus determines the color of the puparium: a black puparium (b) is recessive to the normal brown puparium (b^+). These loci are located close together on the second chromosome. Suppose we test cross a fly that is heterozygous at both loci with a fly that is homozygous recessive at both. Because these genes are linked, there are two possible arrangements on the chromosomes of the heterozygous fly. The dominant alleles for green thorax (p^+) and brown puparium (b^+) might reside on the same chromosome, and the recessive alleles for purple thorax (p) and black puparium (b) might reside on the other homologous chromosome:

$$\frac{p^+ \qquad b^+}{p \qquad b}$$

This arrangement, in which wild-type alleles are found on one chromosome and mutant alleles are found on the other

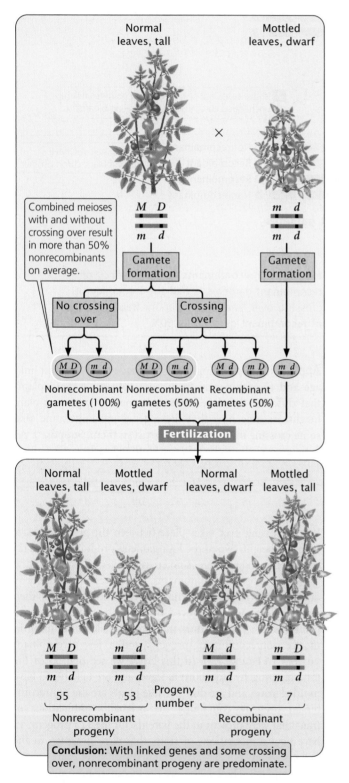

Conclusion: With linked genes and some crossing over, nonrecombinant progeny are predominate.

◄ **7.7 Crossing over between linked genes produces nonrecombinant and recombinant offspring.** In this testcross, genes are linked and there is some crossing over. For comparison, this cross is the same as that illustrated in Figure 7.5.

chromosome, is referred to as **coupling,** or the **cis configuration.** Alternatively, one chromosome might bear the alleles for green thorax (p^+) and black puparium (b), and the other chromosome would carry the alleles for purple thorax (p) and brown puparium (b^+):

$$\frac{p^+ \qquad b}{p \qquad b^+}$$

This arrangement, in which each chromosome contains one wild-type and one mutant allele, is called the **repulsion** or **trans configuration.** Whether the alleles in the heterozygous parent are in coupling or repulsion determines which phenotypes will be most common among the progeny of a testcross.

When the alleles are in the coupling configuration, the most numerous progeny types are those with green thorax and brown puparium and those with purple thorax and black puparium (◀FIGURE 7.8a); but, when the alleles of the heterozygous parent are in repulsion, the most numerous progeny types are those with green thorax and black puparium and those with purple thorax and brown puparium (◀FIGURE 7.8b). Notice that the genotypes of the parents in Figure 7.8a and b are the same ($p^+p\ b^+b \times pp\ bb$) and that the dramatic difference in the phenotypic ratios of the progeny in the two crosses results entirely from the configuration—coupling or repulsion—of the chromosomes. It is essential to know the arrangement of the alleles on the chromosomes to accurately predict the outcome of crosses in which genes are linked.

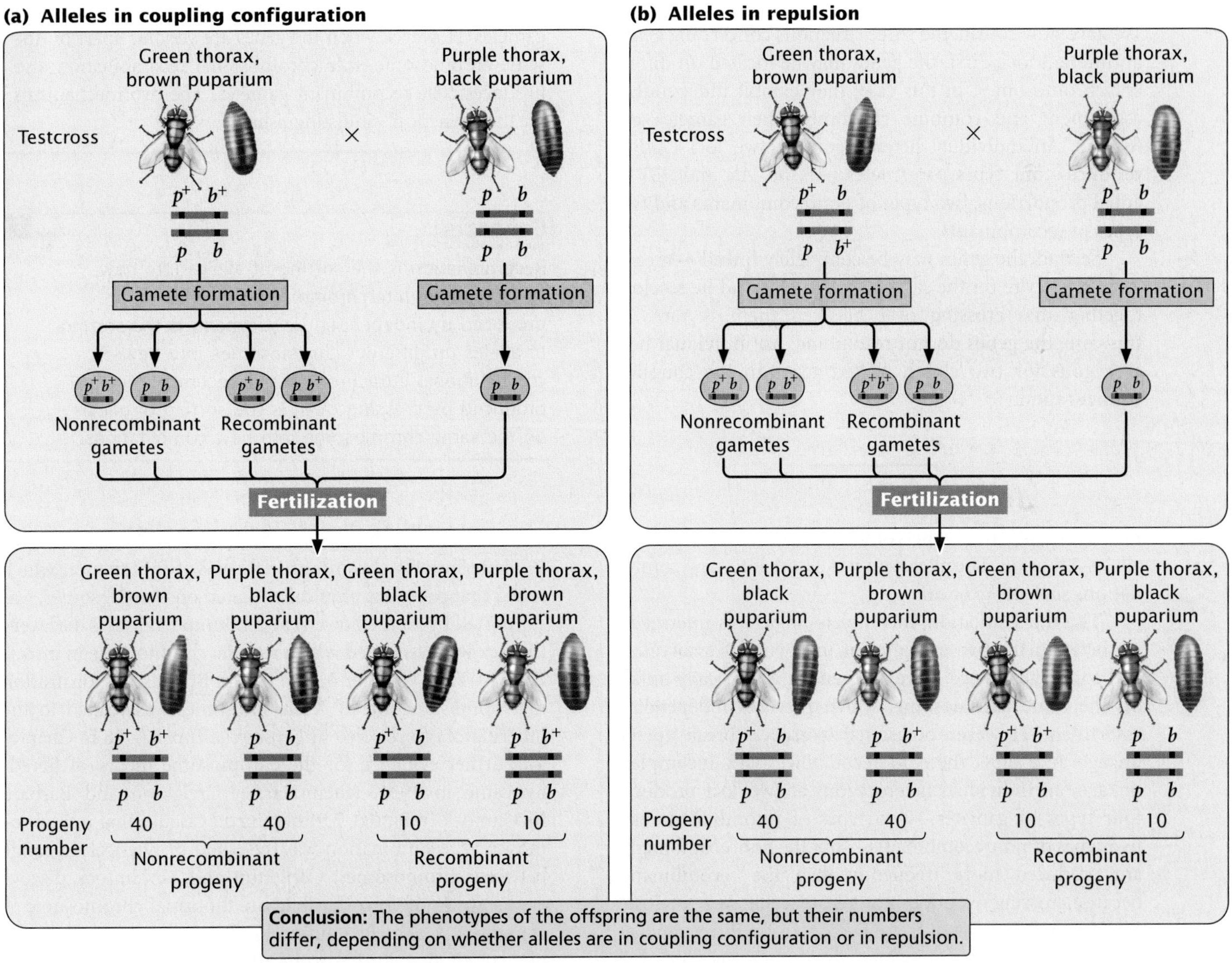

(a) Alleles in coupling configuration

(b) Alleles in repulsion

Conclusion: The phenotypes of the offspring are the same, but their numbers differ, depending on whether alleles are in coupling configuration or in repulsion.

◀7.8 **The arrangement of linked genes on a chromosome (coupling or repulsion) affects the results of a testcross.** Linked loci in the Australian blowfly, *Lucilia cuprina*, determine the color of the thorax and that of the puparium.

Concepts

In a cross, the arrangement of linked alleles on the chromosomes is critical for determining the outcome. When two wild-type alleles are on one homologous chromosome and two mutant alleles are on the other, they are in the coupling configuration; when each chromosome contains one wild-type allele and one mutant allele, the alleles are in repulsion.

Connecting Concepts

Relating Independent Assortment, Linkage, and Crossing Over

We have now considered three situations concerning genes at different loci. First, the genes may be located on different chromosomes; in this case, they exhibit independent assortment and combine randomly when gametes are formed. An individual heterozygous at two loci (*AaBb*) produces four types of gametes (*AB, ab, Ab,* and *aB*) in equal proportions: two types of nonrecombinants and two types of recombinants.

Second, the genes may be completely linked—meaning that they're on the same chromosome and lie so close together that crossing over between them is rare. In this case, the genes do not recombine. An individual heterozygous for two closely linked genes in the coupling configuration:

$$\frac{A \qquad B}{a \qquad b}$$

produces only the nonrecombinant gametes containing alleles *AB* or *ab.* The alleles do not assort into new combinations such as *Ab* or *aB*.

The third situation, incomplete linkage, is intermediate between the two extremes of independent assortment and complete linkage. Here, the genes are physically linked on the same chromosome, which prevents independent assortment. However, occasional crossovers break up the linkage and allow them to recombine. With incomplete linkage, an individual heterozygous at two loci produces four types of gametes—two types of recombinants and two types of nonrecombinants—but the nonrecombinants are produced more frequently than the recombinants because crossing over does not take place in every meiosis. Linkage and crossing over are two opposing forces: linkage binds alleles at different loci together, restricting their ability to associate freely, whereas crossing over breaks the linkage and allows alleles to assort into new combinations.

Earlier in the chapter, the term recombination was defined as the sorting of alleles into new combinations. We can now distinguish between two types of recombination that differ in the mechanism that generates these new combinations of alleles.

Interchromosomal recombination is between genes on *different* chromosomes. It arises from independent assortment—the random segregation of chromosomes in anaphase I of meiosis. **Intrachromosomal recombination** is between genes located on the *same* chromosome. It arises from crossing over—the exchange of genetic material in prophase I of meiosis. Both types of recombination produce new allele combinations in the gametes; so they cannot be distinguished by examining the types of gametes produced. Nevertheless, they can often be distinguished by the *frequencies* of types of gametes: interchromosomal recombination produces 50% nonrecombinant gametes and 50% recombinant gametes, whereas intrachromosomal recombination frequently produces less than 50% recombinant gametes. However, when the genes are very far apart on the same chromosome, intrachromosomal recombination also produces 50% recombinant gametes. The two mechanisms are then genetically indistinguishable.

Concepts

Recombination is the sorting of alleles into new combinations. Interchromosomal recombination, produced by independent assortment, is the sorting of alleles on different chromosomes into new combinations. Intrachromosomal recombination, produced by crossing over, is the sorting of alleles on the same chromosome into new combinations.

The Physical Basis of Recombination

William Sutton's chromosome theory of inheritance, which stated that genes are physically located on chromosomes, was supported by Nettie Stevens and Edmund Wilson's discovery that sex was associated with a specific chromosome in insects (pp. 78–79 in Chapter 4) and Calvin Bridges' demonstration that nondisjunction of X chromosomes was related to the inheritance of eye color in *Drosophila* (pp. 87–88 in Chapter 4). Further evidence for the chromosome theory of heredity came in 1931, when Harriet Creighton and Barbara McClintock (◄FIGURE 7.9) obtained evidence that intrachromosomal recombination was the result of physical exchange between chromosomes. Creighton and McClintock discovered a strain of corn that had an abnormal chromosome 9, containing a densely staining knob at one end and a small piece of another chromosome attached to the other end. This aberrant chromosome allowed them to visually distinguish the two members of a homologous pair.

They studied the inheritance of two traits in corn determined by genes on chromosome 9: at one locus, a dom-

◀7.9 **Barbara McClintock (left) and Harriet Creighton (right) provided evidence that genes are located on chromosomes.** (Karl Maramorosch/Cold Spring Harbor Laboratory Archives.)

inant allele (*C*) produced colored kernels, whereas a recessive allele (*c*) produced colorless kernels; at another, linked locus, a dominant allele (*Wx*) produced starchy kernels, whereas a recessive allele (*wx*) produced waxy kernels. Creighton and McClintock obtained a plant that was heterozygous at both loci in repulsion, with the alleles for colored and waxy on the aberrant chromosome and the alleles for colorless and starchy on a normal chromosome:

They crossed this heterozygous plant with a plant that was homozygous for colorless and heterozygous for waxy:

$$\frac{C}{c} \quad \frac{wx}{Wx} \times \frac{c}{c} \quad \frac{Wx}{wx}$$

This cross will produce different combinations of traits in the progeny, but the only way that colorless and waxy progeny can arise is through crossing over in the doubly heterozygous parent:

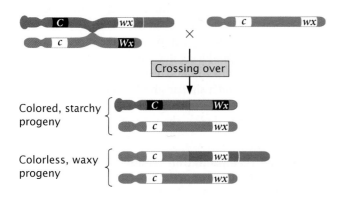

Notice that, if crossing over entails physical exchange between the chromosomes, then the colorless, waxy progeny resulting from recombination should have a chromosome with an extra piece, but not a knob. Furthermore, some of the colored, starchy progeny should possess a knob but not the extra piece. This outcome is precisely what Creighton and McClintock observed, confirming the chromosomal theory of inheritance. Curt Stern provided a similar demonstration by using chromosomal markers in *Drosophila* at about the same time. We will examine the molecular basis of recombination in more detail in Chapter 12.

Predicting the Outcomes of Crosses with Linked Genes

Knowing the arrangement of alleles on a chromosome allows us to predict the types of progeny that will result from a cross entailing linked genes and to determine which of these types will be the most numerous. Determining the *proportions* of the types of offspring requires an additional piece of information—the recombination frequency. The recombination frequency provides us with information about how often the alleles in the gametes appear in new combinations and allows us to predict the proportions of offspring phenotypes that will result from a specific cross entailing linked genes.

In cucumbers, smooth fruit (*t*) is recessive to warty fruit (*T*) and glossy fruit (*d*) is recessive to dull fruit (*D*). Geneticists have determined that these two genes exhibit a recombination frequency of 16%. Suppose we cross a plant homozygous for warty and dull fruit with a plant homozygous for smooth and glossy fruit and then carry out a testcross by using the F_1:

$$\frac{T}{t} \quad \frac{D}{d} \times \frac{t}{t} \quad \frac{d}{d}$$

What types and proportions of progeny will result from this testcross?

Four types of gametes will be produced by the heterozygous parent, as shown in (◀ FIGURE 7.10): two types of nonrecombinant gametes (_T____D_ and _t____d_) and two types of recombinant gametes (_T____d_ and _t____D_). The recombination frequency tells us that

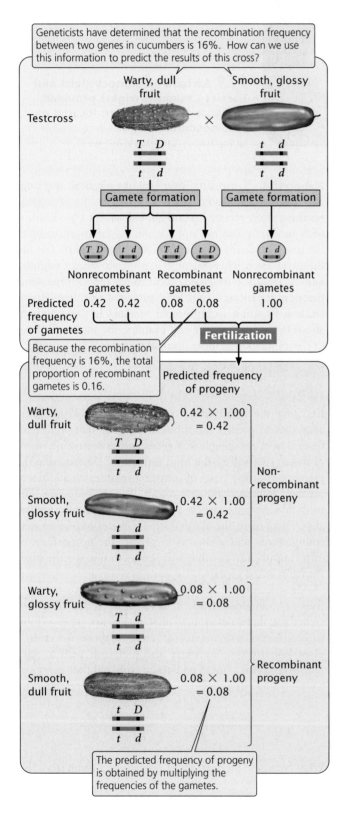

16% of the gametes produced by the heterozygous parent will be recombinants. Because there are two types of recombinant gametes, each should arise with a frequency of $^{16\%}/_2 = 8\%$. All the other gametes will be nonrecombinants; so they should arise with a frequency of 100% − 16% = 84%. Because there are two types of nonrecombinant gametes, each should arise with a frequency of $^{84\%}/_2 = 42\%$. The other parent in the testcross is homozygous and therefore produces only a single type of gamete (_t____d_) with a probability of 1.00.

The progeny of the cross result from the union of two gametes, producing four types of progeny (see Figure 7.10). The expected proportion of each type can be determined by using the multiplication rule, multiplying together the probability of each uniting gamete. Testcross progeny with warty and dull fruit

$$\frac{T \qquad D}{t \qquad d}$$

appear with a frequency of 0.42 (the probability of inheriting a gamete with chromosome _T____D_ from the heterozygous parent) × 1.00 (the probability of inheriting a gamete with chromosome _t____d_ from the recessive parent) = 0.42. The proportions of the other types of F_2 progeny can be calculated in a similar manner (see Figure 7.10). This method can be used for predicting the outcome of any cross with linked genes for which the recombination frequency is known.

Testing for Independent Assortment

In some crosses, the genes are obviously linked because there are clearly more nonrecombinants than recombinants. In other crosses, the difference between independent assortment and linkage is not so obvious. For example, suppose we did a testcross for two pairs of genes, such as AaBb × aabb, and observed the following numbers of progeny: 54 AaBb, 56 aabb, 42 Aabb, and 48 aaBb. Is this outcome a 1:1:1:1 ratio? Not exactly, but it's pretty close. Perhaps these genes are assorting independently and chance produced the slight deviations between the observed numbers and the expected 1:1:1:1 ratio. Alternatively, the genes might be linked, with considerable crossing over taking place between them, and so the number of nonrecombinants is only slightly greater than the number of recombinants. How do we distinguish between the roles of chance and of linkage in producing deviations from the results expected with independent assortment?

We encountered a similar problem in crosses in which genes were unlinked—the problem of distinguishing between deviations due to chance and those due to other

◀ 7.10 **The recombination frequency allows a prediction of the proportions of offspring expected for a cross entailing linked genes.**

factors. We addressed this problem (in Chapter 3) with the goodness-of-fit chi-square test, which serves to evaluate the likelihood that chance alone is responsible for deviations between observed and expected numbers. The chi-square test can also be used to test the goodness of fit between observed numbers of progeny and the numbers expected with independent assortment.

Testing for independent assortment between two linked genes requires the calculation of a series of three chi-square tests. To illustrate this analysis, we will examine the data from a cross between German cockroaches, in which yellow body (y) is recessive to brown body (y^+) and curved wings (cv) are recessive to straight wings (cv^+). A testcross ($y^+y\ cv^+cv \times yy\ cvcv$) produced the following progeny:

63	$y^+y\ cv^+cv$	brown body, straight wings
77	$yy\ cvcv$	yellow body, curved wings
28	$y^+y\ cvcv$	brown body, curved wings
32	$yy\ cv^+cv$	yellow body, straight wings
200	total progeny	

Testing ratios at each locus To determine if the genes for body color and wing shape are assorting independently, we must examine each locus separately and determine whether the observed numbers differ from the expected (we will consider why this step is necessary at the end of this section). At the first locus (for body color), the cross between heterozygote and homozygote ($y^+y \times yy$) is expected to produce $\frac{1}{2}\ y^+y$ brown and $\frac{1}{2}\ yy$ yellow progeny; so we expect 100 of each. We observe $63 + 28 = 91$ brown progeny and $77 + 32 = 109$ yellow progeny. Applying the chi-square test (see Chapter 3) to these observed and expected numbers, we obtain:

$$\chi^2 = \Sigma\ \frac{(\text{observed} - \text{expected})^2}{\text{expected}}$$

$$\chi^2 = \frac{(91 - 100)^2}{100} + \frac{(109 - 100)^2}{100}$$

$$= \frac{81}{100} + \frac{81}{100} = 0.81 + 0.81 = 1.62$$

The degrees of freedom associated with the chi-square test (Chapter 3) are $n - 1$, where n equals the number of expected classes. Here, there are two expected phenotypes; so the degree of freedom is $2 - 1 = 1$. Looking up our calculated chi-square value in Table 3.4, we find that the probability associated with this chi-square value is between .30 and .20. Because the probability is above .05 (our critical probability for rejecting the hypothesis that chance produces the difference between observed and expected values), we conclude that there is no significant difference between the 1 : 1 ratio that we expect in the progeny of the testcross and the ratio that we observed.

We next compare the observed and expected ratios for the second locus, which determines the type of wing. At this locus, a heterozygote and homozygote also were crossed ($cv^+cv \times cvcv$) and are expected to produce $\frac{1}{2}\ cv^+cv$ straight-winged progeny and $\frac{1}{2}\ cvcv$ curved-wing progeny. We actually observe $63 + 32 = 95$ straight-winged progeny and $77 + 28 = 105$ curved-wing progeny; so the calculated chi-square value is:

$$\chi^2 = \frac{(95 - 100)^2}{100} + \frac{(105 - 100)^2}{100}$$

$$= \frac{25}{100} + \frac{25}{100} = 0.25 + 0.25 = 0.50$$

The degree of freedom associated with this chi-square value also is $2 - 1 = 1$, and the associated probability is between .5 and .3. We again assume that there is no significant difference between what we observed and what we expected at this locus in the testcross.

Testing ratios for independent assortment We are now ready to test for the independent assortment of genes at the two loci. If the genes are assorting independently, we can use the multiplication rule to obtain the probabilities and numbers of progeny inheriting different combinations of phenotypes:

Geno-types	Expected pheno-types	Expected propor-tions	Expected numbers	Observed numbers
$y^+y\ cv^+cv$	brown, straight	$\frac{1}{2} \times \frac{1}{2} = \frac{1}{4}$	50	63
$yy\ cvcv$	yellow, curved	$\frac{1}{2} \times \frac{1}{2} = \frac{1}{4}$	50	77
$y^+y\ cvcv$	brown, curved	$\frac{1}{2} \times \frac{1}{2} = \frac{1}{4}$	50	28
$yy\ cv^+cv$	yellow, straight	$\frac{1}{2} \times \frac{1}{2} = \frac{1}{4}$	50	32

The observed and expected numbers of progeny can now be compared by using the chi-square test:

$$\chi^2 = \frac{(63 - 50)^2}{50} + \frac{(77 - 50)^2}{50} + \frac{(28 - 50)^2}{50} + \frac{(32 - 50)^2}{50} = 34.12$$

Here, we have four expected classes of phenotypes; so the degrees of freedom equal $4 - 1 = 3$ and the associated probability is considerably less than .001. This very small probability indicates that the phenotypes are not in the proportions that we would expect if independent assortment were taking place. Our conclusion, then, is that these genes are not assorting independently and must be linked.

In summary, testing for linkage between two genes requires a series of chi-square tests: a chi-square test for the segregation of alleles at each individual locus, followed by a test for independent assortment between alleles at the different loci. The chi-square tests for segregation at individual loci should always be carried out before testing for independent assortment, because the probabilities expected with independent assortment are based on the probabilities expected at the separate loci. Suppose that the alleles in the cockroach example were assorting independently and that some of the cockroaches with curved wings died in embryonic development; the observed proportion with curved wings was then $\frac{1}{3}$ instead of $\frac{1}{2}$. In this case, the proportion of offspring with yellow body and curved wings expected under independent assortment should be $\frac{1}{3} \times \frac{1}{2} = \frac{1}{6}$ instead of $\frac{1}{4}$. Without the initial chi-square test for segregation at the curved-wing locus, we would have no way of knowing that what we expected with independent assortment was $\frac{1}{6}$ instead of $\frac{1}{4}$. If we carried out only the final test for independent assortment and assumed an expected 1:1:1:1 ratio, we would obtain a high chi-square value. We might conclude, erroneously, that the genes were linked.

If a significant chi-square (one that has a probability less than 0.05) is obtained in either of the first two tests for segregation, then the final chi-square for independent assortment should not be carried out, because the true expected values are unknown.

Gene Mapping with Recombination Frequencies

Morgan and his students developed the idea that physical distances between genes on a chromosome are related to the rates of recombination. They hypothesized that crossover events occur more or less at random up and down the chromosome and that two genes that lie far apart are more likely to undergo a crossover than are two genes that lie close together. They proposed that recombination frequencies could provide a convenient way to determine the order of genes along a chromosome and would give estimates of the relative distances between the genes. Chromosome maps calculated by using recombination frequencies are called **genetic maps.** In contrast, chromosome maps based on physical distances along the chromosome (often expressed in terms of numbers of base pairs) are called **physical maps.**

Distances on genetic maps are measured in **map units** (abbreviated m.u.); one map unit equals 1% recombination. Map units are also called **centimorgans** (cM), in honor of Thomas Hunt Morgan; one **morgan** equals 100 m.u. Genetic distances measured with recombination rates are approximately additive: if the distance from gene A to gene B is 5 m.u., the distance from gene B to gene C 10 m.u., and the distance from gene A to gene C is 15 m.u., gene B must be located between genes A and C. On the

basis of the map distances just given, we could draw a simple genetic map for genes A, B, and C, as shown here:

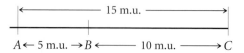

We could just as plausibly draw this map with C on the left and A on the right:

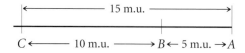

Both maps are correct and equivalent because, with information about the relative positions of only three genes, the most that we can determine is which gene lies in the middle. If we obtained distances to an additional gene, then we could position A and C relative to that gene. An additional gene D, examined through genetic crosses, might yield the following recombination frequencies:

Gene pair	Recombination frequency (%)
A and D	8
B and D	13
C and D	23

Notice that C and D exhibit the greatest amount of recombination; therefore, C and D must be farthest apart, with genes A and B between them. Using the recombination frequencies and remembering that 1 m.u. = 1% recombination, we can now add D to our map:

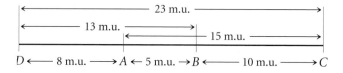

By doing a series of crosses between pairs of genes, we can construct genetic maps showing the linkage arrangements of a number of genes.

Two points should be emphasized about constructing chromosome maps from recombination frequencies. First, recall that the recombination frequency between two genes cannot exceed 50% and that 50% is also the rate of recombination for genes located on different chromosomes. Consequently, one cannot distinguish between genes on different chromosomes and genes located far apart on the same chromosome. If genes exhibit 50% recombination, the most that can be said about them is that they belong to different groups of linked genes (different linkage groups), either on different chromosomes or far apart on the same chromosome.

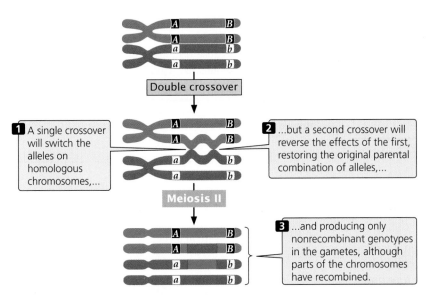

1 A single crossover will switch the alleles on homologous chromosomes,...

2 ...but a second crossover will reverse the effects of the first, restoring the original parental combination of alleles,...

Meiosis II

3 ...and producing only nonrecombinant genotypes in the gametes, although parts of the chromosomes have recombined.

◀ **7.11 A double crossover between two linked genes produces only nonrecombinant gametes.**

A second point is that a testcross for two genes that are relatively far apart on the same chromosome tends to underestimate the true physical distance, because the cross does not reveal double crossovers that might take place between the two genes (◀ FIGURE 7.11). A double crossover arises when two separate crossover events take place between the same two loci. Whereas a single crossover switches the alleles on the homologous chromosomes—producing combinations of alleles that were not present on the original parental chromosomes—a second crossover between the same two genes reverses the effects of the first, thus restoring the original parental combination of alleles (see Figure 7.11). Double crossovers produce only nonrecombinant gametes, and we cannot distinguish between the progeny produced by double crossovers and the progeny produced when there is no crossing over. However, as we shall see in the next section, it is possible to detect double crossovers if we examine a third gene that lies between the two crossovers. Because double crossovers between two genes go undetected, map distances will be underestimated whenever double crossovers take place. Double crossovers are more frequent between genes that are far apart; therefore genetic maps based on short distances are always more accurate than those based on longer distances.

Concepts

A genetic map provides the order of the genes on a chromosome and the approximate distances among the genes based on recombination frequencies. In genetic maps, 1% recombination equals 1 map unit, or 1 centimorgan. Double crossovers between two genes go undetected; so map distances between distant genes tend to underestimate genetic distances.

Constructing a Genetic Map with Two-Point Testcrosses

Genetic maps can be constructed by conducting a series of testcrosses between pairs of genes and examining the recombination frequencies between them. A testcross between two genes is called a **two-point testcross** or a two-point cross for short. Suppose that we carried out a series of two-point crosses for four genes, *a, b, c,* and *d,* and obtained the following recombination frequencies:

Gene loci in testcross	Recombination frequency (%)
a and *b*	50
a and *c*	50
a and *d*	50
b and *c*	20
b and *d*	10
c and *d*	28

We can begin constructing a genetic map for these genes by considering the recombination frequencies for each pair of genes. The recombination frequency between *a* and *b* is 50%, which is the recombination frequency expected with independent assortment. Genes *a* and *b* may therefore either be on different chromosomes or be very far apart on the same chromosome; so we will place them in different linkage groups with the understanding that they may or may not be on the same chromosome:

Linkage group 1

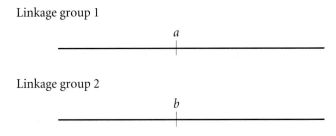

The recombination frequency between *a* and *c* is 50%, indicating that they, too, are in different linkage groups. The recombination frequency between *b* and *c* is 20%; so these genes are linked and separated by 20 map units:

Linkage group 1

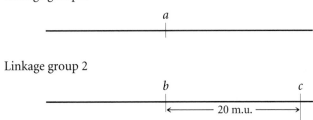

Linkage group 2

The recombination frequency between *a* and *d* is 50%, indicating that these genes belong to different linkage groups, whereas genes *b* and *d* are linked, with a recombination frequency of 10%. To decide whether gene *d* is 10 map units to the left or right of gene *b*, we must consult the *c*-to-*d* distance. If gene *d* is 10 map units to the left of gene *b*, then the distance between *d* and *c* should be 20 m.u. + 10 m.u. = 30 m.u. This distance will be only approximate because any double crossovers between the two genes will be missed and the map distance will be underestimated. If, on the other hand, gene *d* lies to the right of gene *b*, then the distance between gene *d* and *c* will be much shorter, approximately 20 m.u. − 10 m.u. = 10 m.u.

By examining the recombination frequency between *c* and *d*, we can distinguish between these two possibilities. The recombination frequency between *c* and *d* is 28%; so gene *d* must lie to the left of gene *b*. Notice that the sum of the recombination between *d* and *b* (10%) and between *b* and *c* (20%) is greater than the recombination between

d and *c* (28%). (This is what was meant by saying that recombination rates are *approximately* additive.) This discrepancy arises because double crossovers between the two outer genes go undetected, causing an underestimation of the true map distance. The genetic map of these genes is now complete:

Linkage group 1

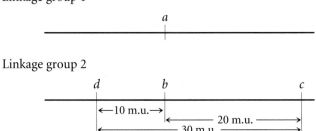

Linkage group 2

Linkage and Recombination Between Three Genes

Genetic maps can be constructed from a series of testcrosses for pairs of genes, but this approach is not particularly efficient, because numerous two-point crosses must be carried out to establish the order of the genes and because double crossovers are missed. A more efficient mapping technique is a testcross for three genes (a **three-point testcross,** or three-point cross). With a three-point cross, the order of the three genes can be established in a single set of progeny and some double crossovers can usually be detected, providing more accurate map distances.

Consider what happens when crossing over takes place among three hypothetical linked genes. ◀FIGURE 7.12 illustrates a pair of homologous chromosomes from an

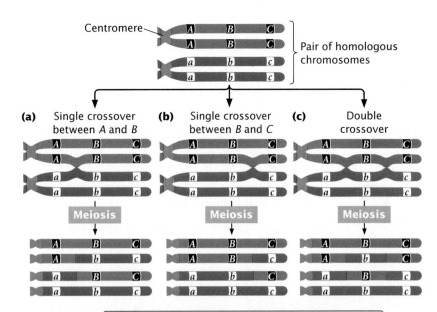

Conclusion: Recombinant chromosomes resulting from the double crossover have only the middle gene altered.

◀7.12 **Three types of crossovers can take place among three linked loci.**

individual that is heterozygous at three loci ($AaBbCc$). Notice that the genes are in the coupling configuration; that is, all the dominant alleles are on one chromosome ($\underline{A \qquad B \qquad C}$) and all the recessive alleles are on the other chromosome ($\underline{a \qquad b \qquad c}$). Three types of crossover events can take place between these three genes: two types of single crossovers (see Figure 7.12a and b) and a double crossover (see Figure 7.12c). In each type of crossover, two of the resulting chromosomes are recombinants and two are nonrecombinants.

Notice that, in the recombinant chromosomes resulting from the double crossover, the outer two alleles are the same as in the nonrecombinants, but the middle allele is different. This result provides us with an important clue about the order of the genes. In progeny that result from a double crossover, only the middle allele should differ from the alleles present in the nonrecombinant progeny.

Gene Mapping with the Three-Point Testcross

To examine gene mapping with a three-point testcross, we will consider three recessive mutations in the fruit fly *Drosophila melanogaster*. In this species, scarlet eyes (*st*) are recessive to red eyes (*st$^+$*), ebony body color (*e*) is recessive to gray body color (*e$^+$*), and spineless (*ss*)—that is, the presence of small bristles—is recessive to normal bristles (*ss$^+$*). All three mutations are linked and located on the third chromosome.

We will refer to these three loci as *st*, *e*, and *ss*, but keep in mind that either recessive alleles (*st*, *e*, and *ss*) or the dominant alleles (*st$^+$*, *e$^+$*, and *ss$^+$*) may be present at each locus. So, when we say that there are 10 m.u. between *st* and *ss*, we mean that there are 10 m.u. between the loci at which these mutations occur; we could just as easily say that there are 10 m.u. between *st$^+$* and *ss$^+$*.

To map these genes, we need to determine their order on the chromosome and the genetic distances between them. First, we must set up a three-point testcross, a cross between a fly heterozygous at all three loci and a fly homozygous for recessive alleles at all three loci. To produce flies heterozygous for all three loci, we might cross a stock of flies that are homozygous for normal alleles at all three loci with flies that are homozygous for recessive alleles at all three loci:

$$P \qquad \frac{st^+ \quad e^+ \quad ss^+}{st^+ \quad e^+ \quad ss^+} \qquad \times \qquad \frac{st \quad e \quad ss}{st \quad e \quad ss}$$

$$\downarrow$$

$$F_1 \qquad \frac{st^+ \quad e^+ \quad ss^+}{st \quad e \quad ss}$$

The order of the genes has been arbitrarily assigned because at this point we do not know which is the middle gene.

Additionally, the alleles in these heterozygotes are in coupling configuration (because all the wild-type dominant alleles were inherited from one parent and all the recessive mutations from the other parent), although the testcross can also be done with alleles in repulsion.

In the three-point testcross, we cross the F_1 heterozygotes with flies that are homozygous for all three recessive mutations. In many organisms, it makes no difference whether the heterozygous parent in the testcross is male or female (provided that the genes are autosomal) but, in *Drosophila*, no crossing over takes place in males. Because crossing over in the heterozygous parent is essential for determining recombination frequencies, the heterozygous flies in our testcross must be female. So we mate female F_1 flies that are heterozygous for all three traits with male flies that are homozygous for all the recessive traits:

$$\frac{st^+ \quad e^+ \quad ss^+}{st \quad e \quad ss} \text{ female} \quad \times \quad \frac{st \quad e \quad ss}{st \quad e \quad ss} \text{ male}$$

The progeny produced from this cross are listed in ◀FIGURE 7.13. For each locus, two classes of progeny are produced: progeny that are heterozygous, displaying the dominant trait, and progeny that are homozygous, displaying the recessive trait. With two classes of progeny possible for each of the three loci, there will be $2^3 = 8$ classes of phenotypes possible in the progeny. In this example, all eight phenotypic classes are present but, in some three-point crosses, one or more of the phenotypes may be missing if the number of progeny is limited. Nevertheless, the absence of a particular class can provide important information about which combination of traits is least frequent and ultimately the order of the genes, as we will see.

To map the genes, we need information about where and how often crossing over has occurred. In the homozygous recessive parent, the two alleles at each locus are the same; and so crossing over will have no effect on the types of gametes produced; with or without crossing over, all gametes from this parent have a chromosome with three recessive alleles ($\underline{st \qquad e \qquad ss}$). In contrast, the heterozygous parent has different alleles on its two chromosomes; so crossing over can be detected. The information that we need for mapping, therefore, comes entirely from the gametes produced by the heterozygous parent. Because chromosomes contributed by the homozygous parent carry only recessive alleles, whatever alleles are present on the chromosome contributed by the heterozygous parent will be expressed in the progeny.

As a shortcut, we usually do not write out the complete genotypes of the testcross progeny, listing instead only the alleles expressed in the phenotype (as shown in Figure 7.13), which are the alleles inherited from the heterozygous parent.

Concepts

To map genes, information about the location and number of crossovers in the gametes that produced the progeny of a cross is needed. An efficient way to obtain this information is to use a three-point testcross, in which an individual heterozygous at three linked loci is crossed with an individual that is homozygous recessive at the three loci.

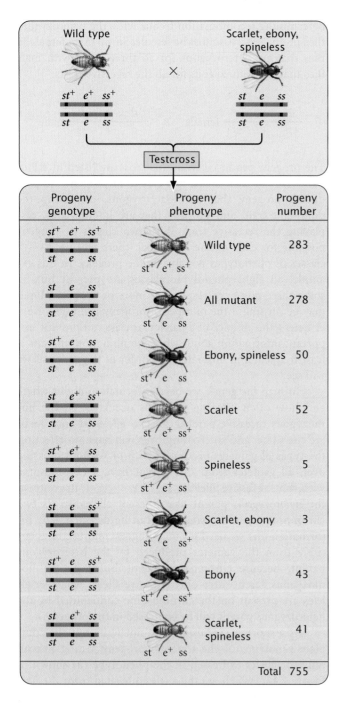

Determining the gene order The first task in mapping the genes is to determine their order on the chromosome. In Figure 7.13, we arbitrarily listed the loci in the order *st, e, ss,* but we had no way of knowing which of the three loci was between the other two. We can now identify the middle locus by examining the double-crossover progeny.

First, determine which progeny are the nonrecombinants—they will be the two most-numerous classes of progeny. (Even if crossing over takes place in every meiosis, the nonrecombinants will comprise at least 50% of the progeny.) Among the progeny of the testcross in Figure 7.13, the most numerous are those with all three dominant traits (st^+ e^+ ss^+) and those with all three recessive traits (st e ss).

Next, identify the double-crossover progeny. These should always be the two least-numerous phenotypes, because the probability of a double crossover is always less than the probability of a single crossover. The least-common progeny among those listed in Figure 7.13 are progeny with spineless bristles (st^+ e^+ ss) and progeny with scarlet eyes and ebony body (st e ss^+); so they are the double-crossover progeny.

Three orders of genes are possible: the eye-color locus could be in the middle (e st ss), the body-color locus could be in the middle (st e ss), or the bristle locus could be in the middle (st ss e). To determine which gene is in the middle, we can draw the chromosomes of the heterozygous parent with all three possible gene orders and then see if a double crossover produces the combination of genes observed in the double-crossover progeny. The three possible gene orders and the types of progeny produced by their double crossovers are:

Original chromosomes		**Chromosomes after crossing over**
e^+ st^+ ss^+	e^+ st^+ ss^+	e^+ st ss^+
1. ————————— → ✕✕ → —————————		
e st ss	e st ss	e st^+ ss
st^+ e^+ ss^+	st^+ e^+ ss^+	st^+ e ss^+
2. ————————— → ✕✕ → —————————		
st e ss	st e ss	st e^+ ss
st^+ ss^+ e^+	st^+ ss^+ e^+	st^+ ss e^+
3. ————————— → ✕✕ → —————————		
st ss e	st ss e	st ss^+ e

◀ **7.13 The results of a three-point testcross can be used to map linked genes.** In this three-point testcross in *Drosophila melanogaster,* the recessive mutations scarlet eyes (*st*), ebony body color (*e*), and spineless bristles (*ss*) are at three linked loci. The order of the loci has been designated arbitrarily, as has the sex of the progeny flies.

The only gene order that produces chromosomes with alleles for the traits observed in the double crossovers (st^+ e^+ ss and st e ss^+) is the third one, where the locus for bristle shape lies in the middle. Therefore, this order (st ss e) must be the correct sequence of genes on the chromosome.

With a little practice, it's possible to quickly determine which locus is in the middle without writing out all the gene orders. The phenotypes of the progeny are expressions of the alleles inherited from the heterozygous parent. Recall that, when we looked at the results of double crossovers (see Figure 7.13), only the alleles at the middle locus differed from the nonrecombinants. If we compare the nonrecombinant progeny with double-crossover progeny, they should differ only in alleles of the middle locus.

Let's compare the alleles in the double-crossover progeny st^+ e^+ ss with those in the nonrecombinant progeny st^+ e^+ ss^+. We see that both have an allele for red eyes (st^+) and both have an allele for gray body (e^+), but the nonrecombinants have an allele for normal bristles (ss^+), whereas the double crossovers have an allele for spineless bristles (ss). Because the bristle locus is the only one that differs, it must lie in the middle. We would obtain the same results if we compared the other class of double-crossover progeny (st e ss^+) with other nonrecombinant progeny (st e ss). Again the only trait that differs is the one for bristles. Don't forget that the nonrecombinants and the double crossovers should differ only at one locus; if they differ in two loci, the wrong classes of progeny are being compared.

Concepts

To determine the middle locus in a three-point cross, compare the double-crossover progeny with the nonrecombinant progeny. The double crossovers will be the two least-common classes of phenotypes; the nonrecombinants will be the two most-common classes of phenotypes. The double-crossover progeny should have the same alleles as the nonrecombinant types at two loci and different alleles at the locus in the middle.

Determining the locations of crossovers When we know the correct order of the loci on the chromosome, we should rewrite the phenotypes of the testcross progeny in Figure 7.13 with the loci in the correct order so that we can determine where crossovers have taken place (◀FIGURE 7.14).

Among the eight classes of progeny, we have already identified two classes as nonrecombinants (st^+ ss^+ e^+ and st ss e) and two classes as double crossovers (st^+ ss e^+ and st ss^+ e). The other four classes include progeny that resulted from a chromosome that underwent a single crossover: two underwent single

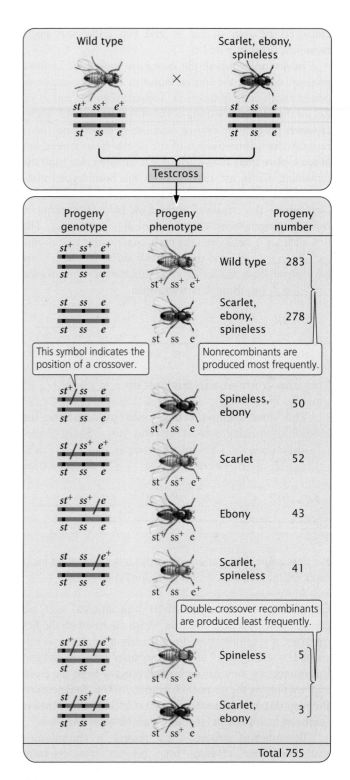

◀7.14 Writing the results of a three-point testcross with the loci in the correct order allows the locations of crossovers to be determined. These results are from the testcross illustrated in Figure 7.13, with the loci shown in the correct order. The location of a crossover is indicated with a slash (/). The sex of the progeny flies has been designated arbitrarily.

crossovers between *st* and *ss,* and two underwent single crossovers between *ss* and *e.*

To determine where the crossovers took place in these progeny, compare the alleles found in the single-crossover progeny with those found in the nonrecombinants, just as we did for the double crossovers. Some of the alleles in the single-crossover progeny are derived from one of the original (non-recombinant) chromosomes of the heterozygous parent, but at some place there is a switch (due to crossing over) and the remaining alleles are derived from the homologous non-recombinant chromosome. The position of the switch indicates where the crossover event took place. For example, consider progeny with chromosome $\underline{st^+ \quad ss \quad e}$. The first allele (st^+) came from the nonrecombinant chromosome $\underline{st^+ \quad ss^+ \quad e^+}$ and the other two alleles (*ss* and *e*) must have come from the other nonrecombinant chromosome $\underline{st \quad ss \quad e}$ through crossing over:

$$\underset{\underline{st \quad ss \quad e}}{\overline{st^+ \quad ss^+ \quad e^+}} \rightarrow \underset{\underline{st \quad ss \quad e}}{\overline{st^+ \quad ss^+ \quad e^+}} \times \rightarrow \underset{\underline{st \quad ss^+ \quad e^+}}{\overline{st^+ \quad ss \quad e}}$$

This same crossover also produces the $\underline{st \quad ss^+ \quad e^+}$ progeny.

This same method can be used to determine the location of crossing over in the other two types of single-crossover progeny. Crossing between *ss* and *e* produces $\underline{st^+ \quad ss^+ \quad e}$ and $\underline{st \quad ss \quad e^+}$ chromosomes:

$$\underset{\underline{st \quad ss \quad e}}{\overline{st^+ \quad ss^+ \quad e^+}} \rightarrow \underset{\underline{st \quad ss \quad e}}{\overline{st^+ \quad ss^+ \quad e^+}} \times \rightarrow \underset{\underline{st \quad ss \quad e^+}}{\overline{st^+ \quad ss^+ \quad e}}$$

We now know the locations of all the crossovers; their locations are marked with a slash in Figure 7.14.

Calculating the recombination frequencies

Next, we can determine the map distances, which are based on the frequencies of recombination. Recombination frequency is calculated by adding up all of the recombinant progeny, dividing this number by the total number of progeny from the cross, and multiplying the number obtained by 100%. To determine the map distances accurately, we must include all crossovers (both single and double) that take place between two genes.

Recombinant progeny that possess a chromosome that underwent crossing over between the eye-color locus (*st*) and the bristle locus (*ss*) include the single crossovers ($\underline{st^+ \quad / \quad ss \quad e}$ and $\underline{st \quad / \quad ss^+ \quad e^+}$) and the two double crossovers ($\underline{st^+ \quad / \quad ss \quad / \quad e^+}$ and $\underline{st \quad / \quad ss^+ \quad / \quad e}$); see Figure 7.14. There are a total of 755 progeny; so the recombination frequency between *ss* and *st* is:

$$st\text{–}ss \text{ recombination frequency} = \frac{(50 + 52 + 5 + 3)}{755} \times 100\% = 14.6\%$$

The distance between the *st* and *ss* loci can be expressed as 14.6 m.u.

The map distance between the bristle locus (*ss*) and the body locus (*e*) is determined in the same manner. The recombinant progeny that possess a crossover between *ss* and *e* are the single crossovers $\underline{st^+ \quad ss^+ \quad / \quad e}$ and $\underline{st \quad ss \quad / \quad e^+}$, and the double crossovers $\underline{st^+ \quad / \quad ss \quad / \quad e^+}$ and $\underline{st \quad / \quad ss^+ \quad / \quad e}$. The recombination frequency is:

$$ss\text{–}e \text{ recombination frequency} =$$
$$\frac{(43 + 41 + 5 + 3)}{755} \times 100\% = 12.2\%$$

Thus, the map distance between *ss* and *e* is 12.2 m.u.

Finally, calculate the map distance between the outer two loci, *st* and *e*. This map distance can be obtained by summing the map distances between *st* and *ss* and between *ss* and *e* (14.6 m.u. + 12.2 m.u. = 26.8 m.u.). Alternatively, it can be calculated by adding up all the progeny with crossovers between the two loci. These progeny include those with a single crossover between *st* and *ss*, those with a single crossover between *ss* and *e*, and the double crossovers ($\underline{st^+ \quad / \quad ss \quad / \quad e^+}$ and $\underline{st \quad / \quad ss^+ \quad / \quad e}$). Because the double crossovers have two crossovers between *st* and *e*, we must add the double crossovers *twice*:

$$st\text{–}e \text{ map distance} =$$
$$\frac{(50 + 52 + 43 + 41 + (2 \times 5) + (2 \times 3)}{755} \times 100\% = 26.8\%$$

Notice that the distances between *st* and *ss* (14.6 m.u.) and between *ss* and *e* (12.2 m.u.) add up to the distance between *st* and *e* (26.8 m.u.). We can now use the map distances to draw a map of the three genes on the chromosome:

A genetic map of *D. melanogaster* is illustrated in ◀ FIGURE 7.15.

Interference and coefficient of coincidence

Map distances give us information not only about the physical distances that separate genes, but also about the proportions of recombinant and nonrecombinant gametes that will be produced in a cross. For example, knowing that genes *st* and *ss* on the third chromosome of *D. melanogaster* are separated by a distance of 14.6 m.u. tells us that 14.6% of the gametes produced by a fly heterozygous at these two loci will be recombinants. Similarly, 12.2% of the gametes from a fly heterozygous for *ss* and *e* will be recombinants.

Theoretically, we should be able to calculate the proportion of double-recombinant gametes by using the

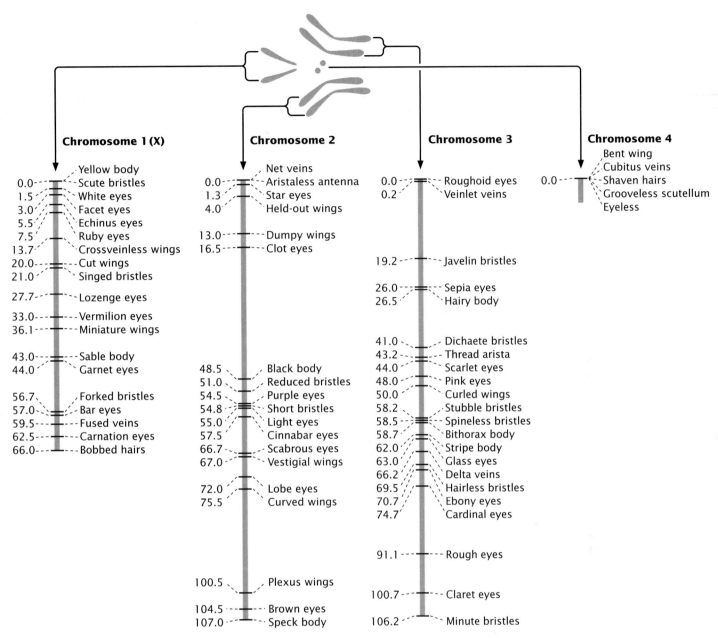

◀7.15 *Drosophila melanogaster* has four linkage groups corresponding to its four pairs of chromosomes. Distances between genes within a linkage group are in map distances.

multiplication rule of probability (Chapter 3), which states that the probability of two independent events occurring together is the multiplication of their independent probabilities. Applying this principle, we should find that the proportion (probability) of gametes with double crossovers between *st* and *e* is equal to the probability of recombination between *st* and *ss*, multiplied by the probability of recombination between *ss* and *e*, or 0.146 × 0.122 = 0.0178. Multiplying this probability by the total number of progeny gives us the *expected* number of double-crossover progeny from the cross: 0.0178 × 755 = 13.4. Only 8 dou-

ble crossovers—considerably fewer than the 13 expected—were observed in the progeny of the cross (see Figure 7.13).

This phenomenon is common in eukaryotic organisms. The calculation assumes that each crossover event is independent and that the occurrence of one crossover does not influence the occurrence of another. But crossovers are frequently *not* independent events: the occurrence of one tends to inhibit additional crossovers in the same region of the chromosome, and so double crossovers are less frequent than expected.

The degree to which one crossover interferes with additional crossovers in the same region is termed the

interference. To calculate the interference, we first determine the **coefficient of coincidence,** which is the ratio of observed double crossovers to expected double crossovers:

coefficient of coincidence =
$$\frac{\text{number of observed double crossovers}}{\text{number of expected double crossovers}}$$

For the loci that we mapped on the third chromosome of *D. melanogaster* (see Figure 7.14), we find that:

coefficient of coincidence =
$$\frac{5+3}{0.146 \times 0.122 \times 755} = \frac{8}{13.4} = 0.6$$

which indicates that we are actually observing only 60% of the double crossovers that we expected on the basis of the single-crossover frequencies. The interference is calculated as:

$$\text{interference} = 1 - \text{coefficient of coincidence}$$

So the interference for our three-point cross is:

$$\text{interference} = 1 - 0.6 = 0.4$$

This value of interference tells us that 40% of the double-crossover progeny expected will not be observed because of interference. When interference is complete and no double-crossover progeny are observed, the coefficient of coincidence is 0 and the interference is 1.

Sometimes *more* double-crossover progeny appear than expected, which happens when a crossover increases the probability of another crossover occurring nearby. In this case, the coefficient of coincidence is greater than 1 and the interference will be negative.

Concepts

The coefficient of coincidence equals the number of double crossovers observed, divided by the number of double crossovers expected on the basis of the single-crossover frequencies. The interference equals 1 − the coefficient of coincidence; it indicates the degree to which one crossover interferes with additional crossovers.

Connecting Concepts

Stepping Through the Three-Point Cross

We have now examined the three-point cross in considerable detail, seeing how the information derived from the cross can be used to map a series of three linked genes. Let's briefly review the steps required to map genes from a three-point cross.

1. Write out the phenotypes and numbers of progeny produced in the three-point cross. The progeny phenotypes will be easier to interpret if you use allelic symbols for the traits (such as $st^+ e^+ ss$).

2. Write out the genotypes of the original parents used to produce the triply heterozygous individual in the testcross and, if known, the arrangement of the alleles on their chromosomes (coupling or repulsion).

3. Determine which phenotypic classes among the progeny are the nonrecombinants and which are the double crossovers. The nonrecombinants will be the two most-common phenotypes; the double crossovers will be the two least-common phenotypes.

4. Determine which locus lies in the middle. Compare the alleles present in the double crossovers with those present in the nonrecombinants; each class of double crossovers should be like one of the nonrecombinants for two loci and should differ for one locus. The locus that differs is the middle one.

5. Rewrite the phenotypes with genes in correct order.

6. Determine where crossovers must have taken place to give rise to the progeny phenotypes by comparing each phenotype with the phenotype of the nonrecombinant progeny.

7. Determine the recombination frequencies. Add the numbers of the progeny that possess a chromosome with a crossover between a pair of loci. Add the double crossovers to this number. Divide this sum by the total number of progeny from the cross, and multiply by 100%; the result is the recombination frequency between the loci, which is the same as the map distance.

8. Draw a map of the three loci, indicating which locus lies in the middle, and label the distances between them.

9. Determine the coefficient of coincidence and the interference. The coefficient of coincidence is the number of observed double-crossover progeny divided by the number of expected double-crossover progeny. The expected number can be obtained by multiplying the product of the two single-recombination probabilities by the total number of progeny in the cross.

Worked Problem

In *D. melanogaster*, cherub wings (*ch*), black body (*b*), and cinnabar eyes (*cn*) result from recessive alleles that are all located on chromosome 2. A homozygous wild-type fly was mated with a cherub, black, and cinnabar fly, and the resulting F_1 females were test-crossed with cherub, black, and cinnabar males. The following progeny were produced from the testcross:

ch	*b*⁺	*cn*	105
ch⁺	*b*⁺	*cn*⁺	750
ch⁺	*b*	*cn*	40
ch⁺	*b*⁺	*cn*	4
ch	*b*	*cn*	753
ch	*b*⁺	*cn*⁺	41
ch⁺	*b*	*cn*⁺	102
ch	*b*	*cn*⁺	5
total			1800

(a) Determine the linear order of the genes on the chromosome (which gene is in the middle).

(b) Calculate the recombinant distances between the three loci.

(c) Determine the coefficient of coincidence and the interference for these three loci.

• Solution

(a) We can represent the crosses in this problem as follows:

P $\quad \dfrac{ch^+ \quad b^+ \quad cn^+}{ch^+ \quad b^+ \quad cn^+} \times \dfrac{ch \quad b \quad cn}{ch \quad b \quad cn}$

$\downarrow$

F₁ $\quad \dfrac{ch^+ \quad b^+ \quad cn^+}{ch \quad b \quad cn}$

Testcross $\quad \dfrac{ch^+ \quad b^+ \quad cn^+}{ch \quad b \quad cn} \times \dfrac{ch \quad b \quad cn}{ch \quad b \quad cn}$

Note that we do not know, at this point, the order of the genes; we have arbitrarily put *b* in the middle.

The next step is to determine which of the testcross progeny are nonrecombinants and which are double crossovers. The nonrecombinants should be the most-frequent phenotype; so they must be the progeny with phenotypes encoded by *ch*⁺   *b*⁺   *cn*⁺ and *ch   b   cn* . These genotypes are consistent with the genotypes of the parents, which we outlined earlier. The double crossovers are the least-frequent phenotypes and are encoded by *ch*⁺   *b*⁺   *cn* and *ch   b   cn*⁺.

We can determine the gene order by comparing the alleles present in the double crossovers with those present in the nonrecombinants. The double-crossover progeny should be like one of the nonrecombinants at two loci and unlike it at one; the allele that differs should be in the middle. Compare the double-crossover progeny *ch   b   cn*⁺ with the nonrecombinant *ch   b   cn* . Both have cherub wings (*ch*) and black body (*b*), but the double-crossover progeny have wild-type eyes (*cn*⁺), whereas the nonrecombinants have cinnabar eyes (*cn*). The locus that determines cinnabar eyes must be in the middle.

(b) To calculate the recombination frequencies among the genes, we first write the phenotypes of the progeny with the genes encoding them in the correct order. We have already identified the nonrecombinant and double-crossover progeny; so the other four progeny types must have resulted from single crossovers. To determine *where* single crossovers took place, we compare the alleles found in the single-crossover progeny with those in the nonrecombinants. Crossing over must have taken place where the alleles switch from those found in one nonrecombinant to those found in the other nonrecombinant. The locations of the crossovers are indicated with a slash:

ch	*cn* /	*b*⁺	105	single crossover
ch⁺	*cn*⁺	*b*⁺	750	nonrecombinant
ch⁺ /	*cn*	*b*	40	single crossover
ch⁺ /	*cn* /	*b*⁺	4	double crossover
ch	*cn*	*b*	753	nonrecombinant
ch /	*cn*⁺	*b*⁺	41	single crossover
ch⁺	*cn*⁺ /	*b*	102	single crossover
ch /	*cn*⁺ /	*b*	5	double crossover
total			1800	

Next, we determine the recombination frequencies and draw a genetic map:

ch–cn recombination frequency = $\dfrac{40 + 4 + 41 + 5}{1800} \times 100\% = 5\%$

cn–b recombination frequency = $\dfrac{105 + 4 + 102 + 5}{1800} \times 100\% = 12\%$

ch–b map distance = $\dfrac{105 + 40 + (2 \times 4) + 41 + 102 + (2 \times 5)}{1800} \times 100\% = 17\%$

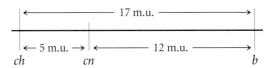

(c) The coefficient of coincidence is the number of observed double crossovers, divided by the number of expected double crossovers. The number of expected double crossovers is obtained by multiplying the probability of a crossover between *ch* and *cn* (0.05) × the probability of a crossover between *cn* and *b* (0.12) × the total number of progeny in the cross (1800):

coefficient of coincidence = $\dfrac{4 + 5}{0.05 \times 0.12 \times 1800} = 0.83$

Finally, the interference is equal to 1 − the coefficient of coincidence:

interference = 1 − 0.83 = 0.17

Gene Mapping in Humans

Efforts in mapping the human genome are hampered by the inability to perform desired crosses and the small number of progeny in most human families. Geneticists are restricted to analyses of pedigrees, which are often incomplete and provide limited information. Nevertheless, techniques have been developed that use pedigree data to analyze linkage, and a large number of human traits have been successfully mapped with the use of these methods. Because the number of progeny from any one mating is usually small, data from several families and pedigrees are usually combined to test for independent assortment. The methods used in these types of analysis are beyond the scope of this book, but an example will illustrate how linkage can be detected from pedigree data.

One of the first documented demonstrations of linkage in humans was between the locus for nail–patella syndrome and the locus that determines the ABO blood types. Nail–patella syndrome is an autosomal dominant disorder characterized by abnormal fingernails and absent or rudimentary kneecaps. The ABO blood types are determined by an autosomal locus with multiple alleles (Chapter 5). Linkage between the genes encoding these traits was established in families in which both traits segregate. Part of one such family is illustrated in ◀ FIGURE 7.16.

Nail–patella syndrome is relatively rare; so we can assume that people having this trait are heterozygous (*Nn*); unaffected people are homozygous (*nn*). The ABO genotypes can be inferred from the phenotypes and the types of offspring produced. Person I-2 in Figure 7.16, for example,

has blood type B, which has two possible genotypes: $I^B I^B$ or $I^B i$ (see Figure 5.5). Because some of her offspring are blood type O (genotype *ii*) and must therefore have inherited an *i* allele from each parent, female I-2 must have genotype $I^B i$. Similarly, the presence of blood type O offspring in generation II indicates that male I-1, with blood type A, also must carry an *i* allele and therefore has genotype $I^A i$. The ABO and nail–patella genotypes for all persons in the pedigree are given below the squares and circles.

From generation II, we can see that the genes for nail–patella syndrome and the blood types do not appear to assort independently. The parents of this family are:

$$I^A i \, Nn \times I^B i \, nn$$

If the genes coding for nail–patella syndrome and the ABO blood types assorted independently, we would expect that some children in generation II would have blood type A and nail–patella syndrome, inheriting both the I^A and *N* genes from their father. However, all children in generation II with nail–patella syndrome have either blood type B or blood type O; all those with blood type A have normal nails and kneecaps. This outcome indicates that the arrangements of the alleles on the chromosomes of the crossed parents are:

$$\frac{I^A \qquad n}{i \qquad N} \times \frac{I^B \qquad n}{i \qquad n}$$

There is no recombination among the offspring of these parents (generation II), but there are two instances of

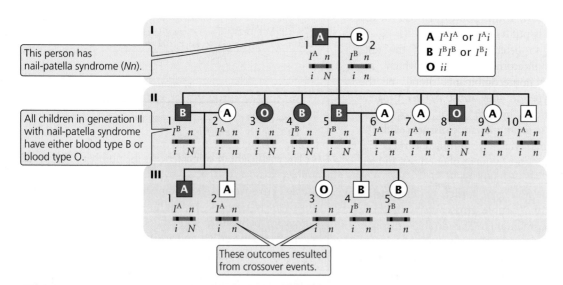

◀ 7.16 **Linkage between ABO blood types and nail–patella syndrome was established by examining families in whom both traits segregate.** The pedigree shown here is for one such family. Solid circles and squares represent the presence of nail–patella syndrome; the ABO blood type is indicated in each circle or square. The genotype, inferred from phenotype, is given below each square or circle.

recombination among the persons in generation III. Individuals II-1 and II-2 have the following genotypes:

$$\frac{I^B \qquad n}{i \qquad N} \times \frac{I^A \qquad n}{i \qquad n}$$

Their child III-2 has blood type A and does not have nail–patella syndrome; so he must have genotype:

$$\frac{I^A \qquad n}{i \qquad n}$$

and must have inherited both the *i* and the *n* alleles from his father. These alleles are on different chromosomes in the father; so crossing over must have taken place. Crossing over also must have taken place to produce child III-3.

In the pedigree of Figure 7.16, there are 13 children from matings in which the genes encoding nail–patella syndrome and ABO blood types segregate; 2 of them are recombinants. On this basis, we might assume that the loci for nail–patella syndrome and ABO blood types are linked, with a recombination frequency of $^2/_{13} = 0.154$. However, it is possible that the genes *are* assorting independently and that the small number of children just makes it seem as though the genes are linked. To determine the probability that genes are actually linked, geneticists often calculate **lod** (logarithm of odds) **scores.**

To obtain a lod score, one calculates both the probability of obtaining the observations with a specified degree of linkage and the probability of obtaining the observations with independent assortment. One then determines the ratio of these two probabilities, and the logarithm of this ratio is the lod score. Suppose that the probability of obtaining a particular set of observations with linkage and a certain recombination frequency is 0.1 and that the probability of obtaining the same observations with independent assortment is 0.0001. The ratio of these two probabilities is $^{0.1}/_{0.0001} = 1000$, the logarithm of which (the lod score) is 3. Thus linkage with the specified recombination is 1000 times as likely to produce what was observed as independent assortment. A lod score of 3 or higher is usually considered convincing evidence for linkage.

Mapping with Molecular Markers

For many years, gene mapping was limited in most organisms by the availability of **genetic markers,** variable genes with easily observable phenotypes whose inheritance could be studied. Traditional genetic markers include genes that encode easily observable characteristics such as flower color, seed shape, blood types, and biochemical differences. The paucity of these types of characteristics in many organisms limited mapping efforts.

In the 1980s, new molecular techniques made it possible to examine variations in DNA itself, providing an almost unlimited number of genetic markers that can be used for creating genetic maps and studying linkage relations. The earliest of these molecular markers consisted of restriction fragment length polymorphisms (RFLPs), variations in DNA sequence detected by cutting the DNA with restriction enzymes (see Chapter 18). Later, methods were developed for detecting variable numbers of short DNA sequences repeated in tandem, called variable number of tandem repeats (VNTRs). More recently, DNA sequencing allows the direct detection of individual variations in the DNA nucleotides, called single nucleotide polymorphisms (SNPs; see Chapter 19). All of these methods have expanded the availability of genetic markers and greatly facilitated the creation of genetic maps.

Gene mapping with molecular markers is done essentially in the same manner as mapping performed with traditional phenotypic markers: the cosegregation of two or more markers is studied and map distances are based on the rates of recombination between markers. These methods and their use in mapping are presented in more detail in Chapters 18 and 19.

Physical Chromosome Mapping

Genetic maps reveal the relative positions of genes on a chromosome on the basis of frequencies of crossing over, but they do not provide information that can allow us to place groups of linked genes on particular chromosomes. Furthermore, the units of a genetic map do not always precisely correspond to physical distances on the chromosome, because a number of factors other than physical distances between genes (such as the type and sex of the organism) can influence rates of crossing over. Because of these limitations, physical-mapping methods that do not rely on rates of crossing over have been developed.

Deletion Mapping

One method for determining the chromosomal location of a gene is **deletion mapping.** Special staining methods have been developed that make it possible to detect chromosome deletions, mutations in which a part of a chromosome is missing. Genes are assigned to regions of particular chromosomes by studying the association of a gene's phenotype or product and particular chromosome deletions.

In deletion mapping, an individual that is homozygous for a recessive mutation in the gene of interest is crossed with an individual that is heterozygous for a deletion (◀ FIGURE 7.17). If the gene of interest is in the region of the chromosome represented by the deletion (the red part of chromosome in Figure 7.17), approximately half of the progeny will display the mutant phenotype (see Figure 7.17a). If the gene is not within the deleted region, all of the progeny will be wild type (see Figure 7.17b).

Deletion mapping has been used to reveal the chromosomal locations of a number of human genes. For example,

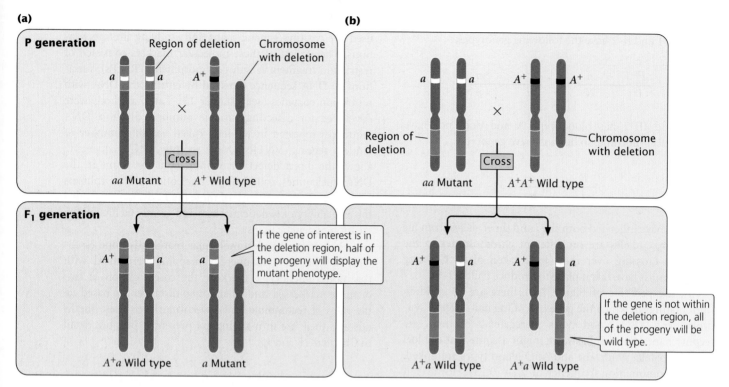

(a)

P generation

Region of deletion   Chromosome with deletion

a   *a*   *A⁺*

×

Cross

aa Mutant   *A⁺* Wild type

F₁ generation

A⁺   *a*   *a*

If the gene of interest is in the deletion region, half of the progeny will display the mutant phenotype.

A⁺a Wild type   *a* Mutant

(b)

a   *a*   *A⁺*   *A⁺*

×

Region of deletion   Chromosome with deletion

Cross

aa Mutant   *A⁺A⁺* Wild type

A⁺   *a*   *A⁺*   *a*

If the gene is not within the deletion region, all of the progeny will be wild type.

A⁺a Wild type   *A⁺a* Wild type

◀ 7.17 **Deletion mapping can be used to determine the chromosomal location of a gene.** An individual homozygous for a recessive mutation in the gene of interest (*aa*) is crossed with an individual heterozygous for a deletion.

Duchenne muscular dystrophy is a disease that causes progressive weakening and degeneration of the muscles. From its X-linked pattern of inheritance, the mutated allele causing this disorder was known to be on the X chromosome, but its precise location was uncertain. Examination of a number of patients having Duchenne muscular dystrophy, who also possessed small deletions, allowed researchers to position the gene to a small segment of the short arm of the X chromosome.

Somatic-Cell Hybridization

Another method used for positioning genes on chromosomes is **somatic cell hybridization,** which requires the fusion of different types of cells. Most mature somatic (nonsex) cells can undergo only a limited number of divisions and therefore cannot be grown continuously. However, cells that have been altered by viruses or derived from tumors that have lost the normal constraints on cell division will divide indefinitely; these types of cells can be cultured in the laboratory and are referred to as a **cell line.**

Cells from two different cell lines can be fused by treating them with polyethylene glycol or other agents that alter their plasma membranes. After fusion, the cell possesses two nuclei and is called a **heterokaryon.** The two nuclei of a heterokaryon eventually also fuse, generating a hybrid cell that contains chromosomes from both cell lines. If human

and mouse cells are mixed in the presence of polyethylene glycol, fusion results in human–mouse somatic-cell hybrids (◀ FIGURE 7.18). The hybrid cells tend to lose chromosomes as they divide and, for reasons that are not understood, chromosomes from one of the species are lost preferentially. In human–mouse somatic-cell hybrids, the human chromosomes tend to be lost, whereas the mouse chromosomes are retained. Eventually, the chromosome number stabilizes when all but a few of the human chromosomes have been lost. Chromosome loss is random and differs among cell lines. The presence of these "extra" human chromosomes in the mouse genome makes it possible to assign human genes to specific chromosomes.

In the first step of this procedure, hybrid cells must be separated from original parental cells that have not undergone hybridization. This separation is accomplished by using a selection method that allows hybrid cells to grow while suppressing the growth of parental cells. The most commonly used method is called HAT selection (◀ FIGURE 7.19), which stands for *h*ypoxanthine, *a*minopterin, and *t*hymidine, three chemicals that are used to select for hybrid cells. In the presence of HAT medium, a cell must possess two enzymes to synthesize DNA: thymidine kinase (TK) and hypoxanthine-guanine phosphoribosyl transferase (HPRT). Cells that are *tk⁻* or *hprt⁻* cannot synthesize DNA and will not grow on HAT medium. The mouse cells used in the hy-

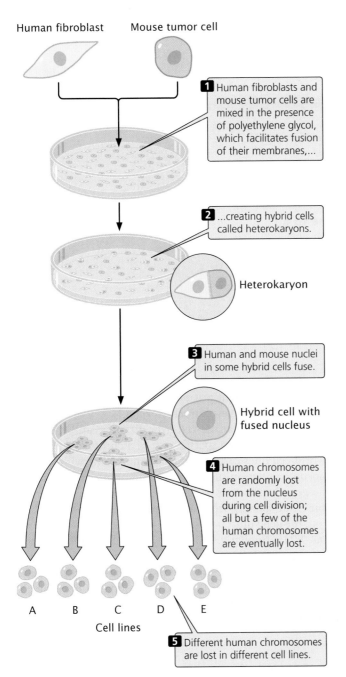

1 Human fibroblasts and mouse tumor cells are mixed in the presence of polyethylene glycol, which facilitates fusion of their membranes,...

2 ...creating hybrid cells called heterokaryons.

Heterokaryon

3 Human and mouse nuclei in some hybrid cells fuse.

Hybrid cell with fused nucleus

4 Human chromosomes are randomly lost from the nucleus during cell division; all but a few of the human chromosomes are eventually lost.

A   B   C   D   E
Cell lines

5 Different human chromosomes are lost in different cell lines.

◀ **7.18 Somatic-cell hybridization can be used to determine which chromosome contains a gene of interest.**

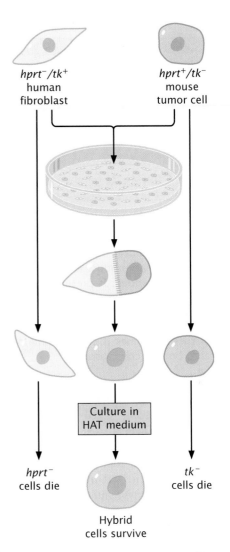

$hprt^-/tk^+$ human fibroblast

$hprt^+/tk^-$ mouse tumor cell

Culture in HAT medium

$hprt^-$ cells die

tk^- cells die

Hybrid cells survive

◀ **7.19 HAT medium can be used to separate human–mouse hybrid cells from the original hybridized cells.**

bridization procedure are deficient in TK, but can produce HPRT (the cells are $tk^-\ hprt^+$); the human cells can produce TK but are deficient for HPRT (they are $tk^+\ hprt^-$). On HAT medium, the mouse cells do not survive, because they are tk^-; the human cells do not survive, because they are $hprt^-$. Hybrid cells, on the other hand, inherit the ability to make HPRT from the mouse cell and the ability to make TK from the human cell; thus, they produce both enzymes (the cells are $tk^+\ hprt^+$) and will grow on HAT medium.

To map genes using somatic-cell hybridization requires the use of a panel of different hybrid cell lines. The cell lines of the panel differ in the human chromosomes that they have retained. For example, one cell line might possess human chromosomes 2, 4, 7, and 8, whereas another might possess chromosomes 4, 19, and 20. Each cell line in the panel is examined for evidence of a particular human gene. The human gene can be detected either by looking for the protein that it produces or by looking for the gene itself with the use of molecular probes (discussed in Chapter 18). Correlation of the presence of the gene with the presence of specific human chromosomes often allows the gene to be assigned to the correct chromosome. For example, if a gene was detected in both of the aforementioned cell lines, the gene must be on chromosome 4, because it is the only human chromosome common to both cell lines (◀ FIGURE 7.20).

Human chromosomes present

Cell line	Gene product present	1	2	3	4	5	6	7	8	9	10	11	12	13	14	15	16	17	18	19	20	21	22	X
A	+		+		+			+	+															
B	+		+	+		+				+	+	+	+	+	+									
C	−															+		+		+			+	
D	+				+		+	+	+															
E	−												+								+			
F	+				+															+	+		

7.20 **Somatic-cell hybridization is used to assign a gene to a particular human chromosome.** A panel of six cell lines, each line containing a different subset of human chromosomes, is examined for the presence of the gene product (such as an enzyme). A plus sign means that the gene product is present; a minus sign means that the gene product is missing. Four of the cell lines (A, B, D, and F) have the gene product, indicating that the gene is present on one of the chromosomes found in these cell lines. The only chromosome common to all four of these cell lines is chromosome 4, indicating that the gene is located on this chromosome.

Two genes determined to be on the same chromosome with the use of somatic-cell hybridization are said to be **syntenic genes.** This term is used because syntenic genes may or may not exhibit linkage in the traditional genetic sense—remember that two genes can be located on the same chromosome but may be so far apart that they assort independently. Syntenic refers to genes that are physically linked, regardless of whether they exhibit genetic linkage. (Synteny is sometimes also used to refer to gene loci in different organisms located on a chromosome region of common evolutionary origin.)

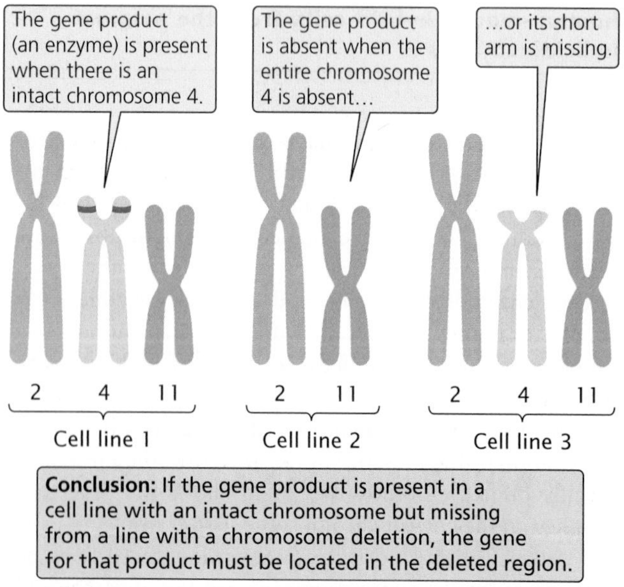

Sometimes somatic-cell hybridization can be used to position a gene on a specific part of a chromosome. Some hybrid cell lines carry a human chromosome with a chromosome mutation such as a deletion or a translocation. If the gene is present in a cell line with the intact chromosome but missing from a line with a chromosome deletion, the gene must be located in the deleted region (FIG-URE 7.21). Similarly, if a gene is usually absent from a chromosome but consistently appears whenever a translocation (a piece of another chromosome that has broken off and attached itself to the chromosome in question) is present, it must be present on the translocated part of the chromosome.

In Situ Hybridization

Described in more detail in Chapter 18, in situ hybridization is another method for determining the chromosomal location of a particular gene. This method requires a DNA copy of the gene or its RNA product, which is used to make a molecule (called a probe) that is complementary to the gene of interest. The probe is made radioactive or is attached to a special molecule that fluoresces under ultraviolet (UV) light and is added to chromosomes from specially treated cells that have been spread on a microscope slide. The probe binds to the complementary DNA sequence of the gene on the chromosome. The presence of radioactivity or fluorescence from the bound probe reveals the location of the gene on a particular chromosome (FIGURE 7.22a). The use of fluorescence in situ hybridiza-

The gene product (an enzyme) is present when there is an intact chromosome 4.

The gene product is absent when the entire chromosome 4 is absent…

…or its short arm is missing.

2	4	11

Cell line 1

2	11

Cell line 2

2	4	11

Cell line 3

Conclusion: If the gene product is present in a cell line with an intact chromosome but missing from a line with a chromosome deletion, the gene for that product must be located in the deleted region.

7.21 **Genes can be localized to a specific part of a chromosome by using somatic-cell hybridization.**

(a)

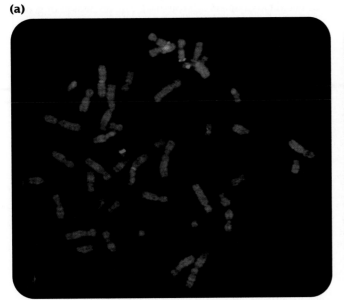

(b)

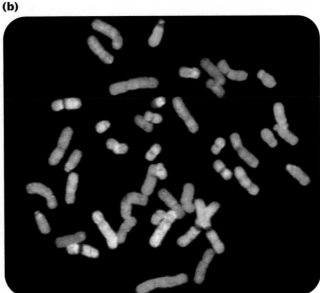

◀ 7.22 **In situ hybridization is another technique for determining the chromosomal location of a gene.** (a) FISH technique. The red fluorescence is produced by a probe for sequences on chromosome 9; the green fluorescence is produced by a probe for sequences on chromosome 22. (b) SKY technique: 24 different probes, each specific for a different human chromosome and producing a different color, identify the different human chromosomes. (Courtesy of Dr. Hesed Padilla-Nash and Dr. Thomas Ried, NIH.)

tion (FISH) has been widely used to identify the chromosomal location of human genes. In spectral karyotyping (SKY) (◀ FIGURE 7.22b), a set of 24 FISH probes, each specific to a different human chromosome and attached to a molecule that fluoresces a different color, allows each chromosome in a karyotype to be identified.

<u>www.whfreeman.com/pierce</u>   More on fluorescence in situ hybridization (FISH)

Mapping by DNA Sequencing

Another means of physically mapping genes is to determine the sequence of nucleotides in the DNA (DNA sequencing, Chapter 19). With this technique, physical distances between genes are measured in numbers of base pairs. Continuous sequences can be determined for only relatively small fragments of DNA; so, after sequencing, some method is still required to map the individual fragments. This mapping is often done by using the traditional gene mapping that examines rates of crossing over between molecular markers located on the fragments. It can also be accomplished by generating a set of overlapping fragments, sequencing each fragment, and then aligning the fragments by using a computer program that identifies the overlap in the sequence of adjacent fragments. With these methods, complete physical maps of entire genomes have been produced (Chapter 19).

Concepts

Physical-mapping methods determine the physical locations of genes on chromosomes and include deletion mapping, somatic-cell hybridization, in situ hybridization, and direct DNA sequencing.

Connecting Concepts Across Chapters

The principle of independent assortment states that alleles at different loci assort (separate) independently in meiosis but only if the genes are located on different chromosomes or are far apart on the same chromosome. This chapter has focused on the inheritance of genes that are physically linked on the same chromosome and do not assort independently. To predict the outcome of crosses entailing linked genes, we must consider not only the genotypes of the parents but also the physical arrangement of the alleles on the chromosomes.

An important principle learned in this chapter is that rates of recombination are related to the physical distances between genes. Crossing over is more frequent between genes that are far apart than between genes that lie close together. This fact provides the foundation for gene mapping in eukaryotic organisms: recombination frequencies are used to determine the relative order and distances between linked genes. Gene mapping therefore requires the setting up of crosses in which recombinant progeny can be detected.

This chapter also examined several methods of physical mapping that do not rely on recombination rates but use methods to directly observe the association between genes and particular chromosomes or to position genes by determining the nucleotide sequences. Although genetic and physical distances are correlated, they are not identical, because factors other than the distances between genes can influence rates of crossing over.

Gene mapping requires a firm understanding of the behavior of chromosomes (Chapter 2) and basic principles of heredity (Chapters 3 through 5). The discussion of gene mapping with pedigrees assumes knowledge of how families are displayed in pedigrees (Chapter 6). In Chapter 8, we will consider specialized mapping techniques used in bacteria and viruses; in Chapter 18, techniques for detecting molecular markers used in gene mapping are examined in more detail. Techniques for mapping whole genomes are discussed in Chapter 19. Chromosome mutations that play a role in deletion mapping and somatic-cell hybridization are considered in more detail in Chapter 9.

CONCEPTS SUMMARY

- Soon after Mendel's principles were rediscovered, examples of genes that did not assort independently were discovered. These genes were subsequently shown to be linked on the same chromosome.

- In a testcross for two completely linked genes (which exhibit no crossing over), only nonrecombinant progeny containing the original combinations of alleles present in the parents are produced. When two genes assort independently, recombinant progeny and nonrecombinant progeny are produced in equal proportions. When two genes are linked with some crossing over between them, more nonrecombinant progeny than recombinant progeny are produced.

- Because a single crossover between two linked genes produces two recombinant gametes and two nonrecombinant gametes, crossing over and independent assortment produce the same results.

- Recombination frequency is calculated by summing the number of recombinant progeny, dividing by the total number of progeny produced in the cross, and multiplying by 100%.

- The recombination frequency is half the frequency of crossing over, and the maximum frequency of recombinant gametes is 50%.

- When two wild-type alleles are found on one homologous chromosome and their mutant alleles are found on the other chromosome, the genes are said to be in coupling configuration. When one wild-type allele and one mutant allele are found on each homologous chromosome, the genes are said to be in repulsion. Whether genes are in coupling configuration or in repulsion determines which combination of phenotypes will be most frequent in the progeny of a testcross.

- Linkage and crossing over are two opposing forces: linkage keeps alleles at different loci together, whereas crossing over

breaks up linkage and allows alleles to recombine into new associations.

- Interchromosomal recombination takes place among genes located on different chromosomes and occurs through the random segregation of chromosomes in meiosis. Intrachromosomal recombination takes place among genes located on the same chromosome and occurs through crossing over.

- Testing for independent assortment between genes requires a series of chi-square tests, in which segregation is first tested at each locus individually, followed by testing for independent assortment among genes at the different loci.

- Recombination rates can be used to determine the relative order of genes and distances between them on a chromosome. Maps based on recombination rates are called genetic maps; maps based on physical distances are called physical maps.

- One percent recombination equals one map unit, which is also a centiMorgan.

- When genes exhibit 50% recombination, they belong to different linkage groups, which may be either on different chromosomes or far apart on the same chromosome.

- Recombination rates between two genes will underestimate the true distance between them because double crossovers cannot be detected.

- Genetic maps can be constructed by examining recombination rates from a series of two-point crosses or by examining the progeny of a three-point testcross.

- Gene mapping in humans can be accomplished by examining the cosegregation of traits in pedigrees, although the inability to control crosses and the small number of progeny in many families limit mapping with this technique.

- A lod score is obtained by calculating the logarithm of the ratio of the probability of obtaining the observed progeny with a specified degree of linkage to the probability of obtaining the observed progeny with independent assortment. A lod score of 3 or higher is usually considered evidence for linkage.

- Molecular techniques that allow the detection of variable differences in DNA sequence have greatly facilitated gene mapping.

- In deletion mapping, genes are physically associated with particular chromosomes by studying the expression of recessive mutations in heterozygotes that possess chromosome deletions.

- In somatic-cell hybridization, cells from two different cell lines (human and rodent) are fused. The resulting hybrid

cells initially contain chromosomes from both species but randomly lose different human chromosomes. The hybrid cells are examined for the presence of specific genes; if a human gene is present in the hybrid cell, it must be present on one of the human chromosomes in that the cell line.

- With in situ hybridization, a radioactive or fluorescence label is added to a fragment of DNA that is complementary to a specific gene. This probe is then added to specially prepared chromosomes, where it pairs with the gene of interest. The presence of the label on a particular chromosome reveals the physical location of the gene.

- Nucleotide sequencing is another method of physically mapping genes.

IMPORTANT TERMS

linked genes (p. 161)
linkage group (p. 161)
nonrecombinant (parental) gamete (p. 163)
nonrecombinant (parental) progeny (p. 163)
recombinant gamete (p. 163)
recombinant progeny (p. 163)
recombination frequency (p. 166)

coupling (cis) configuration (p. 167)
repulsion (trans) configuration (p. 167)
interchromosomal recombination (p. 168)
intrachromosomal recombination (p. 168)
genetic map (p. 172)
physical map (p. 172)

map unit (m.u.) (p. 172)
centimorgan (p. 172)
morgan (p. 172)
two-point testcross (p. 173)
three-point testcross (p. 174)
interference (p. 179)
coefficient of coincidence (p. 180)
lod score (p. 183)
genetic marker (p. 183)

deletion mapping (p. 183)
somatic-cell hybridization (p. 184)
cell line (p. 184)
heterokaryon (p. 184)
syntenic gene (p. 186)

Worked Problems

1. In guinea pigs, white coat (w) is recessive to black coat (W) and wavy hair (v) is recessive to straight hair (V). A breeder crosses a guinea pig that is homozygous for white coat and wavy hair with a guinea pig that is black with straight hair. The F_1 are then crossed with guinea pigs having white coats and wavy hair in a series of testcrosses. The following progeny are produced from these testcrosses:

black, straight	30
black, wavy	10
white, straight	12
white, wavy	31
total	83

(a) Are the genes that determine coat color and hair type assorting independently? Carry out chi-square tests to test your hypothesis.

(b) If the genes are not assorting independently, what is the recombination frequency between them?

• Solution

(a) Assuming independent assortment, outline the crosses conducted by the breeder:

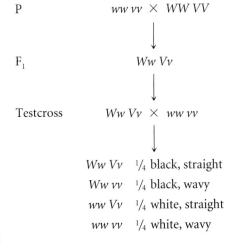

Because a total of 83 progeny were produced in the testcrosses, we expect $\frac{1}{4} \times 83 = 20.75$ of each. The observed numbers of progeny from the testcross (30, 10, 12, 31) do not appear to fit the expected numbers (20.75, 20.75, 20.75, 20.75) well; so independent assortment may not have occurred.

To test the hypothesis, carry out a series of three chi-square tests. First, look at each locus separately and determine if the observed numbers fit those expected from the testcross. For the locus determining coat color, crossing $Ww \times ww$ is expected to produce $\frac{1}{2}$ Ww (black) and $\frac{1}{2}$ ww (white) progeny, or 41.5 of a total of 83 progeny. Ignoring the hair type, we find that $30 + 10 = 40$ black progeny and $12 + 31 = 43$ white progeny were observed. Thus, the observed and expected values for this chi-square test are:

Phenotype	Observed	Expected
black	40	41.5
white	43	41.5

The chi-square value is:

$$\chi^2 = \sum \frac{(\text{observed} - \text{expected})^2}{\text{expected}}$$
$$= \frac{(40 - 41.5)^2}{41.5} + \frac{(43 - 41.5)^2}{41.5} = 0.108$$

The degrees of freedom for the chi-square goodness-of-fit test are $n - 1$, where n equals the number of expected classes. There are two expected classes (black and white) so the degree of freedom is $2 - 1 = 1$. On the basis of the calculated chi-square value in Table 3.4, the probability associated with this chi-square is greater than .05 (the critical probability for rejecting the hypothesis that chance might be responsible for the differences between observed and expected numbers); so the black and white progeny appear to be in the $1:1$ ratio expected in a testcross.

Next, compute a second chi-square value comparing the number of straight and wavy progeny with the numbers expected from the testcross. From the $Vv \times vv$, $\frac{1}{2}$ Vv (straight) and $\frac{1}{2}$ vv (wavy) progeny are expected:

Phenotype	Observed	Expected
straight	42	41.5
wavy	41	41.5

$$\chi^2 = \frac{(42 - 41.5)^2}{41.5} + \frac{(41 - 41.5)^2}{41.5} = 0.012$$

$$\text{degrees of freedom} = n - 1 = 2 - 1 = 1$$

In Table 3.4, the probability associated with this chi-square value is much greater than .05; so straight and wavy progeny are in a $1:1$ ratio.

Having established that the observed numbers for each trait do not differ from the numbers expected from the testcross, we next test for independent assortment. With independent assortment, 20.75 of each phenotype are expected; so

the observed and expected numbers and the associated chi-square value are:

Phenotype	Observed	Expected
black, straight	30	20.75
black, wavy	10	20.75
white, straight	12	20.75
white, wavy	31	20.75

$$\chi^2 = \frac{(30 - 20.75)^-}{20.75} + \frac{(10 - 20.75)^2}{20.75} + \frac{(12 - 20.75)^2}{20.75}$$
$$+ \frac{(31 - 20.75)^2}{20.75}$$
$$= 118.44$$

$$\text{degrees of freedom} = n - 1 = 4 - 1 = 3$$

In Table 3.4, the associated probability is much less than .05, indicating that chance is very unlikely to be responsible for the differences between the observed numbers and the numbers expected with independent assortment. The genes for coat color and hair type have therefore not assorted independently.

(b) To determine the recombination frequencies, identify the recombinant progeny. Using the notation for linked genes, write the crosses:

The recombination frequency is:

$$\frac{\text{number of recombinant progeny}}{\text{total number progeny}} \times 100\%$$

or

$$\text{recombination frequency} = \frac{10 + 12}{30 + 10 + 12 + 31} \times 100\%$$

$$= \frac{22}{83} \times 100 = 26.5\%$$

2. A series of two-point crosses entailed seven loci (a, b, c, d, e, f, and g), producing the following recombination frequencies. Using these recombination frequencies, map the seven loci, showing their linkage groups and the order and distances between the loci of each linkage group:

Loci	Recombination frequency (%)	Loci	Recombination frequency (%)
a and b	10	c and d	50
a and c	50	c and e	8
a and d	14	c and f	50
a and e	50	c and g	12
a and f	50	d and e	50
a and g	50	d and f	50
b and c	50	d and g	50
b and d	4	e and f	50
b and e	50	e and g	18
b and f	50	f and g	50
b and g	50		

• **Solution**

To work this problem, remember that 1% recombination equals 1 map unit and a recombination frequency of 50% means that genes at the two loci are assorting independently (located in different linkage groups).

The recombination frequency between a and b is 10%; so these two loci are in the same linkage group, approximately 10 m.u. apart.

Linkage group 1

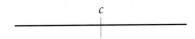

The recombination frequency between a and c is 50%; so c must lie in a second linkage group.

Linkage group 1

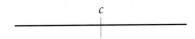

Linkage group 2

c

The recombination frequency between a and d is 14%; so d is located in linkage group 1. Is locus d 14 m.u. to the right or to the left of gene a? If d is 14 m.u. to the left of a, then the b-to-d distance should be 10 m.u. + 14 m.u. = 24 m.u. On the other hand, if d is to the right of a, then the distance between b and d should be 14 m.u. − 10 m.u. = 4 m.u. The b−d recombination frequency is 4%; so d is 14 m.u. to the right of a. The updated map is:

Linkage group 1

Linkage group 2

The recombination frequencies between each of loci a, b, and d, and locus e are all 50%; so e is not in linkage group 1 with a, b, and d. The recombination frequency between e and c is 8 m.u.; so e is in linkage group 2:

Linkage group 1

Linkage group 2

There is 50% recombination between f and all the other genes; so f must belong to a third linkage group:

Linkage group 1

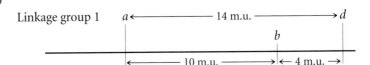

Linkage group 2

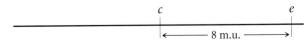

Linkage group 3

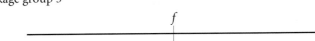

Finally, position locus g with respect to the other genes. The recombination frequencies between g and loci a, b, and d are all 50%; so g is not in linkage group 1. The recombination frequency between *g* and *c* is 12 m.u.; so *g* is a part of linkage group 2. To determine whether *g* is 12 map units to the right or left of *c*, consult the *g*–*e* recombination frequency. Because this recombination frequency is 18%, *g* must lie to the left of *c*:

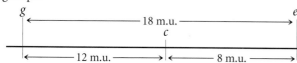

Linkage group 3

Note that the *g*-to-*e* distance (18 m.u.) is shorter than the sum of the *g*-to-*c* (12 m.u.) and *c*-to-*e* distances (8 m.u.), because of undetectable double crossovers between *g* and *e*.

3. Ebony body color (*e*), rough eyes (*ro*), and brevis bristles (*bv*) are three recessive mutations that occur in fruit flies. The loci for these mutations have been mapped and are separated by the following map distances:

The interference between these genes is 0.4.

A fly with ebony body, rough eyes, and brevis bristles is crossed with a fly that is homozygous for the wild-type traits. The resulting F₁ females are test-crossed with males that have ebony body, rough eyes, and brevis bristles; 1800 progeny are produced. Give the phenotypes and expected numbers of phenotypes in the progeny of the testcross.

• **Solution**

The crosses are:

P $\quad \dfrac{e^+ \quad ro^+ \quad bv^+}{e^+ \quad ro^+ \quad bv^+} \quad \times \quad \dfrac{e \quad ro \quad bv}{e \quad ro \quad bv}$

$\downarrow$

F₁ $\quad \dfrac{e^+ \quad ro^+ \quad bv^+}{e \quad ro \quad bv}$

Testcross $\quad \dfrac{e^+ \quad ro^+ \quad bv^+}{e \quad ro \quad bv} \quad \times \quad \dfrac{e \quad ro \quad bv}{e \quad ro \quad bv}$

In this case, we know that *ro* is the middle locus because the genes have been mapped. Eight classes of progeny will be produced from this cross:

e^+	ro^+	bv^+	nonrecombinant
e	ro	bv	nonrecombinant
e^+ /	ro	bv	single crossover between *e* and *ro*
e /	ro^+	bv^+	single crossover between *e* and *ro*
e^+	ro^+ /	bv	single crossover between *ro* and *bv*
e	ro /	bv^+	single crossover between *ro* and *bv*
e^+ /	ro /	bv^+	double crossover
e /	ro^+ /	bv	double crossover

To determine the numbers of each type, use the map distances, starting with the double crossovers. The expected number of double crossovers is equal to the product of the single-crossover probabilities:

$$\text{expected number of double crossovers} = 0.20 \times 0.12 \times 1800$$
$$= 43.2$$

However, some interference occurs; so the observed number of double crossovers will be less than the expected. The interference is 1 − coefficient of coincidence; so the coefficient of coincidence is:

$$\text{coefficient of coincidence} = 1 - \text{interference}$$

The interference is given as 0.4; so the coefficient of coincidence equals 1 − 0.4 = 0.6. Recall that the coefficient of coincidence is:

$$\text{coefficient of coincidence} =$$
$$\frac{\text{number of observed double crossovers}}{\text{number of expected double crossovers}}$$

Rearranging this equation, we obtain:

$$\text{number of observed double crossovers} =$$
$$\text{coefficient of coincidence} \times \text{number of expected double crossovers}$$

$$\text{number of observed double crossovers} = 0.6 \times 43.2 = 26$$

A total of 26 double crossovers should be observed. Because there are two classes of double crossovers ($\underline{e^+ \quad / \quad ro \quad / \quad bv^+}$ and $\underline{e \quad / \quad ro^+ \quad / \quad bv}$), we should observe 13 of each.

Next, we determine the number of single-crossover progeny. The genetic map indicates that there are 20 m.u. between *e* and *ro*; so 360 progeny (20% of 1800) are expected to have resulted from recombination between these two loci. Some of them will be single-crossover progeny and some will be double-crossover progeny. We have already determined that the number of double-crossover progeny is 26; so the number of progeny resulting from a single crossover between *e* and *ro* is 360 − 26 = 334, which will be divided equally between the two single-crossover phenotypes (e / ro^+ bv^+ and e^+ / ro bv).

There are 12 map units between ro and bv; so the number of progeny resulting from recombination between these two genes is $0.12 \times 1800 = 216$. Again, some of these recombinants will be single-crossover progeny and some will be double-crossover progeny. To determine the number of progeny resulting from a single crossover, subtract the double crossovers: $216 - 26 = 190$. These single-crossover progeny will be divided between the two single-crossover phenotypes (e^+ ro^+ / bv and e ro / bv^+); so there will be $^{190}/_2 = 95$ of each. The remaining progeny will be nonrecombinants, and they can be obtained by subtraction: $1800 - 26 - 334 - 190 = 1250$; there are two nonrecombinants (e^+ ro^+ bv^+ and e ro bv); so there will be $^{1250}/_2 = 625$ of each. The numbers of the various phenotypes are listed here:

e^+	ro^+	bv^+	625	nonrecombinant
e	ro	bv	625	nonrecombinant
e^+ / ro		bv	167	single crossover between e and ro
e / ro^+		bv^+	167	single crossover between e and ro
e^+	ro^+ / bv		95	single crossover between ro and bv
e	ro / bv^+		95	single crossover between ro and bv
e^+ / ro / bv^+			13	double crossover
e / ro^+ / bv			13	double crossover
total			1800	

4. The locations of six deletions have been mapped to the *Drosophila* chromosome as shown in the following diagram. Recessive mutations a, b, c, d, e, f, and g are known to be located in the same regions as the deletions, but the order of the mutations on the chromosome is not known. When flies homozygous for the recessive mutations are crossed with flies homozygous for the deletions, the following results are obtained, where the letter "m" represents a mutant phenotype and a plus sign ($+$) represents the wild type. On the basis of these data, determine the relative order of the seven mutant genes on the chromosome:

Chromosome

Deletion 1 ——————
　　Deletion 2 ————————
　　　Deletion 3 ——————
　　　Deletion 4 ————————————
　　　　　　Deletion 5 ————
　　　Deletion 6 ————————————

				Mutations			
Deletion	a	b	c	d	e	f	g
1	$+$	m	m	m	$+$	$+$	$+$
2	$+$	$+$	m	m	$+$	$+$	$+$
3	$+$	$+$	$+$	m	m	$+$	$+$
4	m	$+$	$+$	m	m	$+$	$+$
5	m	$+$	$+$	$+$	$+$	m	m
6	m	$+$	$+$	m	m	m	$+$

• Solution

The offspring of the cross will be heterozygous, possessing one chromosome with the deletion and wild-type alleles and one chromosome without the deletion and recessive mutant alleles. For loci within the deleted region, only the recessive mutations will be present in the offspring, which will exhibit the mutant phenotype. The presence of a mutant trait in the offspring therefore indicates that the locus for that trait is within the region covered by the deletion. We can map the genes by examining the expression of the recessive mutations in the flies with different deletions.

Mutation a is expressed in flies with deletions 4, 5, and 6 but not in flies with other deletions; so a must be in the area that is unique to deletions 4, 5, and 6:

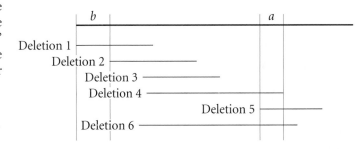

Mutation b is expressed only when deletion 1 is present; so it must be located in the region of the chromosome covered by deletion 1 and none of the other deletions:

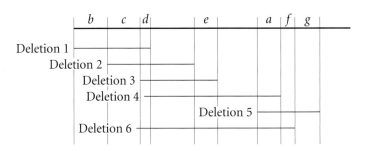

Using this procedure, we can map the remaining mutations. For each mutation, we look for the area of overlap among deletions that express the mutations and exclude any areas of overlap that are covered by other deletions that do not express the mutation:

5. A panel of cell lines was created from mouse–human somatic-cell fusions. Each line was examined for the presence of human chromosomes and for the production of human haptoglobin (a protein). The following results were obtained:

Cell line	Human haptoglobin	Human chromosomes							
		1	2	3	14	15	16	21
A	−	+	−	+	−	+	−	−
B	+	+	−	+	−	−	+	−
C	+	+	−	−	−	+	+	−
D	−	+	+	−	−	+	−	−

On the basis of these results, which human chromosome carries the gene for haptoglobin?

• Solution

Examine those cell lines that are positive for human haptoglobin and see what chromosomes they have in common. Lines B and C produce human haptoglobin; the chromosomes that they have in common are 1 and 16. Next, examine all lines that possess chromosomes 1 and 16 and determine whether they produce haptoglobin. Chromosome 1 is found in cell lines A, B, C, and D. If the gene for human haptoglobin were found on chromosome 1, human haptoglobin would be present in all of these cell lines. However, lines A and D do not produce human haptoglobin; so the gene cannot be on chromosome 1. Chromosome 16 is found only in cell lines B and C, and only these lines produce human haptoglobin; so the gene for human haptoglobin lies on chromosome 16.

COMPREHENSION QUESTIONS

* 1. What does the term recombination mean? What are two causes of recombination?

* 2. In a testcross for two genes, what types of gametes are produced with **(a)** complete linkage, **(b)** independent assortment, and **(c)** incomplete linkage?

3. What effect does crossing over have on linkage?

4. Why is the frequency of recombinant gametes always half the frequency of crossing over?

* 5. What is the difference between genes in coupling configuration and genes in repulsion? What effect does the arrangement of linked genes (whether they are in coupling configuration or in repulsion) have on the results of a cross?

6. How does one test to see if two genes are linked?

7. What is the difference between a genetic map and a physical map?

* 8. Why do calculated recombination frequencies between pairs of loci that are located relatively far apart underestimate the true genetic distances between loci?

9. Explain how one can determine which of three linked loci is the middle locus from the progeny of a three-point testcross.

*10. What does the interference tell us about the effect of one crossover on another?

11. List some of the methods for physically mapping genes and explain how they are used to position genes on chromosomes.

12. What is a lod score and how is it calculated?

APPLICATION QUESTIONS AND PROBLEMS

*13. In the snail *Cepaea nemoralis,* an autosomal allele causing a banded shell (B^B) is recessive to the allele for unbanded shell (B^O). Genes at a different locus determine the background color of the shell; here, yellow (C^Y) is recessive to brown (C^{Bw}). A banded, yellow snail is crossed with a homozygous brown, unbanded snail. The F_1 are then crossed with banded, yellow snails (a testcross).

(a) What will be the results of the testcross if the loci that control banding and color are linked with no crossing over?

(b) What will be the results of the testcross if the loci assort independently?

(c) What will be the results of the testcross if the loci are linked and 20 map units apart?

*14. In silkmoths (*Bombyx mori*) red eyes (*re*) and white-banded wing (*wb*) are encoded by two mutant alleles that are recessive to those that produce wild-type traits (*re*$^+$ and *wb*$^+$); these two genes are on the same chromosome. A moth homozygous for red eyes and white-banded wings is crossed with a moth homozygous for the wild-type traits. The F_1 have normal eyes and normal wings. The F_1 are crossed with moths that have red eyes and white-banded wings in a testcross. The progeny of this testcross are:

wild-type eyes, wild-type wings	418
red eyes, wild-type wings	19
wild-type eyes, white-banded wings	16
red eyes, white-banded wings	426

(a) What phenotypic proportions would be expected if the genes for red eyes and white-banded wings were located on different chromosomes?

(b) What is the genetic distance between the genes for red eyes and white-banded wings?

*15. A geneticist discovers a new mutation in *Drosophila melanogaster* that causes the flies to shake and quiver. She calls this mutation spastic *(sps)* and determines that spastic is due to an autosomal recessive gene. She wants to determine if the spastic gene is linked to the recessive gene for vestigial wings *(vg)*. She crosses a fly homozygous for spastic and vestigial traits with a fly homozygous for the wild-type traits and then uses the resulting F_1 females in a testcross. She obtains the following flies from this testcross.

vg^+	sps^+	230
vg	sps	224
vg	sps^+	97
vg^+	sps	99
total		650

Are the genes that cause vestigial wings and the spastic mutation linked? Do a series of chi-square tests to determine if the genes have assorted independently.

16. In cucumbers, heart-shaped leaves *(hl)* are recessive to normal leaves *(Hl)* and having many fruit spines *(ns)* is recessive to having few fruit spines *(Nl)*. The genes for leaf shape and number of spines are located on the same chromosome; mapping experiments indicate that they are 32.6 map units apart. A cucumber plant having heart-shaped leaves and many spines is crossed with a plant that is homozygous for normal leaves and few spines. The F_1 are crossed with plants that have heart-shaped leaves and many spines. What phenotypes and proportions are expected in the progeny of this cross?

*17. In tomatoes, tall *(D)* is dominant over dwarf *(d)* and smooth fruit *(P)* is dominant over pubescent fruit *(p)*, which is covered with fine hairs. A farmer has two tall and smooth tomato plants, which we will call plant A and plant B. The farmer crosses plants A and B with the same dwarf and pubescent plant and obtains the following numbers of progeny:

	Progeny of	
	Plant A	**Plant B**
Dd Pp	122	2
Dd pp	6	82
dd Pp	4	82
dd pp	124	4

(a) What are the genotypes of plant A and plant B?

(b) Are the loci that determine height of the plant and pubescence linked? If so, what is the map distance between them?

(c) Explain why different proportions of progeny are produced when plant A and plant B are crossed with the same dwarf pubescent plant.

18. A cross between individuals with genotypes $a^+ a\ b^+b$ × $aa\ bb$ produces the following progeny:

$a^+a\ b^+b$	83
$a^+a\ bb$	21
$aa\ b^+b$	19
$aa\ bb$	77

(a) Does the evidence indicate that the *a* and *b* loci are linked?

(b) What is the map distance between *a* and *b*?

(c) Are the alleles in the parent with genotype $a^+a\ b^+b$ in coupling configuration or repulsion? How do you know?

19. In tomatoes, dwarf *(d)* is recessive to tall *(D)* and opaque (light green) leaves *(op)* are recessive to green leaves *(Op)*. The loci that determine the height and leaf color are linked and separated by a distance of 7 m.u. For each of the following crosses, determine the phenotypes and proportions of progeny produced.

(a) $\dfrac{D}{d}\ \dfrac{Op}{op}$ × $\dfrac{d}{d}\ \dfrac{op}{op}$

(b) $\dfrac{D}{d}\ \dfrac{op}{Op}$ × $\dfrac{d}{d}\ \dfrac{op}{op}$

(c) $\dfrac{D}{d}\ \dfrac{Op}{op}$ × $\dfrac{D}{d}\ \dfrac{Op}{op}$

(d) $\dfrac{D}{d}\ \dfrac{op}{Op}$ × $\dfrac{D}{d}\ \dfrac{op}{Op}$

*20. In *Drosophila melanogaster*, ebony body *(e)* and rough eyes *(ro)* are encoded by autosomal recessive genes found on chromosome 3; they are separated by 20 map units. The gene that encodes forked bristles *(f)* is X-linked recessive and assorts independently of *e* and *ro*. Give the phenotypes of progeny and their expected proportions when each of the following genotypes is test-crossed.

(a) $\dfrac{e^+}{e}\ \dfrac{ro^+}{ro}\ \dfrac{f^+}{f}$

(b) $\dfrac{e^+}{e}\ \dfrac{ro}{ro^+}\ \dfrac{f}{f}$

*21. A series of two-point crosses were carried out among seven loci (*a, b, c, d, e, f,* and *g*), producing the following recombination frequencies. Map the seven loci, showing their linkage groups, the order of the loci in each linkage group, and the distances between the loci of each group:

Loci	Recombination frequency (%)	Loci	Recombination frequency (%)
a and b	50	c and d	50
a and c	50	c and e	26
a and d	12	c and f	50
a and e	50	c and g	50
a and f	50	d and e	50
a and g	4	d and f	50
b and c	10	d and g	8
b and d	50	e and f	50
b and e	18	e and g	50
b and f	50	f and g	50
b and g	50		

*22. Waxy endosperm (wx), shrunken endosperm (sh), and yellow seedling (v) are encoded by three recessive genes in corn that are linked on chromosome 5. A corn plant homozygous for all three recessive alleles is crossed with a plant homozygous for all the dominant alleles. The resulting F_1 are then crossed with a plant homozygous for the recessive alleles in a three-point testcross. The progeny of the testcross are:

wx	sh	V	87
Wx	Sh	v	94
Wx	Sh	V	3,479
wx	sh	v	3,478
Wx	sh	V	1,515
wx	Sh	v	1,531
wx	Sh	V	292
Wx	sh	v	280
total			10,756

(a) Determine order of these genes on the chromosome.

(b) Calculate the map distances between the genes.

(c) Determine the coefficient of coincidence and the interference among these genes.

23. Fine spines (s), smooth fruit (tu), and uniform fruit color (u) are three recessive traits in cucumbers whose genes are linked on the same chromosome. A cucumber plant heterozygous for all three traits is used in a testcross, and the following progeny are produced from this testcross:

S	U	Tu	2
s	u	Tu	70
S	u	Tu	21
s	u	tu	4
S	U	tu	82
s	U	tu	21
s	U	Tu	13
S	u	tu	17
total			230

(a) Determine the order of these genes on the chromosome.

(b) Calculate the map distances between the genes.

(c) Determine the coefficient of coincidence and the interference among these genes.

(d) List the genes found on each chromosome in the parents used in the testcross.

*24. In *Drosophila melanogaster*, black body (b) is recessive to gray body (b^+), purple eyes (pr) are recessive to red eyes (pr^+), and vestigial wings (vg) are recessive to normal wings (vg^+). The loci coding for these traits are linked, with the following map distances:

The interference among these genes is 0.5. A fly with black body, purple eyes, and vestigial wings is crossed with a fly homozygous for gray body, red eyes, and normal wings. The female progeny are then crossed with males that have black body, purple eyes, and vestigial wings. If 1000 progeny are produced from this testcross, what will be the phenotypes and proportions of the progeny?

*25. The locations of six deletions have been mapped to the *Drosophila* chromosome shown here. Recessive mutations a, b, c, d, e, and f are known to be located in the same region as the deletions, but the order of the mutations on the chromosome is not known. When flies homozygous for the recessive mutations are crossed with flies homozygous for the deletions, the following results are obtained, where "m" represents a mutant phenotype and a plus sign (+) represents the wild type. On the basis of these data, determine the relative order of the seven mutant genes on the chromosome:

```
                     Chromosome
                     _____

Deletion 1  ———————————————
   Deletion 2  ————
             Deletion 3  ————————————————————
             Deletion 4  ——————————————
                Deletion 5  ——————
                   Deletion 6  ——————————————
```

Deletion	Mutations					
	a	b	c	d	e	f
1	m	+	m	+	+	m
2	m	+	+	+	+	+
3	+	m	m	m	m	+
4	+	+	m	m	m	+
5	+	+	+	m	m	+
6	+	m	+	m	+	+

26. A panel of cell lines was created from mouse–human somatic-cell fusions. Each line was examined for the presence of human chromosomes and for the production of an enzyme. The following results were obtained:

Cell line	Enzyme	1	2	3	4	5	6	7	8	9	10	17	22
A	–	+	–	–	–	+	–	–	–	–	–	+	–
B	+	+	+	–	–	–	–	–	+	–	–	+	+
C	–	+	–	–	–	+	–	–	–	–	–	–	+
D	–	–	–	–	+	–	–	–	–	–	–	–	–
E	+	+	–	–	–	–	–	–	+	–	+	+	–

On the basis of these results, which chromosome has the gene that codes for the enzyme?

*27. A panel of cell lines was created from mouse–human somatic-cell fusions. Each line was examined for the presence of human chromosomes and for the production of three enzymes. The following results were obtained.

Cell line	Enzyme			Human chromosomes								
	1	2	3	4	8	9	12	15	16	17	22	X
A	+	–	+	–	–	+	–	+	+	–	–	+
B	+	–	–	–	–	+	–	–	+	+	–	–
C	–	+	+	+	–	–	–	–	–	+	–	+
D	–	+	+	+	+	–	–	–	+	–	–	+

On the basis of these results, give the chromosome location of enzyme 1, enzyme 2, and enzyme 3.

CHALLENGE QUESTION

28. In calculating map distances, we did not concern ourselves with whether double crossovers were two stranded, three stranded, or four stranded; yet, these different types of double crossovers produce different types of gametes. Can you explain why we do not need to determine how many strands take part in double crossovers in diploid organisms? (Hint: Draw out the types of gametes produced by the different types of double crossovers and see how they contribute to the determination of map distances.)

SUGGESTED READINGS

Creighton, H. B., and B. McClintock. 1931. A correlation of cytological and genetical crossing over in *Zea mays. Proceedings of the National Academy of Science U. S. A.* 17:492–497.
Paper reporting Creighton and McClintock's finding that crossing over is associated with exchange of chromosome segments.

Crow, J. 1988. A diamond anniversary: the first genetic map. *Genetics* 118:1–3.
A brief review of the history of Sturtevant's first genetic map.

Ruddle, F. H., and R. S. Kucherlapati. 1974. Hybrid cells and human genes. *Scientific American* 231(1):36–44.
A readable review of somatic-cell hybridization.

Stern, C. 1936. Somatic crossing over and segregation in *Drosophila melanogaster. Genetics* 21:625–631.
Stern's finding, similar to Creighton and McClintock's, of a correlation between crossing over and physical exchange of chromosome segments.

Sturtevant, A. H. 1913. The linear arrangement of six sex-linked factors in *Drosophila*, as shown by their mode of association. *Journal of Experimental Zoology* 14:43–59.
Sturtevant's report of the first genetic map.

8

Bacterial and Viral Genetic Systems

London experienced severe epidemics of cholera in the 1850s. Shown here is a map of London in 1856. (Mylne, Robert W., *Map of the Contours of London and Its Environs*, Published by Edward Stanford, Charing Cross, London. Engraved and Printed from Stone by Waterlow and Sons, 1856.)

Pump Handles and Cholera Genes

On the night of August 31 in 1854, a terrible epidemic of cholera broke out in the Soho neighborhood of London. Hundreds of residents were stricken with severe diarrhea and vomiting and, in the next three days, 127 people living on or near Broad Street died. By September 10, the number of fatalities had climbed to more than 500. It was the worst outbreak of cholera ever seen in England. Residents of Soho fled their homes in terror, leaving businesses closed, homes locked, and streets deserted.

The Soho epidemic was witnessed firsthand by Dr. John Snow, a physician who lived on Sackville Street and saw the devastating effects and rapid spread of the disease. He had conducted research on cholera and suspected that it was spread through the water supply, but most medical authorities dismissed his suspicion.

Snow conducted a thorough survey of the Soho neighborhood, identifying all those who were sick with cholera. He plotted the locations of the cases on a map and observed that they clustered around one particular water pump located on Broad Street. Cholera cases did not cluster around other nearby water pumps. Snow contacted the parish officials and convinced them to remove the handle to the Broad Street pump and, with this simple action, the spread of cholera stopped dramatically. Snow later conducted additional studies of cholera outbreaks in London and established that the disease was spread in water contaminated with sewage.

Cholera has existed in Asia for at least 1000 years but, at the time of the Soho epidemic, it was a relatively new disease in England. Today, cholera is recognized as a severe infection of the intestine caused by *Vibrio cholerae* (◀ FIGURE 8.1). This bacterium produces a potent endotoxin that induces

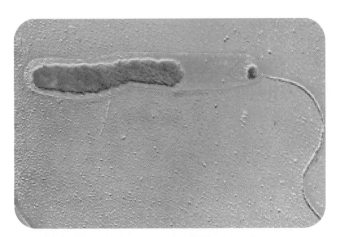

◀ 8.1 *Vibrio cholerae* is the bacterium that causes cholera. (CNRI/Science Photo Library/Photo Researchers.)

copious diarrhea and vomiting. If untreated, the condition can lead to serious dehydration and death. Although the number of cholera deaths has dropped dramatically since the advent of oral rehydration treatment and antibiotics, the disease continues to be a serious public-health problem, particularly in areas that lack modern water-supply systems. A recent epidemic in South Africa infected more than 25,000 people in a 6-month period.

Many of cholera's secrets have now been revealed through the sequencing of *V. cholerae*'s genome. Most bacteria have a single circular chromosome, but *V. cholerae* has two. One of the most significant findings to emerge from the sequencing of the *V. cholerae* genome is that many of the bacterium's genes for pathogenesis were acquired from other bacteria. *Vibrio cholerae* apparently has a long history

Table 8.1	Advantages of using bacteria and viruses for genetic studies

1. Reproduction is rapid.
2. Many progeny are produced.
3. Haploid genome allows all mutations to be expressed directly.
4. Asexual reproduction simplifies the isolation of genetically pure strains.
5. Growth in the laboratory is easy and requires little space.
6. Genomes are small.
7. Techniques are available for isolating and manipulating their genes.
8. They have medical importance.
9. They can be genetically engineered to produce substances of commercial value.

of exchanging genetic material with other bacteria and viruses, and it seems that much of this acquired DNA is responsible for its virulence. The gene that encodes the cholera toxin, for example, is found imbedded in a viral genome that infected the bacterium and became a permanent part of its genome long ago. *Vibrio cholerae* illustrates the importance of gene exchange between bacteria and viruses, a major theme of this chapter.

Since the 1940s, the genetic systems of bacteria and viruses have contributed to the discovery of many important concepts in genetics. The study of molecular genetics initially focused almost entirely on their genes; today, bacteria and viruses are still essential tools for probing the nature of genes in more-complex organisms, in part because they possess a number of characteristics that make them suitable for genetic studies (Table 8.1).

The genetic systems of bacteria and viruses are also studied because these organisms play important roles in human society. They have been harnessed to produce a number of economically important substances, and they are of immense medical significance, causing many human diseases. In this chapter, we focus on several unique aspects of bacterial and viral genetic systems. Important processes of gene transfer and recombination, like those that contributed to the pathogenesis of the cholera bacterium, will be described, and we will see how these processes can be used to map bacterial and viral genes.

www.whfreeman.com/pierce   Information about John Snow and his contributions to the study of cholera

Bacterial Genetics

Heredity in bacteria is fundamentally similar to heredity in more-complex organisms, but the bacterial haploid genome and the small size of bacteria (which makes observation of their phenotypes difficult) require different approaches and methods. First, we will consider how bacteria are studied and then examine several processes that transfer genes from one bacterium to another.

Techniques for the Study of Bacteria

Microbiologists have defined the nutritional needs of a number of bacteria and developed culture media for growing them in the laboratory. Culture media typically contain a carbon source, essential elements such as nitrogen and phosphorus, certain vitamins, and other required ions and nutrients. Wild-type (prototrophic) bacteria can use these simple ingredients to synthesize all the compounds that they need for growth and reproduction. A medium that contains only the nutrients required by prototrophic bacteria is termed **minimal medium.** Mutant strains called auxotrophs lack one or more enzymes necessary for metabolizing nutrients or synthesizing essential molecules and will grow only on medium supplemented with one or more

(a) Culturing bacteria in liquid medium   **(b) Culturing bacteria on petri plates**

Inoculating loop

Sterile liquid medium

Inoculate medium with bacteria.

Bacteria grow and divide.

Pipet

Lid

Glass rod

Dilute soluion of bacterial cells

Petri plate

A growth medium is suspended in gelatin-like agar.

Add a dilute solution of bacteria to petri plate.

Spread bacterial solution evenly with glass rod.

After incubation for 1 to 2 days, bacteria multiply, forming visible colonies.

◀ 8.2 **Bacteria can be grown in liquid medium.**

nutrients. For example, auxotrophic strains that are unable to synthesize the amino acid leucine will not grow on minimal medium but *will* grow on medium to which leucine has been added. **Complete medium** contains all the substances required by bacteria for growth and reproduction.

Cultures of bacteria are often grown in test tubes that contain sterile liquid medium (◀ FIGURE 8.2a). A few bacteria are added to the tube, and they grow and divide until all the nutrients are used up or—more commonly—until the concentration of their waste products becomes toxic. Bacteria are also grown in petri plates (◀ FIGURE 8.2b). Growth medium suspended in agar is poured into the bottom half of the petri plate, providing a solid, gel-like base

for bacterial growth. The chief advantage of this method is that it allows one to isolate and count bacteria, which individually are too small to see without a microscope. In a process called plating, a dilute solution of bacteria is spread over the surface of an agar-filled petri plate. As each bacterium grows and divides, it gives rise to a visible clump of genetically identical cells (a **colony**). Genetically pure strains of the bacteria can be isolated by collecting bacteria from a single colony and transferring them to a new test tube or petri plate.

Because individual bacteria are too small to be seen directly, it is often easier to study phenotypes that affect the appearance of the colony (◀ FIGURE 8.3) or can be de-

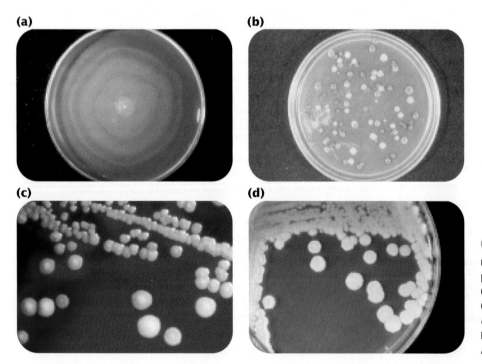

(a)   **(b)**

(c)   **(d)**

◀ 8.3 **Bacteria can be grown on solid media and show a variety of phenotypes.** (a) *Proteus mirabilis.* (b) *Serratia marcescens* with color variation. (c) *Staphylococcus aureus.* (d) *Bacillus cereus.* (Parts a and d, Biophoto Associates/ Photo Researchers; part b, Dr. E. Bottone/Peter Arnold; part c, Larry Jensen/Visuals Unlimited.)

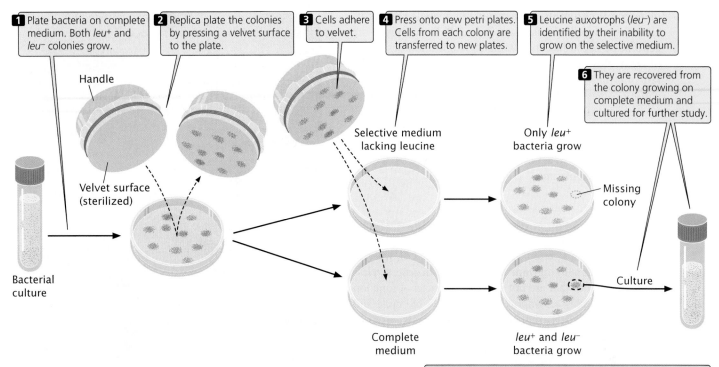

1 Plate bacteria on complete medium. Both *leu+* and *leu−* colonies grow.

2 Replica plate the colonies by pressing a velvet surface to the plate.

3 Cells adhere to velvet.

4 Press onto new petri plates. Cells from each colony are transferred to new plates.

5 Leucine auxotrophs (*leu−*) are identified by their inability to grow on the selective medium.

6 They are recovered from the colony growing on complete medium and cultured for further study.

Handle

Velvet surface (sterilized)

Bacterial culture

Selective medium lacking leucine

Only *leu+* bacteria grow

Missing colony

Complete medium

leu+ and *leu−* bacteria grow

Culture

Conclusion: A colony that grows on the complete medium but not on the selective medium has a mutation in a gene that encodes the synthesis of an essential nutrient.

◀8.4 Mutant bacterial strains can be isolated on the basis of their nutritional requirements.

tected by simple chemical tests. Nutritional requirements of the bacteria are used to detect some commonly studied phenotypes. Suppose we want to detect auxotrophic mutants that cannot synthesize leucine (*leu−* mutants). We first spread the bacteria on a petri plate containing complete medium that includes leucine; so both prototrophs that have the *leu+* allele and auxotrophs that have *leu−* alleles will grow on it (◀ FIGURE 8.4). Next, using a technique called replica plating, we transfer a few cells from each of the colonies on the original plate to two new replica plates, one containing complete medium and the other containing *selective medium* that lacks leucine. The *leu+* bacteria will grow on both media, but the *leu−* mutants will grow only on the complete medium, because they cannot synthesize the leucine that is absent from the selective medium. Any colony that grows on complete medium but not on this selective medium consists of *leu−* bacteria. The auxotrophic mutants growing on complete medium can then be cultured for further study.

The Bacterial Genome

Bacteria are unicellular organisms that lack a nuclear membrane (◀ FIGURE 8.5). Most bacterial genomes consist

◀8.5 Most bacterial cells possess a single, circular chromosome, shown here emerging from a ruptured bacterial cell. (K. G. Murti/Visuals Unlimited.)

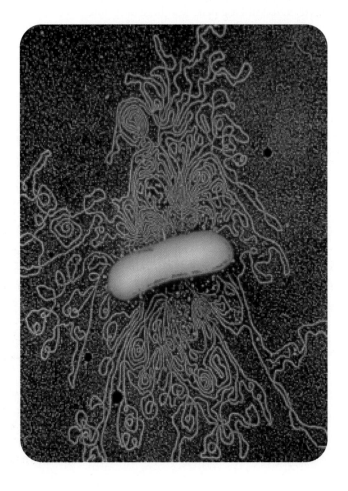

of a circular chromosome that contains a single DNA molecule several million base pairs in length. For example, the genome of *E. coli* has approximately 4.6 million base pairs of DNA. However, some bacteria (such as *V. cholerae*) contain multiple chromosomes, and a few even have linear chromosomes. Bacterial chromosomes are usually organized efficiently, with little DNA between genes.

www.whfreeman.com/pierce   General information on bacteria, bacterial structure, and major groups of bacteria, with some great pictures

Plasmids

In addition to having a chromosome, many bacteria possess **plasmids,** small, circular DNA molecules (◀FIGURE 8.6). Some plasmids are present in many copies per cell, whereas others are present in only one or two copies. In general, plasmids carry genes that are not essential to bacterial function but that may play an important role in the life cycle and growth of their bacterial hosts. Some plasmids promote mating between bacteria; others contain genes that kill other bacteria. Of great importance, plasmids are used extensively in genetic engineering (Chapter 18) and some of them play a role in the spread of antibiotic resistance among bacteria.

Most plasmids are circular and several thousand base pairs in length, although plasmids consisting of several hundred thousand base pairs also have been found. Possessing its own origin of replication, a plasmid replicates independently of the bacterial chromosome. Replication proceeds from the origin in one or two directions until the entire plasmid is copied. In ◀FIGURE 8.7, the origin of replication is *oriV*. A few plasmids have multiple replication origins.

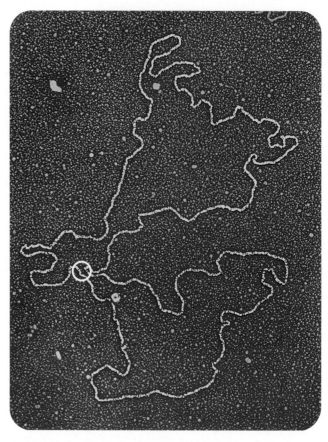

◀8.6 **Many bacteria contain plasmids, small, circular molecules of DNA.** An electron micrograph of bacterial plasmid. The connected plasmids (indicated by the circle) have just replicated. (A. B. Dowsett/Science Photo Library/Photo Researchers.)

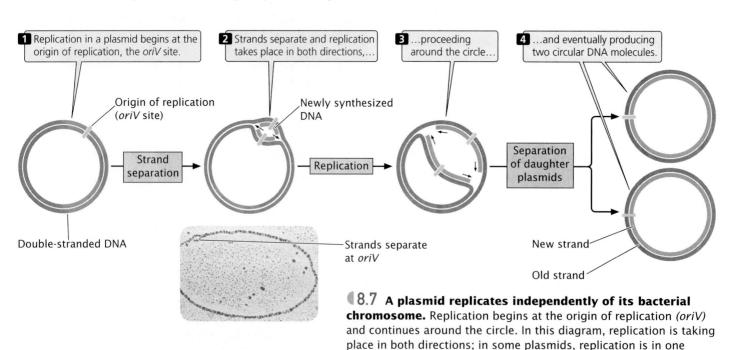

◀8.7 **A plasmid replicates independently of its bacterial chromosome.** Replication begins at the origin of replication (*oriV*) and continues around the circle. In this diagram, replication is taking place in both directions; in some plasmids, replication is in one direction only. (Photo from Photo Researchers.)

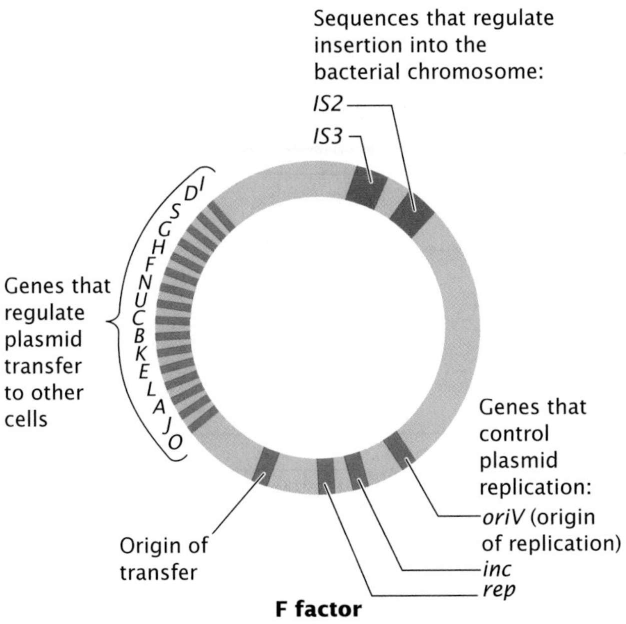

Sequences that regulate insertion into the bacterial chromosome:
IS2
IS3

Genes that regulate plasmid transfer to other cells

Genes that control plasmid replication:
oriV (origin of replication)
inc
rep

Origin of transfer

F factor

◀ **8.8   The F factor, a circular episome of *E. coli*, contains a number of genes that regulate transfer into the bacterial cell, replication, and insertion into the bacterial chromosome.** Replication is initiated at *oriV*. Insertion sequences (Chapter 11) *IS3* and *IS2* control insertion into the bacterial chromosome and excision from it.

Episomes are plasmids that are capable of either freely replicating or integrating into the bacterial chromosomes. The **F (fertility) factor** of *E. coli* (◀ FIGURE 8.8) is an episome that controls mating and gene exchange between *E. coli* cells, as will be discussed in the next section.

⬡ **Concepts**

The typical bacterial genome consists of a single circular chromosome that contains several million base pairs. Some bacterial genes may be present on plasmids, which are small, circular DNA molecules that replicate independently of the bacterial chromosome.

Gene Transfer in Bacteria

For many years, bacteria were thought to reproduce only by simple binary fission, in which one cell splits into two identical cells without any exchange or recombination of genetic material. In 1946, Joshua Lederberg and Edward Tatum demonstrated that bacteria can transfer and recombine genetic information. This finding paved the way for the use of bacteria as model genetic organisms. Bacteria exchange genetic material by three different mechanisms,

all entailing some type of DNA transfer and recombination between the transferred DNA and the bacterial chromosome.

1. **Conjugation** (◀ FIGURE 8.9a) is the direct transfer of genetic material from one bacterium to another. In conjugation, two bacteria lie close together and a connection forms between them. A plasmid or a part of the bacterial chromosome passes from one cell (the donor) to the other (the recipient). Subsequent to conjugation, crossing over takes place between homologous sequences in the transferred DNA and the chromosome of the recipient cell. In conjugation, DNA is transferred only from donor to recipient, with no reciprocal exchange of genetic material.

2. In **transformation** (◀ FIGURE 8.9b), DNA in the medium surrounding a bacterium is taken up. After transformation, recombination may take place between the introduced genes and those of the bacterial chromosome.

3. In **transduction** (◀ FIGURE 8.9c), bacterial viruses (bacteriophages) carry DNA from one bacterium to another. Once inside the bacterium, the newly introduced DNA may undergo recombination with the bacterial chromosome.

Not all bacterial species exhibit all three types of genetic transfer. Conjugation is more frequent for some bacteria than for others. Transformation takes place to a limited extent in many bacteria, but laboratory techniques have been developed that increase the rate of DNA uptake. Most bacteriophages have a limited host range; so transduction is normally between bacteria of the same or closely related species only.

These processes of genetic exchange in bacteria differ from the sexual reproduction of diploid eukaryotes in two important ways. First, DNA exchange and reproduction are not coupled in bacteria. Second, donated genetic material that is not recombined into the host DNA is usually degraded and so the recipient cell remains haploid. Each type of genetic transfer can be used to map genes, as will be discussed in the following sections.

⬡ **Concepts**

DNA may be transferred between bacterial cells through conjugation, transformation, or transduction. Each type of genetic transfer consists of a one-way movement of genetic information to the recipient cell, sometimes followed by recombination. These processes are not connected to cellular reproduction in bacteria.

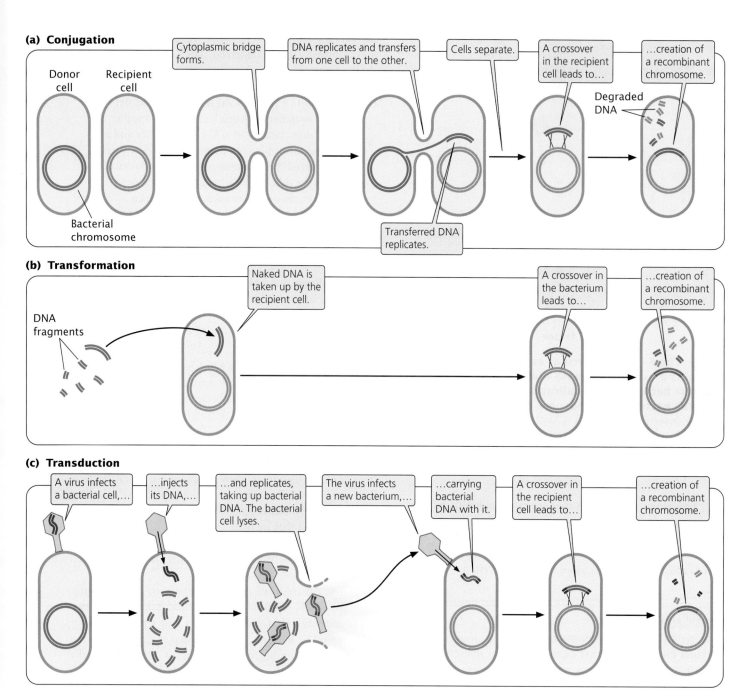

8.9 Conjugation, transformation, and transduction are three processes of gene transfer in bacteria. All three processes require transferred DNA to undergo recombination with the bacterial chromosome for the transferred DNA to be stably inherited.

Conjugation

In the course of their research, Lederberg and Tatum studied strains of *E. coli* possessing auxotrophic mutations. The Y10 strain required the amino acids threonine (and genotypically was *thr⁻*) and leucine (*leu⁻*) and the vitamin thiamine (*thi⁻*) for growth but did not require the vitamin biotin (*bio⁺*) or the amino acids phenylalanine (*phe⁺*) and cysteine (*cys⁺*); the genotype of this strain can be written

as: *thr⁻ leu⁻ thi⁻ bio⁺ phe⁺ cys⁺*. The Y24 strain required biotin, phenylalanine, and cysteine in its medium, but it did not require threonine, leucine, or thiamine; its genotype was: *thr⁺ leu⁺ thi⁺ bio⁻ phe⁻ cys⁻*. In one experiment, Lederberg and Tatum mixed Y10 and Y24 bacteria together and plated them on minimal medium (◀FIGURE 8.10). Each strain was also plated separately on minimal medium.

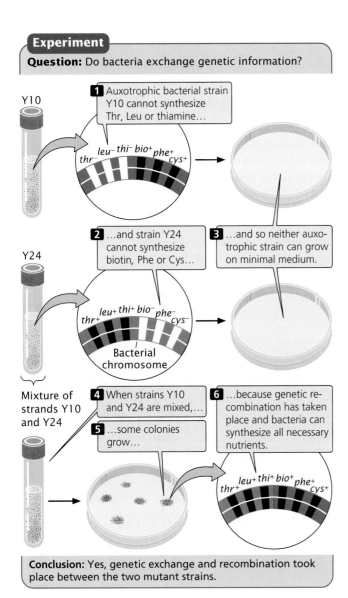

Question: Do bacteria exchange genetic information?

Y10

1 Auxotrophic bacterial strain Y10 cannot synthesize Thr, Leu or thiamine...

leu^- thi^- bio^+ phe^+ thr^- cys^+

Y24

2 ...and strain Y24 cannot synthesize biotin, Phe or Cys...

3 ...and so neither auxotrophic strain can grow on minimal medium.

leu^+ thi^+ bio^- thr^+ phe^- cys^-

Bacterial chromosome

Mixture of strands Y10 and Y24

4 When strains Y10 and Y24 are mixed,...

5 ...some colonies grow...

6 ...because genetic recombination has taken place and bacteria can synthesize all necessary nutrients.

leu^+ thi^+ bio^+ thr^+ phe^+ cys^+

Conclusion: Yes, genetic exchange and recombination took place between the two mutant strains.

◀ **8.10 Lederberg and Tatum's experiment demonstrated that bacteria undergo genetic exchange.**

Alone, neither Y10 nor Y24 grew on minimal medium. Strain Y10 was unable to grow, because it required threonine, leucine, and thiamine, which were absent in the minimal medium; strain Y24 was unable to grow, because it required biotin, phenylalanine, and cysteine, which also were absent from the minimal medium. When Lederberg and Tatum mixed the two strains, however, a few colonies did grow on the minimal medium. These prototrophic bacteria must have had genotype thr^+ leu^+ thi^+ bio^+ phe^+ cys^+. Where had they come from?

If mutations were responsible for the prototrophic colonies, then some colonies should also have grown on the plates containing Y10 or Y24 alone, but no bacteria grew on these plates. Multiple simultaneous mutations ($thr^- \rightarrow thr^+$,

$leu^- \rightarrow leu^+$, and $thi^- \rightarrow thi^+$ in strain Y10 or $bio^- \rightarrow bio^+$, $phe^- \rightarrow phe^+$, and $cys^- \rightarrow cys^+$ in strain Y24) would have been required for either strain to become prototrophic by mutation, which was very improbable. Lederberg and Tatum concluded that some type of genetic transfer and recombination had taken place:

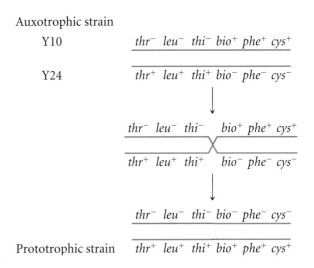

Auxotrophic strain

Y10 thr^- leu^- thi^- bio^+ phe^+ cys^+

Y24 thr^+ leu^+ thi^+ bio^- phe^- cys^-

thr^- leu^- thi^- bio^+ phe^+ cys^+

thr^+ leu^+ thi^+ bio^- phe^- cys^-

thr^- leu^- thi^- bio^- phe^- cys^-

Prototrophic strain thr^+ leu^+ thi^+ bio^+ phe^+ cys^+

What they did not know was *how* it had taken place.

To study this problem, Bernard Davis constructed a U-shaped tube (◀ FIGURE 8.11) that was divided into two compartments by a filter having fine pores. This filter allowed liquid medium to pass from one side of the tube to the other, but the pores of the filter were too small to allow passage of bacteria. Two auxotrophic strains of bacteria were placed on opposite sides of the filter, and suction was applied alternately to the ends of the U-tube, causing the medium to flow back and forth between the two compartments. Despite hours of incubation in the U-tube, bacteria plated out on minimal medium did not grow; there had been no genetic exchange between the strains. The exchange of bacterial genes clearly required direct contact between the bacterial cells. This type of genetic exchange entailing cell-to-cell contact in bacteria is called conjugation.

F^+ and F^- cells In most bacteria, conjugation depends on a fertility (F) factor that is present in the donor cell and absent in the recipient cell. Cells that contain F are referred to as F^+, and cells lacking F are F^-.

The F factor contains an origin of replication and a number of genes required for conjugation (see Figure 8.8). For example, some of these genes encode sex **pili** (singular, pilus), slender extensions of the cell membrane. A cell containing F produces the sex pili, which makes contact with a receptor on an F^- cell (◀ FIGURE 8.12) and pulls the two cells together. DNA is then transferred from the F^+ cell to the F^- cell. Conjugation can take place only between a cell that possesses F and a cell that lacks F.

Experiment

Question: How did the genetic exchange seen in Lederberg and Tatum's experiment take place?

Auxotrophic strain A

Auxotrophic strain B

Airflow

Strain A ——

—— Strain B

When two auxotrophic strains were separated by a filter that allowed mixing of medium but not bacteria,...

...no prototrophic bacteria were produced

Minimal medium	Minimal medium	Minimal medium	Minimal medium
No growth	No growth	No growth	No growth

Conclusion: Genetic exchange requires direct contact between bacterial cells.

◀8.11   **Davis's U-tube experiment.**

In most cases, the only genes transferred during conjugation between an F$^+$ and F$^-$ cell are those on the F factor (◀FIGURE 8.13a and b). Transfer is initiated when one of the DNA strands on the F factor is nicked at an origin (*oriT*). One end of the nicked DNA separates from the circle and passes into the recipient cell (◀FIGURE 8.13c). Replication takes place on the nicked strand, proceeding around the circular plasmid and replacing the transferred strand (◀FIGURE 8.13d). Because the plasmid in the F$^+$ cell is always nicked at the *oriT* site, this site always enters the recipient cell first, followed by the rest of the plasmid. Thus, the transfer of genetic material has a defined direction. Once inside the recipient cell, the single strand

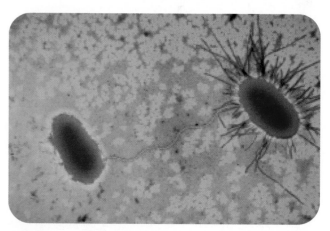

◀8.12   **A sex pilus connects F$^+$ and F$^-$ cells during bacterial conjugation.** *E. Coli* cells in conjugation. (Dr Dennis Kunkel/Phototake.)

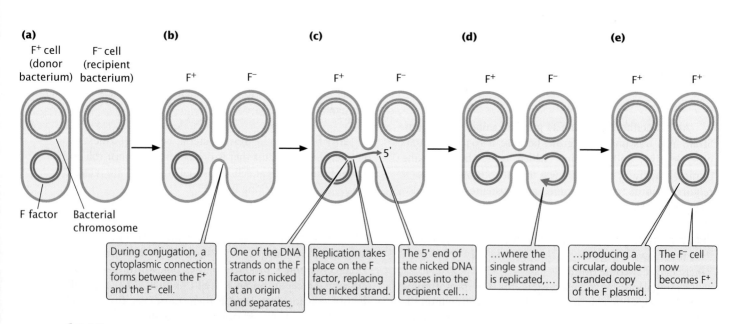

(a)

F$^+$ cell (donor bacterium)   F$^-$ cell (recipient bacterium)

F factor   Bacterial chromosome

(b) F$^+$ F$^-$

(c) F$^+$ F$^-$ 5'

(d) F$^+$ F$^-$

(e) F$^+$ F$^+$

During conjugation, a cytoplasmic connection forms between the F$^+$ and the F$^-$ cell.

One of the DNA strands on the F factor is nicked at an origin and separates.

Replication takes place on the F factor, replacing the nicked strand.

The 5' end of the nicked DNA passes into the recipient cell...

...where the single strand is replicated,...

...producing a circular, double-stranded copy of the F plasmid.

The F$^-$ cell now becomes F$^+$.

◀8.13   **The F factor is transferred during conjugation between an F$^+$ and F$^-$ cell.**

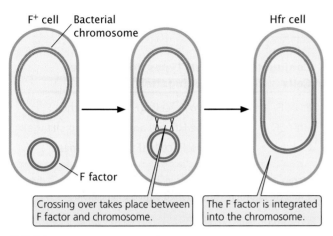

8.14 **The F factor is integrated into the bacterial chromosome in an Hfr cell.**

is replicated, producing a circular, double-stranded copy of the F plasmid (◀FIGURE 8.13e). If the entire F factor is transferred to the recipient F⁻ cell, that cell becomes an F⁺ cell.

Hfr cells Conjugation transfers genetic material in the F plasmid from F⁺ to F⁻ cells but does not account for the transfer of chromosomal genes observed by Lederberg and Tatum. In Hfr (high-frequency) strains, the F factor is integrated into the bacterial chromosome (◀FIGURE 8.14). Hfr cells behave as F⁺ cells, forming sex pili and undergoing conjugation with F⁻ cells.

In conjugation between Hfr and F⁻ cells (◀FIGURE 8.15a), the integrated F factor is nicked, and the end of the nicked strand moves into the F⁻ cell (◀FIGURE 8.15b), just as it does in conjugation between F⁺ and F⁻ cells. In the Hfr

cells, the F factor is linked to the bacterial chromosome, so the chromosome follows it into the recipient cell. How much of the bacterial chromosome is transferred depends on the length of time that the two cells remain in conjugation.

Once inside the recipient cell, the donor DNA strand is replicated (◀FIGURE 8.15c), and crossing over between it and the original chromosome of the F⁻ cell (◀FIGURE 8.15d) may take place. This gene transfer between Hfr and F⁻ cells is how the recombinant prototrophic cells observed by Lederberg and Tatum were produced. When crossing over has taken place in the recipient cell, the donated chromosome is degraded, and the recombinant recipient chromosome remains (◀FIGURE 8.15e) to be replicated and passed to later generations by binary fission.

In a mating of Hfr × F⁻, the F⁻ cell almost never becomes F⁺ or Hfr, because the F factor is nicked in the middle during the initiation of strand transfer, placing part of F at the beginning and part at the end of the strand to be transferred. To become F⁺ or Hfr, the recipient cell must receive the entire F factor, requiring that the entire bacterial chromosome is transferred. This event happens rarely, because most conjugating cells break apart before the entire chromosome has been transferred.

The F plasmid in F⁺ cells integrates into the bacterial chromosome, causing an F⁺ cell to become Hfr, at a frequency of only about $^1/_{10,000}$. This low frequency accounts for the low rate of recombination observed by Lederberg and Tatum in their F⁺ cells. The F factor is excised from the bacterial chromosome at a similarly low rate, causing a few Hfr cells to become F⁺.

F′ cells When an F factor does excise from the bacterial chromosome, a small amount of the bacterial chromosome

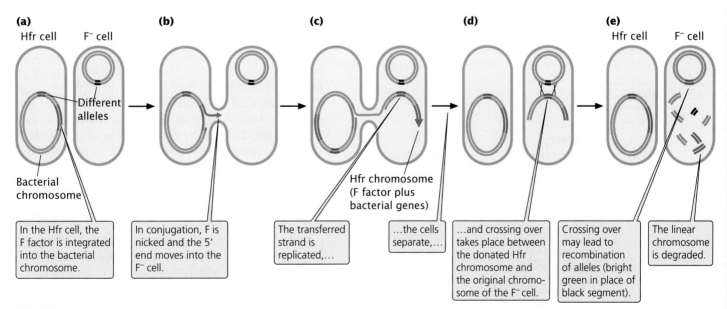

8.15 **Bacterial genes may be transferred from an Hfr cell to an F⁻ cell in conjugation.**

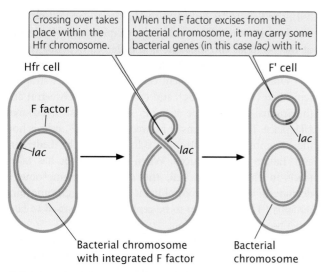

Crossing over takes place within the Hfr chromosome.

When the F factor excises from the bacterial chromosome, it may carry some bacterial genes (in this case *lac*) with it.

Hfr cell

F factor

lac

lac

F' cell

lac

Bacterial chromosome with integrated F factor

Bacterial chromosome

8.16 **An Hfr cell may be converted into an F′ cell when the F factor excises from the bacterial chromosome and carries bacterial genes with it.**

may be removed with it, and these chromosomal genes will then be carried with the F plasmid (◀ FIGURE 8.16). Cells containing an F plasmid with some bacterial genes are called F prime (F′). For example, if an F factor integrates into a chromosome adjacent to the chromosome's *lac* operon, the F factor may pick up *lac* genes when it excises, becoming F′*lac*. F′ cells can conjugate with F⁻ cells, given that they possess the F plasmid with all the genetic information necessary for conjugation and gene transfer. Characteristics of different mating types of *E. coli* (cells with different types of F) are summarized in Table 8.2.

During conjugation between an F′*lac* cell and an F⁻ cell, the F plasmid is transferred to the F⁻ cell, which means that any genes on the F plasmid, including those from the bacterial chromosome, may be transferred to F⁻ recipient cells. This process is called sexduction. It produces partial diploids, or *merozygotes*, which are cells with two copies of some genes,

Table 8.2	Characteristics of *E. coli* cells with different types of F factor	
Type	**F Factor Characteristics**	**Role in Conjugation**
F⁺	Present as separate circular DNA	Donor
F⁻	Absent	Recipient
Hfr	Present, integrated into bacterial chromosome	High-frequency donor
F′	Present as separate circular DNA, carrying some bacterial genes	Donor

Table 8.3	Results of conjugation between cells with different F factors
Conjugating Cells	**Cell Types Present After Conjugation**
F⁺ × F⁻	Two F⁺ cells (F⁻ cell becomes F⁺)
Hfr × F⁻	One Hfr cell and one F⁻ (no change)*
F′ × F⁻	Two F′ cells (F⁻ cell becomes F′)

*Rarely, the F⁻ cell becomes F⁺ in an Hfr × F⁻ conjugation if the entire chromosome is transferred during conjugation.

one on the bacterial chromosome and one on the newly introduced F plasmid. The outcomes of conjugation between different mating types of *E. coli* are summarized in Table 8.3.

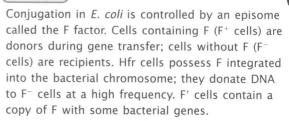

Concepts

Conjugation in *E. coli* is controlled by an episome called the F factor. Cells containing F (F⁺ cells) are donors during gene transfer; cells without F (F⁻ cells) are recipients. Hfr cells possess F integrated into the bacterial chromosome; they donate DNA to F⁻ cells at a high frequency. F′ cells contain a copy of F with some bacterial genes.

Mapping bacterial genes with interrupted conjugation

The transfer of DNA that takes place during conjugation between Hfr and F⁻ cells allows bacterial genes to be mapped. During conjugation, the chromosome of the Hfr cell is transferred to the F⁻ cell. Transfer of the entire *E. coli* chromosome requires about 100 minutes; if conjugation is interrupted before 100 minutes have elapsed, only part of the chromosome will pass into the F⁻ cell and have an opportunity to recombine with the recipient chromosome. Chromosome transfer always begins within the integrated F factor and proceeds in a continuous direction; so genes are transferred according to their sequence on the chromosome. The time required for individual genes to be transferred indicates their relative positions on the chromosome. In most genetic maps, distances are expressed as percent recombination; but, in bacterial maps constructed with interrupted conjugation, the basic unit of distance is a minute.

Worked Problem

To illustrate the method of mapping genes with interrupted conjugation, let's look at a cross analyzed by François Jacob and Elie Wollman, who first developed this method of gene mapping (◀ FIGURE 8.17a). They used donor Hfr cells that were sensitive to the antibiotic streptomycin (genotype *str*ˢ); resistant to sodium azide (*azi*ʳ) and infection by bacteriophage T1 (*ton*ʳ); prototrophic for threonine (*thr*⁺) and leucine (leu^+); and able to break down lactose (lac^+) and galactose (gal^+). They used F⁻ recipient cells that were resistant to streptomycin (str^r); sensitive to sodium azide (azi^s) and to infection by bacteriophage T1 (ton^s); auxotrophic for threonine (thr^-) and leucine (leu^-); and unable to breakdown lactose (lac^-) and galactose (gal^-). Thus, the genotypes of the donor and recipient cells were:

Donor Hfr cells: Hfr str^s thr^+ leu^+ azi^r ton^r lac^+ gal^+
Recipient F⁻ cells: F⁻ str^r thr^- leu^- azi^s ton^s lac^- gal^-

The two strains were mixed in nutrient medium and allowed to conjugate. After a few minutes, the medium was diluted to prevent any new pairings. At regular intervals, a sample of cells was removed and agitated vigorously in a kitchen blender to halt all conjugation and DNA transfer. The cells were plated on a selective medium that contained streptomycin and lacked leucine and threonine. The original donor cells were streptomycin sensitive (str^s) and would not grow on this medium. The F⁻ recipient cells were auxotrophic for leucine and threonine and also failed to grow on this medium. Only cells that underwent conjugation and received at least the leu^+ and thr^+ genes from the Hfr donors could grow on the selective medium. All str^r leu^+ thr^+ cells were then tested for the presence of other genes that might have been transferred from the donor Hfr strain.

All of the cells that grow on the selective medium are str^r leu^+ thr^+; so we know that these genes were transferred. The percentage of str^r leu^+ thr^+ exconjugates receiving specific alleles (azi^r, ton^r, lac^+, and gal^+) from the Hfr strain are plotted against the duration of conjugation (◀FIGURE 8.17b). What is the order and distances among the genes?

• Solution

The first donor gene to appear in all of these exconjugates (at about 9 minutes) was azi^r. Gene ton^r appeared next (after about 10 minutes), followed by lac^+ (at about 18 minutes) and by gal^+ (after 25 minutes). These transfer times indicate the order and relative distances among the genes (◀FIGURE 8.17b).

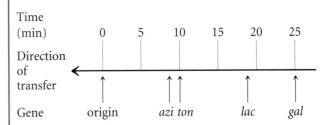

Notice that the maximum frequency of exconjugates decreased for the more distant genes. For example, about 90% of the exconjugates received the azi^r allele, but only about 30% received the gal^+ allele. The lower percentage for gal^+ is due to the fact that some conjugating cells spontaneously broke apart before they were disrupted by the blender. The probability of spontaneous disruption increases with time; so fewer cells had an opportunity to receive genes that were transferred later.

Experiment

Question: How can interrupted conjugation be used to map bacterial genes?

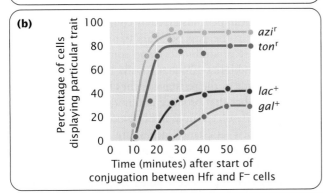

Conclusion: The transfer times indicate the order and relative distances between genes and can be used to construct a genetic map.

◀8.17 **Jacob and Wollman used interrupted conjugation to map bacterial genes.**

Directional transfer and mapping Different Hfr strains have the F factor integrated into the bacterial chromosome at different sites and in different orientations. Gene transfer always begins within F, and the orientation and position of F determine the direction and starting point of gene transfer. In ◀ FIGURE 8.18a, strain Hfr1 has F integrated between *leu* and *azi*; the orientation of F at this site dictates that gene transfer will proceed in a counterclockwise direction around the circular chromosome. Genes from this strain will be transferred in the order of:

$$\leftarrow leu - thr - thi - his - gal - lac - pro - azi$$

Strain Hfr5 has F integrated between the *thi* and *his* genes (◀ FIGURE 8.18b) and in the opposite orientation. Here gene transfer will proceed in a clockwise direction:

$$\leftarrow thi - thr - leu - azi - pro - lac - gal - his$$

Although the starting point and direction of transfer may differ between two strains, the relative distance in time between any two pairs of genes is constant.

Notice that the order of gene transfer is not the same for different Hfr strains (◀ FIGURE 8.19a). For example, *azi* is transferred just after *leu* in strain HfrH, but long after *leu* in strain Hfr1. Aligning the sequences (◀ FIGURE 8.19b) shows that the two genes on either side of *azi* are always the same: *leu* and *pro*. That they are the same makes sense when one recognizes that the bacterial chromosome is circular and the starting point of transfer varies from strain to strain. These data provided the first evidence that the bacterial chromosome is circular (◀ FIGURE 8.19c).

> **Concepts**
>
> Conjugation can be used to map bacterial genes by mixing Hfr and F⁻ cells that differ in genotype and interrupting conjugation at regular intervals. The amount of time required for individual genes to be transferred from the Hfr to the F⁻ cells indicates the relative positions of the genes on the bacterial chromosome.

Natural Gene Transfer and Antibiotic Resistance

Many pathogenic bacteria have developed resistance to antibiotics, particularly in environments where antibiotics are routinely used, such as hospitals and fish farms. (Massive amounts of antibiotics are often used in aquaculture to prevent infection in the fish and enhance their growth.) The continual presence of antibiotics in these environments selects for resistant bacteria, which reduces the effectiveness of antibiotic treatment for medically important infections.

Antibiotic resistance in bacteria frequently results from the action of genes located on *R plasmids,* small circular

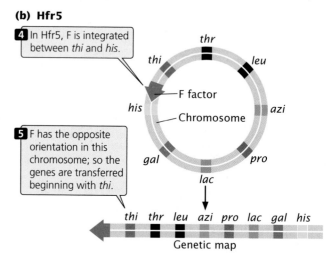

(a) Hfr1

1 Transfer always begins within F, and the orientation of F determines the direction of transfer.

2 In Hfr1, F is integrated between the *leu* and *azi* genes;....

3 ...so the genes are transferred beginning with *leu*.

leu thr thi his gal lac pro azi
Genetic map

(b) Hfr5

4 In Hfr5, F is integrated between *thi* and *his*.

5 F has the opposite orientation in this chromosome; so the genes are transferred beginning with *thi*.

thi thr leu azi pro lac gal his
Genetic map

◀ **8.18 The orientation of the F factor in an Hfr strain determines the direction of gene transfer.** Arrowheads indicate the origin and direction of transfer.

plasmids that can be transferred by conjugation. R plasmids have evolved in the past 50 years (since the beginning of widespread use of antibiotics), and some convey resistance to several antibiotics simultaneously. Ironic but plausible sources of some of the resistance genes found in R plasmids are the microbes that produce antibiotics in the first place.

The results of recent studies demonstrate that R plasmids and their resistance genes are transferred among bacteria in a variety of natural environments. In one study, plasmids carrying genes for resistance to multiple antibiotics were transferred from a cow udder infected with *E. coli* to a human strain of *E. coli* on a hand towel: a farmer wiping his hands after milking an infected cow might unwittingly transfer antibiotic resistance from bovine- to human-inhabiting microbes. Conjugation taking place in minced meat on a cutting board allowed R plasmids to be passed from porcine

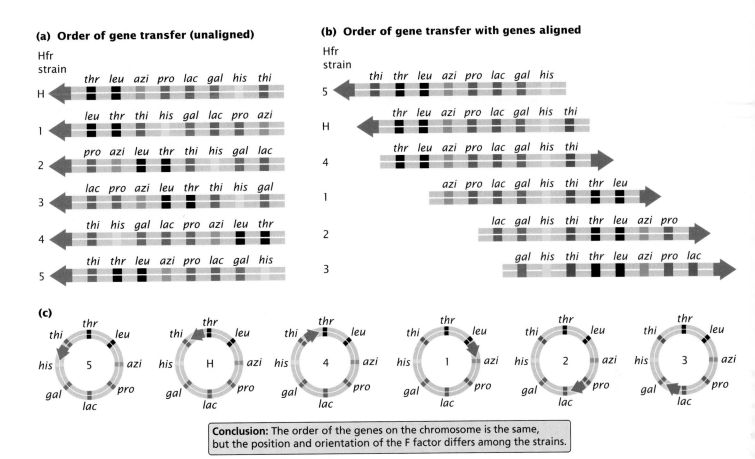

Conclusion: The order of the genes on the chromosome is the same, but the position and orientation of the F factor differs among the strains.

◀8.19 The order of gene transfer in a series of different Hfr strains indicates that the *E. coli* chromosome is circular.

(pig) to human *E. coli*. The transfer of R plasmids also occurs in sewage, soil, lake water, and marine sediments.

Perhaps most significantly, the transfer of R plasmids is not restricted to bacteria of the same or even related species. R plasmids with multiple antibiotic resistances have been transferred in marine waters from *E. coli* and other human-inhabiting bacteria (in sewage) to the fish bacterium *Aeromona salmonicida* and then back to *E. coli* through raw salmon chopped on a cutting board. These results indicate that R plasmids can spread easily through the environment, passing among related and unrelated bacteria in a variety of common situations. That they can do so underscores both the importance of limiting antibiotic use to treating medically important infections and the importance of hygiene in everyday life.

Transformation in Bacteria

A second way that DNA can be transferred between bacteria is through transformation (see Figure 8.9b). Transformation played an important role in the initial identification of DNA as the genetic material, which will be discussed in Chapter 10.

Transformation requires both the uptake of DNA from the surrounding medium and its incorporation into the bacterial chromosome or a plasmid. It may occur naturally when dead bacteria break up and release DNA fragments into the environment. In soil and marine environments, this means may be an important route of genetic exchange for some bacteria.

Cells that take up DNA are said to be **competent.** Some species of bacteria take up DNA more easily than do others; competence is influenced by growth stage, the concentration of available DNA, and the composition of the medium. The uptake of DNA fragments into a competent bacterial cell appears to be a random process. The DNA need not even be bacterial: virtually any type of DNA (bacterial or otherwise) can be transferred to competent cells under the appropriate conditions.

As a DNA fragment enters the cell in the course of transformation (◀FIGURE 8.20), one of the strands is hydrolyzed, whereas the other strand associates with proteins as it moves across the membrane. Once inside the cell, this single strand may pair with a homologous region and become integrated into the bacterial chromosome. This integration requires two crossover events, after which the remaining single-stranded DNA is degraded by bacterial enzymes.

Bacterial geneticists have developed techniques to increase the frequency of transformation in the laboratory

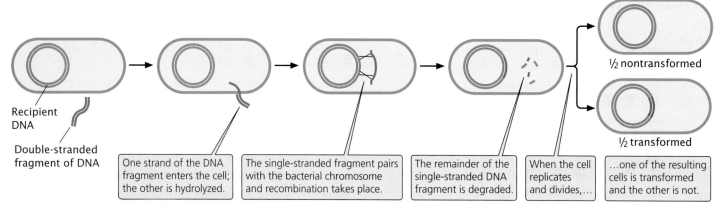

Recipient
DNA

Double-stranded
fragment of DNA

| One strand of the DNA fragment enters the cell; the other is hydrolyzed. | The single-stranded fragment pairs with the bacterial chromosome and recombination takes place. | The remainder of the single-stranded DNA fragment is degraded. | When the cell replicates and divides,… | …one of the resulting cells is transformed and the other is not. |

½ nontransformed

½ transformed

◀ 8.20 **Genes can be transferred between bacteria through transformation.**

in order to introduce particular DNA fragments into cells. They have developed strains of bacteria that are more competent than wild-type cells. Treatment with calcium chloride, heat shock, or an electrical field makes bacterial membranes more porous and permeable to DNA, and the efficiency of transformation can also be increased by using high concentrations of DNA. These techniques make it possible to transform bacteria such as *E. coli*, which are not naturally competent.

Transformation, like conjugation, is used to map bacterial genes, especially in those species that do not undergo conjugation or transduction (see Figure 8.9a and c). Transformation mapping requires two strains of bacteria that differ in several genetic traits; for example, the recipient strain might be $a^- b^- c^-$ (auxotrophic for three nutrients), with the donor cell being prototrophic with alleles $a^+ b^+ c^+$.

DNA from the donor strain is isolated and purified. The recipient strain is treated to increase competency, and DNA from the donor strain is added to the medium. Fragments of the donor DNA enter the recipient cells and undergo recombination with homologous DNA sequences on the bacterial chromosome. Cells that receive genetic material through transformation are called **transformants.**

Genes can be mapped by observing the rate at which two or more genes are transferred together (**cotransformed**) in transformation. When the DNA is fragmented during isolation, genes that are physically close on the chromosome are more likely to be present on the same DNA fragment and transferred together, as shown for genes a^+ and b^+ in ◀ FIGURE 8.21. Genes that are far apart are unlikely to be present on the same DNA fragment and rarely will be transferred together. Once inside the cell, DNA becomes

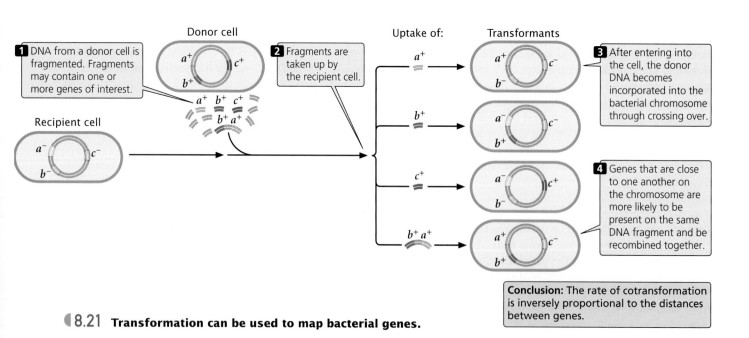

1 DNA from a donor cell is fragmented. Fragments may contain one or more genes of interest.

Donor cell

Uptake of: Transformants

2 Fragments are taken up by the recipient cell.

Recipient cell

3 After entering into the cell, the donor DNA becomes incorporated into the bacterial chromosome through crossing over.

4 Genes that are close to one another on the chromosome are more likely to be present on the same DNA fragment and be recombined together.

Conclusion: The rate of cotransformation is inversely proportional to the distances between genes.

◀ 8.21 **Transformation can be used to map bacterial genes.**

incorporated into the bacterial chromosome through recombination. If two genes are close together on the same fragment, any two crossovers are likely to occur on either side of the two genes, allowing both to become part of the recipient chromosome. If the two genes are far apart, there may be one crossover between them, allowing one gene but not the other to recombine with the bacterial chromosome. Thus, two genes are more likely to be transferred together when they are close together on the chromosome, and genes located far apart are rarely cotransformed. Therefore, the frequency of cotransformation can be used to map bacterial genes. If genes *a* and *b* are frequently cotransformed, and genes *b* and *c* are frequently cotransformed, but genes *a* and *c* are rarely cotransformed, then gene *b* must be between *a* and *c*—the gene order is *a b c*.

> **Concepts**
>
> Genes can be mapped in bacteria by taking advantage of transformation, the ability of cells to take up DNA from the environment and incorporate it into their chromosomes through crossing over. The relative rate at which pairs of genes are cotransformed indicates the distance between them: the higher the rate of cotransformation, the closer the genes are on the bacterial chromosome.

Bacterial Genome Sequences

Genetic maps serve as the foundation for more detailed information provided by DNA sequencing, such as gene content and organization (see Chapter 19 for a discussion of gene sequencing).

Geneticists have now determined the complete nucleotide sequence of a number of bacterial genomes. The genome of *E. coli*, one of the most widely studied of all bacteria, is a single circular DNA molecule approximately 1 mm in length. It consists of 4,638,858 nucleotides and an estimated 4300 genes, more than half of which have no known function. These "orphan genes" may play important roles in adapting to unusual environments, coordinating metabolic pathways, organizing the chromosome, or communicating with other bacterial cells.

A number of other bacterial genomes have been completely sequenced (see Table 19.2), and many additional microbial sequencing projects are underway. A substantial proportion of genes in all bacteria have no known function. Certain genes, particularly those with related functions, tend to reside next to one another, but these clusters are in very different locations in different species, suggesting that bacterial genomes are constantly being reshuffled. Comparisons of the gene sequences of pathogenic and nonpathogenic bacteria are helping to identify genes implicated in disease and may suggest new targets for antibiotics and other antimicrobial agents.

www.whfreeman.com/pierce For a current list of completed and partial microbial genome projects and a list of microbial genome projects funded by the U.S. Department of Energy

Viral Genetics

All organisms—plants, animals, fungi, and bacteria—are infected by viruses. A **virus** is a simple replicating structure made up of nucleic acid surrounded by a protein coat (see Figure 2.4). Viruses come in a great variety of shapes and sizes (◀FIGURE 8.22). Some have DNA as their genetic material, whereas others have RNA; the nucleic acid may be double stranded or single stranded, linear or circular. Not surprisingly, viruses reproduce in a number of different ways.

Bacteriophages (phages) have played a central role in genetic research since the late 1940s. They are ideal for many types of genetic research because they have small and easily manageable genomes, reproduce rapidly, and produce large numbers of progeny. Bacteriophages have two alternative life cycles: the lytic and the lysogenic cycles. In the lytic cycle, a phage attaches to a receptor on the bacterial cell wall and

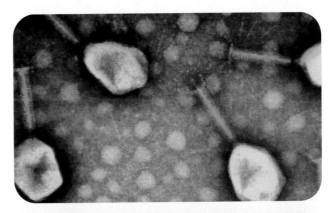

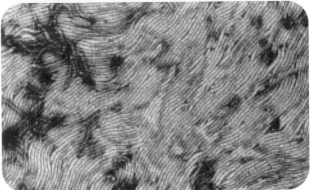

◀ **8.22 Viruses occur in different strutures and sizes.** Top, T4 bacteriophage. Bottom, potato virus. (Top, Dr. Dennis Kunkel/Phototake; bottom, R. W. Horne/Photo Researchers.)

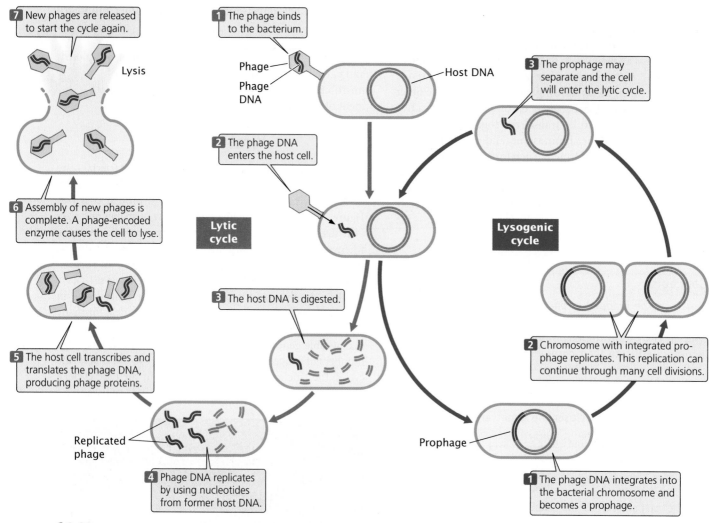

7 New phages are released to start the cycle again.

Lysis

1 The phage binds to the bacterium.

Phage

Phage DNA

Host DNA

3 The prophage may separate and the cell will enter the lytic cycle.

2 The phage DNA enters the host cell.

6 Assembly of new phages is complete. A phage-encoded enzyme causes the cell to lyse.

Lytic cycle

Lysogenic cycle

5 The host cell transcribes and translates the phage DNA, producing phage proteins.

3 The host DNA is digested.

2 Chromosome with integrated pro-phage replicates. This replication can continue through many cell divisions.

Replicated phage

Prophage

4 Phage DNA replicates by using nucleotides from former host DNA.

1 The phage DNA integrates into the bacterial chromosome and becomes a prophage.

◀**8.23 Bacteriophages have two alternating life cycles—lytic and lysogenic.**

injects its DNA into the cell (◀FIGURE 8.23). Once inside the cell, the phage DNA is replicated, transcribed, and translated, producing more phage DNA and phage proteins. New phage particles are assembled from these components. The phages then produce an enzyme that breaks open the cell, releasing the new phages. **Virulent phages** reproduce strictly through the lytic cycle and always kill their host cells.

Temperate phage can utilize either the lytic or the lysogenic cycle. The lysogenic cycle begins like the lytic cycle (see Figure 8.23) but, inside the cell, the phage DNA integrates into the bacterial chromosome, where it remains as an inactive **prophage.** The prophage is replicated along with the bacterial DNA and is passed on when the bacterium divides. Certain stimuli cause the prophage to dissociate from the bacterial chromosome and enter into the lytic cycle, producing new phage particles and lysing the cell.

www.whfreeman.com/pierce   For additional information on viruses

Techniques for the Study of Bacteriophages

Viruses reproduce only within host cells; so bacteriophages must be cultured in bacterial cells. To do so, phages and bacteria are mixed together and plated on solid medium in a petri plate. A high concentration of bacteria is used so that the colonies grow into one another and produce a continuous layer of bacteria, or "lawn," on the agar. An individual phage infects a single bacterial cell and goes through its lytic cycle. Many new phages are released from the lysed cell and infect additional cells; the cycle is then repeated. The bacteria grow on solid medium; so the diffusion of the phages is restricted and only nearby cells are infected. After several rounds of phage reproduction, a clear patch of lysed cells (a **plaque**) appears on the plate (◀FIGURE 8.24). Each plaque represents a single phage that multiplied and lysed many cells. Plating a known volume of a dilute solution of phages on a bacterial lawn and counting the number of plaques that appear can be used to determine the original concentration of phage in the solution.

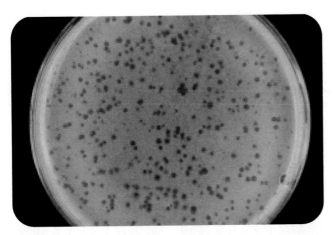

8.24 Plaques are clear patches of lysed cells on a lawn of bacteria. (E. C. S. Chan/Visuals Unlimited.)

(Concepts)

Viral genomes may be DNA or RNA, circular or linear, and double or single stranded. Bacteriophages are used in many types of genetic research.

Gene Mapping in Phages

Mapping genes in bacteriophage requires the application of the same principles as those applied to mapping genes in eukaryotic organisms (Chapter 7). Crosses are made between viruses that differ in two or more genes, and recombinant progeny phage are identified and counted. The proportion of recombinant progeny is then used to estimate the distances between the genes and their linear order on the chromosome.

In 1949, Alfred Hershey and Raquel Rotman examined rates of recombination between genes in two strains of the T2 bacteriophage that differed in plaque appearance and host range (the bacterial strains that the phages could infect). One strain was able to infect and lyse type B *E. coli* cells but not B/2 cells (normal host range, h^+) and produced an abnormal plaque that was large with distinct borders (r^-). The second strain was able to infect and lyse *both* B *and* B/2 cells (mutant host range, h^-) and produced normal plaques that were small with fuzzy borders (r^+).

Hershey and Rotman crossed the $h^+ r^-$ and $h^- r^+$ strains of T2 by infecting type B *E. coli* cells with a mixture of the two strains. They used a high concentration of phages so that most cells could be simultaneously infected by both strains (◀FIGURE 8.25). Homologous recombination occasionally took place between the chromosomes of the different strains, producing $h^+ r^+$ and $h^- r^-$ chromosomes, which

8.25 Hershey and Rotman developed a technique for mapping viral genes. (Photo from G. S. Stent, *Molecular Biology of Bacterial Viruses.* © 1963 by W. H. Freeman and Company.)

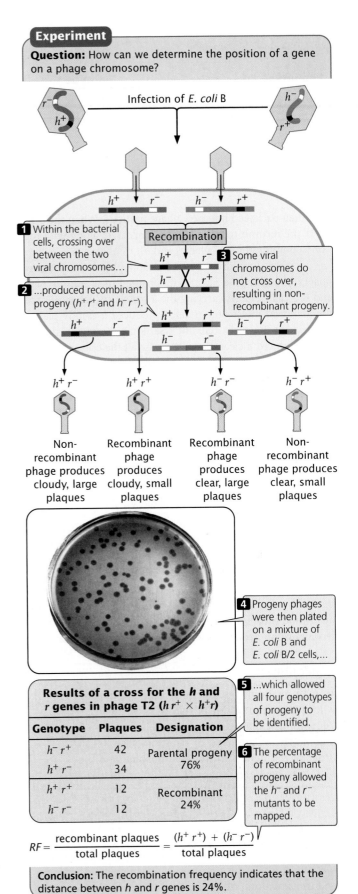

Question: How can we determine the position of a gene on a phage chromosome?

Genotype	Plaques	Designation
$h^- r^+$	42	Parental progeny 76%
$h^+ r^-$	34	
$h^+ r^+$	12	Recombinant 24%
$h^- r^-$	12	

$$RF = \frac{\text{recombinant plaques}}{\text{total plaques}} = \frac{(h^+ r^+) + (h^- r^-)}{\text{total plaques}}$$

Conclusion: The recombination frequency indicates that the distance between *h* and *r* genes is 24%.

Table 8.4	Progeny phage produced from $h^-r^+ \times h^+r^-$	
Phenotype		**Genotype**
Clear and small		h^-r^+
Cloudy and large		h^+r^-
Cloudy and small		h^+r^+
Clear and large		h^-r^-

were then packaged into new phage particles. When the cells lysed, the recombinant phages were released, along with the nonrecombinant $h^+\ r^-$ phages and $h^-\ r^+$ phages.

Hershey and Rotman diluted and plated the progeny phages on a bacterial lawn that consisted of a *mixture* of B and B/2 cells. Phages carrying the h^+ allele (which conferred the ability to infect only B cells) produced a cloudy plaque because the B/2 cells did not lyse. Phages carrying the h^- allele produced a clear plaque because all the cells within the plaque were lysed. The r^+ phages produced small plaques, whereas the r^- phages produced large plaques. The genotypes of these progeny phages could therefore be determined by the appearance of the plaque (see Figure 8.25 and Table 8.4).

In this type of phage cross, the recombination frequency (*RF*) between the two genes can be calculated by using the following formula:

$$RF = \frac{\text{recombinant plaques}}{\text{total plaques}}$$

In Hershey and Rotman's cross, the recombinant plaques were $h^+\ r^+$ and $h^-\ r^-$; so the recombination frequency was

$$RF = \frac{(h^+\ r^+) + (h^-\ r^-)}{\text{total plaques}}$$

Recombination frequencies can be used to determine the distances and orders of genes on the phage chromosome, just as recombination frequencies are used to map genes in eukaryotes.

> **Concepts**
>
> To map phage genes, bacterial cells are infected with viruses that differ in two or more genes. Recombinant plaques are counted, and rates of recombination are used to determine the linear order of the genes on the chromosome and the distances between them.

Transduction: Using Phages to Map Bacterial Genes

In the discussion of bacterial genetics, we identified three mechanisms of gene transfer: conjugation, transformation, and transduction (see Figure 8.9). Let's take a closer look at transduction, in which genes are transferred between bacte-

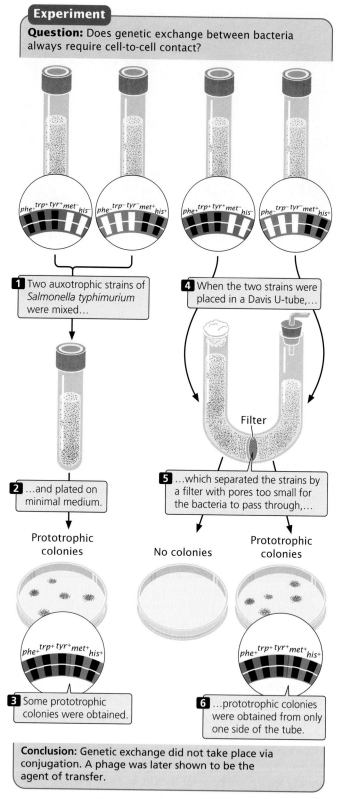

Experiment

Question: Does genetic exchange between bacteria always require cell-to-cell contact?

1 Two auxotrophic strains of *Salmonella typhimurium* were mixed...

2 ...and plated on minimal medium.

Prototrophic colonies

3 Some prototrophic colonies were obtained.

4 When the two strains were placed in a Davis U-tube,...

Filter

5 ...which separated the strains by a filter with pores too small for the bacteria to pass through,...

No colonies

Prototrophic colonies

6 ...prototrophic colonies were obtained from only one side of the tube.

Conclusion: Genetic exchange did not take place via conjugation. A phage was later shown to be the agent of transfer.

◀ **8.26 The Lederberg and Zinder experiment.**

ria by viruses. In **generalized transduction,** any gene may be transferred. In **specialized transduction,** only a few genes are transferred.

Generalized transduction Joshua Lederberg and Norton Zinder discovered generalized transduction in 1952. They were trying to produce recombination in the bacterium *Salmonella typhimurium* by conjugation. They mixed a strain of *S. typhimurium* that was *phe⁺ trp⁺ tyr⁺ met⁻ his⁻* (◀FIGURE 8.26) and plated them on minimal medium. A few prototrophic recombinants (*phe⁺ trp⁺ tyr⁺ met⁺ his⁺*) appeared, suggesting that conjugation had taken place. However, when they tested the two strains in a U-shaped tube similar to the one used by Davis, some *phe⁺ trp⁺ tyr⁺ met⁺ his⁺* prototrophs were obtained on one side of the tube (compare Figure 8.26 with Figure 8.11). This apparatus separated the two strains by a filter with pores too small for the passage of bacteria; so how were genes being transferred between bacteria in the absence of conjugation? The results of subsequent studies revealed that the agent of transfer was a bacteriophage.

In the lytic cycle of phage reproduction, the bacterial chromosome is broken into random fragments (◀FIGURE 8.27). For some types of bacteriophage, a piece of the bacterial chromosome occasionally gets packaged into a phage coat instead of phage DNA; these phage particles are called **transducing phages.** The transducing phage infects a new cell, releasing the bacterial DNA, and the introduced genes may then become integrated into the bacterial chromosome by a double crossover. Bacterial genes can, by this process, be moved from one bacterial strain to another, producing recombinant bacteria called **transductants.**

Not all phages are capable of transduction, a rare event that requires (1) that the phage degrade the bacterial chromosome; (2) that the process of packaging DNA into the phage protein not be specific for phage DNA; and (3) that the bacterial genes transferred by the virus recombine with the chromosome in the recipient cell.

Because of the limited size of a phage particle, only about 1% of the bacterial chromosome can be transduced. Only genes located close together on the bacterial chromosome will be transferred together (**cotransduced**). The overall rate of transduction ranges from only about 1 in 100,000 to 1 in 1,000,000. Because the chance of a cell being transduced by two separate phages is exceedingly small, any cotransduced genes are usually located close together on the bacterial chromosome. Thus, rates of cotransduction, like rates of cotransformation, give an indication of the physical distances between genes on a bacterial chromosome.

To map genes by using transduction, two bacterial strains with different alleles at several loci are used. The donor strain is infected with phages (◀FIGURE 8.28), which reproduce within the cell. When the phages have lysed the donor cells, a suspension of the progeny phage is mixed with a recipient strain of bacteria, which are then plated on several different kinds of media to determine the phenotypes of the transducing progeny phages.

> **Concepts**
>
> In transduction, bacterial genes become packaged into a viral coat, are transferred to another bacterium by the virus, and become incorporated into the bacterial chromosome by crossing over. Bacterial genes can be mapped with the use of generalized transduction.

Specialized transduction Like generalized transduction, specialized transduction requires gene transfer from one bacterium to another through phages, but here only genes near particular sites on the bacterial chromosome are transferred. This process requires lysogenic bacteriophages. The

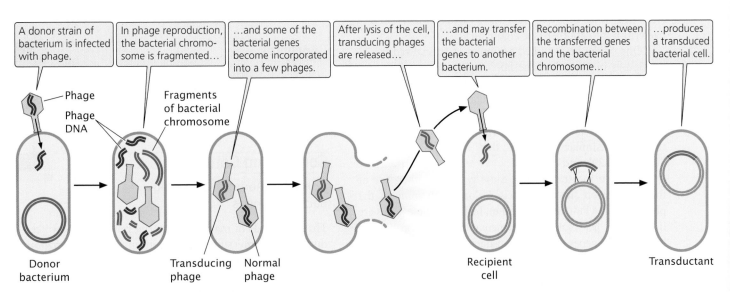

| A donor strain of bacterium is infected with phage. | In phage reproduction, the bacterial chromosome is fragmented... | ...and some of the bacterial genes become incorporated into a few phages. | After lysis of the cell, transducing phages are released... | ...and may transfer the bacterial genes to another bacterium. | Recombination between the transferred genes and the bacterial chromosome... | ...produces a transduced bacterial cell. |

Phage
Phage DNA
Fragments of bacterial chromosome

Donor bacterium   Transducing phage   Normal phage   Recipient cell   Transductant

◀8.27 **Genes can be transferred from one bacterium to another through generalized transduction.**

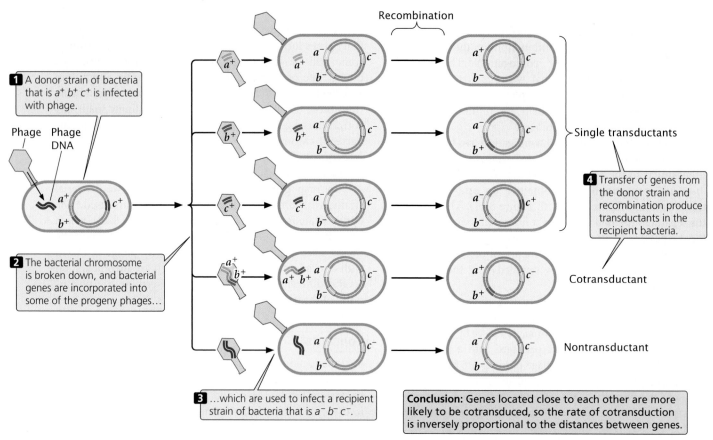

8.28   Generalized transduction can be used to map genes.

prophage may imperfectly excise from the bacterial chromosome, carrying with it a small part of the bacterial DNA adjacent to the site of prophage integration. A phage carrying this DNA will then inject it into another bacterial cell in the next round of infection. This process resembles the situation in F′ cells, where the F plasmid carries genes from one bacterium into another (see Figure 8.16).

One of the best-studied examples of specialized transduction is in bacteriophage lambda (λ), which integrates into the *E. coli* chromosome at the **attachment (*att*) site.** The phage DNA contains a site similar to the *att* site; a single crossover integrates the phage DNA into the bacterial chromosome (◀ FIGURE 8.29a). The λ prophage is excised through a similar crossover that reverses the process (◀ FIGURE 8.29b and c).

An error in excision may cause genes on either side of the bacterial *att* site to be excised along with some of the phage DNA (◀ FIGURE 8.29d and e). In *E. coli*, these genes are usually the *gal* (galactose fermentation) and *bio* (biotin biosynthesis) genes. When a transducing phage carrying the *gal* gene infects another bacterium, the gene may integrate into the bacterial chromosome along with the prophage (◀ FIGURE 8.29f), giving the bacterial chromosome two copies of the *gal* gene (◀ FIGURE 8.29g). These transductants are

unstable, because the prophage DNA may excise from the chromosome, carrying the introduced gene with it. Stable transductants are produced when the *gal* gene in the phage is exchanged for the *gal* gene in the chromosome through a double crossover (◀ FIGURE 8.29h).

Concepts

Specialized transduction transfers only those bacterial genes located near the site of prophage insertion.

Connecting Concepts

Three Methods for Mapping Bacterial Genes

Three methods of mapping bacterial genes have now been outlined: (1) interrupted conjugation; (2) transformation; and (3) transduction. These methods have important similarities and differences.

Mapping with interrupted conjugation is based on the time required for genes to be transferred from one

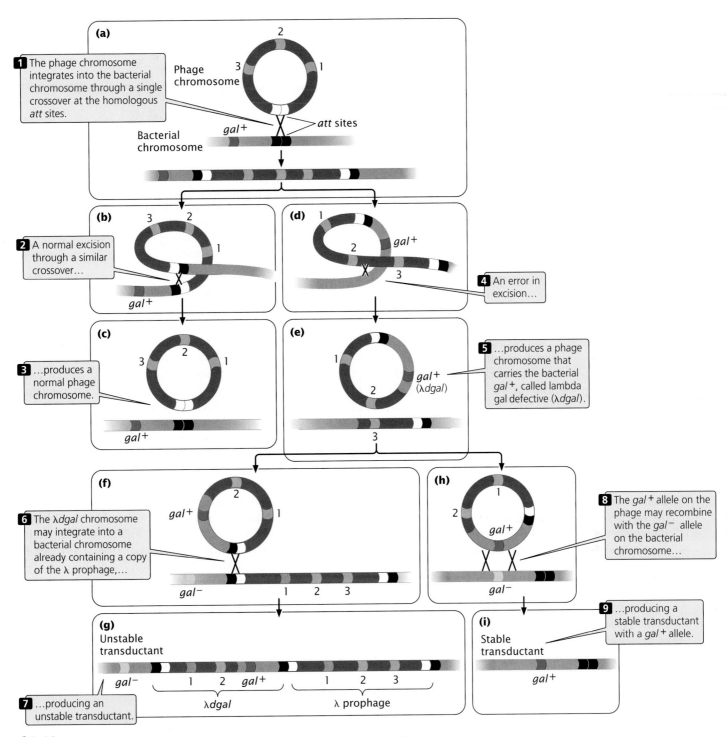

◀ 8.29 Bacteria can exchange genes through specialized transduction.
Segments 1, 2, and 3 represent genes on the phage chromosome.

bacterium to another by means of cell-to-cell contact. The key to this technique is that the bacterial chromosome itself is transferred, and the order of genes and the time required for their transfer provide information about the positions of the genes on the chromosome. In contrast with other mapping methods, the distance between genes is measured

not in recombination frequencies but units of time required for genes to be transferred. Here, the basic unit of conjugation mapping is a minute.

In gene mapping with transformation, DNA from the donor strain is isolated, broken up, and mixed with the recipient strain. Some fragments pass into the recipient cells, where the transformed DNA may recombine with the bacterial chromosome. The unit of transfer here is a random fragment of the chromosome. Loci that are close together on the donor chromosome tend to be on the same DNA fragment; so the rates of cotransformation provide information about the relative positions of genes on the chromosome.

Transduction mapping also relies on the transfer of genes between bacteria that differ in two or more traits, but here the vehicle of gene transfer is a bacteriophage. In a number of respects, transduction mapping is similar to transformation mapping. Small fragments of DNA are carried by the phage from donor to recipient bacteria, and the rates of cotransduction, like the rates of cotransformation, provide information about the relative distances between the genes.

All of the methods use a common strategy for mapping bacterial genes. The movement of genes from donor to recipient is detected by using strains that differ in two or more traits, and the transfer of one gene relative to the transfer of others is examined. Additionally, all three methods rely on recombination between the transferred DNA and the bacterial chromosome. In mapping with interrupted conjugation, the relative order and timing of gene transfer provide the information necessary to map the genes; in transformation and transduction, the rate of cotransfer provides this information.

In conclusion, the same basic strategies are used for mapping with interrupted conjugation, transformation, and transduction. The methods differ principally in their mechanisms of transfer: in conjugation mapping, DNA is transferred though contact between bacteria; in transformation, DNA is transferred as small naked fragments; and, in transduction, DNA is transferred by bacteriophages.

Fine-Structure Analysis of Bacteriophage Genes

In the 1950s and 1960s, Seymour Benzer conducted a series of experiments to examine the structure of a gene. Because no molecular techniques were available at the time for directly examining nucleotide sequences, Benzer was forced to infer gene structure from analyses of mutations and their effects. The results of his studies showed that different mutational sites *within* a single gene could be mapped (**intragenic mapping**) by using techniques similar to those just described. Different sites within a single gene are very close together; so recombination between them takes place at a very low frequency. Because large numbers of progeny are required to detect these recombination events, Benzer used the bacteriophage T4, which reproduces rapidly and produces large numbers of progeny.

Benzer's mapping techniques Wild-type T4 phages normally produce small plaques with rough edges when grown on

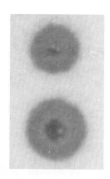

◀8.30 **T4 phage *rII* mutants produce distinct plaques when grown on *E. coli* B cells.** Plaque produced by wild type phage at top; plaque produced by rII mutant below. (Dr. D. P. Snustad, College of Biological Sciences, University of Minnesota.)

a lawn of *E. coli* bacteria. Certain mutants, called *r* for rapid lysis, produce larger plaques with sharply defined edges. Benzer isolated phages with a number of different *r* mutations, concentrating on one particular subgroup called *rII* mutants.

Wild-type T4 phages produce typical plaques (◀FIGURE 8.30) on *E. coli* strains B and K. In contrast, the *rII* mutants produce *r* plaques on strain B and do not form plaques at all on strain K. Benzer recognized the *r* mutants by their distinctive plaques when grown on *E. coli* B. He then collected lysate from these plaques and used it to infect *E. coli* K. Phages that did not produce plaques on *E. coli* K were defined as the *rII* type.

Benzer collected thousands of *rII* mutations. He simultaneously infected bacterial cells with two different mutants and looked for recombinant progeny (◀FIGURE 8.31). Consider two *rII* mutations, a^- and b^-, and their wild-type alleles, a^+ and b^+. Benzer infected *E. coli* B cells with two different strains of phages, one $a^- b^+$ and the other $a^+ b^-$ (Figure 8.31, step 1). While reproducing within the B cells, a few phages of the two strains recombined (Figure 8.31, step 2). A single crossover produces two recombinant chromosomes; one with genotype $a^+ b^+$ and the other with genotype $a^- b^-$:

Phage 1 $\underline{a^-\quad b^+}$

Phage 2 $\overline{a^+\quad b^-}$

$\downarrow$

$a^-\qquad b^+$

$a^+\qquad b^-$

$\downarrow$

$a^-\qquad b^-$

$\overline{a^+\quad b^+}$

The resulting recombinant chromosomes, along with the nonrecombinant (parental) chromosomes, were incorporated

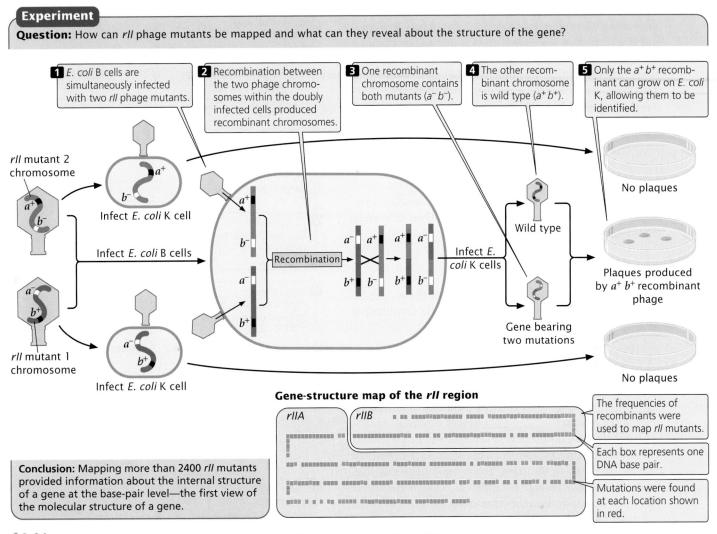

Experiment

Question: How can *rII* phage mutants be mapped and what can they reveal about the structure of the gene?

1 *E. coli* B cells are simultaneously infected with two *rII* phage mutants.

2 Recombination between the two phage chromosomes within the doubly infected cells produced recombinant chromosomes.

3 One recombinant chromosome contains both mutants ($a^- b^-$).

4 The other recombinant chromosome is wild type ($a^+ b^+$).

5 Only the $a^+ b^+$ recombinant can grow on *E. coli* K, allowing them to be identified.

rII mutant 2 chromosome — Infect *E. coli* K cell

Infect *E. coli* B cells

rII mutant 1 chromosome — Infect *E. coli* K cell

Recombination — Infect *E. coli* K cells

Wild type — No plaques

Gene bearing two mutations — Plaques produced by $a^+ b^+$ recombinant phage — No plaques

Gene-structure map of the *rII* region

rIIA   *rIIB*

The frequencies of recombinants were used to map *rII* mutants.

Each box represents one DNA base pair.

Mutations were found at each location shown in red.

Conclusion: Mapping more than 2400 *rII* mutants provided information about the internal structure of a gene at the base-pair level—the first view of the molecular structure of a gene.

8.31 Benzer developed a procedure for mapping *rII* mutants. Two different *rII* mutants ($a^- b^+$ and $a^+ b^-$) are isolated on *E. coli* B cells. Neither will grow on *E. coli* K cells. Only the $a^+ b^+$ recombinant can grow on *E. coli* K, allowing these recombinants to be identified. *rIIA* and *rIIB* refer to different parts of the gene.

into progeny phages (Figure 8.31, steps 3 and 4), which were then used to infect *E. coli* K cells. The resulting plaques were examined to determine the genotype of the infecting phage.

The *rII* mutants would not grow on *E. coli* K, but wild-type phages could; so progeny phages with the recombinant genotype $a^+ b^+$ produced plaques on *E. coli* K (Figure 8.31, step 5). Each recombination event produces an equal number of double mutants ($a^- b^-$) and wild-type chromosomes ($a^+ b^+$); so the number of recombinant progeny should be twice the number of wild-type plaques that appeared on *E. coli* K. The recombination frequency between the two *rII* mutants would be:

$$\text{recombination frequency} = \frac{2 \times \text{number of plaques on } E.\ coli\ K}{\text{total number of plaques on } E.\ coli\ B}$$

Benzer was able to detect a single recombinant among billions of progeny phages, allowing very low rates of recombination to be detected. Recombination frequencies are proportional to physical distances along the chromosome (p. 172 in Chapter 7), revealing the positions of the different mutations within the *rII* region of the phage chromosome. In this way, Benzer eventually mapped more than 2400 *rII* mutations, many corresponding to single base pairs in the viral DNA. His work provided the first molecular view of a gene.

Concepts

In a series of experiments with the bacteriophage T4, Seymour Benzer showed that recombination could occur within a single gene and created the first molecular map of a gene.

Complementation experiments At the time Benzer was conducting his experiments, the relationship between genes and DNA structure was unknown. A gene had been defined as a functional unit of heredity that coded for a phenotype. To test whether different *rII* mutations belonged to different functional genes, Benzer used the complementation (cis-trans) test (see pp. 114–115 in Chapter 5).

Individuals heterozygous for two mutations may have the mutations in trans,

$$\frac{a^+ \qquad b^-}{a^- \qquad b^+}$$

meaning that they are located on different chromosomes, or in cis, meaning that they are located on the same chromosome

$$\frac{a^- \qquad a^-}{b^+ \qquad b^+}$$

(see p. 166–167 in Chapter 7). Suppose that the a^- and b^- mutations occur at different loci, which code for different proteins. In the trans heterozygote

$$\frac{a^+ \qquad b^-}{a^- \qquad b^+}$$

one chromosome has a functional allele at the *a* locus ($\underline{a^+ \qquad b^-}$) and the other chromosome has a functional allele at the *b* locus ($\underline{a^- \qquad b^+}$); since a^- and b^- are recessive mutations, both A and B proteins will be produced. The two mutations complement each other, so the presence of wild type trait in the trans heterozygote indicates that these mutations belong to different complementation groups—they come from different loci.

Suppose the two mutations occur within a single locus that codes for one protein. In the trans heterozygote, one chromosome fails to produce a functional protein because it has a defect at the *b* site ($\underline{a^+ \qquad b^-}$) and the other chromosome fails to produce a functional protein because it has a defect at the *a* site ($\underline{a^- \qquad b^+}$). No functional protein is produced by either chromosome, and the trans heterozygote has a mutant phenotype—the mutations are unable to complement each other.

The heterozygous individual used in complementation testing must have the mutations in the trans configuration. When the mutations are in the cis configuration:

$$\frac{a^- \qquad b^-}{a^+ \qquad b^+}$$

heterozygotes will have a wild-type phenotype regardless of whether the two mutations occur at the same locus or at different loci, because one chromosome ($\underline{a^+ \qquad b^+}$) is mutation free.

To carry out the complementation test in bacteriophage, Benzer infected cells of *E. coli* K with large numbers of two mutant strains of phage (◀FIGURE 8.32, step 1). We will refer to the two mutations as *rIIa* ($\underline{a^- \qquad b^+}$) and *rIIb* ($\underline{a^+ \qquad b^-}$). Cells infected with both mutants:

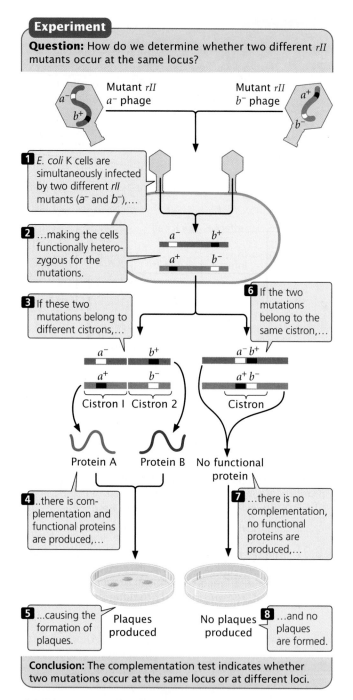

Experiment

Question: How do we determine whether two different *rII* mutants occur at the same locus?

Mutant *rII* a^- phage   Mutant *rII* b^- phage

1 *E. coli* K cells are simultaneously infected by two different *rII* mutants (a^- and b^-),…

2 …making the cells functionally heterozygous for the mutations.

3 If these two mutations belong to different cistrons,…

6 If the two mutations belong to the same cistron,…

Cistron 1   Cistron 2   Cistron

Protein A   Protein B   No functional protein

4 …there is complementation and functional proteins are produced,…

7 …there is no complementation, no functional proteins are produced,…

5 …causing the formation of plaques.   Plaques produced   No plaques produced   **8** …and no plaques are formed.

Conclusion: The complementation test indicates whether two mutations occur at the same locus or at different loci.

◀8.32 **Complementation tests are used to determine whether different mutations are at the same functional gene.**

$$\frac{a^- \qquad b^+}{a^+ \qquad b^-}$$

were effectively heterozygous for the phage genes, with the mutations in the trans configuration (◀FIGURE 8.32, step 2). In the complementation testing, the phenotypes of progeny phages were examined on the K strain, rather than the B strain as illustrated in Figure 8.31.

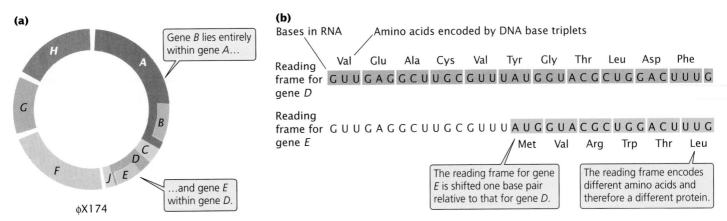

8.33 The genome of bacteriophage φX174 contains overlapping genes. The genome contains nine genes (*A* through *J*).

If the *rIIa* and *rIIb* mutations occur at different loci that code for different proteins then, in bacterial cells infected by both mutants, the wild-type sequences on the chromosome opposite each mutation will overcome the effects of the recessive mutations; the phages will produce normal plaques on *E. coli* K cells (◀ FIGURE 8.32, steps 3, 4, and 5). (Benzer coined the term *cistron* to designate a functional gene defined by the complementation test.) If, on the other hand, the mutations occur at the same locus, no functional protein is produced by either chromosome, and no plaques develop in the *E. coli* K cells (◀ FIGURE 8.32 steps 6, 7, and 8). Thus, the absence of plaques indicates that the two mutations occur at the same locus.

In the complementation test, the cis heterozygote is used as a control. Benzer simultaneously infected bacteria with wild-type phage (a^+ b^+) and with phage carrying both mutations (a^- b^-). This test also produced cells that were heterozygous and cis for the phage genes:

$$\frac{a^+ \qquad b^+}{a^- \qquad b^-}$$

Regardless of whether the *rIIa* and *rIIb* mutations are in the same functional unit, these cells contain a copy of the wild-type phage chromosome (a^+ b^+) and will produce normal plaques in *E. coli* K.

Benzer carried out complementation testing on many pairs of *rII* mutants. He found that the *rII* region consists of two loci, designated *rIIA* and *rIIB*. Mutations belonging to the *rIIA* and *rIIB* groups complemented each other, but mutations in the *rIIA* group did not complement others in *rIIA;* nor did mutations in the *rIIB* group complement others in *rIIB*.

Concepts

Benzer used the complementation test to distinguish between functional genes (loci).

At the time of Benzer's research, many geneticists believed that genes were indivisible and that recombination could not take place within them. Benzer demonstrated that intragenic recombination did indeed take place (although at a very low rate) and gave geneticists their first glimpse at the structure of an individual gene.

Overlapping Genes

The first viral genome to be completely sequenced, that of bacteriophage φX174, revealed surprising information: the nucleotide sequences of several genes overlapped. This genome encodes nine proteins (◀ FIGURE 8.33). Two of the genes are nested within other genes; in both cases, the same DNA sequence codes for two different proteins by using different reading frames (see p. 415 in Chapter 15). In five of the φX174 genes, the initiation codon of one gene overlaps the termination codon of another.

The results of subsequent studies revealed that overlapping genes are found in a number of viruses and bacteria. Viral genome size is strictly limited by the capacity of the viral protein coat; so there is strong selective pressure for economic use of the DNA.

Concepts

Some viruses contain overlapping genes, in which the same base sequence specifies more than one protein.

RNA Viruses

Viral genomes may be encoded in either DNA or RNA. Some medically important human viruses have RNA as their genetic material, including those that cause influenza, common colds, polio, and AIDS. Almost all viruses that infect plants have RNA genomes. The medical and economic importance of RNA viruses has encouraged their study.

RNA viruses, like bacteriophages, reproduce by infecting cells and making copies of themselves. Most use RNA-dependent RNA polymerases encoded by their own genes.

In **positive-strand RNA viruses,** the genomic RNA molecule carried inside the viral particle codes directly for viral proteins (◀ FIGURE **8.34a**). In **negative-strand RNA viruses,** the virus first makes a complementary copy of its RNA genome, which is then translated into viral proteins (◀ FIGURE **8.34b**).

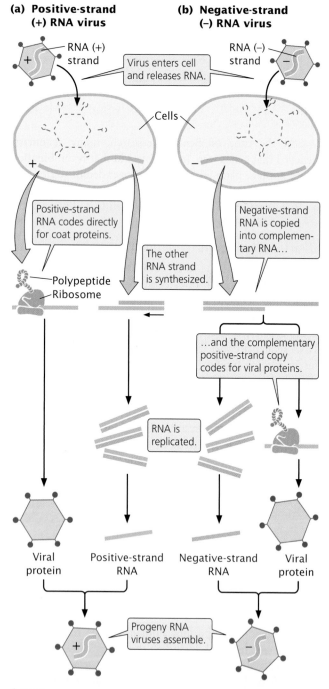

(a) Positive-strand (+) RNA virus

(b) Negative-strand (–) RNA virus

RNA (+) strand

RNA (–) strand

Virus enters cell and releases RNA.

Cells

Positive-strand RNA codes directly for coat proteins.

Negative-strand RNA is copied into complementary RNA...

The other RNA strand is synthesized.

Polypeptide
Ribosome

...and the complementary positive-strand copy codes for viral proteins.

RNA is replicated.

Viral protein

Positive-strand RNA

Negative-strand RNA

Viral protein

Progeny RNA viruses assemble.

◀8.34 **The process of reproduction differs in positive-strand RNA viruses and negative-strand RNA viruses.**

RNA viruses capable of integrating into the genome of their hosts, much as temperate phages insert themselves into bacterial chromosomes, are called **retroviruses** (◀ FIGURE **8.35a**). Because the retroviral genome is RNA, whereas that of the host is DNA, a retrovirus must produce **reverse transcriptase,** an enzyme that synthesizes complementary DNA (cDNA) from either an RNA or a DNA template. A retrovirus uses reverse transcriptase to make a double-stranded DNA copy from its single-stranded RNA genome. The DNA copy then integrates into the host chromosome to form a **provirus,** which is replicated by host enzymes when the host chromosome is duplicated (◀ FIGURE **8.35b**).

When conditions are appropriate, the provirus undergoes transcription to produce numerous copies of the original RNA genome. This RNA codes for viral proteins and serves as genomic RNA for new viral particles. As these viruses escape the cell, they collect patches of the cell membrane to use as their envelopes.

All known retroviral genomes have in common three genes: *gag, pol,* and *env* (◀ FIGURE **8.36**), each encoding a precursor protein that is cleaved into two or more functional proteins. The *gag* gene encodes the three or four proteins that make up the viral capsid. The *pol* gene codes for reverse transcriptase and an enzyme, called **integrase,** that inserts the viral DNA into the host chromosome. The *env* gene codes for the glycoproteins, which appear on the viral envelope that surrounds the viral capsid.

Some retroviruses contain **oncogenes** (Chapter 21) that may stimulate cell division and cause the formation of tumors. The first retrovirus to be isolated, the Rous sarcoma virus, was originally recognized by its ability to produce connective-tissue tumors (sarcomas) in chickens.

The human immunodeficiency virus (HIV) is a retrovirus that causes acquired immune deficiency syndrome. AIDS was first recognized in 1982, when a number of homosexual males in the United States began to exhibit symptoms of a new immune-system-deficiency disease. In that year, Robert Gallo proposed that AIDS was caused by a retrovirus. Between 1983 and 1984, as the AIDS epidemic became widespread, the HIV retrovirus was isolated from AIDS patients.

HIV is thought to have appeared first in Africa in the 1950s or 1960s. It is closely related to several retroviruses found in monkeys and may have evolved when a monkey retrovirus mutated and infected humans. HIV is transmitted by sexual contact between humans and through any type of blood-to-blood contact, such as that caused by the sharing of dirty needles by drug addicts. Until screening tests could identify HIV-infected blood, transfusions and clotting factors used by hemophiliacs also were sources of infection.

HIV principally attacks a class of blood cells called helper T lymphocytes (◀ FIGURE **8.37**). HIV enters a helper T cell, undergoes reverse transcription, and integrates into the chromosome. The virus reproduces rapidly, destroying the T cell as new virus particles escape from the cell. Because helper T cells are central to immune function and

(a)

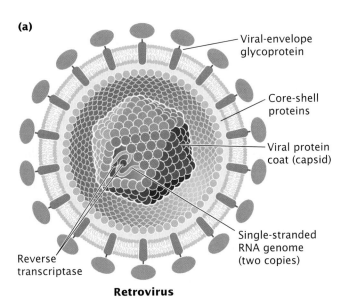

Viral-envelope glycoprotein

Core-shell proteins

Viral protein coat (capsid)

Single-stranded RNA genome (two copies)

Reverse transcriptase

Retrovirus

◀ **8.35 A retrovirus uses reverse transcription to incorporate its RNA into the host DNA.** (a) Structure of a typical retrovirus. Two copies of the single-stranded RNA genome and reverse transcriptase enzyme are shown enclosed within a protein capsid. The capsid is surrounded by a viral envelope that is studded with viral glycoproteins. (b) The retrovirus life cycle.

are destroyed in the infection, AIDS patients have a diminished immune response—most AIDS patients die of secondary infections that develop because they have lost the ability to fight off pathogens.

The HIV genome is 9749 nucleotides long and carries *gag*, *pol*, *env*, and six other genes that regulate the life cycle of the virus. HIV's reverse transcriptase is very error prone, giving the virus a high mutation rate and allowing it to evolve rapidly, even within a single host. This rapid evolution makes the development of an effective vaccine against HIV particularly difficult.

Concepts

Retrovirus is an RNA virus that integrates into its host chromosome by making a DNA copy of its RNA genome through the process of reverse transcription. Human immunodeficiency virus, the causative agent of AIDS, is a retrovirus.

Prions: Pathogens Without Genes

In 1997, Stanley B. Prusiner was awarded the Nobel Prize in Physiology or Medicine for his discovery and characterization of **prions,** a novel class of pathogens that cause several rare neurodegenerative diseases and that appear to replicate without any genes. Initially, Prusiner's proposal that prions were composed entirely of protein and lacked any trace of nucleic acid was met with skepticism. One of the foundations of modern biology is that all living things possess

(b)

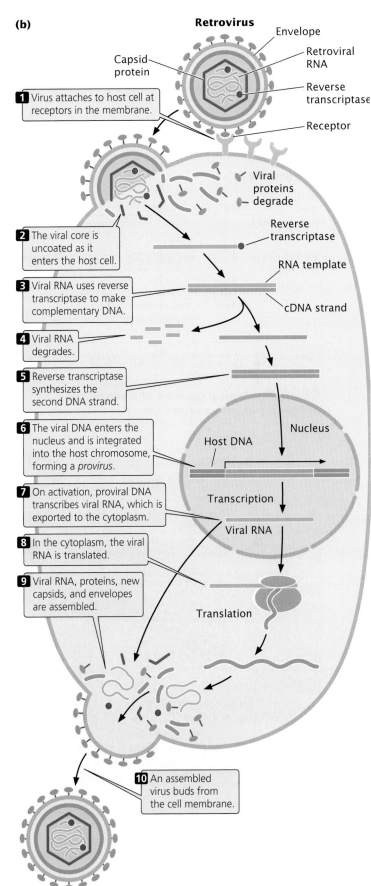

Retrovirus

Envelope

Capsid protein

Retroviral RNA

Reverse transcriptase

1 Virus attaches to host cell at receptors in the membrane.

Receptor

Viral proteins degrade

2 The viral core is uncoated as it enters the host cell.

Reverse transcriptase

3 Viral RNA uses reverse transcriptase to make complementary DNA.

RNA template

cDNA strand

4 Viral RNA degrades.

5 Reverse transcriptase synthesizes the second DNA strand.

6 The viral DNA enters the nucleus and is integrated into the host chromosome, forming a *provirus*.

Nucleus

Host DNA

7 On activation, proviral DNA transcribes viral RNA, which is exported to the cytoplasm.

Transcription

Viral RNA

8 In the cytoplasm, the viral RNA is translated.

9 Viral RNA, proteins, new capsids, and envelopes are assembled.

Translation

10 An assembled virus buds from the cell membrane.

RNA genome

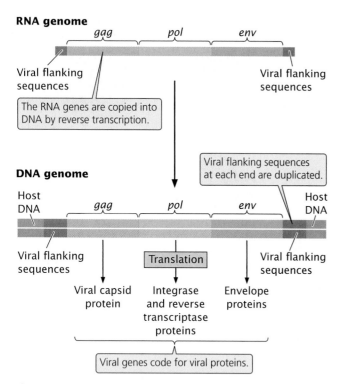

gag   *pol*   *env*

Viral flanking sequences

Viral flanking sequences

The RNA genes are copied into DNA by reverse transcription.

Viral flanking sequences at each end are duplicated.

DNA genome

Host DNA   *gag*   *pol*   *env*   Host DNA

Viral flanking sequences

Translation

Viral flanking sequences

Viral capsid protein

Integrase and reverse transcriptase proteins

Envelope proteins

Viral genes code for viral proteins.

◀ **8.36 The typical genome of a retrovirus contains *gag*, *pol*, and *env* genes.**

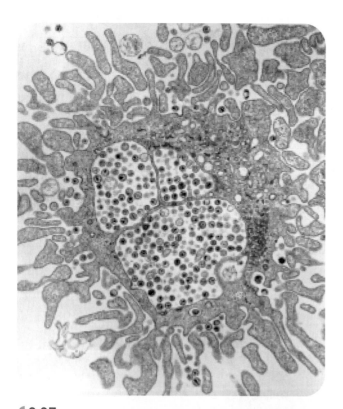

◀ **8.37 HIV principally attacks T lymphocytes.**
Electron micrograph showing a T cell infected with HIV, visible as small circles with dark centers. (Courtesy of Dr. Hans Gelderblom.)

hereditary information in the form of DNA or RNA, and so how are prions able to reproduce without nucleic acid?

Prions were first recognized as unusual infectious agents that cause scrapie, a disease of sheep that destroys the brain. In 1982, Prusiner purified the scrapie pathogen and reported that it consisted entirely of protein. Prusiner and his colleagues eventually showed that the prion protein (PrP) is derived from a normal protein that is encoded by a gene found throughout eukaryotes, including yeast. Normal PrP (PrP^C) is folded into a helical shape, but the protein can also fold into a flattened β sheet that causes scrapie (PrP^{Sc}) (◀ FIGURE 8.38). When PrP^{Sc} is present, it interacts with and causes PrP^C to fold into the disease-causing form of the protein; infection with PrP^{Sc} converts an individual's normal PrP protein into abnormal PrP that forms prions. Accumulation of the PrP^{Sc} in the brain appears to be responsible for the neurological degeneration associated with diseases caused by prions. This explanation for prion diseases, called the "protein only" hypothesis, is not universally accepted; some scientists still believe that these diseases are caused by an as-of-yet unisolated virus.

Prions cause scrapie, bovine spongiform encephalopathy (BSE, or "mad cow" disease), and kuru, an exotic disorder spread among New Guinea aborigines by ritualistic cannibalism. They also play a role in some inherited human neurodegenerative disorders, including Creutzfeldt-Jakob disease and Gerstmann-Scheinker disease. In these inherited diseases, the PrP gene is mutated and produces a type of PrP that is more susceptible to folding into PrP^{Sc}. Nearly all those who carry such a mutated gene eventually produce prions and get the disease.

Some cases of human prion diseases have been traced to injections of growth hormone, which until recently was obtained from the brains of human cadavers infected with prions. In England, an epidemic of mad cow disease erupted in the late 1980s, the origin of which was traced to cattle feed containing the remains of sheep infected with scrapie.

www.whfreeman.com/pierce   For more information on prions, prion-caused diseases, and Stanley Prusiner's account of his hunt for the secret of prions.

Connecting Concepts Across Chapters

Bacteria and viruses have been used extensively in the study of genetics: their rapid reproduction, large numbers of progeny, small haploid genomes, and medical importance make them ideal organisms for many types of genetic investigations.

This chapter examined some of the techniques used to study and map bacterial and viral genomes. Some of these methods are an extension of the principles of recombination and gene mapping explored in Chapter 7.

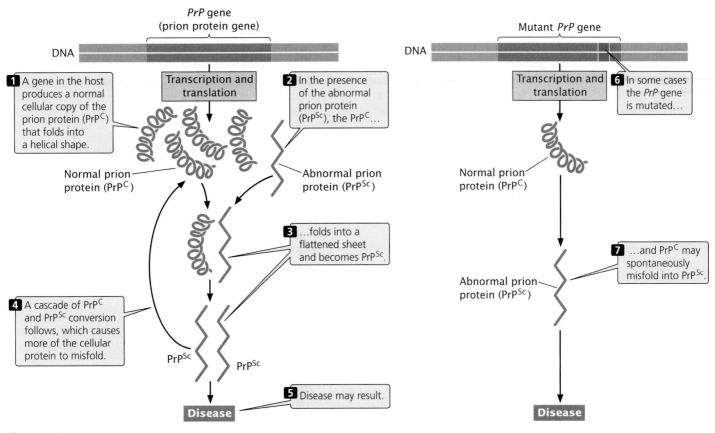

◀ **8.38** **The protein-only hypothesis describes a method for the replication of prions.**

Bacterial reproduction was discussed in Chapter 2, and a number of the principles and techniques covered in this chapter are linked to topics in future chapters. Bacterial chromosomes will be considered in more detail in Chapter 11, and bacterial replication, transcription, translation, and gene regulation will be the topics of Chapters 12 through 16. Bacteria are central to recombinant DNA technology, the topic of Chapter 18, where they are often used in mass producing specific DNA fragments. Many of the tools of recombinant DNA technology, including plasmids, restriction enzymes, DNA polymerases, and many other enzymes, have been isolated and engineered from natural components of bacterial cells. Engineered viruses are common vehicles for delivering genes to host cells.

Some transposable genetic elements (discussed in Chapter 11) are closely related to viruses, and considerable evidence suggests that viruses evolved from such elements. Because their mutations are easily isolated, bacteria also play an important role in the study of gene mutations, a topic examined in Chapter 17. Chapter 20 deals with mitochondrial and chloroplast DNA, which in many respects are more similar to bacterial DNA than to the nuclear DNA of the cells in which these organelles are found. Finally, viruses cause some cancers, and the role of viral genes in cancer development is studied in Chapter 21.

CONCEPTS SUMMARY

- Bacteria and viruses are well suited to genetic studies: they are small, have a small haploid genome, undergo rapid reproduction, and produce large numbers of progeny through asexual reproduction. When spread on a petri plate, individual bacteria grow into colonies of identical cells that can be easily seen.

- The bacterial genome normally consists of a single, circular molecule of double-stranded DNA.

- Plasmids are small pieces of bacterial DNA that can replicate independently of the large chromosome. Episomes are plasmids that can exist either in a freely replicating state or can integrate into the bacterial chromosome.

- DNA may be transferred between bacteria by means of conjugation, transformation, and transduction.

- Conjugation is the union and the transfer of genetic material between two bacterial cells and is controlled by a fertility factor called F, which is an episome. F^+ cells are donors, and F^- cells are recipients during conjugation. An Hfr cell has F incorporated into the bacterial chromosome. An F′ cell has an F factor that has excised from the bacterial genome and carries some bacterial genes.

- The rate at which individual genes are transferred from Hfr to F^- cells during conjugation provides information about the order and distance between the genes on the bacterial chromosome.

- In transformation, bacteria take up DNA from their environment. Frequencies of cotransformation provide information about the physical distances between chromosomal genes.

- Viruses are replicating structures with DNA or RNA genomes that may be double stranded or single stranded, linear or circular. Bacteriophages are viruses that infect bacteria. An individual phage can be identified when it enters a bacterial cell, multiplies, and eventually produces a patch of lysed bacterial cells (a plaque) on an agar plate.

- Phage genes can be mapped by infecting bacterial cells with two different strains of phage. The numbers of recombinant plaques produced by the progeny phages are used to estimate recombination rates between phage genes.

- In generalized transduction, bacterial genes become incorporated into phage coats and are transferred to other bacteria during phage infection. Rates of cotransduction can be used to determine the order and distance between genes on the bacterial chromosome.

- In specialized transduction, DNA near the site of phage integration on the bacterial chromosome is transferred from one bacterium to another.

- Benzer mapped a large number of mutations that occurred within the *rII* region of phage T4 and showed that intragenic recombination takes place. The results of his complementation studies demonstrated that the *rII* region consists of two functional units (cistrons).

- A number of viruses have RNA genomes. In positive-strand viruses, the RNA genome codes directly for viral proteins; in negative-strand viruses, a complementary copy of the genome is translated to form viral proteins. Retroviruses encode a reverse transcriptase enzyme used to make a DNA copy of the viral genome, which then integrates into the host genome as a provirus.

- Prions are infectious agents consisting only of protein; they are thought to cause disease by altering the shape of proteins encoded by the host genome.

IMPORTANT TERMS

minimal medium (p. 199)	competent cell (p. 211)	specialized transduction (p. 216)	negative-strand RNA virus (p. 224)
complete medium (p. 200)	transformant (p. 212)	transducing phage (p. 217)	retrovirus (p. 224)
colony (p. 200)	cotransformation (p. 212)	transductants (p. 217)	reverse transcriptase (p. 224)
plasmid (p. 202)	virus (p. 213)	cotransduction (p. 217)	provirus (p. 224)
episome (p. 203)	virulent phage (p. 214)	attachment site (p. 218)	integrase (p. 224)
F factor (p. 203)	temperate phage (p. 214)	intragenic mapping (p. 220)	oncogene (p. 224)
conjugation (p. 203)	prophage (p. 214)	positive-strand RNA virus (p. 224)	prion (p. 225)
transformation (p. 203)	plaque (p. 214)		
transduction (p. 203)	generalized transduction (p. 216)		
pili (p. 205)			

Worked Problems

1. DNA from a strain of bacteria with genotype $a^+ b^+ c^+ d^+ e^+$ was isolated and used to transform a strain of bacteria that was $a^- b^- c^- d^- e^-$. The transformed cells were tested for the presence of donated genes. The following genes were cotransformed:

a^+ and d^+

b^+ and e^+

c^+ and d^+

c^+ and e^+

What is the order of genes *a*, *b*, *c*, *d*, and *e* on the bacterial chromosome?

• Solution

The rate at which genes are cotransformed is inversely proportional to the distance between them: genes that are close together are frequently cotransformed, whereas genes that are far apart are rarely cotransformed. In this transformation experiment, gene c^+ is cotransformed with both genes e^+ and d^+, but genes e^+ and d^+

are not cotransformed; therefore the *c* locus must be between the *d* and *e* loci:

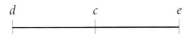

Gene *e*⁺ is also cotransformed with gene *b*⁺; so the *e* and *b* loci must be located close together. Locus *b* could be on either side of locus *e*. To determine whether locus *b* is on the same side of *e* as locus *c*, we look to see whether genes *b*⁺ and *c*⁺ are cotransformed. They are not; so locus *b* must be on the opposite side of *e* from *c*:

Gene *a*⁺ is cotransformed with gene *d*⁺; so they must be located close together. If locus *a* were located on the same side of *d* as locus *c*, then genes *a*⁺ and *c*⁺ would be cotransformed. Because these genes display no cotransformation, locus *a* must be on the opposite side of locus *d*:

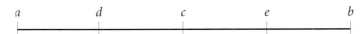

2. Consider three genes in *E. coli*: *thr*⁺ (the ability to synthesize threonine), *ara*⁺ (the ability to metabolize arabinose), and *leu*⁺ (the ability to synthesize leucine). All three of these genes are close together on the *E. coli* chromosome. Phages are grown in a *thr*⁺ *ara*⁺ *leu*⁺ strain of bacteria (the donor strain). The phage lysate is collected and used to infect a strain of bacteria that is *thr*⁻ *ara*⁻ *leu*⁻. The recipient bacteria are then tested on medium lacking leucine. Bacteria that grow and form colonies on this medium (*leu*⁺ transductants) are then replica plated onto medium lacking threonine and medium lacking arabinose to see which are *thr*⁺ and which are *ara*⁺.

Another group of recipient bacteria are tested on medium lacking threonine. Bacteria that grew and formed colonies on this medium (*thr*⁺ transductants) were then replica plated onto medium lacking leucine and medium lacking arabinose to see

which are *ara*⁺ and which are *leu*⁺. Results from these experiments are as follows:

Selected marker	Cells with cotransduced genes (3%)
leu⁺	3 *thr*⁺
	76 *ara*⁺
thr⁺	3 *leu*⁺
	0 *ara*⁺

How are the loci arranged on the chromosome?

• Solution

Notice that, when we select for *leu*⁺ (the top half of the table), most of the selected cells also are *ara*⁺. This finding indicates that the *leu* and *ara* genes are located close together, because they are usually cotransduced. In contrast, *thr*⁺ is only rarely cotransduced with *leu*⁺, indicating that *leu* and *thr* are much farther apart. On the basis of these observations, we know that *leu* and *ara* are closer together than are *leu* and *thr*, but we don't yet know the order of three genes—whether *thr* is on the same side of *ara* as *leu* or on the opposite side, as shown here:

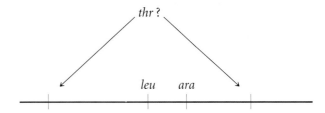

We can determine the position of *thr* with respect to the other two genes by looking at the cotransduction frequencies when *thr*⁺ is selected (the bottom half of the table). Notice that, although the cotransduction frequency for *thr* and *leu* also is 3%, no *thr*⁺ *ara*⁺ cotransductants are observed. This finding indicates that *thr* is closer to *leu* than to *ara*, and therefore *thr* must be to the left of *leu*, as shown here:

<hr>

COMPREHENSION QUESTIONS

* 1. List some of the characteristics that make bacteria and viruses ideal organisms for many types of genetic studies.

2. Explain how auxotrophic bacteria are isolated.

3. Briefly explain the differences between F⁺, F⁻, Hfr, and F′ cells.

* 4. What types of matings are possible between F⁺, F⁻, Hfr, and F′ cells? What outcomes do these matings produce? What is the role of F factor in conjugation?

* 5. Explain how interrupted conjugation, transformation, and transduction can be used to map bacterial genes. How are these methods similar and how are they different?

6. What types of genomes do viruses have?

7. Briefly describe the differences between the lytic cycle of virulent phages and the lysogenic cycle of temperate phages.

8. Briefly explain how genes in phages are mapped.

* 9. How does specialized transduction differ from generalized transduction?

*10. Briefly explain the method used by Benzer to determine whether two different mutations occurred at the same locus.

 11. What is the difference between a positive-strand RNA virus and a negative-strand RNA virus?

*12. Explain how a retrovirus, which has an RNA genome, is able to integrate its genetic material into that of a host having a DNA genome.

 13. Briefly describe the genetic structure of a typical retrovirus.

APPLICATION QUESTIONS AND PROBLEMS

*14. John Smith is a pig farmer. For the past 5 years, Smith has been adding vitamins and low doses of antibiotics to his pig food; he says that these supplements enhance the growth of the pigs. Within the past year, however, several of his pigs died from infections of common bacteria, which failed to respond to large doses of antibiotics. Can you offer an explanation for the increased rate of mortality due to infection in Smith's pigs? What advice might you offer Smith to prevent this problem in the future?

 15. Rarely, conjugation of Hfr and F^- cells produces two Hfr cells. Explain how this occurs.

*16. A strain of Hfr cells that is sensitive to the antibiotic streptomycin (str^s) has the genotype gal^+ his^+ bio^+ pur^+ gly^+. These cells were mixed with an F^- strain that is resistant to streptomycin (str^r) and has genotype gal^- his^- bio^- pur^- gly^-. The cells were allowed to undergo conjugation. At regular intervals, a sample of cells was removed and conjugation was interrupted by placing the sample in a blender. The cells were then plated on medium that contains streptomycin. The cells that grew on this medium were then tested for the presence of genes transferred from the Hfr strain. Genes from the donor Hfr strain first appeared in the recipient F^- strain at the times listed here. On the basis of these data, give the order of the genes on the bacterial chromosome and indicate the minimum distances between them:

gly^+	3 minutes
his^+	14 minutes
bio^+	35 minutes
gal^+	36 minutes
pur^+	38 minutes

*17. A series of Hfr strains that have genotype m^+ n^+ o^+ p^+ q^+ r^+ are mixed with an F^- strain that has genotype m^- n^- o^- p^- q^- r^-. Conjugation is interrupted at regular intervals and the order of appearance of genes from the Hfr strain is determined in the recipient cells. The order of gene transfer for each Hfr strain is:

Hfr5	m^+ q^+ p^+ n^+ r^+ o^+
Hfr4	n^+ r^+ o^+ m^+ q^+ p^+
Hfr1	o^+ m^+ q^+ p^+ n^+ r^+
Hfr9	q^+ m^+ o^+ r^+ n^+ p^+

What is the order of genes on the circular bacterial chromosome? For each Hfr strain, give the location of the F factor in the chromosome and its polarity.

*18. Crosses of three different Hfr strains with separate samples of an F^- strain are carried out, and the following mapping data are provided from studies of interrupted conjugation:

Appearance of Genes in F^- cells

Hfr1:	Genes	b^+	d^+	c^+	f^+	g^+
	Time	3	5	16	27	59
Hfr2:	Genes	e^+	f^+	c^+	d^+	b^+
	Time	6	24	35	46	48
Hfr3:	Genes	d^+	c^+	f^+	e^+	g^+
	Time	4	15	26	44	58

Construct a genetic map for these genes, indicating their order on the bacterial chromosome and the distances between them.

 19. DNA from a strain of *Bacillus subtilis* with the genotype trp^+ tyr^+ is used to transform a recipient strain with the genotype trp^- tyr^-. The following numbers of transformed cells were recovered:

Genotype	Number of transformed cells
trp^+ tyr^-	154
trp^- tyr^+	312
trp^+ tyr^+	354

What do these results suggest about the linkage of the *trp* and *tyr* genes?

 20. DNA from a strain of *Bacillus subtilis* with genotype a^+ b^+ c^+ d^+ e^+ is used to transform a strain with genotype a^- b^- c^- d^- e^-. Pairs of genes are checked for cotransformation and the following results are obtained:

Pair of genes	Cotransformation
a^+ and b^+	no
a^+ and c^+	no
a^+ and d^+	yes
a^+ and e^+	yes
b^+ and c^+	yes
b^+ and d^+	no
b^+ and e^+	yes
c^+ and d^+	no
c^+ and e^+	yes
d^+ and e^+	no

On the basis of these results, what is the order of the genes on the bacterial chromosome?

21. DNA from a bacterial strain that is his^+ leu^+ lac^+ is used to transform a strain that is his^- leu^- lac^-. The following percentages of cells were transformed:

Donor Strain	Recipient strain	Genotype of transformed cells	Percentage
his^+ leu^+ lac^+	his^- leu^- lac^-	his^+ leu^+ lac^+	0.02
		his^+ leu^+ lac^-	0.00
		his^+ leu^- lac^+	2.00
		his^+ leu^- lac^-	4.00
		his^- leu^+ lac^+	0.10
		his^- leu^- lac^+	3.00
		his^- leu^+ lac^-	1.50

(a) What conclusions can you make about that order of these three genes on the chromosome?

(b) Which two genes are closest?

22. Two mutations that affect plaque morphology in phages (a^- and b^-) have been isolated. Phages carrying both mutations (a^- b^-) are mixed with wild-type phages (a^+ b^+) and added to a culture of bacterial cells. Subsequent to infection and lysis, samples of the phage lysate are collected and cultured on bacterial cells. The following numbers of plaques are observed:

Plaque phenotype	Number
a^+ b^+	2043
a^+ b^-	320
a^- b^+	357
a^- b^-	2134

What is the frequency of recombination between the a and b genes?

*23. A geneticist isolates two mutations in bacteriophage. One mutation causes the clear plaques (c) and the other produces minute plaques (m). Previous mapping experiments have established that the genes responsible for these two mutations are 8 map units apart. The geneticist mixes phages with genotype c^+ m^+ and genotype c^- m^- and uses the mixture to infect bacterial cells. She collects the progeny phages and cultures a sample of them on plated bacteria. A total of 1000 plaques are observed. What numbers of the different types of plaques (c^+ m^+, c^- m^-, c^+ m^-, c^- m^+) should she expect to see?

24. The geneticist carries out the same experiment described in Problem 23, but this time she mixes phages with genotypes c^+ m^- and c^- m^+. What results are expected with this cross?

*25. A geneticist isolates two r mutants (r_{13} and r_2) that cause rapid lysis. He carries out the following crosses and counts the number of plaques listed here:

Genotype of parental phage	Progeny	Number of plaques
h^+ $r_{13}^- \times h^-$ r_{13}^+	h^+ r_{13}^+	1
	h^- r_{13}^+	104
	h^+ r_{13}^-	110
	h^- r_{13}^-	2
	total	216
h^+ $r_2^- \times h^-$ r_2^+	h^+ r_2^+	6
	h^- r_2^+	86
	h^+ r_2^-	81
	h^- r_2^-	7
	total	180

(a) Calculate the recombination frequencies between r_2 and h and between r_{13} and h.

(b) Draw all possible linkage maps for these three genes.

*26. E. coli cells are simultaneously infected with two strains of phage λ. One strain has a mutant host range, is temperature sensitive, and produces clear plaques (genotype = h st c); another strain carries the wild-type alleles (genotype = h^+ st^+ c^+). Progeny phage are collected from the lysed cells and are plated on bacteria. The genotypes of the progeny phage are:

Progeny phage genotype	Number of plaques
h^+ c^+ t^+	321
h c t	338
h^+ c t	26
h c^+ t^+	30
h^+ c t^+	106
h c^+ t	110
h^+ c^+ t	5
h c t^+	6

(a) Determine the order of the three genes on the phage chromosome.

(b) Determine the map distances between the genes.

(c) Determine the coefficient of coincidence and the interference (see pp. 178–180 in Chapter 7).

27. A donor strain of bacteria with genes a^+ b^+ c^+ is infected with phages to map the donor chromosome with generalized transduction. The phage lysate from the bacterial cells is collected and used to infect a second strain of bacteria that are a^- b^- c^-. Bacteria with the a^+ gene are selected and the percentage of cells with cotransduced b^+ and c^+ genes are recorded.

Donor	Recipient	Selected gene	Cells with cotransduced gene (%)
a^+ b^+ c^+	a^- b^- c^-	a^+	25 b^+
		a^+	3 c^+

Is the b or c gene closer to a? Explain your reasoning.

28. A donor strain of bacteria with genotype *leu*⁺ *gal*⁻ *pro*⁺ is infected with phages. The phage lysate from the bacterial cells is collected and used to infect a second strain of bacteria that are *leu*⁻ *gal*⁺ *pro*⁻. The second strain is selected for *leu*⁺, and the following cotransduction data are obtained:

Donor	Recipient	Selected gene	Cells with cotransduced gene (%)
leu⁺ *gal*⁻ *pro*⁺	*leu*⁻ *gal*⁺ *pro*⁻	*leu*⁺	47 *pro*⁺
		leu⁺	26 *gal*⁻

Which genes are closest, *leu* and *gal* or *leu* and *pro*?

29. A geneticist isolates two new mutations from the *rII* region of bacteriophage T4, called *rII*ₓ and *rII*ᵧ. *E. coli* B cells are simultaneously infected with phages carrying the *rII*ₓ mutation *and* with phages carrying the *rII*ᵧ mutation. After the cells have lysed, samples of the phage lysate are collected. One sample is grown on *E. coli* K cells and a second sample on *E. coli* B cells. There are 8322 plaques on *E. coli* B and 3 plaques on *E. coli* K. What is the recombination frequency between these two mutations?

30. A geneticist is working with a new bacteriophage called phage Y3 that infects *E. coli*. He has isolated eight mutant phages that fail to produce plaques when grown on *E. coli* strain K. To determine whether these mutations occur at the same functional gene, he simultaneously infects *E. coli* K cells with paired combinations of the mutants and looks to see whether plaques are formed. He obtains the following results. (A plus sign means that plaques were formed on *E. coli* K; a minus sign means that no plaques were formed on *E. coli* K.)

Mutant	1	2	3	4	5	6	7	8
1								
2	+							
3	+	+						
4	+	−	+					
5	−	+	+	+				
6	−	+	+	+	−			
7	+	−	+	−	+	+		
8	−	+	+	+	−	−	+	

(a) To how many functional genes (cistrons) do these mutations belong?
(b) Which mutations belong to the same functional gene?

CHALLENGE QUESTIONS

31. As a summer project, a microbiology student independently isolates two mutations in *E. coli* that are auxotrophic for glycine (*gly*⁻). The student wants to know whether these two mutants occur at the same cistron. Outline a procedure that the student could use to determine whether these two *gly*⁻ mutations occur within the same cistron.

32. A group of genetics students mix two auxotrophic strains of bacteria: one is *leu*⁺ *trp*⁺ *his*⁻ *met*⁻ and the other is *leu*⁻ *trp*⁻ *his*⁺ *met*⁺. After mixing the two strains, they plate the bacteria on minimal medium and observe a few prototrophic colonies (*leu*⁺ *trp*⁺ *his*⁺ *met*⁺). They assume that some gene transfer has taken place between the two strains. How can they determine whether the transfer of genes is due to conjugation, transduction, or transformation?

SUGGESTED READINGS

Aguzzi, A., and C. Weissman. 1997. Prion research: the next frontiers. *Nature* 389:795–798.
A review of research into the nature of prions.

Benzer, S. 1962. The fine structure of the gene. *Scientific American* 206(1):70–84.
A good summary of Benzer's methodology for intragenic mapping, written by Benzer.

Birge, E. A. 2000. *Bacterial and Bacteriophage Genetics,* 4th ed. New York: Springer-Verlag.
An excellent textbook on the genetics of bacteria and bacteriophage.

Cole, L. A. 1996. The specter of biological weapons. *Scientific American* 275(6):60–65.
Reviews germ warfare and what can be done to discourage it.

Dale, J. 1998. *Molecular Genetics of Bacteria,* 3rd ed. New York: Wiley.
A concise summary of basic and molecular genetics of bacteria and bacteriophage.

Davies, J. 1994. Inactivation of antibiotics and the dissemination of resistance genes. *Science* 264:375–382.
Reviews the crisis of antibiotic resistance in bacteria, with particular emphasis on the physiology and genetics of resistance.

Doolittle, R. F. 1998. Microbial genomes opened up. *Nature* 392:339–342.
Discussion of sequence data on bacterial genomes and what this information provides.

Fraser, C. M., J. A. Eisen, and S. L. Salzberg. 2000. Microbial genome sequencing. *Nature* 406:799–803.
A short review of DNA sequencing of bacterial genomes.

Heidelberg, J. F., et al. 2000. DNA sequence of both chromosomes of the cholera pathogen *Vibrio cholera*. *Nature* 406:477–483.
 Report of the sequencing and analysis of the genome of the bacterium that causes chorea.

Hershey, A. D., and R. Rotman. 1942. Genetic recombination between host-range and plaque-type mutants of bacteriophage in single bacterial cells. *Genetics* 34:44–71.
 Original report of Hershey and Rotman's mapping experiments with phage.

Ippen-Ihler, K. A., and E. G. Minkley, Jr. 1986. The conjugation system of F, the fertility factor of *Escherichia coli*. *Annual Review of Genetics* 20:593–624.
 A detailed review of the F factor.

Kruse, H., and H. Sørum. 1994. Transfer of multiple drug resistance plasmids between bacteria of diverse origins in natural microenvironments. *Applied and Environmental Microbiology* 60:4015–4021.
 Reports experiments demonstrating the transfer of R plasmids between diverse bacteria under natural conditions.

Lederberg, J., and E. L. Tatum. 1946. Gene recombination in *Escherichia coli*. *Nature* 158:558.
 One of the original descriptions of Lederberg and Tatum's discovery of gene transfer in bacteria. A slightly different set of experiments showing the same result were published in 1946 in *Cold Spring Harbor Symposium on Quantitative Biology* 11:113–114.

Miller, R. V. 1998. Bacterial gene swapping in nature. *Scientific American* 278(1):66–71.
 Discusses the importance of gene transfer by conjugation, transformation, and transduction in nature.

Novick, R. P. 1980. Plasmids. *Scientific American* 243(6):103–124.
 A good summary of plasmids and their importance in drug resistance.

Pace, N. R. 1997. A molecular view of microbial diversity and the biosphere. *Science* 276:734–740.
 Good review of the diversity and classification of bacteria based on DNA sequence data.

Scientific American. 1998. Volume 279, issue 1.
 This issue contains a special report with a number of articles on HIV and AIDS.

Walsh, C. 2000. Molecular mechanisms that confer antibacterial drug resistance. *Nature* 406:775–781.
 A very good review of how antibiotic resistance develops and how antibiotics can be developed that are less likely to be resisted by bacteria.

Wollman, E. L., F. Jacob, and W. Hayes. 1962. Conjugation and genetic recombination in *Echerichia coli* K-12. *Cold Spring Harbor Symposium on Quantitative Biology* 21:141–162.
 Original work on the use of interrupted conjugation to map genes in *E. coli*.

9 Chromosome Variation

A cross between a female horse with 64 chromosomes and a male donkey with 62 chromosomes results in a mule with 63 chromosomes. Most mules are sterile but occasionally female mules give birth to viable offspring. (Charles Palek/Animals Animals.)

Once in a Blue Moon

One of the best-known facts of genetics is that a cross between a horse and a donkey produces a mule. Actually, it's a cross between a *female* horse and a *male* donkey that produces the mule; the reciprocal cross, between a male horse and a female donkey, produces a hinny, which has smaller ears and a bushy tail, like a horse (FIGURE 9.1). Both mules and hinnies are sterile because horses and donkeys are different species with different numbers of chromosomes: a horse has 64 chromosomes, whereas a donkey has only 62. There are also considerable differences in the sizes and shapes of the chromosomes that horses and donkeys have in common. A mule inherits 32 chromosomes from its horse mother and 31 chromosomes from its donkey father, giving the mule a chromosome number of 63. The maternal and paternal chromosomes of a mule are not homologous, and so they do not pair and separate properly in meiosis; consequently, a mule's gametes are abnormal and the animal is sterile.

In spite of the conventional wisdom that mules are sterile, reports of female mules with foals have surfaced over the years, although many of them can be attributed to mistaken identification. In several instances, a chromosome check of the alleged fertile mule has demonstrated that she is actually a donkey. In other instances, analyses of genetic markers in both mule and foal demonstrated that the foal was not the offspring of the mule; female mules are capable of lactation and sometimes they adopt the foal of a nearby horse or donkey.

In the summer of 1985, a female mule named Krause, who was pastured with a male donkey, was observed with a newborn foal. There were no other female horses or donkeys in the pasture; so it seemed unlikely that the mule had adopted the foal. Blood samples were collected from Krause, her horse and donkey parents, and her male foal, which was appropriately named Blue Moon. A team of geneticists led by Oliver Ryder of the San Diego Zoo examined their chromosomal makeup and analyzed 17 genetic markers from the blood samples.

Parents

Horse
(2n = 64)

Donkey
(2n = 62)

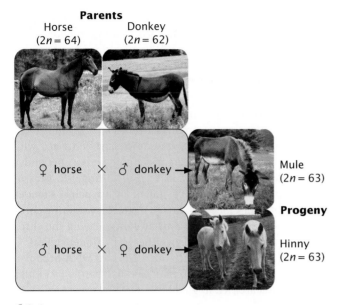

♀ horse × ♂ donkey →

Mule
(2n = 63)

Progeny

♂ horse × ♀ donkey →

Hinny
(2n = 63)

9.1 A cross between a female horse and a male donkey produces a mule; a cross between a male horse and a female donkey produces a hinny.
(Clockwise from top left, Bonnie Rauch/Photo Researchers; R. J. Erwin/Photo Researchers; Bruce Gaylord/Visuals Unlimited; Bill Kamin/Visuals Unlimited).

Krause's karyotype revealed that she was indeed a mule, with 63 chromosomes and blood type genes that were a mixture of those found in donkeys and horses. Blue Moon also had 63 chromosomes and, like his mother, he possessed both donkey and horse genes (◀FIGURE 9.2). Remarkably, he seemed to have inherited the entire set of horse chromosomes that were present in his mother. A mule's horse and donkey chromosomes would be expected to segregate randomly when the mule produces its own gametes; so Blue Moon *should* have inherited a mixture of horse and donkey chromosomes from his mother. The genetic markers that Ryder and his colleagues studied suggested that random segregation had not occurred. Krause and Blue Moon were therefore not only mother and son, but also sister and brother because they have the same father and they inherited the same maternal genes. The mechanism that allowed Krause to pass only horse chromosomes to

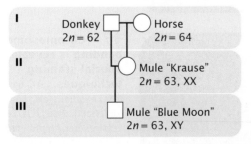

I   Donkey
2n = 62   Horse
2n = 64

II   Mule "Krause"
2n = 63, XX

III   Mule "Blue Moon"
2n = 63, XY

9.2 Blue Moon resulted from a cross between a fertile mule and a donkey. The probable pedigree of Blue Moon, the foal of a fertile mule, is shown.

her son is not known; possibly all Krause's donkey chromosomes passed into the polar body during the first division of meiosis (see Figure 2.22), leaving the oocyte with only horse chromosomes.

Krause later gave birth to another male foal named White Lightning. Like his brother, White Lightning possessed mule chromosomes and appeared to have inherited only horse chromosomes from his mother. Additional reports of fertile female mules support the idea that their offspring inherit only horse chromosomes from their mother. When the father of a mule's offspring is a horse, the offspring is horselike in appearance, because it apparently inherits horse chromosomes from both of its parents. When the father of a mule's offspring is a donkey, however, the offspring is mulelike in appearance, because it inherits horse chromosomes from its mule mother and donkey chromosomes from its father.

Most species have a characteristic number of chromosomes, each with a distinct size and structure, and all the tissues of an organism (except for gametes) generally have the same set of chromosomes. Nevertheless, variations in chromosome number and structure do periodically arise. Individual chromosomes may lose or gain parts; the sequence of genes within a chromosome may become altered; whole chromosomes can even be lost or gained. These variations in the number and structure of chromosomes are termed **chromosome mutations,** and they frequently play an important role in evolution.

We begin this chapter by briefly reviewing some basic concepts of chromosome structure, which we learned in Chapter 2. We then consider the different types of chromosome mutations, their definitions, their features, and their phenotypic effects. Finally, we examine the role of chromosome mutations in cancer.

www.whfreeman.com/pierce   More information on mules

Chromosome Variation

Before we consider the different types of chromosome mutations, their effects, and how they arise, we will review the basics of chromosome structure.

Chromosome Morphology

Each functional chromosome has a centromere, where spindle fibers attach, and two telomeres that stabilize the chromosome (see Figure 2.7). Chromosomes are classified into four basic types: **metacentric,** in which the centromere is located approximately in the middle, and so the chromosome has two arms of equal length; **submetacentric,** in which the centromere is displaced toward one end, creating a long arm and a short arm; **acrocentric,** in which the centromere is near one end, producing a long arm and a knob, or satellite, at the other; and **telocentric,** in which the centromere is at or very near the end of the chromosome (see

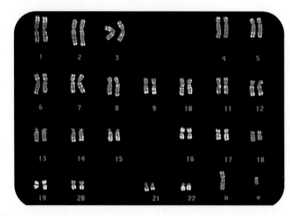

◀ 9.3 **A human karyotype consists of 46 chromosomes.** A karyotype for a male is shown here; a karyotype for a female would have two X chromosomes. (ISM/Phototake).

Figure 2.8). On human chromosomes, the short arm is designated by the letter p and the long arm by the letter q.

The complete set of chromosomes that an organism possesses is called its *karyotype* and is usually presented as a picture of metaphase chromosomes lined up in descending order of their size (◀ FIGURE 9.3). Karyotypes are prepared from actively dividing cells, such as white blood cells, bone marrow cells, or cells from meristematic tissues of plants. After treatment with a chemical (such as colchicine) that prevents them from entering anaphase, the cells are chemically preserved, spread on a microscope slide, stained, and photographed. The photograph is then enlarged, and the individual chromosomes are cut out and arranged in a karyotype. For human chromosomes, karyotypes are often

routinely prepared by automated machines, which scan a slide with a video camera attached to a microscope, looking for chromosome spreads. When a spread has been located, the camera takes a picture of the chromosomes, the image is digitized, and the chromosomes are sorted and arranged electronically by a computer.

Preparation and staining techniques have been developed to help distinguish among chromosomes of similar size and shape. For instance, chromosomes may be treated with enzymes that partly digest them; staining with a special dye called Giemsa reveals G bands, which distinguish areas of DNA that are rich in adenine–thymine base pairs (◀ FIGURE 9.4a). Q bands (◀ FIGURE 9.4b) are revealed by staining chromosomes with quinacrine mustard and viewing the chromosomes under UV light. Other techniques reveal C bands (◀ FIGURE 9.4c), which are regions of DNA occupied by centromeric heterochromatin, and R bands (◀ FIGURE 9.4d), which are rich in guanine–cytosine base pairs.

www.whfreeman.com/pierce   Pictures of karyotypes, including specific chromosome abnormalities, the analysis of human karyotypes, and links to a number of Web sites on chromosomes

Types of Chromosome Mutations

Chromosome mutations can be grouped into three basic categories: chromosome rearrangements, aneuploids, and polyploids. **Chromosome rearrangements** alter the structure of chromosomes; for example, a piece of a chromosome might be duplicated, deleted, or inverted. In **aneuploidy,** the *number* of chromosomes is altered: one or more individual chromosomes are added or deleted. In **polyploidy,** one or more

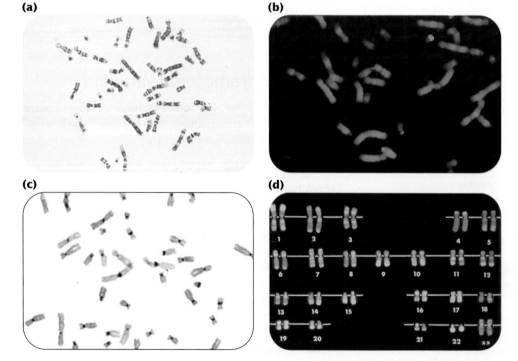

(a)

(b)

(c)

(d)

◀ 9.4 **Chromosome banding is revealed by special staining techniques.** (a) G banding. (b) Q banding. (c) C banding. (d) R banding. (Part a, Leonard Lessin/Peter Arnold; parts b and c, Dr. Dorothy Warburton, HICCC, Columbia University; part d, Dr. Ram Verma/ Phototake).

complete *sets* of chromosomes are added. Some organisms (such as yeast) possess a single chromosome set (1*n*) for most of their life cycles and are referred to as haploid, whereas others possess two chromosome sets and are referred to as diploid (2*n*). A polyploid is any organism that has more than two sets of chromosomes (3*n*, 4*n*, 5*n*, or more).

Chromosome Rearrangements

Chromosome rearrangements are mutations that change the structure of individual chromosomes. The four basic types of rearrangements are duplications, deletions, inversions, and translocations (◀ FIGURE 9.5).

(a) Duplication

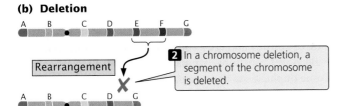

(b) Deletion

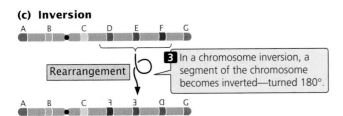

(c) Inversion

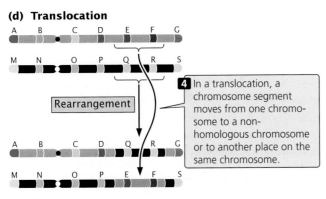

(d) Translocation

◀ 9.5 **The four basic types of chromosome rearrangements are duplication, deletion, inversion, and translocation.**

Duplications

A **chromosome duplication** is a mutation in which part of the chromosome has been doubled (see Figure 9.5a). Consider a chromosome with segments AB•CDEFG, in which • represents the centromere. A duplication might include the EF segments, giving rise to a chromosome with segments AB•CDEFEFG. This type of duplication, in which the duplicated region is immediately adjacent to the original segment, is called a **tandem duplication.** If the duplicated segment is located some distance from the original segment, either on the same chromosome or on a different one, this type is called a **displaced duplication.** An example of a displaced duplication would be AB•CDEFGEF. A duplication can either be in the same orientation as the original sequence, as in the two preceding examples, or be inverted: AB•CDEFFEG. When the duplication is inverted, it is called a **reverse duplication.**

An individual homozygous for a rearrangement carries the rearrangement (the mutated sequence) on both homologous chromosomes, and an individual heterozygous for a rearrangement has one unmutated chromosome and one chromosome with the rearrangement. In the heterozygotes (◀ FIGURE 9.6a), problems arise in chromosome pairing at prophase I of meiosis, because the two chromosomes are not homologous throughout their length. The pairing and synapsis of homologous regions require that one or both

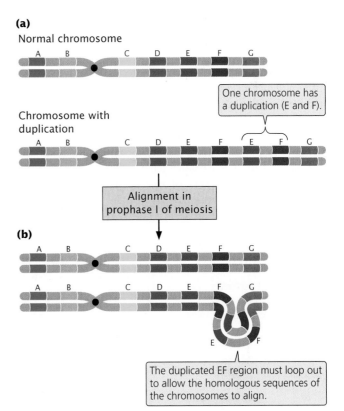

◀ 9.6 **In an individual heterozygous for a duplication, the duplicated chromosome loops out during pairing in prophase I.**

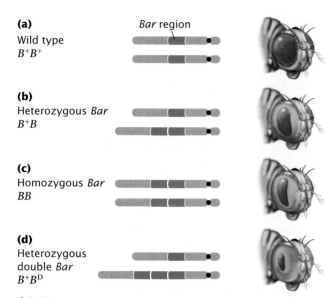

(a)
Wild type
B^+B^+

Bar region

(b)
Heterozygous *Bar*
B^+B

(c)
Homozygous *Bar*
BB

(d)
Heterozygous
double *Bar*
B^+B^D

◀9.7 The Bar phenotype in *Drosophila melanogaster* results from an X-linked duplication. (a) Wild-type fruit flies have normal-size eyes. (b) Flies heterozygous and (c) homozygous for the *Bar* mutation have smaller, bar-shaped eyes. (d) Flies with double *Bar* have three copies of the duplication and much smaller bar-shaped eyes.

chromosomes loop and twist so that these regions are able to line up (◀Figure 9.6b). The appearance of this characteristic loop structure during meiosis is one way to detect duplications.

Duplications may have major effects on the phenotype. In *Drosophila melanogaster*, for example, a *Bar* mutant has a reduced number of facets in the eye, making the eye smaller and bar shaped instead of oval (◀Figure 9.7). The *Bar* mutant results from a small duplication on the X chromosome, which is inherited as an incompletely dominant, X-linked trait: heterozygous female flies have somewhat smaller eyes (the number of facets is reduced; see Figure 9.7b), whereas, in homozygous female and hemizygous male flies, the number of facets is greatly reduced (see Figure

9.7c). Occasionally, a fly carries three copies of the *Bar* duplication on its X chromosome; in such mutants, which are termed *double Bar,* the number of facets is extremely reduced (see Figure 9.7d). *Bar* arises from unequal crossing over, a duplication-generating process (◀Figure 9.8; see also Figure 17.15).

How does a chromosome duplication alter the phenotype? After all, gene sequences are not altered by duplications, and no genetic information is missing; the only change is the presence of additional copies of normal sequences. The answer to this question is not well understood, but the effects are most likely due to imbalances in the amounts of gene products (abnormal gene dosage). The amount of a particular protein synthesized by a cell is often directly related to the number of copies of its corresponding gene: an individual with three functional copies of a gene often produces 1.5 times as much of the protein encoded by that gene as that produced by an individual with two copies. Because developmental processes often require the interaction of many proteins, they may critically depend on the relative amounts of the proteins. If the amount of one protein increases while the amounts of others remain constant, problems can result (◀Figure 9.9). Although duplications can have severe consequences when the precise balance of a gene product is critical to cell function, duplications have arisen frequently throughout the evolution of many eukaryotic organisms and are a source of new genes that may provide novel functions. Human phenotypes associated with some duplications are summarized in Table 9.1.

Concepts

A chromosome duplication is a mutation that doubles part of a chromosome. In individuals heterozygous for a chromosome duplication, the duplicated region of the chromosome loops out when homologous chromosomes pair in prophase I of meiosis. Duplications often have major effects on the phenotype, possibly by altering gene dosage.

Wild-type chromosomes

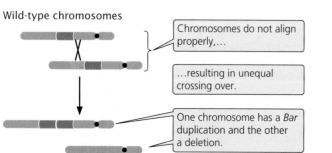

Chromosomes do not align properly,…

…resulting in unequal crossing over.

One chromosome has a *Bar* duplication and the other a deletion.

Bar chromosomes

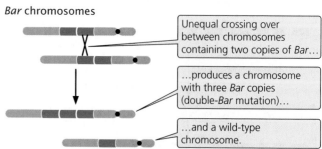

Unequal crossing over between chromosomes containing two copies of *Bar*…

…produces a chromosome with three *Bar* copies (double-*Bar* mutation)…

…and a wild-type chromosome.

◀9.8 Unequal crossing over produces *Bar* and double-*Bar* mutations.

Table 9.1	Effects of some chromosome rearrangements in humans		
Type of Rearrangement	Chromosome	Disorder	Symptoms
Duplication	4, short arm	—	Small head, short neck, low hairline, growth and mental retardation
Duplication	4, long arm	—	Small head, sloping forehead, hand abnormalities
Duplication	7, long arm	—	Delayed development, asymmetry of the head, fuzzy scalp, small nose, low-set ears
Duplication	9, short arm	—	Characteristic face, variable mental retardation, high and broad forehead, hand abnormalities
Deletion	5, short arm	Cri-du-chat syndrome	Small head, distinctive cry, widely spaced eyes, a round face, mental retardation
Deletion	4, short arm	Wolf-Hirschhorn syndrome	Small head with high forehead, wide nose, cleft lip and palate, severe mental retardation
Deletion	4, long arm	—	Small head, mild to moderate mental retardation, cleft lip and palate, hand and foot abnormalities
Deletion	15, long arm	Prader-Willi syndrome	Feeding difficulty at early age, but becoming obese after 1 year of age, mild to moderate mental retardation
Deletion	18, short arm	—	Round face, large low set-ears, mild to moderate mental retardation
Deletion	18, long arm	—	Distinctive mouth shape, small hands, small head, mental retardation

Deletions

A second type of chromosome rearrangement is a **deletion,** the loss of a chromosome segment (see Figure 9.5b). A chromosome with segments AB•CDEFG that undergoes a deletion of segment EF would generate the mutated chromosome AB•CDG.

A large deletion can be easily detected because the chromosome is noticeably shortened. In individuals heterozygous for deletions, the normal chromosome must loop out during the pairing of homologs in prophase I of meiosis (FIGURE 9.10) to allow the homologous regions of the two chromosomes to align and undergo synapsis. This looping out generates a structure that looks very much like that seen in individuals heterozygous for duplications.

The phenotypic consequences of a deletion depend on which genes are located in the deleted region. If the deletion includes the centromere, the chromosome will not segregate in meiosis or mitosis and will usually be lost. Many deletions are lethal in the homozygous state because all copies of any essential genes located in the deleted region are missing. Even individuals heterozygous for a deletion may have multiple defects for three reasons.

First, the heterozygous condition may produce imbalances in the amounts of gene products, similar to the imbalances produced by extra gene copies. Second, deletions may

allow recessive mutations on the undeleted chromosome to be expressed (because there is no wild-type allele to mask their expression). This phenomenon is referred to as **pseudodominance.** The appearance of pseudodominance in otherwise recessive alleles is an indication that a deletion is present on one of the chromosomes. Third, some genes must be present in two copies for normal function. Such a gene is said to be **haploinsufficient;** loss of function mutations in haploinsufficient genes are dominant. *Notch* is a series of X-linked wing mutations in *Drosophila* that often result from chromosome deletions. *Notch* deletions behave as dominant mutations: when heterozygous for the *Notch* deletion, a fly has wings that are notched at the tips and along the edges (FIGURE 9.11). The *Notch* locus is therefore haploinsufficient—a single copy of the gene is not sufficient to produce a wild-type phenotype. Females that are homozygous for a *Notch* deletion (or males that are hemizygous) die early in embryonic development. The *Notch* gene codes for a receptor that normally transmits signals received from outside the cell to the cell's interior and is important in fly development. The deletion acts as a recessive lethal because loss of all copies of the *Notch* gene prevents normal development.

In humans, a deletion on the short arm of chromosome 5 is responsible for *cri-du-chat* syndrome. The name

(a)

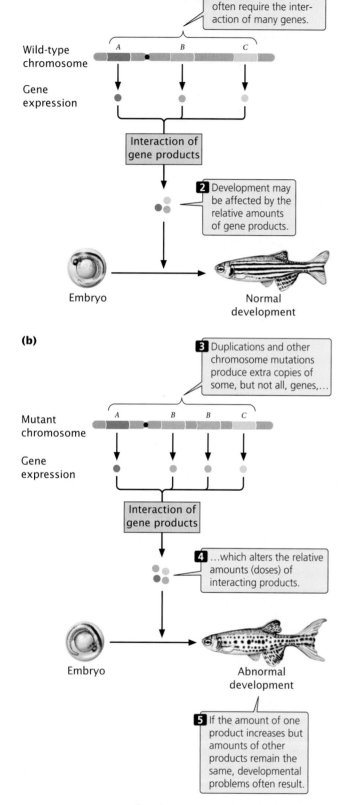

1 Developmental processes often require the interaction of many genes.

Wild-type chromosome

Gene expression

Interaction of gene products

2 Development may be affected by the relative amounts of gene products.

Embryo → Normal development

(b)

3 Duplications and other chromosome mutations produce extra copies of some, but not all, genes,...

Mutant chromosome

Gene expression

Interaction of gene products

4 ...which alters the relative amounts (doses) of interacting products.

Embryo → Abnormal development

5 If the amount of one product increases but amounts of other products remain the same, developmental problems often result.

◀ **9.9 Unbalanced gene dosage leads to developmental abnormalities.**

(French for "cry of the cat") derives from the peculiar, cat-like cry of infants with this syndrome. A child who is heterozygous for this deletion has a small head, widely spaced eyes, a round face, and mental retardation. Deletion of part of the short arm of chromosome 4 results in another human disorder—Wolf-Hirschhorn syndrome, which is characterized by seizures and by severe mental and growth retardation.

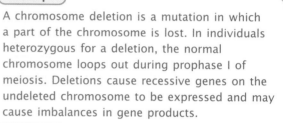

Concepts

A chromosome deletion is a mutation in which a part of the chromosome is lost. In individuals heterozygous for a deletion, the normal chromosome loops out during prophase I of meiosis. Deletions cause recessive genes on the undeleted chromosome to be expressed and may cause imbalances in gene products.

www.whfreeman.com/pierce   Information on rare chromosome disorders

Inversions

A third type of chromosome rearrangement is a **chromosome inversion,** in which a chromosome segment is inverted—turned 180 degrees (see Figure 9.5c). If a chromosome originally had segments AB•CDEFG, then chromosome AB•CFEDG represents an inversion that includes segments DEF. For an inversion to take place, the chromosome must break in two places. Inversions that do not include the centromere, such as AB•CFEDG, are termed **paracentric inversions** (*para* meaning "next to"), whereas inversions that include the centromere, such as ADC•BEFG, are termed **pericentric inversions** (*peri* meaning "around").

Individuals with inversions have neither lost nor gained any genetic material; just the gene order has been altered. Nevertheless, these mutations often have pronounced phenotypic effects. An inversion may break a gene into two parts, with one part moving to a new location and destroying the function of that gene. Even when the chromosome breaks are between genes, phenotypic effects may arise from the inverted gene order in an inversion. Many genes are regulated in a position-dependent manner; if their positions are altered by an inversion, they may be expressed at inappropriate times or in inappropriate tissues. This outcome is referred to as a **position effect**.

When an individual is homozygous for a particular inversion, no special problems arise in meiosis, and the two homologous chromosomes can pair and separate normally. When an individual is heterozygous for an inversion, however, the gene order of the two homologs differs, and the homologous sequences can align and pair only if

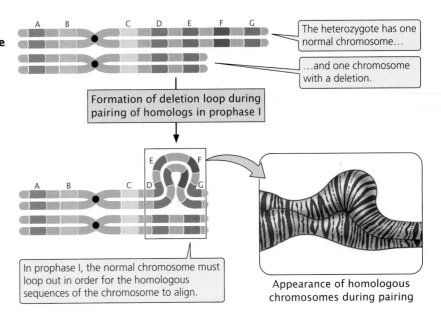

9.10 In an individual heterozygous for a deletion, the normal chromosome loops out during chromosome pairing in prophase I.

The heterozygote has one normal chromosome…

…and one chromosome with a deletion.

Formation of deletion loop during pairing of homologs in prophase I

In prophase I, the normal chromosome must loop out in order for the homologous sequences of the chromosome to align.

Appearance of homologous chromosomes during pairing

9.11 The Notch phenotype is produced by a chromosome deletion that includes the *Notch* gene. (Top) Normal wing veination. (Bottom) Wing veination produced by *Notch* mutation. (Spyros Artavanis-Tsakonas, Kenji Matsuno, and Mark E. Fortini).

the two chromosomes form an inversion loop (◀FIGURE 9.12). The presence of an inversion loop in meiosis indicates that an inversion is present.

Individuals heterozygous for inversions also exhibit reduced recombination among genes located in the inverted region. The frequency of crossing over within the inversion is not actually diminished but, when crossing over does take place, the result is a tendency to produce gametes that are

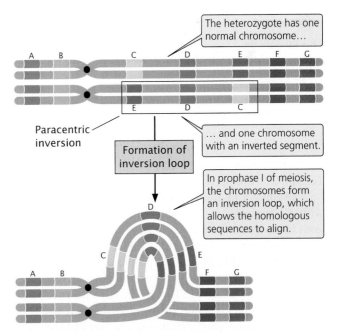

The heterozygote has one normal chromosome…

Paracentric inversion

… and one chromosome with an inverted segment.

Formation of inversion loop

In prophase I of meiosis, the chromosomes form an inversion loop, which allows the homologous sequences to align.

9.12 In an individual heterozygous for a paracentric inversion, the chromosomes form an inversion loop during pairing in prophase I.

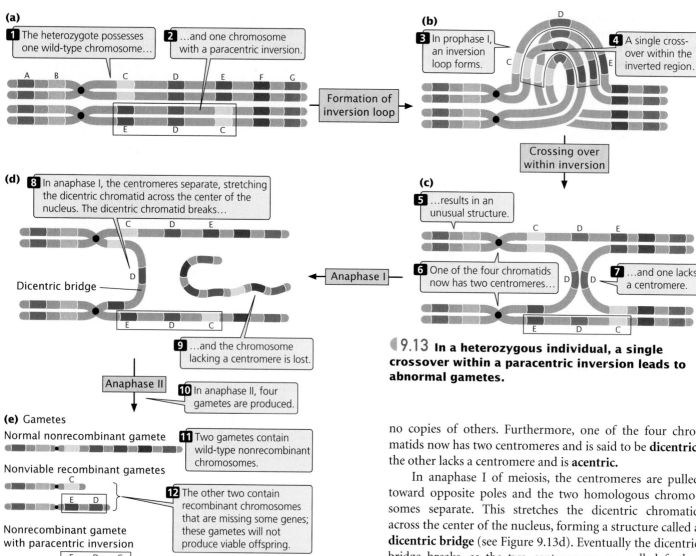

9.13 In a heterozygous individual, a single crossover within a paracentric inversion leads to abnormal gametes.

no copies of others. Furthermore, one of the four chromatids now has two centromeres and is said to be **dicentric**; the other lacks a centromere and is **acentric**.

In anaphase I of meiosis, the centromeres are pulled toward opposite poles and the two homologous chromosomes separate. This stretches the dicentric chromatid across the center of the nucleus, forming a structure called a **dicentric bridge** (see Figure 9.13d). Eventually the dicentric bridge breaks, as the two centromeres are pulled farther apart. The acentric fragment has no centromere. Spindle fibers do not attach to it, and so this fragment does not segregate into a nucleus in meiosis and is usually lost.

In the second division of meiosis, the chromatids separate and four gametes are produced (see Figure 9.11e). Two of the gametes contain the original, nonrecombinant chromosomes (AB•CDEFG and AB•EDCFG). The other two gametes contain recombinant chromosomes that are missing some genes; these gametes will not produce viable offspring. Thus, no recombinant progeny result when crossing over takes place within a paracentric inversion.

Recombination is also reduced within a pericentric inversion (◀FIGURE 9.14). No dicentric bridges or acentric fragments are produced, but the recombinant chromosomes have too many copies of some genes and no copies of others; so gametes that receive the recombinant chromosomes cannot produce viable progeny.

Figures 9.13 and 9.14 illustrate the results of single crossovers within inversions. Double crossovers, in which both crossovers are on the same two strands (two-strand,

not viable and thus no recombinant progeny are observed. Let's see why this occurs.

◀FIGURE 9.13 illustrates the results of crossing over within a paracentric inversion. The individual is heterozygous for an inversion (see Figure 9.13a), with one wild-type, unmutated chromosome (AB•CDEFG) and one inverted chromosome (AB•EDCFG). In prophase I of meiosis, an inversion loop forms, allowing the homologous sequences to pair up (see Figure 9.13b). If a single crossover takes place in the inverted region (between segments C and D in Figure 9.13), an unusual structure results (see Figure 9.13c). The two outer chromatids, which did not participate in crossing over, contain original, nonrecombinant gene sequences. The two inner chromatids, which did cross over, are highly abnormal: each has two copies of some genes and

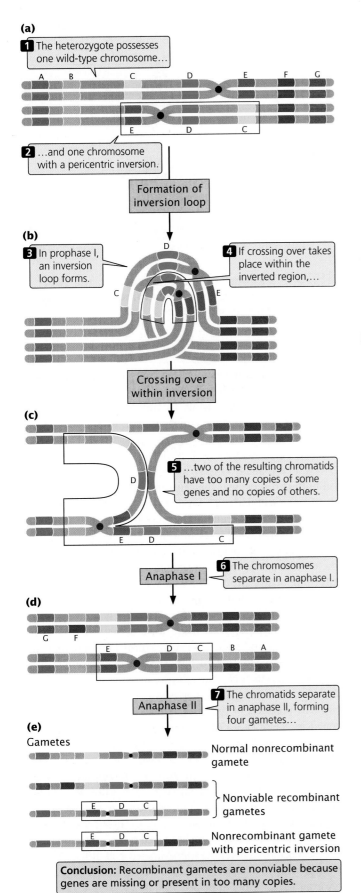

(a)

1 The heterozygote possesses one wild-type chromosome…

A B C D E F G

2 …and one chromosome with a pericentric inversion.

E D C

Formation of inversion loop

(b)

3 In prophase I, an inversion loop forms.

D

C E

4 If crossing over takes place within the inverted region,…

Crossing over within inversion

(c)

D

5 …two of the resulting chromatids have too many copies of some genes and no copies of others.

E D C

Anaphase I

6 The chromosomes separate in anaphase I.

(d)

G F

E D C B A

Anaphase II

7 The chromatids separate in anaphase II, forming four gametes…

(e)

Gametes

Normal nonrecombinant gamete

E D C

Nonviable recombinant gametes

E D C

Nonrecombinant gamete with pericentric inversion

Conclusion: Recombinant gametes are nonviable because genes are missing or present in too many copies.

9.14 In a heterozygous individual, a single crossover within a pericentric inversion leads to abnormal gametes.

double crossovers), result in functional, recombinant chromosomes. (Try drawing out the results of a double crossover.) Thus, even though the overall rate of recombination is reduced within an inversion, some viable recombinant progeny may still be produced through two-stranded double crossovers.

Inversion heterozygotes are common in many organisms, including a number of plants, some species of *Drosophila*, mosquitoes, and grasshoppers. Inversions may have played an important role in human evolution: G-banding patterns reveal that several human chromosomes differ from those of chimpanzees by only a pericentric inversion (◄ FIGURE 9.15).

Concepts

In an inversion, a segment of a chromosome is inverted. Inversions cause breaks in some genes and may move others to new locations. In heterozygotes for a chromosome inversion, the chromosomes form loops in prophase I of meiosis. When crossing over takes place within the inverted region, nonviable gametes are usually produced, resulting in a depression in observed recombination frequencies.

Translocations

A **translocation** entails the movement of genetic material between nonhomologous chromosomes (see Figure 9.5d) or within the same chromosome. Translocation should not be confused with crossing over, in which there is an exchange of genetic material between *homologous* chromosomes.

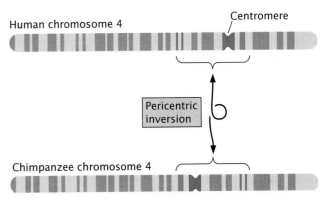

Human chromosome 4

Centromere

Pericentric inversion

Chimpanzee chromosome 4

9.15 Chromosome 4 differs in humans and chimpanzees in a pericentric inversion.

In **nonreciprocal translocations,** genetic material moves from one chromosome to another without any reciprocal exchange. Consider the following two nonhomologous chromosomes: AB•CDEFG and MN•OPQRS. If chromosome segment EF moves from the first chromosome to the second without any transfer of segments from the second chromosome to the first, a nonreciprocal translocation has taken place, producing chromosomes AB•CDG and MN•OPEFQRS. More commonly, there is a two-way exchange of segments between the chromosomes, resulting in a **reciprocal translocation.** A reciprocal translocation between chromosomes AB•CDEFG and MN•OPQRS might give rise to chromosomes AB•CDQRG and MN•OPEFS.

Translocations can affect a phenotype in several ways. First, they may create new linkage relations that affect gene expression (a position effect): genes translocated to new locations may come under the control of different regulatory sequences or other genes that affect their expression—an example is found in Burkitt lymphoma, to be discussed later in this chapter.

Second, the chromosomal breaks that bring about translocations may take place within a gene and disrupt its function. Molecular geneticists have used these types of effects to map human genes. Neurofibromatosis is a genetic disease characterized by numerous fibrous tumors of the skin and nervous tissue; it results from an autosomal dominant mutation. Linkage studies first placed the locus for neurofibromatosis on chromosome 17. Geneticists later identified two patients with neurofibromatosis who possessed a translocation affecting chromosome 17. These patients were assumed to have developed neurofibromatosis because one of the chromosome breaks that occurred in the translocation disrupted a particular gene that causes neurofibromatosis. DNA from the regions around the breaks was sequenced and eventually led to the identification of the gene responsible for neurofibromatosis.

Deletions frequently accompany translocations. In a **Robertsonian translocation,** for example, the long arms of two acrocentric chromosomes become joined to a common centromere through a translocation, generating a metacentric chromosome with two long arms and another chromosome with two very short arms (◀FIGURE 9.16). The smaller chromosome often fails to segregate, leading to an overall reduction in chromosome number. As we will see, Robertsonian translocations are the cause of some cases of Down syndrome.

The effects of a translocation on chromosome segregation in meiosis depend on the nature of the translocation. Let us consider what happens in an individual heterozygous for a reciprocal translocation. Suppose that the original chromosomes were AB•CDEFG and MN•OPQRS (designated N₁ and N₂), and a reciprocal translocation takes place, producing chromosomes AB•CDQRS and MN•OPEFG (designated T₁ and T₂). An individual heterozygous for this translocation would possess one normal

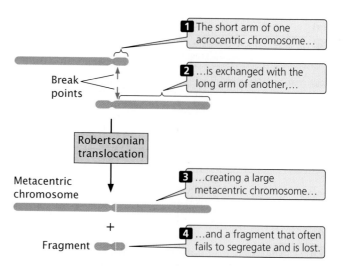

◀9.16 **In a Robertsonian translocation, the short arm of one acrocentric chromosome is exchanged with the long arm of another.**

copy of each chromosome and one translocated copy (◀FIGURE 9.17a). Each of these chromosomes contains segments that are homologous to *two* other chromosomes. When the homologous sequences pair in prophase I of meiosis, crosslike configurations consisting of all four chromosomes (◀FIGURE 9.17b) form.

Notice that N₁ and T₁ have homologous centromeres (in both chromosomes the centromere is between segments B and C); similarly, N₂ and T₂ have homologous centromeres (between segments N and O). Normally, homologous centromeres separate and move toward opposite poles in anaphase I of meiosis. With a reciprocal translocation, the chromosomes may segregate in three different ways. In **alternate segregation** (◀FIGURE 9.17c), N₁ and N₂ move toward one pole and T₁ and T₂ move toward the opposite pole. In **adjacent-1 segregation,** N₁ and T₂ move toward one pole and T₁ and N₂ move toward the other pole. In both alternate and adjacent-1 segregation, homologous centromeres segregate toward opposite poles. **Adjacent-2 segregation,** in which N₁ and T₁ move toward one pole and T₂ and N₂ move toward the other, is rare.

The products of the three segregation patterns are illustrated in ◀FIGURE 9.17d. As you can see, the gametes produced by alternate segregation possess one complete set of the chromosome segments. These gametes are therefore functional and can produce viable progeny. In contrast, gametes produced by adjacent-1 and adjacent-2 segregation are not viable, because some chromosome segments are present in two copies, whereas others are missing. Adjacent-2 segregation is rare, and so most gametes are produced by alternate and adjacent segregation. Therefore, approximately half of the gametes from an individual heterozygous for a reciprocal translocation are expected to be functional.

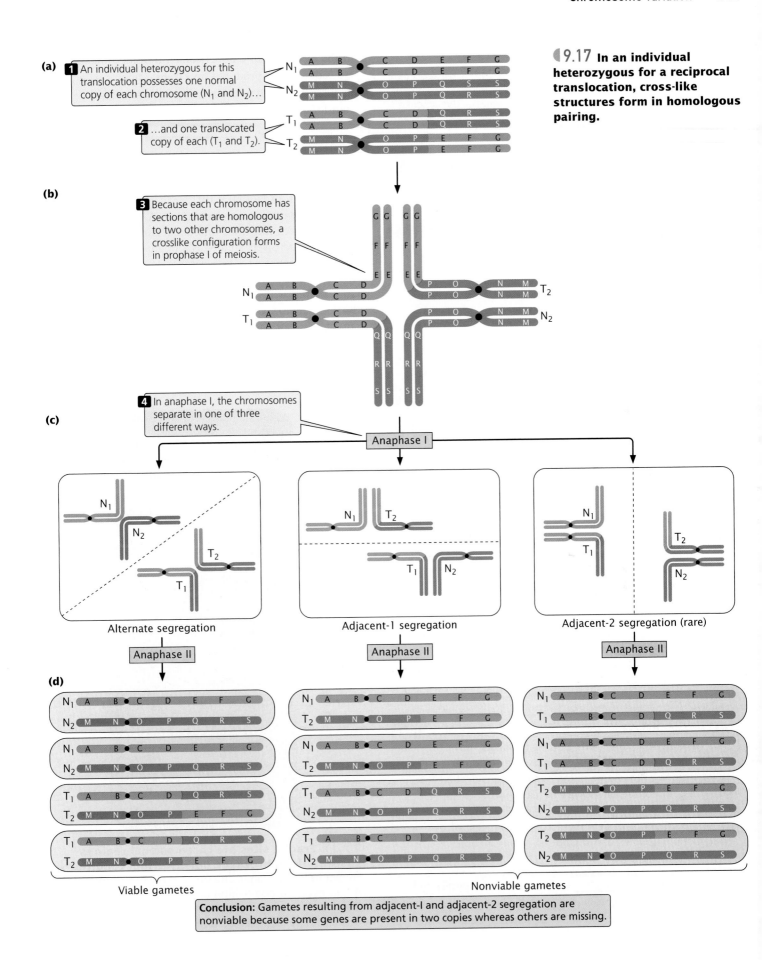

9.17 In an individual heterozygous for a reciprocal translocation, cross-like structures form in homologous pairing.

(a)

1 An individual heterozygous for this translocation possesses one normal copy of each chromosome (N_1 and N_2)…

2 …and one translocated copy of each (T_1 and T_2).

(b)

3 Because each chromosome has sections that are homologous to two other chromosomes, a crosslike configuration forms in prophase I of meiosis.

(c)

4 In anaphase I, the chromosomes separate in one of three different ways.

Anaphase I

Alternate segregation

Adjacent-1 segregation

Adjacent-2 segregation (rare)

Anaphase II

Anaphase II

Anaphase II

(d)

Viable gametes

Nonviable gametes

Conclusion: Gametes resulting from adjacent-I and adjacent-2 segregation are nonviable because some genes are present in two copies whereas others are missing.

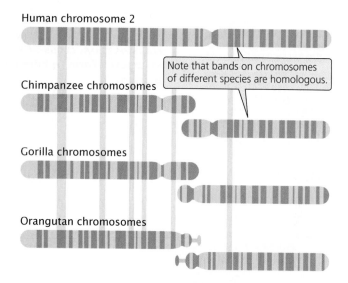

Human chromosome 2

Note that bands on chromosomes of different species are homologous.

Chimpanzee chromosomes

Gorilla chromosomes

Orangutan chromosomes

◀9.18 **Human chromosome 2 contains a Robertsonian translocation that is not present in chimps, gorillas, or orangutans.** G-banding reveals that a Robertsonian translocation in a human ancestor switched the long and short arms of the two acrocentric chromosomes that are still found in the other three primates. This translocation created the large metacentric human chromosome 2.

Translocations can play an important role in the evolution of karyotypes. Chimpanzees, gorillas, and orangutans all have 48 chromosomes, whereas humans have 46. Human chromosome 2 is a large, metacentric chromosome with G-banding patterns that match those found on two different acrocentric chromosomes of the apes (◀FIGURE 9.18). Apparently, a Robertsonian translocation took place in a human ancestor, creating a large metacentric chromosome from the two long arms of the ancestral acrocentric chromosomes and a small chromosome consisting of the two short arms. The small chromosome was subsequently lost, leading to the reduced human chromosome number.

Concepts

In translocations, parts of chromosomes move to other, nonhomologous chromosomes or other regions of the same chromosome. Translocations may affect the phenotype by causing genes to move to new locations, where they come under the influence of new regulatory sequences, or by breaking genes and disrupting their function.

www.whfreeman.com/pierce Information on gorilla and other great-ape chromosomes with a comparison of the human karyotype and great-ape karyotypes; animation

of the formation of a Robertsonian translocation and types of gametes produced by a translocation carrier; and pictures of karyotypes containing Robertsonian translocations

Fragile Sites

Chromosomes of cells grown in culture sometimes develop constrictions or gaps at particular locations called **fragile sites** (◀FIGURE 9.19) because they are prone to breakage under certain conditions. A number of fragile sites have been identified on human chromosomes. One of the most intensively studied is a fragile site on the human X chromosome that is associated with mental retardation, the fragile-X syndrome. Exhibiting X-linked inheritance and arising with a frequency of about 1 in 1250 male births, fragile-X syndrome has been shown to result from an increase in the number of repeats of a CGG trinucleotide (see Chapter 17). However, other common fragile sites do not consist of trinucleotide repeats, and their nature is still incompletely understood.

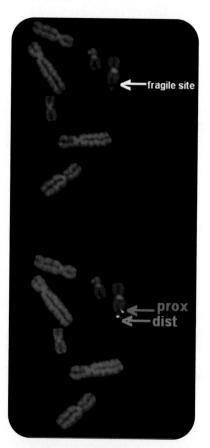

◀9.19 **Fragile sites are chromosomal regions susceptible to breakage under certain conditions.** Shown here is a fragile site on human chromosome 16. A fluorescein probe (white) identifies the proximal and distal ends of the fragile site. (Erica Woollatt, Women's and Children's Hospital, Adelaide, Australia).

The New Genetics
ETHICS · SCIENCE · TECHNOLOGY

Fragile-X Syndrome

Ron Green

Ryan, age 4, is brought to the medical genetics clinic by his 27-year-old mother, Janet. Ryan is developmentally delayed and hyperactive and has undergone many tests, but all the results were normal. Janet and her husband, Terry, very much want another child. The family history is unremarkable with the exception of the 6-year-old son of one of Janet's cousins, who is apparently "slow." However, Janet reports that she does not get along with her siblings and that in fact she has little contact with any of the rest of her family. Both her parents are deceased. A friend told her recently that Janet's 25-year-old sister just found out that she is pregnant with her first child.

The cause of Ryan's delay is determined to be fragile-X syndrome. As its name suggests, this condition is an X-linked disorder carried by females and most seriously affecting males, in whom it can cause severe mental retardation. After describing the genetics of fragile-X syndrome and its hereditary risks, the medical geneticist asks Janet to notify her sister of the information and alert her to the availability of prenatal testing. The next week, the geneticist calls Janet, who states that she has not called her sister and does not intend to.

Are there ways that the geneticist can persuade Janet to inform her sister? Failing that, can the physician alert Janet's sister to her risk without compromising the obligation to preserve the confidentiality of the relationship with Janet? If there is no other recourse, does the genetic professional have an ethical or legal right to breach confidentiality and inform Janet's sister—and perhaps others in the family—of the risk? Does the professional have a *duty* to do so? By law, a professional's ethical duty to a patient or client can be overridden only if (1) reasonable efforts to gain consent to disclosure have failed; (2) there is a high probability of harm if information is withheld, and the information will be used to avert harm; (3) the harm that person would suffer is serious; and (4) precautions are taken to ensure that only the genetic information needed for diagnosis or treatment or both is disclosed. However, who determines which genetic risks are among the "serious harms" that would permit breaching confidentiality in medical contexts?

Some might argue that all these questions miss the point: the familiar duties of doctor and patient don't apply in this case because, where genes are concerned, the patient is not the *individual*, but the *entire family* to which that patient belongs. Thus, the physician must do whatever best meets the needs of all members of the family. This line of reasoning may be increasingly popular as the powers of genetic medicine grow and physicians are more frequently asked to utilize and interpret genetic tests in the course of routine care. Nevertheless, we should be careful not to hastily discard the traditional ethical principle that the doctor's and medical team's first responsibility is to the presenting patient. Replacing it with a generalized responsibility to the whole family takes medical practice into uncharted territory and can impose serious new burdens on medical professionals.

In this case, the patient refuses to inform other family members who might benefit from prenatal testing—which would allow them to decide to continue or terminate a pregnancy or prepare for the birth of a child with a genetic disorder. But the same type of problem can arise in many other ways where genetic medicine or research is concerned. In some conditions, testing is aimed at determining whether an individual or family has a genetic susceptibility to a disease, the knowledge of which can help them pursue preventative strategies. Current testing for known breast cancer mutations is an example. In these cases, whenever the specific genes involved have not yet been identified, researchers or clinicians must conduct extensive family linkage studies to determine the pattern of inheritance. One or more family members can block progress by refusing to participate in the study. The principle of respect for autonomy certainly supports such refusals, but should this principle trump research that is needed to improve the health of other members of the family? Sometimes, the reverse problem arises: some members demand participation by other relatives in ways that exert pressure on them.

Alternatively, in some cases one person's use of a genetic test may harm others in the family. This problem has arisen in connection with Huntington disease, a fatal, later-onset neurological disorder for which no treatment exists. There have been instances when one twin of a pair of identical twins has insisted on testing and the other twin, unwilling to be subjected to the fearful psychosocial harms that testing can bring, has objected.

Social workers, psychologists, genetic counselors, and others who work closely with families know that such disputes often reveal deep fault lines and sources of conflict within a family. When faced with these cases, they also recognize that it is not a matter of just solving an ethical problem, but of understanding and addressing the underlying problems that give rise to these conflicts. The familial nature of genetic information will undoubtedly increase the number and intensity of conflicts that come before caregivers or counseling professionals.

Aneuploidy

In addition to chromosome rearrangements, chromosome mutations also include changes in the *number* of chromosomes. Variations in chromosome number can be classified into two basic types: changes in the number of individual chromosomes (aneuploidy) and changes in the number of chromosome sets (polyploidy).

Aneuploidy can arise in several ways. First, a chromosome may be lost in the course of mitosis or meiosis if, for example, its centromere is deleted. Loss of the centromere prevents the spindle fibers from attaching; so the chromosome fails to move to the spindle pole and does not become incorporated into a nucleus after cell division. Second, the small chromosome generated by a Robertsonian translocation may be lost in mitosis or meiosis. Third, aneuploids may arise through nondisjunction, the failure of homolo-

gous chromosomes or sister chromatids to separate in meiosis or mitosis (see p. 87 in Chapter 4). Nondisjunction leads to some gametes or cells that contain an extra chromosome and others that are missing a chromosome (◀FIGURE 9.20).

Types of Aneuploidy

We will consider four types of relatively common aneuploid conditions in diploid individuals: nullisomy, monosomy, trisomy, and tetrasomy.

1. **Nullisomy** is the loss of both members of a homologous pair of chromosomes. It is represented as $2n - 2$, where n refers to the haploid number of chromosomes. Thus, among humans, who normally possess $2n = 46$ chromosomes, a nullisomic person has 44 chromosomes.

2. **Monosomy** is the loss of a single chromosome, represented as $2n - 1$. A monosomic person has 45 chromosomes.

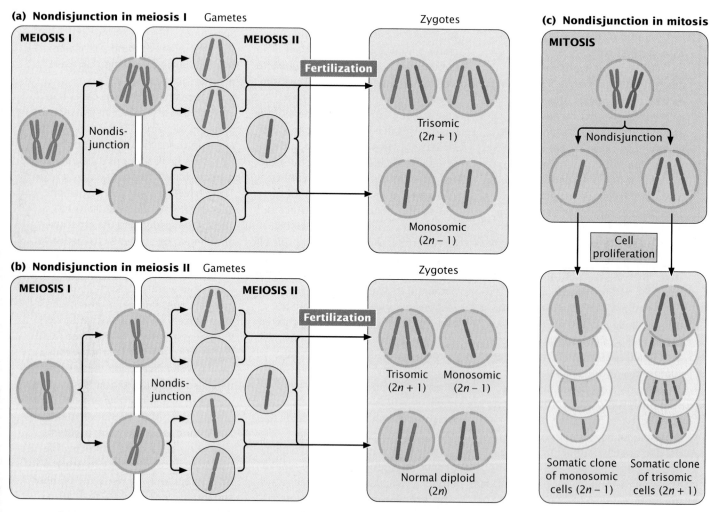

◀9.20 Aneuploids can be produced through nondisjunction in (a) meiosis I, (b) meiosis II, and (c) mitosis. The gametes that result from meioses with nondisjunction combine with a gamete (with blue chromosome) that results from normal meiosis to produce the zygotes.

3. **Trisomy** is the gain of a single chromosome, represented as $2n + 1$. A trisomic person has 47 chromosomes. The gain of a chromosome means that there are three homologous copies of one chromosome.

4. **Tetrasomy** is the gain of two homologous chromosomes, represented as $2n + 2$. A tetrasomic person has 48 chromosomes. Tetrasomy is not the gain of *any* two extra chromosomes, but rather the gain of two homologous chromosomes; so there will be four homologous copies of a particular chromosome.

More than one aneuploid mutation may occur in the same individual. An individual that has an extra copy of two different (nonhomologous) chromosomes is referred to as being double trisomic and represented as $2n + 1 + 1$. Similarly, a double monosomic has two fewer nonhomologous chromosomes ($2n - 1 - 1$), and a double tetrasomic has two extra pairs of homologous chromosomes ($2n + 2 + 2$).

Effects of Aneuploidy

One of the first aneuploids to be recognized was a fruit fly with a single X chromosome and no Y chromosome, which was discovered by Calvin Bridges in 1913 (see pp. 87–88 in Chapter 4). Another early study of aneuploidy focused on mutants in the Jimson weed, *Datura stramonium*. A. Francis Blakeslee began breeding this plant in 1913, and he observed that crosses with several Jimson mutants produced unusual ratios of progeny. For example, the *globe* mutant (having a seedcase globular in shape) was dominant but was inherited primarily from the female parent. When *globe* plants were self-fertilized, only 25% of the progeny had the globe phenotype, an unusual ratio for a dominant trait. Blakeslee isolated 12 different mutants (FIGURE 9.21) that also exhibited peculiar patterns of inheritance. Eventually, John Belling demonstrated that these 12 mutants are in fact trisomics. *Datura stramonium* has 12 pairs of chromosomes ($2n = 24$), and each of the 12 mutants is trisomic for a different chromosome pair. The aneuploid nature of the mutants explained the unusual ratios that Blakeslee had observed in the progeny. Many of the extra chromosomes in the trisomics were lost in meiosis, so fewer than 50% of the gametes carried the extra chromosome, and the proportion of trisomics in the progeny was low. Furthermore, the pollen containing an extra chromosome was not as successful in fertilization, and trisomic zygotes were less viable.

Aneuploidy usually alters the phenotype drastically. In most animals and many plants, aneuploid mutations are lethal. Because aneuploidy affects the number of gene copies but not their nucleotide sequences, the effects of aneuploidy are most likely due to abnormal gene dosage. Aneuploidy alters the dosage for some, but not all, genes, disrupting the relative concentrations of gene products and often interfering with normal development.

A major exception to the relation between gene number and protein dosage pertains to genes on the mammalian

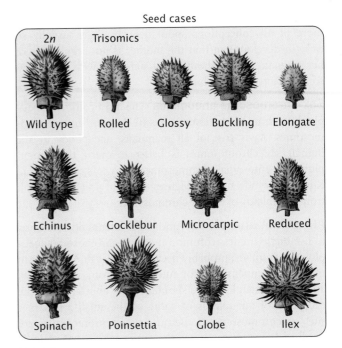

Seed cases

9.21 Mutant capsules in Jimson weed (*Datura stramonium*) result from different trisomies. Each type of capsule is a phenotype that is trisomic for a different chromosome.

X chromosome. In mammals, X-chromosome inactivation ensures that males (who have a single X chromosome) and females (who have two X chromosomes) receive the same functional dosage for X-linked genes (see pp. 90–91 in Chapter 4 for further discussion of X-chromosome inactivation). Extra X chromosomes in mammals are inactivated; so we might expect that aneuploidy of the sex chromosomes would be less detrimental in these animals. Indeed, this is the case for mice and humans, for whom aneuploids of the sex chromosomes are the most common form of aneuploidy seen in living organisms. Y-chromosome aneuploids are probably common because there is so little information on the Y-chromosome.

> **Concepts**
>
> Aneuploidy, the loss or gain of one or more individual chromosomes, may arise from the loss of a chromosome subsequent to translocation or from nondisjunction in meiosis or mitosis. It disrupts gene dosage and often has severe phenotypic effects.

Aneuploidy in Humans

Aneuploidy in humans usually produces serious developmental problems that lead to spontaneous abortion (miscarriage). In fact, as many as 50% of all spontaneously aborted fetuses carry chromosome defects, and a third or more of all conceptions spontaneously abort, usually so early in development that the mother is not even aware of her pregnancy. Only about 2% of all fetuses with a chromosome defect survive to birth.

Sex-chromosome aneuploids The most common aneuploidy seen in living humans has to do with the sex chromosomes. As is true of all mammals, aneuploidy of the human sex chromosomes is better tolerated than aneuploidy of autosomal chromosomes. Turner syndrome and Klinefelter syndrome (see Figures 4.10 and 4.11) both result from aneuploidy of the sex chromosomes.

Autosomal aneuploids Autosomal aneuploids resulting in live births are less common than sex-chromosome aneuploids in humans, probably because there is no mechanism of dosage compensation for autosomal chromosomes. Most autosomal aneuploids are spontaneously aborted, with the exception of aneuploids of some of the small autosomes such as chromosome 21. Because these chromosomes are small and carry fewer genes, the presence of extra copies is less detrimental than for larger chromosomes. For example, the most common autosomal aneuploidy in humans is **trisomy 21,** also called **Down syndrome.** The number of genes on different human chromosomes is not precisely known at the present time, but DNA sequence data indicate that chromosome 21 has fewer genes than any other autosome, with perhaps only 300 genes out of a total of 30,000 to 35,000 for the entire genome.

The incidence of Down syndrome in the United States is about 1 in 700 human births, although the incidence is higher among children born to older mothers. People with Down syndrome (◀ FIGURE 9.22a) show variable degrees of mental retardation, with an average IQ of about 50 (compared with an average IQ of 100 in the general population). Many people with Down syndrome also have characteristic facial features, some retardation of growth and development, and an increased incidence of heart defects, leukemia, and other abnormalities.

Approximately 92% of those who have Down syndrome have three full copies of chromosome 21 (and therefore a total of 47 chromosomes), a condition termed **primary Down syndrome** (◀ FIGURE 9.22b). Primary Down syndrome usually arises from random nondisjunction in egg formation: about 75% of the nondisjunction events that cause Down syndrome are maternal in origin, and most arise in meiosis I. Most children with Down syndrome are born to normal parents, and the failure of the chromosomes to divide has little hereditary tendency. A couple who has conceived one child with primary Down syndrome has only a slightly higher risk of conceiving a second child with Down syndrome (compared with other couples of similar age who have not had any Down-syndrome children). Similarly, the couple's relatives are not more likely to have a child with primary Down syndrome.

Most cases of primary Down syndrome arise from maternal nondisjunction, and the frequency of this occurring correlates with maternal age (◀ FIGURE 9.23). Although the underlying cause of the association between maternal age and nondisjunction remains obscure, recent studies have indicated a strong correlation between nondisjunction and aberrant meiotic recombination. Most chromosomes that failed to separate in meiosis I do not show any evidence of having recombined with one another. Conversely, chromosomes that failed to separate in meiosis II often show evidence of recombination in regions that do not normally

(a)

(b)

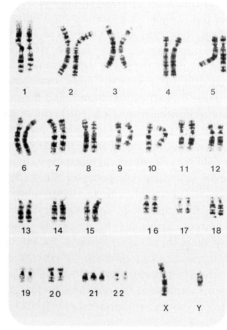

◀ 9.22 **Primary Down syndrome is caused by the presence of three copies of chromosome 21.** (a) A child who has Down syndrome. (b) Karyotype of a person who has primary Down syndrome. (Part a, Hattie Young/Science Photo Library/Photo Researchers; part b, L. Willatt. East Anglian Regional Genetics Service/Science Photo Library/Photo Researchers).

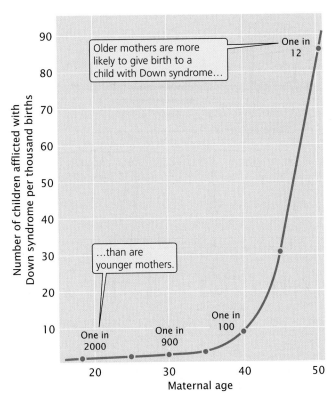

◀9.23 The incidence of primary Down syndrome increases with maternal age.

are produced continually after puberty, with no long suspension of the meiotic divisions. This fundamental difference between the meiotic process in females and males may explain why most chromosome aneuploidy in humans is maternal in origin.

About 4% of people with Down syndrome have 46 chromosomes, but an extra copy of part of chromosome 21 is attached to another chromosome through a translocation (◀FIGURE 9.24). This condition is termed **familial Down syndrome** because it has a tendency to run in families. The phenotypic characteristics of familial Down syndrome are the same as those for primary Down syndrome.

Familial Down syndrome arises in offspring whose parents are carriers of chromosomes that have undergone a Robertsonian translocation, most commonly between chromosome 21 and chromosome 14: the long arm of 21 and the short arm of 14 exchange places. This exchange produces a chromosome that includes the long arms of chromosomes 14 and 21, and a very small chromosome that consists of the short arms of chromosomes 21 and 14. The small chromosome is generally lost after several cell divisions.

Persons with the translocation, called **translocation carriers,** do not have Down syndrome. Although they possess only 45 chromosomes, their phenotypes are normal because they have two copies of the long arms of chromosomes 14 and 21, and apparently the short arms of these chromosomes (which are lost) carry no essential genetic information. Although translocation carriers are completely healthy, they have an increased chance of producing children with Down syndrome.

When a translocation carrier produces gametes, the translocation chromosome may segregate in three different

recombine, most notably near the centromere. Although aberrant recombination appears to play a role in nondisjunction, the maternal age effect is more complex. In female mammals, prophase I begins in all oogonia during fetal development, and recombination is completed prior to birth. Meisosis then arrests in diplotene, and the primary oocytes remain suspended until just before ovulation. As each primary oocyte is ovulated, meiosis resumes and the first division is completed, producing a secondary oocyte. At this point, meiosis is suspended again, and remains so until the secondary oocyte is penetrated by a sperm. The second meiotic division takes place immediately before the nuclei of egg and sperm unite to form a zygote.

An explanation of the maternal age effect must take into account the aberrant recombination that occurs prenatally and the long suspension in prophase I. One theory is that the "best" oocytes are ovulated first, leaving those oocytes that had aberrant recombination to be used later in life. However, evidence indicates that the frequency of aberrant recombination is similar between oocytes that are ovulated in young women and those ovulated in older women. Another possible explanation is that aging of the cellular components needed for meiosis results in nondisjunction of chromosomes that are "at risk," because they have failed to recombine or had some recombination defect. In younger oocytes, these chromosomes can still be segregated from one another, but in older oocytes, they are sensitive to other perturbations in the meiotic machinery. In contrast, sperm

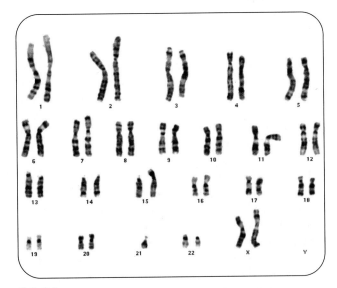

◀9.24 The translocation of chromosome 21 onto another chromosome results in familial Down syndrome. Here, the long arm of chromosome 21 is attached to chromosome 15. (Dr. Dorothy Warburton, HICCC, Columbia University).

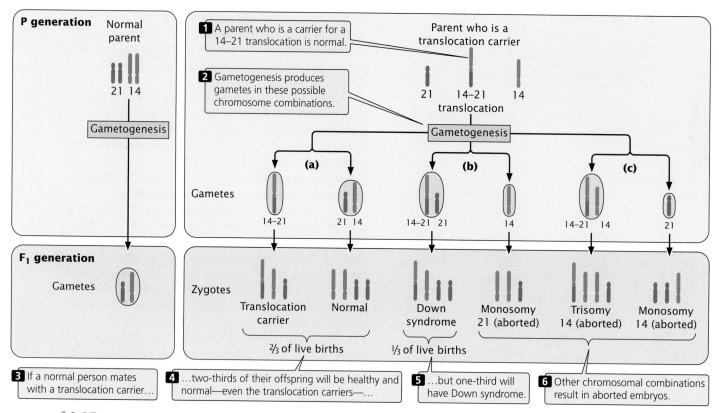

◀9.25 **Translocation carriers are at increased risk for producing children with Down syndrome.**

ways. First, it may separate from the normal chromosomes 14 and 21 in anaphase I of meiosis (◀FIGURE 9.25a). In this type of segregation, half of the gametes will have the translocation chromosome and no other copies of chromosomes 21 and 14; the fusion of such a gamete with a normal gamete will give rise to a translocation carrier. The other half of the gametes produced by this first type of segregation will be normal, each with a single copy of chromosomes 21 and 14, and will result in normal offspring.

Alternatively, the translocation chromosome may separate from chromosome 14 and pass into the same cell with the normal chromosome 21 (◀FIGURE 9.25b). This type of segregation produces all abnormal gametes; half will have two functional copies of chromosome 21 (one normal and one attached to chromosome 14) and the other half will lack chromosome 21. The gametes with the two functional copies of chromosome 21 will produce children with familial Down syndrome; the gametes lacking chromosome 21 will result in zygotes with monosomy 21 and will be spontaneously aborted.

In the third type of segregation, the translocation chromosome and the normal copy of chromosome 14 segregate together, and the normal chromosome 21 segregates by itself (◀FIGURE 9.25c). This pattern is presumably rare, because the two centromeres are both derived from chromosome 14 and usually separate from each other. In any case, all the gametes produced by this process are abnormal:

half result in monosomy 14 and the other half result in trisomy 14—all are spontaneously aborted. Thus, only three of the six types of gametes that can be produced by a translocation carrier will result in the birth of a baby and, theoretically, these gametes should arise with equal frequency. One-third of the offspring of a translocation carrier should be translocation carriers like their parent, one-third should have familial Down syndrome, and one-third should be normal. In reality, however, fewer than one-third of the children born to translocation carriers have Down syndrome, which suggests that some of the embryos with Down syndrome are spontaneously aborted.

www.whfreeman.com/pierce   Additional information on Down syndrome

Few autosomal aneuploids besides trisomy 21 result in human live births. **Trisomy 18,** also known as **Edward syndrome,** arises with a frequency of approximately 1 in 8000 live births. Babies with Edward syndrome are severely retarded and have low-set ears, a short neck, deformed feet, clenched fingers, heart problems, and other disabilities. Few live for more than a year after birth. **Trisomy 13** has a frequency of about 1 in 15,000 live births and produces features that are collectively known as **Patau syndrome.** Characteristics of this condition include severe mental retardation, a small head, sloping forehead, small eyes, cleft lip and palate, extra fingers and toes, and numerous other problems. About half of children with trisomy 13 die within the first month of life, and 95% die by the age of 3. Rarer still is **trisomy 8,** which arises with a frequency of about 1 in 25,000 to 50,000 live births. This aneuploid is characterized by mental retardation, contracted fingers and toes, low-set malformed ears, and a prominent forehead. Many who have this condition have normal life expectancy.

> **Concepts**
>
> In humans, sex-chromosome aneuploids are more common than are autosomal aneuploids. X-chromosome inactivation prevents problems of gene dosage for X-linked genes. Down syndrome results from three functional copies of chromosome 21, either through trisomy (primary Down syndrome) or a Robertsonian translocation (familial Down syndrome).

www.whfreeman.com/pierce   Additional information on trisomy 13 and trisomy 18

Uniparental Disomy

Normally, the two chromosomes of a homologous pair are inherited from different parents—one from the father and one from the mother. The development of molecular techniques that facilitate the identification of specific DNA sequences (see Chapter 18), has made it possible to determine the parental origins of chromosomes. Surprisingly, sometimes both chromosomes are inherited from the same parent, a condition termed **uniparental disomy.**

Uniparental disomy violates the rule that children affected with a recessive disorder appear only in families where both parents are carriers. For example, cystic fibrosis is an autosomal recessive disease; typically, both parents of an affected child are heterozygous for the cystic fibrosis mutation on chromosome 7. However, a small proportion of people with cystic fibrosis have only a single parent who is heterozygous for the cystic fibrosis gene. How can this be? These people must have inherited from the heterozygous parent two copies of the chromosome 7 that carries the defective cystic fibrosis allele and no copy of the normal allele from the other parent. Uniparental disomy has also been observed in Prader-Willi syndrome, a rare condition that arises when a paternal copy of a gene on chromosome 15 is missing. Although most cases of Prader-Willi syndrome result from a chromosome deletion that removes the paternal copy of the gene (see p. 120 in Chapter 5), from 20% to 30% arise when both copies of chromosome 15 are inherited from the mother and no copy is inherited from the father.

Many cases of uniparental disomy probably originate as a trisomy. Although most autosomal trisomies are lethal, a trisomic embryo can survive if one of the three chromosomes is lost early in development. If, just by chance, the two remaining chromosomes are both from the same parent, uniparental disomy results.

www.whfreeman.com/pierce   More on uniparental disomy and links to information on Prader-Willi syndrome

Mosaicism

Nondisjunction in a mitotic division may generate patches of cells in which every cell has a chromosome abnormality and other patches in which every cell has a normal karyotype. This type of nondisjunction leads to regions of tissue with different chromosome constitutions, a condition known as **mosaicism.** Growing evidence suggests that mosaicism is relatively common.

Only about 50% of those diagnosed with Turner syndrome have the 45,X karyotype (presence of a single X chromosome) in all their cells; most others are mosaics, possessing some 45,X cells and some normal 46,XX cells. A few may even be mosaics for two or more types of abnormal karyotypes. The 45,X/46,XX mosaic usually arises when an X chromosome is lost soon after fertilization in an XX embryo.

Fruit flies that are XX/XO mosaics (O designates the absence of a homologous chromosome; XO means the cell has a single X chromosome and no Y chromosome) develop a mixture of male and female traits, because the presence of two X chromosomes in fruit flies produces female traits and the presence of a single X chromosome produces male traits (◀ FIGURE 9.26). Sex determination in fruit flies occurs

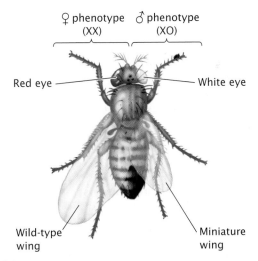

◀ **9.26 Mosaicism for the sex chromosomes produces a gynandromorph.** This XX/XO gynandromorph fruit fly carries one wild-type X chromosome and one X chromosome with recessive alleles for white eyes and miniature wings. The left side of the fly has a normal female phenotype, because the cells are XX and the recessive alleles on one X chromosome are masked by the presence of wild-type alleles on the other. The right side of the fly has a male phenotype with white eyes and miniature wing, because the cells are missing the wild-type X chromosome (are XO), allowing the white and miniature alleles to be expressed.

independently in each cell during development. Those cells that are XX express female traits; those that are XO express male traits. Such sexual mosaics are called **gynandromorphs.** Normally, X-linked recessive genes are masked in heterozygous females but, in XX/XO mosaics, any X-linked recessive genes present in the cells with a single X chromosome will be expressed.

> ### Concepts
>
> In uniparental disomy, an individual has two copies of a chromosome from one parent and no copy from the other. It may arise when a trisomic embryo loses one of the triplicate chromosomes early in development. In mosaicism, different cells within the same individual have different chromosome constitutions.

Polyploidy

Most eukaryotic organisms are diploid ($2n$) for most of their life cycles, possessing two sets of chromosomes. Occasionally, whole sets of chromosomes fail to separate in meiosis or mitosis, leading to polyploidy, the presence of more than two genomic sets of chromosomes. Polyploids include *triploids* ($3n$) *tetraploids* ($4n$), *pentaploids* ($5n$), and even higher numbers of chromosome sets.

Polyploidy is common in plants and is a major mechanism by which new plant species have evolved. Approximately 40% of all flowering-plant species and from 70% to 80% of grasses are polyploids. They include a number of agriculturally important plants, such as wheat, oats, cotton, potatoes, and sugar cane. Polyploidy is less common in animals, but is found in some invertebrates, fishes, salamanders, frogs, and lizards. No naturally occurring, viable polyploids are known in birds, but at least one polyploid mammal—a rat from Argentina—has been reported.

We will consider two major types of polyploidy: **autopolyploidy,** in which all chromosome sets are from a single species; and **allopolyploidy,** in which chromosome sets are from two or more species.

Autopolyploidy

Autopolyploidy results when accidents of meiosis or mitosis produce extra sets of chromosomes, all derived from a single species. Nondisjunction of all chromosomes in mitosis in an early $2n$ embryo, for example, doubles the chromosome number and produces an autotetraploid ($4n$) (◀FIGURE 9.27a). An autotriploid may arise when nondisjunction in meiosis produces a diploid gamete that then fuses with a normal haploid gamete to produce a triploid zygote (◀FIGURE 9.27b). Alternatively, triploids may arise from a cross between an autotetraploid that produces $2n$ gametes and a diploid that produces $1n$ gametes.

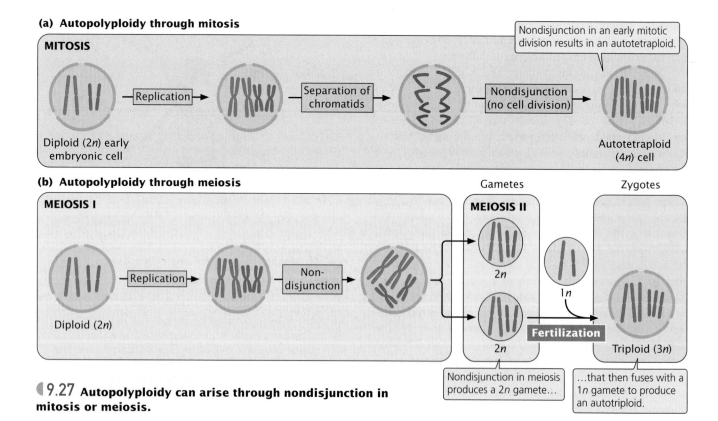

(a) Autopolyploidy through mitosis

MITOSIS

Diploid ($2n$) early embryonic cell → Replication → Separation of chromatids → Nondisjunction (no cell division) → Autotetraploid ($4n$) cell

Nondisjunction in an early mitotic division results in an autotetraploid.

(b) Autopolyploidy through meiosis

MEIOSIS I

Diploid ($2n$) → Replication → Non-disjunction →

MEIOSIS II

Gametes   Zygotes

$2n$

$2n$ → Fertilization → $1n$ → Triploid ($3n$)

Nondisjunction in meiosis produces a $2n$ gamete…

…that then fuses with a $1n$ gamete to produce an autotriploid.

◀9.27 **Autopolyploidy can arise through nondisjunction in mitosis or meiosis.**

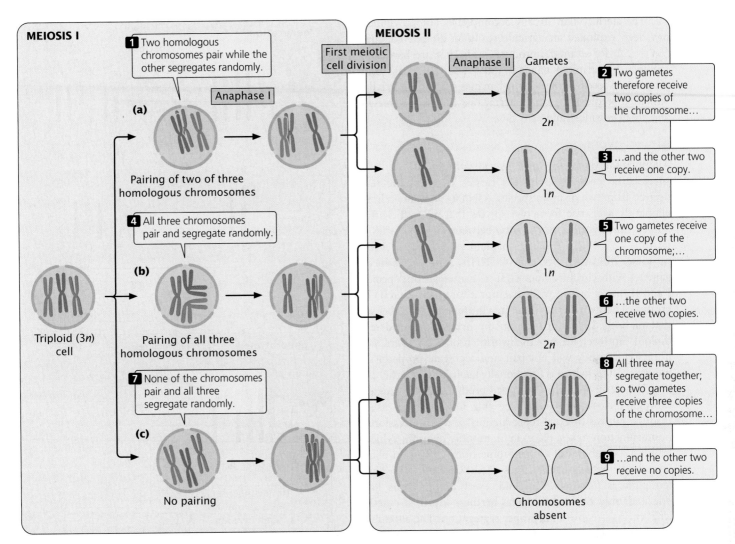

◀9.28 In meiosis of an autotriploid, homologous chromosomes can pair or not pair in three ways. In this example, pairing and segregation of a single homologous set of chromosomes is illustrated.

Because all the chromosome sets in autopolyploids are from the same species, they are homologous and attempt to align in prophase I of meiosis, which usually results in sterility. Consider meiosis in an autotriploid (◀FIGURE 9.28). In meiosis in a diploid cell, two chromosome homologs pair and align, but, in autotriploids, three homologs are present. One of the three homologs may fail to align with the other two, and this unaligned chromosome will segregate randomly (see Figure 9.28a). Which gamete gets the extra chromosome will be determined by chance and will differ for each homologous group of chromosomes. The resulting gametes will have two copies of some chromosomes and one copy of others. Even if all three chromosomes do align, two chromosomes must segregate to one gamete and one chromosome to the other (see Figure 9.28b). Occasionally, the presence of a third chromosome interferes with normal alignment, and all three chromosomes segregate to the same gamete (see Figure 9.28c).

No matter how the three homologous chromosomes align, their random segregation will create **unbalanced gametes,** with various numbers of chromosomes. A gamete produced by meiosis in such an autotriploid might receive, say, two copies of chromosome 1, one copy of chromosome 2, three copies of chromosome 3, and no copies of chromosome 4. When the unbalanced gamete fuses with a normal gamete (or with another unbalanced gamete), the resulting zygote has different numbers of the four types of chromosomes. This difference in number creates unbalanced gene dosage in the zygote, which is often lethal. For this reason, triploids do not usually produce viable offspring.

In even-numbered autopolyploids, such as autotetraploids, it is theoretically possible for the homologous chromosomes to form pairs and divide equally. However, this event rarely happens; so these types of autotetraploids also produce unbalanced gametes.

The sterility that usually accompanies autopolyploidy has been exploited in agriculture. Wild diploid bananas ($2n = 22$), for example, produce seeds that are hard and inedible, but triploid bananas ($3n = 33$) are sterile, and produce no seeds—they are the bananas sold commercially. Similarly, seedless triploid watermelons have been created and are now widely sold.

Allopolyploidy

Allopolyploidy arises from hybridization between two species; the resulting polyploid carries chromosome sets derived from two or more species. ◀ FIGURE 9.29 shows how alloploidy can arise from two species that are sufficiently related that hybridization occurs between them. Species I (AABBCC, $2n = 6$) produces haploid gametes with chromosomes ABC, and species II (GGHHII, $2n = 6$) produces gametes with chromosomes GHI. If gametes from species I and II fuse, a hybrid with six chromosomes (ABCGHI) is created. The hybrid has the same chromosome number as that of both diploid species; so the hybrid is considered diploid. However, because the hybrid chromosomes are not homologous, they will not pair and segregate properly in meiosis; so this hybrid is functionally haploid and sterile.

The sterile hybrid is unable to produce viable gametes through meiosis, but it may be able to perpetuate itself through mitosis (asexual reproduction). On rare occasions, nondisjunction takes place in a mitotic division, which leads to a doubling of chromosome number and an allo-tetraploid, with chromosomes AABBCCGGHHII. This tetraploid is *functionally* diploid: every chromosome has one and only one homologous partner, which is exactly what meiosis requires for proper segregation. The allopolyploid can now undergo normal meiosis to produce balanced gametes having six chromosomes.

George Karpechenko created polyploids experimentally in the 1920s. Today, as well as in the early twentieth century, cabbage (*Brassica oleracea*, $2n = 18$) and radishes (*Raphanus sativa*, $2n = 18$) are agriculturally important plants, but only the leaves of the cabbage and the roots of the radish are normally consumed. Karpechenko wanted to produce a plant that had cabbage leaves and radish roots so that no part of the plant would go to waste. Because both cabbage and radish possess 18 chromosomes, Karpechenko was able to successfully cross them, producing a hybrid with $2n = 18$, but, unfortunately, the hybrid was sterile. After several crosses, Karpechenko noticed that one of his hybrid plants produced a few seeds. When planted, these seeds grew into plants that were viable and fertile. Analysis of their chromosomes revealed that the plants were allotetraploids, with $2n = 36$ chromo-

◀ 9.29 **Allopolyploids usually arise from hybridization between two species followed by chromosome doubling.**

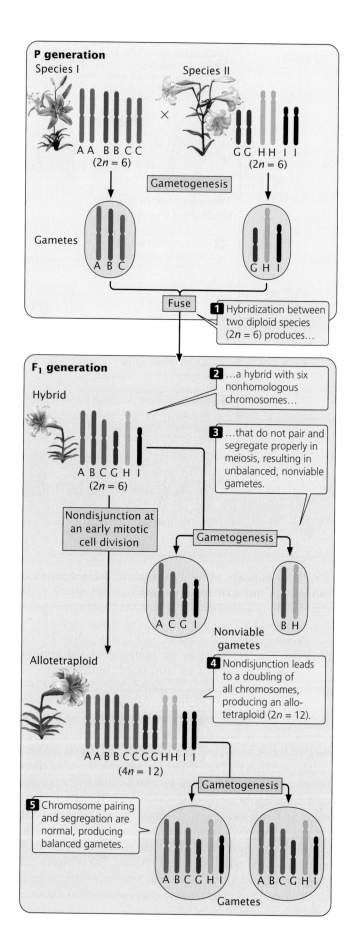

somes. To Karpechencko's great disappointment, however, the new plants possessed the roots of a cabbage and the leaves of a radish.

The Significance of Polyploidy

In many organisms, cell volume is correlated with nuclear volume, which, in turn, is determined by genome size. Thus, the increase in chromosome number in polyploidy is often associated with an increase in cell size, and many polyploids are physically larger than diploids. Breeders have used this effect to produce plants with larger leaves, flowers, fruits, and seeds. The hexaploid ($6n = 42$) genome of wheat probably contains chromosomes derived from three different wild species (◀ FIGURE 9.30). Many other cultivated plants also are polyploid (Table 9.2).

Polyploidy is less common in animals than in plants for several reasons. As discussed, allopolyploids require hybridization between different species, which occurs less frequently in animals than in plants. Animal behavior often prevents interbreeding, and the complexity of animal development causes most interspecific hybrids to be nonviable. Many of the polyploid animals that do arise are in groups that reproduce through parthenogenesis (a type of reproduction in which individuals develop from unfertilized eggs). Thus asexual reproduction may facilitate the development of polyploids, perhaps because the perpetuation of hybrids through asexual reproduction provides greater opportunities for nondisjunction. Only a few human polyploid babies have been reported, and most died within a few days of birth. Polyploidy—usually triploidy—is seen in about 10% of all spontaneously aborted human fetuses. Different types of chromosome mutations are summarized in Table 9.3.

Concepts

Polyploidy is the presence of extra chromosome sets: autopolyploids possess extra chromosome sets from the same species; allopolyploids possess extra chromosome sets from two or more species. Problems in chromosome pairing and segregation often lead to sterility in autopolyploids, but many allopolyploids are fertile.

◀ 9.30 **Modern bread wheat, *Triticum aestivum*, is a hexaploid with genes derived from three different species.** Two diploids species *T. monococcum* ($n = 14$) and probably *T. searsii* ($n = 14$) originally crossed to produce a diploid hybrid ($2n = 14$) that underwent chromosome doubling to create *T. turgidum* ($4n = 28$). A cross between *T. turgidum* and *T. tauschi* ($2n = 14$) produced a triploid hybrid ($3n = 21$) that then underwent chromosome doubling to produce *T. aestivum*, which is a hexaploid ($6n = 42$).

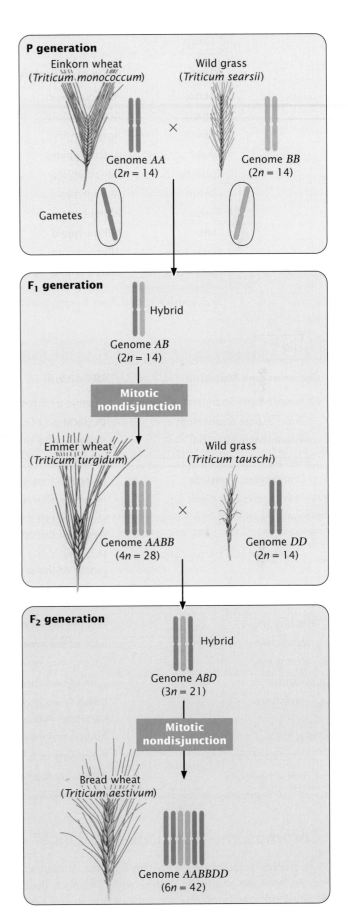

Table 9.2	Examples of polyploid crop plants		
Plant	**Type of Polyploidy**	**Ploidy**	**Chromosome Number**
Potato	Autopolyploid	$4n$	48
Banana	Autopolyploid	$3n$	33
Peanut	Autopolyploid	$4n$	40
Sweet potato	Autopolyploid	$6n$	90
Tobacco	Allopolyploid	$4n$	48
Cotton	Allopolyploid	$4n$	52
Wheat	Allopolyploid	$6n$	42
Oats	Allopolyploid	$6n$	42
Sugar cane	Allopolyploid	$8n$	80
Strawberry	Allopolyploid	$8n$	56

Source: After F. C. Elliot, *Plant Breeding and Cytogenetics* (New York: McGraw-Hill, 1958).

Table 9.3	Different types of chromosome mutations
Chromosome Mutation	**Definition**
Chromosome rearrangement	Change in chromosome structure
Chromosome duplication	Duplication of a chromosome segment
Chromosome deletion	Deletion of a chromosome segment
Inversion	Chromosome segment inverted 180 degrees
Paracentric inversion	Inversion that does not include the centromere in the inverted region
Pericentric inversion	Inversion that includes the centromere in the inverted region
Translocation	Movement of a chromosome segment to a nonhomologous chromosome or region of the same chromosome
Nonreciprocal translocation	Movement of a chromosome segment to a nonhomologous chromosome or region of the same chromosome without reciprocal exchange
Reciprocal translocation	Exchange between segments of nonhomologous chromosomes or regions of the same chromosome
Aneuploidy	Change in number of individual chromosomes
Nullisomy	Loss of both members of a homologous pair
Monosomy	Loss of one member of a homologous pair
Trisomy	Gain of one chromosome, resulting in three homologous chromosomes
Tetrasomy	Gain of two homologous chromosomes, resulting in four homologous chromosomes
Polyploidy	Addition of entire chromosome sets
Autopolyploidy	Polyploidy in which extra chromosome sets are derived from the same species
Allopolyploidy	Polyploidy in which extra chromosome sets are derived from two or more species

Chromosome Mutations and Cancer

Most tumors contain cells with chromosome mutations. For many years, geneticists argued about whether these chromosome mutations were the cause or the result of cancer. Some types of tumors are consistently associated with *specific* chromosome mutations, suggesting that in these cases the specific chromosome mutation played a pivotal role in the development of the cancer. However, many cancers are not associated with specific types of chromosome abnormalities, and individual *gene* mutations are now known to contribute to many types of cancer. Nevertheless, chromosome instability is a general feature of cancer cells, causing them to accumulate chromosome mutations, which then affect individual genes that contribute to the cancer process. Thus, chromosome mutations appear to both *cause* and *be a result* of cancer.

At least three types of chromosome rearrangements—deletions, inversions, and translocations—are associated with certain types of cancer. Deletions may result in the loss of one or more genes that normally hold cell division in check. When these so-called tumor-suppressor genes are lost, cell division is not regulated and cancer may result.

Inversions and translocations contribute to cancer in several ways. First, the chromosomal breakpoints that accompany these mutations may lie within tumor-suppressor genes, disrupting their function and leading to cell proliferation. Second, translocations and inversions may bring together sequences from two different genes, generating a fused protein that stimulates some aspect of the cancer process. Such fusions are seen in most cases of chronic myeloid leukemia, a form of leukemia affecting bone-marrow cells. About 90% of patients with chronic myeloid leukemia have a reciprocal translocation between the long arm of chromosome 22 and the tip of the long arm of chromosome 9 (◀ FIGURE 9.31). This translocation produces a shortened

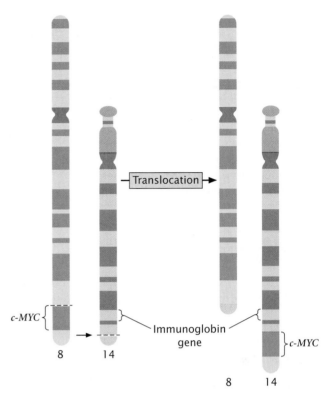

◀ 9.32 **A reciprocal translocation between chromosomes 8 and 14 causes Burkitt lymphoma.**

chromosome 22, called the Philadelphia chromosome because it was first discovered in Philadelphia. At the end of a normal chromosome 9 is a potential cancer-causing gene called *c-ABL*. As a result of the translocation, part of the *c-ABL* gene is fused with the *BCR* gene from chromosome 22. The protein produced by this *BCR–c-ABL* fusion gene is much more active than the protein produced by the normal *c-ABL* gene; the fusion protein stimulates increased, unregulated cell division and eventually leads to leukemia.

A third mechanism by which chromosome rearrangements may produce cancer is by the transfer of a potential cancer-causing gene to a new location, where it is activated by different regulatory sequences. Burkitt lymphoma is a cancer of the B cells, the lymphocytes that produce antibodies. Many people having Burkitt lymphoma possess a reciprocal translocation between chromosome 8 and chromosome 2, 14, or 22, each of which carries genes for immunological proteins (◀ FIGURE 9.32). This translocation relocates a gene called *c-MYC* from the tip of chromosome 8 to a position in one of the aforementioned chromosomes that is next to a gene for one of the immunoglobulin proteins. At this new location, *c-MYC* comes under the control of regulatory sequences that normally activate the production of immunoglobulins, and *c-MYC* is expressed in B cells. The c-MYC protein stimulates the division of the B cells and leads to Burkitt lymphoma.

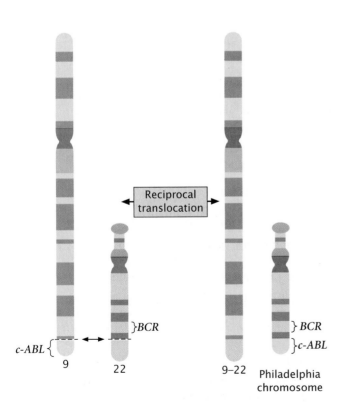

◀ 9.31 **A reciprocal translocation between chromosomes 9 and 22 causes chronic myeloid leukemia.**

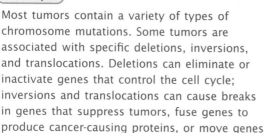

Concepts

Most tumors contain a variety of types of chromosome mutations. Some tumors are associated with specific deletions, inversions, and translocations. Deletions can eliminate or inactivate genes that control the cell cycle; inversions and translocations can cause breaks in genes that suppress tumors, fuse genes to produce cancer-causing proteins, or move genes to new locations, where they are under the influence of different regulatory sequences.

www.whfreeman.com/pierce   More information on chronic myeloid leukemia

Connecting Concepts Across Chapters

This chapter has focused on variations in the number and structure of chromosomes. Because these chromosome mutations affect many genes simultaneously, they have major effects on the phenotypes and often are not compatible with development. A major theme of this chapter has been that, even when the structure of a gene is not disrupted, changes in gene number and position produced by chromosome mutations can have severe effects on gene expression.

Chromosome mutations most frequently arise through errors in mitosis and meiosis, and so a thorough understanding of these processes and chromosome structure (covered in Chapter 2) is essential for grasping the material in this chapter. The process of crossing over, discussed in Chapters 2 and 7, also is helpful for understanding the consequences of recombination in individuals heterozygous for chromosome rearrangements.

This chapter has provided a foundation for understanding the molecular nature of chromosome structure (discussed in Chapter 11). The movement of genes through a process called transposition often produces chromosome mutations, and so the current chapter is also relevant to the discussion of transposition in Chapter 11. The discussion in this chapter of chromosomes and cancer is closely linked to the more extended discussion of cancer genetics found in Chapter 21. Variation produced by chromosome mutations, along with gene mutations and recombination, provides the raw material for evolutionary change, which is covered in Chapters 22 and 23.

CONCEPTS SUMMARY

- Three basic types of chromosome mutations are: (1) chromosome rearrangements, which are changes in the structure of chromosomes; (2) aneuploidy, which is an increase or decrease in chromosome number; and (3) polyploidy, which is the presence of extra chromosome sets.

- Chromosome rearrangements include duplications, deletions, inversions, and translocations.

- Chromosome duplications arise when a chromosome segment is doubled. The segment may be adjacent to the original segment (a tandem duplication) or distant from the original segment (a displaced duplication). Reverse duplications have the duplicated sequence in the reverse order. In individuals heterozygous for a duplication, the duplicated region will form a loop when homologous chromosomes pair in meiosis. Duplications often have pronounced effects on the phenotype owing to unbalanced gene dosage.

- Chromosome deletion is the loss of part of a chromosome. In individuals heterozygous for a deletion, one of the chromosomes will loop out during pairing in meiosis. Many chromosome deletions are lethal in the homozygous state and cause deleterious effects in the heterozygous state, because of unbalanced gene dosage. Deletions may cause recessive alleles to be expressed.

- A chromosome inversion is the inversion of a chromosome segment. Pericentric inversions include the centromere; paracentric inversions do not. The phenotypic effects caused by inversions are due to the breaking of genes and their movement to new locations, where they may be influenced by different regulatory sequences. In individuals heterozygous for an inversion, the chromosomes form inversion loops in meiosis, with reduced recombination taking place within the inverted region.

- A translocation is the attachment of part of one chromosome to a nonhomologous chromosome. In translocation heterozygotes, the chromosomes form crosslike structures in meiosis, and the segregation of chromosomes produces unbalanced gametes.

- Fragile sites are constrictions or gaps that appear at particular regions on the chromosomes of cells grown in culture and are prone to breakage under certain conditions.

- Aneuploidy is the addition or loss of individual chromosomes. Nullisomy refers to the loss of two homologous chromosomes; monosomy is the loss of one homologous chromosome; trisomy is the addition of one homologous chromosome; tetrasomy is the addition of two homologous chromosomes.

- Aneuploidy usually causes drastic phenotypic effects because it leads to unbalanced gene dosage. In humans, sex-chromosome aneuploids are less detrimental than autosomal aneuploids because X-chromosome inactivation reduces the problems of unbalanced gene dosage.

- The most common autosomal aneuploid in living humans is trisomy 21, which results in Down syndrome. Primary Down syndrome is caused by the presence of three full copies of chromosome 21, whereas familial Down syndrome is caused by the presence of two normal copies of chromosome 21 and a third copy that is attached to another chromosome through a translocation.

- Uniparental disomy is the presence of two copies of a chromosome from one parent and no copy from the other.

- Mosaicism is caused by nondisjunction in an early mitotic division that leads to different chromosome constitutions in different cells of a single individual.

- Polyploidy is the presence of more than two full chromosome sets. In autopolyploidy, all the chromosomes derive from one species; in allopolyploidy, they come from two or more species.

- Autopolyploidy arises from nondisjunction in meiosis or mitosis. Here, problems with chromosome alignment and segregation frequently lead to the production of nonviable gametes.

- Allopolyploidy arises from nondisjunction that follows hybridization between two species. Allopolyploids are frequently fertile.

- Some types of cancer are associated with specific chromosome deletions, inversions, and translocations. Deletions may cause cancer by removing or disrupting genes that suppress tumors; inversions and translocations may break tumor-suppressing genes or they may move genes to positions next to different regulatory sequences, which alters their expression.

IMPORTANT TERMS

chromosome mutation (p. 235)
metacentric chromosome (p. 235)
submetacentric chromosome (p. 235)
acrocentric chromosome (p. 235)
telocentric chromosome (p. 235)
chromosome rearrangement (p. 236)
aneuploidy (p. 236)
polyploidy (p. 236)
chromosome duplication (p. 237)
tandem duplication (p. 237)

displaced duplication (p. 237)
reverse duplication (p. 237)
chromosome deletion (p. 239)
pseudodominance (p. 239)
haploinsufficient gene (p. 239)
chromosome inversion (p. 240)
paracentric inversion (p. 240)
pericentric inversion (p. 240)
position effect (p. 240)
dicentric chromatid (p. 242)
acentric chromatid (p. 242)
dicentric bridge (p. 242)
translocation (p. 243)
nonreciprocal translocation (p. 244)

reciprocal translocation (p. 244)
Robertsonian translocation (p. 244)
alternate segregation (p. 244)
adjacent-1 segregation (p. 244)
adjacent-2 segregation (p. 244)
fragile site (p. 246)
nullisomy (p. 248)
monosomy (p. 248)
trisomy (p. 249)
tetrasomy (p. 249)
Down syndrome (trisomy 21) (p. 250)
primary Down syndrome (p. 250)

familial Down syndrome (p. 251)
translocation carrier (p. 251)
Edward syndrome (trisomy 18) (p. 252)
Patau syndrome (trisomy 13) (p. 252)
trisomy 8 (p. 253)
uniparental disomy (p. 253)
mosaicism (p. 253)
gynandromorph (p. 254)
autopolyploidy (p. 254)
allopolyploidy (p. 254)
unbalanced gametes (p. 255)

Worked Problems

1. A chromosome has the following segments, where • represents the centromere.

$$\underline{A \quad B \quad C \quad D \quad E \quad • \quad F \quad G}$$

What types of chromosome mutations are required to change this chromosome into each of the following chromosomes? (In some cases, more than one chromosome mutation may be required.)

(a) $\underline{A \quad B \quad E \quad • \quad F \quad G}$

(b) $\underline{A \quad E \quad D \quad C \quad B \quad • \quad F \quad G}$

(c) $\underline{A \quad B \quad A \quad B \quad C \quad D \quad E \quad • \quad F \quad G}$

(d) $\underline{A \quad F \quad • \quad E \quad D \quad C \quad B \quad G}$

(e) $\underline{A \quad B \quad C \quad D \quad E \quad E \quad D \quad C \quad • \quad F \quad G}$

• Solution

The types of chromosome mutations are identified by comparing the mutated chromosome with the original, wild-type chromosome.

(a) The mutated chromosome ($\underline{A \quad B \quad E \quad • \quad F \quad G}$) is missing segment $\underline{C \quad D}$; so this mutation is a deletion.

(b) The mutated chromosome ($\underline{A \quad E \quad D \quad C \quad B \quad • \quad F \quad G}$) has one and only one copy of all the gene segments, but segment

B C D E has been inverted 180 degrees. Because the centromere has not changed location and is not in the inverted region, this chromosome mutation is a paracentric inversion.

(c) The mutated chromosome (A B A B C D E • F G) is longer than normal, and we see that segment A B has been duplicated. This mutation is a tandem duplication.

(d) The mutated chromosome (A F • E D C B G) is normal length, but the gene order and the location of the centromere have changed; this mutation is therefore a pericentric inversion of region (B C D E • F).

(e) The mutated chromosome (A B C D E E D C • F G) contains a duplication (C D E) that is also inverted; so this chromosome has undergone a duplication and a paracentric inversion.

2. Species I is diploid ($2n = 4$) with chromosomes AABB; related species II is diploid ($2n = 6$) with chromosomes MMNNOO. Give the chromosomes that would be found in individuals with the following chromosome mutations.

(a) Autotriploid for species I

(b) Allotetraploid including species I and II

(c) Monosomic in species I

(d) Trisomic in species II for chromosome M

(e) Tetrasomic in species I for chromosome A

(f) Allotriploid including species I and II

(g) Nullisomic in species II for chromosome N

• Solution

To work this problem, we should first determine the haploid genome complement for each species. For species I, $n = 2$ with chromosomes AB and, for species II, $n = 3$ with chromosomes MNO.

(a) An autotriploid is $3n$, with all the chromosomes coming from a single species; so an autotriploid of species I will have chromosomes AAABBB ($3n = 6$).

(b) An allotetraploid is $4n$, with the chromosomes coming from more than one species. An allotetraploid could consist of $2n$ from species I and $2n$ from species II, giving the allotetraploid ($4n = 2 + 2 + 3 + 3 = 10$) chromosomes AABBMMNNOO. An allotetraploid could also possess $3n$ from species I and $1n$ from species II ($4n = 2 + 2 + 2 + 3 = 9$; AAABBBMNO) or $1n$ from species I and $3n$ from species II ($4n = 2 + 3 + 3 + 3$; ABMMMNNNOOO).

(c) A monosomic is missing a single chromosome; so a monosomic for species 1 would be $2n - 1 = 4 - 1 = 3$. The monosomy might include either of the two chromosome pairs, with chromosomes ABB or AAB.

(d) Trisomy requires an extra chromosome; so a trisomic of species II for chromosome M would be $2n + 1 = 6 + 1 = 7$ (MMMNNOO).

(e) A tetrasomic has two extra homologous chromosomes; so a tetrasomic of species I for chromosome A would be $2n + 2 = 4 + 2 = 6$ (AAAABB).

(f) An allotriploid is $3n$ with the chromosomes coming from two different species; so an allotriploid could be $3n = 2 + 2 + 3 = 7$ (AABBMNO) or $3n = 2 + 3 + 3 = 8$ (ABMMNNOO).

(g) A nullisomic is missing both chromosomes of a homologous pair; so a nullisomic of species II for chromosome N would be $2n - 2 = 6 - 2 = 4$ (MMOO).

The New Genetics
MINING GENOMES

THE HUMAN GENOME PROJECT

The first successful efforts to clone single genes and to determine the sequence of DNA molecules began in the 1970s. Less than two decades later, techniques and strategies were being developed to organize and sequence clones that covered the entire human genome. Today, the sequence of our genome is freshly available. This exercise will introduce you to some of the tools that are used to organize, retrieve, and understand information derived from the Human Genome Project.

COMPREHENSION QUESTIONS

* 1. List the different types of chromosome mutations and define each one.

* 2. Why do extra copies of genes sometimes cause drastic phenotypic effects?

  3. Draw a pair of chromosomes as they would appear during synapsis in prophase I of meiosis in an individual heterozygous for a chromosome duplication.

  4. How does a deletion cause pseudodominance?

* 5. What is the difference between a paracentric and a pericentric inversion?

  6. How do inversions cause phenotypic effects?

* 7. Draw a pair of chromosomes as they would appear during synapsis in prophase I of meiosis in an individual heterozygous for a paracentric inversion.

8. Explain why recombination is suppressed in individuals heterozygous for paracentric and pericentric inversions.

* 9. How do translocations produce phenotypic effects?

10. Sketch the chromosome pairing and the different segregation patterns that can arise in an individual heterozygous for a reciprocal translocation.

11. What is a Robertsonian translocation?

12. List four major types of aneuploidy.

*13. Why are sex-chromosome aneuploids more common in humans than autosomal aneuploids?

*14. What is the difference between primary Down syndrome and familial Down syndrome? How does each arise?

*15. What is uniparental disomy and how does it arise?

16. What is mosaicism and how does it arise?

*17. What is the difference between autopolyploidy and allopolyploidy? How does each arise?

APPLICATION QUESTIONS AND PROBLEMS

*18. Which types of chromosome mutations

(a) increase the amount of genetic material on a particular chromosome?

(b) increase the amount of genetic material for all chromosomes?

(c) decrease the amount of genetic material on a particular chromosome?

(d) change the position of DNA sequences on a single chromosome without changing the amount of genetic material?

(e) move DNA from one chromosome to a nonhomologous chromosome?

*19. A chromosome has the following segments, where • represents the centromere:

$$\underline{A \quad B \quad \bullet \quad C \quad D \quad E \quad F \quad G}$$

What types of chromosome mutations are required to change this chromosome into each of the following chromosomes? (In some cases, more than one chromosome mutation may be required.)

(a) $\underline{A \quad B \quad A \quad B \quad \bullet \quad C \quad D \quad E \quad F \quad G}$
(b) $\underline{A \quad B \quad \bullet \quad C \quad D \quad E \quad A \quad B \quad F \quad G}$
(c) $\underline{A \quad B \quad \bullet \quad C \quad F \quad E \quad D \quad G}$
(d) $\underline{A \quad \bullet \quad C \quad D \quad E \quad F \quad G}$
(e) $\underline{A \quad B \quad \bullet \quad C \quad D \quad E}$
(f) $\underline{A \quad B \quad \bullet \quad E \quad D \quad C \quad F \quad G}$
(g) $\underline{C \quad \bullet \quad B \quad A \quad D \quad E \quad F \quad G}$
(h) $\underline{A \quad B \quad \bullet \quad C \quad F \quad E \quad D \quad F \quad E \quad D \quad G}$
(i) $\underline{A \quad B \quad \bullet \quad C \quad D \quad E \quad F \quad C \quad D \quad F \quad E \quad G}$

20. A chromosome initially has the following segments:

$$\underline{A \quad B \quad \bullet \quad C \quad D \quad E \quad F \quad G}$$

Draw and label the chromosome that would result from each of the following mutations.

(a) Tandem duplication of DEF

(b) Displaced duplication of DEF

(c) Deletion of FG

(e) Paracentric inversion that includes DEFG

(f) Pericentric inversion of BCDE

21. The following diagram represents two nonhomologous chromosomes:

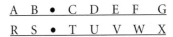

$$\underline{A \quad B \quad \bullet \quad C \quad D \quad E \quad F \quad G}$$
$$\underline{R \quad S \quad \bullet \quad T \quad U \quad V \quad W \quad X}$$

What type of chromosome mutation would produce the following chromosomes?

(a) $\underline{A \quad B \quad \bullet \quad C \quad D}$
$\underline{R \quad S \quad \bullet \quad T \quad U \quad V \quad W \quad X \quad E \quad F \quad G}$

(b) $\underline{A \quad U \quad V \quad B \quad \bullet \quad C \quad D \quad E \quad F \quad G}$
$\underline{R \quad S \quad \bullet \quad T \quad W \quad X}$

(c) $\underline{A \quad B \quad \bullet \quad T \quad U \quad V \quad F \quad G}$
$\underline{R \quad S \quad \bullet \quad C \quad D \quad E \quad W \quad X}$

(d) $\underline{A \quad B \quad \bullet \quad C \quad W \quad G}$
$\underline{R \quad S \quad \bullet \quad T \quad U \quad V \quad D \quad E \quad F \quad X}$

*22. A species has $2n = 16$ chromosomes. How many chromosomes will be found per cell in each of the following mutants in this species?

(a) Monosomic

(b) Autotriploid

(c) Autotetraploid

(d) Trisomic

(e) Double monosomic

(f) Nullisomic

(g) Autopentaploid

(h) Tetrasomic

*23. The *Notch* mutation is a deletion on the X chromosome of *Drosophila melanogaster*. Female flies heterozygous for *Notch* have an indentation on the margin of their wings; *Notch* is lethal in the homozygous and hemizygous conditions. The *Notch* deletion covers the region of the X chromosome that contains the locus for white eyes, an X-linked recessive trait. Give the phenotypes and proportions of progeny produced in the following crosses.

(a) A red-eyed, Notch female is mated with white-eyed male.

(b) A white-eyed, Notch female is mated with a red-eyed male.

(c) A white-eyed, Notch female is mated with a white-eyed male.

24. The green nose fly normally has six chromosomes, two metacentric and four acrocentric. A geneticist examines the chromosomes of an odd-looking green nose fly and discovers that it has only five chromosomes; three of them are metacentric and two are acrocentric. Explain how this change in chromosome number might have occurred.

25. Species I is diploid ($2n = 8$) with chromosomes AABBCCDD; related species II is diploid ($2n = 8$) with chromosomes MMNNOOPP. Individuals with the following sets of chromosomes represent what types of chromosome mutations?

(a) AAABBCCDD

(b) MMNNOOOOPP

(c) AABBCDD

(d) AAABBBCCCDDD

(e) AAABBCCDDD

(f) AABBDD

(g) AABBCCDDMMNNOOPP

(h) AABBCCDDMNOP

*26. A wild-type chromosome has the following segments:

A B C • D E F G H I

An individual is heterozygous for the following chromosome mutations. For each mutation, sketch how the wild-type and mutated chromosomes would pair in prophase I of meiosis, showing all chromosome strands.

(a) A B C • D E F D E F G H I

(b) A B C • D H I

(c) A B C • D G F E H I

(d) A B E D • C F G H I

27. An individual that is heterozygous for a pericentric inversion has the following two chromosomes:

A B C D • E F G H I
A B C F E • D G H I

(a) Sketch the pairing of these two chromosomes in prophase I of meiosis, showing all four strands.

(b) Draw the chromatids that would result from a single crossover between the E and F segments.

(c) What will happen when the chromosomes separate in anaphase I of meiosis?

28. Answer part *b* of problem 28 for a two-strand double crossover between E and F.

*29. An individual heterozygous for a reciprocal translocation possesses the following chromosomes:

A B • C D E F G
A B • C D V W X
R S • T U E F G
R S • T U V W X

(a) Draw the pairing arrangement of these chromosomes in prophase I of meiosis.

(b) Diagram the alternate, adjacent-1, and adjacent-2 segregation patterns in anaphase I of meiosis.

(c) Give the products that result from alternate, adjacent-1, and adjacent-2 segregation.

30. Red–green color blindness is a human X-linked recessive disorder. A young man with a 47,XXY karyotype (Klinefelter syndrome) is color-blind. His 46,XY brother also is color-blind. Both parents have normal color vision. Where did the nondisjunction occur that gave rise to the young man with Klinefelter syndrome?

31. Some people with Turner syndrome are 45,X/46,XY mosaics. Explain how this mosaicism could arise.

*32. Bill and Betty have had two children with Down syndrome. Bill's brother has Down syndrome and his sister has two children with Down syndrome. On the basis of these observations, which of the following statements is most likely correct? Explain your reasoning.

(a) Bill has 47 chromosomes.

(b) Betty has 47 chromosomes.

(c) Bill and Betty's children each have 47 chromosomes.

(d) Bill's sister has 45 chromosomes.

(e) Bill has 46 chromosomes.

(f) Betty has 45 chromosomes.

(g) Bill's brother has 45 chromosomes.

*33. Tay-Sachs disease is an autosomal recessive disease that causes blindness, deafness, brain enlargement, and premature death in children. It is possible to identify carriers for Tay-Sachs disease by means of a blood test. Mike and Sue have both been tested for the Tay-Sachs gene; Mike is a heterozygous carrier for Tay-Sachs, but Sue is homozygous for the normal allele. Mike and Sue's baby boy is completely normal at birth, but at age 2 develops Tay-Sachs disease. Assuming that a new mutation has not occurred, how could Mike and Sue's baby have inherited Tay-Sach's disease?

34. In mammals, sex-chromosome aneuploids are more common than autosomal aneuploids but, in fishes, sex-chromosome aneuploids and autosomal aneuploids are found with equal

frequency. Offer an explanation for these differences in mammals and fishes.

*35. A young couple is planning to have children. Knowing that there have been a substantial number of stillbirths, miscarriages, and fertility problems on the husband's side of the family, they see a genetic counselor. A chromosome analysis reveals that, whereas the woman has a normal karyotype, the man possesses only 45 chromosomes and is a carrier for a Robertsonian translocation between chromosomes 22 and 13.

(a) List all the different types of gametes that might be produced by the man.

(b) What types of zygotes will develop when each of gametes produced by the man fuses with a normal gamete from the woman?

(c) If trisomies and monosomies entailing chromosome 13 and 22 are lethal, what proportion of the surviving offspring will be carriers of the translocation?

CHALLENGE QUESTION

36. Red–green color blindness is a human X-linked recessive disorder. Jill has normal color vision, but her father is color-blind. Jill marries Tom, who also has normal color vision. Jill and Tom have a daughter who has Turner syndrome and is color-blind.

(a) How did the daughter inherit color blindness?

(b) Did the daughter inherit her X chromosome from Jill or from Tom?

SUGGESTED READINGS

Boue, A. 1985. Cytogenetics of pregnancy wastage. *Advances in Human Genetics* 14:1–58.
A study showing that many human spontaneously aborted fetuses contain chromosome mutations.

Brewer, C., S. Holloway, P. Zawalnyski, A. Schinzel, and D. FitzPatrick. 1998. A chromosomal deletion map of human malformations. *American Journal of Human Genetics* 63:1153–1159.
A study of human malformations associated with specific chromosome deletions.

Epstein, C. J. 1988. Mechanisms of the effects of aneuploidy in mammals. *Annual Review of Genetics* 22:51–75.
A review of how aneuploidy produces phenotypic effects in mammals.

Feldman, M., and E. R. Sears. 1981. The wild resources of wheat. *Scientific American* 244 (1):98.
An account of how polyploidy has led to the evolution of modern wheat.

Gardner, R. J. M., and G. R. Sunderland. 1996. *Chromosome Abnormalities and Genetic Counseling.* Oxford: Oxford University Press.
A guide to chromosome abnormalies for genetic counselors.

Goodman, R. M., and R. J. Gorlin. 1983. *The Malformed Infant and Child: An Illustrated Guide.* New York: Oxford University Press.
A pictorial compendium of genetic and chromosomal syndromes in humans.

Hall, J. C. 1988. Review and hypothesis: somatic mosaicism—observations related to clinical genetics. *American Journal of Human Genetics* 43:355–363.
A review of the significance of mosaicism in human genetics.

Hieter, P., and T. Griffiths. 1999. Polyploidy: more is more or less. *Science* 285:210–211.
Discusses current research that shows that there is some unbalanced gene expression in polyploid cells.

Patterson, D. 1987. The causes of Down syndrome. *Scientific American* 257(2):52–60.
An excellent review of research concerning the genes on chromosome 21 that cause Down syndrome.

Rabbitts, T. H. 1994. Chromosomal translocations in human cancers. *Nature* 372:143–149.
Reviews the association of some chromosomal translocations with specific human cancers.

Rowley, J. D. 1998. The critical role of chromosome translocations in human leukemias. *Annual Review of Genetics* 32:495–519.
A review of molecular analyses of chromosome translocations in leukemias.

Ryder, O. A., L. G. Chemnick, A. T. Bowling, and K. Benirschke. 1985. Male mule foal qualifies as the offspring of a female mule and jack donkey. *Journal of Heredity* 76:379–381.
A study of a male foal (Blue Moon) born to a mule, which was discussed at the beginning of the chapter.

Sánchez-García, I. 1997. Consequences of chromosome abnormalities in tumor development. *Annual Review of Genetics* 31:429–453.
Reviews the nature of fusion proteins produced by chromosome translocations that play a role in tumor development.

Schulz-Schaeffer, J. 1980. *Cytogenetics: Plants, Animals, Humans.* New York: Springer Verlag.
A detailed treatment of chromosomal variation.

10 DNA: The Chemical Nature of the Gene

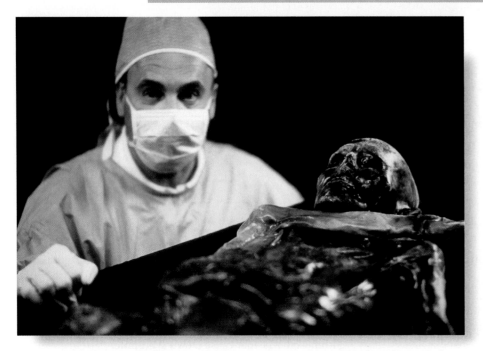

Ice Man is a 5300-year-old frozen corpse found in the Alps. Analysis of his mitochondrial DNA has revealed that he was a Neolithic hunter related to present-day Europeans living north of the Alps. (Brando Quilici.)

The Elegantly Stable Double Helix: Ice Man's DNA

DNA, with its gentle double-stranded spiral, is among the most elegant of all biological molecules. But the double helix is not just a beautiful structure; it also gives DNA incredible stability and permanence, as illustrated by the story of Ice Man.

On September 19, 1991, German tourists hiking in the Tyrolean Alps near the border between Austria and Italy spotted a corpse trapped in glacial ice. A copper ax, dagger, bow, and quiver with 14 arrows were found alongside the body. Not realizing its antiquity, local residents made several crude and unsuccessful attempts to free the body from the ice. After 4 days, a team of forensic experts arrived to recover the body and transport it to the University of Innsbruck. There the mummified corpse, known as Ice Man, was refrozen and subjected to scientific study.

Radiocarbon dating indicates that Ice Man is approximately 5000 years old. Recent evidence from the South Tyrol Museum of Archeology has led to the conclusion that Ice Man was shot in the chest with an arrow and died soon thereafter. The body became dehydrated in the cold high-altitude air, was covered with snow that turned into ice, and remained frozen for the next 5000 years.

Some experts challenged Ice Man's origin, suggesting that he was a South American mummy who had been planted at the glacier site in an elaborate hoax. To establish his authenticity and ethnic origin, scientists removed eight samples of muscle, connective tissue, and bone from his left hip. Under sterile conditions, the investigators extracted DNA from the samples and used the polymerase chain reaction (see Chapter 18) to amplify a very small region of his mitochondrial DNA a millionfold. They determined the base sequence of this amplified DNA and compared it with mitochondrial sequences from present-day humans.

This analysis revealed that Ice Man's mitochondrial DNA sequences resemble those found in present-day Europeans living north of the Alps and are quite different from those of sub-Saharan Africans, Siberians, and Native Americans. Together, radiocarbon dating, the artifacts, and the DNA analysis all indicate that Ice Man was a Neolithic hunter who died while attempting to cross the Alps 5000 years ago. That some of Ice Man's DNA persists and faithfully carries his genetic instructions even after the passage of 5000 years is testimony to the remarkable stability of the double helix. Even more ancient DNA has been isolated from the fossilized bones of Neanderthals that are at least 30,000 years old.

This chapter focuses on how DNA was identified as the source of genetic information and how this elegant molecule encodes the genetic instructions. We begin by considering the basic requirements of the genetic material and the history of our understanding of DNA—how its relation to genes was uncovered and how its structure was determined. The history of DNA illustrates several important points about the nature of scientific research. As with so many important scientific advances, DNA's structure and its role as the genetic material were not discovered by any single person but were gradually revealed over a period of almost 100 years, thanks to the work of many investigators. Our understanding of the relation between DNA and genes was enormously enhanced in 1953, when James Watson and Francis Crick proposed a three-dimensional structure for DNA that brilliantly illuminated its role in genetics. As illustrated by Watson and Crick's discovery, major scientific advances are often achieved not through the collection of new data but through the interpretation of old data in new ways.

After reviewing the history of DNA, we will examine DNA structure. DNA structure is important in its own right, but the key genetic concept is the relation between the structure and the function of DNA—how its structure allows it to serve as the genetic material.

Characteristics of Genetic Material

Life is characterized by tremendous diversity, but the coding instructions of all living organisms are written in the same genetic language—that of nucleic acids. Surprisingly, the idea that genes are made of nucleic acids was not widely accepted until after 1950. This late recognition of the role of nucleic acids in genetics resulted principally from a lack of knowledge about the structure of deoxyribonucleic acid (DNA). Until the structure of DNA was fully elucidated, it wasn't clear how DNA could store and transmit genetic information. Even before nucleic acids were identified as the genetic material, biologists recognized that, whatever the nature of genetic material, it must possess three important characteristics.

1. **Genetic material must contain complex information.** First and foremost, the genetic material must be capable of storing large amounts of information—instructions for all the traits and functions of an organism. This information must have the capacity to vary, because different species and even individual members of a species differ in their genetic makeup. At the same time, the genetic material must be stable, because most alterations to the genetic instructions (mutations) are likely to be detrimental.

2. **Genetic material must replicate faithfully.** A second necessary feature is that genetic material must have the capacity to be copied accurately. Every organism begins life as a single cell, which must undergo billions of cell divisions to produce a complex, multicellular creature like yourself. At each cell division, the genetic instructions must be transmitted to descendent cells with great accuracy. When organisms reproduce and pass genes to their progeny, the coding instructions must be copied with fidelity.

3. **Genetic material must encode phenotype.** The genetic material (the genotype) must have the capacity to "code for" (determine) traits (the phenotype). The product of a gene is often a protein; so there must be a mechanism for genetic instructions to be translated into the amino acid sequence of a protein.

Concepts

The genetic material must be capable of carrying large amounts information, replicating faithfully, and translating its coding instructions into phenotypes.

The Molecular Basis of Heredity

Although our understanding of how DNA encodes genetic information is relatively recent, the study of DNA structure stretches back 100 years.

Early Studies of DNA

In 1868, Johann Friedrich Miescher (◀ FIGURE 10.1) graduated from medical school in Switzerland. Influenced by an uncle who believed that the key to understanding disease lay in the chemistry of tissues, Miescher traveled to Tubingen, Germany, to study under Ernst Felix Hoppe-Seyler, an early leader in the emerging field of biochemistry. Under Hoppe-Seyler's direction, Miescher turned his attention to the chemistry of pus, a substance of clear medical importance. Pus contains white blood cells with large nuclei; Miescher developed a method of isolating

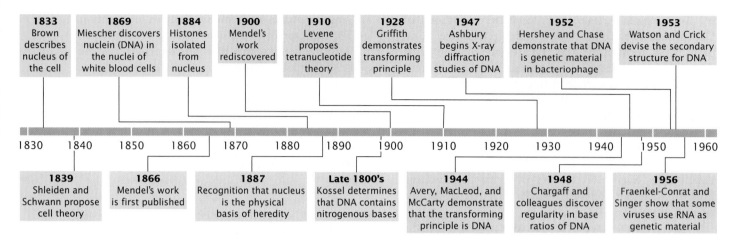

| 1833 Brown describes nucleus of the cell | 1869 Miescher discovers nuclein (DNA) in the nuclei of white blood cells | 1884 Histones isolated from nucleus | 1900 Mendel's work rediscovered | 1910 Levene proposes tetranucleotide theory | 1928 Griffith demonstrates transforming principle | 1947 Ashbury begins X-ray diffraction studies of DNA | 1952 Hershey and Chase demonstrate that DNA is genetic material in bacteriophage | 1953 Watson and Crick devise the secondary structure for DNA |

1830 1840 1850 1860 1870 1880 1890 1900 1910 1920 1930 1940 1950 1960

| 1839 Shleiden and Schwann propose cell theory | 1866 Mendel's work is first published | 1887 Recognition that nucleus is the physical basis of heredity | Late 1800's Kossel determines that DNA contains nitrogenous bases | 1944 Avery, MacLeod, and McCarty demonstrate that the transforming principle is DNA | 1948 Chargaff and colleagues discover regularity in base ratios of DNA | 1956 Fraenkel-Conrat and Singer show that some viruses use RNA as genetic material |

◀10.1 **Many people have contributed to our understanding of the structure and function of DNA.**

these nuclei. The minute amounts of nuclear material that he obtained were insufficient for a thorough chemical analysis, but he did establish that it contained a novel substance that was slightly acidic and high in phosphorus. This material, which consisted of DNA and protein, Miescher called *nuclein*. The substance was later renamed *nucleic acid* by one of his students.

By 1887, researchers had concluded that the physical basis of heredity lies in the nucleus. Chromatin was shown to consist of nucleic acid and proteins, but which of these substances is actually the genetic information was not clear. In the late 1800s, further work on the chemistry of DNA was carried out by Albrecht Kossel, who determined that DNA contains four nitrogenous bases: adenine, cytosine, guanine, and thymine (abbreviated A, C, G, and T).

In the early twentieth century, the Rockefeller Institute in New York City became a center for nucleic acid research. Phoebus Aaron Levene joined the Institute in 1905 and spent the next 40 years studying the chemistry of DNA. He discovered that DNA consists of a large number of linked, repeating units, each containing a sugar, a phosphate, and a base (together forming a **nucleotide**).

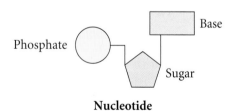

Nucleotide

He incorrectly proposed that DNA consists of a series of four-nucleotide units, each unit containing all four bases—adenine, guanine, cytosine, and thymine—in a fixed sequence. This concept, known as the tetranucleotide theory, implied that the structure of DNA is too regular to serve as the genetic material. The tetranucleotide theory

contributed to the idea that *protein* is the genetic material because, with its 20 different amino acids, protein structure could be highly variable.

As additional studies of the chemistry of DNA were completed in the 1940s and 1950s, this notion of DNA as a simple, invariant molecule began to change. Erwin Chargaff and his colleagues carefully measured the amounts of the four bases in DNA from a variety of organisms and found that DNA from different organisms varies greatly in base composition. This finding disproved the tetranucleotide theory. They discovered that, within each species, there is some regularity in the ratios of the bases: the total amount of adenine is always equal to the amount of thymine (A = T), and the amount of guanine is always equal to the amount of cytosine (G = C; Table 10.1). These findings became known as **Chargaff's rules.**

> **Concepts**
>
> Details of the structure of DNA were worked out by a number of scientists. At first, DNA was interpreted as being too regular in structure to carry genetic information but, by the 1940s, DNA from different organisms was shown to vary in its base composition.

DNA As the Source of Genetic Information

While chemists were working out the structure of DNA, biologists were attempting to identify the source of genetic information. Two sets of experiments, one conducted on bacteria and the other on viruses, provided pivotal evidence that DNA, rather than protein, was the genetic material.

					Ratio		
Source of DNA	**A**	**T**	**G**	**C**	**A/T**	**G/C**	**A + G/T + C**
E. coli	26.0	23.9	24.9	25.2	1.09	0.99	1.04
Yeast	31.3	32.9	18.7	17.1	.95	1.09	1.00
Sea urchin	32.8	32.1	17.7	18.4	1.02	.96	1.00
Rat	28.6	28.4	21.4	21.5	1.01	1.00	1.00
Human	30.3	30.3	19.5	19.9	1.00	0.98	0.99

Table 10.1 Base composition of DNA from different sources and rations of bases

The discovery of the transforming principle The first clue that DNA was the carrier of hereditary information came with the demonstration that DNA was responsible for a phenomenon called *transformation*. The phenomenon was first observed in 1928 by Fred Griffith, an English physician whose special interest was the bacterium that causes pneumonia, *Streptococcus pneumonia*. Griffith had succeeded in isolating several different strains of *S. pneumonia* (type I, II, III, and so forth). In the virulent (disease-causing) forms of a strain, each bacterium is surrounded by a polysaccharide coat, which makes the bacterial colony appear smooth when grown on an agar plate; these forms are referred to as S, for smooth. Griffith found that these virulent forms occasionally mutated to nonvirulent forms, which lack a polysaccharide coat and produce a rough-appearing colony on an agar plate; these forms are referred to as R, for rough.

Griffith was interested in the origins of the different strains of *S. pneumonia* and why some types were virulent, whereas others were not. He observed that small amounts of living type IIIS bacteria injected into mice caused the mice to develop pneumonia and die; on autopsy, he found large amounts of type IIIS bacteria in the blood of the mice (◀FIGURE 10.2a). When Griffith injected type IIR bacteria into mice, the mice lived, and no bacteria were recovered from their blood (◀FIGURE 10.2b). Griffith knew that boiling killed all the bacteria and destroyed their virulence; when he injected large amounts of heat-killed type IIIS bacteria into mice, the mice lived and no type IIIS bacteria were recovered from their blood (◀FIGURE 10.2c).

The results of these experiments were not unusual. However, Griffith got a surprise when he infected his mice with a small amount of living type IIR bacteria, along with a large amount of heat-killed type IIIS bacteria. Because both the type IIR bacteria and the heat-killed type IIIS bacteria were nonvirulent, he expected these mice to live. Surprisingly, 5 days after the injections, the mice became infected with pneumonia and died (◀FIGURE 10.2d). When Griffith examined blood from the hearts of these mice, he observed live type IIIS bacteria. Furthermore, these bacteria retained their type IIIS characteristics through several generations; so the infectivity was heritable.

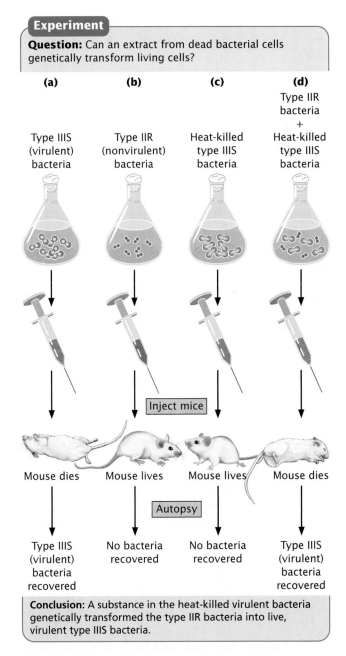

Experiment

Question: Can an extract from dead bacterial cells genetically transform living cells?

(a)	(b)	(c)	(d)
Type IIIS (virulent) bacteria	Type IIR (nonvirulent) bacteria	Heat-killed type IIIS bacteria	Type IIR bacteria + Heat-killed type IIIS bacteria

Inject mice

Mouse dies	Mouse lives	Mouse lives	Mouse dies

Autopsy

Type IIIS (virulent) bacteria recovered	No bacteria recovered	No bacteria recovered	Type IIIS (virulent) bacteria recovered

Conclusion: A substance in the heat-killed virulent bacteria genetically transformed the type IIR bacteria into live, virulent type IIIS bacteria.

◀10.2 **Griffith's experiments demonstrated transformation in bacteria.**

Griffith's results had several possible interpretations, all of which he considered. First, it could have been the case that he had not sufficiently sterilized the type IIIS bacteria and thus a few live bacteria remained in the culture. Any live bacteria injected into the mice would have multiplied and caused pneumonia. Griffith knew that this possibility was unlikely, because he had used only heat-killed type IIIS bacteria in the control experiment, and they never produced pneumonia in the mice.

A second interpretation was that the live, type IIR bacteria had mutated to the virulent S form. Such a mutation would cause pneumonia in the mice, but it would produce type IIS bacteria, not the type IIIS that Griffith found in the dead mice. Many mutations would be required for type II bacteria to mutate to type III bacteria, and the chance of all the mutations occurring simultaneously was impossibly low.

Griffith finally concluded that the type IIR bacteria had somehow been *transformed*, acquiring the genetic virulence of the dead type IIIS bacteria. This transformation had produced a permanent, genetic change in the bacteria; though Griffith didn't understand the nature of transformation, he theorized that some substance in the polysaccharide coat of the dead bacteria might be responsible. He called this substance the **transforming principle.**

Identification of the transforming principle At the time of Griffith's report, Oswald Avery (see Figure 10.1) was a microbiologist at the Rockefeller Institute. At first Avery was skeptical but, after other microbiologists successfully repeated Griffith's experiments using other bacteria and showed that transformation took place, Avery set out to identify the nature of the transforming substance.

After 10 years of research, Avery, Colin MacLeod, and Maclyn McCarty succeeded in isolating and purifying the transforming substance. They showed that it had a chemical composition closely matching that of DNA and quite different from that of proteins. Enzymes such as trypsin and chymotrypsin, known to break down proteins, had no effect on the transforming substance. Ribonuclease, an enzyme that destroys RNA, also had no effect. Enzymes capable of destroying DNA, however, eliminated the biological activity of the transforming substance (◀FIGURE 10.3). Avery, MacLeod, and McCarty showed that purified transforming substance precipitated at about the same rate as purified DNA and that it absorbed ultraviolet light at the same wavelengths as does DNA. These results, published in 1944, provided compelling evidence that the transforming principle—and therefore genetic information—resides in DNA. Many biologists still refused to accept the idea, however, still preferring the hypothesis that the genetic material is protein.

◀10.3 **Avery, MacLeod, and McCarty's experiment revealed the nature of the transforming principle.**

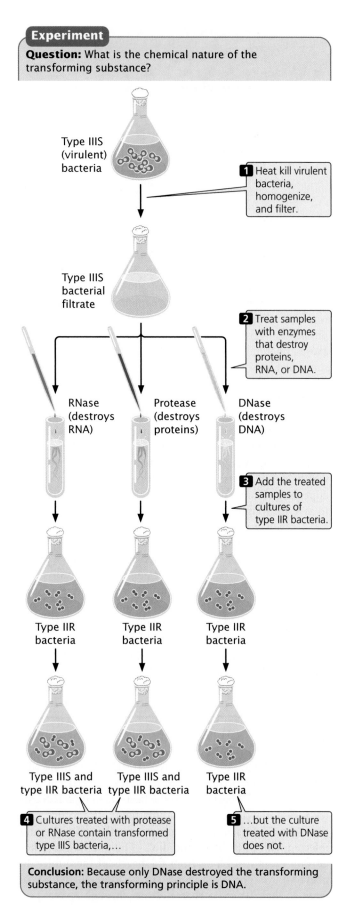

Experiment

Question: What is the chemical nature of the transforming substance?

Type IIIS (virulent) bacteria

1 Heat kill virulent bacteria, homogenize, and filter.

Type IIIS bacterial filtrate

2 Treat samples with enzymes that destroy proteins, RNA, or DNA.

RNase (destroys RNA)   Protease (destroys proteins)   DNase (destroys DNA)

3 Add the treated samples to cultures of type IIR bacteria.

Type IIR bacteria   Type IIR bacteria   Type IIR bacteria

Type IIIS and type IIR bacteria   Type IIIS and type IIR bacteria   Type IIR bacteria

4 Cultures treated with protease or RNase contain transformed type IIIS bacteria,...

5 ...but the culture treated with DNase does not.

Conclusion: Because only DNase destroyed the transforming substance, the transforming principle is DNA.

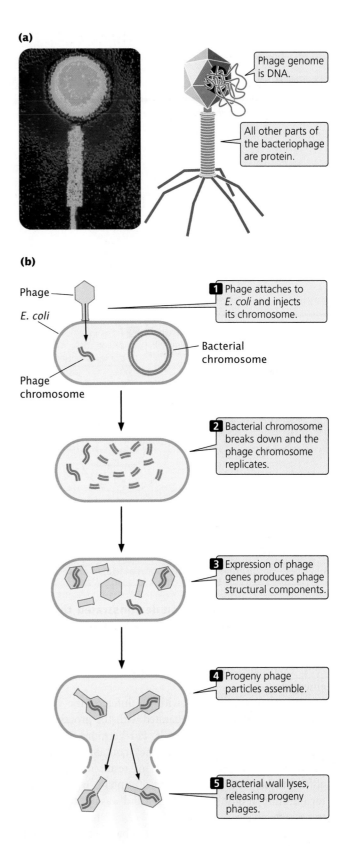

(a)

Phage genome is DNA.

All other parts of the bacteriophage are protein.

(b)

Phage

E. coli

Phage chromosome

Bacterial chromosome

1 Phage attaches to E. coli and injects its chromosome.

2 Bacterial chromosome breaks down and the phage chromosome replicates.

3 Expression of phage genes produces phage structural components.

4 Progeny phage particles assemble.

5 Bacterial wall lyses, releasing progeny phages.

10.4 T2 is a bacteriophage that infects *E. coli.* (a) T2 phage. (b) Its life cycle. (Micrograph from George Musil/Visuals Unlimited.)

Concepts

The process of transformation indicates that some substance—the transforming principle—is capable of genetically altering bacteria. Avery, MacLeod, and McCarty demonstrated that the transforming principle is DNA, providing the first evidence that DNA is the genetic material.

The Hershey-Chase experiment A second piece of evidence implicating DNA as the genetic material resulted from a study of the T2 virus conducted by Alfred Hershey and Martha Chase. T2 is a *bacteriophage* (phage) that infects the bacterium *Escherichia coli* (◀FIGURE 10.4a). As stated in Chapter 8, a phage reproduces by attaching to the outer wall of a bacterial cell and injecting its DNA into the cell, where it replicates and directs the cell to synthesize phage protein. The phage DNA becomes encapsulated within the proteins, producing progeny phages that lyse (break open) the cell and escape (◀FIGURE 10.4b).

At the time of the Hershey-Chase study (their paper was published in 1952), biologists did not understand exactly how phages reproduce. What they did know was that the T2 phage consists of approximately 50% protein and 50% nucleic acid, that a phage infects a cell by first attaching to the cell wall, and that progeny phages are ultimately produced within the cell. Because the progeny carried the same traits as the infecting phage, genetic material from the infecting phage must be transmitted to the progeny, but how this occurs was unknown.

Hershey and Chase designed a series of experiments to determine whether the phage *protein* or the phage *DNA* was transmitted in phage reproduction. To follow the fate of protein and DNA, they used radioactive forms (**isotopes**) of phosphorus and sulfur. A radioactive isotope can be used as a tracer to identify the location of a specific molecule, because any molecule containing the isotope will be radioactive and therefore easily detected. DNA contains phosphorus but not sulfur; so Hershey and Chase used ^{32}P to follow phage DNA during reproduction. Protein contains sulfur but not phosphorus; so they used ^{35}S to follow the protein.

First, Hershey and Chase grew *E. coli* in a medium containing ^{32}P and infected the bacteria with T2 so that all the new phages would have DNA labeled with ^{32}P (◀FIGURE 10.5). They grew a second batch of *E. coli* in a medium containing ^{35}S and infected these bacteria with T2 so that all these new phages would have protein labeled with ^{35}S. Hershey and Chase then infected separate batches of unlabeled *E. coli* with the ^{35}S- and ^{32}P-labeled phages. After allowing time for the phages to infect the cells, they placed the *E. coli* cells in a blender and sheared off the now-empty protein coats (ghosts) from the cell walls. They separated out the protein coats and cultured the infected bacterial

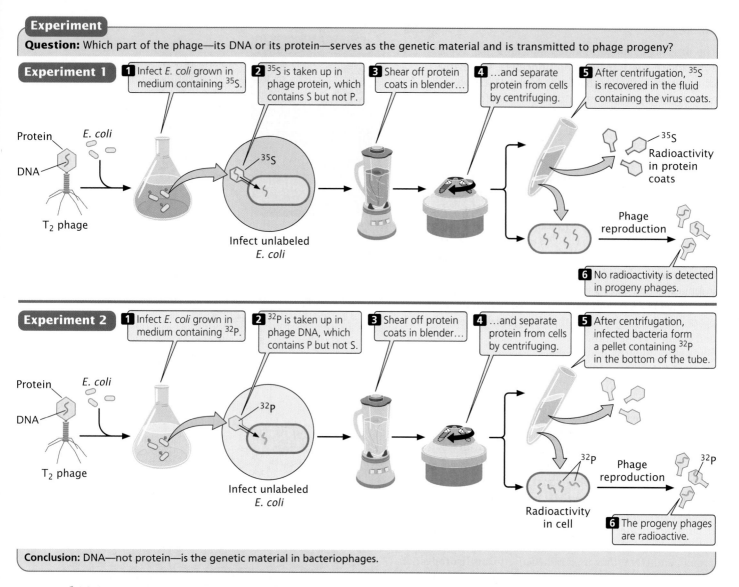

Experiment

Question: Which part of the phage—its DNA or its protein—serves as the genetic material and is transmitted to phage progeny?

Experiment 1

1 Infect *E. coli* grown in medium containing ^{35}S.

2 ^{35}S is taken up in phage protein, which contains S but not P.

3 Shear off protein coats in blender…

4 …and separate protein from cells by centrifuging.

5 After centrifugation, ^{35}S is recovered in the fluid containing the virus coats.

Protein

E. coli

DNA

T₂ phage

Infect unlabeled *E. coli*

^{35}S

Radioactivity in protein coats

Phage reproduction

6 No radioactivity is detected in progeny phages.

Experiment 2

1 Infect *E. coli* grown in medium containing ^{32}P.

2 ^{32}P is taken up in phage DNA, which contains P but not S.

3 Shear off protein coats in blender…

4 …and separate protein from cells by centrifuging.

5 After centrifugation, infected bacteria form a pellet containing ^{32}P in the bottom of the tube.

Protein

E. coli

DNA

T₂ phage

Infect unlabeled *E. coli*

^{32}P

^{32}P

Phage reproduction

^{32}P

Radioactivity in cell

6 The progeny phages are radioactive.

Conclusion: DNA—not protein—is the genetic material in bacteriophages.

◀ **10.5 Hershey and Chase demonstrated that DNA carries the genetic information in bacteriophages.**

cells. Eventually, the cells burst and new phage particles emerged.

When phages labeled with ^{35}S infected the bacteria, most of the radioactivity separated with the protein ghosts and little remained in the cells. Furthermore, when new phages emerged from the cell, they contained almost no radioactivity (see Figure 10.5). This result indicated that, although the protein component of a phage was necessary for infection, it didn't enter the cell and was not transmitted to progeny phages.

In contrast, when Hershey and Chase infected bacteria with ^{32}P-labeled phages and removed the protein ghosts, the bacteria were still radioactive. Most significantly, after the cells lysed and new progeny phages emerged, many of these phages emitted radioactivity from ^{32}P, demonstrating that DNA from the infecting phages had been passed on to the

progeny (see Figure 10.5). These results confirmed that DNA, not protein, is the genetic material of phages.

Concepts

Using radioactive isotopes, Hershey and Chase traced the movement of DNA and protein during phage infection. They demonstrated that DNA, not protein, enters the bacterial cell during phage reproduction and that only DNA is passed on to progeny phages.

www.whfreeman.com/pierce   A discussion of the requirements of the genetic material and the history of our understanding of DNA structure and function.

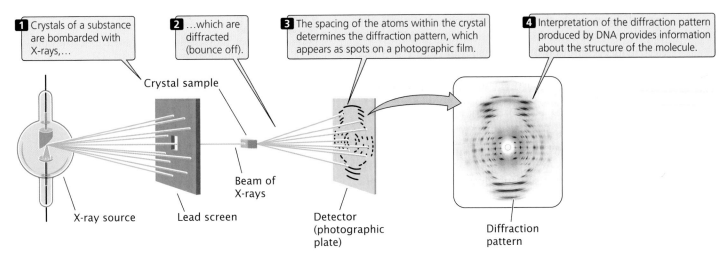

1 Crystals of a substance are bombarded with X-rays,...

2 ...which are diffracted (bounce off).

3 The spacing of the atoms within the crystal determines the diffraction pattern, which appears as spots on a photographic film.

4 Interpretation of the diffraction pattern produced by DNA provides information about the structure of the molecule.

Crystal sample

X-ray source

Lead screen

Beam of X-rays

Detector (photographic plate)

Diffraction pattern

◀ 10.6 **X-ray diffraction provides information about the structures of molecules.** (Photo from M. H. F. Wilkins, Department of Biophysics, King's College, University of London.)

Watson and Crick's Discovery of the Three-Dimensional Structure of DNA

The experiments on the nature of the genetic material set the stage for one of the most important advances in the history of biology—the discovery of the three-dimensional structure of DNA by James Watson and Francis Crick in 1953.

Watson had studied bacteriophage for his Ph.D.; he was familiar with Avery's work and thus understood the tremendous importance of DNA to genetics. Shortly after receiving his Ph.D., Watson went to the Cavendish Laboratory at Cambridge University in England, where a number of researchers were studying the three-dimensional structure of large molecules. Among these researchers was Francis Crick, who was still working on his Ph.D. Watson and Crick immediately became friends and colleagues.

Much of the basic chemistry of DNA had already been determined by Miescher, Kossel, Levene, Chargaff, and others, who had established that DNA consisted of nucleotides, and that each nucleotide contained a sugar, base, and phosphate group. However, how the nucleotides fit together in the three-dimensional structure of the molecule was not at all clear.

In 1947, William Ashbury began studying the three-dimensional structure of DNA by using a technique called **X-ray diffraction** (◀ FIGURE 10.6), but his diffraction pictures did not provide enough resolution to reveal the structure. A research group at King's College in London, led by Maurice Wilkins and Rosalind Franklin, also was studying the structure of DNA by using X-ray diffraction and obtained strikingly better pictures of the molecule. Wilkins and Franklin, however, were unable to develop a complete structure of the molecule; their progress was impeded by personal discord that existed between them.

Watson and Crick investigated the structure of DNA, not by collecting new data but by using all available information about the chemistry of DNA to construct molecular models (◀ FIGURE 10.7). By applying the laws of structural chemistry, they were able to limit the number of possible structures that DNA could assume. Watson and Crick tested various structures by building models made of wire and metal plates. With their models, they were able to see whether a structure was compatible with chemical principles and with the X-ray images.

The key to solving the structure came when Watson recognized that an adenine base could bond with a thymine base and that a guanine base could bond with a cytosine base; these pairings accounted for the base ratios that Chargaff had discovered earlier. The model developed by Watson and Crick showed that DNA consists of two strands of nucleotides wound around each other to form a right-

◀ 10.7 **Watson and Crick provided a three-dimensional model of the structure of DNA.**

(A. Barrington Brown/Science Photo Library/Photo Researchers.)

handed helix, with the sugars and phosphates on the outside and the bases in the interior. They published an electrifying description of their model in *Nature* in 1953. At the same time, Wilkins and Franklin published their X-ray diffraction data, which demonstrated experimentally the theory that DNA was helical in structure.

Many have called the solving of DNA's structure the most important biological discovery of the twentieth century. For their discovery, Watson and Crick, along with Maurice Wilkins, were awarded a Nobel Prize in 1962. (Rosalind Franklin had died of cancer in 1957 and, thus, could not be considered a candidate for the shared prize.)

> **Concepts**
>
> By collecting existing information about the chemistry of DNA and building molecular models, Watson and Crick were able to discover the three-dimensional structure of the DNA molecule.

www.whfreeman.com/pierce   A commentary on Watson and Crick's original paper describing the structure of DNA and more information about some of the key players in the discovery of DNA

RNA As Genetic Material

In most organisms, DNA carries the genetic information. However, a few viruses utilize RNA, not DNA, as their genetic material. This fact was demonstrated in 1956 by Heinz Fraenkel-Conrat and Bea Singer, who worked with tobacco mosaic virus (TMV), a virus that infects and causes disease in tobacco plants. The tobacco mosaic virus possesses a single molecule of RNA surrounded by a helically arranged cylinder of protein molecules. Fraenkal-Conrat found that, after separating the RNA and protein of TMV, he could remix them and obtain intact, infectious viral particles.

With Singer, Fraenkal-Conrat then created hybrid viruses by mixing RNA and protein from different strains of TMV (◀ Figure 10.8). When these hybrid viruses infected tobacco leaves, new viral particles were produced. The new viral progeny were identical to the strain from which the RNA had been isolated and did not exhibit the characteristics of the strain that donated the protein. These results showed that RNA carries the genetic information in TMV.

Also in 1956, Alfred Gierer and Gerhard Schramm demonstrated that RNA isolated from TMV is sufficient to infect tobacco plants and direct the production of new TMV particles, confirming that RNA carries genetic instructions.

> **Concepts**
>
> RNA serves as the genetic material in some viruses.

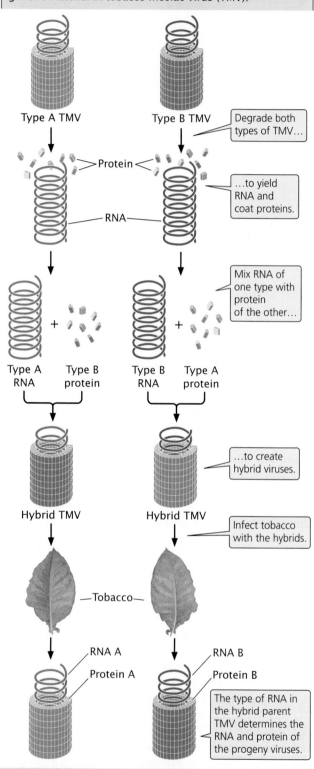

Experiment

Question: What substance—RNA or protein—carries the genetic material in tobacco mosaic virus (TMV)?

Type A TMV   Type B TMV

Degrade both types of TMV…

Protein

…to yield RNA and coat proteins.

RNA

Mix RNA of one type with protein of the other…

Type A RNA   Type B protein   Type B RNA   Type A protein

…to create hybrid viruses.

Hybrid TMV   Hybrid TMV

Infect tobacco with the hybrids.

Tobacco

RNA A   RNA B
Protein A   Protein B

The type of RNA in the hybrid parent TMV determines the RNA and protein of the progeny viruses.

Conclusion: RNA is the genetic material of TMV.

◀ **10.8 Fraenkal-Conrat and Singer's experiment demonstrated that, in the tobacco mosaic virus, RNA carries the genetic information.**

The Structure of DNA

DNA, though relatively simple in structure, has an elegance and beauty unsurpassed by other large molecules. It is useful to consider the structure of DNA at three levels of increasing complexity, known as the primary, secondary, and tertiary structures of DNA. The primary structure of DNA refers to its nucleotide structure and how the nucleotides are joined together. The secondary structure refers to DNA's stable three-dimensional configuration, the helical structure worked out by Watson and Crick. In Chapter 11, we will consider DNA's tertiary structures, which are the complex packing arrangements of double-stranded DNA in chromosomes.

The Primary Structure of DNA

The primary structure of DNA consists of a string of nucleotides joined together by phosphodiester linkages.

Nucleotides DNA is typically a very long molecule and is therefore termed a macromolecule. For example, within each human chromosome is a single DNA molecule that, if stretched out straight, would be several centimeters in length. In spite of its large size, DNA has a relatively simple structure: it is a polymer, a chain made up of many repeating units linked together. As already mentioned, the repeating units of DNA are *nucleotides*, each comprising three parts: (1) a sugar, (2) a phosphate, and (3) a nitrogen-containing base.

The sugars of nucleic acids—called pentose sugars—have five carbon atoms, numbered $1'$, $2'$, $3'$, and so forth (◀ FIGURE 10.9). Four of the carbon atoms are joined by an oxygen atom to form a five-sided ring; the fifth ($5'$) carbon atom projects upward from the ring. Hydrogen atoms or hydroxyl groups (OH) are attached to each carbon atom.

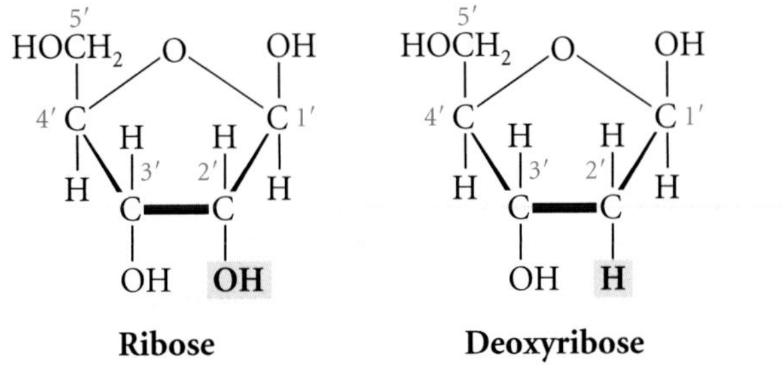

Ribose   Deoxyribose

◀ 10.9 **A nucleotide contains either a ribose sugar (in RNA) or a deoxyribose sugar (in DNA).** The carbons are assigned primed numbers.

The sugars of DNA and RNA are slightly different in structure. RNA's **ribose** sugar has a hydroxyl group attached to the $2'$-carbon atom, whereas DNA's sugar, called **deoxyribose,** has a hydrogen atom at this position and contains one oxygen atom fewer overall. This difference gives rise to the names ribonucleic acid (RNA) and *deoxy*ribonucleic acid (DNA). This minor chemical difference is recognized by all the cellular enzymes that interact with DNA or RNA, thus yielding specific functions for each nucleic acid. Further, the additional oxygen atom in the RNA nucleotide makes it more reactive and less chemically stable than DNA. For this reason, DNA is better suited to serve as the long-term repository of genetic information.

The second component of a nucleotide is its **nitrogenous base,** which may be of two types—a **purine** or a **pyrimidine** (◀ FIGURE 10.10). Each purine consists of a six-sided ring attached to a five-sided ring, whereas each pyrimidine consists of a six-sided ring only. DNA and RNA both contain two purines, **adenine** and **guanine** (A

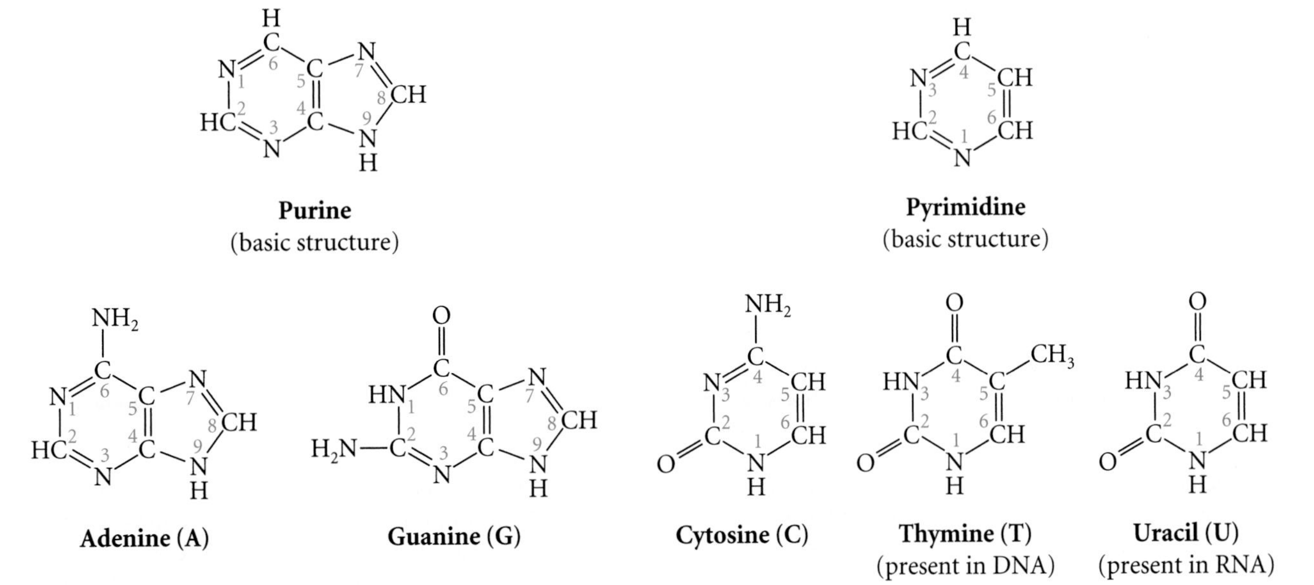

Purine
(basic structure)

Pyrimidine
(basic structure)

Adenine (A)   Guanine (G)   Cytosine (C)   Thymine (T)   Uracil (U)
                                                                (present in DNA)   (present in RNA)

◀ 10.10 **A nucleotide contains either a purine or a pyrimidine base.** The atoms of the rings in the bases are assigned unprimed numbers.

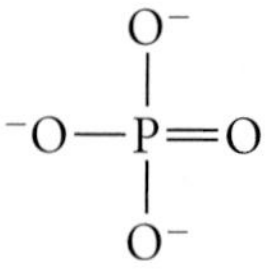

$$O^-$$
$$^-O—P{=}O$$
$$O^-$$

Phosphate

◀ **10.11 A nucleotide contains a phosphate group.**

and G), which differ in the positions of their double bonds and in the groups attached to the six-sided ring. There are three pyrimidines found in nucleic acids: **cytosine** (C), **thymine** (T), and **uracil** (U). Cytosine is present in both DNA and RNA; however, thymine is restricted to DNA, and uracil is found only in RNA. The three pyrimidines differ in the groups or atoms attached to the carbon atoms of the ring and in the number of double bonds in the ring. In a nucleotide, the nitrogenous base always forms a covalent bond with the 1′-carbon atom of the sugar (see Figure 10.9). A deoxyribose (or ribose) sugar and a base together are referred to as a **nucleoside.**

The third component of a nucleotide is the **phosphate group,** which consists of a phosphorus atom bonded to four oxygen atoms (◀ FIGURE 10.11). Phosphate groups are found in every nucleotide and frequently carry a negative charge, which makes DNA acidic. The phosphate is always bonded to the 5′-carbon atom of the sugar in a nucleotide (see Figure 10.9).

The DNA nucleotides are properly known as **deoxyribonucleotides** or deoxyribonucleoside 5′-monophosphates. Because there are four types of bases, there are four different kinds of DNA nucleotides (◀ FIGURE 10.12). The equivalent RNA nucleotides are termed **ribonucleotides** or ribonucleoside 5′-monophosphates. RNA molecules sometimes contain additional rare bases, which

are modified forms of the four common bases. These modified bases will be discussed in more detail when we examine the function of RNA molecules in Chapter 14.

Concepts

The primary structure of DNA consists of a string of nucleotides. Each nucleotide consists of a five-carbon sugar, a phosphate, and a base. There are two types of DNA bases: purines (adenine and guanine) and pyrimidines (thymine and cytosine).

Polynucleotide strands DNA is made up of many nucleotides connected by covalent bonds, which join the 5′-phosphate group of one nucleotide to the 3′-carbon atom of the next nucleotide (◀ FIGURE 10.13). These bonds, called **phosphodiester linkages,** are relatively strong covalent bonds; a series of nucleotides linked in this way constitutes a **polynucleotide strand.** The backbone of the polynucleotide strand is composed of alternating sugars and phosphates; the bases project away from the long axis of the strand. The negative charges of the phosphate groups are frequently neutralized by the association of positive charges on proteins, metals, or other molecules.

An important characteristic of the polynucleotide strand is its direction, or polarity. At one end of the strand a phosphate group is attached only to the 5′-carbon atom of the sugar in the nucleotide. This end of the strand is therefore referred to as the **5′ end.** The other end of the strand, referred to as the **3′ end,** has an OH group attached to the 3′-carbon atom of the sugar.

RNA nucleotides also are connected by phosphodiester linkages to form similar polynucleotide strands (see Figure 10.13).

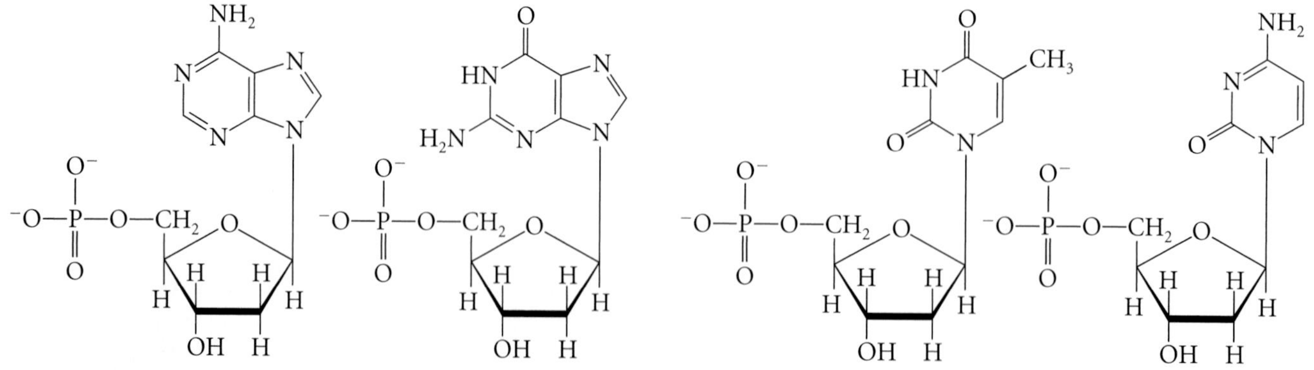

Deoxyadenosine 5′-monophosphate (dAMP)   **Deoxyguanosine 5′-monophosphate (dGMP)**   **Deoxythymidine 5′-monophosphate (dTMP)**   **Deoxycytidine 5′-monophosphate (dCMP)**

◀ 10.12 **There are four types of DNA nucleotides.**

DNA polynucleotide strand

RNA polynucleotide strand

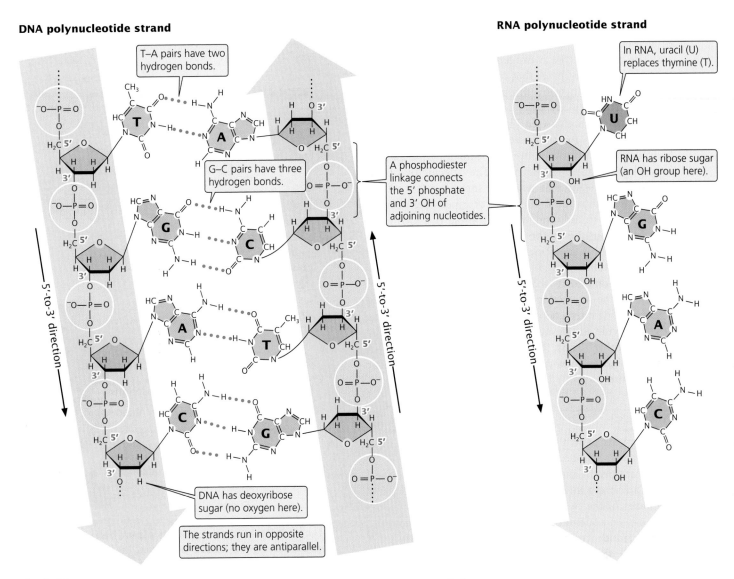

T–A pairs have two hydrogen bonds.

G–C pairs have three hydrogen bonds.

A phosphodiester linkage connects the 5' phosphate and 3' OH of adjoining nucleotides.

In RNA, uracil (U) replaces thymine (T).

RNA has ribose sugar (an OH group here).

DNA has deoxyribose sugar (no oxygen here).

The strands run in opposite directions; they are antiparallel.

5'-to-3' direction

10.13 DNA consists of two polynucleotide chains that are antiparallel and complementary, and RNA consists of a single nucleotide chain.

Concepts

The nucleotides of DNA are joined in polynucleotide strands by phosphodiester bonds that connect the 3' carbon atom of one nucleotide to the 5' phosphate group of the next. Each polynucleotide strand has polarity, with a 5' end and a 3' end.

Secondary Structures of DNA

The secondary structure of DNA refers to its three-dimensional configuration—its fundamental helical structure. DNA's secondary structure can assume a variety of configurations, depending on its base sequence and the conditions in which it is placed.

The double helix A fundamental characteristic of DNA's secondary structure is that it consists of two polynucleotide strands wound around each other—it's a double helix. The sugar–phosphate linkages are on the outside of the helix, and the bases are stacked in the interior of the molecule (see Figure 10.13). The two polynucleotide strands run in opposite directions—they are **antiparallel,** which means that the 5' end of one strand is opposite the 3' end of the second.

The strands are held together by two types of molecular forces. Hydrogen bonds link the bases on opposite strands (see Figure 10.13). These bonds are relatively weak compared with the covalent phosphodiester bonds that connect the sugar and phosphate groups of adjoining nucleotides. As we will see, several important functions of DNA require the separation of its two nucleotide strands, and this separation can be readily accomplished because of the relative ease of breaking and reestablishing the hydrogen bonds.

The nature of the hydrogen bond imposes a limitation on the types of bases that can pair. Adenine normally pairs only with thymine through two hydrogen bonds, and cytosine normally pairs only with guanine through three hydrogen bonds (see Figure 10.13). Because three hydrogen bonds form between C and G and only two hydrogen bonds form between A and T, C–G pairing is stronger than A–T pairing. The specificity of the base pairing means that wherever there is an A on one strand, there must be a T in the corresponding position on the other strand, and wherever there is a G on one strand, a C must be on the other. The two polynucleotide strands of a DNA molecule are therefore not identical but are **complementary.**

The second force that holds the two DNA strands together is the interaction between the stacked base pairs. These stacking interactions contribute to the stability of the DNA molecule and do not require that any particular base follow another. Thus, the base sequence of the DNA molecule is free to vary, allowing DNA to carry genetic information.

> ### Concepts
>
> DNA consists of two polynucleotide strands. The sugar–phosphate groups of each polynucleotide strand are on the outside of the molecule, and the bases are in the interior. Hydrogen bonding joins the bases of the two strands: guanine pairs with cytosine, and adenine pairs with thymine. The two polynucleotide strands of a DNA molecule are complementary and antiparallel.

Different secondary structures As we have seen, DNA normally consists of two polynucleotide strands that are antiparallel and complementary (exceptions are single-stranded DNA molecules in a few viruses). The precise three-dimensional shape of the molecule can vary, however, depending on the conditions in which the DNA is placed and, in some cases, on the base sequence itself.

The three-dimensional structure of DNA that Watson and Crick described is termed the **B-DNA** structure (◀ Figure 10.14). This structure exists when plenty of water surrounds the molecule and there is no unusual base sequence in the DNA—conditions that are likely to be present in cells. The B-DNA structure is the most stable configuration for a random sequence of nucleotides under physiological conditions, and most evidence suggests that it is the predominate structure in the cell.

B-DNA is an alpha helix, meaning that it has a right-handed, or clockwise, spiral. It possesses approximately 10 base pairs (bp) per 360-degree rotation of the helix; so

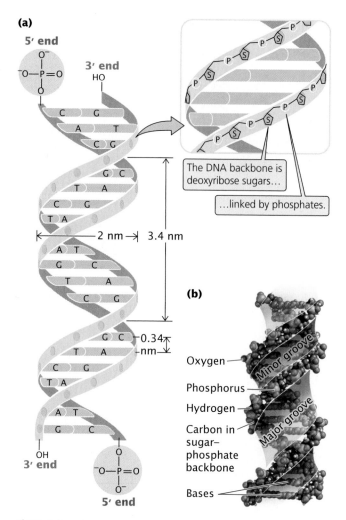

◀ 10.14 B-DNA consists of an alpha helix with approximately 10 bases per turn. (a) Diagrammatic representation showing that the bases are 0.34 nanometer (nm) apart, that each rotation encompasses 3.4 nm, and that the diameter of the helix is 2 nm. (b) Space-filling model of B-DNA showing major and minor grooves.

each base pair is twisted 36 degrees relative to the adjacent bases (see Figure 10.14a). The base pairs are 0.34 nanometer (nm) apart; so each complete rotation of the molecule encompasses 3.4 nm. The diameter of the helix is 2 nm, and the bases are perpendicular to the long axis of the DNA molecule. A space-filling model shows that B-DNA has a relatively slim and elongated structure (see Figure 10.14b). Spiraling of the nucleotide strands creates major and minor grooves in the helix, features that are important for the binding of some DNA-binding proteins that regulate the expression of genetic information (Chapter 16). Some characteristics of the B-DNA structure, along with characteristics of other secondary structures that exist under certain conditions or with unusual base sequences, are given in Table 10.2.

Table 10.2	Characteristics of DNA secondary structures		
Characteristic	**A-DNA**	**B-DNA**	**Z-DNA**
Conditions required to produce structure	75% H_2O	92% H_2O	Alternating purine and pyrimidine bases
Helix direction	Right-handed	Right-handed	Left-handed
Average base pairs per turn	11	10	12
Rotation per base pair	32.7°	36°	−30°
Distance between adjacent bases	0.26 nm	0.34 nm	0.37 nm
Diameter	2.3 nm	1.9 nm	1.8 nm
Overall shape	Short and wide	Long and narrow	Elongated and narrow

Note: Within each structure, the parameters may vary somewhat owing to local variation and method of analysis.

Another secondary structure that DNA can assume is the **A-DNA** structure, which exists when less water is present. Like B-DNA, A-DNA is an alpha (right-handed) helix (◀ FIGURE 10.15a), but it is shorter and wider than B-DNA (◀ FIGURE 10.15b) and its bases are tilted away from the main axis of the molecule. There is little evidence that A-DNA exists under physiological conditions.

A radically different secondary structure called **Z-DNA** (◀ FIGURE 10.15c) forms a left-handed helix. In this form, the sugar–phosphate backbones zigzag back and forth, giving rise to the name Z-DNA (for zigzag). Z-DNA structures can arise under physiological conditions when particular base sequences are present, such as stretches of alternating C and G sequences. Parts of some active genes form Z-DNA, suggesting that Z-DNA may play a role in regulating gene transcription.

Other secondary structures may exist under special conditions or with special base sequences, and characteristics of some of these structures are given in Table 10.2. Structures other than B-DNA exist rarely, if ever, within cells.

Local variation in secondary structures   DNA is frequently presented as a static, rigid structure that is invariant in its secondary structure. In reality, the numbers describing the parameters for B-DNA in Figure 10.14 are average values, and the actual measurements vary slightly from one part of the molecule to another. The twist between base pairs within a single molecule of B-DNA, for example, can vary from 27 degrees to as high as 42 degrees. This **local variation** in DNA structure arises because of differences in local environmental conditions, such as the presence of proteins, metals, and ions that may bind to the DNA. The base sequence also influences DNA structure locally.

28Å

(a) A form   (b) B form   (c) Z form

◀ 10.15 **DNA can assume several different secondary structures.** These structures depend on the base sequence of the DNA and the conditions under which it is placed.

Concepts

DNA can assume different secondary structures, depending on the conditions in which it is placed and on its base sequence. B-DNA is thought to be the most common configuration in the cell. Local variation in DNA arises as a result of environmental factors and base sequence.

www.whfreeman.com/pierce   More on DNA structure and some interesting images of DNA

Connecting Concepts

Genetic Implications of DNA Structure

After Oswald Avery and his colleagues demonstrated that the transforming principle is DNA, it was clear that the genotype resides within the chemical structure of DNA. Watson and Crick's great contribution was their elucidation of the genotype's chemical structure, making it possible for geneticists to begin to examine genes directly, instead of looking only at the phenotypic consequences of gene action. Determining the structure of DNA permitted the birth of molecular genetics—the study of the chemical and molecular nature of genetic information.

Watson and Crick's structure did more than just create the potential for molecular genetic studies; it was an immediate source of insight into key genetic processes. At the beginning of this chapter, three fundamental properties of the genetic material were identified. First, it must be capable of carrying large amounts of information; so it must vary in structure. Watson and Crick's model suggested that genetic instructions are encoded in the base sequence, the only variable part of the molecule. The sequence of the four bases—adenine, guanine, cytosine, and thymine—along the helix encodes the information that ultimately determines the phenotype. Watson and Crick were not sure *how* the base sequence of DNA determined the phenotype, but their structure clearly indicated that the genetic instructions were encoded in the bases.

A second necessary property of genetic material is its ability to replicate faithfully. The complementary polynucleotide strands of DNA make this replication possible. Watson and Crick wrote, "It has not escaped our attention that the specific base pairing we have postulated immediately suggests a possible copying mechanism for the genetic material." They proposed that, in replication, the two polynucleotide strands unzip, breaking the weak hydrogen bonds between the two strands, and each strand serves as a template on which a new strand is synthesized. The specificity of the base pairing means that only one possible sequence of bases—the complementary sequence—can be synthesized from each template. Newly replicated double-stranded DNA molecules will therefore be identical with the original double-stranded DNA molecule (see Chapter 12 on DNA replication).

The third essential property of genetic material is the ability to translate its instructions into the phenotype. For most traits, the immediate phenotype is production of a protein; so the genetic material must be capable of encoding proteins. Proteins, like DNA, are polymers, but their repeating units are amino acids, not nucleotides. A protein's function depends on its amino acid sequence; so the genetic material must be able to specify that sequence in a form that can be transferred in the course of protein synthesis.

DNA expresses its genetic instructions by first transferring its information to an RNA molecule, in a process termed **transcription** (see Chapter 13). The term *transcription* is appropriate because, although the information is transferred from DNA to RNA, the information remains in the language of nucleic acids. The RNA molecule then transfers the genetic information to a protein by specifying its amino acid sequence. This process is termed **translation** (see Chapter 15) because the information must be *translated* from the language of nucleotides into the language of amino acids.

We can now identify three major pathways of information flow in the cell (◀ Figure 10.16): in **replication**, information passes from one DNA molecule to other DNA molecules; in transcription, information passes from DNA to RNA; and, in translation, information passes from RNA to protein. This concept of information flow was formalized by Francis Crick in a concept that he called the

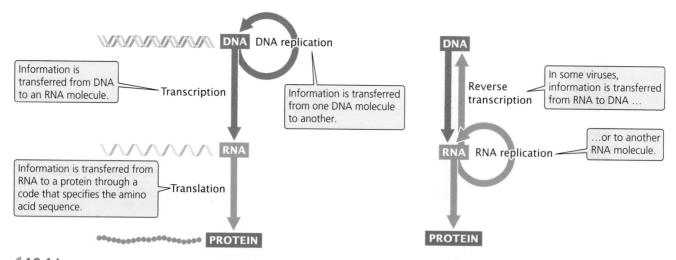

◀10.16 **The three major pathways of information transfer within the cell are DNA replication, transcription, and translation.**

central **dogma** of molecular biology. The central dogma states that genetic information passes from DNA to protein in a one-way information pathway. It indicates that genotype codes for phenotype but phenotype cannot code for genotype. We now realize, however, that the central dogma is an oversimplification. In addition to the three general information pathways of replication, transcription, and translation, other transfers may take place in certain organisms or under special circumstances, including the transfer of information from RNA to DNA, (in **reverse transcription**) and the transfer of information from RNA to RNA (in **RNA replication;** see Figure 10.16). Reverse transcription takes place in retroviruses and in some transposable elements; RNA replication takes place in some RNA viruses (see Chapter 8).

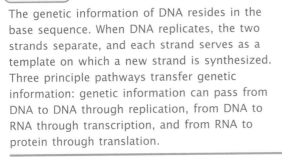

Concepts

The genetic information of DNA resides in the base sequence. When DNA replicates, the two strands separate, and each strand serves as a template on which a new strand is synthesized. Three principle pathways transfer genetic information: genetic information can pass from DNA to DNA through replication, from DNA to RNA through transcription, and from RNA to protein through translation.

www.whfreeman.com/pierce   More information on the central dogma

Special Structures in DNA and RNA

In double-stranded DNA, the pairing of bases on opposite nucleotide strands provides stability and produces the helical secondary structure of the molecule. Single-stranded DNA and RNA (the latter of which is almost always single stranded) lack the stabilizing influence of the paired nucleotide strands; so they exhibit no common secondary structure. Sequences *within* a single strand of nucleotides may be complementary to each other and can pair by forming hydrogen bonds, producing double-stranded regions (◀FIGURE 10.17). This internal base pairing imparts a secondary structure to a single-stranded molecule. In fact, internal base pairing within single strands of nucleotides can result in a great variety of secondary structures.

One common type of secondary structure found in single strands of nucleotides is a **hairpin,** which forms when sequences of nucleotides on the same strand are inverted complements. The sequence 5′ TGCGAT 3′ and 5′-ATCGCA-3′ are examples of inverted complements and, when these sequences are on the same nucleotide strand, they can pair

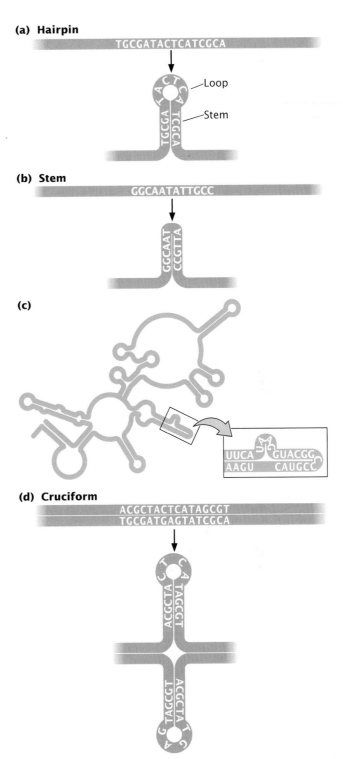

◀10.17 **Both DNA and RNA can form special secondary structures.** (a) A hairpin, consisting of a region of paired bases (which forms the stem) and a region of unpaired bases between the complementary sequences (which form a loop at the end of the stem). (b) A stem with no loop. (c) Secondary structure of RNA component of RNase P of *E. coli.* RNA molecules often have complex secondary structures. (d) A cruciform structure

and form a hairpin (see Figure 10.17a). A hairpin consists of a region of paired bases (the stem) and sometimes includes intervening unpaired bases (the loop). When the complementary sequences are contiguous, the hairpin has a stem but no loop (see Figure 10.17b). Hairpins frequently control aspects of information transfer. RNA molecules may contain numerous hairpins, allowing them to fold up into complex structures (see Figure 10.17c).

In double-stranded DNA, sequences that are inverted replicas of each other are called **inverted repeats.** The following double-stranded sequence is an example of inverted repeats:

$$5'-AAAG \ldots CTTT-3'$$
$$3'-TTTC \ldots GAAA-5'$$

Notice that the sequences on the two strands are the same when read from 5' to 3' but, because the polarities of the two strands are opposite, their sequences are reversed from left to right. An inverted repeat that is complementary to itself, such as:

$$5'-ATCGAT-3'$$
$$3'-TAGCTA-5'$$

is also a **palindrome,** defined as a word or sentence that reads the same forward and backward, such as "rotator." Inverted repeats are palindromes because the sequences on the two strands are the same but in reverse orientation. When an inverted repeat forms a perfect palindrome, the double-stranded sequence reads the same forward and backward.

Another secondary structured, called a **cruciform,** can be made from an inverted repeat when a hairpin forms within each of the two single-stranded sequences. (see Figure 10.17d).

> **Concepts**
>
> In DNA and RNA, base pairing between nucleotides on the same strand produces special secondary structures such as hairpins and cruciforms.

DNA Methylation

The primary structure of DNA can be modified in various ways. These modifications are important in the expression of the genetic material, as we will see in the chapters to come. One such modification is **DNA methylation,** a process in which methyl groups ($-CH_3$) are added (by specific enzymes) to certain positions on the nucleotide bases.

In bacteria, adenine and cytosine are commonly methylated, whereas, in eukaryotes, cytosine is the most

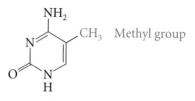

5-Methylcytosine

◀ **10.18 In eukaryotic DNA, cytosine bases are often methylated to form 5-methylcytosine.**

commonly methylated base. Bacterial DNA is frequently methylated to distinguish it from foreign, unmethylated DNA that may be introduced by viruses; bacteria use proteins called restriction enzymes to cut up any unmethylated viral DNA (see Chapter 18).

In eukaryotic DNA, cytosine bases are often methylated to form **5-methylcytosine** (◀ FIGURE 10.18). The extent of cytosine methylation varies; in most animal cells, about 5% of the cytosine bases are methylated, but more than 50% of the cytosine bases in some plants are methylated. On the other hand, no methylation of cytosine has been detected in yeast cells, and only very low levels of methylation (about 1 methylated cytosine base per 12,500 nucleotides) are found in *Drosophila.* Why eukaryotic organisms differ so widely in their degree of methylation is not clear.

Methylation is most frequent on cytosine nucleotides that sit next to guanine nucleotides on the same strand:

$$\ldots GC \ldots$$
$$\ldots CG \ldots$$

In eukaryotic cells, methylation is often related to gene expression. Sequences that are methylated typically show low levels of transcription while sequences lacking methylation are actively being transcribed (see Chapter 16). Methylation can also affect the three-dimensional structure of the DNA molecule.

> **Concepts**
>
> Methyl groups may be added to certain bases in DNA, depending on their positions in the molecule. Both prokaryotic and eukaryotic DNA can be methylated. In eukaryotes, cytosine bases are most often methylated to form 5-methylcytosine, and methylation is often related to gene expression.

www.whfreeman.com/pierce   The latest on DNA methylation

Bends in DNA

Some specific base sequences—such as a series of four or more adenine–thymine base pairs—cause the DNA double helix to bend. Bending affects how the DNA binds to certain proteins and may be important in controlling the transcription of some genes.

The DNA helix can also be made to bend by the binding of proteins to specific DNA sequences (◀FIGURE 10.19). The SRY protein, which is encoded by a Y-linked gene and is responsible for sex determination in mammals (see Chapter 4), binds to certain DNA sequences (along the minor groove) and activates nearby genes that encode male traits. When the SRY protein grips the DNA, it bends the molecule about 80 degrees. This distortion of the DNA helix apparently facilitates the binding of other proteins that activate the transcription of genes that encode male characteristics.

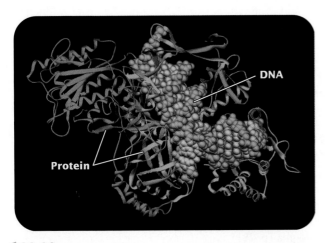

◀10.19 **The DNA helix can be bent by the binding of proteins to the DNA molecule.** The protein shown here is the Ku heterodimer, which helps carry out DNA repair.

Connecting Concepts Across Chapters

This chapter has shifted the focus of our study to molecular genetics. The first nine chapters of this book examined various aspects of transmission genetics. In these chapters, the focus was on the individual: which phenotype was produced by an individual genotype, how the genes of an individual were transmitted to the next generation, and what types of offspring were produced when two individuals were crossed. In molecular genetics, our focus now shifts to genes: how they are encoded in DNA, how they are replicated, and how they are expressed.

Much of what follows in this book will depend on your knowledge of DNA. An understanding of all the major processes of information transfer—replication, transcription, and translation—requires an understanding of nucleic acid structure; discussions of recombinant DNA, mutation, gene expression, cancer genetics, and even population genetics are based on the assumption that you understand the basic structure and function of DNA. Thus the information in this chapter provides a critical foundation for much of the remainder of the book.

In this chapter, the history of how DNA's structure and function were unraveled has been strongly emphasized, because the DNA story illustrates how pivotal scientific discoveries are often made. No one scientist discovered the structure of DNA; rather, numerous persons, over a long period of time, made important contributions to our understanding of its structure. Watson and Crick's proposal for DNA's double-helical structure stands out as a singularly important contribution, because it combined many known facts about the structure into a new model that allowed important inferences about the fundamental nature of genes. The DNA story also illustrates the important lesson that science is a human enterprise, influenced by personalities, relations, and motivation.

CONCEPTS SUMMARY

- Genetic material must contain complex information, be replicated accurately, and have the capacity to be translated into the phenotype.
- Evidence that DNA is the source of genetic information came from the finding by Avery, MacLeod, and McCarty that transformation—the genetic alteration of bacteria—was dependent on DNA and from the demonstration by Hershey and Chase that viral DNA is passed on to progeny phages. The results of experiments with tobacco mosaic virus showed that RNA carries genetic information in some viruses.
- James Watson and Francis Crick proposed a new model for the three-dimensional structure of DNA in 1953.
- A DNA nucleotide consists of a deoxyribose sugar, a phosphate group, and a nitrogenous base. RNA consists of a ribose sugar, a phosphate group, and a nitrogenous base.
- The bases of a DNA nucleotide are of two types: purines (adenine and guanine) and pyrimidines (cytosine and thymine). RNA contains the pyrimidine uracil instead of thymine.
- Nucleotides are joined by phosphodiester linkages in a polynucleotide strand. Each polynucleotide strand has a 5′ end with a phosphate and a 3′ end with a hydroxyl group.
- DNA consists of two nucleotide strands that wind around each other to form a double helix. The sugars and phosphates lie on the outside of the helix, and the bases are stacked in

the interior. Bases from the two strands are joined by hydrogen bonding. The two strands are antiparallel and complementary.

- DNA molecules can form a number of different secondary structures, depending on the conditions in which the DNA is placed and on its base sequence. B-DNA, which consists of a right-handed helix with approximately 10 bases per turn, is the most common form of DNA in cells.

- The structure of DNA has several important genetic implications. Genetic information resides in the base sequence of DNA, which ultimately specifies the amino acid sequence of proteins. Complementarity of the bases on DNA's two strands allows genetic information to be replicated.

- Important pathways by which information passes from DNA to other molecules include: (1) replication, in which one molecule of DNA serves as a template for the synthesis of two new DNA molecules; (2) transcription, in which DNA serves as a template for the synthesis of an RNA molecule; and (3) translation, in which RNA codes for protein.

- The central dogma of molecular biology proposes that information flows in a one-way direction, from DNA to RNA to protein. Clear exceptions to the central dogma are now known.

- Pairing between bases on the same nucleotide strand can lead to hairpins and other secondary structures. Inverted repeats are sequences on the same strand that are inverted and complementary; they can lead to cruciform structures.

- DNA methylation is the addition of methyl groups to the nucleotide bases. In bacteria, adenine and cytosine are commonly methylated. Among eukaryotes, cytosine bases are most commonly methylated to form 5-methylcytosine.

- Some sequences, such as a series of four or more adenine–thymine base pairs, can cause DNA to bend, which may affect gene expression.

IMPORTANT TERMS

nucleotide (p. 268)
Chargaff's rules (p. 268)
transforming principle (p. 270)
isotopes (p. 271)
X-ray diffraction (p. 273)
ribose (p. 275)
deoxyribose (p. 275)
nitrogenous base (p. 275)
purine (p. 275)
pyrimidine (p. 275)
adenine (A) (p. 275)

guanine (G) (p. 275)
cytosine (C) (p. 276)
thymine (T) (p. 276)
uracil (U) (p. 276)
nucleoside (p. 276)
phosphate group (p. 276)
deoxyribonucleotide (p. 276)
ribonucleotide (p. 276)
phosphodiester linkage (p. 276)
polynucleotide strand (p. 276)
5′ end (p. 276)

3′ end (p. 276)
antiparallel (p. 277)
complementary (p. 278)
B-DNA (p. 278)
A-DNA (p. 279)
Z-DNA (p. 279)
local variation (p. 279)
transcription (p. 280)
translation (p. 280)
replication (p. 280)
central dogma (p. 281)

reverse transcription (p. 281)
RNA replication (p. 281)
hairpin (p. 281)
inverted repeats (p. 282)
palindrome (p. 282)
cruciform (p. 282)
DNA methylation (p. 282)
5-methylcytosine (p. 282)

Worked Problems

1. The percentage of cytosine in a double-stranded DNA molecule is 40%. What is the percentage of thymine?

- **Solution**

In double-stranded DNA, A pairs with T, whereas G pairs with C; so the percentage of A equals the percentage of T, and the percentage of G equals the percentage of C. If C = 40%, then G also must be 40%. The total percentage of C + G is therefore 40% + 40% = 80%. All the remaining bases must be either A or T; so the total percentage of A + T = 100% − 80% = 20%; because the percentage of A equals the percentage of T, the percentage of T is $^{20\%}/_2$ = 10%.

2. Which of the following relations will be true for the percentage of bases in double-stranded DNA?

(a) C + T = A + G   (b) $\dfrac{C}{A} = \dfrac{T}{G}$

- **Solution**

An easy way to determine whether the relations are true is to arbitrarily assign percentages to the bases, remembering that, in double-stranded DNA, A = T and G = C. For example, if the percentages of A and T are each 30%, then the percentages of G and C are each 20%. We can substitute these values into the equations to see if the relations are true.

(a) 20 + 30 = 30 + 20, so this relation is true.

(b) $\dfrac{20}{30} \neq \dfrac{30}{20}$; so this relation is not true.

The New Genetics
MINING GENOMES

INTRODUCTION TO GENBANK AND PUBMED

This exercise introduces you to some of the genetics databases that are most frequently used by contemporary researchers. You will explore some of the tools available at the National Center for Biotechology Information (NCBI), which is managed by the National Library of Medicine of the United States.

COMPREHENSION QUESTIONS

* 1. What three general characteristics must the genetic material possess?

  2. Briefly outline the history of our knowledge of the structure of DNA until the time of Watson and Crick. Which do you think were the principle contributions and developments?

* 3. What experiments demonstrated that DNA is the genetic material?

  4. What is transformation? How did Avery and his colleagues demonstrate that the transforming principle is DNA?

* 5. How did Hershey and Chase show that DNA is passed to new phages in phage reproduction?

  6. Why was Watson and Crick's discovery so important?

* 7. Draw and label the three parts of a DNA nucleotide.

  8. How does an RNA nucleotide differ from a DNA nucleotide?

  9. How does a purine differ from a pyrimidine? What purines and pyrimidines are found in DNA and RNA?

*10. Draw a short segment of a single polynucleotide strand, including at least three nucleotides. Indicate the polarity of the strand by labeling the 5′ end and the 3′ end.

 11. Which bases are capable of forming hydrogen bonds with each other?

*12. What is local variation in DNA structure and what causes it?

 13. What are some of the important genetic implications of the DNA structure?

*14. What are the major transfers of genetic information?

 15. What are hairpins and how do they form?

 16. What is DNA methylation?

APPLICATION QUESTIONS AND PROBLEMS

 17. A student mixes some heat-killed type IIS *Streptococcus pneumonia* bacteria with live type IIR bacteria and injects the mixture into a mouse. The mouse develops pneumonia and dies. The student recovers some type IIS bacteria from the dead mouse. It is the only experiment conducted by the student. Has the student demonstrated that transformation has taken place? What other explanations might explain the presence of the type IIS bacteria in the dead mouse?

*18. (a) Why did Hershey and Chase choose ^{32}P and ^{35}S for use in their experiment? (b) Could they have used radioactive isotopes of carbon (C) and oxygen (O) instead? Why or why not?

 19. What results would you expect if the Hershey and Chase experiment were conducted on tobacco mosaic virus?

*20. Each nucleotide pair of a DNA double helix weighs about 1×10^{-21} g. The human body contains approximately 0.5 g of DNA. How many nucleotide pairs of DNA are in the human body? If you assume that all the DNA in human cells is in the B-DNA form, how far would the DNA reach if stretched end to end?

 21. What aspects of its structure contribute to the stability of the DNA molecule? Why is RNA less stable than DNA?

*22. Which of the following relations will be found in the percentages of bases of a double-stranded DNA molecule?

  (a) $A + T = G + C$ (d) $\dfrac{A + T}{C + G} = 1.0$ (g) $\dfrac{A}{G} = \dfrac{T}{C}$

  (b) $A + G = T + C$ (e) $\dfrac{A + G}{C + T} = 1.0$ (h) $\dfrac{A}{T} = \dfrac{G}{C}$

  (c) $A + C = G + T$ (f) $\dfrac{A}{C} = \dfrac{G}{T}$

*23. If a double-stranded DNA molecule is 15% thymine, what are the percentages of all the other bases?

 24. A virus contains 10% adenine, 24% thymine, 30% guanine, and 36% cytosine. Is the genetic material in this virus double-stranded DNA, single-stranded DNA, double-stranded RNA, or single-stranded RNA? Support your answer.

*25. A B-DNA molecule has 1 million nucleotide pairs.

  (a) How many complete turns are there in this molecule?

  (b) If this same molecule were in the Z-DNA configuration, how many complete turns would it have?

26. For entertainment on a Friday night, a genetics professor proposed that his children diagram a polynucleotide strand of DNA. Having learned about DNA in preschool, his 5-year-old daughter was able to draw a polynucleotide strand, but she made a few mistakes. The daughter's diagram (represented here) contained at least 10 mistakes.

    (a) Make a list of all the mistakes in the structure of this DNA polynucleotide strand.

    (b) Draw the correct structure for the polynucleotide strand.

*27. Chapter 1 considered the theory of the inheritance of acquired characteristics and noted that this theory is no longer accepted. Is the central dogma consistent with the theory of the inheritance of acquired characteristics? Why or why not?

28. Write a sequence of bases in an RNA molecule that would produce a hairpin structure.

29. The following sequence is present in one strand of a DNA molecule:

$$5'-CATTGACCGA-3'$$

Write the sequence on the same strand that produces an inverted repeat and the sequence on the complementary strand.

CHALLENGE QUESTIONS

30. Suppose that an automated, unmanned probe is sent into deep space to search for extraterrestrial life. After wandering for many light-years among the far reaches of the universe, this probe arrives on a distant planet and detects life. The chemical composition of life on this planet is completely different from that of life on Earth, and its genetic material is not composed of nucleic acids. What predictions can you make about the chemical properties of the genetic material on this planet?

31. How might ^{32}P and ^{35}S be used to demonstrate that the transforming principle is DNA? Briefly outline an experiment that would show that DNA and not protein is the transforming principle.

32. Scientists have reportedly isolated short fragments of DNA from fossilized dinosaur bones hundreds of millions of years old. The technique used to isolate this DNA is the polymerase chain reaction (PCR), which is capable of amplifying very small amounts of DNA a millionfold (see Chapter 18). Critics have claimed that the DNA isolated from dinosaur bones is not of ancient origin but instead represents contamination of the samples with DNA from present-day organisms such as bacteria, mold, or humans. What precautions, analyses, and control experiments could be carried out to ensure that DNA recovered from fossils is truly of ancient origin?

SUGGESTED READINGS

Avery, O. T., C. M. MacLeod, and M. McCarty. 1944. Studies on the chemical nature of the substance inducing transformation of pneumococcal types. *Journal of Experimental Medicine* 79:137–158.
    Avery, MacLeod, and McCarty's paper describing their demonstration that the transforming principle is DNA.

Crick, F. 1988. *What Mad Pursuit: A Personal View of Scientific Discovery.* New York: Basic Books.

Francis Crick's personal account of the discovery of the structure of DNA.

Dickerson, R. E., H. R. Drew, B. N. Conner, R. M Wing, A. V. Fratini, and M. L. Kopka. 1982. The anatomy of A-, B-, and Z-DNA. *Science* 216:475–485.
    A review of differences in secondary structures of DNA.

Fraenkal-Conrat, H., and B. Singer. 1957. Virus reconstitution II: combination of protein and nucleic acid from different strains.

Biochimica et Biophysica Acta 24:540–548.
Report of Fraenkal-Conrat and Singer's well-known experiment showing that RNA is the genetic material in tobacco mosaic virus.

Griffith, F. 1928. The significance of pneumoncoccal types. *Journal of Hygiene* 27:113–159.
Griffith's original report of the transforming principle.

Handt, O., M. Richards, M. Trommsdorff, et al. 1994. Molecular genetic analysis of the Tyrolean Ice Man. *Science* 264:1775–1778.
Describes the isolation and analysis of DNA from a 5000-year-old frozen man found on a glacier in the Alps.

Hershey, A. D., and M. Chase. 1952. Independent functions of viral protein and nucleic acid in growth of bacteriophage. *Journal of General Physiology* 36:39–56.
Original report of Hershey and Chase's well-known experiment with T2 bacteriophage.

Judson, H. F. 1996. *The Eighth Day of Creation: Makers of the Revolution in Biology,* expanded edition. Cold Spring Harbor, NY: Cold Spring Harbor Laboratory Press.
A comprehensive account of the early years of molecular genetics.

Miescher, F. 1871. On the chemical composition of pus cells. *Hoppe-Seyler's Med.-Chem. Untersuch.* 4:441–460. Abridged and translated in *Great Experiments in Biology,* M. L. Gabriel, and S. Fogel (Eds.). Englewood Cliffs, NJ: Prentice-Hall, 1955.
An abridged and translated version of Miescher's original paper chemically characterizing DNA.

Mirsky, A. E. 1968 The discovery of DNA. *Scientific American* 218 (6):78–88.
A good account of the discovery of DNA structure.

Rich, A., A. Nordheim, and A. H.-J. Wang. 1984. The chemistry and biology of left-handed Z-DNA. *Annual Review of Biochemistry* 53:791–846.
Good review article on the structure and possible function of Z-DNA.

Watson, J. D. 1968. *The Double Helix.* New York: Atheneum.
An excellent account of Watson and Crick's discovery of DNA.

Watson, J. D., and F. C. Crick. 1953. Molecular structure of nucleic acids: a structure for deoxyribose nucleic acids. *Nature* 171:737–738.
Original paper in which Watson and Crick first presented their new structure for DNA.

Zimmerman, S. B. 1982. The three-dimensional structure of DNA. *Annual Review of Biochemistry* 51:395–427.
Review of the different secondary structures that DNA can assume.

11

Chromosome Structure and Transposable Elements

The house mouse, *Mus musculus,* is one of the oldest and most important organisms used for genetic studies. Molecular techniques allow genes to be introduced into mice on yeast artificial chromosomes (YACs). (Carolyn A. McKeone/Photo Researchers.)

YACs and the Common Mouse

The common house mouse, *Mus musculus,* is among the oldest and most valuable subjects for genetic study. It's an excellent genetic organism—small, prolific, and easy to keep, with a short generation time (about 3 months). It tolerates inbreeding well; so a large number of inbred strains have been developed through the years. Finally, being a mammal, the mouse is genetically and physiologically more similar to humans than are other organisms used in genetics studies, such as bacteria, yeast, corn, and fruit flies.

Powerful tools of molecular biology have enhanced the mouse's role in probing fundamental questions of heredity. New and altered genes can be added to the mouse genome

by injecting DNA directly into embryos that are implanted into surrogate mothers. The resulting **transgenic mice** can be bred to produce offspring carrying the new genes.

Today, it is possible to introduce not just individual genes, but entire chromosomes into mouse cells. In 1983, the first artificial chromosomes, made of parts culled from yeast and protozoans, were created for studying chromosome structure and segregation. In 1987, David Burke and Maynard Olson (at Washington University, St. Louis) used yeast to create much larger artificial chromosomes called yeast artificial chromosomes or YACs. Each YAC includes the three essential elements of a chromosome: a centromere, a pair of telomeres, and an origin of replication. These elements ensure that artificial chromosomes will segregate in mitosis and meiosis, will not be degraded, and will replicate successfully. Large chunks of extra DNA from any source can be added to a YAC, and the new artificial chromosome can be inserted into a cell. Eukaryotic centromeres, telomeres, and origins of replication are similar in different organisms; so YACs function well in almost any eukaryotic cell.

In 1993, molecular geneticists successfully modified YACs so that they could be transferred to mouse cells. Previously, transgenic mice could carry only relatively small pieces of DNA, usually no more than 50,000 bp. Now, large genes as well as the surrounding DNA, which may be important in the regulation of those genes, can be added to mouse-cell nuclei. Artificial chromosomes have also been made from chromosomal components of bacteria (BACs) and mammals (MACs).

The successful construction of YACs, BACs, and MACs illustrates the fundamental nature of eukaryotic chromosomes: huge amounts of DNA complexed with proteins and possessing telomeres, centromeres, and origins of replication. In this chapter, we explore the molecular nature of chromosomes, including details of the DNA–protein complex and the structure of telomeres and centromeres; origins of replication will be discussed in Chapter 12.

Much of this chapter focuses on a storage problem: how to cram tremendous amounts of DNA into the limited confines of a cell. Even in those organisms having the smallest amounts of DNA, the length of genetic material far exceeds the length of the cell. Thus, cellular DNA must be highly folded and tightly packed, but this packing creates problems—it renders the DNA inaccessible, unable to be copied or read. Functional DNA must be capable of partly unfolding and expanding so that individual genes can undergo replication and transcription. The flexible, dynamic nature of DNA packing will be a central theme of this chapter.

We begin this chapter by considering supercoiling, an important tertiary structure of DNA found in both prokaryotic and eukaryotic cells. After a brief look at the bacterial chromosome, we examine the structure of eukaryotic chromosomes. After considering chromosome structure, we pay special attention to the working parts of a chromosome, specifically centromeres and telomeres. We also consider the types of DNA sequences present in many eukaryotic chromosomes and how DNA sequences are analyzed.

The second part of this chapter focuses on genes that move. For many years, biologists viewed genes as static entities that occupied fixed positions on chromosomes. But we now recognize that many genetic elements do not occupy fixed positions. Genes that can move have been given a variety of names, including transposons, transposable genetic elements, mobile DNA, movable genes, controlling elements, and jumping genes. We will refer to mobile DNA sequences as **transposable elements,** and by this term we mean any DNA sequence that is capable of moving from one place to another place within the genome.

We begin the second part of the chapter by outlining some of the general features of transposable elements and the processes by which they move from place to place. We then consider several different types of transposable elements found in prokaryotic and eukaryotic genomes. Finally, we consider the evolutionary significance of transposable elements.

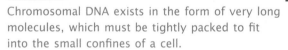 More information about YACs, how genes are cloned into YACs, and mouse genetics

Packing DNA into Small Spaces

The packaging of tremendous amounts of genetic information into the small volume of a cell has been called the ultimate storage problem. Consider the chromosome of the bacterium *E. coli*, a single molecule of DNA with approximately 4.64 million base pairs. Stretched out straight, this DNA would be about 1000 times as long as the cell within which it resides. ◀FIGURE 11.1, encircling pp. 290–291, suggests the molecular shape and length. Human cells contain 6 billion base pairs of DNA, which would measure some 1.8 meters stretched end to end. Even DNA in the smallest human chromosome would stretch 14,000 times the length of the nucleus. Clearly, DNA molecules must be tightly packed to fit into such small spaces.

The structure of DNA can be considered at three hierarchical levels: the primary structure of DNA is its nucleotide sequence; the secondary structure is the double-stranded helix; and the tertiary structure refers to higher-order folding that allows DNA to be packed into the confined space of a cell.

> **Concepts**
>
> Chromosomal DNA exists in the form of very long molecules, which must be tightly packed to fit into the small confines of a cell.

One type of DNA tertiary structure is **supercoiling,** which occurs when the DNA helix is subjected to strain by being overwound or underwound. The lowest energy state for B-DNA is when it has approximately 10 bp per turn of its helix. In this **relaxed state,** a stretch of 100 bp of DNA would assume about 10 complete turns (◀FIGURE 11.2a). If energy is used to add or remove any turns by rotating one strand around the other, strain is placed on the molecule, causing the helix to supercoil, or twist, on itself (◀FIGURE 11.2b and c).

Supercoiling is a natural consequence of the overrotating or underrotating of the helix; it occurs only when the molecule is placed under strain. Molecules that are overrotated exhibit **positive supercoiling** (see Figure 11.2b). Underrotated molecules exhibit **negative supercoiling** (see Figure 11.2c), in which the direction of the supercoil is opposite that of the right-handed coil of the DNA helix.

E. coli bacterium

Bacterial chromosome

◀11.1 The DNA in *E. coli* is about 1000 times as long as the cell itself.

Supercoiling occurs only if the two polynucleotide strands of the DNA double helix are unable to rotate about each other freely. If the chains *can* turn freely, their ends will simply turn as extra rotations are added or removed, and the molecule will spontaneously revert to the relaxed state. Supercoiling takes place when the strain of overrotating or underrotating cannot be compensated for by the turning of the ends of the double helix, which is the case if the DNA is circular—that is, there are no free ends. Some viral chromosomes are in the form of simple circles and readily undergo supercoiling. Large molecules of bacterial DNA are typically a series of large loops, the ends of which are held together by proteins. Eukaryotic DNA is normally linear but also tends to fold into loops stabilized by proteins. In these chromosomes, the anchoring proteins prevent free rotation of the ends of the DNA; so supercoiling does take place.

Supercoiling relies on **topoisomerases,** enzymes that add or remove rotations from the DNA helix by temporarily breaking the nucleotide strands, rotating the ends around each other, and then rejoining the broken ends. The two classes of topoisomerases are: type I, which breaks only one of the nucleotide strands and reduces supercoiling by removing rotations; and type II, which adds or removes rotations by breaking both nucleotide strands.

Most DNA found in cells is negatively supercoiled, which has two advantages for the cell. First, supercoiling makes the separation of the two strands of DNA easier during replication and transcription. Negatively supercoiled DNA is underrotated; so separation of the two strands during replication and transcription is more rapid and requires less energy. Second, supercoiled DNA can be packed into a smaller space because it occupies less volume than relaxed DNA.

Concepts

Overrotation or underrotation of a DNA double helix places strain on the molecule, causing it to supercoil. Supercoiling is controlled by topoisomerase enzymes. Most cellular DNA is negatively supercoiled, which eases the separation of nucleotide strands during replication and transcription and allows DNA to be packed into small spaces.

The Bacterial Chromosome

Most bacterial genomes consist of a single, circular DNA molecule, although linear DNA molecules have been found in a few species. In circular bacterial chromosomes, the

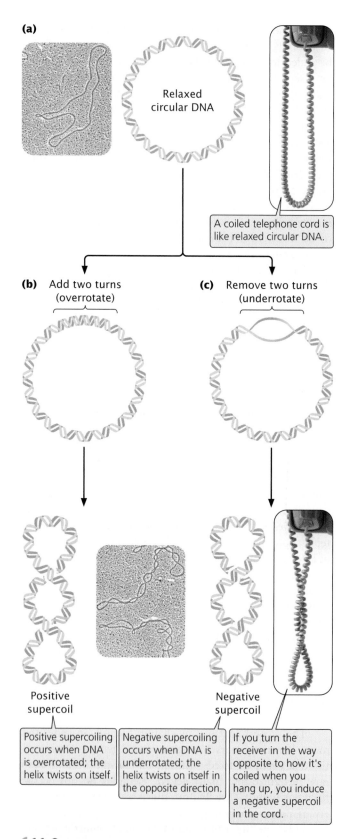

(a)

Relaxed circular DNA

A coiled telephone cord is like relaxed circular DNA.

(b) Add two turns (overrotate)

(c) Remove two turns (underrotate)

Positive supercoil

Negative supercoil

Positive supercoiling occurs when DNA is overrotated; the helix twists on itself.

Negative supercoiling occurs when DNA is underrotated; the helix twists on itself in the opposite direction.

If you turn the receiver in the way opposite to how it's coiled when you hang up, you induce a negative supercoil in the cord.

◀11.2 **Supercoiled DNA is overwound or underwound, causing it to twist on itself.** Electron micrographs are of relaxed DNA (top) and supercoiled DNA (bottom). (Dr. Gopal Murti/Phototake.)

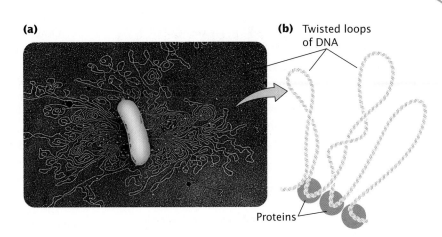

(a)

(b) Twisted loops of DNA

Proteins

◀11.3 **Bacterial DNA is highly folded into a series of twisted loops.**
(Part a, Dr. Gopal Murti/Photo Researchers.)

DNA does not exist in an open, relaxed circle; the 3 million to 4 million base pairs of DNA found in a typical bacterial genome would be much too large to fit into a bacterial cell (see Figure 11.1). Bacterial DNA is not attached to histone proteins (as is eukaryotic DNA, discussed later in the chapter). Consequently, for many years bacterial DNA was called "naked DNA." However, this term is inaccurate, because bacterial DNA is complexed to a number of proteins that help compact it.

When a bacterial cell is viewed with the electron microscope, its DNA frequently appears as a distinct clump, the **nucleoid,** which is confined to a definite region of the cytoplasm. If a bacterial cell is broken open gently, its DNA spills out in a series of twisted loops (◀ FIGURE 11.3a). The ends of the loops are most likely held in place by proteins (◀ FIGURE 11.3b). Many bacteria contain additional DNA in the form of small circular molecules called plasmids, which replicate independently of the chromosome (see Chapter 8).

> **Concepts**
>
> The typical bacterial chromosome consists of a large, circular molecule of DNA that is a series of twisted loops. Bacterial DNA appears as a distinct clump, the nucleoid, within the bacterial cell.

www.whfreeman.com/pierce   Information about the genome of the common bacterium *E. coli*

The Eukaryotic Chromosome

Individual eukaryotic chromosomes contain enormous amounts of DNA. Like bacterial chromosomes, each eukaryotic chromosome consists of a single, extremely long molecule of DNA. For all of this DNA to fit into the nucleus, tremendous packing and folding are required, the extent of which must change through time. The chromosomes are in an elongated, relatively uncondensed state during interphase of the cell cycle (see p. 23 in Chapter 2), but the term *relatively* is an important qualification here.

Although the DNA of interphase chromosomes is less tightly packed than DNA in mitotic chromosomes, it is still highly condensed; it's just *less* condensed. In the course of the cell cycle, the level of DNA packaging changes—chromosomes progress from a highly packed state to a state of extreme condensation. DNA packaging also changes locally in replication and transcription, when the two nucleotide strands must unwind so that particular base sequences are exposed. Thus, the packaging of eukaryotic DNA (its tertiary, chromosomal structure) is not static but changes regularly in response to cellular processes.

Chromatin Structure

As mentioned in Chapter 2, eukaryotic DNA is closely associated with proteins, creating *chromatin.* The two basic types of chromatin are: **euchromatin,** which undergoes the normal process of condensation and decondensation in the cell cycle, and **heterochromatin,** which remains in a highly condensed state throughout the cell cycle, even during interphase. Euchromatin constitutes the majority of the chromosomal material, whereas heterochromatin is found at the centromeres and telomeres of all chromosomes, at other specific places on some chromosomes, and along the entire inactive X chromosome in female mammals (see p. 90 in Chapter 4).

The most abundant proteins in chromatin are the *histones,* which are relatively small, positively charged proteins of five major types: H1, H2A, H2B, H3, and H4 (Table 11.1). All histones have a high percentage of arginine and lysine, positively charged amino acids that give them a net positive charge. The positive charges attract the negative charges on the phosphates of DNA and holds the DNA in contact with the histones.

A heterogeneous assortment of **nonhistone chromosomal proteins** make up about half of the protein mass of the chromosome. A fundamental problem in the study of these proteins is that the nucleus is full of all sorts of proteins; so, whenever chromatin is isolated from the nucleus, it may be contaminated by nonchromatin proteins. On the other hand, isolation procedures may also remove proteins that

Table 11.1	Characteristics of histone proteins	
Histone Protein	**Molecular Weight**	**Number of Amino Acids**
H1	21,130	223
H2A	13,960	129
H2B	13,774	125
H3	15,273	135
H4	11,236	102

Note: The sizes of H1, H2A, and H2B histones vary somewhat from species to species. The values given are for bovine histones. Source: Data are from A.L. Lehninger, D. L. Nelson, and M. M. Cox, *Principles of Biochemistry*, 3d ed. (New York: Worth Publishers, 1993), p. 924.

are associated with chromatin. In spite of these difficulties, we know that some groups of nonhistone proteins are clearly associated with chromatin.

Nonhistone chromosomal proteins may be broadly divided into those that serve structural roles and those that take part in genetic processes such as transcription and replication. **Chromosomal scaffold proteins** (◀FIGURE 11.4) are revealed when chromatin is treated with a concentrated salt solution, which removes histones and most other chromosomal proteins, leaving a chromosomal protein "skeleton" to which the DNA is attached. These scaffold proteins may play a role in the folding and packing of the chromosome. Other structural proteins make up the kinetochore, cap the chromosome ends by attaching to telomeres, and constitute the molecular motors that move chromosomes in mitosis and meiosis.

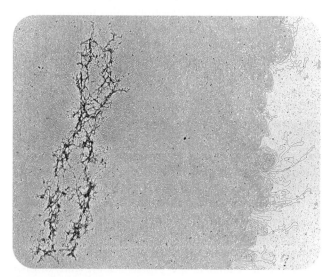

◀11.4 **Scaffold proteins play a role in the folding and packing of chromosomes.** (Professor U. Laemmli/ Photo Researchers.)

Other types of nonhistone chromosomal proteins play a role in genetic processes. They are components of the replication machinery (DNA polymerases, helicases, primases; see Chapter 12) and proteins that carry out and regulate transcription (RNA polymerases, transcription factors, acetylases; see Chapter 13). **High-mobility-group proteins** are small, highly charged proteins that vary in amount and composition, depending on tissue type and stage of the cell cycle. Several of these proteins may play an important role in altering the packing of chromatin during transcription.

The highly organized structure of chromatin is best viewed from several levels. In the next sections, we will examine these levels of chromatin organization.

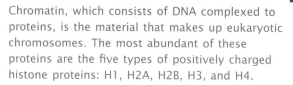

Concepts

Chromatin, which consists of DNA complexed to proteins, is the material that makes up eukaryotic chromosomes. The most abundant of these proteins are the five types of positively charged histone proteins: H1, H2A, H2B, H3, and H4.

The nucleosome Chromatin has a highly complex structure with several levels of organization. The simplest level (◀FIGURE 11.5) is the double helical structure of DNA discussed in Chapter 10. At a more complex level, the DNA molecule is associated with proteins and is highly folded to produce a chromosome.

When chromatin is isolated from the nucleus of a cell and viewed with an electron microscope, it frequently looks like beads on a string (◀FIGURE 11.6a on page 294), If a small amount of nuclease is added to this structure, the enzyme cleaves the string between the beads, leaving individual beads attached to about 200 bp of DNA (◀FIGURE 11.6b). If more nuclease is added, the enzyme chews up all of the DNA between the beads and leaves a core of proteins attached to a fragment of DNA (◀FIGURE 11.6c). Such experiments demonstrated that chromatin is not a random association of proteins and DNA but has a fundamental repeating structure.

The repeating core of protein and DNA produced by digestion with nuclease enzymes is the simplest level of chromatin structure, the **nucleosome** (see Figure 11.5). The nucleosome is a core particle consisting of DNA wrapped about two times around an octamer of eight histone proteins (two copies each of H2A, H2B, H3, and H4), much like thread wound around a spool (◀FIGURE 11.6d). The DNA in direct contact with the histone octamer is between 145 and 147 bp in length, coils around the histones in a left-handed direction, and is supercoiled. It does not wrap around the octamer smoothly; there are four bends,

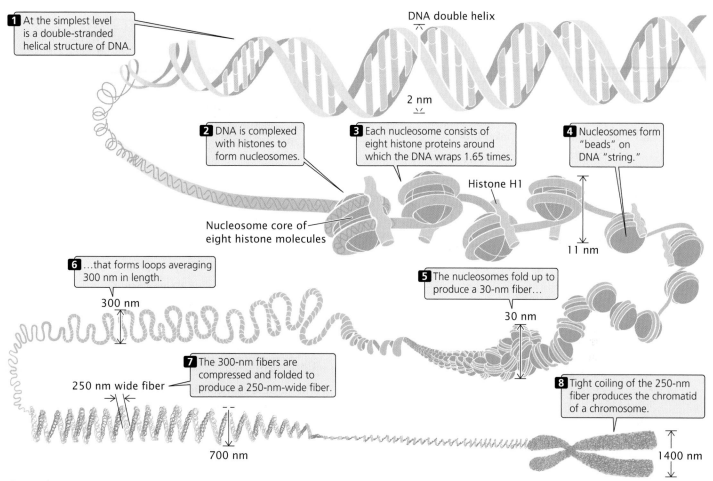

1 At the simplest level is a double-stranded helical structure of DNA.

DNA double helix

2 nm

2 DNA is complexed with histones to form nucleosomes.

3 Each nucleosome consists of eight histone proteins around which the DNA wraps 1.65 times.

4 Nucleosomes form "beads" on DNA "string."

Histone H1

Nucleosome core of eight histone molecules

11 nm

6 ...that forms loops averaging 300 nm in length.

5 The nucleosomes fold up to produce a 30-nm fiber...

300 nm

30 nm

7 The 300-nm fibers are compressed and folded to produce a 250-nm-wide fiber.

250 nm wide fiber

8 Tight coiling of the 250-nm fiber produces the chromatid of a chromosome.

700 nm

1400 nm

◀ **11.5 Chromatin has a highly complex structure with several levels of organization.**

or kinks, in its helical structure as it winds around the histones.

The fifth type of histone, H1, is not a part of the core particle but plays an important role in the nucleosome structure. The precise location of H1 with respect to the core particle is still uncertain. The traditional view is that H1 sits outside the octamer and binds to the DNA where the DNA joins and leaves the octamer (see Figure 11.5). However, the results of recent experiments suggest that the H1 histone sits *inside* the coils of the nucleosome. Regardless of its position, H1 helps to lock the DNA into place, acting as a clamp around the nucleosome octamer.

Together, the core particle and its associated H1 histone are called the **chromatosome,** the next level of chromatin organization. The H1 protein is attached to between 20 and 22 bp of DNA, and the nucleosome encompasses an additional 145 to 147 bp of DNA; so about 167 bp of DNA are held within the chromatosome. Chromatosomes are located at regular intervals along the DNA molecule and are separated from one another by

linker DNA, which varies in size among cell types—most cells have from about 30 bp to 40 bp of linker DNA. Nonhistone chromosomal proteins may be associated with this linker DNA, and a few also appear to bind directly to the core particle.

Higher-order chromatin structure In chromosomes, adjacent nucleosomes are not separated by space equal to the length of the linker DNA; rather, nucleosomes fold on themselves to form a dense, tightly packed structure (see Figure 11.5). This structure is revealed when nuclei are gently broken open and their contents are examined with the use of an electron microscope; much of the chromatin that spills out appears as a fiber with a diameter of about 30 nm (◀FIGURE 11.7a). A model of how this 30-nm fiber forms is shown in ◀FIGURE 11.7b.

The next-higher level of chromatin structure is a series of loops of 30-nm fibers, each anchored at its base by proteins in the nuclear scaffold (see Figure 11.5). On average, each loop encompasses some 20,000 to 100,000 bp of

DNA and is about 300 nm in length, but the individual loops vary considerably. The 300-nm fibers are packed and folded to produce a 250-nm-wide fiber. Tight helical coiling of the 250-nm fiber, in turn, produces the structure that appears in metaphase: an individual chromatid approximately 700 nm in width.

(a) Core histones of nucleosome   Linker DNA

"Beads-on-a-string" view of chromatin

Nuclease

(b)

1 A small amount of nuclease cleaves the "string" between the beads,…

2 …releasing individual beads attached to about 200 bp of DNA.

Nuclease

(c)

3 More nuclease destroys all of the unprotected DNA between the beads,…

11 nm

4 …leaving a core of proteins attached to 145–147 bp of DNA.

(d)

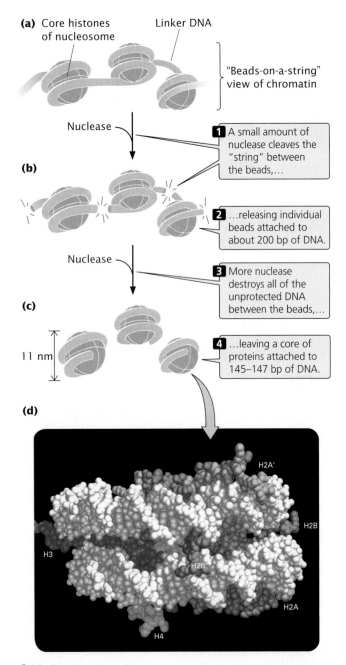

H2A′

H2B

H3

H2B

H2A

H4

◀11.6 **The nucleosome is the fundamental repeating unit of chromatin.** The space-filling model shows that the nucleosome core particle consists of two copies each of H2A, H2B, H3, and H4, around which DNA (white) coils. (Part d, from K. Luger et al., 1997, *Nature* 389:251; courtesy of T. H. Richmond.)

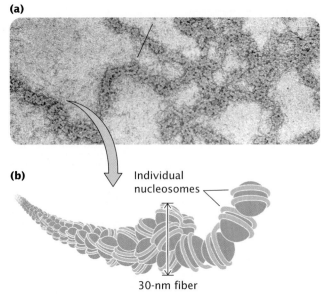

(a)

(b)   Individual nucleosomes

30-nm fiber

◀11.7 **Adjacent nucleosomes pack together to form a 30-nm fiber.** (Part a, Barbara Hamkalo, Molecular Biology and Biochemistry, University of California at Irvine.)

Concepts

The nucleosome consists of a core particle of eight histone proteins and DNA, about 146 bp in length, that wraps around the core. Chromatosomes, each including the core particle plus an H1 histone, are separated by linker DNA. Nucleosomes fold up to form a 30-nm chromatin fiber, which appears as a series of loops that pack to create a 250-nm-wide fiber. Helical coiling of the 250-nm fiber produces a 700-nm-wide chromatid.

www.whfreeman.com/pierce   A virtual tour of nucleosome and chromatin structure and links to additional sites on chromatin structure and research

Changes in chromatin structure Although eukaryotic DNA must be tightly packed to fit into the cell nucleus, it must also periodically unwind to undergo transcription and replication. Evidence of the changing nature of chromatin structure is seen in the puffs of polytene chromosomes and in the sensitivity of genes to digestion by DNase I.

Polytene chromosomes are giant chromosomes found in certain tissues of *Drosophila* and some other organisms (◀FIGURE 11.8). These large, unusual chromosomes arise when repeated rounds of DNA replication take place without accompanying cell divisions, producing thousands of copies of DNA that lie side by side. When polytene chromosomes are stained with dyes, numerous bands are revealed. Under certain conditions, the bands may exhibit **chromosomal puffs**—localized swellings of the chromosome. Each puff is a region of the chromatin that has relaxed its

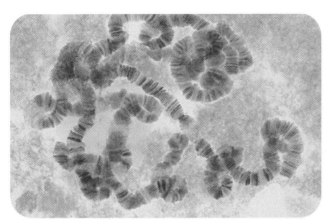

◀ 11.8 **Polytene chromosomes are giant chromosomes isolated from the salivary glands of larval Drosophila.** (Andrew Syred/Science Photo Library/Photo Researchers.

structure, assuming a more open state. If radioactively labeled uridine (a precursor to RNA) is briefly added to a *Drosophila* larva, radioactivity accumulates in chromosomal puffs, indicating that they are regions of active transcription. Additionally, the appearance of puffs at particular locations on the chromosome can be stimulated by exposure to hormones and other compounds that are known to induce the transcription of genes at those locations. This correlation between the occurrence of transcription and the relaxation of chromatin at a puff site indicates that chromatin structure undergoes dynamic change associated with gene activity.

A second piece of evidence indicating that chromatin structure changes with gene activity is sensitivity to DNase I, an enzyme that digests DNA. The ability of this enzyme to digest DNA depends on chromatin structure: when DNA is tightly bound to histone proteins, it is less sensitive to DNase I, whereas unbound DNA is more sensitive to digestion by DNase I. The results of experiments that examine the effect of DNase I on specific genes show that DNase sensitivity is correlated with gene activity. For example, globin genes code for hemoglobin in the erythroblasts (precursors of red blood cells) of chickens. The forms of hemoglobin produced in chick embryos and chickens are different and are encoded by different genes (◀ FIGURE 11.9a). However, no hemoglobin is synthesized in chick embryos in the first 24 hours after fertilization. If DNase I is applied to chromatin from chick erythroblasts in this first 24-hour period, all the globin genes are insensitive to digestion (◀ FIGURE 11.9b). From day 2 to day 6 after fertilization, after hemoglobin synthesis has begun, the globin genes become sensitive to DNase I, and the genes that code for embryonic hemoglobin are the most sensitive (◀ FIGURE 11.9c). After 14 days of development, embryonic hemoglobin is replaced by the adult forms of hemoglobin. The most

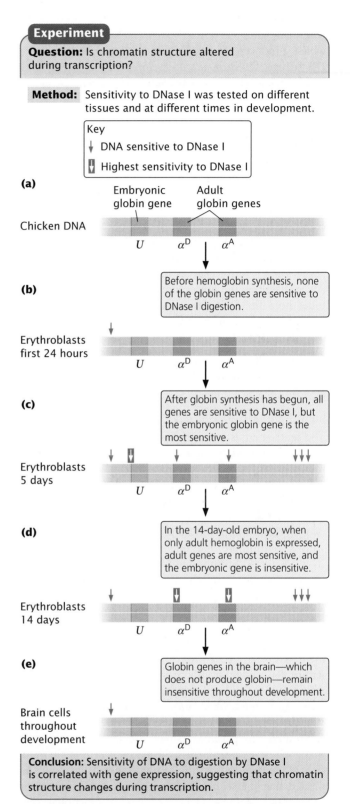

◀ 11.9 **DNase I sensitivity is correlated with the transcription of globin genes in erythroblasts of chick embryos.** The *U* gene codes for embryonic hemoglobin; the α^D and α^A genes code for adult hemoglobin.

sensitive regions now lie near the genes that produce the adult hemoglobins (◀ FIGURE 11.9d). DNA from brain cells, which produce no hemoglobin, remains insensitive to DNase digestion throughout development (◀ FIGURE 11.9e). In summary, when genes become transcriptionally active, they also become sensitive to DNase I, indicating that the chromatin structure is more exposed during transcription.

What is the nature of the change in chromatin structure that produces chromosome puffs and DNase I sensitivity? In both cases, the chromatin relaxes; presumably the histones loosen their grip on the DNA. One process that appears to be implicated in changing chromatin structure is acetylation, a reaction that adds chemical groups called acetyls to the histone proteins. Enzymes called acetyltransferases attach acetyl groups to lysine amino acids at one end (called a tail) of the histone protein. This modification reduces the positive charges that normally exist on lysine and destabilizes the nucleosome structure, and so the histones hold the DNA less tightly. Proteins taking part in transcription can then bind more easily to the DNA and carry out transcription.

www.whfreeman.com/pierce   Images of polytene chromosomes

Centromere Structure

The centromere is a constricted region of the chromosome where spindle fibers attach and is essential for proper movement of the chromosome in mitosis and meiosis (Chapter 2). The essential role of the centromere in chromosome movement was recognized by early geneticists, who observed what happens when a chromosome breaks in two. A chromosome break produces two fragments, one with a centromere and one without (◀ FIGURE 11.10a). In mitosis, the chromosome fragment containing the centromere attaches to spindle fibers and moves to the spindle pole, whereas the fragment lacking a centromere never connects to a spindle fiber and is usually lost because it fails to move into the nucleus of a daughter cell (◀ FIGURE 11.10b).

Although the centromere's role in chromosome movement has been recognized for some time, its molecular nature has only recently been revealed. The first centromeres to be isolated and studied at the molecular level came from yeast, which have small, linear chromosomes. When molecular biologists attached sequences from yeast centromeres to plasmids (small circular DNA molecules that don't have centromeres), the plasmids behaved in mitosis as if they were eukaryotic chromosomes. This finding indicated that the sequences from yeast, called **centromeric sequences** (◀ FIGURE 11.11), contain a functional centromere that allows segregation to take place. Centromeric sequences are the binding sites for proteins that function as the *kinetochore*, a complex that assembles on the centromere and to which the spindle fibers attach.

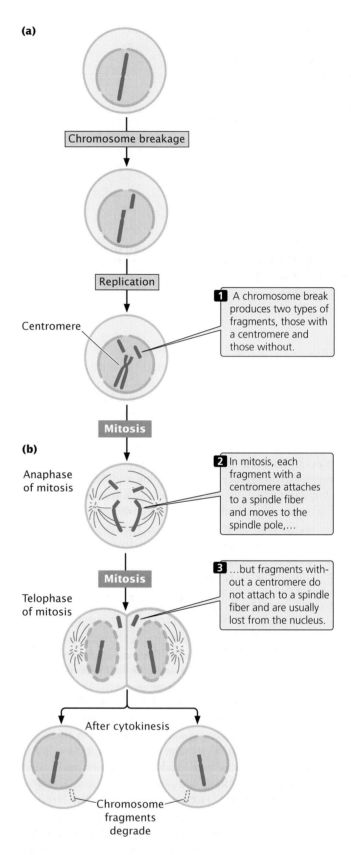

(a)

Chromosome breakage

Replication

Centromere

1 A chromosome break produces two types of fragments, those with a centromere and those without.

Mitosis

(b)

Anaphase of mitosis

2 In mitosis, each fragment with a centromere attaches to a spindle fiber and moves to the spindle pole,...

Mitosis

Telophase of mitosis

3 ...but fragments without a centromere do not attach to a spindle fiber and are usually lost from the nucleus.

After cytokinesis

Chromosome fragments degrade

◀ 11.10 **Chromosome fragments that lack a centromere are lost in mitosis.**

TCACATGATGATATTTGATTTTATTATATTTTTAAAAAAAGTAAAAAAATAAAAAGTAGTTTATTTTTAAAAAATAAAATTTAAAATATTTCACAAAATGATTTCCGAA
AGTGTACTACTATAAACTAAAATAATATAAAAATTTTTTTCATTTTTTATTTTTCATCAAATAAAAATTTTTTATTTTAAATTTTATAAAGTGTTTTACTAAAGGCTT

Region I   Region II   Region III

80–90 bp, more than 90% A + T

◀ **11.11 Centromeres consist of particular sequences repeated many times.** This nucleotide sequence is found in the point centromere of *Saccharomyces cerevisiae*. It is repeated many times in the centromeric region. Each copy of the sequence has approximately 110 bp and possesses three regions. Region I (9 bp) and region III (11 bp) are located at the ends of the sequence. Region II, consisting of about 80 to 90 mostly A–T base pairs, is in the middle. No part of the centromeric sequence codes for a protein; specific centromere proteins bind to centromeric sequences and provide anchor sites for spindle fibers.

The centromeres of different organisms exhibit considerable variation in centromeric sequences. Some organisms have chromosomes with diffuse centromeres, and spindle fibers attach along the entire length of the chromosome. Most have chromosomes with localized centromeres; in these organisms, spindle fibers attach at a specific place on the chromosome. Localized centromeres appear constricted, but there also can be secondary constrictions at places that do not have centromeric functions.

Two major classes of localized centromeres are point centromeres and regional centromeres. Point centromeres are relatively small; the point centromere of budding yeast (*Saccharomyces cerevisiae*) encompasses 125 bp of DNA.

Regional centromeres are found on the chromosomes of fission yeast (*Schizosaccharomyces pombe*) and most plants and animals. In fission yeast, centromeres consist of a central core of 4000–7000 bp. This core is flanked by blocks of centromere-specific sequences that may be repeated several times. Some of these blocks have specialized functions, such as chromosome movement during meiosis. In *Drosophila*, *Arabidopsis*, and humans, centromeres span hundreds of thousands of base pairs. Most of the centromere is made up of short sequences of DNA that are repeated thousands of times in tandem. Within these repeats are "islands" of more complex sequence, primarily transposable element sequences. However, there do not appear to be any sequences that are unique to the centromere, which raises the question of what exactly determines where the centromere is. One possibility is that centromeres are defined not by a specific sequence but by a specific chromatin structure. In support of this idea, some nucleosomes at centromeres contain variant forms of certain histone proteins.

In addition to their roles in the attachment of the spindle fibers and the movement of chromosomes, centromeres also help control the cell cycle (see pp. 26–28 in Chapter 2). In mitosis, the spindle fibers make contact with the kinetochore of the centromere and orient the chromosomes on the metaphase plate. If anaphase is initiated before each chromosome is attached to the spindle fibers, chromosomes will not move toward the spindle pole and will be lost.

Research findings indicate that the commencement of anaphase is inhibited by a signal from the centromere. This inhibitory signal disappears only after the centromere of each chromosome is attached to spindle fibers from opposite poles.

Concepts

The centromere is a region of the chromosome to which spindle fibers attach. Centromeres display considerable variation in structure. In addition to their role in chromosome movement, centromeres also help control the cell cycle by inhibiting anaphase until chromosomes are attached to spindle fibers from both poles.

Telomere Structure

Telomeres are the natural ends of a chromosome (see p. 21 in Chapter 2). Pioneering work by Hermann Muller (on fruit flies) and Barbara McClintock (on corn) showed that chromosome breaks produce unstable ends that have a tendency to stick together and allow the chromosome to be degraded. Because attachment and degradation don't happen to the ends of a chromosome that has telomeres, each telomere must serve as a cap that stabilizes the chromosome, much like the plastic tips on the ends of a shoelace that prevent the lace from unraveling.

Telomeres also provide a means of replicating the ends of the chromosome. The enzymes that synthesize DNA are unable to replicate the last few nucleotides at the end of each newly synthesized DNA strand (discussed in Chapter 12). Consequently, a chromosome should get shorter each time its DNA is synthesized, and this progressive shortening would eventually damage genes on the chromosome. Indeed, such chromosome shortening does occur in somatic cells, which are capable of only a limited number of divisions. Germ cells and cells in single-celled organisms, however, must divide continually.

Chromosomes in these cells don't progressively shorten and self-destruct, because the cells possess an enzyme called telomerase that replicates the telomeres. The ability of telomerase to replicate a chromosome end depends on the unique molecular structure of the telomere. We will examine this mechanism of replication in Chapter 12.

Telomeres were first isolated from the protozoan *Tetrahymena thermophila* and were found to possess multiple copies of the sequence:

$$5'-CCCCAA-3'$$
$$3'-GGGGTT-5'$$

Telomeres have now been isolated from protozoans, plants, humans, and other organisms; most are similar in structure (Table 11.2). These **telomeric sequences** usually consist of a series of cytosine nucleotides followed by several adenine or thymine nucleotides or both, taking the form $5'-C_n(A \text{ or } T)_m-3'$, where n is 2 or greater and m is from 1 to 4. For example, the repeating unit in human telomeres is CCCTAA, which may be repeated from 250 to 1500 times. The sequence is always oriented with the string of Cs and Gs toward the end of the chromosome, as shown here:

end of $5'-CCCTAA$ toward
chromosome   ← $3'-GGGATT$ →   centromere

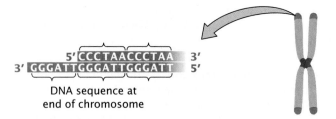

$5'$ CCCTAACCCTAA $3'$
$3'$ GGGATTGGGATTGGGATT $5'$

DNA sequence at
end of chromosome

◄ **11.12 DNA at the ends of eukaryotic chromosomes consists of telomeric sequences.**

The G-rich strand often protrudes beyond the complementary C-rich strand at the end of the chromosome (◄FIGURE 11.12). The length of the telomeric sequence varies from chromosome to chromosome and from cell to cell, suggesting that each telomere is a dynamic structure that actively grows and shrinks. The telomeres of *Drosophila* chromosomes are different in structure. They consist of multiple copies of the two different retrotransposons (discussed later in this chapter), *Het-A* and *Tart*, arranged in tandem repeats. Apparently, in *Drosophila*, loss of telomere sequences during replication is balanced by transposition of additional copies of the *Het-A* and *Tart* elements.

Farther away from the end of the chromosome, from several thousand to hundreds of thousands of base pairs form **telomere-associated sequences.** They, too, contain repeated sequences, but the repeats are longer, more varied, and more complex than those found in telomeric sequences.

Table 11.2 | DNA sequences typically found in telomeres of various organisms

Organism	Sequence
Tetrahymena (protozoan)	$5'-CCCCAA-3'$ $3'-GGGGTT-5'$
Oxytricha (protozoan)	$5'-CCCCAAAA-3'$ $3'-GGGGTTTT-5'$
Trypanosoma (protozoan)	$5'-CCCTAA-3'$ $3'-GGGATT-5'$
Saccharomyces (yeast)	$5'-C_{2-3} ACA_{1-6}-3'$ $3'-G_{2-3} TGT_{1-6}-5'$
Neurospora (fungus)	$5'-CCCTAA-3'$ $3'-GGGATT-5'$
Caenorhabditis (nematode)	$5'-GCCTAA-3'$ $3'-CGGATT-5'$
Bombyx (insect)	$5'-CCTAA-3'$ $3'-GGATT-5'$
Vertebrate	$5'-CCCTAA-3'$ $3'-GGGATT-5'$
Arabidopsis (plant)	$5'-CCCTAAA-3'$ $3'-GGGATTT-5'$

Source: V. A. Zakian, *Science* 270(1995): 1602.

Concepts

A telomere is the stabilizing end of a chromosome. At the end of each telomere are many short telomeric sequences. Longer, more complex telomere-associated sequences are found adjacent to the telomeric sequences.

www.whfreeman.com/pierce   More detailed information on telomeres

Variation in Eukaryotic DNA Sequences

Prokaryotic and eukaryotic cells differ dramatically in the amount of DNA per cell, a quantity termed an organism's **C value** (Table 11.3). Each cell of a fruit fly, for example, contains 35 times the amount of DNA found in a cell of the bacterium *E. coli*. In general, eukaryotic cells contain more DNA than that of prokaryotes, but variability in the C values of different eukaryotes is huge. Human cells contain more than 10 times the amount of DNA found in *Drosophila* cells, whereas some salamander cells contain 20

Table 11.3	Genome sizes of various organisms

Organism	Approximate Genome Size (bp)
λ (bacteriophage)	50,000
E. coli (bacterium)	4,640,000
Saccharomyces cerevisiae (yeast)	12,000,000
Arabidopsis thaliana (plant)	167,000,000
Drosophila melanogaster (insect)	180,000,000
Homo sapiens (human)	3,400,000,000
Zea mays (corn)	4,500,000,000
Amphiuma (salamander)	765,000,000,000

times as much DNA as that of human cells. Clearly, these differences in C value cannot be explained simply by differences in organismal complexity. So what is all this extra DNA in eukaryotic cells doing? We do not yet have a complete answer to this question, but examination of DNA sequences has revealed that eukaryotic DNA has complexity that is absent from prokaryotic DNA.

Denaturation and Renaturation of DNA

The first clue that the DNA of eukaryotes contains several types of sequences came from the results of studies in which double-stranded DNA was separated and then allowed to reassociate. When double-stranded DNA in solution is heated, the hydrogen bonds that hold the two strands together are weakened and, with enough heat, the two nucleotide strands separate completely, a process called **denaturation** or melting (◀FIGURE 11.13). DNA is typically denatured within a narrow temperature range. The midpoint of this range, the **melting temperature** (T_m), depends on the base sequence of a particular sample of DNA: G–C base pairs have three hydrogen bonds, whereas A–T base

pairs only have two; so the separation of G–C pairs requires more energy than does the separation of A–T pairs. A DNA molecule with a higher percentage of G–C pairs will therefore have a higher T_m than that of DNA with more A–T pairs.

The denaturation of DNA by heating is reversible; if single-stranded DNA is slowly cooled, single strands will collide and hydrogen bonds will again form between complementary base pairs, producing double-stranded DNA (see Figure 11.13). This reaction, called **renaturation** or reannealing, takes place in two steps. First, single strands in solution collide randomly with their complementary strands. Second, hydrogen bonds form between complementary bases.

Two single-stranded molecules of DNA from different sources will anneal if they are complementary, a process termed **hybridization.** For hybridization to take place, the two strands do not have to be complementary at all their bases—just at enough bases to hold the two strands together. The extent of hybridization can be used to measure the similarity of nucleic acids from two different sources and is a common tool for assessing evolutionary relationships. The rate at which hybridization takes place also provides information about the sequence complexity of DNA (see next subsection).

Renaturation Reactions and C_0t Curves

In a typical renaturation reaction, DNA molecules are first sheared into fragments several hundred base pairs in length. Next, the fragments are heated to about 100°C, which causes the DNA to denature. The solution is then cooled slowly, and the amount of renaturation is measured by observing optical absorbance. Double-stranded DNA absorbs less UV light than does single-stranded DNA; so the amount of renaturation can be monitored by shining a UV light through the solution and measuring the amount of the light absorbed.

The amount of renaturation depends on two critical factors: (1) initial concentration of single-stranded DNA (C_0) and (2) amount of time allowed for renaturation (t). Other things being equal, there will be more renaturation

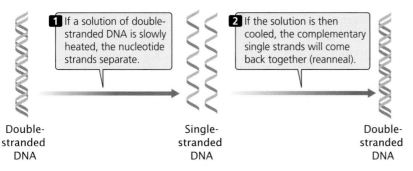

1 If a solution of double-stranded DNA is slowly heated, the nucleotide strands separate.

2 If the solution is then cooled, the complementary single strands will come back together (reanneal).

Double-stranded DNA   Single-stranded DNA   Double-stranded DNA

◀11.13 **The slow heating of DNA causes the two strands to separate (denature).**

at higher concentrations of DNA, because high concentrations increase the likelihood that the two complementary strands will collide. There will also be more renaturation with increasing time, because there are more opportunities for two complementary sequences to collide. These two factors together form a parameter called C_0t, which equals the initial concentration multiplied by the renaturation time ($C_0 \times t = C_0t$).

A plot of the fraction of single-stranded DNA as a function of C_0t during a renaturation reaction is called a C_0t curve. A typical C_0t curve for a prokaryotic organism is shown in ◀ FIGURE 11.14. The upper left-hand side of the curve represents the start of the renaturation reaction, when all of the DNA is single stranded, and so the proportion of single-stranded DNA is 1. As the reaction proceeds, single-stranded DNA pairs to form double-stranded DNA, represented by the decreasing fraction of single-stranded DNA. At the end of the reaction, the proportion of single-stranded DNA is 0, because all of the DNA is now double stranded. The value at which half of the DNA is reannealed is called $C_0t\ ^1/_2$.

The rate of renaturation also depends on the size and complexity of the DNA molecules used. Consider the following analogy. Suppose we distribute 100 cards equally among the students in a class. We ask each student to write his or her name on the cards, and we put all the cards in a hat. We then randomly draw two cards from the hat and see if the names on the two cards match. If they don't match, we put them back in the hat; if they do match, we remove them, and we continue drawing until all the cards have been removed. If there are only four students in the class, each student will receive 25 cards. Because each student's name is on 25 cards, the chance of drawing two cards that match is high, and we will quickly empty the hat. If we do the same exercise in another class with 50 students, again using 100 cards, each student's name will appear on only two cards, and the chance of removing two cards with the same name is much lower. Thus, it will take longer to empty the hat.

This exercise resembles what occurs in the renaturation reaction. If we start with the same total amount of DNA, but there are only a few different sequences in the DNA, a chance collision between two complementary fragments is more likely to occur than if there were many different sequences. Therefore DNA from organisms with larger genomes will have a larger $C_0t\ ^1/_2$ value.

Thus far, we have considered renaturation reactions in which each DNA sequence is present only once in each molecule. If some sequences are present in multiple copies, these sequences will be more likely to collide with a complementary copy, and renaturation of these sequences will be rapid. Think about our analogy of drawing names from a hat. Imagine that we have 50 students and 100 cards; each student gets two cards. This time, the students write only their first names on the cards. Again, we place the cards in the hat and draw out two cards at random. If

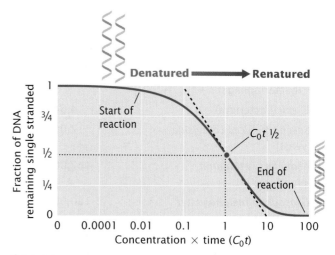

◀ 11.14 A C_0t curve represents the fraction of DNA remaining single stranded in a renaturation reaction, plotted as a function of DNA concentration × time (C_0t). This graph is a typical C_0t curve for a prokaryotic organism.

there are five students in the class named Scott, this name will appear on ten cards; so the chance of drawing out two cards at random bearing the name Scott is fairly high. On the other hand, if there is only one Susan in the class, this name will appear on only two cards, and the chance of drawing out two cards with the name Susan is low. The cards with Scott match up more quickly than the cards with Susan, because there are more copies with the name Scott. Similarly, in a renaturation reaction, if some sequences of DNA are present in multiple copies, they will renature more quickly.

Concepts

When double-stranded DNA is heated, it denatures, separating into single-stranded molecules. On cooling, these single-stranded molecules pair and re-form double-stranded DNA, a process called renaturation. A C_0t curve is a plot of a renaturation reaction.

Types of DNA Sequences in Eukaryotes

For most eukaryotic organisms, C_0t curves similar to the one presented in ◀ FIGURE 11.15 are produced and indicate that eukaryotic DNA consists of at least three types of sequences. Slowly renaturing DNA consists of sequences that are present only once, or at most a few times, in the genome. This nonrepetitive, **unique-sequence DNA** includes sequences that code for proteins, as well as a great deal of DNA whose function is unknown. The more rapidly renaturing DNA represents two kinds of **repetitive DNA**—

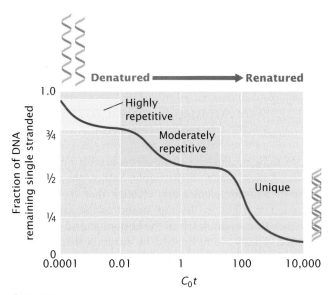

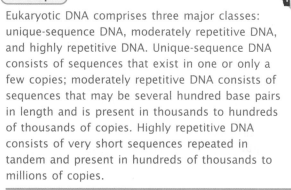

◀ 11.15 A typical C_0t curve for a eukaryotic organism contains several steps. The first step in the curve represents DNA renaturing at very low C_0t values, because these sequences are present in many copies (highly repetitive). The second step represents DNA renaturing at intermediate C_0t values; these sequences are present in an intermediate number of copies (moderately repetitive). The last step represents DNA that renatures slowly; these sequences are present singly or in few copies (unique).

DNA sequences that exist in multiple copies. Although not identical, these copies are similar enough to reanneal.

Moderately repetitive DNA typically consists of sequences from 150 to 300 bp in length (although they may be longer) that are repeated many thousands of times. Some of these sequences perform important functions for the cell; for example, the genes for ribosomal RNAs (rRNAs) and transfer RNAs (tRNAs) make up a part of the moderately repetitive DNA. However, much of the moderately repetitive DNA has no known function in the cell. Moderately repetitive DNA itself is of two types of repeats. **Tandem repeat sequences** appear one after another and tend to be clustered at a few locations on the chromosomes. **Interspersed repeat sequences** are scattered throughout the genome. An example of an interspersed repeat is the *Alu* sequence, each of which consists of about 200 bp. The *Alu* sequence is present more than a million times in the human genome and makes up about 11% of each person's DNA. Short repeats, such as the *Alu* sequences, are called **SINEs (short interspersed elements)**. Longer interspersed repeats consisting of several thousand base pairs are called **LINEs (long interspersed elements)**. Most interspersed repeats are transposable genetic elements, sequences that can multiply and move (see next section).

The other major class of repetitive DNA is **highly repetitive DNA.** These short sequences, often less than 10 bp in

length, are present in hundreds of thousands to millions of copies that are repeated in tandem and clustered in certain regions of the chromosome, especially at centromeres and telomeres. Highly repetitive DNA is sometimes called satellite DNA, because it has a different base composition from those of the other DNA sequences and separates as a satellite fraction when centrifuged at high speeds. Highly repetitive DNA is rarely transcribed into RNA. Although these sequences may contribute to centromere and telomere function, most highly repetitive DNA has no known function.

Concepts

Eukaryotic DNA comprises three major classes: unique-sequence DNA, moderately repetitive DNA, and highly repetitive DNA. Unique-sequence DNA consists of sequences that exist in one or only a few copies; moderately repetitive DNA consists of sequences that may be several hundred base pairs in length and is present in thousands to hundreds of thousands of copies. Highly repetitive DNA consists of very short sequences repeated in tandem and present in hundreds of thousands to millions of copies.

The Nature of Transposable Elements

Transposable elements are mobile DNA sequences found in the genomes of all organisms. In many genomes, they are quite abundant: for example, they make up at least 50% of human DNA. Most transposable elements are able to insert at many different locations, relying on mechanisms that are distinct from homologous recombination. They often cause mutations, either by inserting into another gene and disrupting it or by promoting DNA rearrangements such as deletions, duplications, and inversions (see Chapter 9).

General Characteristics of Transposable Elements

There are many different types of transposable elements: some have simple structures, encompassing only those sequences necessary for their own transposition (movement), whereas others have complex structures and encode a number of functions not directly related to transposition. Despite this variation, many transposable elements have certain features in common.

Short, **flanking direct repeats** of 3 to 12 base pairs are present on both sides of most transposable elements. They are not a part of a transposable element and do not travel with it. Rather, they are generated in the process of transposition, at the point of insertion. The sequences of these repeats vary, but the length is constant for each type of transposable element.

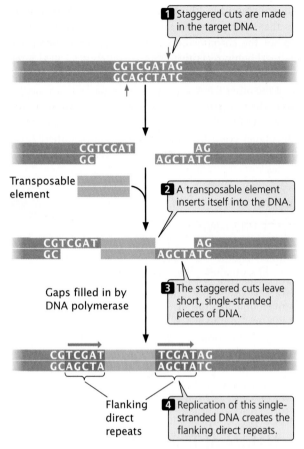

1 Staggered cuts are made in the target DNA.

```
CGTCGATAG
GCAGCTATC
```

```
CGTCGAT          AG
GC          AGCTATC
```

Transposable element

2 A transposable element inserts itself into the DNA.

```
CGTCGAT          AG
GC          AGCTATC
```

Gaps filled in by DNA polymerase

3 The staggered cuts leave short, single-stranded pieces of DNA.

```
CGTCGAT     TCGATAG
GCAGCTA     AGCTATC
```

Flanking direct repeats

4 Replication of this single-stranded DNA creates the flanking direct repeats.

◀11.16 Flanking direct repeats are generated when a transposable element inserts into DNA.

The presence of flanking direct repeats indicates that staggered cuts are made in the target DNA when a transposable element inserts itself, as shown in ◀ FIGURE 11.16. The staggered cuts leave short, single-stranded pieces of DNA on either side of the transposable element. Replication of the single-stranded DNA then creates the flanking direct repeats.

At the ends of many, but not all, transposable elements are **terminal inverted repeats**, which are sequences from 9 to 40 bp in length that are inverted complements of one another. For example, the following sequences are inverted repeats:

$$5'-ACAGTTCAG\dots CTGAACTGT-3'$$
$$3'-TGTCAAGTC\dots GACTTGACA-5'$$

On the same strand, the two sequences are not simple inversions, as their name might imply; rather, they are both inverted and complementary. (Notice that the sequence from left to right in the top strand is the same as the sequence from right to left in the bottom strand.) Terminal inverted repeats are recognized by enzymes that carry out transposition and are required for transposition to take place. ◀ FIGURE 11.17 summarizes the general characteristics of transposable elements.

Concepts

Transposable elements are mobile DNA sequences that often cause mutations. There are many different types of transposable elements; most generate short, flanking direct repeats at the target site as they insert. Many transposable elements also possess short terminal inverted repeats.

Transposition

Transposition is the movement of a transposable element from one location to another. Although our understanding of transposition is still incomplete, it's clear that, rather than a single mechanism, several different mechanisms are used for transposition in both prokaryotic and eukaryotic cells. Nevertheless, all types of transposition have several features in common: (1) staggered breaks are made in the target DNA (see Figure 11.16); (2) the

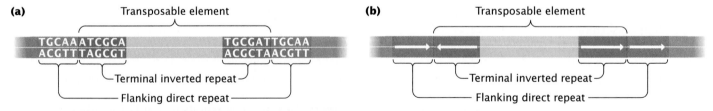

(a) Transposable element

```
TGCAAATCGCA          TGCGATTGCAA
ACGTTTAGCGT          ACGCTAACGTT
```

Terminal inverted repeat

Flanking direct repeat

(b) Transposable element

Terminal inverted repeat

Flanking direct repeat

◀11.17 Many transposable elements have common characteristics. (a) Most transposable elements generate flanking direct repeats on each side of the point of insertion into target DNA. Many transposable elements also possess terminal inverted repeats. (b) These representations of direct and indirect repeats are used in illustrations throughout this chapter.

transposable element is joined to single-stranded ends of the target DNA; and (3) DNA is replicated at the single-strand gaps.

Mechanisms of Transposition

Some transposable elements transpose through DNA intermediates, whereas others use RNA intermediates. Among those that transpose through DNA, transposition may be replicative or nonreplicative. In **replicative transposition,** a new copy of the transposable element is introduced at a new site while the old copy remains behind at the original site; the number of copies of the transposable element increases. In **nonreplicative transposition,** the transposable element excises from the old site and inserts at a new site without any increase in the number of its copies. Nonreplicative transposition requires replication of only the few nucleotides that constitute the direct repeats.

Replicative transposition Replicative transposition, sometimes called copy-and-paste transposition, can be either between two different DNA molecules or between two parts of the same DNA molecule. ◀FIGURE 11.18 summarizes the steps of transposition between two circular DNA molecules. Before transposition (see Figure 11.18a), the transposable element is on one molecule. In the first step, the two DNA molecules are joined, and the transposable element is replicated, producing the **cointegrate structure** that consists of molecules A + B fused together with two copies of the transposable element (see Figure 11.18b). In a moment, we'll see how the copy is produced, but let's first look at the second step of the replicative transposition process. After the cointegrate has formed, crossing over at regions within the transposable elements produces two molecules, each with a copy of the transposable element

(see Figure 11.18c). This second step is known as resolution of the cointegrate.

How are the steps of replicative transposition (cointegrate formation and resolution) brought about? Cointegrate formation requires four events. First, a **transposase** enzyme (often encoded by the transposable element) makes single-strand breaks at each end of the transposable element and on either side of the target sequence where the element inserts (◀FIGURE 11.19 a and b). Second, the free ends of the transposable element attach to the free ends of the target sequence (◀FIGURE 11.19c). Third, replication takes place on the single-stranded templates, beginning at the 3′ ends of the single strands and proceeding through the transposable element (◀FIGURE 11.19d and e). This replication creates the cointegrate, with its two copies of both the transposable element and the sequence at the target site, which is now on one side of each copy (◀FIGURE 11.19f). The enzymes that perform the replication and ligation functions are cellular enzymes that function in replication and DNA repair. Fourth, after the cointegrate has formed, it undergoes resolution, which requires crossing over between sites located within the transposon. Resolution gives rise to two copies of the transposable element (◀FIGURE 11.19g). The resolution step is brought about by **resolvase** enzymes (encoded in some cases by the transposable element and in other cases by a cellular gene) that function in homologous recombination.

Nonreplicative transposition In nonreplicative transposition, the transposable element moves from one site to another without replication of the entire transposable element, although short sequences in the target DNA are replicated, generating flanking direct repeats. Sometimes referred to as cut-and-paste transposition, nonreplicative transposition requires only that the transposable element

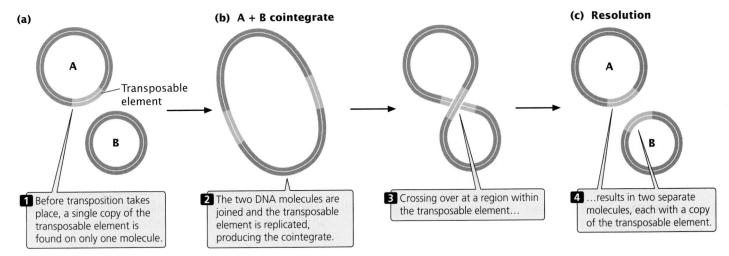

(a)

A

Transposable element

B

1 Before transposition takes place, a single copy of the transposable element is found on only one molecule.

(b) A + B cointegrate

2 The two DNA molecules are joined and the transposable element is replicated, producing the cointegrate.

3 Crossing over at a region within the transposable element...

(c) Resolution

A

B

4 ...results in two separate molecules, each with a copy of the transposable element.

◀**11.18 Replicative transposition increases the number of copies of the transposable element.**

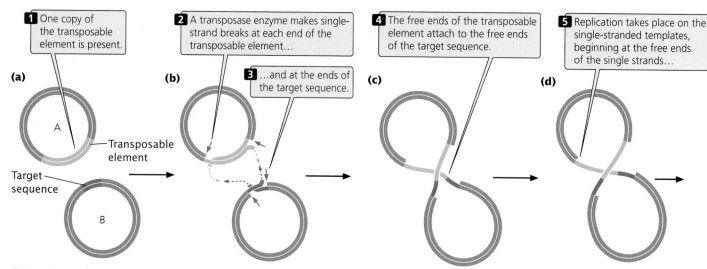

◀ 11.19 **Replicative transposition requires single-strand breaks, replication, and resolution.**

and the target DNA be cleaved and joined together. Cleavage requires a transposase enzyme produced by the transposable element. The joining of the transposable element and target DNA is probably carried out by normal replication and repair enzymes. If a transposable element moves by nonreplicative transposition, how does it increase in copy number in the genome? The answer comes from examining the fate of the original site of the element. After excision, a break will be left at the original insertion site. Such breaks are harmful to the cell, and so they are repaired efficiently (see Chapter 17). One common method of repair is to copy sequence information from a homologous template; the sister chromatid is the preferred template for this type of repair. Before transposition, both sister chromatids will have a copy of the transposable element. After excision from one chromatid, repair of the break can result in copying the transposable element sequence off the sister chromatid. Thus, the transposable element is moved from the original site to a new site, but a copy is restored to the original site by DNA repair mechanisms.

Transposition through an RNA intermediate Eukaryotic transposable elements that transpose through RNA intermediates are called **retrotransposons.** A retrotransposon in DNA (◀ FIGURE 11.20a) is first transcribed into an RNA sequence (◀ FIGURE 11.20b), which may be processed. The processed RNA undergoes reverse transcription by a reverse transcriptase enzyme to produce a double-stranded DNA copy of the RNA (◀ FIGURE 11.20c). Staggered cuts are made in the target DNA (◀ FIGURE 11.20d), and the DNA copy of the retrotransposon inserts into the genome (◀ FIGURE 11.20e). Replication fills in the short gaps produced by the staggered cuts, generating flanking direct repeats on both sides of the retrotransposon.

Concepts

Transposition may be through either a DNA or an RNA intermediate. In replicative transposition, a new copy of the transposable element inserts in a new location and the old copy stays behind; in nonreplicative transposition, the old copy excises from the old site and moves to a new site. Transposition through an RNA intermediate requires reverse transcription, in which a retrotransposon is transcribed into RNA, the RNA is copied into DNA, and the new DNA copy is integrated into the target site.

The Mutagenic Effects of Transposition

Because transposable elements may insert into other genes and disrupt their function, transposition is generally mutagenic. In fact, more than half of all spontaneously occurring mutations in *Drosophila* result from the insertion of a transposable element in or near a functional gene. Although most of these mutations are detrimental, transposition may occasionally activate a gene or change the phenotype of the cell in a beneficial way. Additionally, a transposable element may carry information that benefits the cell, such as antibiotic resistance conferred by genes carried on bacterial transposable elements.

In 1991, Francis Collins and his colleagues discovered a 31-year-old man with neurofibromatosis caused by a transposition of the *Alu* sequence. Neurofibromatosis is a disease

◀ 11.20 (facing page) **Retrotransposons transpose through RNA intermediates.**

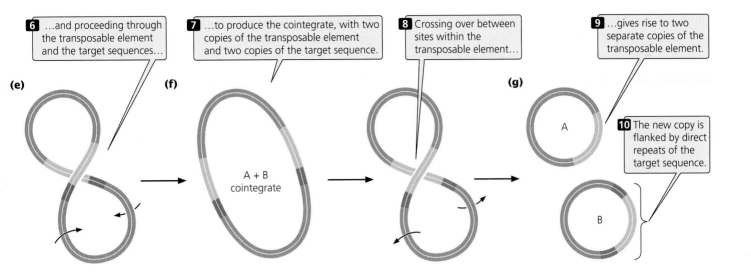

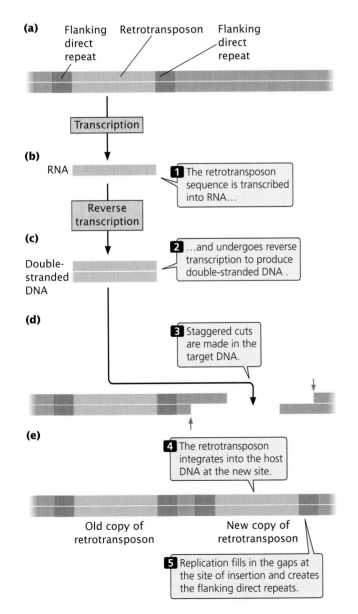

that produces numerous tumors of the skin and nerves; it results from mutations in a gene called *NF1*. Collins and his colleagues found a copy of *Alu* in one of the introns of this man's *NF1* gene. The *Alu* had caused an RNA splicing error, with the result that one of the exons was left out of the *NF1* mRNA. The absence of the exon caused a shift in the reading frame and resulted in an abnormal protein, which eventually caused the neurofibromatosis. Examination of DNA from the man's mother and father revealed that the *Alu* sequence was not present in their *NF1* genes—the insertion was new. Cases of hemophilia and muscular dystrophy also have been traced to mutations caused by transposition.

Because transposition entails the exchange of DNA sequences and recombination, it often leads to DNA rearrangements. Homologous recombination between multiple copies of transposons also leads to duplications, deletions, and inversions, as shown in ◀ FIGURE 11.21. The *Bar* mutation in *Drosophila* (see Figures 9.7 and 9.8) is a tandem duplication thought to have arisen through homologous recombination between two copies of a transposable element present in different locations on the X chromosome.

DNA rearrangements can also be caused by excision of transposable elements in a cut-and-paste transposition. If the broken DNA is not repaired properly, a chromosome rearrangement can be generated. If it is not repaired at all, the acentric fragment will be last, resulting in a deletion. This type of chromosome breakage led to the first discovery of transposable elements by Barbara McClintock (described below). She named the gene that appeared at these sites *Dissociation* because of the tendency for it to cause chromosome breakage and loss of a fragment.

The Regulation of Transposition

Many transposable elements move through replicative transposition and increase in number with each transposition. As the number of copies of the transposon increases, the rate of transposition increases because the concentration

(a)

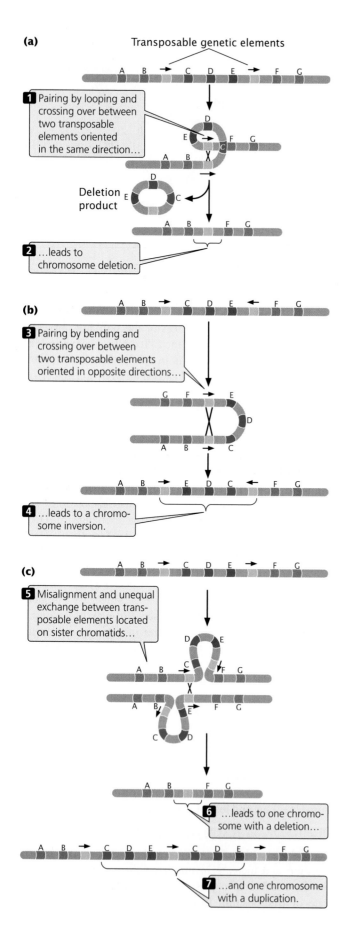

Transposable genetic elements

1 Pairing by looping and crossing over between two transposable elements oriented in the same direction...

Deletion product

2 ...leads to chromosome deletion.

(b)

3 Pairing by bending and crossing over between two transposable elements oriented in opposite directions...

4 ...leads to a chromosome inversion.

(c)

5 Misalignment and unequal exchange between transposable elements located on sister chromatids...

6 ...leads to one chromosome with a deletion...

7 ...and one chromosome with a duplication.

◀ **11.21 Chromosome rearrangements are often generated by transposition.**

of transposase in the cell becomes greater (remember that transposase is produced by the transposon). In the absence of mechanisms to restrict transposition, the number of copies of transposable elements would increase continuously, and the host DNA would be harmed by the resulting high rate of mutation (caused by frequent insertion of transposable elements). Furthermore, large amounts of energy and resources would be required to replicate the "extra" DNA in the proliferating transposable elements. For these reasons, it isn't surprising that cells have evolved mechanisms to regulate transposition, just as they have mechanisms to regulate gene expression (see Chapter 16).

When a transposable element first enters a cell that possesses no other copies of that element, transposition is frequent. As the number of copies of the transposable element increases, the frequency of transposition diminishes until a steady-state number of transposable elements is reached. This regulation of transposition means that most cells have a characteristic number of copies of a particular transposable element.

Many transposable elements regulate transposition by limiting the production of the transposase enzyme required for movement. In some cases, *transcription* of the transposase gene is regulated but, more frequently, *translation* of the transposase mRNA is controlled (see p. 463 in Chapter 16). Other regulatory mechanisms do not affect the level of transposase; rather, they directly inhibit the transposition event.

Concepts

Transposable elements frequently cause mutations and DNA rearrangements. Many transposable elements regulate their own transposition, either by controlling the amount of transposase produced or by direct inhibition of the transposition event.

The Structure of Transposable Elements

Bacteria and eukaryotic organisms possess a number of different types of transposable elements, the structures of which vary extensively. In this section, we consider the structures of representative types of transposable elements.

Transposable Elements in Bacteria

The two major groups of bacterial transposable elements are: (1) simple transposable elements that carry only the information required for movement and (2) more-complex

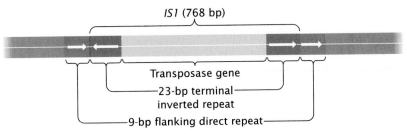

IS1 (768 bp)

Transposase gene
—23-bp terminal—
inverted repeat
—9-bp flanking direct repeat—

◀11.22 **Insertion sequences are simple transposable elements found in bacteria.**

transposable elements that contain DNA sequences not directly related to transposition.

Insertion sequences The simplest type of transposable element in bacterial chromosomes and plasmids is an **insertion sequence** (IS). This type of element carries only the genetic information necessary for its movement. Insertion sequences are common constituents of bacteria and plasmids. They are designated by IS, followed by an identifying number. For example, IS1 is a common insertion sequence found in *E. coli*.

Insertion sequences are typically from 800 to 2000 bp in length and possess the two hallmarks of transposable elements: terminal inverted repeats and the generation of flanking direct repeats at the site of insertion. Most insertion sequences contain one or two genes that code for transposase. IS1, a typical insertion sequence, is 768 nucleotide pairs long and has terminal inverted repeats of 23 bp at each end (◀FIGURE 11.22). The flanking direct repeats created by IS1 are each 9 bp long—the most common length for flanking direct repeats. Table 11.4 summarizes these features for several bacterial insertion sequences.

Composite transposons Any segment of DNA that becomes flanked by two copies of an insertion sequence may itself transpose and is called a **composite transposon**. Each type of composite transposon is designated by the abbreviation *Tn*, followed by a number. *Tn10* is a

composite transposon of about 9300 bp that carries a gene (about 6500 bp) for tetracycline resistance between two IS10 insertion sequences (◀FIGURE 11.23). The insertion sequences have terminal inverted repeats; so the composite transposon also ends in inverted repeats. Composite transposons also generate flanking direct repeats at their sites of insertion (see Figure 11.23). The insertion sequences at the ends of a composite transposon may be in the same orientation or they may be inverted relative to one other (as in *Tn10*).

The insertion sequences at the ends of a composite transposon are responsible for transposition. The DNA between the insertion sequences is not required for movement and may carry additional information (such as antibiotic resistance). Presumably, composite transposons evolve when one insertion sequence transposes to a location close to another of the same type. The transposase produced by one of the IS sequences catalyzes the transposition of both insertions sequences, allowing them to move together and carry along the DNA that lies between

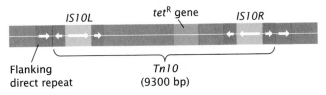

IS10L   tet^R gene   IS10R

Flanking   Tn10
direct repeat   (9300 bp)

◀11.23 ***Tn10* is a composite transposon in bacteria.**

Table 11.4	Structures of some common insertion sequences		
		Length of	
Insertion Sequence	**Total Length (bp)**	**Inverted Repeats (bp)**	**Flanking Direct Repeats (bp)**
IS1	768	23	9
IS2	1327	41	5
IS4	1428	18	11 or 12
IS5	1195	16	4

Source: B. Lewin, *Genes*, 3d ed. (New York: Wiley, 1987), p. 591.

Table 11.5	Characteristics of several composite transposons		
Composite Transposon	**Total Length (bp)**	**Associated IS Elements**	**Other Genes Within the Transposon**
Tn9	2500	IS1	Chloramphenicol resistance
Tn10	9300	IS10	Tetracycline resistance
Tn5	5700	IS50	Kanamycin resistance
Tn903	3100	IS903	Kanamycin resistance

them. In some composite transposons (such as *Tn10*), one of the insertion sequences may be defective; so its movement depends on the transposase produced by the other. Characteristics of several composite transposons are listed in Table 11.5.

Noncomposite transposons As already stated, insertion sequences carry only information for their own movement, whereas bacterial transposons are more complex. Some transposable elements in bacteria lack insertion sequences and are referred to as noncomposite transposons. For instance, *Tn3* is a noncomposite transposon that is about 5000 bp long, possesses terminal inverted repeats of 38 bp, and generates flanking direct repeats that are 5 bp in length. *Tn3* carries genes for transposase and resolvase (mentioned earlier in this chapter), plus a gene that codes for the enzyme β-lactamase, which provides resistance to ampicillin.

A few bacteriophage genomes reproduce by transposition and use transposition to insert themselves into a bacterial chromosome in their lysogenic cycle; the best studied of these transposing bacteriophages is Mu (◀ FIGURE 11.24). Although Mu does not possess terminal inverted repeats, it does generate short (5-bp) flanking direct repeats when it inserts randomly into DNA. Mu replicates through transposition and causes mutations at the site of insertion, properties characteristic of transposable elements.

Concepts

Insertion sequences are prokaryotic transposable elements that carry only the information needed for transposition. A composite transposon is a more complex element that consists of two insertion sequences plus intervening DNA. Noncomposite transposons in bacteria lack insertion sequences but have terminal inverted repeats and carry information not related to transposition. All of these transposable elements generate flanking direct repeats at their points of insertion.

Transposable Elements in Eukaryotes

Eukaryotic transposable elements can be divided into two groups. One group is structurally similar to transposable elements found in bacteria, typically ending in short inverted repeats and transposing through DNA intermediates. The other group comprises retrotransposons (see Figure 11.20); they use RNA intermediates and many are similar in structure and movement to retroviruses (see p. 224 in Chapter 8). On the basis of their structure, function, and genomic sequences, it is clear that some retrotransposons are evolutionarily related to retroviruses. Although their mechanism of movement is fundamentally different from that of other transposable elements, retrotransposons

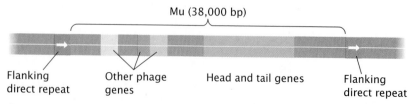

Mu (38,000 bp)

Flanking direct repeat   Other phage genes   Head and tail genes   Flanking direct repeat

◀ 11.24 **Mu is a transposing bacteriophage.**

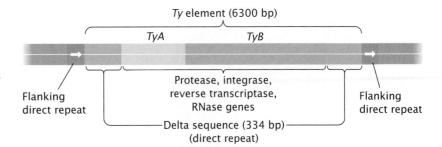

11.25 *Ty* **is a transposable element of yeast.**

Ty element (6300 bp)

TyA   TyB

Flanking direct repeat

Protease, integrase, reverse transcriptase, RNase genes

Flanking direct repeat

Delta sequence (334 bp) (direct repeat)

also generate direct repeats at the point of insertion. Retrotransposons include the *Ty* elements in yeast, the *copia* elements in *Drosophila*, and the *Alu* sequences in humans.

Ty elements in yeast *Ty* (for transposon yeast) elements are a family of common transposable elements found in yeast; many yeast cells have 30 copies of *Ty* elements. These elements are retrotransposons that are about 6300 nucleotide pairs in length and generate 5-bp flanking direct repeats when they insert into DNA (◀ FIGURE 11.25). At each end of a *Ty* element are direct repeats called **delta sequences,** which are 334 bp long. The delta sequences are analogous to the long terminal repeats found in retroviruses (see p. 224 in Chapter 8). These delta sequences contain promoters required for the transcription of *Ty* genes, and the promoters may also stimulate the transcription of genes that lie downstream of the *Ty* element. Between the delta sequences at each end of a *Ty* element are two genes (*TyA* and *TyB*, which encode several enzymes) that are related to the *gag* and *pol* genes found in retroviruses (see p. 224 in Chapter 8). Many *Ty* elements are defective and no longer capable of undergoing transposition.

Ac and Ds elements in maize Transposable elements were first identified in maize (corn), more than 50 years ago by Barbara McClintock (◀ FIGURE 11.26). McClintock spent much of her long career studying their properties, and her work stands among the landmark discoveries of genetics. Her results, however, were misunderstood and ignored for many years. Not until molecular techniques were developed in the late 1960s and 1970s did the importance of transposable elements become widely accepted.

Born in 1902, Barbara McClintock attended Cornell University as an undergraduate and, later, as a graduate student. She was especially interested in genetics, but the subject was taught in the department of plant breeding, which did not accept women students. So she registered for botany instead and studied maize chromosomes for her Ph.D. dissertation.

After receiving her degree, McClintock remained at Cornell, continuing her cytogenetic analysis of maize chromosomes. Her discoveries in the next 10 years included the identification of all the chromosomes in maize, the assignment of linkage groups to chromosomes, proof of crossing over, mapping genes to chromosomes by using rearrangements, and associating chromosome elements with the nucleolus.

11.26 **Barbara McClintock was the first to discover transposable elements.** (CSHL Archives/Peter Arnold.)

McClintock's discovery of transposable elements had its genesis in the early work of Rollins A. Emerson on the maize genes that caused variegated (multicolored) kernels. Most corn kernels are either wholly pigmented or colorless (yellow), but Emerson noted that some yellow kernels had spots or streaks of color (◀ FIGURE 11.27). He proposed that these kernels resulted from an unstable mutation: a mutation in the wild-type gene for pigment produced a colorless kernel; but, in some cells, the mutation reverted back to the wild type, causing a spot of pigment. However, Emerson didn't know why these mutations were unstable.

McClintock discovered that the cause of the unstable mutation was a gene that moved. She noticed that chromosome breakage in maize often occurred at a gene that she called *Dissociation* (*Ds*) but only if another gene, the *Activator* (*Ac*), also was present. *Ds* and *Ac* exhibited unusual patterns of inheritance; occasionally, the genes moved together. McClintock called these moving genes controlling elements, because they controlled the expression of other genes.

McClintock published her conclusion that controlling elements moved in 1948. Although her results were not disputed, they were neither understood nor recognized by most

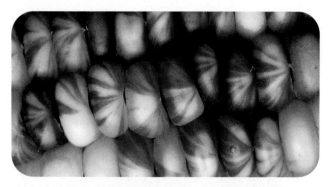

 11.27 **Variegated (multicolored) kernels in corn are caused by mobile genes.** The study of variegated corn led Barbara McClintock to discover transposable elements. (Matt Meadows/Peter Arnold.)

geneticists. Of her work, Alfred Sturtevant, then a prominent geneticist remarked, "I didn't understand one word she said, but if she says it is so, it must be so!" He expressed what seems to have been the attitude of many geneticists at the time. McClintock was frustrated by the genetics community's reaction to her research, but she continued to pursue it nonetheless. In the 1960s, bacteria and bacteriophages were shown to possess transposable elements, and the development of recombinant DNA techniques in the 1970s and 1980s demonstrated that transposable elements exist in all organisms. The significance of McClintock's early discoveries was finally recognized in 1983, when she was awarded the Nobel Prize in Physiology or Medicine.

www.whfreeman.com/pierce   A series of links to Barbara McClintock and her work on transposable elements

Ac and *Ds* elements in maize have now been examined in detail, and their structure and function are similar to those of transposable elements found in bacteria: they possess terminal inverted repeats and generate flanking direct repeats at the points of insertion. *Ac* elements are about 4500 bp long, including terminal inverted repeats of 11 bp, and the flanking direct repeats that they generate are 8 bp in length (◀FIGURE 11.28a). Each *Ac* element contains a single gene that encodes a transposase enzyme. Thus *Ac* elements are *autonomous*—that is, able to transpose. *Ds* elements are *Ac* elements with one or more deletions that have inactivated the transposase gene (◀FIGURE 11.28b). Unable to transpose on their own (*nonautonomous*), *Ds* elements can transpose in the presence of *Ac* elements because they still possess terminal inverted repeats recognized by *Ac* transposase.

Each kernel in an ear of corn is a separate individual, originating as an ovule fertilized by a pollen grain. A kernel's pigment pattern is determined by several loci. A pigment-encoding allele at one of these loci can be designated *C*, and an allele at the same locus that does not confer pigment is designated as *c*. A kernel with genotype *cc* will be colorless—that is, yellow or white (◀FIGURE 11.29a); a kernel with genotype *CC* or *Cc* will produce pigment and be purple (◀FIGURE 11.29b).

A *Ds* element, transposing under the influence of a nearby *Ac* element, may insert into the *C* allele, destroying its ability to produce pigment (◀FIGURE 11.29c). An allele inactivated by a transposable element is designated with a subscript "t"; so in this case it would be designated C_t. After the transposition of *Ds* into the *C* allele, the kernel cell has genotype $C_t c$. This kernel will be colorless (white or yellow), because neither the C_t nor the *c* allele confers pigment.

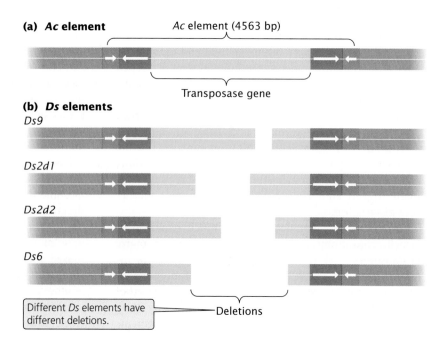

(a) *Ac* element

Ac element (4563 bp)

Transposase gene

(b) *Ds* elements

Ds9

Ds2d1

Ds2d2

Ds6

Different *Ds* elements have different deletions. ——— Deletions

◀**11.28** ***Ac* and *Ds* are transposable elements in maize.**

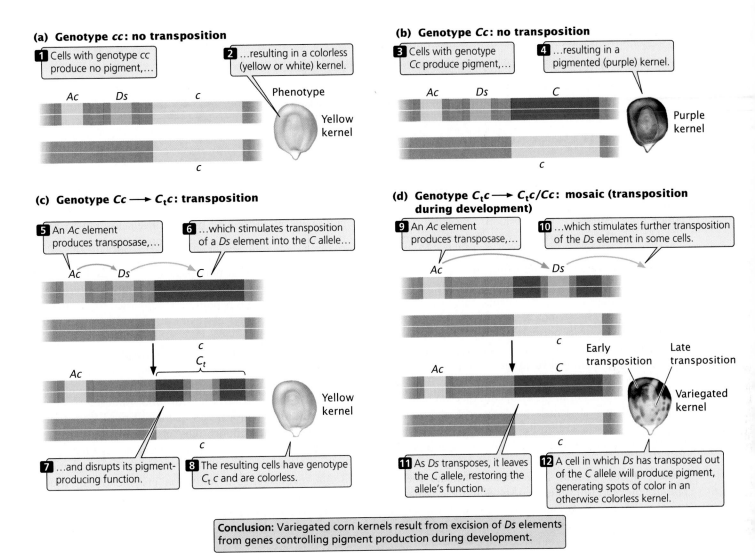

(a) Genotype cc: no transposition

1 Cells with genotype cc produce no pigment,...

2 ...resulting in a colorless (yellow or white) kernel.

Ac   Ds   c

Phenotype

Yellow kernel

c

(b) Genotype Cc: no transposition

3 Cells with genotype Cc produce pigment,...

4 ...resulting in a pigmented (purple) kernel.

Ac   Ds   C

Purple kernel

c

(c) Genotype Cc ⟶ Ctc: transposition

5 An Ac element produces transposase,...

6 ...which stimulates transposition of a Ds element into the C allele...

Ac   Ds   C

c

Ct

Ac

Yellow kernel

c

7 ...and disrupts its pigment-producing function.

8 The resulting cells have genotype Ct c and are colorless.

(d) Genotype Ctc ⟶ Ctc/Cc: mosaic (transposition during development)

9 An Ac element produces transposase,...

10 ...which stimulates further transposition of the Ds element in some cells.

Ac   Ds

c   Early transposition   Late transposition

Ac   C

Variegated kernel

c

11 As Ds transposes, it leaves the C allele, restoring the allele's function.

12 A cell in which Ds has transposed out of the C allele will produce pigment, generating spots of color in an otherwise colorless kernel.

Conclusion: Variegated corn kernels result from excision of Ds elements from genes controlling pigment production during development.

◀ **11.29 Transposition results in variegated maize kernels.**

As development takes place and the original one-celled maize embryo divides by mitosis, additional transpositions may take place in some cells. In any cell in which the transposable element excises from the C_t allele and moves to a new location, the C allele is rendered functional again: all cells derived from those in which this event has taken place will have the genotype Cc and be purple. The presence of these pigmented cells, surrounded by the colorless (C_tc) cells, produces a purple spot or streak (called a sector) in the otherwise yellow kernel (◀ FIGURE 11.29d). The size of the sector varies, depending on when the excision of the transposable element from the C_t allele occurred. If excision occurred early in development, then many cells will contain the functional C allele and the pigmented sector will be large; if excision occurred late in development, few cells will have the functional C allele and the pigmented sector will be small.

Transposable elements in *Drosophila* A number of different transposable elements are found in *Drosophila*. One of the best studied is *copia*, a retrotransposon about 5000 bp long (◀ FIGURE 11.30). *Copia* has direct (i.e., *not* inverted) repeats of 276 bp at each end, and within each direct repeat are terminal inverted repeats. When *copia* transposes, it generates flanking direct repeats that are 5 bp long at the site of insertion. Like *Ty* elements, *copia* contains sequences similar to those found in the *gag* and *pol* genes of retroviruses (see Figure 8.36). The number of *copia* elements in a typical fruit fly genome varies from 20 to 60.

Another family of transposable elements found in *Drosophila* are the *P* elements. Most functional *P* elements are about 2900 bp long, although shorter *P* elements with deletions also exist. Each *P* element possesses terminal inverted repeats that are 31 bp long and generates flanking direct repeats of 8 bp at the site of insertion. Like transpos-

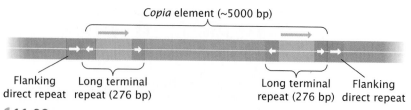

11.30 *Copia* is a transposable element of *Drosophila*.

able elements in bacteria, *P* elements transpose through DNA intermediates. Each element encodes both a transposase and a repressor of transposition.

The role of this repressor in controlling transposition is demonstrated dramatically in **hybrid dysgenesis,** which is the sudden appearance of numerous mutations, chromosome aberrations, and sterility in the offspring of a cross between a P^+ male fly (*with P* elements) and a P^- female fly (*without* them). The reciprocal cross between a P^+ female and a P^- male produces normal offspring.

Hybrid dysgenesis arises from a burst of transposition that takes place when *P* elements are introduced into a cell that does not possess them. A cell that contains *P* elements produces the repressor in the cytoplasm that inhibits transposition. When a P^+ female produces eggs, the repressor protein is incorporated into the egg cytoplasm, which prevents further transposition in the embryo and thus prevents mutations from arising. The resulting offspring are fertile as adults (◀ FIGURE 11.31a). However, a P^- females does not produce the repressor; so none is stored in the cytoplasm of her eggs. When her eggs are fertilized by sperm from a P^+ male, the absence of repression allows the *P* elements contributed by the sperm to undergo rapid transposition in the embryo, causing hybrid dysgenesis (◀ FIGURE 11.31b).

P elements appear to have invaded *D. melanogaster* within the past 50 years. Today, almost all *D. melanogaster* caught in the wild possess *P* elements, but these transposable elements are uncommon in laboratory colonies of flies that were established more than 30 years ago. In fact, no strain of *D. melanogaster* collected before 1945 possesses them, suggesting that *P* elements have recently invaded *D. melanogaster* and have spread rapidly throughout the species.

Because *P* elements are not present in most laboratory stocks, they have been useful experimentally as vectors for introducing modified or foreign DNA into the *Drosophila* genome. *P* elements have been extensively manipulated and engineered for a variety of uses.

If *P* elements are a recent addition to the genome of *D. melanogaster*, where did they come from? A likely source is *Drosophila willistoni*, another fruit fly species. *D. willistoni* appears to have long possessed *P* elements that are virtually identical with those now found in *D. melanogaster*. Researchers Marilyn Houck and Margaret Kidwell proposed that the *P* elements made the leap from *D. willistoni* to *D. melanogaster* by hitching a ride on a mite.

All fruit flies are infected with a variety of mites. One mite species, *Proctolaelaps regalis*, infests both *D. willistoni* and *D. melanogaster*. This mite has needlelike mouth parts that allow it to pierce and feed on the eggs and larvae of the flies. Houck and Kidwell suggest that, while feeding on *D. willistoni*, a mite picked up fruit fly DNA with *P* elements, which it later injected into a developing *D. melanogaster*. This hypothesis is supported by the finding that mites do pick up *P* element DNA from P^+ fruit flies.

www.whfreeman.com/pierce   More information on *copia* and *P* elements

Transposable elements in humans Almost 50% of the human genome consists of sequences derived from transposable elements, although most of these elements are now inactive and no longer capable of transposing. One of the most common transposable elements in the human genome is *Alu*, named after a restriction enzyme (*Alu*I), which cleaves the element into two parts. Every human cell contains more than 1 million related, but not identical, copies of *Alu* in its chromosomes. Unlike the retrotransposons we have described earlier (*Ty* elements from yeast and *copia* elements from *Drosophila*), *Alu* sequences are not similar to retroviruses. They do not have genes resembling *gag* and *pol*, and are therefore nonautonomous. Rather, *Alu* sequences are similar to the gene that encodes the 7S RNA molecule, which transports newly synthesized proteins across the endoplasmic reticulum. *Alu* sequences create short flanking direct repeats when they insert into DNA and have characteristics that suggest that they have transposed through an RNA intermediate.

Alu belongs to a class of repetitive sequences found frequently in mammalian and some other genomes. These sequences are collectively referred to as SINEs, (short interspersed sequences). The human genome also has many LINEs (long interspersed sequences), which are somewhat more similar in structure to retroviruses, but not as similar as *Ty* or *copia*.

The human genome contains evidence for several classes of transposable elements that transpose through a DNA intermediate, by the cut-and-paste mechanism. However, these all appear to have been inactive for about 50 million years; the nonfunctional sequences that remain have been referred to as DNA fossils.

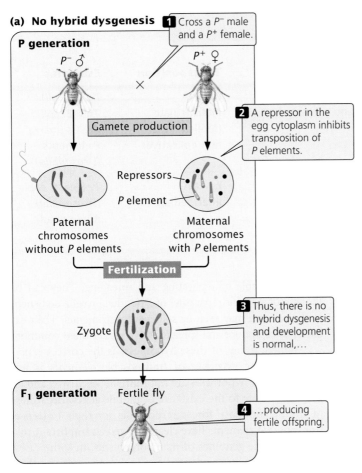

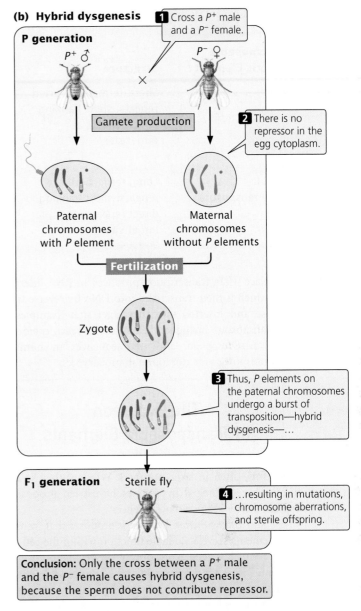

◄11.31 Hybrid dysgenesis in *Drosophila* is caused by the transposition of *P* elements.

Conclusion: Only the cross between a P^+ male and the P^- female causes hybrid dysgenesis, because the sperm does not contribute repressor.

Concepts

A great variety of transposable elements exist in eukaryotes. Some resemble transposable elements in prokaryotes, having terminal inverted repeats, and transpose through a DNA intermediate. Others are retrotransposons with long direct repeats at their ends and transpose through an RNA intermediate.

Connecting Concepts

Classes of Transposable Elements

Now that we have examined the process of transposition, let us review the major classes of transposable elements (Table 11.6).

Transposable elements can be divided into two major classes on the basis of structure and movement. The first class consists of elements that possess terminal inverted repeats and transpose through DNA intermediates. They all generate flanking direct repeats at their points of insertion into DNA. All active forms of these transposable elements encode transposase, which is required for their movement. Some also encode resolvase, repressors, and other proteins. Their transposition may be replicative or nonreplicative, but they never use RNA intermediates. Examples of transposable elements in this first class include insertion sequences and all complex transposons in bacteria, the *Ac* and *Ds* elements of maize, and the *P* elements of *Drosophila*.

The second class of transposable elements are the retrotransposons, which transpose through RNA intermediates. They generate flanking direct repeats at their points of insertion when they transpose into DNA. Retrotransposons do not encode transposase, but some types are similar in structure to retroviruses and carry sequences that produce reverse transcriptase. Transposition

Table 11.6	Characteristics of two major classes of transposable genetic elements			
Transposable Genetic Element	**Structure**	**Genes Encoded**	**Transposition**	**Examples**
Class I	Short, terminal inverted repeats; short flanking direct repeats at target site	Transposase gene (and sometimes others)	By DNA intermediate (replicative or nonreplicative)	*IS1 (E. coli)* *Tn3 (E. coli)* *Ac, Ds* (maize) *P* elements *(Drosophila)*
Class II (retrotransposon)	Long, terminal direct repeats; short flanking direct repeats at target site	Reverse transcriptase gene (and sometimes others)	By RNA intermediate	*Ty* (yeast) *copia (Drosophila)* *Alu* (human)

takes place when transcription produces an RNA intermediate, which is then transcribed into DNA by reverse transcriptase and inserted into the target site. Examples of retrotransposons include *Ty* elements in yeast, *copia* elements in *Drosophila,* and *Alu* sequences in humans. Retrotransposons are not found in prokaryotes.

The Evolution of Transposable Elements

As mentioned earlier, transposable elements exist in all organisms, often in large numbers. Why are they so common? Three principal hypotheses have been proposed to explain their widespread occurrence.

The *cellular function hypothesis* proposes that transposable elements serve a valuable function within the cell, such as the control of gene expression or the regulation of development. Although the insertion of a transposable element can alter gene expression, there are few data to suggest that transposition plays a routine role in either of these or any other cellular processes.

The *genetic variation hypothesis* proposes that transposable elements exist because of their mutagenic activity. It suggests that a certain amount of genetic variation is useful because it allows a species to adapt to environmental change. Although some mutations caused by transposable elements may allow species to evolve beneficial traits, the vast majority of mutations generated by random transposition have deleterious effects. Thus, although mutations produced by transposable elements may be useful in the future, their immediate effect is usually deleterious and they will be rejected. The fact that many organisms have evolved mechanisms to regulate transposition suggests that there is selective pressure to limit the extent of transposition. In fact, if their only effect were to generate mutations, transposable genetic elements could be expected to disappear in time.

The *selfish DNA hypothesis* asserts that transposable elements serve no purpose for the cell; they exist simply because they are capable of replicating and spreading. They can be thought of as "selfish" parasites of DNA that provide no benefit to the cell and may even be somewhat detrimental. Their capacity to reproduce and spread is what makes them common.

Which, if any, of these hypotheses is the correct explanation for the existence of transposable elements is not known. These hypotheses are not mutually exclusive, and all may contribute to the existence of mobile genes. Regardless of the evolutionary forces responsible for their existence, transposable elements have clearly played an important role in shaping the genomes of many organisms. In some cases, they have even been adopted for useful purposes by their host cells. One example is the mechanism that generates antibody diversity in the immune systems of vertebrates.

As will be discussed in Chapter 21, the ability of the immune system to recognize and attack foreign substances (antigens) depends on a mechanism whereby lymphocytes join several DNA segments that code for antigen-recognition proteins. Three DNA segments, called V, D, and J, exist in multiple forms within each cell. In the development of a lymphocyte, particular V, D, and J segments are randomly joined to produce a protein that recognizes a specific antigen. Within different lymphocytes, different V, D, and J segments are joined together in different combinations. The variety of combinations provides a large array of cells, each of which recognizes a particular antigen. Close examination of the V, D, and J joining process reveals that its mechanism is the same as that for transposition. The genes—designated *RAG1* and *RAG2*—participating in bringing about V, D, and J joining may have at one time been transposable elements that inserted into the germ line of a vertebrate ancestor, some 450 million years ago.

Another cellular function that may have originated as the result of a transposable element is the process that maintains the ends of chromosomes in eukaryotic organisms. As mentioned earlier in this chapter, DNA polymerases are unable to replicate the ends of chromosomes. In germ cells and single-celled eukaryotic organisms, chromosome length is maintained by telomerase, an enzyme that extends the chromosome ends by copying repeated DNA sequences from an RNA template that is a part of the telomerase enzyme. The mechanism used by the telomerase enzyme is similar to the reverse transcription process used in retrotransposition, and telomerase is evolutionarily related to the reverse transcriptases encoded by certain retrotransposons.

These findings suggest that an invading retrotransposon in an ancestral eukaryotic cell may have provided the ability to copy the ends of chromosomes and eventually evolved into the gene that encodes the modern telomerase enzyme. *Drosophila* lacks the telomerase enzyme; retrotransposons appear to have resumed the role of telomere maintenance in this case.

Connecting Concepts Across Chapters

The material covered in this chapter has important connections to several topics already covered and to others in chapters yet to come. In Chapter 2, the gross structure of chromosomes and their behavior during mitosis and meiosis were introduced. The present chapter has built on that introduction by examining the molecular details of chromosome structure and the higher-level folding and packing of DNA that allows these very large molecules to maintain their functionality and still fit into the confined space of the cell. The solution to this cellular storage problem and the essential elements of eukaryotic chromosomes have been major themes of this chapter, completing the story of DNA structure introduced in Chapter 10.

Transposable genetic elements, DNA sequences that move, are a part of chromosome structure. Earlier chapters dealt with crossing over, in which homologous DNA sequences switch places, and chromosome rearrangements, in which the breakage and rejoining of chromosome segments moves blocks of genes to new locations. The movement of transposable elements is fundamentally different from these other mechanisms of gene movement because transposable elements possess sequences that facilitate their movement. Understanding the structure of transposable genetic elements requires a basic knowledge of DNA structure and sequence, topics covered in Chapter 10.

Transposable elements violate a basic premise of classical genetics—that genes have a particular fixed location on a chromosome. This departure from a long-held view helps to explain why the discovery of transposable elements by Barbara McClintock was ignored for many years. A common theme in the history of genetics is that fundamental discoveries are often overlooked or unrecognized, because they require a radical rethinking of basic principles. Transposable elements today are recognized as ubiquitous DNA sequences with important implications for medicine, recombinant DNA technology, and evolution, but the reason for their widespread occurrence is still not completely understood.

This chapter has provided a foundation for topics introduced in several later chapters of the book. Transposition requires the replication of DNA (Chapter 12) or reverse transcription (Chapter 14) and generates gene mutations (Chapter 17). In Chapter 16, we explore the control of gene expression, which requires changes in chromatin structure. Condensed chromatin structure tends to inhibit the transcription of genetic information; some of the proteins that take part in activating and repressing transcription are known to affect the binding of DNA to histones. The regulation of transposition is by some of the same mechanisms that regulate the expression of other genes, also discussed in Chapter 16. Additional topics covered in more detail in later chapters include the origins of replication (Chapter 12) and the application of repetitive sequences to DNA fingerprinting (Chapter 18). Transposable elements are important in the generation of immune-system diversity (Chapter 21) and in molecular evolution (Chapter 23).

CONCEPTS SUMMARY

- Chromosomes contain very long DNA molecules that are tightly packed. Packing is accomplished through tertiary structures and the binding of DNA to proteins.

- Supercoiling results from strain produced when rotations are added or removed from a relaxed DNA molecule. Overrotation produces positive supercoiling; underrotation produces negative supercoiling.

- Topoisomerases control the degree of supercoiling by adding or removing rotations to DNA.

- A bacterial chromosome consists of a single, circular DNA molecule that is bound to proteins and exists as a series of large loops. It usually appears in the cell as a distinct clump known as the nucleoid.

- Each eukaryotic chromosome contains a single, very long linear DNA molecule that is bound to histone and nonhistone chromosomal proteins. Euchromatin undergoes the normal cycle of decondensation and condensation in the cell cycle. Heterochromatin remains highly condensed throughout the cell cycle.

- The nucleosome is a core of eight histone proteins (two each of H2A, H2B, H3, and H4) and DNA (145–147 bp) that wraps around it. The H1 protein holds DNA onto the histone core.

- Nucleosomes are folded into a 30-nm fiber that forms a series of 300-nm-long loops; these loops are anchored at their bases by proteins associated with the nuclear scaffold. The 300-nm loops are condensed to form a fiber that is 250 nm in diameter, which is itself tightly coiled to produce a 700-nm-wide chromatid.

- Chromosomal puffs are regions of localized unpacking of the DNA that are associated with regions of active transcription. Chromosome regions that are undergoing active transcription are relatively sensitive to digestion by DNase I, indicating that DNA unfolds during transcription.

- Centromeres are chromosomal regions where spindle fibers attach; chromosomes without centromeres are usually lost in the course of cell division. Centromeres play an important role in the regulation of the cell cycle.

- Telomeres stabilize the ends of chromosomes. Telomeric sequences consist of many copies of short sequences, which usually consist of a series of cytosine nucleotides followed by several adenine nucleotides. Longer telomere-associated sequences are found adjacent to the telomeric sequences.

- The C value is the amount of DNA in an organism's genome. Eukaryotic organisms exhibit much variation in C value owing to differences in sequence complexity, which can be measured by observing the time required for denatured DNA to reanneal in a hybridization reaction, as plotted by a C_0t curve.

- Eukaryotic DNA exhibits three classes of sequences. Unique-sequence DNA exists in very few copies. Moderately repetitive DNA consists of moderately long sequences that are repeated from hundreds to thousands of times. Highly repetitive DNA consists of very short sequences that are repeated in tandem from many thousands to millions of times.

- Transposable elements are mobile DNA sequences that insert into many locations within a genome and often cause mutations and DNA rearrangements.

- Most transposable elements have two common characteristics: terminal inverted repeats and the generation of short direct repeats in DNA at the point of insertion.

- Transposition may take place through a DNA molecule or through the production of an RNA molecule that is then reverse transcribed into DNA. Transposition may be replicative, in which the transposable element is copied and the copy moves to a new site, or nonreplicative, in which the transposable element excises from the old site and moves to a new site.

- Retrotransposons transpose through RNA molecules that undergo reverse transcription to produce DNA.

- In many transposable elements, transposition is tightly regulated.

- Insertion sequences are small bacterial transposable elements that carry only the information needed for their own movement. Composite transposons in bacteria are more complex elements that consist of DNA between two insertion sequences. Some complex transposable elements in bacteria do not contain insertion sequences.

- Some transposable elements in eukaryotic cells are similar to those found in bacteria, ending in short inverted repeats and producing flanking direct repeats at the point of insertion. Others are retrotransposons, similar in structure to retroviruses and transposing through RNA intermediates.

- Hybrid dysgenesis is the appearance of numerous mutations, chromosome rearrangements, and sterility when transposable P elements undergo a burst of transposition in *Drosophila*.

- The evolutionary significance of transposable elements is unknown, but three hypotheses have been proposed to explain their common occurrence. The cellular function hypothesis suggests that transposable elements provide some important function for the cell; the genetic variation hypothesis proposes that transposable elements provide evolutionary flexibility by inducing mutations; and the selfish DNA hypothesis suggests that transposable elements do not benefit the cell but are widespread because they can replicate and spread.

IMPORTANT TERMS

transgenic mouse (p. 288)
transposable element (p. 289)
supercoiling (p. 289)
relaxed state of DNA (p. 289)
positive supercoiling (p. 290)
negative supercoiling (p. 290)
topoisomerase (p. 290)
nucleoid (p. 291)
euchromatin (p. 291)
heterochromatin (p. 291)
nonhistone chromosomal proteins (p. 291)
chromosomal scaffold protein (p. 292)
high-mobility-group proteins (p. 292)

nucleosome (p. 292)
chromatosome (p. 293)
linker DNA (p. 293)
polytene chromosome (p. 294)
chromosomal puff (p. 294)
centromeric sequence (p. 296)
telomeric sequence (p. 298)
telomere-associated sequence (p. 298)
C value (p. 298)
denaturation (melting) (p. 299)
melting temperature (T_m) (p. 299)
renaturation (reannealing) (p. 299)
hybridization (p. 299)

unique-sequence DNA (p. 300)
repetitive DNA (p. 300)
moderately repetitive DNA (p. 301)
tandem repeat sequences (p. 301)
interspersed repeat sequences (p. 301)
short interspersed element (SINE) (p. 301)
long interspersed element (LINE) (p. 301)
highly repetitive DNA (p. 301)
flanking direct repeats (p. 301)
terminal inverted repeats (p. 302)

transposition (p. 302)
replicative transposition (p. 303)
nonreplicative transposition (p. 303)
cointegrate structure (p. 303)
transposase (p. 303)
resolvase (p. 303)
retrotransposon (p. 304)
insertion sequence (p. 307)
composite transposon (p. 307)
delta sequence (p. 309)
hybrid dysgenesis (p. 312)

Worked Problems

1. A diploid plant cell contains 2 billion base pairs of DNA.

(a) How many nucleosomes are present in the cell?

(b) Give the numbers of molecules of each type of histone protein associated with the genomic DNA.

• **Solution**

Each nucleosome encompasses about 200 bp of DNA: from 144 to 147 bp of DNA wrapped twice around the histone core, from 20 to 22 bp of DNA associated with the H1 protein, and another 30 to 40 bp of linker DNA.

(a) To determine how many nucleosomes are present in the cell, we simply divide the total number of base pairs of DNA (2×10^9 bp) by the number of base pairs per nucleosome:

$$\frac{2 \times 10^9 \text{ nucleotides}}{2 \times 10^2 \text{ nucleotides per nucleosome}} = 1 \times 10^7 \text{ nuclesomes}$$

Thus, there are approximately 10 million nucleosomes in the cell.

(b) Each nucleosome includes two molecules each of H2A, H2B, H3, and H4 histones. Therefore, there are 2×10^7 molecules each of H2A, H2B, H3, and H4 histones. Each nucleosome has associated with it one copy of the H1 histone; so there are 1×10^7 molecules of H1.

2. A renaturation reaction is carried out on the genomic DNA from three different bacterial species. Species I has a genome size of 2×10^6 bp, species II has a genome size of 1×10^8 bp, and species III has a genome size of 1×10^6 bp. Assume that the same total amount of DNA is used in each renaturation reaction and draw a C_0t curve for each species, showing the relative positions of each species on the same graph.

• **Solution**

Because this DNA is from bacteria, which contain only unique-sequence DNA, the complexity of renaturation with repetitive DNA can be ignored. If the total amount of DNA is the same for all three bacterial species, the number of copies of each sequence will depend on the genome size: there are fewer different sequences in organisms with small genomes, and so a chance collision is more likely to be between two complementary sequences that will anneal. Consequently, renaturation will proceed more rapidly in organisms with smaller genomes. (Recall the analogy of drawing cards from a hat; when there are only a few different names on the cards, the hat empties more quickly.)

At the start of the reaction, all the DNA is single stranded; so the proportion of single-stranded DNA is 1. As the reaction proceeds, single-stranded DNA pairs to form double-stranded DNA; so the proportion of single-stranded DNA decreases. This decrease will occur at a low C_0t in the organisms with a smaller genome, as shown in the following graph.

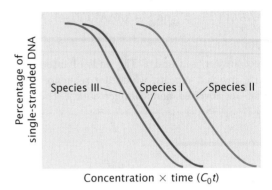

3. Genomic DNAs from species I, II, and III have the following base compositions:

			%	
Species	**A**	**G**	**T**	**C**
I	10	40	10	40
II	27	23	27	23
III	46	4	46	4

DNA from which species has a higher T_m? Explain your reasoning.

• **Solution**

The melting temperature (T_m) of DNA depends on its base sequence. The three hydrogen bonds of a G–C base pair require more energy to break than do the two hydrogen bonds of an A–T pair; so a molecule with a higher percentage of G–C pairs will have a higher T_m. Species I has the highest G–C content of the three species; so it should exhibit the highest T_m.

4. Certain repeated sequences in eukaryotes are flanked by short direct repeats, suggesting that they originated as transposable elements. These same sequences lack introns and possess a string of thymine nucleotides at their 3′ ends. Have these elements transposed through DNA or RNA sequences? Explain your reasoning.

• **Solution**

The absence of introns and the string of thymine nucleotides (which would be complementary to adenine nucleotides in RNA) at the 3′ end are characteristics of processed RNA. These similarities to RNA suggest that the element was originally transcribed into mRNA, processed to remove the introns and to add a poly(A) tail, and then reverse transcribed into a complementary DNA that was inserted into the chromosome.

5. Which of the following pairs of sequences might be found at the ends of an insertion sequence?

(a) 5′–TAAGGCCG–3′ and 5′–TAAGGCCG–3′

(b) 5′–AAAGGGCTA–3′ and 5′–ATCGGGAAA–3′

(c) 5′–GATCCCAGTT–3′ and 5′–CTAGGGTCAA–3′

(d) 5′–GATCCAGGT–3′ and 5′–ACCTGGATC–3′

(e) 5′–AAAATTTT–3′ and 5′–TTTTAAAA–3′

(f) 5′–AAAATTTT–3′ and 5′–AAAATTTT–3′

• Solution

The correct answer is *d* and *f*. The ends of insertion sequences always have inverted repeats, which are sequences on the same strand that are inverted and complementary. The sequences in part *a* are direct repeats, which are generated on the outside of an insertion sequence but are not part of the transposable element itself. The sequences in part *b* are inverted but not complementary. The sequences in part *c* are complementary but not inverted. The sequences in part *d* are both inverted and complementary. The sequences in part *e* are complementary but not inverted. Interestingly, the sequences in part *f* are both inverted complements and direct repeats.

The New Genetics
MINING GENOMES

INTRODUCTION TO BLAST AND BLAST SEARCHING

This exercise casts you in the role of biological detective, trying to figure out the functions of newly discovered genes. The simplest way to determine what is encoded by new sequences is to compare them with information already in the databases by using BLAST (Basic Local Alignment Search Tools). You will use the National Center for Biotechnology Information (NCBI) Web site to explore some of the strengths and weaknesses of this powerful approach.

COMPREHENSION QUESTIONS

* 1. How does supercoiling arise? What is the difference between positive and negative supercoiling?

2. What functions does supercoiling serve for the cell?

* 3. Describe the composition and structure of the nucleosome. How do core particles differ from chromatosomes?

4. Describe in steps how the double helix of DNA, which is 2 nm in width, gives rise to a chromosome that is 700 nm in width.

5. What are polytene chromosomes and chromosomal puffs?

* 6. Describe the function and molecular structure of the centromere.

* 7. Describe the function and molecular structure of a telomere.

8. What is the C value of an organism?

9. What is a C_0t curve? Explain how C_0t curves of DNA provide evidence for the existence of repetitive DNA in eukaryotic cells.

*10. Describe the different types of DNA sequences that exist in eukaryotes.

*11. What general characteristics are found in many transposable elements? Describe the differences between replicative and nonreplicative transposition.

*12. What is a retrotransposon and how does it move?

*13. Describe the process of replicative transposition through DNA intermediates. What enzymes are required?

*14. Draw and label the structure of a typical insertion sequence.

15. Draw and label the structure of a typical composite transposon in bacteria.

16. How are composite transposons and retrotransposons alike and how are they different?

17. Explain how *Ac* and *Ds* elements produce variegated corn kernels.

18. Briefly explain hybrid dysgenesis and how *P* elements lead to hybrid dysgenesis.

*19. Briefly summarize three hypotheses for the widespread occurrence of transposable elements.

APPLICATION QUESTIONS AND PROBLEMS

*20. Compare and contrast prokaryotic and eukaryotic chromosomes. How are they alike and how do they differ?

21. (a) In a typical eukaryotic cell, would you expect to find more molecules of the H1 histone or more molecules of the H2A histone? Explain your reasoning. (b) Would you expect to find more molecules of H2A or more molecules of H3? Explain your reasoning.

22. Suppose you examined polytene chromosomes from the salivary glands of fruit fly larvae and counted the number of chromosomal puffs observed in different regions of DNA.

(a) Would you expect to observe more puffs from euchromatin or from heterochromatin? Explain your answer.

(b) Would you expect to observe more puffs in unique- sequence DNA, moderately repetitive DNA, or repetitive DNA? Why?

*23. A diploid human cell contains approximately 6 billion base pairs of DNA.

   (a) How many nucleosomes are present in such a cell? (Assume that the linker DNA encompasses 40 bp.)

   (b) How many histone proteins are complexed to this DNA?

*24. Would you expect to see more or less acetylation in regions of DNA that are sensitive to digestion by DNase I? Why?

25. A YAC that contains only highly repetitive, nonessential DNA is added to mouse cells that are growing culture. The cells are divided into two groups, A and B. A laser is then used to damage the centromere on the YACs in cells of group A. The centromeres on the YACs of group B are not damaged. In spite of the fact that the YACs contain no essential DNA, the cells in group A divide more slowly than those in group B. Provide a possible explanation.

26. Species A possesses only unique-sequence DNA. Species B possesses unique-sequence DNA and highly repetitive DNA. Species C possesses only moderately repetitive DNA. The genomes of all three species are similar in size. A student performs typical renaturation reactions with DNA from each species and plots a $C_0 t$ curve for each. Draw a $C_0 t$ curve for the renaturation reaction of each species.

*27. Which of the following two molecules of DNA has the lower melting temperature? Why?

   AGTTACTAAAGCAATACATC
   TCAATGATTTCGTTATGTAG

   AGGCGGGTAGGCACCCTTA
   TCCGCCCATCCGTGGGAAT

28. DNA was isolated from a newly discovered worm collected near a deep-sea vent in the Pacific Ocean. This DNA was sheared into pieces, heated to melting, and then cooled slowly. The amount of renaturation was measured with optical absorbance, and the following results were obtained.

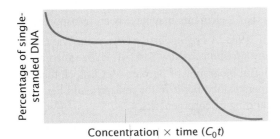

What conclusions can you draw about the type of sequences found in this DNA?

29. Which of the following pairs of sequences might be found at the ends of an insertion sequence?

   (a) 5′–GGGCCAATT–3′ and 5′–CCCGGTTAA–3′

   (b) 5′–AAACCCTTT–3′ and 5′–AAAGGGTTT–3′

   (c) 5′–TTTCGAC–3′ and 5′–CAGCTTT–3′

   (d) 5′–ACGTACG–3′ and 5′–CGTACGT–3′

   (e) 5′–GCCCCAT–3′ and 5′–GCCCAT–3′

*30. A particular transposable element generates flanking direct repeats that are 4 bp long. Give the sequence that will be found on both sides of the transposable element if this transposable element inserts at the position indicated on each of the following sequences.

   (a) Transposable element

   5′ – ATTCGAACTGACCGATCA – 3′

   (b) Transposable element

   5′ – ATTCGAACTGACCGATCA – 3′

*31. White eyes in *Drosophila melanogaster* result from an X-linked recessive mutation. Occasionally, white-eyed mutants give rise to offspring that possess white eyes with small red spots. The number, distribution, and size of the red spots are variable. Explain how a transposable element could be responsible for this spotting phenomenon.

*32. An insertion sequence contains a large deletion in its transposase gene. Under what circumstances would this insertion sequence be able to transpose?

*33. What factor do you think determines the length of the flanking direct repeats that are produced in transposition?

34. A transposable element is found to encode a transposase enzyme. On the basis of this information, what conclusions can you make about the likely structure and method of transposition of this element?

35. A transposable element is found to encode a reverse transcriptase enzyme. On the basis of this information, what conclusions can you make about the likely structure and method of transposition of this element?

36. Transposition often produces chromosome rearrangements, such as deletions, inversions, and translocations. Can you suggest a reason why transposition leads to these chromosome mutations?

37. A geneticist studying the DNA of the Japanese bottle fly finds many copies of a particular sequence that appears similar to the *copia* transposable element in *Drosophila*. Using recombinant DNA techniques, the geneticist places an intron into a copy of this DNA sequence and inserts it into the genome of a Japanese bottle fly. If the sequence is a transposable element similar to *copia*, what prediction would you make concerning the fate of the introduced sequence in the genomes of offspring of the fly receiving it?

CHALLENGE QUESTIONS

38. An explorer discovers a strange new species of plant and sends some of the plant tissue to a geneticist to study. The geneticist isolates chromatin from the plant and examines it with the electron microscope. She observes what appear to be beads on a string. She then adds a small amount of nuclease, which cleaves the string into individual beads that each contain 280 bp of DNA. After digestion with more nuclease, she finds that a 120-bp fragment of DNA remains attached to a core of histone proteins. Analysis of the histone core reveals histones in the following proportions:

H1	12.5%
H2A	25%
H2B	25%
H3	0%
H4	25%
H7 (a new histone)	12.5%

On the basis of these observations, what conclusions could the geneticist make about the probable structure of the nucleosome in the chromatin of this plant?

39. Although highly repetitive DNA is common in eukaryotic chromosomes, it does not code for proteins; in fact, it is probably never transcribed into RNA. If highly repetitive DNA does not code for RNA or proteins, why is it present in eukaryotic genomes? Suggest some possible reasons for the widespread presence of highly repetitive DNA.

40. As discussed in the chapter, *Alu* sequences are retrotransposons that are common in the human genome. *Alu* sequences are thought to have evolved from the 7S RNA gene, which encodes an RNA molecule that takes part in transporting newly synthesized proteins across the endoplasmic reticulum. The 7S RNA gene is transcribed by RNA polymerase III, which uses an internal promoter (see Chapter 13). How might this observation explain the large number of copies of *Alu* sequences?

41. Houck and Kidwell proposed that *P* elements were carried from *Drosophila willistoni* to *D. melanogaster* by mites that fed on fruit flies. What evidence do you think would be required to demonstrate that *D. melanogaster* acquired *P* elements in this way? Propose a series of experiments to provide such evidence.

SUGGESTED READINGS

Beermann, W., and U. Clever. 1964. Chromosome puffs. *Scientific American* 210(4):50–58.
  Describes early research on chromosome puffs that led to the conclusion that puffs are areas of active transcription.

Blackburn, E. H. 2000. Telomere states and cell fates. *Nature* 408:53–56.
  Proposal that telomere length is less important to cell aging than telomere capping.

Burlingame, R. W., W. E. Love, B. C. Wang, R. Hamlin, H. X. Nguyen, and E. N. Moudrianakis. 1985. Crystallographic structure of the octameric histone core of the nucleosome at a resolution of 3.3 Å. *Science* 228:546–553.
  A detailed description of the histone octamer based on X-ray crystallographic data.

Cohen, S. N., and J. A. Shapiro. 1980. Transposable genetic elements. *Scientific American* 242(2):40–49.
  A readable review of transposable elements.

Fedoroff, N. V. 1993. Barbara McClintock (June 16, 1902–September 2, 1992). Genetics 136:1–10.
  An excellent summary of Barbara McClintock's life and her influence in genetics.

Grindley, N. D. F., and R. R. Reed. 1985. Transpositional recombination in prokaryotes. *Annual Review of Biochemistry* 54:863–896.
  A thorough review of mechanisms of transposition.

Greider, C. W., and E. H. Blackburn. 1996. Telomeres, telomerase, and cancer. *Scientific American* 274(2):92–97.
  An excellent review of telomeres and telomerase.

Hagmann, M. 1999. How chromatin changes its shape. *Science* 285:1200–1203.
  A report of research on how chromatin changes its shape in response to various genetic functions.

Houck, M. A., J. B. Clark, K. R. Peterson, and M. G. Kidwell. 1991. Possible horizontal transfer of *Drosophila* genes by the mite *Proctolaelaps regalis*. *Science* 253:1125–1129.
  Reports that *P* elements may have been transported by mites.

Keller, E. F. 1983. *A Feeling for the Organism: The Life and Work of Barbara McClintock*. New York: W. H. Freeman and Company.
  A wonderful biography of Barbara McClintock that captures her unique personality, her love for research, and her deep understanding of corn genetics.

Kornberg, R. D., and A. Klug. 1981. The nucleosome. *Scientific American* 244(2):52–64.
  A good review of basic chromatin structure and how it was discovered.

Luger, K., A. W. Mäder, R. K. Richmond, D. F. Sargent, and T. J. Richmond. 1997. Crystal structure of the nucleosome core particle at 2.8 Å resolution. *Nature* 389:251–260.
  A report of the detailed structure of the nucleosome as revealed by X-ray crystallography.

McEachern, M. J., A. Krauskopf, and E. H. Blackburn. 2000. Telomeres and their control. *Annual Review of Genetics* 34:331–358.
A review of telomere structure and control.

Murray, A. W., and J. W. Szostak. 1987. Artificial chromosomes. *Scientific American* 257(5):62–68.
Describes how YACs are created and gives the history of their discovery.

Ng, H. H., and A. Bird. 2000. Histone deacetylases: silencers for hire. *Trends in Biochemical Science* 25:121–126.
Reviews the role of acetylation in the control of chromatin structure and gene expression.

Pluta, A. F., A. M. Mackay, A. M. Ainsztein, I. G. Goldberg, and W. C. Earnshaw. 1995. The centromere: hub of chromosomal activities. *Science* 270:1591–1594.
An excellent review of centromere structure and function.

Syvanen, M. 1984. The evolutionary implications of mobile genetic elements. *Annual Review of Genetics* 18:271–293.
A review that discusses the evolutionary significance of transposable elements.

Travers, A. 1999. The location of the linker histone on the nucleosome. *Trends in Biochemical Science* 24:4–7.
Reviews and discusses different models concerning the location of the linker H1 histone in the chromatosome.

Voytas, D. F. 1996. Retroelements in genome organization. *Science* 274:737–738.
A discussion of the organization of transposable elements in maize.

Weiner, A. M., P. L. Deininger, and A. Efstatiadis. 1986. Nonviral retroposon: genes, pseudogenes, and transposable elements generated by the reverse flow of genetic information. *Annual Review of Biochemistry* 55:631–661.
A review of retrotransposons and related genetic elements.

Wolffe, A. P. 1998. *Chromatin: Structure and Function,* 3d ed. San Diego: Academic Press.
A detailed and current review of chromatin structure and function.

Zakian, V. A. 1995. Telomeres: beginning to understand the end. *Science* 270:1601–1606.
An excellent review of telomere structure and function.

12

DNA Replication and Recombination

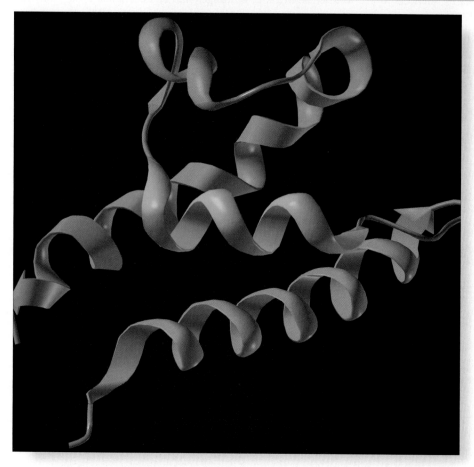

Molecular model of Recq helicase, an enzyme that unwinds DNA during replication and repair. Bloom syndrome is caused by mutation of a gene that encodes Recq helicase.

The Cause of Bloom Syndrome

Tommy was a full-term baby but weighed only 4.5 pounds (2 kg) at birth. At about 9 months of age, an unusual and persistent rash appeared on his face, and he frequently caught colds and infections. The illnesses caused no serious problems; so his parents were not concerned. Throughout childhood, Tommy remained small; by age 18, he was only 4 feet 6 inches (137 cm) in height.

Tommy's first major health problem arose shortly after he turned 22—he was diagnosed with intestinal cancer. The tumor was surgically removed but additional, unrelated tumors appeared spontaneously over the next 10 years. Their appearance startled Tommy's doctors; the chance of multiple, independent cancers arising in the same person is generally remote. The propensity of Tommy's cells to become cancerous hinted at a high mutation rate in his genes. Indeed, when pathologists studied Tommy's chromosomes, they observed a wide range of abnormalities. Tommy had inherited Bloom syndrome.

Bloom syndrome is a rare autosomal recessive condition characterized by short stature, a facial rash induced by sun exposure, a small narrow head, and a predisposition to cancers of all types. The disorder is extremely rare; only several hundred cases have been reported worldwide. Cells from persons with Bloom syndrome exhibit excessive mutations in all genes, and numerous gaps and breaks occur in chromosomes that lead to extensive genetic exchange in cell division. Rates of DNA synthesis are retarded.

The characteristics of Bloom syndrome suggest that its underlying cause is a defect in DNA synthesis. In 1995, researchers at the New York Blood Center traced Bloom syndrome to a gene on chromosome 15 that encodes an enzyme called DNA helicase. A variety of helicase enzymes are responsible for unwinding double-stranded DNA during replication and repair. The cells of a person with Bloom syndrome carry two mutated copies of the gene and possess little or no activity for a particular helicase. Normal DNA synthesis is disrupted, leading to chromosome breaks and numerous mutations. The genetic damage resulting from faulty DNA synthesis leads to tumors. It is not yet clear whether the basic defect in DNA synthesis is associated with replication or DNA repair or both.

Rapid and accurate DNA replication is fundamental to normal cell function and health. Replication is a complex process in which dozens of proteins, enzymes, and DNA structures take part; a single defective component, such as DNA helicase, can disrupt the whole process.

This chapter deals with DNA replication, the process whereby a cell doubles its DNA before division. We begin with the basic mechanism of replication that emerged from the Watson and Crick structure of DNA. We then examine several different modes of replication, the requirements of replication, and the universal direction of DNA synthesis. We examine the enzymes and proteins that participate in the process and conclude the chapter by considering the molecular details of recombination, which is closely related to replication and is essential for the segregation of homologous chromosomes in meiosis, production of genetic variation, and for DNA repair.

www.whfreeman.com/pierce   More information about the symptoms and genetics of Bloom syndrome

The Central Problem of Replication

In a schoolyard game, a verbal message, such as "John's brown dog ran away from home," is whispered to a child, who runs to a second child and repeats the message. The message is relayed from child to child around the schoolyard until it returns to the original sender. Inevitably, the last child returns with an amazingly transformed message, such as "Joe Brown has a pig living under his porch." The more children playing the game, the more garbled the message becomes. This game illustrates an important principle: errors arise whenever information is copied; the more times it is copied, the greater the number of errors.

A complex, multicellular organism faces a problem similar to that of the children in the schoolyard game: how to faithfully transmit genetic instructions each time its cells divide. The solution to this problem is central to replication. A huge amount of genetic information and an enormous number of cell divisions are required to produce a multicellular adult organism; even a low rate of error during copying would be catastrophic. A single-celled human zygote contains 6 billion base pairs of DNA. If a copying error occurred only once per million base pairs, 6000 mistakes would be made every time a cell divided—errors that would be compounded at each of the millions of cell divisions that take place in human development.

Not only must the copying of DNA be astoundingly accurate, it must also take place at breakneck speed. The single, circular chromosome of *E. coli* contains about 4.7 million base pairs. At a rate of more than 1000 nucleotides per minute, replication of the entire chromosome would require almost 3 days. Yet, these bacteria are capable of dividing every 20 minutes. *E. coli* actually replicates its DNA at a rate of 1000 nucleotides per *second*, with fewer than one in a billion errors. How is this extraordinarily accurate and rapid process accomplished?

Semiconservative Replication

From the three-dimensional structure of DNA that Watson and Crick proposed in 1953 (see Figure 10.7), several important genetic implications were immediately apparent. The complementary nature of the two nucleotide strands in a DNA molecule suggested that, during replication, each strand can serve as a template for the synthesis of a new strand. The specificity of base pairing (adenine with thymine; guanine with cytosine) implied that only one sequence of bases can be specified by each template, and so two DNA molecules built on the pair of templates will be identical with the original. This process is called **semiconservative replication,** because each of the original nucleotide strands remains intact (conserved), despite no longer being combined in the same molecule; the original DNA molecule is half (semi) conserved during replication.

Initially, three alternative models were proposed for DNA replication. In conservative replication (◀ FIGURE 12.1a), the entire double-stranded DNA molecule serves as a template for a whole new molecule of DNA, and the original DNA molecule is *fully* conserved during replication. In dispersive replication (◀ FIGURE 12.1b), both nucleotide strands break down (disperse) into fragments, which serve as templates for the synthesis of new DNA fragments, and then somehow reassemble into two complete DNA molecules. In this model, each resulting DNA molecule is interspersed with fragments of old and new DNA; none of the original molecule is conserved. Semiconservative replication (◀ FIGURE 12.1c) is intermediate between these two models; the two nucleotide strands unwind and each serves as a template for a new DNA molecule.

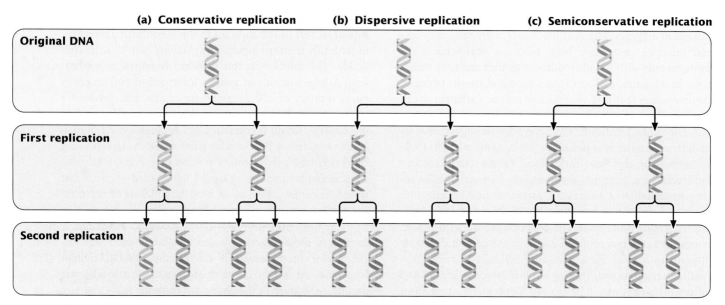

◄12.1 Three proposed models of replication are conservative replication, dispersive replication, and semiconservative replication.

These three models allow different predictions to be made about the distribution of original DNA and newly synthesized DNA after replication. With conservative replication, after one round of replication, 50% of the molecules would consist entirely of the original DNA and 50% would consist entirely of new DNA. After a second round of replication, 25% of the molecules would consist entirely of the original DNA and 75% would consist entirely of new DNA. With each additional round of replication, the proportion of molecules with new DNA would increase, although the number of molecules with the original DNA would remain constant. Dispersive replication would always produce hybrid molecules, containing some original and some new DNA, but the proportion of new DNA within the molecules would increase with each replication event. In contrast, with semiconservative replication, one round of replication would produce two hybrid molecules, each consisting of half original DNA and half new DNA. After a second round of replication, half the molecules would be hybrid, and the other half would consist of new DNA only. Additional rounds of replication would produce more and more molecules consisting entirely of new DNA, and a few hybrid molecules would persist.

Meselson and Stahl's Experiment

To determine which of the three models of replication applied to *E. coli* cells, Matthew Meselson and Franklin Stahl needed a way to distinguish old and new DNA. They did so by using two isotopes of nitrogen, ^{14}N (the common form) and ^{15}N (a rare, heavy form). Meselson and Stahl grew a culture of *E. coli* in a medium that contained ^{15}N as the sole nitrogen source; after many generations, all

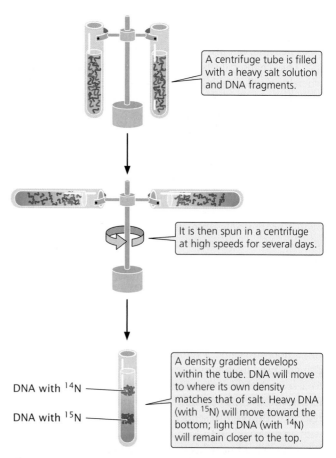

A centrifuge tube is filled with a heavy salt solution and DNA fragments.

It is then spun in a centrifuge at high speeds for several days.

DNA with ^{14}N

DNA with ^{15}N

A density gradient develops within the tube. DNA will move to where its own density matches that of salt. Heavy DNA (with ^{15}N) will move toward the bottom; light DNA (with ^{14}N) will remain closer to the top.

◄12.2 Meselson and Stahl used equilibrium density gradient centrifugation to distinguish between heavy, ^{15}N-laden DNA and lighter, ^{14}N-laden DNA.

the *E. coli* cells had ^{15}N incorporated into the purine and pyrimidine bases of DNA (see Figure 10.10). Meselson and Stahl took a sample of these bacteria, switched the rest of the bacteria to a medium that contained only ^{14}N, and then took additional samples of bacteria over the next few cellular generations. In each sample, the bacterial DNA that was synthesized before the change in medium contained ^{15}N and was relatively heavy, whereas any DNA synthesized after the switch contained ^{14}N and was relatively light.

Meselson and Stahl distinguished between the heavy ^{15}N-laden DNA and the light ^{14}N-containing DNA with the use of **equilibrium density gradient centrifugation** (FIGURE 12.2). In this technique, a centrifuge tube is filled with a heavy salt solution and a substance whose density is to be measured—in this case, DNA fragments. The tube is then spun in a centrifuge at high speeds. After several days of spinning, a gradient of density develops within the tube, with high density at the bottom and low density at the top.

The density of the DNA fragments matches that of the salt: light molecules rise and heavy molecules sink.

Meselson and Stahl found that DNA from bacteria grown only on medium containing ^{15}N produced a single band at the position expected of DNA containing only ^{15}N (FIGURE 12.3a). DNA from bacteria transferred to the medium with ^{14}N and allowed one round of replication also produced a single band, but at a position intermediate between that expected of DNA with only ^{15}N and that expected of DNA with only ^{14}N (FIGURE 12.3b). This result is inconsistent with the conservative replication model, which predicts one heavy band (the original DNA molecules) and one light band (the new DNA molecules). A single band of intermediate density is predicted by both the semiconservative and dispersive models.

To distinguish between these two models, Meselson and Stahl grew the bacteria in medium containing ^{14}N for a second generation. After a second round of replication in medium with ^{14}N, two bands of equal intensity appeared, one

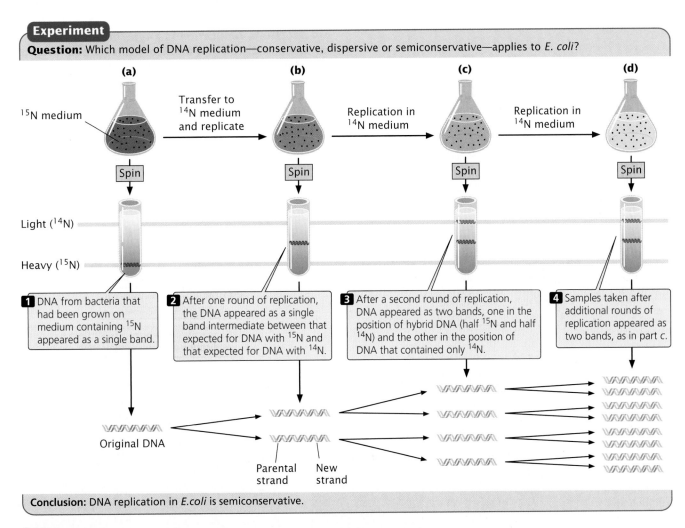

Experiment

Question: Which model of DNA replication—conservative, dispersive or semiconservative—applies to *E. coli*?

(a)   (b)   (c)   (d)

^{15}N medium   Transfer to ^{14}N medium and replicate   Replication in ^{14}N medium   Replication in ^{14}N medium

Spin   Spin   Spin   Spin

Light (^{14}N)

Heavy (^{15}N)

1 DNA from bacteria that had been grown on medium containing ^{15}N appeared as a single band.

2 After one round of replication, the DNA appeared as a single band intermediate between that expected for DNA with ^{15}N and that expected for DNA with ^{14}N.

3 After a second round of replication, DNA appeared as two bands, one in the position of hybrid DNA (half ^{15}N and half ^{14}N) and the other in the position of DNA that contained only ^{14}N.

4 Samples taken after additional rounds of replication appeared as two bands, as in part c.

Original DNA

Parental strand   New strand

Conclusion: DNA replication in *E.coli* is semiconservative.

 12.3 **Meselson and Stahl demonstrated that DNA replication is semiconservative.**

in the intermediate position and the other at the position expected of DNA that contained only ^{14}N (◀FIGURE 12.3c). All samples taken after additional rounds of replication produced two bands, and the band representing light DNA became progressively stronger (◀FIGURE 12.3d). Meselson and Stahl's results were exactly as expected for semiconservative replication and are incompatible with those predicated for both conservative and dispersive replication.

Concepts

Replication is semiconservative: each DNA strand serves as a template for the synthesis of a new DNA molecule. Meselson and Stahl convincingly demonstrated that replication in *E. coli* is semiconservative.

www.whfreeman.com/pierce   A summary of Meselson and Stahl's experiment

Modes of Replication

Following Meselson and Stahl's work, investigators confirmed that other organisms also use semiconservative replication. No evidence was found for conservative or dispersive replication. There are, however, several different ways that semiconservative replication can take place, differing principally in the nature of the template DNA—whether it is linear or circular—and in the number of replication forks.

Individual units of replication are called **replicons,** each of which contains a **replication origin.** Replication starts at the origin and continues until the entire replicon has been replicated. Bacterial chromosomes have a single replication origin, whereas eukaryotic chromosomes contain many.

Theta replication A common type of replication that takes place in circular DNA, such as that found in *E. coli* and other bacteria, is called **theta replication** (◀FIGURE 12.4a), because it generates a structure that resembles the Greek letter theta (θ). In theta replication, double-stranded DNA begins to unwind at the replication origin, producing single-stranded

◀12.4 **Theta replication is a type of replication common in *E. coli* and other organisms possessing circular DNA.** (Electron micrographs from Bernard Hint, Institut Suisse de Richerdies Experimentals sur le Cancer.)

(a)

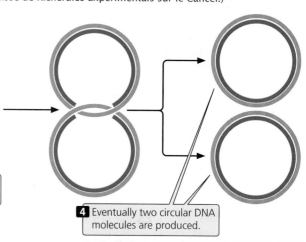

1 Double-stranded DNA unwinds at the replication origin,...

Origin of replication

Replication fork

Newly synthesized DNA

Replication bubble

2 ...producing single-stranded templates for the synthesis of new DNA. A replication bubble forms, usually having a replication fork at each end.

3 The forks proceed around the circle.

4 Eventually two circular DNA molecules are produced.

Conclusion: The products of theta replication are two circular DNA molecules.

(b)

Replication Fork

Origin of replication

Replication bubble

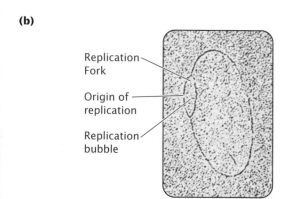
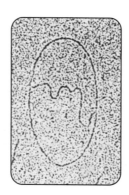
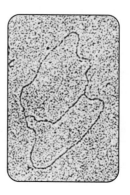
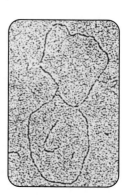

nucleotide strands that then serve as templates on which new DNA can be synthesized. The unwinding of the double helix generates a loop, termed a **replication bubble.** Unwinding may be at one or both ends of the bubble, making it progressively larger. DNA replication on both of the template strands is simultaneous with unwinding. The point of unwinding, where the two single nucleotide strands separate from the double-stranded DNA helix, is called a **replication fork.**

If there are two replication forks, one at each end of the replication bubble, the forks proceed outward in both directions in a process called **bidirectional replication,** simultaneously unwinding and replicating the DNA until they eventually meet. If a single replication fork is present, it proceeds around the entire circle to produce two complete circular DNA molecules, each consisting of one old and one new nucleotide strand.

John Cairns provided the first visible evidence of theta replication in 1963 by growing bacteria in the presence of radioactive nucleotides. After replication, each DNA molecule consisted of one "hot" (radioactive) strand and one "cold" (nonradioactive) strand. Cairns isolated DNA from the bacteria after replication and placed it on an electron microscope grid, which was then covered with a photographic emulsion. Radioactivity present in the sample exposes the emulsion and produces a picture of the molecule (called an autoradiograph), similar to the way that light exposes a photographic film. Because the newly synthesized DNA contained radioactive nucleotides, Cairns was able to produce an electron micrograph of the replication process, similar to those shown in ◀ FIGURE 12.4b.

Rolling-circle replication Another form of replication, called **rolling-circle replication** (◀ FIGURE 12.5), takes place in some viruses and in the F factor (a small circle of extrachromosomal DNA that controls mating, discussed in Chapter 8) of *E. coli*. This form of replication is initiated by a break in one of the nucleotide strands that creates a 3′-OH group and a 5′-phosphate group. New nucleotides are added to the 3′ end of the broken strand, with the inner (unbroken) strand used as a template. As new nucleotides are added to the 3′ end, the 5′ end of the broken strand is displaced from the template, rolling out like thread being pulled off a spool. The 3′ end grows around the circle, giving rise to the name rolling-circle model.

The replication fork may continue around the circle a number of times, producing several linked copies of the same sequence. With each revolution around the circle, the growing 3′ end displaces the nucleotide strand synthesized in the preceding revolution. Eventually, the linear DNA molecule is cleaved from the circle, resulting in a double-stranded circular DNA molecule and a single-stranded linear DNA molecule. The linear molecule circularizes either before or after serving as a template for the synthesis of a complementary strand.

Linear eukaryotic replication Circular DNA molecules that undergo theta or rolling-circle replication have a single origin of replication. Because of the limited size of these DNA molecules, replication starting from one origin can traverse the entire chromosome in a reasonable amount of time. The large linear chromosomes in eukaryotic cells, however, contain far too much DNA to be replicated speedily from a single origin. Eukaryotic replication proceeds at a rate ranging from 500 to 5000 nucleotides per minute at each replication fork (considerably slower than bacterial replication). Even at 5000 nucleotides per minute at each fork, DNA synthesis starting from a single origin would require 7 days to replicate a typical human chromosome consisting of 100 million base pairs of DNA. The replication of eukaryotic chromosomes actually occurs in a matter of minutes or hours, not

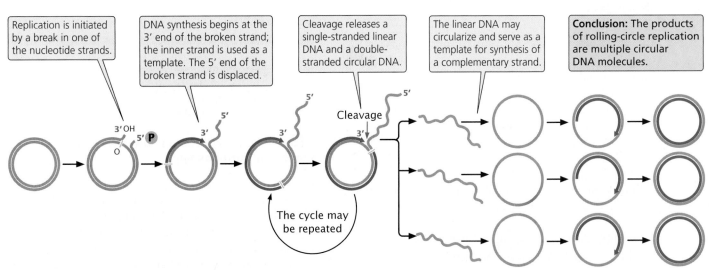

◀ 12.5 **Rolling-circle replication takes place in some viruses and in the F factor of *E. coli*.**

Table 12.1	Number and length of replicons	
Organism	**Number of Replication Origins**	**Average Length of Replicon (bp)**
Escherichia coli (bacterium)	1	4,200,000
Saccharomyces cerevisiae (yeast)	500	40,000
Drosophila melanogaster (fruit fly)	3,500	40,000
Xenopus laevis (toad)	15,000	200,000
Mus musculus (mouse)	25,000	150,000

Source: Data from B. L. Lewin, *Genes V* (Oxford: Oxford University Press, 1994), p. 536.

days. This rate is possible because replication takes place simultaneously from thousands of origins.

Typical eukaryotic replicons are from 20,000 to 300,000 base pairs in length (Table 12.1). At each replication origin, the DNA unwinds and produces a replication bubble. Replication takes place on both strands at each end of the bubble, with the two replication forks spreading outward. Eventually, replication forks of adjacent replicons run into each other, and the replicons fuse to form long stretches of newly synthesized DNA (◀FIGURE 12.6). Replication and fusion of all the replicons leads to two identical DNA molecules. Important features of theta replication, rolling-circle replication, and linear eukaryotic replication are summarized in Table 12.2.

> **Concepts**
>
> Theta replication, rolling-circle replication, and linear replication differ with respect to the initiation and progression of replication, but all produce new DNA molecules by semiconservative replication.

Requirements of Replication

Although the process of replication includes many components, they can be combined into three major groups:

1. a template consisting of single-stranded DNA,

2. raw materials (substrates) to be assembled into a new nucleotide strand, and

3. enzymes and other proteins that "read" the template and assemble the substrates into a DNA molecule.

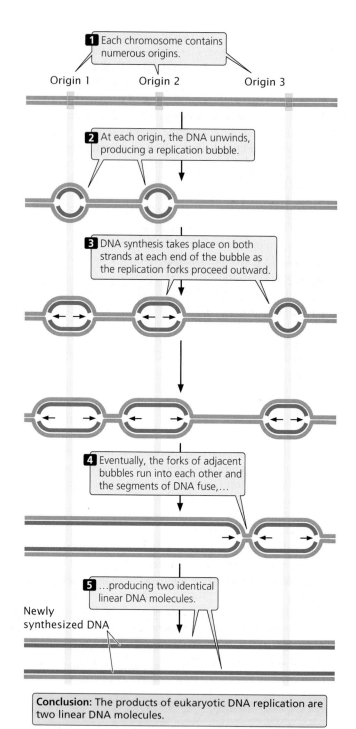

1 Each chromosome contains numerous origins.

2 At each origin, the DNA unwinds, producing a replication bubble.

3 DNA synthesis takes place on both strands at each end of the bubble as the replication forks proceed outward.

4 Eventually, the forks of adjacent bubbles run into each other and the segments of DNA fuse,...

5 ...producing two identical linear DNA molecules.

Newly synthesized DNA

Conclusion: The products of eukaryotic DNA replication are two linear DNA molecules.

◀ **12.6 Linear DNA replication takes place in eukaryotic chromosomes.**

Because of the semiconservative nature of DNA replication, a double-stranded DNA molecule must unwind to expose the bases that act as a template for the assembly of new polynucleotide strands, which are made complementary and antiparallel to the template strands. The raw materials from which new DNA molecules are synthesized are deoxyribonucleoside triphosphates (dNTPs), each consisting of a deoxyribose sugar and a base (a nucleoside)

Table 12.2	Characteristics of theta, rolling-circle, and linear eukaryotic replication				
Replication Model	DNA Template	Breakage of Nucleotide Strand	Number of Replicons	Unidirectional or Bidirectional	Products
Theta	Circular	No	1	Unidirectional or bidirectional	Two circular molecules
Rolling circle	Circular	Yes	1	Unidirectional	One circular molecule and one linear molecule that may circularize
Linear eukaryotic	Linear	No	Many	Bidirectional	Two linear molecules

attached to three phosphates (◀ FIGURE 12.7a). In DNA synthesis, nucleotides are added to the 3′-OH group of the growing nucleotide strand (◀ FIGURE 12.7b). The 3′-OH group of the last nucleotide on the strand attacks the 5′-phosphate group of the incoming dNTP. Two phosphates are cleaved from the incoming dNTP, and a phosphodiester bond is created between the two nucleotides.

DNA synthesis does not happen spontaneously. Rather, it requires a host of enzymes and proteins that function in a coordinated manner. We will examine this complex array of

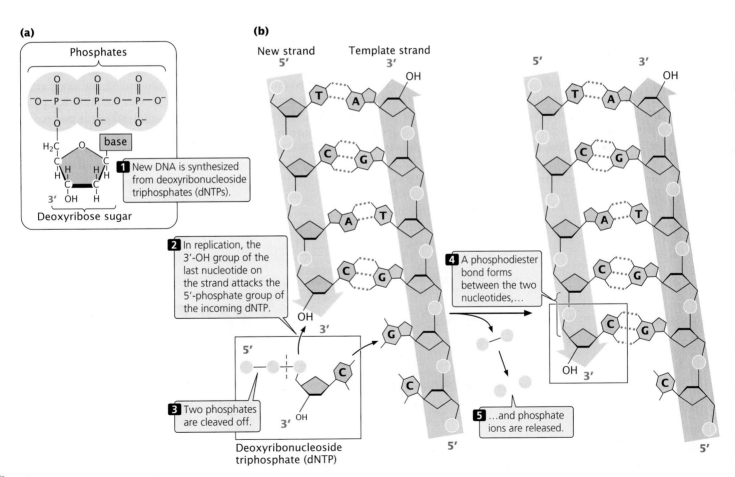

(a)

Phosphates

1 New DNA is synthesized from deoxyribonucleoside triphosphates (dNTPs).

Deoxyribose sugar

(b)

New strand 5′   Template strand 3′

2 In replication, the 3′-OH group of the last nucleotide on the strand attacks the 5′-phosphate group of the incoming dNTP.

3 Two phosphates are cleaved off.

Deoxyribonucleoside triphosphate (dNTP)

4 A phosphodiester bond forms between the two nucleotides,...

5 ...and phosphate ions are released.

◀ **12.7** In replication, new DNA is synthesized from deoxyribonucleoside triphosphates (dNTPs).

proteins and enzymes as we consider the replication process in more detail.

> **Concepts**
>
> DNA synthesis requires a single-stranded DNA template, deoxyribonucleoside triphosphates, a growing nucleotide strand, and a group of enzymes and proteins.

Direction of Replication

In DNA synthesis, new nucleotides are joined one at a time to the 3′ end of the newly synthesized strand. **DNA polymerases,** the enzymes that synthesize DNA, can add nucleotides *only* to the 3′ end of the growing strand (not the 5′ end), so new DNA strands always elongate in the same 5′-to-3′ direction (5′→3′). Because the two single-stranded DNA templates are antiparallel and strand elongation is always 5′→3′, if synthesis on one template proceeds from, say, right to left, then synthesis on the other template must proceed in the opposite direction, from left to right (◀FIGURE 12.8). As DNA unwinds during replication, the antiparallel nature of the two DNA strands means that one template is exposed in the 5′→3′ direction and the other template is exposed in the 3′→5′ direction (see Figure 12.8); so how can synthesis take place simultaneously on both strands at the fork?

As the DNA unwinds, the template strand that is exposed in the 3′→5′ direction (the lower strand in Figures 12.8 and 12.9) allows the new strand to be synthesized continuously, in the 5′→3′ direction. This new strand, which undergoes **continuous replication,** is called the **leading strand.**

The other template strand is exposed in the 5′→3′ direction (the upper strand in Figures 12.8 and 12.9). After

a short length of the DNA has been unwound, synthesis must proceed 5′→3′; that is, in the direction *opposite* that of unwinding (◀FIGURE 12.9). Because only a short length of DNA needs to be unwound before synthesis on this strand gets started, the replication machinery soon runs out of template. By that time, more DNA has unwound, providing new template at the 5′ end of the new strand. DNA synthesis must start anew at the replication fork and proceed in the direction opposite that of the movement of the fork until it runs into the previously replicated segment of DNA. This process is repeated again and again, so synthesis of this strand is in short, discontinuous bursts. The newly made strand that undergoes **discontinuous replication** is called the **lagging strand.**

The short lengths of DNA produced by discontinuous replication of the lagging strand are called **Okazaki fragments,** after Reiji Okazaki, who discovered them. In bacterial cells, each Okazaki fragment ranges in length from about 1000 to 2000 nucleotides; in eukaryotic cells, they are about 100 to 200 nucleotides long. Okazaki fragments on the lagging strand are linked together to create a continuous new DNA molecule.

Let's relate the direction of DNA synthesis to the modes of replication examined earlier. In the theta model (◀FIGURE 12.10a), the DNA unwinds at one particular location, the origin, and a replication bubble is formed. If the bubble has two forks, one at each end, synthesis takes place simultaneously at both forks (bidirectional replication). At each fork, synthesis on one of the template strands proceeds in the same direction as that of unwinding; the newly replicated strand is the leading strand with continuous replication. On the other template strand, synthesis is proceeding in the direction opposite that of unwinding; this newly synthesized strand is the lagging strand with discontinuous replication. Focus on just one of the template strands within the bubble. Notice that synthesis on this template

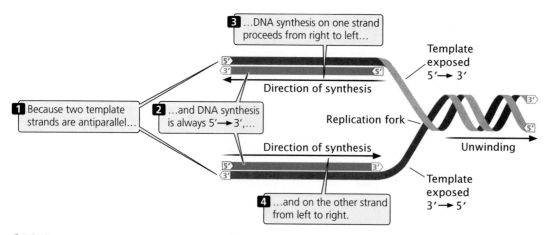

 ◀12.8 **DNA synthesis takes place simultaneously but in opposite directions on the two DNA template strands.** DNA replication at a single replication fork begins when a double-stranded DNA molecule unwinds to provide two single-strand templates.

strand is continuous at one fork but discontinuous at the other. This difference arises because DNA synthesis is always in the same direction (5′→3′), but the two forks are moving in opposite directions.

Replication in the rolling-circle model (◀FIGURE 12.10b) is somewhat different, because there is no replication bubble. Replication begins at the 3′ end of the broken nucleotide

strand. Continuous replication takes place on the circular template as new nucleotides are added to this 3′ end.

The replication of linear molecules of DNA, such as those found in eukaryotic cells, produces a series of replication bubbles (◀FIGURE 12.10c). DNA synthesis in these

1 On the lower template strand, DNA synthesis proceeds continuously in the 5′→3′ direction, the same as that of unwinding.

Template strands

Unwinding →
and replication

Newly
synthesized DNA

2 On the upper template strand, DNA synthesis begins at the fork and proceeds in the direction opposite that of unwinding; so it soon runs out of template.

3 DNA synthesis starts again on the upper strand, at the fork, each time proceeding away from the fork until it runs out of template.

4 DNA synthesis on this strand is discontinuous; short fragments of DNA produced by discontinuous synthesis are called Okazaki fragments.

Okazaki fragments

Lagging strand
Discontinuous
DNA synthesis

Leading strand
Continuous
DNA synthesis

 ◀12.9 **DNA synthesis is continuous on one template strand of DNA and discontinuous on the other.**

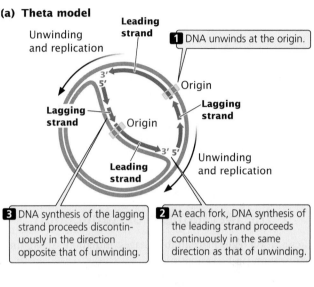

(a) **Theta model**

1 DNA unwinds at the origin.

2 At each fork, DNA synthesis of the leading strand proceeds continuously in the same direction as that of unwinding.

3 DNA synthesis of the lagging strand proceeds discontinuously in the direction opposite that of unwinding.

(b) **Rolling-circle model**

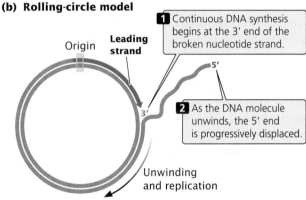

1 Continuous DNA synthesis begins at the 3′ end of the broken nucleotide strand.

2 As the DNA molecule unwinds, the 5′ end is progressively displaced.

(c) **Linear eukaryotic replication**

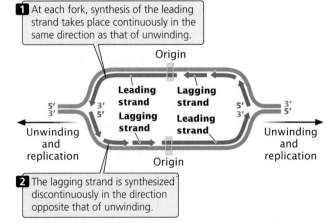

1 At each fork, synthesis of the leading strand takes place continuously in the same direction as that of unwinding.

2 The lagging strand is synthesized discontinuously in the direction opposite that of unwinding.

◀12.10 **The process of replication differs in theta replication, rolling-circle replication, and linear replication.**

bubbles is the same as that in the single replication bubble of the theta model; it begins at the center of each replication bubble and proceeds at two forks, one at each end of the bubble. At both forks, synthesis of the leading strand proceeds in the same direction as that of unwinding, whereas synthesis of the lagging strand proceeds in the direction opposite that of unwinding.

> **Concepts**
>
> All DNA synthesis is 5′→3′, meaning that new nucleotides are always added to the 3′ end of the growing nucleotide strand. At each replication fork, synthesis of the leading strand proceeds continuously and that of the lagging strand proceeds discontinuously.

The Mechanism of Replication

Replication takes place in four stages: initiation, unwinding, elongation, and termination.

Bacterial DNA Replication

The following discussion of the process of replication will focus on bacterial systems, where replication has been most thoroughly studied and is best understood. Although many aspects of replication in eukaryotic cells are similar to those of prokaryotic cells, there are some important differences. We will compare bacterial and eukaryotic replication later in the chapter.

Initiation The circular chromosome of *E. coli* has a single replication origin (*oriC*). The minimal sequence required for *oriC* to function consists of 245 bp that contain several critical sites. **Initiator proteins** bind to *oriC* and cause a short section of DNA to unwind. This unwinding allows helicase and other single-strand-binding proteins to attach to the polynucleotide strand (◀ FIGURE 12.11).

Unwinding Because DNA synthesis requires a single-stranded template and double-stranded DNA must be unwound before DNA synthesis can take place, the cell relies on several proteins and enzymes to accomplish the unwinding. **DNA helicases** break the hydrogen bonds that exist between the bases of the two nucleotide strands of a DNA molecule. Helicases cannot *initiate* the unwinding of double-stranded DNA; the initiator proteins first separate DNA strands at the origin, providing a short stretch of single-stranded DNA to which a helicase binds. Helicases bind to the lagging-strand template at each replication fork and move in the 5′→3′ direction along this strand, thus also moving the replication fork (◀ FIGURE 12.12).

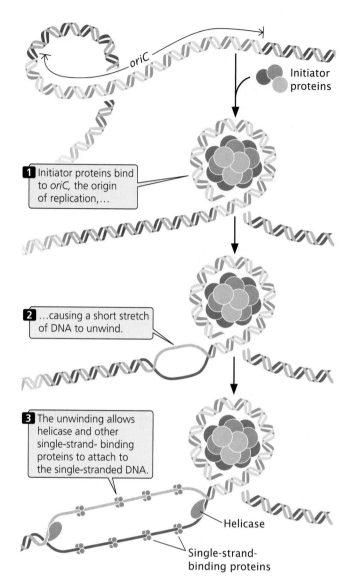

1 Initiator proteins bind to *oriC*, the origin of replication,...

2 ...causing a short stretch of DNA to unwind.

3 The unwinding allows helicase and other single-strand- binding proteins to attach to the single-stranded DNA.

Helicase

Single-strand-binding proteins

◀ **12.11** *E. coli* **DNA replication begins when initiator proteins bind to *oriC*, the origin of replication, causing a short stretch of DNA to unwind.**

After DNA has been unwound by helicase, the single-stranded nucleotide chains have a tendency to form hydrogen bonds and reanneal (stick back together). Secondary structures, such as hairpins (see Figure 10.17), also may form between complementary nucleotides on the same strand. To stabilize the single-stranded DNA long enough for replication to take place, **single-strand-binding** (SSB) **proteins** attach tightly to the exposed single-stranded DNA (see Figure 12.12). Unlike many DNA-binding proteins, SSBs are indifferent to base sequence—they will bind to any single-stranded DNA. Single-strand-binding proteins form tetramers (groups of four); each tetramer covers from 35 to 65 nucleotides.

Another protein essential for the unwinding process is the enzyme **DNA gyrase,** a topoisomerase. As discussed in

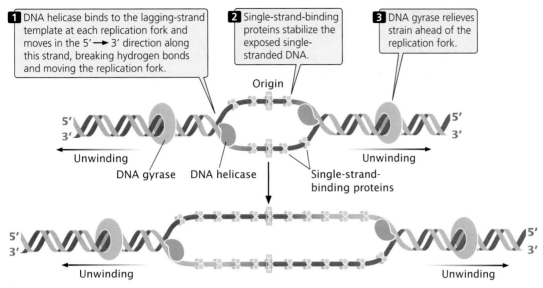

1 DNA helicase binds to the lagging-strand template at each replication fork and moves in the 5'→3' direction along this strand, breaking hydrogen bonds and moving the replication fork.

2 Single-strand-binding proteins stabilize the exposed single-stranded DNA.

3 DNA gyrase relieves strain ahead of the replication fork.

Origin

Unwinding · DNA gyrase · DNA helicase · Single-strand-binding proteins · Unwinding

Unwinding · Unwinding

12.12 **DNA helicase unwinds DNA by binding to the lagging-strand template at each replication fork and moving in the 5'→3' direction along the strand.**

Chapter 11, topoisomerases control the supercoiling of DNA. In replication, DNA gyrase reduces torsional strain (torque) that builds up ahead of the replication fork as a result of unwinding (see Figure 12.12). It reduces torque by making a double-stranded break in one segment of the DNA helix, passing another segment of the helix through the break, and then resealing the broken ends of the DNA. This action removes a twist in the DNA and reduces the supercoiling.

Concepts

Replication is initiated at a replication origin, where an initiator protein binds and causes a short stretch of DNA to unwind. DNA helicase breaks hydrogen bonds at a replication fork, and single-strand-binding proteins stabilize the separated strands. DNA gyrase reduces torsional strain that develops as the two strands of double-helical DNA unwind.

Primers All DNA polymerases require a nucleotide with a 3'-OH group to which a new nucleotide can be added. Because of this requirement, DNA polymerases cannot initiate DNA synthesis on a bare template; rather, they require a primer—an existing 3'-OH group—to get started. How, then, does DNA synthesis begin?

An enzyme called **primase** synthesizes short stretches of nucleotides (**primers**) to get DNA replication started. Primase synthesizes a short stretch of RNA nucleotides (about 10–12 nucleotides long), which provides a 3'-OH

group to which DNA polymerase can attach DNA nucleotides. (Because primase is an RNA polymerase, it does not require an existing 3'-OH group to which nucleotides can be added.) All DNA molecules initially have short RNA primers imbedded within them; these primers are later removed and replaced by DNA nucleotides.

On the leading strand, where DNA synthesis is continuous, a primer is required only at the 5' end of the newly synthesized strand. On the lagging strand, where replication is discontinuous, a new primer must be generated at the beginning of each Okazaki fragment (◄FIGURE 12.13). Primase forms a complex with helicase at the replication fork and moves along the template of the lagging strand. The single primer on the leading strand is probably synthesized by the primase–helicase complex on the template of the lagging strand of the *other* replication fork, at the opposite end of the replication bubble.

Concepts

Primase synthesizes a short stretch of RNA nucleotides (primers), which provides a 3'-OH group for the attachment of DNA nucleotides to start DNA synthesis.

Elongation After DNA is unwound and a primer has been added, DNA polymerases elongate the polynucleotide strand by catalyzing DNA polymerization. The best-studied polymerases are those of *E. coli*, which has at least five different DNA polymerases. Two of them, DNA polymerase I and DNA polymerase III, carry out DNA synthesis associated

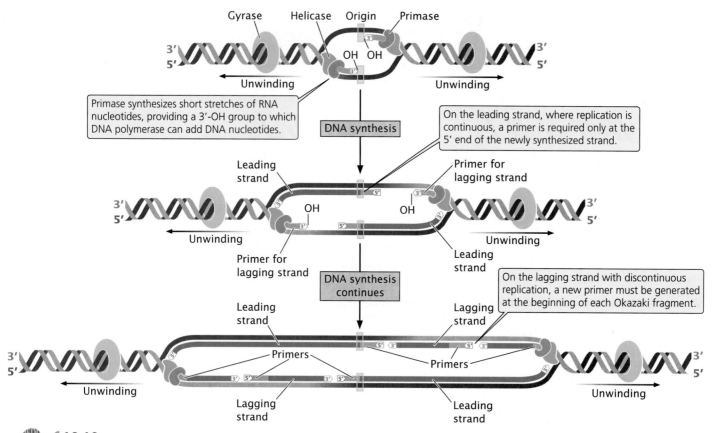

Gyrase   Helicase   Origin   Primase

Primase synthesizes short stretches of RNA nucleotides, providing a 3'-OH group to which DNA polymerase can add DNA nucleotides.

DNA synthesis

On the leading strand, where replication is continuous, a primer is required only at the 5' end of the newly synthesized strand.

Leading strand

Primer for lagging strand

Primer for lagging strand

Leading strand

DNA synthesis continues

On the lagging strand with discontinuous replication, a new primer must be generated at the beginning of each Okazaki fragment.

Leading strand

Lagging strand

Leading strand

Primers

Primers

Lagging strand

Leading strand

12.13   Primase synthesizes short stretches of RNA nucleotides, providing a 3'-OH group to which DNA polymerase can add DNA nucleotides.

with replication; the other three have specialized functions in DNA repair (Table 12.3).

DNA polymerase III is a large multiprotein complex that acts as the main workhorse of replication. DNA polymerase III synthesizes nucleotide strands by adding new nucleotides to the 3' end of growing DNA molecules. This enzyme has two enzymatic activities (Table 12.3). Its 5'→3'

polymerase activity allows it to add new nucleotides in the 5'→3' direction. Its 3'→5' exonuclease activity allows it to remove nucleotides in the 3'→5' direction, enabling it to correct errors. If a nucleotide having an incorrect base is inserted into the growing DNA molecule, DNA polymerase III uses its 3'→5' exonuclease activity to back up and remove the incorrect nucleotide. It then resumes its 5'→3'

Table 12.3	Characteristics of DNA Polymerases in *E. coli*			
DNA Polymerase	**5'→3' Polymerization**	**3'→5' Exonuclease**	**5'→3' Exonuclease**	**Function**
I	Yes	Yes	Yes	Removes and replaces primers
II	Yes	Yes	No	DNA repair; restarts replication after damaged DNA halts synthesis
III	Yes	Yes	No	Elongates DNA
IV	Yes	No	No	DNA repair
V	Yes	No	No	DNA repair; translesion DNA synthesis

polymerase activity. These two functions together allow DNA polymerase III to efficiently and accurately synthesize new DNA molecules.

The first *E. coli* polymerase to be discovered, **DNA polymerase I,** also has $5' \rightarrow 3'$ polymerase and $3' \rightarrow 5'$ exonuclease activities (see Table 12.3), permitting the enzyme to synthesize DNA and to correct errors. Unlike DNA polymerase III, however, DNA polymerase I also possesses $5' \rightarrow 3'$ exonuclease activity, which is used to remove the primers laid down by primase and to replace them with DNA nucleotides by moving in a $5' \rightarrow 3'$ direction. The removal and replacement of primers appear to constitute the main function of DNA polymerase I. DNA polymerases II, IV, and V function in DNA repair.

Despite their differences, all of *E. coli*'s DNA polymerases

1. synthesize any sequence specified by the template strand;
2. synthesize in the $5' \rightarrow 3'$ direction by adding nucleotides to a $3'$-OH group;
3. use dNTPs to synthesize new DNA;
4. require a primer to initiate synthesis;
5. catalyze the formation of a phosphodiester bond by joining the $5'$ phosphate group of the incoming nucleotide to the $3'$-OH group of the preceding nucleotide on the growing strand, cleaving off two phosphates in the process;
6. produce newly synthesized strands that are complementary and antiparallel to the template strands; and
7. are associated with a number of other proteins.

Concepts

DNA polymerases synthesize DNA in the $5' \rightarrow 3'$ direction by adding new nucleotides to the $3'$ end of a growing nucleotide strand.

DNA ligase After DNA polymerase III attaches a DNA nucleotide to the $3'$-OH group on the last nucleotide of the RNA primer, each new DNA nucleotide then provides the $3'$-OH group needed for the next DNA nucleotide to be added. This process continues as long as template is available (◀ FIGURE 12.14a). DNA polymerase I follows DNA polymerase III and, using its $5' \rightarrow 3'$ exonuclease activity, removes the RNA primer. It then uses its $5' \rightarrow 3'$ polymerase activity to replace the RNA nucleotides with DNA nucleotides. DNA polymerase I attaches the first nucleotide to the OH group at the $3'$ end of the preceding Okazaki fragment and then continues, in the $5' \rightarrow 3'$ direction along the nucleotide strand, removing and replacing, one at a time, the RNA nucleotides of the primer (◀ FIGURE 12.14b).

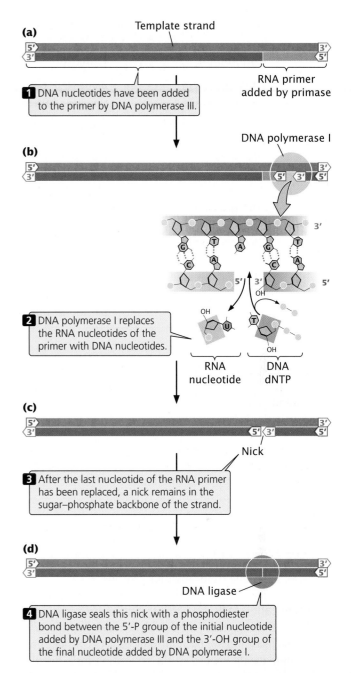

12.14 DNA ligase seals the nick left by DNA polymerase I in the sugar–phosphate backbone after the polymerase has added the final nucleotide.

After polymerase I has replaced the last nucleotide of the RNA primer with a DNA nucleotide, a nick remains in the sugar–phosphate backbone of the new DNA strand. The $3'$-OH group of the last nucleotide to have been added by DNA polymerase I is not attached to the $5'$-phosphate group of the first nucleotide added by DNA polymerase III (◀ FIGURE 12.14c). This nick is sealed by the enzyme **DNA ligase,** which catalyzes the formation of a

Table 12.4	Components required for replication in bacterial cells
Component	**Function**
Initiator protein	Binds to origin and separates strands of DNA to initiate replication
DNA helicase	Unwinds DNA at replication fork
Single-strand-binding proteins	Attach to single-stranded DNA and prevent reannealing
DNA gyrase	Moves ahead of the replication fork, making and resealing breaks in the double-helical DNA to release torque that builds up as a result of unwinding at the replication fork
DNA primase	Synthesizes short RNA primers to provide a 3'-OH group for attachment of DNA nucleotides
DNA polymerase III	Elongates a new nucleotide strand from the 3'-OH group provided by the primer
DNA polymerase I	Removes RNA primers and replaces them with DNA
DNA ligase	Joins Okazaki fragments by sealing nicks in the sugar–phosphate backbone of newly synthesized DNA

phosphodiester bond without adding another nucleotide to the strand (◀ FIGURE 12.14d). Some of the major enzymes and proteins required for replication are summarized in Table 12.4.

Concepts

After primers are removed and replaced, the nick in the sugar–phosphate linkage is sealed by DNA ligase.

www.whfreeman.com/pierce   More information on helicase, primase, and single-strand-binding proteins

The replication fork   Now that the major enzymatic components of elongation—DNA polymerases, helicase, primase, and ligase—have been introduced, let's consider how these components interact at the replication fork. Because the synthesis of both strands takes place simultaneously, two units of DNA polymerase III must be present at the replication fork, one for each strand. In one model of the replication process (◀ FIGURE 12.15), the two units of DNA polymerase III are connected, and the lagging-strand template loops around so that, as the DNA polymerase III complex moves along the helix, the two antiparallel strands can undergo 5'→3' replication simultaneously.

In summary, each active replication fork requires five basic components:

1. helicase to unwind the DNA,

2. single-strand-binding proteins to keep the nucleotide strands separate long enough to allow replication,

3. the topoisomerase gyrase to remove strain ahead of the replication fork,

4. primase to synthesize primers with a 3'-OH group at the beginning of each DNA fragment, and

5. DNA polymerase to synthesize the leading and lagging nucleotide strands.

www.whfreeman.com/pierce   Additional information about the mechanism of replication and an animation of a replication fork

Termination   In some DNA molecules, replication is terminated whenever two replication forks meet. In others, specific termination sequences block further replication. A termination protein, called Tus in *E. coli*, binds to these sequences. Tus blocks the movement of helicase, thus stalling the replication fork and preventing further DNA replication.

The fidelity of DNA replication   Overall, replication results in an error rate of less than one mistake per billion nucleotides. How is this incredible accuracy achieved? No single process could produce this level of accuracy; a series of processes are required, each catching errors missed by the preceding ones (◀ FIGURE 12.16).

DNA polymerases are very particular in pairing nucleotides with their complements on the template strand. Errors in nucleotide selection by DNA polymerase arise only about once per 100,000 nucleotides. Most of the errors that do arise in nucleotide selection are corrected in a second process called **proofreading.** When a DNA polymerase inserts an incorrect nucleotide into the growing strand, the 3'-OH group of the mispaired nucleotide is not correctly positioned for accepting the next nucleotide. The incorrect positioning stalls the poly-

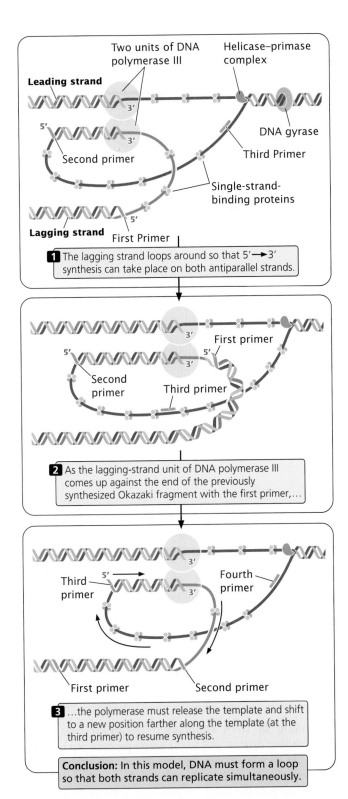

Leading strand

Two units of DNA polymerase III

Helicase–primase complex

DNA gyrase

Second primer

Third Primer

Single-strand-binding proteins

Lagging strand First Primer

1 The lagging strand loops around so that 5′→3′ synthesis can take place on both antiparallel strands.

First primer

Second primer

Third primer

2 As the lagging-strand unit of DNA polymerase III comes up against the end of the previously synthesized Okazaki fragment with the first primer,…

Third primer

Fourth primer

First primer

Second primer

3 …the polymerase must release the template and shift to a new position farther along the template (at the third primer) to resume synthesis.

Conclusion: In this model, DNA must form a loop so that both strands can replicate simultaneously.

◄ **12.15** **In one model of DNA replication in *E. coli*, the two units of DNA polymerase III are connected, and the lagging-strand template forms a loop so that replication can take place on the two anti-parallel DNA strands. Components of the replication machinery at the replication fork are shown at the top.**

merization reaction, and the 3′→5′ exonuclease activity of DNA polymerase removes the incorrectly paired nucleotide. DNA polymerase then inserts the correct nucleotide. Together, proofreading and nucleotide selection result in an error rate of only one in 10 million nucleotides.

A third process, called **mismatch repair** (discussed further in Chapter 17), corrects errors after replication is complete. Any incorrectly paired nucleotides remaining after replication produce a deformity in the secondary structure of the DNA; the deformity is recognized by enzymes that excise an incorrectly paired nucleotide and use the original nucleotide strand as a template to replace the incorrect nucleotide. Mismatch repair requires the ability to distinguish between the old and the new strands of DNA, because the enzymes need some way of determining which of the two incorrectly paired bases to remove. In *E. coli*, methyl groups ($-CH_3$) are added to particular nucleotide sequences, but only *after* replication. Thus, methylation lags behind replication: so, immediately after DNA synthesis, only the old DNA strand is methylated. Therefore it can be distinguished from the newly synthesized strand, and mismatch repair takes place preferentially on the unmethylated nucleotide strand.

Concepts

Replication is extremely accurate, with less than one error per billion nucleotides. This accuracy results from the processes of nucleotide selection, proofreading, and mismatch repair.

Connecting Concepts

The Basic Rules of Replication

Bacterial replication requires a number of enzymes, proteins, and DNA sequences that function together to synthesize a new DNA molecule. These components are important, but it is critical that we not become so immersed in the details of the process that we lose sight of general principles of replication.

1. Replication is always semiconservative.

2. Replication begins at sequences called origins.

3. DNA synthesis is initiated by short segments of RNA called primers.

4. The elongation of DNA strands is always in the 5′→3′ direction.

5. New DNA is synthesized from dNTPs; in the polymerization of DNA, two phosphates are cleaved from a dNTP and the resulting nucleotide is added to the 3′-OH group of the growing nucleotide strand.

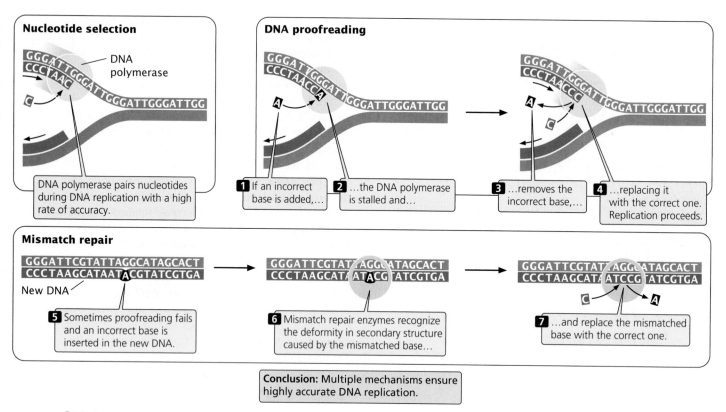

Nucleotide selection

DNA polymerase

DNA polymerase pairs nucleotides during DNA replication with a high rate of accuracy.

DNA proofreading

1 If an incorrect base is added,…

2 …the DNA polymerase is stalled and…

3 …removes the incorrect base,…

4 …replacing it with the correct one. Replication proceeds.

Mismatch repair

New DNA

5 Sometimes proofreading fails and an incorrect base is inserted in the new DNA.

6 Mismatch repair enzymes recognize the deformity in secondary structure caused by the mismatched base…

7 …and replace the mismatched base with the correct one.

Conclusion: Multiple mechanisms ensure highly accurate DNA replication.

◀ **12.16 A series of processes are required to ensure the incredible accuracy of DNA replication.** Among these processes are DNA selection, proofreading, and mismatch repair.

6. Replication is continuous on the leading strand and discontinuous on the lagging strand.

7. New nucleotide strands are made complementary and antiparallel to their template strands.

8. Replication takes place at very high rates and is astonishingly accurate, thanks to precise nucleotide selection, proofreading, and repair mechanisms.

Eukaryotic DNA Replication

Although not as well understood, eukaryotic replication resembles bacterial replication in many respects. The most obvious differences are that eukaryotes have: (1) multiple replication origins in their chromosomes; (2) more types of DNA polymerases, with different functions; and (3) nucleosome assembly immediately following DNA replication.

Eukaryotic origins Researchers first isolated eukaryotic origins of replication from yeast cells by demonstrating that certain DNA sequences confer the ability to replicate when transferred from a yeast chromosome to small circular pieces of DNA (plasmids). These **autonomously replicating sequences** (ARSs) enabled any DNA to which they were attached to replicate. They were subsequently shown to be the origins of replication in yeast chromosomes.

Yeast ARSs typically consist of 100 to 120 bp of DNA. A multiprotein complex, the origin recognition complex (ORC), binds to the ARS and probably unwinds the DNA in this region. Interestingly, ORCs also function in regulating transcription.

Concepts

Eukaryotic DNA contains many origins of replication. At each origin, a multiprotein origin recognition complex binds to initiate the unwinding of the DNA.

Licensing of DNA replication Eukaryotic cells utilize thousands of origins, and so the entire genome can be replicated in a timely manner. The use of multiple origins, however, creates a special problem in the timing of replication: the entire genome must be precisely replicated once and only once in each cell cycle so that no genes are left unreplicated and no genes are replicated more than once. How does a cell ensure that replication is initiated at thousands of origins only once per cell cycle?

The precise replication of DNA is accomplished by the separation of the initiation of replication into two distinct steps. In the first step, the origins are licensed, meaning that they are approved for replication. This step is early in the cell cycle when a **replication licensing factor** attaches to an origin. In the second step, initiator proteins cause the separation of DNA strands and the initiation of replication at each licensed origin. The key is that initiator proteins function only at *licensed* origins. As the replication forks move away from the origin, the licensing factor is removed, leaving the origin in an unlicensed state, where replication cannot be initiated again until the license is renewed. To ensure that replication takes place only once each cell cycle, the licensing factor is active only after the cell has completed mitosis and before the initiator proteins become active.

Unwinding Several helicases that separate double-stranded DNA have been isolated from eukaryotic cells, as have single-strand-binding proteins and topoisomerases (which have a function equivalent to the DNA gyrase in bacterial cells). These enzymes and proteins are assumed to function in unwinding eukaryotic DNA in much the same way as unwinding in bacterial cells.

Eukaryotic DNA polymerases A significant difference in the processes of bacterial and eukaryotic replication is in the number and functions of DNA polymerases. Eukaryotic cells contain a number of different DNA polymerases that function in replication, recombination, and DNA repair

(Table 12.5). **DNA polymerase α,** which contains primase activity, initiates nuclear DNA synthesis by synthesizing an RNA primer, followed by a short string of DNA nucleotides. After DNA polymerase α has laid down from 30 to 40 nucleotides, **DNA polymerase δ** completes replication on the leading and lagging strands. **DNA polymerase β** does not participate in replication but is associated with the repair and recombination of nuclear DNA. **DNA polymerase γ** replicates mitochondrial DNA; a γ-like polymerase also replicates chloroplast DNA. Similar in structure and function to DNA polymerase δ, **DNA polymerase ε** appears to take part in nuclear replication of both the leading and the lagging strands, but its precise role is not yet clear. Other DNA polymerases (ζ, η, θ, κ, λ, μ) allow replication to bypass damaged DNA (called translesion replication) or play a role in DNA repair. Many of the DNA polymerases have multiple roles in replication and DNA repair (see Table 12.5).

Concepts

There are at least thirteen different DNA polymerases in eukaryotic cells. DNA polymerases α and δ carry out replication on the leading and lagging strands.

Table 12.5	DNA polymerases in eukaryotic cells		
DNA Polymerase	**5′→3′ Polymerase Activity**	**3′→5′ Exonuclease Activity**	**Cellular Function**
α (alpha)	Yes	No	Initiation of nuclear DNA synthesis and DNA repair
β (beta)	Yes	No	DNA repair and recombination of nuclear DNA
γ (gamma)	Yes	Yes	Replication of mitochondrial DNA
δ (delta)	Yes	Yes	Leading- and lagging-strand synthesis of nuclear DNA, DNA repair, and translesion DNA synthesis
ε (epsilon)	Yes	Yes	Unknown; probably repair and replication of nuclear DNA
ζ (zeta)	Yes	No	Translesion DNA synthesis
η (eta)	Yes	No	Translesion DNA synthesis
θ (theta)	Yes	No	DNA repair
ι (iota)	Yes	No	Translesion DNA synthesis
κ (kappa)	Yes	No	Translesion DNA synthesis
λ (lambda)	Yes	No	DNA repair
μ (mu)	Yes	No	DNA repair
σ (sigma)	Yes	No	Nuclear DNA replication (possibly), DNA repair, and sister-chromatid cohesion

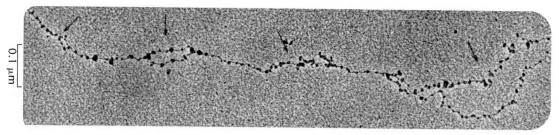

◀ **12.17 This electron micrograph of eukaryotic DNA in the process of replication clearly shows that newly replicated DNA is already covered with nucleosomes (dark circles).** (Victoria Foe.)

Nucleosome assembly Eukaryotic DNA is complexed to histone proteins in nucleosome structures that contribute to the stability and packing of the DNA molecule (see Figure 11.6). The disassembly and reassembly of nucleosomes on newly synthesized DNA probably takes place in replication, but the precise mechanism for these processes has not yet been determined. The unwinding of double-stranded DNA and the assembly of the replication enzymes on the single-stranded templates probably require the disassembly of the nucleosome structure. Electron micrographs of eukaryotic DNA show recently replicated DNA already covered with nucleosomes (◀FIGURE 12.17), indicating that nucleosome structure is reassembled quickly.

Before replication, a single DNA molecule is associated with histone proteins. After replication and nucleosome assembly, two DNA molecules are associated with histone proteins. Do the original histones remain together, attached to one of the new DNA molecules, or do they disassemble and mix with new histones on both DNA molecules?

Techniques similar to those employed by Meselson and Stahl to determine the mode of DNA replication were used to address this question. Cells were cultivated for several generations in a medium containing amino acids labeled with a heavy isotope. The histone proteins incorporated these heavy amino acids and were relatively dense (◀FIGURE 12.18). The cells were then transferred to a culture medium that contained amino acids with a light isotope. Histones assembled after the transfer possessed the new, relatively light amino acids and were less dense.

After allowing replication to take place, the histone octamers were isolated and centrifuged in a density gradient. Results show that, after replication, the octamers were in a continuous band between high density (representing old octamers) and low density (representing new octamers). This finding suggests that newly assembled octamers consist of a random mixture of old and new histones.

Location of replication within the nucleus The DNA polymerases that carry out replication are frequently depicted as moving down the DNA template, much as a locomotive travels along a train track. Recent evidence suggests that this view is incorrect. A more accurate view is that the polymerase is fixed in location, and template DNA is threaded through it, with newly synthesized DNA molecules emerging from the other end.

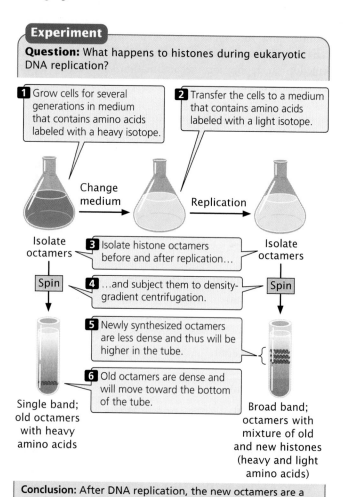

Experiment

Question: What happens to histones during eukaryotic DNA replication?

1 Grow cells for several generations in medium that contains amino acids labeled with a heavy isotope.

2 Transfer the cells to a medium that contains amino acids labeled with a light isotope.

Change medium → Replication →

Isolate octamers

3 Isolate histone octamers before and after replication...

Isolate octamers

Spin

4 ...and subject them to density-gradient centrifugation.

Spin

5 Newly synthesized octamers are less dense and thus will be higher in the tube.

6 Old octamers are dense and will move toward the bottom of the tube.

Single band; old octamers with heavy amino acids

Broad band; octamers with mixture of old and new histones (heavy and light amino acids)

Conclusion: After DNA replication, the new octamers are a random mixture of old and new histones.

◀ **12.18 Experimental procedure for studying how nucleosomes dissociate and reassociate during replication.**

Concepts

After DNA replication, new nucleosomes quickly reassemble on the molecules of DNA.
Nucleosomes apparently break down in the course of replication and reassemble from a random mixture of old and new histones.

Techniques of fluorescence microscopy, which are capable of revealing active sites of DNA synthesis, show that most replication in the nucleus of a eukaryotic cell takes place at a limited number of fixed sites, often referred to as replication factories. Time-lapse micrographs reveal that newly duplicated DNA is extruded from these particular sites. Similar results have also been obtained with bacterial cells.

DNA synthesis at the ends of chromosomes

A fundamental difference between eukaryotic and bacterial replication arises because eukaryotic chromosomes are linear and thus have ends. As already stated, the 3'-OH group needed for replication by DNA polymerases is provided at the initiation of replication by RNA primers that are synthesized by primase. This solution is temporary, because eventually the primers must be removed and replaced by DNA nucleotides. In a circular DNA molecule, elongation around the circle eventually provides a 3'-OH group immediately in front of the primer (◀ FIGURE 12.19a). After the primer has been removed, the replacement DNA nucleotides can be added to this 3'-OH group.

In linear chromosomes with multiple origins, the elongation of DNA in adjacent replicons also provides a 3'-OH group preceding each primer (◀ FIGURE 12.19b). At the very end of a linear chromosome, however, there is no adjacent stretch of replicated DNA to provide this crucial 3'-OH group. Once the primer at the end of the chromosome has been removed, it cannot be replaced with DNA nucleotides, which produces a gap at the end of the chromosome (◀ FIGURE 12.19c), suggesting that the chromosome should become progressively shorter with each round of replication. The chromosome would be shortened each generation, leading to the eventual elimination of the entire telomere, destabilization of the chromosome, and cell death. But chromosomes don't become shorter each generation and destabilize; so how are the ends of linear chromosomes replicated?

The ends of chromosomes—the telomeres—possess several unique features, one of which is the presence of many copies of a short repeated sequence. In the protozoan *Tetrahymena*, this telomeric repeat is CCCCAA (see Table 11.2), with the G-rich strand typically protruding beyond the C-rich strand (◀ FIGURE 12.20a):

$$\text{end of chromosome} \xleftarrow{} \begin{array}{l} 5'\text{–CCCCAA} \\ 3'\text{–GGGGTTGGGGTT} \end{array} \xrightarrow{} \text{toward centromere}$$

The single-stranded protruding end of the telomere can be extended by **telomerase,** an enzyme with both a protein and an RNA component (also known as a ribonucleoprotein). The RNA part of the enzyme contains from 15 to 22 nucleotides that are complementary to the sequence on the G-rich strand. This sequence pairs with the overhanging 3' end of the DNA (◀ FIGURE 12.20b) and provides a template for the synthesis of additional DNA copies of the repeats. DNA nucleotides are added to the 3' end of the

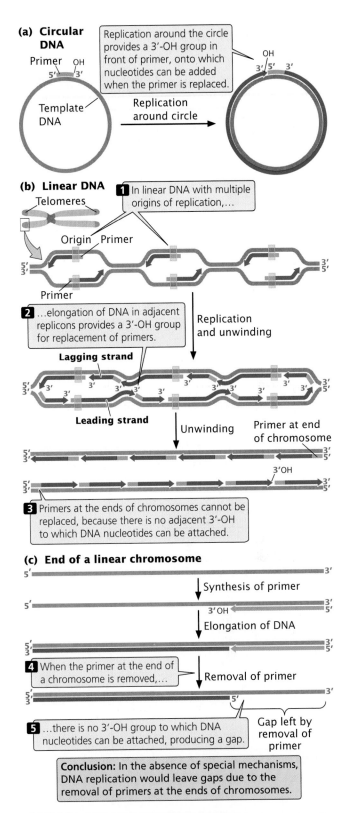

(a) Circular DNA

Primer OH

Replication around the circle provides a 3'-OH group in front of primer, onto which nucleotides can be added when the primer is replaced.

Template DNA

Replication around circle

(b) Linear DNA

Telomeres

1 In linear DNA with multiple origins of replication,...

Origin Primer

Primer

2 ...elongation of DNA in adjacent replicons provides a 3'-OH group for replacement of primers.

Replication and unwinding

Lagging strand

Leading strand

Unwinding

Primer at end of chromosome

3'OH

3 Primers at the ends of chromosomes cannot be replaced, because there is no adjacent 3'-OH to which DNA nucleotides can be attached.

(c) End of a linear chromosome

Synthesis of primer

3'OH

Elongation of DNA

4 When the primer at the end of a chromosome is removed,...

Removal of primer

5 ...there is no 3'-OH group to which DNA nucleotides can be attached, producing a gap.

Gap left by removal of primer

Conclusion: In the absence of special mechanisms, DNA replication would leave gaps due to the removal of primers at the ends of chromosomes.

◀ 12.19 **DNA synthesis must differ at the ends of circular and linear chromosomes.**

strand one at a time (◀ FIGURE 12.20c) and, after several nucleotides have been added, the RNA template moves down the DNA and more nucleotides are added to the 3' end

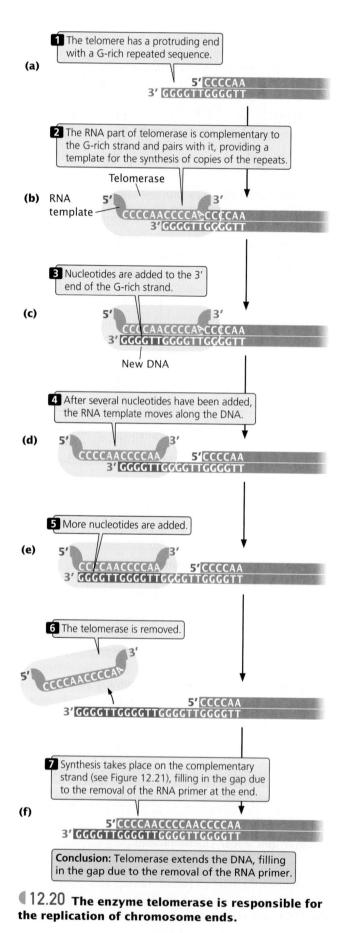

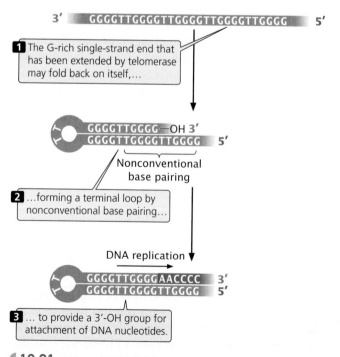

◀12.20 The enzyme telomerase is responsible for the replication of chromosome ends.

(◀FIGURE 12.20d). Usually, from 14 to 16 nucleotides are added to the 3′ end of the G-rich strand.

In this way, the telomerase can extend the 3′ end of the chromosome without the use of a complementary DNA template (◀FIGURE 12.20e). How the complementary C-rich strand is synthesized (◀FIGURE 12.20f) is not yet clear. It may be synthesized by conventional replication, with primase synthesizing an RNA primer on the 5′ end of the extended (G-rich) template. The removal of this primer once again leaves a gap at the 5′ end of the chromosome, but this gap does not matter, because the end of the chromosome is extended at each replication by telomerase; no genetic information is lost, and the chromosome does not become shorter overall. The extended single-strand end may fold back on itself, forming a terminal loop by nonconventional pairing of bases (◀FIGURE 12.21). This loop could provide a 3′-OH group for the attachment of DNA nucleotides along the C-rich strand.

Telomerase is present in single-celled organisms, germ cells, early embryonic cells, and certain proliferative somatic cells (such as bone-marrow cells and cells lining the intestine), all of which must undergo continuous cell division. Most somatic cells have little or no telomerase activity, and chromosomes in these cells progressively shorten with each cell division. These cells are capable of only a limited number of divisions; once the telomeres shorten beyond a critical point, a chromosome becomes unstable, has a tendency to undergo rearrangements, and is degraded. These events lead to cell death.

The shortening of telomeres may contribute to the process of aging. Genetically engineered mice that lack a functional telomerase gene (and therefore do not express

◀12.21 The complementary G-rich strand at the end of the telomere must be primed before the extension of the 3′ end of the chromosome by telomerase.

telomerase in somatic or germ cells) experience progressive shortening of their telomeres in successive generations. After several generations, these mice show some signs of premature aging, such as graying, hair loss, and delayed wound healing. Through genetic engineering, it is also possible to create somatic cells that express telomerase. In these cells, telomeres do not shorten, cell aging is inhibited, and the cells will divide indefinitely.

Telomerase also appears to play a role in cancer. Cancer tumor cells have the capacity to divide indefinitely, and many tumor cells express the telomerase enzyme. As will be discussed in Chapter 21, cancer is a complex, multistep process that usually requires mutations in at least several genes. Telomerase activation alone does not lead to cancerous growth in most cells, but it does appear to be required along with other mutations for cancer to develop.

> **Concepts**
>
> The ends of eukaryotic chromosomes are replicated by an RNA–protein enzyme called telomerase. This enzyme adds extra nucleotides to the G-rich DNA strand of the telomere.

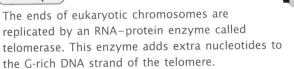 **www.whfreeman.com/pierce** More on telomerase, including an animated cartoon that illustrates the process of replication by telomerase

Replication in Archaea

The process of replication in archaebacteria has a number of features in common with replication in eukaryotic cells—many of the proteins taking part are more similar to those in eukaryotic cells than to those in to those in eubacteria. Although some archaea have a single origin of replication, as do eubacteria, this origin does not contain the typical sequences recognized by bacterial initiator proteins but instead has sequences that are similar to those found in eukaryotic origins. These similarities in replication between archaeal and eukaryotic cells reinforce the conclusion that the archaea are more closely related to eukaryotic cells than to the prokaryotic eubacteria.

The Molecular Basis of Recombination

Recombination is the exchange of genetic information between DNA molecules; when the exchange is between homologous DNA molecules, it is called **homologous recombination**. This process takes place in crossing over, in which homologous regions of chromosomes are exchanged (◄FIGURE 12.22) and genes are shuffled into new combinations. Recombination is an extremely important genetic process because it increases genetic variation. Rates of recombination provide important information about linkage relations among genes, which is used to create genetic maps (see Figures 7.12 through 7.14). Recombination is also essen-

tial for some types of DNA repair (as will be discussed in Chapter 17).

Homologous recombination is a remarkable process: a nucleotide strand of one chromosome aligns precisely with a nucleotide strand of the homologous chromosome, breaks arise in corresponding regions of different DNA molecules, parts of the molecules precisely change place, and then the pieces are correctly joined. In this complicated series of events, no genetic information is lost or gained. Although the precise

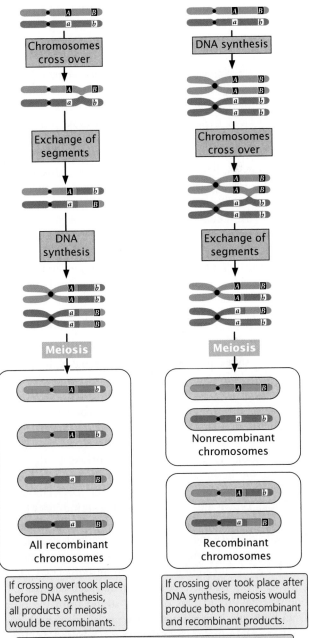

Conclusion: Because crossing over results in recombinant and nonrecombinant products, it must take place after DNA synthesis.

◄ **12.22 Genetic evidence suggests that crossing over takes place after DNA synthesis.**

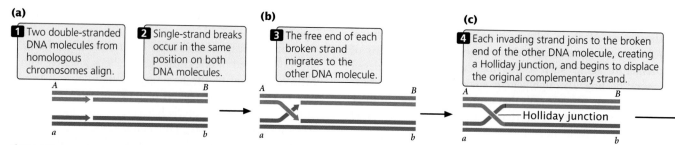

◄12.23 In the Holliday model, homologous recombination is accomplished through a single-strand break in each DNA duplex, strand displacement, branch migration, and resolution of a single Holliday junction.

molecular mechanism of homologous recombination is still poorly known, the exchange is probably accomplished through the pairing of complementary bases. A single-stranded DNA molecule of one chromosome pairs with a single-stranded DNA molecule of another, forming **heteroduplex DNA.**

In meiosis, homologous recombination (crossing over) could theoretically take place before, during, or after DNA synthesis. Cytological, biochemical, and genetic evidence indicates that it takes place in prophase I of meiosis, whereas DNA replication takes place earlier, in interphase. Thus, crossing over must entail the breaking and rejoining of chromatids when homologous chromosomes are at the four-strand stage (see Figure 12.22). This section explores some theories about how the process of recombination takes place.

The Holliday Model

One model of homologous recombination, the **Holliday junction,** states that the process is initiated by single-strand breaks in the DNA molecule. This model begins with double-stranded DNA molecules from two homologous chromosomes that carry identical (or nearly identical) nucleotide sequences. These two DNA molecules align precisely, and so their homologous sequences sit side by side (◄FIGURE 12.23a). Single-strand breaks in the same position on both DNA molecules allow the free ends of the strands to move to the other DNA molecule (◄FIGURE 12.23b and c). Each invading strand joins to the broken end of the other homologous DNA molecule and begins to displace the original complementary strand, taking its place by hydrogen bonding to the original strand. The invasion and joining take place on both DNA molecules, creating two heteroduplex DNAs, each consisting of one original strand plus one new strand from the other DNA molecule. The point at which nucleotide strands pass from one DNA molecule to the other is the Holliday junction. In the Holliday model of recombination, there is a single junction. As the two nucleotide strands exchange positions, the junction moves along the molecules in a process called **branch migration** (◄FIGURE 12.23d). The

exchange of nucleotide strands and branch migration create two duplex molecules connected by the Holliday junction. This structure is termed the **Holliday intermediate.** Holliday intermediates in *E. coli* and yeast have been observed with electron microscopy.

If the ends of the two interconnected duplexes illustrated in Figure 12.23d are pulled away from one another, we obtain the structure illustrated in ◄FIGURE 12.23e. If you carefully compare parts *d* and *e*, you will see that the structures in each are the same; the only difference is that, in part *e*, the ends of the molecules have been pulled apart.

The next step in the Holliday model is easier to visualize if we rotate the bottom half of the Holliday intermediate by 180 degrees, producing the structure shown in ◄FIGURE 12.23f. These interconnected DNA duplexes are then separated by additional cleavage and reunion of the nucleotide strands. The duplexes can be cleaved in one of two ways, as shown by two different pathways in Figure 12.23. Cleavage may be in the horizontal plane (◄FIGURE 12.23g), in which case the nucleotide strands are rejoined as shown in ◄FIGURE 12.23h, and two DNA molecules are produced. Although both resulting DNA molecules contain a patch of heteroduplex DNA, the genes on either end of the molecules are identical with those originally present (gene *A* with *B*, and gene *a* with *b*). These DNA molecules are called noncrossover (patched) recombinants (◄FIGURE 12.23i).

On the other hand, cleavage of the Holliday structure in the vertical plane and rejoining of the nucleotide strands (◄FIGURE 12.23j) produces crossover (spliced) recombinants (◄FIGURE 12.23k). In these recombinants, *both* resulting DNA molecules are heteroduplex, and recombination has taken place between loci at the ends of the molecules; now gene *A* is paired with *b*, and gene *a* is paired with *B*. Recombination is equally likely to produce noncrossover and crossover recombinants.

www.whfreeman.com/pierce   An animated cartoon that will help you visualize the Holliday structure and its resolution

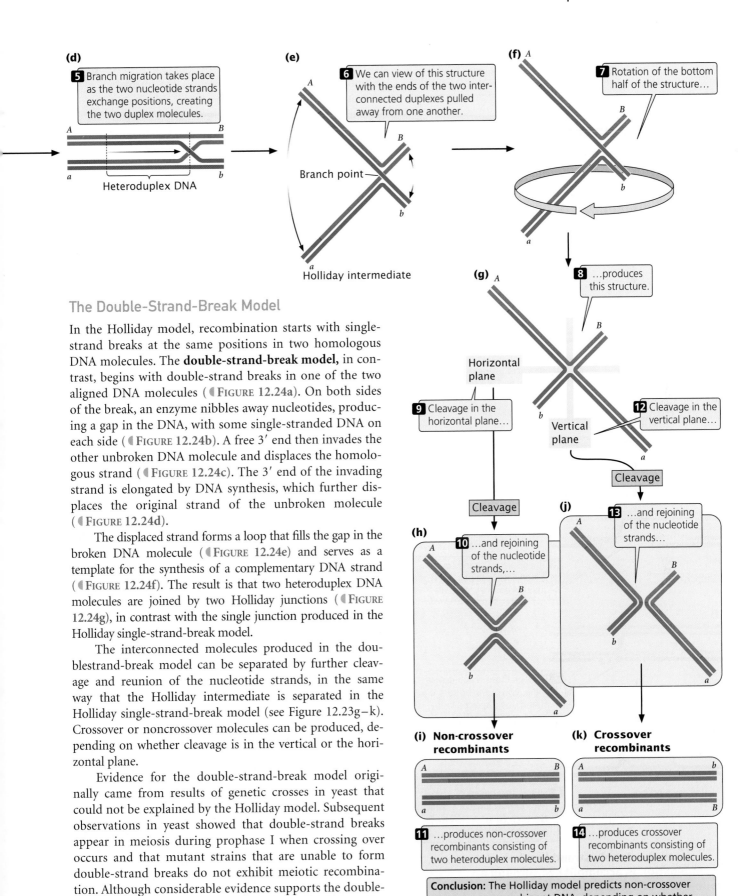

(d)

5 Branch migration takes place as the two nucleotide strands exchange positions, creating the two duplex molecules.

Heteroduplex DNA

(e)

6 We can view of this structure with the ends of the two interconnected duplexes pulled away from one another.

Branch point

Holliday intermediate

(f)

7 Rotation of the bottom half of the structure...

8 ...produces this structure.

(g)

Horizontal plane

9 Cleavage in the horizontal plane...

Vertical plane

12 Cleavage in the vertical plane...

Cleavage

Cleavage

(h)

10 ...and rejoining of the nucleotide strands,...

(j)

13 ...and rejoining of the nucleotide strands...

(i) Non-crossover recombinants

11 ...produces non-crossover recombinants consisting of two heteroduplex molecules.

(k) Crossover recombinants

14 ...produces crossover recombinants consisting of two heteroduplex molecules.

Conclusion: The Holliday model predicts non-crossover or crossover recombinant DNA, depending on whether cleavage is in the horizontal or the vertical plane.

The Double-Strand-Break Model

In the Holliday model, recombination starts with single-strand breaks at the same positions in two homologous DNA molecules. The **double-strand-break model,** in contrast, begins with double-strand breaks in one of the two aligned DNA molecules (◀ FIGURE 12.24a). On both sides of the break, an enzyme nibbles away nucleotides, producing a gap in the DNA, with some single-stranded DNA on each side (◀ FIGURE 12.24b). A free 3′ end then invades the other unbroken DNA molecule and displaces the homologous strand (◀ FIGURE 12.24c). The 3′ end of the invading strand is elongated by DNA synthesis, which further displaces the original strand of the unbroken molecule (◀ FIGURE 12.24d).

The displaced strand forms a loop that fills the gap in the broken DNA molecule (◀ FIGURE 12.24e) and serves as a template for the synthesis of a complementary DNA strand (◀ FIGURE 12.24f). The result is that two heteroduplex DNA molecules are joined by two Holliday junctions (◀ FIGURE 12.24g), in contrast with the single junction produced in the Holliday single-strand-break model.

The interconnected molecules produced in the double-strand-break model can be separated by further cleavage and reunion of the nucleotide strands, in the same way that the Holliday intermediate is separated in the Holliday single-strand-break model (see Figure 12.23g–k). Crossover or noncrossover molecules can be produced, depending on whether cleavage is in the vertical or the horizontal plane.

Evidence for the double-strand-break model originally came from results of genetic crosses in yeast that could not be explained by the Holliday model. Subsequent observations in yeast showed that double-strand breaks appear in meiosis during prophase I when crossing over occurs and that mutant strains that are unable to form double-strand breaks do not exhibit meiotic recombination. Although considerable evidence supports the double-strand-break model in yeast, the extent to which it applies to other organisms is not known.

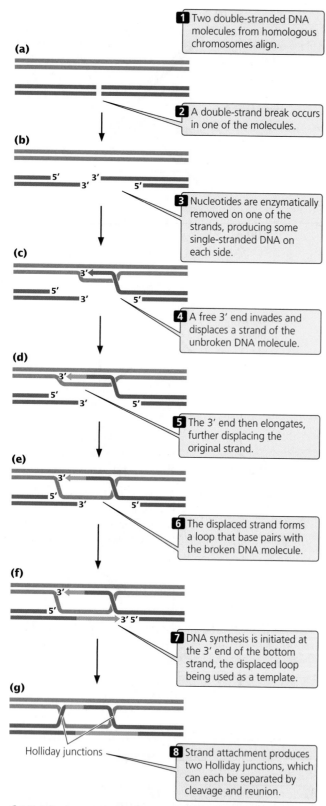

(a)

1 Two double-stranded DNA molecules from homologous chromosomes align.

2 A double-strand break occurs in one of the molecules.

(b)

3 Nucleotides are enzymatically removed on one of the strands, producing some single-stranded DNA on each side.

(c)

4 A free 3′ end invades and displaces a strand of the unbroken DNA molecule.

(d)

5 The 3′ end then elongates, further displacing the original strand.

(e)

6 The displaced strand forms a loop that base pairs with the broken DNA molecule.

(f)

7 DNA synthesis is initiated at the 3′ end of the bottom strand, the displaced loop being used as a template.

(g)

Holliday junctions

8 Strand attachment produces two Holliday junctions, which can each be separated by cleavage and reunion.

◀ **12.24 In the double-strand-break model, recombination is accomplished through a double-strand break in one DNA duplex, strand displacement, DNA synthesis, and resolution of two Holliday junctions.**

Concepts

Homologous recombination requires the formation of heteroduplex DNA consisting of one nucleotide strand from each of two different chromosomes. In the Holliday model, homologous recombination is accomplished through a single-strand break in the DNA, strand displacement, and branch migration. In the double-strand-break model, recombination is accomplished through double-strand breaks, strand displacement, and branch migration.

Enzymes Required for Recombination

Recombination between DNA molecules requires the unwinding of DNA helices, the cleavage of nucleotide strands, strand invasion, and branch migration, followed by further strand cleavage and union to remove Holliday junctions. Much of what we know about these processes arises from studies of gene exchange in *E. coli*. Although bacteria do not undergo meiosis, they do have a type of sexual reproduction (conjugation), in which one bacterium donates its chromosome to another (discussed more fully in Chapter 8). Subsequent to conjugation, the recipient bacterium has two chromosomes, which may undergo homologous recombination. Geneticists have isolated mutant strains of *E. coli* that are deficient in recombination; the study of these strains has resulted in the identification of genes and proteins that play a role in bacterial recombination, revealing several different pathways by which it can take place.

Three genes that play a pivotal role in *E. coli* recombination are *recB*, *recC*, and *recD*, which encode three polypeptides that together form the RecBCD protein. This protein unwinds double-stranded DNA and is capable of cleaving nucleotide strands. The *recA* gene encodes the RecA protein that allows a single strand to invade a DNA helix and the subsequent displacement of one of the original strands. Thus invasion and displacement are necessary for both the single-strand- and the double-strand-break models of homologous recombination.

The *ruvA* and *ruvB* genes encode proteins that catalyze branch migration, and the *ruvC* gene produces a protein, called resolvase, that cleaves Holliday structures. Single-strand-binding proteins, DNA ligase, DNA polymerases, and DNA gyrase also play roles in various types of recombination, in addition to their functions in DNA replication.

Concepts

A number of proteins have roles in recombination, including RecA, RecBCD, RuvA, RuvB, resolvase, single-strand-binding proteins, ligase, DNA polymerases, and gyrase.

Connecting Concepts Across Chapters

This chapter has built on a central concept introduced in Chapter 2, that cell division is preceded by replication of the genetic material. In Chapter 2, we saw that DNA replication takes place in the S phase of the cell cycle and that several checkpoints ensure that division does not take place in the absence of DNA replication. The current chapter examined the process of DNA synthesis.

DNA is sometimes said to be a self-replicating molecule, but nothing could be farther from the truth. Replication requires much more than a DNA template; a large number of proteins and enzymes also are necessary. Despite this complexity, a few rules summarize the process: (1) all replication is semiconservative, (2) new DNA molecules always elongate at the 3′ end (replication is 5′→3′), (3) replication begins at sequences called origins and requires RNA primers for initiation, (4) DNA synthesis takes place continuously on one strand and discontinuously on the other, and (5) newly synthesized nucleotide strands are antiparallel and complementary to their template strands.

As we have seen, replication takes place with a high degree of accuracy; this accuracy is essential to maintain the integrity of genetic information as DNA molecules are copied again and again. The accuracy of replication is maintained by several different mechanisms, including precision in nucleotide selection, the ability of DNA polymerases to proofread and correct mistakes, and the detection and repair of residual mismatches after replication (mismatch repair).

An understanding of DNA replication provides a foundation for several topics that will be introduced in later chapters of this book. Chapter 18 (on recombinant DNA technology) examines the polymerase chain reaction and other techniques (DNA sequence analysis and cloning) that require an understanding of DNA synthesis. In Chapter 17 (on gene mutation and DNA repair), we learn that, in spite of the accuracy of DNA synthesis, errors do arise and sometimes lead to mutations. These errors are addressed by mechanisms of DNA repair, many of which require DNA synthesis. The movement of transposable genetic elements (Chapter 11) also requires DNA synthesis.

CONCEPTS SUMMARY

- Replication is semiconservative: DNA's two nucleotide strands separate and each serves as a template on which a new strand is synthesized

- A replicon is a unit of replication that contains an origin of replication.

- In theta replication of DNA, the two nucleotide strands of a circular DNA molecule unwind, creating a replication bubble; within each replication bubble, DNA is normally synthesized on both strands and at both replication forks, producing two circular DNA molecules.

- Rolling-circle replication is initiated by a nick in one strand of circular DNA, which produces a 3′-OH group to which new nucleotides are added while the 5′ end of the broken strand is displaced from the circle. Replication proceeds around the circle, producing a circular DNA molecule and a single-stranded linear molecule.

- Linear eukaryotic DNA contains many origins of replication. At each origin, the DNA unwinds, producing two nucleotide strands that serve as templates. Unwinding and replication take place on both templates at both ends of the replication bubble until adjacent replicons meet, resulting in two linear DNA molecules.

- DNA synthesis requires a single-stranded DNA template, deoxyribonucleoside triphosphates; and a group of enzymes and proteins that carry out replication.

- All DNA synthesis is in the 5′→3′ direction. Because the two nucleotide strands of DNA are antiparallel, replication takes place continuously on one strand (the leading strand) and discontinuously on the other (the lagging strand).

- Replication begins when an initiator protein binds to a replication origin and unwinds a short stretch of DNA, to which DNA helicase attaches. DNA helicase unwinds the DNA at the replication fork, single-strand-binding proteins bind to single nucleotide strands to prevent them from reannealing, and DNA gyrase (a topoisomerase) removes the strain ahead of the replication fork that is generated by unwinding.

- During replication, primase synthesizes short primers of RNA nucleotides, providing a 3′-OH group to which DNA polymerase can add DNA nucleotides.

- DNA polymerase adds new nucleotides to the 3′ end of a growing polynucleotide strand. Bacteria have two DNA polymerases that have primary roles in replication: DNA polymerase III, which synthesizes new DNA on the leading and lagging strands; and DNA polymerase I, which removes and replaces primers.

- DNA ligase seals nicks that remain in the sugar–phosphate backbones when the RNA primers are replaced by DNA nucleotides.

- Several mechanisms ensure the high rate of accuracy in replication, including precise nucleotide selection, proofreading, and mismatch repair.

- Eukaryotic replication is similar to bacterial replication, although eukaryotes have multiple origins of replication and different DNA polymerases.

- Precise replication at multiple origins is ensured by a licensing factor that must attach to an origin before replication can begin. The licensing factor is removed after replication is initiated and renewed after cell division.

- Eukaryotic nucleosomes are quickly assembled on new molecules of DNA; newly assembled nucleosomes consist of a random mixture of old and new histone proteins.

- The ends of linear eukaryotic DNA molecules are replicated by the enzyme telomerase.

- Replication in archaeal bacteria has a number of features in common with eukaryotic replication.

- Homologous recombination takes place through the exchange of genetic material between homologous DNA molecules. In the Holliday model, homologous recombination begins with single-strand breaks in both DNA molecules, followed by strand displacement, branch migration, and Holliday junction resolution. In the double-strand break model, it begins with a double-strand-break, followed by strand displacement, DNA synthesis, and resolution of two Holliday junctions.

- Homologous recombination in *E. coli* requires a number of enzymes, including RecA, RecBCD, resolvase, single-strand-binding proteins, ligase, DNA polymerases, and gyrase.

IMPORTANT TERMS

semiconservative replication (p. 323)
equilibrium density gradient centrifugation (p. 325)
replicon (p. 326)
replication origin (p. 326)
theta replication (p. 326)
replication bubble (p. 327)
replication fork (p. 327)
bidirectional replication (p. 327)
rolling-circle replication (p. 327)
DNA polymerase (p. 330)

continuous replication (p. 330)
leading strand (p. 330)
discontinuous replication (p. 330)
lagging strand (p. 330)
Okazaki fragments (p. 330)
initiator protein (p. 332)
DNA helicase (p. 332)
single-strand-binding protein (SSB) (p. 332)
DNA gyrase (p. 332)
primase (p. 333)

primer (p. 333)
DNA polymerase III (p. 334)
DNA polymerase I (p. 335)
DNA ligase (p. 335)
proofreading (p. 336)
mismatch repair (p. 337)
autonomously replicating sequence (p. 338)
replication licensing factor (p. 339)
DNA polymerase α (p. 339)
DNA polymerase δ (p. 339)

DNA polymerase β (p. 339)
DNA polymerase γ (p. 339)
DNA polymerase ε (p. 339)
telomerase (p. 341)
homologous recombination (p. 343)
heteroduplex DNA (p. 344)
Holliday junction (p. 344)
branch migration (p. 344)
Holliday intermediate (p. 344)
double-strand-break model (p. 345)

Worked Problems

1. The diagram below represents the template strands of a replication bubble in a DNA molecule. Draw in the newly synthesized strands and label the leading and lagging strands.

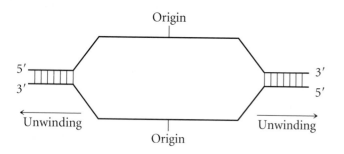

• Solution

To determine the leading and lagging strands, first note which end of each template strands is 5′ and which end is 3′. With a pencil, draw in the strands being synthesized on these templates, and label their 5′ and 3′ ends, recalling that the newly synthesized strands must be antiparallel to the templates.

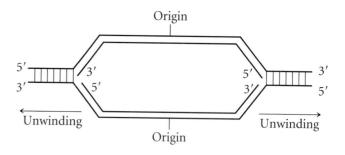

Next, determine the direction of replication for each new strand, which must be 5′→3′. You might draw arrows on the new strands to indicate the direction of replication. After you have established the direction of replication for each strand, look at each fork and determine whether the direction of replication for a strand is the same as the direction of unwinding. The strand on which replication is in the same direction as unwinding is the leading strand. The strand on which replication is in the direction opposite that of unwinding is the lagging strand. Make sure that you have one leading strand and one lagging strand for each fork.

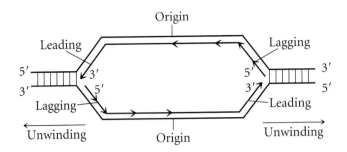

2. Consider the experiment conducted by Meselson and Stahl in which they used ^{14}N and ^{15}N in cultures of *E. coli* and equilibrium density gradient centrifugation. Draw pictures to represent the bands produced by bacterial DNA in the density-gradient tube before the switch to medium containing ^{14}N and after one, two, and three rounds of replication after the switch to the medium containing ^{14}N. Use a separate set of drawings to show the bands that would appear if replication were **(a)** semiconservative; **(b)** conservative; **(c)** dispersive.

• Solution

DNA labeled with ^{15}N will be denser than DNA labeled with ^{14}N; therefore ^{15}N-labeled DNA will sink lower in the density-gradient tube. Before the switch to medium containing ^{14}N, all DNA in the bacteria will contain ^{15}N and will produce a single band in the lower end of the tube.

(a) With semiconservative replication, the two strands separate, and each serves as a template on which a new strand is synthesized. After one round of replication, the original template strand of each molecule will contain ^{15}N and the new strand of each molecule will contain ^{14}N; so a single band will appear in the density gradient halfway between the positions expected of DNA with ^{15}N and of DNA with ^{14}N. In the next round of replication, the two strands again separate and serve as templates for new strands. Each of the new strands contains only ^{14}N, thus some DNA molecules will contain one strand with the original ^{15}N and one strand with new ^{14}N, whereas the other molecules will contain two strands with ^{14}N. This labeling will produce two bands, one at the intermediate position and one at a higher position in the tube. Additional rounds of replication should produce increasing amounts of DNA that contains only ^{14}N; so the higher band will get darker.

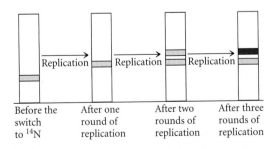

(b) With conservative replication, the entire molecule serves as a template. After one round of replication, some molecules will consist entirely of ^{15}N, and others will consist entirely of ^{14}N, so

two bands should be present. Subsequent rounds of replication will increase the fraction of DNA consisting entirely of new ^{14}N, thus the upper band will get darker. However, the original DNA with ^{15}N will remain, so two bands will be present.

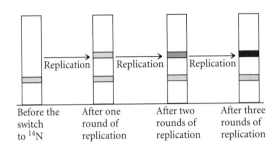

(c) In dispersive replication, both nucleotide strands break down into fragments that serve as templates for the synthesis of new DNA. The fragments then reassemble into DNA molecules. After one round of replication, all DNA should contain approximately half ^{15}N and half ^{14}N, producing a single band that is halfway between the positions expected of DNA labeled with ^{15}N and of DNA labeled with ^{14}N. With further rounds of replication, the proportion of ^{14}N in each molecule increase; so a single hybrid band remains, but its position in the density gradient will move upward. The band is also expected to get darker as the total amount of DNA increases.

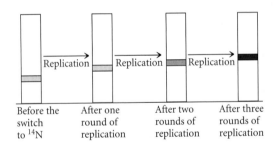

3. The *E. coli* chromosome contains 4.7 million base pairs of DNA. If synthesis at each replication fork occurs at a rate of 1000 nucleotides per second, how long will it take to completely replicate the *E. coli* chromosome with theta replication?

• Solution

Bacterial chromosomes contain a single origin of replication, and theta replication usually employs two replication forks, which proceed around the chromosome in opposite directions. Thus, the overall rate of replication for the whole chromosome is 2000 nucleotides per second. With a total of 4.7 million base pairs of DNA, the entire chromosome will be replicated in:

$$4{,}700{,}000 \text{ bp} \times \frac{1 \text{ second}}{2000 \text{ bp}} = 2350 \text{ seconds} \times \frac{1 \text{ minute}}{60 \text{ seconds}}$$

$$= 39.17 \text{ minutes}$$

At the beginning of this chapter it was stated that *E. coli* is capable of dividing every 20 minutes. How is this possible if it takes almost twice as long to replicate its genome? The answer is that a second round of replication begins before the first round has finished. Thus, when an *E. coli* cell divides, the chromosomes that are passed on to the daughter cells are already partially replicated. This is in contrast to eukaryotic cells, which replicate their entire genome once, and only once, during each cell cycle.

COMPREHENSION QUESTIONS

1. What is semiconservative replication?

* 2. How did Meselson and Stahl demonstrate that replication in *E. coli* takes place in a semiconservative manner?

* 3. Draw a molecule of DNA undergoing theta replication. On your drawing, identify (1) origin, (2) polarity (5′ and 3′ ends) of all template strands and newly synthesized strands, (3) leading and lagging strands, (4) Okazaki fragments, and (5) location of primers.

4. Draw a molecule of DNA undergoing rolling-circle replication. On your drawing, identify (1) origin, (2) polarity (5′ and 3′ ends) of all template and newly synthesized strands, (3) leading and lagging strands, (4) Okazaki fragments, and (5) location of primers.

5. Draw a molecule of DNA undergoing eukaryotic linear replication. On your drawing, identify (1) origin; (2) polarity (5′ and 3′ ends) of all template and newly synthesized strands, (3) leading and lagging strands, (4) Okazaki fragments, and (5) location of primers.

6. What are three major requirements of replication?

* 7. What substrates are used in the DNA synthesis reaction?

8. List the different proteins and enzymes taking part in bacterial replication. Give the function of each in the replication process.

9. What similarities and differences exist in the enzymatic activities of DNA polymerases I, II, and III? What is the function of each type of DNA polymerase in bacterial cells?

*10. Why is primase required for replication?

11. What three mechanisms ensure the accuracy of replication in bacteria?

12. How does replication licensing ensure that DNA is replicated only once at each origin per cell cycle?

*13. In what ways is eukaryotic replication similar to bacterial replication, and in what ways is it different?

14. Outline in words and pictures how telomeres at the end of eukaryotic chromosomes are replicated.

15. Briefly outline with diagrams the Holliday model of homologous recombination.

*16. What are some of the enzymes taking part in recombination in *E. coli* and what roles do they play?

APPLICATION QUESTIONS AND PROBLEMS

*17. Suppose a future scientist explores a distant planet and discovers a novel form of double-stranded nucleic acid. When this nucleic acid is exposed to DNA polymerases from *E. coli*, replication takes place continuously on both strands. What conclusion can you make about the structure of this novel nucleic acid?

*18. Phosphorus is required to synthesize the deoxyribonucleoside triphosphates used in DNA replication. A geneticist grows some *E. coli* in a medium containing nonradioactive phosphorous for many generations. A sample of the bacteria is then transferred to a medium that contains a radioactive isotope of phosphorus (^{32}P). Samples of the bacteria are removed immediately after the transfer and after one and two rounds of replication. What will be the distribution of radioactivity in the DNA of the bacteria in each sample? Will radioactivity be detected in neither, one, or both strands of the DNA?

19. A line of mouse cells is grown for many generations in a medium with ^{15}N. Cells in G_1 are then switched to a new medium that contains ^{14}N. Draw a pair of homologous chromosomes from these cells at the following stages, showing the two strands of DNA molecules found in the chromosomes. Use different colors to represent strands with ^{14}N and ^{15}N.

(a) Cells in G_1, before switching to medium with ^{14}N

(b) Cells in G_2, after switching to medium with ^{14}N

(c) Cells in anaphase of mitosis, after switching to medium with ^{14}N

(d) Cells in metaphase I of meiosis, after switching to medium with ^{14}N

(e) Cells in anaphase II of meiosis, after switching to medium with ^{14}N

*20. A circular molecule of DNA contains 1 million base pairs. If DNA synthesis at a replication fork occurs at a rate of 100,000 nucleotides per minute, how long will theta replication require to completely replicate the molecule, assuming that theta replication is bidirectional? How long will replication of this circular chromosome take by rolling-circle replication? Ignore replication of the displaced strand in rolling-circle replication.

21. A bacterium synthesizes DNA at each replication fork at a rate of 1000 nucleotides per second. If this bacterium completely replicates its circular chromosome by theta replication in 30 minutes, how many base pairs of DNA will its chromosome contain?

*22. The following diagram represents a DNA molecule that is undergoing replication. Draw in the strands of newly synthesized DNA and identify the following:

(a) Polarity of newly synthesized strands

(b) Leading and lagging strands

(c) Okazaki fragments

(d) RNA primers

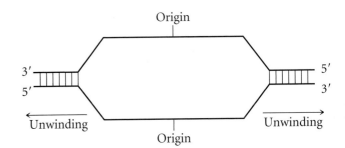

*23. What would be the effect on DNA replication of mutations that destroyed each of the following activities in DNA polymerase I?

(a) $3' \rightarrow 5'$ exonuclease activity

(b) $5' \rightarrow 3'$ exonuclease activity

(c) $5' \rightarrow 3'$ polymerase activity

CHALLENGE QUESTIONS

24. Conditional mutations express their mutant phenotype only under certain conditions (the restrictive conditions) and express the normal phenotype under other conditions (the permissive conditions). One type of conditional mutation is a temperature-sensitive mutation, which expresses the mutant phenotype only at certain temperatures.

Strains of *E. coli* have been isolated that contain temperature-sensitive mutations in the genes encoding different components of the replication machinery. In each of these strains, the protein produced by the mutated gene is nonfunctional under the restrictive conditions. These strains are grown under permissive conditions and then abruptly switched to the restrictive condition. After one round of replication under the restrictive condition, the DNA from each strain is isolated and analyzed. What would you predict to see in the DNA isolated from each strain in the following list?

Temperature-sensitive mutation in gene encoding:

(a) DNA ligase

(b) DNA polymerase I

(c) DNA polymerase III

(d) Primase

(e) Initiator protein

SUGGESTED READINGS

Baker, T. A., and S. H. Wickner. 1992. Genetics and enzymology of DNA replication in *Escherichia coli*. *Annual Review of Genetics* 26:447–477.
A detailed review of replication in bacteria.

Bell, S. P., R. Kobayashi, and B. Stillman. 1993. Yeast origin recognition complex functions in transcription silencing and DNA replication. *Science* 262:1844–1849.
A research article on the role of eukaryotic origins of replication in replication and transcription.

Blow, J. J., and S. Tada. 2000. A new check on issuing the license. *Nature* 404:560–561.
A short review of the molecular basis of replication licensing.

Cairns, J. 1966. The bacterial chromosome. *Scientific American* 214(1):36–44.

Classical research that verified semiconservative replication and the theta model in bacteria.

Campbell, J. L. 1986. Eukaryotic DNA replication. *Annual Review of Biochemisty* 55:733–771.
A detailed review of replication in eukaryotic cells.

Cook. P. R. 1999. The organization of replication and transcription. *Science* 284:1790–1795.
A review of the location of DNA and RNA polymerase enzymes that carry out replication and transcription. Provides evidence that the polymerases are immobilized and the DNA template is threaded through the enzymes.

Echols, H., and M. F. Goodman. 1991. Fidelity mechanisms in DNA replication. *Annual Review of Biochemistry* 60:477–511.
A review of error avoidance mechanisms in replication.

Ellis, N., J. Groden, T. Ye, J. Straughen, D. J. Lennon, S. Ciocci, M. Proytcheva, and J. German. 1995. The Blooms's syndrome gene product is homologous to RecQ helicases. *Cell* 83:655–666.
Report of the isolation of the gene causing Bloom syndrome and the identification of its biochemical function.

Frick, D. N., and C. C. Richardson. 2000. DNA primases. *Annual Review of Biochemistry* 70:39–80.
An excellent and detailed review of DNA primases, which are essential to the replication process.

Greider, C. W., and E. H. Blackburn. 1996. Telomeres, telomerase, and cancer. *Scientific American* 274(2):92–97.
A readable account of telomeres, how they are replicated, and their role in cancer.

Haber, J. E. 1999. DNA recombination: the replication connection. *Trends in Biochemical Science* 24:271–276.
The role of the establishment of replication forks in recombination.

Huberman, J. A. 1998. Choosing a place to begin. *Science* 281:929–930.
A short review on evidence regarding replication origins in eukaryotic cells.

Hübscher, U., H. Nasheuer, and J. E. Syväoja. 2000. Eukaryotic DNA polymerases: a growing family. *Trends in Biochemical Science* 25:143–147.
An excellent review of the increasing number of different DNA polymerases found in eukaryotic cells and their functions.

Keck, J. L. , D. D. Roche, A. S. Lynch, and J. M. Berger. 2000. Structure of the RNA polymerase domain of *E. coli* primase. *Science* 287:2482–2492.
Report of the detailed structure of bacterial primase.

Kornberg, A., and T. A. Baker. 1992. DNA Replication, 2d ed. New York: W. H. Freeman and Company.
The "Bible" of DNA replication by one of the world's foremost authorities on the subject.

Kowalczykowski, S. C. 2000. Initiation of genetic recombination and recombination-dependent replication. *Trends in Biochemical Science* 25:156–164.
A review of the role of replication processes in recombination.

Lee, H., M. A. Blasco, G. J. Gottlieb, J. W. Horner, II, G. W. Greider, and R. A. DePinho. 1998. Essential role of mouse telomerase in highly proliferative organs. *Nature* 392:569–574.
Describes the role of telomerase, investigated by creating knockout mice that lack the gene for telomerase.

Matson, S. W., and K. A. Kaiser-Rogers. 1990. DNA helicases. *Annual Review of Biochemistry* 59:289–329.
A detailed review of the enzymes that unwind DNA.

Newton, C. S. 1993. Two jobs for the origin of replication. *Science* 262:1830–1831.
Discusses findings about the molecular structure and functioning of origins of replication and their role in transcription.

Nossal, N. C. 1983. Prokaryotic DNA replication systems. *Annual Review of Biochemistry* 53:581–615.
A good review of replication in bacteria.

Radman, M., and R. Wagner. 1988. The high fidelity of DNA duplication. *Scientific American* 259(2):40–46.
A very readable account of how the accuracy of DNA replication is ensured.

Stahl, F. W. 1987 Genetic recombination. *Scientific American* 256(2):90–101.
An interesting discussion of research examining the molecular mechanism of homologous recombination.

Stahl, F. W. 1994. The Holliday junction on its thirtieth anniversary. *Genetics* 138:241–246.
A brief history of the Holliday model of recombination and an update on its relevance today.

West, S. C. 1992. Enzymes and molecular mechanisms of genetic recombination. *Annual Review of Biochemistry* 61:603–640.
An excellent but detailed review of recombination at the molecular level.

Waga, S., and B. Stillman. 1998. The DNA replication fork in eukaryotic cells. *Annual Review of Biochemistry* 67:721–751.
Summarizes the components of the replication machinery and the process of DNA synthesis that takes place at the replication fork in eukaryotic cells.

Zakian, V. A. 1995. Telomeres: beginning to understand the ends. *Science* 270:1601–1606.
A review article that discusses telomeres and how they are replicated.

13

Transcription

Molecular image of the hammerhead ribozyme (in blue) bound to RNA (in orange). Ribozymes are catalytic RNA molecules that may have been the first carriers of genetic information. (K. Eward/Biografx/Photo Researchers.)

RNA in the Primeval World

Life requires two basic functions. First, living organisms must be able to store and faithfully transmit genetic information during reproduction. Second, they must have the ability to catalyze chemical transformations, to fire the reactions that drive life processes. It was long believed that the functions of information storage and chemical transformation are handled by two entirely different types of molecules. Genetic information is stored in nucleic acids. Catalysis of chemical transformations was held to be the exclusive domain of certain proteins that serve as biological catalysts or enzymes, making reactions take place rapidly within the cell. This biochemical dichotomy—nucleic acid

for information, proteins for catalysts—revealed a dilemma in our understanding of the early stages in the evolution of life. Which came first: proteins or nucleic acids? If nucleic acids carry the coding instructions for proteins, how could proteins be generated without them? Because nucleic acids are unable to copy themselves, how could they be generated without proteins? If DNA and proteins each require the other, how could life begin?

This apparent paradox disappeared in 1981 when Thomas Cech and his colleagues discovered that RNA can serve as a biological catalyst. They found that RNA from the protozoan *Tetrahymena thermophila* can excise 400 nucleotides from its RNA in the absence of any protein. Other examples of catalytic RNAs have now been discovered in different types of cells. Called **ribozymes,** these RNA molecules can cut out parts of their own sequences, connect some RNA molecules together, replicate others, and even catalyze the formation of peptide bonds between amino acids. The discovery of ribozymes complements other evidence suggesting that the original genetic material was RNA.

Ribozymes that were self-replicating probably first arose between 3.5 billion and 4 billion years ago and may have begun the evolution of life on Earth. Early life was an RNA world, with RNA molecules serving both as carriers of genetic information and as catalysts that drove the chemical reactions needed to sustain and perpetuate life. These catalytic RNAs may have acquired the ability to synthesize protein-based enzymes, which are more efficient catalysts; with enzymes taking over more and more of the catalytic functions, RNA probably became relegated to the role of information storage and transfer. DNA, with its chemical stability and faithful replication, eventually replaced RNA as the primary carrier of genetic information. In modern cells, RNA still plays a vital role in both DNA replication and protein synthesis.

Transcription is the synthesis of RNA molecules, with DNA as a template, and it is the first step in the transfer of genetic information from genotype to phenotype. The process is complex, and requires a number of protein components. As we examine the stages of transcription, try to keep all the detail in perspective; focus on understanding how the details relate to the overall purpose of transcription—the selective synthesis of an RNA molecule.

This chapter begins with a brief review of RNA structure and a discussion of the different classes of RNA. We then consider the major components required for transcription. Finally, we explore the process of transcription in eubacteria and eukaryotic cells. At several points in the text, we'll pause to absorb some general principles that emerge.

www.whfreeman.com/pierce   Current research on ribozymes

RNA Molecules

Before we begin our study of transcription, let's review the structure of RNA and consider the different types of RNA molecules.

The Structure of RNA

RNA, like DNA, is a polymer consisting of nucleotides joined together by phosphodiester bonds (see Chapter 10 for a discussion of RNA structure). However, there are several important differences in the structures of DNA and RNA. Whereas DNA nucleotides contain deoxyribose sugars, RNA nucleotides have ribose sugars (◀Figure 13.1a). With a free hydroxyl group on the 2′-carbon atom of the ribose sugar, RNA is degraded rapidly under alkaline condi-

tions. The deoxyribose sugar of DNA lacks this free hydroxyl group; so DNA is a more stable molecule. Another important difference is that thymine, one of the two pyrimidines found in DNA, is replaced by uracil in RNA.

A final difference in the structures of DNA and RNA is that RNA is usually single stranded, consisting of a single polynucleotide strand (◀Figure 13.1b), whereas DNA normally consists of two polynucleotide strands joined by hydrogen bonding between complementary bases. Some viruses contain double-stranded RNA genomes, as discussed in Chapter 8. Although RNA is usually single stranded, short complementary regions within a nucleotide strand can pair and form secondary structures (see Figure 13.1b). These RNA secondary structures are often called hairpin-loops or stem-loop structures. When two regions within a single RNA molecule pair up, the strands in those regions must be antiparallel, with pairing between cytosine and guanine and between adenine and uracil (although occasionally guanine pairs with uracil).

The formation of secondary structures plays an important role in RNA function. Secondary structure is determined by the base sequence of the nucleotide strand; so different RNA molecules can assume different structures. Because their structure determines their function, RNA molecules have the potential for tremendous variation in function. With its two complementary strands forming a helix, DNA is much more restricted in the range of secondary structures that it can assume, and it serves fewer functional roles in the cell. Similarities and differences in DNA and RNA structures are summarized in Table 13.1.

Table 13.1	The structures of DNA and RNA compared	
Characteristic	**DNA**	**RNA**
Composed of nucleotides	Yes	Yes
Type of sugar	Deoxyribose	Ribose
Presence of 2′-OH group	No	Yes
Bases	A, G, C, T	A, G, C, U
Nucleotides joined by phosphodiester bonds	Yes	Yes
Double or single stranded	Usually double	Usually single
Secondary structure	Double helix	Many types
Stability	Quite stable	Easily degraded

(a)

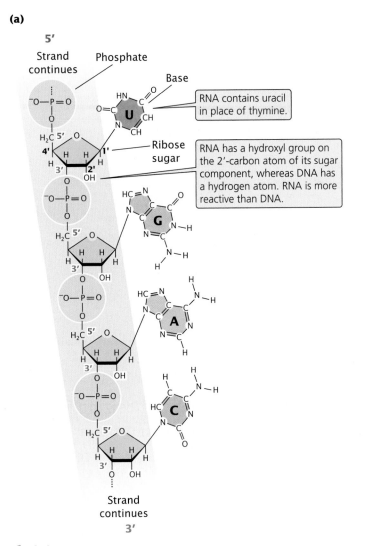

RNA contains uracil in place of thymine.

RNA has a hydroxyl group on the 2'-carbon atom of its sugar component, whereas DNA has a hydrogen atom. RNA is more reactive than DNA.

(b) Primary structure

5' AUGCGGCUACGUAACGAGCUUAGCGCGUAUACCGAAAGGGUAGAAC 3'

An RNA molecule folds to form secondary structures…   Folding

…owing to hydrogen bonding between complementary bases on the same strand.

Secondary structure

◀ 13.1 **RNA has a primary and a secondary structure.**

Classes of RNA

RNA molecules perform a variety of functions in the cell. **Ribosomal RNA** (rRNA), along with ribosomal protein subunits, makes up the ribosome, the site of protein assembly. We'll take a more detailed look at the ribosome in Chapter 14. **Messenger RNA** (mRNA) carries the coding instructions for polypeptide chains from DNA to the ribosome. After attaching to a ribosome, an mRNA molecule specifies the sequence of the amino acids in a polypeptide chain and provides a template for joining amino acids. Large precursor molecules, which are termed **pre-messenger RNAs** (pre-mRNAs), are the immediate products of transcription in eukaryotic cells. Pre-mRNAs are modified extensively before they exit the nucleus for translation into protein. Bacterial cells do not possess pre-mRNA; in these cells, transcription takes place concurrently with translation.

Transfer RNA (tRNA) serves as the link between the coding sequence of nucleotides in the mRNA and the amino acid sequence of a polypeptide chain. Each tRNA attaches to one particular type of amino acid and helps to incorporate that amino acid into a polypeptide chain (discussed in Chapter 15).

Additional classes of RNA molecules are found in the nuclei of eukaryotic cells. **Small nuclear RNAs** (snRNAs) combine with small nuclear protein subunits to form **small nuclear ribonucleoproteins** (snRNPs, affectionately known as "snurps"). The snRNPs are analogous to ribosomes in structure, only smaller, and they typically contain a single RNA molecule combined with approximately 10 small nuclear protein subunits. Some snRNAs participate in the processing of RNA, converting pre-mRNA into mRNA. **Small nucleolar RNAs** (snoRNAs) take part in the processing of rRNA. Small

Table 13.2 Locations and functions of different classes of RNA molecules

Class of RNA	Cell Type	Location of Function* in Eukaryotic Cells	Function
Ribosomal RNA (rRNA)	Bacterial and eukaryotic	Cytoplasm	Structural and functional components of the ribosome
Messenger RNA (mRNA)	Bacterial and eukaryotic	Nucleus and cytoplasm	Carries genetic code for proteins
Transfer RNA (tRNA)	Bacterial and eukaryotic	Cytoplasm	Helps incorporate amino acids into polypeptide chain
Small nuclear RNA (snRNA)	Eukaryotic	Nucleus	Processing of pre-mRNA
Small nucleolar RNA (snoRNA)	Eukaryotic	Nucleus	Processing and assembly of rRNA
Small cytoplasmic RNA (scRNA)	Eukaryotic	Cytoplasm	Variable

*All eukaryotic RNAs are transcribed in the nucleus.

RNA molecules also are found in the cytoplasm of eukaryotic cells; these molecules are called **small cytoplasmic RNAs** (scRNAs). The different classes of RNA molecules are summarized in Table 13.2.

Concepts

RNA differs from DNA in that it possesses a hydroxyl group on the 2′-carbon atom of its sugar, contains uracil instead of thymine, and is normally single stranded. Several classes of RNA exist within bacterial and eukaryotic cells.

Transcription: Synthesizing RNA from a DNA Template

All cellular RNAs are synthesized from a DNA template through the process of transcription (◀FIGURE 13.2). Transcription is in many ways similar to the process of replication, but one fundamental difference relates to the length of the template used. During replication, all the nucleotides in the DNA template are copied, but, during transcription, only small parts of the DNA molecule—usually a single gene or, at most, a few genes—are transcribed into RNA. Because not all gene products are needed at the same time or in the same cell, it would be highly

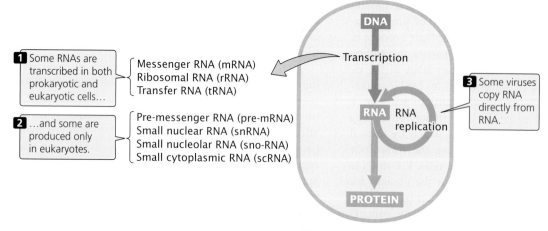

1 Some RNAs are transcribed in both prokaryotic and eukaryotic cells...

Messenger RNA (mRNA)
Ribosomal RNA (rRNA)
Transfer RNA (tRNA)

2 ...and some are produced only in eukaryotes.

Pre-messenger RNA (pre-mRNA)
Small nuclear RNA (snRNA)
Small nucleolar RNA (sno-RNA)
Small cytoplasmic RNA (scRNA)

DNA
Transcription
RNA   RNA replication
PROTEIN

3 Some viruses copy RNA directly from RNA.

◀13.2 **All cellular types of RNA are transcribed from DNA.**

Transcription 357

inefficient for a cell to constantly transcribe all of its genes. Furthermore, much of the DNA does not code for a functional product, and transcription of such sequences would be pointless. Transcription is, in fact, a highly selective process—individual genes are transcribed only as their products are needed. But this selectivity imposes a fundamental problem on the cell—the problem of how to recognize individual genes and transcribe them at the proper time and place.

Like replication, transcription requires three major components:

1. a DNA template;
2. the raw materials (substrates) needed to build a new RNA molecule; and
3. the transcription apparatus, consisting of the proteins necessary to catalyze the synthesis of RNA.

The Template

In 1970, Oscar Miller, Jr., Barbara Hamkalo, and Charles Thomas used electron microscopy to examine cellular contents and demonstrate that RNA is transcribed from a DNA template. The results of this study revealed within the cell the presence of Christmas-tree-like structures: thin central fibers (the trunk of the tree), to which were attached strings (the branches) with granules (◀FIGURE 13.3). The addition of deoxyribonuclease (an enzyme that degrades DNA) caused the central fibers to disappear, indicating that the "tree trunks" were DNA molecules. Ribonuclease (an enzyme that degrades RNA) removed the granular strings, indicating that the branches were RNA. Their conclusion was that each Christmas tree represented a gene undergoing transcription. The transcription of each gene begins at the top of the tree; there, little of the DNA has been transcribed and the RNA branches are short. As the transcription apparatus moves down the tree, transcribing more of the template, the RNA molecules lengthen, producing the long branches at the bottom.

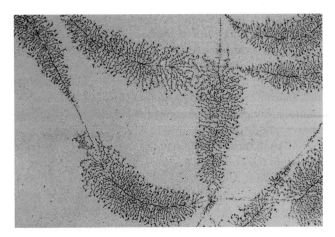

◀13.3 **Under the electron microscope, DNA molecules undergoing transcription exhibit Christmas-tree-like structures.** The trunk of each "Christmas tree" (a transcription unit) represents a DNA molecule; the tree branches (granular strings attached to the DNA) are RNA molecules that have been transcribed from the DNA. As the transcription apparatus moves down the DNA, transcribing more of the template, the RNA molecules become longer and longer. (O. L. Miller, B. R. Beatty, D. W. Fawcett/Visuals Unlimited.)

The transcribed strand The template for RNA synthesis, as for DNA synthesis, is a single strand of the DNA double helix. Unlike replication, however, transcription typically takes place on only one of the two nucleotide strands of DNA (◀FIGURE 13.4). The nucleotide strand used for transcription is termed the **template strand.** The other strand, called the **nontemplate strand,** is not ordinarily transcribed. Thus, in any one section of DNA, only one of the nucleotide strands normally carries the genetic information that is transcribed into RNA (there *are* some exceptions to this rule).

Evidence that only one DNA strand serves as a template came from several experiments carried out by Julius

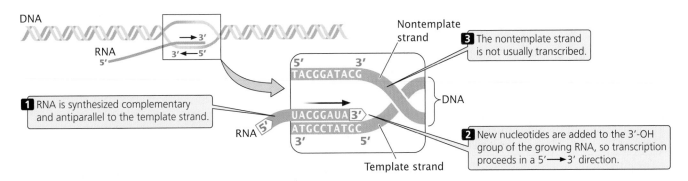

◀13.4 **RNA molecules are synthesized that are complementary and antiparallel to one of the two nucleotide strands of DNA, the template strand.**

Marmur and his colleagues in 1963 on the DNA of bacteriophage SP8, which infects the bacterium *Bacillus subtilus*. This phage carries its genetic information in the form of a double-stranded DNA molecule. The two strands have different base compositions and therefore different densities, which permits the separation of the strands by equilibrium density gradient centrifugation (see Figure 12.2) into "heavy" and "light" DNA strands.

Marmur and his colleagues placed some *B. subtilis* in a medium that contained a radioactively labeled precursor of RNA (◀ FIGURE 13.5). They infected the bacteria with SP8, and the phage injected their DNA into the bacterial cells. Transcription of the phage DNA within the cells incorporated the radioactive precursor into the newly synthesized RNA, producing radioactively labeled RNA complementary to the phage DNA (step 2), which was then isolated from the cells (step 3).

The DNA of another culture of SP8 phage was isolated (step 4) and the heavy and light strands of the DNA were separated (step 5). When the radioactively labeled RNA (obtained in steps 1 through 3 of Figure 13.5) was com-

bined with the heavy strand (step 6), the RNA hybridized to it, indicating that the RNA and DNA were complementary (step 7). However, when radioactively labeled RNA was added to the light strand (step 8), no hybridization took place. These findings led Marmur and his colleagues to conclude that RNA is transcribed from only one of the DNA strands in SP8—in this case, the heavy strand.

SP8 is unusual in that all of its genes are transcribed from the same strand. In most organisms, each gene is transcribed from a single strand, but different genes may be transcribed from different strands (◀ FIGURE 13.6). Notice that one of the strands in Figure 13.6 is identified as plus (+) and the other as minus (−). The plus strand is the template for genes *a* and *c*, and the minus strand is the template for gene *b*.

During transcription, an RNA molecule is synthesized that is complementary and antiparallel to the DNA template strand (see Figure 13.4). The RNA transcript has the same polarity and base sequence as does the nontemplate strand, with the exception that U in RNA substitutes for T in DNA.

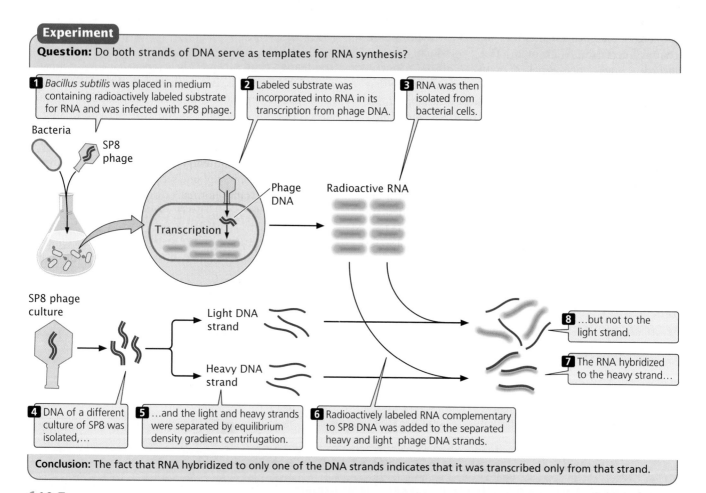

Experiment

Question: Do both strands of DNA serve as templates for RNA synthesis?

1 *Bacillus subtilis* was placed in medium containing radioactively labeled substrate for RNA and was infected with SP8 phage.

2 Labeled substrate was incorporated into RNA in its transcription from phage DNA.

3 RNA was then isolated from bacterial cells.

Bacteria

SP8 phage

Transcription

Phage DNA

Radioactive RNA

SP8 phage culture

Light DNA strand

Heavy DNA strand

8 ...but not to the light strand.

7 The RNA hybridized to the heavy strand...

4 DNA of a different culture of SP8 was isolated,...

5 ...and the light and heavy strands were separated by equilibrium density gradient centrifugation.

6 Radioactively labeled RNA complementary to SP8 DNA was added to the separated heavy and light phage DNA strands.

Conclusion: The fact that RNA hybridized to only one of the DNA strands indicates that it was transcribed only from that strand.

◀ **13.5   Marmur and colleagues showed that only one DNA strand serves as template during transcription.**

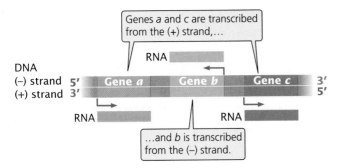

13.6 RNA is transcribed from one DNA strand. In most organisms, each gene is transcribed from a single DNA strand, but different genes may be transcribed from one or the other of the two DNA strands.

Concepts

Within a single gene, only one of the two DNA strands, the template strand, is generally transcribed into RNA.

The transcription unit A **transcription unit** is a stretch of DNA that codes for an RNA molecule and the sequences necessary for its transcription. In eukaryotes, as discussed in Chapter 14, alternative RNA molecules can be produced from each transcription unit. How does the complex of enzymes and proteins that performs transcription—the transcription apparatus—recognize a transcription unit? How does it know which DNA strand to read, and where to start and stop? This information is encoded by the DNA sequence.

Included within a transcription unit are three critical regions: a promoter, an RNA coding sequence, and a terminator (FIGURE 13.7). The **promoter** is a DNA sequence that the transcription apparatus recognizes and binds. It indicates which of the two DNA strands is to be read as the template and the direction of transcription. The promoter also determines the transcription start site, the first nucleotide that will be transcribed into RNA. In most transcription units, the promoter is located next to the transcription start site but is not, itself, transcribed.

The second critical region of the transcription unit is the **RNA-coding region,** a sequence of DNA nucleotides

that is copied into an RNA molecule. A third component of the transcription unit is the **terminator,** a sequence of nucleotides that signals where transcription is to end. Terminators are usually part of the coding sequence; that is, transcription stops only after the terminator has been copied into RNA.

Molecular biologists often use the terms *upstream* and *downstream* to refer to the direction of transcription and the location of nucleotide sequences surrounding the RNA coding sequence. The transcription apparatus is said to move downstream during transcription: it binds to the promoter (which is usually upstream of the start site) and moves toward the terminator (which is downstream of the start site).

When DNA sequences are written out, often the sequence of only one of the two strands is listed. Molecular biologists typically write the sequence of the nontemplate strand, because it will be the same as the sequence of the RNA transcribed from the template (with the exception that U in RNA replaces T in DNA). By convention, the sequence on the nontemplate strand is written with the 5′ end on the left and the 3′ end on the right. The first nucleotide transcribed (the transcription start site) is numbered +1; nucleotides downstream of the start site are assigned positive numbers, and nucleotides upstream of the start site are assigned negative numbers. So, nucleotide +34 would be 34 nucleotides downstream of the start site, whereas nucleotide −75 would be 75 nucleotides upstream of the start site.

Concepts

A transcription unit is a piece of DNA that encodes an RNA molecule and the sequences necessary for its proper transcription. Each transcription unit includes a promoter, an RNA-coding region, and a terminator.

The Substrate for Transcription

RNA is synthesized from **ribonucleoside triphosphates** (rNTPs) (FIGURE 13.8). In synthesis, nucleotides are added one at a time to the 3′-OH group of the growing RNA molecule. Two phosphates are cleaved from the

 13.7 A transcription unit includes a promoter, an RNA-coding region, and a terminator.

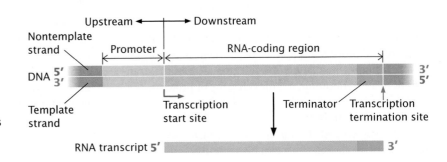

Triphosphate

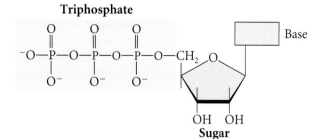

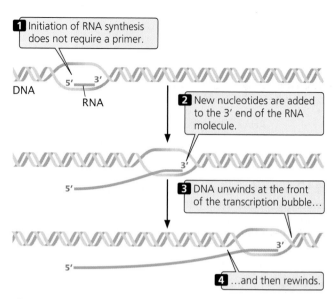

1 Initiation of RNA synthesis does not require a primer.

2 New nucleotides are added to the 3′ end of the RNA molecule.

3 DNA unwinds at the front of the transcription bubble…

4 …and then rewinds.

◀ 13.8 **Ribonucleoside triphosphates are substrates used in RNA synthesis.**

◀ 13.9 **In transcription, nucleotides are always added to the 3′ end of the RNA molecule.**

incoming ribonucleoside triphosphate; the remaining phosphate participates in a phosphodiester bond that connects the nucleotide to the growing RNA molecule. The overall chemical reaction for the addition of each nucleotide is:

$$RNA_n + rNTP \longrightarrow RNA_{n+1} + PP_i$$

where PP_i represents two atoms of inorganic phosphorus. Nucleotides are always added to the 3′ end of the RNA molecule, and the direction of transcription is therefore $5' \rightarrow 3'$ (◀ FIGURE 13.9), the same as the direction of DNA synthesis during replication. RNA is made complementary and antiparallel to one of the DNA strands (the template strand).

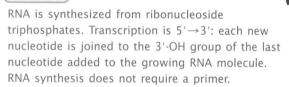

Concepts

RNA is synthesized from ribonucleoside triphosphates. Transcription is $5' \rightarrow 3'$: each new nucleotide is joined to the 3′-OH group of the last nucleotide added to the growing RNA molecule. RNA synthesis does not require a primer.

The Transcription Apparatus

Recall that, in replication, a number of different enzymes and proteins are required to bring about DNA synthesis. Although transcription might initially appear to be quite different, because a single enzyme—**RNA polymerase**—carries out all the required steps of transcription, on closer inspection, the processes are actually similar. The action of RNA polymerase is enhanced by a number of accessory proteins that join and leave the polymerase at different stages of the process. Each accessory protein is responsible for providing or regulating a special function. Thus, transcription, like replication, requires an array of proteins.

Bacterial RNA polymerase Bacterial cells typically possess only one type of RNA polymerase, which catalyzes the synthesis of all classes of bacterial RNA: mRNA, tRNA, and

rRNA. Bacterial RNA polymerase is a large, multimeric enzyme (meaning that it consists of several polypeptide chains).

At the heart of bacterial RNA polymerase are four subunits (individual polypeptide chains) that make up the **core enzyme:** two copies of a subunit called alpha (α), a single copy of beta (β), and single copy of beta prime (β′) (◀ FIGURE 13.10). The core enzyme catalyzes the elongation of the RNA molecule by the addition of RNA nucleotides. Other functional subunits join and leave the core enzyme at particular stages of the transcription process. The **sigma (σ) factor** controls the binding of the RNA polymerase to the promoter. Without sigma, RNA polymerase will initiate transcription at a random point along the DNA. After sigma has associated with the core enzyme (forming a **holoenzyme**), RNA polymerase binds stably only to the promoter region and initiates transcription at the proper start site. Sigma is required only for promoter binding and initiation; when a few RNA nucleotides have been joined together, sigma detaches from the core enzyme.

Many bacteria possess multiple types of sigma. *E. coli*, for example, possesses sigma 28 (σ^{28}), sigma 32 (σ^{32}), sigma 54 (σ^{54}), and sigma 70 (σ^{70}), named on the basis of their molecular weights. Each type of sigma initiates the binding of RNA polymerase to a particular set of promoters. For example, σ^{32} binds to promoters of genes that protect against environmental stress, σ^{54} binds to promoters of genes used during nitrogen starvation, and σ^{70} binds to many different promoters.

Other subunits provide the core RNA polymerase with additional functions. Rho (ρ) and NusA, for example, facilitate the termination of transcription.

(a)

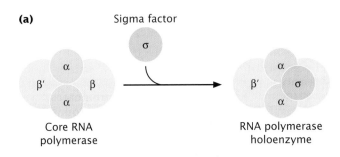

Sigma factor

σ

Core RNA polymerase → RNA polymerase holoenzyme

(b)

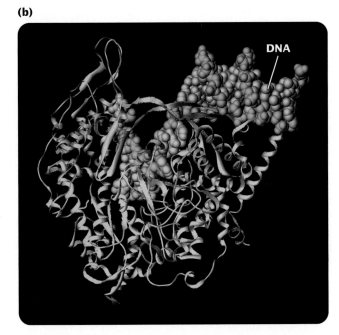

DNA

◀ 13.10 **In bacterial RNA polymerase, the core enzyme consists of four subunits: two copies of alpha (α), a single copy of beta (β), and single copy of beta prime (β′).** The core enzyme catalyzes the elongation of the RNA molecule by the addition of RNA nucleotides. (a) The sigma factor (σ) joins the core to form the holoenzyme, which is capable of binding to a promoter and initiating transcription. (b) The molecular model shows RNA polymerase (shown in yellow) binding DNA.

Eukaryotic RNA polymerases Eukaryotic cells possess three distinct types of RNA polymerase, each of which is responsible for transcribing a different class of RNA: **RNA polymerase I** transcribes rRNA; **RNA polymerase II** transcribes pre-mRNAs, snoRNAs, and some snRNAs; and **RNA polymerase III** transcribes small RNA molecules—specifically tRNAs, small rRNA, and some snRNAs (Table 13.3). All three eukaryotic polymerases are large, multimeric enzymes, typically consisting of more than a dozen subunits. Some subunits are common to all three RNA polymerases, whereas others are limited to one of the polymerases. As in bacterial cells, a number of accessory proteins bind to the core enzyme and affect its function.

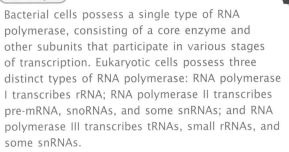

Table 13.3	Eukaryotic RNA polymerases
Type	**Transcribes**
RNA polymerase I	Large rRNAs
RNA polymerase II	Pre-mRNA, some snRNAs, snoRNAs
RNA polymerase III	tRNAs, small rRNA, snRNAs

Concepts

Bacterial cells possess a single type of RNA polymerase, consisting of a core enzyme and other subunits that participate in various stages of transcription. Eukaryotic cells possess three distinct types of RNA polymerase: RNA polymerase I transcribes rRNA; RNA polymerase II transcribes pre-mRNA, snoRNAs, and some snRNAs; and RNA polymerase III transcribes tRNAs, small rRNAs, and some snRNAs.

The Process of Bacterial Transcription

Now that we've considered some of the major components of transcription, we're ready to take a detailed look at the process. Transcription can be conveniently divided into three stages:

1. initiation, in which the transcription apparatus assembles on the promoter and begins the synthesis of RNA;

2. elongation, in which RNA polymerase moves along the DNA, unwinding it and adding new nucleotides, one at a time, to the 3′ end of the growing RNA strand; and

3. termination, the recognition of the end of the transcription unit and the separation of the RNA molecule from the DNA template.

We will first examine each of these steps in bacterial cells, where the process is best understood; then we will consider eukaryotic transcription.

Initiation

Initiation includes all the steps necessary to begin RNA synthesis, including (1) promoter recognition, (2) formation of the transcription bubble, (3) creation of the first bonds between rNTPs, and (4) escape of the transcription apparatus from the promoter.

Transcription initiation requires that the transcription apparatus recognize and bind to the promoter. At this step, the selectivity of transcription is enforced; the binding of RNA polymerase to the promoter determines which parts of the DNA template are to be transcribed and how often.

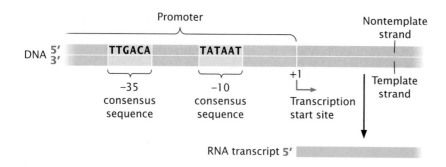

13.11 In bacterial promoters, consensus sequences are found upstream of the start site, approximately at positions −10 and −35.

Different genes are transcribed with different frequencies, and promoter binding is primarily responsible for determining the frequency of transcription for a particular gene. Promoters also have different affinities for RNA polymerase. Even within a single promoter, the affinity can vary over time, depending on its interaction with RNA polymerase and a number of other factors.

Bacterial promoters Essential information for the transcription unit—where it will start transcribing, which strand is to be read, and in what direction the RNA polymerase will move—is imbedded in the nucleotide sequence of the promoter. Promoters are sequences in the DNA that are recognized by the transcription apparatus and are required for transcription to take place. In bacterial cells, promoters are usually adjacent to an RNA coding sequence. The examination of many promoters in *E. coli* and other bacteria reveals a general feature: although most of the nucleotides within the promoters vary in sequence, short stretches of nucleotides are common to many. Furthermore, the spacing and location of these nucleotides relative to the transcription start site are similar in most promoters. These short stretches of common nucleotides are called **consensus sequences.**

The term "consensus sequence" refers to sequences that possess considerable similarity or consensus. By definition, the consensus sequence comprises the most commonly encountered nucleotides found at a specific location. For example, consider the following nucleotides found near the transcription start site of four prokaryotic genes.

$$5' - A A T A A A - 3'$$
$$5' - T T T A A T - 3'$$
$$5' - T A T T T T - 3'$$
$$\underline{5' - T A A A A T - 3'}$$
Consensus sequence = $5' - T A T A A T - 3'$

If two bases are equally frequent, they are designated by listing both bases separated by a line or a slash, as in 5′–T A T A A A/T–3′. Purines can be indicated by the abbreviation R, pyrimidines by Y, and any nucleotide by N. For example, the consensus sequence 5′–T A Y A R N A–3′

means that the third nucleotide in the consensus sequence (Y) is usually a pyrimidine, but either pyrimidine is equally likely. Similarly, the fifth nucleotide in the sequence (R) is most likely one of the purines, but both are equally frequent. In the sixth position (N), no particular base is more common than any other. The presence of consensus in a set of nucleotides usually implies that the sequence is associated with an important function. Consensus exists in a sequence because natural selection has favored a restricted set of nucleotides in that position.

The most commonly encountered consensus sequence, found in almost all bacterial promoters, is located just upstream of the start site, centered on position −10. Called the **−10 consensus sequence** or, sometimes, the Pribnow box, its sequence is

$$5' \ TATAAT \ 3'$$
$$3' \ ATATTA \ 5'$$

often written simply as TATAAT (◀FIGURE 13.11). Remember that TATAAT is just the *consensus* sequence—representing the most commonly encountered nucleotides at each of these positions. In most prokaryotic promoters, the actual sequence is not TATAAT (◀FIGURE 13.12).

Another consensus sequence common to most bacterial promoters is TTGACA, which lies approximately 35 nucleotides upstream of the start site and is termed the **−35 consensus sequence** (see Figure 13.11). The nucleotides on either side of the −10 and −35 consensus sequences and those between them vary greatly from promoter to promoter, suggesting that they are relatively unimportant in promoter recognition.

The function of these consensus sequences in bacterial promoters has been studied by inducing mutations at various positions within the consensus sequences and observing the effect of the changes on transcription. The results of these studies reveal that most base substitutions within the −10 and −35 consensus sequences reduce the rate of transcription; these substitutions are termed *down mutations* because they slow down the rate of transcription. Occasionally, a particular change in a consensus sequence increases the rate of transcription; such a change is called an *up mutation.*

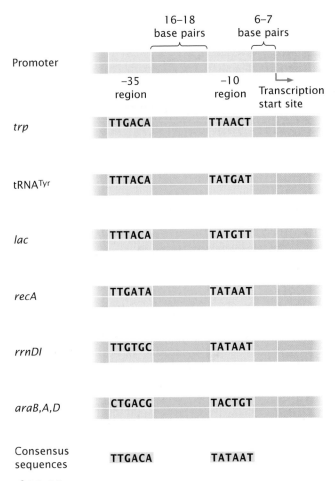

13.12 In most prokaryotic promoters, the actual sequence is not TATAAT. The sequences shown are found in five *E. coli* promoters, including those of genes for tryptophan biosynthesis (*trp*), tyrosine tRNA (tRNA^Tyr), lactose metabolism (*lac*), a recombination protein (rec*A*), and arabinose metabolism (*araB, A, D*). These sequences are on the nontemplate strand and read 5′→3′, left to right.

The sigma factor associates with the core enzyme (◀FIGURE 13.13a) to form a holoenzyme, which binds to the −35 and −10 consensus sequences in the DNA promoter (◀FIGURE 13.13b). Although it binds only the nucleotides of consensus sequences, the enzyme extends from −50 to +20 when bound to the promoter. The holoenzyme initially binds weakly to the promoter but then undergoes a change in structure that allows it to bind more tightly and unwind the double-stranded DNA (◀FIGURE 13.13c). Unwinding begins within the −10 consensus sequence and extends downstream for about 17 nucleotides, including the start site.

Some bacterial promoters contain a third consensus sequence that also takes part in the initiation of transcription. Called the **upstream element,** this sequence contains a number of A−T pairs and is found at about −40 to −60. The alpha subunit of the RNA polymerase interacts directly

with this upstream element, greatly enhancing the rate of transcription in those bacterial promoters that possess it. A number of other proteins may bind to sequences in and near the promoter; some stimulate the rate of transcription and others repress it; we will consider the proteins that regulate gene expression in Chapter 16.

Concepts

A promoter is a DNA sequence that is adjacent to a gene and required for transcription. Promoters contain short consensus sequences that are important in the initiation of transcription.

Initial RNA synthesis After the holoenzyme has attached to the promoter, RNA polymerase is positioned over the start site for transcription (at position +1) and has unwound the DNA to produce a single-stranded template. The orientation and spacing of consensus sequences on a DNA strand determine which strand will be the template for transcription, and thereby determine the direction of transcription.

The start site itself is not marked by a consensus sequence but often has the sequence CAT, with the start site at the A. The position of the start site is determined not by the sequences located there but by the location of the consensus sequences, which positions RNA polymerase so that the enzyme's active site is aligned for initiation of transcription at +1. If the consensus sequences are artificially moved upstream or downstream, the location of the starting point of transcription correspondingly changes.

To begin the synthesis of an RNA molecule, RNA polymerase pairs the base on a ribonucleoside triphosphate with its complementary base at the start site on the DNA template strand (◀FIGURE 13.13d). No primer is required to initiate the synthesis of the 5′ end of the RNA molecule. Two of the three phosphates are cleaved from the ribonucleoside triphosphate as the nucleotide is added to the 3′ end of the growing RNA molecule. However, because the 5′ end of the first ribonucleoside triphosphate does not take part in the formation of a phosphodiester bond, all three of its phosphates remain. An RNA molecule therefore possesses, at least initially, three phosphates at its 5′ end (◀FIGURE 13.13e).

Elongation

After initiation, RNA polymerase moves downstream along the template, progressively unwinding the DNA at the leading (downstream) edge of the transcription bubble, joining nucleotides to the RNA molecule according to the sequence on the template, and rewinding the DNA at the trailing (upstream) edge of the bubble. In bacterial

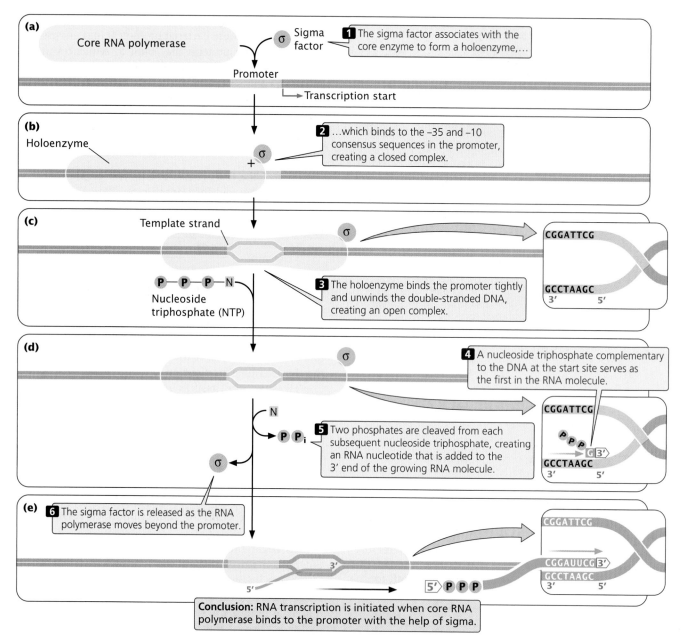

(a)

Core RNA polymerase

Sigma factor

σ

1 The sigma factor associates with the core enzyme to form a holoenzyme,…

Promoter

Transcription start

(b)

Holoenzyme

+ σ

2 …which binds to the −35 and −10 consensus sequences in the promoter, creating a closed complex.

(c)

Template strand

σ

CGGATTCG

GCCTAAGC
3'   5'

P─P─P─N

Nucleoside triphosphate (NTP)

3 The holoenzyme binds the promoter tightly and unwinds the double-stranded DNA, creating an open complex.

(d)

σ

4 A nucleoside triphosphate complementary to the DNA at the start site serves as the first in the RNA molecule.

CGGATTCG

N

P P i

5 Two phosphates are cleaved from each subsequent nucleoside triphosphate, creating an RNA nucleotide that is added to the 3' end of the growing RNA molecule.

P P P

G 3'

GCCTAAGC
3'   5'

σ

(e)

σ

6 The sigma factor is released as the RNA polymerase moves beyond the promoter.

CGGATTCG

3'

CGGAUUCG 3'
GCCTAAGC
3'   5'

5'

5' P P P

Conclusion: RNA transcription is initiated when core RNA polymerase binds to the promoter with the help of sigma.

◀13.13   Transcription in bacteria is carried out by RNA polymerase, which must bind to the sigma factor to initiate transcription.

cells at 37°C, about 40 nucleotides are added per second. This rate of RNA synthesis is much lower than that of DNA synthesis, which is more than 1500 nucleotides per second in bacterial cells.

Transcription takes place within a short stretch of about 18 nucleotides of unwound DNA—the transcription bubble. Within this region, RNA is continuously synthesized, with single-stranded DNA used as a template. About 8 nucleotides of newly synthesized RNA are paired with the DNA-template nucleotides at any one time. As the transcription apparatus moves down the DNA template, it

generates positive supercoiling ahead of the transcription bubble and negative supercoiling behind it. Topoisomerase enzymes probably relieve the stress associated with the unwinding and rewinding of DNA in transcription, as they do in DNA replication.

Concepts

Transcription is initiated at the start site, which, in bacterial cells, is set by the binding of RNA polymerase to the consensus sequences of the promoter. Transcription takes place within the transcription bubble. DNA is unwound ahead of the bubble and rewound behind it.

Termination

RNA polymerase moves along the template, adding nucleotides to the 3' end of the growing RNA molecule until it transcribes a terminator. Most terminators are found upstream of the point of termination. Transcription therefore does not suddenly end when polymerase reaches a terminator, like a car stopping in front of a stop sign. Rather, transcription ends after the terminator has been transcribed, like a car that stops only after running over a speed bump. At the terminator, several overlapping events are needed to bring an end to transcription: RNA polymerase must stop synthesizing RNA, the RNA molecule must be released from RNA polymerase, the newly made RNA molecule must dissociate fully from the DNA, and RNA polymerase must detach from the DNA template.

Bacterial cells possess two major types of terminators. **Rho-dependent terminators** are able to cause the termination of transcription only in the presence of an ancillary protein called the **rho factor. Rho-independent terminators** are able to cause the end of transcription in the absence of rho.

Rho-independent terminators have two common features. First, they contain inverted repeats (sequences of nucleotides on one strand that are inverted and complementary). When inverted repeats have been transcribed into RNA, a hairpin secondary structure forms (◀FIGURE 13.14). Second, in rho-independent terminators, a string of approximately six adenine nucleotides follows the second inverted repeat in the template DNA. Their transcription produces a string of uracil nucleotides after the hairpin in the transcribed RNA.

The presence of a hairpin in an RNA transcript causes RNA polymerase to slow down or pause, which creates an opportunity for termination. The adenine–uracil base pairings downstream of the hairpin are relatively unstable compared with other base pairings, and the formation of the hairpin may itself destablize the DNA–RNA pairing, causing the RNA molecule to separate from its DNA template. When the RNA transcript has separated from the template, RNA synthesis can no longer continue (see Figure 13.14).

Rho-dependent terminators have two features: (1) DNA sequences that produce a pause in transcription; and (2) a DNA sequence that encodes a stretch of RNA upstream of the terminator that is devoid of any secondary structures. This unstructured RNA serves as binding site for the rho protein, which binds the RNA and moves toward its 3' end, following the RNA polymerase (◀FIGURE 13.15). When RNA polymerase encounters the terminator, it pauses, allowing

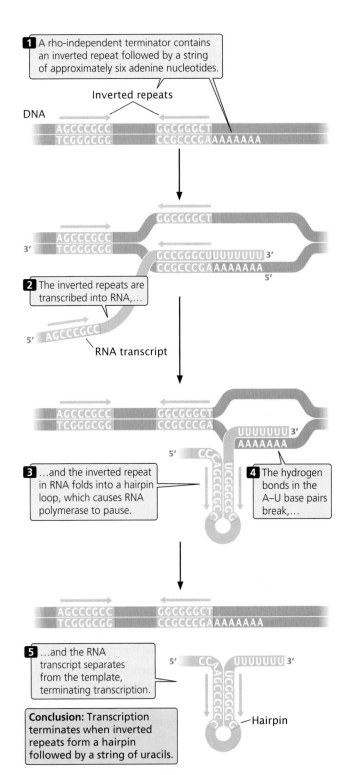

13.14 **Termination by bacterial rho-independent terminators is a multistep process.**

rho to catch up. The rho protein has helicase activity, which it uses to unwind the RNA–DNA hybrid in the transcription bubble, bringing an end to transcription.

Concepts

Transcription ends after RNA polymerase transcribes a terminator. Bacterial cells possess two types of terminator: a rho-independent terminator, which RNA polymerase can recognize by itself; and a rho-dependent terminator, which RNA polymerase can recognize only with the help of the rho protein.

Connecting Concepts

The Basic Rules of Transcription

Before we examine the process of eukaryotic transcription, let's pause to summarize some of the general principles of bacterial transcription.

The Basic Rules of Transcription

1. Transcription is a selective process; only certain parts of the DNA are transcribed.

2. RNA is transcribed from single-stranded DNA. Normally, only one of the two DNA strands—the template strand—is copied into RNA.

3. Ribonucleoside triphosphates are used as the substrates in RNA synthesis. Two phosphates are cleaved from a ribonucleoside triphosphate, and the resulting nucleotide is joined to the 3'-OH group of the growing RNA strand.

4. RNA molecules are antiparallel and complementary to the DNA template strand. Transcription is always in the 5'→3' direction, meaning that the RNA molecule grows at the 3' end.

5. Transcription depends on RNA polymerase—a complex, multimeric enzyme. RNA polymerase consists of a core enzyme, which is capable of synthesizing RNA, and other subunits that may join transiently to perform additional functions.

6. The core enzyme of RNA polymerase requires a sigma factor in order to bind to a promoter and initiate transcription.

7. Promoters contain short sequences crucial in the binding of RNA polymerase to DNA; these consensus sequences are interspersed with nucleotides that play no known role in transcription.

8. RNA polymerase binds to DNA at a promoter, begins transcribing at the start site of the gene, and ends transcription after a terminator has been transcribed.

www.whfreeman.com/pierce   A brief overview of the process of transcription

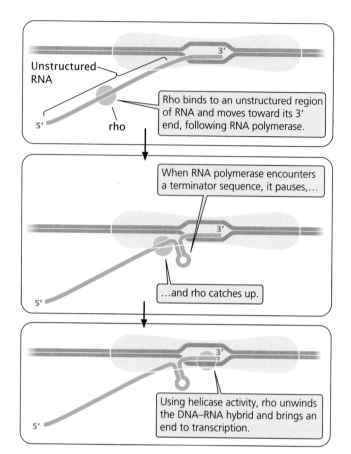

Rho binds to an unstructured region of RNA and moves toward its 3' end, following RNA polymerase.

When RNA polymerase encounters a terminator sequence, it pauses,…

…and rho catches up.

Using helicase activity, rho unwinds the DNA–RNA hybrid and brings an end to transcription.

◀13.15 The termination of transcription in some bacterial genes requires the presence of the rho protein.

The Process of Eukaryotic Transcription

The process of eukaryotic transcription is similar to that of bacterial transcription. Eukaryotic transcription also includes initiation, elongation, and termination, and the basic principles of transcription already outlined apply to eukaryotic transcription. However, there are some important differences. Eukaryotic cells possess three different RNA polymerases, each of which transcribes a different class of RNA and recognizes a different type of promoter. Thus, a generic promoter cannot be described for eukaryotic cells, as was done for bacterial cells; rather, a promoter's description depends on whether the promoter is recognized by RNA polymerase I, II, or III. Another difference is in the nature of promoter recognition and initiation. Many proteins take part in the binding of eukaryotic RNA polymerases to DNA templates, and the different types of promoters require different proteins.

Transcription and Nucleosome Structure

Transcription requires that sequences on DNA are accessible to RNA polymerase and other proteins. However, in eukaryotic cells, DNA is complexed with histone proteins in highly compressed chromatin (see Figure 11.5). How can the proteins necessary for transcription gain access to eukaryotic DNA when it is complexed with histones?

The answer to this question is that, before transcription, the chromatin structure is modified so that the DNA is in a more open configuration and is more accessible to the transcription machinery. Several types of proteins have roles in chromatin modification. Acetyltransferases add acetyl groups to amino acids at the ends of the histone proteins, which destabilizes the nucleosome structure and makes the DNA more accessible. Other types of histone modification also can affect chromatin packing. In addition, proteins called chromatin- remodeling proteins may bind to the chromatin and displace nucleosomes from promoters and other regions important for transcription. We will take a closer look at the role of changes to chromatin structure associated with gene expression in Chapter 16.

Transcription Initiation

The initiation of transcription is a complex processes in eukaryotic cells because of the variety of initiation sequences and because numerous proteins bind to these sequences. Two broad classes of DNA sequences are important for the initiation of transcription: promoters and enhancers. A promoter is always found adjacent to (or sometimes within) the gene that it regulates and has a fixed location with regard to the transcription start point. An enhancer, in contrast, need not be adjacent to the gene; enhancers can affect the transcription of genes that are thousands of nucleotides away, and their positions relative to start sites can vary.

A significant difference between bacterial and eukaryotic transcription is the existence of three different eukaryotic RNA polymerases, which recognize different types of promoters. In bacterial cells, the holoenzyme (RNA polymerase plus sigma) recognizes and binds directly to sequences in the promoter. In eukaryotic cells, promoter recognition is carried out by accessory proteins that bind to the promoter and then recruit a specific RNA polymerase (I, II, or III) to the promoter.

One class of accessory proteins comprises **general transcription factors,** which, along with RNA polymerase, form the **basal transcription apparatus** that assembles near the

start site and is sufficient to initiate minimal levels of transcription. Another class of accessory proteins consists of **transcriptional activator proteins,** which bind to specific DNA sequences and bring about higher levels of transcription by stimulating the assembly of the basal transcription apparatus at the start site.

> **Concepts**
>
> Two classes of DNA sequences in eukaryotic cells affect transcription: enhancers and promoters. A promoter is near the gene and has a fixed position relative to the start site of transcription. An enhancer can be distant from the gene and variable in location.

RNA Polymerase II Promoters

We will focus most of our attention on promoters recognized by RNA polymerase II, which transcribes the genes that encode proteins. A promoter for a gene transcribed by RNA polymerase II typically consists of two primary parts: the core promoter and the regulatory promoter.

Core promoter The **core promoter** is located immediately upstream of the gene (◀ FIGURE 13.16) and typically includes one or more consensus sequences. The most common of these consensus sequences is the **TATA box,** which has the consensus sequence TATAAA and is located from -25 to -30 bp upstream of the start site. Mutations in the sequence of the TATA box affect the rate of transcription, and changing its position alters the location of the transcription start site.

Another common consensus sequence in the core promoter is the TFIIB recognition element (BRE), which has the consensus sequence G/C G/C G/C C G C C and is located from -32 to -38 bp upstream of the start site. (TFIIB is the abbreviation for a transcription factor that binds to this element; see next subsection). Instead of a TATA box, some core promoters have an initiator element (Inr) that directly overlaps the start site and has the consensus Y Y A N T/A Y Y. Another consensus sequence called the downstream core promoter element (DPE) is found

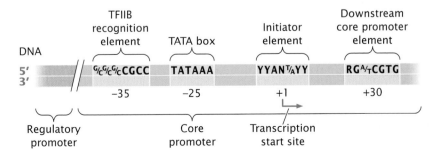

◀ 13.16 **The promoters of genes transcribed by RNA polymerase II consist of a core promoter and a regulatory promoter that contain consensus sequences.** Not all the consensus sequences shown are found in all promoters.

approximately +30 bp downstream of the start site in many promoters that also have Inr; the consensus sequence of DPE is R G A/T C G T G. All of these consensus sequences in the core promoter are recognized by transcription factors that bind to them and serve as a platform for the assembly of the basal transcription apparatus.

Assembly of the basal transcription apparatus The basic transcriptional machinery that binds to DNA at the start site is called the basal transcription apparatus and is required to initiate minimal levels of transcription. It consists of RNA polymerase, a series of general transcription factors, and a complex of proteins known as the mediator

(◀FIGURE 13.17). The general transcription factors include TFIIA, TFIIB, TFIID, TFIIE, TFIIF, and TFIIH, in which TFII stands for transcription factor for RNA polymerase II and the letter designates the individual factor.

TFIID binds to the TATA box and positions the active site of RNA polymerase II so that it begins transcription at the correct place. TFIID consists of at least nine polypeptides. One of them is the **TATA-binding protein** (TBP), which recognizes and binds to the TATA box on the DNA template. The TATA-binding protein binds to the minor groove and straddles the DNA as a molecular saddle (◀FIGURE 13.18), bending the DNA and partly unwinding it. Other proteins, called TBP-associated factors (TAFs),

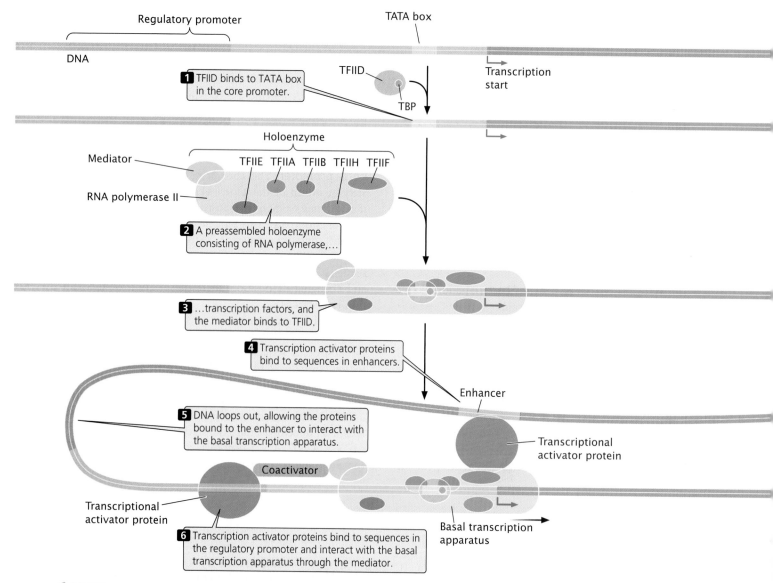

◀13.17 **Transcription is initiated at RNA polymerase II promoters when the TFIID transcription factor binds to the TATA box, followed by the binding of a preassembled holoenzyme containing general transcription factors, RNA polymerase II, and the mediator.**

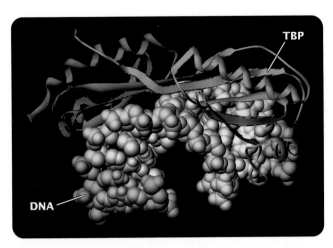

13.18 The TATA-binding protein (TBP) binds to the minor groove of DNA, straddling the double helix of DNA like a saddle.

combine with TBP to form the complete TFIID transcription factor.

The large holoenzyme consisting of RNA polymerase, additional transcription factors, and the mediator are thought to preassemble and bind as a unit to TFIID. The other transcription factors provide additional functions: TFIIA helps to stabilize the interaction between TBP and DNA, TFIIB plays a role in the selection of the start site, and TFIIH has helicase activity and unwinds the DNA during transcription. The mediator plays a role in communication between the basal transcription apparatus and transcriptional activator proteins (see next subsection).

Regulatory promoter The **regulatory promoter** is located immediately upstream of the core promoter. A variety of different consensus sequences may be found in the regulatory promoters, and they can be mixed and matched in different combinations (**FIGURE 13.19**). Transcriptional activator proteins bind to these sequences and, either directly or indirectly (through the mediation of coactivator proteins), make contact with the mediator in the basal transcription apparatus and affect the rate at which transcription is initiated. Some regulatory promoters also contain repressing sequences, which are bound by proteins that lower the rate of transcription through inhibitory inactions with the mediator.

Enhancers DNA sequences that increase the rate of transcription at distant genes are called **enhancers.** Furthermore, the precise position of an enhancer relative to a gene's transcriptional start site is not critical; most enhancers can stimulate any promoter in their vicinities, and an enhancer may be upstream or downstream from the affected gene or, in some cases, within an intron of the gene itself.

Enhancers also contain sequences that are recognized by transcriptional activator proteins. How does the binding of a transcriptional activator protein to an enhancer affect the initiation of transcription at a gene thousands of nucleotides away? The answer is that the DNA between the enhancer and the promoter loops out, allowing the enhancer and the promoter to lie close to each other. Transcriptional activator proteins bound to the enhancer interact with proteins bound to the promoter and stimulate the transcription of the adjacent gene (see Figure 13.17). The looping of DNA between the enhancer and the promoter explains how the position of an enhancer can vary with regard to the start site—enhancers that are farther from the start site simply cause a longer length of DNA to loop out.

Sequences having many of the properties possessed by enhancers sometimes take part in *repressing* transcription instead of enhancing it; such sequences are called **silencers.** Although enhancers and silencers are characteristic of eukaryotic DNA, some enhancer-like sequences have been found in bacterial cells.

Concepts

General transcription factors assemble into the basal transcription apparatus, which binds to DNA near the start site and is necessary for transcription to take place at minimal levels. Additional proteins called transcriptional activators bind to other consensus sequences in promoters and enhancers, and affect the rate of transcription.

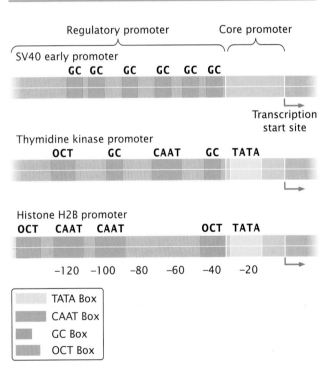

13.19 The consensus sequences in promoters of three eukaryotic genes illustrate the principle that different sequences can be mixed and matched to yield a functional promoter.

RNA Polymerase I Promoters

RNA polymerase I promoters have two functional sequences near the start site (◄FIGURE 13.20). A core element surrounds the start site, extending from −45 to +20, and is needed to initiate transcription. An upstream control element extends from −180 to −107 and increases the efficiency of the core element. The DNA sequences of the core element and the upstream control element are rich in guanine and cytosine nucleotides and are similar in sequence.

RNA polymerase I requires two proteins to initiate transcription: SL1 and UBF. SL1 is made up of four subunits, one of which is TBP—the same protein that binds the TATA box in RNA polymerase II promoters (see Figure 13.17). The other protein, UBF, binds to both the core element and the upstream control element (see Figure 13.20) and enables SL1 to bind to the promoter. SL1 then recruits RNA polymerase I to the promoter. Thus, RNA polymerase I promoters function much as RNA polymerase II promoters do: transcription factors bind to a consensus sequence in the promoter and recruit RNA polymerase to the start site.

> **Concepts**
>
> RNA polymerase I promoters have two key components: (1) the core element, which surrounds the start site and is sufficient to initiate transcription, and (2) the upstream control sequence, which increases the efficiency of the core promoter.

RNA Polymerase III Promoters

RNA polymerase III transcribes small rRNA, tRNAs, and some snRNAs (see Table 13.3) and recognizes several distinct types of promoters. The promoters of snRNA genes transcribed by RNA polymerase III contain several consensus sequences that are also found in some promoters transcribed by RNA polymerase II (◄FIGURE 13.21a). These consensus sequences include the TATA box, which is recognized by a transcription factor that contains TBP. As in other types of eukaryotic promoters, TBP positions the active site of RNA polymerase over the start site for transcription in these promoters.

Promoters for small rRNA and tRNA genes, also transcribed by RNA polymerase III, contain **internal promoters** that are *downstream* of the start site and are actually transcribed into the RNA (◄FIGURE 13.21 b AND c). These promoters contain critical sets of nucleotides (boxes A, B, and C) that also are recognized by transcription factors. One of these transcription factors includes TBP, which, again, positions the active site of RNA polymerase III over the upstream start site and ensures that the enzyme initiates transcription at the correct location. Additional transcription factors then bind to the DNA-binding factor and recruit RNA polymerase to the initiation complex.

> **Concepts**
>
> Some RNA polymerase III promoters are upstream of the start site and contain a TATA box. Others are internal, imbedded within the transcribed sequence downstream of the start site.

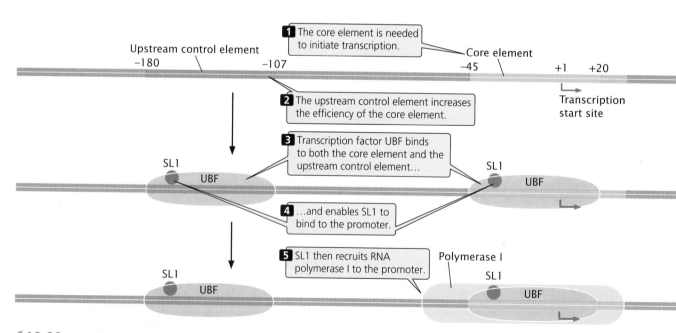

◄13.20 The basal transcription apparatus assembles at RNA polymerase I promoters.

(a)

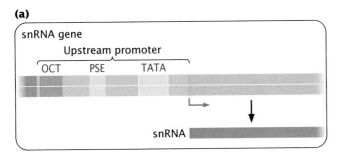

snRNA gene

Upstream promoter

OCT   PSE   TATA

snRNA

(b)

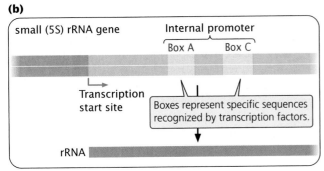

small (5S) rRNA gene   Internal promoter

Box A   Box C

Transcription start site

Boxes represent specific sequences recognized by transcription factors.

rRNA

(c)

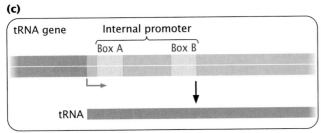

tRNA gene   Internal promoter

Box A   Box B

tRNA

Conclusion: Promoters for RNA polymerase III vary in their sequences and positions relative to the gene.

13.21 RNA polymerase III recognizes several different types of promoters. OCT and PSE are consensus sequences that may also be present in RNA polymerase II promoters.

Connecting Concepts

Characteristics of Eukaryotic Promoters and Transcription Factors

Mastering the details of eukaryotic promoters and their associated transcription factors is a daunting task even for experienced researchers, never mind the beginning genetics student. Let's step back from the detail for a moment and identify some general principles of eukaryotic promoters and transcription factors:

1. Several types of DNA sequences take part in the initiation of transcription in eukaryotic cells. These sequences generally serve as the binding sites for proteins that interact with RNA polymerase and influence the initiation of transcription.

2. Some sequences that affect transcription, called promoters, are adjacent to or within the RNA coding region and are relatively fixed with regard to the start site of transcription. Promoters consist of a core promoter located adjacent to the gene and a regulatory promoter located farther upstream.

3. Other sequences, called enhancers, are distant from the gene and function independently of position and direction. Enhancers stimulate transcription.

4. General transcription factors bind to the core promoter near the start site and, with RNA polymerase, assemble into a basal transcription apparatus. The TATA-binding protein (TBP) is a critical transcription factor that positions the active site of RNA polymerase over the start site.

5. Transcriptional activator proteins bind to sequences in the regulatory promoter and enhancers and affect transcription by interacting with the basal transcription apparatus.

6. Proteins binding to enhancers interact with the basal transcription apparatus by causing the DNA between the promoter and the enhancer to loop out, bringing the enhancer into close proximity to the promoter.

Evolutionary Relationships and the TATA-Binding Protein

Some 2 billion to 3 billion years ago, life diverged into three lines of evolutionary descent: the eubacteria, the archaea, and the eukaryotes (see Chapter 2). Although eubacteria and archaea are superficially similar—both are unicellular and lack a nucleus—the results of studies of their DNA sequences and other biochemical properties indicate that they are distantly related. The evolutionary distinction between archaea, eubacteria, and eukaryotes is clear: however, did eukaryotes first diverge from an ancestral prokaryote, with the later separation of prokaryotes into eubacteria and archaea, or did the archaea and the eubacteria split first, with the eukaryotes later evolving from one of these groups?

Studies of transcription in eubacteria, archaea, and eukaryotes have yielded important findings about the evolutionary relationships of these organisms. The results of studies in 1994 demonstrated that archaea possess a TATA-binding protein, a critical transcription factor in all three of the eukaryotic polymerases. The binding of TBP to DNA is the first step in the assembly of the eukaryotic transcription apparatus. In earlier studies, TATA-like sequences were found in eukaryotic cells, but no such sequences have been found in eubacteria. TBP binds the TATA box in archaea with the help of another transcription factor, TFIIB, which is also found in eukaryotes but not in eubacteria.

Together these findings indicate that transcription, one of the most basic of life processes, has strong similarities in eukaryotes and archaea, suggesting that these two groups are more closely related to each other than either is to the eubacteria. This conclusion is supported by other data, including those obtained from a comparison of gene sequences.

Termination

The termination of transcription in eukaryotic genes is less well understood than in bacterial genes. The three eukaryotic RNA polymerases use different mechanisms for termination. RNA polymerase I requires a termination factor, like the rho factor utilized in termination of some bacterial genes. Unlike rho, which binds to the newly transcribed RNA molecule, the termination factor for RNA polymerase I binds to a DNA sequence downstream of the termination site.

RNA polymerase III ends transcription after transcribing a terminator sequence that produces a string of Us in the RNA molecule, like that produced by the rho-independent terminators of bacteria. Unlike rho-independent terminators in bacterial cells, RNA polymerase III does not require that a hairpin structure precede the string of Us.

In many of the genes transcribed by RNA polymerase II, transcription can end at multiple sites located within a span of hundreds or thousands of base pairs. As we will see in Chapter 14, the transcription of these genes continues well beyond the coding sequence necessary to produce the mRNA. After transcription, the 3′ end of pre-mRNA is cleaved at a specific site, designated by a consensus sequence, producing the mature mRNA. Research findings suggest that termination is coupled to cleavage, which is carried out by a cleavage complex that probably associates with the RNA polymerase. This complex may suppress termination until the consensus sequence that marks the cleavage site is encountered. The 3′ end of the pre-mRNA is then cleaved by the complex, and transcription is terminated downstream.

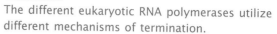

Concepts

The different eukaryotic RNA polymerases utilize different mechanisms of termination.

Connecting Concepts Across Chapters

This chapter has focused on the process of transcription, during which an RNA molecule that is complementary and antiparallel to a DNA template is synthesized. Transcription is the first step in gene expression, the transfer of genetic information from genotype to phenotype and, as we will see in Chapter 16, is an important point at which gene expression is regulated. Transcription is similar in many respects to replication—it utilizes a DNA template, takes place in the 5′→3′ direction, synthesizes a molecule that is antiparallel and complementary to the template, and utilizes nucleoside triphosphates as substrates. But there are important differences as well: only one strand is typically transcribed, each gene is transcribed separately, and the process is subject to numerous regulatory mechanisms.

This chapter has provided important links to topics discussed in several other chapters of the book. Transcription is the first step in the molecular transfer of genetic information from the genotype to the phenotype and is therefore the starting point for discussions of RNA processing in Chapter 14 and translation in Chapter 15. Knowledge of the details of transcription is also essential for understanding gene regulation (Chapter 16), because transcription is an important point at which the expression of many genes is controlled. Additionally, because transcription factors play an important role in some types of cancer, the information in this chapter will be useful when we consider the molecular basis of cancer in Chapter 21.

CONCEPTS SUMMARY

- RNA molecules can function as biological catalysts and may have been the first carriers of genetic information.

- RNA is a polymer, consisting of nucleotides joined together by phosphodiester bonds. Each RNA nucleotide consists of a ribose sugar, a phosphate, and a base. RNA contains the base uracil; it is usually single stranded, which allows it to form secondary structures.

- Ribosomal RNA is a component of the ribosome, messenger RNA carries coding instructions for proteins, and transfer RNA helps incorporate the amino acids into a polypeptide chain. Other RNA molecules found in eukaryotic cells include pre-mRNAs, the precursor of mRNA; snRNAs, which function in the processing of pre-mRNAs; snoRNAs, which process rRNA; and scRNAs, which exist in the cytoplasm.

- The template for RNA synthesis is single-stranded DNA. In transcription, RNA synthesis is complementary and antiparallel to the DNA template strand.

- A transcription unit consists of a promoter, an RNA-coding region, and a terminator.

- The substrates for RNA synthesis are ribonucleoside triphosphates. In transcription, two phosphates are cleaved from a ribonucleoside triphosphate and the remaining phosphate takes part in a phosphodiester bond with the 3′-OH group at the growing end of the RNA molecule.

- RNA polymerase in bacterial cells consists of a core enzyme, which catalyzes the addition of nucleotides to an RNA molecule, and other subunits, which join the core enzyme to provide additional functions. The sigma factor controls the

- binding of the core enzyme to the promoter; rho and NusA assist in the termination of transcription.

- Eukaryotic cells contain three RNA polymerases: RNA polymerase I, which transcribes rRNA; RNA polymerase II, which transcribes pre-mRNA and some snRNAs; and RNA polymerase III, which transcribes tRNAs, small rRNA, and some snRNAs.

- The process of transcription consists of three stages: initiation, elongation, and termination.

- Promoters are recognized by the transcription apparatus and are required for transcription. They contain short consensus sequences imbedded within longer stretches of DNA.

- Transcription begins at the start site, which is determined by the consensus sequences. A short stretch of DNA is unwound near the start site, RNA is synthesized from a single strand of DNA as a template, and the DNA is rewound at the lagging end of the transcription bubble.

- Terminators consist of sequences within the RNA coding region; RNA synthesis ceases after the terminator has been transcribed. Bacterial cells have two types of terminators: rho-independent terminators, which RNA polymerase can recognize by itself, and rho-dependent terminators, which RNA polymerase can recognize only with the help of the rho protein.

- In eukaryotic cells, DNA is complexed to histone proteins, which interfere with the binding of transcription factors and RNA polymerase. Chromatin may be modified by acetylation, chromatin-remodeling proteins, and other factors, allowing transcription factors and RNA polymerase to bind to the DNA.

- Two classes of sequences affect transcription in eukaryotic cells: promoters, which are adjacent to genes, and enhancers, which may be distant to the genes that they affect.

- A promoter for RNA polymerase II consists of a core promoter, which is required for minimal levels of transcription, and a regulatory promoter, which affects the rate of transcription.

- General transcription factors bind to the core promoter and are part of the basal transcription apparatus. Transcriptional activator proteins bind to sequences in regulatory promoters and enhancers and interact with the basal transcription apparatus at the core promoter.

- The three types of RNA polymerase in eukaryotic cells recognize different types of promoters, all of which have consensus sequences that serve as binding sites for transcription factors.

- The three RNA polymerases found in eukaryotic cells use different mechanisms of termination.

IMPORTANT TERMS

ribozyme (p. 354)
ribosomal RNA (rRNA) (p. 355)
messenger RNA (mRNA) (p. 355)
pre-messenger RNA (pre-mRNA) (p. 355)
transfer RNA (tRNA) (p. 355)
small nuclear RNA (snRNA) (p. 355)
small nuclear ribonucleoprotein (snRNP) (p. 355)
small nucleolar RNA (snoRNA) (p. 355)

small cytoplasmic RNA (scRNA) (p. 356)
template strand (p. 357)
nontemplate strand (p. 357)
transcription unit (p. 359)
promoter (p. 359)
RNA-coding region (p. 359)
terminator (p. 359)
ribonucleoside triphosphate (rNTP) (p. 359)
RNA polymerase (p. 360)
core enzyme (p. 360)
sigma factor (p. 360)
holoenzyme (p. 360)
RNA polymerase I (p. 361)

RNA polymerase II (p. 361)
RNA polymerase III (p. 361)
consensus sequence (p. 362)
−10 consensus sequence (Pribnow box) (p. 362)
−35 consensus sequence (p. 362)
upstream element (p. 363)
rho-dependent terminator (p. 365)
rho factor (p. 365)
rho-independent terminator (p. 365)
general transcription factor (p. 367)

basal transcription apparatus (p. 367)
transcriptional activator protein (p. 367)
core promoter (p. 367)
TATA box (p. 367)
TATA-binding protein (TBP) (p. 368)
regulatory promoter (p. 369)
enhancer (p. 369)
silencer (p. 369)
internal promoter (p. 370)

Worked Problems

1. The following diagram represents a sequence of nucleotides surrounding an RNA coding sequence.

```
5'-CATGTT...TTGATGT-[RNA coding sequence]-GACGA...TTTATA...GGCGCGC-3'
3'-GTACAA...AACTACA-[RNA coding sequence]-CTGCT...AAATAT...CCGCGCG-5'
```

(a) Is the RNA coding sequence likely to be from a bacterial cell or from a eukaryotic cell? How can you tell?

(b) Which DNA strand will serve as the template strand during transcription of the RNA coding sequence?

• Solution

(a) Bacterial and eukaryotic cells use the same DNA bases (A, T, G, and C); so the bases themselves provide no clue to the origin of the sequence. The RNA coding sequence must have a promoter, and bacterial and eukaryotic cells do differ in the consensus sequences found in their promoters; so we should examine the sequences for the presence of familiar consensus sequences. On the bottom strand to the right of the RNA coding sequence we find AAATAT, which, written in the conventional manner (5′ on the left), is 5′–TATAAA–3′. This sequence is the TATA box found in most eukaryotic promoters. However, the sequence is also quite similar to the −10 consensus sequence (5′–TATAAT–3′) found in bacterial promoters.

Farther to the right on the bottom strand, we also see 5′–GCGCGCC–3′, which is the TFIIB recognition element (BRE) in eukaryotic RNA polymerase II promoters. No similar consensus sequence is found in bacterial promoters; so we can be fairly certain that this sequence is a eukaryotic promoter and RNA coding sequence.

(b) The TATA box and BRE of RNA polymerase II promoters are upstream of the RNA coding sequences; so RNA polymerase must bind to these sequences and then proceed downstream, transcribing the RNA coding sequence. Thus RNA polymerase must proceed from right (upstream) to left (downstream). The RNA molecule is always synthesized in the 5′→3′ direction and is antiparallel to the DNA template strand; so the template strand must be read 3′→5′. If the enzyme proceeds from right to left and reads the template in the 3′→5′ direction, the upper strand must be the template, as shown here.

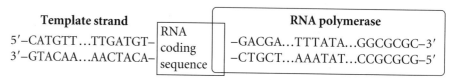

2. Suppose that a consensus sequence in the regulatory promoter of a gene that encodes enzyme A were deleted. Which of the following effects would result from this deletion?

(a) Enzyme A would have a different amino acid sequence.

(b) The mRNA for enzyme A would be abnormally short.

(c) Enzyme A would be missing some amino acids.

(d) The mRNA for enzyme A would be transcribed but not translated.

(e) The amount of mRNA transcribed would be affected.
Explain your reasoning.

• Solution

The correct answer is part e. The regulatory protein contains binding sites for transcriptional activator proteins. These sequences are not part of the RNA coding sequence for enzyme A; so the mutation would have no effect on the length or amino acid sequence of the enzyme, eliminating answers a, b, and c. The TATA box is the binding site for the basal transcription apparatus. Transcriptional activator proteins bind to the regulatory promoter and affect the amount of transcription that takes place through interactions with the basal transcription apparatus at the core promoter.

COMPREHENSION QUESTIONS

* 1. Draw an RNA nucleotide and a DNA nucleotide, highlighting the differences. How is the structure of RNA similar to that of DNA? How is it different?

2. What are the major classes of cellular RNA? Where would you expect to find each class of RNA within eukaryotic cells?

* 3. What parts of DNA make up a transcription unit? Draw and label a typical transcription unit in a bacterial cell.

4. What is the substrate for RNA synthesis? How is this substrate modified and joined together to produce an RNA molecule?

5. Describe the structure of bacterial RNA polymerase.

* 6. Give the names of the three RNA polymerases found in eukaryotic cells and the types of RNA that they transcribe.

7. What are the four basic stages of transcription? Describe what happens at each stage.

* 8. Draw and label a typical bacterial promoter. Include any common consensus sequences.

9. What are the two basic types of terminators found in bacterial cells? Describe the structure of each.

10. How is the process of transcription in eukaryotic cells different from that in bacterial cells?

*11. How are promoters and enhancers similar? How are they different?

12. How can an enhancer affect the transcription of a gene that is thousands of nucleotides away?

13. Compare the roles of general transcription factors and transcriptional activator proteins.

14. What are some of the common consensus sequences found in RNA polymerase II promoters?

*15. What protein associated with a transcription factor is common to all eukaryotic promoters? What is its function in transcription?

*16. Compare and contrast transcription and replication. How are these processes similar and how are they different?

APPLICATION QUESTIONS AND PROBLEMS

17. Write the consensus sequence for the following set of nucleotide sequences.

$$AGGAGTT$$
$$AGCTATT$$
$$TGCAATA$$
$$ACGAAAA$$
$$TCCTAAT$$
$$TGCAATT$$

*18. List at least five properties that DNA polymerases and RNA polymerases have in common. List at least three differences.

19. RNA molecules have *three* phosphates at the 5′ end, but DNA molecules never do. Explain this difference.

20. An RNA molecule has the following percentages of bases: A = 23%, U = 42%, C = 21%, and G = 14%.

(a) Is this RNA single stranded or double stranded? How can you tell?

(b) What would be the percentages of bases in the template strand of the DNA that contains the gene for this RNA?

*21. The following diagram represents DNA that is part of the RNA-coding sequence of a transcription unit. The bottom strand is the template strand. Give the sequence found on the RNA molecule transcribed from this DNA and label the 5′ and 3′ ends of the RNA.

5′–ATAGGCGATGCCA–3′
3′–TATCCGCTACGGT–5′ ←—— Template strand

22. The following sequence of nucleotides is found in a single-stranded DNA template:

ATTGCCAGATCATCCCAATAGAT

Assume that RNA polymerase proceeds along this template from left to right.

(a) Which end of the DNA template is 5′ and which end is 3′?

(b) Give the sequence and label the 5′ and 3′ ends of the RNA copied from this template.

23. Write out a hypothetical sequence of bases that might be found in the first 20 nucleotides of a promoter of a bacterial gene. Include both strands of DNA and label the 5′ and 3′ ends of both strands. Be sure to include the start site for transcription and any consensus sequences found in the promoter.

24. The following diagram represents a transcription unit in a hypothetical DNA molecule.

5′...TTGACA...TATAAT...3′
3′...AACTGT...ATATTA...5′

(a) On the basis of the information given, is this DNA from a bacterium or from a eukaryotic organism?

(b) If this DNA molecule is transcribed, which strand will be the template strand and which will be the nontemplate strand?

(c) Where, approximately, will the start site of transcription be?

*25. What would be the most likely effect of a mutation at the following locations in *E. coli* gene?

(a) −8   (c) −20
(b) −35   (d) Start site

26. A strain of bacteria possesses a temperature-sensitive mutation in the gene that encodes the sigma factor. At elevated temperatures, the mutant bacteria produce a sigma factor that is unable to bind to RNA polymerase. What effect will this mutation have on the process of transcription when the bacteria are raised at elevated temperatures?

*27. The following diagram represents a transcription unit on a DNA molecule.

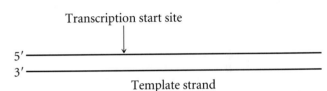

(a) Assume that this DNA molecule is from a bacterial cell. Draw in the approximate location of the promoter and terminator for this transcription unit.

(b) Assume that this DNA molecule is from a eukaryotic cell. Draw in the approximate location of an RNA polymerase II promoter.

(c) Assume that this DNA molecule is from a eukaryotic cell. Draw in the approximate location of an internal RNA polymerase III promoter.

28. The following DNA nucleotides are found near the end of a bacterial transcription unit. Find the terminator in this sequence.

3′–AGCATACAGCAGACCGTTGGTCTGAAAAAAGCATACA–5′

(a) Mark the point at which transcription will terminate.

(b) Is this terminator rho independent or rho dependent?

(c) Draw a diagram of the RNA that will be transcribed from this DNA, including its nucleotide sequence and any secondary structures that form.

*29. A strain of bacteria possesses a temperature-sensitive mutation in the gene that encodes the rho subunit of RNA polymerase. At high temperatures, rho is not functional. When these bacteria are raised at elevated temperatures, which of the following effects would you expect to see?

(a) Transcription does not take place.

(b) All RNA molecules are shorter than normal.

(c) All RNA molecules are longer than normal.

(d) Some RNA molecules are longer than normal.

(e) RNA is copied from both DNA strands.

Explain your reasoning for accepting or rejecting each of these five options.

30. Suppose that the string of As following the inverted repeat in a rho-independent terminator were deleted, but the inverted repeat were left intact. How would this deletion affect termination? What would happen when RNA polymerase reached this region?

*31. Through genetic engineering, a geneticist mutates the gene that encodes TBP in cultured human cells. This mutation destroys the ability of TBP to bind to the TATA box. Predict the effect of this mutation on cells that possess it.

32. Elaborate repair mechanisms are associated with replication to prevent permanent mutations in DNA, yet no similar repair is associated with transcription. Can you think of a reason for these differences in replication and transcription? (Hint: Think about the relative effects of a permanent mutation in a DNA molecule compared with one in an RNA molecule.)

CHALLENGE QUESTIONS

33. Enhancers are sequences that affect the initiation of the transcription of genes that are hundreds or thousands of nucleotides away. Enhancer-binding proteins usually interact directly with transcription factors at promoters by causing the intervening DNA to loop out. An enhancer of bacteriophage T4 does not function by looping of the DNA (D. R. Herendeen, et al., 1992, *Science* 256:1298–1303). Propose some additional mechanisms (other than DNA looping) by which this enhancer might affect transcription at a gene thousands of nucleotides away.

*34. The location of the TATA box in two species of yeast, *Saccharomyces pombe* and *S. cerevisiae*, differs dramatically. The TATA box of *S. pombe* is about 30 nucleotides upstream of the start site, similar to the location for most other eukaryotic cells. However, the TATA box of *S. cerevisiae* can be as many as 120 nucleotides upstream of the start site. To understand how the TATA box functions in these two species, a series of experiments was conducted to determine which components of the transcription apparatus of these two species could be interchanged. In these experiments, different components of the transcription apparatus were switched in *S. pombe* and *S. cerevisiae*, and the effects of the switch on the level of RNA synthesis and on the start point of transcription were observed. TFIID from *S. pombe* could be used in *S. cerevisiae* cells and vice versa, without any effect on the transcription start site in either cell type. Switching TFIIB, TFIIE, or RNA polymerase did alter the level of transcription. However, the following pairs of components could be exchanged without affecting transcription: TFIIE together with TFIIH; and TFIIB together with RNA polymerase. The exchange of TFIIE–TFIIH did not alter the start point, but the exchange of TFIIB–RNA polymerase did shift it. (Y. Li, P. M. Flanagan, H. Tschochner, and R. D. Kornberg, 1994, *Science* 263:805–807.)

On the basis of these results, what conclusions can you draw about how the different components of the transcription apparatus interact and which components are

responsible for setting the start site? Propose a mechanism for the determination of the start site in eukaryotic RNA polymerase II promoters.

35. The relation between chromatin structure and transcription activity has been the focus of recent research. In one set of experiments, the level of in vitro transcription of a *Drosophila* gene by RNA polymerase II was studied with the use of DNA and various combinations of histone proteins.

First, the level of transcription was measured for naked DNA with no associated histone proteins. Then, the level of transcription was measured after nucleosome octamers (without H1) were added to the DNA. The addition of the octamers caused the level of transcription to drop by 50%. When both nucleosome octamers and H1 proteins were added to the DNA, transcription was greatly repressed, dropping to less than 1% of that obtained with naked DNA (see the table below).

GAL4-VP16 is a protein that binds to the DNA of certain eukaryotic genes. When GAL4-VP16 is added to DNA, the level of RNA polymerase II transcription is greatly elevated. Even in the presence of the H1 protein, GAL4-VP16 stimulates high levels of transcription.

Propose a mechanism for how the H1 protein represses transcription and how GAL4-VP16 overcomes this repression. Explain how your proposed mechanism would produce the results obtained in these experiments.

Treatment	Relative amount of transcription
Naked DNA	100
DNA + octamers	50
DNA + octamers + H1	<1
DNA + GAL4-VP16	1000
DNA + octamers + GAL4-VP16	1000
DNA + octamers + H1 + GAL4-VP16	1000

(Based on experiments reported in an article by G. E. Croston et al., 1991, *Science* 251:643–649.)

SUGGESTED READINGS

Atchinson, M. L. 1988. Enhancers: mechanisms of action and cell specificity. *Annual Review of Cell Biology* 4:127–154.
A good review of the mechanism by which enhancers influence the transcription of distant genes.

Baumann, P., S. A. Qreshi, and S. P. Jackson. 1995. Transcription: new insights from studies on archaea. *Trends in Genetics* 11:279–283.
A discussion of how the transcription in archaea is similar to that of eukaryotes.

Cramer, P., D. A. Bushnell, J. Fu, A. L. Gnatt, B. Maier-Davis, N. E. Thompson, R. R. Burgess, A. M. Edwards, P. R. David, and R. D. Kornberg. 2000. Architecture of RNA polymerase II and implications for the transcription mechanism. *Science* 288:640–649.
Report of the detailed structure of RNA polymerase II.

Gesteland, R. F., and J. F. Atkins. 1993. *The RNA World.* Cold Spring Harbor, NY: Cold Spring Harbor Laboratory Press.
Contains a number of chapters on ribozymes and their possible role in the early evolution of life.

Helmann, J. D., and M. J. Chamberlin. 1988. Structure and function of bacterial sigma factors. *Annual Review of Biochemistry* 57:839–872.
Review of sigma factors and their role in bacterial transcription initiation.

Kim, Y., J. H. Geiger, S. Hahn, and P. B. Sigler. 1993. Crystal structure of a yeast TBP/TATA-box complex. *Nature* 365:512–527.
Report of the three-dimensional structure of the TBP protein binding to the TATA box.

Korzheva, N., A. Mustaev, M. Kozlov, A. Malhotra, V. Nikiforov, A. Goldfarb, and S. A. Darst. 2000. A structural model of transcription elongation. *Science* 289:619–625.
Presentation of a model of the transcription apparatus.

Lee, T. I., and R. A. Young. 2000. Transcription of eukaryotic protein-encoding genes. *Annual Review of Genetics* 34:77–138.
A good review of how eukaryotic genes are transcribed by RNA polymerase II.

Nikolov, D. B. 1992. Crystal structure of TFIID TATA-box binding protein. *Nature* 360:40–45.
A look at the three-dimensional structure of TFIID.

Ptashne, M., and A. Gann. 1997. Transcriptional activation by recruitment. *Nature* 386:569–577.
Excellent summary of how prokaryotic and eukaryotic proteins that bind to promoters affect transcription.

Rowlands, R., P. Baumann, and S. P. Jackson. 1994. The TATA-binding protein: a general transcription factor in eukaryotes and archaebacteria. *Science* 264:1326–1329.
Report of a TATA-binding protein in archaea.

von Hippel, P. H. 1998. An integrated model of the transcription complex in elongation, termination, and editing. *Science* 281:660–665.
Review of how the transcription apparatus elongates, terminates, and edits during transcription.

Young, R. A. 1991. RNA polymerase II. *Annual Review of Biochemistry* 60:689–716.
A review of RNA polymerase II.

14 RNA Molecules and RNA Processing

For almost 50 years, entertainer Jerry Lewis has served as national chairman of the Muscular Dystrophy Association, a partnership between scientists and citizens aimed at fighting neuromuscular diseases. (Courtesy of the Muscular Dystrophy Association.)

The Immense Dystrophin Gene

The most common and devastating of the muscular dystrophies is Duchenne muscular dystrophy, a fatal disease that strikes nearly 1 in 3500 males. At birth, affected boys appear normal. The first symptom is mild muscle weakness appearing between 3 and 5 years of age: the child stumbles frequently, has difficulty climbing stairs, and is unable to rise from a sitting position. In time, the arm and leg muscles become progressively weaker. By age 11, those affected are usually confined to a wheel chair and, by age 20, most persons with Duchenne muscular dystrophy have died. At present, there is no cure for the disease.

Duchenne muscular dystrophy was first recognized in 1852, and the disease was fully described in 1861 by Benjamin A. Duchenne, a Paris physician. Even before Mendel's laws were discovered, physicians noticed its X-linked

pattern of inheritance, remarking that the disease developed almost exclusively in males and seemed to be inherited through unaffected mothers. In spite of this early recognition of its hereditary basis, the biochemical cause of Duchenne muscular dystrophy remained a mystery until 1987.

In 1985, Louis Kunkel and his colleagues at Harvard Medical School observed a boy with Duchenne muscular dystrophy whose X chromosome had a visible deletion on the short arm. Reasoning that this boy's disease was caused by the absence of a gene within the deletion, they recognized that the deletion pointed to the location on the X chromosome of the gene responsible for Duchenne muscular dystrophy. Kunkel and his colleagues located and cloned the piece of DNA responsible for the disease. Shortly thereafter, the sequence of the gene was determined, and the protein that it encodes was isolated. This large protein, called dystrophin, consists of nearly 4000 amino acids and is an integral component of muscle cells. Persons with Duchenne muscular dystrophy lack functional dystrophin.

The dystrophin gene is among the most remarkable of all genes yet examined. It's *huge,* encompassing more than 2 million nucleotides of DNA. However, only about 12,000 of its nucleotides encode its amino acids. Why is the dystrophin gene so large? What are all those other nucleotides doing?

The unusual properties of the dystrophin gene make sense only in the context of RNA processing—the alteration of RNA after it has been transcribed. Dystrophin mRNA, like many eukaryotic RNAs, undergoes extensive processing after transcription, including the removal of large sections that are not required for translation. Chapter 13 focused on transcription—the process of RNA synthesis. In this chapter, we will examine the function and processing of RNA.

We begin by taking a careful look at the nature of the gene. Next, we examine messenger RNA, its structure, and how it is modified in eukaryotes after transcription. We'll also see how, through alternative pathways of RNA modification, one gene can produce several different proteins. Then, we turn to transfer RNA, the adapter molecule that forms the interface between amino acids and mRNA in protein synthesis. Finally, we examine ribosomal RNA, the structure and organization of rRNA genes, and how rRNAs are processed.

As we explore the world of RNA and its role in gene function, we will see evidence of two important characteristics of this nucleic acid. First, RNA is extremely versatile, both structurally and biochemically. It can assume a number of different secondary structures, which provide the basis for its functional diversity. Second, RNA processing and function frequently include interactions between two or more RNA molecules.

www.whfreeman.com/pierce   More information about Duchenne muscular dystrophy

Gene Structure

What is a gene? In Chapter 3, it was noted that the definition of *gene* would appear to change as we explored different aspects of heredity. A gene was defined as an inherited factor that determined a trait. This definition may have seemed vague, because it says nothing about what a gene is, only what it does. Nevertheless, this definition was appropriate for our purposes at the time, because our focus was on how genes influence the inheritance of traits. It wasn't necessary to consider the physical nature of the gene in learning the rules of inheritance.

Knowing something about the chemical structure of DNA and the process of transcription enables us to be more precise about what a gene is. Chapter 10 described how genetic information is encoded in the base sequence of DNA; so a gene consists of a set of DNA nucleotides. But how many nucleotides are encompassed in a gene, and how is the information in these nucleotides organized? In 1902, Archibald Garrod suggested, correctly, that genes code for proteins (see pp. 45–46). Proteins are made of amino acids; so a gene contains the nucleotides that specify the amino acids of a protein. We could, then, define a gene as a set of nucleotides that specifies the amino acid sequence of a protein, which indeed was for many years the working definition of a gene. As geneticists learned more about the structure of genes, however, it became clear that this concept of a gene was an oversimplification.

Gene Organization

Early work on gene structure was carried out largely through the examination of mutations in bacteria and viruses. This research led Francis Crick in 1958 to propose that genes and proteins are **colinear**—that there is a direct correspondence between the nucleotide sequence of DNA and the amino acid sequence of a protein (◀ FIGURE 14.1).

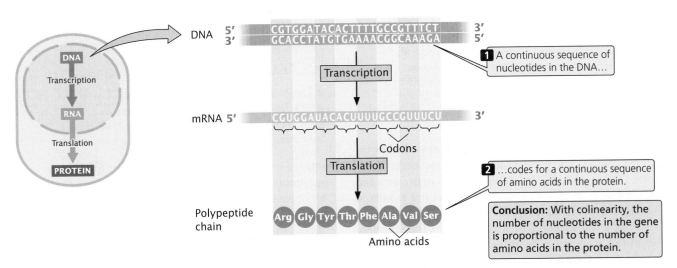

◀ 14.1 **The concept of colinearity suggests that a continuous sequence of nucleotides in DNA encodes a continuous sequence of amino acids in a protein.**

The concept of colinearity suggests that the number of nucleotides in a gene should be proportional to the number of amino acids in the protein encoded by that gene. In a general sense, this concept is true for genes found in bacterial cells and many viruses, although these genes are slightly longer than expected if colinearity is strictly applied (the mRNAs encoded by the genes contain sequences at their ends that do not specify amino acids). At first, eukaryotic genes and proteins also were generally assumed to be colinear, but there were hints that eukaryotic gene structure was fundamentally different. Eukaryotic cells contain far more DNA than is required to encode proteins (see Chapter 11). Furthermore, many large RNA molecules observed in the nucleus were absent from the cytoplasm, suggesting that nuclear RNAs undergo some type of change before they are exported to the cytoplasm.

Most geneticists were nevertheless surprised by the announcement in the 1970s that four coding sequences in a gene from a eukaryotic virus were interrupted by nucleotides that did not specify amino acids. This discovery was made when the viral DNA was hybridized with the mRNA transcribed from it, and the hybridized structure was examined with the use of an electron microscope (◀ FIGURE 14.2). The DNA was clearly much longer than the mRNA, because regions of DNA looped out from the hybridized molecules. These regions contained nucleotides in the DNA that were absent from the coding nucleotides in the mRNA. Many other examples of interrupted genes were subsequently discovered; it quickly became apparent that most eukaryotic genes comprise stretches of coding and noncoding nucleotides.

Concepts

When a continuous sequence of nucleotides in DNA encodes a continuous sequence of amino acids in a protein, the two are said to be colinear. The discovery of coding and noncoding regions within eukaryotic genes shows that not all genes are colinear with the proteins that they encode.

Introns

Many eukaryotic genes contain coding regions called **exons** and noncoding regions called intervening sequences or **introns**. For example, the ovalbumin gene has eight exons and seven introns; the gene for cytochrome *b* has five exons and four introns (◀ FIGURE 14.3). All the introns and the exons are initially transcribed into RNA but, after transcription, the introns are removed by splicing and the exons are joined to yield the mature RNA.

Introns are common in eukaryotic genes but are rare in bacterial genes. For a number of years after their discovery,

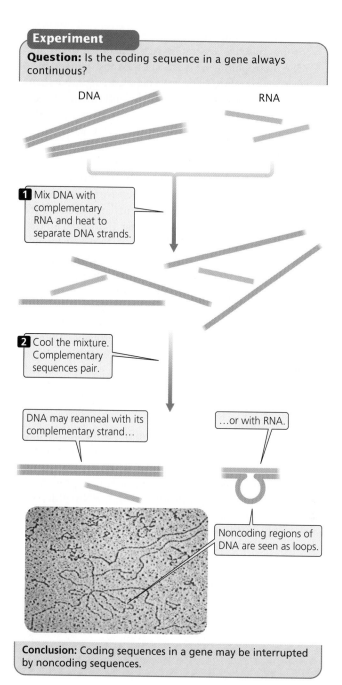

Experiment

Question: Is the coding sequence in a gene always continuous?

1 Mix DNA with complementary RNA and heat to separate DNA strands.

2 Cool the mixture. Complementary sequences pair.

DNA may reanneal with its complementary strand…

…or with RNA.

Noncoding regions of DNA are seen as loops.

Conclusion: Coding sequences in a gene may be interrupted by noncoding sequences.

◀ 14.2 **The noncolinearity of eukaryotic genes was discovered by hybridizing DNA and mRNA.** (Electromicrograph from O.L. Miller, B.R. Beatty, D.W. Fawcett/Visuals Unlimited.)

introns were thought to be entirely absent from prokaryotic genomes, but they have now been observed in archaea, bacteriophages, and even some eubacteria. Introns are present in mitochondrial and chloroplast genes, as well as nuclear genes. In eukaryotic genomes, the size and number of introns appear to be directly related to increasing organismal complexity. Yeast genes contain only a few short introns; *Drosophila* introns are longer and more numerous; and most vertebrate genes are interrupted by long introns. All

Ovalbumin gene

Cytochrome _b_ gene

◀14.3 **The coding sequences of many eukaryotic genes are disrupted by noncoding introns.**

classes of genes—those that code for rRNA, tRNA, and proteins—may contain introns. The number and size of introns vary widely: some eukaryotic genes have no introns, whereas others may have more than 60; intron length varies from fewer than 200 nucleotides to more than 50,000. Introns tend to be longer than exons, and most eukaryotic genes contain more noncoding nucleotides than coding nucleotides. Finally, most introns do not encode proteins (an intron of one gene is not usually an exon for another), although there are exceptions.

There are four major types of introns (Table 14.1). **Group I introns,** found in some rRNA genes, are self-splicing—they can catalyze their own removal. **Group II introns** are present in some protein-encoding genes of mitochondria, chloroplasts, and a few eubacteria; they also are self-splicing, but their mechanism of splicing differs from that of the group I introns. **Nuclear pre-mRNA introns** are the best studied; they include introns located in the protein-encoding genes of the nucleus. The splicing mechanism by which these introns are removed is similar to that of the group II introns, but nuclear introns are not self-splicing; their removal requires snRNAs (discussed later) and a number of proteins. **Transfer RNA introns,** found in tRNA genes, utilize yet another splicing mechanism that relies on enzymes to cut and reseal the RNA. In addition to these major groups, there are several other types of introns.

We'll take a detailed look at the chemistry and mechanics of RNA splicing later in the chapter. For now, we should keep in mind two general characteristics of the splicing process: (1) the splicing of all pre-mRNA introns takes place in the nucleus and is probably required for RNA to move to the cytoplasm; and (2) the order of exons in DNA is usually maintained in the spliced RNA—the coding sequences of a gene may be split up, but they are not usually jumbled up.

Concepts

Many eukaryotic genes contain exons and introns, both of which are transcribed into RNA, but introns are later removed by RNA processing. The number and size of introns vary from gene to gene; they are common in many eukaryotic genes but uncommon in bacterial genes.

The Concept of the Gene Revisited

How does the presence of introns affect our concept of a gene? It no longer seems appropriate to define a gene as a sequence of nucleotides that codes for amino acids in a protein, because this definition excludes from the gene those sequences in introns that don't specify amino acids. This definition also excludes nucleotides that code for the 5′ and 3′ ends of a mRNA molecule, which are required for translation but do not code for amino acids. And defining a gene

Table 14.1	Major types of introns	
Type of Intron	**Location**	**Type of Splicing**
Group I	Some rRNA genes	Self-splicing
Group II	Protein-encoding genes in mitochondria and chloroplasts	Self-splicing
Nuclear pre-mRNA	Protein-encoding genes in the nucleus	Spliceosomal
tRNA	tRNA genes	Enzymatic

Note: There are also several types of minor introns, including group III introns, twintrons, and archaeal introns.

in these terms also excludes sequences that encode rRNA, tRNA, and other RNAs that do not encode proteins. In view of our current understanding of DNA structure and function, we need a more satisfactory definition of gene.

Many geneticists have broadened the concept of a gene to include all sequences in the DNA that are transcribed into a single RNA molecule. Defined in this way, a gene includes all exons, introns, and those sequences at the beginning and end of the RNA that are not translated into a protein. This definition also includes DNA sequences that code for rRNAs, tRNAs, and other types of non-messenger RNA. Many geneticists have expanded the definition of a gene even further, to include the entire transcription unit—the promoter, the RNA coding sequence, and the terminator.

> **Concepts**
>
> The discovery of introns forced a reevaluation of the definition of the gene. Today, a gene is often defined as a DNA sequence that codes for an RNA molecule.

Messenger RNA

As soon as DNA was identified as the source of genetic information, it became clear that DNA could not directly encode proteins. In eukaryotic cells, DNA resides in the nucleus, yet most protein synthesis takes place in the cytoplasm. Geneticists recognized that an additional molecule must take part in the transfer of genetic information.

The results of studies of bacteriophage infection conducted in the late 1950s and early 1960s pointed to RNA as a likely candidate for this transport function. Bacteriophages inject their DNA into bacterial cells, where the DNA is replicated, and large amounts of phage protein are produced on the bacterial ribosomes. As early as 1953, Alfred Hershey discovered a type of RNA that was synthesized rapidly after bacteriophage infection. Findings from later studies showed that the bacteriophage T2 produced short-lived RNA having a nucleotide composition similar to that of phage DNA but quite different from that of the bacterial RNA. These observations were consistent with the idea that RNA was copied from DNA and that this RNA then directed the synthesis of proteins.

At the time, ribosomes were known to be *somehow* implicated in protein synthesis, and much of the RNA in a cell was known to be in the form of ribosomes. Each gene was thought to direct the synthesis of a special type of ribosome in the nucleus, which then moved to the cytoplasm and produced a specific protein. Using equilibrium density-gradient centrifugation (see Figure 12.2), Sydney Brenner, François Jacob, and Matthew Meselson demonstrated in 1961 that new ribosomes are *not* produced during the burst of protein synthesis that accompanies phage infection

(◄ Figure 14.4). The genetic information needed to produce new phage proteins was not carried by the ribosomes.

In a related experiment, François Gros and his colleagues infected *E. coli* cells with bacteriophages while radioactively

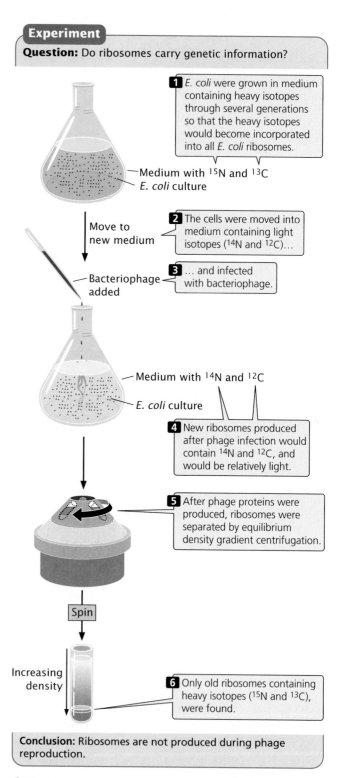

Experiment

Question: Do ribosomes carry genetic information?

1 *E. coli* were grown in medium containing heavy isotopes through several generations so that the heavy isotopes would become incorporated into all *E. coli* ribosomes.

Medium with ^{15}N and ^{13}C
E. coli culture

Move to new medium

2 The cells were moved into medium containing light isotopes (^{14}N and ^{12}C)…

Bacteriophage added

3 … and infected with bacteriophage.

Medium with ^{14}N and ^{12}C
E. coli culture

4 New ribosomes produced after phage infection would contain ^{14}N and ^{12}C, and would be relatively light.

5 After phage proteins were produced, ribosomes were separated by equilibrium density gradient centrifugation.

Spin

Increasing density

6 Only old ribosomes containing heavy isotopes (^{15}N and ^{13}C), were found.

Conclusion: Ribosomes are not produced during phage reproduction.

◄ **14.4 Brenner, Jacob, and Meselson demonstrated that ribosomes do not carry genetic information.**

labeled ("hot") uracil was added to the medium (which would become incorporated into newly produced phage RNA). After a few minutes, they transferred the cells to a medium that contained unlabeled ("cold") uracil. This type of experiment is called a pulse–chase experiment: the cells are exposed to a brief pulse of label, which is then "chased" by cold, unlabeled precursor. Pulse–chase experiments make it possible to follow, by tracking the presence of the radioactivity, products of short-term biochemical events, such as RNA synthesis immediately following phage infection.

Gros and his coworkers found that the newly produced phage RNA was short lived, lasting only a few minutes, and was associated with ribosomes but was distinct from them. They concluded that newly synthesized, short-lived RNA carries the genetic information for protein structure to the ribosome. The term *messenger RNA* was coined for this carrier.

The Structure of Messenger RNA

Messenger RNA functions as the template for protein synthesis; it carries genetic information from DNA to a ribosome and helps to assemble amino acids in their correct order. Each amino acid in a protein is specified by a set of three nucleotides in the mRNA, called a **codon**. Both prokaryotic and eukaryotic mRNAs contain three primary regions (◀ FIGURE 14.5). The **5′ untranslated region** (5′ UTR; sometimes call the leader) is a sequence of nucleotides at the 5′ end of the mRNA that does not code for the amino acid sequence of a protein. In bacterial mRNA, this region contains a consensus sequence called the **Shine-Dalgarno sequence,** which serves as the ribosome-binding site during translation; it is found approximately seven nucleotides upstream of the first codon translated into an amino acid (called the start codon). Eukaryotic mRNA has no equivalent consensus sequence in its 5′ untranslated region. In eukaryotic cells, ribosomes bind to a modified 5′ end of mRNA, as discussed later in the chapter.

The next section of mRNA is the **protein-coding region,** which comprises the codons that specify the amino acid sequence of the protein. The protein-coding region begins with a start codon and ends with a stop codon. The last region of mRNA is the **3′ untranslated region** (3′ UTR), a sequence of nucleotides at the 3′ end of mRNA that is not translated into protein. The 3′ untranslated region affects the stability of mRNA and the translation of the mRNA protein-coding sequence.

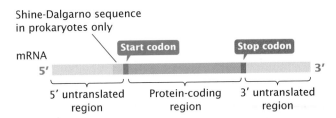

◀ 14.5 **Three primary regions of mature mRNA are the 5′ untranslated region, the protein-coding region, and the 3′ untranslated region.**

Pre-mRNA Processing

In bacterial cells, transcription and translation take place simultaneously; while the 3′ end of an mRNA is undergoing transcription, ribosomes attach to the Shine-Dalgarno sequence near the 5′ end and begin translation. Because transcription and translation are coupled, there is little opportunity for the bacterial mRNA to be modified before protein synthesis. In contrast, transcription and translation are separated in both time and space in eukaryotic cells. Transcription takes place in the nucleus, whereas most translation takes place in the cytoplasm; this separation provides an opportunity for eukaryotic RNA to be modified before it is translated. Indeed, eukaryotic mRNA is extensively altered after transcription. Changes are made to the 5′ end, the 3′ end, and the protein-coding section of the RNA molecule. The initial transcript of protein-encoding genes of eukaryotic cells is called pre-mRNA, whereas the mature, processed transcript is mRNA. We will reserve the term mRNA for RNA molecules that have been completely processed and are ready to undergo translation.

The Addition of the 5′ Cap

Almost all eukaryotic pre-mRNAs are modified at their 5′ ends by the addition of a structure called a **5′ cap.** This capping consists of the addition of an extra nucleotide at the 5′ end of the mRNA and methylation by the addition of a methyl group (CH_3) to the base in the newly added neucleotide and to the 2′−OH group of the sugar of one or more nucleotides at the 5′ end (◀ FIGURE 14.6). Capping takes place rapidly after the initiation of transcription and, as will be discussed in more depth in Chapter 15, the 5′ cap functions in the initiation of translation. Cap-binding proteins recognize the cap and attach to it; a ribosome then binds to these proteins and moves downstream along the mRNA until the start codon is reached and translation begins. The presence of a 5′ cap also increases the stability of mRNA and influences the removal of introns.

In the discussion of transcription in Chapter 13, it was noted that three phosphates are present at the 5′ end of all RNA molecules, because phosphates are not cleaved

---

Concepts

Messenger RNA molecules contain three main regions: a 5′ untranslated region, a protein-coding region, and a 3′ untranslated region. The 5′ and 3′ untranslated regions do not code for the amino acids of a protein.

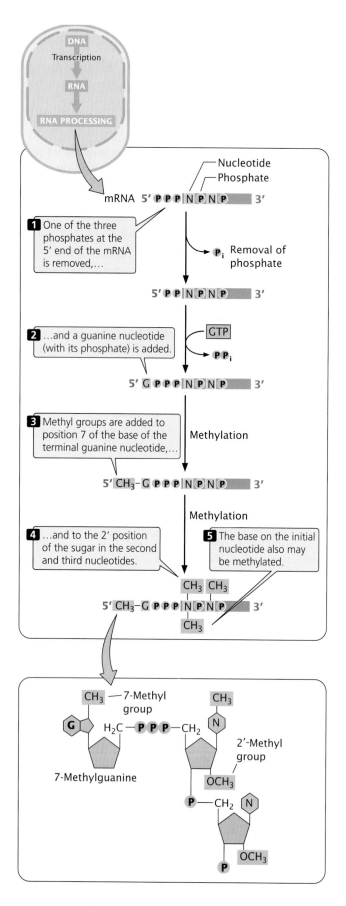

mRNA 5′ P P P N P N P ▨ 3′
— Nucleotide
— Phosphate

1 One of the three phosphates at the 5′ end of the mRNA is removed,…

→ Pᵢ Removal of phosphate

5′ P P N P N P ▨ 3′

2 …and a guanine nucleotide (with its phosphate) is added.

GTP
→ PPᵢ

5′ G P P P N P N P ▨ 3′

3 Methyl groups are added to position 7 of the base of the terminal guanine nucleotide,…

Methylation

5′ CH₃–G P P P N P N P ▨ 3′

Methylation

4 …and to the 2′ position of the sugar in the second and third nucleotides.

5 The base on the initial nucleotide also may be methylated.

CH₃ CH₃
5′ CH₃–G P P P N P N P ▨ 3′
CH₃

CH₃ —7-Methyl group
CH₃
G H₂C–P P P–CH₂ N
7-Methylguanine
2′-Methyl group
OCH₃
P–CH₂ N
OCH₃
P

14.6 Most eukaryotic mRNAs have a 5′ cap. The cap consists of a nucleotide with 7-methyl guanine attached to the pre-mRNA by a unique 5′–5′ bond (shown in detail in the bottom box). The cap is added shortly after the initiation of transcription. A methyl group is added to position 7 of the guanine base of the newly added (now the terminal) nucleotide and to the 2′ position of each sugar of the next two nucleotides.

from the first ribonucleoside triphosphate in the transcription reaction. The 5′ end of pre-mRNA can be represented as 5′–pppNpNpN…, in which the letter N represents a ribonucleotide and p represents a phosphate. Shortly after the initiation of transcription, one of these phosphates is removed and a guanine nucleotide is added (see Figure 14.6). This guanine nucleotide is attached to the pre-mRNA by a unique 5′–5′ bond, which is quite different from the usual 5′–3′ phosphodiester bond that joins all the other nucleotides in RNA. One or more methyl groups are then added to the 5′ end; the first of these methyl groups is added to position 7 of the base of the terminal guanine nucleotide, making the base 7-methylguanine. Next, a methyl group may be added to the 2′ position of the sugar in the second and third nucleotides, as shown in Figure 14.6. Rarely, additional methyl groups may be attached to the bases of the second and third nucleotides of the pre-mRNA.

The Addition of the Poly(A) Tail

Most mature eukaryotic mRNAs have from 50 to 250 adenine nucleotides at the 3′ end (a **poly(A) tail**). These nucleotides are not encoded in the DNA but are added after transcription (◀FIGURE 14.7) in a process termed polyadenylation. Many eukaryotic genes transcribed by RNA polymerase II are transcribed well beyond the end of the coding sequence (see Chapter 13); the extra material at the 3′ end is then cleaved and the poly(A) tail is added. For some pre-mRNA molecules, more than 1000 nucleotides may be cleaved from the 3′ end.

Processing of the 3′ end of pre-mRNA requires sequences both upstream and downstream of the cleavage site (◀FIGURE 14.8a). The consensus sequence AAUAAA is usually from 11 to 30 nucleotides upstream of the cleavage site (see Figure 14.7) and determines the point at which cleavage will take place. A sequence rich in Us (or Gs and Us) is typically downstream of the cleavage site.

In mammals, 3′ cleavage and the addition of the poly(A) tail requires a complex consisting of several proteins: cleavage and polyadenylation specificity factor (CPSF); cleavage stimulation factor (CstF); at least two cleavage factors (CFI and CFII); and polyadenylate polymerase (PAP). CPSF binds to the upstream AAUAAA consensus sequence, whereas CstF binds to the downstream sequence (◀FIGURE 14.8b).

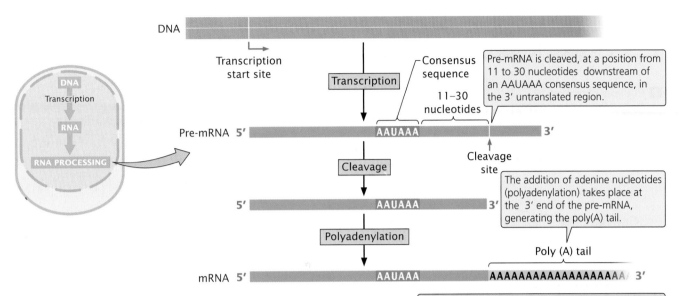

DNA

start site

Transcription

Consensus
sequence

11–30
nucleotides

Pre-mRNA is cleaved, at a position from
11 to 30 nucleotides downstream of
an AAUAAA consensus sequence, in
the 3' untranslated region.

Pre-mRNA 5' — AAUAAA — 3'

Cleavage
site

Cleavage

5' — AAUAAA — 3'

The addition of adenine nucleotides
(polyadenylation) takes place at
the 3' end of the pre-mRNA,
generating the poly(A) tail.

Polyadenylation

Poly (A) tail

mRNA 5' — AAUAAA — AAAAAAAAAAAAAAAAAAAA 3'

Conclusion: In pre-mRNA processing, a poly(A) tail
is added through cleavage and polyadenylation.

 14.7 Most eukaryotic mRNAs have a 3' poly(A) tail.

The pre-mRNA is cleaved, and CstF and the cleavage factors leave the complex; the cleaved 3' end of the pre-mRNA is then degraded (◀ FIGURE 14.8c). CFSF and PAP remain bound to the pre-mRNA and carry out polyadenylation (◀ FIGURE 14.8d). After the addition of approximately 10 adenine nucleotides, a poly(A)-binding protein (PABII) attaches to the poly(A) tail and increases the rate of polyadenylation (◀ FIGURE 14.8e). As more of the tail is synthesized, additional molecules of PABII attach to it(◀ FIGURE 14.8f).

The poly(A) tail confers stability on many mRNAs, increasing the time during which the mRNA remains intact and available for translation before it is degraded by cellular enzymes. The stability conferred by the poly(A) tail is dependent on the proteins that attached to the tail.

Eukaryotic mRNAs that lack a poly(A) tail depend on a different mechanism for 3' cleavage that requires the formation of a hairpin structure in the pre-mRNA and a small ribonucleoprotein particle (snRNP) called U7 (◀ FIGURE 14.9). U7 contains an snRNA with nucleotides that are complementary to a sequence on the pre-mRNA just downstream of the cleavage site, and U7 most likely binds to this sequence. A hairpin-binding protein binds to the hairpin structure and stabilizes the binding of U7 to the complementary sequence on the pre-mRNA.

Concepts

Eukaryotic pre-mRNAs are processed at their 5' and 3' ends. A cap, consisting of a modified nucleotide and several methyl groups, is added to the 5' end. The cap facilitates the binding of

a ribosome, increases the stability of the mRNA, and may affect the removal of introns. Processing at the 3' end includes cleavage downstream of an AAUAAA consensus sequence and the addition of a poly(A) tail.

RNA Splicing

The other major type of modification that takes place in eukaryotic pre-mRNA is the removal of introns by **RNA splicing.** This occurs in the nucleus following transcription but before the RNA moves to the cytoplasm.

Consensus sequences and the spliceosome Splicing requires the presence of three sequences in the intron. One end of the intron is referred to as the **5' splice site,** and the other end is the **3' splice site** (◀ FIGURE 14.10); these splice sites possess short consensus sequences. Most introns in pre-mRNA begin with GU and end with AG, suggesting that these sequences play a crucial role in splicing. Changing a single nucleotide at either of these sites does indeed prevent splicing. A few introns in pre-mRNA begin with AU and end with AC. These introns are spliced by a process that is similar to that seen in GU…AG introns but utilizes a different set of splicing factors. This discussion will focus on splicing of the more common GU…AG introns.

The third sequence important for splicing is at the so-called **branch point,** which is an adenine nucleotide that lies from 18 to 40 nucleotides upstream of the 3' splice site (see Figure 14.10). The sequence surrounding the branch point does not have a strong consensus but usually takes the

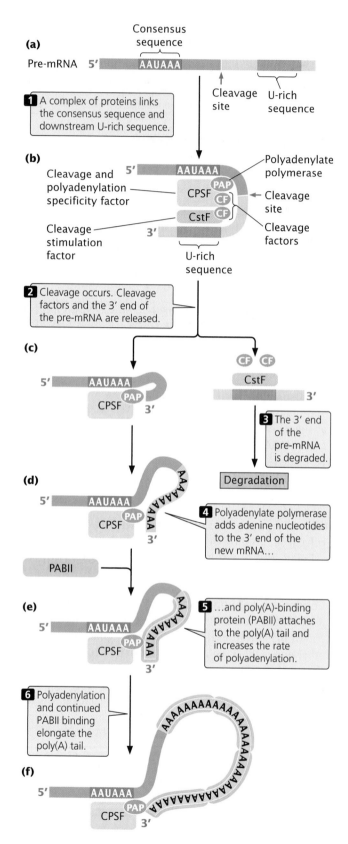

(a)

Consensus sequence

Pre-mRNA 5′ AAUAAA

Cleavage site

U-rich sequence

1 A complex of proteins links the consensus sequence and downstream U-rich sequence.

(b)

Cleavage and polyadenylation specificity factor

5′ AAUAAA

CPSF

CstF

Cleavage stimulation factor

3′

Polyadenylate polymerase

PAP

CF

CF

Cleavage site

Cleavage factors

U-rich sequence

2 Cleavage occurs. Cleavage factors and the 3′ end of the pre-mRNA are released.

(c)

5′ AAUAAA

CPSF PAP

3′

CF CF

CstF

3′

3 The 3′ end of the pre-mRNA is degraded.

Degradation

(d)

5′ AAUAAA

CPSF PAP

AAAAAA

3′

4 Polyadenylate polymerase adds adenine nucleotides to the 3′ end of the new mRNA…

PABII

(e)

5′ AAUAAA

CPSF PAP

AAAAAAA

3′

5 …and poly(A)-binding protein (PABII) attaches to the poly(A) tail and increases the rate of polyadenylation.

6 Polyadenylation and continued PABII binding elongate the poly(A) tail.

(f)

5′ AAUAAA

CPSF PAP

AAAAAAAAAAAAAA

3′

◀ 14.8   Processing of the 3′ end of pre-mRNA requires a consensus sequence and several factors.

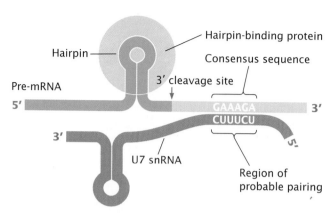

Hairpin

Hairpin-binding protein

Consensus sequence

Pre-mRNA

3′ cleavage site

5′ GAAAGA 3′

CUUUCU

3′

U7 snRNA

5′

Region of probable pairing

◀ 14.9   Eukaryotic mRNAs that lack a poly(A) tail depend on a different mechanism for 3′ cleavage. Cleavage requires the presence of U7 snRNA, which has bases complementary to a consensus sequence downstream of the 3′ cleavage site. Cleavage depends on the formation of a hairpin structure near the 3′ end of the pre-mRNA; base pairing probably takes place between the complementary regions of the pre-mRNA and the U7 snRNA.

form YNYYRAY (Y is any pyrimidine, N is any base, R is any purine, and A is adenine). The deletion or mutation of the adenine nucleotide at the branch point prevents splicing.

Splicing takes place within a large complex called the **spliceosome,** which consists of several RNA molecules and many proteins. The RNA components are small nuclear RNAs (Chapter 13); these snRNAs associate with proteins to form small ribonucleoprotein particles. Each snRNP contains a single snRNA molecule and multiple proteins. The spliceosome is composed of five snRNPs, named for the snRNAs that they contain (U1, U2, U4, U5, and U6), and some proteins not associated with an snRNA.

Concepts

Introns in nuclear genes contain three consensus sequences critical to splicing: a 5′ splice site, a 3′ splice site, and a branch point. Splicing of pre-mRNA takes place within a large complex called the spliceosome, which consists of snRNAs and proteins.

The process of splicing   To illustrate the process of RNA splicing, we'll first consider the chemical reactions that take place. Then we'll see how these splicing reactions constitute a set of coordinated processes within the context of the spliceosome.

Before splicing takes place, an upstream exon (exon 1) and a downstream exon (exon 2) are separated by an intron (◀ FIGURE 14.11). Pre-mRNA is spliced in two distinct steps.

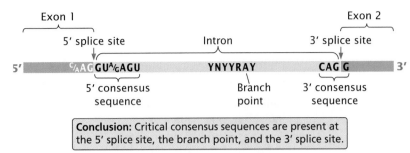

14.10 Splicing of pre-mRNA requires consensus sequences. In the consensus sequence surrounding the branch point (YNYYRAY) Y is any pyrimidine, R is any purine, A is adenine, and N is any base.

Conclusion: Critical consensus sequences are present at the 5' splice site, the branch point, and the 3' splice site.

In the first step, the pre-mRNA is cut at the 5' splice site. This cut frees exon 1 from the intron, and the 5' end of the intron attaches to the branch point; that is, the intron folds back on itself, forming a structure called a **lariat.** The guanine nucleotide in the consensus sequence at the 5' splice site bonds with the adenine nucleotide at the branch point. This bonding is accomplished through **transesterification,** a chemical reaction in which the OH group on the 2'-carbon atom of the adenine nucleotide at the branch point attacks the 5' phosphodiester bond of the guanine nucleotide at the 5' splice site, cleaving it and forming a new 5'−2' phosphodiester bond between the guanine and adenine nucleotides.

In the second step of RNA splicing, a cut is made at the 3' splice site and, simultaneously, the 3' end of exon 1 becomes covalently attached (spliced) to the 5' end of exon 2. This bond also forms through a transesterification reaction, in which the 3'-OH group attached to the end of exon 1 attacks the phosphodiester bond at the 3' splice site, cleaving it and forming a new phosphodiester bond between the 3' end of exon 1 and the 5' end of exon 2; the intron is released as a lariat. The intron becomes linear when the bond breaks at the branch point and is then rapidly degraded by nuclear enzymes. The mature mRNA consisting of the exons spliced together is exported to the cytoplasm where it is translated.

Although splicing is illustrated in Figure 14.11 as a two-step process, the reactions are in fact coordinated within the spliceosome. A key feature of the spliceosome is a series of interactions between the mRNA and snRNAs and

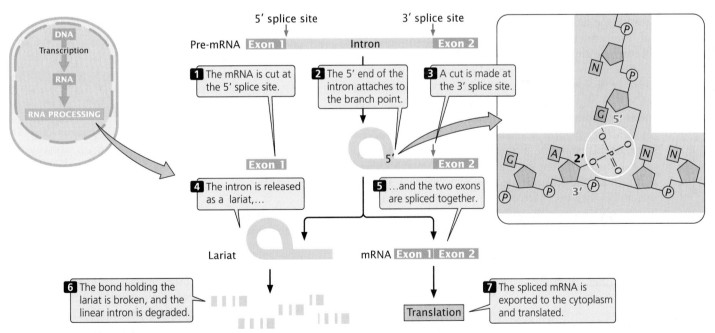

 14.11 The splicing of nuclear introns requires a two-step process. First, cleavage takes place at the 5' splice site, and a lariat is formed by the attachment of the 5' end of the intron to the branch point. Second, cleavage takes place at the 3' splice site, and two exons are spliced together.

Table 14.2	RNA–RNA interactions in pre-mRNA splicing
Interaction	**Function**
U1 with 5′ splice site	U1 attaches to 5′ end of intron; commits intron to splicing; no direct role in splicing
U2 with branch point	Positions 5′ end of intron near branch point for lariat formation
U2 with U6	Holds 5′ end of intron near branch point
U6 with 5′ splice site	Positions 5′ end of intron near branch point
U5 with 3′ end of first exon	Anchors first exon to spliceosome subsequent to cleavage; juxtaposes two ends of exon for splicing
U5 with 3′ end of one exon and 5′ end of the other	Juxtaposes two ends of exon for splicing
U4 with U6	Delivers U6 to intron; no direct role in splicing

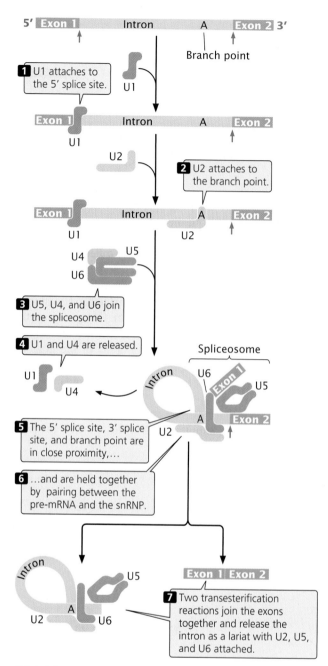

14.12 RNA splicing takes place within the spliceosome.

between different snRNAs (summarized in Table 14.2). These interactions depend on complementary base pairing between the different RNA molecules and bring the essential components of the pre-mRNA transcript and the spliceosome close together, which makes splicing possible.

The spliceosome is assembled on the pre-mRNA transcript in a step-by-step fashion (◀FIGURE 14.12). First, snRNP U1 attaches to the 5′ splice site, and then U2 attaches to the branch point. A complex consisting of U5 and U4–U6 (which form a single snRNP) joins the spliceosome. At this point, the intron loops over and the 5′ splice site is brought close to the branch point. U1 and U4 dissociate from the spliceosome. The 5′ splice site, 3′ splice site, and branch point are in close proximity, held together by the spliceosome. The two transesterification reactions take place, joining the two exons together and releasing the intron as a lariat.

www.whfreeman.com/pierce An animation of the splicing process

Nuclear organization RNA splicing takes place in the nucleus and must occur before the RNA can move into the cytoplasm. For many years, the nucleus was viewed as a biochemical soup, in which components such as the spliceosome diffused and reacted randomly. Now, the nucleus is believed to have a highly ordered internal structure, with transcription and RNA processing taking place at particular locations within it. By attaching fluorescent tags to pre-mRNA and using special imaging techniques, researchers have been able to observe the location of pre-mRNA as it is transcribed and processed. The results of these studies revealed that intron removal and other processing reactions take place at the same sites as those of transcription (◀FIGURE 14.13), suggesting that these processes may be physically coupled. This suggestion is supported by the observation that part of RNA polymerase II is also required for the splicing and 3′ processing of pre-mRNA.

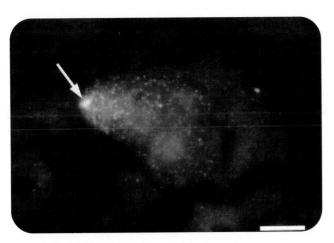

14.13 Intron removal, processing, and transcription take place at the same site. RNA tracks can be seen in the nucleus of a eukaryotic cell. Fluorescent tags were attached to DNA (red) and RNA (green). Transcribed RNA does not disperse; rather, it accumulates near the site of synthesis and follows a defined track during processing. (R. W. Dirks, K. C. Daniël, and A. K. Raap.)

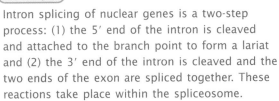

Concepts

Intron splicing of nuclear genes is a two-step process: (1) the 5′ end of the intron is cleaved and attached to the branch point to form a lariat and (2) the 3′ end of the intron is cleaved and the two ends of the exon are spliced together. These reactions take place within the spliceosome.

Self-splicing introns Some introns are self-splicing, meaning that they possess the ability to remove themselves from an RNA molecule. These self-splicing introns fall into two major categories. Group I introns are found in a variety of genes, including some rRNA genes in protists, some mitochondrial genes in fungi, and even some bacteriophage genes. Although the lengths of group I introns vary, all of them fold into a common secondary structure with nine looped stems (◄FIGURE 14.14a), which are necessary for splicing. Transesterification reactions are required for the splicing of group I introns (◄FIGURE 14.14b).

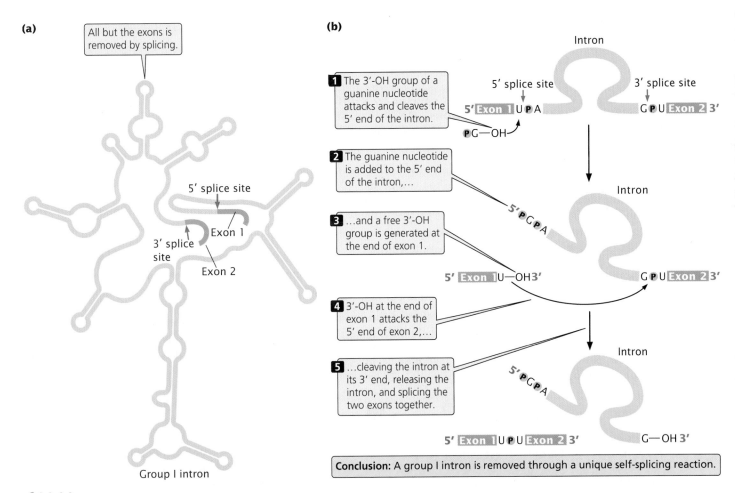

14.14 Group I introns undergo self-splicing. (a) Secondary structure of a group I intron. (b) Self-splicing of a group I intron.

(a)

(b)

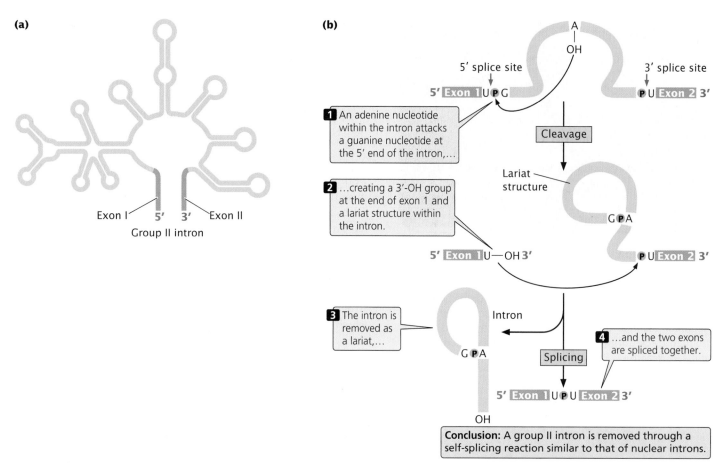

1 An adenine nucleotide within the intron attacks a guanine nucleotide at the 5′ end of the intron,…

2 …creating a 3′-OH group at the end of exon 1 and a lariat structure within the intron.

3 The intron is removed as a lariat,…

4 …and the two exons are spliced together.

5′ splice site

3′ splice site

Cleavage

Lariat structure

Splicing

Intron

Conclusion: A group II intron is removed through a self-splicing reaction similar to that of nuclear introns.

14.15 Group II introns undergo self-splicing by a different mechanism from that for group I introns. (a) Secondary structure of a group II intron. (b) Self-splicing of group II introns, which is similar to the splicing of nuclear introns.

Group II introns, present in some mitochondrial genes, also have the ability to self-splice. All group II introns fold into similar secondary structures (◀FIGURE 14.15a). The splicing of group II introns is accomplished by a mechanism that has some similarities to the spliceosomal-mediated splicing of nuclear genes; splicing takes place through two transesterification reactions that generate a lariat structure (◀FIGURE 14.15b). Because of these similarities, group II introns and nuclear pre-mRNA introns have been suggested to be evolutionarily related—perhaps the nuclear introns evolved from self-splicing group II introns and later adopted the proteins and snRNAs of the spliceosome to carry out the splicing reaction.

Concepts

Self-splicing introns are of two types: group I introns and group II introns. These introns have complex secondary structures that enable them to catalyze their excision from RNA molecules without the aid of enzymes or other proteins.

Alternative Processing Pathways

Another finding that complicates the view of a gene as a sequence of nucleotides that specifies the amino acid sequence of a protein is the existence of **alternative processing pathways,** in which a single pre-mRNA is processed in different ways to produce alternative types of mRNA, resulting in the production of different proteins from the same DNA sequence.

One type of alternative processing is **alternative splicing,** in which the same pre-mRNA can be spliced in more than one way to yield multiple mRNAs that are translated into proteins with different amino acid sequences (◀FIGURE 14.16a). Another type of alternative processing requires the use of **multiple 3′ cleavage sites** (◀FIGURE 14.16b); two or more potential sites for cleavage and polyadenylation are present in the pre-mRNA. In our example, cleavage at the first site produces a relatively short mRNA, compared with the mRNAs produced through cleavage at other sites.

Both alternative splicing and multiple 3′ cleavage sites can exist in the same pre-mRNA transcript; an example is seen in the mammalian calcitonin gene, which contains six exons and five introns (◀FIGURE 14.17a). The entire gene is transcribed into pre-mRNA (◀FIGURE 14.17b). There are two possible 3′ cleavage sites. In cells of the thyroid gland, 3′ cleavage and polyadenylation take place after the fourth exon, and the first three introns are then removed to produce a mature mRNA consisting of exons 1, 2, 3, and 4 (◀FIGURE 14.17c). This mRNA is translated into the hormone calcitonin. In brain cells, the *identical* pre-RNA is transcribed from DNA, but it is processed differently. Cleavage and polyadenylation take place after the sixth exon, yielding an initial transcript that includes all six exons. During splicing, exon 4 (part of the calcitonin mRNA) is removed, along with all the introns; so only exons 1, 2, 3, 5, and 6 are present in the mature mRNA (◀FIGURE 14.17d). When translated, this mRNA produces a protein called calcitonin-gene-related peptide (CGRP), which has an amino acid sequence quite different from that of calcitonin. Alternative splicing may produce different combinations of

exons in the mRNA, but the order of the exons is not usually changed. Different processing pathways contribute to gene regulation, as discussed in Chapter 16.

Concepts

Alternative splicing enables exons to be spliced together in different combinations to yield mRNAs that encode different proteins. Alternative 3′ cleavage sites allow pre-mRNA to be cleaved at different sites to produce mRNAs of different lengths.

RNA Editing

A long-standing principle of molecular genetics is that genetic information ultimately resides in the nucleotide sequence of DNA, except in RNA viruses (Chapter 10). This

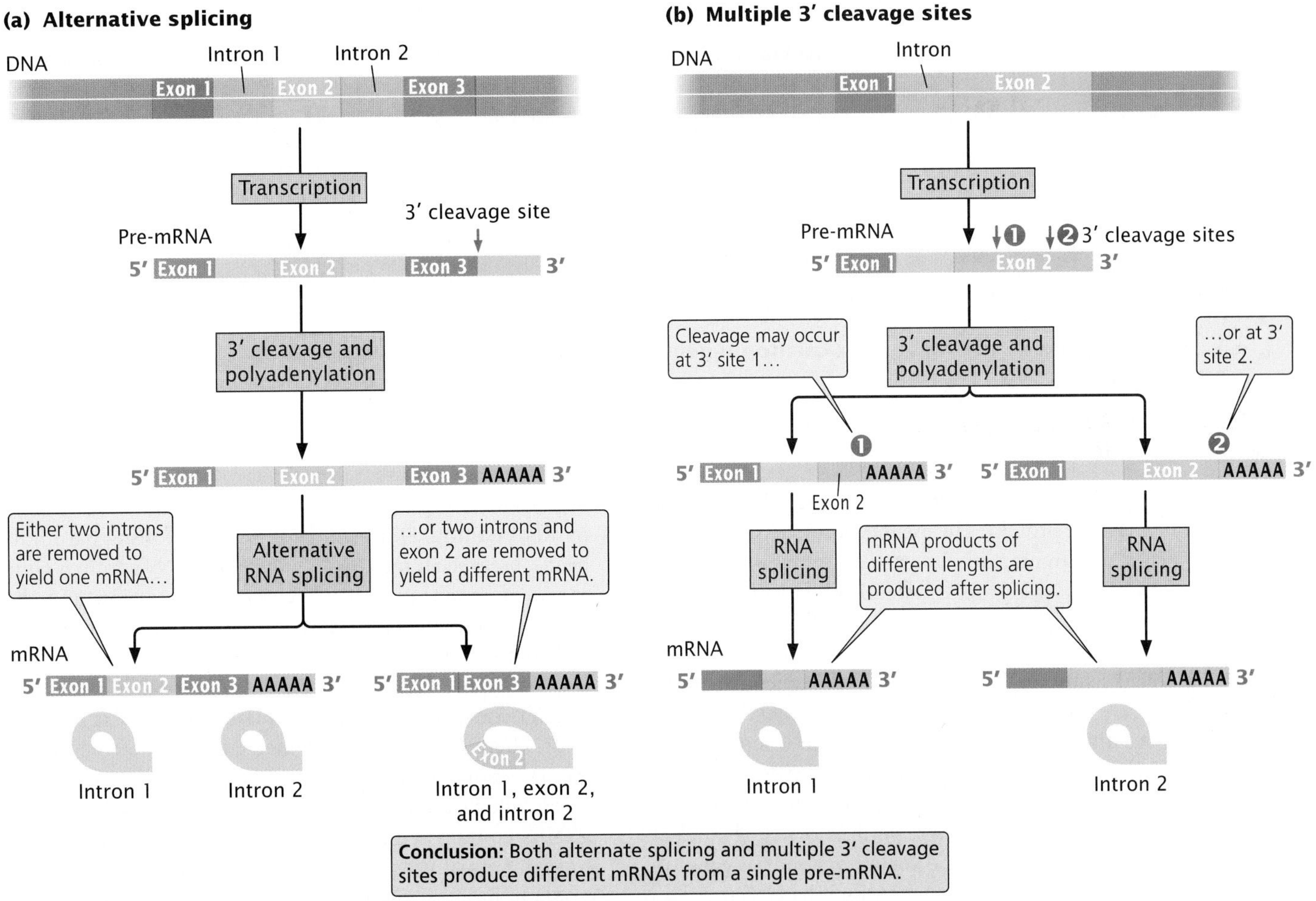

14.16 Eukaryotic cells have alternative pathways for processing pre-mRNA. (a) With alternative splicing; pre-mRNA can be spliced in different ways to produce different mRNAs. (b) With multiple 3′ cleavage sites, there are two or more potential sites for cleavage and polyadenylation; use of the different sites produces mRNAs of different lengths.

(a)

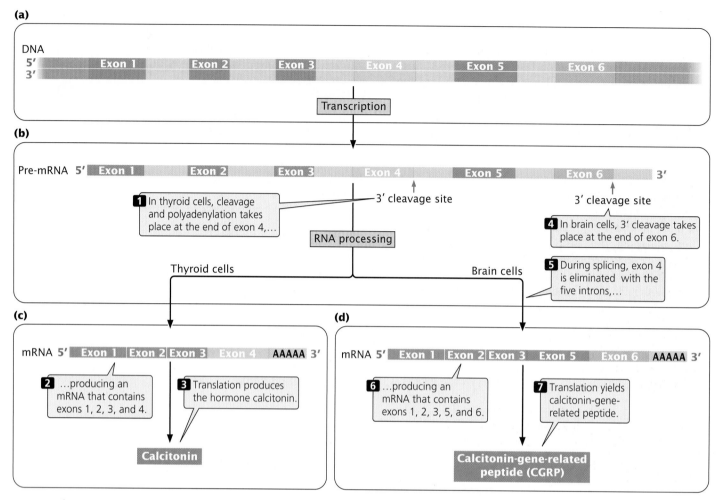

14.17 **Pre-mRNA encoded by the calcitonin gene undergoes alternative processing.**

information is transcribed into mRNA, and mRNA is then translated into a protein. The assumption that all information about the amino acid sequence of a protein resides in DNA is violated by a process called **RNA editing.** In RNA editing, the coding sequence of an mRNA molecule is altered after transcription, and so the protein has an amino acid sequence that differs from that encoded by the gene.

RNA editing was first detected in 1986 when the coding sequences of mRNAs were compared with the coding sequences of the DNAs from which they had been transcribed. Discrepancies were found for some nuclear genes in mammalian cells and for mitochondrial genes in plant cells. In these cases, substitutions had occurred in some of the nucleotides of the mRNA. More extensive RNA editing has been found in the mRNA for some mitochondrial genes in trypanosome parasites (which cause African sleeping sickness). In some mRNAs of these

organisms, more than 60% of the sequence is determined by RNA editing. Different types of RNA editing have now been observed in mRNAs, tRNAs, and rRNAs from a wide range of organisms; they include the insertion and the deletion of nucleotides and the conversion of one base into another.

If the modified sequence in edited RNA molecules doesn't come from a DNA template, then how is it specified? There are a variety of mechanisms that may bring about changes in RNA sequences. In some cases, molecules called **guide RNAs** (gRNAs) play a crucial role. The gRNAs contain sequences that are partly complementary to segments of the preedited RNA, and the two molecules undergo base pairing in these sequences (FIGURE 14.18). After the mRNA is anchored to the gRNA, the mRNA undergoes cleavage and nucleotides are added, deleted, or altered according to the template provided by gRNA. The ends of the mRNA are then joined together.

www.whfreeman.com/pierce More information on RNA editing and a database of guide RNA sequences

Connecting Concepts

Eukaryotic Gene Structure and Pre-mRNA Processing

Chapters 13 and 14 have introduced a number of different components of genes and RNA molecules, including promoters, 5′ untranslated regions, coding sequences, introns, 3′ untranslated regions, poly(A) tails, and caps. Let's see how some of these components are combined to create a typical eukaryotic gene and how a mature mRNA is produced from them.

The promoter, which typically encompasses about 100 nucleotides upstream of the transcription start site, is necessary for transcription to take place but is itself not usually transcribed when protein-encoding genes are transcribed by RNA polymerase II (◀FIGURE 14.19a). Farther upstream or downstream of the start site, there may be enhancers that also regulate transcription.

In transcription, all the nucleotides between the transcription start site and the stop site are transcribed into pre-mRNA, including exons, introns, and a long 3′ end that is later cleaved from the transcript (◀FIGURE 14.19b). Notice that the 5′ end of the first exon contains the sequence that codes for the 5′ untranslated region, and the 3′ end of the last exon contains the sequence that codes for the 3′ untranslated region.

The pre-mRNA is then processed to yield a mature mRNA. The first step in this processing is the addition of a cap to the 5′ end of the pre-mRNA (◀FIGURE 14.19c). Next, the 3′ end is cleaved at a site downstream of the AAUAAA consensus sequence in the last exon (◀FIGURE 14.19d). Immediately after cleavage, a poly(A) tail is added to the 3′ end (◀FIGURE 14.19e). Finally, the introns are removed to yield the mature mRNA (◀FIGURE 14.19f). The mRNA now contains 5′ and 3′ untranslated regions, which are not translated into amino acids, and the nucleotides that carry the protein-coding sequences. The nucleotide sequence of a small gene, with these components labeled, is presented in (◀FIGURE 14.20).

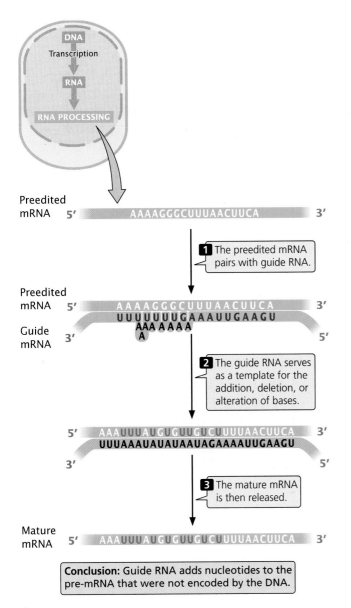

Conclusion: Guide RNA adds nucleotides to the pre-mRNA that were not encoded by the DNA.

◀14.18 **RNA editing is carried out by guide RNAs.**

In other cases, enzymes bring about base conversion. In humans, for example, a gene is transcribed into mRNA that codes for a lipid-transporting polypeptide called apolipoprotein-B100, which has 4563 amino acids and is synthesized in liver cells. A truncated form of the protein called apolipoprotein-B48—with only 2153 amino acids—is synthesized in intestinal cells. The truncated protein is produced from an edited version of the same mRNA that codes for apolipoprotein-B100. In editing, an enzyme deaminates a cytosine base, converting it into uracil. This conversion changes a codon that specifies the amino acid glutamine into a stop codon that prematurely terminates translation, resulting in the shortened protein.

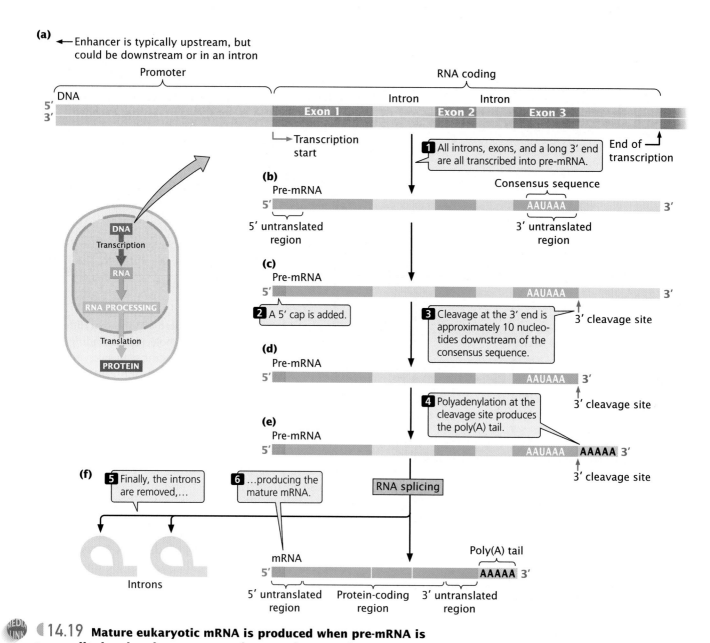

◀14.19 **Mature eukaryotic mRNA is produced when pre-mRNA is transcribed and undergoes several types of processing.**

Transfer RNA

In 1956, Francis Crick proposed the idea of a molecule that transported amino acids to the ribosome and interacted with codons in mRNA, placing amino acids in their proper order in protein synthesis. By 1963, the existence of such an adapter molecule, called transfer RNA, had been confirmed. Transfer RNA (tRNA) serves as a link between the genetic code in mRNA and the amino acids that make up a protein. Each tRNA attaches to a particular amino acid and carries it to the ribosome, where the tRNA adds its amino acid to the growing polypeptide chain at the position specified by the genetic instructions in the mRNA.

We'll take a closer look at the mechanism of this process in Chapter 15.

Each tRNA is capable of attaching to only one type of amino acid. The complex of tRNA plus its amino acid can be written in abbreviated form by adding a three-letter superscript representing the amino acid to the term tRNA. For example, a tRNA that attaches to the amino acid alanine is written as tRNA[Ala]. Because 20 different amino acids are found in proteins, there must be a minimum of 20 different types of tRNA. In fact, most organisms possess from at least 30 to 40 different types of tRNA, each encoded by a different gene (or, in some cases, multiple copies of a gene) in DNA.

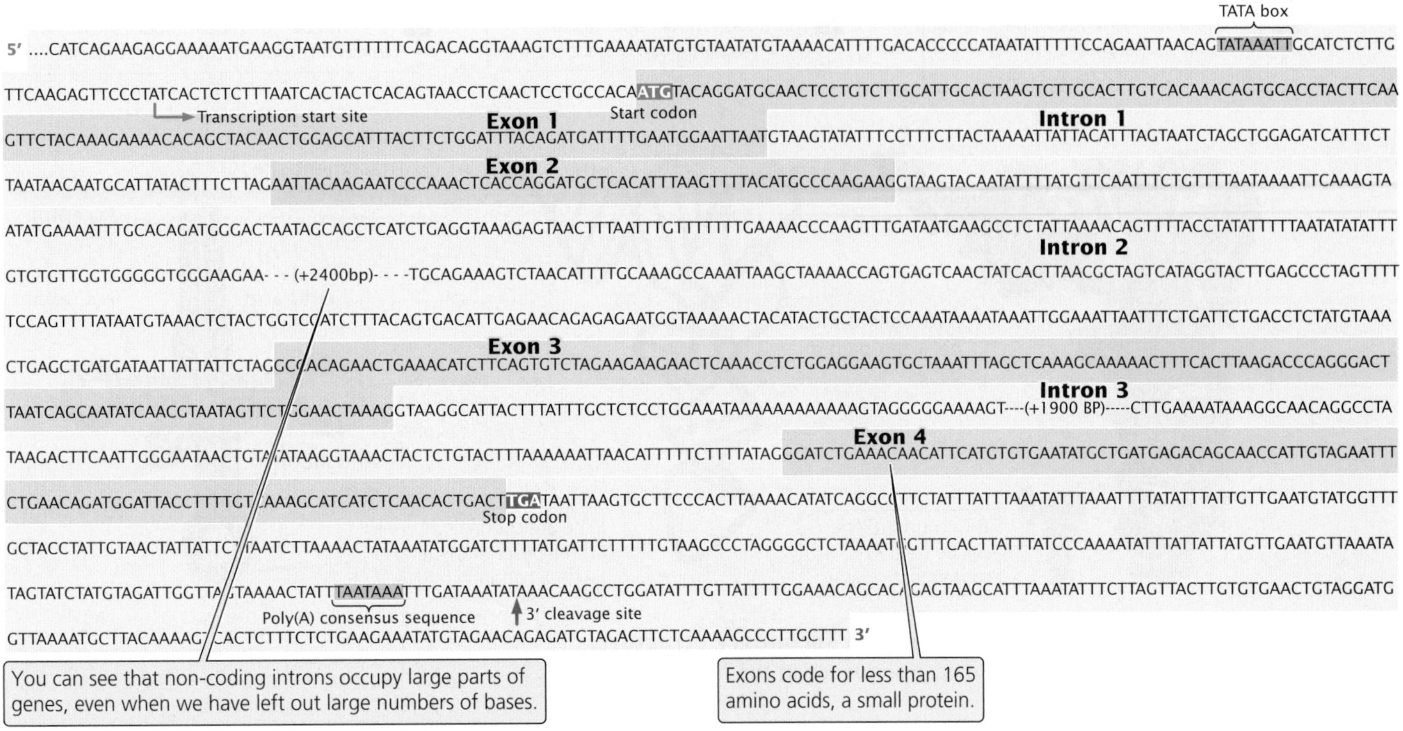

TATA box

5′ ....CATCAGAAGAGGAAAAATGAAGGTAATGTTTTTTCAGACAGGTAAAGTCTTTGAAAATATGTGTAATATGTAAAACATTTTGACACCCCCATAATATTTTTCCAGAATTAACAGTATAAATTGCATCTCTTG

TTCAAGAGTTCCCTATCACTCTCTTTAATCACTACTCACAGTAACCTCAACTCCTGCCACAATGTACAGGATGCAACTCCTGTCTTGCATTGCACTAAGTCTTGCACTTGTCACAAACAGTGCACCTACTTCAA
└──► Transcription start site   **Exon 1**   Start codon
GTTCTACAAAGAAAACACAGCTACAACTGGAGCATTTACTTCTGGATTTACAGATGATTTTGAATGGAATTAATGTAAGTATATTTCCTTTCTTACTAAAATTATTACATTTAGTAATCTAGCTGGAGATCATTTCT
                                                                                                              Intron 1

Exon 2
TAATAACAATGCATTATACTTTCTTAGAATTACAAGAATCCCAAACTCACCAGGATGCTCACATTTAAGTTTTACATGCCCAAGAAGGTAAGTACAATATTTTATGTTCAATTTCTGTTTTAATAAAATTCAAAGTA

ATATGAAAATTTGCACAGATGGGACTAATAGCAGCTCATCTGAGGTAAAGAGTAACTTTAATTTGTTTTTTTGAAAACCCAAGTTTGATAATGAAGCCTCTATTAAAACAGTTTTACCTATATTTTTAATATATATTT
                                                                                                    Intron 2

GTGTGTTGGTGGGGGTGGGAAGAA- - - (+2400bp)- - - -TGCAGAAAGTCTAACATTTTGCAAAGCCAAATTAAGCTAAAACCAGTGAGTCAACTATCACTTAACGCTAGTCATAGGTACTTGAGCCCTAGTTTT

TCCAGTTTTATAATGTAAACTCTACTGGTCCATCTTTACAGTGACATTGAGAACAGAGAGAATGGTAAAAACTACATACTGCTACTCCAAATAAAATAAATTGGAAATTAATTTCTGATTCTGACCTCTATGTAAA

Exon 3
CTGAGCTGATGATAATTATTATTCTAGGCCACAGAACTGAAACATCTTCAGTGTCTAGAAGAAGAACTCAAACCTCTGGAGGAAGTGCTAAATTTAGCTCAAAGCAAAAACTTTCACTTAAGACCCAGGGACT

                                                                                        Intron 3
TAATCAGCAATATCAACGTAATAGTTCTGGAACTAAAGGTAAGGCATTACTTTATTTGCTCTCCTGGAAATAAAAAAAAAAAAAGTAGGGGGAAAAGT----(+1900 BP)-----CTTGAAAATAAAGGCAACAGGCCTA

                                                                    Exon 4
TAAGACTTCAATTGGGAATAACTGTAATAAGGTAAACTACTCTGTACTTTAAAAAAATTAACATTTTTCTTTTATAGGGATCTGAAACAACATTCATGTGTGAATATGCTGATGAGACAGCAACCATTGTAGAATTT

CTGAACAGATGGATTACCTTTTGTAAAAGCATCATCTCAACACTGACTTGATAATTAAGTGCTTCCCACTTAAAACATATCAGGCCTTCTATTTATTTAAATATTTAAATTTTATATTTATTGTTGAATGTATGGTTT
                                    Stop codon

GCTACCTATTGTAACTATTATTCTAATCTTAAAAACTATAAATATGGATCTTTTATGATTCTTTTTGTAAGCCCTAGGGGGCTCTAAAATGTTTCACTTATTTATCCCAAAATATTTATTATTATGTTGAATGTTAAATA

TAGTATCTATGTAGATTGGTTAGTAAAACTATTTAATAAATTTGATAAATATAAACAAGCCTGGATATTTGTTATTTTGGAAACAGCACAGAGTAAGCATTTAAATATTTCTTAGTTACTTGTGTGAACTGTAGGATG
                    Poly(A) consensus sequence   ↑3′ cleavage site
GTTAAAATGCTTACAAAAGTCACTCTTTCTCTGAAGAAATATGTAGAACAGAGATGTAGACTTCTCAAAAGCCCTTGCTTT 3′

┌───┐
│ You can see that non-coding introns occupy large parts of │
│ genes, even when we have left out large numbers of bases. │
└───┘

┌────────────────────────────┐
│ Exons code for less than 165 │
│ amino acids, a small protein. │
└────────────────────────────┘

◄14.20 **This representation of the nucleotide sequence of the human interleukin 2 gene includes the TATA box, transcription start site, start and stop codons, introns, exons, poly(A) consensus sequence, and the 3′ cleavage site.**

The Structure of Transfer RNA

A unique feature of tRNA is the occurrence of rare, **modified bases**. All RNAs have the four standard bases (adenine, cytosine, guanine, and uracil) specified by DNA, but tRNAs have additional bases, including ribothymine, pseudourasil (which is also occasionally present in snRNAs and rRNA), and dozens of others. The structures of some of these modified bases are shown in (◄FIGURE 14.21).

If there are only four bases in DNA, and all RNA molecules are transcribed from DNA, how do tRNAs acquire these additional bases? Modified bases arise from chemical changes made to the four standard bases after transcription. These changes are carried out by special **tRNA-modifying enzymes**. For example, the addition of a methyl group to uracil creates the modified base ribothymine.

The structures of all tRNAs are similar, a feature critical to tRNA function. Most tRNAs contain between 74 and 95 nucleotides, some of which are complementary to each other and form intramolecular hydrogen bonds. As a result, each tRNA has a **cloverleaf structure** (◄FIGURE 14.22). The cloverleaf has four major arms, each consisting of a stem and a loop. The stem is formed by the paring of complementary nucleotides, and the loop lies at the terminus of the stem, where there is no nucleotide pairing. If we start at

the top and proceed clockwise around the tRNA shown at the right in Figure 14.22, the four major arms are the acceptor arm, the TψC arm, the anticodon arm, and the DHU arm.

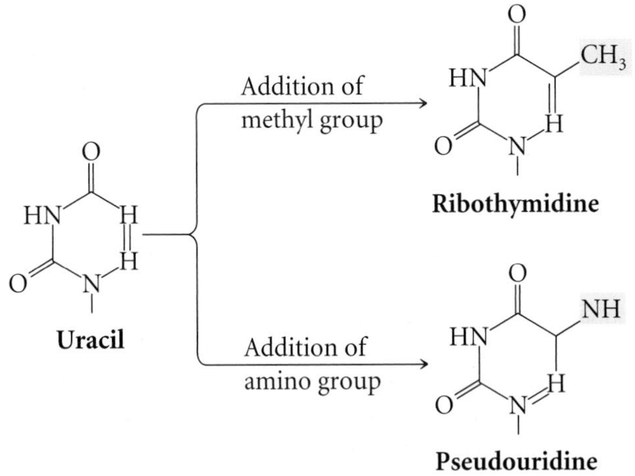

Uracil → Addition of methyl group → Ribothymidine
Uracil → Addition of amino group → Pseudouridine

◄14.21 **Modified bases are found in tRNAs.** All the modified bases are produced by the chemical alteration of the four standard RNA bases.

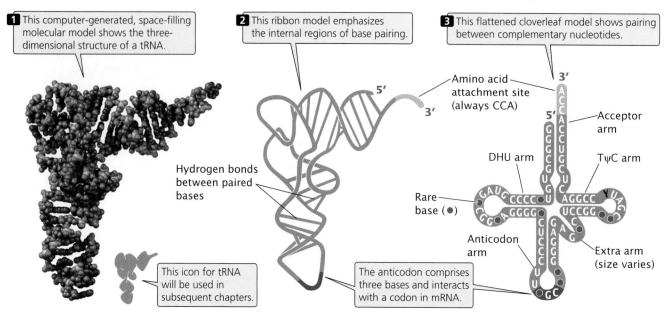

◀14.22 All tRNAs possess a common secondary structure, the cloverleaf structure. The base sequence in the flattened model is for tRNA^Ala.

The acceptor arm has no loop but contains the 5′ and 3′ ends of the tRNA molecule. All tRNAs have the same sequence (CCA) at the 3′ end, where the amino acid attaches to the tRNA; so clearly this sequence is not responsible for specifying which amino acid will attach to the tRNA.

The TψC arm is named for the bases of three nucleotides in the loop of this arm: thymine (T), pseudouracil (ψ), and cytosine (C). The anticodon arm lies at the bottom of the tRNA. Three nucleotides at the end of this arm make up the **anticodon,** which pairs with the corresponding codon on mRNA to ensure that the amino acids link in the correct order. The DHU arm is so named because it often contains the modified base dihydrouridine.

Although each tRNA molecule folds into a cloverleaf owing to the complementary paring of bases, the cloverleaf is not the three-dimensional (tertiary) structure of tRNAs found in the cell. The results of X-ray crystallographic studies have shown that the cloverleaf folds upon itself to from an L-shaped structure, as illustrated by the space-filling and ribbon models in Figure 14.22. Notice that the acceptor stem is at one end of the tertiary structure and the anticodon is at the other end.

Transfer RNA Gene Structure and Processing

The genes that produce tRNAs may be scattered about the genome or may be in clusters. In *E. coli*, the genes for some tRNAs are present in a single copy, whereas the genes for other tRNAs are present in several copies; eukaryotic cells usually have many copies of each tRNA gene. All tRNA molecules in both bacterial and eukaryotic cells undergo processing after transcription.

In *E. coli*, several tRNAs are usually transcribed together as one large precursor tRNA, which is then cut up into pieces, each containing a single tRNA. Additional nucleotides may then be removed one at a time from the 5′ and 3′ ends of the tRNA in a process known as trimming. Base-modifying enzymes may then change some of the standard bases into modified bases, and additional bases (such as CCA at the 3′ end) may be added (◀FIGURE **14.23**). Different tRNAs are processed in different ways; so it is not possible to outline a generic processing pathway for all tRNAs. Eukaryotic tRNAs are processed in a manner similar to that for bacterial tRNAs: most are transcribed as larger precursors that are then cleaved, trimmed, and modified to produce mature tRNAs.

Some eukaryotic tRNA genes possess introns of variable length that must be removed in processing. For example, about 40 of the 400 tRNA genes in yeast contain a single intron that is always found adjacent to the 3′ side of the anticodon. The splicing process for tRNA genes (see Figure 14.23) is quite different from the spliceosome-mediated reactions that remove introns from protein-encoding genes. The intron in the precursor tRNA is cut at both ends by an endonuclease enzyme, which releases the linear intron from the rest of the tRNA. The two pieces of tRNA, which are held together by intramolecular bonding, are then folded and ligated to produce the mature tRNA.

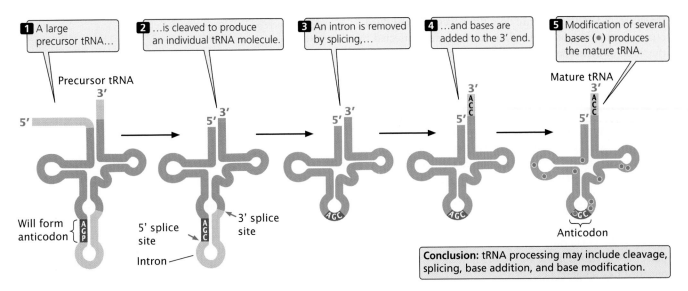

1 A large precursor tRNA...

2 ...is cleaved to produce an individual tRNA molecule.

3 An intron is removed by splicing,...

4 ...and bases are added to the 3' end.

5 Modification of several bases (●) produces the mature tRNA.

Precursor tRNA

Mature tRNA

Will form anticodon

5' splice site

3' splice site

Intron

Anticodon

Conclusion: tRNA processing may include cleavage, splicing, base addition, and base modification.

◀ **14.23 Transfer RNAs are processed in both bacterial and eukaryotic cells.** Different tRNAs are modified in different ways. One example is shown here.

Concepts

All tRNAs are similar in size and have a common secondary structure known as the cloverleaf. Transfer RNAs contain modified bases and are extensively processed after transcription in both bacterial and eukaryotic cells.

Ribosomal RNA

Within ribosomes, the genetic instructions contained in mRNA are translated into the amino acid sequences of polypeptides. Thus, ribosomes play an integral part in the transfer of genetic information from genotype to phenotype. We will examine the role of ribosomes in the process of translation in Chapter 15. Here, we will consider ribosome structure and examine how ribosomes are processed before becoming functional.

The Structure of the Ribosome

The ribosome is one of the most abundant organelles in the cell: a single bacterial cell may contain as many as 20,000 ribosomes, and eukaryotic cells possess even more. Ribosomes typically contain about 80% of the total cellular RNA. They are complex organelles, each consisting of more than 50 different proteins and RNA molecules (Table 14.3). A functional ribosome consists of two subunits, a **large subunit** and a **small subunit,** each of which consists of one or more pieces of RNA and a number of proteins. The sizes of the ribosomes and their RNA components are given in Svedberg (S) units (a measure of how rapidly an object sediments in a centrifugal field). It is important to note that Svedberg units are not additive; in other words, combining a 10S

Table 14.3	Composition of ribosomes in bacterial and eukaryotic cells			
Cell Type	**Ribosome Size**	**Subunit**	**rRNA Component**	**Proteins**
Bacterial	70S	Large (50S)	23S (2900 nucleotides) 5S (120 nucleotides)	31
		Small (30S)	16S (1500 nucleotides)	21
Eukaryotic	80S	Large (60S)	28S (4700 nucleotides) 5.8S (160 nucleotides) 5S (120 nucleotides)	49
		Small (40S)	18S (1900 nucleotides)	33

Note: The letter S stands for Svedberg unit.

structure and a 20S structure does not necessarily produce a 30S structure, because the sedimentation rate is affected by the three-dimensional structure as well as the mass.

Ribosomal RNA Gene Structure and Processing

The genes for rRNA, like those for tRNA, can be present in multiple copies, and the numbers vary among species (Table 14.4); all copies of the rRNA gene in a species are identical or nearly identical. In bacteria, rRNA genes are dispersed, but, in eukaryotic cells, they are clustered, with the genes arrayed in tandem, one after another.

Eukaryotic cells possess two types of rRNA genes: a large one that encodes 18S rRNA, 28S rRNA, and 5.8S rRNA, and a small one that encodes the 5S rRNA. All three bacterial rRNAs (23S rRNA, 16S rRNA, and 5S rRNA) are encoded by a single type of gene.

Ribosomal RNA is processed in both bacterial and eukaryotic cells. In *E. coli*, the immediate product of transcription is a 30S rRNA precursor (◀FIGURE 14.24a). Methyl groups (CH_3) are added to specific bases and to the 2′ carbon of some of the ribose sugars of this 30S precursor, which is then cleaved into several pieces and trimmed to produce 16S rRNA, 23S rRNA, and 5S rRNA, along with one or more tRNAs. All rRNA genes in *E. coli* produce the same three rRNA molecules, but the number and location of these rRNAs within the 30S rRNA transcript differ among genes.

Eukaryotic rRNAs undergo similar processing (◀FIGURE 14.24b). Small nucleolar RNAs help to cleave and

Table 14.4	Number of rRNA genes in different organisms	
Species	**Copies of rRNA Genes or Genome**	
Escherichia coli	1	
Yeast	100–200	
Human	280	
Frog	450	

modify eukaryotic rRNAs (as well as some archaeal rRNAs), and help to assemble the processed rRNAs into mature ribosomes. The snoRNAs have extensive complementarity to the rRNA sequences where modification takes place. Interestingly, some snoRNAs are encoded by sequences in the introns of other protein-encoding genes.

Concepts

A ribosome is a complex organelle consisting of several rRNA molecules and many proteins. Each functional ribosome consists of a large and a small subunit. rRNAs in both bacterial and eukaryotic cells are modified after transcription. In eukaryotes, rRNA processing is carried out by small nucleolar RNAs (snoRNAs).

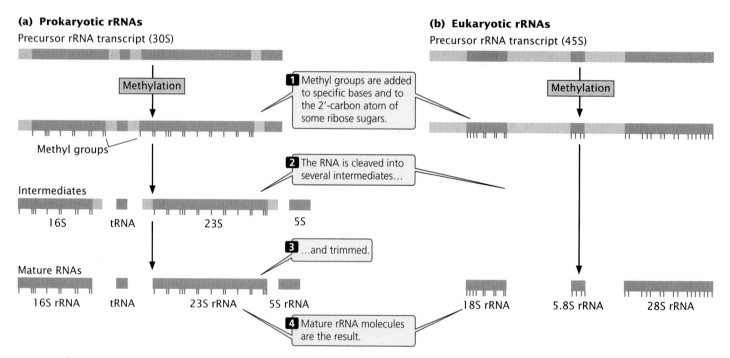

(a) Prokaryotic rRNAs
Precursor rRNA transcript (30S)

Methylation

1 Methyl groups are added to specific bases and to the 2′-carbon atom of some ribose sugars.

Methyl groups

2 The RNA is cleaved into several intermediates…

Intermediates
16S   tRNA   23S   5S

3 …and trimmed.

Mature RNAs
16S rRNA   tRNA   23S rRNA   5S rRNA

4 Mature rRNA molecules are the result.

(b) Eukaryotic rRNAs
Precursor rRNA transcript (45S)

Methylation

18S rRNA   5.8S rRNA   28S rRNA

◀14.24 **Ribosomal RNA is processed after transcription.** Note that eukaryotic 5S rRNA is transcribed separately from the small eukaryotic rRNA gene.

Connecting Concepts Across Chapters

Because it is single stranded and can form hydrogen bonds between complementary bases on the same strand, RNA is capable of assuming a number of secondary structures. This ability gives RNA functional flexibility, and it assumes a number of important roles in information transfer within the cell.

A central theme in this chapter has been the nature of the gene. The concept of a gene has changed with time and even today depends on the particular question that is being addressed. A modern definition used by many geneticists is: a gene is a sequence of nucleotides in DNA that is transcribed into a single RNA molecule.

The details of RNA function and processing covered in this chapter are important for understanding the process of protein synthesis, which is the focus of Chapter 15. Knowledge of the structure of the ribosome and tRNAs will be important for understanding how amino acids are assembled into a protein. In eukaryotic cells, features [such as the 5′ cap and the poly(A) tail] that are added to pre-mRNA and those removed (introns) from it are essential for translation to proceed properly. These features of processed mRNA also play an important role in eukaryotic gene regulation, a subject to be addressed in Chapter 16.

CONCEPTS SUMMARY

- The discovery of introns in eukaryotic genes forced the redefinition of the gene at the molecular level. Today, a gene is often defined as a sequence of DNA nucleotides that is transcribed into a single RNA molecule.

- Introns are noncoding sequences that interrupt the coding sequences (exons) of genes. Common in eukaryotic cells but rare in bacterial cells, introns exist in all types of genes and vary in size and number. They comprise four major types: group I introns, group II introns, nuclear pre-mRNA introns, and tRNA introns.

- The results of experiments in the late 1950s and early 1960s suggested that genetic information is carried from DNA to ribosomes by short-lived RNA molecules called messenger RNA. An mRNA molecule has three primary parts: a 5′ untranslated region, a protein-coding sequence, and a 3′ untranslated region.

- Bacterial mRNA is translated immediately after transcription and undergoes little processing.

- The primary transcript (pre-mRNA) of a eukaryotic protein-encoding gene is extensively processed: a modified nucleotide and methyl groups, collectively termed the cap, are added to the 5′ end of pre-mRNA; the 3′ end is cleaved and a poly(A) tail is added; and introns are removed.

- The process of RNA splicing takes place within a structure called the spliceosome, which is composed of several small nuclear RNAs and proteins. RNA splicing takes place in a two-step process that entails RNA–RNA interactions among snRNAs of the spliceosome and the pre-mRNA.

- Some introns found in rRNA genes and mitochondrial genes are self-splicing.

- Some pre-mRNAs undergo alternative splicing, in which different combinations of exons are spliced together or different 3′ cleavage sites are used.

- Messenger RNAs may also be altered by the addition, deletion, or modification of nucleotides in the coding sequence, a process called RNA editing.

- Transfer RNA serves as a bridge between amino acids and the genetic information carried in mRNA. Transfer RNAs are relatively short molecules that assume a common secondary structure and contain modified bases. Most organisms have multiple copies of tRNA genes; the tRNAs transcribed from these genes are extensively processed in bacterial and eukaryotic cells.

- Ribosomes are the sites of protein synthesis in the cell. Each ribosome is composed of several rRNA molecules and a number of proteins that form a large and a small subunit. Genes for rRNA exist in multiple copies; the primary transcripts from these genes are extensively modified after transcription in bacterial and eukaryotic cells. In eukaryotic cells, rRNA processing is carried out by small nucleolar RNAs.

IMPORTANT TERMS

colinearity (p. 379)
exon (p. 380)
intron (p. 380)
group I intron (p. 381)
group II intron (p. 381)
nuclear pre-mRNA intron
(p. 381)
tRNA intron (p. 381)
codon (p. 383)
5' untranslated region (p. 383)

Shine-Dalgarno sequence
(p. 383)
protein-coding region (p. 383)
3' untranslated region (p. 383)
5' cap (p. 383)
poly(A) tail (p. 384)
RNA splicing (p. 385)
5' splice site (p. 385)
3' splice site (p. 385)
branch point (p. 385)

spliceosome (p. 386)
lariat (p. 387)
transesterification (p. 387)
alternative processing pathway
(p. 390)
alternative splicing (p. 390)
multiple 3' cleavage site
(p. 390)
RNA editing (p. 392)
guide RNA (p. 392)

modified base (p. 395)
tRNA-modifying enzyme
(p. 395)
cloverleaf structure (p. 395)
anticodon (p. 396)
large ribosomal subunit
(p. 397)
small ribosomal subunit
(p. 397)

Worked Problems

1. DNA from a eukaryotic gene was isolated, denatured, and hybridized to the mRNA transcribed from the gene; the hybridized structure was then observed with the use of an electron microscope. The following structure was observed.

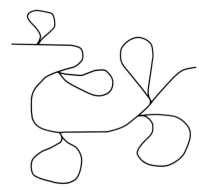

(a) How many introns and exons are there in this gene? Explain your answer.

(b) Identify the exons and introns in this hybridized structure.

• Solution

(a) Each of the loops represents a region where there are sequences in the DNA that do not have corresponding sequences in the RNA; these regions are introns. There are five loops in the hybridized structure; so there must be five introns in the DNA.

(b)

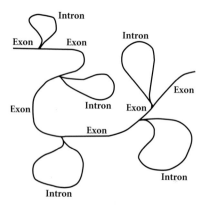

2. Draw a typical bacterial mRNA and the gene from which it was transcribed. Label the 5' and 3' ends of the RNA and DNA molecules, and identify the following regions or sequences:

(a) Promoter   (e) Transcription start site

(b) 5' untranslated region   (f) Terminator

(c) 3' untranslated region   (g) Shine-Dalgarno sequence

(d) Protein-coding sequence

• Solution

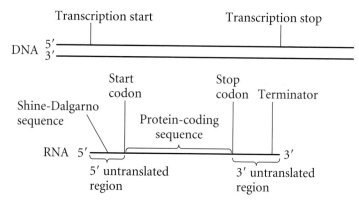

3. A test-tube splicing system has been developed that contains all the components (snRNAs, proteins, splicing factors) necessary for the splicing of nuclear genes. When a piece of RNA containing an intron and two exons is added to the system, the intron is removed as a lariat and the exons are spliced together. If the RNA molecule added to the system has the following mutations, what intermediate products of the splicing reactions will accumulate? Explain your answer.

(a) GT at the 5' splice site is deleted.

(b) A at the branch point is deleted.

(c) AG at the 3' splice site is deleted.

• Solution

(a) The GT sequence at the 5' splice site is required for the attachment of the U1 snRNP and the first cleavage reaction. If this sequence is mutated, cleavage will not take place. Thus, the original pre-mRNA with the intron will accumulate.

(b) After cleavage at the 5′ splice site, the 5′ end of the intron attaches to the A at the branch point in a transesterification reaction. If the A at the branch point is deleted, no lariat structure will form. The separated first exon and the intron attached to the second exon will accumulate as intermediate products.

(c) The AG sequence at the 3′ splice site is required for cleavage at the 3′ splice site. If this sequence is mutated, accumulated intermediate products will be: (1) the separated first exon and (2) the intron attached to the second exon, with the 5′ end of the intron attached to the branch point to form a lariat structure.

The New Genetics
MINING GENOMES

RIBOSOMAL RNA STUDIES

Ribosomal RNA is the most plentiful nucleic acid in cells and is widely exploited for a variety of genetic studies. This exercise introduces you to some of the ways that ribosomal RNA

sequences are used and to the collection of tools at the Ribosomal Database Project II, managed by Michigan State University's Center for Microbial Ecology.

COMPREHENSION QUESTIONS

* 1. What is the concept of colinearity? In what way is this concept fulfilled in bacterial and eukaryotic cells?
  2. What are some characteristics of introns?
* 3. What are the four basic types of introns? In which genes are they found?
* 4. What are the three principal elements in mRNA sequences in bacterial cells?
  5. What is the function of the Shine-Dalgarno consensus sequence?
* 6. (a) What is the 5′ cap? (b) How is the 5′ cap added to eukaryotic pre-mRNA? (c) What is the function of the 5′ cap?
  7. How is the poly(A) tail added to pre-mRNA? What is the purpose of the poly(A) tail?

* 8. What makes up the spliceosome? What is the function of the spliceosome?
  9. Explain the process of pre-mRNA splicing in nuclear genes.
 10. Describe two types of alternative processing pathways. How do they lead to the production of multiple proteins from a single gene?
*11. What is RNA editing? Explain the role of guide RNAs in RNA editing.
*12. Summarize the different types of processing that can take place in pre-mRNA.
*13. What are some of the modifications in tRNA processing?
 14. Describe the basic structure of ribosomes in bacterial and eukaryotic cells
*15. Explain how rRNA is processed.

APPLICATION QUESTIONS AND PROBLEMS

*16. At the beginning of the chapter, we considered Duchenne muscular dystrophy and the dystrophin gene. We learned that the gene causing Duchenne muscular dystrophy encompasses more than 2 million nucleotides, but less than 1% of the gene encodes the protein dystrophin. On the basis of what you now know about gene structure and RNA processing in eukaryotic cells, provide a possible explanation for the large size of the dystrophin gene.

17. How do the mRNA of bacterial cells and the pre-mRNA of eukaryotic cells differ? How do the mature mRNAs of bacterial and eukaryotic cells differ?

*18. Draw a typical eukaryotic gene and the pre-mRNA and mRNA derived from it. Assume that the gene contains three exons. Identify the following items and, for each item, give a brief description of its function:

(a) 5′ untranslated region   (e) 3′ untranslated region
(b) Promoter   (f) Introns
(c) AAUAAA consensus   (g) Exons
    sequence   (h) Poly(A) tail
(d) Transcription start site   (i) 5′ cap

19. How would the deletion of the Shine-Dalgarno sequence affect a bacterial mRNA?

*20. How would the deletion of the following sequences or features most likely affect a eukaryotic pre-mRNA?

(a) AAUAAA consensus sequence
(b) 5′ cap
(c) Poly(A) tail

21. What would be the most likely effect on the amino acid sequence of a protein of a mutation that occurred in an intron of the gene encoding the protein? Explain your answer.

22. A geneticist induces a mutation in the gene that codes for cleavage and polyadenylation specificity factor (CPSF) in a line of cells growing in the laboratory. What would be the immediate effect of this mutation on RNA molecules in the cultured cells?

*23. A geneticist mutates the gene for proteins that bind to the poly(A) tail in a line of cells growing in the laboratory. What would be the immediate effect of this mutation in the cultured cells?

*24. An in vitro (within a test tube) splicing system has been developed that contains all the components (snRNAs, proteins, splicing factors) necessary for the splicing of nuclear pre-mRNA genes. When a piece of RNA containing an intron and two exons is added to the system, the intron is removed as a lariat and the exons are spliced together. What intermediate products of the splicing reaction would accumulate if the following components were omitted from the splicing system? Explain your reasoning.

(a) U1   (c) U6   (e) U4
(b) U2   (d) U5

25. The splicing system introduced in Problem 24 is used to splice an RNA molecule containing two exons and one intron. This time, however, the U2 snRNA used in the splicing reaction contains several mutations in the sequence that pairs with the U6 snRNA. What would be the effect of these mutations on the splicing process?

26. A geneticist isolates a gene that contains five exons. He then isolates the mature mRNA produced by this gene. After making the DNA single stranded, he mixes the single-stranded DNA and RNA. Some of the single-stranded DNA hybridizes (pairs) with the complementary mRNA. Draw a picture of what the DNA–RNA hybrids would look like under the electron microscope.

27. The chemical reagent psoralen can be used to elucidate nucleic acid structure. This chemical attaches itself to nucleic acids and, on exposure to UV light, forms covalent bonds between closely associated nucleotide sequences. Such cross-links provide information about the proximity of RNA molecules to one another in complex structures.

Psoralen cross-linking has been used to examine the structure of the spliceosome. In one study, the following cross-linked structures were obtained during splicing. U1, U2, U5, and U6 became cross-linked to pre-mRNA. U2 was cross-linked to U6 and to pre-mRNA. The U1, U5, and U6 cross-links with pre-mRNA were mapped to sequences near the 5′ splice site, whereas the U2 snRNA cross-links with pre-mRNA were mapped to the branch site. After splicing, U2, U5, and U6 were cross-linked to the excised lariat.

Explain these results in regard to what is known about the structure of the spliceosome and how it functions in RNA splicing. (Based on D. A. Wassarman, and J. A. Steitz, 1992, Interactions of small nuclear RNAs with precursor messenger RNA during in vitro splicing, *Science* 257:1918–1925.)

CHALLENGE QUESTIONS

28. In addition to snRNAs, the spliceosome contains a number of proteins. Some of these proteins are associated with the snRNAs to form snRNPs. Other proteins are associated with the spliceosome but are not associated with any specific snRNA.

One group of spliceosomal proteins comprises the precursor RNA-processing (PRP) proteins. Three PRP proteins that directly take part in splicing are PRP2, PRP16, and PRP22. The results of studies have shown that PRP2 is required for the first step of the splicing reaction, PRP16 acts at the second step, and PRP22 is required for the release of the mRNA from the spliceosome. Other studies have found that these PRP proteins have amino acid sequences similar to the sequences found in RNA helicase enzymes—enzymes that are capable of unwinding two paired RNA molecules. On the basis of this information, propose a functional role for PRP2, PRP16, and PRP22 in RNA splicing.

29. Propose a scenario by which spliceosomal-mediated splicing might have evolved from the splicing of group II introns.

SUGGESTED READINGS

Bjork, G. R., J. U. Erikson, C. E. D. Gustafsson, T. G. Hagervall, Y. H. Jonsson, and P. M. Wikstrom. 1987. Transfer RNA modification. *Annual Review of Biochemistry* 56:263–288.
A review of how tRNA is processed.

Broker, T. R., L. T. Chow, A. R. Dunn, R. E. Gelinas, J. A. Hassel, D. F. Klessig, J. B. Lewis, R. J. Roberts, and B. S. Zain. 1978. Adenovirus-2 messengers: an example of baroque molecular architecture. *Cold Spring Harbor Symposium on Quantatative Biology* 42:531–534.
One of the first reports of introns in eukaryotic genes.

Gott, J. M., and R. B. Emerson. 2000. Functions and mechanisms of RNA editing. *Annual Review of Genetics* 34:499–531.
An extensive review of the different types of RNA editing and their mechanisms.

Hurst, L. D. 1994. The uncertain origin of introns. *Nature* 371:381–382.
A discussion of some of the ideas about when and how introns first arose.

Keller, W. 1995. No end yet to messenger RNA 3′ processing. *Cell* 81:829–832.
An excellent review of processing at the 3′ end of eukaryotic pre-mRNA.

Lake, J. A. 1981. The ribosome. *Scientific American* 245(2):84–97.
A review of the structure of ribosomes.

Landweber, L. F., P. J. Simon, and T. A. Wagner. 1998. Ribozyme engineering and early evolution. *Bioscience* 48:94–103.
A nice review of the idea that early life may have consisted of an RNA world.

McKeown, M. 1992. Alternative mRNA splicing. *Annual Review of Cell Biology* 8:133–155.
An extensive review that discusses the different types of alternative splicing with specific examples of each type.

Misteli, T., J. F. Caceres, and D. L. Spector. 1997. The dynamics of a pre-mRNA splicing factor in living cells. *Nature* 387:523–527.
Reports that pre-mRNA splicing and transcription take place at the same sites in the nucleus.

Nilsen, T. W. 1994. RNA–RNA interactions in the spliceosome: unraveling the ties that bind. *Cell* 78: 1–4.
An excellent summary of how RNA–RNA interactions play an important role in the splicing of nuclear pre-mRNAs.

Noller, H. F. 1984. Structure of ribosomal RNA. *Annual Review of Biochemistry* 53:119–162.
A review of rRNA.

Rich, A., and S. H. Kim. 1978. The three-dimensional structure of transfer RNA. *Scientific American* 238(1):52–62.
Discusses the structure of tRNA.

Scott, J. 1995. A place in the world for RNA editing. *Cell* 81:833–836.
A good, succinct review of RNA editing.

Smith, C. M., and J. A. Steitz. 1997. Sno storm in the nucleolus: new roles for myriad small RNPs. *Cell* 89:669–672.
A good review of the role of snoRNAs in rRNA processing.

Volkin, E., and L. Astrachan. 1956. Phosphorous incorporation in *Escherichia coli* ribonucleic acid after infections with bacteriophage T2. *Virology* 2:149–161.
Reports the discovery of short-lived RNA after phage infection.

15 The Genetic Code and Translation

Uric acid in polarized light. Large amounts of uric acid are produced in the blood and urine of people who have Lesch-Nyhan disease, and X-linked disorder of purine metabolism. (Phil A. Harrington/Peter Arnold.)

Lesch-Nyhan Syndrome and the Relation Between Genotype and Phenotype

In 1962, Dr. William Nyhan and his student Michael Lesch examined a seriously ill boy with a strange combination of symptoms. The boy had blood in his urine, high concentrations of uric acid in his blood, and uncontrollable spasms in his arms and legs. He was mentally retarded and self-destructively bit his fingers and lips. After carefully studying the boy, Nyhan and Lesch came to the conclusion that he was afflicted by an undescribed disease. Soon other patients with similar symptoms were reported, and the disease became known as the Lesch-Nyhan syndrome.

One of the earliest symptoms of Lesche-Nyhan syndrome is the appearance of orange "sand"—actually uric acid crystals—in diapers a few weeks after birth. Within a year, the child begins to exhibit writhing movements of the hands and feet and involuntary spasms. About half of the children have seizures. After 2 or 3 years, some of the children exhibit compulsive self-mutilation—biting fingers, lips, the tongue, and the insides of the mouth.

Lesch-Nyhan syndrome develops almost exclusively in boys, and the trait is inherited as an X-linked recessive disorder; the presence of a defective gene on a male's single X chromosome causes the disease. In 1967, a team of scientists at the National Institutes of Health determined that the disease results from a defective copy of the gene that normally encodes the enzyme hypoxanthine-guanine phosphoribosyl transferase (HGPRT). As DNA and RNA are degraded in the cell, purines are liberated, and HGPRT salvages these purines and uses them to synthesize new RNA and DNA nucleotides. In people who have Lesch-Nyhan syndrome, a mutation in the gene for HGPRT changes the amino acid sequence of the enzyme, rendering it nonfunctional. The result is that purines are not recycled; they accumulate and are converted into uric acid. High levels of uric acid produce the symptoms of the disease.

Lesch-Nyhan syndrome illustrates the link between genotype and phenotype: a mutation in a gene affects a protein, which then produces the symptoms of the disease. Preceding chapters described how DNA encodes genetic information and how that information is transferred from DNA to RNA. In this chapter, we examine translation, the process by which the nucleotide sequence in mRNA specifies the amino acid sequence of a protein.

We begin by examining the molecular relation between genotype and phenotype. As with Lesch-Nyhan syndrome, the final phenotype may be complex, including biochemical, physiological, and behavioral traits, but it is ultimately caused by the protein that the gene encodes. We next study the genetic code—the instructions that specify the amino acid sequence of a protein—and then examine the mechanism of protein synthesis. Our primary focus will be on protein synthesis in bacterial cells, but we will highlight some of the differences in eukaryotic cells.

www.whfreeman.com/pierce   More information on Lesch-Nyhan syndrome

The Molecular Relation Between Genotype and Phenotype

The first person to suggest the existence of a relation between genotype and proteins was Archibald Garrod. In 1908, Garrod correctly proposed that genes encode enzymes (see pp. 45–46), but, unfortunately, his theory made little impression on his contemporaries. Not until the 1940s, when George Beadle and Edward Tatum examined the genetic basis of biochemical pathways in *Neurospora*, did the relation between genes and proteins become widely accepted.

The One Gene, One Enzyme Hypothesis

Beadle and Tatum used the bread mold *Neurospora* to study the biochemical result of mutations. *Neurospora* is a good model organism because it is easy to cultivate in the laboratory and because the main vegetative part of the fungus is haploid; the haploid state allows the effects of recessive mutations to be easily observed (FIGURE 15.1).

Wild-type *Neurospora* grows on minimal medium, which contains only inorganic salts, nitrogen, a carbon source such as sucrose, and the vitamin biotin. The fungus can synthesize all the biological molecules that it needs from these basic compounds. However, mutations may arise that disrupt fungal growth by destroying the fungus's ability to synthesize one or more essential biological molecules.

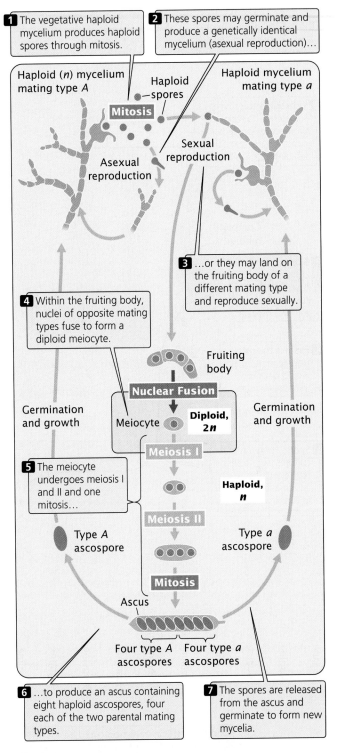

1 The vegetative haploid mycelium produces haploid spores through mitosis.

2 These spores may germinate and produce a genetically identical mycelium (asexual reproduction)…

Haploid (*n*) mycelium mating type *A*

Haploid mycelium mating type *a*

Haploid spores

Mitosis

Asexual reproduction

Sexual reproduction

3 …or they may land on the fruiting body of a different mating type and reproduce sexually.

4 Within the fruiting body, nuclei of opposite mating types fuse to form a diploid meiocyte.

Fruiting body

Nuclear Fusion

Germination and growth

Meiocyte

Diploid, 2*n*

Germination and growth

Meiosis I

5 The meiocyte undergoes meiosis I and II and one mitosis…

Haploid, *n*

Meiosis II

Type *A* ascospore

Type *a* ascospore

Mitosis

Ascus

Four type *A* ascospores

Four type *a* ascospores

6 …to produce an ascus containing eight haploid ascospores, four each of the two parental mating types.

7 The spores are released from the ascus and germinate to form new mycelia.

 15.1 **Beadle and Tatum used the fungus** *Neurospora*, **which has a complex life cycle, to work out the relation of genes to proteins.**

These nutritionally deficient mutants, termed **auxotrophs,** will not grow on minimal medium, but they *can* grow on medium that contains the substance that they are no longer able to synthesize.

Beadle and Tatum first irradiated spores of *Neurospora* to induce mutations (◀FIGURE 15.2). After irradiation, each spore was placed into a different culture tube with complete medium (medium containing all the biological substances needed for growth (see Figure 15.2). Next, they transferred spores from each culture to tubes containing minimal medium. Fungi containing auxotrophic mutations grew on complete medium but would not grow on minimal medium, which allowed Beadle and Tatum to identify cultures that contained mutations.

After they had determined that a particular culture had an auxotrophic mutation, Beadle and Tatum set out to determine the specific *effect* of the mutation. They transferred spores of each mutant strain from complete medium to a series of tubes (see Figure 15.2), each of which possessed minimal medium plus one of a variety of essential biological molecules, such as an amino acid. If the spores in a tube grew, Beadle and Tatum were able to identify the added substance as the biological molecule whose synthesis had been affected by the mutation. For example, an auxotrophic mutant that would grow only on minimal medium to which arginine had been added must have possessed a mutation that disrupts the synthesis of arginine.

Patient application of this procedure allowed the genetic dissection of multistep biochemical pathways. Adrian Srb and Norman H. Horowitz used this method to investigate genes that control arginine synthesis (◀FIGURE 15.3).

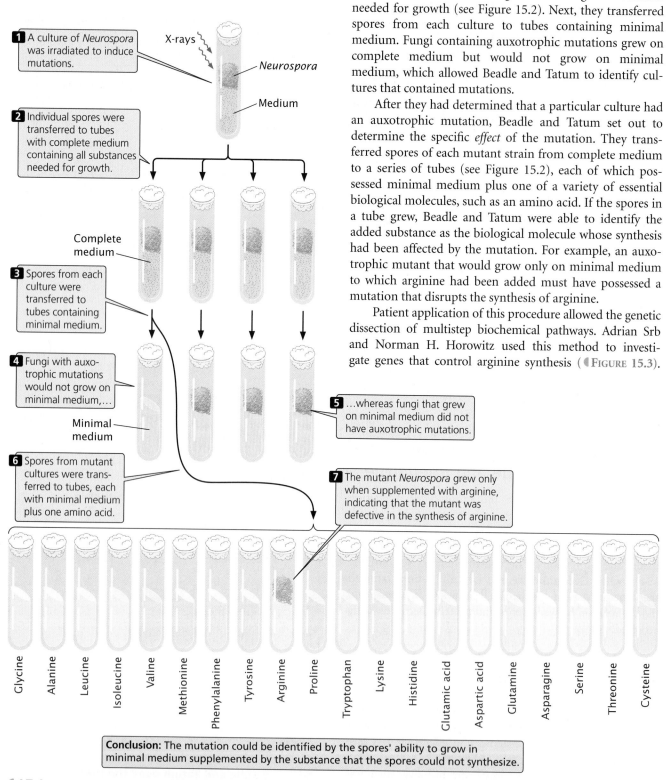

1 A culture of *Neurospora* was irradiated to induce mutations.

X-rays

Neurospora

Medium

2 Individual spores were transferred to tubes with complete medium containing all substances needed for growth.

Complete medium

3 Spores from each culture were transferred to tubes containing minimal medium.

4 Fungi with auxotrophic mutations would not grow on minimal medium,…

Minimal medium

5 …whereas fungi that grew on minimal medium did not have auxotrophic mutations.

6 Spores from mutant cultures were transferred to tubes, each with minimal medium plus one amino acid.

7 The mutant *Neurospora* grew only when supplemented with arginine, indicating that the mutant was defective in the synthesis of arginine.

Glycine · Alanine · Leucine · Isoleucine · Valine · Methionine · Phenylalanine · Tyrosine · Arginine · Proline · Tryptophan · Lysine · Histidine · Glutamic acid · Aspartic acid · Glutamine · Asparagine · Serine · Threonine · Cysteine

Conclusion: The mutation could be identified by the spores' ability to grow in minimal medium supplemented by the substance that the spores could not synthesize.

◀15.2 **Beadle and Tatum developed a method for isolating auxotrophic mutants in *Neurospora*.**

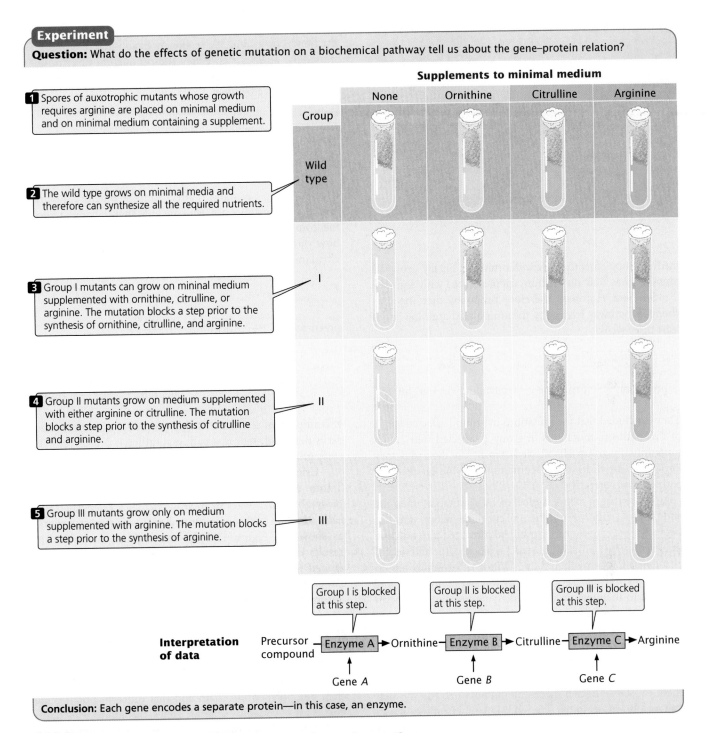

Experiment

Question: What do the effects of genetic mutation on a biochemical pathway tell us about the gene–protein relation?

Supplements to minimal medium

| | None | Ornithine | Citrulline | Arginine |

1 Spores of auxotrophic mutants whose growth requires arginine are placed on minimal medium and on minimal medium containing a supplement.

Group — Wild type

2 The wild type grows on minimal media and therefore can synthesize all the required nutrients.

3 Group I mutants can grow on mininal medium supplemented with ornithine, citrulline, or arginine. The mutation blocks a step prior to the synthesis of ornithine, citrulline, and arginine.

I

4 Group II mutants grow on medium supplemented with either arginine or citrulline. The mutation blocks a step prior to the synthesis of citrulline and arginine.

II

5 Group III mutants grow only on medium supplemented with arginine. The mutation blocks a step prior to the synthesis of arginine.

III

Group I is blocked at this step.

Group II is blocked at this step.

Group III is blocked at this step.

Interpretation of data

Precursor compound — Enzyme A → Ornithine — Enzyme B → Citrulline — Enzyme C → Arginine

Gene *A*   Gene *B*   Gene *C*

Conclusion: Each gene encodes a separate protein—in this case, an enzyme.

◄15.3   Beadle and Tatum established that each step in a pathway is controlled by a different enzyme. This biochemical pathway leads to the synthesis of arginine in *Neurospora*. Steps in the pathway are catalyzed by enzymes affected by mutants.

They first isolated a series of auxotrophic mutants whose growth required arginine. They then tested these mutants for their ability to grow on minimal medium supplemented with three compounds: ornithine, citrulline, and arginine. From the results, they were able to place the mutants into three groups (Table 15.1) based on which of the substances allowed growth. Group I mutants grew on minimal medium supplemented with ornithine, citrulline, or arginine. Group II mutants grew on minimal medium supplemented with either arginine or citrulline but did not grow on

Table 15.1	Growth of arginine auxotrophic mutants on minimal medium with various supplements		
Mutant Strain Number	Ornithine	Citrulline	Arginine
Group I	+	+	+
Group II	−	+	+
Group III	−	−	+

Note: + indicates growth; − indicates no growth.

medium supplemented only with ornithine. Finally, group III mutants grew only on medium supplemented with arginine.

Srb and Horowitz therefore proposed that the biochemical pathway leading to the amino acid arginine has at least three steps:

$$\text{precursor} \xrightarrow[1]{\text{Step}} \text{ornithine} \xrightarrow[2]{\text{Step}} \text{citrulline} \xrightarrow[3]{\text{Step}} \text{arginine}$$

They concluded that the mutations in group I affected step 1 of this pathway, mutations in group II affected step 2, and mutations in group III affected step 3. But how did they know that the order of the compounds in the biochemical pathway was correct?

Notice that, if step 1 is blocked by a mutation, then the addition of either ornithine or citrulline allows growth, because these compounds can still be converted into arginine (see Figure 15.3). Similarly, if step 2 is blocked, the addition of citrulline allows growth, but the addition of ornithine has no effect. If step 3 is blocked, the spores will grow only if arginine is added to the medium. The underlying principle is that an auxotrophic mutant cannot synthesize any compound that comes after the step blocked by a mutation.

Using this reasoning with the information in Table 15.1, we can see that the addition of arginine to the medium allows all three groups of mutants to grow. Therefore, biochemical steps affected by all the mutants precede the step that results in arginine. The addition of citrulline allows group I and group II mutants to grow but not group III mutants; therefore, group III mutations must affect a biochemical step that takes place after the production of citrulline but before the production of arginine.

$$\text{citrulline} \xrightarrow[\text{mutations}]{\substack{\text{Group}\\ \text{III}}} \text{arginine}$$

The addition of ornithine allows the growth of group I mutants but not group II or group III mutants; thus, muta-

tions in groups II and III affect steps that come after the production of ornithine. We've already established that group II mutations affect a step *before* the production of citrulline; so group II mutations must block the conversion of ornithine into citrulline.

$$\text{ornithine} \xrightarrow[\text{mutations}]{\substack{\text{Group}\\ \text{II}}} \text{citrulline} \xrightarrow[\text{mutations}]{\substack{\text{Group}\\ \text{III}}} \text{arginine}$$

Because group I mutations affect some step before the production of ornithine, we can conclude that they must affect the conversion of some precursor into ornithine. We can now outline the biochemical pathway yielding ornithine, citrulline, and arginine.

$$\text{precursor} \xrightarrow[\text{mutations}]{\substack{\text{Group}\\ \text{I}}} \text{ornithine} \xrightarrow[\text{mutations}]{\substack{\text{Group}\\ \text{II}}}$$

$$\text{citrulline} \xrightarrow[\text{mutations}]{\substack{\text{Group}\\ \text{III}}} \text{arginine}$$

It is important to note that this procedure does not necessarily detect all steps in a pathway; rather, it detects only the steps producing the compounds tested.

Using mutations and this type of reasoning, Beadle and Tatum were able to identify genes that control several biosynthetic pathways in *Neurospora*. They established that each step in a pathway is controlled by a different enzyme, as shown in Figure 15.3 for the arginine pathway. The results of genetic crosses and mapping studies demonstrated that mutations affecting any one step in a pathway always map to the same chromosomal location. Beadle and Tatum reasoned that mutations affecting a particular biochemical step occurred at a single locus that encoded a particular enzyme. This idea became known as the **one gene, one enzyme hypothesis:** genes function by encoding enzymes, and each gene encodes a separate enzyme. When research showed that some proteins are composed of more than one polypeptide chain and that different polypeptide chains are encoded by separate genes, this model was modified to become the **one gene, one polypeptide hypothesis.**

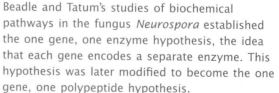

Concepts

Beadle and Tatum's studies of biochemical pathways in the fungus *Neurospora* established the one gene, one enzyme hypothesis, the idea that each gene encodes a separate enzyme. This hypothesis was later modified to become the one gene, one polypeptide hypothesis.

(a)

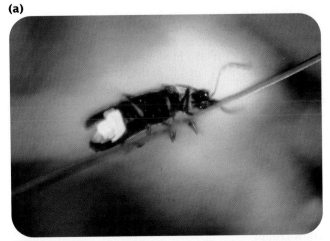

(b)

(c)

◀15.4 **Proteins serve a number of biological functions and are central to all living processes.**
(a) The light produced by fireflies is the result of a light-producing reaction between luciferin and ATP catalyzed by the enzyme luciferase. (b) The protein fibroin is the major structural component of spider webs.
(c) Castor beans contain a highly toxic protein called ricin.
(Part a, Gregory K. Scott/Photo Researchers; part b, A. Shay/ Animals Animals; part c, Gerald & Buff Corsi/Visuals Unlimited.)

www.whfreeman.com/pierce   Further information about the use of *Neurospora* in genetics

The Structure and Function of Proteins

Proteins are central to all living processes (◀FIGURE 15.4). Many proteins are enzymes, the biological catalysts that drive the chemical reactions of the cell; others are structural components, providing scaffolding and support for membranes, filaments, bone, and hair. Some proteins help transport substances; others have a regulatory, communication, or defense function.

All proteins are composed of **amino acids,** linked end to end. There are 20 common amino acids found in proteins; these amino acids are shown in ◀FIGURE 15.5 with both their three- and one-letter abbreviations. (Other amino acids that are sometimes found in proteins are modified forms of the common amino acids.) The 20 common amino acids are similar in structure, differing only in the structures of the R (radical) groups. The amino acids in proteins are joined by **peptide bonds** (◀FIGURE 15.6) to form **polypeptide** chains, and a protein consists of one or more polypeptide chains. Like nucleic acids, polypeptides have polarity with one end having a free amino group (NH_3^+) and the other end possessing a free carboxyl group (COO^-). Some proteins consist of only a few amino acids, whereas others may have thousands.

Like that of nucleic acids, the molecular structure of proteins has several levels of organization. The *primary structure* of a protein is its sequence of amino acids (◀FIGURE 15.7a). Through interactions between neighboring amino acids, a polypeptide chain folds and twists into a *secondary structure* (◀FIGURE 15.7b); two common secondary structures found in proteins are the beta (β) pleated sheet and the alpha (α) helix. Secondary structures interact and fold further to form a *tertiary structure* (◀FIGURE 15.7c), which is the overall, three-dimensional shape of the protein. The secondary and tertiary structures of a protein are ultimately determined by the primary structure—the amino acid sequence—of the protein. Finally, some proteins consist of two or more polypeptide chains that associate to produce a *quaternary structure* (◀FIGURE 15.7d).

Concepts

The product of many genes is a protein, whose action produces the trait encoded by that gene. Proteins are polymers, consisting of amino acids linked by peptide bonds. The amino acid sequence of a protein is its primary structure. This structure folds to create the secondary and tertiary structures; two or more polypeptide chains may associate to create a quaternary structure.

www.whfreeman.com/pierce   More information on protein structure

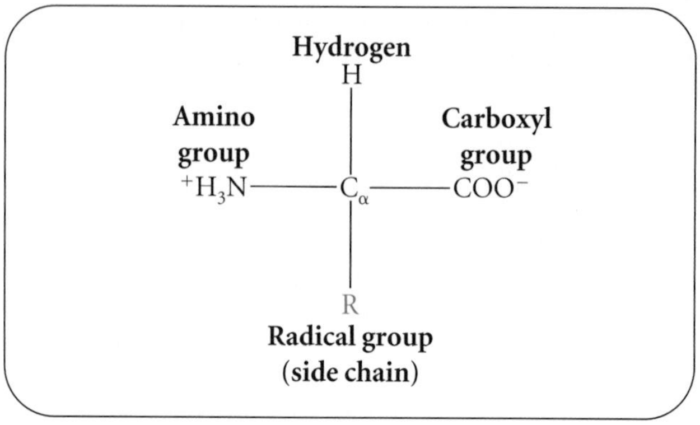

Hydrogen
H

Amino group
^+H_3N

Carboxyl group
COO^-

C_α

R
Radical group
(side chain)

15.5 The common amino acids have similar structures. Each amino acid consists of a central (α) carbon atom attached to: (1) an amino group (NH_3^+); (2) a carboxyl group (COO^-); (3) a hydrogen atom (H); and (4) a radical group, designated R. In the structures of the 20 common amino acids, the parts in black are common to all amino acids and the parts in red are the R groups.

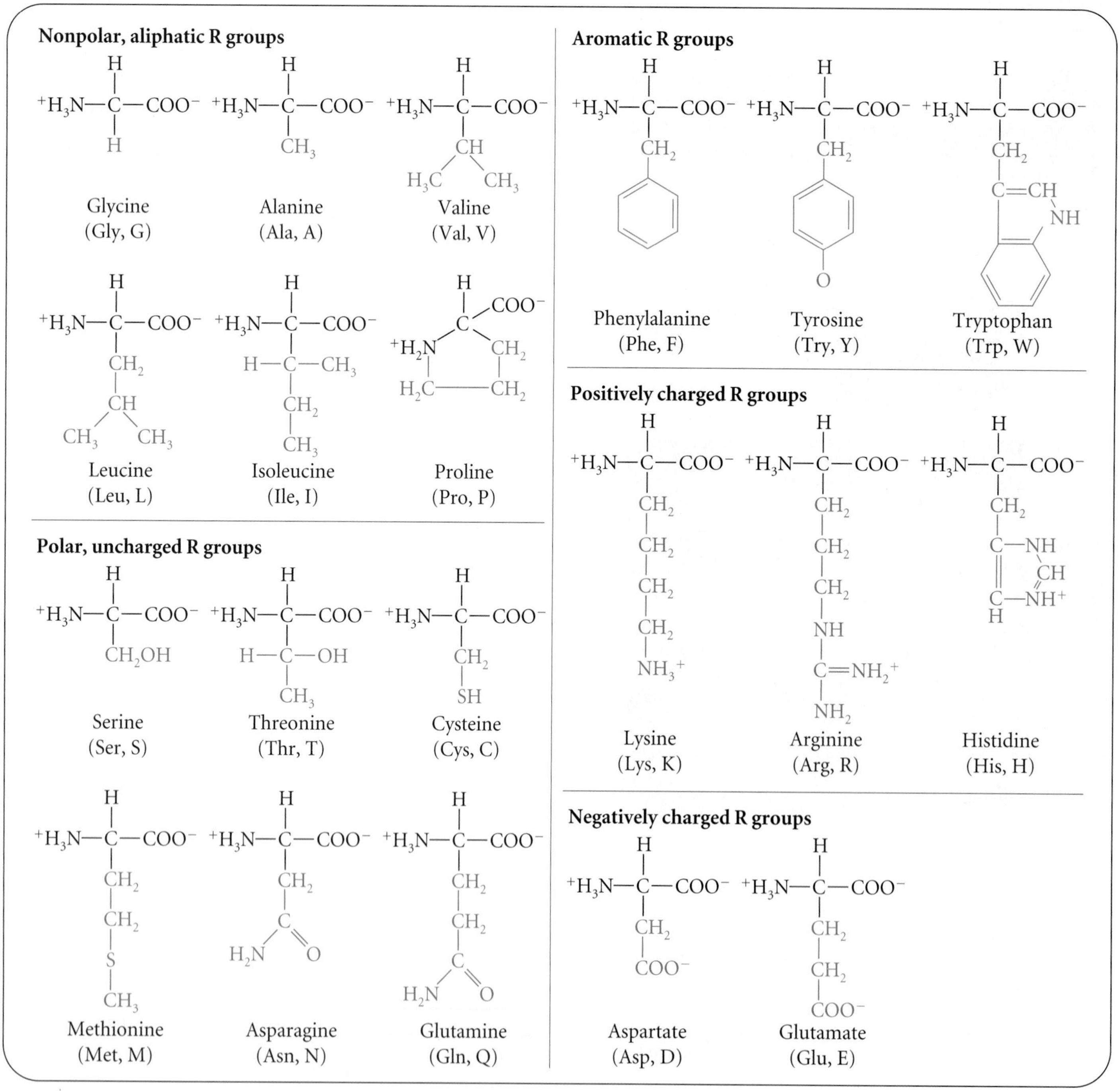

Nonpolar, aliphatic R groups

Glycine (Gly, G)

Alanine (Ala, A)

Valine (Val, V)

Leucine (Leu, L)

Isoleucine (Ile, I)

Proline (Pro, P)

Polar, uncharged R groups

Serine (Ser, S)

Threonine (Thr, T)

Cysteine (Cys, C)

Methionine (Met, M)

Asparagine (Asn, N)

Glutamine (Gln, Q)

Aromatic R groups

Phenylalanine (Phe, F)

Tyrosine (Try, Y)

Tryptophan (Trp, W)

Positively charged R groups

Lysine (Lys, K)

Arginine (Arg, R)

Histidine (His, H)

Negatively charged R groups

Aspartate (Asp, D)

Glutamate (Glu, E)

15.6 Amino acids are joined together by peptide bonds. In a peptide bond, the carboxyl group of one amino acid is covalently attached to the amino group of another amino acid.

$$^+H_3N-CH-C-O^- \quad {}^+H-N-CH-COO^-$$

with substituents R^1 and O; H, R^2

$$\downarrow H_2O$$

$$^+H_3N-CH-C-N-CH-COO^-$$

Peptide bond

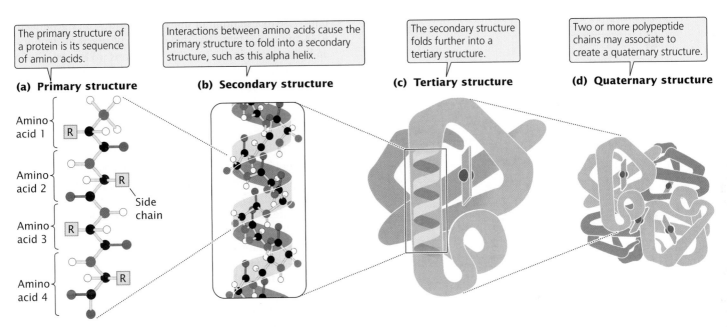

The primary structure of a protein is its sequence of amino acids.

Interactions between amino acids cause the primary structure to fold into a secondary structure, such as this alpha helix.

The secondary structure folds further into a tertiary structure.

Two or more polypeptide chains may associate to create a quaternary structure.

(a) Primary structure   **(b) Secondary structure**   **(c) Tertiary structure**   **(d) Quaternary structure**

Amino acid 1

Amino acid 2 — Side chain

Amino acid 3

Amino acid 4

15.7 Proteins have several levels of structural organization.

The Genetic Code

In 1953, Watson and Crick solved the structure of DNA and identified the base sequence as the carrier of genetic information. However, the way in which the base sequence of DNA specified the amino acid sequences of proteins (the genetic code) was not immediately obvious and remained elusive for another 10 years.

One of the first questions about the genetic code to be addressed was: *How many nucleotides are necessary to specify a single amino acid?* This basic unit of the genetic code—the set of bases that encode a single amino acid—is a *codon* (Chapter 14). Many early investigators recognized that codons must contain a minimum of three nucleotides. Each nucleotide position in mRNA can be occupied by one of four bases: A, G, C, or U. If a codon consisted of a single nucleotide, only four different codons (A, G, C, and U) would be possible, which is not enough to code for the 20 different

amino acids commonly found in proteins. If codons were made up of two nucleotides each (i.e., GU, AC, etc.) there would be $4 \times 4 = 16$ possible codons—still not enough to code for all 20 amino acids. With three nucleotides per codon, there are $4 \times 4 \times 4 = 64$ possible codons, which is more than enough to specify 20 different amino acids. Therefore, a *triplet code* requiring three nucleotides per codon is the most efficient way to encode all 20 amino acids. Using mutations in bacteriophage, Francis Crick and his colleagues confirmed in 1961 that the genetic code is indeed a triplet code.

Concepts

The genetic code is a triplet code, in which three nucleotides code for each amino acid in a protein.

www.whfreeman.com/pierce   An electronic table of codons and the amino acids they specify

Breaking the Genetic Code

When it had been firmly established that the genetic code consists of codons that are three nucleotides in length, the next step was to determine which groups of three nucleotides specify which amino acids. This task required the development of a cell-free system for protein synthesis (◀ FIGURE 15.8), which would make it possible to study the translation of a known mRNA.

Logically, the easiest way to break the code would have been to determine the base sequence of a piece of RNA, add it to a cell-free protein-synthesizing system, and allow it to direct the synthesis of a protein. The amino acid sequence of the newly synthesized protein could then be determined, and its sequence could be compared with that of the RNA. Unfortunately, there was no way at that time to determine the nucleotide sequence of a piece of RNA; so indirect methods were necessary to break the code.

The first clues to the genetic code came in 1961, from the work of Marshall Nirenberg and Johann Heinrich Matthaei. These investigators created synthetic RNAs by using an enzyme called polynucleotide phosphorylase. Unlike RNA polymerase, polynucleotide phosphorylase does not require a template; it randomly links together any RNA nucleotides that happen to be available. The first synthetic mRNAs used by Nirenberg and Matthaei were homopolymers, RNA molecules consisting of a single type of nucleotide. For example, by adding polynucleotide phosphorylase to a solution of uracil nucleotides, they generated RNA molecules that consisted entirely of uracil nucleotides and thus contained only UUU codons (◀ FIGURE 15.9). These poly(U) RNAs were then added to 20 tubes, each containing a cell-free protein-synthesizing system and the 20 different amino acids, one of which was radioactively labeled. Translation took place in all 20 tubes, but radioactive protein appeared in only one of the tubes—the one containing labeled phenylalanine (see Figure 15.9). This result showed that the codon UUU specifies the amino acid phenylalanine. The results of similar experiments using poly(C) and poly(A) RNA demonstrated that CCC codes for proline and AAA codes for lysine; for technical reasons, the results from poly(G) were uninterpretable.

To gain information about additional codons, Nirenberg and his colleagues created synthetic RNAs containing two or three different bases. Because polynucleotide phosphorylase incorporates nucleotides randomly, these RNAs contained random mixtures of the bases and are thus called random copolymers. For example, when adenine and cytosine nucleotides are mixed with polynucleotide phosphorylase, the RNA molecules produced have eight different codons: AAA, AAC, ACC, ACA, CAA, CCA, CAC, and CCC. In cell-free protein-synthesizing systems, these poly(AC) RNAs produced proteins containing six different amino acids: asparagine, glutamine, histidine, lysine, proline, and threonine.

The proportions of the different amino acids in the proteins depended on the ratio of the two nucleotides used in creating the synthetic mRNA, and the theoretical probability of finding a particular codon could be calculated from the ratios of the bases. If a $4 : 1$ ratio of C to A were used in making the RNA, then the probability of C occurring at any given position in a codon is $\frac{4}{5}$ and the probability of A being in it is $\frac{1}{5}$. With random incorporation of bases, the probability of any one of the codons with two Cs and one A (CCA, CAC, or ACC) should be $\frac{4}{5} \times \frac{4}{5} \times \frac{1}{5} = \frac{16}{125} = 0.13$, or 13%, and the probability of any codon with two As and one C (AAC, ACA, or CAA) should be $\frac{1}{5} \times \frac{1}{5} \times \frac{4}{5} = \frac{4}{125} = 0.032$, or about 3%. Therefore, an amino acid encoded by

Prepairing a cell-free synthesizing system

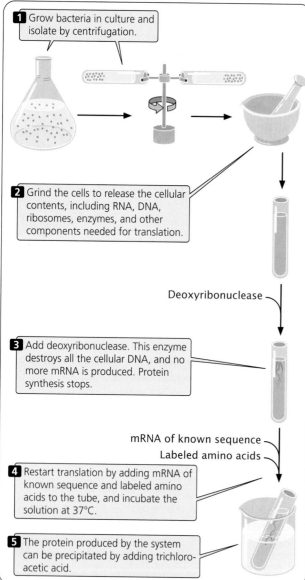

1 Grow bacteria in culture and isolate by centrifugation.

2 Grind the cells to release the cellular contents, including RNA, DNA, ribosomes, enzymes, and other components needed for translation.

Deoxyribonuclease

3 Add deoxyribonuclease. This enzyme destroys all the cellular DNA, and no more mRNA is produced. Protein synthesis stops.

mRNA of known sequence

Labeled amino acids

4 Restart translation by adding mRNA of known sequence and labeled amino acids to the tube, and incubate the solution at 37°C.

5 The protein produced by the system can be precipitated by adding trichloroacetic acid.

◀ **15.8 Breaking the genetic code required a cell-free protein-synthesizing system.**

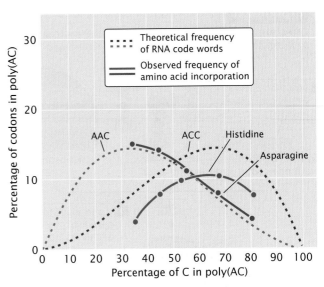

15.10 Nirenberg and Matthaei's use of random copolymers provided information about the genetic code.

The theoretical percentage of codons (vertical axis) is plotted against various percentages of cytosine in random AC copolymers. Notice that the distribution of histidine approximates the theoretical percentage of a codon with two Cs and one A, whereas the distribution of asparagine approximates the percentage expected of a codon with two As and one C. (Modified from Nirenberg, M. W. et al., 1963. Cold Spring Harbor Symposium on Quantitative Biology 28:549–557.)

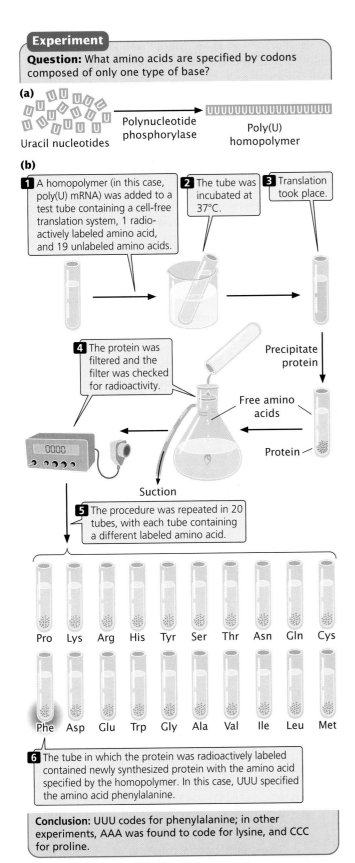

Experiment

Question: What amino acids are specified by codons composed of only one type of base?

(a) Uracil nucleotides → Polynucleotide phosphorylase → Poly(U) homopolymer

(b)

1 A homopolymer (in this case, poly(U) mRNA) was added to a test tube containing a cell-free translation system, 1 radioactively labeled amino acid, and 19 unlabeled amino acids.

2 The tube was incubated at 37°C.

3 Translation took place.

4 The protein was filtered and the filter was checked for radioactivity.

Precipitate protein

Free amino acids

Protein

Suction

5 The procedure was repeated in 20 tubes, with each tube containing a different labeled amino acid.

Pro Lys Arg His Tyr Ser Thr Asn Gln Cys

Phe Asp Glu Trp Gly Ala Val Ile Leu Met

6 The tube in which the protein was radioactively labeled contained newly synthesized protein with the amino acid specified by the homopolymer. In this case, UUU specified the amino acid phenylalanine.

Conclusion: UUU codes for phenylalanine; in other experiments, AAA was found to code for lysine, and CCC for proline.

15.9 Nirenberg and Matthaei developed a method for identifying the amino acid specified by a homopolymer.

two Cs and one A should be more common than an amino acid encoded by two As and one C. By comparing the percentages of amino acids in proteins produced by random copolymers with the theoretical frequencies expected for the codons (**◄ FIGURE 15.10**), information about the base *composition* of the codons was derived. Findings from these experiments revealed nothing, however, about the codon base *sequence*; histidine was clearly encoded by a codon with two Cs and one A (see Figure 15.10), but whether that codon was ACC, CAC, or CCA was unknown. There were other problems with this method: the theoretical calculations depended on the random incorporation of bases, which did not always occur, and, because the genetic code is redundant, sometimes several different codons specify the same amino acid.

To overcome the limitations of random copolymers, Nirenberg and Philip Leder developed another technique in 1964 that used ribosome-bound tRNAs. They found that a very short sequence of mRNA—even one consisting of a single codon—would bind to a ribosome. The codon on the short mRNA would then base pair with the matching anticodon on a tRNA that carries the amino acid specified by the codon (**◄ FIGURE 15.11**). The ribosome-bound mRNA was mixed with tRNAs and amino acids, and this mixture was passed through a nitrocellulose filter. The tRNAs paired with the ribosome-bound mRNA stuck to the filter, whereas unbound tRNAs passed through. The

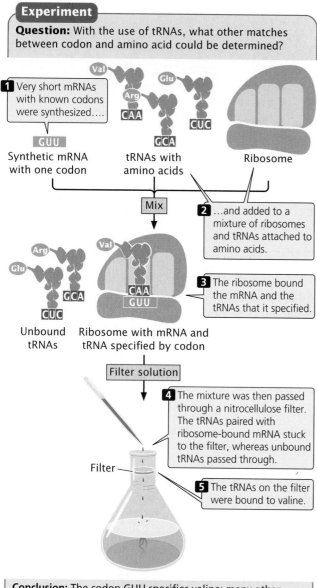

Experiment

Question: With the use of tRNAs, what other matches between codon and amino acid could be determined?

1 Very short mRNAs with known codons were synthesized....

Synthetic mRNA with one codon | tRNAs with amino acids | Ribosome

Mix

2 ...and added to a mixture of ribosomes and tRNAs attached to amino acids.

Unbound tRNAs | Ribosome with mRNA and tRNA specified by codon

3 The ribosome bound the mRNA and the tRNAs that it specified.

Filter solution

4 The mixture was then passed through a nitrocellulose filter. The tRNAs paired with ribosome-bound mRNA stuck to the filter, whereas unbound tRNAs passed through.

Filter

5 The tRNAs on the filter were bound to valine.

Conclusion: The codon GUU specifies valine; many other codons were determined by using this method.

◄ **15.11 Nirenberg and Leder developed a technique for using ribosome-bound tRNAs to provide additional information about the genetic code.**

advantage of this system was that it could be used with very short synthetic mRNA molecules that could be synthesized with a known sequence. Nirenberg and Leder synthesized over 50 short mRNAs with known codons and added them individually to a mixture of ribosomes and tRNAs. They then isolated the ribosome-bound tRNAs and determined which amino acids were present on the bound tRNAs. For example, synthetic RNA with the codon GUU retained a tRNA to which valine was attached, whereas RNAs with the codons UGU and UUG did not. Using this method, Nirenberg and his colleagues were able to determine the amino acids encoded by more than 50 codons.

A third method provided additional information about the genetic code. Gobind Khorana and his colleagues used chemical techniques to synthesize RNA molecules that contained known repeating sequences. They hypothesized that an mRNA that contained, for instance, alternating uracil and guanine nucleotides (UGUG UGUG UGUG) would be read during translation as two alternating codons, UGU GUG UGU GUG, producing a protein composed of two alternating amino acids. When Khorana and his colleagues placed this synthetic mRNA in a cell-free protein-synthesizing system, it produced a protein made of alternating cysteine and valine residues. This technique could not determine which of the two codons (UGU or GUG) specified cysteine, but, combined with other methods, it made a crucial contribution to cracking the genetic code. The genetic code was fully understood by 1968 (◄ FIGURE 15.12). In the next section, we will examine some of the features of the code, which is so important to modern biology that Francis Crick has compared its place to that of the periodic table of the elements in chemistry.

www.whfreeman.com/pierce   A brief biography of Marshall Nirenberg

Second base

First base	U	C	A	G	Third base
U	UUU Phe / UUC Phe / UUA Leu / UUG Leu	UCU / UCC / UCA / UCG Ser	UAU Tyr / UAC Tyr / UAA Stop / UAG Stop	UGU Cys / UGC Cys / UGA Stop / UGG Trp	U C A G
C	CUU / CUC / CUA / CUG Leu	CCU / CCC / CCA / CCG Pro	CAU His / CAC His / CAA Gln / CAG Gln	CGU / CGC / CGA / CGG Arg	U C A G
A	AUU / AUC Ile / AUA / AUG Met	ACU / ACC / ACA / ACG Thr	AAU Asn / AAC Asn / AAA Lys / AAG Lys	AGU Ser / AGC Ser / AGA Arg / AGG Arg	U C A G
G	GUU / GUC / GUA / GUG Val	GCU / GCC / GCA / GCG Ala	GAU Asp / GAC Asp / GAA Glu / GAG Glu	GGU / GGC / GGA / GGG Gly	U C A G

◄ **15.12 The genetic code consists of 64 codons and the amino acids specified by these codons.** The codons are written 5'→3', as they appear in the mRNA. AUG is an initiation codon; UAA, UAG, and UGA are termination codons.

The Degeneracy of the Code

One amino acid is encoded by three consecutive nucleotides in mRNA, and each nucleotide can have one of four possible bases (A, G, C, and U) at each nucleotide position, thus permitting $4^3 = 64$ possible codons (see Figure 15.12). Three of these codons are stop codons, specifying the end of translation. Thus, 61 codons, called **sense codons,** code for amino acids. Because there are 61 sense codons and only 20 different amino acids commonly found in proteins, the code contains more information than is needed to specify the amino acids and is said to be a **degenerate code.** This expression does not mean that the genetic code is depraved; *degenerate* is a term that Francis Crick borrowed from quantum physics, where it describes multiple physical states that have equivalent meaning. The degeneracy of the genetic code means that amino acids may be specified by more than one codon. Only tryptophan and methionine are encoded by a single codon (see Figure 15.12). Others amino acids are specified by two codons, and some, such as leucine, are specified by six different codons. Codons that specify the same amino acid are said to be **synonymous,** just as synonymous words are different words that have the same meaning.

Isoaccepting tRNAs As we learned in Chapter 14, tRNAs serve as adapter molecules, binding particular amino acids and delivering them to a ribosome, where the amino acids are then assembled into polypeptide chains. Each type of tRNA attaches to a single type of amino acid. The cells of most organisms possess from about 30 to 50 different tRNAs, and yet there are only 20 different amino acids in proteins. Thus, some amino acids are carried by more than one tRNA. Different tRNAs that accept the same amino acid but have different anticodons are called **isoaccepting tRNAs.** Some synonymous codons code for different isoacceptors.

Wobble Many synonymous codons differ only in the third position (see Figure 15.12). For example, alanine is encoded by the codons GCU, GCC, GCA, and GCG, all of which begin with GC. When the codon on the mRNA and the anticodon of the tRNA join (◀ FIGURE 15.13), the first (5′) base of the codon pairs with the third base (3′) of the anticodon, strictly according to Watson and Crick rules: A with U; C with G. Next, the middle bases of codon and anticodon pair, also strictly following the Watson and Crick rules. After these pairs have hydrogen bonded, the third bases pair weakly— there may be flexibility, or **wobble,** in their pairing.

In 1966, Francis Crick developed the wobble hypothesis, which proposed that some nonstandard pairings of bases could occur at the third position of a codon. For example, a G in the anticodon may pair with either a C or a U in the third position of the codon (Table 15.2). The important thing to remember about wobble is that it allows

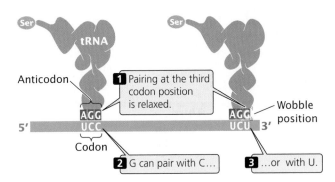

◀ **15.13** **Wobble may exist in the pairing of a codon on mRNA with an anticodon on tRNA.** The mRNA and tRNA pair in an antiparallel fashion. Pairing at the first and second codon positions is in accord with the Watson and Crick pairing rules (A with T, G with C); however, pairing rules are relaxed at the third position of the codon, and G on the anticodon can pair with either U or C on the codon in this example.

some tRNAs to pair with more than one codon on an mRNA; thus from 30 to 50 tRNAs can pair with 61 sense codons. Some codons are synonymous through wobble.

Concepts

> The genetic code consists of 61 sense codons that specify the 20 common amino acids; the code is degenerate and some amino acids are encoded by more than one codon. Isoaccepting tRNAs are different tRNAs with different anticodons that specify the same amino acid. Wobble exists when more than one codon can pair with the same anticodon.

The Reading Frame and Initiation Codons

Findings from early studies of the genetic code indicated that it is generally **nonoverlapping.** An overlapping code is one in which a single nucleotide is included in more than one codon, as shown in ◀ FIGURE 15.14. Usually, however, each nucleotide sequence of an mRNA specifies a single amino acid. A few overlapping codes are found in viruses; in these cases, two different proteins may be encoded within the same sequence of mRNA.

For any sequence of nucleotides, there are three potential sets of codons—three ways that the sequence can be read in groups of three. Each different way of reading the sequence is called a **reading frame,** and any sequence of nucleotides has three potential reading frames. The three reading frames have completely different sets of codons and therefore will specify proteins with entirely different amino acid sequences. Thus, it is essential for the

Table 15.2	The wobble rules, indicating which bases in the third position (3′ end) of the mRNA codon can pair with bases at the first (5′ end) of the anticodon of the tRNA

First Position of Anticodon	Third Position of Codon	Pairing
C	G	Anticodon 3′–X—Y—C–5′ │ │ │ 5′–Y—X—G–3′ Codon
G	U or C	Anticodon 3′–X—Y—G–5′ │ │ │ 5′–Y—X—U–3′ C Codon
A	U	Anticodon 3′–X—Y—A–5′ │ │ │ 5′–Y—X—U–3′ Codon
U	A or G	Anticodon 3′–X—Y—U–5′ │ │ │ 5′–Y—X—A–3′ G Codon

I (inosine)	A, U, or C	Anticodon
		3′–X—Y—I–5′
		 │ │ │
		5′–Y—X—A–3′
		          U
		          C
		Codon

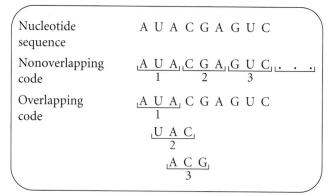

Nucleotide sequence	A U A C G A G U C
Nonoverlapping code	⌐A U A⌐⌐C G A⌐⌐G U C⌐. . . 1 2 3

Overlapping code	⌐A U A⌐C G A G U C
	   1
	   ⌐U A C⌐
	      2
	      ⌐A C G⌐
	         3

◀ **15.14 The genetic code is generally nonoverlapping.** In a nonoverlapping code, each nucleotide belongs to only one codon. In an overlapping code, some nucleotides belong to more than one codon. The genetic code used in almost all living organisms is nonoverlapping.

position in a gene, it codes for unformylated methionine. In archaeal and eukaryotic cells, AUG specifies unformylated methionine both at the initiation position and at internal positions.

Termination Codons

Three codons—UAA, UAG, and UGA—do not encode amino acids. These codons signal the end of the protein in both bacterial and eukaryotic cells and are called **stop codons, termination codons,** or **nonsense codons.** No tRNA molecules have anticodons that pair with termination codons.

The Universality of the Code

For many years the genetic code was assumed to be **universal,** meaning that each codon specifies the same amino acid in all organisms. We now know that the genetic code is almost, but not completely, universal; a few exceptions have been found. Most of these exceptions are termination codons, but there are a few cases in which one sense codon substitutes for another. The majority of exceptions are found in mitochondrial genes; a few nonuniversal codons have also been detected in nuclear genes of protozoans and in bacterial DNA (Table 15.3).

Concepts

Each sequence of nucleotides possesses three potential reading frames. The correct reading frame is set by the initiation codon. The end of a protein-encoding sequence is marked by a termination codon. With a few exceptions, all organisms use the same genetic code.

translational machinery to use the correct reading frame. How is the correct reading frame established? The reading frame is set by the **initiation codon,** which is the first codon of the mRNA to specify an amino acid. After the initiation codon, the other codons are read as successive groups of three nucleotides. No bases are skipped between the codons; so there are no punctuation marks to separate the codons.

The initiation codon is usually AUG, although GUG and UUG are used on rare occasions. The initiation codon is not just a punctuation mark; it specifies an amino acid. In bacterial cells, AUG encodes a modified type of methionine, *N*-formylmethionine; all proteins in bacteria begin with this amino acid, but the formyl group (or, in some cases, the entire amino acid) may be removed after the protein has been synthesized. When the codon AUG is at an internal

Connecting Concepts

Characteristics of the Genetic Code

We have now considered a number of characteristics of the genetic code. Let's pause for a moment and review these characteristics.

1. The genetic code consists of a sequence of nucleotides in DNA or RNA. There are four letters in the code, corresponding to the four bases—A, G, C, and U (T in DNA).

2. The genetic code is a triplet code. Each amino acid is encoded by a sequence of three consecutive nucleotides, called a codon.

3. The genetic code is degenerate—there are 64 codons but only 20 amino acids in proteins. Some codons are synonymous, specifying the same amino acid.

4. Isoaccepting tRNAs are tRNAs with different anticodons that accept the same amino acid; wobble allows the anticodon on one type of tRNA to pair with more than one type of codon on mRNA.

5. The code is generally nonoverlapping; each nucleotide in an mRNA sequence belongs to a single reading frame.

6. The reading frame is set by an initiation codon, which is usually AUG.

7. When a reading frame has been set, codons are read as successive groups of three nucleotides.

8. Any one of three termination codons (UAA, UAG, and UGA) can signal the end of a protein; no amino acids are encoded by the termination codons.

9. The code is almost universal.

Table 15.3	Some exceptions to the universal genetic code		
Genome	**Codon**	**Universal Code**	**Altered Code**
Bacterial DNA			
Mycoplasma capricolum	UGA	Stop	Trp
Mitochondrial DNA			
Human	UGA	Stop	Trp
Human	AUA	Ile	Met
Human	AGA, AGG	Arg	Stop
Yeast	UGA	Stop	Trp
Trypanosomes	UGA	Stop	Trp
Plants	CGG	Arg	Trp
Nuclear DNA			
Tetrahymena	UAA	Stop	Gln
Paramecium	UAG	Stop	Gln

The Process of Translation

Now that we are familiar with the genetic code, we can begin to study the mechanism by which amino acids are assembled into proteins. Because more is known about translation in bacteria, we will focus primarily on bacterial translation. In most respects, eukaryotic translation is similar, although there are some significant differences that will be noted as we proceed through the stages of translation.

Translation takes place on ribosomes; indeed, ribosomes can be thought of as moving protein-synthesizing machines. Through a variety of techniques, a detailed view of the structure of the ribosome has been produced in recent years, which has greatly improved our understanding of the translational process. A ribosome attaches near the 5′ end of an mRNA strand and moves toward the 3′ end, translating the codons as it goes (◀ FIGURE 15.15). Synthesis begins at the amino end of the protein, and the protein is elongated by the addition of new amino acids to the carboxyl end.

Protein synthesis can be conveniently divided into four stages: (1) the binding of amino acids to the tRNAs; (2) initiation, in which the components necessary for translation

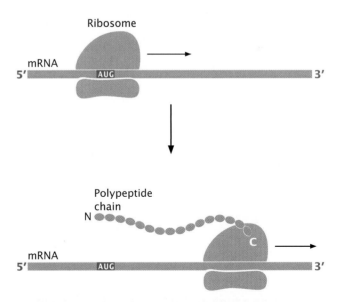

◀ 15.15 **The translation of an mRNA molecule takes place on a ribosome.** N represents the amino end of the protein; C represents the carboxyl end.

are assembled at the ribosome; (3) elongation, in which amino acids are joined, one at a time, to the growing polypeptide chain; and (4) termination, in which protein synthesis halts at the termination codon and the translation components are released from the ribosome.

The Binding of Amino Acids to Transfer RNAs

The first stage of translation is the binding of tRNA molecules to their appropriate amino acids. When linked to its amino acid, a tRNA delivers that amino acid to the ribosome, where the tRNA's anticodon pairs with a codon on mRNA. This process enables the amino acids to be joined in the order specified by the mRNA. Proper translation, then, first requires the correct binding of tRNA and amino acid.

As already mentioned, a cell typically possesses from 30 to 50 different tRNAs, and, collectively, these tRNAs are attached to the 20 different amino acids. Each tRNA is specific for a particular kind of amino acid. All tRNAs have the sequence CCA at the 3′ end, and the carboxyl group (COO⁻) of the amino acid is attached to the 2′- or 3′-hydroxyl group of the adenine nucleotide at the end of the tRNA, (FIGURE 15.16). If each tRNA is specific for a particular amino acid but all amino acids are attached to the same nucleotide (A) at the 3′ end of a tRNA, how does a tRNA link up with its appropriate amino acid?

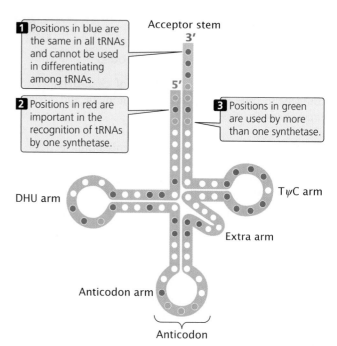

15.17 Certain positions on tRNA molecules are recognized by the appropriate aminoacyl-tRNA synthetase.

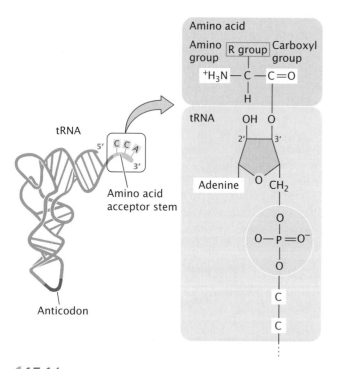

15.16 An amino acid attaches to the 3′ end of a tRNA. The carboxyl group (COO⁻) of the amino acid attaches to the hydroxyl group of the 2′- or 3′- carbon atom of the final nucleotide at the 3′ end of the tRNA, in which the base is always an adenine.

The key to specificity between an amino acid and its tRNA is a set of enzymes called **aminoacyl-tRNA synthetases.** A cell has 20 different aminoacyl-tRNA synthetases, one for each of the 20 amino acids. Each synthetase recognizes a particular amino acid, as well as all the tRNAs that accept that amino acid. Recognition of the appropriate amino acid by a synthetase is based on the different sizes, charges, and R groups of the amino acids. The tRNAs, however, are all similar in tertiary structure. How does a synthetase distinguish among tRNAs?

The recognition of tRNAs by a synthetase depends on the differing nucleotide sequences of tRNAs. Researchers have identified which nucleotides are important in recognition by altering different nucleotides in a particular tRNA and determining whether the altered tRNA is still recognized by its synthetase. The results of these studies revealed that the anticodon loop, the DHU-loop, and the acceptor stem are particularly critical for the identification of most tRNAs (FIGURE 15.17).

The attachment of a tRNA to its appropriate amino acid (termed **tRNA charging**) requires energy, which is supplied by adenosine triphosphate (ATP):

$$\text{amino acid} + \text{tRNA} + \text{ATP} \longrightarrow$$
$$\text{aminoacyl-tRNA} + \text{AMP} + \text{PP}_i$$

Two phosphates are cleaved from ATP, producing adenosine monophosphate (AMP) and pyrophosphate (PP$_i$), as well as the aminoacylated tRNA (the tRNA with its attached

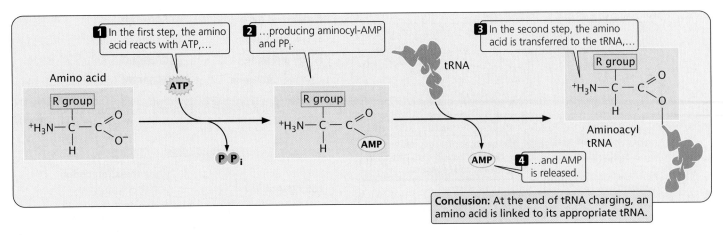

◀15.18 An amino acid becomes attached to the appropriate tRNA in a two-step reaction.

amino acid). This reaction takes place in two steps (◀FIGURE 15.18). To identify the resulting aminoacylated tRNA, we write the three-letter abbreviation for the amino acid in front of the tRNA; for example, the amino acid alanine (Ala) attaches to its tRNA (tRNA^{Ala}), giving rise to its aminoacyl-tRNA (Ala-tRNA^{Ala}).

Errors in tRNA charging are rare; they occur in only about 1 in 10,000 to 1 in 100,000 reactions. This fidelity is due to the presence of proofreading activity in the synthetases, which detects and removes incorrectly paired amino acids from the tRNAs.

Concepts

Amino acids are attached to specific tRNAs by aminoacyl-tRNA synthetases in a two-step reaction that requires ATP.

The Initiation of Translation

The second stage in the process of protein synthesis is initiation. During initiation, all the components necessary for protein synthesis assemble: (1) mRNA; (2) the small and large subunits of the ribosome; (3) a set of three proteins called initiation factors; (4) initiator tRNA with *N*-formylmethionine attached (fMet-tRNA^{fMet}); and (5) guanosine triphosphate (GTP). Initiation comprises three major steps. First, mRNA binds to the small subunit of the ribosome. Second, initiator tRNA binds to the mRNA through base pairing between the codon and anticodon. Third, the large ribosome joins the initiation complex. Let's look at each of these steps more closely.

A functional ribosome exists as two subunits, the small 30S subunit and the large 50S subunit (in bacterial cells). When not actively translating, the two subunits exist in dynamic equilibrium, in which they are constantly joining

and separating (◀FIGURE 15.19). An mRNA molecule can bind to the small ribosome subunit only when the subunits are separate. **Initiation factor 3** (IF-3) binds to the small subunit of the ribosome and prevents the large subunit from binding during initiation (see Figure 15.19b).

Key sequences on the mRNA required for ribosome binding have been identified in experiments in which the ribosome is allowed to bind to mRNA under conditions that allow initiation but prevent later stages of protein synthesis, thereby stalling the ribosome at the initiation site. After the ribosome has attached to the mRNA in these experiments, ribonuclease is added, which degrades all the mRNA except the region covered by the ribosome. The remaining mRNA can be separated from the ribosome and studied. The sequence covered by the ribosome during initiation is from 30 to 40 nucleotides long and includes the AUG initiation codon. Within the ribosome-binding site is the Shine-Dalgarno consensus sequence (◀FIGURE 15.20) (see Chapter 14), which is complementary to a sequence of nucleotides at the 3′ end of 16S rRNA (part of the small subunit of the ribosome). During initiation, the nucleotides in the Shine-Dalgarno sequence pair with their complementary nucleotides in the 16S rRNA, allowing the small subunit of the ribosome to attach to the mRNA and positioning the ribosome directly over the initiation codon.

Next, the initiator fMet-tRNA^{fMet} attaches to the initiation codon (see Figure 15.19c). This step requires **initiation factor 2** (IF-2), which forms a complex with GTP. A third factor, **initiation factor 1** (IF-1), enhances the dissociation of the large and small ribosomal subunits.

At this point, the initiation complex consists of (1) the small subunit of the ribosome; (2) the mRNA; (3) the initiator tRNA with its amino acid (fMet-tRNA^{fMet}); (4) one molecule of GTP; and (5) IF-3, IF-2, and IF-1. These components are collectively known as the **30S initiation complex** (see Figure 15.19c). In the final step of initiation, IF-3 dissociates from the small subunit, allowing the large

(a)

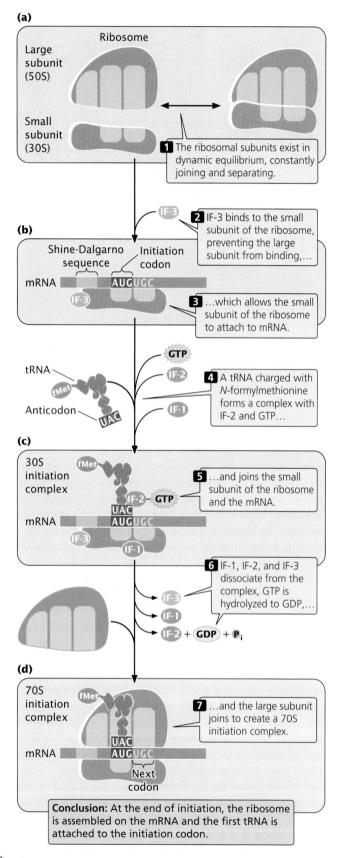

Ribosome

Large subunit (50S)

Small subunit (30S)

1 The ribosomal subunits exist in dynamic equilibrium, constantly joining and separating.

(b)

Shine-Dalgarno sequence   Initiation codon

mRNA   AUGUGC

IF-3

2 IF-3 binds to the small subunit of the ribosome, preventing the large subunit from binding,...

3 ...which allows the small subunit of the ribosome to attach to mRNA.

tRNA

fMet

Anticodon   UAC

GTP

IF-2

IF-1

4 A tRNA charged with N-formylmethionine forms a complex with IF-2 and GTP...

(c)

30S initiation complex

fMet

IF-2   GTP

UAC

mRNA   AUGUGC

IF-3

IF-1

5 ...and joins the small subunit of the ribosome and the mRNA.

6 IF-1, IF-2, and IF-3 dissociate from the complex, GTP is hydrolyzed to GDP,...

IF-3

IF-1

IF-2 + GDP + Pᵢ

(d)

70S initiation complex

fMet

UAC

mRNA   AUGUGC

Next codon

7 ...and the large subunit joins to create a 70S initiation complex.

Conclusion: At the end of initiation, the ribosome is assembled on the mRNA and the first tRNA is attached to the initiation codon.

◀15.19   The initiation of translation in bacterial cells requires several initiation factors and GTP.

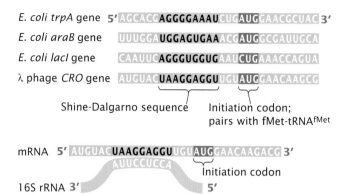

E. coli trpA gene   5′ AGCACC AGGGGAAAU CUG AUG GAACGCUAC 3′

E. coli araB gene   UUUGGA UGGAGUGAA ACC AUG GCGAUUGCA

E. coli lacI gene   CAAUUC AGGGUGGUG AAU CUG AAACCAGUA

λ phage CRO gene   AUGUAC UAAGGAGGU UGU AUG GAACAAGCG

Shine-Dalgarno sequence   Initiation codon; pairs with fMet-tRNAᶠᴹᵉᵗ

mRNA   5′ AUGUAC UAAGGAGGU UGU AUG GAACAAGACG 3′

AUUCCUCCA

Initiation codon

16S rRNA   3′   5′

◀15.20   Shine-Dalgarno consensus sequences in mRNA are required for the attachment of the small subunit of the ribosome. The Shine-Dalgarno sequences are complementary to a sequence of nucleotides found near the 3′ end of 16S rRNA in the small subunit of the ribosome. These complementary nucleotides base pair during the initiation of translation.

subunit of the ribosome to join the initiation complex. The molecule of GTP (provided by IF-2) is hydrolyzed to guanosine diphosphate (GDP), and IF-1 and IF-2 depart (see Figure 15.19d). When the large subunit has joined the initiation complex, it is called the **70S initiation complex.**

Similar events take place in the initiation of translation in eukaryotic cells, but there are some important differences. In bacterial cells, sequences in 16S rRNA of the small subunit of the ribosome bind to the Shine-Dalgarno sequence in mRNA; this binding positions the ribosome over the start codon. No analogous consensus sequence exists in eukaryotic mRNA. Instead, the cap at the 5′ end of eukaryotic mRNA plays a critical role in the initiation of translation. The small subunit of the eukaryotic ribosome, with the help of initiation factors, recognizes the cap and binds there; the small subunit then migrates along (scans) the mRNA until it locates the first AUG codon. The identification of the start codon is facilitated by the presence of a consensus sequence (called the Kozak sequence) that surrounds the start codon:

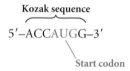

Kozak sequence

5′–ACCAUGG–3′

Start codon

Another important difference is that eukaryotic initiation requires more initiation factors. Some factors keep the ribosomal subunits separated, just as IF-3 does in bacterial cells. Others recognize the 5′ cap on mRNA and allow the small subunit of the ribosome to bind there. Still others possess RNA helicase activity, which is used to unwind secondary structures that may exist in the 5′ untranslated region of mRNA, allowing the small subunit to move down the mRNA until the initiation codon is reached. Other initiation factors help bring the initiator tRNA and methionine (Met-tRNA^{fMet}) to the initiation complex.

The poly(A) tail at the 3′ end of eukaryotic mRNA also plays a role in the initiation of translation. Proteins that attach to the poly(A) tail interact with proteins that bind to the 5′ cap, enhancing the binding of the small subunit of the ribosome to the 5′ end of the mRNA. This interaction between the 5′ cap and the 3′ tail suggests that the mRNA bends backward during the initiation of translation, forming a circular structure (◀ FIGURE 15.21). A few eukaryotic mRNAs contain internal ribosome entry sites, where ribosomes can bind directly without first attaching to the 5′ cap.

> ### Concepts
>
> In the initiation of translation in bacterial cells, the small ribosomal subunit attaches to mRNA, and initiator tRNA attaches to the initiation codon. This process requires several initiation factors (IF-1, IF-2, and IF-3) and GTP. In the final step, the large ribosomal subunit joins the initiation complex.

Elongation

The next stage in protein synthesis is elongation, in which amino acids are joined to create a polypeptide chain. Elongation requires (1) the 70S complex just described; (2) tRNAs charged with their amino acids; (3) several elongation factors (EF-Ts, EF-Tu, and EF-G); and (4) GTP.

A ribosome has three sites that can be occupied by tRNAs; the **aminoacyl, or A, site,** the **peptidyl, or P, site,** and the **exit, or E, site** (◀ FIGURE 15.22a). The initiator tRNA immediately occupies the P site (the only site to which the fMet-tRNA^{fMet} is capable of binding), but all other tRNAs first enter the A site. After initiation, the ribosome is attached to the mRNA, and fMet-tRNA^{fMet} is positioned over the AUG start codon in the P site; the adjacent A site is unoccupied (see Figure 15.22a).

Elongation occurs in three steps. The first step (◀ FIGURE 15.22b) is the delivery of a charged tRNA (tRNA with its amino acid attached) to the A site. This requires the presence of **elongation factor Tu (EF-Tu), elongation factor Ts (EF-Ts),** and GTP. EF-Tu first joins with GTP and then binds to a charged tRNA to form a three-part complex. This three-part complex enters the A site of the ribosome, where the anticodon on the tRNA pairs with the codon on the mRNA. After the charged tRNA is in the A site, GTP is cleaved to GDP, and the EF-Tu–GDP complex is released (◀ FIGURE 15.22c). Factor EF-Ts regenerates EF-Tu–GDP to EF-Tu–GTP. In eukaryotic cells, a similar set of reactions delivers the charged tRNA to the A site.

The second step of elongation is the creation of a peptide bond between the amino acids that are attached to

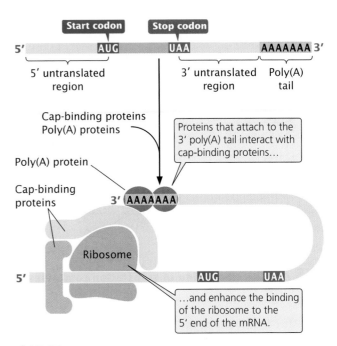

Start codon   **Stop codon**

5′ untranslated region

3′ untranslated region   Poly(A) tail

Cap-binding proteins
Poly(A) proteins

Proteins that attach to the 3′ poly(A) tail interact with cap-binding proteins…

Poly(A) protein

Cap-binding proteins

Ribosome

…and enhance the binding of the ribosome to the 5′ end of the mRNA.

◀ **15.21** **The poly(A) tail at the 3′ end of eukaryotic mRNA plays a role in the initiation of translation.**

tRNAs in the P and A sites (◀ FIGURE 15.22d). The formation of this peptide bond releases the amino acid in the P site from its tRNA. The activity responsible for peptide-bond formation in the ribosome is referred to as **peptidyl transferase.** For many years, the assumption was that this activity is carried out by one of the proteins in the large subunit of the ribosome. Evidence, however, now indicates that the catalytic activity is a property of the rRNA in the large subunit of the ribosome; this rRNA acts as a ribozyme (see p. 354 in Chapter 13).

The third step in elongation is **translocation,** (◀ FIGURE 15.22e), the movement of the ribosome down the mRNA in the 5′→3′ direction. This step positions the ribosome over the next codon and requires **elongation factor G** (EF-G) and the hydrolysis of GTP to GDP. Because the tRNAs in the P and A site are still attached to the mRNA through codon–anticodon pairing, they do not move with the ribosome as it translocates. Consequently, the ribosome shifts so that the tRNA that previously occupied the P site now occupies the E site, from which it moves into the cytoplasm where it may be recharged with another amino acid. Translocation also causes the tRNA that occupied the A site (which is attached to the growing polypeptide chain) to be in the P site, leaving the A site open. Thus, the progress of each tRNA through the ribosome during elongation can be summarized as follows: cytoplasm → A site → P site → E site → cytoplasm. As discussed earlier, the initiator tRNA is an exception: it attaches directly to the P site and never occupies the A site.

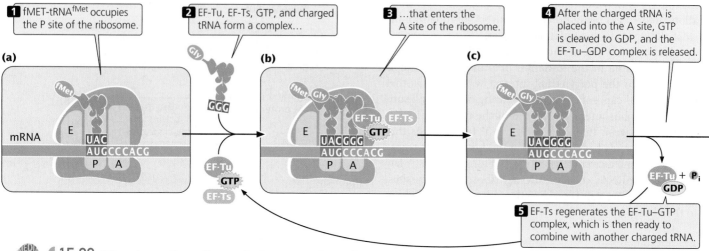

◄ 15.22 The elongation of translation comprises three steps.

After translocation, the A site of the ribosome is empty and ready to receive the tRNA specified by the next codon. The elongation cycle (Figure 15.22a through d) repeats itself: a charged tRNA and its amino acid occupy the A site, a peptide bond is formed between the amino acids in the A and P sites, and the ribosome translocates to the next codon. Throughout the cycle, the polypeptide chain remains attached to the tRNA in the P site. The ribosome moves down the mRNA in the 5′→3′ direction, adding amino acids one at a time according to the order specified by the mRNA's codon sequence. Elongation in eukaryotic cells takes place in a similar manner.

Concepts

Elongation consists of three steps: (1) a charged tRNA enters the A site, (2) a peptide bond is created between amino acids in the A and P sites, and (3) the ribosome translocates to the next codon. Elongation requires several elongation factors (EF-Tu, EF-Ts, and EF-G) and GTP.

Termination

Protein synthesis terminates when the ribosome translocates to a termination codon. Because there are no tRNAs with anticodons complementary to the termination codons, no tRNA enters the A site of the ribosome when a termination codon is encountered (◄ FIGURE 15.23a). Instead, proteins called **release factors** bind to the ribosome (◄ FIGURE 15.23b). *E. coli* has three release factors—RF₁, RF₂, and RF₃. Release factor 1 recognizes the termination codons UAA and UAG, and RF₂ recognizes UGA and UAA. Release factor 3 forms a complex with GTP and binds to the ribosome. The release factors then promote the cleavage of the tRNA in the P site from the polypeptide chain; in the

process, the GTP that is complexed to RF₃ is hydrolyzed to GDP. Additional factors help bring about the release of the tRNA from the P site, the release of the mRNA from the ribosome, and the dissociation of the ribosome (◄ FIGURE 15.23c). Translation in eukaryotic cells terminates in a similar way, except that there are two release factors: eRF1, which recognizes all three termination codons, and eRF2, which binds GTP and stimulates the release of the polypeptide from the ribosome.

Findings from recent studies suggest that the release factors bring about the termination of translation by completing a final elongation cycle of protein synthesis. In this model, RF₁ and RF₂ are similar in size and shape to tRNAs and occupy the A site of the ribosome, just as the amino acid–tRNA–EF–Tu–GTP complex does during an elongation cycle. Release factor 3 is structurally similar to EF-G; it then translocates RF₁ or RF₂ to the P site, as well as the last tRNA to the E site, in a way similar to that in which EF-G brings about translocation. When both the A site and the P site of the ribosome are cleared of tRNAs, the ribosome can dissociate. Research findings also indicate that some of the sequences in the rRNA play a role in the recognition of termination codons.

Concepts

Termination takes place when the ribosome reaches a termination codon. Release factors bind to the termination codon, causing the release of the polypeptide from the last tRNA, the tRNA from the ribosome, and the mRNA from the ribosome.

The overall process of protein synthesis, including tRNA charging, initiation, elongation, and termination, is summarized in ◄ FIGURE 15.24, and the components taking part in this process are listed in Table 15.4 (see page 425).

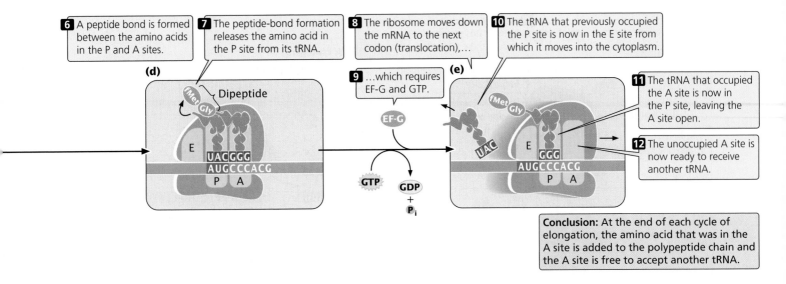

6 A peptide bond is formed between the amino acids in the P and A sites.

7 The peptide-bond formation releases the amino acid in the P site from its tRNA.

8 The ribosome moves down the mRNA to the next codon (translocation),...

9 ...which requires EF-G and GTP.

10 The tRNA that previously occupied the P site is now in the E site from which it moves into the cytoplasm.

11 The tRNA that occupied the A site is now in the P site, leaving the A site open.

12 The unoccupied A site is now ready to receive another tRNA.

(d) Dipeptide

(e)

Conclusion: At the end of each cycle of elongation, the amino acid that was in the A site is added to the polypeptide chain and the A site is free to accept another tRNA.

www.whfreeman.com/pierce A brief overview of translation and how it fits into the central dogma of genetics

RNA–RNA Interactions in Translation

The process of translation is rich in RNA–RNA interactions (which were discussed in Chapter 14 in the context of RNA processing). For example, in bacterial translation, the Shine-Dalgarno consensus sequence at the 5′ end of the mRNA pairs with the 3′ end of the 16S rRNA (see Figure 15.20), which ensures the binding of the ribosome to mRNA. Mutations that alter the Shine-Dalgarno sequence, so that the mRNA and rRNA are no longer complementary, inhibit translation. Corresponding mutations affecting the rRNA that restore complementarity allow translation to proceed. RNA–RNA interactions also take place between the tRNAs in the A and P sites and the rRNAs found in both the large and the small subunits of the ribosome. Furthermore, association of the large and small subunits of the ribosome may require interactions between the 16S rRNA and the 23S rRNA, although whether ribosomal proteins are implicated is not yet clear. Finally, tRNAs and mRNAs interact through their codon–anticodon pairing.

Polyribosomes

In both prokaryotic and eukaryotic cells, mRNA molecules are translated simultaneously by multiple ribosomes; see page 426 (◀ FIGURE 15.25). The resulting structure—an mRNA with several ribosomes attached—is called a **polyribosome.** Each ribosome successively attaches to the ribosome-binding site at the 5′ end of the mRNA and moves toward the 3′ end; the polypeptide associated with each ribosome becomes progressively longer as the ribosome moves along the mRNA.

In prokaryotic cells, transcription and translation are simultaneous; so multiple ribosomes may be attached to the 5′ end of the mRNA while transcription is still taking place at the 3′ end, as shown in ◀ FIGURE 15.26; see page 426. Until recently, transcription and translation were thought *not* to be simultaneous in eukaryotes, because transcription takes place in the nucleus and all translation was assumed to take place in the cytoplasm. However, research findings have now demonstrated that some translation takes place within the eukaryotic nucleus, and evidence suggests that, when the nucleus is the site of translation, transcription and translation may be simultaneous, much as in prokaryotes.

Concepts

In both prokaryotic and eukaryotic cells, multiple ribosomes may be attached to a single mRNA, generating a structure called a polyribosome.

Connecting Concepts

A Comparison of Bacterial and Eukaryotic Translation

We have now considered the process of translation in bacterial cells and noted some distinctive differences that exist in eukaryotic cells. Let's take a few minutes to reflect on some of the important similarities and differences of protein synthesis in bacterial and eukaryotic cells.

First, we should emphasize that the genetic code of bacterial and eukaryotic cells is virtually identical; the only difference is in the amino acid specified by the initiation codon. In bacterial cells, AUG codes for a modified type of methionine, *N*-formylmethionine, whereas, in eukaryotic

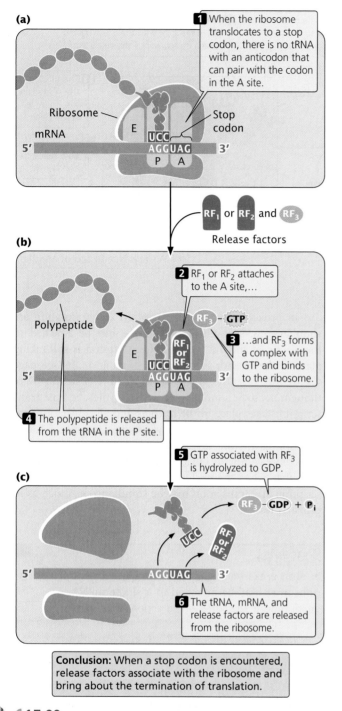

(a)

Ribosome

mRNA

E   UCC
    AGGUAG
    P   A

Stop codon

1 When the ribosome translocates to a stop codon, there is no tRNA with an anticodon that can pair with the codon in the A site.

RF₁ or RF₂ and RF₃

Release factors

(b)

Polypeptide

2 RF₁ or RF₂ attaches to the A site,...

RF₃ – GTP

3 ...and RF₃ forms a complex with GTP and binds to the ribosome.

E   UCC
    AGGUAG
    P   A

RF₁ or RF₂

4 The polypeptide is released from the tRNA in the P site.

5 GTP associated with RF₃ is hydrolyzed to GDP.

(c)

RF₃ – GDP + Pᵢ

UCC

RF₁ or RF₂

AGGUAG

6 The tRNA, mRNA, and release factors are released from the ribosome.

Conclusion: When a stop codon is encountered, release factors associate with the ribosome and bring about the termination of translation.

15.23 Translation ends when a stop codon is encountered.

15.24 The four steps involved in translation are tRNA charging (the binding of amino acids to tRNAs), initiation, elongation, and termination. In this process, amino acids are linked together in the order specified by the mRNA to create a polypeptide chain. A number of initiation, elongation, and release factors take part in the process, and energy is supplied by ATP and GTP.

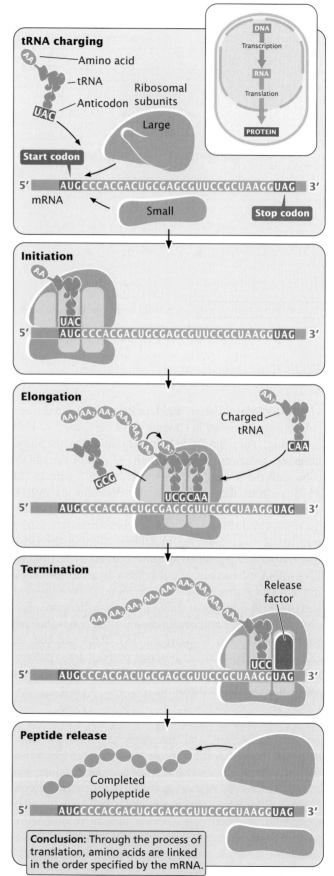

tRNA charging

Amino acid

tRNA

Anticodon

Ribosomal subunits

Large

UAC

Start codon

5′ AUGCCCACGACUGCGAGCGUUCCGCUAAGGUAG 3′

mRNA

Small

Stop codon

DNA
Transcription
RNA
Translation
PROTEIN

Initiation

AA

UAC
5′ AUGCCCACGACUGCGAGCGUUCCGCUAAGGUAG 3′

Elongation

AA₁ AA₂ AA₃ AA₄ AA₅ AA₆ AA₇

Charged tRNA

CAA

GCG

UCGCAA

5′ AUGCCCACGACUGCGAGCGUUCCGCUAAGGUAG 3′

Termination

AA₁ AA₂ AA₃ AA₄ AA₅ AA₆ AA₇ AA₈ AA₉

Release factor

UCC

5′ AUGCCCACGACUGCGAGCGUUCCGCUAAGGUAG 3′

Peptide release

Completed polypeptide

5′ AUGCCCACGACUGCGAGCGUUCCGCUAAGGUAG 3′

Conclusion: Through the process of translation, amino acids are linked in the order specified by the mRNA.

Table 15.4	Components required for protein synthesis in bacterial cells	

Stage	Component	Function
Binding of amino acid to tRNA	Amino acids	Building blocks of proteins
	tRNAs	Deliver amino acids to ribosomes
	aminoacyl-tRNA synthetase	Attaches amino acids to tRNAs
	ATP	Provides energy for binding amino acid to tRNA
Initiation	mRNA	Carries coding instructions
	fMet-tRNA^{fMet}	Provides first amino acid in peptide
	30S ribosomal subunit	Attaches to mRNA
	50S ribosomal subunit	Stabilizes tRNAs and amino acids
	Initiation factor 1	Enhances dissociation of large and small subunits of ribosome
	Initiation factor 2	Binds GTP; delivers fMet-tRNA^{fMet} to initiation codon
	Initiation factor 3	Binds to 30S subunit and prevents association with 50S subunit
Elongation	70S initiation complex	Functional ribosome with A, P, and E sites and peptidyl transferase activity where protein synthesis takes place
	Charged tRNAs	Bring amino acids to ribosome and help assemble them in order specified by mRNA
	Elongation factor Tu	Binds GTP and charged tRNA; delivers charged tRNA to A site
	Elongation factor Ts	Generates active elongation factor Tu
	Elongation factor G	Stimulates movement of ribosome to next codon
	GTP	Provides energy
	Peptidyl transferase	Creates peptide bond between amino acids in A site and P site
Termination	Release factors 1, 2, and 3	Bind to ribosome when stop codon is reached and terminate translation

cells, AUG codes for unformylated methionine. One consequence of the fact that bacteria and eukaryotes use the same code is that eukaryotic genes can be translated in bacterial systems, and vice versa; this feature makes genetic engineering possible, as we will see in Chapter 18.

Another difference is that transcription and translation take place simultaneously in bacterial cells, but the nuclear envelope may separate these processes in eukaryotic cells. The physical separation of transcription and translation has important implications for the control of gene expression, which we will consider in Chapter 16, and it allows for extensive modification of eukaryotic mRNAs, as discussed in Chapter 14. However, it is now evident that some translation does take place in the eukaryotic nucleus and, there, transcription and translation may be simultaneous. The extent of nuclear translation and how it may affect gene regulation are not yet clear.

Yet another difference is that mRNA in bacterial cells is short lived, typically lasting only a few minutes, but the longevity of mRNA in eukaryotic cells is highly variable and is frequently hours or days. Thus the synthesis of a particular bacterial protein ceases very quickly after transcription of the corresponding mRNA stops, but protein synthesis in eukaryotic cells may continue long after transcription has ended.

In both bacterial and eukaryotic cells, aminoacyl-tRNA synthetases attach amino acids to their appropriate tRNAs; the chemical reaction employed is the same. There are significant differences in the sizes and compositions of bacterial and eukaryotic ribosomal subunits. For example, the large subunit of the eukaryotic ribosome contains three rRNAs, whereas the bacterial ribosome contains only two. These differences allow antibiotics and other substances to inhibit bacterial translation while having no effect on the transla-

(a)

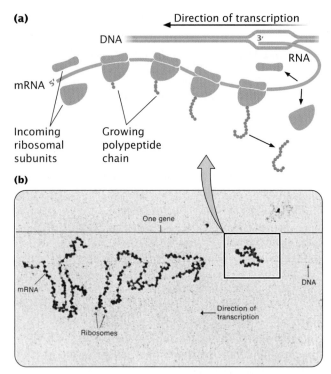

Direction of transcription

DNA

RNA

mRNA

Incoming ribosomal subunits

Growing polypeptide chain

(b)

One gene

mRNA

DNA

Direction of transcription

Ribosomes

◀ **15.25 An mRNA molecule may be transcribed simultaneously by several ribosomes.** (a) Four ribosomes are translating a mRNA molecule, moving from the 5′ end to the 3′ end. (b) In this electron micrograph of a polyribosome, the dark staining spheres are ribosomes, and the long, thin filament connecting the ribosomes is mRNA. The 5′ end of the mRNA is toward the lower left-hand corner of the micrograph. (Part b, O. L. Miller, Jr., and Barbara A. Hamaklo.)

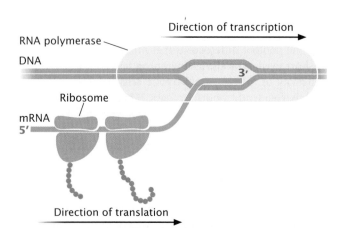

RNA polymerase

DNA

Direction of transcription

3′

Ribosome

mRNA
5′

Direction of translation

◀ **15.26 In prokaryotic cells, transcription and translation take place simultaneously.** While mRNA is being transcribed from the DNA template at mRNA's 3′ end, translation is taking place simultaneously at mRNA's 5′ end.

tion of eukaryotic nuclear genes, as will be discussed near the end of this chapter.

Other fundamental differences lie in the process of initiation. In bacterial cells, the small subunit of the ribosome attaches directly to the region surrounding the start codon through hydrogen bonding between the Shine-Dalgarno consensus sequence in the 5′ untranslated region of the mRNA and a sequence at the 3′ end of the 16S rRNA. In contrast, the small subunit of a eukaryotic ribosome first binds to proteins attached to the 5′ cap on mRNA and then migrates down the mRNA, scanning the sequence until it encounters the first AUG initiation codon. (A few eukaryotic mRNAs have internal ribosome-binding sites that utilize a specialized initiation mechanism similar to that seen in bacterial cells.) Additionally, more initiation factors take part in eukaryotic initiation than in bacterial initiation.

Elongation and termination are similar in bacterial and eukaryotic cells, although different elongation and termination factors are used. In both types of organisms, mRNAs are translated multiple times and are simultaneously attached to several ribosomes, forming polyribosomes.

What about translation in archaea, which are prokaryotic in structure (see Chapter 2) but are similar to eukaryotes in other genetic processes such as transcription? Much less is known about the process of translation in archaea, but available evidence suggests that they possess a mixture of eubacterial and eukaryotic features. Because archaea lack nuclear membranes, transcription and translation take place simultaneously, just as they do in eubacterial cells. As mentioned earlier, archaea utilize unformylated methionine as the initiator amino acid, a characteristic of eukaryotic translation. Findings from recent studies of DNA sequences that code for initiation and elongation factors in archaea suggest that some of them are similar to those found in eubacteria, whereas others are similar to those found in eukaryotes. Finally, some of the antibiotics that inhibit translation in eubacteria have no effect on translation in archaea.

The Posttranslational Modifications of Proteins

After translation, proteins in both prokaryotic and eukaryotic cells may undergo alterations termed posttranslational modifications. A number of different types of modifications are possible. As mentioned earlier, the formyl group or the entire methionine residue may be removed from the amino end of a protein. Some proteins are synthesized as larger precursor proteins and must be cleaved and trimmed by enzymes before the proteins can become functional. For others, the attachment of carbohydrates may be required for activation. The functions of many proteins depend critically on the proper folding of the polypeptide chain; some proteins spontaneously fold into their correct shapes, but, for others, correct folding may initially require the participation of other molecules called **molecular chaperones.**

In eukaryotic cells, the amino end of a protein is often acetylated after translation. Another modification of some proteins is the removal of 15 to 30 amino acids, called the **signal sequence,** at the amino end of the protein. The signal sequence helps direct a protein to a specific location within the cell, after which the sequence is removed by special enzymes. Amino acids within a protein may also be modified: phosphates, carboxyl groups, and methyl groups are added to some amino acids.

Concepts

Many proteins undergo posttranslational modifications after their synthesis.

Translation and Antibiotics

Antibiotics are drugs that kill microorganisms. To make an effective antibiotic—not just any poison will do—the trick is to kill the microbe without harming the patient. Antibiotics must be carefully chosen so that they destroy bacterial cells but not the eukaryotic cells of their host.

Translation is frequently the target of antibiotics because translation is essential to all living organisms and differs significantly between bacterial and eukaryotic cells. For example, bacterial and eukaryotic ribosomes differ in size and composition. A number of antibiotics bind selectively to bacterial ribosomes and inhibit various steps in translation, but they do not affect eukaryotic ribosomes. Tetracyclines, for instance, are a class of antibiotics that bind to the A site of bacterial ribosomes and block the entry of charged tRNAs, yet they have no effect on eukaryotic ribosomes. Neomycin binds to the ribosome near the A site and induces translational errors, probably by causing mistakes in the binding of charged tRNAs to the A site. Chloramphenicol binds to the large subunit of the ribosome and blocks peptide-bond formation. Streptomycin binds to the small subunit of the ribosome and inhibits initiation, and erythromycin blocks translocation. Although chloramphenicol and streptomycin are potent inhibitors of translation in bacteria, they do not inhibit translation in archaebacteria.

The three-dimensional structure of puromycin resembles the 3′ end of a charged tRNA, permitting puromycin to enter the A site of a ribosome efficiently and inhibit the entry of tRNAs. A peptide bond can form between the puromycin molecule in the A site and an amino acid on the tRNA in the P site of the ribosome, but puromycin cannot bind to the P site and translocation does not take place, blocking further elongation of the protein. Because tRNA structure is similar in all organisms, puromycin inhibits translation in both bacterial and eukaryotic cells; consequently, puromycin kills eukaryotic cells along with

bacteria and is sometimes used in cancer therapy to destroy tumor cells.

Many antibiotics act by blocking specific steps in translation, and different antibiotics affect different steps in protein synthesis. Because of this specificity, antibiotics are frequently used to study the process of protein synthesis.

Connecting Concepts Across Chapters

This chapter has focused on the process by which genetic information in an mRNA molecule is transferred to the amino acid sequence of a protein. This process is termed translation because information contained in the language of nucleotides must be "translated" into the language of amino acids.

The link between genotype and phenotype is usually a protein: most genes affect phenotypes by encoding proteins. How the presence of a protein produces a particular anatomical, physiological, or behavioral trait, however, is often far from clear, as was illustrated by the story of Lesch-Nyhan disease. The relation between genes and traits is the subject of much current research and will be explored further in Chapters 16 and 21.

In this chapter, we have examined the nature of the genetic code. It is a very concise code, with each codon consisting of three nucleotides, the minimum number capable of specifying all 20 common amino acids. Breaking the genetic code required great ingenuity and hard work on the part of a number of geneticists.

Much of this chapter has centered on protein synthesis. We learned that translation is a highly complex process: rRNAs, ribosomal proteins, tRNAs, mRNA, initiation factors, elongation factors, release factors, and aminoacyl-tRNA synthetases all help to assemble amino acids into a protein. This complexity might seem surprising, because the peptide bonds that hold amino acids together are simple covalent bonds. Translation is complex not because of any special property of the peptide bond, but rather because the amino acids must be linked in a highly precise order. The amino acid sequence determines the secondary and tertiary structures of a protein, which are critical to its function; so the genetic information in a mRNA molecule must be accurately translated. The complexity of translation has evolved to ensure that few mistakes are made in the course of protein synthesis.

An important theme in protein synthesis is RNA–RNA interaction, which takes place between tRNAs and mRNA, between mRNA and rRNAs, and between tRNAs and rRNAs. The prominence of these RNA–RNA interactions in translation reinforces the proposal that life first evolved in an RNA world, where flexible and versatile RNA molecules carried out many life processes (Chapter 13).

This chapter has built on our understanding of other processes of information transfer covered earlier in the book: replication (Chapter 12), transcription (Chapter 13), and RNA processing (Chapter 14). It also provides a critical foundation for later discussions of gene regulation (Chapter 16), gene mutations (Chapter 17), and the advanced topics of developmental genetics, cancer genetics, and immunological genetics (Chapter 21).

CONCEPTS SUMMARY

- Genes code for phenotypes by specifying the amino acid sequences of proteins.
- The relation between genes and proteins was first suggested by Archibald Garrod.
- Beadle and Tatum developed the one gene, one enzyme hypothesis, which proposed that each gene specifies one enzyme; this hypothesis was later modified to become the one gene, one polypeptide hypothesis.
- Proteins are composed of 20 different amino acids, several or many of which are linked together by peptide bonds. Chains of amino acids fold and associate to produce the secondary, tertiary, and quaternary structures of proteins.
- The genetic code is the way in which genetic information is stored in the nucleotide sequence of a gene.
- Solving the genetic code required several different approaches: the use of synthetic mRNAs with random sequences; short mRNAs that bind tRNAs with their amino acids; and long synthetic mRNAs with regularly repeating sequences.
- The genetic code is a triplet code: three nucleotides specify a single amino acid. It is also degenerate, nonoverlapping, and universal (almost).
- The degeneracy of the code means that more than one codon may specify an amino acid. Different tRNAs (isoaccepting tRNAs) may accept the same amino acid, and different anticodons may pair with the same codon through wobble, which can exist at the third position of the codon and which allows some nonstandard pairing of bases in this position.

- The reading frame is set by the initiation codon.
- The end of the protein-coding section of an mRNA is marked by one of three termination codons.
- Protein synthesis comprises four steps: (1) the binding of amino acids to the appropriate tRNAs, (2) initiation, (3) elongation, and (4) termination.
- The binding of an amino acid to a tRNA requires the presence of a specific aminoacyl-tRNA synthetase and ATP. The amino acid is attached by its carboxyl end to the 3' end of the tRNA.
- In bacterial translation initiation, the small subunit of the ribosome attaches to the mRNA and is positioned over the initiation codon. It is joined by the first tRNA and its associated amino acid (N-formylmethionine in bacterial cells) and, later, by the large subunit of the ribosome. Initiation requires several initiation factors and GTP.
- In elongation, a charged tRNA enters the A site of a ribosome, a peptide-bond is formed between amino acids in the A and P sites, and the ribosome moves (translocates) along the mRNA to the next codon. Elongation requires several elongation factors and GTP.
- Translation is terminated when the ribosome encounters one of the three termination codons. Release factors and GTP are required to bring about termination.
- Like RNA processing, translation requires a number of RNA–RNA interactions.
- Each mRNA may be simultaneously translated by several ribosomes, producing a structure called a polyribosome.
- Many proteins undergo posttranslational modification.

IMPORTANT TERMS

auxotroph (p. 406)
one gene, one enzyme hypothesis (p. 408)
one gene, one polypeptide hypothesis (p. 408)
amino acid (p. 409)
peptide bond (p. 409)
polypeptide (p. 409)
sense codon (p. 415)
degenerate genetic code (p. 415)
synonymous codons (p. 415)

isoaccepting tRNAs (p. 415)
wobble (p. 415)
nonoverlapping genetic code (p. 415)
reading frame (p. 415)
initiation codon (p. 416)
stop (termination or nonsense) codon (p. 416)
universal genetic code (p. 416)
aminoacyl-tRNA synthetase (p. 418)
tRNA charging (p. 418)

initiation factors (IF-1, IF-2, IF-3) (p. 419)
30S initiation complex (p. 419)
70S initiation complex (p. 420)
aminoacyl (A) site (p. 421)
peptidyl (P) site (p. 421)
exit (E) site (p. 421)
elongation factor Tu (EF-Tu) (p. 421)

elongation factor Ts (EF-Ts) (p. 421)
peptidyl transferase (p. 421)
translocation (p. 421)
elongation factor G (EF-G) (p. 421)
release factors (RF$_1$, RF$_2$, RF$_3$) (p. 422)
polyribosome (p. 423)
molecular chaperone (p. 427)
signal sequence (p. 427)

Worked Problems

1. A series of auxotrophic mutants were isolated in *Neurospora*. Examination of fungi containing these mutations revealed that they grew on minimal medium to which various compounds (A, B, C, D) were added; growth responses to each of the four compounds are presented in the following table. Give the order of compounds A, B, C, and D in a biochemical pathway. Outline a biochemical pathway that includes these four compounds and indicate which step in the pathway is affected by each of the mutations.

Mutation number	Compound			
	A	B	C	D
134	+	+	−	+
276	+	+	+	+
987	−	−	−	+
773	+	+	+	+
772	−	−	−	+
146	+	+	−	+
333	+	+	−	+
123	−	+	−	+

• Solution

To solve this problem, we should first group the mutations for which compounds allow growth, as follows.

Mutation		Compound			
Group	Number	A	B	C	D
I	276	+	+	+	+
	773	+	+	+	+
II	134	+	+	−	+
	146	+	+	−	+
	333	+	+	−	+
III	123	−	+	−	+
IV	987	−	−	−	+
	772	−	−	−	+

The underlying principle used to determine the order of the compounds in the pathway is as follows: If a compound is added after the block, it will allow the mutant to grow, whereas, if a compound is added before the block, it will have no effect. Applying this principle to the data in the table, we see that mutants in group I will grow if compound A, B, C, or D is added to the medium; so these mutations must affect a step before the production of all four compounds:

$$\xrightarrow[\text{mutations}]{\text{Group I}} \text{compounds A, B, C, D}$$

Group II mutants will grow if compound A, B, or D is added but not if compound C is added. Thus compound C comes before A, B, and D; and group II mutations affect the conversion of compound C into one of the other compounds:

$$\xrightarrow[\text{mutations}]{\text{Group I}} \text{compound C} \xrightarrow[\text{mutations}]{\text{Group II}} \text{compounds A, B, D}$$

Group III mutants allow growth if compound B or D is added but not if compound A or C is added. Thus group III mutations affect steps that follow the production of A and C; we have already determined that compound C precedes A in the pathway; so A must be the next compound in the pathway:

$$\xrightarrow[\text{mutations}]{\text{Group I}} \text{compound C} \xrightarrow[\text{mutations}]{\text{Group II}}$$

$$\text{compound A} \xrightarrow[\text{mutations}]{\text{Group III}} \text{compounds B, D}$$

Finally, mutants in group IV will grow if compound D is added, but not if compound A, B, or C are added. Thus compound D is the fourth compound in the pathway, and mutations in group IV block the conversion of B into D:

$$\xrightarrow[\text{mutations}]{\text{Group I}} \text{compound C} \xrightarrow[\text{mutations}]{\text{Group II}} \text{compound A} \xrightarrow[\text{mutations}]{\text{Group III}}$$

$$\text{compound B} \xrightarrow[\text{mutations}]{\text{Group IV}} \text{compound D}$$

2. If there were five different types of bases in mRNA instead of four, what would be the minimum codon size (number of nucleotides) required to specify the following numbers of different amino acid types: (a) 4, (b) 20, (c) 30?

• Solution

To answer this question, we must determine the number of combinations (codons) possible when there are different numbers of bases and different codon lengths. In general, the number of different codons possible will be equal to:

$$b^{lg} = \text{number of codons}$$

where, b equals the number of different types of bases and lg equals the number of nucleotides in each codon (codon length). If there are five different types of bases, then:

$$5^1 = 5 \text{ possible codons}$$

$$5^2 = 25 \text{ possible codons}$$

$$5^3 = 125 \text{ possible codons}$$

The number of possible codons must be greater than or equal to the number of amino acids specified. Therefore, a codon length of one nucleotide could specify 4 different amino acids, a codon length of 2 nucleotides could specify 20 different amino acids, and a codon length of 3 nucleotides could specify 30 different amino acids: (a) 1, (b) 2, (c) 3.

3. A template strand in bacterial DNA has the following base sequence:

$$5'-AGGTTTAACGTGCAT-3'$$

What amino acids would be encoded by this sequence?

• **Solution**

To answer this question, we must first work out the mRNA sequence that will be transcribed from this DNA sequence. The mRNA must be antiparallel and complementary to the DNA template strand:

DNA template strand: $5'-AGGTTTAACGTGCAT-3'$
mRNA copied from DNA: $3'-UCCAAAUUGCACGUA-5'$

An mRNA is translated $5' \rightarrow 3'$; so it will be helpful if we turn the RNA molecule around with the 5' end on the left:

mRNA copied from DNA: $5'-AUGCACGUUAAACCU-3'$

The codons consist of groups of three nucleotides that are read successively after the first AUG codon; using Figure 15.12, we can determine that the amino acids are:

$$5'-AUG\text{—}CAC\text{—}GUU\text{—}AAA\text{—}CCU-3'$$

$$fMet\text{——}His\text{——}Val\text{—}Lys\text{——}Pro$$

4. The following triplets constitute anticodons found on a series of tRNAs. Give the amino acid carried by each of these tRNAs.

(a) $5'-UUU-3'$

(b) $5'-GAC-3'$

(c) $5'-UUG-3'$

(d) $5'-CAG-3'$

• **Solution**

To solve this problem, we first determine the codons with which these anticodons pair and then look up the amino acid specified by the codon in Figure 15.12. The codons are antiparallel and complementary to the anticodons. For part *a*, the anticodon is $5'-UUU-3'$. According to the wobble rules in Table 15.2, U in the first position of the anticodon can pair with either A or G in the third position of the codon, so there are two codons that can pair with this anticodon:

Anticodon: $5'-UUU-3'$
Codon: $3'-AAA-5'$
Codon: $3'-GAA-5'$

Listing these codons in the conventional manner, with the 5' end on the right, we have:

Codon: $5'-AAA-3'$
Codon: $5'-AAG-3'$

According to Figure 15.12, both codons specify the amino acid lysine (Lys). Recall that the wobble in the third position allows more than one codon to specify the same amino acid; so any wobble that exists should produce the same amino acid as the standard base pairings would, and we do not need to figure the wobble to answer this question. The answers for parts *b, c,* and *d* are:

(b) Anticodon: $5'-GAC-3'$
    Codon: $3'-CUG-5'$
               $5'-GUC-3'$ codes for Val

(c) Anticodon: $5'-UUG-3'$
    Codon: $3'-AAC-5'$
               $5'-CAA-3'$ codes for Gln

(d) Anticodon: $5'-CAG-3'$
    Codon: $3'-GUC-5'$
               $5'-CUG-3'$ codes for Leu

The New Genetics
MINING GENOMES

THREE DIMENSIONAL PROTEIN STRUCTURE

The function of proteins and other biological macromolecules is directly related to their shape, and understanding the three-dimensional shape of proteins is an important developing field in bioinformatics. This exercise uses tools available at the

Biology Workbench, managed by the San Diego Supercomputing Center at the University of California, San Diego, to visualize the three-dimensional shape of some interesting proteins.

COMPREHENSION QUESTIONS

1. What is the one gene, one enzyme hypothesis? Why was this hypothesis an important advance in our understanding of genetics?

* 2. What three different methods were used to help break the genetic code? What did each reveal and what were the advantages and disadvantages of each?

3. What are isoaccepting tRNAs?

* 4. What is the significance of the fact that many synonymous codons differ only in the third nucleotide position?

* 5. Define the following terms as they apply to the genetic code:

   (a) reading frame
   (b) overlapping code
   (c) nonoverlapping code
   (d) initiation codon
   (e) termination codon
   (f) sense codon
   (g) nonsense codon
   (h) universal code
   (i) nonuniversal codons

6. How is the reading frame of a nucleotide sequence set?

* 7. How are tRNAs linked to their corresponding amino acids?

8. What role do the initiation factors play in protein synthesis?

9. How does the process of initiation differ in bacterial and eukaryotic cells?

*10. Give the elongation factors used in bacterial translation and explain the role played by each factor in translation.

11. What events bring about the termination of translation?

12. Give several examples of RNA–RNA interactions that take place in protein synthesis.

13. What are some types of posttranslational modification of proteins?

*14. Explain how some antibiotics work by affecting the process of protein synthesis.

15. Compare and contrast the process of protein synthesis in bacterial and eukaryotic cells, giving similarities and differences in the process of translation in these two types of cells.

APPLICATION QUESTIONS AND PROBLEMS

*16. Sydney Brenner isolated *Salmonella typhimurium* mutants that were implicated in the biosynthesis of tryptophan and would not grow on minimal medium. When these mutants were tested on minimal medium to which one of four compounds (indole glycerol phosphate, indole, anthranilic acid, and tryptophan) had been added, the growth responses shown in the following table were obtained.

Mutant	Minimal medium	Anthranilic acid	Indole glycerol phosphate	Indole	Trypto- phan
trp-1	−	−	−	−	+
trp-2	−	−	+	+	+
trp-3	−	−	−	+	+
trp-4	−	−	+	+	+
trp-6	−	−	−	−	+
trp-7	−	−	−	−	+
trp-8	−	+	+	+	+
trp-9	−	−	−	−	+
trp-10	−	−	−	−	+
trp-11	−	−	−	−	+

Give the order of indole glycerol phosphate, indole, anthranilic acid, and tryptophan in a biochemical pathway leading to the synthesis of tryptophan. Indicate which step in the pathway is affected by each of the mutations.

17. The addition of a series of compounds yielded in the following biochemical pathway:

$$precursor \xrightarrow[enzyme\ A]{} compound\ I \xrightarrow[enzyme\ B]{}$$

$$compound\ II \xrightarrow[enzyme\ C]{} compound\ III$$

Mutation *a* inactivates enzyme A, mutation *b* inactivates enzyme B, and mutation *c* inactivates enzyme C. Mutants, each having one of these defects, were tested on minimal medium to which compound I, II, or III was added. Fill in the results expected of these tests by placing a plus sign (+) for growth or a minus sign (−) for no growth in the following table:

Strain with mutation	Minimal medium to which is added		
	Compound I	Compound II	Compound III
a			
b			
c			

*18. Assume that the number of different types of bases in RNA is four. What would be the minimum codon size (number of nucleotides) required if the number of different types of amino acids in proteins were:
(a) 2, (b) 8, (c) 17, (d) 45, (e) 75.

19. How many codons would be possible in a triplet code if only three bases (A, C, and U) were used?

*20. Using the genetic code given in Figure 15.12, give the amino acids specified by the following bacterial mRNA sequences, and indicate the amino and carboxyl ends of the polypeptide produced.

(a) 5′–AUGUUUAAAUUUAAAUUUUGA–3′

(b) 5′–AUGUAUAUAUAUAUAUGA–3′

(c) 5′–AUGGAUGAAAGAUUUCUCGCUUGA–3′

(d) 5′–AUGGGUUAGGGGACAUCAUUUUGA–3′

21. A nontemplate strand on DNA has the following base sequence. What amino acid sequence would be encoded by this sequence?

5′–ATGATACTAAGGCCC–3′

*22. The following amino acid sequence is found in a tripeptide: Met-Trp-His. Give all possible nucleotide sequences on the mRNA, on the template strand of DNA, and on the nontemplate strand of DNA that could encode this tripeptide.

23. How many different mRNA sequences can code for a polypeptide chain with the amino acid sequence Met-Leu-Arg? (Be sure to include the stop codon.)

*24. A series of tRNAs have the following anticodons. Consider the wobble rules given in Table 15.2 and give all possible codons with which each tRNA can pair.

(a) 5′–GGC–3′

(b) 5′–AAG–3′

(c) 5′–IAA–3′

(d) 5′–UGG–3′

(e) 5′–CAG–3′

25. An anticodon on a tRNA has the sequence 5′–GCA–3′.

(a) What amino acid is carried by this tRNA?

(b) What would be the effect if the G in the anticodon were mutated to a U?

26. Which of the following amino acid changes could result from a mutation that changed a single base? For each change that could result from the alteration of a single base, determine which position of the codon (first, second, or third nucleotide) in the mRNA must be altered for the change to result.

(a) Leu → Gln

(b) Phe → Ser

(c) Phe → Ile

(d) Pro → Ala

(e) Asn → Lys

(f) Ile → Asn

*27. A synthetic mRNA added to a cell-free protein-synthesizing system produces a peptide with the following amino acid sequence: Met-Pro-Ile-Ser-Ala. What would be the effect on translation if the following components were omitted from the cell-free protein-synthesizing system? What, it any, type of protein would be produced? Explain your reasoning.

(a) initiation factor 1

(b) initiation factor 2

(c) elongation factor Tu

(d) elongation factor G

(f) release factors R_1, R_2, and R_3

(g) ATP

(h) GTP

CHALLENGE QUESTIONS

28. In what ways are spliceosomes and ribosomes similar? In what ways are they different? Can you suggest some possible reasons for their similarities.

*29. Several experiments were conducted to obtain information about how the eukaryotic ribosome recognizes the AUG start codon. In one experiment, the gene that codes for methionine initiator tRNA (tRNA$_i^{Met}$) was located and changed. The nucleotides that specify the anticodon on tRNA$_i^{Met}$ were mutated so that the anticodon in the tRNA was 5′–CCA–3′ instead of 5′–CAU–3′. When this mutated

gene was placed into a eukaryotic cell, protein synthesis took place, but the proteins produced were abnormal. Some of the proteins produced contained extra amino acids, and others contained fewer amino acids.

(a) What do these results indicate about how the ribosome recognizes the starting point for translation in eukaryotic cells? Explain your reasoning.

(b) If the same experiment had been conducted on bacterial cells, what results would you expect?

SUGGESTED READINGS

Agrawal, R. K., P. Penczek, R. A. Grassucci, Y. Li, A. Leith, K. H. Nierhaus, and J. Frank. 1996. Direct visualization of A-, P-, and E-site transfer RNAs in the *Escherichia coli* ribosome. *Science* 271:1000–1002.
A three-dimensional reconstruction of the location of tRNAs in the three sites of the ribosome during translation.

Beadle, G. W., and E. L. Tatum. 1942. Genetic control of biochemical reactions in *Neurospora*. *Proceedings of the National Academy of Sciences* 27:499–506.
Seminal paper in which Beadle and Tatum outline their basic methodology for isolating auxotrophic mutants.

Cech, T. R. 2000. The ribosome is a ribozyme. *Science* 289:878–879.
A brief commentary on research indicating that RNA in the ribosome is responsible for catalyzing peptide-bond formation in protein synthesis.

Dever, T. E. 1999. Translation initiation: adept at adapting. *Trends in Biochemical Science.* 24:398–403.
A discussion of factors that play a role in eukaryotic translation initiation.

Fox, T. D. 1987. Natural variation in the genetic code. *Annual Review of Genetics* 21:67–91.
A review of exceptions to the universal genetic code.

Gualerzi, C. O., and C. L. Pon. 1990. Initiation of mRNA translation in prokaryotes. *Biochemistry* 29:5881–5889.
A good, although fairly technical, review of the process of translational initiation in prokaryotic cells.

Ibba, M., and D. Söll. 1999. Quality control mechanisms during translation. *Science* 286:1893–1897.
A review of mechanisms that prevent errors in translation.

Iborra, F. J., D. A. Jackson, and P. R. Cook. 2001. Coupled transcription and translation within nuclei of mammalian cells. *Science* 293:1139–1142.
A report of experiments demonstrating that some translation takes place within the eukaryotic nucleus.

Khorana, H. G., H. Buchi, H. Ghosh, N. Gupta, T. M. Jacob, H. Kossel, R. Morgan, S. A. Narang, E. Ohtsuka, and R. D. Wells. 1966. Polynucleotide synthesis and the genetic code. *Cold Spring Harbor Symposium on Quantitative Biology* 31:39–49.
The use of repeating RNA polymers in solving the code.

Nirenberg, M., and P. Leder. 1964. RNA code words and protein synthesis I: the effect of trinucleotides upon the binding of sRNA to ribosomes. *Science* 145:1399–1407.
A description of the tRNA binding technique for solving the code.

Nirenberg, M. W., O. W. Jones, P. Leder, B. F. C. Clark, W. S. Sly, and S. Pestka. 1963. On the coding of genetic information. *Cold Spring Harbor Symposium on Quantitative Biology* 28:549–557.
The use of random copolymers in solving the code.

Nakamura, Y., K. Ito, and L. A. Isaksson. 1996. Emerging understanding of translational termination. *Cell* 87:147–150.
A review of translational termination.

Noller, H. F. 1991. Ribosomal RNA and translation. *Annual Review of Biochemistry* 60:191–227.
A good review of RNA–RNA interactions in translation.

Preiss, T., and M. W. Hentze. 1998. Dual function of the messenger RNA cap structure in poly(A)-tail-promoted translation in yeast. *Nature* 392:516–519.
Research article describing evidence that the poly(A) tail plays a role in translational initiation.

Sachs, A. B., P. Sarnow, and M. W. Hentz. 1997. Starting at the beginning, middle, and end: translation initiation in eukaryotes. *Cell* 89:831–838.
A good review of different ways in which translation is initiated in eukaryotic mRNAs.

Wickner, S., M. R. Maurizi, and S. Gottesman. 1999. Postranslational quality control: folding, refolding, and degrading proteins. *Science* 286:1888–1893.
A review of posttranslation modifications of proteins.

Yusupov, M. M., G. Z. Yusupova, A. Baucom, K. Lieberman, T. N. Earnest, J. H. D. Cate, and H. F. Noller. 2001. Crystal structure of the ribosome at 5.5 Å resolution. *Science* 292:883–896.
A report of the structure of the complete ribosome with mRNA and bound tRNAs at high resolution.

16

Control of Gene Expression

The giant transgenic mouse on the left was produced by injecting a rat gene for growth hormone into a mouse embryo; a normal-size mouse is on the right. To ensure expression, the rat gene was linked to a DNA sequence that stimulates the transcription of mouse DNA whenever heavy metals are present. Zinc was provided in the food for the transgenic mouse; some transgenic mice produced 800 times the normal levels of growth hormone. (Courtesy of Dr. Ralph L. Brinster, School of Veterinary Medicine, University of Pennsylvania.)

Creating Giant Mice Through Gene Regulation

In 1982, a group of molecular geneticists led by Richard Palmiter at the University of Washington produced gigantic mice that grew to almost twice the size of normal mice. Palmiter and his colleagues created these large mice through genetic engineering, by injecting the rat gene for growth hormone into the nuclei of fertilized mouse embryos and then implanting these embryos into surrogate mouse mothers. In a few embryos, the rat gene became incorporated into the mouse chromosome and, after birth, these *trans-genic* mice produced growth hormone encoded by the rat gene. Some of the transgenic mice produced from 100 to 800 times the amount of growth hormone found in normal mice, which caused them to grow rapidly into giants.

Inserting foreign genes into bacteria, plants, mice, and even humans is now a routine procedure for molecular geneticists (see Chapter 18). However, simply putting a gene into a cell does not guarantee that the gene will be transcribed or produce a protein; indeed, most foreign genes are never transcribed or translated, which isn't surprising. Organisms have evolved complex systems to ensure that genes are expressed at the appropriate time and in the appropriate amounts, and sequences other than the gene itself are required to ensure transcription and translation. In this chapter, we will learn more about these sequences and other mechanisms that control gene expression.

If foreign genes are rarely expressed, why did the transgenic mice with the gene for rat growth hormone grow so big? Palmiter and his colleagues, aware of the need to provide sequences that control gene expression, linked the rat gene with the mouse metallothionein I promoter sequence, a DNA sequence normally found upstream of the mouse metallothionein I gene. When heavy metals such as zinc are present, they activate the metallothionein promoter sequence, thereby stimulating transcription of the metallothionein I gene. By connecting the rat growth-hormone gene to this promoter, Palmiter and his colleagues provided a means of turning on the transcription of the gene, simply by putting extra zinc in the food for the transgenic mice.

This chapter is about **gene regulation,** the mechanisms and systems that control the expression of genes. We begin by discussing why gene regulation is necessary; the levels at which gene expression is controlled; and the difference between genes and regulatory elements. We then examine gene regulation in bacterial cells. In the second half of the chapter, we turn to gene regulation in eukaryotic cells, which is often more complex than in bacterial cells.

General Principles of Gene Regulation

One of the major themes of molecular genetics is the central dogma, which stated that genetic information flows from DNA to RNA to proteins (see Figure 10.16) and provided a molecular basis for the connection between genotype and phenotype. Although the central dogma brought coherence to early research in molecular genetics, it failed to address a critical issue: How is the flow of information along the molecular pathway *regulated*?

Consider *E. coli*, a bacterium that resides in your large intestine. Your eating habits completely determine the nutrients available to this bacteria: it can't seek out nourishment when nutrients are scarce; nor can it move away when confronted with unpleasant changes. *E. coli* makes up for its inability to alter the external environment by being internally flexible. For example, if glucose is present, *E. coli* uses it to generate ATP; if there's no glucose, it utilizes lactose, arabinose, maltose, xylose, or any of a number of other sugars. When amino acids are available, *E. coli* uses them to synthesize proteins; if a particular amino acid is absent, *E. coli* produces the enzymes needed to synthesize that amino acid. Thus, *E. coli* responds to environmental changes by rapidly altering its biochemistry. This biochemical flexibility, however, has a high price. Producing all the enzymes necessary for every environmental condition would be energetically expensive. So how does *E. coli* maintain biochemical flexibility while optimizing energy efficiency?

The answer is through gene regulation. Bacteria carry the genetic information for many proteins, but only a subset of this genetic information is expressed at any time. When the environment changes, new genes are expressed, and proteins appropriate for the new environment are synthesized. For example, if a carbon source appears in the environment, genes encoding enzymes that take up and metabolize this carbon source are quickly transcribed and translated. When this carbon source disappears, the genes that encode them are shut off. This type of response, the synthesis of an enzyme stimulated by a specific substrate, is called **induction.**

Multicellular eukaryotic organisms face a different dilemma. Individual cells in a multicellular organism are specialized for particular tasks. The proteins produced by a nerve cell, for example, are quite different from those produced by a white blood cell. The problem that a eukaryotic cell faces is how to specialize. Although they are quite different in shape and function, a nerve cell and a blood cell still carry the same genetic instructions.

A multicellular organism's challenge is to bring about the specialization of cells that have a common set of genetic instructions. This challenge is met through gene regulation: all of an organism's cells carry the same genetic information, but only a subset of genes are expressed in each cell type. Genes needed for other cell types are not expressed. Gene regulation is therefore the key to both unicellular flexibility and multicellular specialization, and it is critical to the success of all living organisms.

> **Concepts**
>
> In bacteria, gene regulation maintains internal flexibility, turning genes on and off in response to environmental changes. In multicellular eukaryotic organisms, gene regulation brings about cellular differentiation.

Levels of Gene Control

A gene may be regulated at a number of points along the pathway of information flow from genotype to phenotype (◀FIGURE 16.1). First, regulation may be through the alteration of gene structure. Modifications to DNA or its packaging may influence which sequences are available for transcription or the rate at which sequences are transcribed. DNA methylation and changes in chromatin are two processes that play a pivotal role in gene regulation.

A second point at which a gene can be regulated is at the level of transcription. For the sake of cellular economy, it makes sense to limit protein production early in the transfer of information from DNA to protein, and transcription is an important point of gene regulation in both bacterial and eukaryotic cells. A third potential point of gene regulation is mRNA processing. Eukaryotic mRNA is

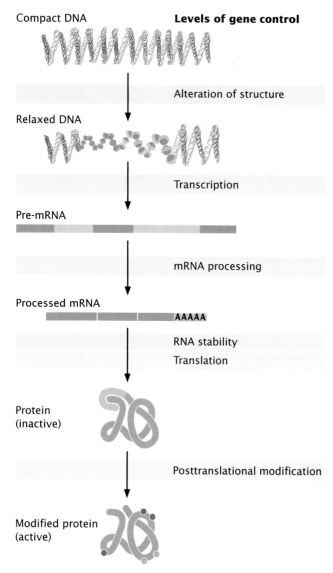

Compact DNA

Levels of gene control

Alteration of structure

Relaxed DNA

Transcription

Pre-mRNA

mRNA processing

Processed mRNA

AAAAA

RNA stability

Translation

Protein
(inactive)

Posttranslational modification

Modified protein
(active)

◀ 16.1 **Gene expression may be controlled at multiple levels.**

extensively modified before it is translated; a 5′ cap is added, the 3′ end is cleaved and polyadenylated, and introns are removed (see Chapter 14). These modifications determine the stability of the mRNA, whether mRNA can be translated, the rate of translation, and the amino acid sequence of the protein produced. There is growing evidence that a number of regulatory mechanisms in eukaryotic cells operate at the level of mRNA processing.

A fourth point for the control of gene expression is the regulation of RNA stability. The amount of protein produced depends not only on the amount of mRNA synthesized, but also on the rate at which the mRNA is degraded; so RNA stability plays an important role in gene expression. A fifth point of gene regulation is at the level of translation, a complex process requiring a large number of enzymes, protein factors, and RNA molecules (Chapter 15).

All of these factors, as well as the availability of amino acids and sequences in mRNA, influence the rate at which proteins are produced and therefore provide points at which gene expression may be controlled.

Finally, many proteins are modified after translation (Chapter 15), and these modifications affect whether the proteins become active; so genes can be regulated through processes that affect posttranscriptional modification. Gene expression may be affected by regulatory activities at any or all of these points.

Concepts

Gene expression may be controlled at any of a number of points along the molecular pathway from DNA to protein, including gene structure, transcription, mRNA processing, RNA stability, translation, and posttranslational modification.

Genes and Regulatory Elements

In our consideration of gene regulation, it will be necessary to distinguish between the DNA sequences that are transcribed and the DNA sequences that regulate the expression of other sequences. We will refer to any DNA sequence that is transcribed into an RNA molecule as a *gene*. According to this definition, genes include DNA sequences that encode proteins, as well as sequences that encode rRNA, tRNA, snRNA, and other types of RNA. **Structural genes** encode proteins that are used in metabolism or biosynthesis or that play a structural role in the cell. **Regulatory genes** are genes whose products, either RNA or proteins, interact with other sequences and affect their transcription or translation. In many cases, the products of regulatory genes are DNA-binding proteins.

We will also encounter DNA sequences that are not transcribed at all but still play a role in regulating other nucleotide sequences. These **regulatory elements** affect the expression of sequences to which they are physically linked. Much of gene regulation takes place through the action of proteins produced by regulatory genes that recognize and bind to regulatory elements.

Concepts

Genes are DNA sequences that are transcribed into RNA. Regulatory elements are DNA sequences that are not transcribed but affect the expression of genes.

DNA-Binding Proteins

Much of gene regulation is accomplished by proteins that bind to DNA sequences and influence their expression. These regulatory proteins generally have discrete functional

(a) Helix-turn-helix

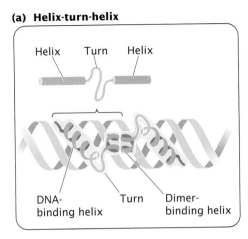

Helix   Turn   Helix

DNA-binding helix   Turn   Dimer-binding helix

(b) Zinc fingers

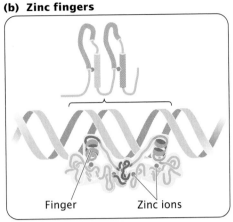

Finger   Zinc ions

(c) Steroid receptor

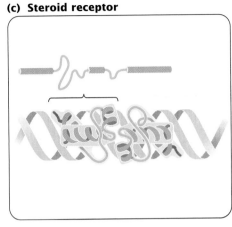

(d) Leucine zipper

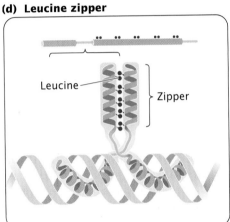

Leucine   Zipper

(e) Helix-loop-helix

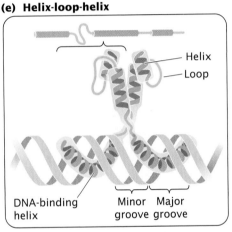

Helix
Loop

DNA-binding helix   Minor groove   Major groove

(f) Homeodomain

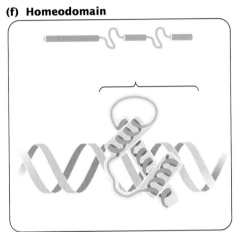

◀ 16.2 **DNA-binding proteins can be grouped into several types on the basis of their structure, or motif.** (a) The helix-turn-helix DNA motif consists of two alpha helices connected by a turn. (b) The zinc-finger motif consists of a loop of amino acids containing a single zinc ion. Most proteins containing zinc fingers have several repeats of the zinc-finger motif. Each zinc finger fits into the major groove of DNA and forms hydrogen bonds with bases in the DNA. (c) The steroid receptor binding motif has two alpha helices, each with a zinc ion surrounded by four cysteine residues. The two alpha helices are perpendicular to one another: one fits into the major groove of the double helix, whereas the other is parallel to the DNA. (d) The leucine-zipper motif consists of a helix of leucine nucleotides and an arm of basic amino acids. DNA-binding proteins usually have two polypeptides; the leucine nucleotides of the two polypeptides face one another, whereas the basic amino acids bind to the DNA. (e) The helix-loop-helix binding motif consists of two alpha helices separated by a loop of amino acids. Two polypeptide chains with this motif join to form a functional DNA-binding protein. A highly basic set of amino acids in one of the helices binds to the DNA. (f) The homeodomain motif consists of three alpha helices; the third helix fits in a major groove of DNA.

parts—called **domains,** typically consisting of 60 to 90 amino acids—that are responsible for binding to DNA. Within a domain, only a few amino acids actually make contact with the DNA. These amino acids (most commonly asparagine, glutamine, glycine, lysine, and arginine) often form hydrogen bonds with the bases or interact with the sugar–phosphate backbone of the DNA. Many regulatory proteins have additional domains that can bind other molecules such as other regulatory proteins.

DNA-binding proteins can be grouped into several distinct types on the basis of a characteristic structure, called a motif, found within the binding domain. Motifs are simple structures, such as alpha helices, that can fit into the major groove of the DNA. Some common DNA-binding motifs are illustrated in ◀ FIGURE 16.2 and are summarized in Table 16.1.

www.whfreeman.com/pierce   Molecular images of several
DNA-binding proteins

Table 16.1 Common DNA-binding motifs

Motif	Location	Characteristics	Binding Site in DNA
Helix-turn-helix	Bacterial regulatory proteins; related motifs in eukaryotic proteins	Two alpha helices	Major groove
Zinc-finger	Eukaryotic regulatory and other proteins	Loop of amino acids with zinc at base	Major groove
Steroid receptor	Eukaryotic proteins	Two perpendicular alpha helices with zinc surrounded by four cysteine residues	Major groove and DNA backbone
Leucine-zipper	Eukaryotic transcription factors	Helix of leucine residues and a basic arm; two leucine residues interdigitate	Two adjacent major grooves
Helix-loop-helix	Eukaryotic proteins	Two alpha helices separated by a loop of amino acids	Major groove
Homeodomain	Eukaryotic regulatory proteins	Three alpha helices	Major groove

Gene Regulation in Bacterial Cells

The mechanisms of gene regulation were first investigated in bacterial cells, where the availability of mutants and the ease of laboratory manipulation made it possible to unravel the mechanisms. When the study of these mechanisms in eukaryotic cells began, it seemed clear that bacterial and eukaryotic gene regulation were quite different. As more and more information has accumulated about gene regulation, however, a number of common themes have emerged, and today many aspects of gene regulation in bacterial and eukaryotic cells are recognized to be similar. Although we will look at gene regulation in these two cell types separately, the emphasis will be on the common themes that apply to all cells.

Operon Structure

One significant difference in bacterial and eukaryotic gene control lies in the organization of functionally related genes. Many bacterial genes that have related functions are clustered and are under the control of a single promoter. These genes are often transcribed together into a single mRNA. Eukaryotic genes, in contrast, are dispersed, and typically, each is transcribed into a separate mRNA. A group of bacterial structural genes that are transcribed together (along with their promoter and additional sequences that control transcription) is called an **operon.**

The organization of a typical operon is illustrated in ◄ FIGURE 16.3. At one end of the operon is a set of structural genes, shown in Figure 16.3 as gene *a*, gene *b*, and gene *c*. These structural genes are transcribed into a single mRNA, which is translated to produce enzymes A, B, and C. These enzymes carry out a series of biochemical reactions that convert precursor molecule X into product Y. The transcription of structural genes *a*, *b*, and *c* is under the control of a promoter, which lies upstream of the first structural gene. RNA polymerase binds to the promoter and then moves downstream, transcribing the structural genes.

A **regulator gene** helps to regulate the transcription of the structural genes of the operon. The regulator gene is not considered part of the operon, although it affects operon function. The regulator gene has its own promoter and is transcribed into a relatively short mRNA, which is translated into a small protein. This **regulator protein** may bind to a region of DNA called the **operator** and affect whether transcription can take place. The operator usually overlaps the 3′ end of the promoter and sometimes the 5′ end of the first structural gene (see Figure 16.3).

Concepts

Functionally related genes in bacterial cells are frequently clustered together as a single transcriptional unit termed an operon. A typical operon includes several structural genes, a promoter for the structural genes, and an operator site where the product of a regulator gene binds.

Negative and Positive Control: Inducible and Repressible Operons

There are two types of transcriptional control: **negative control,** in which a regulatory protein acts as a repressor, binding to DNA and inhibiting transcription; and **positive control,** in

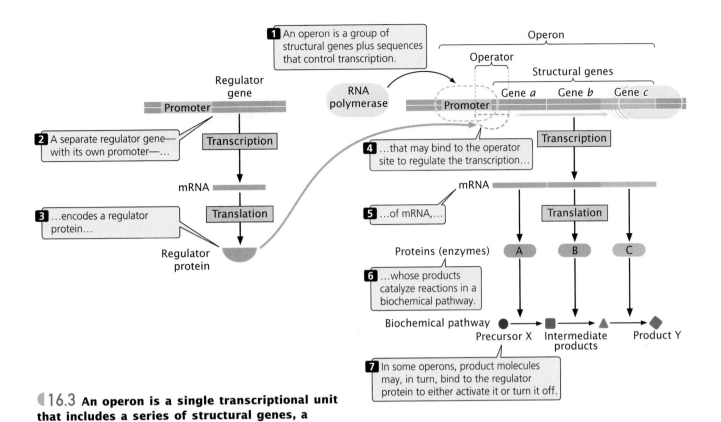

1 An operon is a group of structural genes plus sequences that control transcription.

Regulator gene

Promoter

2 A separate regulator gene—with its own promoter—...

Transcription

mRNA

3 ...encodes a regulator protein...

Translation

Regulator protein

RNA polymerase

Operon

Operator

Structural genes

Promoter   Gene *a*   Gene *b*   Gene *c*

4 ...that may bind to the operator site to regulate the transcription...

Transcription

mRNA

5 ...of mRNA,...

Translation

Proteins (enzymes)   A   B   C

6 ...whose products catalyze reactions in a biochemical pathway.

Biochemical pathway

Precursor X   Intermediate products   Product Y

7 In some operons, product molecules may, in turn, bind to the regulator protein to either activate it or turn it off.

◀ **16.3 An operon is a single transcriptional unit that includes a series of structural genes, a promoter, and an operator.**

which a regulatory protein acts as an activator, stimulating transcription. In the next sections, we will consider several varieties of these two basic control mechanisms.

Negative inducible operons In an operon with negative control at the operator site, the regulatory protein is a repressor—the binding of the regulator protein to the operator inhibits transcription. In a *negative* **inducible operon,** transcription and translation of the regulator gene produce an *active* repressor that readily binds to the operator (◀ FIGURE 16.4a). Because the operator site overlaps with the promoter site, the binding of this protein to the operator physically blocks the binding of RNA polymerase to the promoter and prevents transcription. For transcription to take place, something must happen to prevent the binding of the repressor at the operator site. This type of system is said to be inducible, because transcription is normally off (inhibited) and must be turned on (induced).

Transcription is turned on when a small molecule, an **inducer,** binds to the repressor. ◀ FIGURE 16.4b shows that, when precursor V (acting as the inducer) binds to the repressor, the repressor can no longer bind to the operator. Regulatory proteins frequently have two binding sites: one that binds to DNA and another that binds to a small molecule such as an inducer. Binding of the inducer alters the shape of the repressor, preventing it from binding to DNA. Proteins of this type, which change shape on binding to another molecule, are called **allosteric proteins.**

When the inducer is absent, the repressor binds to the operator, the structural genes are not transcribed, and enzymes D, E, and F (which metabolize precursor V) are not synthesized (see Figure 16.4a). This is an adaptive mechanism: because no precursor V is available, it would be wasteful for the cell to synthesize the enzymes when they have no substrate to metabolize. As soon as precursor V becomes available, some of it binds to the repressor, rendering the repressor inactive and unable to bind to the operator site. Now RNA polymerase can bind to the promoter and transcribe the structural genes. The resulting mRNA is then translated into enzymes D, E, and F, which convert substrate V into product W (see Figure 16.4b). So, an operon with negative inducible control regulates the synthesis of the enzymes economically: the enzymes are synthesized only when their substrate (V) is available.

Negative repressible operons Some operons with negative control are **repressible,** meaning that transcription *normally* takes place and must be turned off, or repressed. The regulator protein in this type of operon also is a repressor but is synthesized in an inactive form that cannot by itself bind to the operator. Because there is no repressor bound to the operator, RNA polymerase readily binds to the promoter and transcription of the structural genes takes place (◀ FIGURE 16.5a).

To turn transcription off, something must happen to make the repressor active. A small molecule called a **core-**

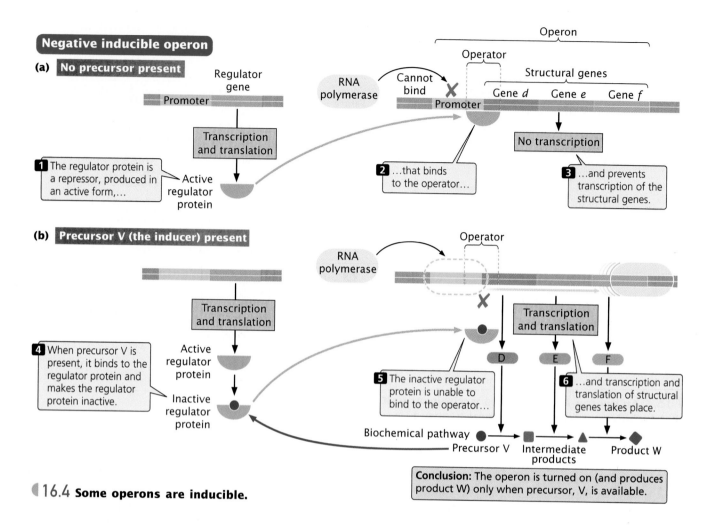

◀16.4 **Some operons are inducible.**

pressor binds to the repressor and makes it capable of binding to the operator. In the example illustrated (see Figure 16.5a), the product (U) of the metabolic reaction is the corepressor. As long as the level of product U is high, it is available to bind to and activate the repressor, preventing transcription (◀FIGURE 16.5b). With the operon repressed, enzymes G, H, and I are not synthesized, and no more U is produced from precursor T. However, when all of product U is used up, the repressor is no longer activated by U and cannot bind to the operator. The inactivation of the repressor allows the transcription of the structural genes and the synthesis of enzymes G, H, and I, resulting in the conversion of precursor T into product U.

As with inducible operons, repressible operons are economical: the enzymes are synthesized only as needed. Note that both the inducible and the repressible systems that we have considered are forms of negative control, in which the regulatory protein is a repressor. We will now consider positive control, in which a regulator protein stimulates transcription.

Positive control With positive control, a regulatory protein binds to DNA (usually at a site other than the operator)

and stimulates transcription. Theoretically, positive control could be inducible or repressible.

In a positive *inducible* operon, transcription would normally be turned off because the regulator protein would be produced in an inactive form. Transcription would take place when an inducer became attached to the regulatory protein, rendering the regulator active. Logically, the inducer should be the precursor of the reaction controlled by the operon so that the necessary enzymes would be synthesized only when the substrate for their reaction was present.

A positive operon could also be repressible; transcription would normally take place and would have to be repressed. In this case, the regulator protein would be produced in a form that readily binds to DNA and stimulates transcription. Transcription would be inhibited when a substance became attached to the activator and rendered it unable to bind to the DNA so that transcription was no longer stimulated. Here, the product (P) of the reaction controlled by the operon would logically be the repressing substance, because it would be economical for the cell to prevent the transcription of genes that allow the synthesis of P when plenty of P is already available.

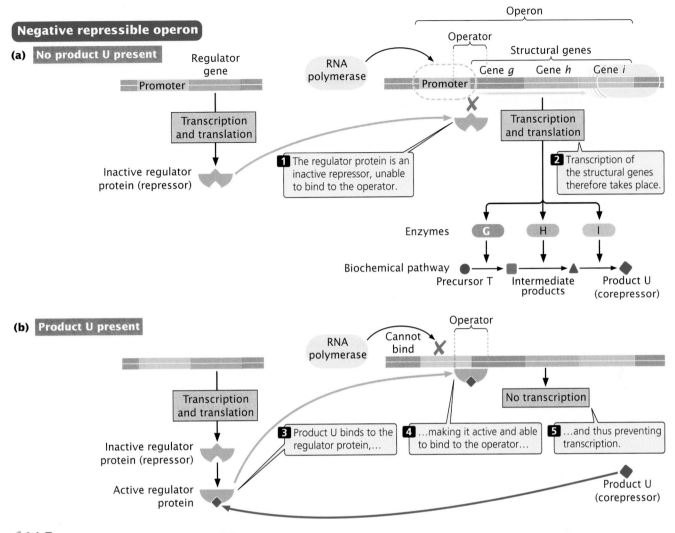

Negative repressible operon

(a) **No product U present**

1 The regulator protein is an inactive repressor, unable to bind to the operator.

2 Transcription of the structural genes therefore takes place.

(b) **Product U present**

3 Product U binds to the regulator protein,...

4 ...making it active and able to bind to the operator...

5 ...and thus preventing transcription.

◀ 16.5 **Some operons are repressible.**

Putting it all together Theoretically, operons might exhibit positive or negative control and be either inducible or repressible. Try sketching out all possible types—negative inducible, negative repressible, positive inducible, and positive repressible. To do so, learn the meanings of *positive* and *negative control* and *inducible* and *repressible*; then use logic to work out the details of whether the regulatory protein is a repressor or an activator and whether it is produced in an active or inactive form. You can check your answers against Table 16.2, where the important features of these four types of operons are summarized. Another useful exercise is to think about the effects of mutations at various sites in different types of operon systems.

Although it is a useful learning device to think of operons as either positive or negative and either inducible or repressive, in reality both positive and negative controls often exist in the same operon.

Concepts

There are two basic types of transcriptional control: negative and positive. In negative control, when a regulatory protein (repressor) binds to DNA, transcription is inhibited; in positive control, when a regulatory protein (activator) binds to DNA, transcription is stimulated. Some operons are inducible; transcription is normally off and must be turned on. Other operons are repressible; transcription is normally on and must be turned off.

The *lac* Operon of *E. coli*

In 1961, François Jacob and Jacques Monod described the "operon model" for the genetic control of lactose metabolism in *E. coli*. This work and subsequent research on the genetics of lactose metabolism established the operon as

Table 16.2	Features of inducible and repressible operons with positive and negative control				
Type of Control	Transcription Normally	Regulator Protein	Effect of Regulatory Protein	Action of Modulator
Negative inducible	Off	Active repressor	Inhibits transcription	Substrate makes repressor inactive
Negative repressible	On	Inactive repressor	Inhibits transcription	Product makes repressor active
Positive inducible	Off	Inactive activator	Stimulates transcription	Substrate makes activator active
Positive repressible	On	Active activator	Stimulates transcription	Product makes activator inactive

the basic unit of transcriptional control in bacteria. Despite the fact that, at the time, no methods were available for determining nucleotide sequences, Jacob and Monod deduced the structure of the operon *genetically* by analyzing the interactions of mutations that interfered with the normal regulation of lactose metabolism. We will examine the effects of some of these mutations after seeing how the *lac* operon regulates lactose metabolism.

Lactose (a disaccharide) is one of the major carbohydrates found in milk; it can be metabolized by *E. coli* bacteria that reside in the gut of mammals. Lactose does not easily diffuse across the *E. coli* cell membrane and must be actively transported into the cell by the enzyme permease (FIGURE 16.6). To utilize lactose as an energy source, *E. coli* must first break it into glucose and galactose, a reaction catalyzed by the enzyme β-galactosidase. This enzyme can also convert lactose into allolactose, a compound that plays an

important role in regulating lactose metabolism. A third enzyme, thiogalactoside transacetylase, also is produced by the *lac* operon, but its function in lactose metabolism is not yet known.

The enzymes β-galactosidase, permease, and transacetylase are encoded by adjacent structural genes in the *lac* operon of *E. coli*. β-Galactosidase is encoded by the *lacZ* gene, permease by the *lacY* gene, and transacetylase by the *lacA* gene (FIGURE 16.7a). When lactose is absent from the medium in which *E. coli* grows, only a few molecules of each enzyme are produced. If lactose is added to the medium and glucose is absent, the rate of synthesis of all three enzymes simultaneously increases about a thousandfold within 2 to 3 minutes. This boost in enzyme synthesis results from transcription of *lacZ*, *lacY*, and *lacA* and examplifies **coordinate induction,** the simultaneous synthesis of several enzymes, stimulated by a specific molecule, the inducer (FIGURE 16.7b). Although

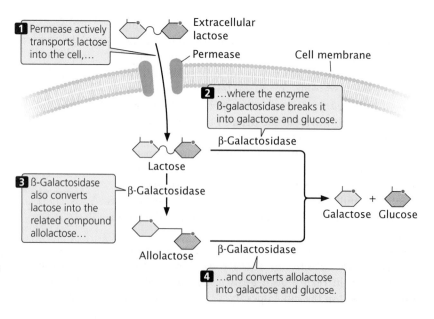

 16.6 **Lactose, a major carbohydrate found in milk, consists of 2 six-carbon sugars linked together.**

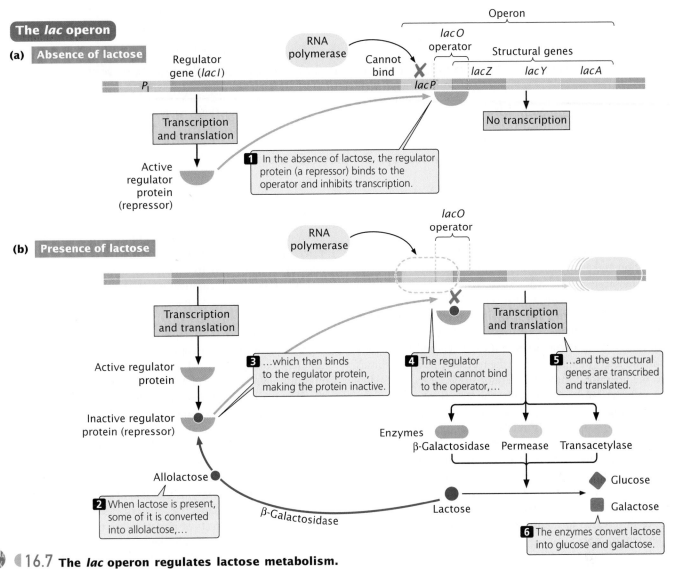

16.7 The *lac* operon regulates lactose metabolism.

lactose appears to be the inducer here, allolactose is actually responsible for induction.

In the *lac* operon, the *lacZ*, *lacY*, and *lacA* genes have a common promoter (*lacP* in Figure 16.7a) and are transcribed together. Upstream of the promoter is a regulator gene, *lacI*, which has its own promoter (P_1). The *lacI* gene is transcribed into a short mRNA that is translated into a repressor. Each repressor consists of four identical polypeptides and has two binding sites; one site binds to allolactose and the other binds to DNA. In the absence of lactose (and, therefore, allolactose), the repressor binds to the *lac* operator site *lacO* (see Figure 16.7a). Jacob and Monod mapped the operator to a position adjacent to the *lacZ* gene; more recent nucleotide sequencing has demonstrated that the operator actually overlaps the 3′ end of the promoter and the 5′ end of *lacZ* (◀FIGURE 16.8).

Immediately upstream of the structural genes is the *lac* promoter. RNA polymerase binds to the promoter and moves

down the DNA molecule, transcribing the structural genes. When the repressor is bound to the operator, the binding of RNA polymerase is blocked, and transcription is prevented. When lactose is present, some of it is converted into allolactose, which binds to the repressor and causes the repressor to be released from the DNA. In the presence of lactose, then, the repressor is inactivated, the binding of RNA polymerase is no longer blocked, the transcription of *lacZ*, *lacY*, and *lacA* takes place, and the *lac* enzymes are produced.

Have you spotted the flaw in the explanation just given for the induction of the *lac* enzymes? You might recall that permease is required to transport lactose into the cell. If the *lac* operon is repressed and no permease is being produced, how does lactose get into the cell to inactivate the repressor and turn on transcription? Furthermore, the inducer is actually allolactose, which must be produced from lactose by β-galactosidase. If β-galactosidase production is repressed, how can lactose metabolism be induced?

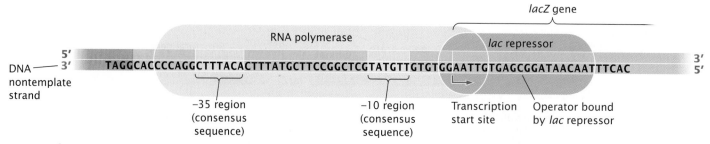

16.8 In the *lac* operon, the operator overlaps the promoter and the 5′ end of the first structural gene.

The answer is that repression never *completely* shuts down transcription of the *lac* operon. Even with active repressor bound to the operator, there is a low level of transcription and a few molecules of β-galactosidase, permease, and transacetylase are synthesized. When lactose appears in the medium, the permease that is present transports a small amount of lactose into the cell. There, the few molecules of β-galactosidase that are present convert some of the lactose into allolactose. The allolactose then attaches to the repressor and alters its shape so that the repressor no longer binds to the operator. When the operator site is clear, RNA polymerase can bind and transcribe the structural genes of the *lac* operon.

Several compounds related to allolactose also can bind to the *lac* repressor and induce transcription of the *lac* operon. One such inducer is isopropylthiogalactoside (IPTG). Although IPTG inactivates the repressor and allows the transcription of *lacZ*, *lacY*, and *lacA*, IPTG is not metabolized by β-galactosidase; for this reason it is often used in research to examine the effects of induction, independent of metabolism.

> **Concepts**
>
> The *lac* operon of *E. coli* controls the transcription of three genes in lactose metabolism: the *lacZ* gene, which encodes β-galactosidase; the *lacY* gene, which encodes permease; and the *lacA* gene, which encodes thiogalactoside transacetylase. The *lac* operon is inducible: a regulator gene produces a repressor that binds to the operator site and prevents the transcription of the structural genes. The presence of allolactose inactivates the repressor and allows the transcription of the *lac* operon.

lac Mutations

Jacob and Monod worked out the structure and function of the *lac* operon by analyzing mutations that affected lactose metabolism. To help define the roles of the different components of the operon, they used **partial diploid** strains of *E. coli*. The cells of these strains possessed two different DNA molecules: the full bacterial chromosome and an extra piece of DNA. Jacob and Monod created these strains by allowing conjugation to take place between two bacteria (see Chapter 8). In conjugation, a small circular piece of DNA (a plasmid) is transferred from one bacterium to another. The plasmid used by Jacob and Monod contained the *lac* operon; so the recipient bacterium became partly diploid, possessing two copies of the *lac* operon. By using different combinations of mutations on the bacterial and plasmid DNA, Jacob and Monod determined that parts of the *lac* operon were cis acting (able to control the expression of genes on the same piece of DNA only) or trans acting (able to control the expression of genes on other DNA molecules).

Structural-gene mutations Jacob and Monod first discovered some mutant strains that had lost the ability to synthesize either β-galactosidase or permease. (They did not study in detail the effects of mutations on the transacetylase enzyme, so it will not be considered here.) These mutations mapped to the *lacZ* or *lacY* structural genes and altered the amino acid sequence of the enzymes encoded by the genes. These mutations clearly affected the *structure* of the enzymes and not the regulation of their synthesis.

Through the use of partial diploids, Jacob and Monod were able to establish that mutations at the *lacZ* and *lacY* genes were independent and usually affected only the product of the gene in which they occurred. Partial diploids with $lacZ^+ \ lacY^-$ on the bacterial chromosome and $lacZ^- \ lacY^+$ on the plasmid functioned normally, producing β-galactosidase and permease in the presence of lactose. (The genotype of a partial diploid is written by separating the genes on each DNA molecule with a slash: $lacZ^+ \ lacY^-/lacZ^- \ lacY^+$.) In this partial diploid, a single functional β-galactosidase gene ($lacZ^+$) is sufficient to produce β-galactosidase; it makes no difference whether the functional β-galactosidase gene is coupled to a functional ($lacY^+$) or a defective ($lacY^-$) permease gene. The same is true of the $lacY^+$ gene.

Regulator-gene mutations Jacob and Monod also isolated mutations that affected the *regulation* of enzyme production.

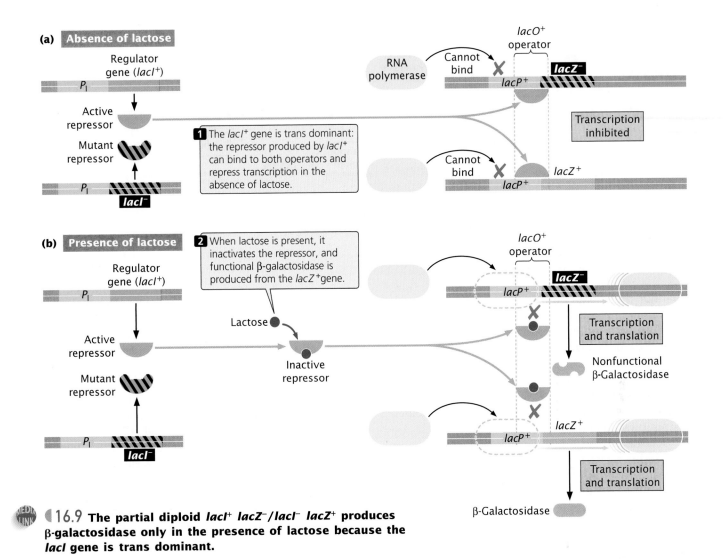

(a) Absence of lactose

Regulator gene (*lacI*⁺)

P_I

Active repressor

Mutant repressor

P_I *lacI*⁻

1 The *lacI*⁺ gene is trans dominant: the repressor produced by *lacI*⁺ can bind to both operators and repress transcription in the absence of lactose.

RNA polymerase

Cannot bind

lacO⁺ operator

lacP⁺   *lacZ*⁻

Transcription inhibited

Cannot bind

lacP⁺   *lacZ*⁺

(b) Presence of lactose

2 When lactose is present, it inactivates the repressor, and functional β-galactosidase is produced from the *lacZ*⁺ gene.

Regulator gene (*lacI*⁺)

P_I

Lactose

Active repressor

Inactive repressor

Mutant repressor

P_I *lacI*⁻

lacO⁺ operator

lacP⁺   *lacZ*⁻

Transcription and translation

Nonfunctional β-Galactosidase

lacP⁺   *lacZ*⁺

Transcription and translation

β-Galactosidase

16.9 The partial diploid *lacI*⁺ *lacZ*⁻/*lacI*⁻ *lacZ*⁺ produces β-galactosidase only in the presence of lactose because the *lacI* gene is trans dominant.

These mutations affected the production of both β-galactosidase and permease, because genes for both enzymes are in the same operon and are regulated coordinately.

Some of these mutations were **constitutive,** causing the *lac* enzymes to be produced all the time, whether lactose was present or not, and these mutations fell into two classes: regulator and operator. Jacob and Monod mapped one class to a site upstream of the structural genes; these mutations occurred in the regulator gene and were designated *lacI*⁻. The construction of partial diploids demonstrated that a *lacI*⁺ gene was dominant over a *lacI*⁻ gene; a single copy of *lacI*⁺ (genotype *lacI*⁺/*lacI*⁻) was sufficient to bring about normal regulation of enzyme production. Furthermore, *lacI*⁺ restored normal control to an operon even if the operon was located on a different DNA molecule, showing that *lacI*⁺ was able to act in trans. A partial diploid with genotype *lacI*⁺ *lacZ*⁻/*lacI*⁻ *lacZ*⁺ functioned normally, synthesizing β-galactosidase only when lactose was present (◄ FIGURE 16.9). In this strain, the *lacI*⁺ gene on the bacterial chromosome was functional, but the *lacZ*⁻ gene was defec-

tive; on the plasmid, the *lacI*⁻ gene was defective, but the *lacZ*⁺ gene was functional. The fact that a *lacI*⁺ gene could regulate a *lacZ*⁺ gene located on a different DNA molecule indicated to Jacob and Monod that the *lacI*⁺ gene product was able to diffuse to either the plasmid or the chromosome.

Some *lacI* mutations isolated by Jacob and Monod prevented transcription from taking place even in the presence of lactose and other inducers such as IPTG. These mutations were referred to as superrepressors (*lacI*ˢ), because they produced repressors that could not be inactivated by an inducer. Recall that the repressor has two binding sites, one for the inducer and one for DNA. The *lacI*ˢ mutations produced a repressor with an altered inducer-binding site, which made the inducer unable to bind to the repressor; consequently, the repressor was always able to attach to the operator site and prevent transcription of the *lac* genes. Superrepressor mutations were dominant over *lacI*⁺; partial diploids with genotype *lacI*ˢ *lacZ*⁺/*lacI*⁺ *lacZ*⁺ were unable to synthesize either β-galactosidase or permease, whether or not lactose was present (◄ FIGURE 16.10).

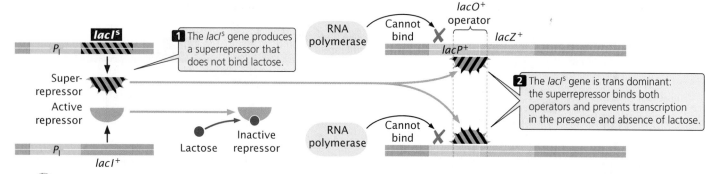

◀ 16.10 **The partial diploid *lacI*ˢ *lacZ*⁺/*lacI*⁺ *lacZ*⁺ fails to produce β-galactosidase in the presence *and* absence of lactose, because the *lacI*ˢ gene encodes a superrepressor.**

Operator mutations Jacob and Monod mapped the other class of constitutive mutants to a site adjacent to *lacZ*. These mutations occurred at the operator site, and were labeled *lacO*ᶜ (O stands for operator and c for constitutive). The *lacO*ᶜ mutations altered the sequence of DNA at the operator so that the repressor protein was no longer able to bind. A partial diploid with genotype *lacI*⁺ *lacO*ᶜ *lacZ*⁺/*lacI*⁺ *lacO*⁺ *lacZ*⁺ exhibited constitutive synthesis of β-galactosidase, indicating that *lacO*ᶜ was dominant over *lacO*⁺.

Analysis of other partial diploids showed that the *lacO* gene was cis acting, affecting only genes on the same DNA molecule. For example, a partial diploid with genotype *lacI*⁺ *lacO*⁺ *lacZ*⁻/*lacI*⁺ *lacO*ᶜ *lacZ*⁺ was constitutive, producing β-galactosidase in the presence or absence of lactose (◀ FIGURE 16.11a), but a partial diploid with genotype *lacI*⁺ *lacO*⁺ *lacZ*⁺/*lacI*⁺ *lacO*ᶜ *lacZ*⁻ produced β-galactosidase only in the presence of lactose (◀ FIGURE 16.11b). In the constitutive partial diploid (*lacI*⁺ *lacO*⁺ *lacZ*⁻/*lacI*⁺ *lacO*ᶜ *lacZ*⁺; see Figure 16.11a), the *lacO*ᶜ mutation and the functional *lacZ*⁺ gene are present on the same DNA molecule; but in *lacI*⁺ *lacO*⁺ *lacZ*⁺/*lacI*⁺ *lacO*ᶜ *lacZ*⁻ (see Figure 16.11b), the *lacO*ᶜ mutation and the functional *lacZ*⁺ gene are on different molecules. The *lacO* mutation affects only genes to which it is physically connected, as is true of all operator mutations. They prevent the binding of a repressor protein to the operator and thereby allow RNA polymerase to transcribe genes on the same DNA molecule. However, they cannot prevent a repressor from binding to normal operators on other DNA molecules.

Promoter mutations Mutations affecting lactose metabolism have also been isolated at the promoter site; these mutations are designated *lacP*⁻, and they interfere with the binding of RNA polymerase to the promoter. Because this binding is essential for the transcription of the structural genes, *E. coli* strains with *lacP*⁻ mutations don't produce *lac* enzymes either in the presence or in the absence of lactose. Like operator mutations, *lacP*⁻ mutations are cis

acting and affect only genes on the same DNA molecule. The partial diploid *lacI*⁺ *lacP*⁺ *lacZ*⁺/*lacI*⁺ *lacP*⁻ *lacZ*⁺ exhibits normal synthesis of β-galactosidase, whereas the *lacI*⁺ *lacP*⁻ *lacZ*⁺/ *lacI*⁺ *lacP*⁺ *lacZ*⁻ fails to produce β-galactosidase whether or not lactose is present.

Positive Control and Catabolite Repression

E. coli and many other bacteria will metabolize glucose preferentially in the presence of lactose and other sugars. They do so because glucose enters glycolysis without further modification and therefore requires less energy to metabolize than do other sugars. When glucose is available, genes that participate in the metabolism of other sugars are repressed, in a phenomenon known as **catabolite repression.** For example, the efficient transcription of the *lac* operon takes place only if lactose is present and glucose is absent. But how is the expression of the *lac* operon influenced by glucose? What brings about catabolite repression?

Catabolite repression results from positive control in response to glucose. (This regulation is in addition to the negative control brought about by the repressor binding at the operator site of the *lac* operon when lactose is absent.) Positive control is accomplished through the binding of a dimeric protein called the **catabolite activator protein** (CAP) to a site that is about 22 nucleotides long and is located within or slightly upstream of the promoter of the *lac* genes (◀ FIGURE 16.12). RNA polymerase does not bind efficiently to many promoters unless CAP is first bound to the DNA. Before CAP can bind to DNA, it must form a complex with a modified nucleotide called **adenosine-3′, 5′-cyclic monophosphate** (cyclic AMP or cAMP), which is important in cellular signaling processes in both bacterial and eukaryotic cells. In *E. coli*, the concentration of cAMP is inversely proportional to the level of available glucose. A high concentration of glucose within the cell lowers the amount of cAMP, and so little cAMP–CAP complex is available to bind to the DNA. Subsequently, RNA polymerase has poor affinity for the *lac* promoter, and little

(a) Partial diploid _lacI+ lacO+ lacZ−/ lacI+ lacOᶜ lacZ+_

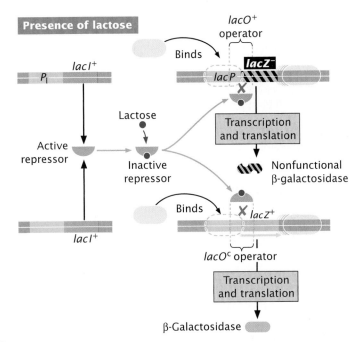

(b) Partial diploid _lacI+ lacO+ lacZ+/ lacI+ lacOᶜ lacZ−_

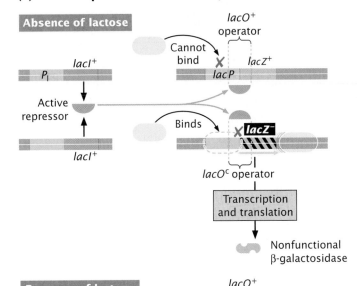

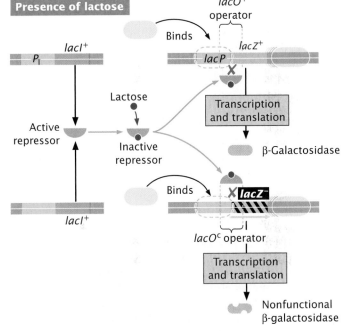

 ◀16.11 **Mutations in _lacO_ are constitutive and cis acting.**
(a) The partial diploid _lacI+ lacO+ lacZ−/lacI+ lacOᶜ lacZ+_ is constitutive,
producing β-galactosidase in the presence and absence of lactose.
(b) The partial diploid _lacI+ lacO+ lacZ+/lacI+ lacOᶜ lacZ−_ is inducible
(produces β-galactosidase only when lactose is present),
demonstrating that the _lacO_ gene is cis acting.

transcription of the _lac_ operon takes place. Low concentrations of glucose stimulate high levels of cAMP, resulting in increased cAMP–CAP binding to DNA. This increase enhances the binding of RNA polymerase to the promoter and increases transcription of the _lac_ genes by some 50-fold.

The catabolite activator protein exerts positive control in more than 20 operons of _E. coli_. The response to

CAP varies among these promoters; some operons are activated by low levels of CAP, whereas others require high levels. CAP contains a helix-turn-helix DNA-binding motif and, when it binds at the CAP site, it causes the DNA helix to bend (◀FIGURE 16.13). The bent helix enables CAP to interact directly with the RNA polymerase enzyme bound to the promoter and facilitate the initiation of transcription.

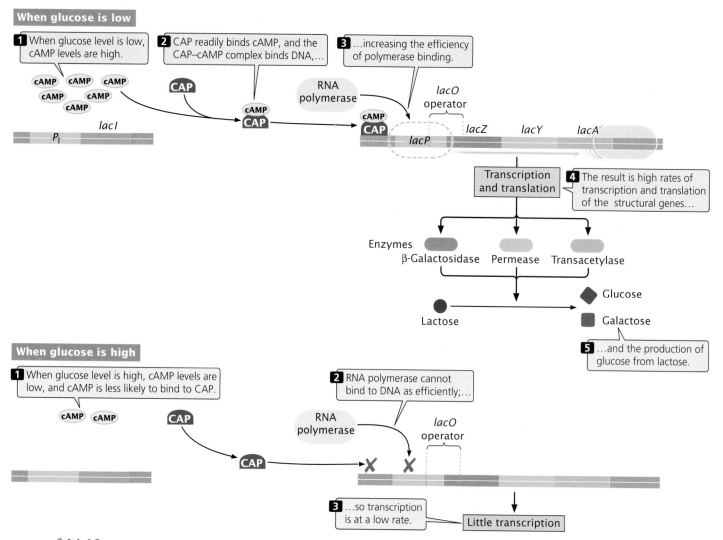

◀16.12 The catabolite activator protein (CAP) binds to the promoter of the *lac* operon and stimulates transcription. CAP must complex with cAMP before binding to the promoter of the *lac* operon. The binding of cAMP–CAP to the promoter activates transcription by facilitating the binding of RNA polymerase. Levels of cAMP are inversely related to glucose: low glucose stimulates high cAMP; high glucose stimulates low cAMP.

Concepts

In spite of its name, catabolite repression is a type of positive control in the *lac* operon. CAP, complexed with cAMP, binds to a site near the promoter and stimulates the binding of RNA polymerase. Cellular levels of cAMP in the cell are controlled by glucose; a low glucose level increases the abundance of cAMP and enhances the transcription of the *lac* structural genes.

www.whfreeman.com/pierce   Information on CAP control and the binding of CAP to DNA

The *trp* Operon of *E. coli*

The *lac* operon just discussed is an inducible operon, one in which transcription does not normally take place and must be turned on. Other operons are repressible; transcription in these operons is normally turned on and must be repressed. The tryptophan (*trp*) operon in *E. coli*, which controls the biosynthesis of the amino acid tryptophan, is an example of a repressible operon.

The *trp* operon contains five structural genes (*trpE, trpD, trpC, trpB,* and *trpA*), which produce components of three enzymes (two of the enzymes consist of two polypeptide chains). These enzymes convert chorismate into tryptophan (◀FIGURE 16.14). The first structural

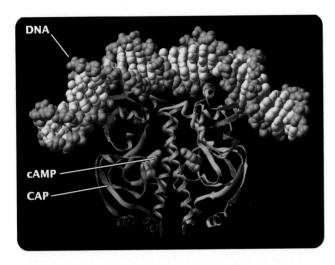

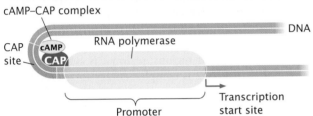

�@16.13 **Binding of the cAMP–CAP complex to DNA produces a sharp bend in DNA that activates transcription.**

gene, *trpE*, contains a long 5′ untranslated region (5′ UTR) that is transcribed but does not encode any of these enzymes. Instead, this 5′ UTR plays an important role in another regulatory mechanism, discussed in the next section. Upstream of the structural genes is the *trp* promoter. When tryptophan levels are low, RNA polymerase binds to the promoter and transcribes the five structural genes into a single mRNA, which is then translated into enzymes that convert chorismate into tryptophan.

Some distance from the *trp* operon is a regulator gene, *trpR*, which encodes a repressor that alone cannot bind DNA (see Figure 16.14). Like the *lac* repressor, the tryptophan repressor has two binding sites, one that binds to DNA at the operator site and another that binds to tryptophan (the activator). Binding with tryptophan causes a conformational change in the repressor that makes it capable of binding to DNA at the operator site, which overlaps the promoter (see Figure 16.14). When the operator is occupied by the tryptophan repressor, RNA polymerase cannot bind to the promoter and the structural genes cannot be transcribed. Thus, when cellular levels of tryptophan are low, transcription of the *trp* operon takes place and more tryptophan is synthesized; when cellular levels of tryptophan are high, transcription of the *trp* operon is inhibited and the synthesis of more tryptophan does not take place.

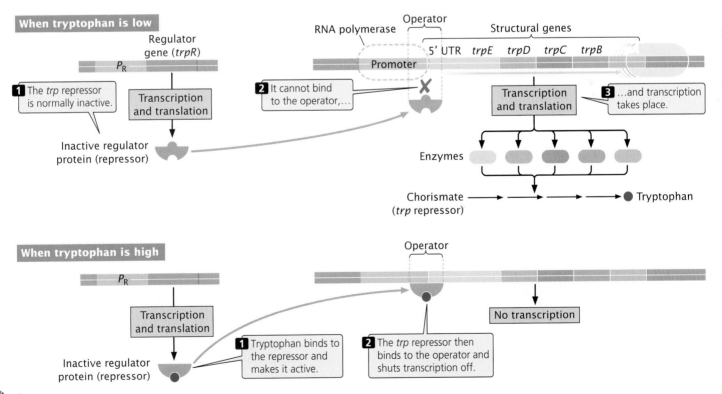

▩ ◀16.14 **The *trp* operon controls the biosynthesis of the amino acid tryptophan in *E. coli*.**

Concepts

The *trp* operon is a repressible operon that controls the biosynthesis of tryptophan. In a repressible operon, transcription is normally turned on and must be repressed. Repression is accomplished through the binding of tryptophan to the repressor, which renders the repressor active. The active repressor binds to the operator and prevents RNA polymerase from transcribing the structural genes.

Attenuation: The Premature Termination of Transcription

We've now seen how both positive and negative control regulate the initiation of transcription in an operon. Some operons have an additional level of control that affects the *continuation* of transcription rather than its initiation. In **attenuation,** transcription begins at the start site, but termination takes place prematurely, before the RNA polymerase even reaches the structural genes. Attenuation occurs in a number of operons that code for enzymes participating in the biosynthesis of amino acids.

We can understand the process of attenuation most easily by looking at one of the best-studied examples,

which is found in the *trp* operon of *E. coli.* Several observations by Charles Yanofsky and his colleagues in the early 1970s indicated that repression at the operator site is not the only method of regulation in the *trp* operon. They isolated a series of mutants that possessed deletions in the transcribed region of the operon. Some of these mutants exhibited increased levels of transcription, yet control at the operator site was unaffected. Furthermore, they observed that two mRNAs of different sizes were transcribed from the *trp* operon: a long mRNA containing sequences for the structural genes and a much shorter mRNA of only 140 nucleotides. These observations led Yanofsky to propose that another mechanism—one that caused premature termination of transcription—also regulates transcription in the *trp* operon.

Close examination of the *trp* operon reveals a region of 162 nucleotides that corresponds to the long 5′ UTR of the mRNA (mentioned earlier) transcribed from the *trp* operon (◀FIGURE 16.15a). The 5′ UTR (also called a leader) contains four regions: region 1 is complementary to region 2, region 2 is complementary to region 3, and region 3 is complementary to region 4. These complementarities allow the 5′ UTR to fold into two different secondary structures (◀FIGURE 16.15b). Which secondary structure is assumed determines whether attenuation will occur.

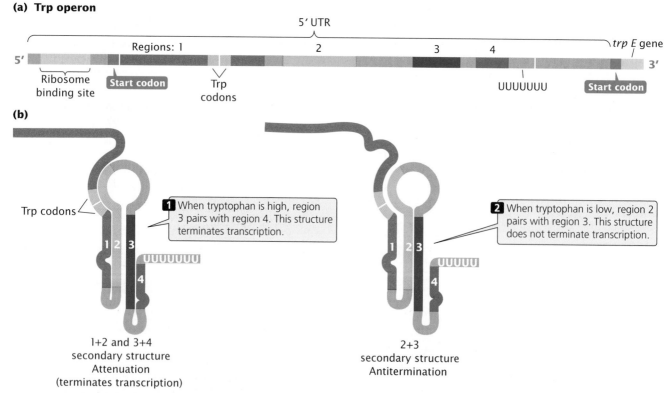

(a) Trp operon

Regions: 1   5′ UTR   2   3   4   *trp E* gene

Ribosome binding site   Start codon   Trp codons   UUUUUUU   Start codon

(b)

Trp codons

1 When tryptophan is high, region 3 pairs with region 4. This structure terminates transcription.

1 2 3   UUUUUUU   4

1+2 and 3+4 secondary structure
Attenuation
(terminates transcription)

2 When tryptophan is low, region 2 pairs with region 3. This structure does not terminate transcription.

1 2 3   UUUUU   4

2+3 secondary structure
Antitermination

 ◀16.15 **Two different secondary structures may be formed by the 5′ UTR of the mRNA transcript of the *trp* operon.**

One of the secondary structures contains one hairpin produced by the base pairing of regions 1 and 2 and another hairpin produced by the base pairing of regions 3 and 4. Notice that a string of uracil nucleotides follows the 3+4 hairpin. Not coincidentally, the structure of a bacterial intrinsic terminator (see Chapter 13) includes a hairpin followed by a string of uracil nucleotides; this secondary structure in the 5′ UTR of the *trp* operon is indeed a terminator and is called an **attenuator.** When cellular levels of tryptophan are high, regions 3 and 4 of the 5′ UTR base pair, to produce the attenuator structure; this base pairing causes transcription to be terminated before the *trp* structural genes can be transcribed.

The alternative secondary structure of the 5′ UTR is produced by the base pairing of regions 2 and 3 (see Figure 16.15b). This base pairing also produces a hairpin, but this hairpin is not followed by a string of uracil nucleotides; so this structure does *not* function as a terminator. When cellular levels of tryptophan are low, regions 2 and 3 base pair, and transcription of the *trp* structural genes is not terminated. RNA polymerase continues past the 5′ UTR into the coding section of the structural genes, and the enzymes that synthesize tryptophan are produced. Because it prevents the termination of transcription, the 2+3 structure is called an **antiterminator.**

To summarize, the 5′ UTR of the *trp* operon can fold into one of two structures. When tryptophan is high, the 3+4 structure forms, transcription is terminated within the 5′ UTR, and no additional tryptophan is synthesized. When tryptophan is low, the 2+3 structure forms, transcription continues through the structural genes, and tryptophan is synthesized. The critical question, then, is, Why does the 3+4 structure arise when tryptophan is high and the 2+3 structure when tryptophan is low?

To answer this question, we must take a closer look at the nucleotide sequence of the 5′ UTR. At the 5′ end, upstream of region 1, is a ribosome-binding site. Region 1 actually encodes a small protein (see Figure 16.15a). Within the coding sequence for this protein are two UGG codons, which specify the amino acid tryptophan; so tryptophan is required for the translation of this 5′ UTR sequence. The protein encoded by the 5′ UTR has not been isolated and is presumed to be unstable; its only apparent function is to control attenuation. Although it was stated in Chapter 14 that a 5′ UTR is not translated into a protein, the 5′ UTR of operons subject to attenuation are exceptions to this rule.

The formation of hairpins in the 5′ UTR of the *trp* operon is controlled by the interplay of transcription and translation that takes place near the 5′ end of the mRNA. Recall that, in prokaryotic cells, transcription and translation are coupled: while transcription is taking place at the 3′ end of the mRNA, translation is initiated at the 5′ end. The precise timing and interaction of these two processes in the 5′ UTR determine whether attenuation occurs.

Transcription when tryptophan levels are high Let's first consider what happens when intracellular levels of tryptophan are high. RNA polymerase begins transcribing the DNA, producing region 1 of the 5′ UTR (◀FIGURE 16.16a). Following RNA polymerase closely, a ribosome binds to the 5′ UTR (at the Shine-Dalgarno sequence, see Chapter 14) and begins to translate the coding region. Meanwhile, RNA polymerase is transcribing region 2 (◀FIGURE 16.16b). Region 2 is complementary to region 1 but, because the ribosome is translating region 1, the nucleotides in regions 1 and 2 cannot base pair. As RNA polymerase begins to transcribe region 3, the ribosome is continuing to translate region 1 (◀FIGURE 16.16c). When the ribosome reaches the two UGG tryptophan codons, it doesn't slow or stall, because tryptophan is abundant and tRNAs charged with tryptophan are readily available. This point is critical to note: because tryptophan is abundant, translation can keep up with transcription.

As it moves past region 1 to the stop codon, the ribosome partly covers region 2; (◀FIGURE 16.16d); meanwhile, RNA polymerase completes the transcription of region 3. Although regions 2 and 3 are complementary, region 2 is partly covered by the ribosome; so it can't base pair with 3.

RNA polymerase continues to move along the DNA, eventually transcribing regions 4 of the 5′ UTR. Region 4 is complementary to region 3, and, because region 3 cannot base pair with region 2, it pairs with region 4. The pairing of regions 3 and 4 (see Figure 16.16d) produces the attenuator—a hairpin followed by a string of uracil nucleotides—and transcription terminates just beyond region 4. The structural genes are not transcribed, no tryptophan-producing enzymes are translated, and no additional tryptophan is synthesized.

Transcription when tryptophan levels are low What happens when tryptophan levels are low? Once again, RNA polymerase begins transcribing region 1 of the 5′ UTR (◀FIGURE 16.16e), and the ribosome binds to the 5′ end of the 5′ UTR and begins to translate region 1 while RNA polymerase continues transcribing region 2 (◀FIGURE 16.16f). When the ribosome reaches the UGG tryptophan codons, it stalls (◀FIGURE 16.16g) because the level of tryptophan is low, and tRNAs charged with tryptophan are scarce or even unavailable. The ribosome sits at the tryptophan codons, awaiting the arrival of a tRNA charged with tryptophan. Stalling of the ribosome does not, however, hinder transcription; RNA polymerase continues to move along the DNA, and transcription gets ahead of translation.

Because the ribosome is stalled at the tryptophan codons in region 1, region 2 is *not* covered by the ribosome when region 3 has been transcribed. Therefore, nucleotides in region 2 and region 3 base pair, forming the 2+3 hairpin (◀FIGURE 16.16h). This hairpin does not cause termination, and so transcription continues. Because region 3 is already paired with region 2, the 3+4 hairpin (the attenuator)

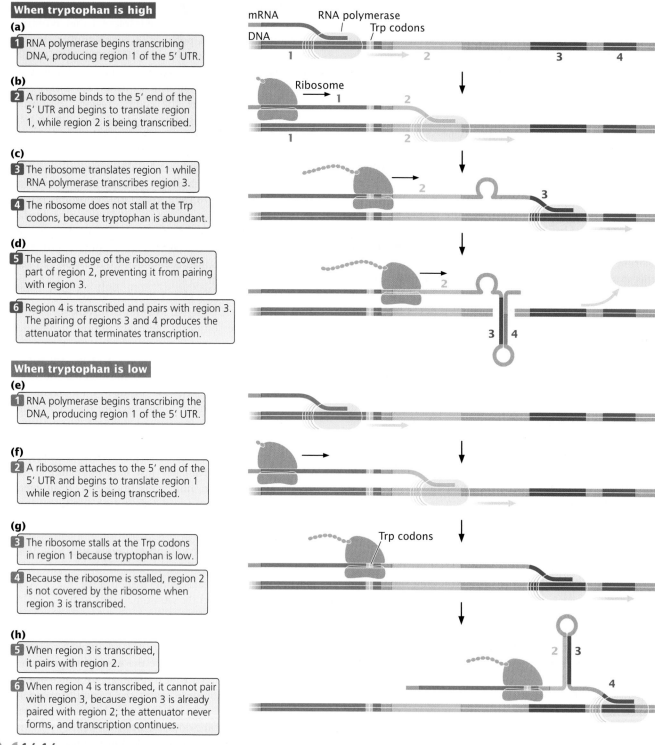

When tryptophan is high

(a)

1 RNA polymerase begins transcribing DNA, producing region 1 of the 5′ UTR.

(b)

2 A ribosome binds to the 5′ end of the 5′ UTR and begins to translate region 1, while region 2 is being transcribed.

(c)

3 The ribosome translates region 1 while RNA polymerase transcribes region 3.

4 The ribosome does not stall at the Trp codons, because tryptophan is abundant.

(d)

5 The leading edge of the ribosome covers part of region 2, preventing it from pairing with region 3.

6 Region 4 is transcribed and pairs with region 3. The pairing of regions 3 and 4 produces the attenuator that terminates transcription.

When tryptophan is low

(e)

1 RNA polymerase begins transcribing the DNA, producing region 1 of the 5′ UTR.

(f)

2 A ribosome attaches to the 5′ end of the 5′ UTR and begins to translate region 1 while region 2 is being transcribed.

(g)

3 The ribosome stalls at the Trp codons in region 1 because tryptophan is low.

4 Because the ribosome is stalled, region 2 is not covered by the ribosome when region 3 is transcribed.

(h)

5 When region 3 is transcribed, it pairs with region 2.

6 When region 4 is transcribed, it cannot pair with region 3, because region 3 is already paired with region 2; the attenuator never forms, and transcription continues.

mRNA
RNA polymerase
DNA
Trp codons

Ribosome

Trp codons

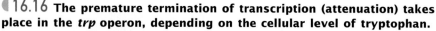
16.16 The premature termination of transcription (attenuation) takes place in the *trp* operon, depending on the cellular level of tryptophan.

never forms, and so attenuation does not occur. RNA polymerase continues along the DNA, past the 5′ UTR, transcribing all the structural genes into mRNA, which is translated into the enzymes encoded by the *trp* operon. These enzymes then synthesize more tryptophan. Important

events in the process of attenuation are summarized in Table 16.3.

Several additional points about attenuation need clarification. The key factor controlling attenuation is the number of tRNA molecules charged with tryptophan, because

Table 16.3	Events in the process of attenuation				
Intracellular Level of Tryptophan	Ribosome Stalls at Trp Codons	Position of Ribosome When Region 3 Is Transcribed	Secondary Structure of 5′ UTR	Termination of Transcription of *trp* Operon	
High	No	Covers region 2	3+4 hairpin	Yes	
Low	Yes	Covers region 1	2+3 hairpin	No	

their availability is what determines whether the ribosome stalls at the tryptophan codons. A second point concerns the synchronization of transcription and translation, which is critical to attenuation. Synchronization is achieved through a pause site located in region 1 of the 5′ UTR. After initiating transcription, RNA polymerase stops temporarily at this site, which allows time for a ribosome to bind to the 5′ end of the mRNA so that translation can closely follow transcription. A third point is that ribosomes do not traverse the convoluted hairpins of the 5′ UTR to translate the structural genes. Ribosomes that attach to the ribosome-binding site at the 5′ end of the mRNA encounter a stop codon at the end of region 1. Ribosomes translating the structural genes attach to a different ribosome-binding site located near the beginning of the *trpE* gene.

Why does attenuation occur? Why do bacteria need attenuation in the *trp* operon? Shouldn't repression at the operator site prevent transcription from taking place when tryptophan levels in the cell are high? Why does the cell have two types of control? Part of the answer is that repression is never complete; some transcription is initiated even when the *trp* repressor is active; repression reduces transcription only as much as 70-fold. Attenuation can further reduce transcription another 8- to 10-fold; so together the two processes are capable of reducing transcription of the *trp* operon more than 600-fold. Both mechanisms provide *E. coli* with a much finer degree of control over tryptophan synthesis than either could achieve alone.

Another reason for the dual control is that attenuation and repression respond to different signals: repression responds to the cellular levels of tryptophan, whereas attenuation responds to the number of tRNAs charged with tryptophan. There may be times when it is advantageous for the cell to be able to respond to these different signals. Finally, the *trp* repressor affects several operons other than the *trp* operon. It's possible that at an earlier stage in the evolution of *E. coli*, the *trp* operon was controlled only by attenuation. The *trp* repressor may have evolved primarily to control the other operons and only incidentally affects the *trp* operon.

Attenuation is a complex process to grasp because you must simultaneously visualize how two dynamic processes—transcription and translation—interact, and it's easy to get the two processes confused. Remember that attenuation

entails the early termination of *transcription*, not translation (although events in translation bring about the termination of transcription). Attenuation often causes confusion because we know that transcription must precede translation. We're comfortable with the idea that transcription might affect translation, but it's harder to imagine that the effects of translation could influence transcription, as it does in attenuation. The reality is that transcription and translation are closely coupled in prokaryotic cells, and events in one process can easily affect the other.

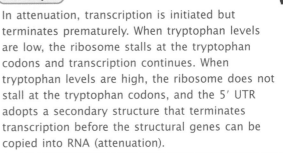

Concepts

In attenuation, transcription is initiated but terminates prematurely. When tryptophan levels are low, the ribosome stalls at the tryptophan codons and transcription continues. When tryptophan levels are high, the ribosome does not stall at the tryptophan codons, and the 5′ UTR adopts a secondary structure that terminates transcription before the structural genes can be copied into RNA (attenuation).

www.whfreeman.com/pierce   More information on attenuation

Antisense RNA in Gene Regulation

All the regulators of gene expression that we have considered so far have been proteins. Several examples of RNA regulators have also been discovered. These small RNA molecules are complementary to particular sequences on mRNAs and are called **antisense RNA.** They control gene expression by binding to sequences on mRNA and inhibiting translation.

Translational control by antisense RNA is seen in the regulation of the *ompF* gene of *E. coli* (◀ FIGURE 16.17a). Two *E. coli* genes, *ompF* and *ompC*, produce outer-membrane proteins that function as diffusion pores, allowing bacteria to adapt to external osmolarities (the tendency of water to move across a membrane owing to different ion concentrations). Under most conditions, both the *ompF* and the *ompC* genes are transcribed and translated. When the osmolarity of the medium increases, a regulator gene named *micF*—for

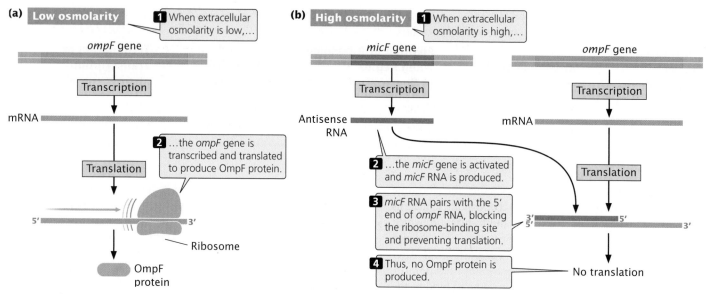

(a) Low osmolarity **1** When extracellular osmolarity is low,...

ompF gene

Transcription

mRNA

2 ...the *ompF* gene is transcribed and translated to produce OmpF protein.

Translation

5' 3' Ribosome

OmpF protein

(b) High osmolarity **1** When extracellular osmolarity is high,...

micF gene

Transcription

Antisense RNA

2 ...the *micF* gene is activated and *micF* RNA is produced.

3 *micF* RNA pairs with the 5' end of *ompF* RNA, blocking the ribosome-binding site and preventing translation.

4 Thus, no OmpF protein is produced.

ompF gene

Transcription

mRNA

Translation

3' 5'
5' 3'

No translation

16.17 Antisense RNA can regulate translation.

mRNA-interfering complementary RNA—is activated and *micF* RNA is produced (Figure 16.17b). The *micF* RNA, an antisense RNA, binds to a complementary sequence in the 5′ UTR of the *ompF* mRNA and inhibits the binding of the ribosome. This inhibition reduces the amount of translation (see Figure 16.17b), which results in fewer OmpF proteins in the outer membrane and thus reduces the detrimental movement of substances across the membrane owing to the changes in osmolarity. A number of examples of antisense RNA controlling gene expression have now been identified in bacteria and bacteriophages.

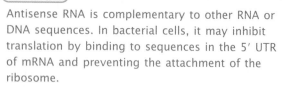

Concepts

Antisense RNA is complementary to other RNA or DNA sequences. In bacterial cells, it may inhibit translation by binding to sequences in the 5′ UTR of mRNA and preventing the attachment of the ribosome.

Transcriptional Control in Bacteriophage Lambda

Bacteriophage λ is a virus that infects the bacterium *E. coli* (Chapter 8). Bacteriophage λ possesses a single DNA chromosome consisting of 48,502 nucleotides surrounded by a protein coat. A bacteriophage infects a bacterial cell by attaching to the cell wall and injecting its DNA into the cell. Inside the cell, λ phage undergoes either of two life cycles.

In the *lytic cycle* (see Chapter 8), phage genes are transcribed and translated to produce phage coat proteins and enzymes that synthesize from 100 to 200 copies of the phage DNA. The viral components are assembled to produce phage particles, and the phage produces a protein that causes the cell to lyse. The released phage can then infect other bacterial cells. In the *lysogenic cycle,* phage genes that encode replication enzymes and phage proteins are not immediately transcribed. Instead, the phage DNA integrates into the bacterial chromosome as a prophage. When the bacterial chromosome replicates, the prophage is duplicated along with the bacterial genes and is passed to the daughter cells in bacterial reproduction. The prophage may later excise from the bacterial chromosome and enter the lytic cycle.

Whether a λ phage enters the lytic or the lysogenic cycle depends on the regulation of the phage genes. In the lytic cycle, the genes that encode replication enzymes, phage proteins, and bacterial cell lysis are transcribed; but, in the lysogenic cycle, these genes are repressed.

Like bacterial genes, functionally related phage genes are clustered together into operons. There are four major operons in the phage λ chromosome (Figure 16.18). The early right operon contains genes that are required for DNA replication and are transcribed early in the lytic cycle. The early left operon contains genes necessary for recombination and the integration of phage DNA into the bacterial chromosome as a part of the lysogenic cycle. A third operon, the late operon, contains genes that encode the protein coat of the phage, produced late in the lytic cycle. The fourth operon is the repressor operon, which produces the λ repressor responsible for maintaining the prophage DNA in a dormant state. Although there are several additional promoters on the λ chromosome that may be activated at special times, here the emphasis is on three general features of transcriptional control in bacteriophage λ.

First, both positive control and negative control are seen in λ gene regulation. Several proteins act as repressors, inhibiting transcription, whereas others act as activators,

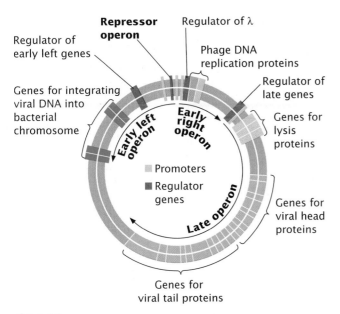

16.18 The bacteriophage λ chromosome contains four major operons: the early left operon, the early right operon, the late operon, and the repressor operon.

this controlled transcription ensures that genes appropriate to each stage of the lytic or lysogenic cycle are expressed.

A third feature of λ gene regulation is the use of **transcriptional antiterminator proteins,** which bind to RNA polymerase and alter its structure, allowing it to ignore certain terminators (◀FIGURE 16.19a). In the absence of the antiterminator protein, RNA polymerase stops at a terminator located early in the operon (◀FIGURE 16.19b), and so only some of the genes in the operon are transcribed and translated.

> **Concepts**
>
> The entry of bacteriophage λ into lysis or lysogeny is controlled by a cascade of reactions, in which the transcription of operons is turned on and off in a specific sequence. The expression of the operons is controlled by the affinity of different promoters for repressor and activator proteins and through transcriptional antiterminators.

Eukaryotic Gene Regulation

Many features of gene regulation are common to both bacterial and eukaryotic cells. For example, in both types of cells, DNA-binding proteins influence the ability of RNA polymerase to initiate transcription. However, there are also some differences, although these differences are often a matter of degree. First, eukaryotic genes are not organized into operons and are rarely transcribed together into a single mRNA molecule; instead, each structural gene typically has its own promoter and is transcribed separately. Second, chromatin structure affects gene expression in eukaryotic cells; DNA must unwind

stimulating transcription. The λ repressor, which plays a major role in λ gene regulation, can act as either an activator or a repressor.

A second feature is that transcription is accomplished through a cascade of reactions. As one operon is transcribed, it produces a protein that regulates the transcription of a second operon, which produces a protein that affects the transcription of a third operon. Thus, the operons are activated and repressed in a particular order, with the use of several different promoters, each with an affinity for specific activators and repressors. As each promoter is activated, only the genes under its control are transcribed;

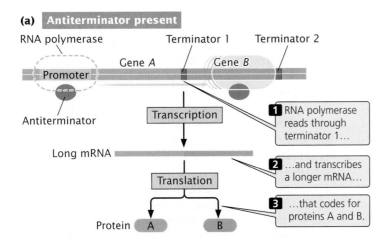

(a) Antiterminator present

(b) Antiterminator absent

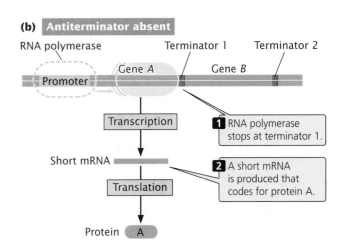

16.19 Antiterminator proteins bind to RNA polymerase and alter its structure so that it ignores certain terminators.

from the histone proteins before transcription can take place. Third, although both repressors and activators function in eukaryotic and bacterial gene regulation, activators seem to be more common in eukaryotic cells. Finally, the regulation of gene expression in eukaryotic cells is characterized by a greater diversity of mechanisms that act at different points in the transfer of information from DNA to protein.

Eukaryotic gene regulation is less well understood than bacterial regulation, partly owing to the larger genomes in eukaryotes, their greater sequence complexity, and the difficulty of isolating and manipulating mutations that can be used in the study of gene regulation. Nevertheless, great advances in our understanding of the regulation of eukaryotic genes have been made in recent years, and eukaryotic regulation continues to be one of the cutting-edge areas of research in genetics.

Chromatin Structure and Gene Regulation

One type of gene control in eukaryotic cells is accomplished through the modification of gene structure. In the nucleus, histone proteins associate to form octamers, around which helical DNA tightly coils to create chromatin (see Figure 11.5). In a general sense, this chromatin structure represses gene expression. For a gene to be transcribed, transcription factors, activators, and RNA polymerase must bind to the DNA. How can these events take place with DNA wrapped tightly around histone proteins? The answer is that before transcription, chromatin structure changes, and the DNA becomes more accessible to the transcriptional machinery.

DNase I hypersensitivity Several types of changes are observed in chromatin structure when genes become transcriptionally active. One type is an increase in the sensitivity of chromatin to degradation by DNase I, an enzyme that digests DNA. When tightly bound by histone proteins, DNA is resistant to DNase I digestion because the enzyme cannot gain access to the DNA. When DNA is less tightly bound by histones, it becomes sensitive to DNase I degradation. Thus, the ability of DNase I to digest DNA provides an indication of the DNA–histone association.

As genes become transcriptionally *active*, regions around the genes become highly sensitive to the action of DNase I (see Chapter 11). These regions, called **DNase I hypersensitive sites,** frequently develop about 1000 nucleotides upstream of the start site of transcription, suggesting that the chromatin in these regions adopts a more open configuration during transcription. This relaxation of the chromatin structure may allow regulatory proteins access to binding sites on the DNA. Indeed, many DNase I hypersensitive sites correspond to known binding sites for regulatory proteins.

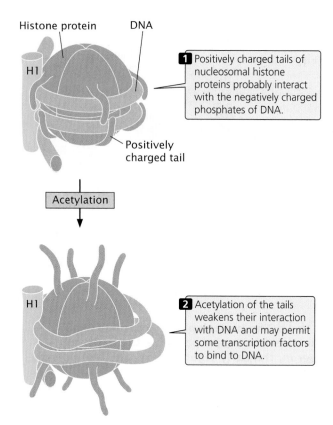

1 Positively charged tails of nucleosomal histone proteins probably interact with the negatively charged phosphates of DNA.

2 Acetylation of the tails weakens their interaction with DNA and may permit some transcription factors to bind to DNA.

◀ **16.20 The acetylation of histone proteins alters chromatin structure and permits some transcription factors to bind to DNA.**

Histone acetylation One factor affecting chromatin structure is acetylation, the addition of acetyl groups (CH_3CO) to histone proteins. Histones in the octamer core of the nucleosome have two domains: (1) a globular domain that associates with other histones and the DNA and (2) a positively charged tail domain that probably interacts with the negatively charged phosphates on the backbone of DNA (◀ FIGURE 16.20).

Acetyl groups are added to histone proteins by acetyltransferase enzymes; the acetyl groups destabilize the nucleosome structure, perhaps by neutralizing the positive charges on the histone tails and allowing the DNA to separate from the histones. Other enzymes called deacetylases strip acetyl groups from histones and restore chromatin repression. Certain transcription factors (see Chapter 13) and other proteins that regulate transcription either have acetyltransferase activity or attract acetyltransferases to the DNA.

Some transcription factors and other regulatory proteins are known to alter chromatin structure without acetylating histone proteins. These **chromatin-remodeling complexes** bind directly to particular sites on DNA and reposition the nucleosomes, allowing transcription factors to bind to promoters and initiate transcription.

DNA methylation Another change in chromatin structure associated with transcription is the methylation of cytosine bases, which yields 5-methylcytosine (see Figure 10.18). Heavily methylated DNA is associated with the repression of transcription in vertebrates and plants, whereas transcriptionally active DNA is usually unmethylated in these organisms.

DNA methylation is most common on cytosine bases adjacent to guanine nucleotides on the same strand (CpG); so two methylated cytosines sit diagonally across from each other on opposing strands:

$$\cdots GC \cdots$$
$$\cdots CG \cdots$$

DNA regions with many CpG sequences are called **CpG islands** and are commonly found near transcription start sites. While genes are not being transcribed, these CpG islands are often methylated, but the methyl groups are removed before the initiation of transcription. CpG methylation is also associated with long-term gene repression, such as on the inactivated X chromosome of female mammals (see Chapter 4).

Recent evidence suggests an association between DNA methylation and the deacetylation of histones, both of which repress transcription. Certain proteins that bind tightly to methylated CpG sequences form complexes with other proteins that act as histone deacetylases. In other words, methylation appears to attract deacetylases, which remove acetyl groups from the histone tails, stabilizing the nucleosome structure and repressing transcription. Demethylation of DNA would allow acetyltransferases to add acetyl groups, disrupting nucleosome structure and permitting transcription.

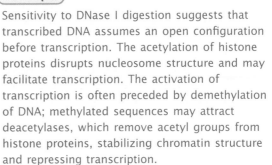

Concepts

Sensitivity to DNase I digestion suggests that transcribed DNA assumes an open configuration before transcription. The acetylation of histone proteins disrupts nucleosome structure and may facilitate transcription. The activation of transcription is often preceded by demethylation of DNA; methylated sequences may attract deacetylases, which remove acetyl groups from histone proteins, stabilizing chromatin structure and repressing transcription.

Transcriptional Control in Eukaryotic Cells

Transcription is an important level of control in eukaryotic cells, and this control requires a number of different types of proteins and regulatory elements. The initiation of eukaryotic transcription was discussed in detail in Chapter 13. Recall that general transcription factors and RNA polymerase assemble into a *basal transcription apparatus*, which binds to a *core promoter* located immediately upstream of a gene. The basal transcription apparatus is capable of minimal levels of transcription; *transcriptional activator proteins* are required to bring about normal levels of transcription. These proteins bind to a regulatory promoter, which is located upstream of the core promoter, and to *enhancers*, which may be located some distance from the gene (◀ FIGURE 16.21).

Transcriptional activators, coactivators and repressors

Transcriptional activator proteins stimulate transcription by facilitating the assembly or action of the basal transcription apparatus at the core promoter; the activators may interact directly with the basal transcription apparatus or indirectly

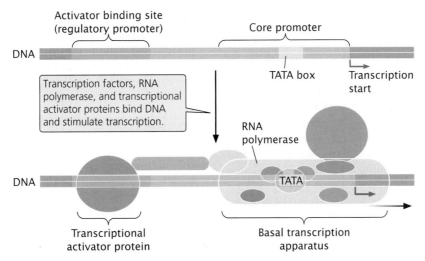

◀ **16.21 Transcriptional activator proteins bind to sites on DNA and stimulate transcription.** Most act by stimulating or stabilizing the assembly of the basal transcription apparatus.

through protein **coactivators.** Some activators and coactivators, as well as the general transcription factors, also have acteyltransferase activity and facilitate transcription further by altering chromatin structure (see earlier subsection on histone acetylation).

Transcriptional activator proteins have two distinct functions (see Figure 16.21). First, they are capable of binding DNA at a specific base sequence, usually a consensus sequence in a regulatory promoter or enhancer; for this function, most transcriptional activator proteins contain one or more of the DNA-binding motifs discussed at the beginning of this chapter. A second function is the ability to interact with other components of the transcriptional apparatus and influence the rate of transcription. Most do so by either stabilizing or stimulating the assembly of the basal transcription apparatus.

GAL4 is a transcription activator protein that regulates the transcription of several yeast genes in galactose metabolism. GAL4 contains several zinc fingers and binds to a DNA sequence called UAS_G (upstream activating sequence for GAL4). UAS_G exhibits the properties of an enhancer—a regulatory sequence that may be some distance from the regulated gene and is independent of the gene in position and orientation (see Chapter 13). When bound to UAS_G, GAL4 stimulates the transcription of yeast genes needed for metabolizing galactose.

A particular region of GAL4 binds another protein called GAL80, which regulates the activity of GAL4 in the presence of galactose. When galactose is absent, GAL80 binds to GAL4 (two molecules of GAL80 bind to each molecule of GAL4), preventing GAL4 from activating transcription (◄FIGURE 16.22). When galactose is present, however, it binds to GAL80, causing a conformational change in the protein so that it can no longer bind GAL4. The GAL4 protein is then available to activate the transcription of the genes whose products metabolize galactose.

GAL4 and a number of other transcriptional activator proteins contain multiple amino acids with negative charges that form an *acidic activation domain.* These acidic activators stimulate transcription by enhancing the ability of TFIIB (see Chapter 13), one of the general transcription factors, to join the basal transcription apparatus. Without the activator, the binding of TFIIB is a slow process; the activator helps "recruit" TFIIB to the initiation complex, thereby stimulating the binding of RNA polymerase and the initiation of transcription. Acidic activators may also enhance other steps in the assembly of the basal transcription apparatus.

Some regulatory proteins in eukaryotic cells act as repressors, inhibiting transcription. These repressors may bind to sequences in the regulatory promoter or to distant sequences called *silencers,* which, like enhancers, are position and orientation independent. Unlike repressors in bacteria, most eukaryotic repressors do not directly block RNA polymerase. These repressors may compete with activators for DNA binding sites: when a site is occupied by an activa-

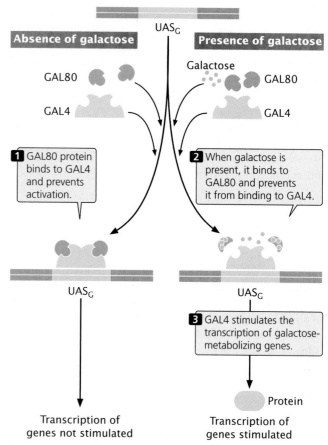

Absence of galactose

GAL80

GAL4

1 GAL80 protein binds to GAL4 and prevents activation.

UAS_G

Transcription of genes not stimulated

Presence of galactose

Galactose

GAL80

GAL4

2 When galactose is present, it binds to GAL80 and prevents it from binding to GAL4.

UAS_G

3 GAL4 stimulates the transcription of galactose-metabolizing genes.

Protein

Transcription of genes stimulated

◄**16.22 Transcription is activated by GAL4 in response to galactose.** GAL4 binds to the UAS_G site and controls the transcription of genes in galactose metabolism.

tor, transcription is stimulated, but, if a repressor occupies that site, no activation occurs. Alternatively, a repressor may bind to sites near an activator site and prevent the activator from contacting the basal transcription apparatus. A third possible mechanism of repressor action is direct interference with the assembly of the basal transcription apparatus, thereby blocking the initiation of transcription.

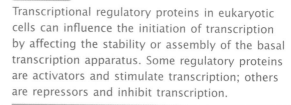

Concepts

Transcriptional regulatory proteins in eukaryotic cells can influence the initiation of transcription by affecting the stability or assembly of the basal transcription apparatus. Some regulatory proteins are activators and stimulate transcription; others are repressors and inhibit transcription.

Enhancers and insulators Enhancers are capable of affecting transcription at distant promoters. For example, an enhancer that regulates the gene encoding the alpha chain of the T-cell receptor is located 69,000 bp down-

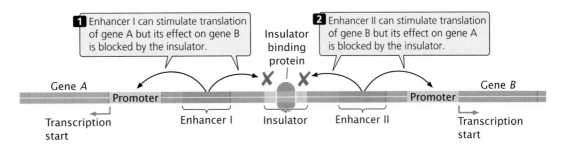

◀ 16.23 An insulator blocks the action of an enhancer on a promoter when the insulator lies between the enhancer and the promoter.

stream of the gene's promoter. Furthermore, the exact position and orientation of an enhancer relative to the promoter can vary. How can an enhancer affect the initiation of transcription taking place at a promoter that is tens of thousands of base pairs away? The mechanism of action of many enhancers is not known, but evidence suggest that, in some cases, activator proteins bind to the enhancer and cause the DNA between the enhancer and the promoter to loop out, bringing the promoter and enhancer close to one another, so that the transcriptional activator proteins are able to directly interact with the basal transcription apparatus at the core promoter.

Most enhancers are capable of stimulating any promoter in their vicinities. Their effects are limited, however, by **insulators** (also called boundary elements), which are DNA sequences that block or insulate the effect of enhancers in a position-dependent manner. If the insulator lies between the enhancer and the promoter, it blocks the action of the enhancer; but, if the insulator lies outside the region between the two, it has no effect (◀ Figure 16.23). Specific proteins bind to insulators and play a role in their blocking activity, but exactly how this takes place is poorly understood. Some insulators also limit the spread of changes in chromatin structure that affect transcription.

Concepts

Some activator proteins bind to enhancers, which are regulatory elements that are distant from the gene whose transcription they stimulate. Insulators are DNA sequences that block the action of enhancers.

Coordinated gene regulation Although eukaryotic cells do not possess operons, several eukaryotic genes may be activated by the same stimulus. For example, many eukaryotic cells respond to extreme heat and other stresses by producing **heat-shock proteins** that help to prevent damage from such stressing agents. Heat-shock proteins are produced by approximately 20 different genes. During times of environmental stress, the transcription of all the heat-shock genes is greatly elevated. Groups of bacterial genes are often coordinately expressed (turned on and off together) because they are physically clustered as an operon and have the same promoter, but coordinately expressed genes in eukaryotic cells are not clustered. How, then, is the transcription of eukaryotic genes coordinately controlled if they are not organized into an operon?

Genes that are coordinately expressed in eukaryotic cells are able to respond to the same stimulus because they have regulatory sequences in common in their promoters or enhancers. For example, different eukaryotic heat-shock genes possess a common regulatory element upstream of their start sites. A transcriptional activator protein binds to this regulatory element during stress and elevates transcription. Such common DNA regulatory sequences are called **response elements;** they typically contain short consensus sequences (Table 16.4) at varying distances from the gene being regulated.

A single eukaryotic gene may be regulated by several different response elements. The metallothionein gene protects cells from the toxicity of heavy metals by encoding a protein that binds to heavy metals and removes them from cells. The basal transcription apparatus assembles around the TATA box, just upstream of the transcription start site for the metallothionein gene, but the apparatus alone is capable of only low rates of transcription. The presence of heavy metals stimulates much higher rates of transcription.

Table 16.4	A few response elements found in eukaryotic cells	
Response Element	**Responds to**	**Consensus Sequence**
Heat-shock element	Heat and other stress	CNNGAANNTCCNNG
Glucocorticoid response element	Glucocorticoids	TGGTACAAATGTTCT
Phorbol ester response element	Phorbal esters	TGACTCA
Serum response element	Serum	CCATATTAGG

Source: Adapted from B. Lewin, *Genes IV* (Oxford: Oxford University Press, 1994), p. 880.

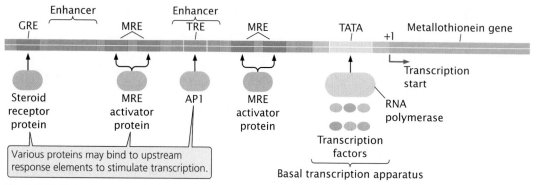

◀ **16.24 Multiple response elements (MREs) are found in the upstream region of the metallothionein gene.** The basal transcription apparatus binds near the TATA box. In response to heavy metals, activator proteins bind to several MRE elements and stimulate transcription. The TRE response element is the binding site for transcription factor AP1. In response to glucocorticoid hormones, steroid receptor proteins bind to the GRE response element located approximately 250 nucleotides upstream of the metallothionein gene and stimulate transcription.

Other response elements found upstream of the metallothionein gene also contribute to increasing its rate of transcription. For example, several copies of a metal response element (MRE) are upstream of the metallothionein gene (◀ FIGURE 16.24). Heavy metals stimulate the binding of an activator protein to MREs, which elevates the rate of transcription of the metallothionein gene. The presence of multiple copies of this response element permits high rates of transcription to be induced by metals. Two enhancers also are located in the upstream region of the metallothionein gene; one enhancer contains a response element known as TRE, which stimulates transcription in the presence of phorbol esters. A third response element called GRE is located approximately 250 nucleotides upstream of the metallothionein gene and stimulates transcription in response to glucocorticoid hormones.

This example illustrates a common feature of eukaryotic transcriptional control: a single gene may be activated by several different response elements, found in both promoters and enhancers. Multiple response elements allow the same gene to be activated by different stimuli. At the same time, the presence of the same response element in different genes allows a single stimulus to activate multiple genes. In this way, response elements allow complex biochemical responses in eukaryotic cells.

Gene Control Through Messenger RNA Processing

Alternative splicing allows a pre-mRNA to be spliced in multiple ways, generating different proteins in different tissues or at different times in development (see Chapter 14). Many eukaryotic genes undergo alternative splicing, and the regulation of splicing is probably an important means of controlling gene expression in eukaryotic cells.

The T-antigen gene of the mammalian virus SV40 serves as a well-studied example of alternative splicing. This gene is capable of encoding two different proteins, the large T and small t antigens. Which of the two proteins is produced depends on which of two alternative 5' splice sites is used during RNA splicing (◀ FIGURE 16.25). The use of one 5' splice site produces mRNA that encodes the large T

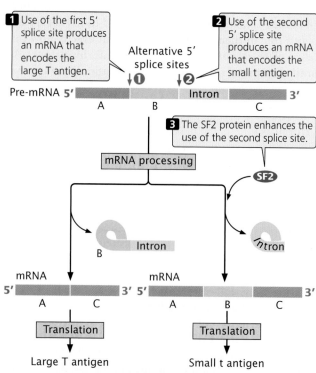

◀ **16.25 Alternative splicing leads to the production of the small t antigen and the large T antigen in the mammalian virus SV40.**

XX genotype

X:A = 1.0

Tra-2 protein

Sxl gene → Sxl protein → *tra* pre-mRNA → Tra protein ⌐→ *dsx* pre-mRNA → Dsx ♀ protein → Female Fly

1 In X:A = 1.0 embryos, the activated *Sxl* gene produces a protein…

2 …that causes *tra* pre-mRNA to be spliced at a downstream 3′ site…

3 …to produce Tra protein.

4 Together, Tra and Tra-2 proteins direct the female-specific splicing of *dsx* pre-mRNA,…

5 …which produces proteins causing the embryo to develop into a female.

XY genotype

X:A = 0.5

Sxl gene → ✗ → No Sxl protein   *tra* pre-mRNA → Nonfunctional Tra protein   *dsx* pre-mRNA → Dsx ♂ protein → Male Fly

1 In X:A = 0.5 embryos, the *Sxl* gene is not activated,…

2 …and no Sxl protein is produced.

3 Thus *tra* pre-mRNA is spliced at an upstream site,…

4 …producing a nonfunctional Tra protein.

5 Without Tra, the male-specific splicing of *dsx* pre-mRNA…

6 …produces male Dsx proteins that cause the embryo to develop into a male.

◀ **16.26 Alternative splicing controls sex determination in Drosophila.**

antigen, whereas the use of the other 5′ splice site (which is farther downstream) produces an mRNA encoding the small t antigen.

A protein called splicing factor 2 (SF2) enhances the production of mRNA encoding the small t antigen (see Figure 16.25). Splicing factor 2 has two binding domains: one is an RNA-binding region and the other has alternating serine and arginine amino acids. These two domains are typical of **SR proteins,** which often play a role in regulating splicing. Splicing factor 2 stimulates the binding of U1 snRNP to the 5′ splice site, one of the earliest steps in RNA splicing (see Chapter 14). The precise mechanism by which SR proteins influence the choice of splice sites is poorly understood. One model suggests that SF2 and other SR proteins bind to specific splice sites on mRNA and stimulate the attachment of snRNPs, which then commit the site to splicing.

Another example of alternative mRNA splicing that regulates the expression of genes controls whether a fruit fly develops as male or female. Sex differentiation in *Drosophila* arises from a cascade of gene regulation (◀ FIGURE 16.26). When the ratio of X chromosomes to the number of haploid sets of autosomes (the X : A ratio; see Chapter 4) is 1, a female-specific promoter is activated early in development and stimulates the transcription of the *sex-lethal* (*Sxl*) gene. The protein encoded by *Sxl* regulates the splicing of the pre-mRNA transcribed from another gene called *transformer* (*tra*). The splicing of *tra* pre-mRNA results in the production of Tra protein. Together with another protein (Tra-2), Tra stimulates the female-specific splicing of

pre-mRNA from yet another gene called *doublesex* (*dsx*). This event produces a female-specific Dsx protein, which causes the embryo to develop female characteristics.

In male embryos, which have an X : A ratio of 0.5 (see Figure 16.26), the promoter that transcribes the *Sxl* gene in females is inactive; so no Sxl protein is produced. In the absence of Sxl protein, Tra pre-mRNA is spliced at a different 3′ splice site to produce a nonfunctional form of Tra protein (◀ FIGURE 16.27). In turn, the presence of this nonfunctional Tra in males causes Dsx pre-mRNAs to be spliced differently (see Figure 16.26), and a male-specific Dsx protein is produced. This event causes the development of male-specific traits.

In summary, the Tra, Tra-2, and Sxl proteins regulate alternative splicing that produces male and female phenotypes in *Drosophila*. Exactly how these proteins regulate alternative splicing is not yet known, but it's possible that the Sxl protein (produced only in females) may block the upstream splice site on the *tra* pre-mRNA. This blockage would force the spliceosome to use the downstream 3′ splice site, which causes the production of Tra protein and eventually results in female traits (see Figure 16.27).

Concepts

Eukaryotic genes may be regulated through the control of mRNA processing. The selection of alternative splice sites leads to the production of different proteins.

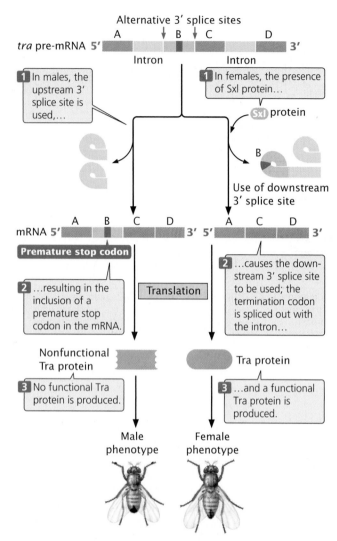

Alternative 3′ splice sites

tra pre-mRNA 5′

Intron   Intron

1 In males, the upstream 3′ splice site is used,…

1 In females, the presence of Sxl protein…

Sxl protein

B

Use of downstream 3′ splice site

mRNA 5′   3′ 5′   3′

Premature stop codon

2 …resulting in the inclusion of a premature stop codon in the mRNA.

Translation

2 …causes the downstream 3′ splice site to be used; the termination codon is spliced out with the intron…

Nonfunctional Tra protein

Tra protein

3 No functional Tra protein is produced.

3 …and a functional Tra protein is produced.

Male phenotype

Female phenotype

◀ **16.27 Alternative splicing of *tra* pre-mRNA.** Two alternative 3′ splice sites are present.

Gene Control Through RNA Stability

The amount of a protein that is synthesized depends on the amount of corresponding mRNA available for translation. The amount of available mRNA, in turn, depends on both the rate of mRNA synthesis and the rate of mRNA degradation. Eukaryotic mRNAs are generally more stable than bacterial mRNAs, which typically last only a few minutes before being degraded, but nonetheless there is great variability in the stability of eukaryotic mRNA: some persist for only a few minutes; others last for hours, days, or even months. These variations can result in large differences in the amount of protein that is synthesized.

Cellular RNA is degraded by ribonucleases, enzymes that specifically break down RNA. Most eukaryotic cells contain 10 or more types of ribonucleases, and there are several different pathways of mRNA degradation. In one pathway, the 5′ cap is first removed, followed by 5′→3′

removal of nucleotides. A second pathway begins at the 3′ end of the mRNA and removes nucleotides in the 3′→5′ direction. In a third pathway, the mRNA can be cleaved at internal sites.

Messenger RNA degradation from the 5′ end is most common and begins with the removal of the 5′ cap. This pathway is usually preceded by the shortening of the poly(A) tail. Poly(A)-binding proteins (PABPs) normally bind to the poly(A) tail and contribute to its stability-enhancing effect. The presence of these proteins at the 3′ end of the mRNA protects the 5′ cap. When the poly(A) tail has been shortened below a critical limit, the 5′ cap is removed, and nucleases then degrade the mRNA by removing nucleotides from the 5′ end. These observations suggest that the 5′ cap and 3′ poly(A) tail of eukaryotic mRNA physically interact with each other, most likely by the poly(A) tail bending around so that the PABPs make contact with the 5′ cap (see Figure 15.21). Other parts of eukaryotic mRNA, including sequences in the 5′ UTR, the coding region, and the 3′ UTR, also affect mRNA stability.

Poly(A) tails are added to the 3′ ends of some bacterial mRNAs, but they are shorter than those typically associated with eukaryotic mRNA and have the opposite effect; they appear to destabilize most prokaryotic mRNAs.

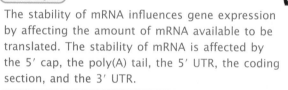

Concepts

The stability of mRNA influences gene expression by affecting the amount of mRNA available to be translated. The stability of mRNA is affected by the 5′ cap, the poly(A) tail, the 5′ UTR, the coding section, and the 3′ UTR.

RNA Silencing

Recent evidence indicates that the expression of some genes may be suppressed through **RNA silencing,** also known as RNA interference and posttranscriptional gene silencing. Although many of the details of this mechanism are still poorly understood, it appears to be widespread, existing in fungi, plants, and animals. It may also prove to be a powerful tool for artificially regulating gene expression in genetically engineered organisms.

RNA silencing is initiated by the presence of double-stranded RNA, which may arise in several ways: by the transcription of inverted repeats in DNA into a single RNA molecule that base pairs with itself; by the simultaneous transcription of two different RNA molecules that are complementary to one another and pair; or by the replication of double-stranded RNA viruses (◀FIGURE 16.28a). In *Drosophila,* an enzyme called Dicer cleaves and processes the double-stranded RNA to produce small pieces of single-stranded RNA that range in length from 21 to 25 nucleotides (◀FIGURE 16.28b). These **small interfering**

RNAs (siRNAs) then pair with complementary sequences in mRNA and attract an RNA–protein complex that cleaves the mRNA approximately in the middle of the bound siRNA. After cleavage, the mRNA is further degraded. In the nucleus, siRNAs serve as guides for the methylation of complementary sequences in DNA, which then affects transcription. Some related RNA molecules produced through the cleavage of double-stranded RNA bind to complementary sequences in the 3′ UTR of mRNA and inhibit their translation.

RNA silencing is thought to have evolved as a defense against RNA viruses and transposable elements that move through an RNA intermediate (see Chapter 11). The extent to which it contributes to normal gene regulation is uncertain, but dramatic phenotypic effects result from some mutations that occur in the enzymes that carry out RNA silencing.

Concepts

RNA silencing is initiated by double-stranded RNA molecules that are cleaved and processed. The resulting small interfering RNAs bind to complementary sequences in mRNA and bring about their cleavage and degradation. Small interfering RNAs may also stimulate the methylation of complementary sequences in DNA.

Translational and Posttranslational Control

Ribosomes, aminoacyl tRNAs, initiation factors, and elongation factors are all required for the translation of mRNA molecules. The availability of these components affects the rate of translation and therefore influences gene expression. The initiation of translation in some mRNAs is regulated by proteins that bind to the mRNA's 5′ UTR and inhibit the binding of ribosomes, similar to the way in which repressor proteins bind to operators and prevent the transcription of structural genes.

Many eukaryotic proteins are extensively modified after translation by the selective cleavage and trimming of amino acids from the ends, by acetylation, or by the addition of phosphates, carboxyl groups, methyl groups, and carbohydrates to the protein. These modifications affect the transport, function, and activity of the proteins and have the capacity to affect gene expression.

Concepts

The initiation of translation may be affected by proteins that bind to specific sequences at the 5′ end of mRNA. The availability of ribosomes, tRNAs, initiation and elongation factors, and other components of the translational apparatus may affect the rate of translation.

16.28 RNA silencing leads to the degradation of mRNA and the methylation of DNA.

Connecting Concepts

A Comparison of Bacterial and Eukaryotic Gene Control

Now that we have considered the major types of gene regulation, let's review some of the similarities and differences of bacterial and eukaryotic gene control.

1. Much of gene regulation in both bacterial and eukaryotic cells is accomplished through proteins that bind to specific sequences in DNA. Regulatory proteins come in a variety of types, but most can be characterized according to a small set of DNA-binding motifs.

2. Regulatory proteins that affect transcription exhibit two basic types of control: *repressors* inhibit transcription (negative control); *activators* stimulate transcription (positive control). Both negative control and positive control are found in bacterial and eukaryotic cells.

3. Complex biochemical and developmental events in bacterial and eukaryotic cells may require a cascade of gene regulation, in which the activation of one set of genes stimulates the activation of another set.

4. Most gene regulation in bacterial cells is at the level of transcription (although it does exist at other levels). Gene regulation in eukaryotic cells often takes place at multiple levels, including chromatin structure, transcription, mRNA processing, and RNA stability.

5. In bacterial cells, genes are often clustered in operons and are coordinately expressed by transcription into a single mRNA molecule. In contrast, each eukaryotic gene typically has its own promoter and is transcribed independently. Coordinate regulation in eukaryotic cells takes place through common response elements, present in the promoters and enhancers of the genes. Different genes that have the same response element in common are influenced by the same regulatory protein.

6. Chromatin structure plays a role in eukaryotic (but not bacterial) gene regulation. In general, condensed chromatin represses gene expression; chromatin structure must be altered before transcription. Acetylation of the histone proteins, which may be influenced by the degree of DNA methylation, appears to be important in bringing about these changes in chromatin structure.

7. The initiation of transcription is a relatively simple process in bacterial cells, and regulatory proteins function by blocking or stimulating the binding of RNA polymerase to DNA. Eukaryotic transcription requires complex machinery that includes RNA polymerase, general transcription factors, and transcriptional activators, which allows transcription to be influenced by multiple factors.

8. Some eukaryotic transcriptional activator proteins function at a distance from the gene by binding to enhancers, causing a loop in the DNA, and bringing the promoter and enhancer into close proximity. Some distant-acting sequences analogous to enhancers have been described in bacterial cells, but they appear to be less common.

9. The greater time lag between transcription and translation in eukaryotic cells than in bacterial cells allows mRNA stability and mRNA processing to play larger roles in eukaryotic gene regulation.

Connecting Concepts Across Chapters

The focus of this chapter has been on how the flow of information from genotype to phenotype is controlled. We have seen that there are a number of potential points of control in this pathway of information flow, including changes in gene structure, transcription, mRNA processing, mRNA stability, translation, and posttranslational modifications.

Gene regulation is critically important from a number of perspectives. It is essential to the survival of cells, which cannot afford to simultaneously transcribe and translate all of their genes. The evolution of complex genomes consisting of thousands of genes would not have been possible without some mechanism to selectively control gene expression. Gene regulation is also important from a practical point of view. A number of human diseases are caused by the breakdown of gene regulation, which produces proteins at inappropriate times or places. Gene regulation is also important to genetic engineering, where the key to success is often not getting genes into a cell, which is relatively easy, but getting them expressed at useful levels. For all of these reasons, there is tremendous interest in how gene expression is controlled, and understanding gene regulation is one of the frontiers of genetic research.

Information presented in this chapter builds on the foundation of molecular genetics developed in Chapters 10 through 15. The mechanisms of gene regulation provide important links to several topics in subsequent chapters. Gene regulation is important to the success of recombinant DNA, which is discussed in Chapter 18. Gene regulation also plays an important role in the genetics of development and cancer, which are discussed in Chapter 21.

CONCEPTS SUMMARY

- Gene expression may be controlled at different levels, including the alteration of gene structure, transcription, mRNA processing, RNA stability, translation, and posttranslational modification. Much of gene regulation is through the action of regulatory proteins binding to specific sequences in DNA.

- Genes in bacterial cells are typically clustered into operons— groups of functionally related structural genes and the sequences that control their transcription. Structural genes in an operon are transcribed together as a single mRNA.

- In negative control, a repressor protein binds to DNA and inhibits transcription. In positive control, an activator protein binds to DNA and stimulates transcription. In inducible operons, transcription is normally off and must be turned on; in repressible operons, transcription is normally on and must be turned off.

- The *lac* operon of *E. coli* is a negative inducible operon that controls the metabolism of lactose. In the absence of lactose, a repressor binds to the operator and prevents transcription of the structural genes that encode β-galactosidase, permease, and transacetylase. When lactose is present, some of it is converted into allolactose, which binds to the repressor and makes it inactive, allowing the structural genes to be transcribed and lactose to be metabolized. When all the lactose has been metabolized, the repressor once again binds to the operator and blocks transcription.

- Positive control in the *lac* operon and other operons is through catabolite repression. When complexed with cAMP, the catabolite activator protein (CAP) binds to a site in or near the promoter and stimulates the transcription of the structural genes. Levels of cAMP are indirectly correlated with glucose; so low levels of glucose stimulate transcription and high levels inhibit transcription.

- The *trp* operon of *E. coli* is a negative repressible operon that controls the biosynthesis of tryptophan.

- Attenuation is another level of control that allows transcription to be stopped before RNA polymerase has reached the structural genes. It takes place through the close coupling of transcription and translation and depends on the secondary structure of the 5′ UTR sequence.

- Small RNA molecules, called antisense RNA, are complementary to sequences in mRNA and may inhibit translation by binding to these sequences, thereby preventing the attachment or progress of the ribosome.

- Transcriptional control regulates the lytic and lysogenic cycles of bacteriophage λ. The transcription of certain operons stimulates the transcription of some operons and represses the transcription of others. Which operons are stimulated and which are repressed depends on the affinity of promoters for repressor and activator proteins.

- Like gene regulation in bacterial cells, much of eukaryotic regulation is accomplished through the binding of regulatory proteins to DNA. However, there are no operons in eukaryotic cells, and gene regulation is characterized by a greater diversity of mechanisms acting at different levels.

- In eukaryotic cells, chromatin structure represses gene expression. During transcription, chromatin structure may be altered by the acetylation of histone proteins and demethylation.

- The initiation of eukaryotic transcription is controlled by general transcription factors that assemble into the basal transcription apparatus and by transcriptional activator proteins that stimulate normal levels of transcription by binding to regulatory promoters and enhancers.

- Some DNA sequences limit the action of enhancers by blocking their action in a position-dependent manner.

- Coordinately controlled genes in eukaryotic cells respond to the same factors because they have common response elements that are stimulated by the same transcriptional activator.

- Gene expression in eukaryotic cells may be influenced by RNA processing.

- Gene expression may be regulated by changes in RNA stability. The 5′ cap, the coding sequence, the 3′ UTR, and the poly(A) tail are important in controlling the stability of eukaryotic mRNAs. Proteins binding to the 5′ end of eukaryotic mRNA may affect its translation.

- RNA silencing takes place when double-stranded RNA is cleaved and processed to produce small interfering RNAs that bind to complementary mRNAs and bring about their cleavage and degradation.

- Control of the posttranslational modification of proteins also may play a role in gene expression.

IMPORTANT TERMS

gene regulation (p. 435)
induction (p. 435)
structural gene (p. 436)
regulatory gene (p. 436)
regulatory element (p. 436)
domain (p. 437)
operon (p. 438)
regulator gene (p. 438)
regulator protein (p. 438)
operator (p. 438)
negative control (p. 438)
positive control (p. 438)

inducible operon (p. 439)
inducer (p. 439)
allosteric protein (p. 439)
repressible operon (p. 439)
corepressor (p. 439)
coordinate induction (p. 442)
partial diploid (p. 444)
constitutive mutation (p. 445)
catabolite repression (p. 446)
catabolite activator protein
  (CAP) (p. 446)

adenosine-3′, 5′-cyclic
  monophosphate
  (cAMP) (p. 446)
attenuation (p. 450)
attenuator (p. 451)
antiterminator (p. 451)
antisense RNA (p. 453)
transcriptional antiterminator
  protein (p. 455)
DNase I hypersensitive
  site (p. 456)

chromatin-remodeling
  complex (p. 456)
CpG island (p. 457)
coactivator (p. 458)
insulator (p. 459)
heat-shock protein (p. 459)
response element (p. 459)
SR protein (p. 461)
RNA silencing (p. 462)
small interfering RNAs
  (siRNAs) (p. 462)

Worked Problems

1. A regulator gene produces a repressor in an inducible operon. A geneticist isolates several constitutive mutations affecting this operon. Where might these constitutive mutations occur? How would the mutations cause the operon to be constitutive?

• Solution

An inducible operon is normally not being transcribed, meaning that the repressor is active and binds to the operator, inhibiting transcription. Transcription takes place when the inducer binds to the repressor, making it unable to bind to the operator. Constitutive mutations cause transcription to take place at all times, whether the inducer is present or not. Constitutive mutations might occur in the regulator gene, altering the repressor so that it is never able to bind to the operator. Alternatively, constitutive mutations might occur in the operator, altering the binding site for the repressor so that the repressor is unable to bind under any conditions.

2. For *E. coli* strains with the *lac* genotypes, use a plus sign (+) to indicate the synthesis of β-galactosidase and permease and a minus sign (−) to indicate no synthesis of the enzymes.

	Lactose absent		Lactose present	
Genotype of strain	**β-Galactosidase**	**Permease**	**β-Galactosidase**	**Permease**
(a) *lacI⁺ lacP⁺ lacO⁺ lacZ⁺ lacY⁺*				
(b) *lacI⁺ lacP⁺ lacOᶜ lacZ⁻ lacY⁺*				
(c) *lacI⁺ lacP⁻ lacO⁺ lacZ⁺ lacY⁻*				
(d) *lacI⁺ lacP⁺ lacO⁺ lacZ⁻ lacY⁻ / lacI⁻ lacP⁺ lacO⁺ lacZ⁺ lacY⁺*				

• Solution

	Lactose absent		Lactose present	
Genotype of strain	**β-Galactosidase**	**Permease**	**β-Galactosidase**	**Permease**
(a) *lacI⁺ lacP⁺ lacO⁺ lacZ⁺ lacY⁺*	−	−	+	+
(b) *lacI⁺ lacP⁺ lacOᶜ lacZ⁻ lacY⁺*	−	+	−	+
(c) *lacI⁺ lacP⁻ lacO⁺ lacZ⁻ lacY⁺*	−	−	−	−
(d) *lacI⁺ lacP⁺ lacO⁺ lacZ⁻ lacY⁻ / lacI⁻ lacP⁺ lacO⁺ lacZ⁺ lacY⁺*	−	−	+	+

(a) All the genes possess normal sequences, and so the *lac* operon functions normally: when lactose is absent, the regulator protein binds to the operator and inhibits the transcription of the structural genes, and so β-galactosidase and permease are not produced. When lactose is present, some of it is converted into allolactose, which binds to the repressor and makes it inactive; the repressor does not bind to the operator, and so the structural genes are transcribed, and β-galactosidase and permease are produced.

(b) The structural *lacZ* gene is mutated; so β-galactosidase will not be produced under any conditions. The *lacO* gene has a constitutive mutation, which means that the repressor is unable to bind to it, and so transcription takes place at all times. Therefore, permease will be produced in both the presence and the absence of lactose.

(c) In this strain, the promoter is mutated, and so RNA polymerase is unable to bind and transcription does not take place. Therefore β-galactosidase and permease are not produced under any conditions.

(d) This strain is a partial diploid, which consists of two copies of the *lac* operon—one on the bacterial chromosome and another on a plasmid. The *lac* operon represented in the upper part of the genotype has mutations in both the *lacZ* and *lacY* genes, and so it is not capable of encoding β-galactosidase or permease under any conditions. The *lac* operon in the lower part of the genotype has a defective regulator gene, but the normal regulator gene in the upper operon produces a diffusible repressor (trans acting) that binds to the lower operon in the absence of lactose and inhibits transcription. Therefore no β-galactosidase or permease is produced when lactose is absent. In the presence of lactose, the repressor cannot bind to the operator, and so the lower operon is transcribed and β-galactosidase and permease are produced.

3. The *fox* operon, which has sequences A, B, C, and D, encodes enzymes 1 and 2. Mutations in sequences A, B, C, and D have the following effects, where a plus sign (+) = enzyme synthesized and a minus sign (−) = enzyme not synthesized.

Mutation in sequence	Fox absent		Fox present	
	Enzyme 1	Enzyme 2	Enzyme 1	Enzyme 2
No mutation	−	−	+	+
A	−	−	−	+
B	−	−	−	−
C	−	−	+	−
D	+	+	+	+

(a) Is the *fox* operon inducible or repressible?

(b) Indicate which sequence (A, B, C, or D) is part of the following components of the operon:

Regulator gene	———
Promoter	———
Structural gene for enzyme 1	———
Structural gene for enzyme 2	———

• Solution

Because the structural genes in an operon are coordinately expressed, mutations that affect only one enzyme are likely to occur in the structural genes; mutations that affect both enzymes must occur in the promoter or regulator.

(a) When no mutations are present, enzymes 1 and 2 are produced in the presence of Fox but not in its absence, indicating that the operon is inducible and Fox is the inducer.

(b) Mutation A allows the production of enzyme 2 in the presence of Fox, but enzyme 1 is not produced in the presence or absence of Fox, and so A must have a mutation in the structural gene for enzyme 1. With B, neither enzyme is produced under any conditions, and so this mutation most likely occurs in the promoter and prevents RNA polymerase from binding. Mutation C affects only enzyme 2, which is not produced in the presence or absence of lactose; enzyme 1 is produced normally (only in the presence of Fox), and so mutation C most likely occurs in the structural gene for enzyme 2. Mutation D is constitutive, allowing the production of enzymes 1 and 2 whether or not Fox is present. This mutation most likely occurs in the regulator gene, producing a defective repressor that is unable to bind to the operator under any conditions.

Regulator gene	D
Promoter	B
Structural gene for enzyme 1	A
Structural gene for enzyme 2	C

4. A mutation occurs in the 5′ UTR of the *trp* operon that reduces the ability of region 2 to pair with region 3. What would be the effect of this mutation when the tryptophan level is high and when the tryptophan level is low?

• Solution

When the tryptophan level is high, regions 2 and 3 do not normally pair, and therefore the mutation will have no effect. When the tryptophan level is low, however, the ribosome normally stalls at the Trp codons in region 1 and does not cover region 2, and so regions 2 and 3 are free to pair, which prevents regions 3 and 4 from pairing and forming a terminator, ending transcription. If regions 2 and 3 cannot pair, then regions 3 and 4 will pair even when tryptophan is low and attenuation will always occur. Therefore, no more tryptophan will be synthesized even in the absence of tryptophan.

The New Genetics
MINING GENOMES

MICROARRAY ANALYSIS AND THE ANALYSIS
OF GENE EXPRESSION

This exercise introduces the powerful technique of microarray analysis, one of the most potent tools in bioinformatics. After a general introduction to microarrays, you will explore the use of microarrays in studies of gene expression. You will use SAGE (Serial Analysis of Gene Expression) to try to identify which genes are important in the development of specific diseases.

COMPREHENSION QUESTIONS

* 1. Name six different levels at which gene expression might be controlled.

* 2. Draw a picture illustrating the general structure of an operon and identify its parts.

3. What is the difference between positive and negative control? What is the difference between inducible and repressible operons?

* 4. Briefly describe the *lac* operon and how it controls the metabolism of lactose.

5. What is catabolite repression? How does it allow a bacterial cell to use glucose in preference to other sugars?

* 6. What is attenuation? What is the mechanism by which the attenuator forms when tryptophan levels are high and the antiterminator forms when tryptophan levels are low?

* 7. What is antisense RNA? How does it control gene expression?

8. What general features of transcriptional control are found in bacteriophage λ?

* 9. What changes take place in chromatin structure and what role do these changes play in eukaryotic gene regulation?

10. Briefly explain how transcriptional activator proteins and repressors affect the level of transcription of eukaryotic genes.

11. What is an insulator?

12. What is a response element? How do response elements bring about the coordinated expression of eukaryotic genes?

13. Outline the role of alternative splicing in the control of sex differentiation in *Drosophila*.

*14. What role does RNA stability play in gene regulation? What controls RNA stability in eukaryotic cells?

15. Define RNA silencing. Explain how siRNAs arise and how they potentially affect gene expression.

*16. Compare and contrast bacterial and eukaryotic gene regulation. How are they similar? How are they different?

APPLICATION QUESTIONS AND PROBLEMS

*17. For each of the following types of transcriptional control, indicate whether the protein produced by the regulator gene will be synthesized initially as an active repressor, inactive repressor, active activator, or inactive activator.

(a) Negative control in a repressible operon

(b) Positive control in a repressible operon

(c) Negative control in an inducible operon

(d) Positive control in an inducible operon

*18. A mutation occurs at the operator site that prevents the regulator protein from binding. What effect will this mutation have in the following types of operons?

(a) Regulator protein is a repressor in a repressible operon.

(b) Regulator protein is a repressor in an inducible operon.

19. The *blob* operon produces enzymes that convert compound A into compound B. The operon is controlled by a regulatory gene S. Normally the enzymes are synthesized only in the absence of compound B. If gene S is mutated, the enzymes are synthesized in the presence *and* in the absence of compound B. Does gene S produce a repressor or an activator? Is this operon inducible or repressible?

*20. A mutation prevents the catabolite activator protein (CAP) from binding to the promoter in the *lac* operon. What will be the effect of this mutation on transcription of the operon?

21. Under which of the following conditions would a *lac* operon produce the greatest amount of β-galactosidase? The least? Explain your reasoning.

	Lactose present	Glucose present
Condition 1	Yes	No
Condition 2	No	Yes
Condition 3	Yes	Yes
Condition 4	No	No

22. A mutant strain of *E. coli* produces β-galactosidase in the presence *and* in the absence of lactose. Where in the operon might the mutation in this strain occur?

*23. For *E. coli* strains with the *lac* genotypes shown below, use a plus sign (+) to indicate the synthesis of β-galactosidase and permease and a minus sign (−) to indicate no synthesis of the enzymes.

	Lactose absent		Lactose present	
Genotype of strain	β-Galactosidase	Permease	β-Galactosidase	Permease
$lacI^+ lacP^+ lacO^+ lacZ^+ lacY^+$				
$lacI^- lacP^+ lacO^+ lacZ^+ lacY^+$				
$lacI^+ lacP^+ lacO^c lacZ^+ lacY^+$				
$lacI^- lacP^+ lacO^+ lacZ^+ lacY^-$				
$lacI^- lacP^- lacO^+ lacZ^+ lacY^+$				
$lacI^+ lacP^+ lacO^+ lacZ^- lacY^+/$ $lacI^- lacP^+ lacO^+ lacZ^+ lacY^-$				
$lacI^- lacP^+ lacO^c lacZ^+ lacY^+/$ $lacI^+ lacP^+ lacO^+ lacZ^- lacY^-$				
$lacI^- lacP^+ lacO^+ lacZ^+ lacY^-/$ $lacI^+ lacP^- lacO^+ lacZ^- lacY^+$				
$lacI^+ lacP^- lacO^c lacZ^- lacY^+/$ $lacI^- lacP^+ lacO^+ lacZ^+ lacY^-$				
$lacI^+ lacP^+ lacO^+ lacZ^+ lacY^+/$ $lacI^+ lacP^+ lacO^+ lacZ^+ lacY^+$				
$lacI^s lacP^+ lacO^+ lacZ^+ lacY^-/$ $lacI^+ lacP^+ lacO^+ lacZ^- lacY^+$				
$lacI^s lacP^- lacO^+ lacZ^- lacY^+/$ $lacI^+ lacP^+ lacO^+ lacZ^+ lacY^+$				

24. Give all possible genotypes of a *lac* operon that produces β-galactosidase and permease under the following conditions. Do not give partial diploid genotypes.

	Lactose absent		Lactose present	
	β-Galactosidase	Permease	β-Galactosidase	Permease
(a)	−	−	+	+
(b)	−	−	−	+
(c)	−	−	+	−
(d)	+	+	+	+
(e)	−	−	−	−
(f)	+	−	+	−
(g)	−	+	−	+

*25. Explain why mutations in the *lacI* gene are trans in their effects, but mutations in the *lacO* gene are cis in their effects.

*26. The *mmm* operon, which has sequences A, B, C, and D, encodes enzymes 1 and 2. Mutations in sequences A, B, C, and D have the following effects, where a plus sign (+) = enzyme synthesized and a minus sign (−) = enzyme not synthesized.

Mutation in sequence	Mmm absent		Mmm present	
	Enzyme 1	Enzyme 2	Enzyme 1	Enzyme 2
No mutation	+	+	−	−
A	−	+	−	−
B	+	+	+	+
C	+	−	−	−
D	−	−	−	−

(a) Is the *mmm* operon inducible or repressible?

(b) Indicate which sequence (A, B, C, or D) is part of the following components of the operon:

    Regulator gene   ____
    Promoter   ____
    Structural gene for enzyme 1   ____
    Structural gene for enzyme 2   ____

*27. Listed in parts *a* through *g* are some mutations that were found in the 5′ UTR region of the *trp* operon of *E. coli*. What would the most likely effect of each of these mutations be on the transcription of the *trp* structural genes?

(a) A mutation that prevented the binding of the ribosome to the 5′ end of the mRNA 5′ UTR

(b) A mutation that changed the tryptophan codons in region 1 of the mRNA 5′ UTR into codons for alanine

(c) A mutation that created a stop codon early in region 1 of the mRNA 5′ UTR

(d) Deletions in region 2 of the mRNA 5′ UTR

(e) Deletions in region 3 of the mRNA 5′ UTR

(f) Deletions in region 4 of the mRNA 5′ UTR

(g) Deletion of the string of adenine nucleotides that follows region 4 in the 5′ UTR

28. Some mutations in the *trp* 5′ UTR region increase termination by the attenuator. Where might these mutations occur and how might they affect the attenuator?

29. Some of the mutations mentioned in Question 28 have an interesting property. They prevent the formation of the antiterminator that normally takes place when the tryptophan level is low. In one of the mutations, the AUG start codon for the 5′ UTR peptide has been deleted. How might this mutation prevent antitermination from occurring?

30. Several examples of antisense RNA regulating translation in bacterial cells have been discovered. Molecular geneticists have also used antisense RNA to artificially control transcription in both bacterial and eukaryotic genes. If you wanted to inhibit the transcription of a bacterial gene with antisense RNA, what sequences might the antisense RNA contain?

*31. What would be the effect of deleting the *Sxl* gene in a newly fertilized *Drosophila* embryo?

32. What would be the effect of a mutation that destroyed the ability of poly(A)-binding protein (PABP) to attach to a poly(A) tail?

CHALLENGE QUESTIONS

33. Would you expect to see attenuation in the *lac* operon and other operons that control the metabolism of sugars? Why or why not?

34. A common feature of many eukaryotic mRNAs is the presence of a rather long 3′ UTR, which often contains consensus sequences. Creatine kinase B (CK-B) is an enzyme important in cellular metabolism. Certain cells—termed U937D cells—have lots of CK-B mRNA, but no CK-B enzyme is present. In these cells, the 5′ end of the CK-B mRNA is bound to ribosomes, but the mRNA is apparently not translated. Something inhibits the translation of the CK-B mRNA in these cells.

In recent experiments, numerous short segments of RNA containing only 3′ UTR sequences were introduced into U937D cells. As a result, the U937D cells began to synthesize the CK-B enzyme, but the total amount of CK-B mRNA did not increase. Short segments of other RNA sequences did not stimulate the synthesis of CK-B; only the 3′ UTR sequences turned on the translation of the enzyme.

On the basis of these experiments, propose a mechanism for how CK-B translation is inhibited in the U937D cells. Explain how the introduction of short segments of RNA containing the 3′ UTR sequences might remove the inhibition.

SUGGESTED READINGS

Beelman, C. A., and R. Parker. 1995. Degradation of mRNA in eukaryotes. *Cell* 81:179–183.

An excellent review of the importance of mRNA stability in eukaryotic gene regulation and of some of the ways in which mRNA is degraded.

Bell, A. C., A. G. West, and G. Felsenfeld. 2001. Insulators and boundaries: versatile regulatory elements in the eukaryotic genome. *Science* 291:447–498.

A good introduction to research on insulators.

Bestor, T. H. 1998. Methylation meets acetylation. *Nature* 393:311–312.

A short review of research demonstrating a connection between DNA methylation and histone acetylation.

Bird, A. P., and A. P. Wolffe. 1999. Methylation-induced repression: belts, braces, and chromatin. *Cell* 99:451–454.

Discusses the role of methylation in gene regulation and development.

Blackwood, E. M., and J. T. Kadonaga. 1998. Going the distance: a current view of enhancer action. *Science* 281:60–63.

Reviews and discusses current models of enhancer action.

Gerasimova, T. I., and V. G. Corces. 2001. Chromatin insulators and boundaries: effects on transcription and nuclear organization. *Annual Review of Genetics* 35:193–208.

Reviews the effects of insulators and chromatin boundaries on the transcription of eukaryotic genes.

Green, P. J., O. Pines, and M. Inouye. 1986. The role of antisense RNA in gene regulation. *Annual Review of Biochemistry* 55:569–597.

A good review of antisense RNA and its role in gene regulation.

Hodgkin, J. 1989. *Drosophila* sex determination: a cascade of regulated splicing. *Cell* 56:905–906.

A good short review of alternative splicing and how it regulates sex differentiation in *Drosophila*.

Jacob, F., and J. Monod. 1961. Genetic regulatory mechanisms in the synthesis of proteins. *Journal of Molecular Biology* 3:318–356.

A classic paper describing Jacob and Monod's work on the *lac* operon, as a well as a review of gene control in several other systems.

Matzke, M., A. J. M. Matzke, and J. M. Kooter. 2001. RNA: guiding gene silencing. *Science* 293:1080–1083.

A review of RNA silencing.

Ng, H. H., and A. Bird. 2000. Histone deacetylases: silencers for hire. *Trends in Biochemical Science* 25:121–126.

Reviews how deacetylation can affect transcription in eukaryotic cells.

Pabo, C. O., and R. T. Sauer. 1992. Transcription factors: structural families and principles. *Annual Review of Biochemistry* 61:1053–1095.

A review of different DNA-binding motifs.

Ptashne, M. 1989. How gene activators work. *Scientific American* 260(1):41–47.

A discussion of some of the similarities in the ways in which gene activators work in prokaryotes and eukaryotes.

Ross, J. 1989. The turnover of messenger RNA. *Scientific American* 260(4):48–55.

A review of factors that control mRNA stability in eukaryotes.

Struhl, K. 1995. Yeast transcriptional regulatory mechanisms. *Annual Review of Genetics* 29:651–674.

A good review of transcriptional control in yeast.

Tuite, M. F. 1996. Death by decapitation for mRNA. *Nature* 382:577–579.

Discusses the interaction of the $3'$ poly(A) tail and the $5'$ cap in mRNA degradation.

Tyler, J. K., and J. T. Kadonaga. 1999. The "dark side" of chromatin remodeling: repressive effects on transcription. *Cell* 99:443–446.

Discusses the role of chromatin-remodeling complexes in eukaryotic gene regulation.

Wolffe, A. P. 1994. Transcription: in tune with histones. *Cell* 77:13–16.

A review of the role of histone proteins in eukaryotic gene regulation.

Wolffe, A. P. 1997. Sinful repression. *Nature* 387:16–17.

A short review of the role of histone acetylation in eukaryotic gene regulation.

Yanofsky, C. 1981. Attenuation in the control of expression of bacterial operons. *Nature* 289:751–758.

A good review of attenuation.

17 Gene Mutations and DNA Repair

Damaged Chernobyl nuclear reactor following a catastrophic explosion on April 26, 1986. Radiation released from the explosion and subsequent fire caused increased rates of somatic and germline mutations in residents of the surrounding area. (Volodymyr Repik/AP.)

The Genetic Legacy of Chernobyl

Early on the morning of April 26, 1986, unit 4 of the Chernobyl nuclear power plant in northern Ukraine exploded, creating the worst nuclear disaster in history. The explosion blew off the 2000-ton metal plate that sealed the top of the reactor and ignited hundreds of tons of graphite, which burned uncontrollably for 10 days. The exact amount of radiation released in the explosion and ensuing fire is still unknown, but a minimum estimate is 100 mil-

lion curies, equal to a medium-sized nuclear strike. A plume of radioactive particles blew west and north from the crippled reactor, raining dangerous levels of radiation down on thousands of square kilometers. Regions as far away as Germany and Norway were affected; even Japan and the United States received measurable increases in radiation.

Immediately after the accident, 31 people, mostly firefighters who heroically battled the blaze, died of acute radiation sickness. More than 400,000 workers later toiled to bury radioactive and chemical wastes from the accident and to entomb the remains of the disabled reactor in a steel and concrete sarcophagus. Many of these workers are now ill, suffering from a variety of problems including immune suppression, increased rates of cancer, and reproductive disorders.

Radiation is a known mutagen, causing damage to DNA. More than 13,000 children in the area surrounding Chernobyl were exposed to the radioactive isotope iodine-131; many had exposures 400 times the maximum annual radiation exposure recommended for workers in the nuclear industry. The rate of thyroid cancer among children in the Ukraine is now 10 times the pre-Chernobyl levels. Chromosome mutations have been detected in the cells of many people who resided near Chernobyl at the time of the accident, and birth defects in the population have increased significantly.

To examine germ-line mutations (those passed on to future generations) resulting from the Chernobyl accident, geneticists collected blood samples from 79 families who resided in heavily contaminated districts. These families included children born in 1994 who had not been exposed to radiation but who might possess mutations acquired from their parents. DNA sequences from these parents and children were analyzed, allowing the researchers to identify possible germ-line mutations. The germ-line mutation rate in these families was found to be twice as high as that in a control group of families in Britain. Furthermore, the mutation rate was correlated with the level of surface radiation: families in which the parents had resided in more-contaminated districts had higher mutation rates than those from less-contaminated districts.

This chapter is about the infidelity of DNA—about how errors arise in genetic instructions and how those errors are sometimes repaired. The Chernobyl catastrophe illustrates one cause of mutations (radiation) and the detrimental effects that DNA damage can have.

We begin with a brief examination of the different types of mutations, including their phenotypic effects, how they may be suppressed, and mutation rates. The next section explores how mutations spontaneously arise in the course of replication and afterward, as well as how chemicals and radiation induce mutations. We then consider the analysis of mutations. Finally, we take a look at DNA repair and some of the diseases that arise when DNA repair is defective. Throughout the chapter, it will be useful to keep in mind that mutations, by definition, are *inherited* changes in the DNA sequence—they must be passed on. Mutation requires both that the structure of a DNA molecule be changed and that this change is replicated.

www.whfreeman.com/pierce   More information about the health effects of radiation released in the Chernobyl accident

The Nature of Mutation

DNA is a highly stable molecule that replicates with amazing accuracy (see Chapters 10 and 12), but changes in DNA structure and errors of replication do occur. A **mutation** is defined as an inherited change in genetic information; the descendants may be cells produced by cell division or individual organisms produced by reproduction.

The Importance of Mutations

Mutations are both the sustainer of life and the cause of great suffering. On the one hand, mutation is the source of all genetic variation, the raw material of evolution. Without mutations and the variation that they generate, organisms could not adapt to changing environments and would risk extinction. On the other hand, most mutations have detrimental effects, and mutation is the source of many human diseases and disorders.

Much of genetics focuses on how variants produced by mutation are inherited; genetic crosses are meaningless if all individuals are identically homozygous for the same alleles. Mutations serve as important tools of genetic analysis; the solution to almost any genetic problem begins with a good set of mutants. Much of Gregor Mendel's success in unraveling the principles of inheritance can be traced to his use of carefully selected variants of the garden pea; similarly, Thomas Hunt Morgan and his students discovered many basic principles of genetics by analyzing mutant fruit flies (◀ FIGURE 17.1).

Mutations are also useful for probing fundamental biological processes. Finding mutations that affect different components of a biological system and studying their effects can often lead to an understanding of the system. This method, referred to as genetic dissection, is analogous to figuring out how an automobile works by breaking different parts of a car and observing the effects—for example, smash the radiator and the engine overheats, revealing that the radiator cools the engine. The disruption of function in individual organisms bearing particular mutations likewise can be a source of insight into biological processes. For example, geneticists have begun to unravel the molecular details of development by studying mutations that interrupt various embryonic stages in *Drosophila* (see Chapter 21). Although this method of breaking "parts" to determine their function might seem like a crude approach to understanding a system, it is actually very powerful and has been used extensively in biochemistry, developmental biology, physiology, and behavioral science (but this method is *not* recommended for learning how your car works).

Concepts

Mutations are heritable changes in the genetic coding instructions of DNA. They are essential to the study of genetics and are useful in many other biological fields.

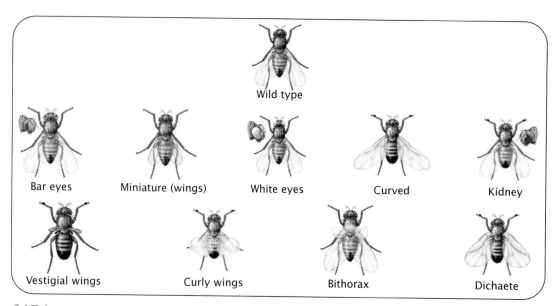

◀17.1 **Morgan and his students discovered many principles of heredity by studying mutation in *Drosophila melanogaster.*** Shown here are several common mutations.

Categories of Mutations

In multicellular organisms, we can distinguish between two broad categories of mutations: somatic mutations and germ-line mutations. **Somatic mutations** arise in somatic tissues, which do not produce gametes (◀FIGURE 17.2). These mutations are passed on to other cells through the process of mitosis, which leads to a population of genetically identical cells (a clone). The earlier in development that a somatic mutation occurs, the larger the clone of cells within that individual organism that will contain the mutation.

Because of the huge number of cells present in a typical eukaryotic organism, somatic mutations must be numerous. For example, there are about 10^{14} cells in the human body. If a mutation arises only once in every million cell divisions (a fairly typical rate of mutation), hundreds of millions of somatic mutations must arise in each person. The effect of these mutations depends on many factors, including the type of cell in which they occur and the developmental stage at which they arise. Many somatic mutations have no obvious effect on the phenotype of the organism, because the function of the mutant cell (even the cell itself) is replaced by that of normal cells. However, cells with a somatic mutation that stimulates cell division can increase in number and spread; this type of mutation can give rise to cells with a selective advantage and is the basis for all cancers (see Chapter 21).

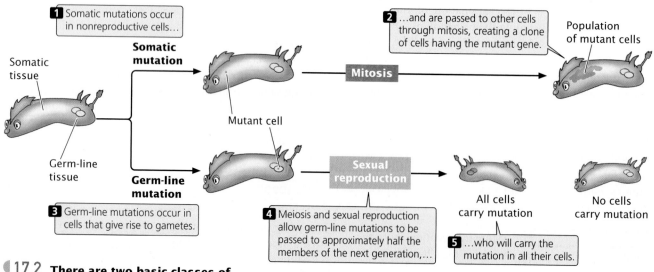

◀17.2 **There are two basic classes of mutations: somatic mutations and germ-line mutations.**

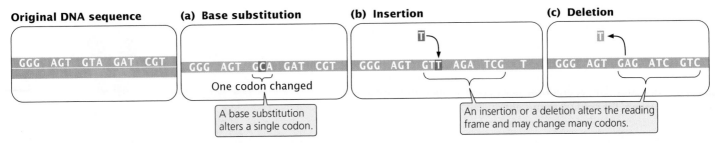

Original DNA sequence

GGG AGT GTA GAT CGT

(a) Base substitution

GGG AGT GCA GAT CGT

One codon changed

A base substitution alters a single codon.

(b) Insertion

T

GGG AGT GTT AGA TCG T

(c) Deletion

T

GGG AGT GAG ATC GTC

An insertion or a deletion alters the reading frame and may change many codons.

◀ **17.3 Three basic types of gene mutations are base substitutions, insertions, and deletions.**

Germ-line mutations arise in cells that ultimately produce gametes. These mutations can be passed to future generations, producing individual organisms that carry the mutation in all their somatic and germ-line cells (see Figure 17.2). When we speak of mutations in multicellular organisms, we're usually talking about germ-line mutations. In single-cell organisms, however, there is no distinction between germ-line and somatic mutations, because cell division results in new individuals.

Historically, mutations have been partitioned into those that affect a single gene, called *gene mutations,* and those that affect the number or structure of chromosomes, called *chromosome mutations.* This distinction arose because chromosome mutations could be observed directly, by looking at chromosomes with a microscope, whereas gene mutations could be detected only by observing their phenotypic effects. Now, with the development of DNA sequencing, gene mutations and chromosome mutations are distinguished somewhat arbitrarily on the basis of the size of the DNA lesion. Nevertheless, it is useful to use the term *chromosome mutation* for a large-scale genetic alteration that affects chromosome structure or the number of chromosomes and the term **gene mutation** for a relatively small DNA lesion that affects a single gene. This chapter focuses on gene mutations; chromosome mutations were discussed in Chapter 9.

Types of Gene Mutations

There are a number of ways to classify gene mutations. Some classification schemes are based on the nature of the phenotypic effect—whether the mutation alters the amino acid sequence of the protein and, if so, how. Other schemes are based on the causative agent of the mutation, and still others focus on the molecular nature of the defect. The most appropriate scheme depends on the reason for studying the mutation. Here, we will categorize mutations primarily on the basis of their molecular nature, but we will also encounter some terms that relate the causes and the phenotypic effects of mutations.

Base substitutions The simplest type of gene mutation is a **base substitution,** the alternation of a single nucleotide in the DNA (◀ FIGURE 17.3a). Because of the complementary nature of the two DNA strands (see Figure 10.13), when the base of one nucleotide is altered, the base of the corresponding nucleotide on the opposite strand also will be altered in the next round of replication. A base substitution therefore usually leads to a base-pair substitution.

Base substitutions are of two types. In a **transition,** a purine is replaced by a different purine or, alternatively, a pyrimidine is replaced by a different pyrimidine (◀ FIGURE 17.4). In a **transversion,** a purine is replaced by a pyrimidine or a pyrimidine is replaced by a purine. The number of possible transversions (see Figure 17.4) is twice the number of possible transitions, but transitions usually arise more frequently.

Insertions and deletions The second major class of gene mutations contains **insertions** and **deletions**—the addition or the removal, respectively, of one or more nucleotide pairs (◀ FIGURE 17.3b and c). Although base substitutions are often assumed to be the most common type of mutation, molecular analysis has revealed that insertions and deletions are more frequent. Insertions and deletions within

◀ **17.4 A transition is the substitution of a purine for a purine or a pyrimidine for a pyrimidine; a transversion is the substitution of a pyrimidine for a purine or a purine for a pyrimidine.**

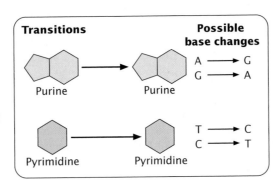

Transitions

Purine → Purine

Pyrimidine → Pyrimidine

Possible base changes

A → G
G → A

T → C
C → T

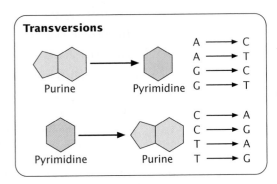

Transversions

Purine → Pyrimidine

Pyrimidine → Purine

A → C
A → T
G → C
G → T

C → A
C → G
T → A
T → G

sequences that encode proteins may lead to **frameshift mutations,** changes in the reading frame (see p. 415 in Chapter 15) of the gene. The initiation codon in mRNA sets the reading frame: after the initiation codon, other codons are read as successive nonoverlapping groups of three nucleotides. The addition or deletion of a nucleotide usually changes the reading frame, altering all amino acids encoded by codons following the mutation (see Figure 17.3b and c). Many amino acids can be affected; so frameshift mutations generally have drastic effects on the phenotype. Not all insertions and deletions lead to frameshifts, however; because codons consist of three nucleotides, insertions and deletions consisting of any multiple of three nucleotides will leave the reading frame intact, although the addition or removal of one or more amino acids may still affect the phenotype. These mutations are called **in-frame insertions** and **deletions,** respectively.

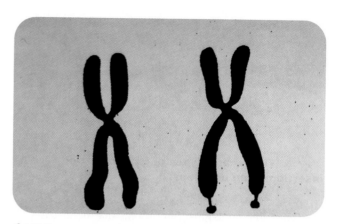

◀ **17.5   The fragile-X chromosome is associated with a characteristic constriction (fragile site) on the long arm.** (Visuals Unlimited.)

Concepts

Gene mutations consist of changes in a single gene and may be base substitutions (a single pair of nucleotides is altered) or insertions or deletions (nucleotides are added or removed). A base substitution may be a transition (substitution of like bases) or a transversion (substitution of unlike bases). Insertions and deletions often lead to a change in the reading frame of a gene.

Expanding trinucleotide repeats   In 1991, an entirely novel type of mutation was discovered. This mutation occurs in a gene called *FMR-1* and causes fragile-X syndrome, the most common hereditary cause of mental retardation. The disorder is so named because, in specially treated cells of persons having the condition, the tip of the X chromosome is attached only by a slender thread (◀ FIGURE 17.5). The *FMR-1* gene contains a number of adjacent copies of the trinucleotide CGG. The normal *FMR-1* allele (not containing the mutation) has 60 or fewer copies of this trinucleotide but, in persons with fragile-X syndrome, the allele may har-

Table 17.1	Examples of genetic diseases caused by expanding trinucleotide repeats			
			Number of Copies of Repeat	
Disease	**Repeated Sequence**	**Normal Range**	**Disease Range**	
Spinal and bulbar muscular atrophy	CAG	11–33	40–62	
Fragile-X syndrome	CGG	6–54	50–1500	
Jacobsen syndrome	CGG	11	100–1000	
Spinocerebellar ataxia (several types)	CAG	4–44	21–130	
Autosomal dominant cerebellar ataxia	CAG	7–19	37–~220	
Myotonic dystrophy	CTG	5–37	44–3000	
Huntington disease	CAG	9–37	37–121	
Friedreich ataxia	GAA	6–29	200–900	
Dentatorubral-pallidoluysian atrophy	CAG	7–25	49–75	
Myoclonus epilepsy of the Unverricht-Lundborg type*	CCCCGCCCCGCG	2–3	12–13	

*Technically not a trinucleotide repeat but does entail a multiple of three nucleotides that expands and contracts in similar fashion to trinucleotide repeats.

bor hundreds or even thousands of copies. Mutations in which copies of a trinucleotide may increase greatly in number are called **expanding trinucleotide repeats.**

Expanding trinucleotide repeats have been found in several other human diseases (Table 17.1). The number of copies of the trinucleotide repeat often correlates with the severity or age of onset of the disease. The number of copies of the repeat also correlates with the instability of trinucleotide repeats—when more repeats are present, the probability of expansion to even more repeats increases. This instability leads to a phenomenon known as anticipation (see p. 121 in Chapter 5), in which diseases caused by trinucleotide-repeat expansions become more severe in each generation. Less commonly, the number of trinucleotide repeats may decrease within a family.

How an increase in the number of trinucleotides produces disease symptoms is not yet clear. In several of the diseases (e.g., Huntington disease), the trinucleotide CAG expands within the coding part of a gene, producing a toxic protein that has extra glutamine residues (the amino acid encoded by CAG). In other diseases (e.g., fragile-X syndrome and myotonic dystrophy), the repeat is outside the coding region of the gene and therefore must have some other mode of action. At least one disease (a rare type of epilepsy) has now been associated with an expanding repeat of a 12-bp sequence. Although this repeat is not a trinucleotide, it is included as a type of expanding trinucleotide because its repeat is a multiple of three.

The mechanism that leads to the expansion of trinucleotide repeats is still unclear. Strand slippage in DNA replication (see Figure 17.14) and crossing over between misaligned repeats (see Figure 17.15) are two possible sources of expansion. Single-stranded regions of some trinucleotide repeats are known to fold into hairpins (◀ FIGURE 17.6) and other special DNA structures. Such structures may promote strand slippage in replication and may prevent these errors from being recognized and corrected, as described later in this chapter in the section on mismatch repair.

Concepts

Expanding trinucleotide repeats are regions of DNA that consist of repeated copies of three nucleotides. Increased numbers of trinucleotide repeats are associated with several genetic diseases.

Phenotypic effects of mutations Mutations have a variety of phenotypic effects. The effect of a mutation must be considered with reference to a phenotype against which the mutant can be compared, which is usually the wild-type phenotype—that is, the most common phenotype in natural populations of the organism. For example, most *Drosophila melanogaster* in nature have red eyes; so

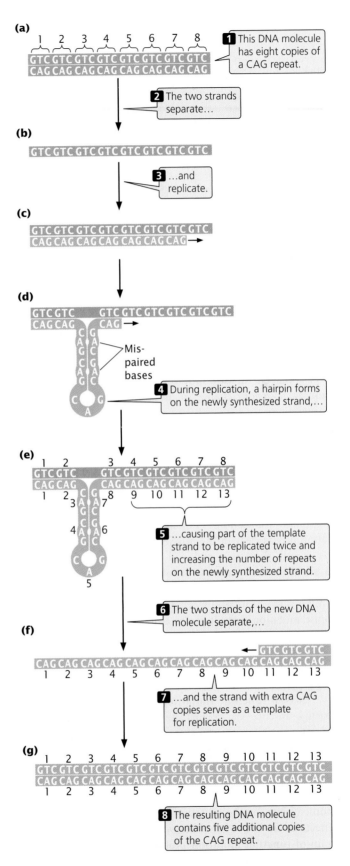

◀ 17.6 **The number of copies of a trinucleotide may increase by strand slippage in replication.**

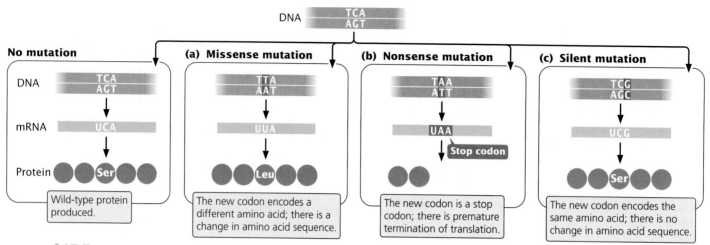

◀ 17.7 **Base substitutions can cause (a) missense, (b) nonsense, and (c) silent mutations.**

red eyes are considered the wild-type eye color; any other genetically determined eye color in fruit flies is considered to be a mutant. A mutation that alters the wild-type phenotype is called a **forward mutation,** where as a **reverse mutation** (a *reversion*) changes a mutant phenotype back into the wild type.

Geneticists use special terms to describe the phenotypic effects of mutations. A base substitution that alters a codon in the mRNA, resulting in a different amino acid in the protein, is referred to as a **missense mutation** (◀ FIGURE 17.7a). A **nonsense mutation** changes a sense codon (one that specifies an amino acid) into a nonsense codon (one that terminates translation; ◀ FIGURE 17.7b). If a nonsense mutation occurs early in the mRNA sequence, the protein will be greatly shortened and will usually be nonfunctional. A **silent mutation** alters a codon but, thanks to the redundancy of the genetic code, the codon still specifies the same amino acid (◀ FIGURE 17.7c). A **neutral mutation** is a missense mutation that alters the amino acid sequence of the protein but does not change its function. Neutral mutations occur when one amino acid is replaced by another that is chemically similar or when the affected amino acid has little influence on protein function.

Loss-of-function mutations cause the complete or partial absence of normal function. A loss-of-function mutation so alters the structure of the protein that the protein no longer works correctly or the mutation can occur in regulatory regions that affect the transcription, translation, or splicing of the protein. Loss-of-function mutations are frequently recessive, and diploid individuals must be homozygous for the mutation before they can exhibit the effects of the loss of the functional protein. In contrast, a **gain-of-function mutation** produces an entirely new trait or it causes a trait to appear in inappropriate tissues or at inappropriate times in development. These mutations are frequently dominant in their expression. Still other types of

mutations are **conditional mutations,** which are expressed only under certain conditions, and **lethal mutations,** which cause premature death.

Suppressor mutations A **suppressor mutation** is a genetic change that hides or suppresses the effect of another mutation. This type of mutation is distinct from a reverse mutation, in which the mutated site changes back into the original wild-type sequence (◀ FIGURE 17.8). A suppressor mutation occurs at a site that is distinct from the site of the original mutation; thus, an individual organism with a suppressor mutation is a double mutant, possessing both the original mutation and the suppressor mutation but exhibiting the phenotype of an unmutated wild type.

Geneticists distinguish between two classes of suppressor mutations: intragenic and intergenic. An **intragenic suppressor** is in the same gene as that containing the mutation being suppressed and may work in several ways. The suppressor may change a second nucleotide in the same codon that was altered by the original mutation, producing a codon that specifies the same amino acid as the original, unmutated codon (◀ FIGURE 17.9). Intragenic suppressors may also work by suppressing a frameshift mutation. If the original mutation is a one-base deletion, then the addition of a single base elsewhere in the gene will restore the former reading frame. Consider the following nucleotide sequence in DNA and the amino acids that it encodes:

DNA	AAA TCA CTT GGC GTA CAA
Amino acids	Phe Ser Glu Pro His Val

Suppose a one-base deletion occurs in the first nucleotide of the second codon. This deletion shifts the reading frame by one nucleotide and alters all the amino acids that follow the mutation.

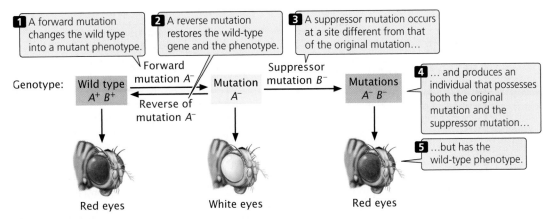

17.8 Relation of forward, reverse, and suppressor mutations.

One-nucleotide deletion

AAA ⃥CAC TTG GCG TAC AA
Phe   Val   Asn   Arg   Met

If a single nucleotide is added to the third codon (the suppressor mutation), the reading frame is restored, although two of the amino acids differ from those specified by the original sequence.

One-nucleotide duplication

AAA CAC TTT GGC GTA CAA
Phe   Val   Lys   Pro   His   Val

Similarly, a mutation due to an insertion may be suppressed by a subsequent deletion in the same gene.

A third way in which an intragenic suppressor may work is by making compensatory changes in the protein. A first missense mutation may alter the folding of a polypeptide chain by changing the way in which amino acids in the protein interact with one another. A second missense mutation at a different site (the suppressor) may recreate the original folding pattern by restoring interactions between the amino acids.

Intergenic suppressors, in contrast, occur in a gene that is different from the one bearing the original mutation. These suppressors sometimes work by changing the way that the mRNA is translated. In the example illustrated in (◀FIGURE 17.10), the original DNA sequence is AAC (UUG in the mRNA) and specifies leucine. This sequence mutates to ATC (UAG in mRNA), a termination codon. The ATC nonsense mutation could be suppressed by a mutation in a gene that encodes a tRNA molecule by changing the anticodon on the tRNA so that it is capable of pairing with the UAG termination codon. For example, the gene that encodes the tRNA for tyrosine (tRNA^Tyr), which has the anticodon AUA, might be mutated to have the anticodon AUC, which will then pair with the UAG stop codon. Instead of translation terminating at the UAG codon, tyrosine would be inserted into the protein and a full-length protein would be produced, although tyrosine would now substitute for leucine. The effect of this change would depend on the role of this amino acid in the overall structure of the protein, but the effect is likely to be less detrimental than the effect of the nonsense mutation, which would halt translation prematurely.

Because cells in many organisms have multiple copies of tRNA genes, other unmutated copies of tRNA^Tyr would

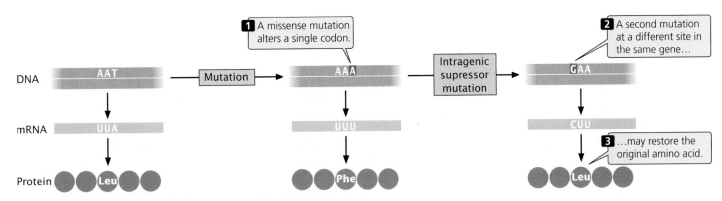

17.9 An intragenic suppressor mutation occurs in the same gene that contains the mutation being suppressed.

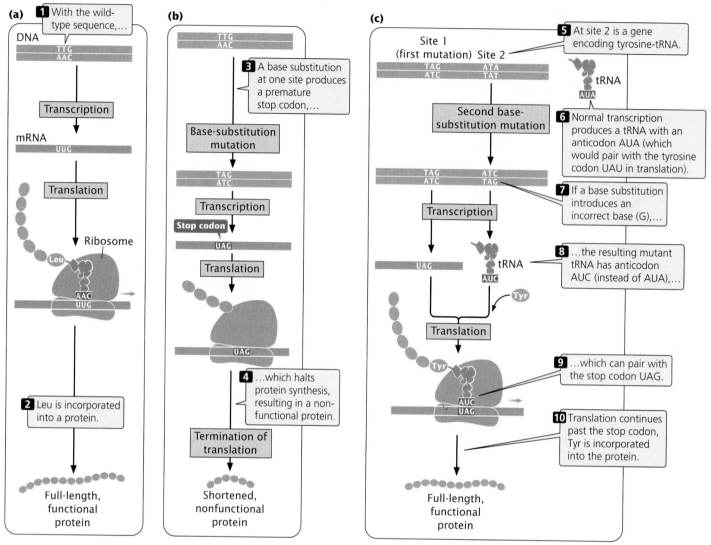

◀17.10 An intergenic suppressor mutation occurs in a different gene from the one bearing the original mutation. (a) The wild-type sequence produces a full-length, functional protein. (b) A base substitution at a site in one gene produces a premature stop codon, resulting in a shortened, nonfunctional protein. (c) A base substitution at a site in another gene, which in this case encodes tRNA, alters the anticodon of tRNA^Tyr so that tRNA^Tyr can pair with the stop codon produced by the original mutation, allowing tyrosine to be incorporated into the protein and translation to continue. Tyrosine replaces the leucine residue present in the original protein.

remain available to recognize the tyrosine codons. However, we might expect that the tRNAs that have undergone a suppressor mutation would also suppress the normal termination codons at the ends of coding sequences, resulting in the production of longer-than-normal proteins, but this event does not usually take place. Mutations in tRNA genes can also suppress missense and frameshift mutations.

Intergenic suppressors can also work through genic interactions (see p. 107 in Chapter 5). Polypeptide chains which are produced by two genes may interact to produce a functional protein. A mutation in one gene may alter the encoded polypeptide so that the interaction is destroyed and then a functional protein is no longer produced. A suppressor mutation in the second gene may produce a compensatory change in its polypeptide therefore restoring the original interaction. Characteristics of some of the different types of mutations are summarized in Table 17.2.

Table 17.2	Characteristics of different types of mutations

Type of Mutation	Definition
Base substitution	Changes the base of a single DNA nucleotide
Transition	Base substitution in which a purine replaces a purine or a pyrimidine replaces a pyrimidine
Transversion	Base substitution in which a purine replaces a pyrimidine or a pyrimidine replaces a purine
Insertion	Addition of one or more nucleotides
Deletion	Deletion of one or more nucleotides
Frameshift mutation	Insertion or deletion that alters the reading frame of a gene
In-frame deletion or insertion	Insertion or deletion of a multiple of three nucleotides that does not alter the reading frame
Expanding trinucleotide repeats	Repeated sequence of three nucleotides (trinucleotide) in which the number of copies of the trinucleotide increases
Forward mutation	Changes the wild-type phenotype to a mutant phenotype
Reverse mutation	Changes a mutant phenotype back to the wild-type phenotype
Missense mutation	Changes a sense codon into a different sense codon, resulting in the incorporation of a different amino acid in the protein
Nonsense mutation	Changes a sense codon into a nonsense codon, causing premature termination of translation
Silent mutation	Changes a sense codon into a synonymous codon, leaving unchanged the amino acid sequence of the protein
Neutral mutation	Changes the amino acid sequence of a protein without altering its ability to function
Loss-of-function mutation	Causes a complete or partial loss of function
Gain-of-function mutation	Causes the appearance of a new trait or function or causes the appearance of a trait in inappropriate tissues or at inappropriate times
Lethal mutation	Causes premature death
Suppressor mutation	Suppresses the effect of an earlier mutation at a different site
Intragenic suppressor mutation	Suppresses the effect of an earlier mutation within the same gene
Intergenic suppressor mutation	Suppresses the effect of an earlier mutation in another gene

Concepts

A suppressor mutation overrides the effect of an earlier mutation at a different site. An intragenic suppressor mutation occurs within the *same* gene, as that containing the original mutation, whereas an intergenic suppressor mutation occurs in a *different* gene.

www.whfreeman.com/pierce Descriptions and illustrations of different types of mutations

Mutation Rates

The frequency with which a gene changes from the wild type to a mutant is referred to as the **mutation rate** and is generally expressed as the number of mutations per biological unit, which may be mutations per cell division, per

The New Genetics
ETHICS · SCIENCE · TECHNOLOGY

Achondroplasia

Ron Green

Achondroplasia is an inherited autosomal dominant condition that causes diminished growth in the long bones of the legs, leading to dwarfism. Several years ago, the gene for achondroplasia was identified and cloned. If two people with achondroplasia marry, and each of them is heterozygous for achondroplasia (has one of two possible copies of the gene), chances are that two of every four children that they have will also be heterozygous and dwarfs. On average, one child in every four born to the couple will not inherit the achondroplasia gene and will be of average height, and one child in four will be homozygous for the gene. Homozygosity for this gene is lethal, and these children usually die in infancy.

A researcher who helped identify the gene understandably felt that he had made a significant contribution by allowing short-statured parents the option of aborting fetuses with the lethal double dose of the gene. To his surprise, shortly after news of the discovery was published, he received a call from one member of an achondroplasia couple, asking whether it was possible to test for both the *presence* and the *absence* of the gene. The couple wanted this information, they said, because they planned to abort not just all fetuses homozygous for the achondroplasia gene, but any completely unaffected ones as well. They were intent on having only short-statured children like themselves.

This case poses at least two major conflicts for genetic professionals. First, there is the conflict between respect for parental autonomy, which would ordinarily encourage acceding to the parents' request for assistance and information, and the medical professional's desire not to visit harm on a child. Children born with achondroplasia frequently must undergo a series of surgical procedures to correct serious bone problems. Throughout life, they also face many social and physical obstacles because of their short stature. Is it right for parents to deliberately bring a child into existence with this condition? Is it appropriate for health professionals to assist such efforts? How do we balance respect for parental autonomy against nonmaleficence?

Matters become more complex when we realize that some people with achondroplasia reject the idea that any harm is being done by the parents in this case. They maintain that most of the problems that they face are socially constructed and are due to society's marginalization and neglect of those who are different. Some also reject medical or genetic "solutions" to their problems. The proper response, they believe, is not to prevent the birth of a child with a genetic condition but to eliminate the social handicaps and discriminatory attitudes. Thus, the parents in this case may be driven not merely by their personal wishes but by a commitment to social justice.

Traditional, nondirective genetic counseling has assumed that people seek prenatal testing to prevent the

A family of three who have achondroplasia. (Gail Burton/AP.)

birth of a child with a genetic disease, to prepare for the birth and treatment of a child with a recognized genetic disorder, or to reconsider their reproductive plans. What this case reveals is that genetics is opening up the possibility of shaping our children's lives in ways that go far beyond what is normally associated with the healing role. Somewhat less dramatic, but perhaps more worrisome, is the fact that the identification of the genetic basis of many traits that are not considered diseases (e.g., height, intelligence, temperament) will offer parents a new range of choices in the "genetic design" of their children. At this moment, research is underway to identify and replace disease-causing genes in human embryos. In the future, such embryonic gene therapy will open up the possibility of *enhancing* children's capabilities. Beginning with genes that improve a child's resistance to cancer or AIDS, genetic interventions may make it possible to increase a child's height, stamina, or IQ. Science could offer parents who yearn for a champion basketball player or world-class swimmer the means to realize their dreams.

As complex as it may seem, the science here is the easy part. Far more difficult are the ethical questions. To begin with, there is the question of whether we will ever have enough knowledge to "play God" in this way. Do we dare alter the course of human evolution? The history of twentieth-century science is littered with well-intended technologies—from DDT to nuclear power—that eventually brought unforeseen harms. Will our genetic interventions follow this path? Will our clumsy attempts to "improve" the human genome unleash an epidemic of new genetic diseases? And what of the child's rights in all this? Is it fair to "engineer" a child into a parent's dream of perfection?

gamete, or per round of replication. For example, the mutation rate for achondroplasia (a type of hereditary dwarfism) is about four mutations per 100,000 gametes, usually expressed more simply as 4×10^{-5}. In contrast, **mutation frequency** is defined as the incidence of a specific type of mutation within a group of individual organisms. For achondroplasia, the mutation frequency in the United States is about 2×10^{-4}, which means that about 1 of every 20,000 persons in the U.S. population carries this mutation.

Mutation rates are affected by three factors. First, they depend on the frequency with which primary changes take place in DNA. Primary change may arise from spontaneous molecular changes in DNA or it may be induced by chemical or physical agents in the environment.

A second factor influencing the mutation rate is the probability that, when a change takes place, it will be repaired. Most cells possess a number of mechanisms to repair altered DNA; so most alterations are corrected before they are replicated. If these repair systems are effective, mutation rates will be low; if they are faulty, mutation rates will be elevated. Some mutations increase the overall rate of mutation at other genes; these mutations usually occur in genes that encode components of the replication machinery or DNA repair enzymes.

A third factor, one that influences our ability to calculate mutation rates, is the probability that a mutation will be recognized and recorded. When DNA is sequenced, all mutations are potentially detectable. In practice, however, sequencing is expensive; so mutations are usually detected by their phenotypic effects. Some mutations may appear to arise at a higher rate simply because they are easier to detect.

Mutation rates vary among organisms and among genes within organisms (Table 17.3), but we can draw several general conclusions about mutation rates. First, spontaneous mutation rates are low for all organisms studied. Typical mutation rates for bacterial genes range from about 1 to 100 mutations per 10 billion cells (1×10^{-8} to 1×10^{-10}). The mutation rates for most eukaryotic genes are a bit higher, from about 1 to 10 mutations per million gametes (1×10^{-5} to 1×10^{-6}). These higher values in eukaryotes may be due to the fact that the rates are calculated *per gamete,* and several cell divisions are required to produce a gamete, whereas mutation rates in prokaryotic cells are calculated *per cell division.*

Within each major class of organisms, mutation rates vary considerably. These differences may be due to differing abilities to repair mutations, unequal exposures to mutagens, or biological differences in rates of spontaneously arising mutations. Even within a single species, spontaneous rates of mutation vary among genes. The reason for this variation is not entirely understood, but some regions of DNA are known to be more susceptible to mutation than others.

Table 17.3	Mutation rates of different genes in different organisms		
Organism	**Mutation**	**Rate**	**Unit**
Bacteriophage T2	Lysis inhibition	1×10^{-8}	Per replication
	Host range	3×10^{-9}	
Escherichia coli	Lactose fermentation	2×10^{-7}	Per cell division
	Histidine requirement	2×10^{-8}	
Neurospora crassa	Inositol requirement	8×10^{-8}	Per asexual spore
	Adenine requirement	4×10^{-8}	
Corn	Kernel color	2.2×10^{-6}	Per gamete
Drosophila	Eye color	4×10^{-5}	Per gamete
	Allozymes	5.14×10^{-6}	
Mouse	Albino coat color	4.5×10^{-5}	Per gamete
	Dilution coat color	3×10^{-5}	
Human	Huntington disease	1×10^{-6}	Per gamete
	Achondroplasia	1×10^{-5}	
	Neurofibromatosis (Michigan)	1×10^{-4}	
	Hemophilia A (Finland)	3.2×10^{-5}	
	Duchenne muscular dystrophy (Wisconsin)	9.2×10^{-5}	

ANT

Concepts

Mutation rate is the frequency with which a specific mutation arises, whereas mutation frequency is the incidence of a mutation within a defined group of individual organisms. Rates of mutations are generally low and are affected by environmental and genetic factors.

Causes of Mutations

Mutations result from both internal and external factors. Those that are a result of natural changes in DNA structure are termed **spontaneous mutations,** whereas those that result from changes caused by environmental chemicals or radiation are **induced mutations.**

Spontaneous Replication Errors

Replication is amazingly accurate: fewer than one in a billion errors are made in the course of DNA synthesis (Chapter 12). However, spontaneous replication errors do occasionally occur.

The primary cause of spontaneous replication errors was formerly thought to be tautomeric shifts, in which the positions of protons in the DNA bases change. Purine and pyrimidine bases exist in different chemical forms called tautomers (◀FIGURE 17.11a). The two tautomeric forms of each base are in dynamic equilibrium, although one form is more common than the other. The standard Watson and Crick base pairings—adenine with thymine, and cytosine with guanine—are between the common forms of the bases, but, if the bases are in their rare tautomeric forms, other base pairings are possible (◀FIGURE 17.11b).

Watson and Crick proposed that tautomeric shifts might produce mutations, and for many years their proposal was the accepted model for spontaneous replication errors, but there has never been convincing evidence that the rare tautomers are the cause of spontaneous mutations. Furthermore, research now shows little evidence of these structures in DNA.

Mispairing can also occur through wobble, in which normal, protonated, and other forms of the bases are

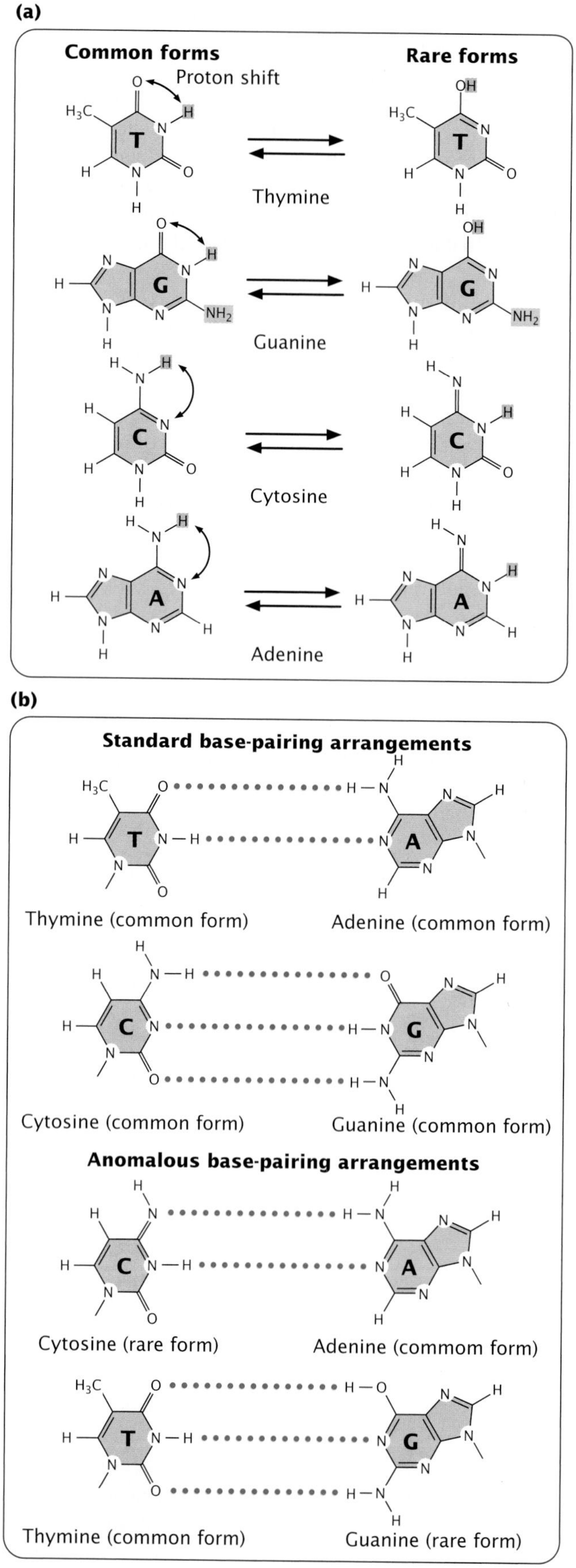

◀**17.11 Purine and pyrimidine bases exist in different forms called tautomers.** (a) A tautomeric shift occurs when a proton changes its position, resulting in a rare tautomeric form. (b) Standard and anomalous base-pairing arrangements that occur if bases are in the rare tautomeric forms. Base mispairings due to tautomeric shifts were originally thought to be a major source of errors in replication, but such structures have not been detected in DNA, and most evidence now suggests that other types of anomalous pairings (see Figure 17.14) are responsible for replication errors.

Non-Watson-Crick base pairing

Thymine–guanine wobble

Cytosine–adenine protonated wobble

◀ **17.12 Nonstandard base pairings can occur as a result of the flexibility in DNA structure.** Thymine and guanine can pair through wobble between normal bases. Cytosine and adenine can pair through wobble when adenine is protonated (has an extra hydrogen).

able to pair because of flexibility in the DNA helical structure (◀ FIGURE 17.12). These structures have been detected in DNA molecules and are now thought to be responsible for many of the mispairings in replication.

When a mismatched base has been incorporated into a newly synthesized nucleotide chain, an **incorporated error** is said to have occurred. Suppose that, in replication, thymine (which normally pairs with adenine) mispairs with guanine through wobble (◀ FIGURE 17.13). In the next round of replication, the two mismatched bases separate, and each serves as template for the synthesis of a new nucleotide strand. This time, thymine pairs with adenine, producing another copy of the original DNA sequence. On the other strand, however, the incorrectly incorporated guanine serves as the template and pairs with cytosine, producing a new DNA molecule that has an error—a C·G pair in

place of the original T·A pair (a T·A→C·G base substitution). The original incorporated error leads to a **replication error,** which creates a permanent mutation, because all the base pairings are correct and there is no mechanism for repair systems to detect the error.

Mutations due to small insertions and deletions also may arise spontaneously in replication and crossing over. **Strand slippage** may occur when one nucleotide strand forms a small loop (◀ FIGURE 17.14). If the looped-out nucleotides are on the newly synthesized strand, an insertion results. At the next round of replication, the insertion will be incorporated into both strands of the DNA molecule. If the looped-out nucleotides are on the template strand, then there is a deletion on the newly replicated strand, and this deletion will be perpetuated in subsequent rounds of replication.

During normal crossing over, the homologous sequences of the two DNA molecules align, and crossing over produces no net change in the number of nucleotides in either molecule. Misaligned pairing may cause **unequal crossing over,** which results in one DNA molecule with an insertion and the other with a deletion (◀ FIGURE 17.15). Some DNA sequences are more likely than others to undergo strand slippage or unequal crossing over. Stretches of repeated sequences, such as trinucleotide repeats or homopolymeric repeats (more than five repeats of the same base in a row), are prone to strand slippage. Stretches with more repeats are more likely to undergo strand slippage. Duplicated or repetitive sequences may misalign during pairing, leading to unequal crossing over. Both strand slippage and unequal crossing over produce duplicated copies of sequences, which in turn promote further strand slippage and unequal crossing over. This chain of events may explain the phenomenon of anticipation often observed for expanding trinucleotide repeats.

Concepts

Spontaneous replication errors arise from altered base structures and from wobble base pairing. Small insertions and deletions may occur through strand slippage in replication and through unequal crossing over.

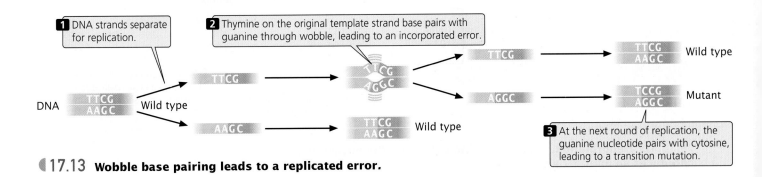

◀ **17.13 Wobble base pairing leads to a replicated error.**

1 DNA strands separate for replication.

2 Thymine on the original template strand base pairs with guanine through wobble, leading to an incorporated error.

3 At the next round of replication, the guanine nucleotide pairs with cytosine, leading to a transition mutation.

DNA   TTCG / AAGC   Wild type

TTCG

AAGC   Wild type

TTCG / AGGC

TTCG

TTCG / AAGC   Wild type

AGGC

TCCG / AGGC   Mutant

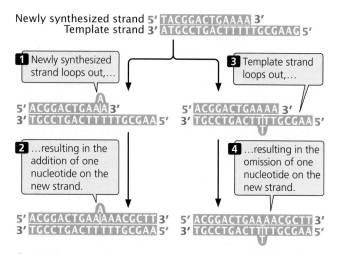

◀17.14 Insertions and deletions may result from strand slippage.

Spontaneous Chemical Changes

In addition to spontaneous mutations that arise in replication, mutations also result from spontaneous chemical changes in DNA. One such change is **depurination,** the loss of a purine base from a nucleotide. Depurination results when the covalent bond connecting the purine to the 1′-carbon atom of the deoxyribose sugar breaks (◀FIGURE 17.16a), producing an apurinic site—a nucleotide that lacks its purine base. An apurinic site cannot act as a template for a complementary base in replication. In the absence of base-pairing constraints, an incorrect nucleotide (most often adenine) is incorporated into the newly synthesized DNA strand opposite the apurinic

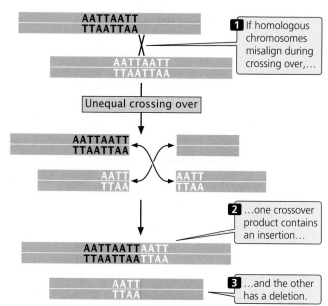

◀17.15 Unequal crossing over produces insertions and deletions.

site (◀FIGURE 17.16b), frequently leading to an incorporated error. The incorporated error is then transformed into a replication error at the next round of replication. Depurination is a common cause of spontaneous mutation; a mammalian cell in culture loses approximately 10,000 purines every day.

Another spontaneously occurring chemical change that takes place in DNA is **deamination,** the loss of an amino group (NH_2) from a base. Deamination may occur spontaneously or be induced by mutagenic chemicals.

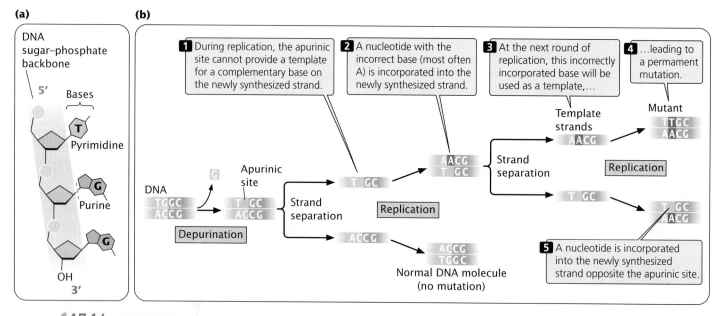

◀17.16 Depurination, loss of a purine base from the nucleotide, produces an apurinic site.

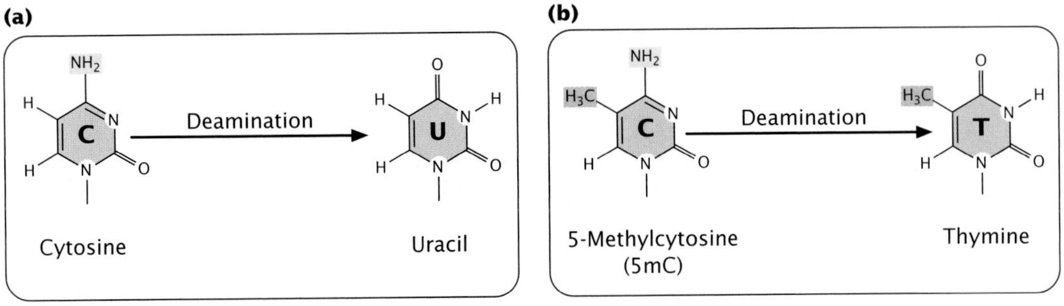

(a)

Cytosine → Deamination → Uracil

(b)

5-Methylcytosine (5mC) → Deamination → Thymine

◀ 17.17 **Deamination alters DNA bases.**

Deamination may alter the pairing properties of a base: the deamination of cytosine, for example, produces uracil (◀ FIGURE 17.17a), which pairs with adenine during replication. After another round of replication, the adenine will pair with thymine, creating a T·A pair in place of the original C·G pair (C·G→U·A→T·A); this chemical change is a transition mutation. This type of mutation is usually repaired by enzymes that remove uracil whenever it is found in DNA. The ability to recognize the product of cytosine deamination may explain why thymine, not uracil, is found in DNA. Some cytosine bases in DNA are naturally methylated and exist in the form of 5-methylcytosine (5mC; see p. 282 in Chapter 10 and Figure 10.18), which when deaminated becomes thymine (◀ FIGURE 17.17b). Because thymine pairs with adenine in replication, the deamination of 5-methylcytosine changes an original C·G pair to T·A (C·G→5mC·A→T·A). This change cannot be detected by DNA repair systems, because it produces a normal base. Consequently, C·G→T·A transitions occur frequently in eukaryotic cells.

> **Concepts**
>
> Some mutations arise from spontaneous alterations to DNA structure, such as depurination and deamination, which may alter the pairing properties of the bases and cause errors in subsequent rounds of replication.

Chemically Induced Mutations

Although many mutations arise spontaneously, a number of environmental agents are capable of damaging DNA, including certain chemicals and radiation. Any environmental agent that significantly increases the rate of mutation above the spontaneous rate is called a **mutagen.**

The first discovery of a chemical mutagen was made by Charlotte Auerbach, who was born in Germany to a Jewish family in 1899. After attending university in Berlin and doing research, she spent several years teaching at various schools in Berlin. Faced with increasing anti-Semitism in Nazi Germany, Auerbach immigrated to Britain, where she conducted research on the development of mutants in *Drosophila*. There she met Herman Muller, who had shown that radiation induces mutations; he suggested that Auerbach try to obtain mutants by treating *Drosophila* with chemicals. Her initial attempts met with little success. Other scientists were conducting top-secret research on mustard gas (used as a chemical weapon in World War I) and noticed that it produced many of the same effects as radiation. Auerbach was asked to determine whether mustard gas was mutagenic.

Collaborating with pharmacologist J. M. Robson, Auerbach studied the effects of mustard gas on *Drosophila melanogaster*. The experimental conditions were crude. They heated liquid mustard gas over a Bunsen burner on the roof of the pharmacology building, and the flies were exposed to the gas in a large chamber. After developing serious burns on her hands from the gas, Auerbach let others carry out the exposures, and she analyzed the flies. Auerbach and Robson showed that mustard gas is indeed a powerful mutagen, reducing the viability of gametes and increasing the numbers of mutations seen in the offspring of exposed flies. Because the research was part of the secret war effort, publication of their findings was delayed until 1947.

www.whfreeman.com/pierce   A brief history of Herman Muller

Base analogs One class of chemical mutagens consists of **base analogs**, chemicals with structures similar to that of any of the four standard bases of DNA. DNA polymerases cannot distinguish these analogs from the standard bases; so, if base analogs are present during replication, they may be incorporated into newly synthesized DNA molecules. For example, 5-bromouracil (5BU) is an analog of thymine; it has the same structure as that of thymine except that it has a bromine (Br) atom on the 5-carbon atom instead of a methyl group (◀ FIGURE 17.18a). Normally, 5-bromouracil pairs with adenine just as thymine does, but it occasionally mispairs with guanine (◀ FIGURE 17.18b), leading to a transition (T·A→5BU·A→5BU·G→C·G), as shown in

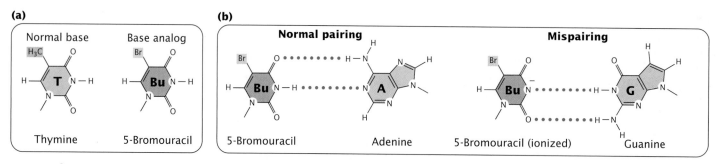

(a)

Normal base

Base analog

Thymine

5-Bromouracil

(b)

Normal pairing

Mispairing

5-Bromouracil   Adenine

5-Bromouracil (ionized)   Guanine

◀ 17.18 **5-Bromouracil (a base analog) resembles thymine, except that it has a bromine atom in place of a methyl group on the 5-carbon atom.** Because of the similarity in their structures, 5-bromouracil may be incorporated into DNA in place of thymine. Like thymine, 5-bromouracil normally pairs with adenine but, when ionized, it may pair with guanine through wobble.

◀ FIGURE 17.19. Through mispairing, 5-bromouracil may also be incorporated into a newly synthesized DNA strand opposite guanine. In the next round of replication, 5-bromouracil may pair with adenine, leading to another transition (G·C→G·5BU→A·5BU→A·T).

Another mutagenic chemical is 2-aminopurine (2AP), which is a base analog of adenine. Normally, 2-aminopurine base pairs with thymine, but it may mispair with cytosine, causing a transition mutation (T·A→T·2AP→C·2AP→ C·G). Alternatively, 2-aminopurine may be incorporated through mispairing into the newly synthesized DNA opposite cytosine and later pair with thymine, leading to a C·G→C·2AP→T·2AP→T·A transition.

Thus, both 5-bromouracil and 2-aminopurine can produce transition mutations. In the laboratory, mutations by base analogs can be reversed by treatment with the same analog or by treatment with a different analog.

Alkylating agents Alkylating agents are chemicals that donate alkyl groups. These agents include methyl (CH_3) and ethyl (CH_3-CH_2) groups, which are added to nucleotide bases by some chemicals. For example, ethylmethanesulfonate (EMS) adds an ethyl group to guanine, producing 6-ethylguanine, which pairs with thymine (see ◀ FIGURE 17.20a). Thus, EMS produces C·G→T·A transitions. EMS is also capable of adding an ethyl group to thymine, producing 4-ethylthymine, which then pairs with guanine, leading to a T·A→C·G transition. Because EMS produces both C·G→T·A and T·A→C·G transitions, mutations produced by EMS can be reversed by additional treatment with EMS. Mustard gas is another alkylating agent.

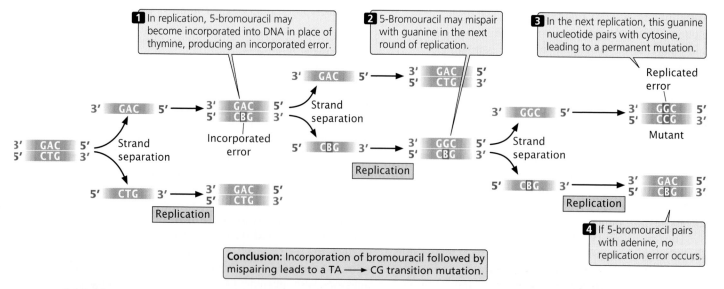

1 In replication, 5-bromouracil may become incorporated into DNA in place of thymine, producing an incorporated error.

2 5-Bromouracil may mispair with guanine in the next round of replication.

3 In the next replication, this guanine nucleotide pairs with cytosine, leading to a permanent mutation.

Strand separation

Strand separation

Incorporated error

Replication

Strand separation

Replication

Replicated error

Mutant

Replication

4 If 5-bromouracil pairs with adenine, no replication error occurs.

Conclusion: Incorporation of bromouracil followed by mispairing leads to a TA ⟶ CG transition mutation.

◀ 17.19 **5-Bromouracil can lead to a replicated error.**

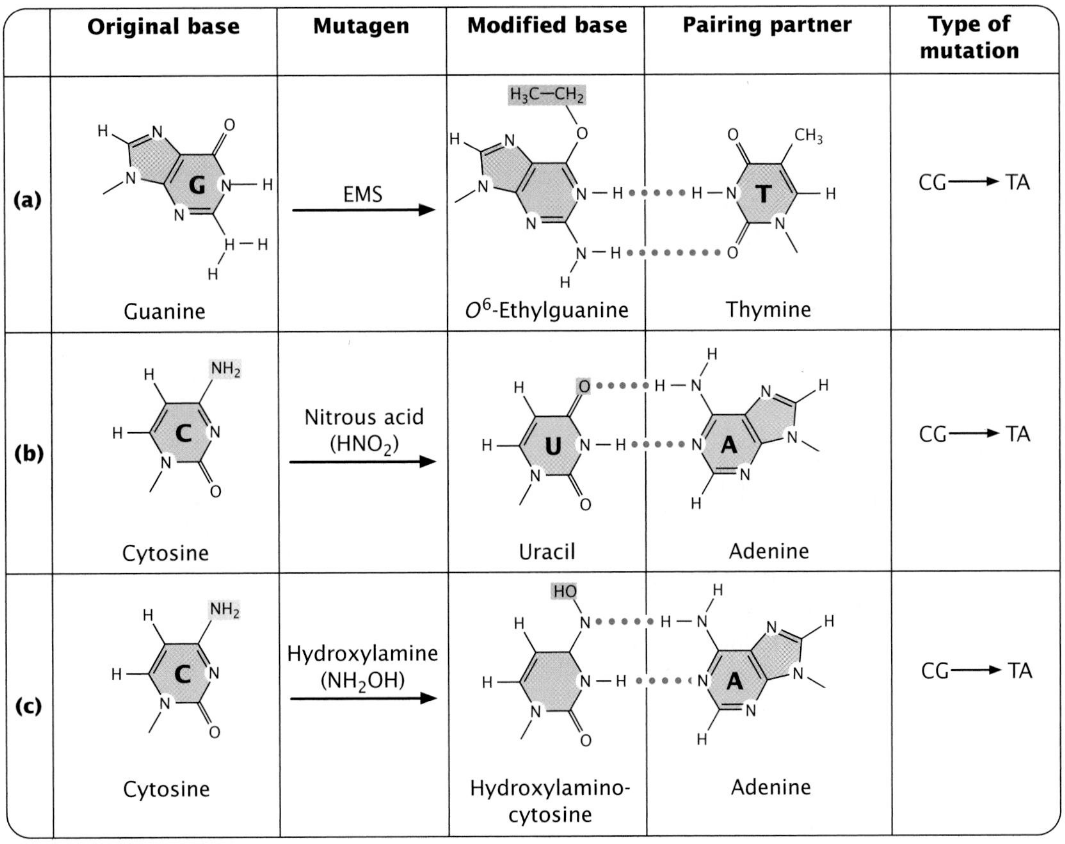

	Original base	Mutagen	Modified base	Pairing partner	Type of mutation
(a)	Guanine	EMS	O⁶-Ethylguanine	Thymine	CG ⟶ TA
(b)	Cytosine	Nitrous acid (HNO₂)	Uracil	Adenine	CG ⟶ TA
(c)	Cytosine	Hydroxylamine (NH₂OH)	Hydroxylamino-cytosine	Adenine	CG ⟶ TA

◀ 17.20 **Chemicals may alter DNA bases.** (a) The alkylating agent ethylmethanesulfonate (EMS) adds an ethyl group to guanine, producing 6-ethylguanine, which pairs with thymine, producing a C·G→T·A transition mutation. (b) Nitrous acid deaminates cytosine to produce uracil, which pairs with adenine, producing a C·G→T·A transition mutation. (c) Hydroxylamine converts cytosine into hydroxylaminocytosine, which frequently pairs with adenine, leading to a C·G→T·A transition mutation.

Deamination In addition to its spontaneous occurrence (see Figure 17.17), deamination can be induced by some chemicals. For instance, nitrous acid deaminates cytosine, creating uracil, which in the next round of replication pairs with adenine (see Figure 17.20b), producing a C·G→T·A transition mutation. Nitrous acid changes adenine into hypoxanthine, which pairs with cytosine, leading to a T·A→C·G transition. Nitrous acid also deaminates guanine, producing xanthine, which pairs with cytosine just as guanine does; however xanthine may also pair with thymine, leading to a C·G→T·A transition. Nitrous acid produces exclusively transition mutations and, because both C·G→T·A and T·A→C·G transitions are produced, these mutations can be reversed with nitrous acid.

Hydroxylamine Hydroxylamine is a very specific base-modifying mutagen that adds a hydroxyl group to cytosine, converting it into hydroxylaminocytosine (see Figure 17.20c). This conversion increases the frequency of a rare tautomer

that pairs with adenine instead of guanine and leads to C·G→T·A transitions. Because hydroxylamine acts only on cytosine, it will *not* generate T·A→C·G transitions; thus, hydroxylamine will not reverse the mutations that it produces.

Oxidative reactions Reactive forms of oxygen (including superoxide radicals, hydrogen peroxide, and hydroxyl radicals) are produced in the course of normal aerobic metabolism, as well as by radiation, ozone, peroxides, and certain drugs. These reactive forms of oxygen damage DNA and induce mutations by bringing about chemical changes to DNA. For example, oxidation converts guanine into 8-oxy-7,8-dihydrodeoxyguanine (◀ FIGURE 17.21), which frequently mispairs with adenine instead of cytosine, causing a G·C→T·A transversion mutation.

Intercalating agents Intercalating agents, such as proflavin, acridine orange, ethidium bromide, and dioxin are

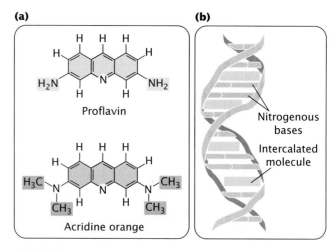

◀ 17.21 Oxidative radicals convert guanine into 8-oxy-7,8-dihydrodeoxyguanine, which frequently mispairs with adenine instead of cytosine, producing a C·G→T·A transversion.

Guanine

8-Oxy-7,8-dihydrodeoxyguanine (may mispair with adenine)

about the same size as a nucleotide (◀ FIGURE 17.22a). They produce mutations by sandwiching themselves (intercalating) between adjacent bases in DNA, distorting the three-dimensional structure of the helix and causing single-nucleotide insertions and deletions in replication (◀ FIGURE 17.22b). These insertions and deletions frequently produce frameshift mutations (which change all amino acids downstream of the mutation), and so the mutagenic effects of intercalating agents are often severe. Because intercalating agents generate both additions and deletions, they can reverse the effects of their own mutations.

(a)

Proflavin

Acridine orange

(b)

Nitrogenous bases

Intercalated molecule

◀ 17.22 Intercalating agents such as proflavin and acridine orange insert themselves between adjacent bases in DNA, distorting the three-dimensional structure of the helix and causing single-nucleotide insertions and deletions in replication.

Concepts

Chemicals can produce mutations by a number of mechanisms. Base analogs are inserted into DNA and frequently pair with the wrong base. Alkylating agents, deaminating chemicals, hydroxylamine, and oxidative radicals change the structure of DNA bases, thereby altering their pairing properties. Intercalating agents wedge between the bases and cause single-base insertions and deletions in replication.

Radiation

In 1927, Herman Muller demonstrated that mutations in fruit flies could be induced by X-rays. The results of subsequent studies showed that X-rays greatly increase mutation rates in all organisms. The high energies of X-rays, gamma rays, and cosmic rays (◀ FIGURE 17.23) are all capable of penetrating tissues and damaging DNA. These forms of radiation, called ionizing radiation, dislodge electrons from the atoms that they encounter, changing stable molecules into free radicals and reactive ions, which then alter the structures of bases and break phosphodiester

◀ 17.23 In the electromagnetic spectrum, as wavelength decreases, energy increases. (Adapted from Purves, et al., *Life* 6th ed., Sinauer Associates).

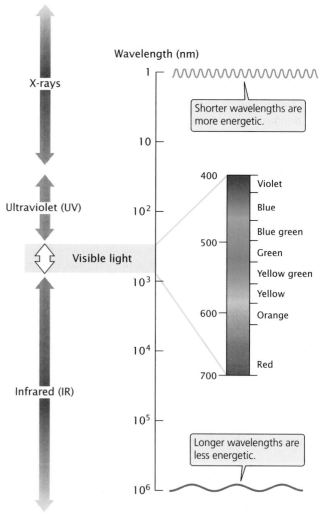

Cosmic rays/Gamma rays

X-rays

Ultraviolet (UV)

Visible light

Infrared (IR)

Microwaves/Radio waves

Wavelength (nm)

Shorter wavelengths are more energetic.

400   Violet
       Blue
       Blue green
500   Green
       Yellow green
       Yellow
600   Orange
       Red
700

Longer wavelengths are less energetic.

(a)

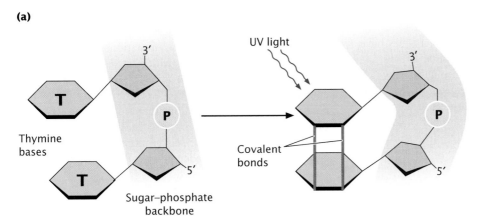

(b)

◄17.24 **Pyrimidine dimers result from ultraviolet light.**
(a) Formation of thymine dimer. (b) Distorted DNA.

bonds in DNA. Ionizing radiation also frequently results in double-strand breaks in DNA. Attempts to repair these breaks can produce chromosome mutations (discussed in Chapter 9).

Ultraviolet light has less energy than that of ionizing radiation and does not eject electrons and cause ionization but is nevertheless highly mutagenic. Purine and pyrimidine bases readily absorb UV light, resulting in the formation of chemical bonds between adjacent pyrimidine molecules on the same strand of DNA and in the creation of structures called **pyrimidine dimers** (◄FIGURE 17.24a). Pyrimidine dimers consisting of two thymine bases (called thymine dimers) are most frequent, but cytosine dimers and thymine–cytosine dimers also can form. These dimers distort the configuration of DNA (◄FIGURE 17.24b) and often block replication. Most pyrimidine dimers are immediately repaired by mechanisms discussed later in this chapter, but some escape repair and inhibit replication and transcription.

When pyrimidine dimers block replication, cell division is inhibited and the cell usually dies; for this reason, UV light kills bacteria and is an effective sterilizing agent. For a mutation—a hereditary error in the genetic instructions—to occur, the replication block must be overcome. How do bacteria and other organisms replicate despite the presence of thymine dimers?

Bacteria can circumvent replication blocks produced by pyrimidine dimers and other types of DNA damage by means of the **SOS system.** This system allows replication blocks to be overcome, but in the process makes numerous mistakes and greatly increases the rate of mutation. Indeed, the very reason that replication can proceed in the presence of a block is that the enzymes in the SOS system do not strictly adhere to the base-pairing rules. The trade-off is that replication may continue and the cell

survives, but only by sacrificing the normal accuracy of DNA synthesis.

The SOS system is complex, including the products of at least 25 genes. A protein called RecA binds to the damaged DNA at the blocked replication fork and becomes activated. This activation promotes the binding of a protein called LexA, which is a repressor of the SOS system. The activated RecA complex induces LexA to undergo self-cleavage, destroying its repressive activity. This inactivation enables other SOS genes to be expressed, and the products of these genes allow replication of the damaged DNA to proceed. The SOS system allows bases to be inserted into a new DNA strand in the absence of bases on the template strand, but these insertions result in numerous errors in the base sequence.

Eukaryotic cells have a specialized DNA polymerase called polymerase η (eta) that bypasses pyrimidine dimers. Polymerase η preferentially inserts AA opposite a pyrimidine dimer. This strategy seems to be reasonable because about two-thirds of pyrimidine dimers are thymine dimers. However, the insertion of AA opposite a CT dimer results in a C·G→A·T transversion. Polymerase η is therefore said to be an error-prone polymerase.

> **Concepts**
>
> Ionizing radiation such as X-rays and gamma rays damage DNA by dislodging electrons from atoms; these electrons then break phosphodiester bonds and alter the structure of bases. Ultraviolet light causes mutations primarily by producing pyrimidine dimers that disrupt replication and transcription. The SOS system enables bacteria to overcome replication blocks but introduces mistakes in replication.

The Study of Mutations

Because mutations often have detrimental effects, they have been the subject of intense study by geneticists. These studies have included the analysis of reverse mutations, which are often sources of important insight into how mutations cause DNA damage; the development of tests to determine the mutagenic properties of chemical compounds; and the investigation of human populations tragically exposed to high levels of radiation.

The Analysis of Reverse Mutations

The study of reverse mutations (reversions) can provide useful information about how mutagens alter DNA structure. For example, any mutagen that produces both A·T→G·C and G·C→A·T transitions should be able to reverse its own mutations. However, if the mutagen produces only G·C→A·T transitions, then reversion by the same mutagen is not possible. Hydroxylamine (see Figure 17.20c) exhibits this type of one-way mutagenic activity; it causes G·C→A·T transitions but is incapable of reversing the mutations that it produces; so we know that it does not produce A·T→G·C transitions. Ethylmethanesulfonate (see Figure 17.20a), on the other hand, produces G·C→A·T transitions and reverses its own mutations; so we know that it also produces T·A→C·G transitions.

Analyses of the ability of different mutagens to cause reverse mutations can be sources of insight into the molecular nature of the mutations. We can use reverse mutations to determine whether a mutation results from a base substitution or a frameshift. Base analogs such as 2-aminopurine cause transitions, and intercalating agents such as acridine orange (see Figure 17.22) produce frameshifts. If a chemical reverses mutations produced by 2-aminopurine but not those produced by acridine orange, we can conclude that

the chemical causes transitions and not frameshifts. If nitrous acid (which produces both G·C→A·T and A·T→G·C transitions) reverses mutations produced by the chemical but hydroxylamine (which causes *only* G·C→A·T transitions) does not, we know that, like hydroxylamine, the chemical produces only G·C→A·T transitions. Table 17.4 illustrates the reverse mutations that are theoretically possible among several mutagenic agents. The actual ability of mutagens to produce reversals is more complex than suggested by Table 17.4 and depends on environmental conditions and the organism tested.

> **Concepts**
>
> The study of the ability of mutagenic agents to produce reverse mutations provides important information about how mutagens alter DNA.

Detecting Mutations with the Ames Test

Humans in industrial societies are surrounded by a multitude of artificially produced chemicals: more than 50,000 different chemicals are in commercial and industrial use today, and from 500 to 1000 new chemicals are introduced each year. Some of these chemicals are potential carcinogens and may cause potential harm to humans. How can we determine which chemicals are hazardous? In a few instances, previous human exposure to a specific chemical is correlated with an increase in cancer incidence, providing good evidence that the chemical is a carcinogen. But, ideally, we would like to know which chemicals are hazardous before we are exposed to them. One method for testing the cancer-causing potential of chemicals is to administer them to laboratory animals (rats or mice) and compare the inci-

Table 17.4	Theoretical reverse mutations possible by various mutagenic agents						
				Reversal of Mutations by			
Mutagen	**Type of Mutation**	**5-Bromo-uracil**	**2-Amino-purine**	**Ethyl methane sulfonate**	**Nitrous acid**	**Hydroxyl-amine**	**Acridine orange**
5-Bromouracil	C·G↔T·A	+	+	+	+	+/−	−
2-Aminopurine	C·G↔T·A	+	+	+	+	+/−	−
Nitrous acid	C·G↔T·A	+	+	+	+	+/−	−
Ethylmethane sulfonate	C·G↔T·A	+	+	+	+	+/−	−
Hydroxylamine	C·G↔T·A	+	+	+	+	−	−
Acridine orange	Frameshift	−	−	−	−	−	+

Note: + indicates that reverse mutations occur, − indicates that reverse mutations do not occur, and +/− indicates that only some mutations are reversed. Not all reverse mutations are equally likely.

dence of cancer in the treated animals with that of control animals. These tests are unfortunately time consuming and expensive. Furthermore, the ability of a substance to cause cancer in rodents is not always indicative of its effect on humans. After all, we aren't rats!

In 1974, Bruce Ames developed a simple test for evaluating the potential of chemicals to cause cancer. The **Ames test** is based on the principle that both cancer and mutations result from damage to DNA, and the results of experiments have demonstrated that 90% of known carcinogens are also mutagens. Ames proposed that mutagenesis in bacteria could serve as an indicator of carcinogenesis in humans.

The Ames test uses four strains of the bacterium *Salmonella typhimurium* that have defects in the lipopolysaccharide coat, which normally protects the bacteria from chemicals in the environment. Furthermore, their DNA repair system has been inactivated, enhancing their susceptibility to mutagens.

One of the four strains used in the Ames test detects base-pair substitutions; the other three detect different types of frameshift mutations. Each strain carries a mutation that renders it unable to synthesize the amino acid histidine (*his⁻*), and the bacteria are plated onto medium that lacks histidine (◀FIGURE 17.25). Only bacteria that have undergone a reverse mutation of the histidine gene (*his⁻ → his⁺*) are able to synthesize histidine and grow on the medium. Different dilutions of a chemical to be tested are added to plates inoculated with the bacteria, and the number of mutant bacterial colonies that appear on each plate is compared with the number that appear on control plates with no chemical (arose through spontaneous mutation). Any chemical that significantly increases the number of colonies appearing on a treated plate is mutagenic and is probably also carcinogenic.

Some compounds are not active carcinogens but may be converted into cancer-causing compounds in the body. To make the Ames test sensitive for such *potential* carcinogens, a compound to be tested is first incubated in mammalian liver extract that contains metabolic enzymes. The Ames test has been applied to thousands of chemicals and commercial products. An early demonstration of its usefulness was the discovery, in 1975, that most hair dyes sold in the United States contained compounds that were mutagenic to bacteria. These compounds were then removed from most hair dyes.

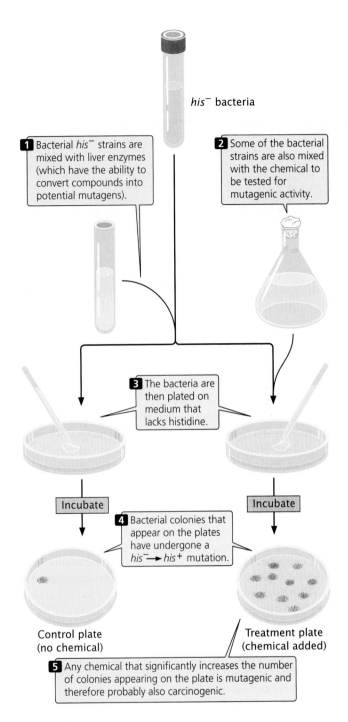

17.25 The Ames test is used to identify chemical mutagens.

www.whfreeman.com/pierce   More information on the Ames test

Radiation Exposure in Humans

People are routinely exposed to low levels of radiation from cosmic, medical, and environmental sources, but there have also been tragic events that produced exposures of much higher degree.

Concepts

The Ames test uses *his⁻* strains of bacteria to test chemicals for their ability to produce *his⁻ → his⁺* mutations. Because mutagenic activity and carcinogenic potential are closely correlated, the Ames test is widely used to screen chemicals for their cancer-causing potential.

 **17.26 Hiroshima was destroyed by an atomic bomb on August 6, 1945.** The atomic explosion produced many somatic mutations among the survivors. (Stanley Troutman/AP.)

On August 6, 1945, a high-flying American plane dropped a single atomic bomb on the city of Hiroshima, Japan. The explosion devastated 4.5 square miles of the city, killed from 90,000 to 140,000 people, and injured almost as many (◀ FIGURE 17.26). Three days later, the United States dropped an atomic bomb on the city of Nagasaki, this time destroying 1.5 square miles of city and killing between 60,000 and 80,000 people. Huge amounts of radiation were released during these explosions and many people were exposed.

After the war, a joint Japanese–U.S. effort was made to study the biological effects of radiation exposure on the survivors of the atomic blasts and their children. Somatic mutations were examined by studying radiation sickness and cancer among the survivors; germ-line mutations were assessed by looking at birth defects, chromosome abnormalities, and gene mutations in children born to people that had been exposed to radiation.

Geneticist James Neel and his colleagues examined almost 19,000 children of parents who were within 2000 meters (1.2 miles) of the center of the atomic blast at Hiroshima or Nagasaki, along with a similar number of children whose parents did not receive radiation exposure. Radiation doses were estimated for the child's parents on the basis of careful assessment of the parents' location, posture, and position at the time of the blast. A blood sample was collected from each child, and gel electrophoresis was used to investigate amino acid substitutions in 28 proteins. When rare variants were detected, blood samples from the child's parents also were analyzed to establish whether the variant was inherited or a new mutation.

Of a total of 289,868 genes examined by Neel and his colleagues, only one mutation was found in the children of exposed parents; no mutations were found in the control group. From these findings, a mutation rate of 3.4×10^{-6} was estimated for the children whose parents were exposed to the blast, which is within the range of spontaneous mutation rates observed for other eukaryotes. Neel and his colleagues also examined the frequency of chromosome mutations, sex ratios of children born to exposed parents, and frequencies of chromosome aneuploidy. There was no evidence in any of these assays for increased mutations among the children of the people who were exposed to radiation from the atomic explosions, suggesting that germ-line mutations were not elevated.

Animal studies clearly show that radiation causes germ-line mutations; so why was there no apparent increase in germ-line mutations among the inhabitants of Hiroshima and Nagasaki? The exposed parents did exhibit an increased incidence of leukemia and other types of cancers; so somatic mutations were clearly induced. The answer to the question is not known, but the lack of germ-line mutations may be due to the fact that those persons who received the largest radiation doses died soon after the blasts.

www.whfreeman.com/pierce Information on studies of the health effects of the nuclear blasts at Hiroshima and Nagasaki

The Techa River in southern Russia is another place where people have been tragically exposed to high levels of radiation. The Mayak nuclear facility, located 60 miles from the city of Chelyabinsk, produced plutonium for nuclear warheads in the early days of the Cold War. Between 1949 and 1956, this plant dumped some 76 million cubic meters of radioactive sludge into the Techa River. People downstream used the river for drinking water and crop irrigation; some received radiation doses 1700 times the annual amount considered safe by today's standards. Radiation in the area was further elevated by a series of nuclear accidents at the Mayak plant; the worst was an explosion of a radioactive liquid storage tank in 1957, which showered radiation over a 27,000-square-kilometer (10,425-square-mile) area.

Although Soviet authorities suppressed information about the radiation problems along the Techa until the 1990s, Russian physicians lead by Mira Kossenko quietly began studying cancer and other radiation-related illnesses among the inhabitants in the 1960s. They found that the overall incidence of cancer was elevated among people who lived on the banks of the Techa River.

Most data on radiation exposure in humans are from the intensive study of the survivors of the atomic bombing of Hiroshima and Nagasaki. However, the inhabitants of Hiroshima and Nagasaki were exposed in one intense burst of radiation, and these data may not be appropriate for understanding the effects of long-term low-dose radiation. Today, U.S. and Russian scientists are studying the people of the Techa River region, as well as those exposed to radiation in the Chernobyl accident (see the story at the beginning of this chapter), in an attempt to better understand the effects of chronic radiation exposure on human populations.

DNA Repair

The integrity of DNA is under constant assault from radiation, chemical mutagens, and spontaneously arising changes. In spite of this onslaught of damaging agents, the rate of mutation remains remarkably low, thanks to the efficiency with which DNA is repaired. It has been estimated that fewer than one in a thousand DNA lesions becomes a mutation; all the others are corrected.

There are a number of complex pathways for repairing DNA, but several general statements can be made about DNA repair. First, most DNA repair mechanisms require two nucleotide strands of DNA because most replace whole nucleotides, and a template strand is needed to specify the base sequence. The complementary, double-stranded nature of DNA not only provides stability and efficiency of replication, but also enables either strand to provide the information necessary for correcting the other.

A second general feature of DNA repair is redundancy, meaning that many types of DNA damage can be corrected by more than one pathway of repair. This redundancy testifies to the extreme importance of DNA repair to the survival of the cell: it ensures that almost all mistakes are corrected. If a mistake escapes one repair system, it's likely to be repaired by another system.

We will consider four general mechanisms of DNA repair: mismatch repair, direct repair, base-excision repair, and nucleotide-excision repair (Table 17.5).

Mismatch Repair

Replication is extremely accurate: each new copy of DNA has only one error per billion nucleotides. However, in the process of replication, mismatched bases are incorporated

Table 17.5	Summary of common DNA repair mechanisms
Repair System	**Type of Damage Repaired**
Mismatch	Replication errors, including mispaired bases and strand slippage
Direct	Pyrimidine dimers; other specific types of alterations
Base-excision	Abnormal bases, modified bases, and pyrimidine dimers
Nucleotide-excision	DNA damage that distorts the double helix, including abnormal bases, modified bases, and pyrimidine dimers

into the new DNA with a frequency of about 10^{-4} to 10^{-5}; so most of the errors that initially arise are corrected and never become permanent mutations. Some of these corrections are made in proofreading (see p. 336 in Chapter 12). DNA polymerases have the capacity to recognize and correct mismatched nucleotides. When a mismatched nucleotide is added to a newly synthesized DNA strand, the polymerase stalls. It then uses its $3' \rightarrow 5'$ exonuclease activity to back up and remove the incorrectly inserted nucleotide before continuing with $5' \rightarrow 3'$ polymerization.

Many incorrectly inserted nucleotides that escape detection by proofreading are corrected by *mismatch repair* (see p. 337 in Chapter 12). Incorrectly paired bases distort the three-dimensional structure of DNA, and mismatch-repair enzymes detect these distortions. In addition to detecting incorrectly paired bases, the mismatch-repair system corrects small unpaired loops in the DNA, such as those caused by strand slippage in replication (see Figure 17.14). Some trinucleotide repeats may form secondary structures on the unpaired strand (see Figure 17.6d), allowing them to escape detection by the mismatch-repair system.

After the incorporation error has been recognized, mismatch-repair enzymes cut out the distorted section of the newly synthesized strand and fill the gap with new nucleotides, by using the original DNA strand as a template. For this strategy to work, mismatch repair must have some way of distinguishing between the old and the new strands of the DNA so that the incorporation error, and not part of the original strand, is removed.

The proteins that carry out mismatch repair in *E. coli* differentiate between old and new strands by the presence of methyl groups on special sequences of the old strand. After replication, adenine nucleotides in the sequence GATC are methylated by an enzyme called Dam methylase. The process of methylation is delayed and so, immediately after replication, the old strand is methylated and the new

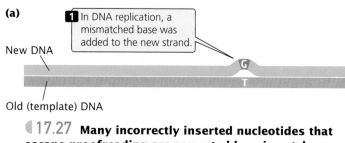

(a)

New DNA

Old (template) DNA

1 In DNA replication, a mismatched base was added to the new strand.

2 Methylation at GATC sequences allows old and newly synthesized nucleotide strands to be differentiated: immediately after replication, the old strand will be methylated but the new strand will not.

◀ 17.27 **Many incorrectly inserted nucleotides that escape proofreading are corrected by mismatch repair.**

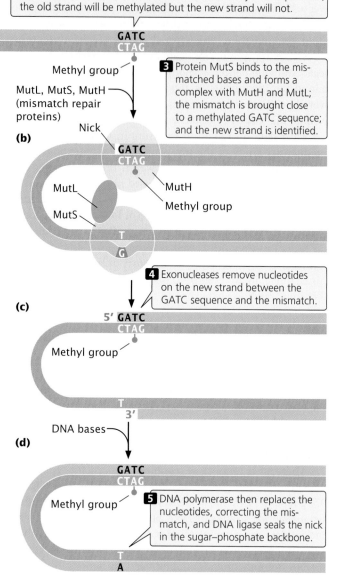

Methyl group

MutL, MutS, MutH (mismatch repair proteins)

Nick

(b)

3 Protein MutS binds to the mismatched bases and forms a complex with MutH and MutL; the mismatch is brought close to a methylated GATC sequence; and the new strand is identified.

MutL

MutS

MutH

Methyl group

4 Exonucleases remove nucleotides on the new strand between the GATC sequence and the mismatch.

(c)

Methyl group

DNA bases

(d)

5 DNA polymerase then replaces the nucleotides, correcting the mismatch, and DNA ligase seals the nick in the sugar–phosphate backbone.

strand is not (◀ FIGURE 17.27a). In *E. coli*, the proteins MutS, MutL, and MutH are required for mismatch repair. MutS binds to the mismatched bases and forms a complex with MutL and MutH; this complex is thought to bring an unmethylated GATC sequence in close proximity to the mismatched bases. MutH nicks the unmethylated strand at the GATC site (◀ FIGURE 17.27b), and exonucleases degrade the unmethylated strand from the nick to the mismatched bases (◀ FIGURE 17.27c). DNA polymerase and DNA ligase fill in the gap on the unmethylated strand with correctly paired nucleotides (◀ FIGURE 17.27d).

Mismatch repair in eukaryotic cells is similar to that in *E. coli*, except that several proteins are related to MutS and several are related to MutL. These proteins function together in different combinations to detect different types of incorporation errors, such as mispaired bases and small unpaired loops. Eukaryotic cells do not have any proteins related to *E. coli* MutH. What enzyme makes the nick in eukaryotic cells is not clear. How the old and new strands are recognized in eukaryotic cells is not known, because in some eukaryotes, such as yeast and fruit flies, there is no detectable methylation of DNA.

Direct Repair

Direct-repair mechanisms do not replace altered nucleotides but instead change them back into their original (correct) structures. One of the best-characterized direct-repair mechanisms is photoreactivation of UV-induced pyrimidine dimers. *E. coli* and some eukaryotic cells possess an enzyme called photolyase, which uses energy captured from light to break the covalent bonds that link the pyrimidines in a dimer.

Direct repair also corrects O^6-methylguanine, an alkylation product of guanine that pairs with adenine, producing G·C→T·A transversions. An enzyme called O^6-methylguanine-DNA methyltransferase removes the methyl group from O^6-methylguanine, restoring the base to guanine (◀ FIGURE 17.28).

Base-Excision Repair

In **base-excision repair,** modified bases are first excised and then the entire nucleotide is replaced. The excision of modified bases is catalyzed by a set of enzymes called DNA

glycosylases, each of which recognizes and removes a specific type of modified base by cleaving the bond that links that base to the 1′-carbon atom of deoxyribose (◀ FIGURE 17.29a). Uracil glycosylase, for example, recognizes and removes uracil produced by the deamination of cytosine. Other glycosylases

O^6-Methylguanine

Methyltransferase

Guanine

◀ 17.28 **Direct repair changes nucleotides back into their original structures.**

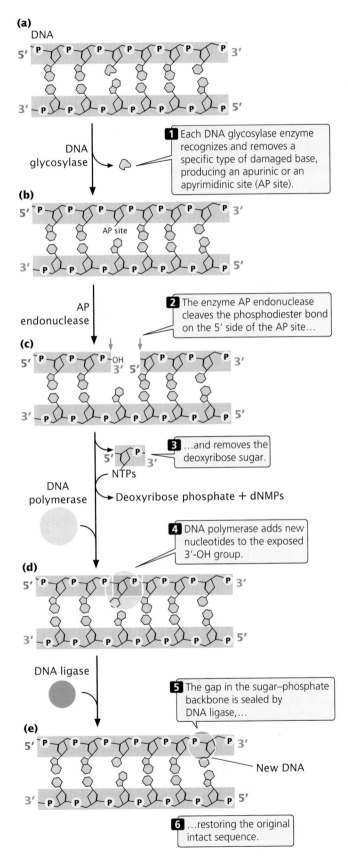

(a)

DNA

1 Each DNA glycosylase enzyme recognizes and removes a specific type of damaged base, producing an apurinic or an apyrimidinic site (AP site).

DNA glycosylase

(b)

AP site

AP endonuclease

2 The enzyme AP endonuclease cleaves the phosphodiester bond on the 5′ side of the AP site...

(c)

3 ...and removes the deoxyribose sugar.

NTPs

DNA polymerase

Deoxyribose phosphate + dNMPs

4 DNA polymerase adds new nucleotides to the exposed 3′-OH group.

(d)

DNA ligase

5 The gap in the sugar–phosphate backbone is sealed by DNA ligase,...

(e)

New DNA

6 ...restoring the original intact sequence.

◀ **17.29** **Base-excision repair excises modified bases and then replaces the entire nucleotide.**

recognize hypoxanthine, 3-methyladenine, 7-methylguanine, and other modified bases.

After the base has been removed, an enzyme called AP (apurinic or apyrimidinic) endonuclease cuts the phosphodiester bond, and other enzymes remove the deoxyribose sugar (◀ FIGURE 17.29b). DNA polymerase then adds new nucleotides to the exposed 3′-OH group (◀ FIGURE 17.29c), replacing a section of nucleotides on the damaged strand. The nick in the phosphodiester backbone is sealed by DNA ligase (◀ FIGURE 17.29d), and the original intact sequence is restored (◀ FIGURE 17.29e).

Nucleotide-Excision Repair

The final repair pathway that we'll consider is **nucleotide-excision repair,** which removes bulky DNA lesions that distort the double helix, such as pyrimidine dimers or large hydrocarbons attached to the DNA. Nucleotide-excision repair is quite versatile and can repair many different types of DNA damage. It is found in cells of all organisms from bacteria to humans and is one of the most important of all repair mechanisms.

The process of nucleotide excision is complex; in humans, a large number of genes take part. First, a complex of enzymes scans DNA, looking for distortions of its three-dimensional configuration (◀ FIGURE 17.30a and b). When a distortion is detected, additional enzymes separate the two nucleotide strands at the damaged region, and single-strand-binding proteins stabilize the separated strands (◀ FIGURE 17.30c). Next, the sugar–phosphate backbone of the damaged strand is cleaved on both sides of the damage. One cut is made 5 nucleotides upstream (on the 3′ side) of the damage, and the other cut is made 8 nucleotides (in prokaryotes) or from 21 to 23 nucleotides (in eukaryotes) downstream (on the 5′ side) of the damage (◀ FIGURE 17.30d). Part of the damaged strand is peeled away (◀ FIGURE 17.30e), and the gap is filled in by DNA polymerase and sealed by DNA ligase (◀ FIGURE 17.30f).

Other Types of DNA Repair

The DNA repair pathways described so far respond to damage that is limited to one strand of a DNA molecule, leaving the other strand to be used as a template for the synthesis of new DNA during the repair process. Some types of DNA damage, however, affect both strands of the molecule and therefore pose a more severe challenge to the DNA repair machinery. Ionizing radiation frequently results in double-strand breaks in DNA. The repair of double-strand breaks is frequently by homologous recombination. Models for homologous recombination were described in Chapter 12.

Another type of damage that affects both strands is an interstrand cross-link, which arises when the two strands of a duplex are connected through covalent bonds. Interstrand cross-links are extremely toxic to cells because they halt replication. Several drugs commonly used in chemotherapy, including cisplatin, mitomycin C, psoralen, and nitrogen mustard, cause interstrand cross-links. Nitrogen mustard, which is structurally related to the mustard gas used by Charlotte Auerbach to induce mutations in *Drosophila*, was the first chemical agent to be used in chemotherapy treat-

ment. Little is known about how interstrand cross-links are repaired. One model proposes that double-strand breaks are made on each side of the cross-link and are subsequently repaired by the pathways that repair double-strand breaks.

(a)

Damaged DNA

1 Damage to the DNA, distorts the configuration of the molecule.

(b)

2 An enzyme complex recognizes the distortion resulting from damage.

(c)

3 The DNA is separated, and single-strand-binding proteins stabilize the single strands.

(d)

4 An enzyme cleaves the strand on both sides of the damage.

(e)

5 A part of the damaged strand is removed,...

3'   5'

(f)

DNA polymerase, ligase

New DNA

6 ...and the gap is filled in by DNA polymerase and sealed by DNA ligase.

◀ **17.30 Nucleotide-excision repair consists of four steps: detection of damage, excision of damage, polymerization, and ligation.**

Connecting Concepts

The Basic Pathway of DNA Repair

We have now examined several different mechanisms of DNA repair. What do these methods have in common? How are they different? Most methods of DNA repair depend on the presence of two strands, because nucleotides in the damaged area are removed and replaced. Nucleotides are replaced in mismatch repair, base excision repair, and nucleotide-excision repair, but are not replaced by direct-repair mechanisms.

Repair mechanisms that include nucleotide removal utilize a common four-step pathway:

1. **Detection:** The damaged section of the DNA is recognized.

2. **Excision:** DNA repair endonucleases nick the phosphodiester backbone on one or both sides of the DNA damage.

3. **Polymerization:** DNA polymerase adds nucleotides to the newly exposed 3'-OH group by using the other strand as a template and replacing damaged (and frequently some undamaged) nucleotides.

4. **Ligation:** DNA ligase seals the nicks in the sugar–phosphate backbone.

The primary differences in the mechanisms of mismatch, base-excision, and nucleotide-excision repair are in the details of detection and excision. In base-excision and mismatch repair, a single nick is made in the sugar–phosphate backbone on one side of the damaged strand; in nucleotide-excision repair, nicks are made on both sides of the DNA lesion. In base-excision repair, DNA polymerase displaces the old nucleotides as it adds new nucleotides to the 3' end of the nick; in mismatch repair, the old nucleotides are degraded; and, in nucleotide-excision repair, nucleotides are displaced by helicase enzymes. All three mechanisms use DNA polymerase and ligase to fill in the gap produced by the excision and removal of damaged nucleotides.

www.whfreeman.com/pierce   Additional information on DNA repair

Genetic Diseases and Faulty DNA Repair

Several human diseases are connected to defects in DNA repair. These diseases are often associated with high incidences of specific cancers, because defects in DNA repair

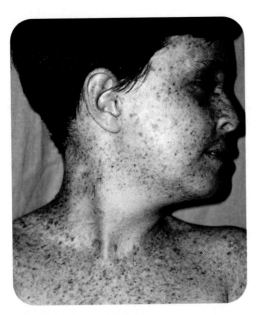

◀17.31 **Xeroderma pigmentosum is a human disease that results from defects in DNA repair.** The disease is characterized by frecklelike spots on the skin (shown here) and predisposition to skin cancer. (Ken Greer/Visuals Unlimited.)

lead to increased rates of mutation. This concept is discussed further in Chapter 21.

Among the best studied of the human DNA repair diseases is xeroderma pigmentosum (◀Figure 17.31), a rare autosomal recessive condition that includes abnormal skin pigmentation and acute sensitivity to sunlight. Persons who have this disease also have a strong predisposition to skin cancer, with an incidence from 1000 to 2000 times that found in unaffected people.

Sunlight includes a strong UV component; so exposure to sunlight produces pyrimidine dimers in the DNA of skin cells. Although human cells lack photolyase (the enzyme that repairs pyrimidine dimers in bacteria), most pyrimidine dimers in humans can be corrected by nucleotide-excision repair (see Figure 17.30). However, the cells of most people with xeroderma pigmentosum are defective in nucleotide-excision repair, and many of their pyrimidine dimers go uncorrected and may lead to cancer.

Xeroderma pigmentosum can result from defects in several different genes; studies have identified at least seven different xeroderma pigmentosum complementation groups, meaning that at least seven genes are required for nucleotide-excision repair in humans. Recent molecular research has led to the identification of genetic defects of nucleotide-excision repair associated with these complementation groups. Some persons with xeroderma pigmentosum have mutations in a gene encoding the protein that recognizes and binds to damaged DNA; others have mutations in a gene encoding helicase. Still others have defects in the genes that play a role in cutting the damaged strand on

the 5′ or 3′ sides of the pyrimidine dimer. Some persons have a slightly different form of the disease (xeroderma pigmentosum variant) owing to mutations in the gene encoding polymerase η, the DNA polymerase that bypasses pyrimidine dimers by inserting AA.

Two other genetic diseases due to defects in nucleotide-excision repair are Cockayne syndrome and trichothiodystrophy (also known as brittle-hair syndrome). Persons who have either of these diseases do not have an increased risk of cancer but do exhibit multiple developmental and neurological problems. Both diseases result from mutations in some of the same genes that cause xeroderma pigmentosum. Several of the genes taking part in nucleotide-excision repair produce proteins that also play a role in recombination and the initiation of transcription. These other functions may account for the developmental symptoms seen in Cockayne syndrome and trichothiodystrophy.

Another genetic disease caused by faulty DNA repair is an inherited form of colon cancer called hereditary nonpolyposis colon cancer (HNPCC). This cancer is one of the most common hereditary cancers, accounting for about 15% of colon cancers. Research indicate that HNPCC arises from mutations in the proteins that carry out mismatch repair (see Figure 17.27).

Li-Fraumeni syndrome is caused by mutations in a gene called *p53*, which plays an important role in regulating the cell cycle. The product encoded by the *p53* gene can halt cell division until damage to DNA has been repaired; it can also directly stimulate DNA repair. The *p53* gene product may actually cause cells with damaged DNA to self-destruct (undergo apoptosis, or controlled cell death; see Chapter 21), preventing their mutated genetic instructions from being passed on. Patients who have Li-Fraumeni syndrome exhibit multiple independent cancers in different tissues. Some additional genetic diseases associated with defective DNA repair are summarized in Table 17.6.

Concepts

Defects in DNA repair are the underlying cause of several genetic diseases. Many of these diseases are characterized by a predisposition to cancer.

www.whfreeman.com/pierce   Additional information about xeroderma pigmentosum

Connecting Concepts Across Chapters

This chapter has been our first comprehensive look at mutations, but we have been considering and using mutations throughout the book.

Mutation is a fact of life. Our DNA is continually assaulted by spontaneously arising and environmentally

Table 17.6	Genetic diseases associated with defects in DNA repair systems	
Disease	**Symptoms**	**Genetic Defect**
Xeroderma pigmentosum	Frecklelike spots on skin, sensitivity to sunlight, predisposition to skin cancer	Defects in nucleotide-excision repair
Cockayne syndrome	Dwarfism, sensitivity to sunlight, premature aging, deafness, mental retardation	Defects in nucleotide-excision repair
Trichothiodystrophy	Brittle hair, skin abnormalities, short stature, immature sexual development, characteristic facial features	Defects in nucleotide-excision repair
Hereditary nonpolyposis colon cancer	Predisposition to colon cancer	Defects in mismatch repair
Fanconi anemia	Increased skin pigmentation, abnormalities of skeleton, heart, and kidneys, predisposition to leukemia	Possibly defects in the repair of interstrand cross-links
Ataxia telangiectasia	Defective muscle coordination, dilation of blood vessels in skin and eyes, immune deficiencies, sensitivity to ionizing radiation, predisposition to cancer	Defects in DNA damage detection and response
Li-Fraumeni syndrome	Predisposition to cancer in many different tissues	Defects in DNA damage response

induced mutations. These mutations are the raw material of evolution and, in the long run, allow organisms to adapt to the environment, a topic that will be taken up in Chapter 23. In spite of their long-term contribution to species evolution, the vast majority of mutations are, in the short term, detrimental to cells. The fact that most are detrimental is evidenced by the number mechanisms that cells possess to reduce the generation of errors in DNA and to repair those that do arise. A dominant theme of this chapter is that cells go to great lengths to prevent mutations.

This chapter has incorporated information presented in a number of earlier chapters, which you might want to review for a better understanding of the processes and structures discussed in the current chapter. Chromosome mutations and transposable elements (which frequently cause mutations) are discussed in Chapters 9 and 11. Although the structural nature of these mutations is different from that of gene mutations, many fundamental aspects of the mutational process that were introduced in this chapter also apply to these other types of mutations. The study of gene mutations is fundamentally about changes in DNA structure; so the discussion of DNA structure in Chapter 10 is critical for understanding the nature of mutations and how they arise. Some mutations spontaneously arise from errors in replication, and many DNA repair mechanisms include some DNA synthesis; hence, the process of replication outlined in Chapter 12 also is important. The relation between the nucleotide sequences of DNA and the amino acid sequences of proteins, which is discussed in Chapter 15, is particularly relevant for understanding the phenotypic effects of mutations and the nature of intra- and intergenic suppressors. Some of the material covered on bacterial and viral genetics in Chapter 8 is helpful for understanding complementation and the Ames test.

The current chapter has provided information that is important for understanding material presented in future chapters. Mutation is the molecular basis of cancer; so the contents of the current chapter will be highly relevant to the discussion of cancer genetics in Chapter 21. The importance of the mutation process to evolution will be revisited in Chapter 23.

CONCEPTS SUMMARY

- Mutations are heritable changes in genetic information. They are important for the study of genetics and can be used to unravel other biological processes.

- Somatic mutations occur in somatic cells; germ-line mutations occur in cells that give rise to gametes. Gene mutations are genetic alterations that affect a single gene; chromosome mutations entail changes in the number or structure of chromosomes.

- The simplest type of mutation is a base substitution, a change in a single base pair of DNA. Transitions are base substitutions in which purines are replaced by purines or pyrimidines are replaced by pyrimidines. Transversions are base substitutions in which a purine replaces a pyrimidine or a pyrimidine replaces a purine.

- Insertions are the addition of nucleotides, and deletions are the removal of nucleotides; these mutations often change the reading frame of the gene.

- Expanding trinucleotide repeats are mutations in which the number of copies of a trinucleotide increases through time; they are responsible for several human genetic diseases.

- A missense mutation alters the coding sequence so that one amino acid substitutes for another. A nonsense mutation changes a codon that specifies an amino acid to a termination codon. A silent mutation produces a synonymous codon that specifies the same amino acid as the original sequence, whereas a neutral mutation alters the amino acid sequence but does not change the functioning of the protein. A suppressor mutation reverses the effect of a previous mutation at a different site and may be intragenic (within the same gene as the original mutation) or intergenic (within a different gene).

- Mutation rate is the frequency with which a particular mutation arises in a population, whereas mutation frequency is the incidence of a mutation in a population. Mutation rates are usually low and are influenced by both genetic and environmental factors.

- Some mutations occur spontaneously. These mutations include the mispairing of bases in replication and spontaneous depurination and deamination.

- Insertions and deletions may arise from strand slippage in replication or from unequal crossing over.

- Base analogs may become incorporated into DNA in replication and pair with the wrong base in subsequent replication events. Alkylating agents and hydroxylamine modify the chemical structure of bases and lead to mutations. Intercalating agents insert into the DNA molecule and cause single-nucleotide additions and deletions. Oxidative reactions alter the chemical structures of bases.

- Ionizing radiation is mutagenic, altering base structures and breaking phosphodiester bonds. Ultraviolet light produces pyrimidine dimers, which block replication. Bacteria use the SOS response to overcome replication blocks produced by pyrimidine dimers and other lesions in DNA, but the SOS response causes the occurrence of more replication errors. Pyrimidine dimers in eukaryotic cells can be bypassed by DNA polymerase η but may result in the placement of incorrect bases opposite the dimer.

- The analysis of reverse mutations provides information about the molecular nature of the original mutation.

- The Ames tests uses bacteria to assess the mutagenic potential of chemical substances.

- Most damage to DNA is corrected by DNA repair mechanisms. These mechanisms include mismatch repair, direct repair, base-excision repair, nucleotide-excision repair, and other repair pathways. Although the details of the different DNA repair mechanisms vary, most require two strands of DNA and exhibit some overlap in the types of damage repaired. Proofreading and mismatch repair correct errors that arise in replication. Direct-repair mechanisms change the altered nucleotides back into their original condition, whereas base-excision and nucleotide-excision repair mechanisms replace nucleotides around the damaged segment of the DNA.

- Defects in DNA repair are the underlying cause of several genetic diseases.

IMPORTANT TERMS

mutation (p. 473)
somatic mutation (p. 474)
germ-line mutation (p. 475)
gene mutation (p. 475)
base substitution (p. 475)
transition (p. 475)
transversion (p. 475)
insertion (p. 475)

deletion (p. 475)
frameshift mutation (p. 476)
in-frame insertion (p. 476)
in-frame deletion (p. 476)
expanding trinucleotide repeat (p. 477)
forward mutation (p. 478)

reverse mutation (reversion) (p. 478)
missense mutation (p. 478)
nonsense mutation (p. 478)
silent mutation (p. 478)
neutral mutation (p. 478)
loss-of-function mutation (p. 478)

gain-of-function mutation (p. 478)
conditional mutation (p. 478)
lethal mutation (p. 478)
suppressor mutation (p. 478)
intragenic suppressor mutation (p. 478)

intergenic suppressor
    mutation (p. 480)
mutation rate (p. 481)
mutation frequency (p. 483)
spontaneous mutation (p. 484)
induced mutation (p. 484)

incorporated error (p. 485)
replicated error (p. 485)
strand slippage (p. 485)
unequal crossing over (p. 485)
depurination (p. 486)
deamination (p. 486)

mutagen (p. 487)
base analog (p. 487)
intercalating agent (p. 489)
pyrimidine dimer (p. 491)
SOS system (p. 491)
Ames test (p. 493)

direct repair (p. 496)
base-excision repair (p. 496)
nucleotide-excision repair
    (p. 497)

Worked Problems

1. A codon that specifies the amino acid Asp undergoes a single-base substitution that yields a codon that specifies Ala. Refer to the genetic code in Figure 15.12 and give all possible DNA sequences for the original and the mutated codon. Is the mutation a transition or a transversion?

• Solution

There are two possible RNA codons for Asp: GAU and GAC. The DNA sequences that encode these codons will be complementary to the RNA codons: CTA and CTG. There are four possible RNA codons for Ala: GCU, GCC, GCA, and GCG, which correspond to DNA sequences CGA, CGG, CGT, and CGC. If we organize the original and mutated sequences as shown in the following table, it is easy to see what type of mutations may have occurred:

Possible original sequence for Asp	Possible mutated sequence for Ala
CTA	CGA
CTG	CGG
	CGT
	CGC

If the mutation is confined to a single-base substitution, then the only mutations possible are that CTA mutated to CGA or that GTG mutated to CGG. In both, there is a T→G transversion in the middle nucleotide of the codon.

2. A gene encodes a protein with the following amino acid sequence:

Met-Arg-Cys-Ile-Lys-Arg

A mutation of a single nucleotide alters the amino acid sequence to:

Met-Asp-Ala-Tyr-Lys-Gly-Glu-Ala-Pro-Val

A second single-nucleotide mutation occurs in the same gene and suppresses the effects of the first mutation (an intragenic suppressor). With the original mutation and the intragenic suppressor present, the protein has the following amino acid sequence:

Met-Asp-Gly-Ile-Lys-Arg

What is the nature and location of the first mutation and the intragenic suppressor mutation?

• Solution

The first mutation alters the reading frame, because all amino acids after Met are changed. Insertions and deletions affect the reading frame; so the original mutation consists of a single-nucleotide insertion or deletion in the second codon. The intragenic suppressor restores the reading frame; so the intragenic suppressor also is most likely a single-nucleotide insertion or deletion: if the first mutation is an insertion, the suppressor must be a deletion; if the first mutation is a deletion, then the suppressor must be an insertion. Notice that the protein produced by the suppressor still differs from the original protein at the second and third amino acids, but the suppressor's second amino acid is the same as that in the protein produced by the original mutation. Thus the suppressor mutation must have occurred in the third codon, because the suppressor does not alter the second amino acid.

3. The mutations produced by the following compounds are reversed by the substances shown. What conclusions can you make about the nature of the mutations originally produced by these compounds?

Mutations produced by compound	Reversed by			
	5-Bromouracil	EMS	Hydroxyl-amine	Acridine orange
(a) 1	Yes	Yes	No	No
(b) 2	Yes	Yes	Some	No
(c) 3	No	No	No	Yes
(d) 4	Yes	Yes	Yes	Yes

• Solution

The ability of various compounds to produce reverse mutations reveals important information about the nature of the original mutation.

(a) Mutations produced by compound 1 are reversed by 5-bromouracil, which produces both A·T→G·C and G·C→A·T transitions. This tells us that compound 1 produces single-base substitutions that *may* include the generation of either A·T or

G·C pairs. The mutations produced by compound 1 are also reversed by EMS, which, like 5-bromouracil, produces both A·T→G·C and G·C→A·T transitions; so no additional information is provided here. Hydroxylamine does not reverse the mutations produced by compound 1. Because hydroxylamine produces only C·G→T·A transitions, we know that compound 1 does not generate C·G base pairs. Acridine orange, an intercalating agent that produces frameshift mutations, also does not reverse the mutations, revealing that compound 1 produces only single-base-pair substitutions, not insertions or deletions. In summary, compound 1 appears to causes single-base substitutions that generate T·A but not G·C base pairs.

(b) Compound 2 generates mutations that are reversed by 5-bromouracil and EMS, indicating that it may produce G·C or A·T base pairs. Some of these mutations are reversed by hydroxylamine, which produces only C·G→T·A transitions. This indicates that some of the mutations produced by compound 2 are C·G base pairs. None of the mutations are reversed by acridine orange; so compound 2 does not induce insertions or deletions. In summary, compound 2 produces single-base substitutions that generate both G·C and A·T base pairs.

(c) Compound 3 produces mutations that are reversed only by acridine orange; so compound 3 appears to produce only insertions and deletions.

(d) Compound 4 is reversed by 5 bromouracil, EMS, hydroxylamine, and acridine orange, indicating that this compound produces single-base substitutions, which include both G·C and A·T base pairs, and insertions and deletions.

The New Genetics
MINING GENOMES

MOLECULAR EVOLUTION

This exercise introduces you to some of the basic principles of molecular evolution, the study of the ways in which molecules evolve, and the reconstruction of the evolutionary history of molecules and organisms. You will use several of the Internet tools most frequently used by contemporary molecular geneticists to analyze analogous sequences from related organisms.

COMPREHENSION QUESTIONS

* 1. What is the difference between somatic mutations and germ-line mutations?

* 2. What is the difference between a transition and a transversion? Which type of base substitution is usually more common?

* 3. Briefly describe expanding trinucleotide repeats. How do they account for the phenomenon of anticipation?

4. What is the difference between a missense mutation and a nonsense mutation? A silent mutation and a neutral mutation?

5. Briefly describe two different ways that intragenic suppressors may reverse the effects of mutations.

* 6. How do intergenic suppressors work?

* 7. What is the difference between mutation frequency and mutation rate?

* 8. What is the cause of errors in DNA replication?

9. How do insertions and deletions arise?

*10. How do base analogs lead to mutations?

11. How do alkylating agents, nitrous acid, and hydroxylamine produce mutations?

12. What types of mutations are produced by ionizing and UV radiation?

*13. What is the SOS system and how does it lead to an increase in mutations?

14. What is the purpose of the Ames test? How are *his⁻* bacteria used in this test?

*15. List at least three different types of DNA repair and briefly explain how each is carried out.

16. What features do mismatch repair, base-excision repair, and nucleotide-excision repair have in common?

APPLICATION QUESTIONS AND PROBLEMS

*17. A codon that specifies the amino acid Gly undergoes a single-base substitution to become a nonsense mutation. In accord with the genetic code given in Figure 15.12, is this mutation a transition or a transversion? At which position of the codon does the mutation occur?

*18. **(a)** If a single transition occurs in a codon that specifies Phe, what amino acids could be specified by the mutated sequence?
(b) If a single transversion occurs in a codon that specifies Phe, what amino acids could be specified by the mutated sequence?

(c) If a single transition occurs in a codon that specifies Leu, what amino acids could be specified by the mutated sequence?

(d) If a single transversion occurs in a codon that specifies Leu, what amino acids could be specified by the mutated sequence?

19. Hemoglobin is a complex protein that contains four polypeptide chains. The normal hemoglobin found in adults—called adult hemoglobin—consists of two α and two β polypeptide chains, which are encoded by different loci. Sickle-cell hemoglobin, which causes sickle-cell anemia, arises from a mutation in the β chain of adult hemoglobin. Adult hemoglobin and sickle-cell hemoglobin differ in a single amino acid: the sixth amino acid from one end in adult hemoglobin is glutamic acid, whereas sickle-cell hemoglobin has valine at this position. After consulting the genetic code provided in Figure 15.12, indicate the type and location of the mutation that gave rise to sickle-cell anemia.

*20. The following nucleotide sequence is found on the template strand of DNA. First, determine the amino acids of the protein encoded by this sequence by using the genetic code provided in Figure 15.12. Then, give the altered amino acid sequence of the protein that will be found in each of the following mutations.

Sequence
of DNA
template
└ 3′–TAC TGG CCG TTA GTT GAT ATA ACT–5′
┌   1   24
Nucleotide
number

(a) Mutant 1: A transition at nucleotide 11.

(b) Mutant 2: A transition at nucleotide 13.

(c) Mutant 3: A one-nucleotide deletion at nucleotide 7.

(d) Mutant 4: A T→A transversion at nucleotide 15.

(e) Mutant 5: An addition of TGG after nucleotide 6.

(f) Mutant 6: A transition at nucleotide 9.

21. A polypeptide has the following amino acid sequence:

Met-Ser-Pro-Arg-Leu-Glu-Gly

The amino acid sequence of this polypeptide was determined in a series of mutants listed in parts a through e. For each mutant, indicate the type of change that occurred in the DNA (single-base substitution, insertion, deletion) and the phenotypic effect of the mutation (nonsense mutation, missense mutation, frameshift, etc.).

(a) Mutant 1: Met-Ser-Ser-Arg-Leu-Glu-Gly

(b) Mutant 2: Met-Ser-Pro

(c) Mutant 3: Met-Ser-Pro-Asp-Trp-Arg-Asp-Lys

(d) Mutant 4: Met-Ser-Pro-Glu-Gly

(e) Mutant 5: Met-Ser-Pro-Arg-Leu-Leu-Glu-Gly

*22. A gene encodes a protein with the following amino acid sequence:

Met-Trp-His-Arg-Ala-Ser-Phe.

A mutation occurs in the gene. The mutant protein has the following amino acid sequence:

Met-Trp-His-Ser-Ala-Ser-Phe.

An intragenic suppressor restores the amino acid sequence to that of the original the protein:

Met-Trp-His-Arg-Ala-Ser-Phe.

Give at least one example of base changes that could produce the original mutation and the intragenic suppressor? (Consult the genetic code in Figure 15.12.)

23. A gene encodes a protein with the following amino acid sequence:

Met-Lys-Ser-Pro-Ala-Thr-Pro

A nonsense mutation from a single-base-pair substitution occurs in this gene, resulting in a protein with the amino acid sequence Met-Lys. An intergenic suppressor mutation allows the gene to produce the full-length protein. With the original mutation and the intergenic suppressor present, the gene now produces a protein with the following amino acid sequence:

Met-Lys-Cys-Pro-Ala-Thr-Pro.

Give the location and nature of the original mutation and the intergenic suppressor.

*24. Can nonsense mutations be reversed by hydroxylamine? Why or why not?

25. XG syndrome is a rare genetic disease that is due to an autosomal dominant gene. A complete census of a small European country reveals that 77,536 babies were born in 2000, of whom 3 had XG syndrome. In the same year, this country had a population of 5,964,321 people, and there were 35 living persons with XG syndrome. What are the mutation rate and mutation frequency of XG syndrome for this country?

*26. The following nucleotide sequence is found in a short stretch of DNA:

5′–ATGT–3′
3′–TACA–5′

If this sequence is treated with hydroxylamine, what sequences will result after replication?

27. The following nucleotide sequence is found in a short stretch of DNA:

5′–AG–3′
3′–TC–5′

(a) Give all the mutant sequences that may result from spontaneous depurination in this stretch of DNA.

(b) Give all the mutant sequences that may result from spontaneous deamination in this stretch of DNA.

28. In many eukaryotic organisms, a significant proportion of cytosine bases are naturally methylated to 5-methylcytosine. Through evolutionary time, the proportion of AT base pairs in the DNA of these organisms increases. Can you suggest a possible mechanism by which this increase occurs?

*29. A chemist synthesizes four new chemical compounds in the laboratory and names them PFI1, PFI2, PFI3, and PFI4. He gives the PFI compounds to a geneticist friend and asks her to determine their mutagenic potential. The geneticist finds that all four are highly mutagenic. She also tests the capacity of mutations produced by the PFI compounds to be reversed by other known mutagens and obtains the following results. What conclusions can you make about the nature of the mutations produced by these compounds?

Mutations produced by	Reversed by			
	2-Amino-purine	Nitrous-acid	Hydroxyl-amine	Acridine orange
PFI1	Yes	Yes	Some	No
PFI2	No	No	No	No
PFI3	Yes	Yes	No	No
PFI4	No	No	No	Yes

*30. A plant breeder wants to isolate mutants in tomatoes that are defective in DNA repair. However, this breeder does not have the expertise or equipment to study enzymes in DNA repair systems. How could the breeder identify tomato plants that are deficient in DNA repair? What are the traits to look for?

31. A genetics instructor designs a laboratory experiment to study the effects of UV radiation on mutation in bacteria. In the experiment, the students expose bacteria plated on petri plates to UV light for different lengths of time, place the plates in an incubator for 48 hours, and then count the number of colonies that appear on each plate. The plates that have received more UV radiation should have more pyrimidine dimers, which block replication; thus, fewer colonies should appear on the plates exposed to UV light for longer periods of time. Before the students carry out the experiment, the instructor warns them that, while the bacteria are in the incubator, the students must not open the incubator door unless the room is darkened. Why should the bacteria not be exposed to light?

CHALLENGE QUESTIONS

32. *Ochre* and *amber* are two types of nonsense mutations. Before the genetic code was worked out, Sydney Brenner, Anthony O. Stretton, and Samuel Kaplan applied different types of mutagens to bacteriophages in an attempt to determine the bases present in the codons responsible for *amber* and *ochre* mutations. They knew that *ochre* and *amber* mutants were suppressed by different types of mutations, demonstrating that each was a different termination codon. They obtained the following results.

    (1) A single-base substitution could convert an *ochre* mutation into an *amber* mutation.

    (2) Hydroxylamine induced both *ochre* and *amber* mutations in wild-type phages.

    (3) 2-Aminopurine caused *ochre* to mutate to *amber*.

    (4) Hydroxylamine did not cause *ochre* to mutate to *amber*.

    These data do not allow the complete nucleotide sequence of the *amber* and *ochre* codons to be worked out, but they do provide some information about the bases found in the nonsense mutations.

    (a) What conclusions about the bases found in the codons of *amber* and *ochre* mutations can be made from these observations?

    (b) Of the three nonsense codons (UAA, UAG, UGA), which represents the *ochre* mutation.

33. To determine whether radiation associated with the atomic bombings of Hiroshima and Nagasaki produced recessive germ-line mutations, scientists examined the sex ratio of the children of the survivors of the blasts. Can you explain why an increase in germ-line mutations might be expected to alter the sex ratio?

34. The results of several studies provide evidence that DNA repair is rapid in genes that are undergoing transcription and that some proteins that play a role in transcription also participate in DNA repair. How are transcription and DNA repair related? Why might a gene that is being transcribed be repaired faster than a gene that is not being transcribed?

SUGGESTED READINGS

Balter, M. 1995. Filtering a river of cancer data. *Science* 267:1084–1086.
Article describing the nuclear disaster on the Techa river in Russia.

Beale, G. 1993. The discovery of mustard gas mutagenesis by Auerbach and Robson in 1941. *Genetics* 134:393–399.
An informative and personal account of Auerbach's life and research.

Dovoret, R. 1979. Bacterial tests for potential carcinogens. *Scientific American* 241(2):40–49.
A discussion of the Ames tests and more recent tests of mutagenesis in bacteria.

Drake, J.W., and R.H. Baltz. 1976. The biochemistry of mutagenesis. *Annual Review of Biochemistry* 45:11–37.
A discussion of how mutations are produced by mutagenic agents.

Dubrova, Y.E., V.N. Nesterov, N.G. Krouchinsky, V.A. Ostapenko, R. Neumann, D.L. Neil, and A.J. Jeffreys. 1996. Human minisatellite mutation rate after the Chernobyl accident. *Nature* 380:683–686.
A report of increased germ-line mutation rate in people exposed to radiation in the Chernobyl accident.

Goodman, M.F. 1995. DNA models: mutations caught in the act. *Nature* 378:237–238.
A review of the role of tautomerization in replication errors.

Hoeijmakers, J.H., and D. Bootsma. 1994. Incisions for excision. *Nature* 371:654–655.
Commentary on the proteins in eukaryotic nucleotide-excision repair.

Martin, J.B. 1993. Molecular genetics of neurological diseases. *Science* 262:674–676.
A discussion of expanding trinucleotide repeats as cause of neurological diseases.

Modrich, P. 1991. Mechanisms and biological effects of mismatch repair. *Annual Review of Genetics* 25:229–253.
A comprehensive review of mismatch repair.

Neel, J.V., C. Satoh, H.B. Hamilton, M. Otake, K. Goriki, T. Kageoka, M. Fujita, S. Neriishi, and J. Asakawa. 1980. Search for mutations affecting protein structure in children of atomic bomb survivors: preliminary report. *Proceedings of the National Academy of Sciences of the United States of America.* 77:4221–4225.
A report of the gene mutations in the children of survivors of the atomic bombings in Japan.

Sancar, A. 1994. Mechanisms of DNA excision repair. *Science* 266:1954–1956.
An excellent review of research on excision repair. This issue of *Science* was about the "molecule of the year" for 1994, which was DNA repair (actually not a molecule).

Schull, W.J., M. Otake, and J.V. Neel. 1981. Genetic effects of the atomic bombs: a reappraisal. *Science* 213:1220–1227.
Research findings concerning the genetic effects of radiation exposure in survivors of the atomic bombings in Japan.

Shcherbak, Y.M. 1996. Ten years of the Chernobyl era. *Scientific American* 274(4):44–49.
Considers the long-term effects of the Chernobyl accident.

Sinden, R.R. 1999. Biological implications of DNA structures associated with disease-causing triplet repeats. *American Journal of Human Genetics* 64:346–353.
A good summary of disease-causing trinucleotide repeats and some models for how they might arise.

Tanaka, K., and R.D. Wood. 1994. Xeroderma pigmentosum and nucleotide excision repair. *Trends in Biochemical Sciences* 19:84–86.
A review of the molecular basis of xeroderma pigmentosum.

Yu, S., J. Mulley, D. Loesch, G. Turner, A. Donnelly, A. Gedeon, D. Hillen, E. Kremer, M. Lynch, M. Pritchard, G.R. Sunderland, and R.I. Richards. 1992. Fragile-X syndrome: unique genetics of the heritable unstable element. *American Journal of Human Genetics* 50:968–980.
A research report describing the expanding trinucleotide repeat that causes fragile-X syndrome.

18 Recombinant DNA Technology

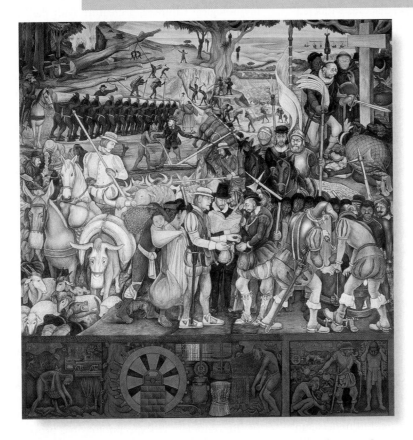

Disembarkation of the Spanish at Veracruz by Mexican artist Diego Rivera. Anthropologists have suggested that Europeans first transmitted tuberculosis to the Native Americans. The polymerase chain reaction—a technique for amplifying very small amounts of DNA—has now demonstrated the presence of the bacterium that causes tuberculosis in a 1000 year old mummy from Peru, demonstrating that the disease was present in South America long before Europeans arrived. (Diego Rivera, *Disembarkation of the Spanish at Veracruz*, 1951. Schalkwijk/Art Resource.)

PCR and the Arrival of Tuberculosis in America

In the early 1600s, soon after Europeans arrived in the New World, devastating epidemics of tuberculosis ravaged many tribes of Native Americans. Anthropologists long argued that this disease was absent from the New World before 1492 and that Europeans first transmitted tuberculosis to the Native Americans. With no prior exposure to the disease and little natural immunity, the indigenous people would, it was argued, have been highly susceptible to tuberculosis.

A few anthropologists challenged this conventional view. On the basis of tuberculosis-like lesions found in a few skeletons and mummified remains that pre-date European contact, they suggested that the disease was present in Native Americans before European contact. On the other hand, many diseases and even bacteria that gain access to the body after death can produce similar marks, and the origin of tuberculosis in America remained controversial.

In 1994, pathologist Arthur C. Aufderheide and molecular biologist Wilmar Salo teamed up to resolve this controversy. Aufderheide obtained access to the remains of a woman who died and was naturally mummified in Peru about 1000 years ago, hundreds of years before Europeans arrived in South America. He removed samples of the woman's right lung and a lymph node and sent them to Salo, who used the newly developed polymerase chain reaction (PCR, which selectively amplifies sequences of DNA) to search the samples for DNA from *Mycobacterium tuberculosis*, the bacterium that causes tuberculosis. When applied to DNA from modern-day *Mycobacterium tuberculosis*, this technique produces copies of a 97-bp fragment of DNA. Salo detected an *identical* 97-bp piece of DNA by applying PCR to DNA samples from the ancient lung and lymph tissue, demonstrating unambiguously the presence of tuberculosis in the tissue of this 1000-year-old woman.

Although Europeans were not the *first* to transmit tuberculosis to Native Americans, they were still the most likely cause of the tuberculosis epidemics that accompanied their arrival in the Americas. European settlement was highly disruptive to many indigenous societies, often causing mass displacements of people and radically altering their traditional life styles. Stressful conditions, accompanied by crowding and malnutrition on reservations, probably lowered the resistance of many Native Americans and contributed to the spread of tuberculosis.

This story of the discovery of tuberculosis in the remains of a 1000-year-old woman illustrates the power of PCR, one of the techniques of molecular biology discussed in this chapter. We begin the chapter with a discussion of recombinant DNA technology and some of its effects. We then examine a number of methods used to isolate, study, alter, and recombine DNA sequences and place them back into cells. Finally, we explore some of the applications of recombinant DNA technology.

In reading this chapter, it will be helpful to understand two things. First, working at the molecular level is quite different from working with whole organisms: different approaches are needed, because the molecular objects of study cannot be seen directly. Second, there are a number of different approaches for isolating DNA sequences, amplifying them, and inserting them into bacteria, each approach with its own strengths and weaknesses. The optimal method depends on the starting materials, how much is known about the sequences to be isolated, and what the final objective is.

www.whfreeman.com/pierce   More information about tuberculosis

Basic Concepts of Recombinant DNA Technology

In 1973, a group of scientists produced the first organisms with recombinant DNA molecules. Stanley Cohen at Stanford University and Herbert Boyer at the University of California School of Medicine at San Francisco and their colleagues inserted a piece of DNA from one plasmid into another, creating an entirely new, recombinant DNA molecule. They then introduced the recombinant plasmid into *E. coli* cells. Within a short time, they used the same methods to stitch together genes from two different types of bacteria, as well as to transfer genes from a frog to a bacterium. They called the hybrid DNA molecules *chimeras,* after the mythological Chimera, a creature with the head of a lion, the body of a goat, and the tail of a serpent. These experiments ushered in one of the most momentous revolutions in the history of science.

Recombinant DNA technology is a set of molecular techniques for locating, isolating, altering, and studying DNA segments. The term *recombinant* is used because frequently the goal is to combine DNA from two distinct sources. Genes from two different bacteria might be joined, for example, or a human gene might be inserted into a viral chromosome. Commonly called **genetic engineering,** recombinant DNA technology now encompasses an array of molecular techniques that can be used to analyze, alter, and recombine virtually any DNA sequences.

The Impact of Recombinant DNA Technology

Recombinant DNA technology has drastically altered the way that genes are studied. Previously, information about the structure and organization of genes was gained by examining their phenotypic effects, but the new technology makes it possible to read the nucleotide sequences themselves. Previously, geneticists had to wait for the appearance of random or induced mutations to analyze the effects of genetic differences; now they can create mutations at precisely defined spots and see how they alter the phenotype.

Recombinant DNA technology has provided new information about the structure and function of genes and has altered many fundamental concepts of genetics. For example, whereas the genetic code was once thought to be entirely universal, we now know that nonuniversal codons exist in mitochondrial DNA. Previously, we thought that the organization of eukaryotic genes was like that of prokaryotes, but we now know that many eukaryotic genes are interrupted by introns. Much of what we know today about replication, transcription, translation, RNA processing, and gene regulation has been learned through the use of recombinant DNA techniques. These techniques are also used in many other fields, including biochemistry, microbiology, developmental biology, neurobiology, evolution, and ecology.

Recombinant DNA technology is also used to create a number of commercial products, including drugs, hormones, enzymes, and crops (◀ FIGURE 18.1). An entirely new industry—**biotechnology**—has grown up around the use of these techniques to develop new products. In medicine, recombinant DNA techniques are used to probe the nature

◀ 18.1 **Recombinant DNA technology has been used to create genetically modified crops.** Genetically engineered corn, which produces a toxin that kills insect pests, now comprises over 30% of all corn grown in the United States. (Chris Knapton/Photo Researchers.)

of cancer, diagnose genetic and infectious diseases, produce drugs, and treat hereditary disorders.

Concepts

Recombinant DNA technology is a set of methods used to locate, analyze, alter, study, and recombine DNA sequences. It is used to probe the structure and function of genes, address questions in many areas of biology, create commercial products, and diagnose and treat diseases.

www.whfreeman.com/pierce Information on genetic engineering and the biotechnology industry

Working at the Molecular Level

The manipulation of genes presents a serious challenge, often requiring strategies that may not, at first, seem obvi-

ous. The basic problem is that genes are minute and there are thousands of them in every cell. Even when viewed with the most powerful microscope, DNA appears as a tiny thread—individual nucleotides cannot be seen, and no physical features mark the beginning or the end of a gene.

To illustrate the problem, let's consider a typical situation faced by a molecular geneticist. Suppose we wanted to isolate a particular human gene, place it inside bacterial cells, and use the bacteria to produce large quantities of the encoded human protein. The first and most formidable problem is to find the desired gene. A haploid human genome consists of 3.3 billion base pairs of DNA. Let's assume that the gene that we want to isolate is 3000 bp long. Our target gene occupies only one-millionth of the genome; so searching for our gene in the huge expanse of genomic DNA is more difficult than looking for a needle in the proverbial haystack. But, even if we are able to locate the gene, how are we to separate it from the rest of the DNA? No forceps are small enough to pick up a single piece of DNA, and no mechanical scissors precise enough to snip out an individual gene.

If we did succeed in locating and isolating the desired gene, we would next need to insert it into a bacterial cell. Linear fragments of DNA are quickly degraded by bacteria; so the gene must be inserted in a stable form. It must also be able to successfully replicate or it will not be passed on when the cell divides.

If we succeed in transferring our gene to bacteria in a stable form, we still must ensure that the gene is properly transcribed and translated. Gene expression is a complex process requiring a number of DNA sequence elements, some of which lie outside the gene itself (Chapters 13 through 16). All of these elements must be present in their proper orientations and positions for the protein to be produced.

Finally, the methods used to isolate and transfer genes are inefficient and, of a million cells that are subjected to these procedures, only *one* cell might successfully take up and express the human gene. So we must search through many bacterial cells to find the one containing the recombinant DNA. We are back to the problem of the needle in the haystack.

Although these problems might seem insurmountable, molecular techniques have been developed to overcome all of them, and human genes are routinely transferred to bacterial cells, where the genes are expressed.

Concepts

Recombinant DNA technology requires special methods because individual genes make up a tiny fraction of the cellular DNA and they cannot be seen.

Recombinant DNA Techniques

In the sections that follow, we will examine some of the following techniques of recombinant DNA technology and see how they are used to create recombinant DNA molecules:

1. Methods for locating specific DNA sequences

2. Techniques for cutting DNA at precise locations

3. Procedures for amplifying a particular DNA sequence billions of times, producing enough copies of a DNA sequence to carry out further manipulations

4. Methods for mutating and joining DNA fragments to produce desired sequences

5. Procedures for transferring DNA sequences into recipient cells

Cutting and Joining DNA Fragments

The key development that made recombinant DNA technology possible was the discovery in the late 1960s of **restriction enzymes** (also called **restriction endonucleases**) that recognize and make double-stranded cuts in the sugar–phosphate backbone of DNA molecules at specific nucleotide sequences. These enzymes are produced naturally by bacteria, where they are used in defense against viruses. In bacteria, restriction enzymes recognize particular sequences in viral DNA and then cut it up. A bacterium protects its own DNA from a restriction enzyme by modifying the recognition sequence, usually by adding methyl groups to its DNA.

Three types of restriction enzymes have been isolated from bacteria (Table 18.1). Type I restriction enzymes recognize specific sequences in the DNA but cut the DNA at random sites that may be some distance (1000 bp or more) from the recognition sequence. Type III restriction enzymes recognize specific sequences and cut the DNA at nearby sites, usually about 25 bp away. Type II restriction enzymes recognize specific sequences and cut the DNA within the recognition sequence. Virtually all work on recombinant DNA is done with type II restriction enzymes; discussions

of restriction enzymes throughout this book, refers to type II enzymes.

More than 800 different restriction enzymes that recognize and cut DNA at more than 100 different sequences have been isolated from bacteria. Many of these enzymes are commercially available; examples of some commonly used restriction enzymes are given in Table 18.2. Each restriction enzyme is referred to by a short abbreviation that signifies its bacterial origin.

The sequences recognized by restriction enzymes are usually from 4 to 8 bp long; most enzymes recognize a sequence of 4 or 6 bp. Most recognition sequences are palindromic—sequences that read the same forward and backward. Notice in Table 18.2 that the sequence on the bottom strand is the same as the sequence on the top strand, only reversed. All type II restriction enzymes recognize palindromic sequences.

Some of the enzymes make staggered cuts in the DNA. For example, *Hin*dIII recognizes the following sequence:

$$
\begin{array}{c}
\quad\quad\downarrow \\
5'-\text{AAGCTT}-3' \\
3'-\text{TTCGAA}-5' \\
\quad\quad\quad\quad\uparrow
\end{array}
$$

*Hin*dIII cuts the sugar–phosphate backbone of each strand at the point indicated by the arrow, generating fragments with short, single-stranded overhanging ends:

$$
\begin{array}{c}
5'-\text{A}\quad\quad\text{AGCTT}-3' \\
3'-\text{TTCGA}\quad\quad\text{A}-5'
\end{array}
$$

Such ends are called **cohesive ends** or sticky ends, because they are complementary to each other and can spontaneously pair to connect the fragments. Thus DNA fragments can be "glued" together: any two fragments cleaved by the same enzyme will have complementary ends and will pair (◀ FIGURE 18.2). When their cohesive ends have paired, two DNA fragments can be joined together permanently by the enzyme DNA ligase, which seals nicks between the sugar–phosphate groups of the fragments.

Not all restriction enzymes produce staggered cuts and sticky ends. *Pvu*II cuts in the middle of its recognition site, producing blunt-ended fragments:

$$
\begin{array}{c}
\quad\quad\quad\downarrow \\
5'-\text{CAGCTG}-3' \\
3'-\text{GTCGAC}-5' \\
\quad\quad\quad\uparrow
\end{array}
$$

$$\downarrow$$

$$
\begin{array}{c}
5'-\text{CAG}\quad\quad\text{CTG}-3' \\
3'-\text{GTC}\quad\quad\text{GAC}-5'
\end{array}
$$

Fragments with blunt ends must be joined together in other ways, which will be discussed later.

Table 18.1	Types of restriction enzymes		
Type	**Activity of Enzyme**	**ATP Required**	**Cleavage Site**
I	Cleavage and methylation	Yes	Random sites distant from recognition site
II	Cleavage only	No	Within recognition site
III	Cleavage and methylation	Yes	Random sites near recognition site

Table 18.2	Characteristics of some common type II restriction enzymes used in recombinant DNA technology		
Enzyme	Microorganism From Which Enzyme Is Isolated	Recognition Sequence	Type of Fragment End Produced
BamHI	Bacillus amyloliquefaciens	5'–GGATCC–3' 3'–CCTAGG–3'	Cohesive
CofI	Clostridium formicoaceticum	5'–GCGC–3' 3'–CGCG–5'	Cohesive
DraI	Deinococcus radiophilus	5'–TTTAAA–3' 3'–AAATTT–5'	Blunt
EcoRI	Escherichia coli	5'–GAATTC–3' 3'–CTTAAG–5'	Cohesive
EcoRII	Escherichia coli	5'–CCAGG–3' 3'–GGTCC–5'	Cohesive
HaeIII	Haemophilus aegyptius	5'–GGCC–3' 3'–CCGG–5'	Blunt
HindIII	Haemophilus influenzae	5'–AAGCTT–3' 3'–TTCGAA–5'	Cohesive
HpaII	Haemophilus parainfluenzae	5'–CCGG–3' 3'–GGCC–5'	Cohesive
NotI	Nocardia otitidis-caviarum	5'–GCGGCCGC–3' 3'–CGCCGGCG–5'	Cohesive
PstI	Providencia stuartii	5'–CTGCAG–3' 3'–GACGTC–5'	Cohesive
PvuII	Proteus vulgaris	5'–CAGCTG–3' 3'–GTCGAC–5'	Blunt
SmaI	Serratia marcescens	5'–CCCGGG–3' 3'–GGGCCC–5'	Blunt

Note: The first three letters of the abbreviation for each restriction enzyme refer to the bacterial species from which the enzyme was isolated (e.g., Eco refers to E. coli). A fourth letter may refer to the strain of bacteria from which the enzyme was isolated (the "R" in EcoRI indicates that this enzyme was isolated from the RY13 strain of E. coli). Roman numerals that follow the letters allow different enzymes from the same species to be identified. For convenience, molecular geneticists have come up with idiosyncratic pronunciations of the names: EcoRI is pronounced "echo-R-one," HindIII is "hin-D-three," and HaeIII is "hay-three." These common pronunciations obey no formal rules and simply have to be learned.

(a)

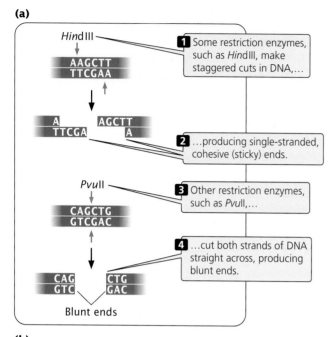

1 Some restriction enzymes, such as *Hind*III, make staggered cuts in DNA,…

2 …producing single-stranded, cohesive (sticky) ends.

3 Other restriction enzymes, such as *Pvu*II,…

4 …cut both strands of DNA straight across, producing blunt ends.

Blunt ends

(b)

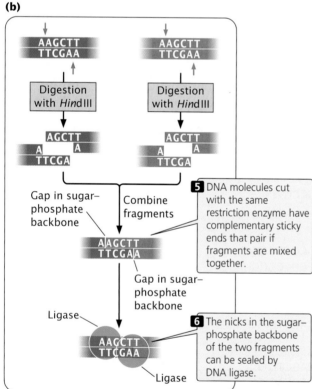

Digestion with *Hind*III

Digestion with *Hind*III

Gap in sugar–phosphate backbone

Combine fragments

5 DNA molecules cut with the same restriction enzyme have complementary sticky ends that pair if fragments are mixed together.

Gap in sugar–phosphate backbone

Ligase

6 The nicks in the sugar–phosphate backbone of the two fragments can be sealed by DNA ligase.

Ligase

◀18.2 **Restriction enzymes make double-stranded cuts in the sugar–phosphate backbone of DNA, producing cohesive, or sticky, ends.**

The sequences recognized by a restriction enzyme occur randomly within genomic DNA. Consequently, there is a relation between the length of the recognition sequence and its frequency of occurrence: there are fewer long recog-

nition sequences than short sequences because the probability of all the bases being in the required order is less.

Restriction enzymes are the workhorses of recombinant DNA technology and are used whenever DNA fragments must be cut or joined. In a typical restriction reaction, a concentrated solution of purified DNA is placed in a small tube with a buffer solution and a small amount of restriction enzyme. The reaction mixture is then heated at the optimal temperature for the enzyme, usually 37°C. Within a few hours, the enzyme cuts all the restriction sites in the DNA, producing a set of DNA fragments (◀FIGURE 18.3).

Concepts

Type II restriction enzymes cut DNA at specific base sequences. Some restriction enzymes make staggered cuts, producing DNA fragments with cohesive ends; others cut both strands straight across, producing blunt-ended fragments. There are fewer long recognition sequences in DNA than short sequences.

(a) Linear DNA

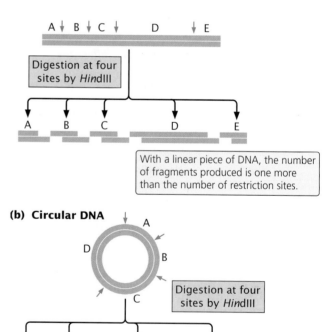

Digestion at four sites by *Hind*III

With a linear piece of DNA, the number of fragments produced is one more than the number of restriction sites.

(b) Circular DNA

Digestion at four sites by *Hind*III

With a circular piece of DNA, the number of fragments produced is equal to the number of restriction sites.

◀18.3 **The number of restriction sites is related to the number of fragments produced when DNA is cut by a restriction enzyme.**

www.whfreeman.com/pierce Information on specific restriction enzymes

Viewing DNA Fragments

After the completion of a restriction reaction, a number of questions arise. Did the restriction enzyme cut the DNA? How many times was the DNA cut? What are the sizes of the resulting fragments? Gel electrophoresis provides us with a means of answering these questions.

Electrophoresis is a standard biochemical technique for separating molecules on the basis of their size and electrical charge. There are a number of different types of electrophoresis; to separate DNA molecules, **gel electrophoresis** is used. A porous gel is often made from agarose (a polysaccharide isolated from seaweed), which is melted in a buffer solution and poured into a plastic mold. As it cools, the agarose solidifies, making a gel that looks something like stiff gelatin.

Small indentions called wells are made at one end of the gel to hold solutions of DNA fragments (◀ Figure 18.4a), and an electrical current is passed through the gel. Because the phosphate group of each DNA nucleotide carries a negative charge, the DNA fragments migrate toward the positive end of the gel (◀ Figure 18.4b). In this migration, the gel acts as a sieve; as the DNA molecules migrate toward the positive pole, they move through the pores between the gel particles. Small DNA fragments migrate more rapidly than do large ones and, with time, the fragments separate on the basis of their size. The distance that each fragment migrates depends on its size. Typically, DNA fragments of known length (a marker sample) are placed in another well. By comparing the migration distance of the unknown fragments with the distance traveled by the marker fragments, one can determine the approximate size of the unknown fragments.

After electrophoresis, the DNA fragments are separated according to size (◀ Figure 18.4c). However, the DNA fragments are still too small to see; so the problem of visualizing the DNA needs to be addressed. Visualization can be accomplished in several ways. The simplest procedure is to stain the gel with a dye specific for nucleic acids, such as ethidium bromide, which wedges itself tightly (intercalates) between the bases of DNA. When exposed to UV light, ethidium bromide fluoresces bright orange; so copies of each DNA fragment appear as a brilliant orange band (◀ Figure 18.4d). The original concentrated sample of purified DNA contained millions of copies of a DNA molecule, and thus each band represents millions of copies of identical DNA fragments.

Alternatively, DNA fragments can be visualized by adding a radioactive or chemical label to the DNA before it is placed in the gel. Nucleotides with radioactively labeled phosphate (^{32}P) can be used as the substrate for DNA synthesis and will be incorporated into the newly synthesized DNA strand. In another method called **end labeling**, the bacteriophage enzyme polynucleotide kinase is used to

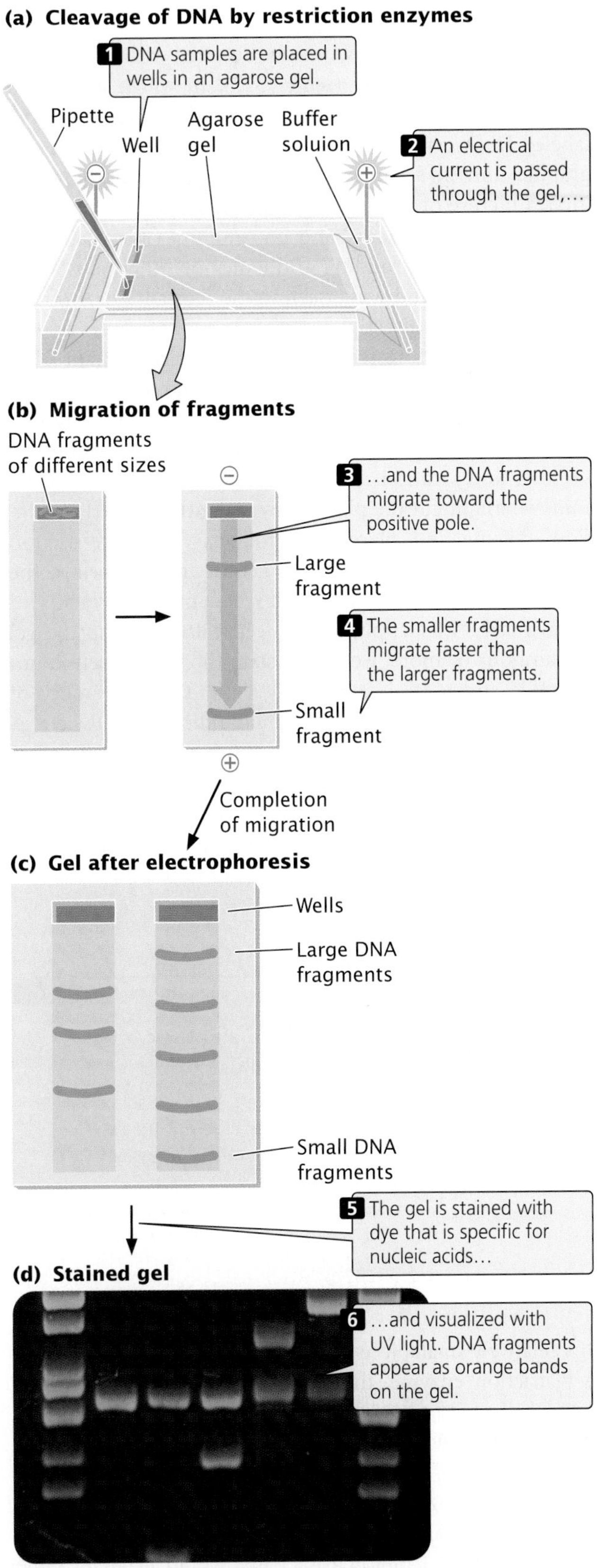

(a) Cleavage of DNA by restriction enzymes

1 DNA samples are placed in wells in an agarose gel.

Pipette   Agarose   Buffer
Well   gel   soluion

2 An electrical current is passed through the gel,...

(b) Migration of fragments

DNA fragments of different sizes

3 ...and the DNA fragments migrate toward the positive pole.

Large fragment

4 The smaller fragments migrate faster than the larger fragments.

Small fragment

Completion of migration

(c) Gel after electrophoresis

Wells

Large DNA fragments

Small DNA fragments

5 The gel is stained with dye that is specific for nucleic acids...

(d) Stained gel

6 ...and visualized with UV light. DNA fragments appear as orange bands on the gel.

◀ **18.4 Gel electrophoresis can be used to separate DNA molecules on the basis of their size and electrical charge.** (Photo courtesy of Carol Eng.)

transfer a single ^{32}P to the 5′ end of each DNA strand. Radioactively labeled DNA can be detected with a technique called **autoradiography** (see p. 327), in which a piece of X-ray film is placed on top of the gel. Radiation from the labeled DNA exposes the film, just as light exposes photographic film in a camera. The developed autoradiograph gives a picture of the fragments in the gel; each DNA fragment appears as a dark band on the film. Chemical labels can be detected by adding antibodies or other substances that carry a dye and will attach to the relevant DNA, which can be visualized directly.

Gel electrophoresis is used widely in recombinant DNA technology; it is often employed when there is a need to determine the number or size of DNA fragments or to isolate DNA fragments by size. For example, to determine the number and location of *Bam*HI restriction sites in a plasmid, we might cut the plasmid by using the *Bam*HI restriction enzyme and place the products of the restriction reaction in a well of an agarose gel. In another well of the same gel, we would place a set of control fragments of known size. After applying an electrical current to the gel for an hour or more, we would stain the gel with ethidium bromide and place it over a UV light. The appearance of three orange bands on the gel would indicate that the circular plasmid had been cut three times and that there are three *Bam*HI restriction sites in the plasmid. A comparison of the migration distance of the plasmid fragments with the migration distance of the standard fragments would reveal the sizes of the fragments and the distances between the *Bam*HI recognition sites.

Concepts

DNA fragments can be separated, and their sizes can be determined with the use of gel electrophoresis. The fragments can be viewed by using a dye that is specific for nucleic acids or by labeling the fragments with a radioactive or chemical tag.

Locating DNA Fragments with Southern Blotting and Probes

If a relatively small piece of DNA, such as a plasmid, is cut by a restriction enzyme, the few fragments produced can be seen as distinct bands on an electrophoretic gel. In contrast, if genomic DNA from a cell is cut by a restriction enzyme, a large number of fragments of different sizes are produced. A restriction enzyme that recognizes a four-base sequence would theoretically cut about once every 256 bp. The human genome, with 3.3 billion base pairs, would generate more than 12 million fragments when cut by this restriction enzyme. Separated by electrophoresis and stained, this large set of fragments would appear as a continuous smear on the gel

because of the presence of so many fragments of differing size. Usually, one is interested in only a few of these fragments, perhaps those carrying a specific gene. How does one locate the desired fragments in such a large pool of DNA?

One approach is to use a **probe,** which is a DNA or RNA molecule with a base sequence complementary to a sequence in the gene of interest. The bases on a probe will pair only with the bases on a complementary sequence and, if suitably tagged with an identifying label, the probe can be used to locate a specific gene or other DNA sequence.

To use a probe, one first cuts the DNA into fragments by using one or more restriction enzymes and then separates the fragments with gel electrophoresis (◀FIGURE 18.5). Next, the separated fragments must be denatured and transferred to a thinner solid medium (such as nitrocellulose or nylon membrane) to prevent diffusion. **Southern blotting** (named after Edwin M. Southern) is one technique used to transfer the denatured, single-stranded fragments from a gel to a thin solid medium.

After the single-stranded DNA fragments have been transferred, the membrane is placed in a hybridization solution of a radioactively or chemically labeled probe (see Figure 18.5). The probe will bind to any DNA fragments on the membrane that bear complementary sequences. The membrane is then washed to remove any unbound probe; bound probe is detected by autoradiography or another method for chemically labeled probes.

RNA can be transferred from a gel to a solid support by a related procedure called **Northern blotting** (not named after anyone but capitalized to match Southern). The hybridization of a probe can reveal the size of a particular mRNA molecule, its relative abundance, or the tissues in which the mRNA is transcribed. **Western blotting** is the transfer of protein from a gel to a membrane. Here, the probe is usually an antibody, used to determine the size of a particular protein and the pattern of the protein's expression.

Concepts

Labeled probes, which are sequences of RNA or DNA that are complementary to the sequence of interest, can be used to locate individual genes or DNA sequences. Southern blotting can be used to transfer DNA fragments from a gel to a membrane such as nitrocellulose.

Cloning Genes

Many recombinant DNA methods require numerous copies of a specific DNA fragment. One way to obtain these copies is to place the fragment in a bacterial cell and allow the cell to replicate the DNA. This procedure is termed **gene cloning,** because identical copies (clones) of the original piece of DNA are produced.

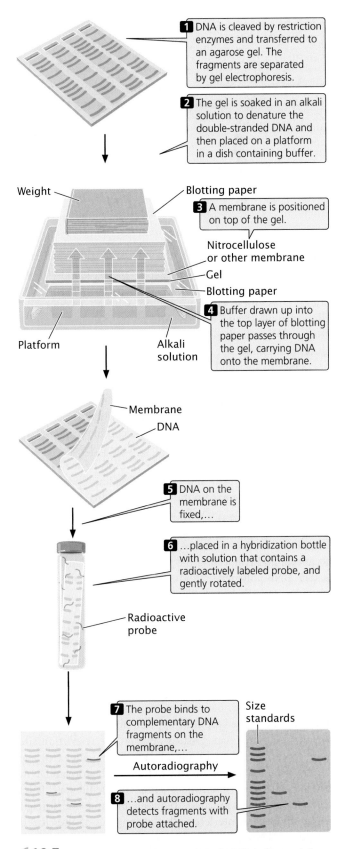

1 DNA is cleaved by restriction enzymes and transferred to an agarose gel. The fragments are separated by gel electrophoresis.

2 The gel is soaked in an alkali solution to denature the double-stranded DNA and then placed on a platform in a dish containing buffer.

Weight

Blotting paper

3 A membrane is positioned on top of the gel.

Nitrocellulose or other membrane

Gel

Blotting paper

4 Buffer drawn up into the top layer of blotting paper passes through the gel, carrying DNA onto the membrane.

Platform

Alkali solution

Membrane

DNA

5 DNA on the membrane is fixed,…

6 …placed in a hybridization bottle with solution that contains a radioactively labeled probe, and gently rotated.

Radioactive probe

7 The probe binds to complementary DNA fragments on the membrane,…

Size standards

Autoradiography

8 …and autoradiography detects fragments with probe attached.

◀18.5 **Southern blotting and hybridization with probes can be used to locate a few specific fragments in a large pool of DNA.**

Cloning vectors A **cloning vector** is a stable, replicating DNA molecule to which a foreign DNA fragment can be attached for introduction into a cell. An effective cloning vector has three important characteristics (◀ FIGURE 18.6): (1) an origin of replication, which ensures that the vector is replicated within the cell; (2) selectable markers, which enable any cells containing the vector to be selected or identified; and (3) one or more unique restriction sites into which a DNA fragment can be inserted. The restriction sites used for cloning must be unique; if a vector is cut at multiple recognition sites, generating several pieces of DNA, there will be no way to get the pieces back together in the correct order. Three types of cloning vectors are commonly used for cloning genes in bacteria: plasmids, bacteriophages, and cosmids.

Plasmid vectors Plasmids are circular DNA molecules that exist naturally in bacteria (see p. 202 in Chapter 8). They contain origins of replication and are therefore able to replicate independently of the bacterial chromosome. The plasmids typically used in cloning have been constructed from the larger, naturally occurring bacterial plasmids.

The pUC19 plasmid is a typical cloning vector (◀ FIGURE 18.7). It has an origin of replication and two selectable markers—an ampicillin-resistance gene and a *lacZ* gene. Ampicillin is an antibiotic that normally kills bacterial cells, but any bacterium that contains a pUC19 plasmid will be resistant to this antibiotic. The *lacZ* gene encodes the enzyme β-galactosidase, which normally cleaves lactose to produce glucose and galactose (see p. 442 in Chapter 16). The enzyme will also cleave a chemical called X-gal, producing a blue substance; when X-gal is placed in the medium, any bacterial colonies that contain intact pUC19 plasmids will turn blue and can be easily identified. (In these experiments, the bacterium's own β-galactosidase gene has been inactivated, and so only bacteria with the plasmid turn blue.) The pUC19 plasmid also possesses a number of different unique restrictions sites grouped together (a polylinker) that allow DNA fragments to be inserted into the plasmid.

The easiest method for inserting a foreign DNA fragment into a plasmid is to use restriction cloning, in which the foreign DNA and the plasmid are cut by the same restriction enzyme. Restriction cloning produces complementary sticky ends on the foreign DNA and the plasmid (◀ FIGURE 18.8a). The DNA and plasmid are then mixed together; some of the foreign DNA fragments will pair with the cut ends of the plasmid. DNA ligase seals the nicks in the sugar–phosphate backbone, creating a recombinant plasmid that contains the foreign DNA fragment.

Although simple, restriction cloning has several disadvantages. First, restriction cloning requires that a single restriction site in the plasmid matches sites on both ends of the foreign sequence to be cloned. If this arrangement of restriction sites is not available, this otherwise relatively

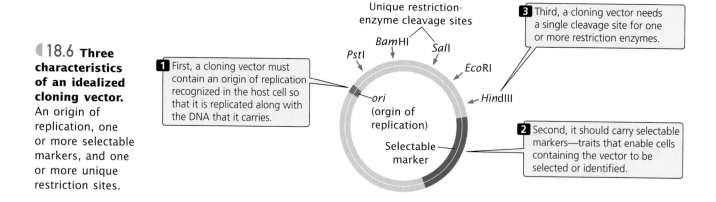

◀ 18.6 **Three characteristics of an idealized cloning vector.** An origin of replication, one or more selectable markers, and one or more unique restriction sites.

Unique restriction-enzyme cleavage sites

*Pst*I   *Bam*HI   *Sal*I   *Eco*RI   *Hind*III

1 First, a cloning vector must contain an origin of replication recognized in the host cell so that it is replicated along with the DNA that it carries.

3 Third, a cloning vector needs a single cleavage site for one or more restriction enzymes.

ori (orgin of replication)

Selectable marker

2 Second, it should carry selectable markers—traits that enable cells containing the vector to be selected or identified.

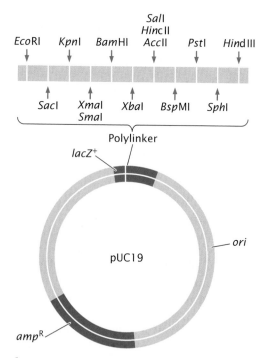

*Eco*RI   *Kpn*I   *Bam*HI   *Sal*I *Hinc*II *Acc*II   *Pst*I   *Hind*III

*Sac*I   *Xma*I *Sma*I   *Xba*I   *Bsp*MI   *Sph*I

Polylinker

lacZ⁺

pUC19

ori

*amp*ᴿ

◀ 18.7 **The pUC19 plasmid is a typical cloning vector.** It contains a cluster of unique restriction sites, an origin of replication, and two selectable markers—an ampicillin-resistance gene and a *lacZ* gene.

straightforward method cannot be used. Second, this technique often leads to undesirable products. The sticky ends of the plasmid are complementary to each other; so the two ends of the plasmid will often simply reanneal, reproducing the intact plasmid. Alternatively, the two complementary ends of the cleaved foreign DNA may anneal; or several pieces of foreign DNA or several plasmids may join. However, these undesirable products do not constitute a serious problem if an efficient method is used for screening bacterial cells for the presence of a recombinant plasmid.

Another method for inserting DNA into a plasmid (a method that gets around the problem of undesired products) is tailing (◀ FIGURE 18.8b). In this procedure, comple-

mentary sticky ends are created on blunt-ended pieces of DNA. The plasmid and the foreign DNA are first cut by any restriction enzyme. If the restriction enzyme produces sticky ends, these ends are removed by an enzyme that digests single-stranded DNA. Alternatively, the plasmid and foreign DNA can be cut by a restriction enzyme that produces blunt ends.

Once the plasmid and the foreign DNA have blunt ends, single-stranded sticky ends are added by an enzyme called terminal transferase, which adds any available nucleotides to the 3′ end of DNA in a template-independent reaction. For example, terminal transferase and deoxyadenosine triphosphate (dATP) might be mixed with the plasmid DNA, creating poly(A) single-stranded tails on the 3′ ends of the plasmid. Terminal transferase and deoxythymidine triphosphate (dTTP) could be mixed with the blunt-ended foreign DNA fragments, creating poly(T) single-stranded tails on their 3′ ends. The poly(A) tail of the plasmid would be complementary to the poly(T) tail of the foreign DNA, allowing them to anneal and connecting the plasmid and foreign DNA together. DNA polymerase can be used to fill in any missing nucleotides, and DNA ligase can be used to seal the nicks in the sugar−phosphate backbone.

One advantage of tailing is that it prevents the production of the undesired products created by restriction cloning: the single-stranded ends of the plasmid are complementary only to the single-stranded ends of the foreign DNA. Another advantage is that identical restriction sites are not required in plasmid and foreign DNA; any restriction site can be used for cleavage. But tailing has several disadvantages of its own. First, it destroys the restriction site used to cut the original molecule, preventing later cleavage by the same restriction enzyme to retrieve the foreign DNA. Second, the new nucleotides (the complementary tails) introduced at the junctions between plasmid and foreign DNA sometimes interfere with the function of the cloned DNA.

A third method of inserting fragments into plasmids is to use the enzyme T4 ligase, which is capable of connecting any two pieces of blunt-ended DNA. Like tailing, this

(a) Restriction cloning

Plasmid   Foreign DNA

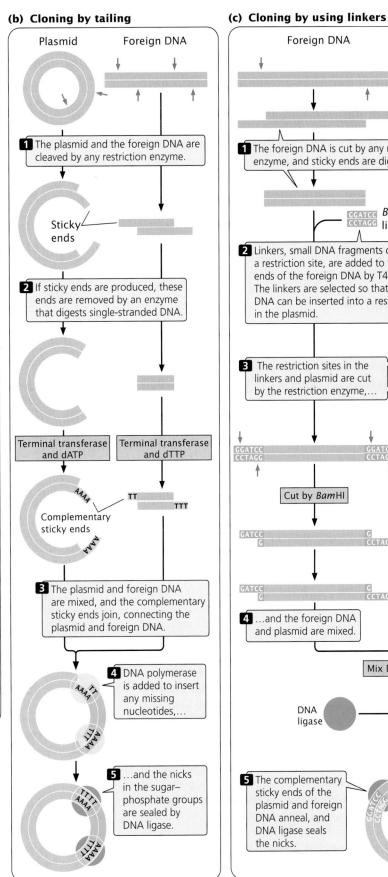

1 The plasmid and the foreign DNA are cut by the same restriction enzyme—in this case, EcoRI.

Complementary sticky ends

2 When mixed, the sticky ends anneal, joining the foreign DNA and plasmid.

DNA ligase

3 Nicks in the sugar–phosphate bonds are sealed by DNA ligase.

(b) Cloning by tailing

Plasmid   Foreign DNA

1 The plasmid and the foreign DNA are cleaved by any restriction enzyme.

Sticky ends

2 If sticky ends are produced, these ends are removed by an enzyme that digests single-stranded DNA.

Terminal transferase and dATP   Terminal transferase and dTTP

Complementary sticky ends

3 The plasmid and foreign DNA are mixed, and the complementary sticky ends join, connecting the plasmid and foreign DNA.

4 DNA polymerase is added to insert any missing nucleotides,...

5 ...and the nicks in the sugar–phosphate groups are sealed by DNA ligase.

(c) Cloning by using linkers

Foreign DNA

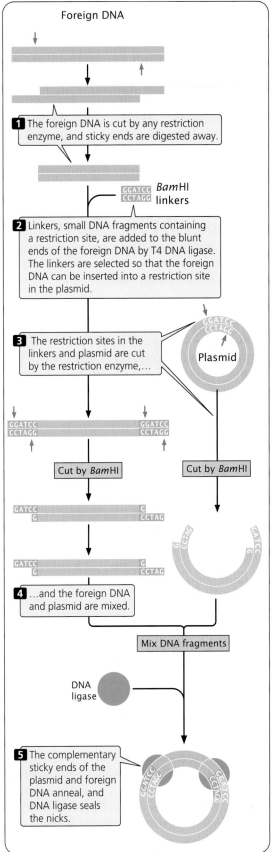

1 The foreign DNA is cut by any restriction enzyme, and sticky ends are digested away.

GGATCC   BamHI
CCTAGG   linkers

2 Linkers, small DNA fragments containing a restriction site, are added to the blunt ends of the foreign DNA by T4 DNA ligase. The linkers are selected so that the foreign DNA can be inserted into a restriction site in the plasmid.

3 The restriction sites in the linkers and plasmid are cut by the restriction enzyme,...

Plasmid

GGATCC   GGATCC
CCTAGG   CCTAGG

Cut by BamHI   Cut by BamHI

GATCC   G
G   CCTAG

GATCC   G
G   CCTAG

4 ...and the foreign DNA and plasmid are mixed.

Mix DNA fragments

DNA ligase

5 The complementary sticky ends of the plasmid and foreign DNA anneal, and DNA ligase seals the nicks.

◄18.8 A foreign DNA fragment can be inserted into a plasmid with the use of (a) restriction cloning, (b) tailing, or (c) linkers.

method requires no specific restriction sites and has great versatility; its chief drawback is that it creates a number of undesired products.

A fourth method, and one commonly used today, is the use of linkers to add complementary ends to DNA molecules (◀FIGURE 18.8c). Linkers are small, synthetic DNA fragments that contain one or more restriction sites. The foreign DNA of interest is cut by any restriction enzyme; if sticky ends are created, they are digested to produce blunt ends. The linkers are then attached to the blunt ends by T4 ligase and are then cut by a restriction enzyme, generating sticky ends that are complementary to sticky ends on the plasmid, which have been generated by using the same restriction enzyme to cut the plasmid. Mixing the plasmid and foreign DNA leads to the formation of recombinant DNA that can be stabilized by ligase. The great advantage of using linkers is that a particular restriction site can be added at almost any desired location; so any two pieces of DNA can be cut and joined.

Transformation When a gene has been placed inside a plasmid, the plasmid must be introduced into bacterial cells. This task is usually accomplished by *transformation,* which is the capacity of bacterial cells to take up DNA from the external environment (see Chapter 10). Some types of cells undergo transformation naturally; others must be treated chemically or physically before they will undergo transformation. Inside the cell, the plasmids replicate and multiply.

The use of selective markers Cells bearing recombinant plasmids can be detected by using the selectable markers on the plasmid. One type of selectable marker commonly used with plasmids is a copy of the *lacZ* gene (◀FIGURE 18.9). The *lacZ* gene contains a series of unique restriction sites into which may be inserted a fragment of DNA to be cloned. In the absence of an inserted fragment, the *lacZ* gene is active and produces β-galactosidase. When foreign DNA is inserted into the restriction site, it disrupts the *lacZ* gene, and β-galactosidase is not produced. The plasmid also usually contains a second selectable marker, which may be a gene that confers resistance to an antibiotic such as ampicillin.

Bacteria that are *lacZ⁻* are transformed by the plasmids and plated on medium that contains ampicillin. Only cells that have been successfully transformed and contain a plasmid with the ampicillin-resistance gene will survive and grow. Some of these cells will contain an intact plasmid, whereas others possess a recombinant plasmid. The medium also contains the chemical X-gal. Bacterial cells with an intact original plasmid—without an inserted fragment—have a functional *lacZ* gene and can synthesize β-galactosidase, which cleaves X-gal and turns the bacteria blue. Bacterial cells with a recombinant plasmid, however, have a β-galactosidase gene that is disrupted by the inserted DNA; they do not synthesize β-galactosidase and remain white. Thus, the color of

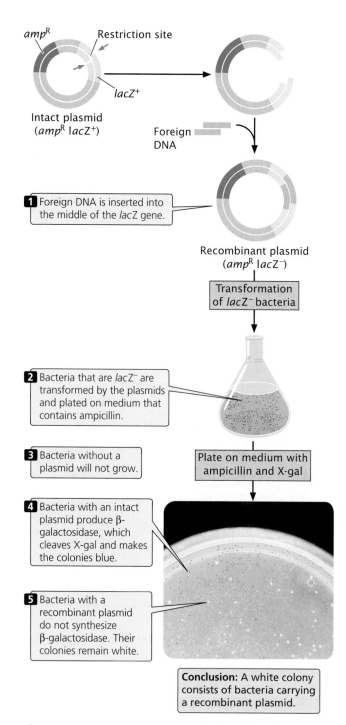

▲18.9 **The *lacZ* gene can be used to screen bacteria containing recombinant plasmids.** A special plasmid carries a copy of the *lacZ* gene and an ampicillin resistance gene. (Photo: Cytographics/Visuals Unlimited.)

the colony allows quick determination of whether a recombinant or intact plasmid is present in the cell.

Plasmids make ideal cloning vectors but can hold only DNA less than about 15 kb in size. When large DNA fragments are inserted into a plasmid vector, the plasmid becomes unstable. Cloning DNA fragments that are longer than 15 kb requires the use of different cloning vectors.

Concepts

DNA fragments can be inserted into cloning vectors, stable pieces of DNA that will replicate within a cell. Cloning vectors must have an origin of replication, one or more unique restriction sites, and selectable markers. Plasmids are commonly used as cloning vectors.

Bacteriophage vectors Bacteriophages offer a number of advantages as cloning vectors. The most widely used bacteriophage vector is bacteriophage λ, which infects *E. coli*. One of its chief advantages is the high efficiency with which it transfers DNA into bacteria cells. A second advantage is that about a third of the λ genome is not essential for infection and reproduction; without these genes, a λ particle will still faithfully inject its DNA into a bacterial cell and reproduce. These nonessential genes, which comprise about 15 kb, can be replaced by as much as about 23 kb of foreign DNA. A third advantage is that DNA will not be packaged into a λ coat unless it is 40 to 50 kb long; so fragments of foreign DNA are not likely to be transferred by the vector unless they are inserted into the λ genome, which ensures that the foreign DNA fragment will be replicated after it enters the cell.

The essential genes of the phage λ genome are located in a cluster. Strains of phage λ, called replacement vectors, have been engineered with unique *Eco*RI sites on either side of the nonessential genes (◀ FIGURE 18.10) so that, by using *Eco*RI, the nonessential genes can be removed. Foreign DNA cut with *Eco*RI will have sticky ends that are complementary to those on the ends of the essential λ genes, to which it can be connected by ligase. The λ chromosome possesses short, single-stranded ends called *cos* sites that are required for packaging λ DNA into a phage head. The recombinant phage chromosomes can then be packaged into protein coats and added to *E. coli*. The phages inject their recombinant DNA into the cell, where it will be replicated. Only DNA fragments of the proper size and containing essential genes will be packaged into the phage coats, providing an automatic selection system for recombinant vectors.

Cosmid vectors Although only about 23 kb of DNA can be cloned in λ vectors, DNA fragments as large as about 44 kb can be cloned in cosmids, which combine the properties of plasmids and phage vectors.

Cosmids are small plasmids that carry phage λ *cos* sites; they can be packaged into viral coats and transferred to bacteria by viral infection. Because all viral genes except the *cos* sites are missing, a cosmid can carry more than twice as much foreign DNA as can a phage vector. Cosmid vectors

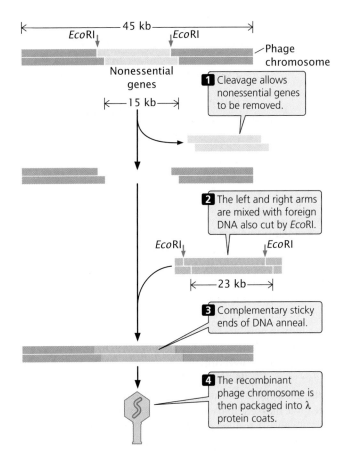

18.10 Phage λ is an effective cloning vector.

have the following components: (1) a plasmid origin of replication (*ori*); (2) one or more unique restriction sites; (3) one or more selectable markers; and (4) *cos* sites to allow the packaging of DNA into phage heads.

Foreign DNA is inserted into cosmids in the same way that DNA is introduced into plasmids: the cosmid and foreign DNA are both cut by a restriction enzyme that produces complementary (sticky) ends, and they are joined by DNA ligase. Recombinant cosmids are incorporated into the coats, and the phage particles are used to infect bacterial cells, where the cosmid replicates as a plasmid. Table 18.3 compares the properties of plasmids, phage λ vectors, and cosmids.

Concepts

Bacteriophage vectors not only hold more DNA than do plasmids, but also transfer foreign DNA into bacterial cells at a relatively high rate. A cosmid vector consists of a plasmid with *cos* sites, which allow DNA to be packaged into phage protein coats. Cosmids hold more DNA than do bacteriophage vectors.

Table 18.3	Comparison of plasmids, phage lambda vectors, and cosmids		
Cloning Vector	**Size of DNA That Can Be Cloned**	**Method of Propagation**	**Introduction to Bacteria**
Plasmid	As large as 15 kb	Plasmid replication	Transformation
Phage lambda	As large as 23 kb	Phage reproduction	Phage infection
Cosmid	As large as 44 kb	Plasmid reproduction	Phage infection

Note: 1 kb = 1000 bp

Expression vectors Sometimes the goal in gene cloning is not just to replicate the gene, but also to produce the protein that it encodes. One of the first commercial products produced by recombinant DNA technology was the protein insulin. The gene for human insulin was isolated and inserted into bacteria, which were then multiplied and used to synthesize human insulin. However, the successful expression of a human gene in a bacterial cell is not a straightforward matter. Although the universality of the genetic code allows human genes to specify the same protein in both human and bacterial cells, the sequences that regulate transcription and translation are quite different in bacteria and eukaryotes.

To ensure transcription and translation, a foreign gene is usually inserted into an **expression vector,** which, in addition to the usual origin of replication, restriction sites, and selectable markers, contains sequences required for transcription and translation in bacterial cells (◀FIGURE 18.11). These additional sequences may include:

1. A bacterial promoter, such as the *lac* promoter. The promoter precedes a restriction site where foreign DNA is to be inserted, allowing transcription of the foreign sequence to be regulated by adding substances that induce the promoter.

2. A DNA sequence that, when transcribed into RNA, produces a bacterial ribosome binding site.

3. Bacterial transcription initiation and termination sequences.

4. Sequences that control transcription initiation, such as regulator genes and operators.

The bacterial promoter and ribosome-binding site are usually placed upstream of the restriction site, which allows the foreign DNA to be inserted just downstream of the initiation codon. When the plasmid is placed in a bacterial cell, RNA polymerase binds to the promoter and transcribes the foreign DNA. Bacterial ribosomes attach to the ribosome-binding site on the RNA and translate the sequence into a foreign protein.

Concepts

An expression vector contains a promoter, a ribosome-binding site, and other sequences that allow a cloned gene to be transcribed and translated in bacteria.

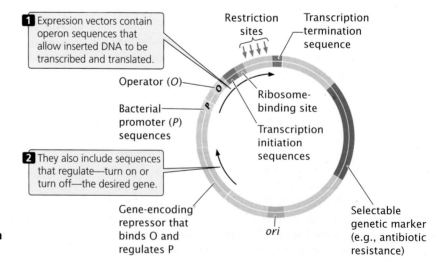

1 Expression vectors contain operon sequences that allow inserted DNA to be transcribed and translated.

2 They also include sequences that regulate—turn on or turn off—the desired gene.

Restriction sites

Transcription termination sequence

Operator (*O*)

Bacterial promoter (*P*) sequences

Ribosome-binding site

Transcription initiation sequences

Gene-encoding repressor that binds O and regulates P

ori

Selectable genetic marker (e.g., antibiotic resistance)

◀18.11 **To ensure transcription and translation, a foreign gene may be inserted into an expression vector — in this example, an *E. coli* expression vector.**

Cloning vectors for eukaryotes The vectors discussed so far allow genes to be cloned in bacterial cells. Other cloning vectors have been developed for transferring genes into eukaryotic cells. Special plasmids, for example, have been developed for cloning in yeast, and retroviral vectors have been developed for cloning in mammals.

Shuttle vectors are used to shuttle genes back and forth between two hosts. For example, plasmids have been engineered that allow gene sequences to be cloned and manipulated in bacteria and then transferred to yeast cells for study. For this reason, they must contain replication origins and selectable markers that work in both hosts.

Yeast artificial chromosomes (YACs) are DNA molecules with a yeast origin of replication, a pair of telomeres, and a centromere. Mitotic spindle fibers attach to the centromere, and YACs segregate in the same way as yeast chromosomes; the telomeres ensure that YACs remain stable within the cell; and the origin of replication allows YACs to be replicated. YACs are particularly useful because they can carry DNA fragments as large as 600 kb, and some special YACs can carry inserts of more than 1000 kb. YACs have been modified so that they can be used in eukaryotic organisms other than yeast (see introduction to Chapter 11). **Bacterial artificial chromosomes** (BACs), constructed from F factors (see Chapter 8), are used to clone large fragments ranging in length from 100 to 500 kb in bacteria.

The soil bacterium *Agrobacterium tumefaciens,* which invades plants through wounds and induces crown galls (tumors), has been used to transfer genes to plants. This bacterium contains a large plasmid called the **Ti plasmid,** part of which is transferred to a plant cell when *A. tumefaciens* infects a plant. In the plant, the Ti plasmid DNA integrates into one of the plant chromosomes where it is transcribed and translated to produce several enzymes that help support the bacterium (FIGURE 18.12a). Transfer of the DNA segment from the Ti plasmid to a plant chromosome requires two 25-bp sequences that flank the Ti DNA, as well as several genes located in the Ti plasmid.

Geneticists have engineered an *Agrobacterium–E. coli* shuttle vector that contains the flanking sequences required

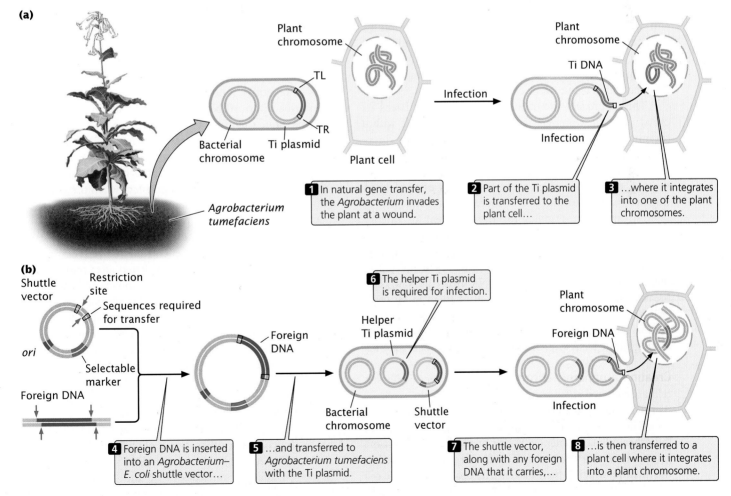

(a)

1 In natural gene transfer, the *Agrobacterium* invades the plant at a wound.

2 Part of the Ti plasmid is transferred to the plant cell...

3 ...where it integrates into one of the plant chromosomes.

(b)

4 Foreign DNA is inserted into an *Agrobacterium–E. coli* shuttle vector...

5 ...and transferred to *Agrobacterium tumefaciens* with the Ti plasmid.

6 The helper Ti plasmid is required for infection.

7 The shuttle vector, along with any foreign DNA that it carries,...

8 ...is then transferred to a plant cell where it integrates into a plant chromosome.

 18.12 **The Ti plasmid can be used to transfer genes into plants.**

to transfer DNA, a selectable marker, and restriction sites into which foreign DNA can be inserted (Figure 18.12b). When placed in *A. tumefaciens* with the Ti plasmid, the shuttle vector will transfer the foreign DNA that it carries into a plant cell, where it will integrate into a plant chromosome. This vector has been used to transfer genes that confer economically significant attributes such as resistances to herbicides, plant viruses, and insect pests.

> **Concepts**
>
> Special cloning vectors are used for introducing genes into eukaryotes; they include shuttle vectors, which can reproduce in two different hosts, yeast and bacterial artificial chromosomes, which hold DNA fragments hundreds of thousands of base pairs in length, and the Ti plasmid, which transfers genes to plants.

Finding Genes

In our consideration of gene cloning, we've glossed over a problem of major significance: How do we find the DNA sequence to be cloned in the first place? In fact, this problem is frequently the most significant one in cloning, because there are often millions or billions of base pairs of DNA in a cell. A discussion of how to solve this problem has been purposely delayed until now because, paradoxically, one must often clone a gene to find it.

This approach—to clone first and search later—is called "shotgun cloning," because it is like hunting with a shotgun: one sprays one's shots widely in the general direction of the quarry, knowing that there is a good chance that one or more of the pellets will hit the intended target. In shotgun cloning, one first clones a large number of DNA fragments, knowing that one or more contains the DNA of interest, and then searches for the fragment of interest among the clones.

A collection of clones containing all the DNA fragments from one source is called a **DNA library**. For example, we might isolate genomic DNA from human cells, break it into fragments, and clone all of them in bacterial cells or phages. The set of bacterial colonies or phages containing these fragments is a human **genomic library**, containing all the DNA sequences found in the human genome.

Creating a genomic library To create a genomic library, cells are collected and disrupted, which causes them to release their DNA and other cellular contents into an aqueous solution. There are several methods for isolating the DNA from the other cellular contents. In one method, phenol (an organic solvent that does not mix well with water) is added to the mixture, which is then shaken. The proteins from the cell associate with phenol, whereas the DNA and RNA remain in the aqueous solution, which is removed with the use of a pipette. The nucleic acids are then precipitated from this solution by adding cold alcohol. RNA can be removed by adding an enzyme that degrades RNA but not DNA.

When DNA has been extracted, it is cut into fragments by using a restriction enzyme to digest it for a limited amount of time only (a partial digestion) so that only *some* of the restriction sites in each DNA molecule are cut. Because which sites are cut is random, different DNA molecules will be cut in different places, and a set of overlapping fragments will be produced (Figure 18.13). The fragments are then joined to plasmid, phage, or cosmid vectors, which can be transferred to bacteria. This technique produces a set of bacterial cells or phage particles containing the overlapping genomic fragments. A few of the clones contain the entire gene of interest, a few contain parts of the gene, but most contain fragments that have no part of the gene of interest.

A genomic library must contain a large number of clones to ensure that all DNA sequences in the genome are represented in the library. A library of the human genome formed by using cosmids, each carrying a random DNA fragment from 35,000 to 44,000 bp long, would require about 350,000 cosmid clones to provide a 99% chance that every sequence is included in the library.

Creating a cDNA library An alternative to creating a genomic library is to create a library consisting only of those DNA sequences that are transcribed into mRNA (called a **cDNA library** because all the DNA in this library is *complementary* to mRNA). Much of eukaryotic DNA consists of repetitive (and other DNA) sequences that are not transcribed into mRNA (see pp. 300–301 in Chapter 11), and the sequences are not represented in a cDNA library.

A cDNA library has two additional advantages. First, it is enriched with fragments from actively transcribed genes. Second, introns do not interrupt the cloned sequences; introns would pose a problem when the goal is to produce a eukaryotic protein in bacteria, because most bacteria have no means of removing the introns.

The disadvantage of a cDNA library is that it contains only sequences that are present in mature mRNA. Introns and any other sequences that are altered after transcription are not present; sequences, such as promoters and enhancers, that are not transcribed into RNA also are not present in a cDNA library. It is also important to note that the cDNA library represents only those gene sequences expressed in the tissue from which the RNA was isolated. Furthermore, the frequency of a particular DNA sequence in a cDNA library depends on the abundance of the corresponding mRNA in the given tissue. In contrast, almost all genes are present at the same frequency in a genomic DNA library.

To create a cDNA library, messenger RNA must first be separated from other types of cellular RNA (tRNA, rRNA, snRNA, etc.). Most eukaryotic mRNAs possess a string of adenine nucleotides at the 3′ end, and this poly(A) tail provides a convenient hook for separating eukaryotic mRNA from the other types. Total cellular RNA is isolated from cells and poured through a column packed

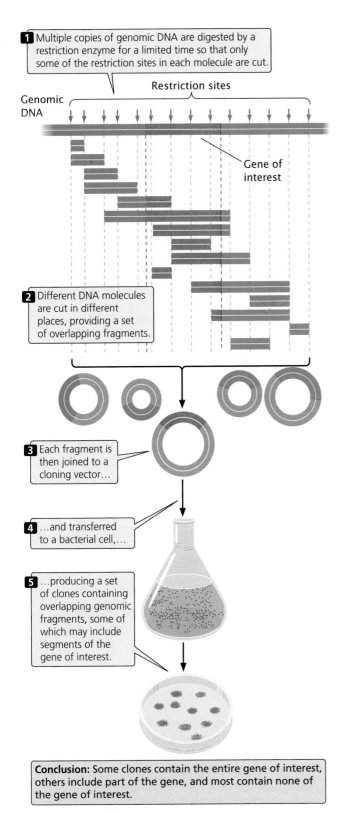

1 Multiple copies of genomic DNA are digested by a restriction enzyme for a limited time so that only some of the restriction sites in each molecule are cut.

Restriction sites

Genomic DNA

Gene of interest

2 Different DNA molecules are cut in different places, providing a set of overlapping fragments.

3 Each fragment is then joined to a cloning vector…

4 …and transferred to a bacterial cell,…

5 …producing a set of clones containing overlapping genomic fragments, some of which may include segments of the gene of interest.

Conclusion: Some clones contain the entire gene of interest, others include part of the gene, and most contain none of the gene of interest.

◀ **18.13 A genomic library contains all of the DNA sequences found in an organism's genome.**

with short fragments of DNA consisting entirely of thymine nucleotides [oligo(dT) chains; ◀ FIGURE 18.14a]. As the RNA moves through the column, the poly(A) tails

of mRNA molecules pair with the oligo(dT) chains and are retained in the column, whereas the rest of the RNA passes through. The mRNA can then be washed from the column by adding a buffer that breaks the hydrogen bonds between poly(A) tails and oligo(dT) chains.

The mRNA molecules are then copied into cDNA by reverse transcription. Short oligo(dT) primers are added to the mRNA. A primer pairs with the poly(A) tail at the 3′ end of the mRNA, providing a 3′-OH group for the initiation of DNA synthesis (◀ FIGURE 18.14b). Reverse transcriptase, an enzyme isolated from retroviruses (see p. 224 in Chapter 8), synthesizes single-stranded complementary DNA from the RNA template by adding DNA nucleotides to the 3′-OH group of the primer.

The resulting RNA–DNA hybrid molecule is then converted into a double-stranded cDNA molecule by one of several methods. One common method is to treat the RNA–DNA hybrid with RNase to partly digest the RNA strand. Partial digestion leaves gaps in the RNA–DNA hybrid, allowing DNA polymerase to synthesize a second DNA strand by using the short undigested RNA pieces as primers and the first DNA strand as a template. DNA polymerase eventually displaces all the RNA fragments, replacing them with DNA nucleotides, and nicks in the sugar–phosphate backbone are sealed by DNA ligase.

Concepts

One method of finding a gene is to create and screen a DNA library. A genomic library is created by cutting genomic DNA into overlapping fragments and cloning each fragment in a separate bacterial cell. A cDNA library is created from mRNA that is converted into cDNA and cloned in bacteria.

Screening DNA libraries Creating a genomic or cDNA library is relatively easy compared with screening the library to find clones that contain the gene of interest. The screening procedure used depends on what is known about the gene.

The first step in screening is to plate out the clones of the library. If a plasmid or cosmid vector was used to construct the library, the cells are diluted and plated so that each bacterium grows into a distinct colony. If a phage vector was used, the phages are allowed to infect a lawn of bacteria on a petri plate. Each plaque or bacterial colony contains a single, cloned DNA fragment that must be screened for the gene of interest.

One common way to screen libraries is with probes. We've seen how probes can be used to find specific fragments of DNA on an electrophoretic gel (see Figure 18.5). In a similar way, probes can be used to find cloned fragments of DNA in bacteria or phages. To use a probe, replicas of the plated colonies or plaques in the library must first be made. ◀ FIGURE 18.15 illustrates this procedure for a cosmid library.

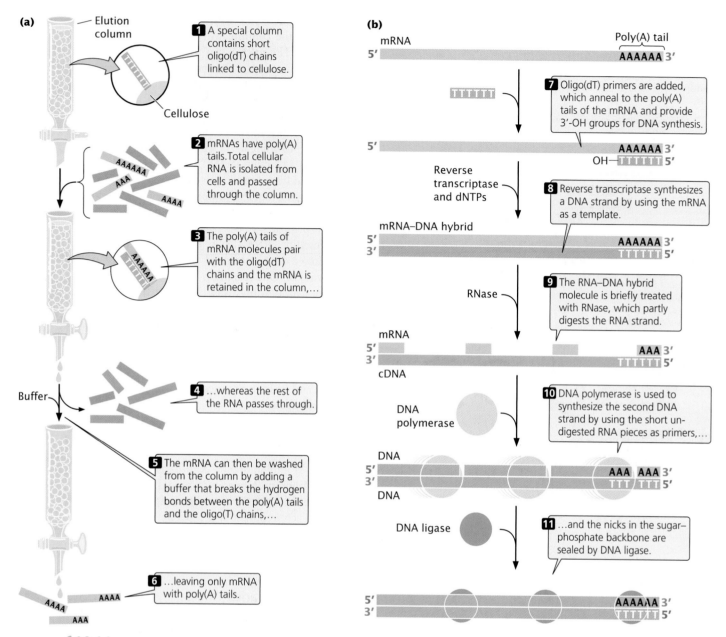

◀18.14 A cDNA library contains only those DNA sequences that are transcribed into mRNA.

How is a probe obtained when the gene has not yet been isolated? One option is to use a similar gene from another organism as the probe. For example, if we wanted to screen a human genomic library for the growth-hormone gene and the gene had already been isolated from rats, we could use a purified rat gene sequence as the probe to find the human gene for growth hormone. Successful hybridization does not require perfect complementarity between the probe and the target sequence; so a related sequence can often be used as a probe. The temperature and salt concentration of the hybridization reaction can be adjusted to regulate the degree of complementarity required for pairing to take place. Alternatively, synthetic probes can be created if the protein pro-

duced by the gene has been isolated and its amino acid sequence has been determined. With the use of the genetic code and the amino acid sequence of the protein, possible nucleotide sequences of a small region of the gene can be deduced. Although only one sequence in the gene encodes a particular protein, the presence of synonymous codons means that the same protein could be produced by several different DNA sequences, and it is impossible to know which is correct. To overcome this problem, a mixture of all the possible DNA sequences is used as a probe. To minimize the number of sequences required in the mixture, a region of the protein is selected with relatively little degeneracy in its codons (◀FIGURE 18.16).

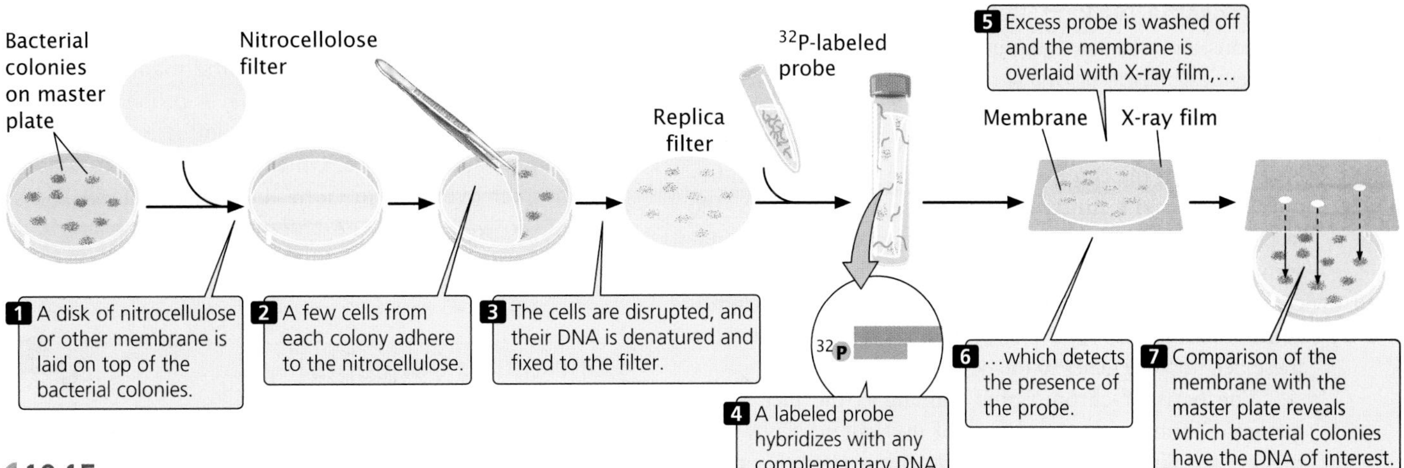

18.15 Genomic and cDNA libraries may be screened with a probe to find the gene of interest.

When part of the DNA sequence of the gene has been determined, a set of DNA probes can be synthesized chemically by using an automated machine known as an oligonucleotide synthesizer. The resulting probes can be used to screen a library for a gene of interest.

Yet another method of screening a library is to look for the protein product of a gene. This method requires that the DNA library be cloned in an expression vector. The clones can be tested for the presence of the protein by using an antibody that recognizes the protein or by using a chemical test for the protein product. This method depends on the existence of a test for the protein produced by the gene.

Almost any method used to screen a library will identify several clones, some of which will be false positives that do not contain the gene of interest; several screening methods may be needed to determine which clones actually contain the gene.

Concepts

A DNA library can be screened for a specific gene by using complementary probes that hybridize to the gene. Alternatively, the library can be cloned into an expression vector, and the gene can be located by examining the clones for the protein product of the gene.

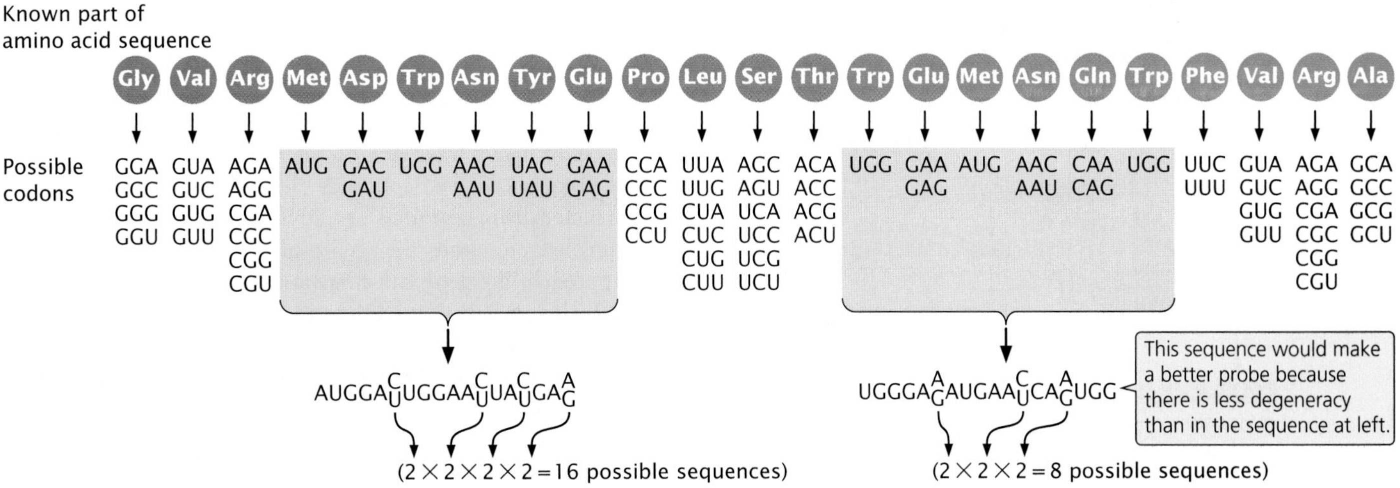

18.16 A synthetic probe can be designed on the basis of the genetic code and the known amino acid sequence of the protein encoded by the gene of interest. Because of ambiguity in the code, the same protein can be encoded by several different DNA sequences, and probes consisting of all the possible DNA sequences must be synthesized. To minimize the number of sequences that must be synthesized, a region of the gene with minimal degeneracy is picked.

Chromosome walking For many genes with important functions, no associated protein product is yet known. The biochemical bases of many human genetic diseases, for example, are still unknown. How could these genes be isolated? One approach is to first determine the general location of the gene on the chromosome by using recombination frequencies derived from crosses or pedigrees (see pp. 172–180 in Chapter 7). After the gene has been placed on a chromosome map, neighboring genes that have already been cloned can be identified. With the use of a technique called **chromosome walking** (◀ FIGURE 18.17), it is possible to move from these neighboring genes to the new gene of interest.

The basis of chromosome walking is the fact that a genomic library consists of a set of *overlapping* DNA fragments (see Figure 18.13). We start with a cloned gene or DNA sequence that is close to the new gene of interest so that the "walk" will be as short as possible. One end of the clone of a neighboring gene (clone A in Figure 18.17) is used to make a complementary probe. This probe is used to screen the genomic library to find a second clone (clone B) that overlaps with the first and extends in the direction of the gene of interest. This second clone is isolated and purified and a probe is prepared from its end. The second probe is used to screen the library for a third clone (clone C) that overlaps with the second. In this way, one can walk systematically toward the gene of interest, one clone at a time. A number of important human genes and genes of other organisms have been found in this way.

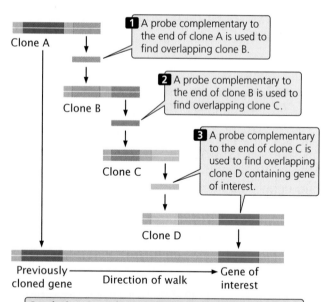

Conclusion: By making probes complementary to areas of overlap between cloned fragments in a genomic library, we can connect a gene of interest to a previously mapped, linked gene.

◀ **18.17 In chromosome walking, neighboring genes are used to locate a gene of interest.**

Concepts

In chromosome walking, a gene is first mapped in relation to a previously cloned gene. A probe made from one end of the cloned gene is used to find an overlapping clone, which is then used to find another overlapping clone. In this way, it is possible to walk down the chromosome to the gene of interest.

Connecting Concepts

Cloning Strategies

All gene-cloning experiments have four basic steps:

1. Isolation of a DNA fragment
2. Joining of the fragment to a cloning vector
3. Introduction of the cloning vector, along with the inserted DNA fragment, into host cells
4. Identification of cells containing the recombinant DNA molecule

We've now considered a number of different methods for carrying out these four steps. There is no single procedure for cloning a gene but rather a variety of methods, each with its strengths and weaknesses. The particular combination of methods chosen for a cloning experiment is termed the **cloning strategy.**

In developing a cloning strategy, a number of factors must be taken into consideration. These factors include how much is known about the gene to be cloned, the size and nature of the gene, and the ultimate purpose of the cloning experiment. The procedure for cloning a small, well-characterized DNA fragment for sequencing would be very different from that for cloning a large, poorly known gene for the commercial production of a protein.

The first step in gene cloning is to find the particular gene or DNA fragment of interest. There are two basic approaches. In one approach, a DNA library can be constructed from genomic or cDNA, and the library can be screened to find the gene of interest. In the other approach, the gene can first be isolated and then cloned. Which approach is used depends largely on what is already known about the gene. Has the gene been mapped? Is there a probe available for screening? Is the amino acid sequence of a protein encoded by the gene known?

If one chooses to make and screen a DNA library, the next problem is to select the best source of DNA. Will it be a genomic library or a cDNA library? If the purpose is to clone the gene in an expression vector and produce a protein, then a cDNA library is ideal. Using a cDNA library means that introns (which bacteria cannot splice out) will be excluded, and fewer colonies will need to be screened. If, on the other hand, the purpose is to examine the regulatory sequences or the introns within a gene, then a genomic library is required.

The next important decision in developing a cloning strategy is to select the cloning vector. The choice depends on a number of factors:

- *The length of the sequence to be cloned.* For a sequence only a few thousand base pairs in length, a plasmid may be the best choice; if one wants to clone a gene that is 35 kb or longer, a cosmid will be required.

- *The organism in which the gene will be cloned.* Some vectors are specific for *E. coli*, whereas others are specific for other bacteria or for eukaryotic cells.

- *The selection methods used to find cells containing a plasmid with the inserted gene.* One may need a vector with selectable markers so that cells containing the gene can be identified.

- *The need for the inserted gene to be expressed.* If the protein product is desired, it may be necessary to use an expression vector that contains a promoter and other sequences that ensure transcription and translation of the inserted gene.

- *The need for efficiency of transfer to host cells.* If selection methods can be used to screen a large number of cells, then a low rate of transfer may be adequate; but, if screening is less efficient or is costly, a higher rate of transfer may be desirable.

A cloning strategy must also take into consideration the best method for joining the DNA fragment and cloning vector. Important points here include the simplicity and ease of the method, the need to retain restriction sites so that the foreign gene can later be retrieved from the vector, and whether the gene sequence must be joined to a promoter and other regulatory sequences to ensure transcription.

The method chosen for moving the vector into the host cell is usually dictated by the type of vector; plasmids are transferred to bacterial cells by transformation, whereas phage vectors and cosmids are transferred by viral infection.

The procedure for screening cells to find those with recombinant molecules depends on how much is known about the cloned fragment, the efficiency of transfer, and the cloning vector used. Considerations used in a developing a cloning strategy are summarized in Table 18.4.

Table 18.4	Considerations in developing a cloning strategy
Step in Gene Cloning	**Considerations**
1. Isolation of DNA fragment	a. The purpose of cloning (is expression required?). Is the entire sequence needed?
	b. What is known about the gene and the protein (if any) that it encodes?
	c. The size of the gene.
	d. Is the chromosomal location of the gene known?
	e. Size of the genome from which the gene is isolated.
2. Joining DNA fragment to vector	a. Type of cloning vector used.
	i. The size of the gene.
	ii. The organism into which the gene will be cloned.
	iii. The need for a selection mechanism.
	iv. Whether expression is required.
	v. Efficiency of transfer to host cell required.
	vi. The purpose of cloning.
	b. Method of joining the gene to vector.
	i. Simplicity of method.
	ii. Availability of restriction sites.
	iii. The need to retrieve the fragment from the vector.
	iv. Whether expression is required.
	v. The purpose of cloning.
3. Transfer of recombinant vector to host cell	a. Type of cloning vector used.
4. Identification of cells carrying recombinant molecule	a. Known information about the gene.
	b. Type of cloning vector used.
	c. Efficiency of transfer.
	d. Purpose of cloning.

Using the Polymerase Chain Reaction to Amplify DNA

A major problem in working at the molecular level is that each gene is a tiny fraction of the total cellular DNA. Because each gene is rare, it must be isolated and amplified before it can be studied. Before mid-1980, the only procedure available for amplifying DNA was gene cloning—placing the gene in a bacterial cell and multiplying the bacteria. Cloning is labor intensive and requires at least several days to grow the bacteria. In 1983, Kary Mullis of the Cetus Corporation conceptualized a new technique for amplifying DNA in a test tube. The **polymerase chain reaction** allows DNA fragments to be amplified a billionfold within just a few hours. It can be used with extremely small amounts of original DNA, even a single molecule. The polymerase chain reaction has revolutionized molecular biology and is now one of the most widely used of all molecular techniques.

The basis of PCR is replication catalyzed by a DNA polymerase enzyme, which has two essential requirements:

(1) a single-stranded DNA template from which a new DNA strand can be copied and (2) a primer with a 3'-OH group to which new nucleotides can be added.

Because a DNA molecule consists of two nucleotide strands, each of which can serve as a template to produce a new molecule of DNA, the amount of DNA doubles with each replication event. The starting point of DNA synthesis on the template is determined by the choice of primers. The primers used in PCR are short fragments of DNA, typically from 17 to 25 nucleotides long, that are complementary to known sequences on the template. A different primer is used for each strand.

To carry out PCR, one begins with a solution that includes the target DNA (the DNA to be amplified), DNA polymerase, all four deoxyribonucleoside triphosphates (dNTPs—the substrates for DNA polymerase), primers that are complementary to short sequences on each strand of the target DNA, and magnesium ions and other salts that are necessary for the reaction to proceed. A typical polymerase chain reaction includes three steps (◀FIGURE 18.18).

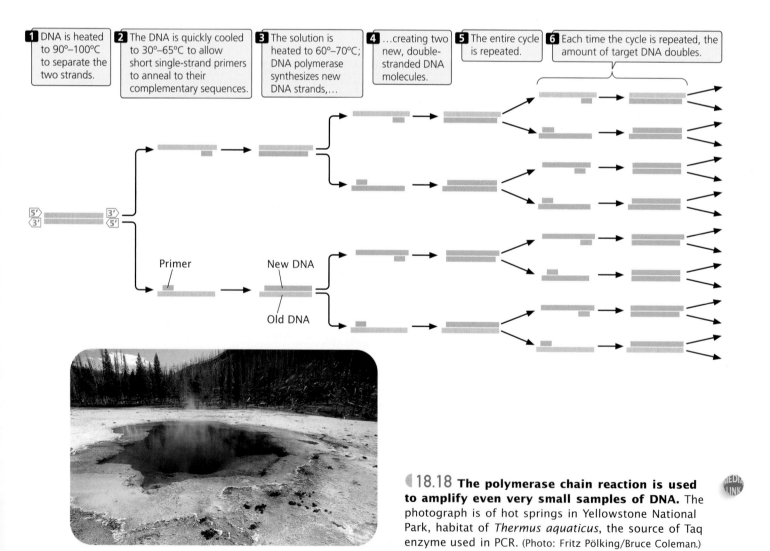

◀18.18 **The polymerase chain reaction is used to amplify even very small samples of DNA.** The photograph is of hot springs in Yellowstone National Park, habitat of *Thermus aquaticus*, the source of Taq enzyme used in PCR. (Photo: Fritz Pölking/Bruce Coleman.)

In step 1, a starting solution of DNA is heated to between 90° and 100°C to break the hydrogen bonds between the two nucleotide strands and thus produce the necessary single-stranded templates. The reaction mixture is held at this temperature for only a minute or two. In step 2, the DNA solution is cooled quickly to between 30° and 65°C and held at this temperature for a minute or less. During this short interval, the DNA strands will not have a chance to reanneal, but the primers will be able to attach to their complementary sequences on the template strands. In step 3, the solution is heated to between 60° and 70°C, the temperature at which DNA polymerase can synthesize new DNA strands by adding nucleotides to the primers. Within a few minutes, two new double-stranded DNA molecules are produced for each original molecule of target DNA.

The whole cycle is then repeated. With each cycle, the amount of target DNA doubles; so the target DNA increases geometrically. One molecule of DNA increases to more than 1000 molecules in 10 PCR cycles, to more than 1 million molecules in 20 cycles, and to more than 1 billion molecules in 30 cycles (Table 18.5). Each cycle is completed within a few minutes; so a large amplification of DNA can be achieved within a few hours.

Two key innovations facilitated the use of PCR in the laboratory. The first was the discovery of a DNA polymerase that is stable at the high temperatures used in step 1 of PCR. The DNA polymerase from *E. coli* that was originally used in PCR denatures at 90°C. For this reason, fresh enzyme had to be added to the reaction mixture during *each* cycle, slowing the process considerably. This obstacle was overcome when DNA polymerase was isolated from the bacterium *Thermus aquaticus,* which lives in the boiling springs of Yellowstone National Park (see Figure 18.18). This enzyme, dubbed ***Taq* polymerase,** is remarkably stable at high temperatures and is not denatured during the strand-separation step of PCR; so it can be added to the reaction mixture at the beginning of the PCR process and will continue to function through many cycles.

The second key innovation was the development of automated thermal cyclers—machines that bring about the rapid temperature changes necessary for the different steps of PCR. Originally, tubes containing reaction mixtures were moved by hand among water baths set at the different temperatures required for the three steps of each cycle. In automated thermal cyclers, the reaction tubes are placed in a metal block that changes temperature rapidly according to a computer program.

The polymerase chain reaction is now often used in place of gene cloning, but it does have several limitations. First, the use of PCR requires prior knowledge of at least part of the sequence of the target DNA to allow construction of the primers. Therefore PCR cannot be used to amplify a gene that has not been at least partly sequenced. Second, the capacity of PCR to amplify extremely small

Table 18.5	Number of copies of DNA fragment in PCR amplification
Number of PCR Cycles (n)	Number of Double-Stranded Copies of Original DNA (2^n)
0	1
1	2
2	4
3	8
4	16
5	32
6	64
7	128
8	256
9	512
10	1,024
20	1,048,576
30	1,073,741,824

amounts of DNA makes contamination a significant problem. Minute amounts of DNA from the skin of laboratory workers and even in small particles in the air can enter a reaction tube and be amplified along with the target DNA. Careful laboratory technique and the use of controls are necessary to circumvent this problem.

A third limitation of PCR is accuracy. Unlike other DNA polymerases, *Taq* polymerase does not have the capacity to proofread (see p. 336–337 in Chapter 12) and, under standard PCR conditions, it incorporates an incorrect nucleotide about once every 20,000 bp. DNA polymerases with proofreading capacity usually incorporate an incorrect nucleotide only about once every billion base pairs. For many applications, the error rate produced by PCR is not a problem, because only a few DNA molecules of the billions produced will contain an error. However, for other applications such as the cloning of PCR products, the relatively high error rate of PCR can pose significant problems. New heat-stable DNA polymerases *with* proofreading capacity have been isolated, giving more accurate PCR results.

A fourth limitation of PCR is that the size of the fragments that can be amplified by standard *Taq* polymerase is usually less than 2000 bp. By using a combination of *Taq* polymerase and a DNA polymerase with proofreading capacity and by modifying the reaction conditions, investigators have been successful in extending PCR amplification to larger fragments. In spite of its limitations, PCR is used routinely in a wide array of molecular applications.

Analyzing DNA Sequences

In addition to cloning and amplifying DNA, molecular techniques are used to analyze DNA molecules through a determination of their sequences and an investigation of their functions.

DNA sequencing A powerful technique to emerge from recombinant DNA technology is the ability to quickly sequence DNA molecules. *DNA sequencing* is the determination of the sequence of bases in DNA. Sequencing allows the genetic information in DNA to be read, providing an enormous amount of information about gene structure and function. Details of DNA sequencing will be covered in Chapter 19.

In situ hybridization DNA probes can be used to determine the chromosomal location of a gene or the cellular location of an mRNA in a process called **in situ hybridization.** The name is derived from the fact that DNA or RNA is visualized while it is in the cell (in situ). This technique requires that the cells be fixed and the chromosomes be spread on a microscope slide. The chromosomes are then briefly exposed to a solution with high pH, which disrupts the pairing of the DNA bases, making them accessible to probes. A labeled probe, which binds to any complementary DNA sequences, is added. Excess probe is washed off, and the location of the bound probe is detected. Originally, probes were radioactively labeled and detected with autoradiography, but now many probes carry attached fluorescent dyes that can be seen directly with the microscope (◀FIGURE 18.19a) Several probes with different colored dyes can be used simultaneously to investigate different sequences or chromosomes.

In situ hybridization can also be used to determine the tissue distribution of specific mRNA molecules, serving as a source of insight into how gene expression differs among cell types (◀FIGURE 18.19b). A labeled DNA or RNA probe complementary to a specific mRNA molecule is added to tissue, and the location of the probe is determined by using either autoradiography or fluorescent tags.

DNA footprinting Many important DNA sequences serve as binding sites for proteins; for example, consensus sequences in promoters are often binding sites for transcription factors (see p. 301 in Chapter 11). A technique called **DNA footprinting** can be used to determine which DNA sequences are bound by such proteins.

In a typical DNA-footprinting experiment, purified DNA fragments are labeled at one end with a radioactive isotope of phosphorus, ^{32}P. An enzyme or chemical that

(a)

(b)

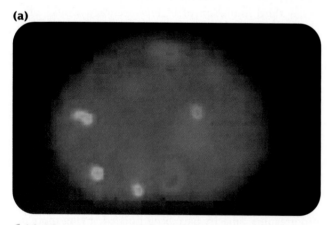
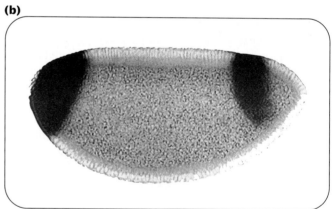

◀18.19 **With in situ hybridization, DNA probes are used to determine the cellular or chromosomal location of a gene or its product.** (a) A probe with green fluorescence is specific to the X chromosome, a probe with red fluorescence is specific to the Y chromosome, and a probe with blue fluorescence is specific to chromosome 18. The probes indicate that this cell has XXXY sex chromosomes and two copies of chromosome 18. (b) In situ hybridization is used to detect the presence of mRNA from the tailless gene in a *Drosophila* embryo. The tailless mRNA is concentrated at both ends of the embryo and is required for proper development. (Part a, Courtesy of Patrick Daniel Storto, Ph.D., Department of Pediatrics and Human Development, Michigan State University; part b, Courtesy of L. Tsuda.)

makes cuts in DNA is used to cleave the DNA randomly into subfragments, which are then denatured and separated by gel electrophoresis. The positions of the subfragments are visualized with autoradiography. This procedure is carried out both in the presence and in the absence of a particular DNA-binding protein. When the protein is absent, cleavage is random along the DNA, producing a continuous "ladder" of bands on the autoradiograph (◀ FIGURE 18.20). When the protein is present, it binds to specific nucleotides and protects their phosphodiester bonds from cleavage. Therefore, there is no cleavage in the area protected by the protein, and no labeled fragments terminating in the binding site appear on the autoradiograph. Their omission leaves a gap, or "footprint", on the ladder of bands (see Figure 18.20), and the position of the footprint identifies those nucleotides bound tightly by the protein.

Concepts

In situ hybridization can be used to visualize the chromosomal location of a gene or to determine the tissue distribution of an mRNA transcribed from a specific gene. DNA footprinting is used to determine the sequences to which DNA-binding proteins attach.

Mutagenesis A powerful way to study gene function is to create mutations at specific locations in a process called **site-directed mutagenesis** and then to study the effects of these mutations on the organism.

A number of different strategies have been developed for site-directed mutagenesis. One strategy is to cut out a short sequence of nucleotides with restriction enzymes and replace it with a short, synthetic oligonulceotide that contains the desired mutated sequence (◀ FIGURE 18.21). The success of this method depends on the availability of restriction sites flanking the sequence to be altered.

If appropriate restriction sites are not available, **oligonucleotide-directed mutagenesis** can be used (◀ FIGURE 18.22). In this method, a single-stranded oligonucleotide is produced that differs from the target sequence by one or a few bases. Because they differ in only a few bases, the target DNA and the oligonucleotide will pair under the appropriate conditions. When successfully paired with the target DNA, the oligonucleotide can act as a primer to initiate DNA synthesis, which produces a double-stranded molecule with a mismatch in the primer region. When this DNA is transferred to bacterial cells, the mismatched bases will be repaired by bacterial enzymes. About half of the time the normal bases will be changed into mutant bases, and about half of the time the mutant bases will be changed into normal bases. The bacteria are then screened for the presence of the mutant gene.

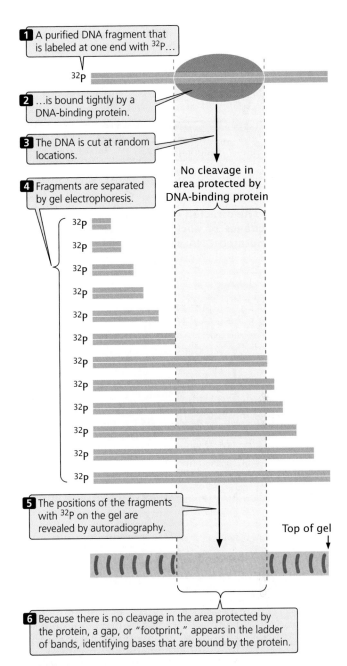

1 A purified DNA fragment that is labeled at one end with ^{32}P…

^{32}P

2 …is bound tightly by a DNA-binding protein.

3 The DNA is cut at random locations.

No cleavage in area protected by DNA-binding protein

4 Fragments are separated by gel electrophoresis.

^{32}P
^{32}P
^{32}P
^{32}P
^{32}P
^{32}P
^{32}P
^{32}P
^{32}P
^{32}P
^{32}P

5 The positions of the fragments with ^{32}P on the gel are revealed by autoradiography.

Top of gel

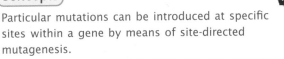

6 Because there is no cleavage in the area protected by the protein, a gap, or "footprint," appears in the ladder of bands, identifying bases that are bound by the protein.

◀18.20 **DNA footprinting can be used to determine which DNA sequences are bound by binding proteins.**

Concepts

Particular mutations can be introduced at specific sites within a gene by means of site-directed mutagenesis.

Transgenic animals The oocytes of mice and other mammals are large enough that DNA can be injected into them directly. Immediately after penetration by sperm,

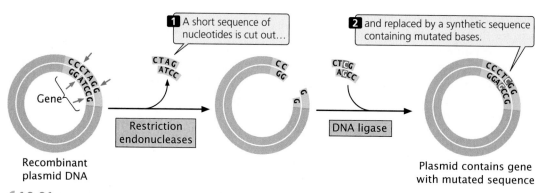

18.21 In site-directed mutagenesis, restriction enzymes cut out a short sequence of nucleotides that is then replaced by a synthetic mutated DNA sequence.

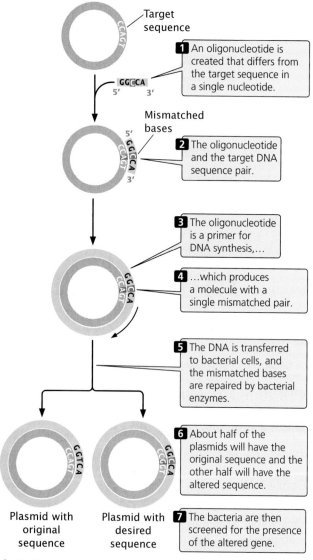

18.22 Oligonucleotide-directed mutagenesis is used to study gene function when appropriate restriction sites are not available.

a fertilized mouse egg contains two pronuclei, one from the sperm and one from the egg; these pronuclei later fuse to form the nucleus of the embryo. Mechanical devices can manipulate extremely fine, hollow glass needles to inject DNA directly into one of the pronuclei of a fertilized egg (◀ FIGURE 18.23). Typically, a few hundred copies of cloned, linear DNA are injected into a pronucleus, and, in a few of the injected eggs, copies of the cloned DNA integrate randomly into one of the chromosomes through a process called nonhomologous recombination. After injection, the embryos are implanted in a pseudopregnant female—a surrogate mother that has been physiologically prepared for pregnancy by mating with a vasectomized male.

Only about 10% to 30% of the eggs survive and, of those that do survive, only a few have a copy of the cloned DNA stably integrated into a chromosome. Nevertheless, if several hundred embryos are injected and implanted, there is a good chance that one or more mice whose chromosomes contain the foreign DNA will be born. Moreover, because the DNA was injected at the one-cell stage of the embryo, these mice usually carry the cloned DNA in every cell of their bodies, including their reproductive cells, and will therefore pass the foreign DNA on to their progeny. Through interbreeding, a strain of mice that is homozygous for the foreign gene can be created. Animals that have been permanently altered in this way are said to be *transgenic*, and the foreign DNA that they carry is called a **transgene.**

Transgenic mice have proved useful in the study of gene function. For example, proof that the *SRY* gene (see pp. 84–85 in Chapter 4) is the male-determining gene in mice was obtained by injecting a copy of the *SRY* gene into XX embryos and observing that these mice developed as males. In addition, a number of transgenic mouse strains that serve as experimental models for human genetic diseases have been created by injecting mutated copies of genes into mouse embryos.

www.whfreeman.com/pierce   More information on
transgenic animals

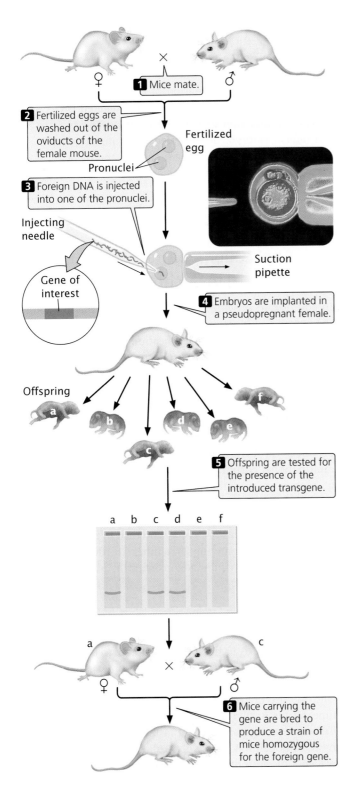

◀ 18.23 Transgenic animals have genomes that have been permanently altered through recombinant DNA technology. In the photograph a mouse embryo (red and blue) is being injected with DNA. (Photo: Jon Gordon/Phototake.)

Knockout mice A particularly useful variant of the transgenic approach is to produce mice in which a normal gene has been disabled. The phenotypes of these animals, called **knockout mice,** help geneticists to determine the function of a gene. The creation of knockout mice begins when a normal gene is cloned in bacteria and then "knocked out", or disabled. There are a number of ways to disable a gene, but a common method is to insert a gene called *neo,* which confers resistance to the antibiotic G418, into the middle of the target gene (◀ FIGURE 18.24). The insertion of *neo* both disrupts (knocks out) the target gene and provides a convenient marker for finding copies of the disabled gene. In addition, a second gene, usually the herpes simplex viral thymidine kinase (*tk*) gene, is cloned adjacent to the disrupted gene. The disabled gene is then transferred to cultured embryonic mouse cells, where it may exchange places with the normal chromosomal copy through homologous recombination.

After the disabled gene has been transferred to the embryonic cells, the cells are screened by adding the antibiotic G418 to the medium. Only cells with the disabled gene containing the *neo* insert will survive. Because the frequency of nonhomologous recombination is higher than that of homologous recombination and because the intact target gene is replaced by the disabled copy only through homologous recombination, a means to select for the rarer homologous recombinants is required. The presence of the viral *tk* gene makes the cells sensitive to gancyclovir. Thus, transfected cells that grow on medium containing G418 and gancyclovir will contain the *neo* gene (disabled target gene) but not the adjacent *tk* gene. These cells contain the desired homologous recombinants. The nonhomologous recombinants (random insertions) will contain both the *neo* and the *tk* genes, and these transfected cells will die on the selection medium owing to the presence of gancyclovir. The surviving cells are injected into an early-stage mouse embryo, which is then implanted into a pseudopregnant mouse. Cells in the embryo carrying the disabled gene and normal embryonic cells carrying the wild-type gene will develop together, producing a chimera—a mouse that is a genetic mixture of the two cell types. The chimeric mice can be identified easily if the injected embryonic cells came from a black mouse and the embryos into which they are injected came from a white mouse; the resulting chimeras will have variegated black and white fur. The chimeras can then be interbred to produce some progeny that are homozygous for the knockout gene. The effects of disabling a particular gene can be observed in these homozygous mice.

Although they are a recent innovation, knockout mice have become important subjects for research in a number of fields. They have been used to study genes implicated in immune function, development, ethology, and human behavior.

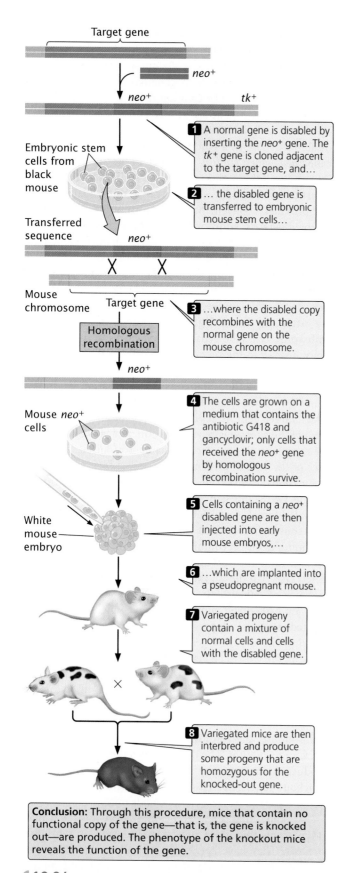

1 A normal gene is disabled by inserting the *neo+* gene. The *tk+* gene is cloned adjacent to the target gene, and...

2 ... the disabled gene is transferred to embryonic mouse stem cells...

3 ...where the disabled copy recombines with the normal gene on the mouse chromosome.

4 The cells are grown on a medium that contains the antibiotic G418 and gancyclovir; only cells that received the *neo+* gene by homologous recombination survive.

5 Cells containing a *neo+* disabled gene are then injected into early mouse embryos,...

6 ...which are implanted into a pseudopregnant mouse.

7 Variegated progeny contain a mixture of normal cells and cells with the disabled gene.

8 Variegated mice are then interbred and produce some progeny that are homozygous for the knocked-out gene.

Conclusion: Through this procedure, mice that contain no functional copy of the gene—that is, the gene is knocked out—are produced. The phenotype of the knockout mice reveals the function of the gene.

◄ **18.24 Knockout mice possess a genome in which a gene has been disabled.**

Concepts

Transgenic mice are produced by the injection of cloned DNA into the pronucleus of a fertilized egg, followed by implantation of the egg into a female mouse. In knockout mice, the injected DNA contains a mutation that disables a gene. Inside the mouse embryo, the disabled copy of the gene can exchange with the normal copy of the gene through homologous recombination.

www.whfreeman.com/pierce   A catalog of knockout mice

Applications of Recombinant DNA Technology

In addition to providing valuable new information about the nature and function of genes, recombinant DNA technology has many practical applications. These applications include the production of pharmaceuticals and other chemicals, specialized bacteria, agriculturally important plants, and genetically engineered farm animals. The technology is also used extensively in medical testing and, in a few cases, is even being used to correct human genetic defects. Hundreds of firms now specialize in developing products through genetic engineering, and many large multinational corporations have invested enormous sums of money in recombinant DNA research. Recombinant DNA technology is also frequently used in criminal investigations and for the identification of human remains. For example, the remains of people who died in the collapse of the World Trade Center on September 11, 2001, were identified through DNA comparisons with the use of recombinant DNA technology.

Pharmaceuticals

The first commercial products to be developed by using recombinant DNA technology were pharmaceuticals used in the treatment of human diseases and disorders. In 1979, the Eli Lilly corporation began selling human insulin produced with the use of recombinant DNA technology. Before this time, all the insulin used in the treatment of diabetics was isolated from the pancreases of farm animals slaughtered for meat. Although this source of insulin worked well for many diabetics, it was not human insulin, and some people suffered allergic reactions to the foreign protein. The human insulin gene was inserted into plasmids and transferred to bacteria that then produced human insulin. Pharmaceuticals produced through recombinant DNA technology include human growth hormone (for children with growth deficiencies), clotting factors (for hemophiliacs), and tissue plasminogen activator (used to dissolve blood clots in heart-attack patients).

Specialized Bacteria

Bacteria play an important role in many industrial processes, including the production of ethanol from plant material, the leaching of minerals from ore, and the treatment of sewage and other wastes. The bacteria engaged in these processes are being modified by genetic engineering so that they work more efficiently. New strains of technologically useful bacteria are being developed that will break down toxic chemicals and pollutants, enhance oil recovery, increase nitrogen uptake by plants, and inhibit the growth of pathogenic bacteria and fungi.

Agricultural Products

Recombinant DNA technology has had a major effect on agriculture, where it is now used to create crop plants and domestic animals with valuable traits. For many years, plant pathologists had recognized that plants infected with mild strains of viruses are resistant to infection by virulent strains. Using this knowledge, geneticists have created viral resistance in plants by transferring genes for viral proteins to the plant cells. A genetically engineered squash, called Freedom II, carries genes from the watermelon mosaic virus 2 and the zucchini yellow mosaic virus that protect the squash against viral infections.

Another objective has been to genetically engineer pest resistance into plants to reduce dependence on chemical pesticides. A protein toxin from the bacterium *Bacillus thuringiensis* selectively kills the larvae of certain insect pests but is harmless to wildlife, humans, and many other insects. The toxin gene has been isolated from the bacteria, linked to active promoters, and transferred into corn, tomato, potato, and cotton plants. The gene produces the insecticidal toxin in the plants, and caterpillars that feed on the plant die.

Recombinant DNA technology has also permitted the development of herbicide resistance in plants. A major problem in agriculture is the control of weeds, which compete with crop plants for water, sunlight, and nutrients. Although herbicides are effective at killing weeds, they can also damage the crop plants. Genes that provide resistance to broad-spectrum herbicides have been transferred into tomato, soybean, cotton, oilseed rape, and other commercially important crops. When the fields containing these crops are sprayed with herbicides, the weeds are killed but the genetically engineered plants are unaffected. In 1999, more than 21 million hectares (1 hectare = 2.471 acres) of genetically engineered soybeans and 11 million hectares of genetically engineered corn was grown throughout the world.

Recombinant DNA techniques are also applied to domestic animals. For example, the gene for growth hormone was isolated from cattle and cloned in *E. coli;* these bacteria produce large quantities of bovine growth hormone, which is administered to dairy cattle to increase milk production. Transgenic animals are being developed to carry genes that encode pharmaceutical products. For example, a gene for human clotting factor VIII has been linked to the regulatory region of the sheep gene for β-lactoglobulin, a milk protein. The fused gene was injected in sheep embryos, creating transgenic sheep that produce in their milk the human clotting factor, which is used to treat hemophiliacs. A similar procedure was used to transfer a gene for α_1-antitrypsin, a protein used to treat patients with hereditary emphysema, into sheep. Female sheep bearing this gene produce as much as 15 grams of α_1-antitrypsin in each liter of their milk, generating $100,000 worth of α_1-antitrypsin per year for each sheep.

The genetic engineering of agricultural products is controversial. One area of concern focuses on the potential effects of releasing novel organisms produced by genetic engineering into the environment. There are many examples in which nonnative organisms released into a new environment have caused ecological disruption because they are free of predators and other natural control mechanisms. Genetic engineering normally transfers only small sequences of DNA, relative to the large genetic differences that often exist between species, but even small genetic differences may alter ecologically important traits that might affect the ecosystem.

Another area of concern is that transgenic organisms may hybridize with native organisms and transfer their genetically engineered traits. For example, herbicide resistance engineered into crop plants might be transferred to weeds, which would then be resistant to the herbicides that are now used for their control. The results of some studies have demonstrated gene transfer between engineered plants and native plants, but the extent and effect of this transfer are uncertain. Other concerns focus on health-safety issues associated with the presence of engineered products in natural foods; some critics have advocated required labeling of all genetically engineered foods that contain transgenic DNA or protein. Such labeling is required in countries of the European Union but not in the United States.

On the other hand, the use of genetically engineered crops and domestic animals has potential benefits. Genetically engineered crops that are pest resistant have the potential to reduce the use of environmentally harmful chemicals, and research findings indicate that lower amounts of pesticides are used in the United States as a result of the adoption of transgenic plants. Transgenic crops also increase yields, providing more food per acre, which reduces the amount of land that must be used for agriculture.

> **Concepts**
>
> Recombinant DNA technology is used to create a wide range of commercial products, including pharmaceuticals, specialized bacteria, genetically engineered crops, and transgenic domestic animals.

www.whfreeman.com/pierce More information about the use of recombinant DNA and biotechnology in agriculture

Oligonucleotide Drugs

A recent application of DNA technology has been the development of oligonucleotide drugs, which are short sequences of synthetic DNA or RNA molecules that can be used to treat diseases. Antisense oligonucleotides are complementary to undesirable RNAs, such as viral RNA. When added to a cell, these antisense DNAs bind to the viral mRNA and inhibit its translation.

Single-stranded DNA oligonucleotides bind tightly to other DNA sequences, forming a triplex DNA molecule (◀FIGURE 18.25). The formation of triplex DNA interferes with the binding of RNA polymerase and other proteins required for transcription. Other oligonucleotides are ribozymes, RNA molecules that function as enzymes (see introduction to Chapter 13). These compounds bind to specific mRNA molecules and cleave them into fragments, destroying their ability to encode proteins. Several oligonucleotide drugs are already being tested for the treatment of AIDS and cancer.

> **Concepts**
>
> Oligonucleotide drugs are short pieces of DNA or RNA that prevent the expression of particular genes.

Genetic Testing

The identification and cloning of many important disease-causing human genes has allowed the development of probes for detecting disease-causing mutations. Prenatal testing is already available for several hundred genetic disorders (see Chapter 6). Additionally, presymptomatic genetic tests for adults and children are available for an increasing number of disorders.

The growing availability of genetic tests raises a number of ethical and social issues. For example, is it ethical to test for genetic diseases for which there is no cure or treatment? Huntington disease, an autosomal dominant disorder that appears in middle age, causes slow physical and mental deterioration and eventually death. No effective treatment is currently available. If one parent is affected, a child has a 50% chance of inheriting the gene for Huntington disease and eventually getting the disorder. Tests are now available that make it possible to determine whether a person carries the Huntington-disease gene, but is it beneficial to tell a young person that he or she has the Huntington-disease gene and will get the disease later in life?

Although learning that you do not have the gene might provide great peace of mind, learning that you *do* have it might lead to despair and depression. Many people at risk for Huntington disease want predictive testing, saying that the uncertainty of not knowing is more debilitating than the certain knowledge that they will get it and, in fact, a number of medical centers now offer predictive testing for Huntington disease. A few people who learned that they have the gene have committed suicide, and others had to be hospitalized for depression, but the results of several studies indicate that most people who undergo predictive testing for Huntington disease are able to cope with the information.

Other ethical and legal questions concern the confidentiality of test results. Who should have access to the results of genetic testing? Should insurance companies be allowed to use results from such tests to deny coverage to healthy people who are at risk for genetic diseases? Should relatives who also might be at risk be informed of the results of genetic testing?

Other concerns focus on whether the cost of genetic testing justifies the benefits. In some cases, genetic tests provide clear benefits because early identification allows for

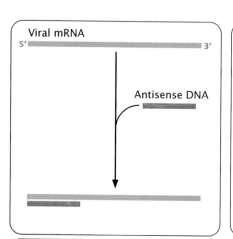

Antisense DNA may bind to the 5' end of viral mRNA and prevent the binding of the ribosome; so no viral protein is produced.

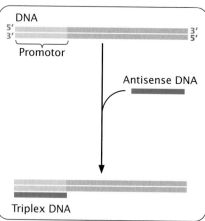

Antisense DNA may bind to a promoter (forming triplex DNA) and prevent the binding of RNA polymerase; so mRNA is not produced.

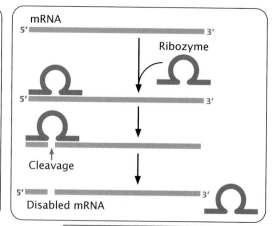

Antisense RNA in the form of a ribozyme may bind and cleave mRNA, destroying it.

◀18.25 **Oligonucleotide drugs are short sequences of DNA or RNA that can be used to treat diseases.**

better treatment. For example, when phenylketonuria (an autosomal recessive disorder that can cause mental retardation) is identified in infants, the administration of a special diet can prevent mental retardation. Because of this obvious benefit and the low cost of testing for this disorder, all states in the United States and many other countries require newborns to be tested for PKU.

Predictive testing for colorectal cancer and breast cancer also may be beneficial for at-risk people, because finding these cancers early improves the chances for successful treatment. Patients with genes that predispose to cancer may require more aggressive treatment than do patients with sporadically arising cancers. In these diseases, genetic testing provides clear benefits.

Another set of concerns is related to the accuracy of genetic tests. For many genetic diseases, the only predictive tests available are those that identify a *predisposing* mutation in DNA, but many genetic diseases may be caused by dozens or hundreds of different mutations. Probes that detect common mutations can be developed, but they won't detect rare mutations and will give a false negative result. Short of sequencing the entire gene—which is expensive and time consuming—there is no way to identify all predisposed persons. These questions and concerns are currently the focus of intense debate by ethicists, physicians, scientists, and patients.

www.whfreeman.com/pierce More information on genetic testing

Gene Therapy

Perhaps the ultimate application of recombinant DNA technology is **gene therapy,** the direct transfer of genes into humans to treat disease. When the first recombinant DNA experiments with bacteria were announced, many researchers recognized the potential for using this new technology in the treatment of patients with genetic diseases. But, before recombinant DNA could be used on humans, a number of difficult obstacles had to be overcome. The genes responsible for particular genetic diseases needed to be located and cloned, and special vectors had to be developed that would reliably and efficiently deliver genes to human cells.

In 1990, gene therapy became reality. W. French Anderson and his colleagues at the U.S. National Institutes of Health (NIH) transferred a functional gene for adenosine deaminase to a young girl with severe combined immunodeficiency disease, an autosomal recessive condition that produces impaired immune function.

Today, thousands of patients have received gene therapy, and many clinical trials are underway. Gene therapy is being used to treat genetic diseases, cancer, heart disease, and even some infectious diseases such as AIDS. All of these

Table 18.6	Vectors used in gene therapy	
Vector	**Advantages**	**Disadvantages**
Retrovirus	Efficient transfer	Transfers DNA only to dividing cells, inserts randomly; risk of producing wild-type viruses
Adenovirus	Transfers to nondividing cells	Causes immune reaction
Adeno-associated virus	Does not cause immune reaction	Holds small amount of DNA; hard to produce
Herpes virus	Can insert into cells of nervous system; does not cause immune reaction	Hard to produce in large quantities
Lentivirus	Can accommodate large genes	Safety concerns
Liposomes and other lipid-coated vectors	No replication; does not stimulate immune reaction	Low efficiency
Direct injection	No replication; directed toward specific tissues	Low efficiency; does not work well within some tissues
Pressure treatment	Safe, because tissues are treated outside the body and then transplanted into the patient	Most efficient with small DNA molecules
Gene gun (DNA coated on small gold particles and shot into tissue)	No vector required	Low efficiency

Source: After E. Marshall, Gene therapy's growing pains, *Science* 269(1995):1050–1055.

therapies depend on an introduced gene's ability to produce a therapeutic protein. A number of different methods for transferring genes into human cells are currently under development. Commonly used vectors include genetically modified retroviruses, adenoviruses, and adeno-associated viruses (Table 18.6). One method of gene transfer is to remove cells (such as white blood cells) from a patient's body, add viruses containing recombinant genes, and then reintroduce the cells back into the patient's body. In other cases, vectors are injected directly into the body.

In spite of the growing number of clinical trials for gene therapy, significant problems remain in transferring foreign genes into human cells, getting them expressed, and limiting immune responses to the gene products and the vectors used to transfer the genes to the cells. There are also heightened concerns about safety, especially after the death in 1999 of a patient participating in a gene-therapy trial who had a fatal immune reaction after he was injected with a viral vector carrying a gene to treat his metabolic disorder. Despite this setback, gene-therapy research has moved ahead. Unequivocal results demonstrating positive benefits from gene therapy for a severe combined immunodeficiency disease and for head and neck cancer were announced in 2000.

Gene therapy conducted to date has targeted only nonreproductive, somatic cells. Correcting a genetic defect in these cells (termed *somatic gene therapy*) may provide positive benefits to patients but will not affect the genes of future generations. Gene therapy that alters reproductive, or germ-line, cells (termed *germ-line gene therapy*) is technically possible but raises a number of significant ethical issues, because it has the capacity to alter the gene pool of future generations.

Concepts

Gene therapy is the direct transfer of genes into humans to treat disease. Gene therapy was first successfully implemented in 1990 and is now being used to treat genetic diseases, cancer, and infectious diseases.

Gene Mapping

A significant contribution of recombinant DNA technology has been to provide numerous genetic markers that can be used in gene mapping. One group of markers used in gene mapping comprises **restriction fragment length polymorphisms** (RFLPs, pronounced rifflips). RFLPs are variations (polymorphisms) in the patterns of fragments produced when DNA molecules are cut with the same restriction enzyme. If DNA from two persons is cut with the same restriction enzyme and different patterns of fragments are produced (◀ FIGURE 18.26), these persons must possess differences in their DNA sequences. These differences are inherited and can be used in mapping, similar to the way in which allelic differences are used to map conventional genes.

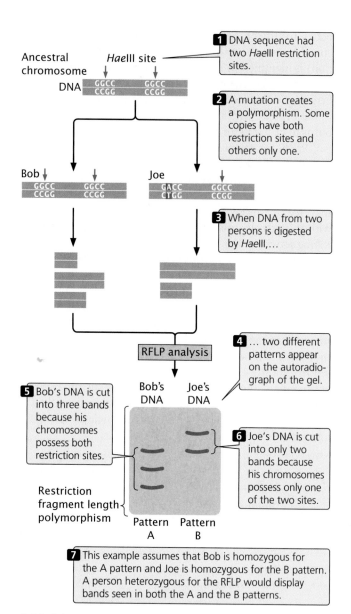

1 DNA sequence had two *Hae*III restriction sites.

2 A mutation creates a polymorphism. Some copies have both restriction sites and others only one.

3 When DNA from two persons is digested by *Hae*III,…

4 … two different patterns appear on the autoradiograph of the gel.

5 Bob's DNA is cut into three bands because his chromosomes possess both restriction sites.

6 Joe's DNA is cut into only two bands because his chromosomes possess only one of the two sites.

7 This example assumes that Bob is homozygous for the A pattern and Joe is homozygous for the B pattern. A person heterozygous for the RFLP would display bands seen in both the A and the B patterns.

◀ **18.26 Restriction fragment length polymorphisms are genetic markers that can be used in mapping.**

Traditionally, gene mapping has relied on the use of genetic differences that produce easily observable phenotypic differences. Unfortunately, because most traits are influenced by multiple genes and the environment, the number of traits with a simple genetic basis suitable for use in mapping is limited. RFLPs provide a large number of genetic markers that can be used in mapping.

To illustrate mapping with RFLPs, let's again consider Huntington disease. As mentioned earlier, this disease is caused by an autosomal dominant gene but, until recently, the chromosomal location of the gene was unknown. A team of scientists led by James Gusella (see introduction to Chapter 5) set out to determine the location of the Huntington gene, in the hope that, when the gene was found, its biochemical basis could be determined and possible treatments might be suggested. DNA was collected from members of the largest known family with Huntington disease, who live near Lake Maracaibo in Venezuela.

The basic strategy employed in the search for the Huntington-disease gene and a number of other human disease-causing genes is to look for coinheritance of the disease-causing gene and an RFLP with a known chromosomal location. If the disease gene and the RFLP have been inherited together, they must be physically linked.

This approach is summarized in ◀ FIGURE 18.27, which illustrates the coinheritance of two traits: (1) the presence or absence of Huntington disease and (2) the type of restriction pattern produced (pattern A or C). In the family shown, the father is heterozygous for Huntington disease (*Hh*) and is also heterozygous for a restriction pattern (*AC*). From the father, each child inherits either a Huntington-

disease allele (*H*) or a normal allele (*h*); any child inheriting the Huntington-disease allele develops the disease, because it is an autosomal dominant disorder. The child also inherits one of the two RFLP alleles from the father, either *A* or *C*, which produces the corresponding RFLP pattern. In ◀ FIGURE 18.27a, there is no correspondence between the inheritance of the RFLP pattern and the inheritance of the disease: children who have inherited Huntington disease (and therefore the *H* allele) from their father are equally likely to have inherited the A or C RFLP pattern. In this case, the *H* allele and the RFLP alleles segregate randomly, so we know that they are not closely linked.

◀ FIGURE 18.27b, on the other hand, shows that every child who inherits the C pattern from the father also inherits Huntington disease (and therefore the *H* allele), because the locus for the RFLP is closely linked to the locus for the disease-causing gene. The chromosomal location of the RFLP provides a general indication of the disease-causing locus. An examination of the cosegregation of other RFLPs from the same region can precisely determine the location of the gene. Actual RFLP patterns and part of the Huntington-disease gene are shown in ◀ FIGURE 18.28.

DNA Fingerprinting

Restriction fragment length polymorphisms are often found in noncoding regions of DNA and are therefore frequently quite variable in humans. Two randomly chosen people will differ at many RFLPs and, if enough RFLPs are examined, no two people (with the exception of identical twins) will be exactly the same. The use of DNA sequences to identify a person, called **DNA fingerprinting,** is a powerful tool for criminal investigations and other forensic applications.

In a typical application, DNA fingerprinting might be used to confirm that a suspect was present at the scene of a crime (◀ FIGURE 18.29). A sample of DNA from blood, semen, hair, or other body tissue is collected from the crime scene. If the sample is very small, PCR can be used to amplify it so that enough DNA is available for testing. Additional DNA samples are collected from one or more suspects.

Each DNA sample is cut with one or more restriction enzymes, and the resulting DNA fragments are separated by gel electrophoresis. The fragments in the gel are denatured and transferred to nitrocellulose paper by Southern blotting. One or more radioactive probes is then hybridized to the nitrocellulose and detected by autoradiography. The pattern of bands produced by DNA from the sample collected at the crime scene is then compared with the patterns produced by DNA from the suspects.

The probes used in DNA fingerprinting detect highly variable regions of the genome; so the chances of DNA from two people producing exactly the same banding pattern is low. When several probes are used in the analysis, the probability that two people have the same set of patterns becomes vanishingly small (unless they are identical twins). A match between the sample from the crime scene and one

(a)

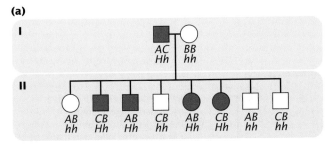

(b)

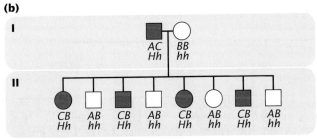

◀ **18.27 Restriction fragment length polymorphisms can be used to detect linkage.** In this hypothetical pedigree, the father and half of the children are affected (red circles and squares) with Huntington disease, an autosomal dominant disease. The father is heterozygous (*Hh*) and will pass the chromosome with the Huntington gene to approximately half of his offspring. The father is also heterozygous for RFLP alleles *A* and *C*; each child receives one of these two alleles from the father. The mother is homozygous for RFLP allele *B*, so all children receive the *B* allele from her. (a) In this case, there is no correspondence between the inheritance of the RFLP allele and inheritance of the disease—children with the disease are just as likely to carry the *A* allele as they are the *C* allele. Thus the disease gene and RFLP alleles segregate independently and are not closely linked. (b) In this case, there is a close correspondence between the inheritance of the RFLP alleles and the presence of the disease—every child who inherits the *C* allele from the father also has the disease. This correspondence indicates that the RFLP is closely linked to the Huntington gene.

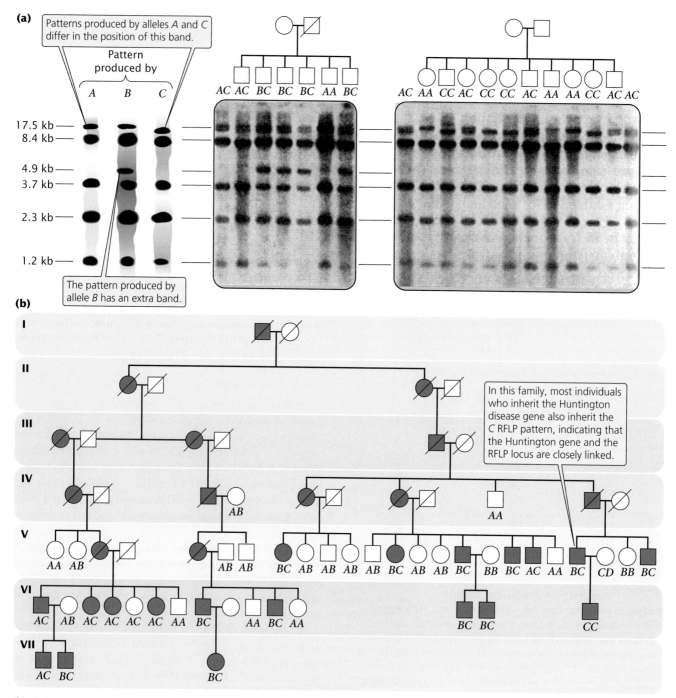

◀18.28 Restriction fragment length polymorphisms were used to map the Huntington-disease gene to chromosome 8.
(a) Autoradiograph showing different banding patterns revealed by cutting the DNA with *Hin*dIII and using a probe to chromosome 8. The RFLP *A* allele produces five bands. The *C* allele also produces five bands, but the first band is just below the first band produced by the *A* allele; *AC* heterozygotes have both bands, which are very close together. The *B* allele has an extra band representing a 4.9-kb fragment. (b) Partial pedigree of large family from Lake Maracaibo. Red symbols represent family members with Huntington disease; the RFLP genotypes are indicated below each person represented in the pedigree. Notice that persons with the disease carry the *C* allele, indicating that the sequences on chromosome 8 revealed by the probe are closely linked to the Huntington-disease gene.

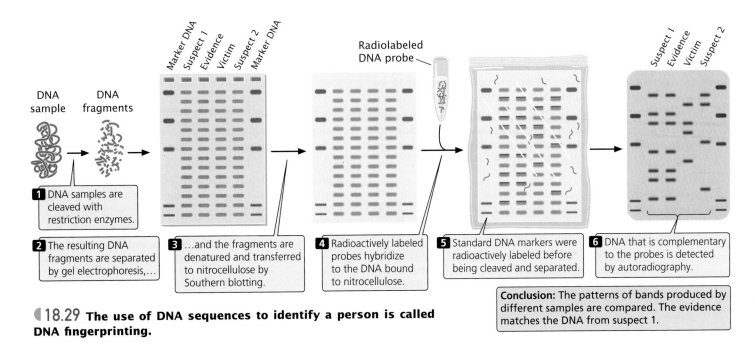

DNA sample

DNA fragments

1 DNA samples are cleaved with restriction enzymes.

2 The resulting DNA fragments are separated by gel electrophoresis,...

3 ...and the fragments are denatured and transferred to nitrocellulose by Southern blotting.

4 Radioactively labeled probes hybridize to the DNA bound to nitrocellulose.

Radiolabeled DNA probe

5 Standard DNA markers were radioactively labeled before being cleaved and separated.

6 DNA that is complementary to the probes is detected by autoradiography.

Conclusion: The patterns of bands produced by different samples are compared. The evidence matches the DNA from suspect 1.

◄ **18.29 The use of DNA sequences to identify a person is called DNA fingerprinting.**

from the suspect can provide evidence that the suspect was present at the scene of the crime.

The probes most commonly used in DNA fingerprinting are complementary to short sequences repeated in tandem that are widely found in the human genome (see pp. 300–301 in Chapter 11). People vary greatly in the number of copies of these repeats; thus, these polymorphisms are termed **variable number of tandem repeats** (VNTRs).

Since its introduction in the 1980s, DNA fingerprinting has helped convict a number of suspects in murder and rape cases. Suspects in other cases have been proved innocent when their DNA failed to match that from the crime scenes. Initially there was some controversy over calculating the odds of a match (the probability that two people could have the same pattern) and concerns about quality control (such as the accidental contamination of samples and the reproducibility of results) in laboratories where DNA analysis is done. In spite of the controversy, DNA fingerprinting has become an important tool in forensic investigations.

DNA fingerprinting has also been used to provide information about the relationships and sources of other organisms. For example, DNA fingerprinting was used to determine that several samples of anthrax mailed to different people in 2001 were all from the same source.

Concepts

RFLPs are variations in the pattern of fragments produced by restriction enzymes, which reveal variations in DNA sequences. They are used extensively in gene mapping. DNA fingerprinting detects genetic differences among people by using probes for highly variable regions of chromosomes.

www.whfreeman.com/pierce   More information on DNA fingerprinting

Concerns About Recombinant DNA Technology

In 1971, as researchers were planning some of the first gene-cloning experiments, in which they planned to transfer genes from tumor viruses to *E. coli*, several scientists raised concerns about the safety of such experiments. *E. coli* is present in the human intestinal tract, and these scientists questioned whether it might be possible for recombinant bacteria to escape from the laboratory and infect people, eventually transferring tumor-causing genes to people. The risks were thought to be small, but the real hazards were quite unknown.

When the first experiments using recombinant DNA were performed in 1973, concerns about risks associated with recombinant technology were heightened. Although no hazard had been demonstrated, a number of potential dangers could be envisioned. In July 1974, leading molecular biologists published a letter in *Science* urging scientists to stop conducting certain types of potentially hazardous recombinant DNA experiments until their risks could be evaluated. In February 1975, a group of more than 100 molecular biologists met and agreed that some restrictions on recombinant DNA research were warranted. They formulated a series of recommendations concerning the types of recombinant DNA experiments that should be prohibited.

The National Institutes of Health then appointed a committee to develop guidelines for recombinant DNA research. Different types of cloning experiments were considered to have different degrees of risk, and more precautions were required for the more "risky" experiments. The

Recombinant DNA Advisory Committee was established to oversee the safety of this work in the United States, and similar committees were established in Europe.

After years of experience with recombinant DNA experiments, the initial concerns about risks turned out to be largely unfounded, and the NIH guidelines have now been significantly relaxed. Current controversy about recombinant DNA technology revolves largely around the release of genetically modified organisms into the environment and the application of recombinant DNA technology to humans.

Connecting Concepts Across Chapters

This chapter has focused on recombinant DNA technology, a set of methods to isolate, study, and manipulate DNA sequences. Before the development of this technology, geneticists were forced to study genes by examining the phenotypes produced by the genes under study. The power of recombinant DNA technology is that it allows geneticists to read and alter genetic information directly, leading to an entirely new approach to the study of heredity in which genes are studied by altering DNA sequences and observing the associated change in phenotype.

A major theme of this chapter has been that working at the molecular level requires special approaches because DNA and other molecules are too small to see and manipulate directly. A number of recombinant DNA techniques are available and can be mixed and matched in different combinations or strategies; the particular set of methods used depends both on the sequences being manipulated and on the ultimate goal of the researcher.

Mastering the information in this chapter requires an understanding of material presented in many of the preceding chapters, particularly those on molecular genetics. A detailed understanding of DNA structure (Chapter 10), replication (Chapter 12), and the genetic code (Chapter 15) are essential for grasping the details of recombinant DNA technology. Knowledge of bacterial and viral genetics (Chapter 8) is helpful, because much of gene cloning takes place in bacteria, and plasmids and viruses are commonly used as cloning vectors. Knowledge of gene regulation (Chapter 16) is useful for understanding expression vectors and recombinant DNA applications where proteins are produced.

The information presented in this chapter will complement and enhance much of the material presented in the remaining chapters of the book. Chapter 19 deals with the use of recombinant DNA technology to compare the organization, content, and expression of genomes of different organisms.

CONCEPTS SUMMARY

- Recombinant DNA technology is a set of molecular techniques for locating, cutting, joining, analyzing, and altering DNA sequences and for inserting the sequences into a cell.

- Restriction endonucleases are enzymes that make double-stranded cuts in DNA at specific base sequences.

- DNA fragments can be separated with the use of gel electrophoresis and visualized by staining the gel with a dye that is specific for nucleic acids or by labeling the fragments with a radioactive or chemical tag.

- Individual genes can be studied by transferring DNA fragments from a gel to nitrocellulose or nylon and applying complementary probes.

- Gene cloning refers to placing a gene or a DNA fragment into a bacterial cell, where it will be multiplied as the cell divides.

- Plasmids, small circular pieces of DNA, are often used as vectors to ensure that a cloned gene is stable and replicated within the recipient cells.

- Bacteriophage λ offers several advantages over plasmids: it can hold larger fragments of foreign DNA and transfers DNA to cells with higher efficiency.

- Cosmids, which combine properties of plasmids and phage vectors, hold even larger amounts of foreign DNA. Yeast and bacterial artificial chromosomes can accommodate large inserts more than 100,000 bp in length.

- Expression vectors contain promoters, ribosome-binding sites, and other sequences necessary for foreign DNA to be transcribed and translated.

- Genes can be isolated by creating a DNA library, a set of bacterial colonies or viral plaques that each contain a different cloned fragment of DNA. A genomic library contains the entire genome of an organism, cloned as a set of overlapping fragments; a cDNA library contains DNA fragments complementary to all the different mRNAs in a cell.

- DNA libraries can be screened with probes complementary to particular genes or DNA fragments in the library can be cloned into an expression vector and screened by looking for the associated protein product.

- Genes can also be located by chromosome walking, in which a neighboring gene is used to make a probe; a genomic library is screened with this probe to find a clone that overlaps the gene. A probe is made from the end of this clone, and the probe is used to screen the library for a second clone that overlaps the first. The process is continued until the gene of interest is reached.

- The cloning strategy depends on the purpose of the cloning experiment, what is known about the gene, the size of the gene to be cloned, the size of the genome from which it is isolated, and the organism into which it will be cloned.

- The polymerase chain reaction is a method for amplifying DNA enzymatically without cloning. A solution containing DNA is heated, so that the two DNA strands separate, and then quickly cooled, allowing primers to attach to the template DNA. The solution is then heated again, and DNA polymerase synthesizes new strands from the primers. Each time the cycle is repeated, the amount of DNA doubles.

- In situ hybridization can be used to determine the chromosomal location of a gene and the distribution of the mRNA produced by a gene. DNA footprinting reveals the nucleotides that are covered by DNA-binding proteins. Site-directed mutagenesis can be used to produce mutations at specific sites in DNA, allowing genes to be tailored for a particular purpose. Transgenic animals, produced by injecting DNA into fertilized eggs, contain foreign DNA that is integrated into a chromosome. Knockout mice are transgenic mice that have a normal gene disabled.

- Recombinant DNA technology has many applications, including not only the production of pharmaceuticals and other biological substances in bacteria but also the creation of bacteria that are genetically engineered for economically or medically important tasks. It is also being used in agriculture to transfer particular traits, such as disease and pest resistance, to crop plants. Transgenic domestic animals can be produced with desirable traits. Oligonucleotide drugs—short nucleotide sequences for treating diseases—are another application of recombinant DNA technology.

- In gene therapy, diseases are being treated by altering the genes of human cells.

- Restriction fragment length polymorphisms and variable number tandem repeats facilitate gene mapping by making available numerous genetic markers and are being used to identify people by their DNA sequences (DNA fingerprinting).

IMPORTANT TERMS

recombinant DNA technology (p. 508)
genetic engineering (p. 508)
biotechnology (p. 508)
restriction enzyme (p. 510)
restriction endonuclease (p. 510)
cohesive end (p. 510)
gel electrophoresis (p. 513)
end labeling (p. 513)
autoradiography (p. 514)
probe (p. 514)
Southern blotting (p. 514)

Northern blotting (p. 514)
Western blotting (p. 514)
gene cloning (p. 514)
cloning vector (p. 515)
cosmid (p. 519)
expression vector (p. 520)
shuttle vector (p. 521)
yeast artificial chromosome (YAC) (p. 521)
bacterial artificial chromosome (BAC) (p. 521)
Ti plasmid (p. 521)
DNA library (p. 522)

genomic library (p. 522)
cDNA library (p. 522)
chromosome walking (p. 526)
cloning strategy (p. 526)
polymerase chain reaction (PCR) (p. 528)
Taq polymerase (p. 529)
in situ hybridization (p. 530)
DNA footprinting (p. 530)
site-directed mutagenesis (p. 531)
oligonucleotide-directed mutagenesis (p. 531)

transgene (p. 532)
knockout mice (p. 533)
gene therapy (p. 537)
restriction fragment length polymorphism (RFLP) (p. 538)
DNA fingerprinting (p. 539)
variable number of tandem repeats (VNTRs) (p. 541)

Worked Problems

1. A molecule of double-stranded DNA that is 5 million base pairs long has a base composition that is 62% G + C. How many times, on average, are the following restriction sites likely to be present in this DNA molecule?

(a) *Bam*HI (recognition sequence = GGATCC)
(b) *Hind*III (recognitions sequence = AAGCTT)
(c) *Hpa*II (recognition sequence = CCGG)

• Solution

The percentages of G and C are equal in double-stranded DNA; so, if G + C = 62%, then %G = %C = 62%/2 = 31%. The percentage of A + T = (100% − G + C) = 48%, and %A = %T = 48%/2 = 24%. To determine the probability of finding a

particular base sequence, we use the multiplicative rule, multiplying together the probably of finding each base at a particular site.

(a) The probability of finding the sequence GGATCC = 0.31 × 0.31 × 0.24 × 0.24 × 0.31 × 0.31 = 0.0005319. To determine the average number of recognition sequences in a 5-million-base-pair piece of DNA, we multiply 5,000,000 bp × 0.00053 = 2659.5 recognition sequences.

(b) The number of AAGCTT recognition sequences is 0.24 × 0.24 × 0.31 × 0.31 × 0.24 × 0.24 × 5,000,000 = 1594 recognition sequences.

(c) The number of CCGG recognition sequences is 0.31 × 0.31 × 0.31 × 0.31 × 5,000,000 = 46,176 recognition sequences.

2. A protein has the following amino acid sequence:

   Met-Leu-Arg-Ser-Arg-Met-Tyr-Trp-Asp-His-Glu-Thr

You wish to make a set of probes to screen a cDNA library for the sequence that encodes this protein. Your probes should be at least 18 nucleotides in length.

(a) Which amino acids in the protein should be used so that the smallest number of probes is required? (Consult the genetic code in Figure 15.12.)

(b) How many different sequences must be synthesized to be certain that you will find the correct cDNA sequence that specifies the protein?

• **Solution**

We first write out all the codons that can specify all the amino acids in the protein, using the genetic code in Figure 15.12 (see table below).

(a) The 18-bp region encoding amino acids 6 through 11 should be used, because this region has the fewest number of possible codons.

(b) For amino acids 6 through 11, there is one possible codon for Met, two for Tyr, one for Trp, two for Asp, two for His, and two for Glu. Thus $1 \times 2 \times 1 \times 2 \times 2 \times 2 = 16$ possible sequences must be synthesized to locate the gene.

1	2	3	4	5	6	7	8	9	10	11	12
Met	Leu	Arg	Ser	Arg	Met	Tyr	Trp	Asp	His	Glu	Thr
AUG	UUA	CGU	UCU	CGU	AUG	UAU	UGG	GAU	CAU	GAA	ACU
	UUG	CGC	UCC	CGC		UAC		GAC	CAC	GAG	ACC
	CUU	CGA	UCA	CGA							ACA
	CUC	CGG	UCG	CGG							ACG
	CUA	AGA	AGU	AGA							
	CUG	AGG	AGC	AGG							

The New Genetics
MINING GENOMES

RECOMBINANT DNA PROJECT

This exercise casts you in the role of research geneticist. Your job is to plan a project to clone a specific gene into a plasmid vector, including the selection of the restriction enzymes and vector that you will use. You will utilize the Web sites of some of the major suppliers of biotchnology reagents in the process.

COMPREHENSION QUESTIONS

1. List some of the effects and applications of recombinant DNA technology.

2. What common feature is seen in the sequences recognized by type II restriction enzymes?

3. What role do restriction enzymes play in bacteria? How do bacteria protect their own DNA from the action of restriction enzymes?

* 4. Explain how gel electrophoresis is used to separate DNA fragments of different lengths.

* 5. After DNA fragments are separated by gel electrophoresis, how can they be visualized?

6. What is the purpose of Southern blotting? How is it carried out?

* 7. What are the differences between Southern, Northern, and Western blotting?

* 8. Give three important characteristics of cloning vectors.

9. Briefly describe four different methods for inserting foreign DNA into plasmids, giving the strengths and weaknesses of each.

10. How are plasmids transferred into bacterial cells?

*11. Briefly explain how an antibiotic-resistance gene and the *lacZ* gene can be used as markers to determine which cells contain a particular plasmid.

12. How are genes inserted into bacteriophage λ vectors? What advantages do λ vectors have over plasmids?

*13. What is a cosmid? What are the advantages of using cosmids as gene vectors?

14. What are yeast artificial chromosomes and shuttle vectors? When are these cloning vectors used?

*15. How does a genomic library differ from a cDNA library? How is each created?

16. How are probes used to screen DNA libraries? Explain how a synthetic probe can be prepared when the protein product of a gene is known.

17. Explain how chromosome walking can be used to find a gene.

18. Discuss some of the considerations that must go into developing an appropriate cloning strategy.

*19. Briefly explain how the polymerase chain reaction is used to amplify a specific DNA sequence. What are some of the limitations of PCR?

*20. Briefly explain in situ hybridization, giving some applications of this technique.

21. What is DNA footprinting?

22. Briefly explain how site-directed mutagenesis is carried out.

*23. What are knockout mice, how are they produced, and for what are they used?

24. Describe how RFLPs can be used in gene mapping.

*25. What is DNA fingerprinting? What types of sequences are examined in DNA fingerprinting?

26. What is gene therapy?

27. As the first recombinant DNA experiments were being carried out, there was concern among some scientists about this research. What were these concerns and how were they addressed?

APPLICATION QUESTIONS AND PROBLEMS

*28. Suppose that a geneticist discovers a new restriction enzyme in the bacterium *Aeromonas ranidae*. This restriction enzyme is the first to be isolated from this bacterial species. Using the standard convention for abbreviating restriction enzymes, give this new restriction enzyme a name (for help, see footnote to Table 18.2).

29. How often, on average, would you expect a type II restriction endonuclease to cut a DNA molecule if the recognition sequence for the enzyme had 5 bp? (Assume that the four types of bases are equally likely to be found in the DNA and that the bases in a recognition sequence are independent.) How often would the endonuclease cut the DNA if the recognition sequence had 8 bp?

*30. A microbiologist discovers a new type II restriction endonuclease. When DNA is digested by this enzyme, fragments that average 1,048,500 bp in length are produced. What is the most likely number of base pairs in the recognition sequence of this enzyme?

31. Will restriction sites for an enzyme that has 4 bp in its restriction site be closer together, farther apart, or similarly spaced, on average, compared with those of an enzyme that has 6 bp in its restriction site? Explain your reasoning.

*32. About 60% of the base pairs in a human DNA molecule are AT. If the human genome has 3 billion base pairs of DNA, about how many times will the following restriction sites be present?

(a) *Bam*HI (restriction site = 5′–GGATCC–3′)

(b) *Eco*RI (restriction site = 5′–GAATTC–3′)

(c) *Hae*III (restriction site = 5′–GGCC–3′)

*33. Restriction mapping of a linear piece of DNA reveals the following *Eco*RI restriction sites.

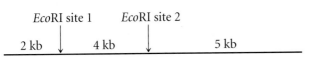

(a) This piece of DNA is cut by *Eco*RI, the resulting fragments are separated by gel electrophoresis, and the gel is stained with ethidium bromide. Draw a picture of the bands that will appear on the gel.

(b) If a mutation that alters *Eco*RI site 1 occurs in this piece of DNA, how will the banding pattern on the gel differ from the one that you drew in part *a*?

(c) If mutations that alter *Eco*RI sites 1 and 2 occur in this piece of DNA, how will the banding pattern on the gel differ from the one that you drew in part *a*?

(d) If a 1000-bp insertion occurred between the two restriction sites, how would the banding pattern on the gel differ from the one that you drew in part *a*?

(e) If a 500-bp deletion occurred between the two restriction sites, how would the banding pattern on the gel differ from the one that you drew in part *a*?

*34. Which vectors (plasmid, phage λ, cosmid) can be used to clone a continuous fragment of DNA with the following lengths?

(a) 4 kb.

(b) 20 kb.

(c) 35 kb.

35. A geneticist uses a plasmid for cloning that has a gene that confers resistance to penicillin and the *lacZ* gene. The geneticist inserts a piece of foreign DNA into a restriction site that is located within the *lacZ* gene and transforms bacteria with the plasmid. Explain how the geneticist can identify bacteria that contain a copy of a plasmid with the foreign DNA.

*36. Suppose that you have just graduated from college and have started working at a biotechnology firm. Your first job assignment is to clone the pig gene for the hormone prolactin. Assume that the pig gene for prolactin has not yet been isolated, sequenced, or mapped; however, the mouse gene for prolactin has been cloned and the amino acid sequence of mouse prolactin is known. Briefly explain two different strategies that you might use to find and clone the pig gene for prolactin.

37. A genetic engineer wants to isolate a gene from a scorpion that encodes the deadly toxin found in its stinger, with the

ultimate purpose of transferring this gene to bacteria and producing the toxin for use as a commercial pesticide. Isolating the gene requires a DNA library. Should the genetic engineer create a genomic library or a cDNA library? Explain your reasoning.

*38. A protein has the following amino acid sequence:

Met-Tyr-Asn-Val-Arg-Val-Tyr-Lys-
                    Ala-Lys-Trp-Leu-Ile-His-Thr-Pro

You wish to make a set of probes to screen a cDNA library for the sequence that encodes this protein. Your probes should be at least 18 nucleotides in length.

(a) Which amino acids in the protein should be used to construct the probes so that the least degeneracy results? (Consult the genetic code in Table 15.12.)

(b) How many different probes must be synthesized to be certain that you will find the correct cDNA sequence that specifies the protein?

*39. A gene in mice is discovered that is similar to a gene in yeast. How might it be determined whether this gene is essential for development in mice?

*40. A hypothetical disorder called G syndrome is an autosomal dominant disease characterized by visual, skeletal, and cardiovascular defects. The disorder appears in middle age. Because the symptoms of the disorder are variable, the disorder is difficult to diagnose. Early diagnosis is important, however, because the cardiovascular symptoms can be treated if the disorder is recognized early. The gene for G syndrome is known to reside on chromosome 7, and

it is closely linked to two RFLPs on the same chromosome, one at the A locus and one at the C locus. The genes at the G, A, and C loci are very close together, and there is little crossing over between them. The following RFLP alleles are found at the A and C loci:

A locus: A1, A2, A3, A4
C locus: C1, C2, C3

Sally, shown in the following pedigree, is concerned that she might have G syndrome. Her deceased mother had G syndrome, and she has a brother with the disorder. A geneticist genotypes Sally and her immediate family for the A and C loci and obtains the genotypes shown on the pedigree.

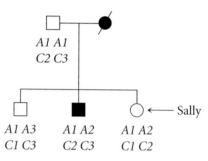

(a) Assume that there is no crossing over between the A, C, and G loci. Does Sally carry the gene that causes G syndrome? Explain why or why not?
(b) Draw the arrangement of the A, C, and G alleles on the chromosomes for all members of the family.

CHALLENGE QUESTIONS

41. Suppose that you are hired by a biotechnology firm to produce a giant strain of fruit flies, by using recombinant DNA technology, so that genetics students will not be forced to strain their eyes when looking at tiny flies. You go to the library and learn that growth in fruit flies is normally inhibited by a hormone called shorty substance P (SSP). You decide that you can produce giant fruit flies if you can somehow turn off the production of SSP. SSP is synthesized from a compound called XSP in a single-step reaction catalyzed by the enzyme *runtase*:

$$XSP \xrightarrow{\text{runtase}} SSP$$

A researcher has already isolated cDNA for runtase and has sequenced it, but the location of the runtase gene in the *Drosophila* genome is unknown.

In attempting to devise a strategy for turning off the production of SSP and producing giant flies by using

standard recombinant DNA techniques, you discover that deleting, inactivating, or otherwise mutating this DNA sequence in *Drosophila* turns out to be extremely difficult. Therefore you must restrict your genetic engineering to gene augmentation (adding new genes to cells). Describe the methods that you will use to turn off SSP and produce giant flies by using recombinant DNA technology.

42. A rare form of polydactyly (extra fingers and toes) in humans is due to an X-linked recessive gene, whose chromosomal location is unknown. Suppose a geneticist studies the family whose pedigree is shown here. She isolates DNA from each member of this family, cuts the DNA with a restriction enzyme, separates the resulting fragments by gel electrophoresis, and transfers the DNA to nitrocellulose by Southern blotting. She then hybridizes the nitrocellulose with a cloned DNA sequence that comes from the X chromosome. The pattern of bands

that appear on the autoradiograph is shown below each person in the pedigree.

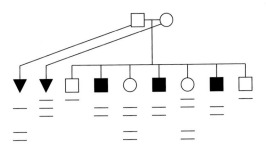

(a) For each person in the pedigree, give his or her genotype for RFLPs revealed by the probe. (Remember that males are hemizygous for X-linked genes, and females can be homozygous or heterozygous.)

(b) Is there evidence for close linkage between the probe sequence and the X-linked gene for polydactyly? Explain your reasoning.

(c) How many of the daughters in the pedigree are likely to be carriers of X-linked polydactyly? Explain your reasoning.

SUGGESTED READINGS

Andrews, L. B., J. E. Fullarton, N. A. Holtzman, and A.G. Molulsky. 1994. *Assessing Genetic Risks: Implications for Health and Social Policy*. Washington, DC: National Academy Press.
Discusses some of the legal, ethical, and social issues surrounding gene testing.

Berg, P., D. Baltimore, H. W. Boyer, S. N. Cohen, R. W. Davis, D. S. Hogness, D. Nathans, R. Roblin, J. D. Watson, S. Weissman, and N. D. Zinder. 1974. Potential biohazards of recombinant DNA molecules. *Science* 185:303.
Well-known letter calling for a moratorium on certain types of recombinant DNA experiments.

Cohen, J. S., and M. E. Hogan. 1994. The new genetic medicine. *Scientific American* 271(6):76–82.
A good review of oligonucleotide drugs.

Cohen, S., A. Chang, H. Boyer, and R. Helling. 1973. Construction of biologically functional bacterial plasmids in vitro. *Proceedings of the National Academy of Sciences of the United States of America* 70:3240–3244.
Description of the first gene-cloning experiments.

Enriquez, J. 1998. Genomics and the world's economy. *Science* 281:925–926.
Discussion of the growing importance of gene sequencing and biotechnology in the world economy.

Friedmann, T. 1997. Overcoming the obstacles to gene therapy. *Scientific American* 276(6):96–101.
Review of some of the current problems in gene therapy.

Gasser, C. S., and R. T. Fraley. 1992. Transgenic crops. *Scientific American* 266(6):62–69.
An excellent and readable account of how genes are put into plants and some of the applications.

Isner, J. M. 2002. Myocardial gene therapy. *Nature* 415:234–239.
Discusses recent research on the use of gene therapy to treat coronary artery disease and heart failure.

Mullis, K. B. 1990. The unusual origin of the polymerase chain reaction. *Scientific American* 262(4):56–65.
Mullis describes his inspiration for the polymerase chain reaction.

Nowak, R. 1994. Forensic DNA goes to court with O. J. *Science* 265:1352–1354.
A report on the use of DNA fingerprinting in the famed O. J. Simpson trial.

Roberts, L. 1992. Science in court: a culture clash. *Science* 257:732–736.
A report on some of the controversy surrounding DNA fingerprinting.

Salo, W. L., A. C. Aufderheide, J. Buikstra, and T. A. Holcomb. 1994. Identification of *Mycobacterium tuberculosis* DNA in a pre-Columbian Peruvian mummy. *Proceedings of the National Academy of Sciences of the United States of America* 91:2091–2094.
Report of tuberculosis-causing bacteria in a 1000-year-old Peruvian mummy.

Stein, C. A., and Y. C. Cheng. 1993. Antisense oligonucleotides as therapeutic agents: is the bullet really magical? *Science* 261:1004–1011.
A review of the development of oligonucleotide drugs.

Verma, I. M., and J. Somia. 1997. Gene therapy: promises, problems, and prospects. *Nature* 389:239–242.
A good review of the state of gene therapy.

Wofenbarger, L. L., and P. R. Phifer. 2000. The ecological risks and benefits of genetically engineered plants. *Science* 290:2088–2093.
A review of scientific evidence of benefits and risks associated with the use of genetically engineered organisms in agriculture.

Yan, H., K. W. Kinzler, and B. Vogelstein. 2000. Genetic testing: present and future. *Science* 289:1890–1892.
A review of some of the problems associated with genetic testing and current techniques being developed to overcome them.

Zanjani, E. D., and W. F. Anderson. 1999. Prospects for in utero human therapy. *Science* 285:2084–2088.
A review of the potential use of gene therapy in utero on unborn fetuses to correct human genetic defects.

19 Genomics

A young girl with leprosy, a disease caused by *Mycobacterium leprae*. Leprosy causes characteristic patches of skin with a loss of sensation, as seen on the face of this young patient; if untreated, leprosy may lead to nerve damage and disfigurement. Genomic studies reveal the genome of *M. leprae* has undergone extensive gene loss, mutation, and rearrangement over evolutionary time. (WHO/OMS.)

The Decaying Genome of *Mycobacterium leprae*

Leprosy, one of the most feared diseases of history, was well known in ancient times and is still a major public health problem today; from 2 million to 3 million people are affected worldwide, and approximately 650,000 new cases are reported each year. In its severest form, leprosy causes paralysis, blindness, and disfigurement. Although human genes play some role in susceptibility to leprosy, the disease is caused by the bacterium *Mycobacterium leprae*, which infects cells of the nervous system and causes nerve damage, sensory loss, and disfigurement. In 1873, Armauer Hansen observed these bacteria in tissue samples taken from people with leprosy, but to this day no one has successfully cultured the bacterium in laboratory media, severely restricting the study of the disease agent.

In 2001, scientists in Britain and France determined the sequence of the entire genome of *M. leprae*. Comparing its genome with that of its close relative *M. tuberculosis* (the pathogen that causes tuberculosis) and other mycobacteria has been a source of important insight into the unique properties of this pathogen.

The genome of *M. leprae* is 3,268,203 bp in size, 1 million base pairs smaller than the genomes of other mycobacteria. In most bacterial genomes, the vast majority of the DNA encodes proteins—there is little noncoding DNA between genes. In contrast, only 50% of the DNA of *M. leprae* encodes proteins (Table 19.1), and *M. leprae* has 2300 fewer genes than *M. tuberculosis*. An incredible 27% of *M. leprae*'s genome consists of pseudogenes—nonfunctional copies of genes that have been inactivated by mutations. *M. leprae* has 1116 pseudogenes, whereas its close relative, *M. tuberculosis*, has just 6.

Table 19.1	Comparison of the genomes of *Mycobacterium leprae*, which causes leprosy, and *Mycobacterium tuberculosis*, which causes tuberculosis		
Characteristics		**M. leprae**	**M. tuberculosis**
Genome size (bp)		3,268,203	4,411,532
Percentage of genome that encodes proteins		49.5%	90.8%
Protein-encoding genes (bp)		1604	3959
Pseudogenes (bp)		1116	6
Gene density (bp/gene)		2037	1114
Average length of gene (bp)		1011	1012

Source: S. T. Cole et al., Massive gene decay in the leprosy bacillus, *Nature* 409 (2001), p. 1007.

The reduced DNA content, fewer functional genes, and the large number of pseudogenes suggest that, evolutionarily, the genome of *M. leprae* has undergone massive decay through time, losing DNA and acquiring mutations that have inactivated many of its genes. Furthermore, the genome of *M. leprae* has undergone extensive rearrangement; comparison with the genome of *M. tuberculosis* has identified at least 65 gene segments that are arranged in different order and distribution.

The mechanisms responsible for gene decay and genomic rearrangement in *M. leprae* are not known, although the loss of proofreading ability in the bacterium's DNA polymerase III (the enzyme responsible for most bacterial DNA replication, see Chapter 12) may contribute to a high rate of mutation and the large number of pseudogenes. Because the leprosy bacterium resides in a highly specialized habitat (human nerve cells), it may have lost the need for many enzymatic functions found in other bacteria. When a function is no longer required for survival, genes encoding that function usually accumulate mutations and deletions.

Regardless of the mechanism for gene inactivation and loss, this genomic decay helps explain some of the bacterium's unique properties. Genes for many metabolic enzymes and structural proteins have been lost, which may explain why the bacterium cannot be cultured on synthetic media containing traditional carbon sources; it may also account for the bacterium's slow growth, with a doubling time of 14 days, compared with a doubling time of 20 minutes for *E. coli.*

A comparison of *M. leprae*'s genome with those of other related bacteria has identified a few unique genes that may contribute to its pathogenesis. The study of these genes has opened the door to an improved understanding of leprosy, better diagnostic tests, and the development of new drugs for the disease.

The information gleaned from sequencing the genome of *M. leprae* illustrates the power of genomics, which is the focus of this chapter. **Genomics** is the field of genetics that attempts to understand the content, organization, function, and evolution of genetic information contained in whole genomes. Genomics consists of two complementary fields: structural genomics and functional genomics. **Structural genomics** determines the organization and sequence of the genetic information contained within a genome, and **functional genomics** characterizes the function of sequences elucidated by structural genomics. A third area, **comparative genomics,** compares the gene content, function, and organization of genomes of different organisms.

The field of genomics is at the cutting edge of modern biology; information resulting from research in this field has made significant contributions to human health, agriculture, and numerous other areas. It has also provided gene sequences necessary for producing medically important proteins through recombinant DNA technology. Comparisons of genome sequences from different organisms are leading to a better understanding of evolution and the history of life.

We begin this chapter by examining genetic and physical maps and methods for sequencing entire genomes. Next, we explore functional genomics—how genes are identified in genomic sequences and how their functions are defined. Some of the genomes that have been sequenced are then examined in detail. We end the chapter by briefly considering the future of genomics.

Concepts

The field of genomics comprises structural genomics, which focuses on the content and organization of genomic information, and functional genomics, which attempts to understand the function of information in genomes. Comparative genomics compares the content and organization of genomes of different organisms. Genomics makes important contributions to human health, agriculture, biotechnology, and our understanding of evolution.

www.whfreeman.com/pierce   Internet sources of information on leprosy

Structural Genomics

Structural genomics is concerned with sequencing and understanding the content of genomes. Often, one of the first steps in characterizing a genome is to prepare genetic and physical maps of its chromosomes. These maps provide information about the relative locations of genes, molecular markers, and chromosome segments, which are often essential for positioning chromosome segments and aligning stretches of sequenced DNA into a whole-genome sequence.

Genetic Maps

Everyone has used a map at one time or another. Maps are indispensable for finding a new friend's house, the way to an unfamiliar city in your state, or the location of a country on the globe. Each of these examples requires a map with a different scale. For finding a friend's house, you would probably use a city street map; for finding your way to an unknown city, you might pick up a state highway map; for finding a country such as Kazakhstan, you would need a world atlas. Similarly, navigating a genome requires maps of different types and scales.

Genetic maps (also called linkage maps) provide a rough approximation of the locations of genes relative to the locations of other known genes (◀ FIGURE 19.1). These maps are based on the *genetic* function of recombination (hence the name genetic map). The basic principles of constructing genetic maps are discussed in detail in Chapter 7. In short, individuals heterozygous at two or more genetic loci are crossed, and the frequency of recombination between loci is determined by examining the progeny. If the recombination frequency between two loci is 50%, then the loci are located on different chromosomes or are far apart on the same chromosome. If the recombination frequency is less than 50%, the loci are located close together on the same chromosome (they belong to the same linkage group). For linked genes, the rate of recombination is proportional to the physical distance between the loci. Distances on genetic maps are measured in percent recombination (centimorgans, cM) or map units. Data from multiple two-point or three-point crosses can be integrated into linkage maps for whole chromosomes.

For many years, genes could be detected only by observing their influence on a trait (the phenotype), and construction of genetic maps was limited by the availability of single-locus traits that could be examined for evidence of recombination. Eventually, this limitation was overcome by the development of molecular techniques such as analysis of restriction fragment length polymorphisms, the polymerase chain reaction, and DNA sequencing (see Chapter 18) that are able to provide molecular markers that can be used to construct and refine genetic maps.

Genetic maps have several limitations, the first of which is resolution or detail. The human genome includes 3.4 billion base pairs of DNA and has a total genetic distance of about 4000 cM, an average of 850,000 bp/cM.

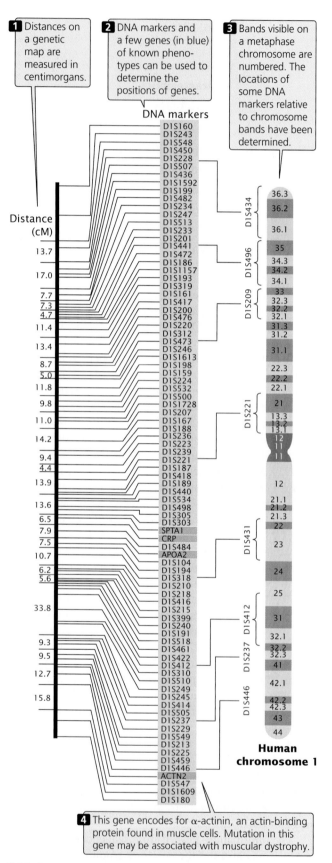

1 Distances on a genetic map are measured in centimorgans.

2 DNA markers and a few genes (in blue) of known phenotypes can be used to determine the positions of genes.

3 Bands visible on a metaphase chromosome are numbered. The locations of some DNA markers relative to chromosome bands have been determined.

4 This gene encodes for α-actinin, an actin-binding protein found in muscle cells. Mutation in this gene may be associated with muscular dystrophy.

Human chromosome 1

◀ **19.1 Genetic maps are based on rates of recombination.** Shown here is a genetic map of human chromosome 1.

Even if a marker occurred every centimorgan (which is unrealistic), the resolution in regard to the physical structure of the DNA would still be quite low. In other words, the detail of the map is very limited. A second problem with genetic maps is that they do not always accurately correspond to physical distances between genes. Genetic maps are based on rates of crossing over, which vary somewhat from one part of a chromosome to another; so the distances on a genetic map are only approximations of real physical distances along a chromosome. ◀ FIGURE 19.2 compares the genetic map of chromosome III of yeast with a physical map determined by DNA sequencing. There are some discrepancies between the distances and even among the positions of some genes. In spite of these limitations, genetic maps have been critical to the development of physical maps and the sequencing of whole genomes.

Physical Maps

Physical maps are based on the direct analysis of DNA, and they place genes in relation to distances measured in number of base pairs, kilobases, or megabases (◀ FIGURE 19.3). A common type of physical map is one that connects isolated pieces of genomic DNA that have been cloned in bacteria or yeast. Physical maps generally have higher resolution and are more accurate than genetic maps. A physical map is analogous to a neighborhood map that shows the location of every house along a street, whereas a genetic map is analogous to a highway map that shows the locations of major towns and cities.

A number of techniques exist for creating physical maps, including restriction mapping, which determines the positions of restriction sites on DNA; sequence-tagged site (STS) mapping, which locates the positions of short unique

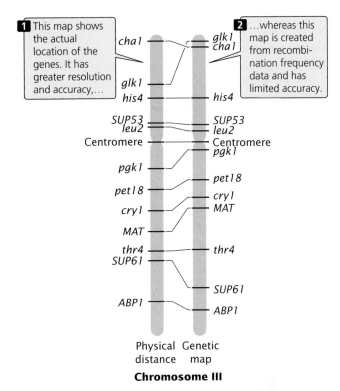

1 This map shows the actual location of the genes. It has greater resolution and accuracy,…

2 …whereas this map is created from recombination frequency data and has limited accuracy.

Physical distance   Genetic map

Chromosome III

◀ 19.2 **Genetic and physical maps may differ in relative distances and even in the position of genes on a chromosome.** Genetic and physical maps of yeast chromosome III reveal such differences.

sequences of DNA on a chromosome; fluorescent in situ hybridization (FISH), by which markers can be visually mapped to locations on chromosomes (see Figure 7.22); and DNA sequencing.

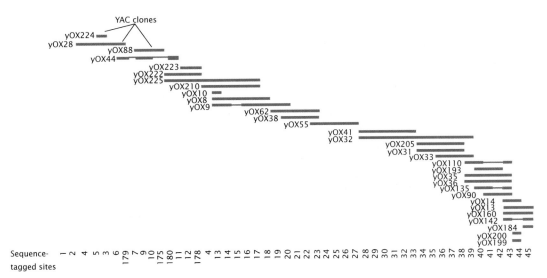

◀ 19.3 **Physical maps are often used to order cloned DNA fragments.** A part of a physical map of a set of overlapping YAC clones from one end of the human Y chromosome.

Both genetic and physical maps provide information about the relative positions and distances between genes, molecular markers, and chromosome segments. Genetic maps are based on rates of recombination and are measured in percent recombination, or centimorgans. Physical maps are based on the physical distances and are measured in base pairs.

Restriction mapping determines the relative positions of restriction sites on a piece of DNA. When a piece of DNA is cut with a restriction enzyme and the fragments are separated by gel electrophoresis, the number of restriction sites in the DNA and the distances between them can be determined by the number and positions of bands on the gel (pp. 513–514 in Chapter 18), but this information does not tell us the order or the precise location of the restriction sites. To map restriction sites, a sample of the DNA is cut with one restriction enzyme, and another sample is cut with a different restriction enzyme. A third sample is cut with both restriction enzymes together (a double digest). The DNA fragments produced by these restriction digests are then separated by gel electrophoresis, and their sizes are compared. Overlap in size of fragments produced by the digests can be used to position the restriction sites on the original DNA molecule.

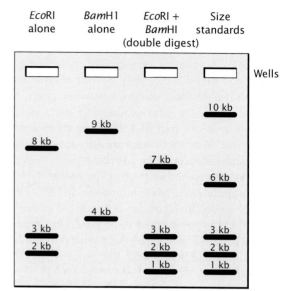

19.4 Restriction sites can be mapped by comparing DNA fragments produced by digestion by restriction enzymes used alone and in various combinations. A sample of a linear piece of DNA was first digested with *Eco*RI alone. Another sample was digested by *Bam*HI alone, and finally a third sample was digested by both *Eco*RI and *Bam*HI. The resulting fragments were separated by gel electrophoresis and stained with ethidium bromide.

Worked Problem

One sample of a linear 13,000-bp (13-kb) DNA fragment is cut with the restriction enzyme *Eco*RI; a second sample of the same DNA is cut with *Bam*HI; and a third sample is cut with *both* *Eco*RI and *Bam*HI together. The resulting fragments are separated and sized by gel electrophoresis (◀ FIGURE 19.4). Determine the positions of the *Eco*RI and *Bam*HI restriction sites on the original 13-kb fragment.

• **Solution**

Using the sizes of the fragments produced from the three digests in Figure 19.4, we can order the positions of the restriction sites on the original 13-kb piece of DNA. First, note that digestion with *Eco*RI alone produced 8-kb, 3-kb, and 2-kb fragments, indicating that there are two *Eco*RI restriction sites in the original linear piece of DNA. Digestion

with *Bam*HI produced 9-kb and 4-kb fragments, indicating that there is only one *Bam*HI site. The *Bam*HI restriction site must be 9 kb from one end and 4 kb from the other end.

The double digest produced four pieces of DNA: 7-kb, 3-kb, 2-kb, and 1-kb fragments. Neither of the fragments generated by *Bam*HI alone is present in the double digest, and so *Eco*RI must have cut both of the *Bam*HI fragments. Consider the 9-kb fragment. How could this fragment be cut by *Eco*RI to produce the fragments found in the double digest? Two of the fragments produced by the double digest, the 7-kb and 2-kb fragments, add up to 9 kb, the length of one fragment produced by digestion by *Bam*HI alone. Similarly, the 3-kb fragment and the 1-kb fragment of the double digest add up to 4 kb, the length of the other fragment produced by *Bam*HI alone. Therefore, *Eco*RI cut the first *Bam*HI fragment into 7-kb and 2-kb fragments and cut the second *Bam*HI fragment into 3-kb and 1-kb fragments:

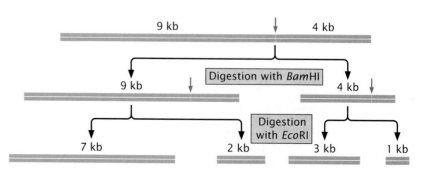

We now know that the *Bam*HI site lies between these two *Eco*RI sites. Considering the four fragments produced by the double digest, there are several possible arrangements by which a *Bam*HI site could fit in between the two *Eco*RI sites. To determine which of the arrangements is correct, compare the results of the *Eco*RI digestion with the double digest. When the original 13-kb DNA fragment was cut by *Eco*RI alone, the three fragments produced were 8 kb, 3 kb, and 2 kb in length. The 2-kb and 3-kb bands are also present in the double digest, indicating that these fragments do not contain a *Bam*HI site. The 8-kb fragment present in the *Eco*RI digest disappears in the double digest and is replaced by the 7-kb fragment and the 1-kb fragment, indicating that the 8-kb fragment has the *Bam*HI site. Thus the 7-kb and 1-kb fragments must lie next to each other, and the 2-kb and 3-kb fragments are on the ends. Thus, the correction arrangement of the restriction sites is:

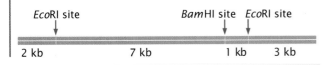

In the example in the Worked Problem, we can map the restriction sites in our heads or with a few simple sketches. Most restriction mapping is done with several restriction enzymes, used alone and in various combinations, producing many restriction fragments. With long pieces of DNA (greater than 30 kb), computer programs are used to determine the restriction maps, and restriction mapping may be facilitated by tagging one end of a large DNA fragment with radioactivity or by identifying the end with the use of a probe.

Physical maps, such as restriction maps of DNA fragments or even whole chromosomes, are often created for genomic analysis. These lengthy maps are often put together by combining maps of shorter, overlapping genomic fragments.

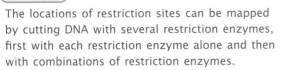

Concepts

The locations of restriction sites can be mapped by cutting DNA with several restriction enzymes, first with each restriction enzyme alone and then with combinations of restriction enzymes.

DNA-Sequencing Methods

The most detailed physical maps are based on direct **DNA sequencing** information. The first methods for quickly sequencing DNA were developed between 1975 and 1977. Frederick Sanger and his colleagues created the dideoxy sequencing method based on the elongation of DNA; Allan

Maxam and Walter Gilbert developed a second method based on the chemical degradation of DNA. The Sanger method quickly became the standard procedure for sequencing any purified fragment of DNA.

The Sanger, or dideoxy, method of DNA sequencing is based on the process of replication. The fragment to be sequenced is used as a template to make a series of new DNA molecules. In the process, replication is sometimes (but not always) terminated when a specific base is encountered, producing DNA strands of different length, each of which ends in the same base.

The method relies on the use of a special substrate for DNA synthesis. Normally, DNA is synthesized from deoxyribonucleoside triphosphates (dNTPs), which have an OH group on the 3′-carbon atom (◀FIGURE 19.5a). In DNA synthesis, two phosphate groups on the 5′-carbon atom of a dNTP are removed, and a phosphodiester bond is formed between the remaining 5′-phosphate group of the dNTP and the 3′-OH group of the last nucleotide on the growing DNA chain (see pp. 328–329 in Chapter 12). In the Sanger method, a special nucleotide, called a **dideoxyribonucleoside triphosphate** (ddNTP; ◀FIGURE 19.5b), is used as substrate. The ddNTPs are identical with dNTPs, except that they lack a 3′-OH group. Like dNTPs, ddNTPs possess three phosphate groups on their 5′ ends, and so they are incorporated into a growing DNA chain. When a ddNTP has been incorporated into a DNA chain, however, no more nucleotides can be added, because there is

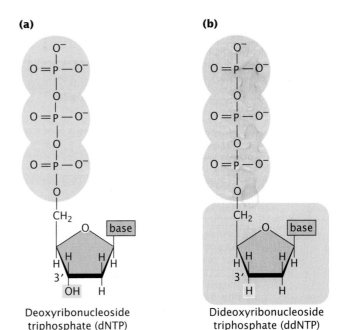

◀19.5 **The dideoxy sequencing reaction requires a special substrate for DNA synthesis.** (a) Structure of deoxyribonucleoside triphosphate, the normal substrate for DNA synthesis. (b) Structure of dideoxyribonucleoside triphosphate, which lacks an OH group on the 3′ carbon atom.

no 3'-OH group to form a phosphodiester bond with an incoming nucleotide. Thus, ddNTPs terminate DNA synthesis.

A single DNA molecule cannot be sequenced; so any DNA fragment to be sequenced must first be amplified by PCR or by cloning in bacteria. Copies of the target DNA are isolated and split into four parts (◀ FIGURE 19.6). Each part is placed in a different tube, to which are added:

1. many copies of a primer that is complementary to one end of the target DNA strand;

2. all four deoxyribonucleoside triphosphates (dCTP, dATP, dGTP, and dTTP), the normal precursors of DNA synthesis;

3. a small amount of *one* of the four types of dideoxyribonucleoside triphosphates (ddCTP, ddATP, ddGTP, *or* ddTTP), which will terminate DNA synthesis as soon as it is incorporated into any growing chain (each of the four tubes received a different ddNTP); and

4. DNA polymerase.

Either the primer or one of the dNTPs is radioactively or chemically labeled so that newly produced DNA can be detected.

Within each of the four tubes, the DNA polymerase enzyme carries out DNA synthesis. Let's consider the reaction in one of the four tubes; the one that received ddATP. Within this tube, each of the single strands of target DNA

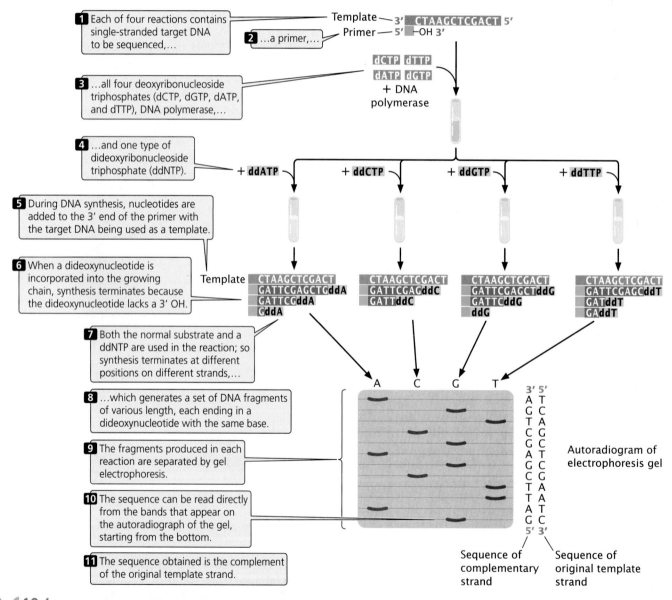

 ◀ 19.6 **The dideoxy method of DNA sequencing is based on the termination of DNA synthesis**.

serves as a template for DNA synthesis. The primer pairs to its complementary sequence at one end of each template strand, providing a 3′-OH group for the initiation of DNA synthesis. DNA polymerase elongates a new strand of DNA from this primer, by using the target DNA strand as a template. Wherever DNA polymerase encounters a T on the template strand, it uses at random either a dATP or a ddATP to introduce an A in the newly synthesized strand. Because there is more dATP than ddATP in the reaction mixture, dATP is incorporated most often, allowing DNA synthesis to continue. Occasionally, however, ddATP is incorporated into the strand and synthesis terminates. The incorporation of ddA into the new strand occurs randomly at different positions in different copies, producing a set of DNA chains of different length (12, 7, and 2 nucleotides long in the example illustrated in Figure 19.6), each ending in a nucleotide with adenine.

Equivalent reactions take place in the other three tubes. In the tube that received ddCTP, all the chains terminate in a nucleotide with cytosine; in the tube that received ddGTP, all the chains terminate in a nucleotide with guanine; and, in the tube that received ddTTP, all the chains terminate in a nucleotide with thymine. After the completion of the polymerization reactions, all the DNA in the tubes is denatured, and the single-strand products of each reaction are separated by gel electrophoresis.

The contents of the four tubes are separated side by side on an acrylamide gel so that DNA strands differing in length by only a single nucleotide can be distinguished. After electrophoresis, the locations of the DNA strands in the gel are revealed by autoradiography. The shortest strands, which terminated at positions early in the DNA sequence, migrate quickly and end up near the bottom of the gel; longer fragments, which terminated late in the sequence, migrate more slowly and end up near the top of the gel.

Reading the DNA sequence is simple and the shortest part of the procedure. In Figure 19.6, you can see that the band closest to the bottom of the gel is from the tube that contained the ddGTP reaction, which means that the first nucleotide synthesized had guanine (G). The next band up is from the tube that contained ddATP; so the next nucleotide in the sequence is adenine (A), and so forth. In this way, the sequence is read from the bottom to the top of the gel, with the nucleotides near the bottom corresponding to the 5′ end of the newly synthesized DNA strand and those near the top corresponding to the 3′ end. Keep in mind that the sequence obtained is not that of the target DNA but that of its *complement*.

You may have wondered how the primers used in dideoxy sequencing are constructed, because the sequence of the target DNA may not be known ahead of time. The trick is to insert a sequence that will be recognized by the primer into the target DNA. This is often done by first cloning the target DNA in a vector that contains sequences

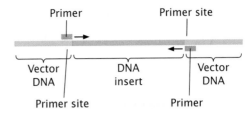

19.7 Sites recognized by sequencing primers are added to the target DNA by cloning the DNA in a vector that contains universal sequencing primer sites on either side of the site where the target DNA will be inserted.

recognized by a common primer (called universal sequencing primer sites) on either side of the site where the target DNA will be inserted. The target DNA is then isolated from the vector and will contain universal sequencing primer sites at each end (◀ FIGURE 19.7).

Sequencing is often carried out by automated machines that use fluorescent dyes and laser scanners to sequence thousands of base pairs in a few hours (◀ FIGURE 19.8). The dideoxy reaction is also used here, but the ddNTPs used in the reaction are labeled with a fluorescent dye, and a different colored dye is used for each type of dideoxynucleotide. For example, a red dye might be used for nucleotides with thymine, a green dye for those with adenine, a black dye for those with guanine, and a blue dye for those with cytosine. In this case, the four sequencing reactions can take place in the same test tube and can be placed in the same well during electrophoresis, given that each ddNTP is distinctively marked. The most recently developed sequencing machines carry out electrophoresis in gel-containing capillary tubes. The different-sized fragments produced by the sequencing reaction separate within a tube and migrate past a laser beam and detector. As the fragments pass the laser, their fluorescent dyes are activated and the resulting fluorescence is detected by an optical scanner. Each colored dye emits fluorescence of a characteristic wavelength, which is read by the optical scanner. The information is fed into a computer for interpretation, and the results are printed out as a set of peaks on a graph (See Figure 19.8). Automated sequencing machines may contain 96 or more capillary tubes, allowing from 50,000 to 60,000 bp of sequence to be read in a few hours.

Concepts

DNA can be rapidly sequenced by the dideoxy method, in which ddNTPs are used to terminate DNA synthesis at specific bases. Automated sequencing methods allow tens of thousands of base pairs to be read in just a few hours.

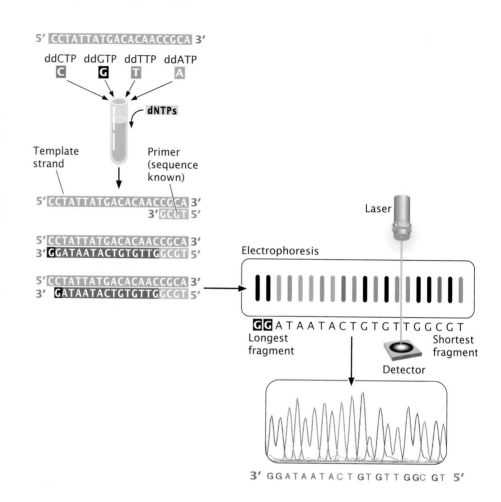

1 A single-stranded DNA fragment whose base sequence is to be determined (the template) is isolated.

2 Each of the four ddNTPs is tagged with a fluorescent dye, and the Sanger sequencing reaction is carried out.

3 The fragments that end in the same base have the same colored dye attached.

4 The products are denatured, and the DNA fragments produced by the four reactions are mixed and loaded into a single well on an electrophoresis gel. The fragments migrate through the gel according to size,...

5 ...and the fluorescent dye on the DNA is detected by using a laser beam and detector.

6 Each fragment appears as a peak on the computer printout; the color of the peak indicates which base the peak represents.

7 The sequence information is read directly into the computer, which converts it into the complementary—target—sequence.

◀ 19.8 The dideoxy sequencing method can be automated.

Sequencing an Entire Genome

The ultimate goal of structural genomics is to determine the ordered nucleotide sequences of entire genomes of organisms. The main obstacle to this task is the immense size of most genomes. Bacterial genomes are usually at least several million base pairs long; many eukaryotic genomes are billions of base pairs long and are distributed among dozens of chromosomes. Furthermore, for technical reasons, it is not possible to begin sequencing at one end of a chromosome and continue straight through to the other end; only small fragments of DNA—usually from 500 to 700 nucleotides—can be sequenced at one time. Therefore, determining the sequence for an entire genome requires that the DNA be broken into thousands or millions of smaller fragments that can then be sequenced. The difficulty lies in putting these short sequences back together in the correct order. As we will see, two different approaches have been used to assemble the short, sequenced fragments into a complete genome.

The first genomes to be sequenced were small genomes of some viruses. The genome of bacteriophage λ, consisting of 49,000 bp, was completed in 1982. In 1995, the first genome of a living organism (*Haemophilus influenzae*) was sequenced by Craig Venter and Claire Fraser of the Institute for Genomic Research (TIGR) and Hamilton Smith of Johns Hopkins University. This bacterium has a relatively small genome of 1.8 million base pairs (◀ FIGURE 19.9). By 1996, the genome the first eukaryotic organism (yeast) had been determined, followed by the genome of *Eschericia coli* (1997), *Caenorhabditis elegans* (1998), and *Drosophila melanogaster* (2000). The first draft of the human genome was completed in June 2000.

Map-based sequencing The first method for assembling short, sequenced fragments into a whole-genome sequence, called a **map-based approach,** requires the initial creation of detailed genetic and physical maps of the genome, which provide known locations of genetic markers (restriction sites, other genes, or known DNA sequences) at regularly spaced intervals along each chromosome. These markers can later be used to help align the short, sequenced fragments into their correct order.

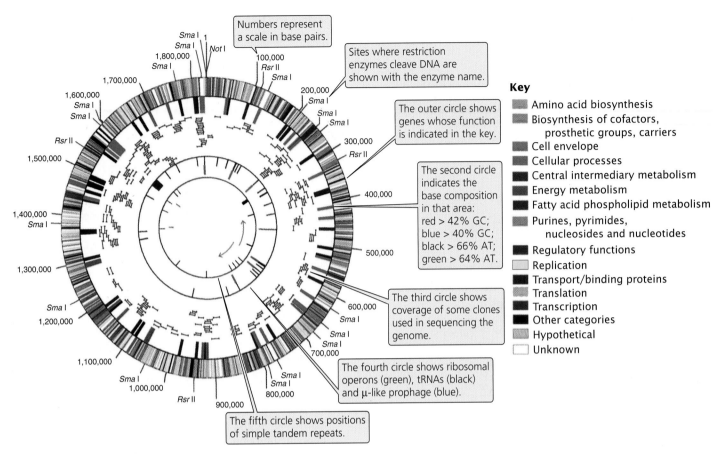

19.9 The bacterium *Haemophilas influenzae* was the first free-living organism to be sequenced. (From R. D. Fleischman et al., 1993, *Science* 269:496; scan courtesy of TIGR.)

After the genetic and physical maps are available, chromosomes or large pieces of chromosomes are separated by pulsed-field gel electrophoresis (PFGE) or by flow cytometry. In pulsed field gel electrophoresis (which is similar to standard gel electrophoresis), large molecules of DNA or whole chromosomes are separated in a gel by periodically alternating the orientation of an electrical current. In flow cytometry, chromosomes are sorted optically by size (◀ FIGURE 19.10).

Each chromosome (or sometimes the entire genome) is then cut up by partial digestion with restriction enzymes (◀ FIGURE 19.11). Partial digestion means that the restriction enzymes are allowed to act only for a limited time so that not all restriction sites in every DNA molecule are cut. Thus partial digestion produces a set of large overlapping DNA fragments, which are then cloned by using cosmids, yeast artificial chromosomes (YACs), or bacterial artificial chromosomes (BACs).

Next, these large-insert clones are put together in their correct order on the chromosome (see Figure 19.11). This assembly can be done in several ways. One method relies on the presence of a high-density map of genetic markers. A complementary DNA probe is made for each genetic marker, and a library of the large-insert clones is screened

with the probe, which will hybridize to any colony containing a clone with the marker. The library is then screened for neighboring markers. Because the clones are much larger than the markers used as probes, some clones will have more than one marker. For example, clone A might have markers M1 and M2, clone B markers M2, M3, and M4, and clone C markers M4 and M5. Such a result would indicate that these clones contain areas of overlap, as shown here.

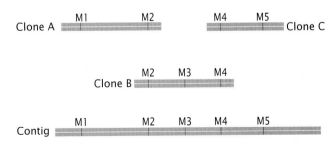

A set of two or more overlapping DNA fragments that form a contiguous stretch of DNA is called a **contig**. This approach was used in 1993 to create a contig of the human Y chromosome consisting of 196 overlapping YAC clones (see Figure 19.3).

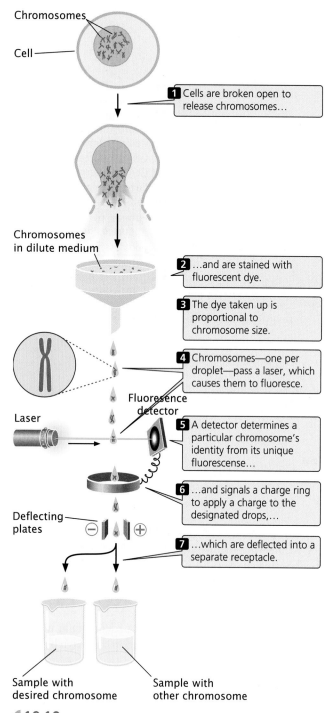

Chromosomes

Cell

1 Cells are broken open to release chromosomes…

Chromosomes in dilute medium

2 …and are stained with fluorescent dye.

3 The dye taken up is proportional to chromosome size.

4 Chromosomes—one per droplet—pass a laser, which causes them to fluoresce.

Fluorescence detector

Laser

5 A detector determines a particular chromosome's identity from its unique fluorescence…

6 …and signals a charge ring to apply a charge to the designated drops,…

Deflecting plates

7 …which are deflected into a separate receptacle.

Sample with desired chromosome

Sample with other chromosome

◀ **19.10 Flow cytometry is used to separate individual chromosomes.**

It is also possible to determine the order of clones without the use of preexisting genetic maps. For example, each clone can be cut with a series of restriction enzymes, and the resulting fragments are then separated by gel electrophoresis. This method generates a unique set of restriction fragments, called a fingerprint, for each clone. The restriction patterns for the clones are stored in a database. A computer program is then used to examine the restriction patterns of all the

clones and look for areas of overlap. The overlap is then used to arrange the clones in order, as shown here:

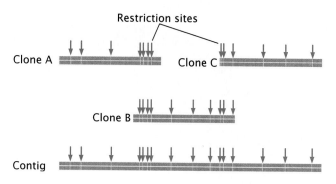

Restriction sites

Clone A   Clone C

Clone B

Contig

Other genetic markers can be used to help position contigs along the chromosome.

When the large-insert clones have been assembled into the correct order on the chromosome, a subset of overlapping clones that efficiently cover the entire chromosome can be chosen for sequencing. Each of the selected large-insert clones is fractured into smaller overlapping fragments, which are themselves cloned with the use of phages or cosmids (Figure 19.11). These smaller clones (called small-insert clones) are then sequenced. The sequences of the small-insert clones are examined for overlap, which allows them to be correctly assembled to give the sequence of the larger insert clones. Enough overlapping small-insert clones are usually sequenced to ensure that the entire genome is sequenced several times. Finally, the whole genome is assembled by putting together the sequences of all overlapping contigs (Figure 19.11). Often, gaps still exist in the genome map that must be filled in by using other methods.

Whole-genome shotgun sequencing The second approach to genome sequencing does not map and assemble the large-insert clones. In this approach, called **whole-genome shotgun sequencing** (◀ FIGURE 19.12), small-insert clones are prepared directly from genomic DNA and sequenced. Powerful computer programs then assemble the entire genome by examining overlap among the small-insert clones. The requirement for overlap means that most of the genome will be sequenced multiple (often from 10 to 15) times.

Concepts

Sequencing a genome requires breaking it up into small overlapping fragments whose DNA sequences can be determined in a sequencing reaction. The sequences can be ordered into the final genome sequence by a map-based approach (large fragments are ordered with the use of genetic and physical maps) or by whole-genome shotgun sequencing (overlap between the sequences of small fragments is compared by computers).

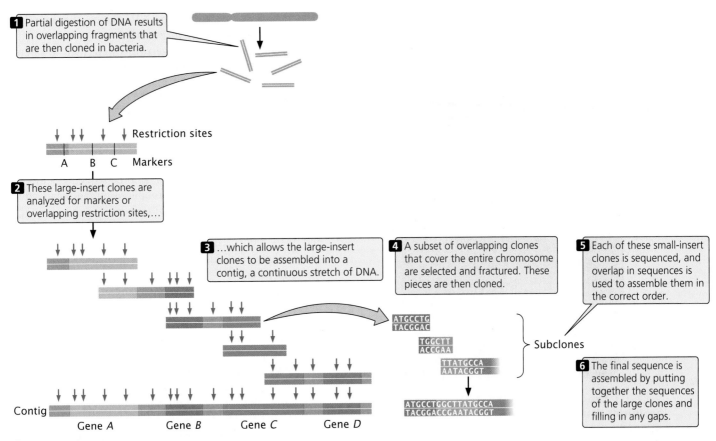

1 Partial digestion of DNA results in overlapping fragments that are then cloned in bacteria.

Restriction sites

A   B   C   Markers

2 These large-insert clones are analyzed for markers or overlapping restriction sites,...

3 ...which allows the large-insert clones to be assembled into a contig, a continuous stretch of DNA.

4 A subset of overlapping clones that cover the entire chromosome are selected and fractured. These pieces are then cloned.

5 Each of these small-insert clones is sequenced, and overlap in sequences is used to assemble them in the correct order.

ATGCCTG
TACGGAC

TGGCTT
ACCGAA

TTATGCCA
AATACGGT

Subclones

6 The final sequence is assembled by putting together the sequences of the large clones and filling in any gaps.

ATGCCTGGCTTATGCCA
TACGGACCGAATACGGT

Contig

Gene *A*   Gene *B*   Gene *C*   Gene *D*

◀ 19.11 **Map-based approaches to whole-genome sequencing rely on detailed genetic and physical maps to align sequenced fragments.**

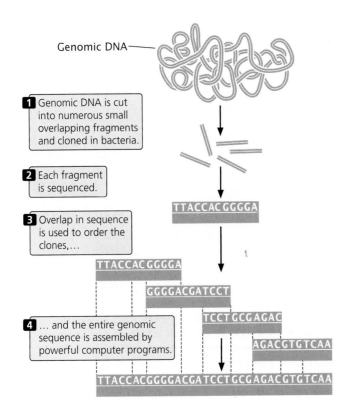

Genomic DNA

1 Genomic DNA is cut into numerous small overlapping fragments and cloned in bacteria.

2 Each fragment is sequenced.

TTACCAC GGGGA

3 Overlap in sequence is used to order the clones,...

TTACCAC GGGGA

GGGGACGATCCT

TCCT GCG AGAC

AGACGTGTCAA

4 ... and the entire genomic sequence is assembled by powerful computer programs.

TTACCACGGGGACGATCCT GCGAGACGTGTCAA

◀ 19.12 **Whole-genome shotgun sequencing utilizes sequence overlap to align sequenced fragments.**

www.whfreeman.com/pierce   More about current developments in DNA sequencing

The Human Genome Project

By 1980, methods for mapping and sequencing DNA fragments had been sufficiently developed that geneticists began seriously proposing that the entire human genome could be sequenced. An international collaboration was planned to undertake the Human Genome Project (◀ FIGURE 19.13); initial estimates suggested that 15 years and $3 billion would be required to accomplish the task. As a part of the effort, the genomes of several model organisms, including *Escherichia coli*, *Saccharomyces cerevisiae* (yeast), *Drosophila melanogaster* (fruit fly), *Arabidopsis thaliana* (a plant), and *Caenorhabditis elegans* (a nematode) were to be sequenced as well. The genomes of these model organisms were sequenced to help develop methods that could then be applied to the sequencing of the human genome and to

◀ **19.13 The Human Genome Project has produced an initial draft of the sequence for the human genome.** (Mario Tama/Getty.)

provide sequenced genomes with which to compare the organization and structure of the human genome.

The Human Genome Project officially got underway in October 1990. Initial efforts focused on developing new and automated methods for cloning and sequencing DNA and on generating detailed physical and genetic maps of the human genome. The methods described earlier for mapping, sequencing, and assembling DNA fragments were pivotal in these early stages of the project. By 1993, large-scale physical maps were completed for all 24 pairs of human chromosomes. At the same time, automated sequencing techniques (◀ FIGURE 19.14) had been developed that made large-scale sequencing feasible.

The initial effort to sequence the genome was a public project consisting of the international collaboration of 20 research groups and hundreds of individual researchers who formed the International Human Genome Sequencing Consortium. In 1998, Craig Venter announced that he would lead a company called Celera Genomics in a private effort to sequence the human genome.

The public and private efforts moved forward simultaneously but used different approaches. The Human Genome Consortium used a map-based approach; many copies of the human genome were cut up into fragments of about 150,000 bp each, which were inserted into bacterial artificial chromosomes. Yeast artificial chromosomes and cosmids had been used in early stages of the project but did

not prove to be as stable as the BAC clones, although YAC clones were instrumental in putting together some of the larger contigs. Restriction fingerprints were used to assemble the BAC clones into contigs, which were positioned on the chromosomes by using genetic markers and probes. The individual BAC clones were sheared into smaller overlapping fragments and sequenced, and the whole genome was assembled by putting together the sequence of the BAC clones.

Celera Genomics used a whole-genome shotgun approach to determine the human genome sequence, although the genetic and physical maps produced by the public effort helped Celera assemble the final sequence. In this approach, small-insert clones were prepared directly from genomic DNA and then sequenced. The overlapping of DNA sequences among these small-insert clones was then used to assemble the entire genome.

Both public and private sequencing projects announced the completion of a rough draft that included most of the sequence of the human genome in the summer of 2000, 5 years ahead of schedule. Analysis of this sequence was published 6 months later.

The availability of the complete sequence of the human genome is proving to be of enormous benefit. It is greatly facilitating the identification and isolation of genes that contribute to many human diseases and is providing probes that can be used in genetic testing, diagnosis, and drug development. The sequence is also providing important information about many basic cellular processes. Comparisons of the human genome with those of other organisms are adding to our understanding of evolution and the history of life.

◀ **19.14 Automated sequencers and powerful computers allowed a rough draft of the human genome sequence to be completed in just 10 years.** (Whitehead/MIT Genome Center, 2001; from *Nature* 409: 860–921.)

The New Genetics
ETHICS · SCIENCE · TECHNOLOGY

Mapping the Human Genome—Where It Leads, What It Means

Arthur L. Caplan and Kelly A. Carroll

In June 2000, scientists from the Human Genome Project and Celera Genomics stood at a podium with President Bill Clinton to announce a stunning achievement—they had successfully constructed a sequence of the entire human genome. Soon this process of identifying and sequencing each and every human gene became characterized as "mapping the human genome." As with maps of the physical world, the map of the human genome provides a picture of locations, terrains, and structures. But, like explorers, scientists must continue to decipher what each location on the map can tell us about diseases, human health, and biology. The map accelerates this process because it allows researchers to identify key structural dimensions of the gene that they are exploring and reminds them where they have been and where they have yet to explore.

What does the map of the human genome depict? When researchers discuss the sequencing of the genome, they are describing the identification of the patterns and order of the 3 billion human DNA base pairs. Although such identification provides valuable information about overall structure and the evolution of humans in relation to other organisms, researchers really want the key information encoded in just 2% of this enormous map—the information that encodes most of the proteins of which you and I are composed. Proteins stand as the link between genes and pharmaceutical drug development, they show which genes are being expressed at any given moment, and they provide information about gene function.

Knowing our genes will lead to a greater understanding and radically improved treatment of many diseases. However, sequencing the entire human genome, in conjunction with sequencing of various nonhuman genomes under the same project, has raised fundamental questions about what it means to be human. After all, fruit flies possess about one-third the number of genes possessed by humans, and an ear of corn has approximately the same number of genes as a human. In addition, the overall DNA sequence of a chimpanzee is about 99% the same as the human genome sequence. As the genomes of other species become available, the similarities to the human genome in both structure and sequence pattern will continue to be identified. At a basic level, the discovery of so many commonalities and links and ancestral trees with other species adds credence to principles of evolution and Darwinism.

Dr. Craig Venter (Celera Genomics), President Clinton, and Dr. Francis Collins (NIH). (Ron Edmonds/AP.)

Some of the most expected developments and potential benefits of the Human Genome Project directly affect human health; researchers, practicing physicians, and the general public eagerly await the development of targeted pharmaceutical agents and more specific diagnostic tests. Pharmacogenomics is at the intersection of genetics and pharmacology; It is the study of how one's genetic makeup will affect one's response to various drugs. In the future, medicine will potentially be safer, cheaper, and more disease specific, all while causing fewer side effects and acting more effectively, the first time around.

There are, however, some hard ethical questions that follow in the wake of new genetic knowledge. Patients will have to undergo genetic testing in order to match drugs to their genetic makeup. Who will have access to these results—just the health care practitioner? Or will the patient's insurance company, employer, school, or family have access? Although the tests may have been administered for one case, will information derived from them be used for other purposes, such as for the identification of other conditions or future diseases or even in research studies?

How should researchers conduct studies in pharmacogenomics? Often, they need to study subjects by some kind of identifiable trait that they believe will assist in separating groups of drugs, and in turn they separate people into populations. The order of almost all the DNA base pairs (99.9 %) is exactly the same in all humans, leaving a small window of difference. The potential for the stigmatization of individuals and groups of people based on race and ethnicity is inherent in genomic research and analysis. As scientists continue drug development, they must be careful not to further such ideas, especially because studies of nuclear DNA indicate that there is often more genetic variation within ethnic groups or cultures than between ethnic groups or cultures.

These are just a few of the ethical issues arising out of one development of the Human Genome Project. The potential applications of genome research are staggering, and the mapping is just the beginning.

www.whfreeman.com/pierce   Information about the Human Genome Project and numerous links to it

Single-Nucleotide Polymorphisms

In addition to the DNA sequence of an entire genome, several other types of data are useful for genomic projects and have been the focus of sequencing efforts. One consists of **single-nucleotide polymorphisms** (SNPs, pronounced "snips"), which are single-base-pair differences in DNA sequence between individual members of a species. Arising through mutation, SNPs are inherited as allelic variants (just like alleles that produce phenotypic differences, such as blood types), although SNPs do not usually produce a phenotypic difference. Single-nucleotide polymorphisms are numerous and are present throughout genomes. In a comparison of the same chromosome from two different people, a SNP can be found approximately every 1000 bp.

Because of their variability and widespread occurrence throughout the genome, SNPs are valuable as markers in linkage studies. For example, human SNPs are being cataloged and mapped for use in identifying genes that contribute to disease. When a SNP is physically close to a disease-causing locus, it will tend to be inherited along with the disease-causing allele. Thus the SNP marks the location of a genetic locus that causes the disease. A SNP can also be useful for determining family relationships—most SNPs are unique within a population, having arisen only once by mutation. Thus the presence of the same SNP in two persons often indicates that they have a common ancestor.

Expressed-Sequence Tags

Another type of data identified by sequencing projects consists of databases of **expressed-sequence tags** (ESTs). In most eukaryotic organisms, only a small percentage of the DNA actually encodes proteins; in humans, less than 2% of human DNA encodes the amino acids of proteins. If only protein-encoding genes are of interest, it is often more efficient to examine RNA than the entire DNA genomic sequence. RNA can be examined by using ESTs—markers associated with DNA sequences that are expressed as RNA. Expressed-sequence tags are obtained by isolating RNA from a cell and subjecting it to reverse transcription, producing a set of cDNA fragments that correspond to RNA

molecules from the cell. Short stretches of these cDNA fragments are then sequenced, and the sequence obtained (called a tag) provides a marker that identifies the DNA fragment. Expressed-sequence tags can be used to find active genes in a particular tissue or at a particular point in development.

Bioinformatics

By the time this book is published, complete genome sequences will have been determined for more than 100 different organisms, with many additional projects underway. These studies are producing tremendous quantities of sequence data. GenBank, one of the major databases of DNA sequence information, now contains more than 19 billion base pairs of sequence, and this number increases in size every month. Cataloging, storing, retrieving, and analyzing this huge data set are a major challenge of modern genetics. **Bioinformatics** is an emerging field consisting of molecular biology and computer science that centers on developing databases, computer-search algorithms, gene-prediction software, and other analytical tools that are used to make sense of DNA, RNA, and protein sequence data. Bioinformatics develops and applies these tools to "mine the data," extracting the useful information from sequencing projects.

Before being sequenced, most genomes contain few genes whose locations have already been determined, which, coupled with the enormous amount of DNA in a genome and the complexities of gene structure, makes finding genes a difficult task. Computer programs have been developed to look for specific sequences in DNA that are associated with certain genes. For example, protein-encoding genes are characterized by an **open reading frame** (ORF), which includes a start codon and a stop codon in the same reading frame. Specific sequences mark the splice sites at the beginning and end of introns; other specific sequences are present in promoters immediately upstream of start codons. Still other sequences are associated with particular functions in certain classes of proteins. Computer programs have been developed that scan DNA databases for these sequences and identify genes on the basis of their presence and position. Some of these programs are capable of examining databases of EST and protein sequences to see if there is evidence that a potential gene is expressed.

It is important to recognize that the programs that have been developed to identify genes on the basis of DNA sequence are not perfect. Therefore, the numbers of genes reported in most genome projects are estimates. The presence of multiple introns, alternative splicing, multiple copies of some genes, and much noncoding DNA between genes makes accurate identification and counting of genes difficult.

www.whfreeman.com/pierce   Information on ESTs, SNPs, and bioinformatics

Functional Genomics

A genomic sequence is, by itself, of limited use. It would be like having a huge set of encyclopedias without being able to read—you could recognize the different letters but the text would be meaningless. Functional genomics is, in essence, probing genome sequences for meaning—identifying genes, recognizing their organization, and understanding their function. The goals of functional genomics include identifying all the RNA molecules transcribed from a genome (the **transcriptome**) and all the proteins encoded by the genome (the **proteome**). Functional genomics exploits both bioinformatics and laboratory-based experimental approaches in its search to define the function of DNA sequences.

Chapter 18 considered several methods for identifying genes and assessing their functions, including in situ hybridization, DNA footprinting, experimental mutagenesis, and the use of transgenic animals and knockouts. These methods can be applied to individual genes and can provide important information about the locations and functions of genetic information. In this section, we will focus primarily on methods that rely on knowing the sequences of other genes or that can be applied to large numbers of genes simultaneously.

Predicting Function from Sequence

The nucleotide sequence of a gene can be used to predict the amino acid sequence of the protein that it encodes. The protein can then be synthesized or isolated and its properties studied to determine its function. However, this biochemical approach to understanding gene function is both time consuming and expensive. A major goal of functional genomics has been to develop computational methods that allow gene function to be identified from DNA sequence alone, bypassing the laborious process of isolating and characterizing individual proteins.

Homology searches One computational method (often the first employed) for determining gene function is to conduct a homology search, which relies on comparing DNA and protein sequences from the same and different organisms. Genes that are evolutionarily related are said to be

homologous. Homologous genes found in different species that evolved from the same gene in a common ancestor are called **orthologs** (◀ FIGURE 19.15). For example, both mouse and human genomes contain a gene that encodes the alpha subunit of hemoglobin; the mouse and human alpha-hemoglobin genes are said to be orthologs, because both genes evolved from an alpha-hemoglobin gene in a mammalian ancestor common to mice and humans. Homologous genes in the same organism (arising by duplication of a single gene in the evolutionary past) are called **paralogs** (see Figure 19.15). Within the human genome is a gene that

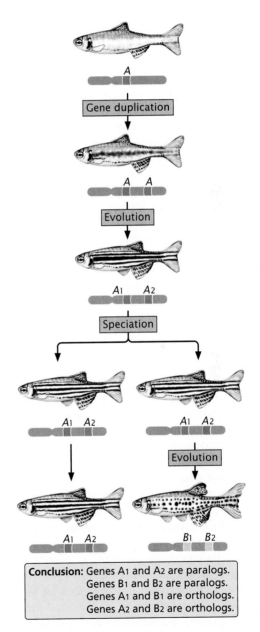

Conclusion: Genes A1 and A2 are paralogs.
Genes B1 and B2 are paralogs.
Genes A1 and B1 are orthologs.
Genes A2 and B2 are orthologs.

◀ 19.15 **Homologous sequences are evolutionarily related.** Orthologs are homologous sequences found in different species; paralogs are homologous genes in the same species that arise from gene duplication.

encodes the alpha subunit of hemoglobin and another homologous gene that encodes the beta subunit of hemoglobin. These two genes arose because an ancestral gene underwent duplication and the resulting two genes diverged through evolutionary time, giving rise to the alpha- and beta-subunit genes; these two genes are paralogs. Homologous genes (both orthologs and paralogs) often have the same or related functions; so, after a function has been assigned to a particular gene, it can provide a clue to the function of a homologous gene.

Databases containing genes and proteins found in a wide array of organisms are available for homology searches. Powerful computer programs have been developed for scanning these databases to look for particular sequences. A commonly used homology search program is BLAST (Basic Local Alignment Search Tool). Suppose a geneticist sequences a genome and locates a gene that encodes a protein of unknown function. A homology search conducted on databases containing the DNA or protein sequences of other organisms may identify one or more orthologous sequences. If a function is known for a protein encoded by one of these sequences, that function may provide information about the function of the newly discovered protein.

In a similar way, computer programs can search a single genome for paralogs. Eukaryotic organisms often contain families of genes that have arisen by duplication of a single gene. If a paralog is found and its function has been previously assigned, this function can provide information about a possible function of the unknown gene. However, paralogs often evolve new functions; so information about their functions must be used cautiously. Of the genes newly identified through genomic-sequencing projects, 50% are significantly similar to orthologs and paralogs whose function has already been described. The 50% of newly identified genes that *cannot* be assigned a function on the basis of homology searches will undoubtedly decrease in number as functions are assigned to more and more genes and as more genomes are sequenced.

Other sequence comparisons

Complex proteins often contain regions that have specific shapes or functions called **protein domains.** For example, certain DNA-binding proteins attach to DNA in the same way; these proteins have in common a domain that provides the DNA-binding function. Each protein domain has an arrangement of amino acids common to that domain. There are probably a limited, though large, number of protein domains, which have mixed and matched through evolutionary time to yield the protein diversity seen in present-day organisms.

Many protein domains have been characterized, and their molecular functions have been determined. The sequence from a newly identified gene can be scanned against a database of known domains. If the gene sequence encodes one or more domains whose functions have been previously determined, the function of the domain can provide important information about a possible function of the new gene.

Another computational method for predicting protein function is a **phylogenetic profile.** In this method, the presence-and-absence pattern of a particular protein is examined across a set of organisms whose genomes have been sequenced. If two proteins are either both present or both absent in all genomes surveyed, the two proteins may be functionally related. For example, the two proteins might function in consecutive steps in a biochemical pathway. The idea is that the two proteins depend on each other and will evolve together. One protein cannot function without the other, and they will either both be present or both be absent.

Consider the following proteins in four bacterial species (◀ FIGURE 19.16a):

E. coli:	protein 1, protein 2, protein 3, protein 4, protein 5, protein 6
Species A:	protein 1, protein 2, protein 3, protein 6
Species B:	protein 1, protein 3, protein 4, protein 6
Species C:	protein 2, protein 4, protein 5

We can create a phylogenetic profile by constructing a table comparing the presence (+) or absence (−) of the proteins in the four bacterial species (◀ FIGURE 19.16b). The phylogenetic profile reveals that proteins 1, 3, and 6 are either all present or all absent in all species; so these proteins might be functionally related.

Examining **fusion patterns** among proteins is another method for predicting functional relations; this technique is sometimes called the Rosetta Stone method. Functionally related, separate proteins in one organism sometimes exist as a single, fused protein in another organism. Thus, the presence of a fused A + B protein in one species suggests that separate proteins A and B in another organism may be functionally related.

Yet another method for determining the function of an unknown gene is **gene neighbor analysis** (◀ FIGURE 19.17). Genes that encode functionally related proteins are often closely linked in bacteria. For example, if two genes are consistently linked in the genomes of several bacteria, they might be functionally related. Functionally related genes are sometimes also linked in eukaryotes; examples are the *hox* genes, which play an important role in embryonic development (Chapter 21).

It is important to recognize that functions suggested by computational methods such as homology searches, phylogenetic profiling, fusion proteins, and neighbor analysis do not define a protein's function; rather these computational methods provide hints about possible

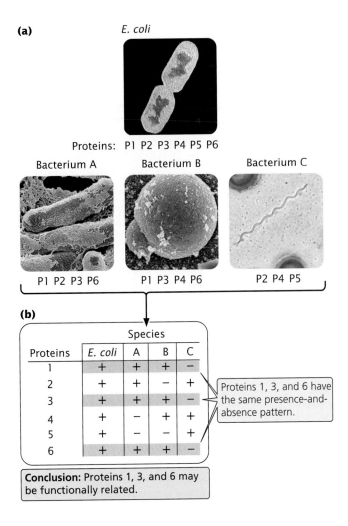

(a)

E. coli

Proteins: P1 P2 P3 P4 P5 P6

Bacterium A   Bacterium B   Bacterium C

P1 P2 P3 P6   P1 P3 P4 P6   P2 P4 P5

(b)

Proteins	Species			
	E. coli	A	B	C
1	+	+	+	−
2	+	+	−	+
3	+	+	+	−
4	+	−	+	+
5	+	−	−	+
6	+	+	+	−

Proteins 1, 3, and 6 have the same presence-and-absence pattern.

Conclusion: Proteins 1, 3, and 6 may be functionally related.

◀19.16 **Phylogenetic profiling can be used to infer protein function.** (Micrographs: top, CNRI/SPL/Photo Researchers; middle left and center, Gary Gaugler/Visuals Unlimited; middle right, M. Abbey/Visuals Unlimited.)

functions that can be pursued through detailed analyses of the biochemistry and cellular location of the protein. Nevertheless, these computational methods and others like them have proved to be invaluable in determining the functions of genes revealed in genomic studies.

⌐**Concepts**⌐

Genes can be identified by computer programs that look for characteristic features of genes, such as start and stop codons in the same reading frame, sequences that mark the beginning and the end of introns, and sequences found within promoters. Clues to the functions of genes can be obtained by homology searches, comparing protein domains, phylogenetic profiling, protein fusion patterns, and gene neighbor analysis.

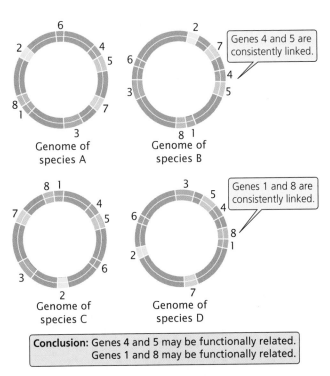

Genome of species A   Genome of species B

Genes 4 and 5 are consistently linked.

Genome of species C   Genome of species D

Genes 1 and 8 are consistently linked.

Conclusion: Genes 4 and 5 may be functionally related.
Genes 1 and 8 may be functionally related.

◀19.17 **The gene neighbor method infers gene function on the basis of the linkage arrangements of the genes.** Genes that are consistently linked in different genomes may be functionally related.

Gene Expression and Microarrays

Many important clues about gene function come from knowing when and where the genes are expressed. The development of microarrays has allowed the expression of thousand of genes to be monitored simultaneously.

Microarrays rely on nucleic acid hybridization (see Chapter 18), in which a known DNA fragment is used as a probe to find complementary sequences (◀FIGURE 19.18). The probe is usually fixed to some type of solid support, such as a nylon filter or a glass slide. A solution containing a mixture of DNA or RNA is applied to the solid support; any nucleic acid that is complementary to the probe will bind to it. Nucleic acids in the mixture are labeled with a radioactive or fluorescent tag so that molecules bound to the probe can be easily detected.

In a microarray (also called a gene chip), numerous known DNA fragments are fixed to a solid support in an orderly pattern or array, usually as a series of dots. These DNA fragments (the probes) usually correspond to known genes.

When the microarray has been constructed, mRNA, DNA, or cDNA isolated from experimental cells is labeled with fluorescent nucleotides and applied to the array. Any of the DNA or RNA molecules that are complementary to probes on the array will hybridize with them and emit

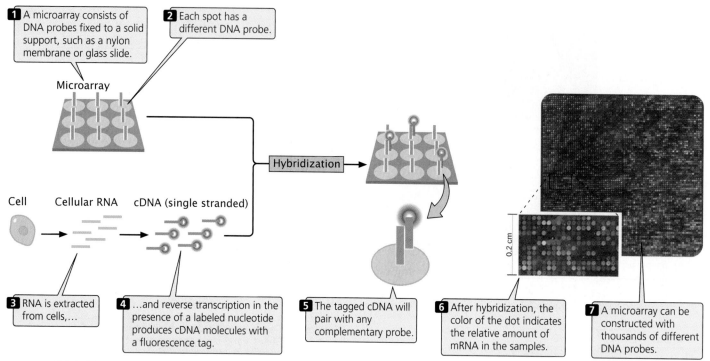

1 A microarray consists of DNA probes fixed to a solid support, such as a nylon membrane or glass slide.

2 Each spot has a different DNA probe.

Microarray

Hybridization

Cell   Cellular RNA   cDNA (single stranded)

3 RNA is extracted from cells,…

4 …and reverse transcription in the presence of a labeled nucleotide produces cDNA molecules with a fluorescence tag.

5 The tagged cDNA will pair with any complementary probe.

6 After hybridization, the color of the dot indicates the relative amount of mRNA in the samples.

7 A microarray can be constructed with thousands of different DNA probes.

0.2 cm

◀ 19.18 **Microarrays are used to simultaneously detect the expression of many genes.** (D. Lockhart and E. Winzeler, 2000, *Nature* 405:827.)

fluorescence, which can be detected by an automated scanner. An array containing tens of thousands of probes can be applied to a glass slide or silicon wafer just a few square centimeters in size.

One type of DNA chip is illustrated in ◀ FIGURE 19.19. For this chip, mRNA from experimental cells is converted into cDNA and labeled with red fluorescent nucleotides. MessengerRNA from control cells is converted into cDNA and labeled with green fluorescent nucleotides. The labeled cDNAs are mixed and hybridized to the DNA chip, which contains DNA probes from different genes. Hybridization of the red (experimental) and green (control) cDNAs is proportional to the relative amounts of mRNA in the samples. The fluorescence of each spot is assessed with microscopic scanning and appears as a single color. Red indicates the overexpression of a gene in the experimental cells relative to that in the control cells (more red-labeled cDNA hybridizes), whereas green indicates the underexpression of a gene in the experimental cells relative to that in the control cells (more green-labeled cDNA hybridizes). Yellow indicates equal expression in experimental and control cells (equal hybridization of red- and green-labeled cDNAs), and no color indicates no expression in either experimental or control cells. Microarrays that allow the detection of specific alleles, SNPs, and even particular proteins also have been created.

Microarrays allow the expression of thousands of genes to be monitored simultaneously, enabling scientists to study which genes are active in particular tissues. They can also be used to investigate how gene expression changes in the course of biological processes such as development or disease progression. In one study, researchers examined gene expression to predict the long-term outcome for women who had undergone treatment for breast cancer. Breast cancer affects 1 of 10 women in the United States, and half of those with the disease die from it. Current treatment depends on a number of factors, including a woman's age, the size of the tumor, the characteristics of tumor cells, and whether the cancer has already spread to nearby lymph nodes. Many women whose cancer has not spread are treated by removal of the tumor and radiation therapy, yet the cancer later reappears in some of the women thus treated. These women might benefit from more-aggressive treatment when the cancer is first detected.

Using microarrays, researchers examined the expression patterns of 25,000 genes from primary tumors of 78 young women who had breast cancer. In 34 of these patients, the cancer later spread to other sites; the other 44 patients remained free of breast cancer for 5 years after their initial diagnoses. The researchers identified a subset of 70 genes whose expression patterns in the initial tumors accurately predicted whether the cancer would later spread (◀ FIGURE 19.20). This degree of prediction was much higher than that of traditional predictive measures, which are based on the size and histology of the tumor. These results, though preliminary and confined to a small sample of cancer patients, suggest that gene-expression data

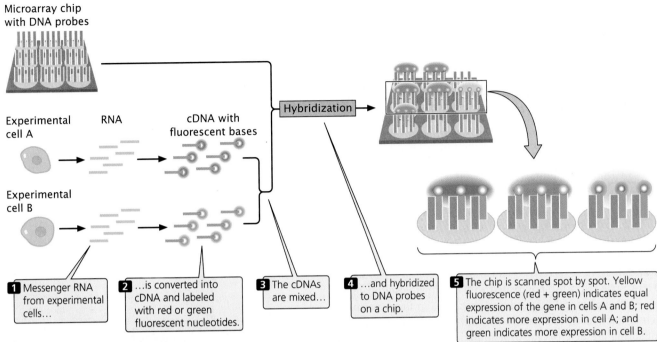

19.19 Microarrays can be used to compare levels of gene expression in different types of cells.

1 Messenger RNA from experimental cells...

2 ...is converted into cDNA and labeled with red or green fluorescent nucleotides.

3 The cDNAs are mixed...

4 ...and hybridized to DNA probes on a chip.

5 The chip is scanned spot by spot. Yellow fluorescence (red + green) indicates equal expression of the gene in cells A and B; red indicates more expression in cell A; and green indicates more expression in cell B.

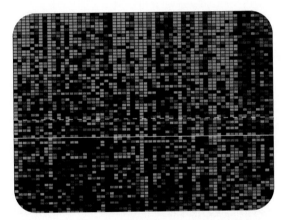

19.20 Microarrays can be used to examine gene expression associated with disease progression. Shown here are expression patterns of 70 genes in the initial tumors from patients whose cancer later spread to other sites and from other patients who remained free of breast cancer for 5 years after their initial diagnosis. Red indicates higher gene expression; green indicates lower gene expression; black indicates no change in gene expression; and gray indicates no data available. Each row represents the primary tumor from a patient and each column represents a different gene. Tumors below the solid yellow line came primarily from patients in whom the cancer spread to distant sites within five years of diagnosis; tumors above the solid line came primarily from patients who remained cancer free for at least five years. (L. J. van't Veer, 2002, *Nature* 405:532.)

obtained from microarrays can be a powerful tool in determining the nature of cancer treatment.

Concepts

Microarrays, consisting of DNA probes attached to a solid support, can be used to determine which RNA and DNA sequences are present in a mixture of nucleic acids. They are capable of determining which RNA molecules are being synthesized and thus can be used to examine changes in gene expression.

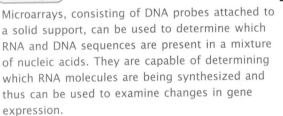 www.whfreeman.com/pierce   More on microarrays

Genomewide Mutagenesis

One of the best methods for determining the function of a gene is to examine the phenotypes of individual organisms that possess a mutation in the gene. Traditionally, genes encoding naturally occurring variations in a phenotype were mapped, the causative genes were isolated, and their products were studied. But this procedure was limited by the number of naturally occurring mutations and the difficulty of mapping genes with a limited number of chromosomal markers. The number of naturally occurring mutations can be increased by exposure to mutagenic agents, and the accuracy of mapping is increased dramatically by the availability of mapped molecular markers, such as RFLPs, microsatellites, STSs, ETSs, and SNPs. These two methods—random inducement of mutations on a genomewide basis and mapping with molecular markers—are coupled and automated in a **mutagenesis screen.**

Mutagenesis screens can be used to search for specific genes encoding a particular function or trait. For example, mutagenesis screens of mice are being used to identify genes having roles in cardiovascular function. When genes that affect cardiovascular function are located in mice, homology searches are carried out to determine if similar genes exist in humans. These genes can then be studied to better understand cardiac disease in humans.

To conduct a mutagenesis screen, random mutations are induced in a population of organisms, creating new phenotypes. The mutations are induced by exposing the organisms to radiation, a chemical mutagen (Chapter 17), or transposable elements (DNA sequences that insert randomly into the DNA; Chapter 11). The procedure for a typical mutagenesis screen is illustrated in ◀FIGURE 19.21. Here, male zebra fish are treated with ethylmethylsulfonate, or EMS, a chemical that induces mutations in their sperm. The treated males are mated with wild-type female fish. The offspring are heterozygous for mutations induced by EMS and are screened for any variant phenotypes that might be the product of dominant mutations expressed in these heterozygous fish.

Recessive mutations will not be expressed in the F_1 progeny but can be revealed with further breeding. The F_1 offspring are mated with wild-type fish, and the offspring from this cross are then backcrossed with their male parents, producing fish that are homozygous for recessive mutations. The offspring of the backcross are then screened for variant phenotypes.

The fish with variant phenotypes undergo further breeding experiments to verify that their variant phenotype is, in fact, due to a single-gene mutation. After the genetic nature of an abnormal phenotype has been verified, the gene that causes the phenotype can be located by **positional cloning.** The first step in positional cloning is to demonstrate linkage between the trait and one or more already mapped genetic markers. The progeny of genetic crosses that include the mutant phenotype are examined for a large number of molecular markers that cover the entire genome. The cosegregation of markers and the mutant phenotype provides evidence of linkage, indicating that the marker and the gene encoding the mutant phenotype are physically linked on the same chromosome. Cosegregating markers provide information about the general chromosome region in which the gene is located.

The next step is to localize the mutated gene to a smaller region of the chromosome, which is usually done by examining a linkage map of the chromosome region to identify other molecular markers in close proximity to the gene of interest. The gene causing the mutant phenotype is then mapped in relation to these markers. Next is the creation of a physical map, which requires a set of overlap-

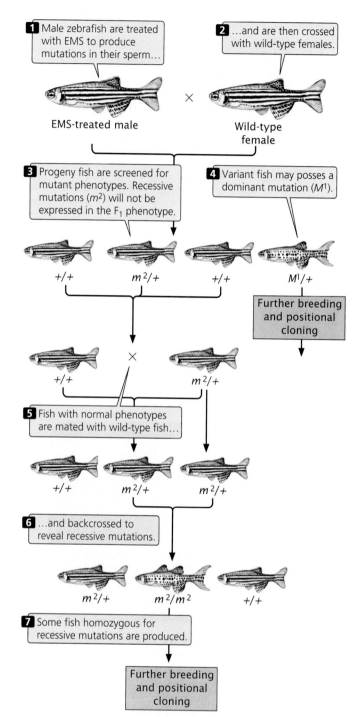

◀ **19.21 Genes affecting a particular characteristic or function can be identified by a genomewide mutagenesis screen.** In this illustration, M^1 represents a dominant mutation and m^2 represents a recessive mutation.

ping clones from the area of interest. A physical map of these overlapping clones that includes information about the molecular markers allows the identification of one or more clones that contain the gene of interest. These clones are then sequenced to find potential candidate genes that might encode the mutant phenotype. Candidate genes are evaluated by studying their expression patterns, protein products, and homology to genes of known function. This information might suggest that one or more of the candidate genes is likely to be the cause of the phenotype. The candidate genes can be examined for the presence of mutations in the gene sequences carried by those individuals having a mutant phenotype. Further proof that a particular gene causes the phenotype can be obtained by mutating a specific gene and observing the phenotype in the offspring.

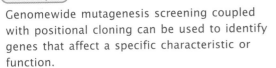

Concepts

Genomewide mutagenesis screening coupled with positional cloning can be used to identify genes that affect a specific characteristic or function.

Comparative Genomics

Genome-sequencing projects provide detailed information about gene content and organization in different species and even in different members of the same species, allowing inferences about how genes function and genomes evolve. They also provide important information about evolutionary relationships among organisms and about factors that influence the speed and direction of evolution.

Prokaryotic Genomes

A large number of bacterial genomes have now been sequenced (Table 19.2). Most prokaryotic genomes consist of a single circular chromosome, but there are exceptions, such as *Vibrio cholerae,* which has two circular chromosomes, and *Borrelia burgdorferi,* which has one large linear chromosome and 21 smaller chromosomes.

The total amount of DNA in prokaryotic genomes ranges from more than 7 million base pairs in *Mesorhizobium loti* to only 580,000 bp in *Mycoplasma genitalium.*

Table 19.2 Characteristics of some completely sequenced representative prokaryotic genomes

Species	Size (Millions of Base Pairs)	Number of Predicted Genes	G + C (%)
Archaea			
Archaeoglobus fulgidus	2.18	2407	49
Methanobacterium thermoautotrophicum	1.75	1869	50
Methanococcus jannaschii	1.66	1715	32
Thermoplasma acidophilum	1.56	1478	46
Eubacteria			
Bacillus subtilis	4.21	4100	44
Bordetella parapertussis	4.75	*	69
Buchnera species	0.64	564	27
Campylobacter jejuni	1.64	1654	31
Escherichia coli	4.64	4289	51
Haemophilus influenzae	1.83	1709	39
Mesorhizobium loti	7.04	6752	63
Mycobacterium tuberculosis	4.41	3918	66
Mycoplasma genitalium	0.58	480	32
Staphylococcus aureus	2.88	2697	33
Treponema pallidum	1.14	1031	53
Ureaplasma urealyticum	0.75	611	26
Vibrio cholerae	4.03	3828	48

Source: Data from the Genome Atlas of the Center for Biological Sequence Analysis, http://www.cbs.dtu.dk/services/GenomeAtlas/
* Data not available.

Escherichia coli, the most widely used bacterium for genetic studies, has 4.6 million base pairs (◀FIGURE 19.22a). The number of genes is usually from 1000 to 2000, but some species have as many as 6700, and others as few as 480. The density of genes is rather constant across all species, with about 1 gene for every 1000 bp. Thus bacteria with larger genomes usually have more genes.

Only about half of the genes identified in prokaryotic genomes can be assigned a function. Almost a quarter of the genes have no significant sequence similarity to any other known genes in bacteria, suggesting that there is considerable genetic diversity among bacteria. The number of genes that encode biological functions such as transcription and translation tends to be similar among species, even when their genomes differ greatly in size. This similarity suggests that these functions are encoded by a basic set of proteins that does not vary among species. On the other hand, the number of genes taking part in biosynthesis, energy metabolism, transport, and regulatory functions varies greatly among species and tends to be higher in larger genomes. The functions of predicted genes (i.e., genes identified by computer programs) and known genes in *E. coli* are presented in ◀FIGURE 19.22b. A substantial part of the "extra" DNA found in the larger bacterial genomes is made up of paralogous genes that have arisen by duplication.

The G + C content (percentage of bases that consist of guanine or cytosine) of prokaryotic genomes varies widely, from 26% to 69%. This more-than-twofold difference in G + C content affects the frequency of particular amino acids in the proteins produced by different bacterial species. For example, glycine, alanine, proline, and arginine are encoded by codons that have G and C nu-

cleotides; so these amino acids are incorporated into proteins with higher frequency in organisms whose genomes have a high G + C content. On the other hand, isoleucine, phenylalanine, tyrosine, and methionine are encoded by codons that tend to have A and T (U in RNA) nucleotides; so these amino acids are found more frequently in proteins encoded by species whose genome has a low G + C content. Which synonymous codons are used is also affected by the G + C content; some synonymous codons have more G and C nucleotides than do others, and these codons tend to be used more frequently in those species with high G + C content.

The results of genomic studies of prokaryotic species support the conclusion that archaea and eubacteria are evolutionarily unique (see Chapter 2). The results also reveal that both closely and distantly related bacterial species periodically exchange genetic information over evolutionary time, a process called **horizontal gene exchange.** Such exchange may take place through bacterial uptake of DNA in the environment (transformation), through the exchange of plasmids, and through viral vectors (see Chapter 8). Horizontal gene exchange has been recognized for some time, but analyses of many microbial genomes now indicate that it is more extensive than was previously recognized. For example, an analysis of two eubacteria species demonstrated that from 20% to 25% of their genes were more similar to genes from archaea than to those from other eubacterial species.

www.whfreeman.com/pierce   Information on prokaryotic and other genomes

(a) *Escherichia coli* (common bacteria)

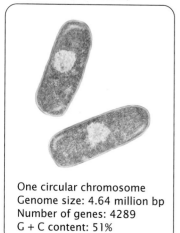

One circular chromosome
Genome size: 4.64 million bp
Number of genes: 4289
G + C content: 51%

(b)

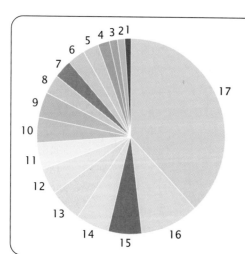

1. Fatty acids, phospholipid metabolism
2. Transcription, RNA metabolism
3. Nucleotide metabolism
4. Phage, transposon, plasmid functions
5. DNA replication, recombination, repair
6. Carbon compounds metabolism
7. Amino acid metabolism
8. Other genes with known functions
9. Regulatory functions
10. Translation, protein metabolism
11. Central intermediary metabolism
12. Adaptation, protection functions
13. Cell, wall, membrane structural components
14. Energy metabolism
15. Putative enzymes
16. Transport proteins
17. Genes with unknown functions

◀19.22 **Genomic characteristics of the bacterium *E. coli*.** (a) Genome size, number of genes, and G + C content. (b) Percentages of genes affecting various known and unknown functions.

Table 19.3	Characteristics of Some Eukaryotic Genomes That Have Been Completely Sequenced		
Species	Genome Size (Millions of Base Pairs)	Number of Predicted Genes	Number of Protein-Domain Families
Saccharomyces cerevisiae (yeast)	12	6,144	851
Arabidopsis thaliana (plant)	125	25,706	1012
Caenorhabditis elegans (roundworm)	100	18,266	1014
Drosophila melanogaster (fruit fly)	180	13,338	1035
Homo sapiens (human)	3400	~32,000	1262

Source: Number of genes and protein-domain families from International Human Genome Sequencing Consortium, Initial sequencing and analysis of the human genome, *Nature* 409 (2001), Table 23.

Eukaryotic Genomes

The genomes of only a few eukaryotic organisms have been completely sequenced, but some tentative statements can be made about the content and organization of eukaryotic genetic information from these organisms.

The genomes of eukaryotic organisms (Table 19.3) are larger than those of prokaryotes, and, in general, multicellular eukaryotes have more DNA than do simple, single-celled eukaryotes such as yeast (see pp. 298–299 in Chapter 11). There is no close relation, however, between genome size and complexity among the multicellular eukaryotes. For example, the roundworm *Caenorhabditis elegans* is structurally more complex than the plant *Arabidopsis* but has considerably less DNA. Eukaryotic genomes also contain more genes than do prokaryotes, and the genomes of multicellular eukaryotes have more genes than do the genomes of single-celled eukaryotes. The *number* of genes among multicellular eukaryotes also is not obviously related to phenotypic complexity: humans have more genes than do invertebrates but only twice as many as fruit flies and only slightly more than the plant *Arabidopsis*. Eukaryotic genomes contain multiple copies of many genes, indicating that gene duplication has been an important process in genome evolution.

A substantial part of the genomes of multicellular organisms consists of moderately and highly repetitive sequences (see Chapter 11), and the percentage of repetitive sequences is usually higher in those species with larger genomes (Table 19.4). Most of these repetitive sequences appear to have arisen through transposition. This is particularly evident in the human genome, where 45% of the DNA is derived from transposable elements, many of which are defective and no longer able to move. The majority of DNA in multicellular organisms is noncoding, and many genes are interrupted by introns. In the more complex eukaryotes, both the number and the length of the introns are greater.

In spite of only a modest increase in gene number, vertebrates have considerably more protein diversity than do invertebrates. The human genome does not encode many new

Table 19.4	Percentage of genome consisting of interspersed repeats derived from transposable elements	
Organism	Percentage of Genome
Plant (*Arabidopsis thaliana*)	10.5
Worm (*Caenorhabditis elegans*)	6.5
Fly (*Drosophila melanogaster*)	3.1
Human (*Homo sapiens*)	44.4

protein domains; there are 1262 domains in humans compared with 1035 in fruit flies (see Table 19.3). However, the existing domains in humans are assembled into more combinations, leading to many more types of proteins. For example, the human genome contains almost two times as many arrangements of protein domains as worms or flies contain and almost six times as many as yeast contains. Humans, worms, and flies have many of the same families of genes in common, but each family in the human genome has a greater number of different genes, suggesting that gene duplication has been an important process in vertebrate evolution.

Concepts

Comparative genomics compares the content and organization of whole genomic sequences from different organisms. Prokaryotic genomes are small, usually ranging from 1 million to 3 million base pairs of DNA, with several thousand genes. Among multicellular eukaryotic organisms, there is no clear relation between organismal complexity and amount of DNA or gene number. A substantial part of the genome in eukaryotic organisms consists of repetitive DNA, much of which is derived from transposable elements.

(a) *Saccharomyces cerevisiae* (yeast)   **(b)**

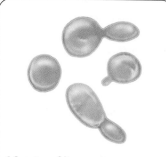

16 pairs of linear chromosomes
Genome size: 12.1 million bp
Number of genes: 6100
G + C content: 38%

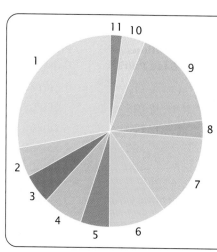

1. Cellular organization and biogenesis
2. Intracellular transport
3. Transport facilitation
4. Protein destination
5. Protein synthesis
6. Transcription
7. Cell growth, cell division, and DNA synthesis
8. Energy
9. Metabolism
10. Cell rescue
11. Signal transduction

◀19.23 **Genomic characteristics of yeast, *Saccharomyces cerevisiae.*** (a) Number of chromosomes, genome size, number of genes, and G + C content. (b) Percentages of genes affecting various known and unknown functions.

Yeast genome As mentioned earlier, *Saccharomyces cerevisiae* (yeast) was the first eukaryotic genome to be completely sequenced. Its genome consists of 12.1 million base pairs of DNA and 6100 potential genes, of which about 5900 encode proteins (◀FIGURE 19.23a), giving a gene density of about one gene for every 2000 bp of DNA. The distribution of gene functions in yeast is displayed in ◀FIGURE 19.23b. The yeast genome contains considerable redundancy; there are a number of blocks of repeated sequences in the genome, and 30% of the genes exist in two or more copies.

Worm genome *Caenorhabditis elegans*, a roundworm, has a genome consisting of 97 million base pairs of DNA (◀FIGURE 19.24). More than 18,000 protein-encoding genes have been identified in the *C. elegans* genome, of which more than 40% are homologous with genes found in other organisms. There is one gene for about every 5000 bp of DNA, and gene density is more uniform across chromosomes than it is in most eukaryotes.

Plant genome The genome of *Arabidopsis thaliana*, a small mustardlike plant, consists of 167 million base pairs of DNA (◀FIGURE 19.25a), encoding 25,706 predicted genes. Although *Arabidopsis* has many proteins in common with yeast, worm, fly, and humans, it has roughly 150 protein families not seen in other eukaryotes, including structural proteins, transcription factors, enzymes, and proteins of unknown function. ◀FIGURE 19.25b shows the distribution of gene functions in *Arabidopsis*.

Gene duplication has played an important role in the evolution of *Arabidopsis*, with 60% of its genome consisting

Caenorhabditis elegans (round worm)

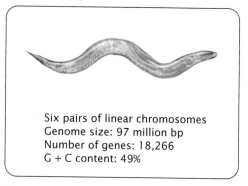

Six pairs of linear chromosomes
Genome size: 97 million bp
Number of genes: 18,266
G + C content: 49%

◀19.24 **Genomic characteristics of the roundworm, *Caenorhabditis elegans.***

of duplicated segments. Seventeen percent of the genes exist in tandem arrays, which are multiple copies of the same gene positioned one after another. One of the processes that produce tandem arrays of duplicated genes is unequal crossing over (see p. 485 in Chapter 17). A number of large duplicated regions, encompassing hundreds of thousands or millions of base pairs of DNA also are present. The large extent of duplication in the *Arabidopsis* genome suggests that this species had a tetraploid (4N) ancestor (see Chapter 9) and that all genes were duplicated in the past, followed by extensive gene rearrangement and divergence. Thus, at least two different mechanisms seem to have led to the large number of duplications seen in the *Arabidopsis* genome: (1) duplication of the whole genome through polyploidy; and

(a) *Arabidopsis thaliana*
(mustard-like weed)

(b)

Five pairs of linear chromosomes
Genome size: 167 million bp
Number of genes: 25,706
G + C content: 47%

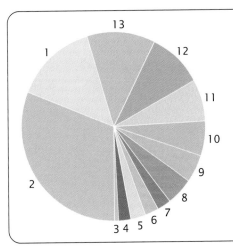

1. Metabolism
2. Unclassified
3. Ionic homeostasis
4. Protein synthesis
5. Energy
6. Transport facilitation
7. Cellular biogenesis
8. Intracellular transport
9. Protein destination
10. Cellular communication and signal transduction
11. Cell rescue, defense, cell death, aging
12. Cell growth, cell division, and DNA synthesis
13. Transcription

◀ **19.25 Genomic characteristics of the mustard plant,**
Arabidopsis thaliana. (a) Number of chromosomes, genome size,
number of genes, and G + C content. (b) Percentages of genes affecting
various known and unknown functions.

(2) duplication of individual genes arrayed in tandem through unequal crossing over.

Transposable elements are common in the *Arabidopsis* genome and make up about 10% of the genome but are much less frequent than in the human genome and in some other plant genomes. Most of these transposable elements are not transcribed, and many are concentrated in the regions surrounding the centromere.

Although *Arabidopsis, C. elegans,* and *Drosophila* have similar numbers of proteins, the *Arabidopsis* genome has more genes. This difference can be explained by the large number of duplicated copies of genes found in the *Arabidopsis* genome.

Fly genome *Drosophila melanogaster,* the fruit fly, has a genome of 180 million base pairs of DNA located on four chromosomes (◀ FIGURE 19.26). A third of its genome is made up of heterochromatin, which contains few genes. This extensive heterochromatin, consisting mainly of short simple repeats, made sequencing the genome of *Drosophila* difficult (because the repeats lead to much overlap in

sequence among cloned fragments, making it difficult to assemble the clones in the correct order). *Drosophila* has more than 13,000 predicted genes. There are 14,113 RNA transcripts produced from these genes, with some genes encoding multiple transcripts through alternative splicing. *Drosophila* genes average four exons per gene, although this number is probably an underestimate. The average RNA molecule encoded by a gene is 3058 nucleotides in length.

Human genome The human genome is 3.4 billion base pairs in length (◀ FIGURE 19.27a). Only about 25% of the DNA is transcribed into RNA, and less than 2% actually encodes proteins (◀ FIGURE 19.27b). Active genes are often separated by vast deserts of noncoding DNA, much of which consists of repeated sequences derived from transposable elements.

The average gene in the human genome is approximately 27,000 bp in length, with about 9 exons. (Table 19.5). (One exceptional gene has 234 exons.) The introns of human genes are much longer, and there are more of them than in other genomes (◀ FIGURE 19.27c). The human genome does not encode substantially more protein domains (see Table 19.3), but the domains are combined in more ways to produce a relatively diverse proteome. Gene functions encoded by the human genome are presented in Figure 19.27b. A single gene often encodes multiple proteins through alternative splicing; each gene may encode, on the average, two or three different mRNAs, meaning that the human genome, with approximately 32,000 genes, might encode as many as 96,000 proteins.

Gene density varies among human chromosomes; chromosomes 17, 19, and 22 have the highest density and

Drosophila melanogaster **(fruit fly)**

Four pairs of linear chromosomes
Genome size: 180 million bp
Number of genes: 13,338
G + C content: 41%

◀ **19.26 Genomic characteristics of *Drosophila***
melanogaster.

(a) *Homo sapiens* **(human)**

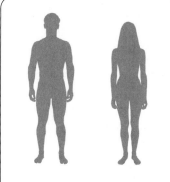

24 pairs of linear chromosomes
Genome size: 3.4 billion bp
Number of genes: ~32,000
G + C content: 41%

(b)

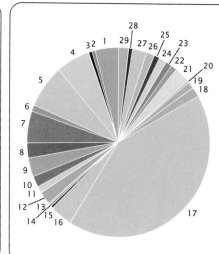

1. Miscellaneous
2. Viral protein
3. Transfer/carrier protein
4. Transcription factor
5. Nucleic acid enzyme
6. Signaling molecule
7. Receptor
8. Kinase
9. Select regulatory molecule
10. Transferase
11. Synthase and synthetase
12. Oxidoreductase
13. Lyase
14. Ligase
15. Isomerase
16. Hydrolase
17. Molecular function unknown
18. Transporter
19. Intracellular transporter
20. Select calcium-binding protein
21. Protooncogene
22. Structural protein of muscle
23. Motor
24. Ion channel
25. Immunoglobulin
26. Extracellular matrix
27. Cytoskeletal structural protein
28. Chaperone
29. Cell adhesion

(c)

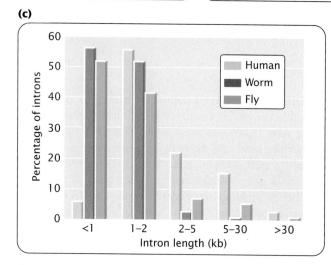

19.27 Genomic characteristics of *Homo sapiens.*
(a) Number of chromosomes, genome size, number of genes, and G + C content. (b) Percentages of genes affecting various known and unknown functions. (c) Intron length of genes in humans, worm, and fly.

Table 19.5	Average characteristics of genes in the human genome	
Characteristic	**Average**
Number of exons	8.8
Size of internal exon	145 bp
Size of intron	3365 bp
Size of 5′ untranslated region	300 bp
Size of 3′ untranslated region	770 bp
Size of coding region	1340 bp
Total length of gene	27,000 bp

chromosomes X, 4, 18, 13, and Y have the lowest density. Some proteins encoded by the human genome that are not found in other animals include those affecting immune function; neural development, structure and function; intercellular and intracellular signaling pathways in development; hemostasis; and apoptosis.

Transposable elements are much more common in the human genome than in worm, plant, and fruit-fly genomes (Table 19.4). The density of transposable elements varies, depending on chromosome location. In one region of the X chromosome, 89% of the DNA is made up of transposable elements, whereas other regions are largely devoid of these elements. There are a variety of types of transposable elements in the human genome, including LINEs, SINEs, retrotransposons, and DNA transposons (see Chapter 11). Most appear to be evolutionarily old and are defective, containing mutations and deletions so that they are no longer capable of transposition.

www.whfreeman.com/pierce   Information on the human genome and other genome-sequencing projects, including animal, plant, protozoan, fungal, and bacterial genomes

The Future of Genomics

The genomes of numerous organisms are in the process of being sequenced. These sequencing efforts, combined with the large amount of known DNA sequence that now exists, provide information that is tremendously useful for agriculture, human health, and biotechnology. The complete genome sequences of the mouse and the chimpanzee will serve as important sources of insight into the function and evolution of the human genome, inasmuch as these organisms are related to humans and are often used in studies of human health. Having complete genome sequences of crop plants and domestic animals will make it easier to identify genes that affect yield, disease and pest resistance, and other agriculturally important traits, which can then be manipulated by traditional breeding or genetic engineering to produce greater quantities and more-nutritious foods.

In the future, whole or partial genomic sequence information will be used in individual patient care. Currently, newborn babies are screened for a few treatable genetic diseases, such as phenylketonuria, which can be identified with the use of simple biochemical tests. In the future, newborns may be screened for a large number of variations in genetic sequence that confer high risk to treatable diseases, such as coronary artery disease, hypertension, asthma, and certain types of cancer. For those persons who are identified as genetically at risk, preventive treatment may be started early. In what has been called "personalized medicine," a person's DNA sequence may be used to predict responses to different treatment regimes, and drug therapy may then be fine-tuned to a person's genetic background. Genetic testing of both patients and pathogens will allow faster and more-precise diagnoses of many diseases.

Along with the many potential benefits of having complete sequence information are concerns about the misuse of this information. With the knowledge gained from genomic sequencing, many more genes for diseases, disorders, and behavioral and physical traits will be identified, increasing the number of genetic tests that can be performed to make predictions about the future phenotype and health of a person. There is concern that information from genetic testing might be used to discriminate against people who are carriers of disease-causing genes or who might be at risk for some future disease. Questions arise about who owns a person's genome sequence. Should employers and insurance companies have access to this information? What about relatives, who have similar genomes and who might also be at risk for some of the same diseases? There are also questions about the use of this information to select for specific traits in future offspring. All of these concerns are legitimate and must be addressed if we are to use the information from genome sequencing responsibly.

www.whfreeman.com/pierce   Ethical issues associated with the Human Genome Project and genomics in general

Connecting Concepts Across Chapters

Genomics, the focus of this chapter, uses many of the techniques described in Chapter 18 for studying individual genes and applies them to the entire genome. What is different about genomics is the tremendous amount of information that is produced by using these techniques, requiring special computational tools. Although the details of many of these methods are beyond the scope of this book, an understanding of the underlying principles of genomics and the general trends emerging from the results of genomic studies is important to a student in a general genetics course. Genomics holds great potential for understanding biological processes and for applications in health, agriculture, and biotechnology. It will undoubtedly be one of the most important areas of future genetic research.

A surprising result to emerge from the study of genomics is the finding that organisms that differ greatly in phenotype and complexity may possess many similar genes and, in fact, may not differ greatly in the total number of genes that they possess. This finding suggests that differences in phenotype are often due more to differing patterns of gene expression than to differences in the protein-coding information of their genomes.

Much of what has already been covered in this book is relevant to the study of genomics. Information on gene mapping (Chapter 7), DNA structure (Chapter 10), chromosome organization (Chapter 11), transcription (Chapter 13), protein synthesis (Chapter 15), and recombinant DNA (Chapter 18) is particularly critical for understanding the concepts presented in this chapter. Comprehension of some of the topics covered in subsequent chapters will be facilitated by an understanding of the information in this chapter; such topics include organelle DNA in Chapter 20 and evolutionary genetics in Chapter 23.

CONCEPTS SUMMARY

- Genomics is the field of genetics that attempts to understand the content, organization, and function of genetic information contained in whole genomes.

- Structural genomics concerns the organization and sequence of the genome. Functional genomics studies the biological function of genomic information. Comparative genomics compares the genomic information in different organisms.

- Genetic maps position genes relative to other genes by determining rates of recombination and are measured in

percent recombination. Physical maps are based on the physical distances between genes and are measured in base pairs.

- The location of sites recognized by restriction enzymes can be determined by cutting the DNA with each restriction enzyme separately and in combinations and then comparing the restriction fragments produced.

- DNA sequencing determines the base sequence of nucleotides along a stretch of DNA. The Sanger (dideoxy) method uses special substrates for DNA synthesis (dideoxynucleoside triphosphates, ddNTPs) that terminate synthesis after they are incorporated into the newly made DNA. Four reactions, each with a different ddNTP, are set up. In each reaction, DNA fragments of varying length are produced, all of which terminate in nucleotides with the same base. The products of the four reactions are separated by gel electrophoresis, and the sequence of the DNA synthesized is read from the pattern of bands on the gel.

- Sequencing a whole genome requires breaking the genome into small overlapping fragments whose DNA sequence can be determined in sequencing reactions. The individual sequences can be ordered into a whole genome sequence with the use of a map-based approach, in which fragments are assembled in order by using previously created genetic and physical maps, or with the use of a whole-genome shotgun approach, in which overlap between fragments is used to assemble them into a whole-genome sequence.

- The Human Genome Project is an effort to determine the entire sequence of the human genome. The project began officially in 1990; rough drafts of the human genome sequence were completed in 2000.

- Single-nucleotide polymorphisms are single-base differences in DNA between individuals and are valuable as markers in linkage studies.

- Expressed-sequence tags are markers associated with expressed (transcribed) DNA sequences. RNA from a cell is subjected to reverse transcription, producing cDNA molecules. A short stretch of the cDNA is then sequenced, which provides a marker that tags (identifies) the DNA fragment. Expressed-sequence tags can be used to find the genes expressed in a genome.

- Bioinformatics is a synthesis of molecular biology and computer science that develops tools to store, retrieve, and analyze DNA, cDNA, and protein sequence data.

- A transcriptome is the set of all RNA molecules transcribed from a genome; a proteome is the set of all the proteins encoded by the genome.

- Computer programs can identify genes by looking for characteristic features of genes within a sequence.

- Homologous genes are evolutionarily related. Orthologs are homologous sequences found in different organisms, whereas paralogs are homologous sequences found in the same organism. Gene function may be determined by looking for homologous sequences (both orthologs and paralogs) whose function has been previously determined.

- Functions of unknown genes may be inferred by searching databases for protein domains in genes that have been previously characterized.

- The functions of unknown genes can be inferred by using methods that compare DNA sequences, including phylogenetic profiling, protein fusion patterns, and linkage arrangements of genes in different organisms.

- A microarray consists of DNA fragments fixed in an orderly pattern to a solid support, such as a nylon filter or glass slide. When a solution containing a mixture of DNA or RNA is applied to the array, any nucleic acid that is complementary to the probe being used will bind to the probe. Microarrays can be used to monitor the expression of thousands of genes simultaneously.

- Genes affecting a particular function or trait can be identified through whole-genome mutagenesis screens. In this process, a group of organisms is screened for abnormal phenotypes subsequent to mutagenesis, and the mutated genes causing the abnormal phenotypes are identified by positional cloning.

- The genomes of many prokaryotic organisms have been determined. Most species have between 1 million and 3 million base pairs of DNA and from 1000 to 2000 genes. Compared with that of eukaryotic genomes, the density of genes in prokaryotic genomes is relatively uniform, with about one gene per 1000 bp. There is relatively little noncoding DNA between prokaryotic genes. Horizontal gene transfer (the movement of genes between different species) has been an important evolutionary process in prokaryotes.

- Eukaryotic genomes are larger and more variable in size than prokaryotic genomes. There is no clear relation between organismal complexity and the amount of DNA or number of genes among multicellular organisms. Much of the genomes of eukaryotic organisms consist of repetitive DNA. Transposable elements are very common in most eukaryotic genomes.

- Genomics is making important contributions to human health, agriculture, biotechnology, and our understanding of evolution.

IMPORTANT TERMS

genomics (p. 549)
structural genomics (p. 549)
functional genomics (p. 549)
comparative genomics (p. 549)
genetic map (p. 550)
physical map (p. 551)

restriction mapping (p. 552)
DNA sequencing (p. 553)
dideoxyribonucleoside
    triphosphate (ddNTP)
    (p. 553)
map-based sequencing (p. 556)

contig (p. 557)
whole-genome shotgun
    sequencing (p. 558)
single-nucleotide
    polymorphism (SNP)
    (p. 562)

expressed-sequence tag (EST)
    (p. 562)
bioinformatics (p. 562)
open reading frame (p. 562)
transcriptome (p. 563)
proteome (p. 563)

Worked Problems

1. A linear piece of DNA that is 30 kb long is first cut with *Bam*HI, then with *Hpa*II, and finally with both *Bam*HI and *Hpa*II together. Fragments of the following sizes were obtained from this reaction.

    *Bam*HI: 20-kb, 6-kb, and 4-kb fragments

    *Hpa*II: 21-kb and 9-kb fragments

    *Bam*HI and *Hpa*II: 20-kb, 5-kb, 4-kb, and 1-kb fragments

Draw a restriction map of the 30-kb piece of DNA, indicating the locations of the *Bam*HI and *Hpa*II restriction sites.

• Solution

This problem can be solved correctly through a variety of approaches; this solution applies one possible approach.

When cut by *Bam*HI alone, the linear piece of DNA is cleaved into three fragments; so there must be two *Bam*HI restriction sites. When cut with *Hpa*II alone, a clone of the same piece of DNA is cleaved into only two fragments; so there is a single *Hpa*II site.

Let's begin to determine the location of these sites by examining the *Hpa*II fragments. Notice that the 21-kb fragment produced when the DNA is cut by *Hpa*II is not present in the fragments produced when the DNA is cut by *Bam*HI and *Hpa*II together (the double digest); this result indicates that the 21-kb *Hpa*II fragment has within it a *Bam*HI site. If we examine the fragments produced by the double digest, we see that the 20-kb and 1-kb fragments sum to 21 kb; so a *Bam*HI site must be 20 kb from one end of the fragment and 1 kb from the other end.

Similarly, we see that the 9-kb *Hpa*II fragment does not appear in the double digest and that the 5-kb and 4-kb fragments in the double digest add up to 9 kb; so another *Bam*HI site must be 5 kb from one end of this fragment and 4 kb from the other end.

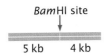

Now, let's examine the fragments produced when the DNA is cut by *Bam*HI alone. The 20-kb and 4-kb fragments are also present in the double digest; so neither of these fragments contains an *Hpa*II site. The 6-kb fragment, however, is not present in the double digest, and the 5-kb and 1-kb fragments in the double digest sum to 6 kb; so this fragment contains an *Hpa*II site that is 5 kb from one end and 1 kb from the other end.

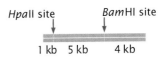

We have accounted for all the restriction sites, but we must still determine the order of the sites on the original 30-kb fragment.

Notice that the 5-kb fragment must be adjacent to both the 1-kb and 4-kb fragments; so it must be in between these two fragments.

We have also established that the 1-kb and 20-kb fragments are adjacent; because the 5-kb fragment is on one side, the 20-kb fragment must be on the other, completing the restriction map:

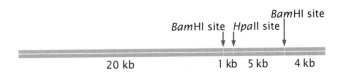

2. You are given the following DNA fragment to sequence: 5′–GCTTAGCATC–3′. You first clone the fragment in bacterial cells to produce sufficient DNA for sequencing. You isolate the DNA from the bacterial cells and carry out the dideoxy sequencing method. You then separate the products of the polymerization reactions by gel electrophoresis. Draw the bands that should appear on the gel from the four sequencing reactions.

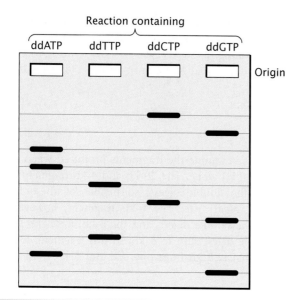

Reaction containing

ddATP   ddTTP   ddCTP   ddGTP

Origin

• Solution

In the dideoxy sequencing reaction, the original fragment is used as a template for the synthesis of a new DNA strand; and it is the sequence of the new strand that is actually determined. The first task, therefore, is to write out the sequence of the newly synthesized fragment, which will be complementary and antiparallel to the original fragment. The sequence of the newly synthesized strand, written $5' \rightarrow 3'$ is: $5'-GATGCTAAGC-3'$. Bands representing this sequence will appear on the gel, with the bands representing nucleotides near the $5'$ end of the molecule at the bottom of the gel.

The New Genetics
MINING GENOMES

GENOME ANALYSIS AND COMPARATIVE GENOMICS

Recent developments in genomics are revolutionizing our understanding of life, evolution, and medicine. This exercise allows you to explore and compare completed genomes at the Comprehensive Microbial Resources site at the Institute for Genomic Research. You will also explore the Human Genome and the Mouse Genome by using tools at the National Center for Biotechnology Information (NCBI).

COMPREHENSION QUESTIONS

1. (a) What is genomics and how does structural genomics differ from functional genomics?
    (b) What is comparative genomics?
* 2. What is the difference between a genetic map and a physical map? Which generally has higher resolution and accuracy and why?
3. What is the purpose of the dideoxynucleoside triphosphate in the dideoxy sequencing reaction?
* 4. What is the difference between a map-based approach to sequencing a whole genome and a whole-genome shotgun approach?
5. How are DNA fragments ordered into a contig by using restriction sites?
* 6. Describe the different approaches to sequencing the human genome that were taken by the international collaboration and Celera Genomics.
7. (a) What is an expressed-sequence tag (EST)?
    (b) How are ESTs created?
    (c) How are ESTs used in genomics studies?

8. What is a single-nucleotide polymorphism (SNP), and how are SNPs used in genomic studies?
9. How are genes recognized within genomic sequences?
*10. What are homologous sequences? What is the difference between orthologs and paralogs?
11. Describe several different methods for inferring the function of a gene by examining its DNA sequence.
12. What is a microarray and how can it be used to obtain information about gene function?
*13. Briefly outline how a mutagenesis screen is carried out.
14. Eukaryotic genomes are typically much larger than prokaryotic genomes. What accounts for the increased amount of DNA seen in eukaryotic genomes?
15. What is one consequence of differences in the G + C content of different genomes?
*16. What is horizontal gene exchange? How might it take place between different species of bacteria?
17. DNA content varies considerably among different multicellular organisms. Is this variation closely related to

the number of genes and the complexity of the organism? If not, what accounts for the differences?

*18. More than half of the genome of *Arabidopsis thaliana* consists of duplicated sequences. What mechanisms are thought to have been responsible for these extensive duplications?

19. The human genome does not encode substantially more protein domains than do invertebrate genomes, and yet it encodes many more proteins. How are more proteins encoded when the number of domains does not differ substantially?

20. What are some of the ethical concerns arising out of the information produced by the Human Genome Project?

APPLICATION QUESTIONS AND PROBLEMS

*21. A 22-kb piece of DNA has the following restriction sites.

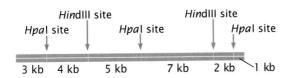

A batch of this DNA is first fully digested by *Hpa*I alone, then another batch is fully digested by *Hind*III alone, and finally a third batch is fully digested by both *Hpa*I and *Hind*III together. The fragments resulting from each of the three digestions are placed in separate wells of an agarose gel, separated by gel electrophoresis, and stained by ethidium bromide. Draw the bands as they would appear on the gel.

*22. A piece of DNA that is 14 kb long is cut first by *Eco*RI alone, then by *Sma*I alone, and finally by both *Eco*RI and *Sma*I together. The following results are obtained.

Digestion by *Eco*RI alone	Digestion by *Sma*I alone	Digestion by both *Eco*RI and *Sma*I
3-kb fragment	7-kb fragment	2-kb fragment
5-kb fragment	7-kb fragment	3-kb fragment
6-kb fragment		4-kb fragment
		5-kb fragment

Draw a map of the *Eco*RI and *Sma*I restriction sites on this 14-kb piece of DNA, indicating the relative positions of the restriction sites and the distances between them.

23. Suppose that you want to sequence the following DNA fragment:

Fragment to be sequenced:
$$5'-TCCCGGGAAA\text{-primer site}-3'$$

You first clone the fragment in bacterial cells to produce sufficient DNA for sequencing. You isolate the DNA from the bacterial cells and carry out the dideoxy sequencing method. You then separate the products of the polymerization reactions by gel electrophoresis. Draw the bands that should appear on the gel from the four sequencing reactions.

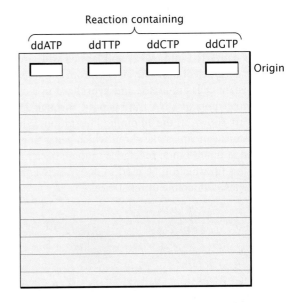

*24. Suppose that you are given a short fragment of DNA to sequence. You clone the fragment, isolate the cloned DNA fragment, and set up a series of four dideoxy reactions. You then separate the products of the reactions by gel electrophoresis and obtain the following banding pattern:

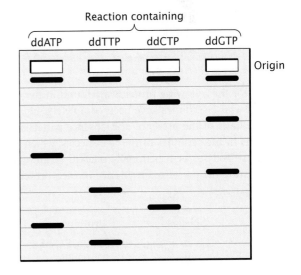

Write out the base sequence of the original fragment that you were given.

Original sequence: 5′ –_____–3′

25. Microarrays can be used to determine the levels of gene expression. In one type of microarray, hybridization of the red (experimental) and green (control) cDNAs is proportional to the relative amounts of mRNA in the samples. Red indicates the overexpression of a gene and green indicates the underexpression of a gene in the experimental cells relative to the control cells, yellow indicates equal expression in experimental and control cells, and no color indicates no expression in either experimental or control cells.

In one experiment, mRNA from a strain of antibiotic-resistant bacteria (experimental cells) is converted into cDNA and labeled with red fluorescent nucleotides; mRNA from a nonresistant strain of the same bacteria (control cells) is converted into cDNA and labeled with green fluorescent nucleotides. The cDNAs from the resistant and nonresistant cells are mixed and hybridized to a chip containing spots of DNA from genes 1 through 25. The results are shown in the adjoining illustration. What conclusions can you make about which genes might be implicated in antibiotic resistance in these bacteria? How might this information be used to design new antibiotics that are less vulnerable to resistance?

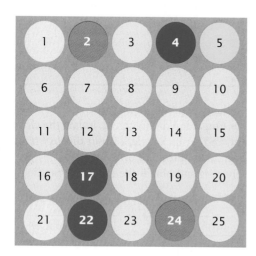

*26. Genes for the following proteins are found in five different species whose genomes have been completely sequenced. On the basis of the presence-and-absence patterns of these proteins in the genomes of the five species, which proteins are most likely to be functionally related? (Hint: Create a table listing the presence or absence of each protein in the five species.)

Species	Proteins
A	P1, P2, P3, P4, P5
B	P1, P2, P3, P5
C	P2, P4
D	P3, P5
E	P1, P3, P4, P5

27. The physical locations of several genes determined from genomic sequences are shown here for three bacterial species. On the basis of this information, which genes might be functionally related?

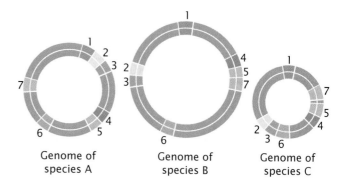

Genome of species A   Genome of species B   Genome of species C

28. The presence (+) or absence (−) of six sequence-tagged sites (STSs) in each of five bacterial artificial chromosome (BAC) clones (A–E) is indicated in the following table. Using these markers, put the BAC clones in their correct order and indicate the locations of the STS sites within them.

	STSs					
BAC clone	1	2	3	4	5	6
A	+	−	−	−	+	−
B	−	−	−	+	−	+
C	−	+	+	−	−	−
D	−	−	+	−	+	−
E	+	−	−	+	−	−

29. How does the density of genes found on chromosome 22 compare with the density of genes found on chromosome 21, two similar-sized chromosomes? How does the number of genes on chromosome 22 compare with the number found on the Y chromosome?

To answer these questions, go to the Ensembl Web site:

http://www.ensembl.org/

Under the heading *Ensembl Species,* click *Human.* On the left-hand side of the next page are pictures of the human chromosomes. Click on chromosome 22. You will be shown a picture of this chromosome and a histogram illustrating the density of total genes (uncolored bars) and known genes (colored bars). The number of known and novel (uncharacterized) genes is given in the upper right-hand side of the page, along with the chromosome length in base pairs.

Now go to chromosome 21 by pulling down the Change Chromosome menu and selecting chromosome 21. Examine the density and total number of genes for chromosome 21. Now do the same for the Y chromosome.

(a) Which chromosome has the highest density and greatest number of genes? Which has the fewest?

(b) Examine in more detail the genes at the tip of the short arm of the Y chromosome by clicking on the top bar in the histogram of genes. A more detailed view will be shown. What known genes are found in this region? How many novel genes are there in this region?

*30. Some researchers have proposed creating an entirely new, free-living organism with a minimal genome, the smallest set of genes that allows for replication of the organism in a particular environment. This organism could be used to design and create, from "scratch," novel organisms that might perform specific tasks such as the breakdown of toxic materials in the environment.

(a) How might the minimal genome required for life be determined?

(b) What, if any, social and ethical concerns might be associated with the creation of novel organisms by constructing an entirely new organism with a minimal genome?

31. What are some of the major differences between the ways in which genetic information is organized in the genomes of prokaryotes versus eukaryotes?

32. How do the following genomic features of prokaryotic organisms compare with those of eukaryotic organisms? How do they compare among eukaryotes?

(a) Genome size

(b) Number of genes

(c) Gene density (bp/gene)

(d) G + C content

(e) Number of exons

SUGGESTED READINGS

Adams, M. D., S. E. Celniker, R. A. Holt, C. A. Evans, J. D. Gocayne, et al. 2000. The genome sequence of *Drosophila melanogaster*. *Science* 287:2185–2195.
Report of the complete sequence of *Drosophila melanogaster*, the fruit fly.

Arabidopsis Genome Initiative. 2000. Analysis of the genome sequence of the flowering plant *Arabidopsis thaliana*. *Nature* 408:796–815.
Analysis of the complete genome of the first plant genome to be published.

C. elegans Sequencing Consortium. 1998. Genome sequence of the nematode *C. elegans*: a platform for investigating biology. *Science* 282: 2012–2018.
Report of the sequence and analysis of the genome of *C. elegans*, the roundworm.

Choe, M. K., D. Magnus, A. L. Caplan, D. McGee, and the Ethics of Genomics Group. 1999. Ethical considerations in synthesizing a minimal genome. *Science* 286:2087–2090.
A discussion of some of the ethical implications of creating novel organisms by constructing a minimal genome.

Cole, S. T., K. Eiglmeier, J. Parkhill, K. D. James, N. R. Thomson, et al. 2001. Massive gene decay in the leprosy bacillus. *Nature* 409:1007–1011.
Report of the genomic sequence of *Mycobacterium leprae*, the bacterium that causes leprosy.

Davies, K. 2001. *Cracking the Genome: Inside the Race to Unlock Human DNA*. New York: Simon & Schuster.
A very readable account of the history of the human genome project, placed within the context of advances in molecular biology.

Dean, P. M., E. D. Zanders, and D. S. Bailey. 2001. Industrial-scale genomics-based drug design and discovery. *Trends in Biotechnology* 19:288–292.
A review of the effect of genomics on drug discovery and design.

Eisenberg, D., E. M. Marcotte, I. Xenarios, and T. O. Yeates. 2000. Protein function in the post-genomic era. *Nature* 405:823–826.
A review of how protein function can be inferred from DNA sequence data.

Fraser, C. M., J. Eisen, R. D. Fleischmann, K. A. Ketchum, and S. Peterson. 2001. Comparative genomics and understanding of microbial biology. *Emerging Infectious Diseases* 6:505–512.
An excellent overview of what has been learned from whole-genome sequences of prokaryotic organisms.

Howard, K. 2000. The bioinformatics gold rush. *Scientific American* 283(1):58–63.
A good overview of bioinformatics and its economic potential. In the same issue, see articles on "The human genome business today" and "Beyond the human genome."

International Human Genome Sequencing Consortium. 2001. Initial sequencing and analysis of the human genome. *Nature* 409:860–921.
A report from the public consortium on its version of the human genome sequence. Many articles in this issue of *Nature* report on various aspects of the human genome.

International SNP Map Working Group. 2001. A map of human genome sequence variation containing 1.42 million single nucleotide polymorphisms. *Nature* 409:928–933.
A report on mapping single nucleotide polymorphisms in the human genome.

Knight, J. 2001. When the chips are down. *Nature* 410:860–861.
News story about progress in using DNA chips to monitor
gene expression.

Mewes, H. W., K. Albermann, M. Bahr, D. Frishman, A.
Gleissner, J. Hani, K. Heumann, K. Kleine, A. Maierl, S. G.
Oliver, F. Pfeiffer, and A. Zollner. 1997. Overview of the yeast
genome. *Nature* 387:7–8.
A broad look at the yeast genome and what can be learned
from its sequence.

Rosamond, J., and A. Allsop. 2000. Harnessing the power of the
genome in the search for new antibiotics. *Science*
287:1973–1976.
Describes how genomic sequences can be useful in the search
for new drugs.

Rubin, G. M., M. D. Yandell, J. R. Wortman, G. L. G. Miklos,
C. R. Nelson, I. K. Hariharan, et al. 2000. Comparative
genomics of the eukaryotes. *Science* 287:2204–2215.
An analysis of the proteins encoded by the genomes of fly,
worm, and yeast.

Sander, C. 2000. Genomic medicine and the future of health
care. *Science* 287:1977–1978.
A discussion of the effect of genomics on the future of
medicine.

Venter, J. C., M. D. Adams, E. W. Myers, P. W. Li, R. J. Mural,
et al. 2001. The sequence of the human genome. *Science*
291:1304–1351.
An analysis of the private draft of the human genome
sequence. Much of this issue of *Science* reports on the human
genome sequence and its analysis.

20 Organelle DNA

Studies of mitochondrial DNA indicate that the genetic integrity of North American wolves is being threatened by hybridization with coyotes. (John Shaw/Bruce Coleman.)

Coyote Genes in Declining Wolves

North America is home to two wild canids: the gray wolf (*Canis lupus*) and the coyote (*Canis latrans*). Before European settlement, gray wolves ranged across much of North America, occupying forest, plains, desert, and tundra habitat (FIGURE 20.1). Coyotes had a more limited distribution, being confined primarily to plains and deserts. With the expansion of European settlement and the development of intensive agriculture in the eighteenth and nineteenth centuries, the distribution of wolves and coyotes changed dramatically (see Figure 20.1). Wolf populations in North America declined precipitously owing to habitat destruction and deliberate extermination. In contrast, coyote populations expanded, probably because competition from wolves was eliminated and because coyotes were better able to

adapt to human disruption of the ecosystem. Today coyotes are found throughout most of North America.

Habitat alternation and changes in their distributions have increased interactions between wolves and coyotes in recent times, providing more opportunities for hybridization between the two species. In captivity, wolves and coyotes will interbreed and produce fertile hybrids. Large coyotes in New England and southeastern Canada may be the result of hybridization between wolves and coyotes in these areas. To what extent is hybridization occurring between coyotes and wolves in nature?

To answer this question, Niles Lehman and his colleagues studied DNA in the mitochondria of wolves and coyotes. **Mitochondrial DNA** (mtDNA) can be helpful in determining hybridization between animals for two reasons: (1) in animals, it is inherited only from the female

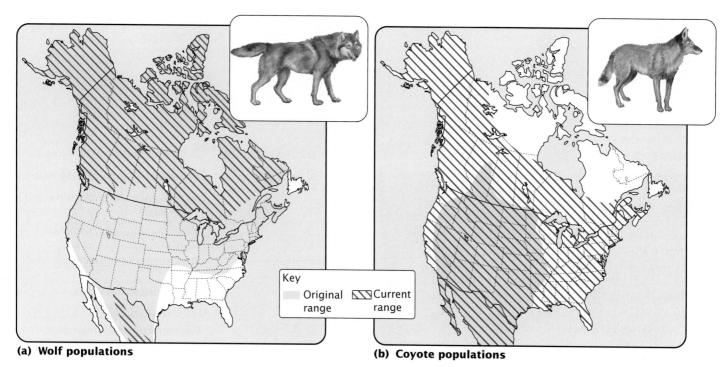

(a) Wolf populations

(b) Coyote populations

Key

▨ Original range   ⧄ Current range

◀ **20.1 Original and current ranges of the gray wolf (Canis lupus) and the coyote (Canis latrans).** Originally, the gray wolf occupied most of North America, but its current range is restricted to northern Minnesota, Canada, and Alaska. The coyote originally occupied the plains and desert habitat in the midwestern United States and Mexico. Today, the coyote is found throughout most of North America. The results of studies of mitochondrial DNA reveal that hybridization is occurring between wolves and coyotes.

parent and (2) it evolves rapidly. Lehman and his colleagues gathered tissue and blood samples from more than 500 gray wolves and coyotes in North America, extracted mtDNA from the samples, and analyzed restriction fragment length polymorphisms (see Chapter 18) in the mtDNA. The results of their study revealed two major clusters of mtDNA among the canids: one consisting of wolf mtDNA and another of coyote-like mtDNA. Surprisingly, the coyote-like mtDNA cluster included several samples that had been obtained from wolves, indicating that some wolves possessed coyote-like mtDNA. The wolves with coyote-like mtDNA were all from the U.S.–Canadian border area, which has recently been invaded by coyotes. No wolves from Alaska or northern Canada possessed coyote-like mtDNA. All the coyotes had only coyote-like mtDNA.

These results indicate that unidirectional hybridization has taken place between coyotes and wolves: coyote mtDNA has entered wolf populations, but wolf mtDNA has not entered coyote populations. The fact that in animals mtDNA is inherited only from the female parent implies that female coyotes are mating successfully with male wolves and the wolf–coyote hybrids are backcrossing with wolves, introducing coyote genes into wolf populations.

These findings have important implications for the future of gray wolves in North America. Hybridization between wolves and coyotes threatens to erode the genetic integrity of wolves. As human activities encroach on areas occupied by wolves, wolves and coyotes will be forced into closer contact and there will be more hybridization; the wolf genome (both mitochondrial and nuclear) will become increasingly diluted by coyote DNA. Current efforts to reintroduce wolves into former territories, which are now occupied by coyotes, may lead to further hybridization, ultimately harming, rather than helping, wolf populations.

DNA sequences found in mitochondria and other organelles possess unique properties that make these sequences useful in the fields of conservation biology, evolution, and genetic diseases. Uniparental inheritance exhibited by genes found in mitochondria and chloroplasts was discussed in Chapter 5; the present chapter examines molecular aspects of organelle DNA. We begin by briefly considering the structures of mitochondria and chloroplasts, the inheritance of traits encoded by their genes, and their evolutionary origin. We then examine the general characteristics of mtDNA, followed by a discussion of the organization and function of different types of mitochondrial genomes.

Mitochondrion

Chloroplast

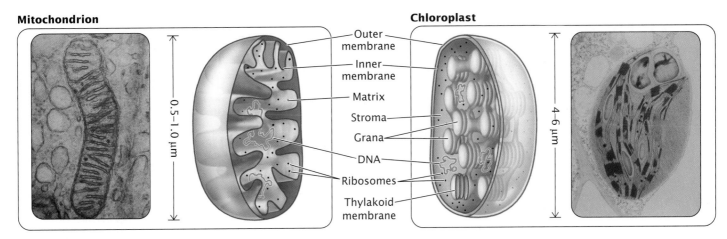

Outer membrane
Inner membrane
Matrix
Stroma
Grana
DNA
Ribosomes
Thylakoid membrane

0.5–1.0 μm

4–6 μm

◀20.2 **Comparison of the structures of mitochondria and chloroplasts.** (Left, Don Fawcett/Visuals Unlimited; right, Biophoto Associates/Photo Researchers.)

Finally, we turn to **chloroplast DNA** (cpDNA), examining its characteristics, organization, and function.

The Biology of Mitochondria and Chloroplasts

Mitochondria and chloroplasts are membrane-bounded organelles located in the cytoplasm of eukaryotic cells (◀FIGURE 20.2). Mitochondria are present in almost all eukaryotic cells, whereas chloroplasts are found in multicellular plants and some algae. Both organelles generate ATP, the universal energy carrier of cells.

Mitochondrion and Chloroplast Structure

Mitochondria are from 0.5 to 1.0 micrometer (μm) in diameter, about the size of a typical bacterium; chloroplasts are typically from about 4 to 6 μm in diameter. Both are surrounded by two membranes that enclose a region (called the matrix in mitochondria and the stroma in chloroplasts) that contains enzymes, ribosomes, RNA, and DNA. In mitochondria, the inner membrane is highly folded; embedded within it are the enzymes that catalyze electron transport and oxidative phosphorylation. Chloroplasts have a third membrane, called the thylakoid membrane, which is highly folded and stacked to form aggregates called grana. This membrane bears the pigments and enzymes required for photophosphorylation. New mitochondria and chloroplasts arise by the division of existing organelles (◀FIGURE 20.3). Mitochondria

◀20.3 **New mitochondria arise by division of existing mitochondria.** (a) DNA molecules within the mitochondria segregate randomly in organelle division. (b) Electron micrograph of a dividing mitochondrion from a liver cell. (T. Kanaseki and D. Fawcett/Visuals Unlimited.)

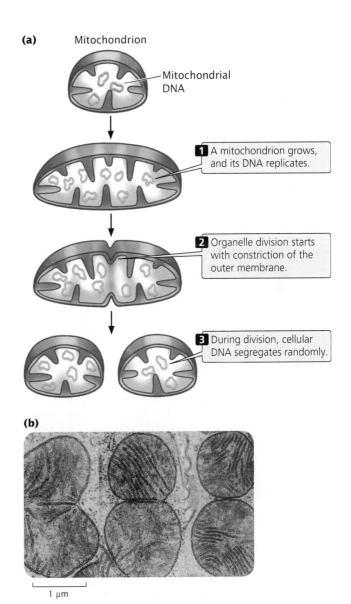

(a)   Mitochondrion

Mitochondrial DNA

1 A mitochondrion grows, and its DNA replicates.

2 Organelle division starts with constriction of the outer membrane.

3 During division, cellular DNA segregates randomly.

(b)

1 μm

and chloroplasts possess DNA that encodes polypeptides used by the organelle, as well as rRNAs and tRNAs needed for the translation of these proteins.

The Genetics of Organelle-Encoded Traits

Mitochondria and chloroplasts are present in the cytoplasm and are usually inherited from a single parent. Thus traits encoded by mtDNA and cpDNA exhibit uniparental inheritance. In animals, mtDNA is inherited almost exclusively from the female parent, although occasional male transmission of mtDNA has been documented. Paternal inheritance of organelles is common in gymnosperms and occurs occasionally in angiosperms as well. Some plants even exhibit biparental inheritance of mtDNA and cpDNA.

Individual cells may contain from dozens to hundreds of organelles, each with numerous copies of the organelle genome; so each cell typically possesses from hundreds to thousands of copies of mitochondrial and chloroplast genomes (◀FIGURE 20.4). A mutation arising within one organellar DNA molecule generates a mixture of mutant and wild-type DNA sequences within that cell. The occurrence of two distinct varieties of DNA within the cytoplasm of a single cell is termed **heteroplasmy**. When a heteroplasmic cell divides, the organelles segregate randomly into the two progeny cells in a process called **replicative segregation** (◀FIGURE 20.5), and chance determines the proportion of mutant organelles in each cell. Although most progeny cells will inherit a mixture of mutant and normal organelles, just by chance some cells may receive organelles with only mutant or only wild-type sequences; this situation is known as **homoplasmy**.

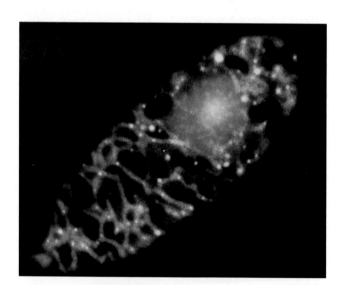

◀20.4 **Individual cells may contain many mitochondria, each with several copies of the mitochondrial genome.** Shown is a cell of *Euglena gracilis*, stained so that the nucleus appears red, mitochondria green, and mtDNA yellow. (From Y. Huyashi and K. Veda, *Journal of Cell Sciences* 93, 1989, 565.)

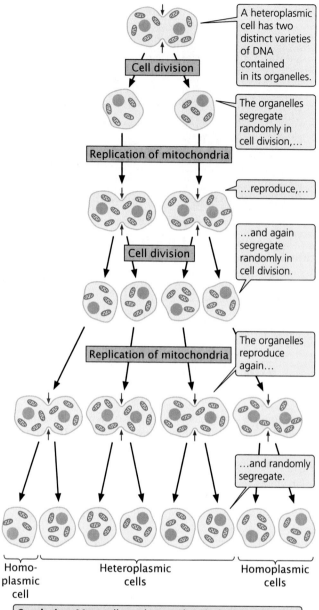

Conclusion: Most cells are heteroplasmic, but, just by chance, some cells may receive only one type of organelle (e.g., they may receive all normal or all mutant).

◀20.5 **Organelles in a heteroplasmic cell divide randomly into the progeny cells.** This diagram illustrates replicative segregation in mitosis; the same process also takes place in meiosis.

When replicative segregation occurs in somatic cells, it may create phenotypic variation within a single organism; different cells of the organism may possess different proportions of mutant and wild-type sequences, resulting in different degrees of phenotypic expression among tissues. When replicative segregation occurs in the germ cells of a heteroplasmic cytoplasmic donor, the offspring may show quite different phenotypes.

The disease known as myoclonic epilepsy and ragged-red fiber disease syndrome (MERRF) is caused by a mutation in an mtDNA gene. A 20-year-old person who carried this mutation in 85% of his mtDNAs displayed a normal phenotype, whereas a cousin who had the mutation in 96% of his mtDNAs was severely affected. In diseases caused by mutations in mtDNA, the severity of the disease is frequently related to the proportion of mutant mtDNA sequences inherited at birth.

A number of traits encoded by organellar DNA have been studied. One of the first to be examined in detail was the phenotype produced by *petite* mutations in yeast (◀FIGURE 20.6). In the late 1940s, Boris Ephrussi and his colleagues noticed that, when grown on solid medium, some colonies of yeast were much smaller than normal. Examination of these *petite* colonies revealed that growth rates of the cells within the colonies were greatly reduced. The results of biochemical studies demonstrated that *petite* mutants were unable to carry out aerobic respiration; they obtained all of their energy from anaerobic respiration (glycolysis), which is much less efficient than aerobic respiration and results in the smaller colony size.

Some *petite* mutations are inherited from both parents and are defects in nuclear DNA. However, most *petite* mutations are inherited from only a single parent; such mutants possess large deletions in mtDNA or, in some cases, are missing mtDNA entirely. Because much of their mtDNA encodes enzymes that catalyze aerobic respiration, the *petite*

mutants are unable to carry out aerobic respiration and therefore cannot produce normal quantities of ATP, which inhibits their growth.

Another known mtDNA mutation occurs in *Neurospora*. Isolated by Mary Mitchell in 1952, *poky* mutants grow slowly, display cytoplasmic inheritance, and have abnormal amounts of cytochromes. Cytochromes are protein components of the electron-transport chain of the mitochondria and play an integral role in the production of ATP. Most organisms have three primary types of cytochromes: cytochrome *a*, cytochrome *b*, and cytochrome *c*. *Poky* mutants have cytochrome *c* but no cytochrome *a* or *b*. Like *petite* mutants, *poky* mutants are defective in ATP synthesis and therefore grow more slowly than normal wild-type cells.

In recent years, a number of genetic diseases that result from mutations in mtDNA have been identified in humans. Leber hereditary optic neuropathy (LHON), which typically leads to sudden loss of vision in middle age, results from mutations in the mtDNA genes that encode electron-transport proteins. Another disease caused by mitochondrial mutations is neurogenic muscle weakness, ataxia, and retinitis pigmentosa (NARP), which is characterized by seizures, dementia, and developmental delay. Other mitochondrial diseases include Kearns-Sayre syndrome (KSS) and chronic external opthalmoplegia (CEOP), both of which result in paralysis of the eye muscles, droopy eyelids, and, in severe cases, vision loss, deafness, and dementia. All of these diseases exhibit cytoplasmic inheritance and variable expression (see Chapter 5).

A trait in plants that is produced by mutations in mitochondrial genes is cytoplasmic male sterility, a mutant phenotype found in more than 140 different plant species and inherited only from the maternal parent. These mutations inhibit pollen development but do not affect female fertility.

A number of cpDNA mutants also have been discovered. One of the first to be recognized was leaf variegation in the *Mirabilis jalapaa*, which was studied by Carl Correns in 1909 (see pp. 118–119 in Chapter 5). In the green alga *Chlamydomonas*, streptomycin-resistant mutations occur in cpDNA, and a number of mutants exhibiting altered pigmentation and growth in higher plants have been traced to defects in cpDNA.

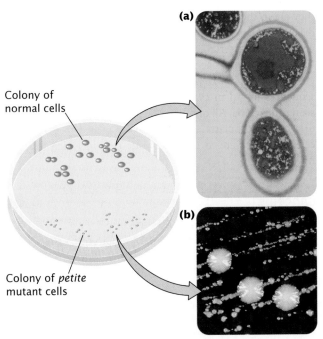

Colony of normal cells

(a)

(b)

Colony of *petite* mutant cells

◀ 20.6 The *petite* mutants have large deletions in their mtDNA and are unable to carry out oxidative phosphorylation. (a) A normal yeast cell and (b) a *petite* mutant. (Part a, David M. Phillips/Visuals Unlimited; part b, Courtesy of Dr. Des Clark-Walker, Research School of Biological Sciences, the Australian National University.)

Concepts

In most organisms, genes encoded by mtDNA and cpDNA are inherited entirely from a single parent. A gamete may contain more than one distinct type of mtDNA or cpDNA; in these cases, random segregation of the organelle DNA may produce phenotypic variation within a single organism or it may produce different degrees of phenotypic expression among progeny of a cross.

www.whfreeman.com/pierce   More information about mitochondria and chloroplasts

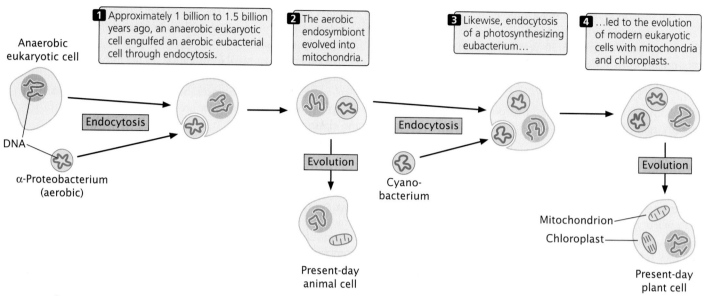

20.7 The endosymbiotic theory proposes that mitochondria and chloroplasts in eukaryotic cells arose from eubacteria.

The Endosymbiotic Theory

Chloroplasts and mitochondria are in many ways similar to bacteria. This resemblance is not superficial; indeed there is compelling evidence that these organelles evolved from eubacteria (see p. 18 in Chapter 2). The **endosymbiotic theory** (▪Figure 20.7) proposes that mitochondria and chloroplasts were once free-living bacteria that became internal inhabitants (endosymbionts) of early eukaryotic cells. According to this theory, between 1 billion and 1.5 billion years ago, a large, anaerobic eukaryotic cell engulfed an aerobic eubacterium, one that possessed the enzymes necessary for oxidative phosphorylation. The eubacterium provided the formerly anaerobic cell with the capacity for oxidative phosphorylation and allowed it to produce more ATP for each organic molecule digested. With time, the endosymbiont became an integral part of the eukaryotic host cell, and its descendants evolved into present-day mitochondria. Sometime later, a similar relation arose between photosynthesizing eubacteria and eukaryotic cells, leading to the evolution of chloroplasts.

A great deal of evidence supports the idea that mitochondria and chloroplasts originated as eubacterial cells. Many modern, single-celled eukaryotes (protists) are hosts to endosymbiotic bacteria. Mitochondria and chloroplasts are similar in size to present-day eubacteria and possess their own DNA, which has many characteristics in common with eubacterial DNA. Mitochondria and chloroplasts possess ribosomes, some of which are similar in size and structure to eubacterial ribosomes. Finally, antibiotics that inhibit protein synthesis in eubacteria but do not affect protein synthesis in eukaryotic cells also inhibit protein synthesis in these organelles.

The strongest evidence for the endosymbiotic theory comes from the study of DNA sequences in organellar DNA. Ribosomal RNA and protein-encoding gene sequences in mitochondria and chloroplasts have been found to be more closely related to sequences in the genes of eubacteria than they are to those found in the eukaryotic nucleus. Mitochondrial DNA sequences are most similar to sequences found in a group of eubacteria called the α-proteobacteria, suggesting that the original bacterial endosymbiont came from this group. Chloroplast DNA sequences are most closely related to sequences found in cyanobacteria, a group of photosynthesizing eubacteria. All of this evidence indicates that mitochondria and chloroplasts are more closely related to eubacterial cells than they are to the eukaryotic cells in which they are now found.

Concepts

Mitochondria and chloroplasts are membrane-bounded organelles of eukaryotic cells that generally possess their own DNA. The well-supported endosymbiotic theory proposes that these organelles began as free-living eubacteria that developed stable endosymbiotic relations with early eukaryotic cells.

Mitochondrial DNA

In animals and most fungi, the mitochondrial genome consists of a single, highly coiled, circular DNA molecule. Plant mitochondrial genomes often exist as a complex collection of multiple circular DNA molecules. Each mitochondrion

Table 20.1 — Sizes of mitochondrial genomes in selected organisms

Organism	Size of mtDNA in Nucleotide Pairs
Ascaris summ (nematode worm)	14,284
Drosophila melanogaster (fruit fly)	19,517
Lumbricus terrestis (earthworm)	14,998
Xenopus laevis (frog)	17,553
Mus musculus (house mouse)	16,295
Canis familiaris (dog)	16,728
Homo sapiens (human)	16,569
Pichia canadensis (fungus)	27,694
Podospora anserina (fungus)	100,314
Schizosaccharomyces pome (fungus)	19,431
Saccharomyces cerevisiae (fungus)	85,779*
Chlamydomonas reinhardtii (green alga)	15,758
Paramecium aurelia (protist)	40,469
Reclinomonas americana (protist)	69,034
Arabidopsis thaliana (plant)	166,924
Brassica hirta (plant)	208,000
Cucumis melo (plant)	2,400,000

*Size varies among strains.

contains multiple copies of the mitochondrial genome, and a cell may contain many mitochondria. A typical rat liver cell, for example, has from 5 to 10 mtDNA molecules in each of about 1000 mitochondria; so each cell possesses from 5000 to 10,000 copies of the mitochondrial genome, and mtDNA constitutes about 1% of the total cellular DNA in a rat liver cell. Like eubacterial chromosomes, mtDNA lacks the histone proteins normally associated with eukaryotic nuclear DNA. The guanine–cytosine (GC) content of mtDNA is often sufficiently different from that of nuclear DNA that mtDNA can be separated from nuclear DNA by density gradient centrifugation.

Mitochondrial genomes are small compared with nuclear genomes and vary greatly in size among different organisms (Table 20.1). Most of this size variation is in noncoding sequences such as introns and intergenic regions.

Gene Structure and Organization of mtDNA

The nucleotide sequence of the mitochondrial genome has been determined for a variety of different organisms, including protists, fungi, plants, and animals. The genes for many of the structural proteins and enzymes found in mitochondria are actually encoded by *nuclear* DNA, translated on cytoplasmic ribosomes, and then transported into the mitochondria; the mitochondrial genome typically encodes only a few rRNA and tRNA molecules needed for mitochondrial protein synthesis. The organization of these mitochondrial genes and how they are expressed is extremely diverse across organisms.

Ancestral and derived mitochondrial genomes Mitochondrial genomes can be divided in two basic types—ancestral genomes and derived genomes—although there is much variation within each type and the mtDNA of some organisms does not fit well into either category. Ancestral mitochondrial genomes are found in some plants and protists and retain many characteristics of their eubacterial ancestors. These mitochondrial genomes contain more genes than do derived genomes, have rRNA genes that encode eubacterial-like ribosomes, and have a complete or almost complete set of tRNA genes. They possess few introns and little noncoding DNA between genes, generally use universal codons, and have their genes organized into clusters similar to those found in eubacteria.

Derived mitochondrial genomes, in contrast, are usually smaller than ancestral genomes and contain fewer genes. Their rRNA genes and ribosomes differ substantially from those found in typical eubacteria. The DNA sequences found in derived mitochondrial genomes differ more from typical eubacterial sequences than do ancestral genomes, and they contain nonuniversal codons. Most animal and fungal mitochondrial genomes fit into this category.

Human mtDNA Human mtDNA is a circular molecule encompassing 16,569 bp that encode two rRNAs, 22 tRNAs, and 13 proteins. The two nucleotide strands of the molecule differ in their base composition: the heavy (H) strand has more guanine nucleotides, whereas the light (L) strand has more cytosine nucleotides. The H strand is the template for both rRNAs, 14 of the 22 tRNAs, and 12 of the 13 proteins, whereas the L strand serves as template for only 8 of the tRNAs and one protein.

The origin of replication for the H strand is within a region known as the **D loop** (◀ FIGURE 20.8), which also contains promoters for both the H and L strands. Human mtDNA is highly economical in its organization: there are few noncoding nucleotides between the genes; almost all the mRNA is translated (there are no 5′ and 3′ untranslated regions); and there are no introns. Each strand has only a single promoter; so transcription produces two very large RNA precursors that are later cleaved into individual RNA molecules. Many of the genes that encode polypeptides even lack a complete termination codon, ending in either U or UA; the addition of a poly(A) tail to the 3′ end of the mRNA provides a UAA termination codon that halts translation. Human mtDNA also contains very little repetitive DNA. The one region of the human mtDNA that does contain some noncoding nucleotides is the D loop.

(a)

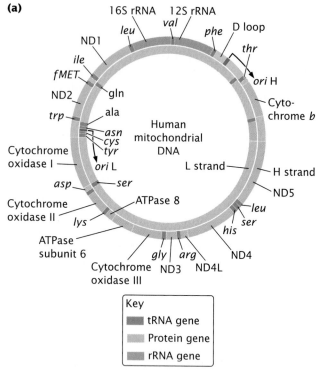

(b)

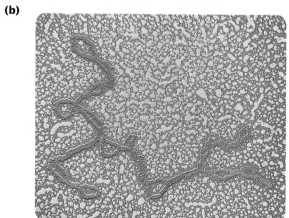

◀20.8 **The human mitochondrial genome, consisting of 16,569 bp, is highly economic in its organization.** (a) The outer circle represents the heavy (H) strand, and the inner circle represents the light (L) strand. The origins of replication for the H and L strands are *ori* H and *ori* L, respectively. (b) Electron micrograph of isolated mtDNA. (Part b, CNRI/Photo Researchers.)

www.whfreeman.com/pierce   Information on genes of the human mitochondrial genome

Yeast mtDNA The organization of yeast mtDNA is quite different from that of human mtDNA. Although the yeast mitochondrial genome with 78,000 bp is nearly five times as large, it encodes only six additional genes, for a total of 2 rRNAs, 25 tRNAs, and 16 polypeptides (◀FIGURE 20.9). Most of the extra DNA in the yeast mitochondrial genome

consists of noncoding sequences. Yeast mitochondrial genes are separated by long intergenic spacer regions that have no known functions. The genes encoding polypeptides often include regions that encode 5′ and 3′ untranslated regions of the mRNA; there are also short repetitive sequences and some duplications.

www.whfreeman.com/pierce   Information on the Fungal Mitochondrial Genome Project (FMGP), whose goals are to sequence and analyze complete mitochondrial genomes from all major groups of fungi

Flowering plant mtDNA Flowering plants (angiosperms) have the largest and most complex mitochondrial genomes known; their mitochondrial genomes range in size from 186,000 bp in white mustard to 2,400,000 bp in muskmelon. Even closely related plant species may differ greatly in the sizes of their mtDNA.

Part of the extensive size variation in the mtDNA of flowering plants can be explained by the presence of large direct repeats, which constitute large parts of the mitochondrial genome. Crossing over between these repeats can generate multiple circular chromosomes of different sizes. The mitochondrial genome in turnip, for example, consists of a "master circle" consisting of 218,000 bp that has direct repeats (◀FIGURE 20.10). Homologous recombination between the repeats can generate two smaller circles of 135,000 bp and 83,000 bp. Other species contain several direct repeats, providing possibilities for complex crossing-over events that may increase or decrease the number and sizes of the circles.

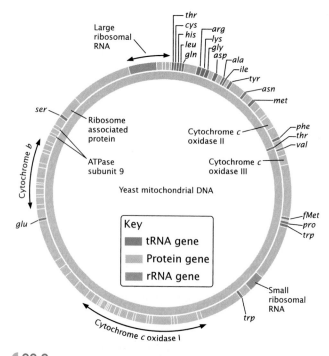

◀20.9 **The yeast mitochondrial genome, consisting of 78,000 bp, contains much noncoding DNA.**

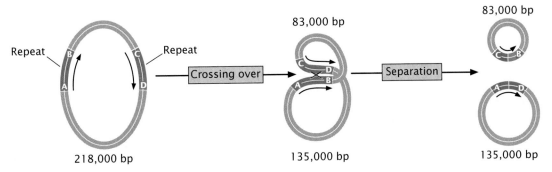

20.10 Size variation in plant mtDNA can be generated through recombination between direct repeats. In turnips, the mitochondrial genome consists of a "master circle" of 218,000 bp, which has direct repeats that are separated by 135,000 bp on one side and 83,000 bp on the other. Crossing over between the direct repeats produces two smaller circles of 135,000 bp and 83,000 nucleotide pairs.

Nonuniversal Codons in mtDNA

In the vast majority of bacterial and eukaryotic DNA, the same codons specify the same amino acids (see p. 416 in Chapter 15). However, there are exceptions to this universal code, and many of these exceptions are in mtDNA (Table 20.2). There is not a "mitochondrial code"; rather, exceptions to the universal code exist in mitochondria, and these exceptions often differ among organisms. For example, AGA specifies arginine in the universal code, but AGA codes for serine in *Drosophila* mtDNA and is a stop codon in mammalian mtDNA.

(Concepts)

The mitochondrial genome consists of circular DNA with no associated histone proteins. The size and structure of mtDNA differ greatly among organisms. Human mtDNA exhibits extreme economy, but mtDNAs found in yeast and flowering plants contain many noncoding nucleotides and repetitive sequences. Mitochondrial DNA in most flowering plants is large and typically has one or more large direct repeats that can recombine to generate smaller or larger molecules.

Replication, Transcription, and Translation of mtDNA

Mitochondrial DNA does not replicate in the orderly, regulated manner of nuclear DNA. Mitochondrial DNA is synthesized throughout the cell cycle and is not coordinated with the synthesis of nuclear DNA. Which mtDNA molecules are replicated at any particular moment appears to be random; within the same mitochondrion, some molecules are replicated two or three times, whereas others are not replicated at all. Furthermore, the two strands in human mtDNA may not replicate synchronously. Mitochondrial DNA is replicated by a special DNA polymerase called DNA polymerase γ (gamma). Presumably, helicases and topoisomerases are required for mitochondrial DNA replication, just as they are in eubacterial and nuclear DNA replication.

The processes of transcription and translation of mitochondrial genes exhibit extensive variation among different organisms. In human mtDNA, eubacterial-like operons are absent, and there are two promoters, one for each nucleotide strand, within the D loop. Transcription of the two strands proceeds in opposite directions, generating two giant precursor RNAs that are then cleaved to yield individual rRNAs, tRNAs, and mRNAs. As the tRNAs are transcribed, they fold up into three-dimensional configurations. These configurations are recognized and cut out by enzymes. The tRNA

Table 20.2	Nonuniversal codons found in mtDNA			
		mtDNA		
Codon	**Universal Code**	**Vertebrate**	***Drosophila***	**Yeast**
UGA	Stop	Tryptophan	Tryptophan	Tryptophan
AUA	Isoleucine	Methionine	Methionine	Methionine
AGA	Arginine	Stop	Serine	Arginine

Source: After T. D. Fox, *Annual Review of Genetics* 21 (1987), p. 69.

genes generally flank the protein and rRNA genes; so cleavage of the tRNAs releases mRNAs and rRNAs. In the mitochondrial genomes of fungi, plants, and protists, there are multiple promoters, although genes are occasionally arranged and transcribed in operons.

Most mRNA molecules produced by the transcription of mtDNA are not capped at their 5′ ends, unlike mRNA transcribed from nuclear genes (See Figure 14.6). Poly(A) tails are added to the 3′ end of some mRNAs encoded by animal mtDNA, but poly(A) tails are missing from those encoded by mtDNA in fungi, plants, and protists. The poly(A) tails added to animal mitochondrial mRNAs are shorter than those attached to nuclear-encoded mRNA and are probably added by an entirely different mechanism.

Some of the genes in yeast and plant mitochondrial DNA contain introns, many of which are self-splicing. RNA encoded by some mitochondrial genomes undergoes extensive editing (see pp. 391–393 in Chapter 14).

Translation in mitochondria has some similarities to eubacterial translation, but there are also important differences. In mitochondria, protein synthesis is initiated at AUG start codons by N-formylmethionine, just as in eubacterial initiation of translation. Mitochondrial translation also employs elongation factors similar to those seen in eubacteria, and the same antibiotics that inhibit translation in eubacteria also inhibit translation in mitochondria. However, mitochondrial ribosomes are variable in structure and are often different from those seen in both eubacterial and eukaryotic cells. Additionally, the initiation of translation in mitochondria must be different from that of both eubacterial and eukaryotic cells, because animal mitochondrial mRNA contains no Shine-Dalgarno ribosome-binding site and no 5′ cap. (A Shine-Dalgarno sequence has been observed in mitochondrial mRNA of the protozoan *Reclinomonas americana,* which has a very primitive, eubacterial-like mitochondrion.)

There is also much diversity in the tRNAs encoded by various mitochondrial genomes. Human mtDNA encodes 22 of the 32 tRNAs required for translation in the cytoplasm. (Only 32 are required in cytoplasmic translation because wobble at the third position of the codon allows tRNAs to pair with more than one codon; see p. 415 in Chapter 15.) In human mitochondrial translation, there is even more wobble than in cytoplasmic translation; many mitochondrial tRNAs will recognize any of the four nucleotides in the third position of the codon, permitting translation to take place with even fewer tRNAs. The increased wobble also means that any change in a DNA nucleotide at the third position of the codon will be a silent mutation (see p. 478 in Chapter 17) and will not alter the amino acid sequence of the protein. Thus more of the changes that occur in mtDNA are silent and accumulate over time, contributing to a higher rate of evolution. In some organisms, fewer than 22 tRNAs are encoded by mtDNA; in these organisms, nuclear-encoded tRNAs are imported from the cytoplasm to help carry out translation. In

yet other organisms, the mitochondrial genome encodes a complete set of all 32 tRNAs.

Concepts

The processes of replication, transcription, and translation vary widely among mitochondrial genomes and exhibit a curious mix of eubacterial, eukaryotic, and unique characteristics.

Evolution of mtDNA

As already mentioned, comparisons of DNA sequences in mitochondrial genomes with homologous sequences in other organisms strongly support a common eubacterial origin for all mtDNA. Nevertheless, patterns of evolution seen in mtDNA vary greatly among different groups of organisms.

The sequences of vertebrate mtDNA exhibit an accelerated rate of change: mammalian mtDNA, for example, typically evolves from 5 to 10 times as fast as mammalian nuclear DNA. The gene content and organization of vertebrate mitochondrial genomes, however, is relatively constant. In contrast, sequences of plant mtDNA evolve slowly at a rate only one-tenth that of the nuclear genome, but their gene content and organization change rapidly. The reason for these basic differences in rates of evolution is not yet known.

One possible reason for the accelerated rate of evolution seen in vertebrate mtDNA is a high mutation rate in mtDNA, which would allow DNA sequences to change quickly. Increased errors associated with replication, the absence of DNA repair functions, and the frequent replication of mtDNA may increase the number of mutations. The large amount of wobble in mitochondrial translation may also allow mutations to accumulate over time, as discussed earlier. The use of mtDNA in evolutionary studies will be described in more detail in Chapter 23.

Concepts

All mtDNA appears to have evolved from a common eubacterial ancestor, but the patterns of evolution seen in different mitochondrial genomes varies greatly. Vertebrate mtDNA exhibits rapid change in sequence but little change in gene content and organization, whereas the mtDNA of plants exhibits little change in sequence but much variation in gene content and organization.

www.whfreeman.com/pierce   Data on mitochondrial genomes that have been completely sequenced, and more information on human diseases and disorders caused by defects in mitochondria

Chloroplast DNA

Geneticists have long recognized that many traits associated with chloroplasts exhibit cytoplasmic inheritance, indicating that these traits are not encoded by nuclear genes. In 1963, chloroplasts were shown to have their own DNA (◄FIGURE 20.11).

Among different plants, the chloroplast genome ranges in size from 80,000 to 600,000 bp, but most chloroplast genomes range from 120,000 to 160,000 bp (Table 20.3). Chloroplast DNA is usually contained on a single, double-stranded DNA molecule that is circular, is highly coiled, and lacks associated histone proteins. As in mtDNA, multiple copies of the chloroplast genome are found in each chloroplast, and there are multiple organelles per cell; so there are several hundred to several thousand copies of cpDNA in a typical plant cell.

Gene Structure and Organization of cpDNA

The chloroplast genomes from a number of plant and algal species have been sequenced, and cpDNA is now recognized to be basically eubacterial in its organization: the order of some groups of genes is the same as that observed in *E. coli*,

Table 20.3	Size of the chloroplast genomes in selected organisms
Organism	**Size of cpDNA in Base Pairs**
Euglena gracilis (protist)	143,172
Porphyra purpurea (red alga)	191,028
Chlorella vulgaris (green alga)	150,613
Marchantia polymorpha (liverwort)	121,024
Nicotiana tabacum (tobacco)	155,939
Zea mays (corn)	140,387
Pinus thunbergii (black pine)	119,707

and many chloroplast genes are organized into operon-like clusters.

Among vascular plants, chloroplast chromosomes are similar in gene content and gene order. A typical chloroplast genome encodes 4 rRNA genes, from 30 to 35 tRNA genes, a number of ribosomal proteins, many proteins engaged in

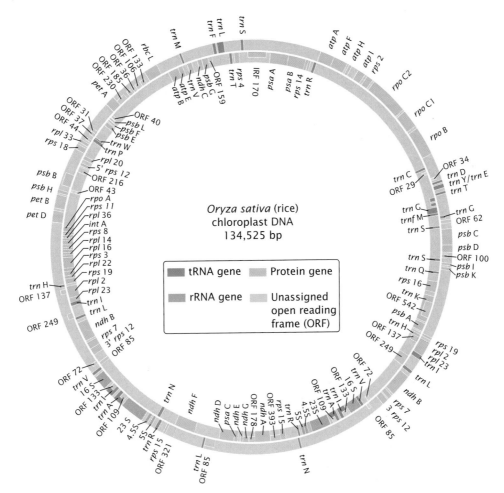

◄20.11 **Chloroplast DNA of rice.**

photosynthesis, and several proteins having roles in nonphotosynthesis processes. A key protein encoded by cpDNA is ribulose-1,5-bisphosphate carboxylase-oxygenase (abbreviated RuBisCO), which participates in carbon fixation of photosynthesis. RuBisCO makes up about 50% of the protein found in green plants and is therefore considered the most abundant protein on earth. It is a complex protein consisting of eight identical large subunits and eight identical small subunits. The large subunit is encoded by chloroplast DNA, whereas the small subunit is encoded by nuclear DNA.

The circular chloroplast genome has genes on both of its strands. Some chloroplast genes have been identified on the basis of a start and stop codon in the same reading frame, but no protein products have yet been isolated for these genes. These sequences are referred to as *open reading frames*. A prominent feature of most chloroplast genomes is the presence of a large inverted repeat. In rice, this repeat includes genes for 23S rRNA, 4.5S rRNA, and 5S rRNA, as well as several genes for tRNAs and proteins (see Figure 20.11). In some plants, these repeats include the majority of the genome, whereas, in others, the repeats are absent entirely. Much of cpDNA consists of noncoding sequences, and introns are found in many chloroplast genes. Finally, many of the sequences in cpDNA are quite similar to those found in equivalent eubacterial genes.

> **Concepts**
>
> Most chloroplast genomes consist of a single, circular DNA molecule not complexed with histone proteins. Although there is considerable size variation, the cpDNAs found in most vascular plants are about 150,000 bp. Genes are scattered in the circular chloroplast genome, and many contain introns. Most cpDNAs contain a large inverted repeat.

Replication, Transcription, and Translation of cpDNA

Little is known about the process of replication of cpDNA. The results of studies viewing cpDNA replication with electron microscopy suggest that replication begins within two D loops and spreads outward to form a theta-like structure (see Figure 12.4). After an initial round of replication, DNA synthesis may switch to a rolling-circle-type mechanism (see Figure 12.5).

The transcription and translation of chloroplast genes are similar in many respects to these processes in eubacteria. For example, promoters found in cpDNA are virtually identical with those found in eubacteria and possess sequences similar to the −10 and −35 consensus sequences of eubacterial promoters. The same antibiotics that inhibit protein synthesis in eubacteria (as well as in mitochondria) inhibit protein synthesis in chloroplasts, indicating that protein synthesis in eubacteria and chloroplasts is similar. Chloroplast translation is initiated by *N*-formylmethionine, just as it is in eubacteria.

Most genes in cpDNA are transcribed in groups; only a few genes have their own promoters and are transcribed as separate mRNA molecules. The RNA polymerase that transcribes cpDNA is more similar to eubacterial RNA polymerase than to any of the RNA polymerases that transcribe eukaryotic nuclear genes. Like eubacterial mRNAs, chloroplast mRNAs are not capped at the 5′ ends, and poly(A) tails are not added to the 3′ ends. However, introns are removed from some RNA molecules after transcription, and the 5′ and 3′ ends may undergo some additional processing before the molecules are translated. Like eubacterial mRNAs, many chloroplast mRNAs have a Shine-Dalgarno sequence in the 5′ untranslated region, which may serve as a ribosome-binding site.

Chloroplasts, like eubacteria, contain 70S ribosomes that consist of two subunits, a large 50S subunit and a smaller 30S subunit. The small subunit includes a single RNA molecule that is 16S in size, similar to that found in the small subunit of eubacterial ribosomes. The larger 50S subunit includes three rRNA molecules: a 23S rRNA, a 5S rRNA, and a 4.5 rRNA. In eubacterial ribosomes, the large subunit possesses only two rRNA molecules, which are 23S and 5S in size. The 4.5S rRNA molecule found in the large subunit of chloroplast ribosomes is homologous to the 3′ end of the 23S rRNA found in eubacteria; so the structure of the chloroplast ribosome is very similar to that of ribosomes found in eubacteria.

Initiation factors, elongation factors, and termination factors function in chloroplast translation and eubacterial translation in similar ways. Most chloroplast chromosomes encode from 30 to 35 different tRNAs, suggesting that the expanded wobble seen in mitochondria does not exist in chloroplast translation. Only universal codons have been found in cpDNA, and translation in chloroplast starts with *N*-formylmethionine as the first amino acid.

Evolution of cpDNA

The DNA sequences of chloroplasts are very similar to those found in cyanobacteria; so chloroplast genomes clearly have a eubacterial ancestry. Overall, cpDNA sequences evolve slowly compared with sequences in nuclear DNA and some mtDNA. For most chloroplast genomes, size and gene organization are similar, although there are some notable exceptions.

> **Concepts**
>
> Many aspects of the transcription and translation of cpDNA are similar to those of eubacteria. Chloroplast DNA sequences are most similar to DNA sequences in cyanobacteria which supports the endosymbiotic theory. Most cpDNA evolves slowly in sequence and structure.

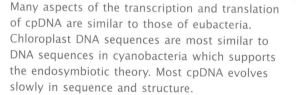 **www.whfreeman.com/pierce**   Information on chloroplast genomes that have been sequenced

Connecting Concepts

Genome Comparisons

A theme running through the preceding discussions of mitochondrial and chloroplast genomes has been a comparison of these genomes with those found in eubacterial and eukaryotic cells (Table 20.4). The endosymbiotic theory indicates that mitochondria and chloroplasts evolved from eubacterial ancestors, and one might therefore assume that mtDNA and cpDNA would be similar to DNA found in eubacterial cells. The actual situation is more complex: mitochondrial DNA and chloroplast DNA possess a mix of eubacterial, eukaryotic, and unique characteristics.

The mitochondrial and chloroplast genomes are similar to those of eubacterial cells in that they are relatively small, lack histone proteins, and are usually on circular DNA molecules. Gene organization and the expression of organelle genomes, however, display some similarities to eubacterial genomes and some similarities to eukaryotic genomes. Introns are present in some organelle genomes but are absent from others. Pre-mRNA introns (see p. 381 in Chapter 14 for a discussion of different types of introns) are absent from mitochondrial and chloroplast genes, as they are from eubacterial genes. Group II introns are present in some organelle and eubacterial genomes but are absent from eukaryotic nuclear genomes. Group I introns are common in some mtDNA and in most cpDNA, and these introns are also found in eubacterial, archaeal, and eukaryotic genomes.

Polycistronic mRNA, which is common in eubacteria but uncommon in eukaryotes, is also found in mitochondria and especially chloroplasts. Human mtDNA, which has little noncoding DNA between genes and little repetitive DNA, is similar in organization to that of typical eubacterial chromosomes, but other mitochondrial and chloroplast genomes possess long noncoding sequences between genes.

Antibiotics that inhibit eubacterial translation also inhibit organelle translation, and the 5′ cap, which is added to eukaryotic mRNA after transcription, is absent from organelle mRNA. A 3′ poly(A) tail, characteristic of most nuclear mRNAs, is present only in some animal mitochondrial mRNA, and it appears to be fundamentally different from that found in nuclear mRNAs. Shine-Dalgarno sequences, the ribosome-binding sites characteristic of eubacterial DNA, are present in some cpDNA but are absent in mtDNA. Finally, some mitochondrial genomes use nonuniversal codons and have extended wobble, which is rare in both eubacterial and eukaryotic DNA.

What conclusions can we draw from these comparisons? Clearly, the genomes of mitochondria and chloroplasts are not typical of the nuclear genomes of the eukaryotic cells in which they reside. In sequence, organelle DNA is most similar to eubacterial DNA, but many aspects of organization and expression in organelle genomes are unique. It is important to remember that the endosymbiotic theory does not propose that mitochondria and chloroplasts are eubacterial in nature but that they arose

Table 20.4 Comparison of nuclear eukaryotic, eubacterial, mitochondrial, and chloroplast genomes

Characteristic	Eukaryotic Genome	Eubacterial Genome	Mitochondrial Genome	Chloroplast Genome
Genome consists of double-stranded DNA	Yes	Yes	Yes	Yes
Circular	No	Yes	Most	Yes
Histone proteins	Yes	No	No	No
Size	Large	Small	Small	Small
Single molecule per genome	No	Yes	Yes in animals No in some plants	Yes
Pre-mRNA introns	Common	Absent	Absent	Absent
Group I introns	Present	Present	Present	Present
Group II introns	Absent	Present	Present	Present
Polycistronic mRNA	Uncommon	Common	Present	Common
5′ cap added to mRNA	Yes	No	No	No
3′ poly(A) tail added to mRNA	Yes	No	Some in animals	No
Shine-Dalgarno sequence in 5′ untranslated region of mRNA	No	Yes	Rare	Some
Nonuniversal codons	Rare	Rare	Yes	No
Extended wobble	No	No	Yes	No
Translation inhibited by tetracycline	No	Yes	Yes	Yes

from eubacterial ancestors more than a billion years ago. Through time, the genomes of the endosymbiont have undergone considerable evolutionary change and have evolved characteristics that distinguish them from contemporary eubacterial and eukaryotic genomes.

The Intergenomic Exchange of Genetic Information

Many proteins found in modern mitochondria and chloroplasts are encoded by nuclear genes, which suggests that much of the original genetic material in the endosymbiont has probably been transferred to the nucleus. This assumption is supported by the observation that some DNA sequences normally found in mtDNA have been detected in nuclear DNA of some strains of yeast and maize. Likewise, chloroplast sequences have been found in the nuclear DNA of spinach. Furthermore, the sequences of nuclear genes that encode organelle proteins are most similar to their eubacterial counterparts.

There is also evidence that genetic material has moved from chloroplasts to mitochondria. For example, DNA fragments from the 16S rRNA gene and two tRNA genes that are normally encoded by cpDNA have been found in the mtDNA of maize. Sequences from the gene that encodes the large subunit of RuBisCO, which is normally encoded by cpDNA, are duplicated in maize mtDNA. And there is even evidence that some nuclear genes have moved into mitochondrial genomes. The exchange of genetic material between the nuclear, mitochondrial, and chloroplast genomes has given rise to the term "promiscuous DNA" to describe this phenomenon. The mechanism by which this exchange takes place is not entirely clear.

Mitochondrial DNA and Aging in Humans

Symptoms of many human genetic diseases caused by defects in mtDNA first appear in middle age or later and increase in severity as people age. One hypothesis to explain the late onset and progressive worsening of mitochondrial diseases is related to the decline in oxidative phosphorylation with aging.

Oxidative phosphorylation is the process that generates ATP, the primary carrier of energy in the cell. This process takes place on the inner membrane of the mitochondrion and requires a number of different proteins, some encoded by mtDNA and others encoded by nuclear genes. Oxidative phosphorylation normally declines with age and, if it falls below some critical threshold, tissues do not make enough ATP to sustain vital functions and disease symptoms appear. Most people start life with an excess capacity for oxidative phosphorylation; this capacity decreases with age, but most people reach old age or die before the critical threshold is passed. Persons born with mitochondrial diseases carry mutations in their mtDNA that lower their oxidative phosphorylation capacity. At birth, their capacity may be sufficient to support their ATP needs but, as their oxidative phosphorylation capacity declines with age, they cross the critical threshold and begin to experience symptoms. These symptoms usually first appear in tissues that are most critically dependent on mitochondrial energy: the central nervous system, heart and skeletal muscle, pancreatic islets, kidneys, and the liver.

Why does oxidative phosphorylation capacity decline with age? One possible explanation is that damage to mtDNA accumulates with age; deletions and base substitutions in mtDNA increase with age. For example, a common 5000-bp deletion in mtDNA is absent in normal heart muscle cells before the age of 40, but afterward this deletion is present with increasing frequency. The same deletion is found at a low frequency in normal brain tissue before age 75 but is found in 11% to 12% of mtDNAs in the basal ganglia by age 80. People with mtDNA genetic diseases may age prematurely because they begin life with damaged mtDNA.

The mechanism of age-related increases in mtDNA damage is not yet known. Oxygen radicals, highly reactive compounds that are natural by-products of oxidative phosphorylation, are known to damage DNA (see p. 489 in Chapter 17). Because mtDNA is physically close to the enzymes taking part in oxidative phosphorylation, it may be more prone to oxidative damage than nuclear DNA. When mtDNA has been damaged, the cell's capacity to produce ATP drops. To produce sufficient ATP to meet the cell's energy needs, even more oxidative phosphorylation must occur, which in turn may stimulate further production of oxygen radicals, leading to a vicious cycle.

Significantly elevated levels of mtDNA defects have been observed in some patients with late-onset degenerative diseases, such as diabetes mellitus, ischemic heart disease, Parkinson disease, Alzheimer disease, and Huntington disease. All of these diseases appear in middle to old age and have symptoms associated with tissues that critically depend on oxidative phosphorylation for ATP production. However, because Huntington disease and some cases of Alzheimer disease are inherited as autosomal dominant conditions, mtDNA defects cannot be the primary cause of these diseases, although they may contribute to their progression.

Connecting Concepts Across Chapters

This chapter is about the unique properties of organelle DNA, which is part of the cytoplasm and usually exhibits uniparental inheritance. A unifying theme has been that mitochondria and chloroplasts evolved from free-living eubacteria that entered into an endosymbiotic relation with the eukaryotic cells in which they are found. Endosymbiosis helps to explain many of the characteristics of mitochondrial DNA and chloroplast DNA, which resemble eubacterial DNA more than they do nuclear eukaryotic DNA. However, not all aspects of mtDNA and cpDNA are similar to eubacterial DNA; organelle DNA has a number of properties that are unique.

Another prominent theme that runs through this chapter is that cpDNA and mtDNA display a bewildering diversity of variation in size and organization. The reason for this variation is unknown, but the variation makes summarizing mitochondrial and chloroplast genomes difficult.

Traits encoded by mitochondrial and chloroplast genes are inherited in a very different manner from those encoded by nuclear genes. Because organelle DNA is located in the cytoplasm, the traits that it encodes exhibit cytoplasmic inheritance and are typically inherited from a single parent, most often the mother. Many traits encoded by mtDNA and cpDNA exhibit phenotypic variation among progeny of a single cross and even among cells and tissues within an individual organism; the latter occurs when there are two or more genetic variants in a single cell and random segregation of the organelles in cell division produces cells with different proportions of the two types of DNA.

Understanding the inheritance of mitochondrial- and chloroplast-encoded traits builds on earlier discussions of uniparental inheritance in Chapter 5 and biparental inheritance (with which it is contrasted) in Chapter 3. Material in the present chapter is closely linked to information on DNA structure and organization found in Chapters 10 and 11 and to discussions of replication, transcription, RNA processing, and translation found in Chapters 12 through 15. Molecular techniques described in this chapter are covered more thoroughly in Chapter 18. The use of mtDNA in evolutionary studies will be discussed in more detail in Chapter 23.

CONCEPTS SUMMARY

- Mitochondria and chloroplasts are eukaryotic organelles that possess their own DNA. Traits encoded by mtDNA and cpDNA exhibit cytoplasmic inheritance and usually are inherited from a single parent, most often the mother. Random segregation of organelles in cell division may produce phenotypic variation among cells within a single individual and among the offspring of a single female.

- The endosymbiotic theory proposes that mitochondria and chloroplasts originated as free-living prokaryotic (specifically eubacterial) organisms that entered into a beneficial association with eukaryotic cells. Similarities in the gene sequences of organelle and eubacterial DNA support a eubacterial origin for mitochondrial and chloroplast DNA.

- The mitochondrial genome usually consists of a single, circular DNA molecule that lacks histone proteins, although plants may have multiple circular molecules. Mitochondrial DNA varies in size among different groups of organisms; most of this variation is due to noncoding DNA. Each cell contains many copies of mtDNA.

- The organization of genes in the mitochondrial genome differs among organisms. Ancestral mitochondrial genomes typically have characteristics of eubacterial genomes, including eubacterial-like ribosomes, a complete or almost complete set of tRNA genes, few introns, little noncoding DNA between genes, genes organized into eubacterial-like clusters, and the use of only universal codons. Derived mitochondrial genomes are smaller and contain fewer genes. Their rRNA genes and ribosomes differ from those found in eubacteria, and they use some nonuniversal codons.

- Human mtDNA is highly economical, with few noncoding nucleotides. Fungal and plant mtDNAs contain much noncoding DNA between genes, introns within genes, and extensive 5' and 3' untranslated regions. Most plant mitochondrial genomes contain one or more large direct repeats, which may recombine to produce smaller or larger DNA molecules.

- Mitochondrial DNA is synthesized throughout the cell cycle, and its synthesis is not coordinated with the replication of nuclear DNA.

- The transcription of mitochondrial genes varies among different organisms. Messenger RNAs produced by the transcription of mtDNA are not capped at their 5' ends; poly(A) tails are added to the 3' ends of some animal mRNAs, but these tails are different from the poly(A) tails found on nuclear-encoded mRNAs.

- Antibiotics that inhibit eubacterial ribosomes also inhibit mitochondrial ribosomes. Protein synthesis in mitochondria is initiated at AUG start codons by N-formylmethionine and employs eubacterial-like elongation factors. Many mitochondrial genomes encode a limited number of tRNAs, with relaxed codon–anticodon pairing rules and extended wobble.

- Comparisons of mtDNA sequences suggest that mitochondria evolved from a eubacterial ancestor. Vertebrate mtDNA exhibits rapid change in sequence but little change in gene content and organization. Plant mtDNA exhibits little change in sequence but much variation in gene content and organization.

- Chloroplast genomes consist of a single, circular DNA molecule that varies little in size and lacks histone proteins. Each plant cell contains multiple copies of cpDNA.

- Most chloroplast chromosomes possess large inverted repeats; some chloroplast genes contain introns.

- Transcription and translation are similar in chloroplasts and eubacteria: most chloroplast genes are transcribed as polycistronic units, their mRNAs are not capped, no poly(A) tails are added, and they possess a Shine-Dalgarno ribosome-binding sequence.

- Chloroplast DNA sequences are most similar to those in cyanobacteria. Chloroplast DNA sequences tend to evolve slowly.

- Through evolutionary time, many mitochondrial and chloroplast genes have moved to nuclear chromosomes. In some plants, there is evidence that copies of chloroplast genes have moved to the mitochondrial genome.

IMPORTANT TERMS

mitochondrial DNA (mtDNA) (p. 583)
chloroplast DNA (cpDNA) (p. 585)

heteroplasmy (p. 586)
replicative segregation (p. 586)
homoplasmy (p. 586)

endosymbiotic theory (p. 588)
D loop (p. 589)

Worked Problems

1. A physician examines a young man who has a progressive muscle disorder and visual abnormalities. A number of the patient's relatives have the same condition, as shown in the adjoining pedigree. The degree of expression of the trait is highly variable among members of the family: some are only slightly affected, whereas others developed severe symptoms at an early age. The physician concludes that this disorder is due to a mutation in the mitochondrial genome. Do you agree with the physician's conclusion? Why or why not? Could the disorder be due to a mutation in a nuclear gene? Explain your reasoning.

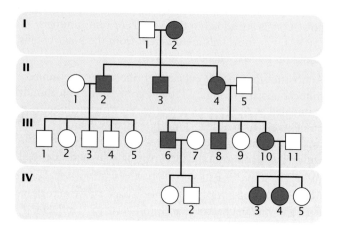

- **Solution**

The conclusion that the disorder is caused by a mutation in the mitochondrial genome is supported by the pedigree and the observation of variable expression in affected members of the same family. The disorder is passed only from affected mothers to offspring; when fathers are affected, none of their children have the trait (as seen in the children of II-2 and III-6). This outcome is expected of traits determined by mutations in mtDNA, because mitochondria are in the cytoplasm and usually inherited only from a single (in humans, the maternal) parent.

The facts that some offspring of affected mothers do not show the trait (III-9 and IV-5) and that expression varies from one person to another suggest that affected persons are heteroplasmic, with both mutant and wild-type mitochondria. Random segregation of mitochondria in meiosis may produce gametes having different proportions of mutant and wild-type sequences, resulting in different degrees of phenotypic expression among the offspring. Most likely, symptoms of the disorder develop when some minimum proportion of the mitochondria are mutant. Just by chance, some of the gametes produced by an affected mother contain few mutant mitochondria and result in offspring that lack the disorder.

Another possible explanation for the disorder is that it results from an autosomal dominant gene. When an affected (heterozygous) person mates with an unaffected (homozygous) person, about half of the offspring are expected to have the trait, but just by chance some affected parents will have no affected offspring. It is possible that individuals II-2 and III-6 in the pedigree just happened to be male and their sex is unrelated to the mode of transmission. The variable expression could be explained by variable expressivity (see p. 68 in Chapter 3).

2. Suppose that a new organelle is discovered in an obscure group of protists. This organelle contains a small DNA genome and some scientists are arguing that, like chloroplasts and mitochondria, this organelle originated as a free-living eubacterium that entered into an endosymbiotic relation with the protist. Outline a research plan to determine if the new organelle evolved from a free-living eubacterium. What kinds of data would you collect and what predictions would you make if the theory is correct?

- **Solution**

We could examine the structure, organization, and sequences of the organelle genome. If the organelle shows only characteristics of eukaryotic DNA, then it most likely has a eukaryotic origin but, if it displays some characteristics of eubacterial DNA, then this finding supports the theory of a eubacterial origin. However, on the basis of our knowledge of mitochondrial and chloroplast genomes, we should not expect the organelle genome to be entirely eubacterial in its characteristics.

We could start by examining the overall characteristics of the organelle DNA. If it has a eubacterial origin, we might expect that the organelle genome will consist of a circular molecule and will lack histone proteins. We might then sequence the organelle DNA to determine its gene content and organization. The presence of any group II introns would suggest a eubacterial origin, because these introns have been found only in eubacterial genomes and genomes derived from

eubacteria. The presence of any pre-mRNA introns, on the other hand, would suggest a eukaryotic origin, because these introns have been found only in nuclear eukaryotic genomes. If the organelle genome has a eubacterial origin, we might expect to see polycistronic mRNA, the absence of a 5′ cap, and inhibition of translation by those antibiotics that typically inhibit eubacterial translation.

Finally, we could compare the DNA sequences found in the organelle genome with homologous sequences from eubacteria and eukaryotic genomes. If the theory of an endosymbiotic origin is correct, then the organelle sequences should be most similar to homologous sequences found in eubacteria.

The New Genetics
MINING GENOMES

EVOLUTIONARY ANALYSIS WITH THE USE OF MITOCHONDRIAL GENOMES

Phylogenetic trees are graphic representations of the relationships between different organisms. Traditionally based on morphological, physiological, and behavioral data, evolutionary analysis has been

substantially changed by the use of molecular information. In this exercise, you will pose and evaluate a question relating to human evolution. You will use mitochondrial DNA sequences and the tools available at the Biology Workbench, managed by the San Diego Supercomputing Center at the University of California, San Diego.

COMPREHENSION QUESTIONS

* 1. Briefly describe the general structures of mtDNA and cpDNA. How are they similar? How do they differ? How do their structures compare with the structures of eubacterial and eukaryotic (nuclear) DNA.

2. Explain why many traits encoded by mtDNA and cpDNA exhibit considerable variation in their expression, even among members of the same family.

* 3. What is the endosymbiotic theory? How does it help to explain some of the characteristics of mitochondria and chloroplasts?

4. What evidence supports the endosymbiotic theory?

5. How are genes organized in the mitochondrial genome? How does this organization differ between ancestral and derived mitochondrial genomes?

* 6. What are nonuniversal codons? Where are they found?

7. How does replication of mtDNA differ from replication of nuclear DNA in eukaryotic cells.

* 8. The human mitochondrial genome encodes only 22 tRNAs, whereas at least 32 tRNAs are required for cytoplasmic translation. Why are fewer tRNAs needed in mitochondria?

9. What are some possible explanations for an accelerated rate of evolution in the sequences of vertebrate mtDNA.

*10. Briefly describe the organization of genes on the chloroplast genome.

11. What is meant by the term "promiscuous DNA"?

APPLICATION QUESTIONS AND PROBLEMS

12. A wheat plant that is light green in color is found growing in a field. Biochemical analysis reveals that chloroplasts in this plant produce only 50% of the chlorophyll normally

found in wheat chloroplasts. Propose a set of crosses to determine whether the light-green phenotype is caused by a mutation in a nuclear gene or a chloroplast gene.

*13. A rare neurological disease is found in the family illustrated in the following pedigree. What is the most likely mode of inheritance for this disorder? Explain your reasoning.

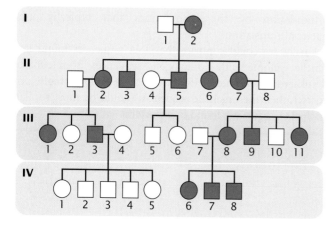

14. In a particular strain of *Neurospora*, a *poky* mutation exhibits biparental inheritance, whereas *poky* mutations in other strains are inherited only from the maternal parent. Explain these results.

15. Antibiotics such as chloramphenicol, tetracycline, and erythromycin inhibit protein synthesis in eubacteria but have no effect on protein synthesis encoded by nuclear genes. Cycloheximide inhibits protein synthesis encoded by nuclear genes but has no effect on eubacterial protein synthesis. How might these compounds be used to determine which proteins are encoded by the mitochondrial and chloroplast genomes?

*16. A scientist collects cells at various points in the cell cycle and isolates DNA from them. Using density gradient centrifugation, she separates the nuclear and mtDNA. She then measures the amount of mtDNA and nuclear DNA present at different points in the cell cycle. On the following graph, draw a line to represent the relative amounts of nuclear DNA that you expect her to find per cell throughout the cell cycle. Then, draw a dotted line on the same graph to indicate the relative amount of mtDNA that you would expect to see at different points throughout the cell cycle.

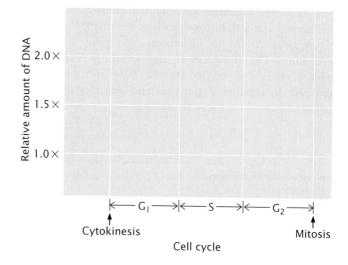

17. The introduction to Chapter 1 described how bones found in 1979 outside Ekaterinburg, Russia, were shown to be those of Tsar Nicholas and his family, who were executed in 1918 by a Bolshevik firing squad in the Russian Revolution. To prove that the skeletons were those of the royal family, mtDNA was extracted from the bone samples, amplified by PCR, and compared with mtDNA from living relatives of the tsar's family. Why was DNA from the mitochondria analyzed instead of nuclear DNA? What are some of the advantages of using mtDNA for this type of study?

18. From Figure 20.8, determine as best you can the percentage of human mtDNA that is coding (transcribed into RNA) and the percentage that is noncoding (not transcribed).

CHALLENGE QUESTIONS

19. Mitochondrial DNA sequences have been detected in the nuclear genomes of many organisms, and cpDNA sequences are sometimes found in the mitochondrial genome. Propose a mechanism for how such "promiscuous DNA" might move between nuclear, mitochondrial, and chloroplast genomes.

20. Steven A. Frank and Laurence D. Hurst argued that a cytoplasmically inherited mutation in humans that has severe effects in males but no effect in females will not be eliminated from a population by natural selection, because only females pass on mtDNA. Using this argument, explain why males with Leber hereditary optic neuropathy are more severely affected than females.

21. Several families have been described that exhibit vision problems, muscle weakness, and deafness. This disorder is inherited as an autosomal dominant trait and the disease-causing gene has been mapped to chromosome 10 in the nucleus. Analysis of the mtDNA from affected persons in these families reveals that large numbers of their mitochondrial genomes possess deletions of varying length. Different members of the same family and even different mitochondria from the same person possess deletions of different sizes; so the underlying defect appears to be a tendency for the mtDNA of affected persons to have deletions. Propose an explanation for how a mutation in a nuclear gene might lead to deletions in mtDNA.

SUGGESTED READINGS

Anderson, S., A. T. Bankier, B. G. Barrell, M. H. L. de Bruijn, A. R. Coulson, et al. 1981. Sequence and organization of the human mitochondrial genome. *Nature* 290:457–465.
Original report of the complete sequencing of the human mtDNA.

Birky, C. W., Jr. 1978. Transmission genetics of mitochondria and chloroplasts. *Annual Review of Genetics* 12:471–512.
A review of how traits encoded by mtDNA and cpDNA are inherited.

Fox, T. D. 1987. Natural variation in the genetic code. *Annual Review of Genetics* 21:67–91.
A review of nonuniversal codons.

Gray, M. W. 1992. The endosymbiotic hypothesis revisited. *International Review of Cytology* 141:233–357.
An excellent review of how data from organelle genomes relate to the endosymbiotic theory.

Gray, M. W. 1998. *Rickettsia,* typhus and the mitochondrial connection. *Nature* 396:109–110.
A short commentary on the DNA sequence of *Rickettsia prowazekii,* the bacterium that causes typhus and is thought to be closely related to the eubacteria that gave rise to mitochondria.

Gray, M. W., G. Burger, and B. Franz Lang. 1999. Mitochondrial evolution. *Science* 283:1476–1481.
A review of the evolution of mitochondria based on DNA sequence data from a number of different species.

Gruissem, W. 1989. Chloroplast RNA: transcription and processing. In A. Marcus, Ed. *The Biochemistry of Plants: A Comprehensive Treatise,* pp. 151–191. Vol. 15, *Molecular Biology.* New York: Academic Press.
A review of the transcription and RNA processing that takes place in chloroplasts.

Lehman, N., A. Eisenhawer, K. Hansen, L. D. Mech, R. O. Peterson, P. J. P. Gogan, and R. K. Wayne. 1991. Introgression of coyote mitochondrial DNA into sympatric North American gray wolf population. *Evolution* 45:104–119.
A report of the introgression of coyote genes into wolf populations as revealed by mtDNA.

Levings, C. S., III, and G. G. Brown. 1989. Molecular biology of plant mitochondria. *Cell* 56:171–179.
A report of some of the unique characteristics and properties of plant mtDNA.

Poulton, J. 1995. Transmission of mtDNA: cracks in the bottleneck. *American Journal of Human Genetics* 57:224–226.
A discussion of the transmission of mtDNA.

Sugiura, M. 1989. The chloroplast genome. In A. Marcus, Ed. *The Biochemistry of Plants: A Comprehensive Treatise,* pp. 133–150. Vol. 15, *Molecular Biology.* New York: Academic Press.
An excellent review of the organization and sequence of cpDNA.

Sugiura, M. 1989. The chloroplast chromosomes in land plants. *Annual Review of Cell Biology* 5:51–70.
A review of the features and evolution of chloroplast DNA among vascular plants.

Sugiura, M., T. Hirose, and M. Sugita. 1998. Evolution and mechanism of translation in chloroplasts. *Annual Review of Genetics* 32:437–459.
A review of the translation of cpDNA.

Wallace, D. C. 1992. Mitochondrial genetics: a paradigm for aging and degenerative diseases? *Science* 256:628–632.
A good review of the role of mtDNA in human genetic disease and in aging.

Wallace, D. C. 1999. Mitochondrial diseases in man and mouse. *Science* 283:1482–1488.
A review of diseases arising from defects in mitochondria, including those due to mutations in mtDNA and nuclear DNA.

Yaffe, M. P. 1999. The machinery of mitochondrial inheritance and behavior. *Science* 283:1493–1497.
A discussion of evidence suggesting that movements of mitochondria in cell division may not be random but regulated by the cell.

Advanced Topics in Genetics: Developmental Genetics, Immunogenetics, and Cancer Genetics

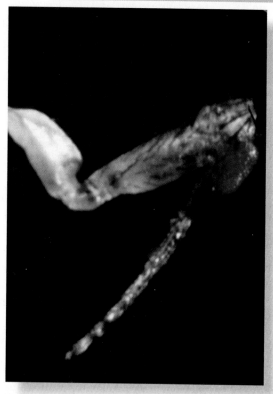

Flies have been genetically engineered to have extra eyes on their legs, wings, and elsewhere. Through genetic engineering, the *eyeless* gene can be expressed in cells of body parts where eyes do not normally appear. (U. Kloter and G. Halder/Biozentrum.)

Flies with Extra Eyes

We can all imagine situations where an extra set of eyes might come in handy: eyeing members of the opposite sex while still paying attention to the professor during lecture, looking both ways at the same time before crossing the street; or watching your backside in a barroom brawl. However useful extra eyes might be, creating them at selected locations is no simple matter. An eye is, after all, an exceedingly complex structure, consisting of photoreceptors, lens, nerves, and other tissues. It would be very unlikely for all of these structures to develop at a site where eyes don't normally exist. Nevertheless, in 1995, a group of geneticists succeeded in genetically engineering fruit flies with extra eyes on their wings, legs, and antennae. How was this amazing feat accomplished?

The story of creating flies with extra eyes began in 1915, when Mildred Hoge discovered a mutant fruit fly with small eyes due to a recessive mutation in a gene called *eyeless*. The product of the normal allele of the *eyeless* locus is required for proper development of the fruit-fly eye.

In 1993, Walter Gehring and his collaborators were investigating *Drosophila* genes that encode transcription factors (see Chapter 13). One of these genes mapped to the same location as that of the *eyeless* gene and, in fact, turned out to *be* the *eyeless* gene. To see what effect *eyeless* might have on development, Gehring's group genetically engineered cells that expressed the *eyeless* gene in parts of the fly where the gene is not normally expressed. When these flies hatched, they had huge eyes on their wings, antennae, and legs. These structures were not just tissue that resembled eyes; they were complete eyes with a cornea, cone cells, and photoreceptors that responded to light, although the flies could not use them to see, because they were not connected to the nervous system. The *eyeless* gene appears to be one of the long-sought master control switches of development: its protein activates a set of other genes that are responsible for making a complete eye.

The *eyeless* gene has counterparts in mice and humans that affect the development of mammalian eyes. There is a striking similarity between the *eyeless* gene of *Drosophila* and the *Small eye* gene that exists in mice. In mice, a mutation in one copy of *Small eye* causes small eyes; a mouse that is homozygous for the *Small eye* mutation has no eyes. There is also a similarity between the *eyeless* gene in *Drosophila* and the *Aniridia* gene in humans; a mutation in *Aniridia* produces a severely malformed human eye. Similarities in the sequences of *eyeless*, *Small eye*, and *Aniridia* suggest that all three genes evolved from a common ancestral sequence. This possibility is surprising, because the eyes of insects and mammals were thought to have evolved independently. Similarities among *eyeless*, *Small eye*, and *Aniridia* suggest that a common pathway underlies eye development in flies, mice, and humans.

This chapter focuses on three specialized topics in genetics: developmental genetics, immunogenetics, and cancer genetics. We begin with a discussion of the genetic control of the early development of *Drosophila* embryos, one of the best-understood developmental systems. We then turn to the genetics of the immune system in vertebrates. This system is capable of generating proteins that recognize virtually any foreign substance in the body. The generation of this huge diversity of proteins relies on a special type of genetic recombination unique to the immune system. Last, we consider the genetic basis of cancer and how mutations in particular types of genes contribute to the growth of tumors in humans.

Developmental Genetics

Every multicellular organism begins life as a unicellular, fertilized egg. This single-celled zygote undergoes repeated cell divisions, eventually producing millions or trillions of cells that constitute a complete adult organism. Initially, each cell in the embryo is **totipotent**—it has the potential to develop into any cell type. Many cells in plants and fungi remain totipotent, but animal cells usually become committed to developing into specific types of cells after just a few early embryonic divisions. This commitment often comes well before a cell begins to exhibit any characteristics of a particular cell type; once the cell becomes committed, it cannot reverse its fate and develop into a different cell type. A cell becomes committed by a process called **determination,** the mechanism of which is still unknown.

For many years, the work of developmental biologists was limited to describing the changes that take place in the course of development, because techniques for probing the intracellular processes behind these changes were unavailable. But, in recent years, powerful genetic and molecular techniques have had a tremendous influence on the study of development. In a few model systems such as *Drosophila*, the molecular mechanisms underlying developmental change are now beginning to be understood.

Cloning Experiments

If all cells in a multicellular organism are derived from the same original cell, how do different cells types arise? One possibility is that, throughout development, genes might be selectively lost or altered, causing different cell types to have different genomes. Alternatively, each cell might contain the same genetic information, but different genes might be expressed in each cell type. Early cloning experiments helped to answer this question.

In the 1950s, Frederick Steward developed methods for cloning plants. He disrupted phloem tissue from the root of a carrot, separating and isolating individual cells. He then placed individual cells in a sterile medium that contained nutrients. Steward was successful in getting the cells to grow and divide, and eventually he obtained whole edible carrots from single cells (◀ FIGURE 21.1). Because all parts of the plant were regenerated from a specialized phloem cell,

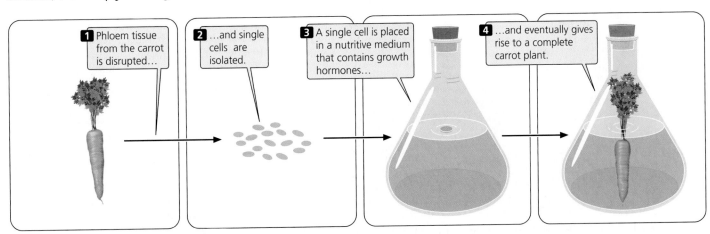

1 Phloem tissue from the carrot is disrupted…

2 …and single cells are isolated.

3 A single cell is placed in a nutritive medium that contains growth hormones…

4 …and eventually gives rise to a complete carrot plant.

◀21.1 **Many plants can be cloned from isolated single cells.**
Thus none of the original genetic material is lost during development.

Steward concluded that each phloem cell contained the genetic potential for a whole plant; none of the original genetic material was lost during determination.

The results of later studies demonstrated that most animal cells also retain a complete set of genetic information during development. In 1952, Robert Briggs and Thomas King removed the nuclei from unfertilized oocytes of the frog *Rana pipiens*. They then isolated nuclei from frog blastulas (an early embryonic stage) and injected these nuclei individually into the oocytes. The eggs were then pricked with a needle to stimulate them to divide. Although most were damaged in the process, a few eggs developed into complete tadpoles that eventually metamorphosed into frogs.

In the late 1960s, John Gurdon used these methods to successfully clone a few frogs with nuclei isolated from intestinal cells of tadpoles. This accomplishment suggested that the differentiated intestinal cells carried the genetic information necessary to encode traits found in all other cells. However, Gurdon's successful clonings may have resulted from the presence of a few undifferentiated stem cells in the intestinal tissue, which were inadvertently used as the nuclei donors.

In 1997, researchers at the Roslin Institute of Scotland announced that they had successfully cloned a sheep by using the genetic material from a differentiated cell of an adult animal. To perform this experiment, they fused an udder cell from a white-faced Finn Dorset ewe with an enucleated egg cell and stimulated the egg electrically to initiate development. After growing it in the laboratory for a week, they implanted the embryo into a Scottish black-faced surrogate mother. Dolly, the first mammal cloned from an adult cell, was born on July 5, 1996 (◀ FIGURE 21.2). Since

the cloning of Dolly, other sheep, mice, and calves have been cloned from differentiated adult cells.

These cloning experiments demonstrated that genetic material is not lost or permanently altered during development—development must require the selective expression of genes. But how do cells regulate their gene expression in a coordinated manner to give rise to a complex, multicellular organism? Research has now begun to provide some answers to this important question.

> **Concepts**
>
> The ability to clone plants and animals from single specialized cells demonstrates that genes are not lost or permanently altered during development.

www.whfreeman.com/pierce   Information about cloning, nuclear transfer research, and the ethics of cloning

The Genetics of Pattern Formation in *Drosophila*

One of the best-studied systems for the genetic control of pattern formation is the early embryonic development of *Drosophila melanogaster*. Geneticists have isolated a large number of mutations in fruit flies that influence all aspects of their development, and these mutations have been subjected to molecular analysis, providing much information about how genes control early development in *Drosophila*.

The development of the fruit fly An adult fruit fly possesses three basic body parts: head, thorax, and abdomen (◀ FIGURE 21.3). The thorax consists of three segments: the first thoracic segment carries a pair of legs; the second thoracic segment carries a pair of legs and a pair of wings; and the third thoracic segment carries a pair of legs and the halteres (rudiments of the second pair of wings found in most other insects). The abdomen contains nine segments.

When a *Drosophila* egg has been fertilized, its diploid nucleus (◀ FIGURE 21.4a) immediately divides nine times without division of the cytoplasm, creating a single, multinucleate cell (◀ FIGURE 21.4b). These nuclei are scattered throughout the cytoplasm but later migrate toward the periphery of the embryo and divide several more times (◀ FIGURE 21.4c). Next, the cell membrane grows inward and around each nucleus, creating a layer of approximately 6000 cells at the outer surface of the embryo (◀ FIGURE 21.4d). Four nuclei at one end of the embryo develop into pole cells, which eventually give rise to germ cells. The early embryo then undergoes further development in three distinct stages: (1) the anterior–posterior axis and the dorsal–ventral axis of the embryo are established (◀ FIGURE 21.5a); (2) the number and orientation of the body segments are determined (◀ FIGURE 21.5b); and (3) the identity of each individual

◀21.2 **In 1996, researchers at the Roslin Institute of Scotland successfully cloned a sheep named Dolly.** They used the genetic material from a differentiated cell of an adult animal. (Paul Clements/AP.)

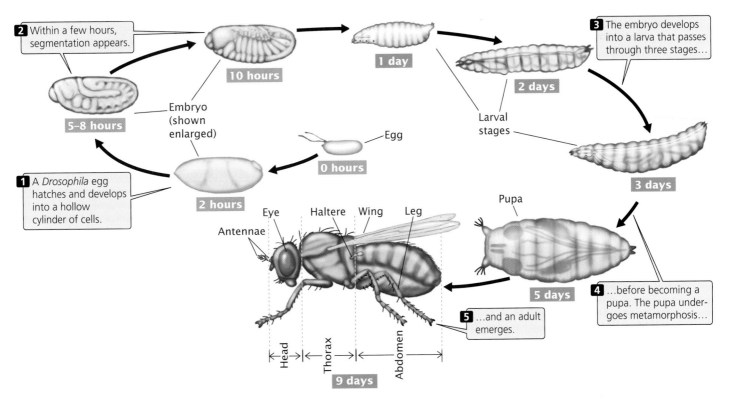

2 Within a few hours, segmentation appears.

3 The embryo develops into a larva that passes through three stages...

10 hours

1 day

2 days

5–8 hours

Embryo (shown enlarged)

Larval stages

Egg

0 hours

3 days

1 A *Drosophila* egg hatches and develops into a hollow cylinder of cells.

2 hours

Pupa

4 ...before becoming a pupa. The pupa undergoes metamorphosis...

Antennae   Eye   Haltere   Wing   Leg

5 days

5 ...and an adult emerges.

Head   Thorax   Abdomen

9 days

◀ 21.3 **The fruit fly,** *Drosophila melanogaster,* **passes through three larval stages and a pupa before developing into an adult fly.** The three major body parts of the adult are head, thorax, and abdomen.

segment is established (◀ FIGURE 21.5c). Different sets of genes control each of these three stages (Table 21.1).

Egg-polarity genes The **egg-polarity genes** play a crucial role in establishing the two main axes of development in fruit flies. You can think of these axes as the longitude and latitude of development: any location in the *Drosophila* embryo can be defined in relation to these two axes. There are two sets of egg-polarity genes: one set determines the anterior–posterior axis and the other determines the dorsal–ventral axis. These genes work by setting up concentration gradients of morphogens within the developing embryo. A **morphogen** is a protein whose concentration gradient affects the developmental fate of the surrounding region.

The egg-polarity genes are transcribed into mRNAs during egg formation in the maternal parent, and these mRNAs become incorporated into the cytoplasm of the egg. After fertilization, the mRNAs are translated into proteins that play an important role in determining the anterior–posterior and dorsal–ventral axes of the embryo. Because the mRNAs of the polarity genes are produced by the female parent and influence the phenotype of their offspring, the traits encoded by them are examples of genetic maternal effects (see pp. 119–120 in Chapter 5).

Egg-polarity genes function by producing proteins that become asymmetrically distributed in the cytoplasm, giving the egg polarity, or direction. This asymmetrical distribution may take place in a couple of ways. The mRNA may be localized to particular regions of the egg cell, leading to an abundance of the protein in those regions when the mRNA is translated. Alternatively, the mRNA may be randomly distributed, but the protein that it encodes may become asymmetrically distributed, either by a transport system that delivers it to particular regions of the cell or by its removal from particular regions by selective degradation.

Determination of the dorsal–ventral axis The dorsal–ventral axis defines the back (dorsum) and belly (ventrum) of a fly (see Figure 21.5). At least 12 different genes determine this axis, one of the most important being a gene called *dorsal*. The *dorsal* gene is transcribed and translated in the maternal ovary, and the resulting mRNA and protein are transferred to the egg during oogenesis. In a newly laid egg, mRNA and protein encoded by the *dorsal* gene are uniformly distributed throughout the cytoplasm but, after the nuclei migrate to the periphery of the embryo (see Figure 21.4c), Dorsal protein becomes redistributed. Along one side of the embryo, Dorsal protein remains in the cytoplasm; this side will become the dorsal surface. Along the other side, Dorsal protein is taken up into the nuclei; this side will become the ventral surface. At this point, there is a smooth gradient of increasing nuclear Dorsal concentration from the dorsal to the ventral side (◀ FIGURE 21.6).

(a) Single-celled diploid zygote

1 Sperm and egg nuclei fuse to create a single-celled diploid zygote.

Single 2*n* nucleus

(b) Multinucleate syncytium

2 Multiple nuclear divisions create a single multinucleate cell, the syncytium.

(c) Syncytial blastoderm

3 The nuclei migrate to the periphery of the embryo and divide several more times, creating the syncytial blastoderm.

Pole nuclei

(d) Cellular blastoderm

4 The cell membrane grows around each nucleus, producing a layer of cells that surrounds the embryo. The resulting structure is the cellular blastoderm.

Pole cells

5 Nuclei at one end of the blastoderm develop into pole cells, which become the primordial germ cells.

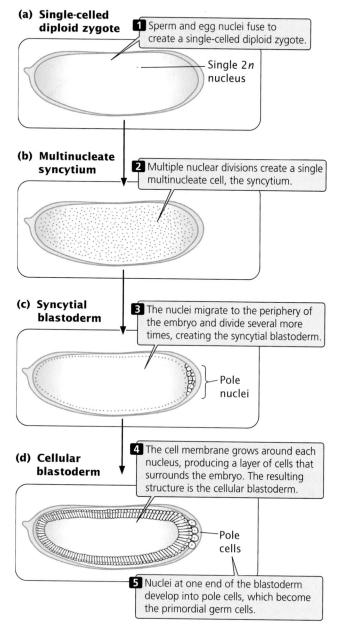

21.4 Early development of a *Drosophila* embryo.

The nuclear uptake of Dorsal protein is thought to be governed by a protein called Cactus, which binds to Dorsal protein and traps it in the cytoplasm. The presence of yet another protein, called Toll, can alter Dorsal, allowing it to dissociate from Cactus and move into the nucleus. Together, Cactus and Toll regulate the nuclear distribution of Dorsal protein, which in turn determines the dorsal–ventral axis of the embryo.

Inside the nucleus, Dorsal protein acts as a transcription factor, binding to regulatory sites on the DNA and activating or repressing the expression of other genes (Table 21.2). High nuclear concentration of Dorsal protein (as on the ventral side of the embryo) activates a gene called *twist*, which causes mesoderm to develop. Low concentrations of

(a) 2-hour embryo

1 The anterior–posterior and dorsal–ventral axes of the embryo are established.

Dorsal

Anterior

Posterior

Ventral

Early embryo

(b) 10-hour embryo

2 The number and orientation of the body segments are established.

Head | Thoracic segments | Abdominal segments

(c) Adult

3 The identity of each individual segment is established.

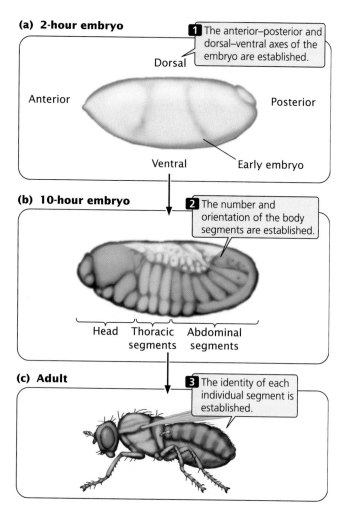

21.5 In an early *Drosophila* embryo, the major body axes are established, the number and orientation of the body segments are determined, and the identity of each individual segment is established. Different sets of genes control each of these three stages.

Dorsal protein (as in cells on the dorsal side of the embryo), activates a gene called *decapentaplegic*, which specifies dorsal structures. In this way, the ventral and dorsal sides of the embryo are determined.

Table 21.1	Stages in the early development of fruit flies and the genes that control each stage
Developmental Stage	**Genes**
Establishment of main body axes	Egg polarity genes
Determination of number and polarity of body segments	Segmentation genes
Establishment of identity of each segment	Homeotic genes

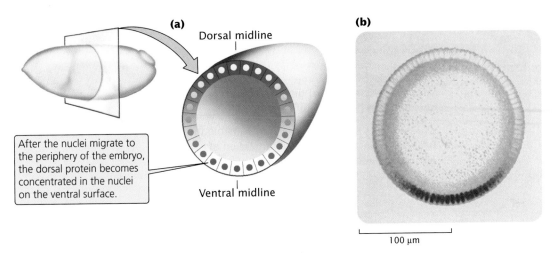

(a) Dorsal midline

After the nuclei migrate to the periphery of the embryo, the dorsal protein becomes concentrated in the nuclei on the ventral surface.

Ventral midline

(b)

100 μm

◀ **21.6 Dorsal protein in the nuclei helps to determine the dorsal–ventral axis of the *Drosophila* embryo.** (a) Relative concentrations of Dorsal protein in the cytoplasm and nuclei of cells in the early *Drosophila* embryo. (b) Micrograph of a cross section of the embryo showing the Dorsal protein, darkly stained, in the nuclei along the ventral surface. (Part b, Max Planck Institute for Developmental Biology.)

Determination of the anterior–posterior axis Establishing the anterior–posterior axis of the embryo is a crucial step in early development. We will consider several genes in this pathway (Table 21.3). One important gene is *bicoid*, which is first transcribed in the ovary of an adult female during oogenesis. *Bicoid* mRNA becomes incorporated into the cytoplasm of the egg and, as it is passes into the egg, *bicoid* mRNA becomes anchored to the anterior end of the egg by part of its 3′ end. This anchoring causes *bicoid* mRNA to become concentrated at the anterior end (FIGURE 21.7a). (A number of other genes that are active in the ovary are required for proper localization of *bicoid* mRNA in the egg.) When the egg has been laid, *bicoid* mRNA is translated into Bicoid protein. Because most of the mRNA is at the anterior end of the egg, Bicoid protein is synthesized there and forms a concentration gradient along the

anterior–posterior axis of the embryo, with a high concentration at the anterior end and a low concentration at the posterior end. This gradient is maintained by the continuous synthesis of Bicoid protein and its short half-life.

The high concentration of Bicoid protein at the anterior end induces the development of anterior structures such as the head of the fruit fly. Bicoid—like Dorsal—is a morphogen. It stimulates the development of anterior structures by binding to regulatory sequences in the DNA and influencing the expression of other genes. One of the most important of the genes stimulated by Bicoid protein is *hunchback*, which is required for the development of the head and thoracic structures of the fruit fly.

The development of the anterior–posterior axis is also greatly influenced by a gene called *nanos*, an egg-polarity

Table 21.2	Key genes that control development of the dorsal–ventral axis in fruit flies and their action	
Gene	**Where Expressed**	**Action of Gene Product**
dorsal	Ovary	Affects expression of genes such as *twist* and *decapentaplegic*
cactus	Ovary	Traps Dorsal protein in cytoplasm
toll	Ovary	Alters Dorsal protein, allowing it to dissociate from Cactus protein and move into nuclei of ventral cells
twist	Embryo	Takes part in development of mesodermal tissues
decapentaplegic	Embryo	Takes part in development of gut structures

Table 21.3	Some key genes that determine the anterior–posterior axis in fruit flies	
Gene	**Where Expressed**	**Action**
bicoid	Ovary	Regulates expression of genes responsible for anterior structures; stimulates *hunchback*
nanos	Ovary	Regulates expression of genes responsible for posterior structures; inhibits translation of *hunchback* mRNA
hunchback	Embryo	Regulates transcription of genes responsible for anterior structures

gene that acts at the posterior end of the axis. The *nanos* gene is transcribed in the adult female, and the resulting mRNA becomes localized at the posterior end of the egg (◀ FIGURE 21.7b). After fertilization, *nanos* mRNA is translated into Nanos protein, which diffuses slowly toward the anterior end. The Nanos protein gradient is opposite that of Bicoid protein: Nanos is most concentrated at the posterior end of the embryo and is least concentrated at the anterior end. Nanos protein inhibits the formation of anterior structures by repressing the translation of *hunchback* mRNA. The synthesis of the Hunchback protein is therefore stimulated at the anterior end of the embryo by Bicoid protein and is repressed at the posterior end by Nanos protein. This combined stimulation and repression results in a Hunchback protein concentration gradient along the anterior–posterior axis that, in turn, affects the expression of other genes and helps determine the anterior and posterior structures.

Concepts

The major axes of development in early fruit-fly embryos are established as a result of initial differences in the distribution of specific mRNAs and proteins encoded by genes in the female parent (genetic maternal effect). These differences in distribution establish concentration gradients of morphogens, which cause different genes to be activated in different parts of the embryo.

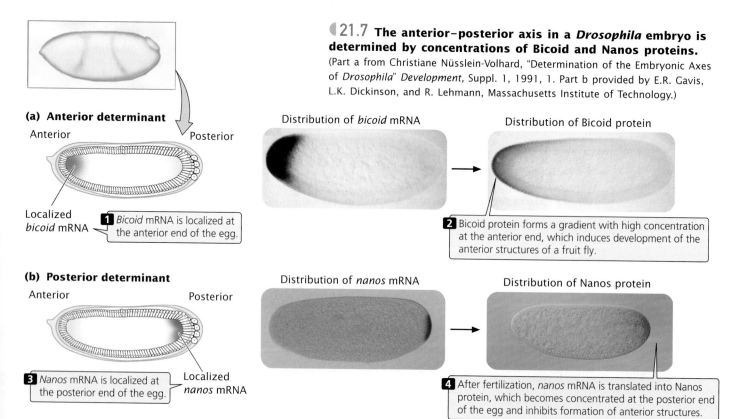

◀ 21.7 **The anterior–posterior axis in a *Drosophila* embryo is determined by concentrations of Bicoid and Nanos proteins.** (Part a from Christiane Nüsslein-Volhard, "Determination of the Embryonic Axes of *Drosophila*" *Development*, Suppl. 1, 1991, 1. Part b provided by E.R. Gavis, L.K. Dickinson, and R. Lehmann, Massachusetts Institute of Technology.)

(a) Anterior determinant

Anterior — Posterior

Localized *bicoid* mRNA

1 *Bicoid* mRNA is localized at the anterior end of the egg.

Distribution of *bicoid* mRNA

Distribution of Bicoid protein

2 Bicoid protein forms a gradient with high concentration at the anterior end, which induces development of the anterior structures of a fruit fly.

(b) Posterior determinant

Anterior — Posterior

3 *Nanos* mRNA is localized at the posterior end of the egg.

Localized *nanos* mRNA

Distribution of *nanos* mRNA

Distribution of Nanos protein

4 After fertilization, *nanos* mRNA is translated into Nanos protein, which becomes concentrated at the posterior end of the egg and inhibits formation of anterior structures.

Table 21.4	Segmentation genes and the effects of mutations in them	
Class of Gene	**Effect of Mutations**	**Examples of Genes**
Gap genes	Delete adjacent segments	*hunchback, Krüppel, knirps, giant, tailless*
Pair-rule genes	Delete same part of pattern in every other segment	*runt, hairy, fushi tarazu, even paired, odd paired, skipped, sloppy, paired, odd skipped*
Segment-polarity genes	Affect polarity of segment; part of segment replaced by mirror image of part of another segment	*engrailed, wingless, gooseberry, cubitus interruptus, patched, hedgehog, disheveled, costal-2, fused*

Segmentation genes Like all insects, the fruit fly has a segmented body plan. When the basic dorsal–ventral and anterior–posterior axes of the fruit-fly embryo have been established, **segmentation genes** control the differentiation of the embryo into individual segments. These genes affect the number and organization of the segments, and mutations in them usually disrupt whole sets of segments. The approximately 25 segmentation genes in *Drosophila* are transcribed after fertilization; so they don't exhibit a genetic maternal effect, and their expression is regulated by the Bicoid and Nanos protein gradients.

The segmentation genes fall into three groups as shown in Table 21.4 and ◀ FIGURE 21.8. **Gap genes** define large sections of the embryo; mutations in these genes eliminate whole groups of adjacent segments. Mutations in the *Krüppel* gene, for example, cause the absence of several adjacent segments. **Pair-rule genes** define regional sections of the embryo and affect alternate segments. Mutations in the *even-skipped* gene cause the deletion of even-numbered segments, whereas mutations in the *fushi tarazu* gene cause the absence of odd-numbered segments. **Segment-polarity genes** affect the organization of segments. Mutations in

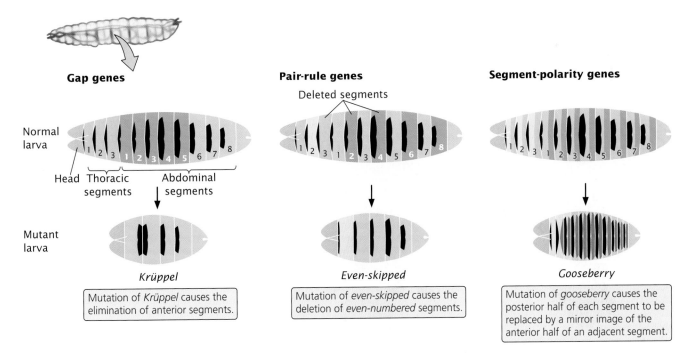

◀ 21.8 Segmentation genes control the differentiation of the *Drosophila* embryo into individual segments. The gap genes affect large sections of the embryo. The pair-rule genes affect alternate segments. The segment-polarity genes affect the polarity of segments.

these genes cause part of each segment to be deleted and replaced by a mirror image of part or all of an adjacent segment. For example, mutations in the *gooseberry* gene cause the posterior half of each segment to be replaced by the anterior half of an adjacent segment.

The gap genes, pair-rule genes, and segment-polarity genes act sequentially, affecting progressively smaller regions of the embryo. First, the egg-polarity genes activate or repress the gap genes, which divide the embryo into broad regions. The gap genes, in turn, regulate the pair-rule genes, which affect the development of pairs of segments. Finally, the pair-rule genes influence the segment-polarity genes, which guide the development of individual segments.

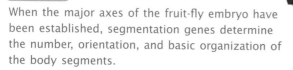

> **Concepts**
>
> When the major axes of the fruit-fly embryo have been established, segmentation genes determine the number, orientation, and basic organization of the body segments.

Homeotic genes After the segmentation genes have established the number and orientation of the segments, **homeotic genes** become active and determine the *identity* of individual segments. Eyes normally arise only on the head segment, whereas legs develop only on the thoracic segments. The products of homeotic genes activate other genes that encode these segment-specific characteristics. Mutations in the homeotic genes cause body parts to appear in the wrong segments.

Homeotic mutations were first identified in 1894, when William Bateson noticed that floral parts of plants occasionally appeared in the wrong place: he found, for example, flowers in which stamens grew in the normal place of

petals. In the late 1940s, Edward Lewis began to study homeotic mutations in *Drosophila*, which caused bizarre rearrangements of body parts. Mutations in the *Antennapedia* gene, for example, cause legs to develop on the head of a fly in place of the antenna (◀ FIGURE 21.9).

Homeotic genes create addresses for the cells of particular segments, telling the cells where they are within the regions defined by the segmentation genes. When a homeotic gene is mutated, the address is wrong and cells in the segment develop as though they were somewhere else in the embryo.

Homeotic genes are expressed after fertilization and are activated by specific concentrations of the proteins produced by the gap, pair-rule, and segment-polarity genes. The homeotic gene *Ultrabithorax (Ubx),* for example, is activated when the concentration of Hunchback protein (a product of a gap gene) is within certain values. These concentrations exist only in the middle region of the embryo; so *Ubx* is expressed only in these segments.

The homeotic genes encode regulatory proteins that bind to DNA; each gene contains a subset of nucleotides, called a **homeobox,** that are similar in all homeotic genes. The homeobox consists of 180 nucleotides and encodes 60 amino acids that serve as a DNA-binding domain; this domain is related to the helix-turn-helix motif (See Figure 16.2a). Homeoboxes are also present in segmentation genes and other genes that play a role in spatial development.

There are two major clusters of homeotic genes in *Drosophila*. One cluster, the **Antennapedia complex,** affects the development of the adult fly's head and anterior thoracic segments. The other cluster consists of the **bithorax complex** and includes genes that influence the adult fly's posterior thoracic and abdominal segments. Together, the *bithorax* and *Antennapedia* genes are termed the **homeotic complex** (HOM-C). In *Drosophila*, the *bithorax* complex contains three

(a)   **(b)**

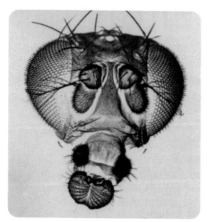
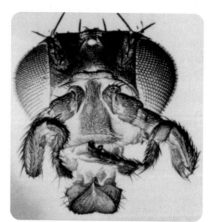

◀ 21.9 **The homeotic mutation *Antennapedia* substitutes legs for the antenna of a fruit fly.** (a) Normal, wild-type antenna. (b) *Antennapedia* mutant. (F.R. Turner/BPS.)

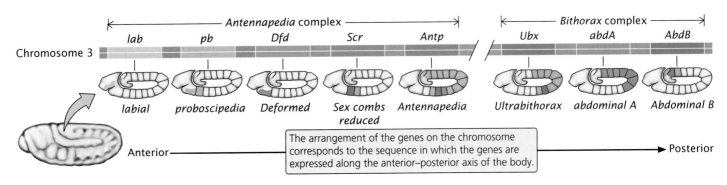

◀21.10 Homeotic genes, which determine the identity of individual segments in *Drosophila*, are present in two complexes. The *Antennapedia* complex has five genes, and the *bithorax* complex has three genes.

genes, and the *Antennapedia* complex has five; they are all located on the same chromosome (◀FIGURE 21.10). In addition to these eight genes, HOM-C contains many sequences that regulate the homeotic genes.

Remarkably, the order of the genes in the HOM-C is the same as the order in which the genes are expressed along the anterior–posterior axis of the body. The genes that are expressed in the more anterior segments are found at the one end of the complex, whereas those expressed in the more posterior end of the embryo are found at the other end of complex (See Figure 21.10). The reason for this correlation is unknown.

> **Concepts**
>
> Homeotic genes help determine the identity of individual segments in *Drosophila* embryos by producing DNA-binding proteins that activate other genes. Each homeotic gene contains a consensus sequence called a homeobox, which encodes the DNA-binding domain.

Homeobox Genes in Other Organisms

After homeotic genes in *Drosophila* had been isolated and cloned, molecular geneticists set out to determine if similar genes exist in other animals; probes complementary to the homeobox of *Drosophila* genes were used to search for homologous genes that might play a role in the development of other animals. The search was hugely successful: homeobox-containing (**Hox**) genes have been found in all animals studied so far, including nematodes, beetles, sea urchins, frogs, birds, and mammals. They have even been discovered in fungi and plants, indicating that *Hox* genes arose early in the evolution of eukaryotes.

In vertebrates, there are four clusters of *Hox* genes, each of which contains from 9 to 11 genes. Interestingly, the *Hox* genes of other organisms exhibit the same relation between order on the chromosome and order of their expression along the anterior–posterior axis of the embryo as that of *Drosophila* (◀FIGURE 21.11). Mammalian *Hox* genes, like those in *Drosophila*, encode transcription factors that help determine the identity of body regions along an anterior–posterior axis.

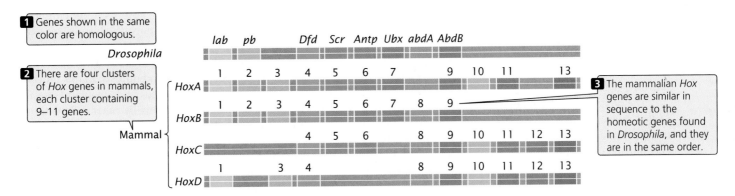

◀21.11 Homeotic genes in mammals are similar to those found in *Drosophila*. The complexes are arranged so that genes with similar sequences lie in the same column. See Figure 21.10 for the full names of the *Drosophila* genes.

Concepts

Homeobox-containing genes are found in many organisms, in which they regulate development.

www.whfreeman.com/pierce   More information about *Hox* genes

Connecting Concepts

The Control of Development

Development is a complex process consisting of numerous events that must take place in a highly specific sequence. The results of studies in fruit flies and other organisms reveal that this process is regulated by a large number of genes. In *Drosophila*, the dorsal–ventral axis and the anterior–posterior axis are established by maternal genes; these genes encode mRNAs and proteins that are localized to specific regions within the egg and cause specific genes to be expressed in different regions of the embryo. The proteins of these genes then stimulate other genes, which in turn stimulate yet other genes in a cascade of control. As might be expected, most of the gene products in the cascade are regulatory proteins, which bind to DNA and activate other genes.

In the course of development, successively smaller regions of the embryo are determined (◀ FIGURE 21.12). In *Drosophila*, first, the major axes and regions of the embryo are established by egg polarity genes. Next, patterns within each region are determined by the action of segmentation genes: the gap genes define large sections; the pair-rule genes define regional sections of the embryo and affect alternate segments; and the segment-polarity genes affect individual segments. Finally, the homeotic genes provide each segment with a unique identity. Initial gradients in proteins and mRNA stimulate localized gene expression, which produces more finely located gradients that stimulate even more localized gene expression. Developmental regulation thus becomes more and more narrowly defined.

The processes by which limbs, organs, and tissues form (called morphogenesis) are less well understood, although this pattern of generalized-to-localized gene expression is encountered frequently.

www.whfreeman.com/pierce   *Drosophila* development, images of fruit-fly anatomy and development, images of mammalian embryos, and many resources on development biology

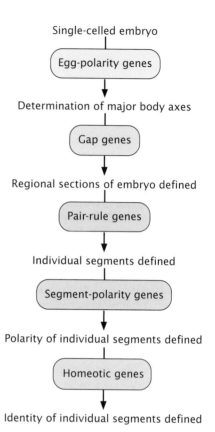

◀ **21.12 A cascade of gene regulation establishes the polarity and identity of individual segments of *Drosophila*.** In development, successively smaller regions of the embryo are determined.

Programmed Cell Death in Development

Cell death is an integral part of multicellular life. Cells in many tissues have a limited life span, and they die and are replaced continually by new cells. Cell death shapes many body parts during development: it is responsible for the disappearance of a tadpole's tail during metamorphosis and causes the removal of tissue between the digits to produce the human hand. Cell death is also used to eliminate dangerous cells that have escaped normal controls (see next section on cancer).

Cell death in animals is often initiated by the cell itself in a kind of cellular suicide termed **apoptosis.** In this process, a cell's DNA is degraded, its nucleus and cytoplasm shrink, and the cell undergoes phagocytosis by other cells without any leakage of its contents (◀ FIGURE 21.13a). Cells that are injured, on the other hand, die in a relatively uncontrolled manner called necrosis. In this process, a cell swells and bursts, spilling its contents over neighboring cells and eliciting an inflammatory response (◀ FIGURE 21.13b). Apoptosis is essential to embryogenesis; most multicellular animals cannot complete development if the process is inhibited.

Surprisingly, most cells are programmed to undergo apoptosis and will survive only if the internal death program is continually held in check. The process of apoptosis is highly regulated and depends on numerous signals inside and outside the cell. Geneticists have identified a number of genes having roles in various stages of the regulation of apoptosis. Some of these genes encode enzymes called **caspases,** which cleave other proteins at specific sites (after aspartic acid). Each caspase is synthesized as a large, inactive precursor (a procaspase) that is activated by cleavage, often by another caspase. When one caspase is activated, it cleaves other procaspases that trigger even more caspase activity. The resulting cascade of caspase activity eventually cleaves proteins essential to cell function, such as those supporting the nuclear membrane and cytoskeleton. Caspases also cleave a protein that normally keeps an enzyme that degrades DNA (DNAse) in an inactive form. Cleavage of this protein activates DNAse and leads to the breakdown of cellular DNA, which eventually leads to cell death.

Procaspases and other proteins required for cell death are continuously produced by healthy cells, so the potential for cell suicide is always present. A number of different signals can trigger apoptosis; for instance, infection by a virus can activate immune cells to secrete substances onto an infected cell, causing that cell to undergo apoptosis. This process is believed to be a defense mechanism designed to prevent the reproduction and spread of viruses. Similarly, DNA damage can induce apoptosis and thus prevent the replication of mutated sequences. Damage to mitochondria and the accumulation of a misfolded protein in the endoplasmic reticulum also stimulate programmed cell death.

Apoptosis in animal development is still poorly understood but is believed to be controlled through cell–cell signaling. The cell death that causes the disappearance of a tadpole's tail, for example, is triggered by thyroxin, a hormone produced by the thyroid gland that increases in concentration during metamorphosis. The elimination of cells between developing fingers in humans is thought to result from localized signals from nearby cells.

The symptoms of many diseases and disorders are caused by apoptosis or, in some cases, its absence. In neurodegenerative diseases such as Parkinson disease and Alzheimer disease, symptoms are caused by a loss of neurons through apoptosis. In heart attacks and stroke, some cells die through necrosis, but many others undergo apoptosis. Cancer is often stimulated by mutations in genes that regulate apoptosis, leading to a failure of apoptosis that would normally eliminate cancer cells.

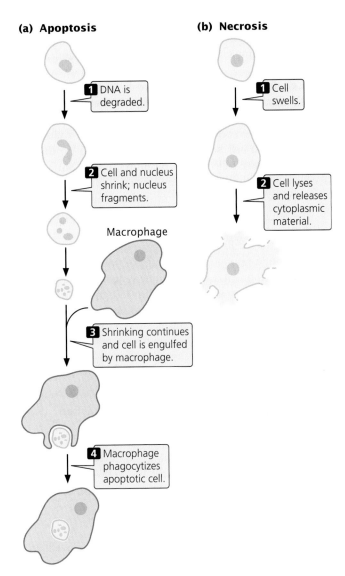

(a) Apoptosis

1 DNA is degraded.

2 Cell and nucleus shrink; nucleus fragments.

Macrophage

3 Shrinking continues and cell is engulfed by macrophage.

4 Macrophage phagocytizes apoptotic cell.

(b) Necrosis

1 Cell swells.

2 Cell lyses and releases cytoplasmic material.

◀ **21.13 Programmed cell death by apoptosis is distinct from uncontrolled cell death through necrosis.**

Concepts

Cells are capable of apoptosis (programmed cell death), a highly regulated process that depends on enzymes called caspases. Apoptosis plays an important role in animal development and is implicated in a number of diseases.

www.whfreeman.com/pierce   Additional information on apoptosis

Evo-Devo: The Study of Evolution and Development

"Ontogeny recapitulates phylogeny" is a familiar phrase that was coined in the 1860s by German zoologist Ernst Haeckel to describe his belief—now considered wrong—that organisms repeat their evolutionary history during development. According to Haeckel's theory, a human embryo passes through fish, amphibian, reptilian, and mammalian stages before developing human traits.

Although ontogeny does not recapitulate phylogeny, many evolutionary biologists today are turning to the study of development for a better understanding of the processes and patterns of evolution. Sometimes called "evo-devo," the study of evolution through the analysis of development is revealing that the same genes often shape developmental pathways in distantly related organisms. In humans and insects, for example, the same gene controls the development of eyes, despite the fact that insect and mammalian eyes are thought to have evolved independently. Similarly, biologists once thought that segmentation in vertebrates and invertebrates was only superficially similar, but we now know that, in both *Drosophila* and amphioxus (a marine organism closely related to vertebrates), a gene called *engrailed* divides the embryo into specific segments. A gene called *distalless*, which creates the legs of a fruit fly, has also been found to play a role in the development of crustacean branched appendages. This same gene also stimulates body outgrowths of many other organisms, from polycheate worms to starfish.

Similar genes may be part of a developmental pathway common to two different species but have quite different effects. For example, a *Hox* gene called *AbdB* helps define the posterior end of a *Drosophila* embryo; a similar group of genes in birds divides the wing into three segments. In another example, the *sog* gene in fruit flies stimulates cells to assume a ventral orientation in the embryo, but the expression of a similar gene called *chordin* in vertebrates causes cells to assume dorsal orientation, exactly the opposite of the situation in fruit flies.

The theme emerging from these studies is that a small, common set of genes may underlie many basic developmental processes in many different organisms. Although Haeckel's euphonious phrase "ontogeny recapitulates phylogeny" was incorrect, evo-devo is proving that development can reveal much about the process of evolution.

Immunogenetics

A basic assumption of developmental biology is that every somatic cell carries an identical set of genetic information and that no genes are lost during development. Although this assumption holds for most cells, there are some important exceptions, one of which concerns genes that encode immune function in vertebrates.

The immune system provides protection against infection by specific bacteria, viruses, fungi, and parasites. The focus of an immune response is an **antigen,** defined as any molecule that elicits an immune reaction. Although any molecule can be an antigen, most are proteins. The immune system is remarkable in its ability to recognize an almost unlimited number of potential antigens.

The body is full of proteins, so it is essential that the immune system be able to distinguish between self-antigens and foreign antigens. Occasionally, the ability to make this distinction breaks down, and the body produces an immune reaction to its own antigens, resulting in an **autoimmune disease** (Table 21.5).

The Organization of the Immune System

The immune system contains a number of different components and uses several mechanisms to provide protection against pathogens, but most immune responses can be grouped into two major classes: humoral immunity and cellular immunity. Although it is convenient to think of these classes as separate systems, they interact and influence each other significantly.

Humoral immunity centers on the production of antibodies by special lymphocytes called **B cells** (◀FIGURE 21.14), which mature in the bone marrow. **Antibodies** are proteins that circulate in the blood and other body fluids, binding to specific antigens and marking them for destruction by phagocytic cells. Antibodies also activate a set of proteins called complement that help to lyse cells and attract macrophages.

Cellular immunity is conferred by **T cells** (see Figure 21.14), which are specialized lymphocytes that mature in the thymus and respond only to antigens found on the surfaces of the body's own cells. After a pathogen such as a virus has infected a host cell, some viral antigens appear on the cell surface. Proteins, called **T-cell receptors,** on the surfaces of T cells bind to these antigens and mark the infected cell for destruction. T-cell receptors must simultaneously bind a foreign antigen and a self-antigen called a **major histocompatibility complex (MHC) antigen** on the cell surface. Not all T cells attack cells having foreign antigens; some help regulate immune responses, providing communication among different components of the immune system.

How can the immune system recognize an almost unlimited number of foreign antigens? Remarkably, each mature lymphocyte is genetically programmed to attack

Table 21.5	Examples of autoimmune diseases
Disease	**Tissues Attacked**
Graves disease, Hashimoto thyroiditis	Thyroid gland
Rheumatic fever	Heart muscle
Systematic lupus erythematosus	Joints, skin, and other organs
Rheumatoid arthritis	Joints
Insulin-dependent diabetes mellitus	Insulin-producing cells in pancreas
Multiple sclerosis	Myelin sheath around nerve cells

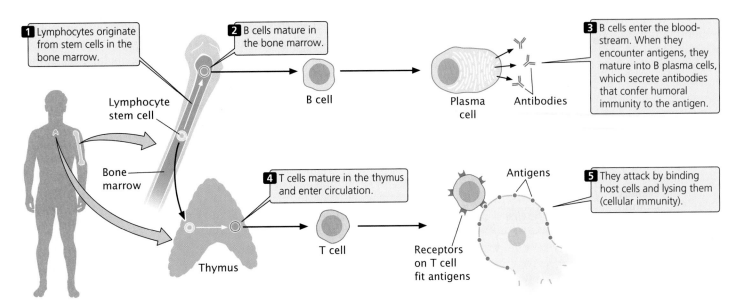

1 Lymphocytes originate from stem cells in the bone marrow.

2 B cells mature in the bone marrow.

3 B cells enter the bloodstream. When they encounter antigens, they mature into B plasma cells, which secrete antibodies that confer humoral immunity to the antigen.

Lymphocyte stem cell

Bone marrow

B cell

Plasma cell   Antibodies

4 T cells mature in the thymus and enter circulation.

5 They attack by binding host cells and lysing them (cellular immunity).

Antigens

Thymus

T cell

Receptors on T cell fit antigens

◀21.14 Immune responses are divided into humoral immunity, in which antibodies are produced by B cells, and cellular immunity produced by T cells.

one and only one specific antigen: each mature B cell produces antibodies against a single antigen, and each T cell is capable of attaching to only one type of foreign antigen.

If each lymphocyte is specific for only one type of antigen, how does an immune response develop? The **theory of clonal selection** proposes that initially there is a large pool of millions of different lymphocytes, each capable of binding only one antigen (◀FIGURE 21.15); so millions of different foreign antigens can be detected. To illustrate clonal selection, let's imagine that a foreign protein enters the body. Only a few lymphocytes in the pool will be specific for this particular foreign antigen. When one of these lymphocytes encounters the foreign antigen and binds to it, that lymphocyte is stimulated to divide. The lymphocyte proliferates rapidly, producing a large population of genetically identical cells—a clone—each of which is specific for that particular antigen.

This initial proliferation of antigen-specific B and T cells is known as a **primary immune response** (see Figure 21.15); in most cases, the primary response destroys the foreign antigen. Subsequent to the primary immune response, most of the lymphocytes in the clone die, but a few continue to circulate in the body. These **memory cells** may remain in circulation for years or even for the rest of one's life. Should the same antigen reappear at some time in the future, memory cells specific to that antigen become activated and quickly give rise to another clone of cells capable of binding the antigen. The rise of this second clone is termed a **secondary immune response** (see Figure 21.15). The ability to quickly produce a second clone of antigen-specific cells permits the long-lasting immunity that often follows recovery from a disease. For example, people who have chicken pox usually have

life-long immunity to the disease. The secondary immune response is also the basis for vaccination, which stimulates a primary immune response to an antigen and results in memory cells that can quickly produce a secondary response if that same antigen appears in the future.

Three sets of proteins are required for immune responses: antibodies, T-cell receptors, and the major histocompatibility antigens. The next section explores how the enormous diversity in these proteins is generated.

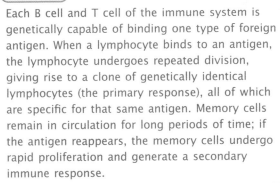

Concepts

Each B cell and T cell of the immune system is genetically capable of binding one type of foreign antigen. When a lymphocyte binds to an antigen, the lymphocyte undergoes repeated division, giving rise to a clone of genetically identical lymphocytes (the primary response), all of which are specific for that same antigen. Memory cells remain in circulation for long periods of time; if the antigen reappears, the memory cells undergo rapid proliferation and generate a secondary immune response.

Immunoglobulin Structure

The principal products of the humoral immune response are antibodies—also called immunoglobulins. Each immunoglobulin (Ig) molecule consists of four polypeptide chains—two identical light chains and two identical heavy chains—which form a Y-shaped structure (◀FIGURE 21.16). Disulfide bonds link the two heavy chains in the stem of the Y

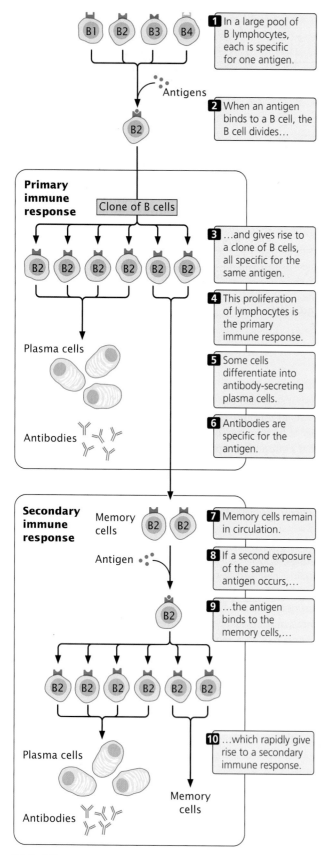

1 In a large pool of B lymphocytes, each is specific for one antigen.

Antigens

2 When an antigen binds to a B cell, the B cell divides…

Primary immune response

Clone of B cells

3 …and gives rise to a clone of B cells, all specific for the same antigen.

4 This proliferation of lymphocytes is the primary immune response.

Plasma cells

5 Some cells differentiate into antibody-secreting plasma cells.

Antibodies

6 Antibodies are specific for the antigen.

Secondary immune response

Memory cells

7 Memory cells remain in circulation.

Antigen

8 If a second exposure of the same antigen occurs,…

9 …the antigen binds to the memory cells,…

Plasma cells

10 …which rapidly give rise to a secondary immune response.

Memory cells

Antibodies

21.15 An immune response to a specific antigen is produced through clonal selection.

and attach a light chain to a heavy chain in each arm of the Y. Binding sites for antigens are at the ends of the two arms.

The light chains of an immunoglobulin come in two basic types, called kappa chains and lambda chains. An immunoglobulin molecule can have two kappa chains or two lambda chains, but it cannot have one of each type. Both the light and the heavy chain has a variable region at one end and a constant region at the other end; the variable regions of different immunoglobulin molecules vary in amino acid sequence, whereas the constant regions of different immunoglobulins are similar in sequence. The variable regions of both light and heavy chains make up the antigen-binding region and specify the type of antigen that the antibody can bind.

Mammals have five basic classes of immunoglobulins, known as IgM, IgD, IgE, IgG, and IgA. Each class is defined

(a)

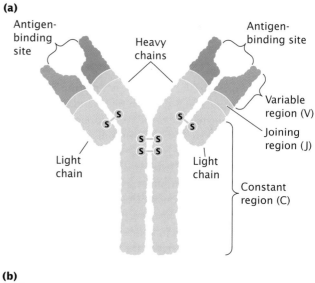

(b)

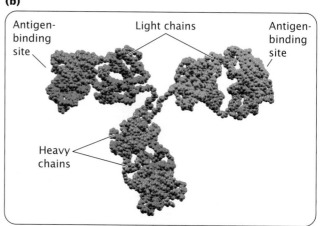

21.16 Each immunoglobulin molecule consists of four polypeptide chains—two light chains and two heavy chains—that combine to form a Y-shaped structure. (a) Structure of an immunoglobulin. (b) Folded, space-filling model.

by the type of heavy chain found in the immunoglobulin. The different classes of antibodies have different functions or they appear at different times during an immune response or both. For example, in a primary response, all B cells initially make IgM but, as the immune response develops, they switch to producing a combination of IgM and IgD. Later, the B cells may switch to one of the other immunoglobulin classes.

The Generation of Antibody Diversity

The immune system is capable of making antibodies against virtually any antigen that might be encountered in one's lifetime: each human is capable of making about 10^{15} different antibody molecules. Antibodies are proteins; so the amino acid sequences of all 10^{15} potential antibodies must be encoded in the human genome. However, there are fewer than 1×10^5 genes in the human genome and, in fact, only 3×10^9 total base pairs; so how can this huge diversity of antibodies be encoded?

The answer lies in the fact that antibody genes are composed of segments. There are a number of copies of each type of segment, each differing slightly from the others. In the maturation of a lymphocyte, the segments are joined to create an immunoglobulin gene. The particular copy of each segment used is random and, because there are multiple copies of each type, there are many possible combinations of the segments. A limited number of segments can therefore encode a huge diversity of antibodies.

To illustrate this process of antibody assembly, let's consider the immunoglobulin light chains. Kappa and lambda chains are encoded by separate genes on different chromosomes. Each gene is composed of three types of segments: V, for variable; J, for joining; and C, for constant. The V segments encode most of the variable region of the light chains, the C segment encodes the constant region of the chain, and the J segments encode a short set of nucleotides that join the V segment and the C segments together.

The number of V, J, and C segments differs among species. For the human kappa gene, there are from 30 to 35

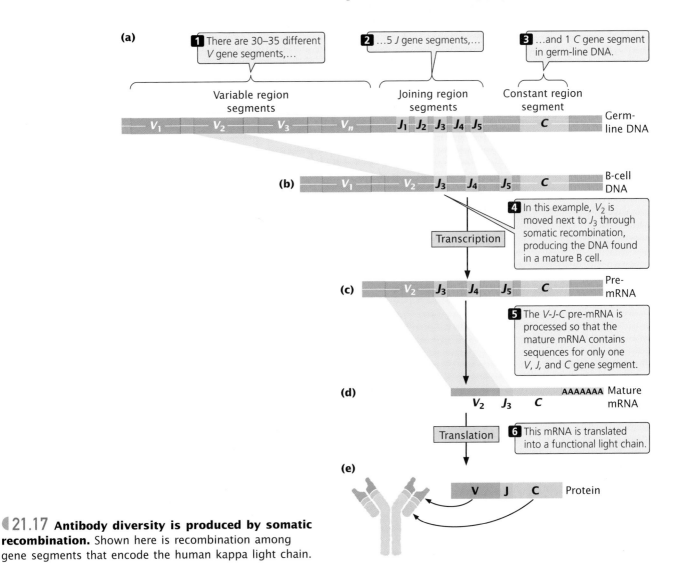

◀21.17 Antibody diversity is produced by somatic recombination. Shown here is recombination among gene segments that encode the human kappa light chain.

different functional *V* gene segments, 5 different *J* genes, and a single *C* gene segment, all of which are present in the germ-line DNA (◀ Figure 21.17a). The *V* gene segments, which are about 400 bp in length, are located on the same chromosome and are separated from one another by about 7000 bp. The *J* gene segments are about 30 bp in length and all together encompass about 1400 bp.

Initially, an immature lymphocyte inherits all of the *V* gene segments and all of the *J* gene segments present in the germ line. In the maturation of the lymphocyte, **somatic recombination** within a single chromosome moves one of the *V* genes to a position next to one of the *J* gene segments (◀ Figure 21.17b). In Figure 21.17b, V_2 (the second of approximately 35 different *V* gene segments) undergoes somatic recombination, which places it next to J_3 (the third of 5 *J* gene segments); the intervening segments are lost.

After somatic recombination has taken place, the combined *V-J-C* gene is transcribed and processed (◀ Figure 21.17c and d). The mature mRNA that results contains only sequences for a single *V*, *J*, and *C* segment; this mRNA is translated into a functional light chain (◀ Figure 21.17e).

In this way, each mature human B cell produces a unique type of kappa light chain, and different B cells produce slightly different kappa chains, depending on the combination of *V* and *J* segments that are joined.

The gene that encodes the lambda light chain is organized in a similar way but differs from the kappa gene in the number of copies of the different segments. In the human gene for the lambda light chain, there are from 29 to 33 different functional *V* gene segments and 4 or 5 different functional *J* and *C* gene segments (each *C* gene segment is attached to a different *J* segment). Somatic recombination takes place among the segments in the same way as that in the kappa gene, generating many possible combinations of lambda light chains.

The gene that encodes the immunoglobulin heavy chain is arranged in *V*, *J*, and *C* segments, but this gene also possesses *D* (for diversity) segments. Somatic recombination taking place in lymphocyte maturation joins one *D* gene segment to one *J* gene segment, and then a *V* gene segment is joined to this combined *D-J* gene segment (◀ Figure 21.18a and b). Transcription and RNA processing of this gene produces a mRNA that encodes only one

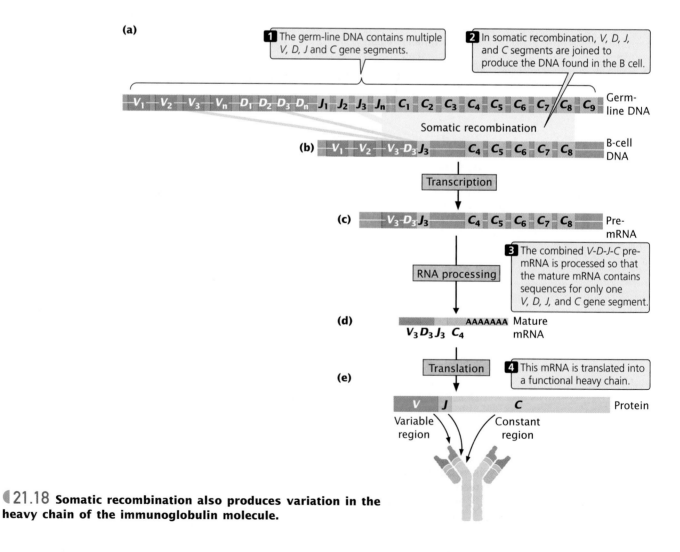

(a)

1 The germ-line DNA contains multiple *V*, *D*, *J* and *C* gene segments.

2 In somatic recombination, *V*, *D*, *J*, and *C* segments are joined to produce the DNA found in the B cell.

V_1—V_2—V_3—V_n—D_1—D_2—D_3—D_n—J_1—J_2—J_3—J_n—C_1—C_2—C_3—C_4—C_5—C_6—C_7—C_8—C_9 Germ-line DNA

Somatic recombination

(b) V_1—V_2—V_3–D_3—J_3—C_4—C_5—C_6—C_7—C_8 B-cell DNA

Transcription

(c) V_3–D_3—J_3—C_4—C_5—C_6—C_7—C_8 Pre-mRNA

3 The combined *V-D-J-C* pre-mRNA is processed so that the mature mRNA contains sequences for only one *V*, *D*, *J*, and *C* gene segment.

RNA processing

(d)   AAAAAAA Mature mRNA
$V_3 D_3 J_3\ C_4$

4 This mRNA is translated into a functional heavy chain.

Translation

(e)

| V | J | C | Protein |

Variable region   Constant region

◀ 21.18 **Somatic recombination also produces variation in the heavy chain of the immunoglobulin molecule.**

particular type of heavy chain (◀Figure 21.18c–e). Thus, many different types of light and heavy chains are possible.

Somatic recombination is brought about by RAG1 and RAG2 proteins, which generate double-strand breaks at specific nucleotide sequences called recombination signal sequences that flank the V, D, J, and C gene segments. DNA repair proteins then process and join the ends of particular segments together (◀Figure 21.19).

In addition to somatic recombination, other mechanisms add to antibody diversity. First, each type of light chain can potentially combine with each type of heavy chain to make a functional immunoglobulin molecule, increasing the amount of possible variation in antibodies. Second, the recombination process that joins V, J, D, and C gene segments in the developing B cell is imprecise, and a few random nucleotides are frequently lost or gained at the junctions of the recombining segments. This **junctional diversity** greatly enhances variation among antibodies. Third, a high rate of mutation, called **somatic hypermutation** (the cause of which is unknown), is characteristic of the immunoglobulin genes.

Concepts

The genes encoding the antibody chains are organized in segments, and germ-line DNA contains multiple versions of each segment. The many possible combinations of V, J, and D segments permit an immense variety of different antibodies to be generated. This diversity is augmented by the different combinations of light and heavy chains,

the random addition and deletion of nucleotides at the junctions of the segments, and the high mutation rates in the immunological genes.

T-Cell-Receptor Diversity

Like B cells, each mature T cell has genetically determined specificity for one type of antigen that is mediated through the cell's receptors. T-cell receptors are structurally similar to immunoglobulins (◀Figure 21.20) and are located on the cell surface; most T-cell receptors are composed of one alpha and one beta polypeptide chain held together by disulfide bonds. One end of each chain is embedded in the cell membrane; the other end projects away from the cell and binds antigens. Like the immunoglobulin chains, each chain of the T-cell receptor possesses a constant region and a variable region (see Figure 21.20); the variable regions of the two chains provide the antigen-binding site.

The genes that encode the alpha and beta chains of the T-cell receptor are organized much like those that encode the heavy and light chains of immunoglobulins: each gene is made up of segments that undergo somatic recombination before the gene is transcribed. For example, the human gene for the alpha chain initially consists of 44 to 46 V gene segments, 50 J gene segments, and a single C gene segment. The organization of the gene for the beta chain is similar, except that it also contains D segments. Random combination of alpha and beta chains and junctional diversity takes place, but there is no evidence for somatic hypermutation in T-cell-receptor genes.

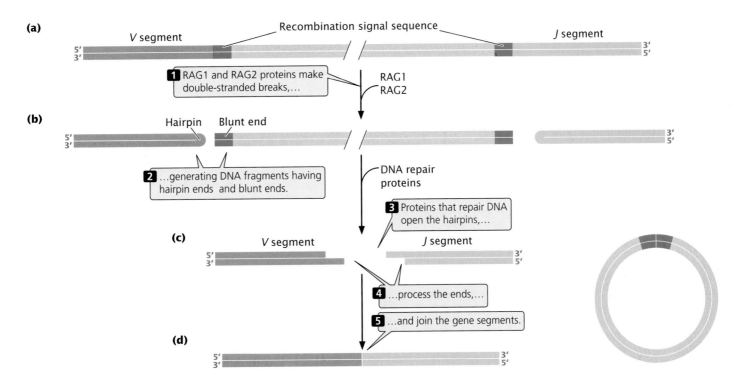

◀21.19 Somatic recombination is brought about by RAG1 and RAG2 proteins and DNA repair proteins.

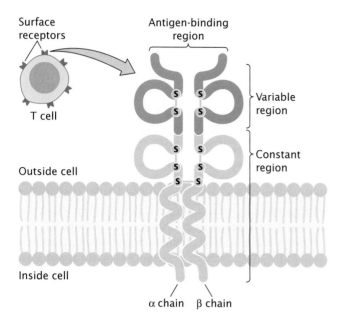

21.20 A T-cell receptor is composed of two polypeptide chains, each having a variable and constant region. Most T-cell receptors are composed of alpha (α) and beta (β) polypeptide chains held together by disulfide bonds. One end of each chain traverses the cell membrane; the other end projects away from the cell and binds antigens.

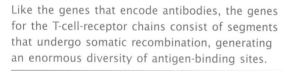

Concepts

Like the genes that encode antibodies, the genes for the T-cell-receptor chains consist of segments that undergo somatic recombination, generating an enormous diversity of antigen-binding sites.

Major Histocompatibility Complex Genes

When tissues are transferred from one species to another or even from one individual member to another within a species, the transplanted tissues are usually rejected by the host animal. The results of early studies demonstrated that this graft rejection is due to an immune response that occurs when antigens on the surface of the grafted tissue are detected and attacked by T cells in the host organism. The antigens that elicit graft rejection are referred to as histocompatibility antigens, and they are encoded by a cluster of genes called the major histocompatibility complex.

T cells are activated only when the T-cell receptor simultaneously binds *both* a foreign antigen *and* the host cell's own histocompatibility antigen. The reason for this requirement is not clear; it may reserve T cells for action against pathogens that have invaded cells. When a foreign body, such as a virus, is ingested by a macrophage or other cell, partly digested pieces of the foreign body containing antigens are displayed on the macrophage's surface (◀FIGURE 21.21). Through their T-cell receptors, T cells bind to both

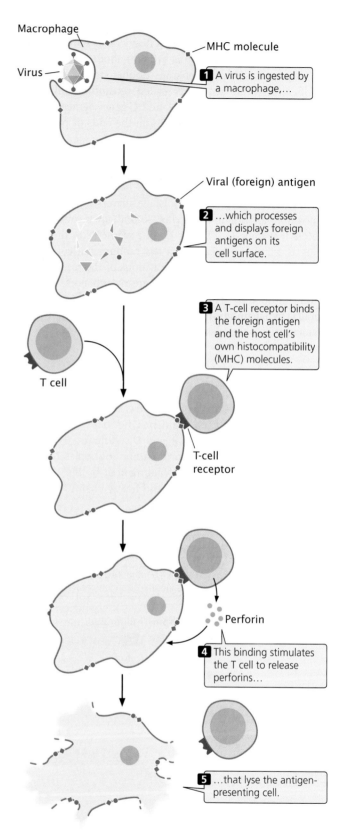

21.21 T cells are activated by binding to a foreign antigen and a histocompatibility antigen on the surface of a self-cell.

the histocompatibility protein and the foreign antigen and secrete substances that either destroy the antigen-containing cell or activate other B and T cells or both.

The MHC genes are among the most variable genes known: there are more than 100 different alleles for some MHC loci. Because each person possesses five or more MHC loci and because many alleles are possible at each locus, no two people (with the exception of identical twins) produce the same set of histocompatibility antigens. The variation in histocompatibility antigens provides each of us with a unique identity for our own cells, which allows our immune systems to distinguish self from nonself. This variation is also the cause of rejection in organ transplants.

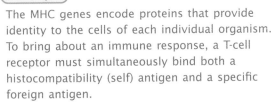

Concepts

The MHC genes encode proteins that provide identity to the cells of each individual organism. To bring about an immune response, a T-cell receptor must simultaneously bind both a histocompatibility (self) antigen and a specific foreign antigen.

www.whfreeman.com/pierce   Additional information about the genetics of the immune system

Genes and Organ Transplants

For a person with a seriously impaired organ, a transplant operation may offer the only hope of survival. Successful transplantation requires more than the skills of a surgeon; it also requires a genetic match between the patient and the person donating the organ.

The fate of transplanted tissue depends largely on the type of antigens present on the surface of its cells. Because foreign tissues are usually rejected by the host, the successful transplantation of tissues between different persons is very difficult. Tissue rejection can be partly inhibited by drugs that interfere with cellular immunity. Unfortunately, this treatment can create serious problems for transplant patients, because they may have difficulty fighting off common pathogens and thus may die of infection. The only other option for controlling the immune reaction is to carefully match the donor and the recipient, maximizing the genetic similarities.

The tissue antigens that elicit the strongest immune reaction are the very ones used by the immune system to mark its own cells, those encoded by the major histocompatibility complex. The MHC spans a region of more than 3 million base pairs on human chromosome 6 and has many alleles, providing different MHC antigens on the cells of different people and allowing the immune system to recognize foreign cells.

The severity of an immune rejection of a transplanted organ depends on the number of mismatched MHC antigens on the cells of the transplanted tissue. The ABO red-blood-cell antigens also are important because they elicit a strong immune reaction. The ideal donor is the patient's own identical twin, who will have exactly the same MHC and ABO antigens. Unfortunately, most patients don't have an identical twin. The next best donor is a sibling with the same major MHC and ABO antigens. If a sibling is not available, donors from the general population are considered. An attempt is made to match as many of the MHC antigens of the donor and recipient as possible, and immunosuppressive drugs are used to control rejection that occurs because of the mismatches. The long-term success of transplants depends on the closeness of the match. Survival rates after kidney transplants (the most successful of the major organ transplants) increase from 63% with zero or one MHC match to 90% with four matches.

Cancer Genetics

Cancer kills one of every five Americans, and cancer treatments cost billions of dollars every year. Cancer is not a single disease; rather, it is a heterogeneous group of disorders characterized by the presence of cells that do not respond to the normal controls on division. Cancer cells divide rapidly and continuously, creating tumors that crowd out normal cells and eventually rob healthy tissues of nutrients. The cells of an advanced tumor can separate from the tumor and travel to distant sites in the body, where they may take up residence and develop into new tumors. The most common cancers in the United States are those of the breast, prostate, lung, colon and rectum, and blood (Table 21.6).

The Nature of Cancer

Normal cells grow, divide, mature, and die in response to a complex set of internal and external signals. A normal cell receives both stimulatory and inhibitory signals, and its growth and division are regulated by a delicate balance between these opposing forces. In a cancer cell, one or more of the signals has been disrupted, which causes the cell to proliferate at an abnormally high rate. As they lose their response to the normal controls, cancer cells gradually lose their regular shape and boundaries, eventually forming a distinct mass of abnormal cells—a tumor. If the cells of the tumor remain localized, the tumor is said to be benign; if the cells invade other tissues, the tumor is said to be **malignant.** Cells that travel to other sites in the body, where they establish secondary tumors, have undergone **metastasis.**

Cancer As a Genetic Disease

Cancer arises as a result of fundamental defects in the regulation of cell division, and its study therefore has significance not only for public health, but also for our basic understanding of cell biology. Through the years, a large number of theories have been put forth to explain cancer, but we now recognize that most, if not all, cancers arise from defects in DNA.

Table 21.6	Estimated incidences of various cancers and cancer mortality in the United States in 2002	
Type of Cancer	**Estimated New Cases per Year**	**Estimated Deaths per Year**
Breast	205,000	40,000
Prostate	189,000	30,200
Lung and bronchus	169,400	154,900
Colon and rectum	148,300	56,600
Lymphoma	60,900	25,800
Bladder	56,500	12,600
Melanoma	53,600	7,400
Uterus	39,300	6,600
Leukemias	30,800	21,700
Oral cavity and pharynx	28,900	7,400
Pancreas	30,300	29,700
Ovary	23,300	13,900
Stomach	21,600	12,400
Brain and nervous system	17,000	13,100
Liver	16,600	14,100
Uterine cervix	13,000	4,100
Cancers of soft tissues including heart	8,300	3,900
All cancers	1,284,900	555,500

Source: American Cancer Society, *Cancer Facts and Figures, 2002* (Atlanta: American Cancer Society, 2001), p. 80

Early observations suggested that cancer might result from genetic damage. First, it was recognized that many agents such as ionizing radiation and chemicals that cause mutations also cause cancer (are carcinogens). Second, some cancers are consistently associated with particular chromosome abnormalities. About 90% of people with chronic myeloid leukemia, for example, have a reciprocal translocation between chromosome 22 and chromosome 9 (see Figure 9.31). Third, some specific types of cancers tend to run in families. Retinoblastoma, a rare childhood cancer of the retina, appears with high frequency in a few families and is inherited as an autosomal dominant trait, suggesting that a single gene is responsible for these cases of the disease.

Although these observations hinted that genes play some role in cancer, the theory of cancer as a genetic disease had several significant problems. If cancer is inherited, every cell in the body should receive the cancer-causing gene, and therefore every cell should become cancerous. In those types of cancer that run in families, however, tumors typically appear only in certain tissues and often only when the person reaches an advanced age. Finally, many cancers do not run in families at all and, even in regard to those cancers that generally do, isolated cases crop up in families with no history of the disease.

In 1971, Alfred Knudson proposed a model to explain the genetic basis of cancer. Knudson was studying retinoblastoma, a cancer that usually develops in only one eye but occasionally appears in both. Knudson found that, when retinoblastoma appears in both eyes, onset is at an early age, and affected children often have close relatives who also have retinoblastoma.

Knudson proposed that retinoblastoma results from two separate genetic defects, both of which are necessary for cancer to develop (◀FIGURE 21.22). He suggested that, in the cases in which the disease affects just one eye, a single cell in one eye undergoes two successive mutations. Because the chance of these two mutations occurring in a single cell is remote, retinoblastoma is rare and typically develops in only one eye. For bilateral cases, Knudson proposed that the child inherited one of the two mutations required for the cancer, and so every cell contains this initial mutation. In these cases, all that is required for cancer to develop is for one eye cell to undergo the second mutation. Because each eye possesses millions of cells, there is a high probability that the second mutation will occur in at least one cell of each eye, producing tumors in both eyes at an early age.

Knudson's hypothesis suggests that cancer is the result of a multistep process that requires several mutations. If one or more of the required mutations is inherited, fewer

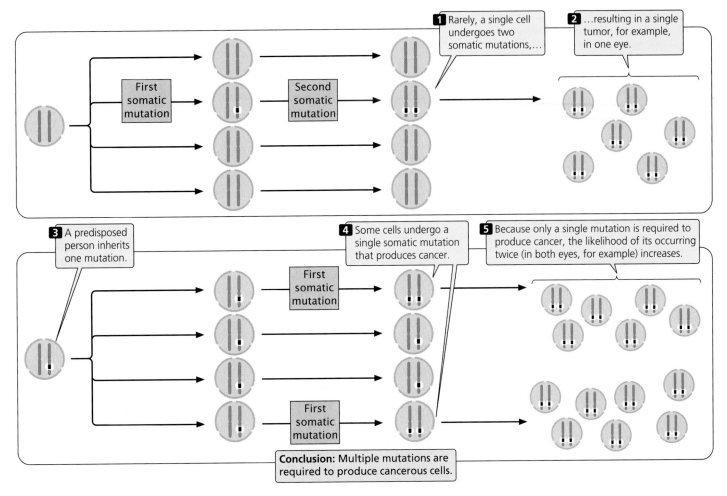

◀ 21.22 **Alfred Knudson proposed that retinoblastoma results from two separate genetic defects, both of which are necessary for cancer to develop.**

additional mutations are required to produce cancer, and the cancer will tend to run in families. The idea that cancer results from multiple mutations turns out to be correct for most cancers.

Knudson's genetic theory for cancer has been confirmed by the identification of genes that, when mutated, cause cancer. Today, we recognize that cancer is fundamentally a genetic disease, although few cancers are actually inherited. Most tumors arise from somatic mutations that accumulate during our life span, either through spontaneous mutation or in response to environmental mutagens.

The clonal evolution of tumors Cancer begins when a single cell undergoes a mutation that causes the cell to divide at an abnormally rapid rate. The cell proliferates, giving rise to a clone of cells, each of which carries the same mutation. Because the cells of the clone divide more rapidly than normal, they soon outgrow other cells. Additional mutations that arise in the clone may further enhance the ability of those cells to proliferate, and cells carrying both mutations soon become dominant in the clone. Eventually, they may be overtaken by cells that contain yet more muta-

tions that enhance proliferation. In this process, called **clonal evolution,** the tumor cells acquire more mutations that allow them to become increasingly more aggressive in their proliferative properties (◀ FIGURE 21.23).

The rate of clonal evolution depends on the frequency with which new mutations arise. Any genetic defect that allows more mutations to arise will accelerate cancer progression. Genes that regulate DNA repair are often found to have been mutated in the cells of advanced cancers, and inherited disorders of DNA repair are usually characterized by increased incidences of cancer. Because DNA repair mechanisms normally eliminate many of the mutations that arise, without DNA repair, mutations are more likely to persist in all genes, including those that regulate cell division. Xeroderma pigmentosum, for example, is a rare disorder caused by a defect in DNA repair (see p. 499 in Chapter 17). People with this condition have elevated rates of skin cancer when exposed to sunlight (which induces mutation).

Mutations in genes that affect chromosome segregation also may contribute to the clonal evolution of tumors. Many cancer cells are aneuploid, and it is clear that chromosome mutations contribute to cancer progression by

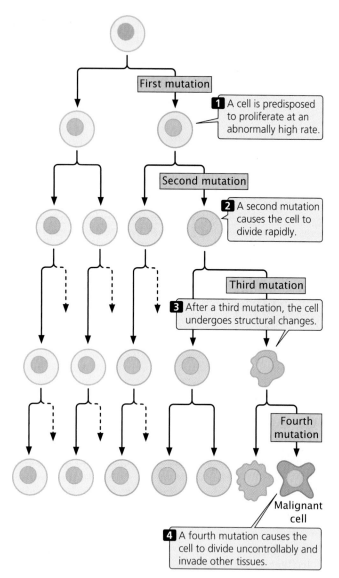

First mutation

1 A cell is predisposed to proliferate at an abnormally high rate.

Second mutation

2 A second mutation causes the cell to divide rapidly.

Third mutation

3 After a third mutation, the cell undergoes structural changes.

Fourth mutation

Malignant cell

4 A fourth mutation causes the cell to divide uncontrollably and invade other tissues.

◀ **21.23 Through clonal evolution, tumor cells acquire multiple mutations that allow them to become increasingly aggressive and proliferative.**

duplicating some genes (those on extra chromosomes) and eliminating others (those on deleted chromosomes). Cellular defects that interfere with chromosome separation increase aneuploidy and therefore may accelerate cancer progression.

Concepts

Cancer is fundamentally a genetic disease. Mutations in several genes are usually required to produce cancer. If one of these mutations is inherited, fewer somatic mutations are necessary for cancer to develop, and the person may have a predisposition to cancer. Clonal evolution is the accumulation of mutations in a clone of cells.

The role of environment in cancer Although cancer is fundamentally a genetic disease, most cancers are not inherited, and there is little doubt that many cancers are influenced by environmental factors. The role of environmental factors in cancer is suggested by differences in the incidence of specific cancers throughout the world (Table 21.7). The results of studies show that migrant populations typically take on the cancer incidence of their host country. For example, the overall rates of cancer are considerably lower in Japan than in Hawaii. However, within a single generation after migration to Hawaii, Japanese people develop cancer at rates similar to those of native Hawaiians. Smoking is a good example of an environmental factor that is strongly associated with cancer. Other environmental factors such as chemicals, ultraviolet light, ionizing radiation, and viruses are known carcinogens and are associated with variation in the incidence of many cancers.

Genes That Contribute to Cancer

The signals that regulate cell division fall into two basic types: molecules that stimulate cell division and those that inhibit it. These control mechanisms are similar to the accelerator and brake of an automobile. In normal cells (but hopefully not your car), both accelerators and brakes are applied at the same time, causing cell division to proceed at the proper speed.

Because cell division is affected by both accelerators and brakes, cancer can arise from mutations in either type of signal, and there are several fundamentally different routes to cancer (◀ FIGURE 21.24). A stimulatory gene can be made hyperactive or active at inappropriate times, analogously to having the accelerator of an automobile stuck in the floored position. Mutations in stimulatory genes are usually dominant, because a mutation in a single copy of the gene is usually sufficient to produce a stimulatory effect. Dominant-acting stimulatory genes that cause cancer are termed **oncogenes.** Cell division may also be stimulated when inhibitory genes are made *inactive,* analogously to having a defective brake in an automobile. Mutated inhibitory genes generally have recessive effects, because both copies must be mutated to remove all inhibition. Inhibitory genes in cancer are termed **tumor-suppressor genes.**

Although oncogenes or mutated tumor-suppressor genes or both are required to produce cancer, mutations in DNA repair genes can increase the likelihood of acquiring mutations in these genes. Having mutated DNA repair genes is analogous to having a lousy car mechanic who does not make the necessary repairs to a broken accelerator or brake.

Oncogenes and tumor-suppressor genes Oncogenes were the first cancer-causing genes to be identified. In 1910, Peyton Rous described a virus that caused connective-tissue tumors (sarcomas) in chickens; this virus became known as the Rous sarcoma virus. A number of other cancer-causing

Table 21.7	Examples of geographic variation in the incidence of cancer	
Type of Cancer	**Location**	**Incidence Rate***
Lip	Canada (Newfoundland)	15.1
	Brazil (Fortaleza)	1.2
Nasopharynx	Hong Kong	30.0
	United States (Utah)	0.5
Colon	United States (Iowa)	30.1
	India (Bombay)	3.4
Lung	United States (New Orleans, African Americans)	110.0
	Costa Rica	17.8
Prostate	United States (Utah)	70.2
	China (Shanghai)	1.8
Bladder	United States (Connecticut, Whites)	25.2
	Philippines (Rizal)	2.8
All cancer	Switzerland (Basel)	383.3
	Kuwait	76.3

Source: C. Muir et al., *Cancer incidence in Five Continents*, vol. 5 (Lyon: International Agency for Research on Cancer, 1987), Table 12-2.

*The incidence rate is the age-standardized rate in males per 100,000 population.

viruses were subsequently isolated from various animal tissues. These viruses were generally assumed to carry a cancer-causing gene that was transferred to the host cell. The first oncogene, called *src*, was isolated from the Rous sarcoma virus in 1970.

In 1975, Michael Bishop, Harold Varmus, and their colleagues began to use probes for viral oncogenes to search for related sequences in normal cells. They discovered that the genomes of all normal cells carry DNA sequences that are closely related to viral oncogenes. These cellular genes are called **proto-oncogenes.** They are responsible for basic cellular functions in normal cells but, when mutated, they become oncogenes that contribute to the development of cancer. When a virus infects a cell, a proto-oncogene may

(a) Oncogenes

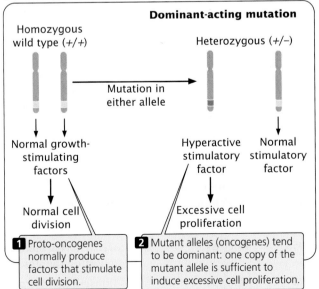

(b) Tumor-suppressor genes

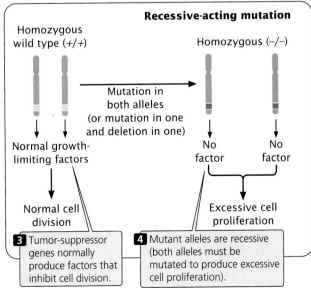

◀ **21.24 Both oncogenes and tumor-suppressor genes contribute to cancer but differ in their modes of action and dominance.**

Table 21.8	Some oncogenes and functions of their corresponding proto-oncogenes	
Oncogene	**Cellular Location of Product**	**Function of Proto-oncogene**
sis	Secreted	Growth factor
erbB	Cell membrane	Part of growth-factor receptor
erbA	Cytoplasm	Thyroid hormone receptor
src	Cell membrane	Protein tyrosine kinase
ras	Cell membrane	GTP binding and GTPase
myc	Nucleus	Transcription factor
fos	Nucleus	Transcription factor
jun	Nucleus	Transcription factor
bcl-1	Nucleus	Cell cycle

become incorporated into the viral genome through recombination. Within the viral genome, the proto-oncogene may mutate to an oncogene that, when inserted back into a cell, causes rapid cell division and cancer. Because the proto-oncogenes are more likely to undergo mutation or recombination within a virus, viral infection is often associated with the cancer.

Proto-oncogenes can be converted into oncogenes in viruses by several different ways. The sequence of the proto-oncogene may be altered or truncated as it is being incorporated into the viral genome. This mutated copy of the gene may then produce an altered protein that causes uncontrolled cell proliferation. Alternatively, through recombination, a proto-oncogene may end up next to a viral promoter or enhancer, which then causes the gene to be overexpressed. Finally, sometimes the function of a proto-oncogene in the host cell may be altered when a virus inserts its own DNA into the gene, disrupting its normal function.

Many oncogenes have been identified by experiments in which selected fragments of DNA are added to cells in culture. Some of the cells take up the DNA and, if these cells become cancerous, then the DNA fragment that was added to the culture must contain an oncogene. The fragments can then be sequenced, and the oncogene can be identified. More than 70 oncogenes have now been discovered (Table 21.8).

Tumor-suppressor genes are more difficult than oncogenes to identify because they *inhibit* cancer and are recessive; both alleles must be mutated before the inhibition of cell division is removed. Because it is the *failure* of their function that promotes cell proliferation, tumor-suppressor genes cannot be identified by adding them to cells and looking for cancer.

One of the first tumor-suppressor genes to be identified was the retinoblastoma gene. In 1985, Raymond White and Webster Cavenne showed that large segments of chromosome 13 were missing in cells of retinoblastoma tumors, and later the tumor-suppressor gene was isolated from these

segments. A number of tumor-suppressor genes have now been discovered in this way (Table 21.9).

Genes controlling the cell cycle Genes that control the cell cycle often serve as proto-oncogenes or tumor-suppressor genes. Let's briefly revisit the regulation of the cell cycle, which was discussed in Chapter 2. The cell cycle is regulated by cyclins, whose concentration oscillates during the cell cycle, and cyclin-dependent kinases (CDKs), which have a relatively constant concentration. Cyclins bind to CDKs, producing activated protein kinases that initiate key events in the cell cycle. Genes that encode cyclins and factors that inhibit or stimulate the formation of activated CDKs are often oncogenes and tumor-suppressor genes, respectively. Mutated cyclin genes have been associated with cancers of the immune system, breast, stomach, and esophagus; genes, such as *p16* and *p21*, that encode inhibitors of CDKs are mutated or missing in many cancer cells.

Some proto-oncogenes and suppressor genes have roles in apoptosis. Cells have the ability to assess themselves and, when they are abnormal or damaged, they normally undergo apoptosis (see pp. 612–613). Cancer cells frequently have chromosome mutations, DNA damage, and other cellular

Table 21.9	Some tumor-suppressor genes and their functions	
Gene	**Cellular Location of Product**	**Function**
NF1	Cytoplasm	GTPase activator
p53	Nucleus	Transcription factor, regulates apoptosis
RB	Nucleus	Transcription factor
WT-1	Nucleus	Transcription factor

Source: J. Marx, Learning how to suppress cancer, *Science* 261(1993):1385.

anomalies that would normally stimulate apoptosis and prevent their proliferation. Often these cells have mutations in genes that regulate apoptosis, and therefore they do not undergo programmed cell death. The ability of a cell to initiate apoptosis in response to DNA damage, for example, depends on a gene called *p53,* which is inactivate in many human cancers.

www.whfreeman.com/pierce   Additional information about *p53* and its role in cancer

DNA repair genes Cancer arises from the accumulation of multiple mutations in a single cell. Some cancer cells have normal rates of mutation, and multiple mutations accumulate because each mutation gives the cell a replicative advantage over cells without the mutations. Other cancer cells may have higher-than-normal rates of mutation in all of their genes, which leads to more frequent mutation of oncogenes and tumor-suppressor genes. What might be the source of these high rates of mutation in some cancer cells?

Two processes control the rate at which mutations arise within a cell: (1) the rate at which errors arise during and after replication; and (2) the efficiency with which these errors are corrected. The error rate during replication is controlled by the fidelity of DNA polymerases and other proteins in the replication process (see Chapter 12). However, defects in genes encoding replication proteins have not been strongly linked to cancer.

The mutation rate is also strongly affected by whether errors are corrected by DNA repair systems (see pp. 495–499 in Chapter 17). Defects in genes that encode components of these repair systems have been consistently associated with a number of cancers. People with xeroderma pigmentosum, for example, are defective in nucleotide-excision repair, an important cellular repair system that normally corrects DNA damage caused by a number of mutagens, including ultraviolet light. Likewise, about 13% of colorectal, endometrial, and stomach cancers have cells that are defective in mismatch repair, another major repair system in the cell.

Some types of colon cancer are inherited as an autosomal dominant trait. In families with this condition, a person can inherit one mutated and one normal allele of a gene that controls mismatch repair. The normal allele provides sufficient levels of the protein for mismatch repair to function, but it is highly likely that this normal allele will become mutated or lost in at least a few cells. If it does so, there is no mismatch repair, and these cells undergo higher-than-normal rates of mutation, leading to defects in oncogenes and tumor-suppressor genes that cause the cells to proliferate.

www.whfreeman.com/pierce   Additional information on DNA repair

Genes affecting chromosome segregation Most advanced tumors contain cells that exhibit a variety of chromosome anomalies, including extra chromosomes, missing chromosomes, and chromosome rearrangements. Aneuploidy in somatic cells usually arises when chromosomes do not segregate properly in mitosis. Normal cells have a checkpoint that monitors the proper assembly of the mitotic spindle; if chromosomes are not properly attached to the microtubules at metaphase, the onset of anaphase is blocked. Some aneuploid cancer cells contain mutant alleles for genes that encode proteins having roles in this checkpoint; in these cells, anaphase is entered into despite the improper or lack of assembly of the spindle, and chromosome abnormalities result.

The tumor-suppressor gene *p53*, in addition to controlling apoptosis, plays a role in the duplication of the centrosome, which is required for proper formation of the spindle and for chromosome segregation. Normally, the centrosome duplicates once per cell cycle. If *p53* is mutated or missing, however, the centrosome may undergo extra duplications, resulting in the unequal segregation of chromosomes. In this way, mutation of the *p53* gene may generate chromosome mutations that contribute to cancer. The *p53* gene is also a tumor-suppressor gene that prevents cell division when the DNA is damaged.

Sequences that regulate telomerase Another factor that may contribute to the progression of cancer is the inappropriate activation of an enzyme called telomerase. Telomeres are special sequences at the ends of eukaryotic chromosomes (see pp. 297–298 in Chapter 11). In DNA replication in somatic cells, DNA polymerases require a 3′-OH group to add new nucleotides. For this reason, the ends of chromosomes cannot be replicated, and telomeres become shorter with each cell division. This shortening eventually leads to the destruction of the chromosome and cell death; so somatic cells are capable of a limited number of cell divisions.

In germ cells, telomerase replicates the chromosome ends (see pp. 341–343 in Chapter 12), thereby maintaining the telomeres, but this enzyme is not normally expressed in somatic cells. In many tumor cells, however, sequences that regulate the expression of the telomerase gene are mutated so that the enzyme *is* expressed, and the cell is capable of unlimited cell division. Although the expression of telomerase appears to contribute to the development of many cancers, its precise role in tumor progression is still being investigated.

Genes that promote vascularization and the spread of tumors A final set of factors that contribute to the progression of cancer includes genes that affect the growth and spread of tumors. Oxygen and nutrients, which are essential to the survival and growth of tumors, are supplied by blood vessels, and the growth of new blood vessels (angiogenesis) is important to tumor progression. Angiogenesis is stimulated by growth factors and others proteins encoded by genes whose expression is carefully regulated in normal cells. In tumor cells, genes encoding these proteins are often overexpressed compared with normal cells, and inhibitors of angiogenesis-promoting factors may be inactivated or underexpressed. At least one inherited cancer syndrome—van Hippel-Lindau disease, in which people develop multiple types of tumors—is caused by the mutation of a gene that affects angiogenesis.

In the development of many cancers, the primary tumor gives rise to cells that spread to distant sites, producing secondary tumors. This process of metastasis is the cause of death in 90% of human cancer cases; it is influenced by cellular changes induced by somatic mutation. By using microarrays to measure levels of gene expression (see Chapter 19), researchers have identified several genes that are transcribed at a significantly higher rate in metastatic cells compared with nonmetastatic cells. These genes encode components of the extracellular matrix and the cytoskeleton, which are thought to affect the migration of cells. Other genes that affect metastasis include adhesion proteins that help hold cells together.

Concepts

Oncogenes are dominant in their action and stimulate cell proliferation. Tumor-suppressor genes are recessive in their action and inhibit cell proliferation. Defects in DNA repair genes allow a higher-than-normal rate of mutation in oncogenes and tumor suppressor genes. Mutations in genes that control chromosome segregation allow chromosome mutations to accumulate, which may then contribute to cancer progression. Mutations that allow telomerase to be expressed in somatic cells and that affect vascularization and metastasis also may contribute to cancer progression.

www.whfreeman.com/pierce   General information about cancer, the genetics of cancer, and telomerase

The Molecular Genetics of Colorectal Cancer

Mutations that contribute to colorectal cancer have received extensive study, and this cancer is an excellent example of how cancer often arises through the accumulation of successive genetic defects.

Colorectal cancers arise in the cells lining the colon and rectum. More than 148,000 new cases of colorectal cancer are diagnosed in the United States each year, where this cancer is responsible for more than 56,000 deaths annually. If detected early, colorectal cancer can be treated successfully; consequently, there has been much interest in identifying the molecular events responsible for the initial stages of colorectal cancer.

Colorectal cancer is thought to originate as benign tumors called adenomatous polyps. Initially, these polyps are microscopic, but in time they enlarge, and the cells of the polyp acquire the abnormal characteristics of cancer cells. In the later stages of the disease, the tumor may invade the mus-

cle layer surrounding the gut and metastasize. The progression of the disease is slow; from 10 to 35 years may be required for a benign tumor to develop into a malignant tumor.

Most cases of colorectal cancer are sporadic, developing in people with no family history of the disease, but a few families display a clear genetic predisposition to this disease. In one form of hereditary colon cancer, known as familial adenomatous polyposis coli, hundreds or thousands of polyps develop in the colon and rectum; if these polyps are not removed, one or more almost invariably becomes malignant.

Because polyps and tumors of the colon and rectum can be easily observed and removed with a colonoscope (a fiber optic instrument that is used to view the interior of the rectum and colon), much is known about the progression of colorectal cancer, and some of the genes responsible for its clonal evolution have been identified. About 75% of colorectal cancers have mutations in tumor-suppressor gene *p53*, and many also have a mutation in the *ras* proto-oncogene. Families with adenomatous polyposis coli carry a defect in a gene called *APC*, and mutations in *APC* are found in the cells of tumors that arise sporadically (in persons without a family history). Additional genes that are frequently mutated in colorectal cancer include the oncogenes *myc* and *neu* and the tumor-suppressor gene *HNPCC*.

Mutations in these genes are responsible for the different steps of colorectal cancer progression. One of the earliest steps is a mutation that inactivates the *APC* gene, which increases the rate of cell division, leading to polyp formation (◀FIGURE 21.25). A person with familial adenomatous polyposis coli inherits one defective copy of the *APC* gene, and defects in this gene are associated with the numerous polyps that appear in those who have this disorder. Mutations in *APC* are also found in the polyps that develop in people who do not have adenomatous polyposis coli.

Mutations of the *ras* oncogene usually occur later, in larger polyps comprising cells that have acquired some genetic mutations. The protein produced by the normal *ras* proto-oncogene sits inside the cell membrane. From there it relays signals from growth factors that stimulate cell division. When *ras* is mutated, the protein that it encodes continually relays a stimulatory signal for cell division, even when growth factor is absent.

Mutations in *p53* and other genes appear still later in tumor progression; these mutations are rare in polyps but common in malignant cells. Because *p53* prevents the replication of cells with genetic damage and controls proper chromosome segregation, mutations in *p53* may allow a cell to rapidly acquire further gene and chromosome mutations, which then contribute to further proliferation and invasion into surrounding tissues.

The sequence of steps just outlined is not the only route to colorectal cancer, and the mutations need not occur in the order presented here, but this sequence is a common pathway by which colon and rectal cells become cancerous.

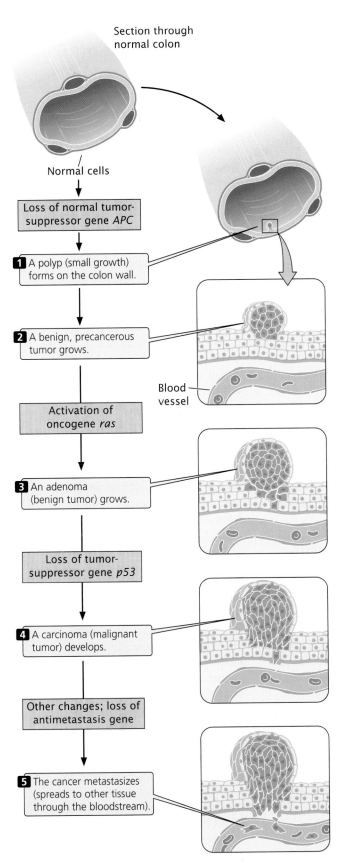

Section through normal colon

Normal cells

Loss of normal tumor-suppressor gene *APC*

1 A polyp (small growth) forms on the colon wall.

2 A benign, precancerous tumor grows.

Blood vessel

Activation of oncogene *ras*

3 An adenoma (benign tumor) grows.

Loss of tumor-suppressor gene *p53*

4 A carcinoma (malignant tumor) develops.

Other changes; loss of antimetastasis gene

5 The cancer metastasizes (spreads to other tissue through the bloodstream).

21.25 Mutations in multiple genes contribute to the progression of colorectal cancer.

Connecting Concepts Across Chapters

This chapter has focused on three specialized but important topics: the genetics of development, the immune system, and cancer. In addition to their relevance to genetics, these topics have obvious medical importance and all are the subject of intense research.

The results of early experiments demonstrated that genes are not usually lost or permanently altered in the course of development; rather, development proceeds through the regulation of gene expression. The basic question for development is how are different sets of genes expressed in different parts of the embryo? Our study of pattern formation in *Drosophila* revealed that many genes take part and that they are regulated in a highly sequential manner. The process is initiated by maternally produced mRNA and proteins that become localized to particular regions of the egg. Sets of genes are successively activated, each set controlling the expression of other sets, so that successively smaller regions of the embryo are determined.

The immune system also is encoded by a complex set of genes whose products interact closely. Unlike those in pattern development, genes encoding antibodies and T-cell receptors are permanently altered in lymphocyte maturation. Lymphocytes violate the general principle that all cells contain the same set of genetic information.

Cancer also is influenced by complex interactions among multiple genes. Paradoxically, cancer is fundamentally a genetic disease, but most cancers are not inherited, because cancer usually requires somatic mutations at multiple genes. Even for those cancers for which a predisposition is clearly inherited, additional somatic mutations are required for cancer to arise. These mutations, each rare, accumulate because they provide the cell with a growth advantage.

This chapter has synthesized much of the information provided in preceding chapters. Gene regulation (Chapter 16) is the basis of development, the understanding of which also requires knowledge of genetic maternal effects (Chapter 5), transcription (Chapter 13), and translation (Chapter 15). The rearrangement of segments in genes of the immune system builds on our understanding of recombination (Chapter 12) and RNA processing (Chapter 14). Chromosome and gene mutations (Chapters 9 and 17) are essential to understanding cancer progression. Many oncogenes and tumor-suppressor genes control the cell cycle (Chapter 2), and predisposition to some cancers may be inherited as single-gene traits (Chapter 3). Cancer may also entail mutations in DNA repair genes (Chapter 17), genes affecting chromosome segregation (Chapter 2), and the regulation of telomerase (Chapter 12). Recombinant DNA techniques (Chapter 18) have contributed tremendously to our understanding of all of these processes.

The New Genetics
ETHICS · SCIENCE · TECHNOLOGY   Breast Cancer   Ron Green

Scientists in a medical genetics program of a major university medical school enroll a 54-year-old woman with metastatic breast cancer into a research protocol. The patient reports that several of her maternal relatives have breast cancer, but no pathological specimens from affected relatives are available for verifying the diagnoses. The patient dies before research studies are completed.

The patient has two daughters who are identical twins. Shortly after her mother's death, one of the twins requests access to her mother's test results. She explains that she wants this information because it will help her learn whether she carries the same mutation that might have contributed to her mother's disease. She has been informed that laboratories will not test for a mutation in a person before the identification of a known cancer-causing mutation in another family member affected by breast cancer. She wishes to learn whether she has a mutation because she is considering a prophylactic mastectomy to reduce her risk of developing breast cancer.

The research team learns from the head of the Institutional Review Board that there are no legal obstacles to this request, because the legal rights of the mother are not being compromised—she is deceased. The deceased are not considered research subjects under existing federal regulations.

On discovering her sister's intentions to request her mother's results, the second twin objects to the release of their mother's genetic information and says that she does not want to know whether she has inherited a greater risk of developing breast cancer. However, she went on to say, there is no way that the information could be kept from her, because she would inevitably learn of

her sister's decision to have surgery. What should the research team do?

The case raises questions that are common in genetic research and clinical care today. Because of the nature and complexity of the questions raised, many of them require interdisciplinary examination. The GenEthics Consortium (GEC) was formed to bring scientists, bioethicists, lawyers, genetic counselors, and consumers together to discuss ethical issues emerging from research associated with the Human Genome Project.

In the late 1990s, the GEC convened to consider this particular case. In the course of their discussion, some members expressed the opinion that the clinical setting should focus on meeting the needs of the individual person, and the research setting should focus on gathering evidence to support a hypothesis. Others disagreed, arguing that the results of genetic testing in research settings often have clinical implications for subjects.

No explicit guidance was given by the mother,—should researchers release this information to a child who requests it? Some argued that the fact that the mother's DNA was being tested for mutations that are markers for breast cancer implies that the mother was not opposed to this type of testing. Some also stated that we can safely presume that the mother expected to share that information with her husband and children.

But whether children have a right to access the mother's test results was not the only issue that merited discussion. What should clinicians and researchers do when family members disagree about whether such information should be made available? Some felt that the sister who first came to the researchers with a request for her mother's

genetic information has a privileged position. Others objected strenuously, saying that one's right to information in complex cases such as this one cannot merely be a matter of who gets to the doctor first.

Still other issues arose in the course of this discussion, most of which fell under the umbrella question of whether the risks of breast cancer for the twins could really be determined with precision. Persons in high-risk families with hereditary breast cancer and cancer-causing mutations face a much greater than average lifetime risk. However, in the absence of a thorough family history, the presence of a certain mutation and breast cancer in the mother alone do not provide the basis for assuming the existence of "hereditary" breast cancer in the family.

A second major area of uncertainty was highlighted when genetic professionals from different institutions disagreed about the statistics regarding risk reduction from prophylactic mastectomy. Some argued that the weighing of benefits and harms must include a comparison of the protection of one twin from the risk of premature death with the risk of psychological distress for the second twin. The research team would be morally justified in choosing the action that is most likely to reduce risk of death by cancer. However, others disagreed on the basis of inconclusive scientific evidence that mastectomy is a valid risk-reduction strategy.

Although there was no agreement on how genetic professionals should respond to this situation, there *was* a sense that, at this stage of genetic research, individual professionals might ethically and responsibly come to different conclusions about what they would do when faced with competing requests of this sort.

CONCEPTS SUMMARY

- Each multicellular organism begins as a single cell that has the potential to develop into any cell type. As development proceeds, cells become committed to particular fates. The results of early cloning experiments demonstrated that this process arises from differential gene expression.

- In the early *Drosophila* embryo, determination is effected through a cascade of gene control.

- The dorsal–ventral and anterior–posterior axes of the *Drosophila* embryo are established by egg-polarity genes. These genes are expressed in the female parent and produce RNA and proteins that are deposited in the egg cytoplasm. Initial differences in the distribution of these molecules regulate gene expression in various parts of the embryo. The dorsal–ventral axis is defined by a concentration gradient of the Dorsal protein, and the anterior–posterior axis is defined by concentration gradients of Bicoid and Nanos proteins.

- After the establishment of the major axes of development, three types of segmentation genes act sequentially to determine the number and organization of the embryonic segments in *Drosophila*. The gap genes establish large sections of the embryo, the pair-rule genes affect alternate segments, and the segment-polarity genes affect the organization of individual segments.

- Homeotic genes then define the identity of individual *Drosophila* segments. All these genes contain a consensus sequence called a homeobox that encodes a DNA-binding domain; the products of homeotic genes are DNA-binding proteins that regulate the expression of other genes. Genes with homeoboxes are found in many other organisms.

- Apoptosis, or programmed cell death, plays an important role in the development of many animals. In apoptosis, DNA is degraded, the nucleus and cytoplasm shrink, and the cell undergoes phagocytosis by other cells. Apoptosis is a highly regulated process that depends on caspases—proteins that cleave proteins. Each caspase is originally synthesized as an inactive precursor that must be activated, often through cleavage by another caspase.

- The immune system is the primary defense network in vertebrates. In humoral immunity, B cells produce antibodies that bind foreign antigens; in cellular immunity, T cells attack cells carrying foreign antigens.

- Each B and T cell is capable of binding only one type of foreign antigen. There are vast numbers of different types of B and T cells, and any potential antigen can be bound. When a lymphocyte binds to an antigen, the lymphocyte divides and gives rise to a clone of cells, each specific for that same antigen. This process is a primary immune response. A few memory cells remain in circulation for long periods of time. If the same antigen is encountered again, memory cells can proliferate rapidly and generate a secondary immune response.

- Immunoglobulins (antibodies) consists of two light chains and two heavy chains, each containing variable and constant regions. Light chains are of two basic types: kappa and lambda chains. The genes that encode the immunoglobulin chains consist of several types of gene segments; germ-line DNA contains multiple copies of these gene segments, which differ slightly in sequence. In B-cell maturation, somatic recombination randomly brings together one version of each segment to produce a single complete gene. Many combinations of the different segments are possible. The potential for diversity of antibodies is further increased by the random addition and deletion of nucleotides at the junctions of the segments. A high mutation rate also increases the potential diversity of antibodies.

- T-cell receptors are composed of alpha and beta chains. The germ-line genes for these proteins consist of segments with multiple varying copies. Somatic recombination allows many different types of T-cell receptors in different cells. Junctional diversity also adds to T-cell receptor variability.

- The major histocompatibility complex encodes a number of histocompatibility antigens. Each T cell simultaneously binds a foreign antigen and a host MHC antigen. The MHC antigen allows the immune system to distinguish self from nonself. Each locus for the MHC contains many alleles.

- Cancer is fundamentally a genetic disorder, arising from somatic mutations in multiple genes that affect cell division and proliferation. If one or more mutations is inherited, then fewer additional mutations are required for cancer to develop.

- A mutation that allows a cell to divide rapidly provides the cell with a growth advantage; this cell gives rise to a clone of cells with the same mutation. Within this clone, other mutations occur that provide additional growth advantages, and cells with these additional mutations become dominant in the clone. In this way, the clone evolves. Environmental factors play an important role in the development of many cancers by increasing the rate of somatic mutations.

- Several types of genes contribute to cancer progression. Oncogenes are dominant mutated copies of genes that normally stimulate cell division. Tumor-suppressor genes normally inhibit cell division; recessive mutations in these genes may contribute to cancer. Oncogenes and tumor-suppressor genes often control the cell cycle or regulate apoptosis.

- Defects in DNA repair genes and genes that control chromosome segregation often increase the overall mutation rate of other genes, leading to defects in proto-oncogenes and tumor-suppressor genes that may contribute to cancer progression.

- Mutations in sequences that regulate telomerase, an enzyme that replicates the ends of chromosomes, are often associated with cancer. Telomerase allows cells to divide indefinitely but is not usually expressed in somatic cells. Mutations in tumor cells allow telomerase to be expressed.

- Tumor progression is also affected by mutations in genes that promote vascularization and the spread of tumors.

- Colorectal cancer offers a model system for understanding tumor progression in humans. Initial mutations stimulate cell division, leading to a small benign polyp. Additional mutations allow the polyp to enlarge, invade the muscle layer of the gut, and eventually spread to other sites. Mutations in particular genes affect different stages of this progression.

IMPORTANT TERMS

totipotent (p. 603)	*bithorax* complex (p. 610)	T cell (p. 614)	somatic recombination (p. 618)
determination (p. 603)	homeotic complex (p. 610)	T-cell receptor (p. 614)	junctional diversity (p. 619)
egg polarity gene (p. 605)	*Hox* gene (p. 611)	major histocompatibility complex (MHC) antigen (p. 614)	somatic hypermutation (p. 619)
morphogen (p. 605)	apoptosis (p. 612)	theory of clonal selection (p. 615)	malignant tumor (p. 621)
segmentation gene (p. 609)	caspase (p. 613)		metastasis (p. 621)
gap gene (p. 609)	antigen (p. 614)	primary immune response (p. 615)	clonal evolution (p. 623)
pair-rule gene (p. 609)	autoimmune disease (p. 614)		oncogene (p. 624)
segment-polarity gene (p. 609)	humoral immunity (p. 614)	memory cell (p. 615)	tumor-suppressor gene (p. 624)
homeotic gene (p. 610)	B cell (p. 614)	secondary immune response (p. 615)	proto-oncogene (p. 625)
homeobox (p. 610)	antibody (p. 614)		
Antennapedia complex (p. 610)	cellular immunity (p. 614)		

Worked Problems

1. If a fertilized *Drosophila* egg is punctured at the anterior end and a small amount of cytoplasm is allowed to leak out, what will be the most likely effect on the development of the fly embryo?

• Solution

The egg-polarity genes determine the major axes of development in the *Drosophila* embryo. One of these genes is *bicoid*, which is transcribed in the maternal parent. As *bicoid* mRNA passes into the egg, the mRNA becomes anchored to the anterior end of the egg. After the egg is laid, *bicoid* mRNA is translated into Bicoid protein, which forms a concentration gradient along the anterior–posterior axis of the embryo. The high concentration of Bicoid protein at the anterior end induces the development of anterior structures such as the head of the fruit fly. If the anterior end of the egg is punctured, cytoplasm containing high concentrations of Bicoid protein will leak out, reducing the concentration of Bicoid protein at the anterior end. The result will be that the embryo fails to develop head and thoracic structures at the anterior end.

2. In some cancer cells, a specific gene has become duplicated many times. Is this gene likely to be an oncogene or a tumor-suppressor gene? Explain your reasoning.

• Solution

The gene is likely to be an oncogene. Oncogenes stimulate cell proliferation and act in a dominant manner. Therefore, extra copies of an oncogene will result in cell proliferation and cancer. Tumor-suppressor genes, on the other hand, suppress cell proliferation and act in a recessive manner; a single copy of a tumor-suppressor gene is sufficient to prevent cell proliferation. Therefore extra copies of the suppressor gene will not lead to cancer.

3. The immunoglobulin molecules of a particular mammalian species has kappa and lambda light chains and heavy chains. The kappa gene consists of 250 V and 8 J segments. The lambda gene contains 200 V and 4 J segments. The gene for the heavy chain consists of 300 V, 8 J, and 4 D segments. Considering just somatic recombination and random combinations of light and heavy chains, how many different types of antibodies can be produced by this species?

• Solution

For the kappa light chain, there are $250 \times 8 = 2000$ combinations; for the lambda light chain, there are $200 \times 4 = 800$ combinations; so a total of 2800 different types of light chains are possible. For the heavy chains, there are $300 \times 8 \times 4 = 9600$ possible types. Any of the 2800 light chains can combine with any of the 9600 heavy chains; so there are $2800 \times 9600 = 26,880,000$ different types of antibodies possible from somatic recombination and random combination alone. Junctional diversity and somatic hypermutation would greatly increase this diversity.

COMPREHENSION QUESTIONS

* 1. What experiments suggested that genes are not lost or permanently altered in development?

2. Briefly explain how the Dorsal protein is redistributed in the formation of the *Drosophila* embryo and how this redistribution helps to establish the dorsal–ventral axis of the early embryo.

* 3. Briefly describe how the *bicoid* and *nanos* genes help to determine the anterior–posterior axis of the fruit fly.

* 4. List the three major classes of segmentation genes and outline the function of each.

5. What role do homeotic genes play in the development of fruit flies?

* 6. What is apoptosis and how is it regulated?

* 7. Explain how each of the following processes contributes to antibody diversity.

    (a) Somatic recombination.

    (b) Junctional diversity.

    (c) Hypermutation.

8. What is the function of the MHC antigens? Why are the genes that encode these antigens so variable?

* 9. Outline Knudson's multistage theory of cancer and describe how it helps to explain unilateral and bilateral cases of retinoblastoma.

10. Briefly explain how cancer arises through clonal evolution.

*11. What is the difference between an oncogene and a tumor-suppressor gene? Give some examples of the functions of proto-oncogenes and tumor suppressers in normal cells.

12. Why do mutations in genes that encode DNA repair enzymes and chromosome segregation often produce a predisposition to cancer?

*13. What role do telomeres and telomerase play in cancer progression?

APPLICATION QUESTIONS AND PROBLEMS

14. If telomeres are normally shortened after each round of replication in somatic cells, what prediction would you make about the length of telomeres in Dolly, the first cloned sheep?

*15. Give examples of genes that affect development in fruit flies by regulating gene expression at the level of
(a) transcription and (b) translation.

16. What would be the most likely effect on development of puncturing the posterior end of a *Drosophila* egg, allowing a small amount of cytoplasm to leak out, and then injecting that cytoplasm into the anterior end of another egg?

*17. What would be the most likely result of injecting *bicoid* mRNA into the posterior end of a *Drosophila* embryo and inhibiting the translation of *nanos* mRNA?

18. What would be the most likely effect of inhibiting the translation of *hunchback* mRNA throughout the embryo?

*19. Molecular geneticists have performed experiments in which they altered the number of copies of the *bicoid* gene in flies, affecting the amount of Bicoid protein produced.

    (a) What would be the effect on development of an increased number of copies of the *bicoid* gene?

    (b) What would be the effect of a decreased number of copies of *bicoid*?

    Justify your answers.

20. What would be the most likely effect on fruit-fly development of a deletion in the *nanos* gene?

21. Give an example of a gene found in each of the categories of genes (egg-polarity, gap, pair-rule, and so forth) listed in Figure 21.12.

*22. In a particular species, the gene for the kappa light chain has 200 *V* gene segments and 4 *J* segments. In the gene for the lambda light chain, this species has 300 *V* segments and 6 *J* segments. Considering only the variability arising from somatic recombination, how many different types of light chains are possible?

23. In the fictional book *Chromosome 6* by Robin Cook, a biotechnology company genetically engineers individual bonobos (a type of chimpanzee) to serve as future organ donors for clients. The genes of the bonobos are altered so that no tissue rejection takes place when their organs are transplanted into a client. What genes would need to be altered for this scenario to work? Explain your answer.

*24. A couple has one child with bilateral retinoblastoma. The mother is free from cancer, but the father had unilateral retinoblastoma and he has a brother who has bilateral retinoblastoma.

    (a) If the couple has another child, what is the probability that this next child will have retinoblastoma?

    (b) If the next child has retinoblastoma, is it likely to be bilateral or unilateral?

    (c) Propose an explanation for why the father's case of retinoblastoma was unilateral, whereas his sons and brother's cases are bilateral.

25. Some cancers are consistently associated with the deletion of a particular part of a chromosome. Does the deleted region contain an oncogene or a tumor-suppressor gene? Explain why.

26. Cells in a tumor contain mutated copies of a particular gene that promotes tumor growth. Gene therapy can be used to introduce a normal copy of this gene into the tumor cells. Would you expect this therapy to be effective if the mutated gene were an oncogene? A tumor-suppressor gene? Explain your reasoning.

CHALLENGE QUESTIONS

27. As we have learned in this chapter, the Nanos protein inhibits the translation of *hunchback* mRNA, thus lowering the concentration of Hunchback protein at the posterior end of a fruit-fly embryo and stimulating the differentiation of posterior characteristics. The results of experiments have demonstrated that the action of Nanos on *hunchback* mRNA depends on the presence of an 11-base sequence that is located in the 3′ untranslated region of *hunchback* mRNA. This sequence has been termed the Nanos response element (NRE). There are two copies of NREs in the trailer of *hunchback* mRNA. If a copy of NRE is added to the 3′ untranslated region of

another mRNA produced by a different gene, the mRNA now becomes repressed by Nanos. The repression is greater if several NREs are added. On the basis of these observations, propose a mechanism for how Nanos inhibits Hunchback translation.

28. Offer a possible explanation for the widespread distribution of *Hox* genes among animals.

29. Many cancer cells are immortal (will divide indefinitely) because they have mutations that allow telomerase to be expressed. How might this knowledge be used to design anticancer drugs?

SUGGESTED READINGS

Developmental Genetics

De Robertis, E. M., G. Oliver, and C. V. E. Wright. 1990. Homeobox genes and the vertebrate body plan. *Scientific American* 264: (1):46–52.
A readable account of how homeobox genes were discovered and how they affect development in vertebrates.

Halder, G., P. Callaerts, and W. J. Gehring. 1995. Induction of ectopic eyes by targeted expression of the *eyeless* gene in *Drosophila*. *Science* 267:1788–1792.
A report of the research that produced extra eyes in *Drosophila*.

Kolata, G. 1998. *Clone: The Road to Dolly and the Path Ahead.* New York: William Morrow.
A readable and accurate account of the cloning of Dolly, the first mammal cloned from an adult cell, and the ethical debate generated by this experiment.

Jan, Y. N., and L. Y. Jan. 1998. Asymmetrical cell division. *Nature* 392:775–778.
A review of the mechanisms by which asymmetrical cell division, which plays a critical role in development, arises.

Meyer, A. 1998. *Hox* gene variation and evolution. *Nature* 391:225–227.
A short review about the evolution of *Hox* gene clusters in vertebrates.

McKinnell, R. G., and M. A. Di Berardino. 1999. The biology of cloning: history and rationale. *Bioscience* 49:875–885.
A good summary of the history of cloning and some of its practical uses.

Pennisi, E., and G. Vogel. 2000. Clones: a hard act to follow. *Science* 288:1722–1727.
A news report on the different organisms that have been successfully cloned.

Raff, M. 1998. Cell suicide for beginners. *Nature* 396:119–122.
An introduction to the process of apoptosis.

Science. 1994. Volume 266(October 28): 513–700.

Deals with the topic of development and contains a number of reviews of research in developmental genetics.

Science. 1998. Volume 281(August 28): 1301–1326.
Contains a number of articles on apoptosis.

Thompson, G. B. 1995. Apoptosis in the pathogenesis and treatment of disease. *Science* 267:1456–1462.
A discussion of the role of apoptosis in disease.

Immunogenetics

Ada, G. L. and G. Nossal. 1987. The clonal-selection theory. *Scientific American* 257(2):62–69.
History and review of the development of the theory of clonal selection.

Gellert, M. 1992. Molecular analysis of *V(D)J* recombination. *Annual Review of Genetics* 22:425–446.
An extensive review of the molecular mechanism of somatic recombination in genes of the immune system.

Gellert, M. 2002. *V(D)J* recombination: RAG proteins, repair factors, and regulation. *Annual Review of Biochemistry* 71:101–132.
A review of the mechanism of recombination that leads to antibody diversity.

Leder, P. 1982. The genetics of antibody diversity. *Scientific American* 247(5):102–115.
A review of the processes that lead to diversity in antibodies.

Weaver, D. T., and F. W. Alt. 1997. From RAGs to stitches. *Nature* 388:428–429.
Reviews findings concerning the mechanism of *V-D-J* joining in the generation of antibody diversity.

Cancer Genetics

Bitttner, M., P. Meltzer, Y. Chen, Y. Jiang, E. Seftor, M. Hendrix, et al. 2000. Molecular classification of cutaneous malignant melanoma by gene expression profiling. *Nature* 406:536–540.

Evidence of genes that affect the spread of cancer.

Fearon, E. R., and B. Vogelstein. 1990. A genetic model for colorectal tumorigenesis. *Cell* 61:759–767.

A review of some of the mutations that led to colorectal cancer.

Hanahan, D., and R. A. Weinberg. 2000. The hallmarks of cancer. *Cell* 100:57–70.

A review of the different types of genes that are associated with cancer.

Knudson, A. G. 2000. Chasing the cancer demon. *Annual Review of Genetics* 34:1–19.

A short history of the search for a genetic cause of cancer, along with a review of hereditary cancers and the genes that cause them.

Lengauer, C., K. W. Kinzler, and B. Vogelstein. 1998. Genetic instabilities in human cancer. *Nature* 396:643–649.

A review of how defects in DNA repair and chromosome segregation genes lead to cancer.

Orr-Weaver, T. L., and R. A. Weinberg. 1998. A checkpoint on the road to cancer. *Nature* 392:223–224.

Discusses how mutations that affect cell-cycle checkpoints may contribute to cancer progression.

Ponder, B. A. 2001. Cancer genetics. *Nature* 411:336–341.

A good review on the types of genetic events that contribute to cancer.

Science. 1997. Volume 278(November 7): 1035–1077.

Contains a number of articles on cancer, including discussions of the genetic basis of human cancer syndromes, genetic testing for cancer risk, genetic approaches to developing drugs for cancer treatment, and environmental influences on cancer.

Weinberg, R. A. 1991. Tumor suppressor genes. *Science* 254:1138–1146.

A review of tumor-suppressor genes.

Weizman, J. B., and M. Yaniv. 1999. Rebuilding the road to cancer. *Nature* 400:401.

Discusses the first successful attempt to convert normal human cells into cancer cells by artificially introducing telomerase-expressing genes, oncogenes, and tumor-suppressor genes into a cell.

22

Quantitative Genetics

Quantitative genetic methods are being used to improve and predict racing speed in thoroughbred horses. (Timothy D. Easley/AP.)

Thoroughbred Winners Through Quantitative Genetics

For more than 300 years, thoroughbred horses have been raised for a single purpose—to win at the racetrack. The origin of these horses can be traced to a small group that was imported to England from North Africa and the Middle East in the 1600s. The population of racing horses remained small until the 1800s, when horse racing became increasing popular; today there are approximately half a million thoroughbred horses worldwide.

Breeding and racing thoroughbred horses is a multibillion-dollar industry that relies on the premise that a horse's speed is inherited. Speed is not, however, a simple genetic characteristic such as seed shape in peas. Numerous genes and nongenetic factors such as diet, training, and the jockey who rides the horse all contribute to a horse's success or failure in a race. The inheritance of racing speed in thoroughbreds is more complex than that of any of the characteristics that we have studied up to this point. Can the inheritance of a complex characteristic such as racing speed be studied? Is it possible to predict the speed of a horse on the basis of its pedigree? The answers are yes—at least in part—but these questions cannot be addressed with the methods that we used for simple genetic characteristics. Instead, we must use statistical procedures that have been developed for analyzing complex characteristics. The genetic analysis of complex characteristics such as racing speed of thoroughbreds is known as **quantitative genetics.**

Although the mathematical methods for analyzing complex characteristics may seem to be imposing at first, most people can intuitively grasp the underlying logic of quantitative genetics. We all recognize family resemblance: we talk about inheriting our father's height or our mother's intelligence. Family resemblance lies at the heart of the statistical methods used in quantitative genetics. When genes influence variation in a characteristic, related individuals resemble one another more than unrelated individuals. Closely related individuals (such as siblings) should resemble one another more than distantly related individuals (such as cousins). Comparing individuals with different degrees of relatedness, then, provides information about the extent to which genes influence a characteristic.

This type of analysis has been applied to the inheritance of racing speed in thoroughbreds. In 1988, Patrick Cunningham and his colleagues examined records of more than 30,000 3-year-old horses that raced between 1961 and 1985. They reasoned that, if genes influence racing success, a horse's racing success should be more similar to that of its parents than to that of unrelated horses. Similarly, the racing speeds of half-brothers and half-sisters should be more similar than the speeds of unrelated horses are. When Cunningham and his colleagues statistically analyzed the racing records for thoroughbreds, they found that a considerable amount of variation in track performance was due to genetic differences—racing speed is heritable. With the use of statistics, it is possible to estimate, with some degree of accuracy, the track performance of a horse from the performance of its relatives.

This chapter is about the genetic analysis of complex characteristics such as racing speed. We begin by considering the differences between quantitative and qualitative characteristics and why the expression of some characteristics varies continuously. We'll see how quantitative characteristics are often influenced by many genes, each of which has a small effect on the phenotype. Next, we will examine statistical procedures for describing and analyzing quantitative characteristics. We will consider the question of how much of phenotypic variation can be attributed to genetic and environmental influences and will conclude by looking at the effects of selection on quantitative characteristics. It's important to recognize that the methods of quantitative genetics are not designed to identify individual genes and genotypes. Rather, the focus is on statistical predictions based on groups of individuals.

www.whfreeman.com/pierce More information on horse genetics, including the genetic basis of coat color, gene mapping, chromosomes, and genetic disorders

Quantitative Characteristics

Qualitative, or discontinuous, characteristics possess only a few distinct phenotypes (FIGURE 22.1a); these characteristics are the types studied by Mendel and have been the focus of our attention thus far. However, many characteristics vary continuously along a scale of measurement with many overlapping phenotypes (FIGURE 22.1b). They are referred to as *continuous characteristics;* they are also called *quantitative characteristics* because any individual's phenotype must be described with a quantitative measurement. Quantitative characteristics might include height, weight, and blood pressure in humans, growth rate in mice, seed weight in plants, and milk production in cattle.

Quantitative characteristics arise from two phenomena. First, many are polygenic—they are influenced by

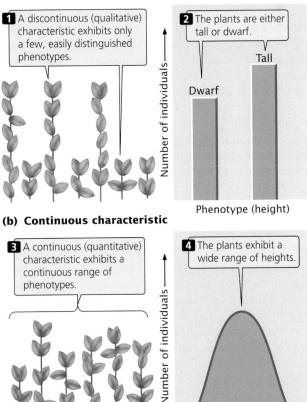

 22.1 **Discontinuous and continuous characteristics differ in the number of phenotypes exhibited.**

genes at many loci. If many loci take part, many genotypes are possible, each producing a slightly different phenotype. Second, quantitative characteristics often arise when environmental factors affect the phenotype, because environmental differences result in a single genotype producing a range of phenotypes. Most continuously varying characteristics are *both* polygenic *and* influenced by environmental factors, and these characteristics are said to be multifactorial.

The Relation Between Genotype and Phenotype

For many discontinuous characteristics, there is a relatively straightforward relation between genotype and phenotype. Each genotype produces a single phenotype, and most phenotypes are encoded by a single genotype. Dominance and epistasis may allow two or three genotypes to produce the same phenotype, but the relation remains relatively simple. This simple relation between genotype and phenotype allowed Mendel to decipher the basic rules of inheritance from his crosses with pea plants; it also permits us both to predict the outcome of genetic crosses and to assign genotypes to individuals.

For quantitative characteristics, the relation between genotype and phenotype is often more complex. If the characteristic is polygenic, many different genotypes are possible, several of which may produce the same phenotype. For instance, consider a plant whose height is determined by three loci (A, B, and C), each of which has two alleles. Assume that one allele at each locus (A^+, B^+, and C^+) encodes a plant hormone that causes the plant to grow 1 cm above its baseline height of 10 cm. The second allele at each locus (A^-, B^-, and C^-) encodes no plant hormone and does not contribute to additional height. Considering only the two alleles at a single locus, 3 genotypes are possible (A^+A^+, A^+A^-, and A^-A^-). If all three loci are taken into account, there are a total of $3^3 = 27$ possible multilocus genotypes ($A^+A^+B^+B^+C^+C^+$, $A^+A^-B^+B^+C^+C^+$, etc.). Although there are 27 genotypes, they produce only seven phenotypes (10 cm, 11 cm, 12 cm, 13 cm, 14 cm, 15 cm, and 16 cm in height). Some of the genotypes produce the same phenotype (Table 22.1); for example, genotypes $A^+A^-B^-B^-C^-C^-$, $A^-A^-B^+B^-C^-C^-$, and $A^-A^-B^-B^-C^+C^-$ all have one gene that encodes plant hormone. These genotypes produce one dose of the hormone and a plant that is 11 cm tall. Even in this simple example of only three loci, the relation between genotype and phenotype is quite complex. The more loci encoding a characteristic, the greater the complexity.

The influence of environment on a characteristic also can complicate the relation between genotype and phenotype. Because of environmental effects, the same genotype may produce a range of potential phenotypes (the norm of reaction; see p. 121 in Chapter 5). The phenotypic ranges of different genotypes may overlap, making it difficult to know whether individuals differ in phenotype because of genetic or environmental differences (◀ FIGURE 22.2).

| Table 22.1 | Hypothetical example of plant height determined by pairs of alleles at each of three loci |

Genotype	Doses of Plant Hormone	Height (cm)
$A^-A^-B^-B^-C^-C^-$	0	10
$A^+A^-B^-B^-C^-C^-$ $A^-A^-B^+B^-C^-C^-$ $A^-A^-B^-B^-C^-C^+$	1	11
$A^+A^+B^-B^-C^-C^-$ $A^-A^-B^+B^+C^-C^-$ $A^-A^-B^-B^-C^+C^+$ $A^+A^-B^+B^-C^-C^-$ $A^+A^-B^-B^-C^+C^-$ $A^-A^-B^+B^-C^+C^-$	2	12
$A^+A^+B^+B^-C^-C^-$ $A^+A^+B^-B^-C^+C^-$ $A^+A^-B^+B^+C^-C^-$ $A^-A^-B^+B^+C^+C^-$ $A^+A^-B^-B^-C^+C^+$ $A^-A^-B^+B^-C^+C^+$ $A^+A^-B^+B^-C^+C^-$	3	13
$A^+A^+B^+B^+C^-C^-$ $A^+A^+B^+B^-C^+C^-$ $A^+A^-B^+B^+C^+C^-$ $A^-A^-B^+B^+C^+C^+$ $A^+A^+B^-B^-C^+C^+$ $A^+A^-B^+B^-C^+C^+$	4	14
$A^+A^+B^+B^+C^+C^-$ $A^+A^-B^+B^+C^+C^+$ $A^+A^+B^+B^-C^+C^+$	5	15
$A^+A^+B^+B^+C^+C^+$	6	16

Note: Each + allele contributes 1 cm in height above a baseline of 10 cm.

In summary, the simple relation between genotype and phenotype that exists for many qualitative (discontinuous) characteristics is absent in quantitative characteristics, and it is impossible to assign a genotype to an individual on the basis of its phenotype alone. The methods used for analyzing qualitative characteristics (examining the phenotypic ratios of progeny from a genetic cross) will not work with quantitative characteristics. Our goal remains the same: we wish to make predictions about the phenotypes of offspring produced in a genetic cross. We may also want to know how much of the variation in a characteristic results from genetic differences and how much results from environmental differences. To answer these questions, we must turn to statistical methods that allow us to make predictions about the inheritance of phenotypes in the absence of information about the underlying genotypes.

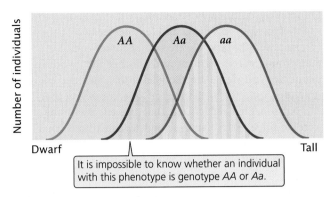

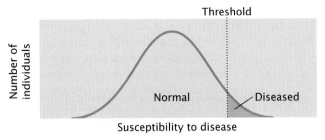

◀22.2 For a quantitative characteristic, each genotype may produce a range of possible phenotypes. In this hypothetical example, the phenotypes produced by genotypes *AA*, *Aa*, and *aa* overlap.

www.whfreeman.com/pierce   Information about some current research in quantitative genetics

Types of Quantitative Characteristics

Before we look more closely at polygenic characteristics and relevant statistical methods, we need to more clearly define what is meant by a quantitative characteristic. Thus far, we have considered only quantitative characteristics that vary continuously in a population. A *continuous characteristic* can theoretically assume any value between two extremes; the number of phenotypes is limited only by our ability to precisely measure the phenotype. Human height is a continuous characteristic because, within certain limits, people can theoretically have any height. Although the number of phenotypes possible with a continuous characteristic is infinite, we often group similar phenotypes together for convenience; we may say that two people are both 5 feet 11 inches tall, but careful measurement may show that one is slightly taller than the other.

Some characteristics are not continuous but are nevertheless considered quantitative because they are determined by multiple genetic and environmental factors. **Meristic characteristics,** for instance, are measured in whole numbers. An example is litter size: a female mouse may have 4, 5, or 6 pups but not 4.13 pups. A meristic characteristic has a limited number of distinct phenotypes, but the underlying determination of the characteristic may still be quantitative. These characteristics must therefore be analyzed with the same techniques that we use to study continuous quantitative characteristics.

Another type of quantitative characteristic is a **threshold characteristic,** which is simply present or absent. Although threshold characteristics exhibit only two phenotypes, they are considered quantitative because they, too, are determined by multiple genetic and environmental factors. The expression of the characteristic depends on an underlying susceptibility (usually referred to as liability or risk) that

◀22.3 Threshold characteristics display only two possible phenotypes—the trait is either present or absent—but they are quantitative because the underlying susceptibility to the characteristic varies continuously. When the susceptibility exceeds a threshold value, the characteristic is expressed.

varies continuously. When the susceptibility is larger than a threshold value, a specific trait is expressed (◀FIGURE 22.3). Diseases are often threshold characteristics because many factors, both genetic and environmental, contribute to disease susceptibility. If enough of the susceptibility factors are present, the disease develops; otherwise, it is absent. Although we focus on the genetics of continuous characteristics in this chapter, the same principles apply to many meristic and threshold characteristics.

It is important to point out that just because a characteristic can be measured on a continuous scale does not mean that it exhibits quantitative variation. One of the characteristics studied by Mendel was height of the pea plant, which can be described by measuring the length of the plant's stem. However, Mendel's particular plants exhibited only two distinct phenotypes (some were tall and others short), and these differences were determined by alleles at a single locus. The differences that Mendel studied were therefore discontinuous in nature.

Concepts

Characteristics whose phenotypes vary continuously are called quantitative characteristics. For most quantitative characteristics, the relation between genotype and phenotype is complex. Some characteristics whose phenotypes do not vary continuously also are considered quantitative because they are influenced by multiple genes and environmental factors.

Polygenic Inheritance

The rediscovery of Mendel's work in 1900 provided a cohesive theory of inheritance, but the characteristics that Mendel studied were all discontinuous. Questions soon arose about the inheritance of continuously varying characteristics. These characteristics had already been the focus of a group of biologists and statisticians, led by Francis Galton, who were known as biometricians. They examined the inheritance of quantitative characteristics such as human height and intelligence by using statistical procedures. The results of these studies showed that quantitative characteristics are inherited, although the mechanism of inheritance was as yet unknown. After Mendel's work was rediscovered, a bitter dispute broke out about whether Mendel's principles applied to quantitative characteristics. Some biometricians argued that the inheritance of quantitative characteristics could not be explained by Mendelian principles, whereas others felt that Mendel's principles acting on numerous genes (polygenes) could adequately account for the inheritance of quantitative characteristics.

This conflict began to be resolved by the work of Wilhelm Johannsen, who showed that continuous variation in the weight of beans was influenced by both genetic and environmental factors. George Udny Yule, a mathematician, proposed in 1906 that several genes acting together could produce continuous characteristics. This hypothesis was later confirmed by Herman Nilsson-Ehle, working on wheat and tobacco, and by Edward East, working on corn. The argument was finally laid to rest in 1918, when Ronald Fisher demonstrated that the inheritance of quantitative characteristics could indeed be explained by the cumulative effects of many genes, each following Mendel's rules.

Kernel Color in Wheat

To illustrate how multiple genes acting on a characteristic can produce a continuous range of phenotypes, let us examine one of the first demonstrations of polygenic inheritance. Nilsson-Ehle studied kernel color in wheat and found that the intensity of red pigmentation was determined by three unlinked loci, each of which had two alleles.

Nilsson-Ehle obtained several homozygous varieties of wheat that differed in color. Like Mendel, he performed crosses between these homozygous varieties and studied the ratios of phenotypes in the progeny. In one experiment, he crossed a variety of wheat that possessed white kernels with a variety that possessed purple (very dark red) kernels and obtained the following results:

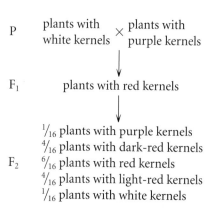

Nilsson-Ehle interpreted this phenotypic ratio as the result of segregation of alleles at two loci. (Although he

found alleles at three loci that affected kernel color, the two varieties used in this cross differed only at two of the loci.) He proposed that there were two alleles at each locus: one that produced red pigment and another that produced no pigment. We'll designate the alleles that encoded pigment A^+ and B^+ and the alleles that encoded no pigment A^- and B^-. Nilsson-Ehle recognized that the effects of the genes were additive. Each gene seemed to contribute equally to color; so the overall phenotype could be determined by adding the effects of all the genes, as shown in this table.

Genotype	Doses of pigment	Phenotype
$A^+A^+B^+B^+$	4	purple
$A^+A^+B^+B^-$ $A^+A^-B^+B^+$	3	dark red
$A^+A^+B^-B^-$ $A^-A^-B^+B^+$ $A^+A^-B^+B^-$	2	red
$A^+A^-B^-B^-$ $A^-A^-B^+B^-$	1	light red
$A^-A^-B^-B^-$	0	white

Notice that the purple and white phenotypes are each encoded by a single genotype, but other phenotypes may result from several different genotypes.

From these results, we see that five phenotypes are possible when alleles at two loci influence the phenotype and the effects of the genes are additive. When alleles at more than two loci influence the phenotype, more phenotypes are possible, and this would make the color appear to vary continuously between white and purple. If environmental factors had influenced the characteristic, individuals of the same genotype would vary somewhat in color, making it even more difficult to distinguish between discrete phenotypic classes. Luckily, environment played little role in determining kernel color in Nilsson-Ehle's crosses, and only a few loci encoded color; so Nilsson-Ehle was able to distinguish among the different phenotypic classes. This ability allowed him to see the Mendelian nature of the characteristic.

Let's now see how Mendel's principles explain the ratio obtained by Nilsson-Ehle in his F_2 progeny. Remember that Nilsson-Ehle crossed a homozygous purple variety $(A^+A^+B^+B^+)$ with the homozygous white variety $(A^-A^-B^-B^-)$, producing F_1 progeny that were heterozygous at both loci $(A^+A^-B^+B^-)$. All the F_1 plants possessed two pigment-producing alleles that allowed two doses of color to make red kernels. The types and proportions of progeny expected in the F_2 can be found by applying Mendel's principles of segregation and independent assortment.

Let's first examine the effects of each locus separately. At the first locus, two heterozygous F_1s are crossed $(A^+A^- \times A^+A^-)$. As we learned in Chapter 3, when two heterozygotes are crossed, we expect progeny in the proportions

$\frac{1}{4} A^+A^+$, $\frac{1}{2} A^+A^-$, and $\frac{1}{4} A^-A^-$. At the second locus, two heterozygotes also are crossed, and again we expect progeny in the proportions $\frac{1}{4} B^+B^+$, $\frac{1}{2} B^+B^-$, and $\frac{1}{4} B^-B^-$.

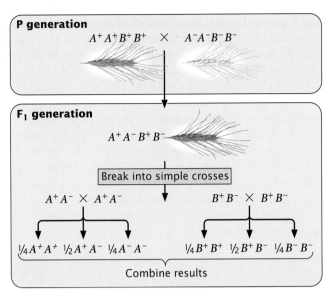

P generation

$A^+A^+B^+B^+$ × $A^-A^-B^-B^-$

F₁ generation

$A^+A^-B^+B^-$

Break into simple crosses

$A^+A^- \times A^+A^-$ $B^+B^- \times B^+B^-$

$\frac{1}{4}A^+A^+$ $\frac{1}{2}A^+A^-$ $\frac{1}{4}A^-A^-$ $\frac{1}{4}B^+B^+$ $\frac{1}{2}B^+B^-$ $\frac{1}{4}B^-B^-$

Combine results

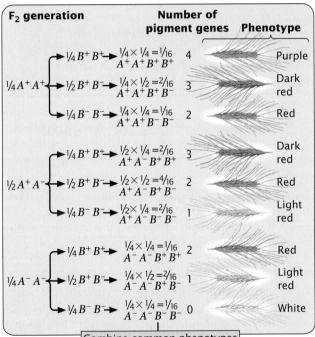

F₂ generation   **Number of pigment genes**   **Phenotype**

$\frac{1}{4}A^+A^+$

$\frac{1}{4}B^+B^+ \rightarrow \frac{1}{4} \times \frac{1}{4} = \frac{1}{16}$ $A^+A^+B^+B^+$ 4   Purple

$\frac{1}{2}B^+B^- \rightarrow \frac{1}{4} \times \frac{1}{2} = \frac{2}{16}$ $A^+A^+B^+B^-$ 3   Dark red

$\frac{1}{4}B^-B^- \rightarrow \frac{1}{4} \times \frac{1}{4} = \frac{1}{16}$ $A^+A^+B^-B^-$ 2   Red

$\frac{1}{2}A^+A^-$

$\frac{1}{4}B^+B^+ \rightarrow \frac{1}{2} \times \frac{1}{4} = \frac{2}{16}$ $A^+A^-B^+B^+$ 3   Dark red

$\frac{1}{2}B^+B^- \rightarrow \frac{1}{2} \times \frac{1}{2} = \frac{4}{16}$ $A^+A^-B^+B^-$ 2   Red

$\frac{1}{4}B^-B^- \rightarrow \frac{1}{2} \times \frac{1}{4} = \frac{2}{16}$ $A^+A^-B^-B^-$ 1   Light red

$\frac{1}{4}A^-A^-$

$\frac{1}{4}B^+B^+ \rightarrow \frac{1}{4} \times \frac{1}{4} = \frac{1}{16}$ $A^-A^-B^+B^+$ 2   Red

$\frac{1}{2}B^+B^- \rightarrow \frac{1}{4} \times \frac{1}{2} = \frac{2}{16}$ $A^-A^-B^+B^-$ 1   Light red

$\frac{1}{4}B^-B^- \rightarrow \frac{1}{4} \times \frac{1}{4} = \frac{1}{16}$ $A^-A^-B^-B^-$ 0   White

Combine common phenotypes

F₂ ratio

Frequency	Number of pigment genes	Phenotype
$\frac{1}{16}$	4	Purple
$\frac{4}{16}$	3	Dark red
$\frac{6}{16}$	2	Red
$\frac{4}{16}$	1	Light red
$\frac{1}{16}$	0	White

Conclusion: Polygenic characteristics are inherited according to Mendel's principles.

To obtain the probability of combinations of genes at both loci, we must use the multiplication rule of probability (pp. 54–55 in Chapter 3), which is based on Mendel's principle of independent assortment. The expected proportion of F₂ progeny with genotype $A^+A^+B^+B^+$ is the product of the probability of obtaining genotype A^+A^+ ($\frac{1}{4}$) and the probability of obtaining genotype B^+B^+ ($\frac{1}{4}$), or $\frac{1}{4} \times \frac{1}{4} = \frac{1}{16}$ (◄ FIGURE 22.4). The probabilities of each of the phenotypes can then be obtained by adding the probabilities of all the genotypes that produce that phenotype. For example, the red phenotype is produced by three genotypes:

Genotype	Probability
$A^+A^+B^-B^-$	$\frac{1}{16}$
$A^-A^-B^+B^+$	$\frac{1}{16}$
$A^+A^-B^+B^-$	$\frac{1}{4}$

Thus, the overall probability of obtaining red kernels in the F₂ progeny is $\frac{1}{16} + \frac{1}{16} + \frac{1}{4} = \frac{6}{16}$. Figure 22.4 shows that the phenotypic ratio expected in the F₂ is $\frac{1}{16}$ purple, $\frac{4}{16}$ dark red, $\frac{6}{16}$ red, $\frac{4}{16}$ light red, and $\frac{1}{16}$ white. This phenotypic ratio is precisely what Nilsson-Ehle observed in his F₂ progeny, demonstrating that the inheritance of a continuously varying characteristic such as kernel color is indeed according to Mendel's basic principles.

Nilsson-Ehle's crosses demonstrated that the difference between the inheritance of genes influencing quantitative characteristics and the inheritance of genes influencing discontinuous characteristics is in the *number* of loci that determine the characteristic. When multiple loci affect a character, more genotypes are possible; so the relation between the genotype and the phenotype is less obvious. As the number of loci affecting a character increases, the number of phenotypic classes in the F₂ increases (◄ FIGURE 22.5).

Several conditions of Nilsson-Ehle's crosses greatly simplified the polygenic inheritance of kernel color and made it possible for him to recognize the Mendelian nature of the characteristic. First, genes affecting color segregated at only two or three loci. If genes at many loci had been segregating, he would have had difficulty in distinguishing the phenotypic classes. Second, the genes affecting kernel color had strictly additive effects, making the relation between genotype and phenotype simple. Third, environment played almost no role in the phenotype; had environmental factors

◄ **22.4 Nilsson-Ehle demonstrated that kernel color in wheat is inherited according to Mendelian principles.** He crossed two varieties of wheat that differed in pairs of alleles at two loci affecting kernel color. A purple strain ($A^+A^+B^+B^+$) was crossed with a white strain ($A^-A^-B^-B^-$), and the F₁ was intercrossed to produce F₂ progeny. The ratio of phenotypes in the F₂ can be determined by breaking the dihybrid cross into two simple single-locus crosses and combining the results with the multiplication rule.

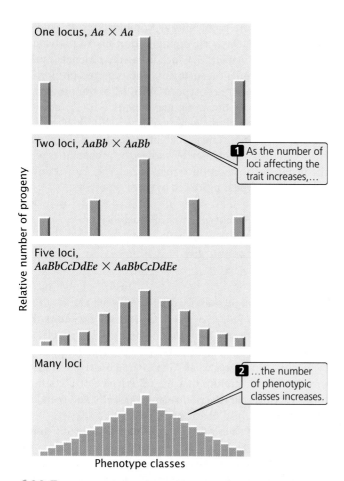

22.5 The results of crossing individuals heterozygous for different numbers of loci affecting a characteristic.

modified the phenotypes, distinguishing between the five phenotypic classes would have been difficult. Finally, the loci that Nilsson-Ehle studied were not linked; so the genes assorted independently. Nilsson-Ehle was fortunate—for many polygenic characteristics, these simplifying conditions are not present and Mendelian inheritance of these characteristics is not obvious.

Determining Gene Number for a Polygenic Characteristic

When two individuals homozygous for different alleles at a single locus are crossed ($A^1A^1 \times A^2A^2$) and the resulting F_1 are interbred ($A^1A^2 \times A^1A^2$), one-fourth of the F_2 should be homozygous like each of the original parents. If the original parents are homozygous for different alleles at *two* loci, as are those in Nilsson-Ehle's crosses, then $1/4 \times 1/4 = 1/16$ of the F_2 should resemble one of the original homozygous parents. Generally, $(1/4)^n$ will be the number of individuals in the F_2 progeny that should resemble each of the original homozygous parents, where n equals the number of loci with a segregating pair of alleles that affects the characteristic. This equation provides us with a possible means of determining the number of loci influencing a quantitative characteristic.

To illustrate the use of this equation, assume that we cross two different homozygous varieties of pea plants that differ in height by 16 cm, interbreed the F_1, and find that approximately $1/256$ of the F_2 are similar to one of the original homozygous parental varieties. This outcome would suggest that 4 loci with segregating pairs of alleles ($1/256 = 1/4^4$) are responsible for the height difference between the two varieties. Because the two homozygous strains differ in height by 16 cm and there are 4 loci each with two alleles (8 alleles in all), each of the alleles contributes 16 cm/8 = 2 cm in height.

This method for determining the number of loci affecting phenotypic differences requires the use of homozygous strains, which may be difficult to obtain in some organisms. It also assumes that all the genes influencing the characteristic have equal effects, that their effects are additive, and that the loci are unlinked. For many polygenic characteristics, these assumptions are not valid, so this method of determining the number of genes affecting a characteristic has limited application.

> **Concepts**
>
> The principles that determine the inheritance of quantitative characteristics are the same as the principles that determine the inheritance of discontinuous characteristics, but more genes take part in the determination of quantitative characteristics.

Statistical Methods for Analyzing Quantitative Characteristics

Because quantitative characteristics are described by a measurement and are influenced by multiple factors, their inheritance must be analyzed statistically. This section will explain the basic concepts of statistics that are used to analyze quantitative characteristics.

Distributions

Understanding the genetic basis of any characteristic begins with a description of the numbers and kinds of phenotypes present in a group of individuals. Phenotypic variation in a group, such as the progeny of a cross, can be conveniently represented by a **frequency distribution,** which is a graph of the frequencies (numbers or proportions) of the different phenotypes (◀ FIGURE 22.6). In a typical frequency distribution, the phenotypic classes are plotted on the horizontal (*x*) axis and the numbers (or proportions) of individuals in each class on the vertical (*y*) axis. Unlike qualitative (discontinuous) characteristics, quantitative (continuous) characteristics often exhibit many phenotypes, so a frequency distribution is a concise method of summarizing them all.

Connecting the points of a frequency distribution with a line creates a curve that is characteristic of the distribution. Many quantitative characteristics exhibit a symmetrical

(a) Qualitative (discontinuous) characteristic

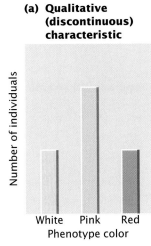

Number of individuals

White   Pink   Red
Phenotype color

(b) Quantitative (continuous) characteristic

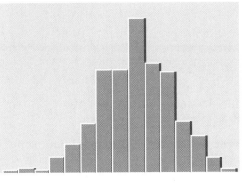

Phenotype (body weight)

◀22.6 A frequency distribution is a graph that displays the number or proportion of different phenotypes. Phenotypic values are plotted on the horizontal axis and the numbers (or proportions) of individuals in each class are plotted on the vertical axis.

(bell-shaped) curve called a **normal distribution** (see Figure 22.7a). Normal distributions arise when a large number of independent factors contribute to a measurement. Quantitative characteristics are frequently affected by numerous genes and environmental factors; so their phenotypes often exhibit normal distributions. Two other common types of distributions (skewed and bimodal) are illustrated in (see Figure 22.7b and c).

Samples and Populations

Biologists frequently need to describe the distribution of phenotypes exhibited by some group of individuals. We might want to describe the height of students at the University of Texas (UT), but there are more than 40,000 students at UT, and measuring every one of them would be impractical. Scientists are constantly confronted with this problem: the group of interest, called the **population,** is too large for a complete census. One solution is to measure a smaller collection of individuals, called a **sample,** and use measurements made on the sample to describe the population.

To provide an accurate description of the population, a good sample must have several characteristics. First, it must

be representative of the whole population. If our sample consisted entirely of members of the UT basketball team, for instance, we would probably overestimate the true height of the students. One way to ensure that a sample is representative of the population is to select the members of the sample randomly. Second, the sample must be large enough that chance differences between individuals in the sample and the overall population do not distort the estimate of the population measurements. If we measured only three students at UT and just by chance all three were short, we would underestimate the true height of the student population. Statistics can provide information about how much confidence to expect from estimates based on random samples.

> **Concepts**
>
> In statistics, the population is the group of interest; a sample is a subset of the population. The sample should be representative of the population and large enough to minimize chance differences between the population and the sample.

(a) Sugar beet percentage of sucrose

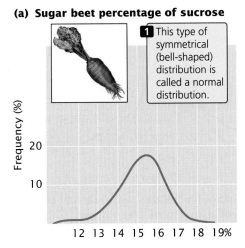

1 This type of symmetrical (bell-shaped) distribution is called a normal distribution.

Frequency (%)

20

10

12 13 14 15 16 17 18 19%

(b) Squash fruit length

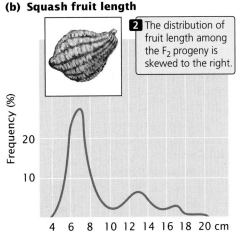

2 The distribution of fruit length among the F₂ progeny is skewed to the right.

Frequency (%)

20

10

4   6   8   10 12 14 16 18 20 cm

(c) Earwig forceps length

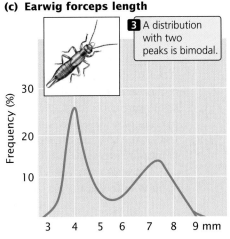

3 A distribution with two peaks is bimodal.

Frequency (%)

30

20

10

3   4   5   6   7   8   9 mm

◀22.7 Distributions of phenotypes may assume several different shapes.

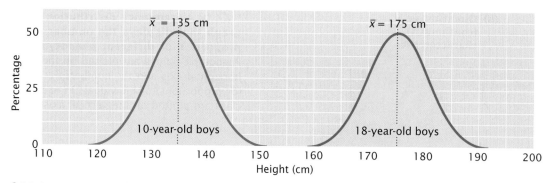

◀ 22.8 **The mean provides information about the center of a distribution.** Both distributions of heights of 10-year-old and 18 year-old boys are normal, but they have different locations along a continuum of height, which makes their means different.

The Mean

The **mean,** also called the average, provides information about the center of the distribution. If we measured the heights of 10-year-old and 18-year-old boys and plotted a frequency distribution for each group, we would find that both distributions are normal, but the two distributions would be centered at different heights, and this difference would be indicated in their different means (◀ Figure 22.8).

If we represent a group of measurements as x_1, x_2, x_3, and so forth, then the mean ($\bar{x}$) is calculated by adding all the individual measurements and dividing by the total number of measurements in the sample (n):

$$\bar{x} = \frac{x_1 + x_2 + x_3 + \cdots + x_n}{n} \quad (22.1)$$

A shorthand way to represent this formula is

$$\bar{x} = \frac{\Sigma x_i}{n} \quad (22.2)$$

or

$$\bar{x} = \frac{1}{n} \Sigma x_i \quad (22.3)$$

where the symbol Σ means "the summation of" and x_i represents individual x values.

The Variance and Standard Deviation

A statistic that provides key information about a distribution is the **variance,** which indicates the variability of a group of measurements (how spread out the distribution is). Distributions may have the same mean but different variances (◀ Figure 22.9). The larger the variance, the greater the spread of measurements in a distribution about its mean.

The variance (s^2) is defined as the average squared deviation from the mean:

$$s^2 = \frac{\Sigma(x_i - \bar{x})^2}{n - 1} \quad (22.4)$$

To calculate the variance, we (1) subtract the mean from each measurement and square the value obtained, (2) add all the squared deviations, and (3) divide this sum by the number of original measurements minus one.

Another statistic that is closely related to the variance is the **standard deviation** (s), which is defined as the square root of the variance:

$$s^2 = \sqrt{s^2} \quad (22.5)$$

Whereas the variance is expressed in units squared, the standard deviation is in the same units as the original measurements; so the standard deviation is often preferred for describing the variability of a measurement.

A normal distribution is symmetrical; so the mean and standard deviation are sufficient to describe its shape. The mean plus or minus one standard deviation ($\bar{x} \pm s$) includes approximately 66% of the measurements in a normal distribution; the mean plus or minus two standard

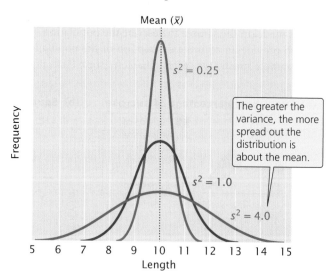

◀ 22.9 **The variance provides information about the variability of a group of phenotypes.** Shown here are three distributions with the same mean but different variances.

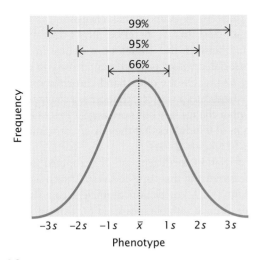

22.10 The proportions of a normal distribution occupied by plus or minus one, two, and three standard deviations from the mean.

deviations $(\bar{x} \pm 2s)$ includes approximately 95% of the measurements, and the mean plus or minus three standard deviations $(\bar{x} \pm 3s)$ includes approximately 99% of the measurements (FIGURE 22.10). Thus, only 1% of a normally distributed population lies outside the range of $\bar{x} \pm 3s$.

Concepts

The mean and variance describe a distribution of measurements: the mean provides information about the location of the center of a distribution, and the variance provides information about its variability.

Worked Problem

The following table lists yearly amounts (in hundreds of pounds) of milk produced by 10 two-year-old Jersey cows. Calculate the mean, variance, and standard deviation of milk production for this sample of 10 cows.

**Annual milk production
(hundreds of pounds)**

60

74

58

61

56

55

54

57

65

42

• Solution

The mean is calculated by using the following formula: $\bar{x} = \Sigma x_i /n$. The value of Σx_i is obtained by summing all the individual measurements, which equals 582; n is the total number of measurements, which equals 10; so $\bar{x} = (582/10) = 58.20$, or 58,200 pounds per year.

The variance is calculated by using the following formula:

$$s_x^2 = \frac{\Sigma(x_i - \bar{x})^2}{n - 1}$$

so we need to determine the deviation of each individual measurement from the mean $(x_i - \bar{x})$, square each value, and sum the squared deviations from the mean.

**Annual milk production
(hundreds of pounds)**

x	$x_i - \bar{x}$	$(x_i - \bar{x})^2$
60	1.80	3.24
74	15.80	249.64
58	−0.20	0.04
61	2.80	7.84
56	−2.20	4.84
55	−3.20	10.24
54	−4.20	17.64
57	−1.20	1.44
65	6.80	46.24
42	−16.20	262.44

$$\Sigma(x_i - \bar{x})^2 = 603.60$$

The variance is therefore:

$$s_x^2 = \frac{\Sigma(x_i - \bar{x})^2}{n - 1} = \frac{603.60}{9} = 67.07$$

The standard deviation is the square root of the variance:

$$s_x = \sqrt{s_x^2} = \sqrt{67.07} = 8.19.$$

Correlation

The mean and the variance can be used to describe an individual characteristic, but geneticists are frequently interested in more than one characteristic. Often, two or more characteristics vary together. For instance, both the number and the weight of eggs produced by hens are important to the poultry industry. These two characteristics are not independent of each other. There is an inverse relation between egg number and weight: hens that lay more eggs produce smaller eggs. This kind of relation between two characteristics is called a **correlation**. When two characteristics are correlated, a change in one characteristic is likely to be associated with a change in the other.

Correlations between characteristics are measured by a **correlation coefficient** (designated r), which measures the strength of their association. Consider two characteristics, such as human height (x) and arm length (y). To determine how these characteristics are correlated, we first obtain the covariance (cov) of x and y:

$$\text{cov}_{xy} = \frac{\Sigma(x_i - \bar{x})(y_i - \bar{y})}{n - 1} \qquad (22.6)$$

The covariance is computed by (1) taking an x value for an individual and subtracting it from the mean of x ($\bar{x}$); (2) taking the y value for the same individual and subtracting it from the mean of y ($\bar{y}$); (3) multiplying the results of these two subtractions; (4) adding the results for all the xy pairs; and (5) dividing this sum by $n - 1$ (where n equals the number of xy pairs).

The correlation coefficient (r) is obtained by dividing the covariance of x and y by the product of the standard deviations of x and y:

$$r = \frac{\text{cov}_{xy}}{s_x s_y} \qquad (22.7)$$

A correlation coefficient can theoretically range from -1 to $+1$. A positive value indicates that there is a direct association between the variables (◀FIGURE 22.11a); as one variable increases, the other variable also tends to increase. A positive correlation exists for human height and weight: tall people tend to weigh more. A negative correlation coefficient indicates that there is an inverse relation between the two variables (◀FIGURE 22.11b); as one variable increases, the other tends to decrease (as is the case for egg number and egg weight in chickens).

The absolute value of the correlation coefficient (the size of the coefficient, ignoring its sign) provides information about the strength of association between the variables. A coefficient of -1 or $+1$ indicates a perfect correlation between the variables, meaning that a change in x is always

accompanied by a proportional change in y. Correlation coefficients close to -1 or close to $+1$ indicate a strong association between the variables—a change in x is almost always associated with a proportional increase in y, as seen in ◀FIGURE 22.11c. On the other hand, a correlation coefficient closer to 0 indicates a weak correlation—a change in x is associated with a change in y but not always (◀FIGURE 22.11d). A correlation of 0 indicates that there is no association between variables (◀FIGURE 22.11e).

A correlation coefficient can be computed for two variables measured for the same individual, such as height (x) and weight (y). A correlation coefficient can also be computed for a single variable measured for pairs of individuals. For example, we can calculate for fish the correlation between the number of vertebrae of a parent (x) and the number of vertebrae of its offspring (y), as shown in ◀FIGURE 22.12. This approach is often used in quantitative genetics.

A correlation between two variables indicates only that the variables are associated; it does not imply a cause and effect relation. Correlation also does not mean that the values of two variables are the same; it means only that a change in one variable is associated with a proportional change in the other variable. For example, the x and y variables in the following list are almost perfectly correlated, with a correlation coefficient of .99.

	x value	y value
	12	123
	14	140
	10	110
	6	61
	3	32
Average:	9	90

A high correlation is found between these x and y variables; larger values of x are always associated with larger values of y. Note that the y values are about 10 times as large as the corresponding x values; so, although x and y are correlated, they are not identical. The distinction between correlation

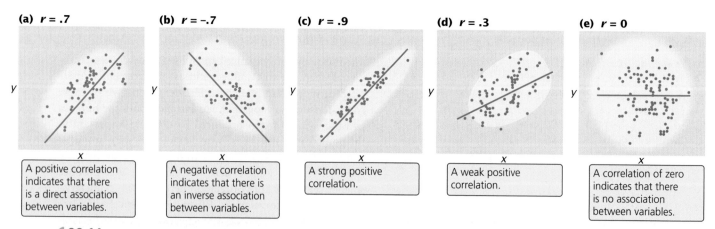

(a) $r = .7$ — A positive correlation indicates that there is a direct association between variables.

(b) $r = -.7$ — A negative correlation indicates that there is an inverse association between variables.

(c) $r = .9$ — A strong positive correlation.

(d) $r = .3$ — A weak positive correlation.

(e) $r = 0$ — A correlation of zero indicates that there is no association between variables.

◀22.11 **The correlation coefficient describes the relation between two or more variables.**

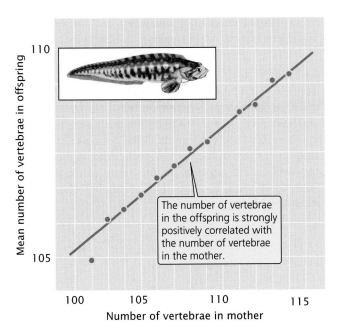

◀ 22.12 A correlation coefficient can be computed for a single variable measured for pairs of individuals. Here, the number of vertebrae in mothers and offspring of the fish *Zoarces viviparus* is compared.

and identity becomes important when we consider the effects of heredity and environment on the correlation of characteristics.

Regression

Correlation provides information only about the strength and direction of association between variables, but often we want to know more than just whether two variables are associated; we want to be able to predict the value of one variable, given a value of the other.

A positive correlation exists between body weight of parents and body weight of their offspring; this correlation exists in part because genes influence body weight, and parents and children share the same genes. Because of this association between phenotypes of parent and offspring, we can predict the weight of an individual on the basis of the weights of its parents. This type of statistical prediction is called **regression**. This technique plays an important role in quantitative genetics because it allows us to predict characteristics of offspring from a given mating, even without knowledge of the genotypes that encode the characteristic.

Regression can be understood by plotting a series of *x* and *y* values. ◀ FIGURE 22.13 illustrates the relation between the weight of fathers (*x*) and the weight of their son (*y*). Each father–son pair is represented by a point on the graph. The overall relation between these two variables is depicted by the regression line, which is the line that best fits all the points on the graph (deviations of the points from the line are minimized). The regression line defines the relation between the *x* and *y* variables and can be represented by

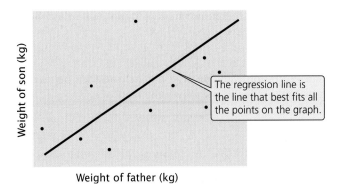

◀ 22.13 A regression line defines the relation between two variables. Illustrated here is a regression of the weights of fathers against the weights of sons. Each father–son pair is represented by a point on the graph: the *x* value of a point is the father's weight and the *y* value of the point is the son's weight.

$$y = a + bx \qquad (22.8)$$

In Equation 22.8, *x* and *y* represent the *x* and *y* variables (in this case, the father's weight and the son's weight, respectively). The variable *a* is the *y* intercept of the line, which is the expected value of *y* when *x* is 0. Variable *b* is the slope of the regression line, also called the **regression coefficient**; it indicates how much *y* increases, on average, per increase in *x*.

Trying to position a regression line by eye is not only very difficult but also inaccurate when there are many points scattered over a wide area. Fortunately, the regression coefficient and *y* intercept can be obtained mathematically. The regression coefficient (*b*) can be computed from the covariance of *x* and *y* (cov_{xy}) and the variance of *x* (s_x^2) by

$$b = \frac{cov_{xy}}{s_x^2} \qquad (22.9)$$

Several regression lines with different regression coefficients are illustrated in ◀ FIGURE 22.14.

After the regression coefficient has been calculated, the *y* intercept can be calculated by substituting the regression

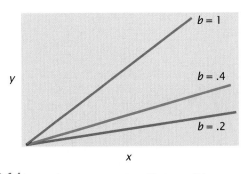

◀ 22.14 The regression coefficient (*b*) represents the change in *y* per unit change in *x*. Shown here are regression lines with different regression coefficients.

coefficient and the mean values of x and y into the following equation:

$$a = \bar{y} - b\bar{x} \qquad (22.10)$$

The regression equation ($y = a + bx$) can then be used to predict the value of any y given the value of x.

Concepts

A correlation coefficient measures the strength of association between two variables. The sign (positive or negative) indicates the direction of the correlation; the absolute value measures the strength of the association. Regression is used to predict the value of one variable on the basis of the value of a correlated variable.

Worked Problem

Body weights of 11 female fish and the numbers of eggs that they produce are:

Weight (mg)	Eggs (thousands)
x	y
14	61
17	37
24	65
25	69
27	54
33	93
34	87
37	89
40	100
41	90
42	97

What are the correlation coefficient and the regression coefficient for body weight and egg number in these 11 fish?

• Solution

The computations needed to answer this question are given in the table below. To calculate the correlation and regression coefficients, we first obtain the sum of all the x_i values (Σx_i) and the sum of all the y_i values (Σy_i); these sums are shown in the last row of the table at the bottom of this page. We can calculate the means of the two variables by dividing the sums by the number of measurements, which is 11:

$$\bar{x} = \frac{\Sigma x_i}{n} = \frac{334}{11} = 30.36$$

$$\bar{y} = \frac{\Sigma y_i}{n} = \frac{842}{11} = 76.55$$

After the means have been calculated, the deviations of each value from the means are computed; these deviations are shown in columns B and E of the table. The deviations are then squared (columns C and F) and summed (last row of columns C and F). Next, the products of the deviation of the x values and the deviation of the y values [$(x_i - \bar{x})(y_i - \bar{y})$] are calculated; these products are shown in column G, and their sum is shown in the last row of column G.

To calculate the covariance, we use Formula 22.6:

$$\text{cov}_{xy} = \frac{\Sigma(x_i - \bar{x})(y_i - \bar{y})}{n - 1} = \frac{1743.84}{10} = 174.38$$

To calculate the covariance and regression requires the variances and standard deviations of x and y:

$$s_x^2 = \frac{(x_i - \bar{x})^2}{n - 1} = \frac{932.55}{10} = 93.26$$

A	B	C	D	E	F	G
Weight (mg)			Eggs (thousands)			
x	$x_i - \bar{x}$	$(x_i - \bar{x})^2$	y	$y_i - \bar{y}$	$(y_i - \bar{y})^2$	$(x_i - \bar{x})(y_i - \bar{y})$
14	−16.36	267.65	61	−15.55	241.80	254.40
17	−13.36	178.49	37	−39.55	1564.20	528.39
24	−6.36	40.45	65	−11.55	133.40	73.46
25	−5.36	28.73	69	−7.55	57.00	40.47
27	−3.36	11.29	54	−22.55	508.50	75.77
33	2.64	6.97	93	16.45	270.60	43.43
34	3.64	13.25	87	10.45	109.20	38.04
37	6.64	44.09	89	12.45	155.00	82.67
40	9.64	92.93	100	23.45	549.90	226.06
41	10.64	113.21	90	13.45	180.90	143.11
42	11.64	135.49	97	20.45	418.20	238.04
$\Sigma x_i =$ 334		$\Sigma(x - \bar{x})^2 =$ 932.55	$\Sigma y_i =$ 842		$\Sigma(y - \bar{y})^2 =$ 4188.70	$\Sigma(x_i - \bar{x})(y_i - \bar{y}) =$ 1743.84

Source: R. R. Sokal and F. J. Rohlf, *Biometry*, 2d ed. (San Francisco: W. H. Freeman and Company, 1981.)

$$s_x = \sqrt{s_x^2} = \sqrt{93.26} = 9.66$$

$$s_y^2 = \frac{(y_i - \bar{y})^2}{n-1} = \frac{4188.70}{10} = 418.87$$

$$s_y = \sqrt{s_y^2} = \sqrt{418.87} = 20.47$$

We can now compute the correlation and regression coefficients.

Correlation coefficient:

$$r = \frac{cov_{xy}}{s_x s_y} = \frac{174.38}{9.66 \times 20.47} = 0.88$$

Regression coefficient:

$$b = \frac{cov_{xy}}{s_x^2} = \frac{174.38}{93.26} = 1.87$$

www.whfreeman.com/pierce   Numerous links to Web sites on statistics

Applying Statistics to the Study of a Polygenic Characteristic

Edward East carried out one early statistical study of polygenic inheritance on the length of flowers in tobacco (*Nicotiana longiflora*). He obtained two varieties of tobacco that differed in flower length: one variety had a mean flower length of 40.5 mm, and the other had a mean flower length of 93.3 mm (◀ FIGURE 22.15). These two varieties had been inbred for many generations and were homozygous at all loci contributing to flower length. Thus, there was no genetic variation in the original parental strains; the small differences in flower length within each strain were due to environmental effects on flower length.

When East crossed the two strains, he found that flower length in the F_1 was about halfway between that in the two parents (see Figure 22.15), as would be expected if the genes determining the differences in the two strains were additive in their effects. The variance of flower length in the F_1 was similar to that seen in the parents, because the F_1 were, like their parents, uniform in genotype (the F_1 were all heterozygous at the genes that differed between the two parental varieties).

East then interbred the F_1 to produce F_2 progeny. The mean flower length of the F_2 was similar to that of the F_1, but the variance of the F_2 was much greater (see Figure 22.15). This greater variability indicates that there were genetic differences within the F_2 progeny.

East selected some F_2 plants and interbred them to produce F_3 progeny. He found that flower length of the F_3 depended on flower length in the plants selected as their

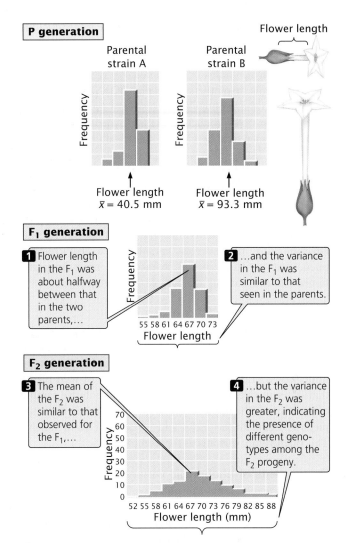

◀ 22.15 **Edward East conducted an early statistical study of the inheritance of flower length in tobacco.**

parents. This finding demonstrated that flower-length differences in the F_2 were partly genetic and thus were passed to the next generation. None of the 444 F_2 plants that East raised exhibited flower lengths similar to those of the two parental strains. This result suggested that more than four loci with pairs of alleles affected flower length in his varieties, because four allelic pairs are expected to produce 1 of 256 progeny ($\frac{1}{4}^4 = \frac{1}{256}$) having one or the other of the original parental phenotypes.

Heritability

In addition to being polygenic, quantitative characteristics are frequently influenced by environmental factors. It is often useful to know how much of the variation in a quantitative characteristic is due to genetic differences and how much is due to environmental differences. That proportion of the total phenotypic variation that is due to genetic differences is known as the **heritability.**

Consider a dairy farmer who owns several hundred milk cows. The farmer notices that some cows consistently produce more milk than others. The *nature* of these differences is important to the profitability of his dairy operation. If the differences in milk production are largely genetic in origin, then the farmer may be able to boost milk production by selectively breeding the cows that produce the most milk. On the other hand, if the differences are largely environmental in origin, selective breeding will have little effect on milk production, and the farmer might better boost milk production by adjusting the environmental factors associated with higher milk production. To determine the extent of genetic and environmental influences on variation in a characteristic, phenotypic variation in the characteristic must be partitioned into components attributable to different factors.

Phenotypic Variance

To determine how much of phenotypic differences in a population is due to genetic and environmental factors, we must first have some quantitative measure of the phenotype under consideration. Consider a population of wild plants that differ in size. We could collect a representative sample of plants from the population, weigh each plant in the sample, and calculate the mean and variance of plant weight. This **phenotypic variance** is represented by V_P.

Components of phenotypic variance Phenotypic variance, which represents the phenotypic differences among individual members of a group, can be attributed to several factors. First, some of the differences in phenotype may be due to differences in genotypes among individual members of the population. These differences are termed the **genetic variance** and are represented by V_G.

Second, some of the differences in phenotype may be due to environmental differences among the plants; these differences are termed the **environmental variance,** V_E. Environmental variance includes differences that can be attributed to specific environmental factors, such as the amount of light or water that the plant receives; it also includes random differences in development that cannot be attributed to any specific factor. Any variation in phenotype that is not inherited is, by definition, a part of the environmental variance.

Third, **genetic–environmental interaction variance** (V_{GE}) arises when the effect of a gene depends on the specific environment in which it is found. An example is shown in ◀FIGURE 22.16. In a dry environment, genotype *AA* produces a plant that averages 12 g in weight, and genotype *aa* produces a smaller plant that averages 10 g. In a wet environment, genotype *aa* produces the larger plant, averaging 24 g in weight, whereas genotype *AA* produces a plant that averages 20 g. In this example, there are clearly differences in the two environments: both genotypes produce heavier plants in the wet environment. There are also differences in the weights of the two genotypes, but the relative perfor-

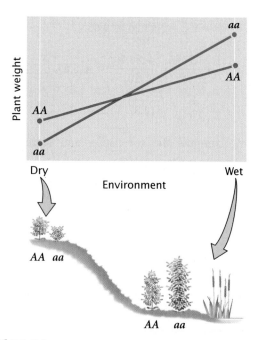

◀22.16 Genetic–environmental interaction variance occurs when the effect of a gene depends on the specific environment in which it is found. In this example, the genotype affects plant weight, but the environmental conditions determine which genotype produces the heavier plant.

mances of the genotypes depend on whether the plants are grown in a wet or dry environment. In this case, the influences on phenotype cannot be neatly allocated into genetic and environmental components, because the expression of the genotype depends on the environment in which the plant grows. The phenotypic variance must therefore include a component that accounts for the way in which genetic and environmental factors interact.

In summary, the total phenotypic variance can be apportioned into three components:

$$V_P = V_G + V_E + V_{GE} \qquad (22.11)$$

Components of genetic variance Genetic variance can be further subdivided into components consisting of different types of genetic effects. First, **additive genetic variance** (V_A) comprises the additive effects of genes on the phenotype, which can be summed to determine the overall effect on the phenotype. For example, suppose that, in a plant, allele A^1 contributes 2 g in weight and allele A^2 contributes 4 g. If the alleles are strictly additive, then the genotypes would have the following weights:

$$A^1A^1 = 2 + 2 = 4\,\text{g}$$

$$A^1A^2 = 2 + 4 = 6\,\text{g}$$

$$A^2A^2 = 4 + 4 = 8\,\text{g}$$

The genes that Nilsson-Ehle studied, which affected kernel color in wheat, were additive in this way.

Second, there is **dominance genetic variance** (V_D) when some genes have a dominance component. In this case, the alleles at a locus are not additive; rather, the effect of an allele depends on the identity of the other allele at that locus. Here, we cannot simply add the effects of the alleles together. Instead, we must add a component (V_D) to the genetic variance to account for the way that alleles interact.

Third, genes at different loci may interact in the same way that alleles at the same locus interact. When this genic interaction occurs, the effects of genes are not additive, and we must include a third component, called **genic interaction variance** (V_I), to the genetic variance:

$$V_G = V_A + V_D + V_I \qquad (22.12)$$

Summary equation We can now integrate these components into one equation to represent all the potential contributions to the phenotypic variance:

$$V_P = V_A + V_D + V_I + V_E + V_{GE} \qquad (22.13)$$

This equation provides us with a model that describes the potential causes of differences that we observe among individual phenotypes. It's important to note that this model deals strictly with the observable *differences* (variance) in phenotypes among individual members of a population; it says nothing about the absolute value of the characteristic or about the underlying genotypes that produce these differences.

Types of Heritability

The model of phenotypic variance that we've just developed can be used to address the question of how much of the phenotypic variance in a characteristic is due to genetic differences. **Broad-sense heritability** (H^2) represents the proportion of phenotypic variance that is due to genetic variance and is calculated by dividing the genetic variance by the phenotypic variance:

$$\text{broad-sense heritability} = H^2 = \frac{V_G}{V_P} \qquad (22.14)$$

It is symbolized H^2 because it is a measure of variance, which is in units squared.

Broad-sense heritability can potentially range from 0 to 1. A value of 0 indicates that none of the phenotypic variance results from differences in genotype and all of the differences in phenotype result from environmental variation. A value of 1 indicates that all of the phenotypic variance results from differences in genotypes. A heritability value between 0 and 1 indicates that both genetic and environmental factors influence the phenotypic variance.

Often, we are more interested in the proportion of the phenotypic variance that results from the additive genetic

variance, because the additive genetic variance primarily determines the resemblance between parents and offspring. **Narrow-sense heritability** (h^2) is equal to the additive genetic variance divided by the phenotypic variance:

$$\text{narrow-sense heritability} = h^2 = \frac{V_A}{V_P} \qquad (22.15)$$

The Calculation of Heritability

Having considered the components that contribute to phenotypic variance and having developed a general concept of heritability, we can ask, How does one go about estimating these different components and calculating heritability? There are several ways to measure the heritability of a characteristic. They include eliminating one or more variance components, comparing the resemblance of parents and offspring, comparing the phenotypic variances of individuals with different degrees of relatedness, and measuring the response to selection. The mathematical theory that underlies these calculations of heritability is complex and beyond the scope of this book. Nevertheless, we can develop a general understanding of how heritability is measured.

Heritability by elimination of variance components One way of calculating the broad-sense heritability is to eliminate one of the variance components. We have seen that $V_P = V_G + V_E + V_{GE}$. If we eliminate all environmental variance ($V_E = 0$), then $V_{GE} = 0$ (because, if either V_G or V_E is zero, no genetic–environmental interaction can take place), and $V_P = V_G$. In theory, we might make V_E equal to 0 by ensuring that all individuals were raised in exactly the same environment but, in practice, it is virtually impossible. Instead, we could make V_G equal to 0 by raising genetically identical individuals, causing V_P to be equal to V_E. In a typical experiment, we might raise cloned or highly inbred, identically homozygous individuals in a defined environment and measure their phenotypic variance to estimate V_E. We could then raise a group of genetically variable individuals and measure their phenotypic variance (V_P). Using V_E calculated on the genetically identical individuals, we could obtain the genetic variance of the variable individuals by subtraction:

$$V_{G[\text{of genetically varying individuals}]} =$$
$$V_{P[\text{of genetically varying individuals}]} - V_{E[\text{of genetically identical individuals}]}$$
$$(22.16)$$

The broad-sense heritability of the genetically variable individuals would then be calculated as follows:

$$H^2 = \frac{V_{G[\text{of genetically varying individuals}]}}{V_{P[\text{of genetically varying individuals}]}} \qquad (22.17)$$

Sewall Wright used this method to estimate the heritability of white spotting in guinea pigs. He first measured the phenotypic variance for white spotting in a genetically variable population and found that $V_P = 573$. Then he inbred the guinea pigs for many generations so that they were essentially homozygous and genetically identical. When he measured their phenotypic variance in white spotting, he obtained V_P equal to 340. Because $V_G = 0$ in this group, their $V_P = V_E$. Wright assumed this value of environmental variance for the original (genetically variable) population and estimated their genetic variance:

$$V_P - V_E = V_G$$

$$573 - 340 = 233$$

He then estimated the broad-sense heritability from the genetic and phenotypic variance:

$$H^2 = \frac{V_G}{V_P}$$

$$H^2 = \frac{233}{573} = .41$$

This value implies that 41% of the variation in spotting of guinea pigs in Wright's population was due to differences in genotype.

Estimating heritability by using this method assumes that the environmental variance of genetically identical individuals is the same as the environmental variance of the genetically variable individuals, which may not be true. Additionally, this approach can be applied only to organisms for which it is possible to create genetically identical individuals.

Heritability by parent–offspring regression Another method for estimating heritability is to compare the phenotypes of parents and offspring. When genetic differences are responsible for phenotypic variance, offspring should resemble their parents more than they resemble unrelated individuals, because offspring and parents have some genes in common that help determine their phenotype. Correlation and regression can be used to analyze the association of phenotypes in different individuals.

To calculate the narrow-sense heritability in this way, we first measure the characteristic on a series of parents and offspring. The data are arranged into families, and the mean parental phenotype is plotted against the mean offspring phenotype (◀FIGURE 22.17). Each data point in the graph represents one family; the value on the x (horizontal) axis is the mean phenotypic value of the parents in a family, and the value on the y (vertical) axis is the mean phenotypic value of the offspring for the family.

Let's assume that there is no narrow-sense heritability for the characteristic ($h^2 = 0$); genetic differences do not contribute to the phenotypic differences among individuals. In this case, offspring will be no more similar to their parents than they are to unrelated individuals, and the data points will be scattered randomly, generating a regression coefficient of zero (see Figure 22.17a). Next, let's assume that all of the phenotypic differences are due to additive genetic differences ($h^2 = 1.0$). In this case, the mean phenotype of the offspring will be equal to the mean phenotype of the parents, and the regression coefficient will be 1 (see Figure 22.17b). If genes and environment both contribute to the differences in phenotype, both heritability and the regression coefficient will lie between 0 and 1 (see Figure 22.17c). The regression coefficient therefore provides information about the *magnitude* of the heritability.

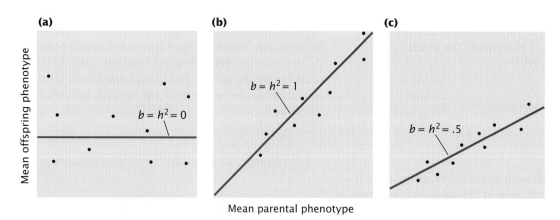

◀ **22.17 The narrow-sense heritability (h^2) equals the regression coefficient (b) in a regression of the mean phenotype of the offspring on the mean phenotype of the parents.** (a) There is no relation between the parental phenotype and the offspring phenotype. (b) The offspring phenotype is the same as the parental phenotypes. (c) Both genes and environment contribute to the differences in phenotype.

A complex mathematical proof (which we will not go into here) demonstrates that, in a regression of the mean phenotype of the offspring against the mean phenotype of the parents, narrow-sense heritability (h^2) equals the regression coefficient (b):

$$h^2 = b_{\text{(regression of mean offspring against mean of both parents)}} \quad (22.18)$$

An example of calculating heritability by regression of the phenotypes of parents and offspring is illustrated in ◀ FIGURE 22.18. In a regression of the mean offspring phenotype against the phenotype of only *one* parent, the narrow-sense heritability equals twice the regression coefficient:

$$h^2 = 2b_{\text{(regression of mean offspring against mean of one parent)}} \quad (22.19)$$

With only one parent, the heritability is twice the regression coefficient because only half the genes of the offspring come from one parent; thus, we must double the regression coefficient to obtain the full heritability.

Heritability and degrees of relatedness A third method for calculating heritability is to compare the phenotypes of individuals having different degrees of relatedness. This method is based on the concept that, the more closely related two individuals are, the more genes they have in common.

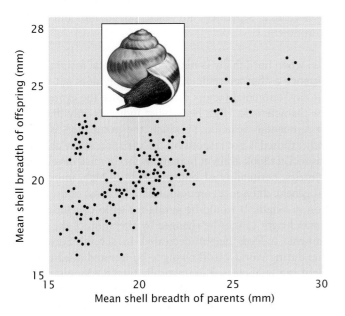

▶ **22.18 The heritability of shell breadth in snails can be determined by regression of the phenotype of offspring against the mean phenotype of the parents.** The regression coefficient, which equals the heritability, is .70. (From L. M. Cook, 1965. *Evolution* 19:86–94.)

Monozygotic (identical) twins have 100% of their genes in common, whereas dizygotic (nonidentical) twins have, on average, 50% of their genes in common. If genes are important in determining variability in a characteristic, then monozygotic twins should be more similar in a particular characteristic than dizygotic twins. By using correlation to compare the phenotypes of monozygotic and dizygotic twins, we can estimate broad-sense heritability. A rough estimate of the broad-sense heritability can be obtained by taking twice the difference of the correlation coefficients for a quantitative characteristic in monozygotic and dizygotic twins:

$$H^2 = 2(r_{MZ} - r_{DZ}) \quad (22.20)$$

where r_{MZ} equals the correlation coefficient among monozygotic twins and r_{DZ} equals the correlation coefficient among dizygotic twins. This calculation assumes that the two individuals of a monozygotic twin pair experience environments that are no more similar to each other than those experienced by the two individuals of a dizygotic twin pair, which is often not the case, unless the twins have been reared apart.

Narrow-sense heritability can also be estimated by comparing the phenotypic variances for a characteristic in full sibs (who have both parents in common, as well as 50% of their genes on the average) and half sibs (who have only one parent in common and thus 25% of their genes on the average).

All estimates of heritability depend on the assumption that the environments of related individuals are not more similar than those of unrelated individuals. This assumption is difficult to meet in human studies, because related people are usually reared together. Heritability estimates for humans should therefore always be viewed with caution.

Concepts

Broad-sense heritability is the proportion of phenotypic variance that is due to genetic variance. Narrow-sense heritability is the proportion of phenotypic variance that is due to additive genetic variance. Heritability can be measured by eliminating one of the variance components, analyzing parent–offspring regression, or comparing individuals having different degrees of relatedness.

The Limitations of Heritability

Knowledge of heritability has great practical value, because it allows us to statistically predict the phenotypes of offspring on the basis of their parent's phenotype. It also provides useful information about how characteristics will respond to selection (see next section). In spite of its importance, heritability is frequently misunderstood. Heritability does not provide information about an individual's genes or the environmental factors that control the development of a characteristic, and it says nothing about the nature of differences between groups. This section outlines some limitations and common misconceptions concerning broad- and narrow-sense heritability.

Heritability does *not* indicate the degree to which a characteristic is genetically determined Heritability is the proportion of the phenotypic variance that is due to genetic variance; it says nothing about the degree to which genes determine a characteristic. Heritability indicates only the degree to which genes determine *variation* in a characteristic. The determination of a characteristic and the determination of variation in a characteristic are two very different things.

Consider polydactyly (the presence of extra digits) in rabbits, which can be caused either by environmental factors or by a dominant gene. Suppose we have a group of rabbits all homozygous for a gene that produces normal numbers of digits. None of the rabbits in this group carries a gene for polydactyly, but a few of the rabbits are polydactylous because of environmental factors. Broad-sense heritability for polydactyly in this group is zero, because there is no genetic variation for polydactyly; all of the variation is due to environmental factors. However, it would be incorrect for us to conclude that genes play no role in determining the number of digits in rabbits. Indeed, we know that there are specific genes that can produce extra digits. Heritability indicates nothing about whether genes control the development of a characteristic; it only provides information about causes of the *variation* in a characteristic within a defined group.

An individual does *not* have heritability Broad- and narrow-sense heritabilities are statistical values based on the genetic and phenotypic variances found in a *group* of individuals. It is impossible to calculate heritability for an individual, and heritability has no meaning for a specific individual. Suppose we calculate the narrow-sense heritability of adult body weight for the students in a biology class and obtain a value of .6. We could conclude that 60% of the variation in adult body weight among the students in this class is determined by additive genetic variation. We should not, however, conclude that 60% of any particular student's body weight is due to additive genes.

There is *no* universal heritability for a characteristic The value of heritability for a characteristic is specific for a given population in a given environment. Recall that broad-sense heritability is genetic variance divided by phenotypic variance. Genetic variance depends on which genes are present, which often differs between populations. In the example of polydactyly in rabbits, there were no genes for polydactyly in the group; so the heritability of the characteristic was zero. A different group of rabbits might contain many genes for polydactyly, and the heritability of the characteristic might be high.

Environmental differences may affect heritability, because V_P is composed of both genetic and environmental variance. When the environmental differences that affect a characteristic differ between two groups, the heritabilities for the two groups also will often differ.

Because heritability is specific to a defined population in a given environment, it is important not to extrapolate heritabilities from one population to another. For example, human height is determined by environmental factors (such as nutrition and health) and by genes. If we measured the heritability of height in a developed country, we might obtain a value of .8, indicating that the variation in height in this population is largely genetic. This population has a high heritability because most people have adequate nutrition and health care (V_E is low); so most of the phenotypic variation in height is genetically determined. It would be incorrect for us to assume that height has a high heritability in all human populations. In developing countries, there may be more variation in a range of environmental factors; some people may enjoy good nutrition and health, whereas others may have a diet deficient in protein and suffer from diseases that affect stature. If we measured the heritability of height in such a country, we would undoubtedly obtain a lower value than we observed in the developed country, because there is more environmental variation and the genetic variance in height constitutes a smaller proportion of the phenotypic variation, making the heritability lower. The important point to remember is that heritability must be calculated separately for each population and each environment.

Even when heritability is high, environmental factors may influence a characteristic High heritability does not mean that environmental factors cannot influence the expression of a characteristic. High heritability indicates only that the environmental variation to which the population is *currently* exposed is not responsible for variation in the characteristic. Let's look again at human height. In most developed countries, heritability of human height is high, indicating that genetic differences are responsible for most of the variation in height. It would be wrong for us to conclude that human height cannot be changed by alteration of the environment. Indeed, height decreased in several European cities during World War II owing to hunger and disease, and height can be increased dramatically by the administration of growth hormone to children. The absence of environmental variation in a characteristic does not mean that the characteristic will not respond to environmental change.

Heritabilities indicate nothing about the nature of population differences in a characteristic A common misconception about heritability is that it provides information about population differences in a characteristic. Heritability is specific for a given population in a given environment, so it cannot be used to draw conclusions about why populations differ in a characteristic.

Suppose we measured heritability for human height in two groups. One group is from a small town in a developed country, where everyone consumes a high-protein diet. Because there is little variation in the environmental factors that affect human height and there is some genetic variation, the heritability of height in this group is high. The second group comprises the inhabitants of a single village in a developing country. The consumption of protein by these people is only 25% of that consumed by those in the first group; so their average adult height is several centimeters less than that in the developed country. Again, there is little variation in the environmental factors that determine height in this group, because everyone in the village eats the same types of food and is exposed to same diseases. Because there is little environmental variation and there is some genetic variation, the heritability of height in this group also is high.

Thus, the heritability of height in both groups is high, and the average height in the two groups is considerably different. We might be tempted to conclude that the difference in height between the two groups is genetically based—that the people in the developed country are genetically taller than the people in the developing country. This conclusion is obviously wrong, however, because the differences in height are due largely to diet—an environmental factor. Heritability provides no information about the causes of differences between populations.

These limitations of heritability have often been ignored, particularly in arguments about possible social implications of genetic differences between humans. Soon after Mendel's principles of heredity were rediscovered, some geneticists began to claim that many human behavioral characteristics are determined entirely by genes. This claim led to debates about whether characteristics such as human intelligence are determined by genes or environment. Many of the early claims of genetically based human behavior were based on poor research; unfortunately, the results of these studies were often accepted at face value and led to a number of eugenic laws that discriminated against certain groups of people. Today, geneticists recognize that many behavioral characteristics are influenced by a complex interaction of genes and environment and that it is very difficult to separate genetic effects from those of the environment.

The results of a number of modern studies indicate that human intelligence as measured by IQ and other intelligence tests has a moderately high heritability (usually from .4 to .8). On the basis of this observation, some people have argued that intelligence is innate and that enhanced educational opportunities cannot boost intelligence. This argument is based on the misconception that, when heritability is high, changing the environment will not alter the characteristic. In addition, because heritabilities of intelligence

range from .4 to .8, a considerable amount of the variance in intelligence originates from environmental differences.

Another argument based on a misconception about heritability is that ethnic differences in measures of intelligence are genetically based. Because the results of some genetic studies show that IQ has moderate heritability and because other studies find differences in the average IQ of ethnic groups, some people have suggested that ethnic differences in IQ are genetically based. As in the example of the effects of diet on nutrition, heritability provides no information about causes of differences among groups; it indicates only the degree to which phenotypic variance within a single group is genetically based. High heritability for a characteristic does not mean that phenotypic differences between ethnic groups are genetic. We should also remember that separating genetic and environmental effects in humans is very difficult; so heritability estimates themselves may be unreliable.

Concepts

Heritability provides information only about the degree to which *variation* in a characteristic is genetically determined. There is no universal heritability for a characteristic; heritability is specific for a given population in a specific environment. Environmental factors can potentially affect characteristics with high heritability, and heritability says nothing about the nature of population differences in a characteristic.

Locating Genes That Affect Quantitative Characteristics

The statistical methods described for use in analyzing quantitative characteristics can be used both to make predictions about the average phenotype expected in offspring and to estimate the overall contribution of genes to variation in the characteristic. These methods do not, however, allow us to identify and determine the influence of individual genes that affect quantitative characteristics. The genes that control polygenic characteristics are referred to as **quantitative trait loci** (QTLs). Although quantitative genetics has made important contributions to basic biology and to plant and animal breeding, the inability to identify QTLs and measure their individual effects has severely limited the application of quantitative genetic methods.

Mapping QTLs In recent years, numerous genetic markers have been identified and mapped with the use of recombinant DNA techniques, making it possible to identify QTLs by linkage analysis. The underlying idea is simple: if the inheritance of a genetic marker is associated consistently with the inheritance of a particular characteristic (such as increased height), then that marker must be linked to a QTL that affects height. The key is to have enough genetic markers so that QTLs can be detected throughout the genome. With the introduction of restriction fragment length polymorphisms and microsatellite variations (see pp. 538–541 in Chapter 18), variable markers are now available for mapping QTLs in a number of different organisms (◀ FIGURE 22.19).

A common procedure for mapping QTLs is to cross two homozygous strains that differ in alleles at many loci. The resulting F_1 progeny are then intercrossed or backcrossed to allow the genes to recombine through independent assortment and crossing over. Genes on different chromosomes and genes that are far apart on the same chromosome will recombine freely; genes that are closely linked will be inherited together. The offspring are measured for one or more quantitative characteristics; at the same time, they are genotyped for numerous genetic markers that span the genome. Any correlation between the inheritance of a particular marker allele and a quantitative phenotype indicates that a QTL is linked to that marker. If enough markers are used, it is theoretically possible to detect all the QTLs affecting a characteristic. This approach has been used to detect genes affecting various characteristics in several plant and animal species (Table 22.2).

Applications of QTL mapping The number of genes affecting a quantitative characteristic can be estimated by

Table 22.2	Quantitative characteristics for which QTLs have been detected	
Organism	**Quantitative Characteristic**	**Number of QTLs Detected**
Tomato	Soluble solids	7
	Fruit mass	13
	Fruit pH	9
	Growth	5
	Leaflet shape	9
	Height	9
Corn	Height	11
	Leaf length	7
	Tiller number	1
	Glume hardness	5
	Grain yield	18
	Number of ears	9
	Thermotolerance	6
Common bean	Number of nodules	4
Mung bean	Seed weight	4
Cow pea	Seed weight	2
Wheat	Preharvest sprout	4
Pig	Growth	2
	Length of small intestine	1
	Average back fat	1
	Abdominal fat	1
Mouse	Epilepsy	2
Rat	Hypertension	2

Source: After S. D. Tanksley, Mapping polygenes, *Annual Review of Genetics* 27 (1993):218.

◀ 22.19 **The availability of molecular markers makes the mapping of QTLs possible in many organisms.** QTL mapping is being used to identify genes that affect yield in corn and other agriculturally important plants. (Inga Spence/Visuals Unlimited.)

locating QTLs with genetic markers and adding up the number of QTLs detected. This method will always be an underestimate, because QTLs that are located close together on the same chromosome will be counted together, and those with small effects are likely to be missed.

QTL mapping also provides information about the magnitude of the effects that individual genes have on a quantitative characteristic. The polygenic model assumes that many genes affect a quantitative characteristic, that the effect of each gene is small, and that the effects of the genes are equal and additive. The results of studies of QTLs in a number of organisms now show that these assumptions are not always valid. Polygenes appear to vary widely in their effects. In many of the characteristics that have been studied, a few QTLs account for much of the phenotypic variation. In some instances, individual QTLs have been mapped that account for more than 20% of the variance in the characteristic.

www.whfreeman.com/pierce   More on QTLs

Response to Selection

Evolution is genetic change among members of a population. Several different forces are potentially capable of producing evolution, and we will explore these forces and the process of evolution more fully in the next chapter. Here, we consider how one of these forces—natural selection—may bring about genetic change in a quantitative characteristic.

Charles Darwin proposed the idea of natural selection in his book *On the Origin of Species* in 1859. **Natural selection** arises through the differential reproduction of individuals with different genotypes, allowing individuals with certain genotypes to produce more offspring than others. Natural selection is one of the most important of the forces that brings about evolutionary change and can be summarized as follows:

Observation 1—Many more individuals are produced each generation than are capable of surviving long enough to reproduce.

Observation 2—There is much phenotypic variation within natural populations.

Observation 3—Some phenotypic variation is heritable. In the terminology of quantitative genetics, some of the phenotypic variation in these characteristics is due to genetic variation, and these characteristics have heritability.

Logical consequence—Individuals with certain characters (called adaptive traits) survive and reproduce better that others. Because the adaptive traits are heritable, offspring will tend to resemble their parents with regard to these traits, and there will be more individuals with these adaptive traits in the next generation. Thus, adaptive traits will tend to increase in the population through time.

In this way, organisms become genetically suited to their environments; as environments change, organisms change in ways that make them better able to survive and reproduce.

For thousands of years, humans have practiced a form of selection by promoting the reproduction of organisms with traits perceived as desirable. This form of selection is **artificial selection,** and it has produced the domestic plants and animals that make modern agriculture possible. The power of artificial selection, the first application of genetic principles by humans, is illustrated by the tremendous diversity of shapes, colors, and behaviors of modern domesticated dogs (◀ FIGURE 22.20).

Predicting the Response to Selection

When a quantitative characteristic is subjected to natural or artificial selection, it will frequently change with the passage of time, provided there is genetic variation for that characteristic in the population. Suppose a dairy farmer breeds only those cows in his herd that have the highest milk production. If there is genetic variation in milk production, the mean milk production in the offspring of the selected cows should be higher than the mean milk production of the original herd. This increased production is due to the fact that the selected cows possess more genes for high milk production than does the average cow, and these genes are passed on to the offspring. The offspring of the selected cows possess a higher proportion of genes for greater milk yield and therefore produce more milk than the average cow in the initial herd.

The extent to which a characteristic subjected to selection changes in one generation is termed the **response to selection.** Suppose that the average cow in a dairy herd produces 80 liters of milk per week. A farmer selects for increased milk production by breeding the highest milk producers, and the progeny of these selected cows produce 100 liters of milk per week on average. The response to selection is calculated by subtracting the mean phenotype of the original population (80 liters) from the mean phenotype of the offspring (100 liters), obtaining a response to selection of $100 - 80 = 20$ liters per week.

The response to selection is determined primarily by two factors. First, it is affected by the narrow-sense heritability, which largely determines the degree of resemblance between parents and offspring. When the narrow-sense heritability is high, offspring will tend to resemble their parents; conversely, when the narrow-sense heritability is low, there will be little resemblance between parents and offspring.

The second factor that determines the response to selection is how much selection there is. If the farmer is very stringent in the choice of parents and breeds only the highest milk producers in the herd (say, the top 2 cows), then all the offspring will receive genes for high-quality milk production. If the farmer is less selective and breeds the top 20 milk producers in the herd, then the offspring will not carry as many superior genes for high milk production, and they will not, on average, produce as much milk as the offspring of the top 2 producers. The response to selection depends on the phenotypic difference of the individuals that are selected as parents; this phenotypic difference is measured by the **selection differential,** defined as the difference between the mean phenotype of the selected parents and the mean phenotype of the original population. If the aver-

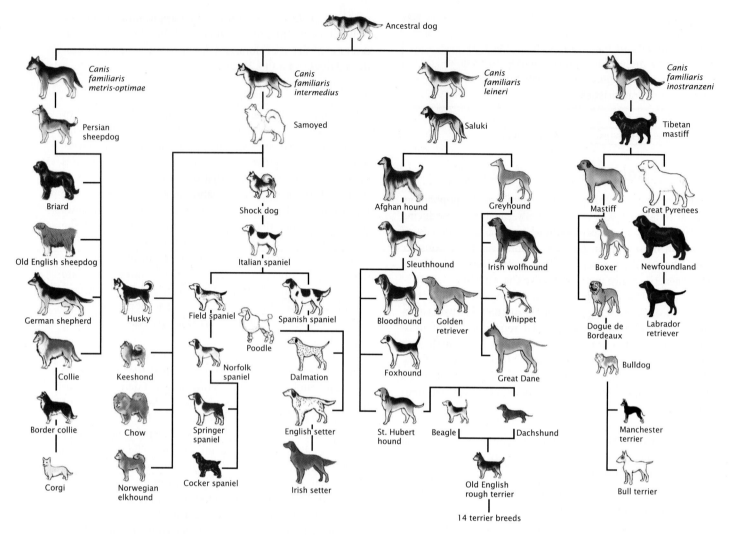

◀ **22.20 Artificial selection has produced the tremendous diversity of shape, size, color, and behavior seen today among breeds of domestic dogs.**

age milk production of the original herd is 80 liters and the farmer breeds cows with an average milk production of 120 liters, then the selection differential is 120 − 80 = 40 liters.

The response to selection (R) depends on the narrow-sense heritability (h^2) and the selection differential (S):

$$R = h^2 \times S \qquad (22.21)$$

This equation can be used to predict the magnitude of change in a characteristic when a given selection differential is applied. G. Clayton and his colleagues estimated the response to selection that would take place in abdominal bristle number of *Drosophila melanogaster*. Using several different methods, including parent–offspring regression, they first estimated the narrow-sense heritability of abdom-

inal bristle number in one population of fruit flies to be .52. The mean number of bristles in the original population was 35.3. They selected individual flies with a mean bristle number of 40.6 and intercrossed them to produce the next generation. The selection differential was 40.6 − 35.3 = 5.3; so they predicted a response to selection to be

$$R = .52 \times 5.3 = 2.8$$

The response to selection of 2.8 represents the expected increase in the characteristic of the offspring above that of the original population. They therefore expected the average number of abdominal bristles in the offspring of their selected flies to be 35.3 + 2.8 = 38.1. Indeed, they found an average bristle number of 37.9 in these flies.

Rearranging Equation 22.21 provides another way to calculate the narrow-sense heritability:

$$h^2 = \frac{R}{S} \qquad (22.22)$$

In this way, h^2 can be calculated by conducting a response-to-selection experiment. First, the selection differential is obtained by subtracting the population mean from the mean of selected parents. The selected parents are then interbred, and the mean phenotype of their offspring is measured. The difference between the mean of the offspring and that of the initial population is the response to selection, which can be used with the selection differential to estimate the heritability. Heritability determined by a response-to-selection experiment is usually termed the **realized heritability.** If certain assumptions are met, the realized heritability is identical with the narrow-sense heritability.

One of the longest selection experiments is a study of oil and protein content in corn seeds (◀FIGURE 22.21). This experiment began at the University of Illinois on 163 ears of corn with an oil content ranging from 4% to 6%. Corn plants having high oil content and those having low oil content were selected and interbreed. Response to selection for increased oil content (the upper line in Figure 22.21) reached about 20%, whereas response to selection for decreased oil content reached a lower limit near zero. Genetic analysis of the high- and low-oil-content strains revealed that at least 20 loci take part in determining oil content.

Concepts

The response to selection is influenced by narrow-sense heritability and the selection differential.

Limits to Selection Response

When a characteristic has been selected for many generations, the response eventually levels off, and the characteristic no longer responds to selection (◀FIGURE 22.22). A potential reason for this leveling off is that the genetic variation in the population may be exhausted; at some point, all individuals in the population have become homozygous for alleles that encode the selected trait. When there is no more additive genetic variation, heritability equals zero, and no further response to selection can occur.

The response to selection may level off even while some genetic variation remains in the population, however, because natural selection opposes further change in the characteristic. Response to selection for small body size in mice, for example, eventually levels off because the smallest animals are sterile and cannot pass on their genes for small body size. In this case, artificial selection for small size is opposed by natural selection for fertility, and the population can no longer respond to the artificial selection.

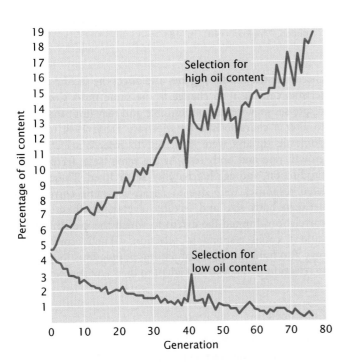

◀**22.21 In a long-term response-to-selection experiment, selection for oil content in corn increased oil content in one line to about 20%, while almost eliminating it altogether in another line.**

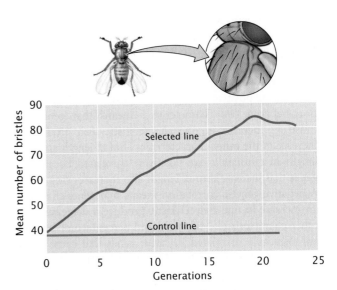

◀**22.22 The response of a population to selection often levels off at some point in time.** In a response-to-selection experiment for increased abdominal chaetae bristle number in female fruit flies, the number of bristles increased steadily for about 20 generations and then leveled off.

Correlated Responses

Two or more characteristics are often correlated. Human height and weight exhibit a positive correlation: tall people, on the average, weigh more than short people. This correlation is a **phenotypic correlation,** because the association is between two phenotypes of the same person. Phenotypic correlations may be due to environmental or genetic correlations. Environmental correlations refer to two or more characteristics that are influenced by the same environmental factor. Moisture availability, for example, may affect both the size of a plant and the number of seeds produced by the plant. Plants growing in environments with lots of water are large and produce many seeds, whereas plants growing in environments with limited water are small and have few seeds.

Alternatively, a phenotypic correlation may result from a **genetic correlation,** which means that the genes affecting two characteristics are associated. The primary genetic cause of phenotypic correlations is pleiotropy, which is due to the effect of one gene on two or more characteristics (see p. 123 in Chapter 5). In humans, for example, many body structures respond to growth hormone, and there are genes that affect the amount of growth hormone secreted by the pituitary gland. People with certain genes produce high levels of growth hormone, which increases both height and hand size. Others possess genes that produce lower levels of growth hormone, which leads to both short stature and small hands. Height and hand size are therefore phenotypically correlated in humans, and this correlation is due to a genetic correlation—the fact that both characteristics are affected by the same genes that control the amount of growth hormone. Genetically speaking, height and hand size are the same characteristic, because they are the phenotypic manifestation of a single set of genes. When two characteristics are influenced by the same genes they are genetically correlated.

Genetic correlations are quite common (Table 22.3) and may be positive or negative. A positive genetic correlation between two characteristics means that genes that cause an increase in one characteristic also produce an increase in the other characteristic. Thorax length and wing length in *Drosophila* are positively correlated because the genes that increase thorax length also increase wing length. A negative genetic correlation means that genes that cause an increase in one characteristic produce a decrease in the other characteristic. Milk yield and percentage of butterfat are negatively correlated in cattle: genes that cause higher milk production result in milk with a lower percentage of butterfat.

Genetic correlations are important in animal and plant breeding because they produce a correlated response to selection, which means that, when one characteristic is selected, genetically correlated characteristics also change. Correlated responses to selection occur because both characteristics are influenced by the same genes; selection for one characteristic causes a change in the genes affecting that

Table 22.3	Genetic correlations in various organisms	
Organism	**Characteristics**	**Genetic Correlation**
Cattle	Milk yield and percentage of butterfat	−.38
Pig	Weight gain and back-fat thickness	.13
	Weight gain and efficiency	.69
Chicken	Body weight and egg weight	.42
	Body weight and egg production	−.17
	Egg weight and egg production	−.31
Mouse	Body weight and tail length	.29
Fruit fly	Abdominal bristle number and sternopleural bristle number	.41

Source: After D. S. Falconer, *Introduction to Quantitative Genetics* (London: Longman, 1981), p. 284.

characteristic, and these genes also affect the second characteristic, causing it to change at the same time. Correlated responses may well be undesirable and may limit the ability to alter a characteristic by selection. From 1944 to 1964, domestic turkeys were subjected to intense selection for growth rate and body size. At the same time, fertility, egg production, and egg hatchability all declined. These correlated responses were due to negative genetic correlations between body size and fertility; eventually, these genetic correlations limited the extent to which the growth rate of turkeys could respond to selection. Genetic correlations may also limit the ability of natural populations to respond to selection in the wild and adapt to their environments.

Concepts

Genetic correlations result from pleiotropy. When two characteristics are genetically correlated, selection for one characteristic will produce a correlated response in the other characteristic.

www.whfreeman.com/pierce Information and links on the use of quantitative genetics in plant and animal breeding

Connecting Concepts Across Chapters

In this chapter, our perspective has shifted from individual genotypes (emphasized in transmission genetics) and the physical nature of the gene (emphasized in molecular genetics) to the genetic properties of groups of individuals. This shift will also be our perspective in Chapter 23, on population genetics.

Many of the most important characteristics in nature are those that display complex phenotypes and vary continuously. Body weight, reproductive output, susceptibility to diseases, and behavioral attributes often have continuous phenotypes. These types of characteristics are important in agriculture and are frequently significant in human health and evolution. An important theme of this chapter has been that such complex characteristics are inherited according to Mendelian principles, but more genes take part and environmental factors modify the phenotype. Because many factors influence the phenotypes of these complex characteristics,

individual genes are difficult to identify, and we cannot predict precise phenotypic ratios among the offspring of a particular cross. Nevertheless, statistical procedures can be used to predict the average offspring phenotype and to assess the extent to which genetic and environmental factors are responsible for phenotypic differences in a characteristic.

Because the genes that influence quantitative characteristics are inherited according to Mendelian principles, the study of quantitative genetics requires a thorough understanding of the basic principles of heredity, which were covered in Chapters 3 through 7. Twin studies, which can be used to calculate heritability, are discussed in detail in Chapter 6; restriction fragment length polymorphisms and microsatellite variants, used to map quantitative trait loci, are explained in Chapter 18. The study of quantitative genetics depends on the genetic composition of populations and how that composition changes with time, which is the focus of Chapter 23.

CONCEPTS SUMMARY

- Quantitative genetics focuses on the inheritance of complex characteristics whose phenotype varies continuously. For many quantitative characteristics, the relation between genotype and phenotype is complex because many genes and environmental factors influence a characteristic.

- Quantitative characteristics also include meristic (counting) characteristics and threshold characteristics whose underlying genetic basis is influenced by multiple factors.

- Many quantitative characteristics are polygenic. The individual genes that influence a polygenic characteristic follow the same Mendelian principles that govern discontinuous characteristics, but, because many genes participate, the expected ratios of phenotypes are obscured.

- A population is the group of interest, and a sample is a subset of the population used to describe it.

- A frequency distribution, in which the phenotypes are represented on one axis and the number of individuals possessing the phenotype is represented on the other, is a convenient means of summarizing phenotypes found in a group of individuals.

- The mean and variance provide key information about a distribution: the mean gives the central location of the distribution, and the variance provides information about how the phenotype varies within a group.

- The correlation coefficient measures the direction and strength of association between two variables. Regression can be used to predict the value of one variable on the basis of the value of a correlated variable.

- Phenotypic variance in a characteristic can be divided into components that are due to additive genetic variance, dominance genetic variance, genic interaction variance,

environmental variance, and genetic–environmental interaction variance.

- Broad-sense heritability is the proportion of the phenotypic variance that is due to genetic variance; narrow-sense heritability is the proportion of the phenotypic variance due to additive genetic variance.

- Broad-sense heritability can be estimated by eliminating the environmental variance component. Narrow-sense heritability can be estimated by comparing the phenotypes of parents and offspring or by comparing phenotypes of individuals with different degrees of relatedness, such as identical twins and nonidentical twins.

- Heritability provides information only about the degree to which variation in a characteristic results from genetic differences. It does not indicate the degree to which a characteristic is genetically determined. Heritability is based on the variances present within a group of individuals, and an individual does not have heritability. Heritability of a characteristic varies among populations and among environments. Even if heritability for a characteristic is high, the characteristic may still be altered by changes in the environment. Heritabilities provide no information about the nature of population differences in a characteristic.

- Quantitative trait loci are genes that control polygenic characteristics. QTLs can be mapped by examining the association between the inheritance of a quantitative characteristic and the inheritance of genetic markers. The mapping of numerous genetic markers with molecular techniques has made QTL mapping feasible for many organisms.

- When selection is applied to a quantitative characteristic, the characteristic will change if additive genetic variation for the characteristic is present. The amount that a quantitative characteristic changes in a single generation when subjected to selection (the response to selection) is directly related to the selection differential and narrow-sense heritability. By applying a selection differential and measuring the response to selection, narrow-sense heritability can calculated.

- After selection has been applied to a quantitative characteristic for a number of generations, the response to

selection may level off because no additive genetic variation in the characteristic remains. Alternatively, the response to selection may level off because of genetic correlations between the selected trait and other traits that affect fitness.

- A genetic correlation may be present when the same gene affects two or more characteristics (pleiotropy). Genetic correlations produce correlated responses to selection.

IMPORTANT TERMS

quantitative genetics (p. 637)
meristic characteristic (p. 639)
threshold characteristic (p. 639)
frequency distribution (p. 642)
normal distribution (p. 643)
population (p. 643)
sample (p. 643)
mean (p. 644)
variance (p. 644)
standard deviation (p. 644)

correlation (p. 645)
correlation coefficient (p. 646)
regression (p. 647)
regression coefficient (p. 647)
heritability (p. 650)
phenotypic variance (p. 650)
genetic variance (p. 650)
environmental variance (p. 650)
genetic–environmental
    interaction variance (p. 650)

additive genetic variance
    (p. 650)
dominance genetic variance
    (p. 651)
genic interaction variance
    (p. 651)
broad-sense heritability
    (p. 651)
narrow-sense heritability
    (p. 651)

quantitative trait locus (QTL)
    (p. 655)
natural selection (p. 657)
artificial selection (p. 657)
response to selection (p. 657)
selection differential (p. 657)
realized heritability (p. 659)
phenotypic correlation
    (p. 660)
genetic correlation (p. 660)

Worked Problems

1. Seed weight in a particular plant species is determined by pairs of alleles at two loci (a^+ a^- and b^+ b^-) that are additive and equal in their effects. Plants with genotype $a^-a^-b^-b^-$ have seeds that average 1 g in weight, whereas plants with genotype $a^+a^+b^+b^+$ have seeds that average 3.4 g in weight. A plant with genotype $a^-a^-b^-b^-$ is crossed with a plant of genotype $a^+a^+b^+b^+$.

(a) What is the predicted weight of seeds from the F_1 progeny of this cross?

(b) If the F_1 plants are intercrossed, what are the expected seed weights and proportions of the F_2 plants?

- **Solution**

The difference in average seed weight of the two parental genotypes is 3.4 g − 1 g = 2.4 g. These two genotypes differ in four genes; so, if the genes have equal and additive effects, each gene difference contributes an additional 2.4 g/4 = .6 g of weight to the 1-g weight of a plant with none of these contributing genes ($a^-a^-b^-b^-$).

The cross between the two homozygous genotypes produces the F_1 and F_2 progeny shown below:

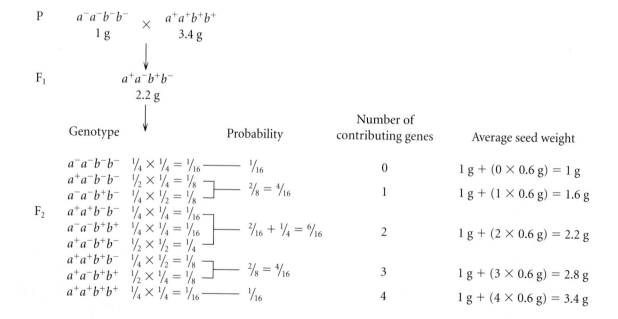

Genotype	Probability		Number of contributing genes	Average seed weight
$a^-a^-b^-b^-$ $\frac{1}{4} \times \frac{1}{4} = \frac{1}{16}$	$\frac{1}{16}$		0	1 g + (0 × 0.6 g) = 1 g
$a^+a^-b^-b^-$ $\frac{1}{2} \times \frac{1}{4} = \frac{1}{8}$				
$a^-a^-b^+b^-$ $\frac{1}{4} \times \frac{1}{2} = \frac{1}{8}$	$\frac{2}{8} = \frac{4}{16}$		1	1 g + (1 × 0.6 g) = 1.6 g
$a^+a^+b^-b^-$ $\frac{1}{4} \times \frac{1}{4} = \frac{1}{16}$				
$a^-a^-b^+b^+$ $\frac{1}{4} \times \frac{1}{4} = \frac{1}{16}$	$\frac{2}{16} + \frac{1}{4} = \frac{6}{16}$		2	1 g + (2 × 0.6 g) = 2.2 g
$a^+a^-b^+b^-$ $\frac{1}{2} \times \frac{1}{2} = \frac{1}{4}$				
$a^+a^+b^+b^-$ $\frac{1}{4} \times \frac{1}{2} = \frac{1}{8}$				
$a^+a^-b^+b^+$ $\frac{1}{2} \times \frac{1}{4} = \frac{1}{8}$	$\frac{2}{8} = \frac{4}{16}$		3	1 g + (3 × 0.6 g) = 2.8 g
$a^+a^+b^+b^+$ $\frac{1}{4} \times \frac{1}{4} = \frac{1}{16}$	$\frac{1}{16}$		4	1 g + (4 × 0.6 g) = 3.4 g

(a) The F_1 are heterozygous at both loci ($a^+a^-b^+b^-$) and possess two genes that contribute an additional 0.6 g each to the 1-g weight of a plant with no contributing genes. Therefore, the seeds of the F_1 should average 1 g + 2(0.6 g) = 2.2 g.

(b) The F_2 will have the following phenotypes and proportions: $\frac{1}{16}$ 1 g; $\frac{4}{16}$ 1.6 g; $\frac{6}{16}$ 2.2 g; $\frac{4}{16}$ 2.8 g; and $\frac{1}{16}$ 3.4 g.

2. Phenotypic variation is analyzed for milk production in a herd of dairy cattle and the following variance components are obtained.

Additive genetic variance (V_A) = .4
Dominance genetic variance (V_D) = .1
Genic interaction variance (V_I) = .2
Environmental variance (V_E) = .5
Genetic-environmental interaction variance (V_{GE}) = .0

(a) What is the narrow-sense heritability of milk production?

(b) What is the broad-sense heritability of milk production?

• Solution

To determine the heritabilities, we first need to calculate V_P and V_G.

$$V_P = V_A + V_D + V_I + V_E + V_{GE}$$
$$= .4 + .1 + .2 + .5 + 0$$
$$= 1.2$$

$$V_G = V_A + V_D + V_I$$
$$= .7$$

(a) The narrow sense heritability is:

$$h^2 = \frac{V_A}{V_P} = \frac{0.4}{1.2} = .33$$

(b) The broad sense heritability is:

$$H^2 = \frac{V_G}{V_P} = \frac{0.7}{1.2} = .58$$

3. The heights of parents and their offspring are measured for 10 families:

Mean height of parents (cm)	Mean height of offspring (cm)
150	152
157	163
188	193
165	163
160	152
142	157
170	183
183	175
152	163
173	180

From these data, determine:

(a) the mean, variance, and standard deviation of height of parents and offspring;

(b) the correlation and regression coefficients for a regression of mean offspring height on mean parental height; and

(c) the narrow-sense heritability of height in these families.

(d) What conclusions can be drawn from the heritability value determined in part c?

• Solution

(a) The best way to begin is by constructing a table, as shown below. To calculate the means, we need to sum the values of x and y, which are shown in the last rows of columns A and D of the table.

For the mean of parental height,

$$\bar{x} = \frac{\Sigma x_i}{n} = \frac{1640}{10} = 164 \text{ cm}$$

A Mean height parents (cm) x_i	B $x_i - \bar{x}$	C $(x_i - \bar{x})^2$	D Mean height offspring (cm) y_i	E $y_i - \bar{y}$	F $(y_i - \bar{y})^2$	G $(x_i - \bar{x})(y_i - \bar{y})$
150	−14	196	152	−16.1	259.21	225.4
157	−7	49	163	−5.1	26.01	35.7
188	24	576	193	24.9	620.01	597.6
165	1	1	163	−5.1	26.01	−5.1
160	−4	16	152	−16.1	259.21	64.4
142	−22	484	157	−11.1	123.21	244.2
170	6	36	183	14.9	222.01	89.4
183	19	361	175	6.9	47.61	131.1
152	−12	144	163	−5.1	26.01	61.2
173	9	81	180	11.9	141.61	107.1
$\Sigma x_i =$ 1640		$\Sigma(x-\bar{x})^2 =$ 1944	$\Sigma y_i =$ 1681		$\Sigma(y-\bar{y})^2 =$ 1750.9	$\Sigma(x_i-\bar{x})(y_i-\bar{y}) =$ 1551

For the mean of the offspring height,

$$\bar{y} = \frac{\Sigma y_i}{n} = \frac{1681}{10} = 168.1 \text{ cm}$$

For the variance, we subtract each x and y value from its mean (columns B and E) and square these differences (columns C and F). The sums of the these squared deviations are shown in the last row of columns C and F. For the regression, we need the covariance, which requires that we take the difference between each x value and its mean and multiply it by the difference between each y value and its mean [$(x_i - \bar{x})(y_i - \bar{y})$, column G] and then sum these products (last row of column G).

The variance is the sum of the squared deviations from the mean divided by $n - 1$, where n is the number of measurements:

$$s_x^2 = \frac{\Sigma(x_i - \bar{x})^2}{n - 1} = \frac{1944}{9} = 216$$

$$s_y^2 = \frac{\Sigma(y_i - \bar{y})^2}{n - 1} = \frac{1750.9}{9} = 194.54$$

The standard deviation is the square root of the variance:

$$s_x = \sqrt{s_x^2} = \sqrt{216} = 14.70$$

$$s_y = \sqrt{s_y^2} = \sqrt{194.54} = 13.95$$

To calculate the correlation coefficient and regression coefficient, we need the covariance:

$$\text{cov}_{xy} = \frac{\Sigma(x_i - \bar{x})(y_i - \bar{y})}{n - 1} = \frac{1551}{9} = 172.33 \text{ cm}$$

The correlation coefficient is the covariance divided by the standard deviation of x and the standard deviation of y:

$$r = \frac{\text{cov}_{xy}}{s_x s_y} = \frac{172.33}{(14.70)(13.95)} = .84$$

The regression coefficient is the covariance divided by the variance of x:

$$r = \frac{\text{cov}_{xy}}{s_x^2} = \frac{172.33}{216} = .80$$

(b) In a regression of the mean phenotype of the offspring against the mean phenotype of the parents, the regression coefficient equals the narrow-sense heritability, which is .80.

(c) We conclude that 80% of the variance in height among the members of these families results from additive genetic variance.

4. A farmer is raising rabbits. The average body weight in his population of rabbits is 3 kg. The farmer selects the 10 largest rabbits in his population, whose average body weight is 4 kg, and interbreeds them. If the heritability of body weight in the rabbit population is .7, what is the expected body weight among offspring of the selected rabbits?

• **Solution**

The farmer has carried out a response-to-selection experiment, in which the response to selection will equal the selection differential times the narrow-sense heritability. The selection differential equals the difference in average weights of the selected rabbits and the entire population: $4 \text{ kg} - 3 \text{ kg} = 1 \text{ kg}$. The narrow-sense heritability is given as .7; so the expected response to selection is: $R = h^2 \times S = .7 \times 1 \text{ kb} = 0.7 \text{ kg}$. This is the increase in weight that is expected in the offspring of the selected parents; so the average weight of the offspring is expected to be: $3 \text{ kg} + 0.7 \text{ kg} = 3.7 \text{ kg}$.

COMPREHENSION QUESTIONS

* 1. How does a quantitative characteristic differ from a discontinuous characteristic?

2. Briefly explain why the relation between genotype and phenotype is frequently complex for quantitative characteristics.

* 3. Why do polygenic characteristics have many phenotypes?

* 4. Explain the relation between a population and a sample. What characteristics should a sample have to be representative of the population?

5. What information do the mean and variance provide about a distribution?

6. How is the standard deviation related to the variance?

* 7. What information does the correlation coefficient provide about the association between two variables?

8. What is regression? How is it used?

* 9. List all the components that contribute to the phenotypic variance and define each component.

*10. How do the broad-sense and narrow-sense heritabilities differ?

11. Briefly outline some of the ways that heritability can be calculated.

12. Briefly discuss common misunderstandings or misapplications of the concept of heritability.

13. Briefly explain how genes affecting a polygenic characteristic are located with the use of QTL mapping.

*14. How is the response to selection related to the narrow-sense heritability and the selection differential? What information does the response to selection provide?

15. Why does the response to selection often level off after many generations of selection?

*16. What is the difference between phenotypic and genetic correlations?

APPLICATION QUESTIONS AND PROBLEMS

*17. For each of the following characteristics, indicate whether it would be considered a discontinuous characteristic or a quantitative characteristic. Briefly justify your answer.

(a) Kernel color in a strain of wheat, in which two codominant alleles segregating at a single locus determine the color. Thus, there are three phenotypes present in this strain: white, light red, and medium red.

(b) Body weight in a family of Labrador retrievers. An autosomal recessive allele that causes dwarfism is present in this family. Two phenotypes are recognized: dwarf (less than 13 kg) and normal (greater than 13 kg).

(c) Presence or absence of leprosy; susceptibility to leprosy is determined by multiple genes and numerous environmental factors.

(d) Number of toes in guinea pigs, which is influenced by genes at many loci.

(e) Number of fingers in humans; extra (more than five) fingers are caused by the presence of an autosomal dominant allele.

*18. The following data are the numbers of digits per foot in 25 guinea pigs. Construct a frequency distribution for these data.

4, 4, 4, 5, 3, 4, 3, 4, 4, 5, 4, 4, 3, 2, 4, 4, 5, 6, 4, 4, 3, 4, 4, 4, 5

19. Ten male Harvard students were weighed in 1916. Their weights are given in the following table. Calculate the mean, variance, and standard deviation for these weights.

Weight (kg) of Harvard students (class of 1920)

51
69
69
57
61
57
75
105
69
63

*20. In Moab, Utah, temperature and rainfall exhibit a correlation of .67. Assuming that this correlation is significant (not due to chance), is there more rainfall, on the average, in the summer or in the winter at this location?

21. Among a population of tadpoles, the correlation coefficient for size at metamorphosis and time required for metamorphosis is −.74. On the basis of this correlation, what conclusions can you make about the relative sizes of tadpoles that metamorphose quickly and those that metamorphose more slowly?

*22. A researcher studying alcohol consumption in North American cities finds a significant, positive correlation between the number of Baptist preachers and alcohol consumption. Is it reasonable for the researcher to conclude that the Baptist preachers are consuming most of the alcohol? Why or why not?

23. Body weight and length were measured on six mosquito fish; these measurements are given in the following table. Calculate the correlation coefficient for weight and length in these fish.

Wet weight (g)	Length (mm)
115	18
130	19
210	22
110	17
140	20
185	21

*24. The heights of mothers and daughters are given in the following table.

(a) Calculate the correlation coefficient for the heights of the mothers and daughters.

(b) Using regression, predict the expected height of a daughter whose mother is 67 inches tall.

Height of mother (in)	Height of daughter (in)
64	66
65	66
66	68
64	65
63	65
63	62
59	62
62	64
61	63
60	62

*25. Assume that plant weight is determined by a pair of alleles at each of two independently assorting loci (A and a, B and b) that are additive in their effects. Further, assume that each allele represented by an uppercase letter contributes 4 g to weight and each allele represented by a lowercase letter contributes 1 g to weight.

(a) If a plant with genotype $AABB$ is crossed with a plant with genotype $aabb$, what weights are expected in the F_1 progeny?

(b) What is the distribution of weight expected in the F_2 progeny?

*26. Assume that three loci, each with two alleles (A and a, B and b, C and c), determine the differences in height between two homozygous strains of a plant. These genes are additive and equal in their effects on plant height. One strain (*aabbcc*) is 10 cm in height. The other strain (*AABBCC*) is 22 cm in height. The two strains are crossed, and the resulting F_1 are interbred to produce F_2 progeny. Give the phenotypes and the expected proportions of the F_2 progeny.

*27. A farmer has two homozygous varieties of tomatoes. One variety, called *Little Pete*, has fruits that average only 2 cm in diameter. The other variety, *Big Boy*, has fruits that average a whopping 14 cm in diameter. The farmer crosses *Little Pete* and *Big Boy*; he then intercrosses the F_1 to produce F_2 progeny. He grows 2000 F_2 tomato plants and doesn't find any F_2 offspring that produce fruits as small as *Little Pete* or as large as *Big Boy*. If we assume that the differences in fruit size of these varieties are produced by genes with equal and additive effects, what conclusion can we make about the minimum number of loci with pairs of alleles determining the differences in fruit size of the two varieties?

28. Seed size in a plant is a polygenic characteristic. A grower crosses two pure-breeding varieties of the plant and measures seed size in the F_1 progeny. He then backcrosses the F_1 plants to one of the parental varieties and measures seed size in the backcross progeny. The grower finds that seed size in the backcross progeny has a higher variance than does seed size in the F_1 progeny. Explain why the backcross progeny are more variable.

*29. Phenotypic variation in tail length of mice has the following components:

Additive genetic variance (V_A)	= .5
Dominance genetic variance (V_D)	= .3
Genic interaction variance (V_I)	= .1
Environmental variance (V_E)	= .4
Genetic–environmental interaction variance (V_{GE})	= .0

(a) What is the narrow-sense heritability of tail length?

(b) What is the broad-sense heritability of tail length?

30. The narrow-sense heritability of ear length in Reno rabbits is .4. The phenotypic variance (V_P) is .8 and the environmental variance (V_E) is .2. What is the additive genetic variance (V_A) for ear length in these rabbits?

*31. Assume that human ear length is influenced by multiple genetic and environmental factors. Suppose you measured ear length on three groups of people, in which group A consists of five unrelated persons, group B consists of five siblings, and group C consists of five first cousins.

(a) Assuming that the environment for each group is similar, which group should have the highest phenotypic variance? Explain why.

(b) Is it realistic to assume that the environmental variance for each group is similar? Explain your answer.

32. A characteristic has a narrow-sense heritability of .6.

(a) If the dominance variance (V_D) increases and all other variance components remain the same, what will happen to the narrow-sense heritability? Will it increase, decrease, or remain the same? Explain.

(b) What will happen to the broad-sense heritability? Explain.

(c) If the environmental variance (V_E) increases and all other variance components remain the same, what will happen to the narrow-sense heritability? Explain.

(d) What will happen to the broad-sense heritability? Explain.

33. Flower color in the pea plants that Mendel studied is controlled by alleles at a single locus. A group of peas homozygous for purple flowers is grown in a garden. Careful study of the plants reveals that all their flowers are purple, but there is some variability in the intensity of the purple color. If heritability were estimated for this variation in flower color, what would it be. Explain your answer.

*34. A graduate student is studying a population of bluebonnets along a roadside. The plants in this population are genetically variable. She counts the seeds produced by 100 plants and measures the mean and variance of seed number. The variance is 20. Selecting one plant, the graduate student takes cuttings from it, and cultivates these cuttings in the greenhouse, eventually producing many genetically identical clones of the same plant. She then transplants these clones into the roadside population, allows them to grow for 1 year, and then counts the number of seeds produced by each of the cloned plants. The graduate student finds that the variance in seed number among these cloned plants is 5. From the phenotypic variance of the genetically variable and genetically identical plants, she calculates the broad-sense heritability.

(a) What is the broad-sense heritability of seed number for the roadside population of bluebonnets?

(b) What might cause this estimate of heritability to be inaccurate?

*35. The length of the middle joint of the right index finger was measured on 10 sets of parents and their adult offspring. The mean parental lengths and the mean offspring lengths for each family are listed in the following table. Calculate the regression coefficient for regression of mean offspring length against mean parental length and estimate the narrow-sense heritability for this characteristic.

Mean parental length (mm)	Mean offspring length (mm)
30	31
35	36
28	31
33	35
26	27
32	30
31	34
29	28
40	38
33	34

*36. Mr. Jones is a pig farmer. For many years, he has fed his pigs the food left over from the local university cafeteria, which is known to be low in protein, deficient in vitamins, and downright untasty. However, the food is free, and his pigs don't complain. One day a salesman from a feed company visits Mr. Jones. The salesman claims that his company sells a new, high-protein, vitamin-enriched feed that enhances weight gain in pigs. Although the food is expensive, the salesman claims that the increased weight gain of the pigs will more than pay for the cost of the feed, increasing Mr. Jones's profit. Mr. Jones responds that he took a genetics class when he went to the university and that he has conducted some genetic experiments on his pigs; specifically, he has calculated the narrow-sense heritability of weight gain for his pigs and found it to be .98. Mr. Jones says that this heritability value indicates that 98% of the variance in weight gain among his pigs is determined by genetic differences, and therefore the new pig feed can have little effect on the growth of his pigs. He concludes that the feed would be a waste of his money. The salesman doesn't dispute Mr. Jones' heritability estimate, but he still claims that the new feed can significantly increase weight gain in Mr. Jones' pigs. Who is correct and why?

37. Joe is breeding cockroaches in his dorm room. He finds that the average wing length in his population of cockroaches is 4 cm. He picks six cockroaches that have the largest wings; the average wing length among these selected cockroaches is 10 cm. Joe interbreeds these selected cockroaches. From previous studies, he knows that the narrow-sense heritability for wing length in his population of cockroaches is .6.

(a) Calculate the selection differential and expected response to selection for wing length in these cockroaches.

(b) What should be the average wing length of the progeny of the selected cockroaches?

38. Three characteristics in beef cattle—body weight, fat content, and tenderness—are measured and the following variance components are estimated.

	Body weight	Fat content	Tenderness
V_A	22	45	12
V_D	10	25	5
V_I	3	8	2
V_E	42	64	8
V_{GE}	0	0	1

In this population, which characteristic would respond best to selection? Explain your reasoning.

*39. A rancher determines that the average amount of wool produced by a sheep in his flock is 22 kg per year. In an attempt to increase the wool production of his flock, the rancher picks five male and five female sheep with the greatest wool production; the average amount of wool produced per sheep by those selected is 30 kg. He interbreeds these selected sheep and finds that the average wool production among the progeny of the selected sheep is 28 kg. What is the narrow-sense heritability for wool production among the sheep in the rancher's flock?

40. The narrow-sense heritability of wing length in a population of Drosophila melanogaster is .8. The narrow-sense heritability of head width in the same population is .9. The genetic correlation between wing length and head width is −.86. If a geneticist selects for increased wing length in these flies, what will happen to head width?

CHALLENGE QUESTIONS

41. We have explored some of the difficulties in separating genetic and environmental components of human behavioral characteristics. Considering these difficulties and what you know about calculating heritability, propose an experimental design for accurately measuring the heritability of musical ability.

42. A student who has just learned about quantitative genetics says, "Heritability estimates are worthless! They don't tell you anything about the genes that affect a characteristic. They don't provide any information about the types of offspring to expect from a cross. Heritability estimates measured in one population can't be used for other populations; so they don't even give you any general information about how much of a characteristic is genetically determined. I can't see that heritabilities do anything other than make undergraduate students sweat during tests." How would you respond to this statement? Is

the student correct? What good are heritabilities, and why do geneticists bother to calculate them?

43. A geneticist selects for increased size in a population of fruit flies that she is raising in her laboratory. She starts with the two largest males and the two largest females and uses them as the parents for the next generation. From the progeny produced by these selected parents, she selects the two largest males and the two largest females and mates them. She repeats this procedure each generation. The average weight of flies in the initial population was 1.1 mg. The flies respond to selection, and their body size steadily increases. After 20 generations of selection, the average weight is 2.3 mg. However, after about 20 generations, the response to selection in subsequent generations levels off, and the average size of the flies no longer increases. At this point, the geneticist takes a long vacation; while she is gone, the fruit flies in her population interbreed randomly. When

she returns from vacation, she finds that the average size of the flies in the population has decreased to 2.0 mg.

(a) Provide an explanation for why the response to selection leveled off after 20 generations.

(b) Why did the average size of the fruit flies decrease when selection was no longer applied during the geneticist's vacation?

44. Manic-depressive illness is a psychiatric disorder that has a strong hereditary basis, but the exact mode of inheritance is not known. Previous research has shown that siblings of patients with manic-depressive illness are more likely also to develop the disorder than are siblings of unaffected persons. A recent study demonstrated that the ratio of manic-depressive brothers to manic-depressive sisters is higher when the patient is male than when the patient is female. In other words, relatively more brothers of manic-depressive patients also have the disease when the patient is male than when the patient is female. What does this new observation suggest about the inheritance of manic-depressive illness?

SUGGESTED READINGS

Barton, N. H. 1989. Evolutionary quantitative genetics: how little do we know? *Annual Review of Genetics* 23:337–3370.
A review of how quantitative genetics is used to study the process of evolution.

Cunningham, P. 1991. The genetics of thoroughbred horses. *Scientific American* 264(5):92–98.
An interesting account of how quantitative genetics is being applied to the breeding of thoroughbred horses.

Dudley, J. W. 1977. 76 generations of selection for oil and protein percentage in maize. In E. Pollak, O. Kempthorne, and T. B. Bailey, Jr., Eds. *Proceedings of the International Conference on Quantitative Genetics*, pp. 459–473. Ames, IA: Iowa State University Press.
A report on the progress of one of the longest running selection experiments.

East, E. M. 1910. A Mendelian interpretation of variation that is apparently continuous. *American Naturalist* 44:65–82.
East's interpretation of how individual genes acting collectively produce continuous variation, including a discussion of Nilsson-Ehle's research on kernel color in wheat.

East, E. M. 1916. Studies on size inheritance in *Nicotiana*. *Genetics* 1:164–176.
East's study of flower length in *Nicotiana*.

Falconer, D. S., and T. F. C. MacKay (Contributor). 1996. *Introduction to Quantitative Genetics*, 4th ed. New York: Addison-Wesley.
An excellent basic text of quantitative genetics.

Frary, A., T. C. Nesbitt, A. Frary, S. Grandillo, E. van der Knaap, et al. 2000. A quantitative trait locus key to the evolution of tomato fruit size. *Science* 289:85–88.
A report of the discovery and cloning of one QTL that is responsible for the quantitative difference in fruit size between wild tomatoes and cultivated varieties.

Gillham, N. W. 2001. Sir Francis Galton and the birth of eugenics. *Annual Review of Genetics* 2001:83–101.
A history of Galton's contributions to the eugenics movement.

Mackay, T. F. C. 2001. The genetic architecture of quantitative traits. *Annual Review of Genetics* 35:303–339.
A review of techniques for QLT mapping and results from current studies on QLTs.

Martienssen, R. 1997. The origin of maize branches out. *Nature* 386:443–445.
Discusses the identification of QTLs that contributed to the domestication of corn.

Moore, K. J., and D. L. Nagle. 2000. Complex trait analysis in the mouse: the strengths, the limitations, and the promise yet to come. *Annual Review of Genetics* 43:653–686.
A review of the genetic analysis of complex characteristics in mice, particularly emphasizing those that are medically important.

Paterson, A. H., E. S. Lander, J. D. Hewitt, S. Peterson, S. E. Lincoln, and S. D. Tanksley. 1988. Resolution of quantitative traits into Mendelian factors by using a complete linkage map of restriction fragment length polymorphisms. *Nature* 335:721–726.
A study identifying QTLs that control fruit mass, pH, and other important characteristics in tomatoes.

Plomin, R. 1999. Genetics and general cognitive ability. *Nature* 402:C25–C29.
A good discussion of the genetics of general intelligence and the search for QTLs that influence it.

Tanksley, S. D. 1993. Mapping polygenes. *Annual Review of Genetics* 27:205–233.
This review article summarizes some of the efforts to map QTLs. It discusses the methodology and some of the findings that are emerging from this research.

23 Population and Evolutionary Genetics

The inhabitants of the island of Tristan da Cuna have one of the highest incidences of asthma in the world due to the population's unique genetic history. (John Eckwall.)

The Genetic History of Tristan da Cuna

In the fall of 1993, geneticist Noé Zamel arrived at Tristan da Cuna, a small remote island in the South Atlantic (◀ FIGURE 23.1). It had taken Zamel 9 days to make the trip from his home in Canada, first by plane from Toronto to South Africa and then aboard a small research vessel to the island. Because of its remote location, the people of Tristan da Cuna call their home "the loneliest island," but isolation was not what attracted Zamel to Tristan da Cuna. Zamel was looking for a gene that causes asthma, and the inhabitants of Tristan da Cuna have one of the world's highest incidences of hereditary asthma: more than half of the islanders display some symptoms of the disease.

The high frequency of asthma on Tristan da Cuna derives from the unique history of the island's gene pool. The population traces its origin to William Glass, a Scot who moved his family there in 1817. They were joined by some shipwrecked sailors and a few women who migrated from the island of St. Helena but, owing to its remote location and lack of a deep harbor, the island population remained largely isolated. The descendants of Glass and the other settlers intermarried, and slowly the island population increased in number; by 1855, about 100 people inhabited the island. However, Tristan da Cuna's population dropped markedly when, after William Glass's death in 1856, many islanders migrated to South America and South Africa. By 1857, only 33 people remained, and the population grew slowly afterward. It was reduced again in 1885 when a small

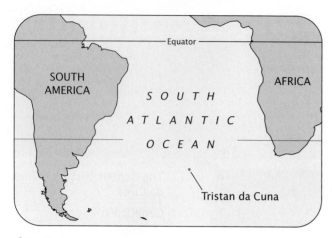

◀ **23.1 Tristan de Cuna is a small island in the South Atlantic.**

boat carrying 15 men was capsized by a huge wave, drowning all on board. Many of the widows and their children left the island, and the population dropped from 106 to 59. In 1961, a volcanic eruption threatened the main village. Fortunately, all of the islanders were rescued and transported to England, where they spent 2 years before returning to Tristan da Cuna.

Today, just a little more than 300 people permanently inhabit the island. These islanders have many genes in common and, in fact, all the island's inhabitants are no less closely related than cousins. Because the founders of the colony were few in number and many were already related, many of the genes in today's population can be traced to just a few original settlers. The population has always been small, which also gives rise to inbreeding and allows chance factors to have a large effect on the frequencies of the alleles in the population. The abrupt population reductions in 1856 and 1885 eliminated some alleles from the population and elevated the frequencies of others. As will be discussed in this chapter, the events affecting these islanders (small number of founders, limited population size, inbreeding, and population reduction) affect the proportions of alleles in a population. All of these factors have contributed to the high proportion of alleles that cause asthma among the inhabitants of Tristan da Cuna.

Tristan da Cuna illustrates how the history of a population shapes its genetic makeup. *Population genetics* is the branch of genetics that studies the genetic makeup of *groups* of individuals and how a group's genetic composition changes with time. Population geneticists usually focus their attention on a **Mendelian population,** which is a group of interbreeding, sexually reproducing individuals that have a common set of genes, the **gene pool.** A population evolves through changes in its gene pool; so population genetics is therefore also the study of evolution. Population geneticists

study the variation in alleles within and between groups and the evolutionary forces responsible for shaping the patterns of genetic variation found in nature. In this chapter, we will learn how the gene pool of a population is measured and what factors are responsible for shaping it. In the later part of the chapter, we will examine molecular studies of genetic variation and evolution.

Genetic Variation

An obvious and pervasive feature of life is variability. Consider a group of students in a typical college class, the members of which vary in eye color, hair color, skin pigmentation, height, weight, facial features, blood type, and susceptibility to numerous diseases and disorders. No two students in the class are likely to be even remotely similar in appearance (◀ FIGURE 23.2a).

Humans are not unique in their extensive variability; almost all organisms exhibit variation in phenotype. For instance, lady beetles are highly variable in their patterns of spots (◀ FIGURE 23.2b), mice vary in body size, snails have different numbers of stripes on their shells, and plants vary in their susceptibility to pests. Much of this phenotypic variation is hereditary. Recognition of the extent of phenotypic variation and its genetic basis led Charles Darwin to the idea of evolution through natural selection.

In fact, even more genetic variation exists in populations than is visible in the phenotype. Much variation exists at the molecular level owing to the redundancy of the genetic code, which allows different codons to specify the same amino acids. Thus two individuals can produce the same protein even if their DNA sequences are different. DNA sequences between the genes and introns within genes do not encode proteins; so much of the variation in these sequences also has little effect on the phenotype.

The amount of genetic variation within natural populations and the forces that limit and shape it are of primary interest to population geneticists. Genetic variation is the basis of all evolution, and the extent of genetic variation within a population affects its potential to adapt to environmental change.

An important, but frequently misunderstood, tool used in population genetics is the mathematical model. Let's take a moment to consider what a model is and how it can be used. A mathematical model usually describes a process in terms of an equation. Factors that may influence the process are represented by variables in the equation; the equation defines the way in which the variables influence the process. Most models are simplified representations of a process, because it is impossible to simultaneously consider all of the influencing factors; some must be ignored in order to examine the effects of others. At first, a model might consider only one or a few factors, but, after their effects are understood, the model can be improved by the addition of

more details. It is important to realize that even a simple model can be a source of valuable insight into how a process is influenced by key variables.

www.whfreeman.com/pierce   More information on genetic diversity within the human species

Before we can explore the evolutionary processes that shape genetic variation, we must be able to describe the genetic structure of a population. The usual way of doing so is to enumerate the types and frequencies of genotypes and alleles in a population.

Calculation of Genotypic Frequencies

A frequency is simply a proportion or a percentage, usually expressed as a decimal fraction. For example, if 20% of the alleles at a particular locus in a population are A, we would say that the frequency of the A allele in the population is .20. For large populations, where it is not practical to determine the genes of all individuals, a sample of individuals from the population is usually taken and the genotypic and allelic frequencies are calculated for this sample (see Chapter 22 for a discussion of samples.) The genotypic and allelic frequencies of the sample are then used to represent the gene pool of the population.

To calculate a **genotypic frequency,** we simply add up the number of individuals possessing the genotype and divide by the total number of individuals in the sample (N). For a locus with three genotypes AA, Aa, and aa, the frequency (f) of each genotype is:

$$f(AA) = \frac{\text{number of } AA \text{ individuals}}{N} \quad (23.1)$$

$$f(Aa) = \frac{\text{number of } Aa \text{ individuals}}{N}$$

$$f(aa) = \frac{\text{number of } aa \text{ individuals}}{N}$$

The sum of all the genotypic frequencies always equals 1.

Calculation of Allelic Frequencies

The gene pool of a population can also be described in terms of the allelic frequencies. There are always fewer alleles than genotypes; so the gene pool of a population can be described in fewer terms when the allelic frequencies are used. In a sexually reproducing population, the genotypes are only temporary assemblages of the alleles: the genotypes break down each generation when individual alleles are passed to the next generation through the gametes, and so it is the types and numbers of alleles, not genotypes, that have real continuity from one generation to the next and that make up the gene pool of a population.

Allelic frequencies can be calculated from (1) the numbers or (2) the frequencies of the genotypes. To calculate the **allelic frequency** from the numbers of genotypes, we count the number of copies of a particular allele present in a sample and divide by the total number of all alleles in the sample:

$$\text{frequency of an allele} = \frac{\text{number of copies of the allele}}{\text{number of copies of all alleles at the locus}} \quad (23.2)$$

For a locus with only two alleles (A and a), the frequencies of the alleles are usually represented by the symbols p and q, and can be calculated as follows:

$$p = f(A) = \frac{2n_{AA} + n_{Aa}}{2N} \quad (23.3)$$

$$q = f(a) = \frac{2n_{aa} + n_{Aa}}{2N}$$

where n_{AA}, n_{Aa}, and n_{aa} represent the numbers of AA, Aa, and aa individuals, and N represents the total number of individuals in the sample. We divide by $2N$ because each diploid individual has two alleles at a locus. The sum of the allelic frequencies always equals 1 ($p + q = 1$); so after p has been obtained, q can be determined by subtraction: $q = 1 - p$.

Alternatively, allelic frequencies can also be calculated from the genotypic frequencies. To do so, we add the frequency of the homozygote for each allele to half the frequency of the heterozygote (because half of the heterozygote's alleles are of each type):

$$p = f(A) = f(AA) + \tfrac{1}{2}f(Aa) \quad (23.4)$$

$$q = f(a) = f(aa) + \tfrac{1}{2}f(Aa)$$

We obtain the same values of p and q whether we calculate the allelic frequencies from the numbers of genotypes (Equation 23.3) or from the genotypic frequencies (Equation 23.4).

Loci with multiple alleles We can use the same principles to determine the frequencies of alleles for loci with more than two alleles. To calculate the allelic frequencies from the numbers of genotypes, we count up the number of copies of an allele by adding twice the number of homozygotes to the number of heterozygotes that possess the allele and divide this sum by twice the number of individuals in the sample. For a locus with three alleles (A^1, A^2, and A^3)

and six genotypes (A^1A^1, A^1A^2, A^2A^2, A^1A^3, A^2A^3, and A^3A^3), the frequencies (p, q, and r) of the alleles are:

$$p = f(A^1) = \frac{2n_{A^1A^1} + n_{A^1A^2} + n_{A^1A^3}}{2N} \quad (23.5)$$

$$q = f(A^2) = \frac{2n_{A^2A^2} + n_{A^1A^2} + n_{A^2A^3}}{2N}$$

$$r = f(A^3) = \frac{2n_{A^3A^3} + n_{A^1A^3} + n_{A^2A^3}}{2N}$$

Alternatively, we can calculate the frequencies of multiple alleles from the genotypic frequencies by extending Equation 23.4. Once again, we add the frequency of the homozygote to half the frequency of each heterozygous genotype that possesses the allele:

$$p = f(A^1) = f(A^1A^1) + \tfrac{1}{2}f(A^1A^2) + \tfrac{1}{2}f(A^1A^3) \quad (23.6)$$

$$q = f(A^2) = f(A^2A^2) + \tfrac{1}{2}f(A^1A^2) + \tfrac{1}{2}f(A^2A^3)$$

$$r = f(A^3) = f(A^3A^3) + \tfrac{1}{2}f(A^1A^3) + \tfrac{1}{2}f(A^2A^3)$$

X-linked loci To calculate allelic frequencies for genes at X-linked loci, we apply these same principles. However, we must remember that a female possesses two X chromosomes and therefore has two X-linked alleles, whereas a male has only a single X chromosome and has one X-linked allele.

Suppose there are two alleles at an X-linked locus, X^A and X^a. Females may be either homozygous ($X^A X^A$ or $X^a X^a$) or heterozygous ($X^A X^a$). All males are hemizygous ($X^A Y$ or $X^a Y$). To determine the frequency of the X^A allele (p), we first count the number of copies of X^A: we multiply the number of $X^A X^A$ females by two and add the number of $X^A X^a$ females and the number of $X^A Y$ males. We then divide the sum by the total number of alleles at the locus, which is twice the total number of females plus the number of males:

$$p = f(X^A) = \frac{2n_{X^AX^A} + n_{X^AX^a} + n_{X^AY}}{2n_{\text{females}} + n_{\text{males}}} \quad (23.7a)$$

Similarly, the frequency of the X^a allele is:

$$q = f(X^a) = \frac{2n_{X^aX^a} + n_{X^AX^a} + n_{X^aY}}{2n_{\text{females}} + n_{\text{males}}} \quad (23.7b)$$

The frequencies of X-linked alleles can also be calculated from genotypic frequencies by adding the frequency of the females that are homozygous for the allele, half the frequency of the females that are heterozygous for the allele, and the frequency of males hemizygous for the allele:

$$p = f(X^A) = f(X^AX^A) + \tfrac{1}{2} f(X^AX^a) + f(X^AY) \quad (23.8)$$

$$q = f(X^a) = f(X^aX^a) + \tfrac{1}{2} f(X^AX^a) + f(X^aY)$$

If you remember the logic behind all of these calculations, you can determine allelic frequencies for any set of genotypes, and it will not be necessary to memorize the formulas.

Concepts

Population genetics is concerned with the genetic composition of a population and how it changes with time. The gene pool of a population can be described by the frequencies of genotypes and alleles in the population.

Worked Problem

The human MN blood type antigens are determined by two codominant alleles, L^M and L^N (see p. 103 in Chapter 5). The MN blood types and corresponding genotypes of 398 Finns from Karjala are tabulated here.

Phenotype	Genotype	Number
MM	L^ML^M	182
MN	L^ML^N	172
NN	L^NL^N	44

Source: W. C. Boyd, *Genetics and the Races of Man* (Boston: Little, Brown, 1950.)

Calculate the allelic and genotypic frequencies at the MN locus for the Karjala population.

• Solution

The genotypic frequencies for the population are calculated with the following formula:

$$\text{genotypic frequency} = \frac{\text{number of individuals with genotype}}{\text{total number of individudals in sample}(N)}$$

$$f(L^ML^M) = \frac{\text{number of } L^ML^M \text{ individuals}}{N} = \frac{182}{398} = .457$$

$$f(L^ML^N) = \frac{\text{number of } L^ML^N \text{ individuals}}{N} = \frac{172}{398} = .432$$

$$f(L^NL^N) = \frac{\text{number of } L^NL^N \text{ individuals}}{N} = \frac{44}{398} = .111$$

The allelic frequencies can be calculated from either the numbers or the frequencies of the genotypes. To calculate allelic frequencies from numbers of genotypes, we add the number of copies of the allele and divide by the number of copies of all alleles at that locus.

$$\text{frequency of an allele} = \frac{\text{number of copies of the allele}}{\text{number of copies of all alleles}}$$

$$p = f(L^M) = \frac{(2n_{L^ML^M}) + (n_{L^ML^N})}{2N} = \frac{2(182) + 172}{2(398)}$$
$$= \frac{536}{796} = .673$$

$$q = f(L^N) = \frac{(2n_{L^NL^N}) + (n_{L^ML^N})}{2N} = \frac{2(44) + 172}{2(398)}$$
$$= \frac{260}{796} = .327$$

To calculate the allelic frequencies from genotypic frequencies, we add the frequency of the homozygote for that genotype to half the frequency of each heterozygote that contains that allele:

$$p = f(L^M) = f(L^ML^M) + \tfrac{1}{2} f(L^ML^N) = .457 + \tfrac{1}{2} (.432)$$
$$= .673$$

$$p = f(L^N) = f(L^NL^N) + \tfrac{1}{2} f(L^ML^N) = .111 + \tfrac{1}{2} (.432)$$
$$= .327$$

The Hardy-Weinberg Law

The primary goal of population genetics is to understand the processes that shape a population's gene pool. First, we must ask what effects reproduction and Mendelian principles have on the genotypic and allelic frequencies: How do the segregation of alleles in gamete formation and the combining of alleles in fertilization influence the gene pool? The answer to this question lies in the **Hardy-Weinberg law,** one of the most important principles of population genetics.

The Hardy-Weinberg law was formulated independently by both Godfrey H. Hardy and Wilhelm Weinberg in 1908. (Similar conclusions were reached by several other geneticists about the same time.) The law is actually a mathematical model that evaluates the effect of reproduction on the genotypic and allelic frequencies of a population. It makes several simplifying assumptions about the population and provides two key predictions if these assumptions are met. For an autosomal locus with two alleles, the Hardy-Weinberg law can be stated as follows:

Assumptions—If a population is large, randomly mating, and not affected by mutation, migration, or natural selection, then:

Prediction 1—the allelic frequencies of a population do not change; and

Prediction 2—the genotypic frequencies stabilize (will not change) after one generation in the proportions p^2 (the frequency of AA), $2pq$ (the frequency of Aa), and q^2 (the frequency of aa), where p equals the frequency of allele A and q equals the frequency of allele a.

The Hardy-Weinberg law indicates that, when the assumptions are met, reproduction alone does not alter allelic or genotypic frequencies and the allelic frequencies determine the frequencies of genotypes.

The statement that genotypic frequencies stabilize after one generation means that they may change in the first generation after random mating, because one generation of random mating is required to produce Hardy-Weinberg proportions of the genotypes. Afterward, the genotypic frequencies, like allelic frequencies, do not change as long as the population continues to meet the assumptions of the Hardy-Weinberg law. When genotypes are in the expected proportions of p^2, $2pq$, and q^2, the population is said to be in **Hardy-Weinberg equilibrium.**

Concepts

The Hardy-Weinberg law describes how reproduction and Mendelian principles affect the allelic and genotypic frequencies of a population.

Closer Examination of the Assumptions of the Hardy-Weinberg Law

Before we consider the implications of the Hardy-Weinberg law, we need to take a closer look at the three assumptions that it makes about a population. First, it assumes that the population is large. How big is "large"? Theoretically, the Hardy-Weinberg law requires that a population be infinitely large in size, but this requirement is obviously unrealistic. In practice, a population need only be large enough that chance deviations from expected ratios do not cause significant changes in allelic frequencies. Later in the chapter, we will examine the effects of small population size on allelic frequencies.

A second assumption of the Hardy-Weinberg law is that individuals in the population mate randomly, which means that each genotype mates in proportion to its frequency. For example, suppose that three genotypes are present in a population in the following proportions: $f(AA) = .6, f(Aa) = .3$, and $f(aa) = .1$. With random mating, the frequency of mating between two AA homozygotes ($AA \times AA$) will be equal to the multiplication of their frequencies: $.6 \times .6 = .36$, whereas the frequency of mating between two aa homozygotes ($aa \times aa$) will be only $.1 \times .1 = .01$.

A third assumption of the Hardy-Weinberg law is that the allelic frequencies of the population are not affected by natural selection, migration, and mutation. Although mutation occurs in every population, its rate is so low that it has

little effect on the predictions of the Hardy-Weinberg law. Although natural selection and migration are significant factors in real populations, we must remember that the purpose of the Hardy-Weinberg law is to examine only the effect of reproduction on the gene pool. When this effect is known, the effects of other factors (such as migration and natural selection) can be examined.

A final point that should be mentioned is that the assumptions of the Hardy-Weinberg law apply to a *single* locus. No real population mates randomly for all traits; nor is a population completely free of natural selection for all traits. The Hardy-Weinberg law, however, does not require random mating and the absence of selection, migration, and mutation for all traits; it requires these conditions only for the locus under consideration. A population may be in Hardy-Weinberg equilibrium for one locus but not for others.

Implications of the Hardy-Weinberg Law

The Hardy-Weinberg law has several important implications for the genetic structure of a population. One implication is that a population cannot evolve if it meets the Hardy-Weinberg assumptions, because evolution consists of change in the allelic frequencies of a population. Therefore the Hardy-Weinberg law tells us that reproduction alone will not bring about evolution. Other processes such as natural selection, mutation, migration, or chance in small populations are required for populations to evolve.

A second important implication is that, when a population is in Hardy-Weinberg equilibrium, the genotypic frequencies are determined by the allelic frequencies. For a locus with two alleles, the frequency of the heterozygote is greatest when allelic frequencies are between .33 and .66 and is at a maximum when allelic frequencies are each .5 (◀FIGURE 23.3). The heterozygote frequency also never exceeds .5. Furthermore, when the frequency of one allele is low, homozygotes for that allele will be rare, and most of the copies of a rare allele will be present in heterozygotes. As you can see from Figure 23.3, when the frequency of allele a is .2, the frequency of the aa homozygote is only $.04$ (q^2), but the frequency of Aa heterozygotes is $.32$ ($2pq$); 80% of the a alleles are in heterozygotes.

A third implication of the Hardy-Weinberg law is that a single generation of random mating produces the equilibrium frequencies of p^2, $2pq$, and q^2. The fact that genotypes are in Hardy-Weinberg proportions does not prove that the population is free from natural selection, mutation, and migration. It means only that these forces have not acted since the last time random mating took place.

Extensions of the Hardy-Weinberg Law

The Hardy-Weinberg law can also be applied to multiple alleles and X-linked alleles. The genotypic frequencies expected under Hardy-Weinberg equilibrium will differ according to the situation.

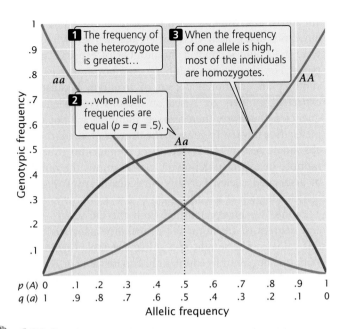

23.3 When a population is in Hardy-Weinberg equilibrium, the proportions of genotypes are determined by frequencies of alleles.

Hardy-Weinberg expectations for loci with multiple alleles

In general, the genotypic frequencies expected at equilibrium are the square of the allelic frequencies. For an autosomal locus with two alleles, these frequencies are $(p + q)^2 = p^2 + 2pq + q^2$. We can also use the square of the allelic frequencies to calculate the equilibrium frequencies for a locus with multiple alleles. An autosomal locus with three alleles, A^1, A^2, and A^3, has six genotypes: A^1A^1, A^1A^2, A^2A^2, A^1A^3, A^2A^3, and A^3A^3. According to the Hardy-Weinberg law, the frequencies of the genotypes at equilibrium depend on the frequencies of the alleles. If the frequencies of alleles A^1, A^2, and A^3 are p, q, and r, respectively, then the equilibrium genotypic frequencies will be the square of the allelic frequencies $(p + q + r)^2 = p^2 + 2pq + q^2 + 2pr + 2qr + r^2$, where:

$$p^2 = f(A^1A^1) \qquad (23.9)$$

$$2pq = f(A^1A^2)$$

$$q^2 = f(A^2A^2)$$

$$2pr = f(A^1A^3)$$

$$2qr = f(A^2A^3)$$

$$r^2 = f(A^3A^3)$$

The square of the allelic frequencies can also be used to calculate the expected genotypic frequencies for loci with four or more alleles.

Hardy-Weinberg expectations for X-linked loci

For an X-linked locus with two alleles, X^A and X^a, there are five possible genotypes: X^AX^A, X^AX^a, X^aX^a, X^AY, and X^aY. Females possess two X-linked alleles, and the expected proportions of the female genotypes can be calculated by using the square of the allelic frequencies. If the frequencies of X^A and X^a are p and q, respectively, then the equilibrium frequencies of the female genotypes are $(p + q)^2 = p^2$ (frequency of X^AX^A) + $2pq$ (frequency of X^AX^a) + q^2 (frequency of X^aX^a). Males have only a single X-linked allele, and so the frequencies of the male genotypes are p (frequency of X^AY) and q (frequency of X^aY). Notice that these expected frequencies are the proportions of the genotypes among males and females rather than the proportions among the entire population. Thus, p^2 is the expected proportion of females with the genotype X^AX^A; if females make up 50% of the population, then the expected proportion of this genotype in the entire population is $.5 \times p^2$.

The frequency of an X-linked recessive trait among males is q, whereas the frequency among females is q^2. When an X-linked allele is uncommon, the trait will therefore be much more frequent in males than in females. Consider hemophilia A, a clotting disorder caused by an X-linked recessive allele with a frequency (q) of approximately 1 in 10,000, or .0001. At Hardy-Weinberg equilibrium, this frequency will also be the frequency of the disease among males. The frequency of the disease among females, however, will be $q^2 = (.0001)^2 = .00000001$, which is only 1 in 10 million. Hemophilia is 1000 times as frequent in males as in females.

Testing for Hardy-Weinberg Proportions

If a population is in equilibrium, then it is randomly mating for the locus in question, and selection, migration, mutation, and small population size have not significantly influenced the genotypic frequencies since random mating last took place. To determine whether these conditions are met, the genotypic proportions expected under the Hardy-Weinberg law must be compared with the observed genotypic frequencies. To do so, we first calculate the allelic frequencies, then find the expected genotypic frequencies by using the square of the allelic frequencies, and finally compare the observed and expected genotypic frequencies by using a chi-square test.

Worked Problem

Jeffrey Mitton and his colleagues found three genotypes (R^2R^2, R^2R^3, and R^3R^3) at a locus encoding the enzyme peroxidase in ponderosa pine trees growing in Colorado. The observed numbers of these genotypes at Glacier Lake, Colorado, were:

Genotypes	Number observed
R^2R^2	135
R^2R^3	44
R^3R^3	11

Are the ponderosa pine trees at Glacier Lake, Colorado, in Hardy-Weinberg equilibrium at the peroxidase locus?

• Solution

If the frequency of the R^2 allele equals p and the frequency of the R^3 allele equals q, the frequency of the R^2 allele is:

$$p = f(R^2) = \frac{(2n_{R^2R^2}) + (n_{R^2R^3})}{2N} = \frac{2(135) + 44}{2(190)} = .826$$

The frequency of the R^3 allele is obtained by subtraction:

$$q = f(R^3) = 1 - p = .174$$

The frequencies of the genotypes expected under Hardy-Weinberg equilibrium are then calculated by using p^2, $2pq$, and q^2:

$$R^2R^2 = p^2 = (.826)^2 = .683$$

$$R^2R^3 = 2pq = 2(.826)(.174) = .287$$

$$R^3R^3 = q^2 = (.174)^2 = .03$$

Multiplying each of these expected genotypic frequencies by the total number of observed individuals in the sample (190), we obtain the *numbers* expected for each genotype:

$$R^2R^2 = .683 \times 190 = 129.8$$

$$R^2R^3 = .287 \times 190 = 54.5$$

$$R^3R^3 = .03 \times 190 = 5.7$$

Comparing these expected numbers with the observed numbers of each genotype, we see that there are more R^2R^2 homozygotes and fewer R^2R^3 heterozygotes and R^3R^3 homozygotes in the population than we expect at equilibrium.

A goodness-of-fit chi-square test is used to determine whether the differences between the observed and the expected numbers of each genotype are due to chance:

$$\chi^2 = \sum \frac{(\text{observed} - \text{expected})^2}{\text{expected}}$$
$$= \frac{(135 - 129.8)^2}{129.8} + \frac{(44 - 54.5)^2}{54.5} + \frac{(11 - 5.7)^2}{5.7}$$
$$= .21 + 2.02 + 4.93 = 7.16$$

The calculated chi-square value is 7.16; to obtain the probability associated with this chi-square value, we determine the appropriate degrees of freedom.

Up to this point, the chi-square test for assessing Hardy-Weinberg equilibrium has been identical with the chi-square tests that we used in Chapter 3 to assess progeny ratios in a genetic cross, where the degrees of freedom were $n - 1$ and n equaled the number of expected genotypes. For the Hardy-Weinberg test, however, we must subtract an additional degree of freedom, because the expected numbers are based on the observed allelic frequencies; therefore, the observed numbers are not completely free to vary. In general, the degrees of freedom for a chi-square test of Hardy-Weinberg equilibrium equal the number of expected genotypic classes minus the number of associated alleles. For this particular Hardy-Weinberg test, the degrees of freedom are $3 - 2 = 1$.

Once we have calculated both the chi-square value and degrees of freedom, the probability associated with this value can be sought in a chi-square table (Table 3.4). With one degree of freedom, a chi-square value of 7.17 has a probability between .01 and .001. It is very unlikely that the peroxidase genotypes observed at Glacier Lake are in Hardy-Weinberg proportions.

Concepts

The observed number of genotypes in a population can be compared to the Hardy-Weinberg expected proportions by using a goodness of fit chi-square test.

Estimating Allelic Frequencies with the Hardy-Weinberg Law

A practical use of the Hardy-Weinberg law is that it allows us to calculate allelic frequencies when dominance is present. For example, cystic fibrosis is an autosomal recessive disorder characterized by respiratory infections, incomplete digestion, and abnormal sweating (see pp. 103–104 in Chapter 5). Among North American Caucasians, the incidence of the disease is approximately 1 person in 2000. The formula for calculating allelic frequency (Equation 23.3) requires that we know the numbers of homozygotes and heterozygotes, but cystic fibrosis is a recessive disease, and so we cannot easily distinguish between homozygous normal persons and heterozygous carriers. Although molecular tests are available for identifying heterozygous carriers of the cystic fibrosis gene, the low frequency of the disease makes widespread screening impractical. In such situations, the Hardy-Weinberg law can be used to estimate the allelic frequencies.

If we assume that a population is in Hardy-Weinberg equilibrium with regard to this locus, then the frequency of the recessive genotype (*aa*) will be q^2, and the allelic frequency is the square root of the genotypic frequency:

$$q = \sqrt{f(aa)} \qquad (23.10)$$

The frequency of cystic fibrosis in North American Caucasians is approximately 1 in 2000, or .0005; so $q = \sqrt{0.0005} = .02$. Thus, about 2% of the alleles in the Caucasian population encode cystic fibrosis. We can calculate the frequency of the normal allele by subtracting: $p = 1 - q = 1 - .02 = .98$. After we have calculated p and q, we can use the Hardy-Weinberg law to determine the frequencies of homozygous normal people and heterozygous carriers of the gene:

$$f(AA) = p^2 = (.98)^2 = .960$$

$$f(Aa) = 2pq = 2(.02)(.98) = .0392$$

Thus about 4% (1 of 25) of Caucasians are heterozygous carriers of the allele that causes cystic fibrosis.

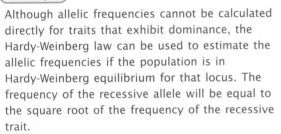

Concepts

Although allelic frequencies cannot be calculated directly for traits that exhibit dominance, the Hardy-Weinberg law can be used to estimate the allelic frequencies if the population is in Hardy-Weinberg equilibrium for that locus. The frequency of the recessive allele will be equal to the square root of the frequency of the recessive trait.

Nonrandom Mating

An assumption of the Hardy-Weinberg law is that mating is random with respect to genotype. Nonrandom mating affects the way in which alleles combine to form genotypes and alters the genotypic frequencies of a population.

We can distinguish between two types of nonrandom mating. **Positive assortative mating** refers to a tendency for like individuals to mate. For example, humans exhibit positive assortative mating for height: tall people mate preferentially with other tall people; short people mate preferentially with other short people. **Negative assortative mating** refers to a tendency for unlike individuals to mate. If people engaged in negative assortative mating for height, tall and short people would preferentially mate.

One form of nonrandom mating is **inbreeding,** which is preferential mating between related individuals. Inbreeding is actually positive assortative mating for relatedness, but it differs from other types of assortative mating because it affects all genes, not just those that determine the trait for which the mating preference occurs. Inbreeding causes a departure from the Hardy-Weinberg equilibrium frequencies of p^2, $2pq$, and q^2. More specifically, it leads to an increase in the proportion of homozygotes and a decrease in the proportion of heterozygotes in a population. **Outcrossing** is the avoidance of mating between related individuals.

Inbreeding is usually measured by the **inbreeding coefficient,** designated F, which is a measure of the probability that two alleles are "identical by descent." In a diploid organism, a homozygous individual has two copies of the same allele. These two copies may be the same in *state*, which means that the two alleles are alike in structure and function but do not have a common origin. Alternatively, the two alleles in a homozygous individual may be the same because they are identical by *descent*—the copies are descended from a single allele that was present in an ancestor (◀ FIGURE 23.4). If we go back far enough in time, many alleles are likely to be identical by descent but, for calculating the effects of inbreeding, we consider identity by descent by going back only a few generations.

Inbreeding coefficients can range from 0 to 1. A value of 0 indicates that mating in a large population is random; a value of 1 indicates that all alleles are identical by descent. Inbreeding coefficients can be calculated from analyses of pedigrees or they can be determined from the reduction in the heterozygosity of a population. Although we will not go into the details of how F is calculated, it's important to understand how inbreeding affects genotypic frequencies.

When inbreeding occurs, the frequency of the genotypes will be:

$$f(AA) = p^2 + Fpq \tag{23.11}$$

$$f(Aa) = 2pq - 2Fpq$$

$$f(aa) = q^2 + Fpq$$

With inbreeding, the proportion of heterozygotes *decreases* by $2Fpq$, and half of this value (Fpq) is *added* to the proportion of each homozygote.

Consider a population that reproduces by self-fertilization (so $F = 1$). We will assume that this population begins with genotypic frequencies in Hardy-Weinberg proportions (p^2, $2pq$, and q^2). With selfing, each homozygote produces

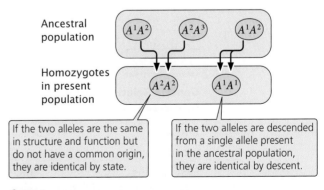

◀23.4 **Individuals may be homozygous by state or by descent.** Inbreeding is a measure of the probability that two alleles are identical by descent.

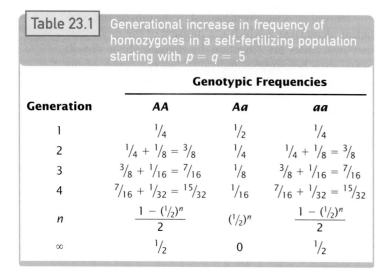

Table 23.1 Generational increase in frequency of homozygotes in a self-fertilizing population starting with $p = q = .5$

	Genotypic Frequencies		
Generation	AA	Aa	aa
1	$\frac{1}{4}$	$\frac{1}{2}$	$\frac{1}{4}$
2	$\frac{1}{4} + \frac{1}{8} = \frac{3}{8}$	$\frac{1}{4}$	$\frac{1}{4} + \frac{1}{8} = \frac{3}{8}$
3	$\frac{3}{8} + \frac{1}{16} = \frac{7}{16}$	$\frac{1}{8}$	$\frac{3}{8} + \frac{1}{16} = \frac{7}{16}$
4	$\frac{7}{16} + \frac{1}{32} = \frac{15}{32}$	$\frac{1}{16}$	$\frac{7}{16} + \frac{1}{32} = \frac{15}{32}$
n	$\dfrac{1 - (\frac{1}{2})^n}{2}$	$(\frac{1}{2})^n$	$\dfrac{1 - (\frac{1}{2})^n}{2}$
∞	$\frac{1}{2}$	0	$\frac{1}{2}$

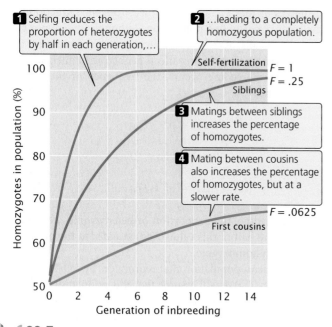

1 Selfing reduces the proportion of heterozygotes by half in each generation,…

2 …leading to a completely homozygous population.

3 Matings between siblings increases the percentage of homozygotes.

4 Mating between cousins also increases the percentage of homozygotes, but at a slower rate.

Self-fertilization $F = 1$
$F = .25$
Siblings
$F = .0625$
First cousins

23.5 Inbreeding increases the percentage of homozygous individuals in a population.

progeny only of the same homozygous genotype ($AA \times AA$ produces all AA; and $aa \times aa$ produces all aa), whereas only half the progeny of a heterozygote will be like the parent ($Aa \times Aa$ produces $\frac{1}{4}$ AA, $\frac{1}{2}$ Aa, and $\frac{1}{4}$ aa). Selfing therefore reduces the proportion of heterozygotes in the population by half with each generation, until all genotypes in the population are homozygous (Table 23.1 and ◀ FIGURE 23.5).

For most outcrossing species, close inbreeding is harmful because it increases the proportion of homozygotes and thereby boosts the probability that deleterious and lethal recessive alleles will combine to produce homozygotes with a harmful trait. Assume that a recessive allele (a) that causes a genetic disease has a frequency (q) of .01. If the population mates randomly ($F = 0$), the frequency of individuals affected with the disease (aa) will be $q^2 = .01^2 = .0001$; so only 1 in 10,000 individuals will have the disease. However, if $F = .25$ (the equivalent of brother–sister mating), then the expected frequency of the homozygote genotype is $q^2 + 2pqF = (.01)^2 + 2(.99)(.01)(.25) = .0026$; thus, the genetic disease is 26 times as frequent at this level of inbreeding. This increased appearance of lethal and deleterious traits with inbreeding is termed **inbreeding depression;** the more intense the inbreeding, the more severe the inbreeding depression.

The harmful effects of inbreeding have been recognized by humans for thousands of years and are the basis of cultural taboos against mating between close relatives. William Schull and James Neel found that, for each 10% increase in F, the mean IQ of Japanese children dropped six points. Child mortality also increases with close inbreeding (Table 23.2); children of first cousins have a 40% increase in mortality over that seen among the children of randomly mated people. Inbreeding also has deleterious effects on crops (◀ FIGURE 23.6) and domestic animals.

Inbreeding depression is most often studied in humans, as well as in plants and animals reared in captivity, but the negative effects of inbreeding may be more severe in natural populations. Julie Jimenez and her colleagues collected wild mice from a natural population in Illinois and bred them in the laboratory for three to four generations. Laboratory matings were chosen so that some mice had no inbreeding,

Table 23.2 Effects of inbreeding on Japanese children

Genetic Relationship of Parents	F	Mortality of Children (Through 12 Years of Age)
Unrelated	0	.082
Second cousins	.016 ($\frac{1}{64}$)	.108
First cousins	.0625 ($\frac{1}{16}$)	.114

Source: After D. L. Hartl, and A. G. Clark, *Principles of Population Genetics.* 2d ed. (Sunderland, MA: Sinauer, 1989), Table 2. Original data from W. J. Schull, and J. V. Neel, *The Effects of Inbreeding on Japanese Children.* (New York: Harper & Row, 1965).

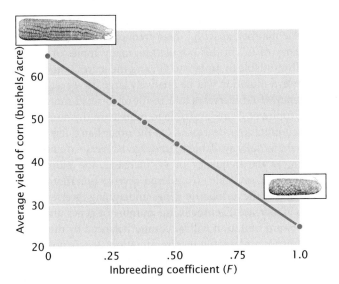

population become homozygous for the same allele. If a species undergoes inbreeding for a number of generations, many deleterious recessive alleles are weeded out by natural or artificial selection so that the population becomes homozygous for beneficial alleles. In this way, the harmful effects of inbreeding may eventually be eliminated, leaving a population that is homozygous for beneficial traits.

Concepts

Nonrandom mating alters the frequencies of the genotypes but not the frequencies of the alleles. Inbreeding is preferential mating between related individuals. With inbreeding, the frequency of homozygotes increases while the frequency of heterozygotes decreases.

◄ 23.6 Inbreeding often has deleterious effects on crops. As inbreeding increases, the average yield of corn, for example, decreases.

whereas others had an inbreeding coefficient of .25. When both types of mice were released back into the wild, the weekly survival of the inbred mice was only 56% of that of the noninbred mice. Inbred male mice also continously lost weight after release into the wild, whereas noninbred male mice regained their body weight within a few days after release.

In spite of the fact that inbreeding is generally harmful for outcrossing species, a number of plants and animals regularly inbreed and are successful (◄ FIGURE 23.7). Inbreeding is commonly used to produce domesticated plants and animals having desirable traits. As stated earlier, inbreeding increases homozygosity, and eventually all individuals in the

Changes in Allelic Frequencies

The Hardy-Weinberg law indicates that allelic frequencies do not change as a result of reproduction; thus, other processes must cause alleles to increase or decrease in frequency. Processes that bring about change in allelic frequency include mutation, migration, genetic drift (random effects due to small population size), and natural selection.

Mutation

Before evolution can occur, genetic variation must exist within a population; consequently, all evolution depends on processes that generate genetic variation. Although new *combinations* of existing genes may arise through recombination in meiosis, all genetic variants ultimately arise through mutation.

The effect of mutation on allelic frequencies Mutation can influence the rate at which one genetic variant increases at the expense of another. Consider a single locus in a population of 25 diploid individuals. Each individual possesses two alleles at the locus under consideration; so the gene pool of the population consists of 50 allelic copies. Let us assume that there are two different alleles, designated G^1 and G^2 with frequencies p and q, respectively. If there are 45 copies of G^1 and 5 copies of G^2 in the population, $p = .90$ and $q = .10$. Now suppose that a mutation changes a G^1 allele into a G^2 allele. After this mutation, there are 44 copies of G^1 and 6 copies of G^2, and the frequency of G^2 has increased from .10 to .12. Mutation has changed the allelic frequency.

If copies of G^1 continue to mutate to G^2, the frequency of G^2 will increase and the frequency of G^1 will decrease (◄ FIGURE 23.8). The amount that G^2 will change (Δq) as a result of mutation depends on: (1) the rate of G^1-to-G^2 mutation (μ); and (2) p, the frequency of G^1 in the population When p is large, there are many copies of G^1 available to mutate to G^2, and the amount of change will be relatively

◄ 23.7 Although inbreeding is generally harmful, a number inbreeding organisms are successful. Shown here is the terrestrial slug *Arion circumscriptos*, an inbreeding species that causes damage in greenhouses and flower gardens. (William Leonard/DRK Photo.)

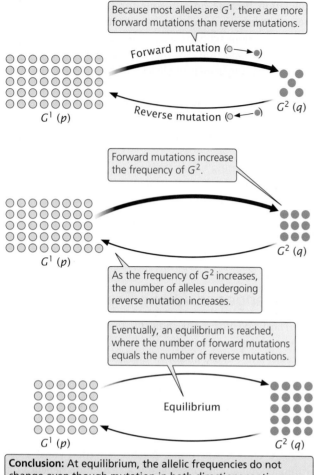

Because most alleles are G^1, there are more forward mutations than reverse mutations.

Forward mutation ($\circ \longrightarrow \bullet$)

Reverse mutation ($\circ \longleftarrow \bullet$)

G^1 (p)

G^2 (q)

Forward mutations increase the frequency of G^2.

G^1 (p)

G^2 (q)

As the frequency of G^2 increases, the number of alleles undergoing reverse mutation increases.

Eventually, an equilibrium is reached, where the number of forward mutations equals the number of reverse mutations.

Equilibrium

G^1 (p)

G^2 (q)

Conclusion: At equilibrium, the allelic frequencies do not change even though mutation in both directions continues.

◀ **23.8 Recurrent mutation changes allelic frequencies.** Forward and reserve mutations eventually lead to a stable equilibrium.

large. As more mutations occur and p decreases, there will be fewer copies of G^1 available to mutate to G^2. The change in G^2 as a result of mutation equals the mutation rate times the allelic frequency:

$$\Delta q = \mu p \qquad (23.12)$$

As the frequency of p decreases as a result of mutation, the change in frequency due to mutation will be less and less.

So far we have considered only the effects of $G^1 \rightarrow G^2$ forward mutations. Reverse $G^2 \rightarrow G^1$ mutations also occur at rate v, which will probably be different from the forward mutation rate, μ. Whenever a reverse mutation occurs, the frequency of G^2 decreases and the frequency of G^1 increases (see Figure 23.8). The rate of change due to reverse mutations equals the reverse mutation rate times the allelic frequency of G^2 ($\Delta q = vq$). The overall change in allelic frequency is a balance between the opposing forces of forward mutation and reverse mutation:

$$\Delta q = \mu p - vq \qquad (23.13)$$

Reaching equilibrium of allelic frequencies

Consider an allele that begins with a high frequency of G^1 and a low frequency of G^2. In this population, many copies of G^1 are initially available to mutate to G^2, and the increase in G^2 due to forward mutation will be relatively large. However, as the frequency of G^2 increases as a result of forward mutations, fewer copies of G^1 are available to mutate; so the number of forward mutations decreases. On the other hand, few copies of G^2 are initially available to undergo a reverse mutation to G^1 but, as the frequency of G^2 increases, the number of copies of G^2 available to undergo reverse mutation to G^1 increases; so the number of genes undergoing reverse mutation will increase. Eventually, the number of genes undergoing forward mutation will be counterbalanced by the number of genes undergoing reverse mutation. At this point, the increase in q due to forward mutation will be equal to the decrease in q due to reverse mutation, and there will be no net change in allelic frequency ($q = 0$), in spite of the fact that forward and reserve mutations continue to occur. The point at which there is no change in the allelic frequency of a population is referred to as **equilibrium** (see Figure 23.8).

Factors determining allelic frequencies at equilibrium

We can determine the allelic frequencies at equilibrium by manipulating Equation 23.13. Recall that $p = 1 - q$. Substituting $1 - q$ for p in Equation 23.13, we get:

$$\Delta q = \mu(1 - q) - vq \qquad (23.14)$$
$$= \mu - \mu q - vq$$
$$= \mu - q(\mu + v)$$

At equilibrium, Δq will be 0; so:

$$0 = \mu - q(\mu + v) \qquad (23.15)$$

$$q(\mu + v) = \mu$$

$$\hat{q} = \frac{\mu}{\mu + v}$$

where $\hat{q}$ equals the frequency of G^2 at equilibrium. This final equation tells us that the allelic frequency at equilibrium is determined solely by the forward and reverse mutation rates.

Summary of effects

When the only evolutionary force acting on a population is mutation, allelic frequencies change with the passage of time because some alleles mutate into others. Eventually, these allelic frequencies reach equilibrium and are determined only by the forward and reverse mutation rates. When the allelic frequencies reach equilibrium, the Hardy-Weinberg law tells us that genotypic frequencies also will remain the same.

The mutation rates for most genes are low; so change in allelic frequency due to mutation in one generation is very

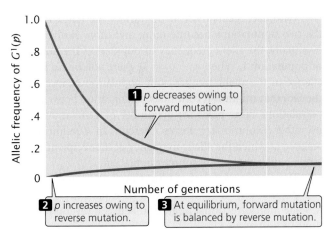

1 p decreases owing to forward mutation.

2 p increases owing to reverse mutation.

3 At equilibrium, forward mutation is balanced by reverse mutation.

◀ **23.9 Change due to recurrent mutation slows as the frequency of p drops.** Allelic frequencies are approaching mutational equilibrium at typical low mutation rates. The allelic frequency of G^1 decreases as a result of forward ($G^1 \rightarrow G^2$) mutation at rate μ (.0001) and increases as a result of reverse ($G^2 \rightarrow G^1$) mutation at rate ν (.00001). Owing to the low rate of mutations, eventual equilibrium takes many generations to be reached.

small, and long periods of time are required for a population to reach mutational equilibrium. For example, if the forward and reverse mutation rates for alleles at a locus are 1×10^{-5} and 0.3×10^{-5} per generation, respectively (rates that have actually been measured at several loci in mice), and the allelic frequencies are $p = .9$ and $q = .1$, then the net change in allelic frequency per generation due to mutation is:

$$\Delta q = \mu p - \nu q$$
$$= (1 \times 10^{-5})(.9) - (.3 \times 10^{-5})(.1)$$
$$= 8.7 \times 10^{-6} = .0000087$$

Therefore, change due to mutation in a single generation is extremely small and, as the frequency of p drops as a result of mutation, the amount of change will become even smaller (◀ FIGURE 23.9). The effect of typical mutation rates on Hardy-Weinberg equilibrium is negligible, and many generations are required for a population to reach mutational equilibrium. Nevertheless, if mutation is the only force acting on a population for long periods of time, mutation rates will determine allelic frequencies.

(Concepts)

Recurrent mutation causes changes in the frequencies of alleles. At equilibrium, the allelic frequencies are determined by the forward and reverse mutation rates. Because mutation rates are low, the effect of mutation per generation is very small.

Migration

Another process that may bring about change in the allelic frequencies is the influx of genes from other populations, commonly called **migration** or **gene flow.** One of the assumptions of the Hardy-Weinberg law is that migration does not take place, but many natural populations do experience migration from other populations. The overall effect of migration is twofold: (1) it prevents genetic divergence *between* populations and (2) it increases genetic variation *within* populations.

The effect of migration on allelic frequencies Let us consider the effects of migration by looking at a simple, unidirectional model of migration between two populations that differ in the frequency of an allele a. Say the frequency of this allele in population I is q_I and in population II is q_{II} (◀ FIGURE 23.10a and b). In each generation, a representative sample of the individuals in population I migrates to population II (◀ FIGURE 23.10c) and reproduces, adding its genes to population II's gene pool. Migration is only from population I to population II (is unidirectional), and all the conditions of the Hardy-Weinberg law apply, except the absence of migration.

After migration, population II consists of two types of individuals (◀ FIGURE 23.10d). Some are migrants; they make up proportion m of population II, and they carry genes from population I; so the frequency of allele a in the migrants is q_I. The other individuals in population II are the original residents. If the migrants make up proportion m of population II, then the residents make up $1 - m$; because the residents originated in population II, the frequency of allele a in this group is q_{II}. After migration, the frequency of allele a in the merged population II (q'_{II}) is:

$$q'_{II} = q_I(m) + q_{II}(1 - m) \tag{23.16}$$

where $q_I(m)$ is the contribution to q made by the copies of allele a in the migrants and $q_{II}(1 - m)$ is the contribution to q made by copies of allele a in the residents. The change in the allelic frequency due to migration (Δq) will be equal to the new frequency of allele a (q'_{II}) minus the original frequency of the allele (q_{II}):

$$\Delta q_{II} = q'_{II} - q_{II}$$

In Equation 23.16, we determined that q'_{II} equals $q_I(m) + q_{II}(1 - m)$. Substituting this value for q'_{II} into the preceding equation, we get:

$$\Delta q = q_I(m) + q_{II}(1 - m) - q_{II}$$

Expanding the term $q_{II}(1 - m)$, we get:

$$\Delta q = q_I m + q_{II} - q_{II}m - q_{II}$$

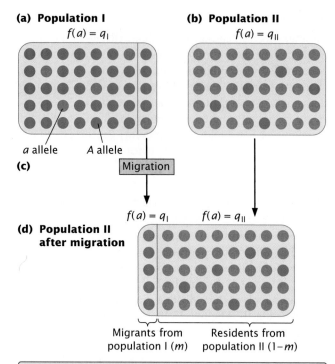

(a) Population I
$f(a) = q_I$

(b) Population II
$f(a) = q_{II}$

a allele   *A* allele

(c)   Migration

$f(a) = q_I$ $f(a) = q_{II}$

(d) Population II after migration

Migrants from population I (*m*)   Residents from population II (1−*m*)

Conclusion: The frequency of allele *a* in population II after migration is $q'_{II} = q_I m + q_{II}(1-m)$.

◀ **23.10 The amount of change in allelic frequency due to migration between populations depends on the difference in allelic frequency and the extent of migration.** Shown here is a model of the effect of unidirectional migration on allelic frequencies. (a) The frequency of allele *a* in the source population (population I) is q_I. (b) The frequency of this allele in the recipient population (population II) is q_{II}. (c) Each generation, a random sample of individuals migrate from population I to population II. (d) After migration, population II consists of migrants and residents. The migrants constitute proportion *m* and have a frequency of *a* equal to q_I; the residents constitute proportion $1 - m$ and have a frequency of *a* equal to q_{II}.

In this last equation, we are subtracting q_{II} from q_{II}, which gives us zero; so the equation simplifies to:

$$\Delta q = q_I m - q_{II} m$$
$$= m(q_I - q_{II}) \qquad (23.17)$$

Equation 23.17 summarizes the factors that determine the amount of change in allelic frequency due to migration. The amount of change in q is directly proportional to the migration (*m*); as the amount of migration increases, the change in allelic frequency increases. The magnitude of change is also affected by the differences in allelic frequencies of the two populations $(q_I - q_{II})$; when the difference is large, the change in allelic frequency will be large.

With each generation of migration, the frequencies of the two populations become more and more similar until, eventually, the allelic frequency of population II equals that of population I. When $q_I - q_{II} = 0$, there will be no further change in the allelic frequency of population II, in spite of the fact that migration continues. If migration between two populations takes place for a number of generations with no other evolutionary forces present, an equilibrium is reached at which the allelic frequency of the recipient population equals that of the source population.

The simple model of unidirectional migration between two populations just outlined can be expanded to accommodate multidirectional migration between several populations (◀ FIGURE 23.11).

The overall effect of migration Migration has two major effects. First, it causes the gene pools of populations to become more similar. Later, we will see how genetic drift and natural selection lead to genetic differences between populations; migration counteracts this tendency and tends to keep populations homogeneous in their allelic frequencies. Second, migration adds genetic variation to populations. Different alleles may arise in different populations owing to rare

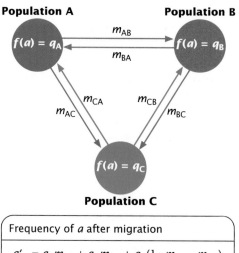

Population A   **Population B**

$f(a) = q_A$ m_{AB} $f(a) = q_B$
m_{BA}

m_{CA} m_{CB}
m_{AC} m_{BC}

$f(a) = q_C$

Population C

Frequency of *a* after migration

$q'_A = q_B m_{BA} + q_C m_{CA} + q_A(1 - m_{BA} - m_{CA})$
$q'_B = q_A m_{AB} + q_C m_{CB} + q_B(1 - m_{AB} - m_{CB})$
$q'_C = q_A m_{AC} + q_B m_{BC} + q_C(1 - m_{AC} - m_{BC})$

◀ **23.11 Model of multidirectional migration among three populations, A, B, and C, with initial frequency of allele *a* equal to q_A, q_B, and q_C, respectively.** The proportion of a population made up of migrants from other populations is designated by *m*, where the subscripts represent the source and recipient populations. For example, m_{AC} represents the proportion of population C that consists of individuals that moved from A to C. The allelic frequencies in populations A, B, and C after migration are represented by q'_A, q'_B, and q'_C.

mutational events, and these alleles can be spread to new populations by migration, increasing the genetic variation within the recipient population.

Concepts

Migration causes changes in the allelic frequency of a population by introducing alleles from other populations. The magnitude of change due to migration depends on both the extent of migration and the difference in allelic frequencies between the source and the recipient populations. Migration decreases genetic differences between populations and increases genetic variation within populations.

Genetic Drift

The Hardy-Weinberg law assumes random mating in an infinitely large population; only when population size is infinite will the gametes carry genes that perfectly represent the parental gene pool. But no real population is infinitely large, and when population size is limited, the gametes that unite to form individuals of the next generation carry a sample of alleles present in the parental gene pool. Just by chance, the composition of this sample may deviate from that of the parental gene pool, and this deviation may cause allelic frequencies to change. The smaller the gametic sample, the greater the chance that its composition will deviate from that of the entire gene pool.

The role of chance in altering allelic frequencies is analogous to flipping a coin. Each time we flip a coin, we have a 50% chance of getting a head and a 50% chance of getting a tail. If we flip a coin 1000 times, the observed ratio of heads to tails will be very close to the expected 50:50 ratio. If, however, we flip a coin only 10 times, there is a good chance that we will obtain not exactly 5 heads and 5 tails, but rather maybe 7 heads and 3 tails or 8 tails and 2 heads. This kind of deviation from an expected ratio due to limited sample size is referred to as **sampling error.**

Sampling error occurs when gametes unite to produce progeny. Many organisms produce a large number of gametes but, when population size is small, a limited number of gametes unite to produce the individuals of the next generation. Chance influences which alleles are present in this limited sample and, in this way, sampling error may lead to changes in allelic frequency, which is called **genetic drift.** Because the deviations from the expected ratios are random, the direction of change is unpredictable. We can nevertheless predict the magnitude of the changes.

The magnitude of genetic drift The amount of sampling error resulting from genetic drift can be estimated from the variance in allelic frequency. Variance is a statistical measure that describes the degree of variability in a trait (see p. 644 in

Chapter 22). Suppose that we observe a large number of separate populations, each with N individuals and allelic frequencies of p and q. After one generation of random mating, genetic drift expressed in terms of the variance in allelic frequency among the populations (s_p^2) will be:

$$s_p^2 = \frac{pq}{2N} \tag{23.18}$$

The amount of change resulting from genetic drift (the variance in allelic frequency) is determined by two parameters: the allelic frequencies (p and q) and the population size (N). Genetic drift will be maximal when p and q are equal (each .5) and when the population size is small.

The effect of population size on genetic drift is illustrated by a study conducted by Luca Cavalli-Sforza and his colleagues. They studied variation in blood types among villagers in the Parm Valley of Italy, where the amount of migration between villages was limited. They found that variation in allelic frequency was greatest between small isolated villages in the upper valley but decreased between larger villages and towns farther down the valley. This result is exactly what we expect with genetic drift: there should be more genetic drift and thus more variation among villages when population size is small.

For ecological and demographic studies, population size is usually defined as the number of individuals in a group. The evolution of a gene pool depends, however, only on those individuals who contribute genes to the next generation. Population geneticists usually define population size as the equivalent number of breeding adults, the **effective population size** (N_e). Several factors determine the equivalent number of breeding adults. One factor is the sex ratio. When the numbers of males and females in the population are equal, the effective population size is simply the sum of reproducing males and females. When they are unequal, then the effective population size is:

$$N_e = \frac{4 \times n_{\text{males}} \times n_{\text{females}}}{n_{\text{males}} + n_{\text{females}}} \tag{23.19}$$

Table 23.3 gives the effective population size for a theoretical population of 100 individuals with different proportions of males and females. Notice that, when the number of males and females is unequal, the effective population size is smaller than it is when the number of males and females is the same. For example, when a population consists of 90 males and 10 females, the effective population size is only 36, and genetic drift will occur as though the actual population consisted of only 36 individuals, equally divided between males and females. A population with 90 males and 10 females has the same effective population size as a population with 10 males and 90 females—it makes no difference which sex is in excess.

Table 23.3	Effective population size (N_e) in theoretical populations of 100 individuals, each with a different sex ratio		
Sex Ratio*	**Number of Males**	**Number of Females**	**N_e**
1.00	50	50	100
3.00	75	25	75
0.33	25	75	75
9.00	90	10	36
0.10	10	90	36
99.00	99	1	3.96
0.01	1	99	3.96

*The sex ratio is the ratio of the number of males to the number of females.

The reason that the sex ratio influences genetic drift is that half the genes in the gene pool come from males and half come from females. When one sex is present in low numbers, genetic drift increases because half of the genes are coming from a small number of individuals. In a population consisting of 10 males and 90 females, the overall population size is relatively large (100), but only 10 males contribute half the genes to the next generation. Sampling error therefore affects the range of genes present in the male gametes, and chance will have a major effect on the allelic frequencies of the next generation.

Other factors that influence effective population size include variation between individuals in reproductive success, fluctuations in population size, the age structure of the population, and whether mating is random.

> ### Concepts
>
> Genetic drift is change in allelic frequency due to chance factors. The amount of change in allelic frequency due to genetic drift is inversely related to the effective population size (the equivalent number of breeding adults in a population). Effective population size decreases when there are unequal numbers of breeding males and females.

Causes of genetic drift All genetic drift arises from sampling error, but there are several different ways in which sampling error can arise. First, a population may be reduced in size for a number of generations because of limitations in space, food, or some other critical resource. Genetic drift in a small population for multiple generations can significantly affect the composition of a population's gene pool.

A second way that sampling error can arise is through the **founder effect,** which is due to the establishment of a population by a small number of individuals; the population of Tristan da Cuna, discussed in the introduction to this chapter, underwent a founder effect. Although a population may increase and become quite large, the genes carried by all its members are derived from the few genes originally present in the founders (assuming no migration or mutation). Chance events affecting which genes were present in the founders will have an important influence on the makeup of the entire population. The small number of founders of Tristan da Cuna included two sisters and a daughter who suffered from asthma; the high incidence of asthma on the island today can be traced to alleles carried by these founders.

A third way that genetic drift arises is through a **genetic bottleneck,** which develops when a population undergoes a drastic reduction in population size. A genetic bottleneck developed in northern elephant seals (◀FIGURE 23.12). Before 1800, thousands of elephant seals were found along the California coast, but the population was devastated by hunting between 1820 and 1880. By 1884, as few as 20 seals survived on a remote beach of Isla de Guadelupe west of Baja, California. Restrictions on hunting enacted by the United States and Mexico allowed the seals to recover, and there are now more than 30,000 seals in the population. All seals in the population today are genetically similar, because they have genes that were carried by the few survivors of the population bottleneck.

The effects of genetic drift Genetic drift has several important effects on the genetic composition of a population. First, it produces change in allelic frequencies within a popu-

◀23.12 **Northern elephant seals underwent a severe genetic bottleneck between 1820 and 1880.** Today, these seals have low levels of genetic variation. (Lisa Husar/DRK Photo.)

lation. Because drift is random, allelic frequency is just as likely to increase as it is to decrease and will wander with the passage of time (hence the name genetic drift). ◀ FIGURE 23.13 illustrates a computer simulation of genetic drift in five populations over 30 generations, starting with $q = .5$ and maintaining a constant population size of 10 males and 10 females. These allelic frequencies change randomly from generation to generation.

A second effect of genetic drift is to reduce genetic variation within populations. Through random change, an allele may eventually reach a frequency of either 1 or 0, at which point all individuals in the population are homozygous for one allele. When an allele has reached a frequency of 1, we say that it has reached **fixation.** Other alleles are lost (reach a frequency of 0) and can be restored only by migration from another population or by mutation. Fixation, then, leads to a loss of genetic variation within a population. This loss can be seen in northern elephant seals. Today, these seals have low levels of genetic variation; a study of 24 protein-encoding genes found no individual or population differences in these genes.

Given enough time, all small populations will become fixed for one allele. Which allele becomes fixed is random and is determined by the initial frequency of the allele. If the population begins with two alleles, each with a frequency of .5, both alleles have an equal probability of fixa-

tion. However, if one allele is initially common, it is more likely to become fixed.

A third effect of genetic drift is that different populations diverge genetically with time. In Figure 23.13, all five populations begin with the same allelic frequency ($q = .5$) but, because drift occurs randomly, the frequencies in different populations do not change in the same way, and so populations gradually acquire genetic differences. Notice that, although the variance in allelic frequency among the populations increases, the average allelic frequency remains basically the same. Eventually, all the populations reach fixation; some will become fixed for one allele and others will become fixed for the alternative allele. This divergence of populations through genetic drift is strikingly illustrated in the results of an experiment carried out by Peter Buri on fruit flies (◀ FIGURE 23.14).

The three results of genetic drift (allelic frequency change, loss of variation within populations, and genetic divergence between populations) occur simultaneously, and all result from sampling error. The first two results occur *within* populations, whereas the third occurs *between* populations.

Concepts

Genetic drift results from continuous small population size, founder effect (establishment of a population by a few founders), and bottleneck effect (population reduction). Genetic drift causes change in allelic frequencies within a population, loss of genetic variation through fixation of alleles, and genetic divergence between populations.

Natural Selection

A final process that brings about changes in allelic frequencies is natural selection, the differential reproduction of genotypes (see p. 657 in Chapter 22). Natural selection takes place when individuals with adaptive traits produce more offspring. If the adaptive traits have a genetic basis, they are inherited by the offspring and appear with greater frequency in the next generation. A trait that provides a reproductive advantage thereby increases over time, enabling populations to become better suited to their environments—to become better adapted. Natural selection is unique among evolutionary forces in that it promotes adaptation (◀ FIGURE 23.15).

Fitness and selection coefficient The effect of natural selection on the gene pool of a population depends on the fitness values of the genotypes in the population. **Fitness** is defined as the relative reproductive success of a genotype. Here the term *relative* is critically important: fitness is the

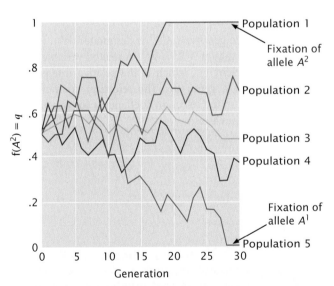

◀ **23.13 Genetic drift changes allelic frequencies within populations, leading to a reduction in genetic variation through fixation and genetic divergence among populations.** Shown here is a computer simulation of changes in the frequency of allele A^2 (q) in five different populations due to random genetic drift. Each population consists of 10 males and 10 females and begins with $q = .5$.

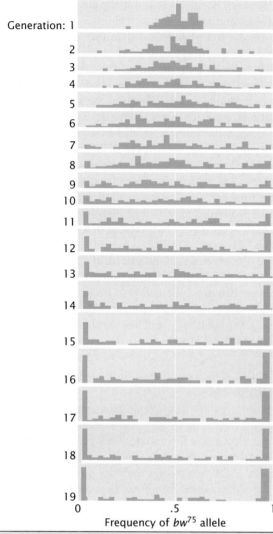

Generation: 1
2
3
4
5
6
7
8
9
10
11
12
13
14
15
16
17
18
19

0   .5   1
Frequency of *bw*[75] allele

Conclusion: As a result of genetic drift, allelic frequencies in the different populations diverged and often became fixed for one allele or the other.

◀23.14 Populations diverge in allelic frequency and become fixed for one allele as a result of genetic drift. In this experiment, Buri examined the frequency of two alleles (*bw*[75] and *bw*) that affect *Drosophila* eye color in 107 replicate populations. Each population consisted of 8 males and 8 females; each population began with the frequency of *bw*[75] equal to .5.

reproductive success of one genotype compared with the reproductive successes of other genotypes in the population.

Fitness (*W*) ranges from 0 to 1. Suppose the number of viable offspring produced by three genotypes is:

Genotypes:	A^1A^1	A^1A^2	A^2A^2
Mean number of offspring produced:	10	5	2

To calculate fitness for each genotype, we take the average number of offspring produced by a genotype and divide it by the mean number of offspring produced by the most prolific genotype:

$$\textbf{Fitness } (W): \quad \overset{\textbf{A}^1\textbf{A}^1}{W_{11} = \frac{10}{10} = 1.0} \quad \overset{\textbf{A}^1\textbf{A}^2}{W_{12} = \frac{5}{10} = .5}$$

$$\overset{\textbf{A}^2\textbf{A}^2}{W_{22} = \frac{2}{10} = .2} \tag{23.20}$$

The fitness of the genotype A^1A^1 is designated W_{11}, that of A^1A^2 is W_{12}, and that of A^2A^2 is W_{22}. A related variable is the **selection coefficient** (*s*), which is the relative intensity of selection against a genotype. The selection coefficient is equal to $1 - W$; so the selection coefficients for the preceding three genotypes are:

◀23.15 Natural selection produces adaptations, such as those seen in polar bears that inhabit the extreme Arctic environment. These bears blend into the snowy background, which helps them in hunting seals. The hairs of their fur stay erect even when wet, and thick layers of blubber provide insulation, which protects against subzero temperatures. Their digestive tracts are adapted to a seal-based carnivorous diet. (Tom and Pat Leeson/DRK Photo.)

$$A^1A^1 \quad A^1A^2 \quad A^2A^2$$

Selection coefficient $(1 - W)$: $s_{11} = 0$ $s_{12} = .5$ $s_{22} = .8$

We usually speak of selection for a particular genotype, but keep in mind that, when selection is *for* one genotype, selection is automatically *against* at least one other genotype.

Concepts

Natural selection is the differential reproduction of genotypes. It is measured as fitness, which is the reproductive success of a genotype compared with other genotypes in a population.

The general selection model Differential fitness among genotypes over time leads to changes in the frequencies of the genotypes, which, in turn, lead to changes in the frequencies of the alleles that make up the genotypes. We can predict the effect of natural selection on allelic frequencies by using a general selection model, which is outlined in Table 23.4. Use of this model requires knowledge of both the initial allelic frequencies and the fitness values of the genotypes. It assumes that mating is random and the only force acting on a population is natural selection.

We have defined fitness in terms of relative reproduction, but it will be easier to understand the logic behind the general selection model if we think of the fitness of the genotypes as differences in survival. It applies equally to fitnesses representing differential reproduction.

Let's apply the general selection model outlined in Table 23.4. Imagine a flock of sparrows overwintering in Rochester, New York. Assume that we can determine the genotypes for a locus that affects the ability of the birds to survive the winter; perhaps the genes at this locus determine the amount of fat that a bird accumulates before the onset of winter. For genotypes A^1A^1, A^1A^2, and A^2A^2, p rep-

resents the frequency of A^1 and q represents the frequency of A^2. On the first line of the table, we record the initial genotypic frequencies before selection has acted, before the onset of winter. If mating has been random (an assumption of the model), the genotypes will have the Hardy-Weinberg equilibrium frequencies of p^2, $2pq$, and q^2. On the second row of the table, we put the fitness values of the corresponding genotypes. Some of the birds die in the winter; so here the fitness values represent the relative survival of the three genotypes. The proportion of the population represented by each genotype after selection is obtained by multiplying the initial genotypic frequency times its fitness (third row of Table 23.4). Now the genotypes are no longer in Hardy-Weinberg equilibrium.

The mean fitness $(\overline{W})$ of the population is the sum of the proportionate contributions of the three genotypes:

$$\overline{W} = p^2W_{11} + 2pqW_{12} + q^2W_{22} \tag{23.21}$$

The mean fitness $\overline{W}$ is the average fitness of all individuals in the population and allows the frequencies of the genotypes after selection to be obtained. In our flock of birds, these frequencies will be those of the three genotypes after the winter mortality. The frequency of a genotype after selection will be equal to its proportionate contribution divided by the mean fitness of the population ($p^2W_{11}/\overline{W}$ for genotype A^1A^1, $2pqW_{12}/\overline{W}$ for genotype A^1A^2, and $q^2W_{22}/\overline{W}$ for genotype A^2A^2), as shown in the fourth line of Table 23.4. When the new genotypic frequencies have been calculated, the new allelic frequency of A^1 (p') can be determined by using the now-familiar formula (Equation 23.4):

$$p' = f(A^1) = f(A^1A^1) + \tfrac{1}{2}f(A^1A^2)$$

and that of q' can be obtained by subtraction:

$$q' = 1 - p'$$

Table 23.4	Method for determining changes in allelic frequency due to selection		
	A^1A^1	**A^1A^2**	**A^2A^2**
Initial genotypic frequencies	p^2	$2pq$	q^2
Fitnesses	W_{11}	W_{12}	W_{22}
Proportionate contribution of genotypes to population	p^2W_{11}	$2pqW_{12}$	q^2W_{22}
Relative genotypic frequency after selection	$\dfrac{p^2W_{11}}{\overline{W}}$	$\dfrac{2pqW_{12}}{\overline{W}}$	$\dfrac{q^2W_{22}}{\overline{W}}$

Note: $\overline{W} = p^2W_{11} + 2pqW_{12} + q^2W_{22}$
Allelic frequencies after selection: $p' = f(A^1) = f(A^1A^1) + \tfrac{1}{2}f(A^1A^2)$
$q' = 1 - p$

The last step is to determine the genotypic frequencies in the *next* generation. In regard to our birds, these genotypic frequencies are those of the offspring of the birds that survived the winter. If the survivors mate randomly, the genotypic frequencies in the next generation will be p'^2, $2p'q'$, and q'^2.

The general selection model can be used to calculate the allelic frequencies after any type of selection. It is also possible to work out formulas for determining the change in allelic frequency when selection is against recessive, dominant, and codominant traits, as well as traits in which the heterozygote has highest fitness (Table 23.5).

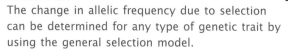

The change in allelic frequency due to selection can be determined for any type of genetic trait by using the general selection model.

The results of selection The results of selection depend on the relative fitnesses of the genotypes. If we have three genotypes (A^1A^1, A^1A^2, and A^2A^2) with fitnesses W_{11}, W_{12}, and W_{22}, we can identify six different types of natural selection (Table 23.6). In type 1 selection, a dominant allele A^1 confers a fitness advantage; in this case, the fitnesses of genotypes A^1A^1 and A^1A^2 are equal and higher than the fitness of A^2A^2 ($W_{11} = W_{12} > W_{22}$). Because the heterozygote and the A^1A^1 homozygote both have copies of the A^1 allele and produce more offspring than the A^2A^2 homozygote does, the frequency of the A^1 allele will increase over time, whereas the frequency of the A^2 allele will decrease. This form of selection, in which one allele or trait is favored over another, is termed **directional selection.**

Type 2 selection (Table 23.6) is directional selection against a dominant allele A^1 ($W_{11} = W_{12} < W_{22}$). In this case, the A^2 allele increases and the A^1 allele decreases. Type

3 and type 4 selection also are directional selection, but in these cases there is incomplete dominance and the heterozygote has a fitness that is intermediate between the two homozygotes ($W_{11} > W_{12} > W_{22}$ for type 3; $W_{11} < W_{12} < W_{22}$ for type 4). When A^1A^1 has the highest fitness (type 3), over time the A^1 allele increases and the A^2 allele decreases. When A^2A^2 has the highest fitness (type 4), over time the A^2 allele increases and the A^1 allele decreases. Eventually, directional selection leads to fixation of the favored allele and elimination of the other allele, as long as no other evolutionary forces act on the population.

Two types of selection (types 5 and 6) are special situations that lead to equilibrium, where there is no further change in allelic frequency. Type 5 selection is referred to as **overdominance** or heterozygote advantage. Here, the heterozygote has higher fitness than the fitnesses of the two homozygotes ($W_{11} < W_{12} > W_{22}$). With overdominance, both alleles are favored in the heterozygote, and neither allele is eliminated from the population. Initially, the allelic frequencies may change because one homozygote has higher fitness than the other; the direction of change will depend on the relative fitness values of the two homozygotes. The allelic frequencies change with overdominant selection until a stable equilibrium is reached, at which point there is no further change. The allelic frequency at equilibrium ($\hat{q}$) depends on the relative fitnesses (usually expressed as selection coefficients) of the two homozygotes:

$$\hat{q} = f(A^2) = \frac{s_{11}}{s_{11} + s_{22}} \tag{23.22}$$

where s_{11} represents the selection coefficient of the A^1A^1 homozygote and s_{22} represents the selection coefficient of the A^2A^2 homozygote.

The last type of selection (type 6) is **underdominance,** in which the heterozygote has lower fitness than both

Table 23.5	Formulas for calculating change in allelic frequencies with different types of selection			
	Fitness Values			
Type of Selection	**A^1A^1**	**A^1A^2**	**A^2A^2**	**Change in q**
Selection against a recessive trait	1	1	$1-s$	$\dfrac{-spq^2}{1-sq^2}$
Selection against a dominant trait	1	$1-s$	$1-s$	$\dfrac{-spq^2}{1-s+sq^2}$
Selection against a trait with no dominance	1	$1-\frac{1}{2}s$	$1-s$	$\dfrac{-\frac{1}{2}spq}{1-sq}$
Selection against both homozygotes (overdominance)	$1-s_{11}$	1	$1-s_{22}$	$\dfrac{pq(s_{11}p-s_{22}q)}{1-s_{11}p^2-s_{22}q^2}$

Table 23.6 Types of natural selection

Type	Fitness Relation	Form of Selection	Result
1	$W_{11} = W_{12} > W_{22}$	Directional selection against recessive allele A^2	A^1 increases, A^2 decreases
2	$W_{11} = W_{12} < W_{22}$	Directional selection against dominant allele A^1	A^2 increases, A^1 decreases
3	$W_{11} > W_{12} > W_{22}$	Directional selection against incompletely dominant allele A^2	A^1 increases, A^2 decreases
4	$W_{11} < W_{12} < W_{22}$	Directional selection against incompletely dominant allele A^1	A^2 increases, A^1 decreases
5	$W_{11} < W_{12} > W_{22}$	Overdominance	Stable equilibrium, both alleles maintained
6	$W_{11} > W_{12} < W_{22}$	Underdominance	Unstable equilibrium

Note: W_{11}, W_{12}, and W_{22} represent the fitnesses of genotypes A^1A^1, A^1A^2, and A^2A^2, respectively.

homozygotes ($W_{11} > W_{12} < W_{22}$). Underdominance leads to an *unstable* equilibrium; here allelic frequencies will not change as long as they are at equilibrium but, if they are disturbed from the equilibrium point by some other evolutionary force, they will move away from equilibrium until one allele eventually becomes fixed.

lower rate than that of dominant alleles. Recessive alleles increase at the lowest rate, because only the homozygote is favored by selection.

The rate at which selection changes allelic frequencies also depends on the allelic frequency itself. If an allele (A^2) is lethal and recessive, $W_{11} = W_{12} = 1$, whereas $W_{22} = 0$. The

Concepts

Natural selection changes allelic frequencies; the direction and magnitude of change depends on the intensity of selection, the dominance relations of the alleles, and the allelic frequencies. Directional selection favors one allele over another and eventually leads to fixation of the favored allele. Overdominance leads to a stable equilibrium with maintenance of both alleles in the population. Underdominance produces an unstable equilibrium because the heterozygote has lower fitness than those of the two homozygotes.

The rate of change in allelic frequency due to natural selection

The rate at which an allele changes in frequency owing to selection depends on the intensity of selection and the dominance relations among the genotypes (◀FIGURE 23.16). Under directional selection, dominant alleles will increase much more rapidly than recessive alleles, because homozygotes and heterozygotes are favored. With incomplete dominance, the heterozygote has a selective advantage, but not as much as the homozygote; so incompletely dominant alleles increase in frequency at a

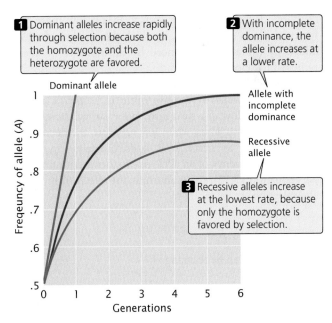

1 Dominant alleles increase rapidly through selection because both the homozygote and the heterozygote are favored.

2 With incomplete dominance, the allele increases at a lower rate.

3 Recessive alleles increase at the lowest rate, because only the homozygote is favored by selection.

◀23.16 The rate of change in allelic frequency due to selection depends on the dominance relations among the genotypes. Here, change in the frequency of an allele is shown for different types of dominance with a constant selection coefficient.

frequency of the A^2 allele will decrease over time (because the A^2A^2 homozygote produces no offspring), and the rate of decrease will be proportional to the frequency of the recessive allele. When the frequency of the allele is high, the change in each generation is relatively large but, as the frequency of the allele drops, a higher proportion of the alleles are in the heterozygous genotypes, where they are immune to the action of natural selection (the heterozygotes have the same phenotype as the favored homozygote). Thus, selection against a rare recessive allele is very inefficient and its removal from the population is slow.

The relation between the frequency of a recessive allele and its rate of change under natural selection has an important implication. Some people believe that the medical treatment of patients with rare recessive diseases will cause the disease gene to increase, eventually leading to degeneration of the human gene pool. This mistaken belief was the basis of eugenic laws that were passed in the early part of the twentieth century prohibiting the marriage of persons with certain genetic conditions and allowing the involuntary sterilization of others. However, most copies of rare recessive alleles are present in heterozygotes, and selection against the homozygotes will have little effect on the frequency of a recessive allele. Thus whether the homozygotes reproduce or not has little effect on the frequency of the disorder.

Mutation and natural selection Recurrent mutation and natural selection act as opposing forces on detrimental alleles; mutation increases their frequency and natural selection decreases their frequency. Eventually, these two forces reach an equilibrium, in which the number of alleles added by mutation is balanced by the number of alleles removed by selection.

Table 23.5 shows that the change in allelic frequency due to selection against a recessive allele is $-spq^2/(1 - sq^2)$. When q is very low, q^2 is near zero; so $1 - sq^2$ will be approximately $1 - 0 = 1$. Thus, when q is very low, the decrease in frequency due to selection is approximately $-spq^2$. The increase in frequency of an allele due to forward mutations is μp (Equation 23.12). At equilibrium, the effects of mutation and selection are balanced; so

$$spq^2 = \mu p$$

This equation can be rearranged:

$$q^2 = \frac{\mu p}{sp} = \frac{\mu}{s}$$

Taking the square root of each side, we get

$$\hat{q} = \sqrt{\frac{\mu}{s}} \qquad (23.23)$$

The frequency of the allele at equilibrium ($\hat{q}$) is therefore equal to the square root of the mutation rate divided by the selection coefficient. With the use of the equation for selection acting on a dominant allele (see Table 23.5) and similar reasoning, the frequency of a dominant allele at equilibrium can be shown to be

$$\hat{q} = \frac{\mu}{s} \qquad (23.24)$$

Achondroplasia (discussed in Chapter 17) is a common type of human dwarfism that results from a dominant gene. People with this condition are fertile, although they produce only about 74% as many children as are produced by people without achondroplasia. The fitness of people with achondroplasia therefore averages .74, and the selection coefficient (s) is $1 - W$, or .26. If we assume that the mutation rate for achondroplasia is about 3×10^{-5} (a typical mutation rate in humans), then we can predict that the equilibrium frequency for the achondroplasia allele will be $\hat{q} = (.00003/.26) = .0001153$. This frequency is close to the actual frequency of the disease.

Connecting Concepts

The General Effects of Evolutionary Forces

You now know that four processes bring about change in the allelic frequencies of a population: mutation, migration, genetic drift, and natural selection. Their short- and long-term effects on allelic frequencies are summarized in Table 23.7. In some cases, these changes continue until one allele is eliminated and the other becomes fixed in the population. Genetic drift and directional selection will eventually result in fixation, provided these forces are the only ones acting on a population. With the other evolutionary forces, allelic frequencies change until an equilibrium point is reached, and then there is no additional change in allelic frequency. Mutation, migration, and some forms of natural selection can lead to stable equilibria (see Table 23.7).

Table 23.7	Effects of different evolutionary forces on allelic frequencies within populations	
Force	**Short-Term Effect**	**Long-Term Effect**
Mutation	Change in allelic frequency	Equilibrium reached between forward and reverse mutations
Migration	Change in allelic frequency	Equilibrium reached when allelic frequencies of source and recipient population are equal
Genetic drift	Change in allelic frequency	Fixation of one allele
Natural selection	Change in allelic frequency	Directional selection: fixation of one allele Overdominant selection: equilibrium reached

The different evolutionary forces affect both genetic variation within populations and genetic divergence between populations. Evolutionary forces that maintain or increase genetic variation within populations are listed in the upper-left quadrant of ◀ FIGURE 23.17. These forces include some types of natural selection, such as overdominance in which both alleles are favored. Mutation and migration also increase genetic variation within populations because they introduce new alleles to the population. Evolutionary forces that decrease genetic variation within populations are listed in the lower-left quadrant of Figure 23.17. These forces include genetic drift, which decreases variation through fixation of alleles, and some forms of natural selection such as directional selection.

The various evolutionary forces also affect the amount of genetic divergence between populations. Natural selection increases divergence among populations if different alleles are favored in the different populations, but it can also *decrease* divergence between populations by favoring the same allele in the different populations. Mutation almost always increases divergence between populations because different mutations arise in each population. Genetic drift also increases divergence between populations because changes in allelic frequencies due to drift are random and are likely to change in different directions in separate populations. Migration, on the other hand, decreases divergence between populations because it makes populations similar in their genetic composition.

Migration and genetic drift act in opposite directions: migration increases genetic variation within populations and decreases divergence between populations, whereas genetic drift decreases genetic variation within populations and increases divergence among populations. Mutation increases both variation within populations and divergence between populations. Natural selection can either increase or decrease variation within populations, and it can increase or decrease divergence between populations.

It is important to keep in mind that real populations are simultaneously affected by many evolutionary forces. This discussion has examined the effects of mutation, migration, genetic drift, and natural selection in isolation so that the influence of each process would be clear. However, in the real world, populations are commonly affected by several evolutionary forces at the same time, and evolution results from the complex interplay of numerous processes.

www.whfreeman.com/pierce Web addresses for a number of resources on evolution

	Within populations	Between populations
Increase genetic variation	**Mutation Migration Some types of natural selection**	**Mutation Genetic drift Some types of natural selection**
Decrease genetic variation	**Genetic drift Some types of natural selection**	**Migration Some types of natural selection**

◀ **23.17** **Mutation, migration, genetic drift, and natural selection have different effects on genetic variation within populations and on genetic divergence between populations.**

Molecular Evolution

For many years, it was not possible to examine genes directly, and evolutionary biology was confined largely to the study of how phenotypes change with the passage of time. The tremendous advances in molecular genetics in recent years have made it possible to investigate evolutionary change directly by analyzing protein and nucleic acid sequences. These molecular data offer a number of advantages for studying the process and pattern of evolution:

1. **Molecular data are genetic.** Evolution results from genetic change over time. Anatomical, behavioral, and physiological traits often have a genetic basis, but the relation between the underlying genes and the trait may be complex. Protein and nucleic acid sequence variation has a clear genetic basis that is easy to interpret.

2. **Molecular methods can be used with all organisms.** Early studies of population genetics relied on simple genetic traits such as human blood types or banding patterns in snails, which are restricted to a small group of organisms. However, all living organisms have proteins and nucleic acids; so molecular data can be collected from any organism.

3. **Molecular methods can be applied to a huge amount of genetic variation.** An enormous amount of data can be accessed by molecular methods. The human genome, for example, contains more than 3 billion base pairs of DNA, which constitutes a large pool of information about our evolution.

4. **All organisms can be compared with the use of some molecular data.** Trying to assess the evolutionary history of distantly related organisms is often difficult because they have few characteristics in common. The evolutionary relationships between angiosperms were traditionally assessed by comparing floral anatomy, whereas the evolutionary relationships of bacteria were determined by their nutritional and staining properties. Because plants and bacteria have so few structural characteristics in common, evaluating how they are related to one another was difficult in the past. All organisms have certain molecular traits in common, such as ribosomal RNA sequences and some fundamental proteins. These molecules offer a valid basis for comparisons among all organisms.

5. **Molecular data are quantifiable.** Protein and nucleic acid sequence data are precise, accurate, and easy to quantify, which facilitates the objective assessment of evolutionary relationships.

6. **Molecular data often provide information about the process of evolution.** Molecular data can reveal important clues about the process of evolution. For example, the results of a study of DNA sequences have revealed that one type of insecticide resistance in mosquitoes probably arose from a single mutation that subsequently spread throughout the world.

7. **The database of molecular information is large and growing.** Today, this database of DNA and protein sequences can be used for making evolutionary comparisons and inferring mechanisms of evolution.

Studies of molecular evolution fall into three primary areas. First, much past research has focused on determining the extent and causes of genetic variation in natural populations. Molecular techniques allow these matters to be addressed directly by examining sequence variation in proteins and DNA. A second area of research examines molecular processes that influence evolutionary events, and the results of these studies have elucidated new mechanisms and processes of evolution that were not suspected before the application of molecular techniques to evolutionary biology. A third area of research in molecular evolution applies molecular techniques to constructing **phylogenies** (evolutionary trees) of various groups of organisms. A detailed evolutionary history is found in the DNA sequences of every organism, and molecular techniques allow this history to be read.

> **Concepts**
>
> Molecular techniques and data offer a number of advantages for evolutionary studies. Molecular data (1) are genetic in nature and can be investigated in all organisms; (2) provide potentially large data sets; (3) allow all organisms to be compared, by using the same characteristics; (4) are easily quantifiable; and (5) provide information about the process of evolution.

Protein Variation

The study of the amounts and kinds of genetic variation in natural populations is central to the study of evolution. For many traits, a complex interaction of many genes and environmental factors determines the phenotype, and assessing the amount of genetic variation by examining phenotypic variation was difficult. Early population geneticists were forced to rely on the phenotypic traits that had a simple genetic basis, such as human blood types or spotting patterns in butterflies (◀ FIGURE 23.18). The initial breakthrough that first allowed the direct examination of molecular evolution was the application of electrophoresis (see Figure 18.4) to population studies. This technique separates macromolecules, such as proteins or nucleic acids, on the basis of their size and charge. In 1966, Richard Lewontin and John Hubby extracted proteins from wild fruit flies, separated the proteins by electrophoresis, and stained for specific enzymes. Examining the pattern of bands on gels enabled them to assign genotypes to individual flies and to quantify the amount of genetic variation in natural populations. In the same year, Harry Harris quantified genetic variation in human populations by using the same technique. Protein variation has now been examined in hundreds of different species by using protein electrophoresis (◀ FIGURE 23.19).

Normal homozygotes

Heterozygotes

Recessive bimacula phenotype

◀ 23.18 **Early population geneticists were forced to rely on the phenotypic traits that had a simple genetic basis.** Variation in the spotting patterns of the butterfly *Panaxia dominula* is an example.

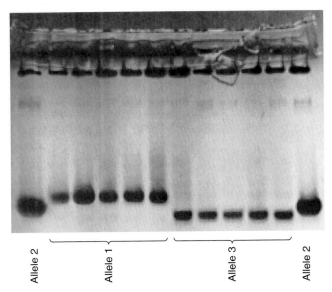

◀ 23.19 **Molecular variation in proteins is revealed by electrophoresis.** Tissue samples from *Drosophila pseudoobscura* have been subjected to electrophoresis and stained for esterase. Esterases encoded by different alleles migrate different distances. Shown on the gel are homozygotes for three different alleles.

Measures of genetic variation

The amount of genetic variation in populations is commonly measured by two parameters. The **proportion of polymorphic loci** is the proportion of examined loci in which more than one allele is present in a population. If we examined 30 different loci and found two or more alleles present at 15 of these loci, the percentage of polymorphic loci would be $15/30 = 0.5$. The **expected heterozygosity** is the proportion of individuals that are expected to be heterozygous at a locus under the Hardy-Weinberg conditions, which is $2pq$ when there are two alleles present in the population. The expected heterozygosity is often preferred to the observed heterozygosity because expected heterozygosity is independent of the breeding system of an organism. For example, if a species self-fertilizes, it may have little or no heterozygosity but still have considerable genetic variation, which will be detected by the expected heterozygosity. Expected heterozygosity is typically calculated for a number of loci and is then averaged over all the loci examined.

The percentage of polymorphic loci and the expected heterozygosity have been determined by protein electrophoresis for a number of species (Table 23.8). About one-third of all protein loci are polymorphic, and expected heterozygosity averages about 10%, although there is considerable diversity among species. These measures actually underestimate the true amount of genetic variation, though, because protein electrophoresis does not detect some amino acid substitutions; nor does it detect genetic variation in DNA that does not alter the amino acids of a protein (synonymous codons and variation in noncoding regions of the DNA).

Explanations for protein variation

By the late 1970s, geneticists recognized that most populations possess large amounts of genetic variation, although the evolutionary significance of this fact was not at all clear. Two opposing hypotheses arose to account for the presence of the extensive molecular variation in proteins. The **neutral-mutation hypothesis** proposed that the molecular variation revealed by protein electrophoresis is adaptively neutral; that is, individuals with different molecular variants have equal fitness. This hypothesis does not propose that the proteins are functionless; rather, it suggests that most variants revealed by protein electrophoresis are functionally equivalent. Because these variants are functionally equivalent, natural selection does not differentiate between them, and their evolution is shaped largely by the random processes of genetic drift and mutation. The neutral-mutation hypothesis accepts that

Table 23.8	Proportion of polymorphic loci and heterozygosity for different organisms, as determined by protein electrophoresis				
Group	Number of Species	Proportion of Polymorphic Loci		Heterozygosity	
		Mean	SD*	Mean	SD*
Plants	15	0.26	0.17	0.07	0.07
Invertebrates (excluding insects)	28	0.40	0.28	0.10	0.07
Insects (excluding *Drosophila*)	23	0.33	0.20	0.07	0.08
Drosophila	32	0.43	0.13	0.14	0.05
Fish	61	0.15	0.01	0.05	0.04
Amphibians	12	0.27	0.13	0.08	0.04
Reptiles	15	0.22	0.13	0.05	0.02
Birds	10	0.15	0.11	0.05	0.04
Mammals	46	0.15	0.10	0.04	0.02

* SD, standard deviation from the mean.
Source: After L. E. Mettler, T. G. Gregg, and H. E. Schaffer, *Population Genetics and Evolution*, 2d ed. (Englewood Cliffs, NJ: Prentice Hall, 1988), Table 9.2. Original data from E. Nevo, Genetic variation in natural populations: patterns and theory, *Theoretical Population Biology* 13(1978):121–177.

natural selection is an important force in evolution, but views selection as a process that favors the "best" allele while eliminating others. It proposes that, when selection is important, there will be *little* genetic variation.

The **balance hypothesis** proposes, on the other hand, that the genetic variation in natural populations is maintained by selection that favors variation (balancing selection). Overdominance, in which the heterozygote has higher fitness than that of either homozygote, is one type of balancing selection. Under this hypothesis, the molecular variants are not physiologically equivalent and do not have the same fitness. Instead, genetic variation within natural populations is shaped largely by selection, and, when selection is important, there will be *much* variation.

Many attempts to prove one hypothesis or the other failed, because precisely how much variation was actually present was not clear (remember that protein electrophoresis detects only *some* genetic variation) and because both hypotheses are capable of explaining many different patterns of genetic variation. The controversy over the forces that control variation revealed by protein electrophoresis continues today, but the results of more-recent studies that provide direct information about DNA sequence variation demonstrate that much variation at the level of DNA has little obvious effect on the phenotype and therefore is likely to be neutral.

Concepts

The application of electrophoresis to the study of protein variation in natural populations revealed that most organisms possess large amounts of genetic variation. The neutral-mutation hypothesis proposes that most molecular variation is neutral with regard to natural selection and is shaped largely by mutation and genetic drift. The balance hypothesis proposes that genetic variation is maintained by balancing selection.

www.whfreeman.com/pierce   Information on genetic variation in natural populations

DNA Sequence Variation

The development of techniques for isolating, restricting, and sequencing DNA in the 1970s and 1980s provided powerful tools for detecting, quantifying, and investigating genetic variation. The application of these techniques has provided a detailed view of molecular variation.

Restriction enzymes are one tool that can be used to detect genetic variation in DNA and examine patterns of genetic variation in nature. Each restriction enzyme recognizes and cuts a particular sequence of DNA nucleotides, known as that enzyme's restriction site (see Chapter 18). Variation in the presence of a restriction site is called a restriction fragment length polymorphism (RFLP; see Figure 18.26). Each restriction enzyme recognizes a limited number of nucleotide sites in a particular piece of DNA but, if a number of different restriction enzymes are used and the sites recognized by the enzymes are assumed to be random sequences, RFLPs can be used to estimate the amount of variation in the DNA and the proportion of nucleotides that differ between organisms.

Methods for determining the complete nucleotide sequences of DNA fragments (see pp. 553–555 in Chapter 19) provide the most detailed evolutionary information, although they are both time consuming and expensive. DNA sequencing in evolutionary studies is therefore usually limited to a few individuals or to short sequences. Nevertheless, the high resolution of information provided by sequencing is often invaluable for understanding molecular processes that influence evolution and for determining phylogenies of closely related organisms. For example, DNA sequencing has been used to study the evolution of human immunodeficiency virus (HIV), the virus that causes AIDS. Like many other RNA viruses, HIV evolves rapidly, often changing its sequences within a single host over a period of several years. Evolutionary comparisons of HIV sequences in a dentist and seven of his patients who had AIDS demonstrated that five of the patients contracted AIDS from the dentist, whereas the other two patients probably acquired their HIV infection elsewhere.

Concepts

Restriction fragment length polymorphisms and DNA sequencing can be used to directly examine genetic variation.

Molecular Evolution of HIV in a Florida Dental Practice

In July 1990, the U.S. Center for Disease Control (CDC) reported that a young woman in Florida (later identified as Kimberly Bergalis) had become HIV positive after undergoing an invasive dental procedure performed by a dentist who had AIDS. Bergalis had no known risk factors for HIV infection and no known contact with other HIV-positive persons. The CDC acknowledged that Bergalis might have acquired the infection from her dentist. Subsequently, the dentist wrote to all of his patients, suggesting that they be tested for HIV infection. By 1992, 7 of the dentist's patients had tested positive for HIV, and this number eventually increased to 10.

Originally diagnosed with HIV infection in 1986, the dentist began to develop symptoms of AIDS in 1987 but continued to practice dentistry for another 2 years. All of his HIV-positive patients had received invasive dental procedures, such as root canals and tooth extractions, in the period when the dentist was infected. Among the seven patients originally studied by the CDC (patients A–G, Table 23.9), two had known risk factors for HIV infection (intravenous drug use, homosexual behavior, or sexual relations with HIV-infected persons), and a third had possible but unconfirmed risk factors.

To determine whether the dentist had infected his patients, the CDC conducted a study of the molecular evolution of HIV isolates from the dentist and the patients. HIV undergoes rapid evolution, making it possible to trace the path of its transmission. This rapid evolution also allows HIV to develop drug resistance quickly, making the development of a treatment for AIDS difficult.

Blood specimens were collected from the dentist, the patients, and a group of 35 local controls (other HIV-infected

Table 23.9			HIV-positive persons included in study of HIV isolates from a Florida dental practice	
			Average Differences in DNA Sequences (%)	
Person	**Sex**	**Known Risk Factors**	**From HIV from Dentist**	**From HIV from Controls**
Dentist	M	Yes		11.0
Patient A	F	No	3.4	10.9
Patient B	F	No	4.4	11.2
Patient C	M	No	3.4	11.1
Patient E	F	No	3.4	10.8
Patient G	M	No	4.9	11.8
Patient D	M	Yes	13.6	13.1
Patient F	M	Yes	10.7	11.9

Source: After C. Ou, et al., *Science* 256(1992):1165–1171, Table 1.

people who lived within 90 miles of the dental practice but who had no known contact with the dentist). DNA was extracted from white blood cells, and a 680-bp fragment of the *envelope* gene of the virus was amplified by PCR (see pp. 528–529 in Chapter 18). The fragments from the dentist, the patients, and the local controls were then sequenced and compared.

The divergence between the viral sequences taken from the dentist, the seven patients, and the controls is shown in Table 23.9. Viral DNA taken from patients with no confirmed risk factors (patients A, B, C, E, and G) differed from the dentist's viral DNA by 3.4% to 4.9%, whereas the viral DNA from the controls differed from the dentist's by an average of 11%. The viral sequences collected from five patients (A, B, C, E, and G) were more closely related to the viral sequences collected from the dentist than to viral sequences from the general population, strongly suggesting that these patients acquired their HIV infection from the dentist. The viral isolates from patients D and F (patients with confirmed risk factors), however, differed from that of the dentist by 10.7% and 13.6%, suggesting that these two patients did not acquire their infection from the dentist.

A phylogenetic tree depicting the evolutionary relationships of the viral sequences (◀FIGURE 23.20) confirmed that the virus taken from the dentist had a close evolutionary relationship to viruses taken from patients A, B, C, E, and G. The viruses from patients D and F, with known risk factors, were no more similar to the virus from the dentist than to viruses from local controls, indicating that the dentist most likely infected five of his patients, whereas the other two patients probably acquired their infections elsewhere. Of three additional HIV-positive patients that have been identified since 1992, only one has viral sequences that are closely related to those from the dentist.

The study of HIV isolates from the dentist and his patients provides an excellent example of the relevance of molecular evolutionary studies to real-world problems. How the dentist infected his patients during their visits to his office remains a mystery, but this case is clearly unusual. A study of almost 16,000 patients treated by HIV-positive health-care workers failed to find a single case of confirmed transmission of HIV from the health-care worker to the patient.

www.whfreeman.com/pierce   Web site of the U.S. Center for Disease Control

Patterns of Molecular Variation

The results of molecular studies of numerous genes have demonstrated that different genes and different parts of the same gene often evolve at different rates. Rates of evolutionary change in nucleotide sequences are usually measured as the rate of nucleotide substitution, which is the number of substitutions taking place per nucleotide site per year. To calculate the rate of nucleotide substitution, we begin by looking at homologous sequences from different organisms. We compare the homologous sequences and

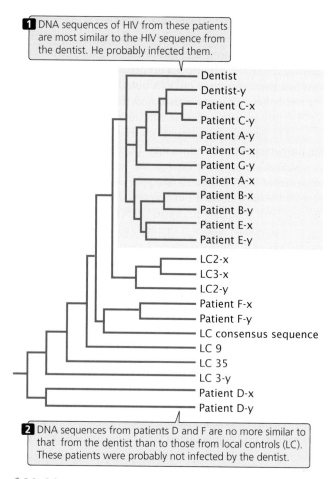

1 DNA sequences of HIV from these patients are most similar to the HIV sequence from the dentist. He probably infected them.

Dentist
Dentist-y
Patient C-x
Patient C-y
Patient A-y
Patient G-x
Patient G-y
Patient A-x
Patient B-x
Patient B-y
Patient E-x
Patient E-y
LC2-x
LC3-x
LC2-y
Patient F-x
Patient F-y
LC consensus sequence
LC 9
LC 35
LC 3-y
Patient D-x
Patient D-y

2 DNA sequences from patients D and F are no more similar to that from the dentist than to those from local controls (LC). These patients were probably not infected by the dentist.

◀ **23.20 Evolutionary tree showing the relationships of HIV isolates from a dentist, seven of his patients (A through G), and other HIV-positive persons from the same region (local controls, LC).** The letters x and y represent different isolates from the same patient. The phylogeny is based on DNA sequences taken from the *envelope* gene of the virus. Viral sequences from patients A, B, C, E, and G cluster with those of the dentist, indicating a close evolutionary relationship. Sequences from patients D and F, along with those of local controls, are more distantly related. [C. Ou, et al. Molecular epidemiology of HIV transmission in a dental practice, *Science* 256(1992): 1167.]

determine the number of nucleotides that differ between the two sequences. We might compare the growth-hormone sequences for mice and rats, which diverged from a common ancestor some 15 million years ago. From the number of different nucleotides in their growth-hormone genes, we compute the number of nucleotide substitutions that must have taken place since they diverged. Because the same site may have mutated more than once, the number of nucleotide substitutions is larger than the number of nucleotide differences in two sequences; so special mathematical methods have been developed for inferring the actual number of substitutions likely to have taken place.

When we have the number of nucleotide substitutions per nucleotide site, we divide by the amount of evolutionary time that separates the two organisms (usually obtained from the fossil record) to obtain an overall rate of nucleotide substitution. For the mouse and rat growth-hormone gene, the overall rate of nucleotide substitution is approximately 8×10^{-9} substitutions per site per year.

Nucleotide changes in a gene that alter the amino acid sequence of a protein are referred to as nonsynonymous substitutions. Nucleotide changes, particularly those at the third position of the codon, that do not alter the amino acid sequence are called synonymous substitutions. The rate of nonsynonymous substitution varies widely among mammalian genes. The rate for the α-actin protein is only 0.01×10^{-9} substitutions per site per year, whereas the rate for interferon γ is 2.79×10^{-9}, about 1000 times as high. The rate of synonymous substitution also varies among genes, but not to the extent of variation in the nonsynonymous rate. For most protein-encoding genes, the synonymous rate of change is considerably higher than the nonsynonymous rate because synonymous mutations are tolerated by natural selection (Table 23.10). Nonsynonymous mutations, on the other hand, alter the amino acid sequence of the protein and in many cases are detrimental to the fitness of the organism, so most of these mutations are eliminated by natural selection.

Different parts of a gene also evolve at different rates, with the highest rates of substitutions in regions of the gene that have the least effect on function, such as the third position of a codon, flanking regions, and introns (◀ FIGURE 23.21). The 5′ and 3′ flanking regions of genes are not transcribed into RNA, and therefore substitutions in these regions do not alter the amino acid sequence of the protein, although they may affect gene expression (see Chapter 16). Rates of substitution in introns are nearly as high. Although these nucleotides do not encode amino acids, introns must be spliced out of the pre-mRNA for a functional protein to be produced, and particular sequences are required at the 5′ splice site, 3′ splice site, and branch point for correct splicing (see Chapter 14).

Substitution rates are somewhat lower in the 5′ and 3′ untranslated regions of a gene. These regions are transcribed into RNA but do not encode amino acids. The 5′ untranslated region contains the ribosome-binding site, which is essential for translation, and the 3′ untranslated region contains sequences that may function in regulating mRNA stability and translation; so substitutions in these regions may have deleterious effects on organismal fitness and will not be tolerated.

Table 23.10	Rates of nonsynonymous and synonymous substitutions in mammalian genes based on human–rodent comparisons	
Gene	**Nonsynonymous Rate (per Site per 10^9 Years)**	**Synonymous Rate (per Site per 10^9 Years)**
α-Actin	0.01	3.68
β-Actin	0.03	3.13
Albumin	0.91	6.63
Aldolase A	0.07	3.59
Apoprotein E	0.98	4.04
Creatine kinase	0.15	3.08
Erythropoietin	0.72	4.34
α-Globin	0.55	5.14
β-Globin	0.80	3.05
Growth hormone	1.23	4.95
Histone 3	0.00	6.38
Immunoglobulin heavy chain (variable region)	1.07	5.66
Insulin	0.13	4.02
Interferon α1	1.41	3.53
Interferon γ	2.79	8.59
Luteinizing hormone	1.02	3.29
Somatostatin-28	0.00	3.97

Source: After W. Li and D. Graur, *Fundamentals of Molecular Evolution* (Sunderland, MA: Sinauer, 1991), p. 69, Table 1.

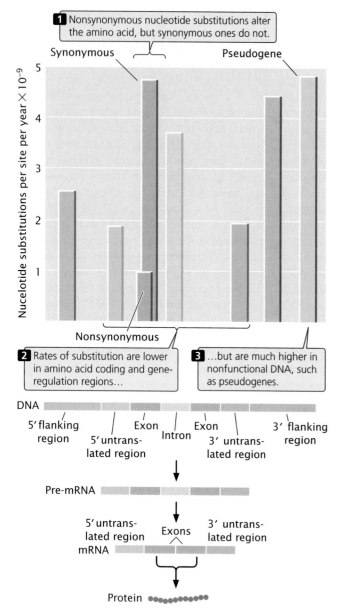

1 Nonsynonymous nucleotide substitutions alter the amino acid, but synonymous ones do not.

2 Rates of substitution are lower in amino acid coding and gene-regulation regions...

3 ...but are much higher in nonfunctional DNA, such as pseudogenes.

◀ **23.21 Different parts of genes evolve at different rates.** The highest rates of nucleotide substitution are in sequences that have the least effect on protein function.

The lowest rates of substitution are seen in nonsynonymous changes in the coding region, because these substitutions always alter the amino acid sequence of the protein and are often deleterious. The highest rates of substitution are in pseudogenes, which are duplicated nonfunctional copies of genes that have acquired mutations. Such a gene no longer produces a functional product; so mutations in pseudogenes have no effect on the fitness of the organism.

In summary, there is a relation between the function of a sequence and its rate of evolution; higher rates are found where they have the least effect on function. This observation

fits with the neutral-mutation hypothesis, which predicts that molecular variation is not affected by natural selection.

The Molecular Clock

The neutral-mutation theory proposes that evolutionary change at the molecular level occurs primarily through the fixation of neutral mutations by genetic drift. The rate at which one neutral mutation replaces another depends only on the mutation rate, which should be fairly constant for any particular gene. If the rate at which a protein evolves is roughly constant over time, the amount of molecular change that a protein has undergone can be used as a **molecular clock** to date evolutionary events.

For example, the enzyme cytochrome c could be examined in two organisms known from fossil evidence to have had a common ancestor 400 million years ago. By determining the number of differences in the cytochrome c amino acid sequences in each organism, we could calculate the number of substitutions that have occurred per amino acid site. The occurrence of 20 amino acid substitutions since the two organisms diverged indicates an average rate of 5 substitutions per 100 million years. Knowing how fast the molecular clock ticks allows us to use molecular changes in cytochrome c to date other evolutionary events: if we found that cytochrome c in two organisms differed by 15 amino acid substitutions, our molecular clock would suggest that they diverged some 300 million years ago. If we assumed some error in our estimate of the rate of amino acid substitution, statistical analysis would show that the true divergence time might range from 160 million to 440 million years. The molecular clock is analogous to geological dating based on the radioactive decay of elements.

The molecular clock was proposed by Emile Zuckerandl and Linus Pauling in 1965 as a possible means of dating evolutionary events on the basis of molecules in present-day organisms. A number of studies have examined the rate of evolutionary change in proteins (◀ FIGURE 23.22), and the molecular clock has been widely used to date evolutionary events when the fossil record is absent or ambiguous. However, the results of several studies have shown that the molecular clock does not always tick at a constant rate, particularly over shorter time periods, and this method remains controversial.

Concepts

Different genes and different parts of the same gene evolve at different rates. Those parts of genes that have the least effect on function tend to evolve at the highest rates. The idea of the molecular clock is that individual proteins and genes evolve at a constant rate and that the differences in the sequences of present-day organisms can be used to date past evolutionary events.

(a)

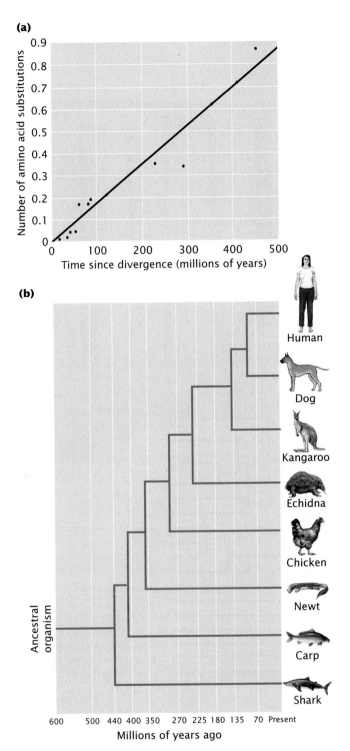

(b)

Molecular Phylogenies

As already mentioned, a phylogeny is an evolutionary history of a group of organisms, usually represented as a tree (◀ FIGURE 23.23). The branches of the phylogenetic tree represent the ancestral relationships between the organisms, and the length of each branch is proportional to the amount of evolutionary change that separates the members of the phylogeny.

Before the rise of molecular biology, phylogenies were based largely on anatomical, morphological, or behavioral traits. Evolutionary biologists attempted to gauge the relationships among organisms by assessing the overall degree of similarity or by tracing the appearance of key characteristics of these traits. The first phylogenies constructed from molecular

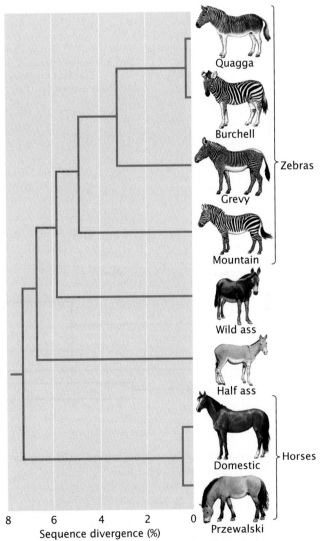

◀ 23.22 **The molecular clock is based on the assumption of a constant rate of change in protein or DNA sequence.** (a) Relation between the rate of amino acid substitution and time since divergence, based in part on amino acid sequences of α hemoglobin from the eight species shown in part *b*. The constant rate of evolution in protein and DNA sequences has been used as a molecular clock to date past evolutionary events. (b) Phylogeny of eight species and their approximate times of divergence, based on the fossil record.

◀ 23.23 **A phylogeny is the evolutionary history—the ancestral relationships—of a group of organisms.** This branching diagram shows the phylogeny of horses based on mitochondrial DNA sequences. DNA of the extinct quagga was extracted from skins from preserved museum specimens.

(a)

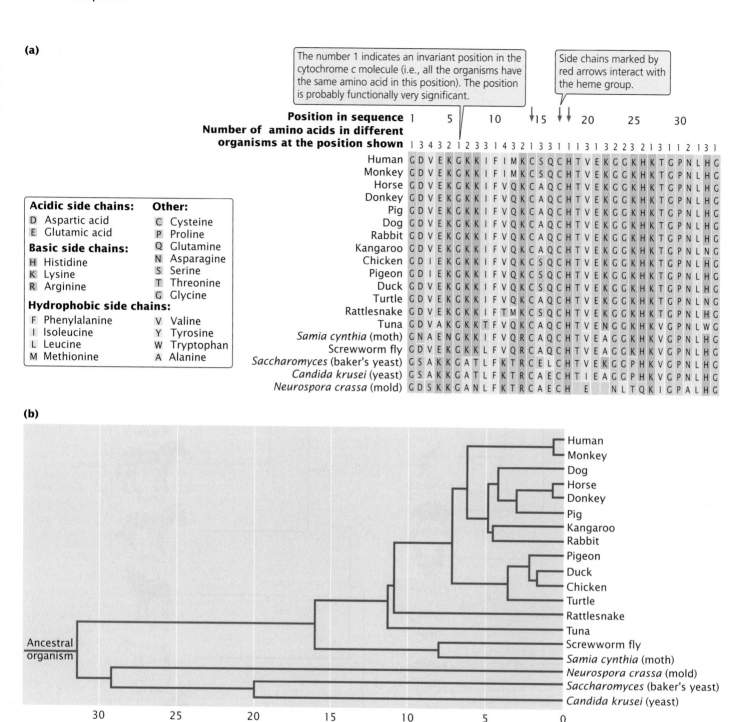

The number 1 indicates an invariant position in the cytochrome c molecule (i.e., all the organisms have the same amino acid in this position). The position is probably functionally very significant.

Side chains marked by red arrows interact with the heme group.

Position in sequence 1   5   10   15   20   25   30

Number of amino acids in different organisms at the position shown 1 3 4 3 2 1 2 3 3 1 4 3 2 1 3 3 1 1 1 3 1 3 2 2 3 2 1 3 1 1 2 1 3 1

	Sequence
Human	G D V E K G K K I F I M K C S Q C H T V E K G G K H K T G P N L H G
Monkey	G D V E K G K K I F I M K C S Q C H T V E K G G K H K T G P N L H G
Horse	G D V E K G K K I F V Q K C A Q C H T V E K G G K H K T G P N L H G
Donkey	G D V E K G K K I F V Q K C A Q C H T V E K G G K H K T G P N L H G
Pig	G D V E K G K K I F V Q K C A Q C H T V E K G G K H K T G P N L H G
Dog	G D V E K G K K I F V Q K C A Q C H T V E K G G K H K T G P N L H G
Rabbit	G D V E K G K K I F V Q K C A Q C H T V E K G G K H K T G P N L H G
Kangaroo	G D V E K G K K I F V Q K C A Q C H T V E K G G K H K T G P N L N G
Chicken	G D I E K G K K I F V Q K C S Q C H T V E K G G K H K T G P N L H G
Pigeon	G D I E K G K K I F V Q K C S Q C H T V E K G G K H K T G P N L H G
Duck	G D V E K G K K I F V Q K C S Q C H T V E K G G K H K T G P N L H G
Turtle	G D V E K G K K I F V Q K C A Q C H T V E K G G K H K T G P N L N G
Rattlesnake	G D V E K G K K I F T M K C S Q C H T V E K G G K H K T G P N L H G
Tuna	G D V A K G K K T F V Q K C A Q C H T V E N G G K H K V G P N L W G
Samia cynthia (moth)	G N A E N G K K I F V Q R C A Q C H T V E A G G K H K V G P N L H G
Screwworm fly	G D V E K G K K L F V Q R C A Q C H T V E A G G K H K V G P N L H G
Saccharomyces (baker's yeast)	G S A K K G A T L F K T R C E L C H T V E K G G P H K V G P N L H G
Candida krusei (yeast)	G S A K K G A T L F K T R C A E C H T I E A G G P H K V G P N L H G
Neurospora crassa (mold)	G D S K K G A N L F K T R C A E C H E N L T Q K I G P A L H G

Acidic side chains:
D Aspartic acid
E Glutamic acid

Basic side chains:
H Histidine
K Lysine
R Arginine

Hydrophobic side chains:
F Phenylalanine
I Isoleucine
L Leucine
M Methionine

Other:
C Cysteine
P Proline
Q Glutamine
N Asparagine
S Serine
T Threonine
G Glycine
V Valine
Y Tyrosine
W Tryptophan
A Alanine

(b)

Ancestral organism

Human
Monkey
Dog
Horse
Donkey
Pig
Kangaroo
Rabbit
Pigeon
Duck
Chicken
Turtle
Rattlesnake
Tuna
Screwworm fly
Samia cynthia (moth)
Neurospora crassa (mold)
Saccharomyces (baker's yeast)
Candida krusei (yeast)

30   25   20   15   10   5   0
Average minimal substitutions

23.24 A phylogeny based on amino acid sequences of the cytochrome c molecule.

data were based on amino acid sequences of proteins such as cytochrome c (◀FIGURE 23.24), but, more recently, phylogenies have been based on DNA sequences. One example is the use of DNA sequences to study the relationship of humans to the other apes. Charles Darwin originally proposed that chimpanzees and gorillas were closely related to humans. However,

subsequent study has placed humans in the family Hominidae and the great apes (chimpanzees, gorilla, orangutan, and gibbon) in the family Pongidae. Some researchers suggested that gibbons belong to a third family; others proposed that humans are most closely related to orangutans. Molecular data support the hypothesis that humans, chimpanzees, and

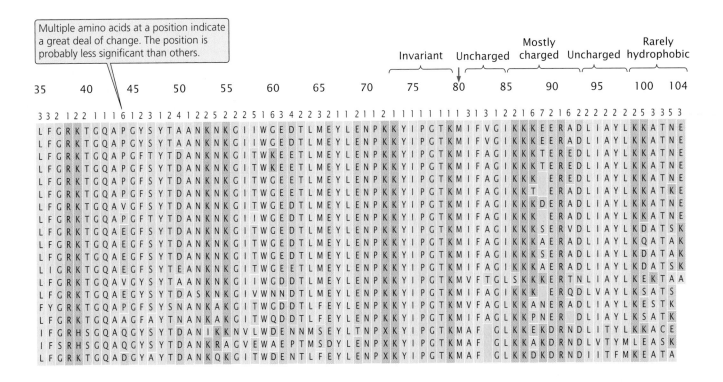

Multiple amino acids at a position indicate a great deal of change. The position is probably less significant than others.

			Invariant	Uncharged	Mostly charged	Uncharged	Rarely hydrophobic

35   40   45   50   55   60   65   70   75   80   85   90   95   100   104

3 3 2 1 2 2 1 1 1 6 1 2 3 1 2 4 1 2 2 5 2 2 2 5 1 6 3 4 2 2 3 2 1 1 2 1 1 2 1 1 1 1 1 1 1 1 1 3 1 3 1 2 2 1 6 7 2 1 6 2 2 2 2 2 2 5 3 3 5 3

L F G R K T G Q A P G Y S Y T A A N K N K G I I W G E D T L M E Y L E N P K K Y I P G T K M I F V G I K K K E E R A D L I A Y L K K A T N E
L F G R K T G Q A P G Y S Y T A A N K N K G I I W G E D T L M E Y L E N P K K Y I P G T K M I F V G I K K K E E R A D L I A Y L K K A T N E
L F G R K T G Q A P G F T Y T D A N K N K G I T W K E E T L M E Y L E N P K K Y I P G T K M I F A G I K K K T E R E D L I A Y L K K A T N E
L F G R K T G Q A P G F S Y T D A N K N K G I T W G E E T L M E Y L E N P K K Y I P G T K M I F A G I K K K   E R E D L I A Y L K K A T N E
L F G R K T G Q A P G F S Y T D A N K N K G I T W G E E T L M E Y L E N P K K Y I P G T K M I F A G I K K T   E R A D L I A Y L K K A T K E
L F G R K T G Q A V G F S Y T D A N K N K G I T W G E D T L M E Y L E N P K K Y I P G T K M I F A G I K K K D E R A D L I A Y L K K A T N E
L F G R K T G Q A P G F T Y T D A N K N K G I I W G E D T L M E Y L E N P K K Y I P G T K M I F A G I K K K   E R A D L I A Y L K K A T N E
L F G R K T G Q A E G F S Y T D A N K N K G I T W G E D T L M E Y L E N P K K Y I P G T K M I F A G I K K K S E R V D L I A Y L K D A T S K
L F G R K T G Q A E G F S Y T D A N K N K G I T W G E D T L M E Y L E N P K K Y I P G T K M I F A G I K K K A E R A D L I A Y L K Q A T A K
L F G R K T G Q A E G F S Y T D A N K N K G I T W G E D T L M E Y L E N P K K Y I P G T K M I F A G I K K K S E R A D L I A Y L K D A T A K
L I G R K T G Q A E G F S Y T E A N K N K G I T W G E E T L M E Y L E N P K K Y I P G T K M I F A G I K K K A E R A D L I A Y L K D A T S K
L F G R K T G Q A V G Y S Y T A A N K N K G I I W G D D T L M E Y L E N P K K Y I P G T K M V F T G L S K K K E R T N L I A Y L K E K T A A
L F G R K T G Q A E G Y S Y T D A S K N K G I V W N N D T L M E Y L E N P K K Y I P G T K M I F A G I K K K   E R Q D L V A Y L K S A T S
F Y G R K T G Q A P G F S Y S N A N K A K G I T W G D D T L F E Y L E N P K K Y I P G T K M V F A G L K K A N E R A D L I A Y L K E S T K
L F G R K T G Q A A G F A Y T N A N K A K G I T W Q D D T L F E Y L E N P K K Y I P G T K M I F A G L K K P N E R   D L I A Y L K S A T K
I F G R H S G Q A Q G Y S Y T D A N I K K N V L W D E N N M S E Y L T N P X K Y I P G T K M A F   G L K K E K D R N D L I T Y L K K A C E
I F S R H S G Q A Q G Y S Y T D A N K R A G V E W A E P T M S D Y L E N P X K Y I P G T K M A F   G L K K A K D R N D L V T Y M L E A S K
L F G R K T G Q A D G Y A Y T D A N K Q K G I T W D E N T L F E Y L E N P X K Y I P G T K M A F   G L K K D K D R N D I I T F M K E A T A

gorillas are most closely related and that orangutans and gibbons diverged from the other apes at a much earlier date. Growing evidence favors a close relationship between humans and chimpanzees (◀ FIGURE 23.25).

Because molecular data can be collected from virtually any organism, comparisons can be made between evolutionary distant organisms. For example, DNA sequences have been used to examine the primary divisions of life and to construct universal phylogenies. On the basis of 16S rRNA, Norman Pace and his colleagues constructed a universal tree of life that included all the major groups of organisms (◀ FIGURE 23.26). The results of their studies

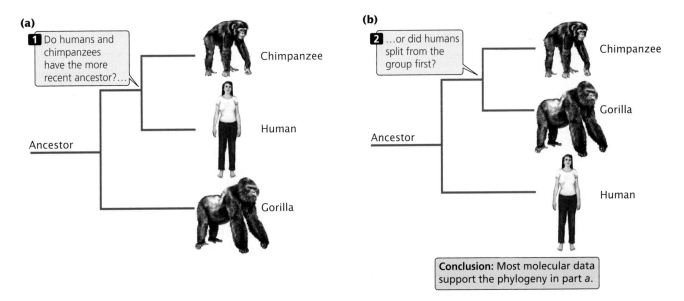

(a)
1 Do humans and chimpanzees have the more recent ancestor?...

Chimpanzee

Human

Ancestor

Gorilla

(b)
2 ...or did humans split from the group first?

Chimpanzee

Gorilla

Ancestor

Human

Conclusion: Most molecular data support the phylogeny in part *a*.

◀ 23.25 **Two possible phylogenies of the human, chimpanzee, and gorilla relationships.** One phylogeny suggests (a) that humans and chimpanzees have the more recent ancestor. The other phylogeny suggests (b) that humans split from the group first and that chimpanzees and gorillas have the more recent ancestor.

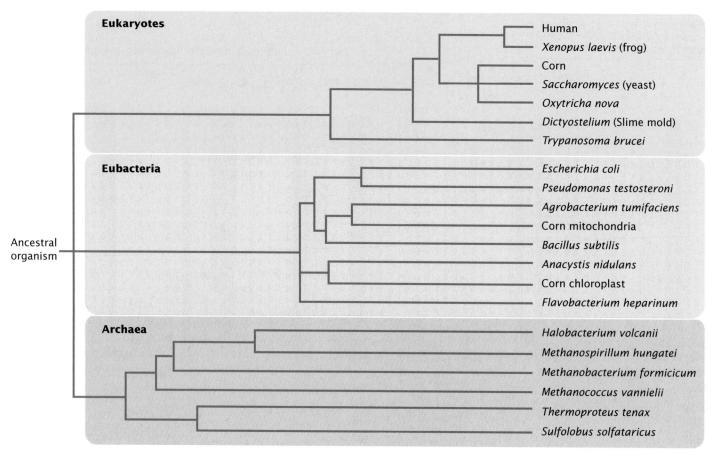

◀ 23.26 A universal tree of life can be constructed from 16S rRNA sequences. Note that sequences from corn mitochondria and chloroplasts are most similar to sequences from eubacteria, confirming the endosymbiotic hypothesis that these eukaryotic organelles evolved from bacteria (see Chapter 20).

revealed that there are three divisions of life: the eubacteria (the common bacteria), the archaea (a distinct group of lesser-known prokaryotes), and the eukaryotes.

Concepts

Molecular data can be used to infer phylogenies (evolutionary histories) of groups of living organisms.

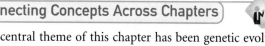 Current research in molecular evolution

Connecting Concepts Across Chapters

The central theme of this chapter has been genetic evolution—how the genetic composition of a population changes with time. Unlike transmission and molecular genetics, which focus on individuals and particular genes, this chapter has focused on the genetic makeup of *groups* of individuals. To describe the genes in these groups, we must rely on mathematics and statistical tools; population genetics is therefore fundamentally quantitative in nature. Mathematical models are commonly used in population genetics to describe processes that bring about change in genotypic and allelic frequencies. These models are, by necessity, simplified representations of the real world, but they nevertheless can be sources of insight into how various factors influence the processes of genetic change.

Our study of population genetics depends on and synthesizes much of the information that we have covered in other parts of this book. Describing the genetic composition of a population requires an understanding of the principles of heredity (Chapters 3 through 5) and how genes are changed by mutation (Chapter 17). Our examination of molecular evolution in the second half of the chapter presupposes an understanding of how genes are encoded in DNA, replicated, and expressed (Chapters 10 through 15). It includes the use of molecular tools, such as restriction enzymes, DNA sequencing, and PCR, which are covered in Chapters 18 and 19.

CONCEPTS SUMMARY

- Population genetics examines the genetic composition of groups of individuals and how this composition changes with time.

- A Mendelian population is a group of interbreeding, sexually reproducing individuals, whose set of genes constitutes the population's gene pool. Evolution occurs through changes in this gene pool.

- Genetic variation and the forces that shape it are important in population genetics. A population's genetic composition can be described by its genotypic and allelic frequencies.

- The Hardy-Weinberg law describes the effect of reproduction and Mendel's laws on the allelic and genotypic frequencies of a population. It assumes that a population is large, randomly mating, and free from the effects of mutation, migration, and natural selection. When these conditions are met, the allelic frequencies do not change and the genotypic frequencies stabilize after one generation in the Hardy-Weinberg equilibrium proportions p^2, $2pq$, and q^3, where p and q equal the frequencies of the alleles.

- Nonrandom mating affects the frequencies of genotypes but not alleles. Positive assortative mating is preferential mating between like individuals; negative assortative mating is preferential mating between unlike individuals.

- Inbreeding, a type of positive assortative mating, increases the frequency of homozygotes while decreasing the frequency of heterozygotes. Inbreeding is frequently detrimental because it increases the appearance of lethal and deleterious recessive traits.

- Mutation, migration, genetic drift, and natural selection can change allelic frequencies.

- Recurrent mutation eventually leads to an equilibrium, with the allelic frequencies being determined by the relative rates of forward and reverse mutation. Change due to mutation in a single generation is usually very small because mutation rates are low.

- Migration, the movement of genes between populations, increases the amount of genetic variation within populations and decreases differences between populations. The magnitude of change depends both on the differences in allelic frequencies between the populations and on the magnitude of migration.

- Genetic drift, the change in allelic frequencies due to chance factors, is important when the effective population size is small. Genetic drift occurs when a population consists of a small number of individuals, is established by a small number of founders, or undergoes a major reduction in size. Genetic drift changes allelic frequencies, reduces genetic variation within populations, and causes genetic divergence among populations.

- Natural selection is the differential reproduction of genotypes; it is measured by the relative reproductive successes of genotypes (fitnesses). The effects of natural selection on allelic frequency can be determined by applying the general selection model. Directional selection leads to the fixation of one allele. The rate of change in allelic frequency due to selection depends on the intensity of selection, the dominance relations, and the initial frequencies of the alleles.

- Mutation and natural selection can produce an equilibrium, in which the number of new alleles introduced by mutation is balanced by the elimination of alleles through natural selection.

- Molecular methods offer a number of advantages for the study of evolution. The use of protein electrophoresis to study genetic variation in natural populations showed that most natural populations have large amounts of genetic variation in their proteins. Two hypotheses arose to explain this variation. The neutral-mutation hypothesis proposed that molecular variation is selectively neutral and is shaped largely by mutation and genetic drift. The balance model proposed that molecular variation is maintained largely by balancing selection.

- Different parts of the genome show different amounts of genetic variation. In general, those that have the least effect on function evolve at the highest rates.

- The molecular-clock hypothesis proposes a constant rate of nucleotide substitution, providing a means of dating evolutionary events by looking at nucleotide differences between organisms.

- Molecular data are often used for constructing phylogenies.

IMPORTANT TERMS

Mendelian population (p. 670)
gene pool (p. 670)
genotypic frequency (p. 671)
allelic frequency (p. 672)
Hardy-Weinberg law (p. 673)
Hardy-Weinberg equilibrium (p. 674)
positive assortative mating (p. 677)

negative assortative mating (p. 677)
inbreeding (p. 677)
outcrossing (p. 677)
inbreeding coefficient (p. 677)
inbreeding depression (p. 678)
equilibrium (p. 680)
migration (gene flow) (p. 681)
sampling error (p. 683)

genetic drift (p. 683)
effective population size (p. 683)
founder effect (p. 684)
genetic bottleneck (p. 684)
fixation (p. 684)
fitness (p. 686)
selection coefficient (p. 686)
directional selection (p. 688)
overdominance (p. 688)

underdominance (p. 688)
phylogeny (p. 692)
proportion of polymorphic loci (p. 693)
expected heterozygosity (p. 693)
neutral-mutation hypothesis (p. 693)
balance hypothesis (p. 694)
molecular clock (p. 698)

Worked Problems

1. The following genotypes were observed in a population:

Genotype	Number
HH	40
Hh	45
hh	50

(a) Calculate the observed genotypic and allelic frequencies for this population.

(b) Calculate the numbers of genotypes expected if this population were in Hardy-Weinberg equilibrium.

(c) Using a chi-square test, determine whether the population is in Hardy-Weinberg equilibrium.

• **Solution**

(a) The observed genotypic and allelic frequencies are calculated by using Equations 23.1 and 23.3:

$$f(HH) = \frac{\text{number of } HH \text{ individuals}}{N} = \frac{40}{135} = .30$$

$$f(Hh) = \frac{\text{number of } Hh \text{ individuals}}{N} = \frac{45}{135} = .33$$

$$f(hh) = \frac{\text{number of } hh \text{ individuals}}{N} = \frac{50}{135} = .37$$

$$p = f(H) = \frac{2n_{HH} + n_{Hh}}{2N} = \frac{2(40) + (45)}{2(135)} = .46$$

$$q = f(h) = (1 - p) = (1 - .46) = .54$$

(b) If the population is in Hardy-Weinberg equilibrium, the expected numbers of genotypes are:

$$HH = p^2 \times N = (.46)^2 \times 135 = 28.57$$

$$Hh = 2pq \times N = 2(.46)(.54) \times 135 = 67.07$$

$$hh = q^2 \times N = (.54)^2 \times 135 = 39.37$$

(c) The observed and expected numbers of the genotypes are:

Genotype	Number observed	Number expected
HH	40	28.57
Hh	45	67.07
hh	50	39.37

These numbers can be compared by using a chi-square test:

$$\chi^2 = \sum \frac{(\text{observed} - \text{expected})^2}{\text{expected}}$$

$$= \sum \frac{(40 - 28.57)^2}{28.57} + \frac{(45 - 67.07)^2}{67.07} + \frac{(50 - 39.37)^2}{39.37}$$

$$= 4.57 + 7.26 + 2.87 = 14.70$$

The degrees of freedom associated with this chi-square value are $n - 2$, where n equals the number of expected genotypes, or $3 - 2 = 1$. By examining Table 3.4, we see that the probability associated with this chi-square and the degrees of freedom is $P < .001$, which means that the difference between the observed and expected values is unlikely to be due to chance. Thus, there is a significant difference between the observed numbers of genotypes and the numbers that we would expect if the population were in Hardy-Weinberg equilibrium. We conclude that the population is not in equilibrium.

2. A recessive allele for red hair (r) has a frequency of .2 in population I and a frequency of .01 in population II. A famine in population I causes a number of people in population I to migrate to population II, where they reproduce randomly with the members of population II. Geneticists estimate that, after migration, 15% of the people in population II consist of people who migrated from population I. What will be the frequency of red hair in population II after the migration?

• **Solution**

From Equation 23.16, the allelic frequency in a population after migration (q'_{II}) is

$$q'_{II} = q_I(m) + q_{II}(1 - m)$$

where q_I and q_{II} are the allelic frequencies in population I (migrants) and population II (residents), respectively, and m is the proportion of population II that consist of migrants. In this problem, the frequency of red hair is .2 in population I and .01 in population II. Because 15% of population II consists of migrants, $m = .15$. Substituting these values into Equation 23.16, we obtain:

$$q'_{II} = .2(.15) + (.01)(1 - .15) = .03 + .0085 = .0385$$

This is the expected frequency of the allele for red hair in population II after migration. Red hair is a recessive trait; if mating is random for hair color, the frequency of red hair in population II after migration will be:

$$f(rr) = q^2 = (.0385)^2 = .0015$$

3. Two populations have the following numbers of breeding adults:

Population A: 60 males, 40 females

Population B: 5 males, 95 females

(a) Calculate the effective population sizes for populations A and B.

(b) What predications can you make about the effects of the different sex ratios of these populations on their gene pools?

• Solution

(a) The effective population size can be calculated by using Equation 23.19:

$$N_e = \frac{4 \times n_{males} \times n_{females}}{n_{males} + n_{females}}$$

For population A:

$$N_e = \frac{4 \times 60 \times 40}{60 + 40} = 96$$

For population B:

$$N_e = \frac{4 \times 5 \times 95}{5 + 95} = 19$$

Although each population has a total of 100 breeding adults, the effective population size of population B is much smaller because it has a greater disparity between the numbers of males and females.

(b) The effective population size determines the amount of genetic drift that will occur. Because the effective population size of B is much smaller than that of population A, we can predict that population B will undergo more genetic drift, leading to greater changes in allelic frequency, greater loss of genetic variation, and greater genetic divergence from other populations.

4. Alcohol is a common substance in rotting fruit, where fruit fly larvae grow and develop; larvae use the enzyme alcohol dehydrogenase (ADH) to detoxify the effects of this alcohol. In some fruit-fly populations, two alleles are present at the locus than encodes ADH: ADH^F, which encodes a form of the enzyme that migrates rapidly (fast) on an electrophoretic gel; and ADH^S, which encodes a form of the enzyme that migrates slowly on an electrophoretic gel. Female fruit flies with different ADH genotypes produce the following numbers of offspring when alcohol is present:

Genotype	Mean number of offspring
$ADH^F ADH^F$	120
$ADH^F ADH^S$	60
$ADH^S ADH^S$	30

(a) Calculate the relative fitnesses of females having these genotypes.

(b) If a population of fruit flies has an initial frequency of ADH^F equal to .2, what will be the frequency in the next generation when alcohol is present?

• Solution

(a) Fitness is the relative reproductive output of a genotype and is calculated by dividing the average number of offspring produced by that genotype by the mean number of offspring produced by the most prolific genotype. The fitnesses of the three ADH genotypes therefore are:

Genotype	Mean number of offspring	Fitness
$ADH^F ADH^F$	120	$W_{FF} = \dfrac{120}{120} = 1$
$ADH^F ADH^S$	60	$W_{FS} = \dfrac{60}{120} = .5$
$ADH^S ADH^S$	30	$W_{SS} = \dfrac{30}{120} = .25$

(b) To calculate the frequency of the ADH^F allele after selection, we can use the table method. The frequencies of the three genotypes before selection are the Hardy-Weinberg equilibrium frequencies of p^2, $2pq$, and q^2. We multiply each of these frequencies by the fitness of each genotype to obtain the frequencies after selection. These products are summed to obtain the mean fitness of the population ($\overline{W}$), and the products are then divided by the mean fitness to obtain the relative genotypic frequencies after selection as shown in the table below:

	$ADH^F ADH^F$	$ADH^F ADH^S$	$ADH^S ADH^S$
Initial genotypic frequencies:	$p^2 = (.2)^2 = .04$	$2pq = 2(.2)(.8) = .32$	$q^2 = (.8)^2 = 0.64$
Fitnesses:	$W_{FF} = 1$	$W_{FS} = .5$	$W_{22} = .25$
Proportionate contribution of genotypes to population:	$p^2 W_{FF} = .04(1) = .04$	$2pq W_{FS} = (.32)(.5) = .16$	$q^2 W_{SS} = (.64)(.25) = .16$
Relative genotypic frequency after selection:	$\dfrac{p^2 W_{FF}}{\overline{W}} = \dfrac{.04}{.36} = .11$	$\dfrac{2pq W_{FS}}{\overline{W}} = \dfrac{.16}{.36} = .44$	$\dfrac{q^2 W_{SS}}{\overline{W}} = \dfrac{.16}{.36} = .44$

$\overline{W} = .04 + .16 + .16 = .36$

To calculate the allelic frequency after selection, we use Equation 23.4:

$$p = f(ADH^F) = f(ADH^F ADH^F) + \tfrac{1}{2} f(ADH^F ADH^S)$$
$$= .11 + \tfrac{1}{2} (.44) = .33$$

We predict that the frequency of ADH^F will increase from .2 to .33.

The New Genetics
MINING GENOMES

POPULATION GENETICS: ANALYSES AND SIMULATIONS

In this exercise, you will analyze real molecular data, primarily generated by high-school and college students, to learn how allele frequencies and genotype distributions can be used to study human populations. To do so, you will use the databases and statistical tools at the Dolan DNA Learning Center of Cold Spring Harbor Laboratory. In addition, you will use simulations to explore how factors such as population size, selection pressure, and genetic drift interact to cause allele frequencies to change.

COMPREHENSION QUESTIONS

1. What is a Mendelian population? How is the gene pool of a Mendelian population usually described? What are the predictions given by the Hardy-Weinberg law?

* 2. What assumptions must be met for a population to be in Hardy-Weinberg equilibrium?

3. What is random mating?

* 4. Give the Hardy-Weinberg expected genotypic frequencies for (a) an autosomal locus with three alleles, and (b) an X-linked locus with two alleles.

5. Define inbreeding and briefly describe its effects on a population.

6. What determines the allelic frequencies at mutational equilibrium?

* 7. What factors affect the magnitude of change in allelic frequencies due to migration?

8. Define genetic drift and give three ways that it can arise. What effect does genetic drift have on a population?

* 9. What is effective population size? How does it affect the amount of genetic drift?

10. Define natural selection and fitness.

11. Briefly discuss the differences between directional selection, overdominance, and underdominance. Describe the effect of each type of selection on the allelic frequencies of a population.

12. What factors affect the rate of change in allelic frequency due to natural selection?

*13. Compare and contrast the effects of mutation, migration, genetic drift, and natural selection on genetic variation within populations and on genetic divergence between populations.

14. Give some of the advantages of using molecular data in evolutionary studies.

*15. What is the key difference between the neutral-mutation hypothesis and the balance hypothesis?

16. Outline the different rates of evolution that are typically seen in different parts of a protein-encoding gene. What might account for these differences?

*17. What is the molecular clock?

APPLICATION QUESTIONS AND PROBLEMS

18. How would you respond to someone who said that models are useless in studying population genetics because they represent oversimplifications of the real world?

*19. Voles (*Microtus ochrogaster*) were trapped in old fields in southern Indiana and were genotyped for a transferrin locus. The following numbers of genotypes were recorded.

$T^E T^E$	$T^E T^F$	$T^F T^F$
407	170	17

Calculate the genotypic and allelic frequencies of the transferrin locus for this population.

20. Orange coat color in cats is due to an X-linked allele (X^O) that is codominant to the allele for black (X^+). Genotypes of the orange locus of cats in Minneapolis and St. Paul, Minnesota, were determined and the following data were obtained.

$X^O X^O$ females	11
$X^O X^+$ females	70
$X^+ X^+$ females	94
$X^O Y$ males	36
$X^+ Y$ males	112

Calculate the frequencies of the X^O and X^+ alleles for this population.

21. A total of 6129 North American Caucasians were blood typed for the MN locus, which is determined by two codominant alleles, L^M and L^N. The following data were obtained:

Blood type	Number
M	1787
MN	3039
N	1303

Carry out a chi-square test to determine whether this population is in Hardy-Weinberg equilibrium at the MN locus.

22. Genotypes of leopard frogs from a population in central Kansas were determined for a locus that encodes the enzyme malate dehydrogenase. The following numbers of genotypes were observed:

Genotype	Number
M^1M^1	20
M^1M^2	45
M^2M^2	42
M^1M^3	4
M^2M^3	8
M^3M^3	6
Total	125

(a) Calculate the genotypic and allelic frequencies for this population.

(b) What would be the expected numbers of genotypes if the population were in Hardy-Weinberg equilibrium?

23. Full color (D) in domestic cats is dominant over dilute color (d). Of 325 cats observed, 194 have full color and 131 have dilute color.

(a) If these cats are in Hardy-Weinberg equilibrium for the dilution locus, what is the frequency of the dilute allele?

(b) How many of the 194 cats with full color are likely to be heterozygous?

24. Tay-Sachs disease is an autosomal recessive disorder. Among Ashkenazi Jews, the frequency of Tay-Sachs disease is 1 in 3600. If the Ashkenazi population is mating randomly for the Tay-Sachs gene, what proportion of the population consists of heterozygous carriers of the Tay-Sachs allele?

25. In the plant *Lotus corniculatus*, cyanogenic glycoside protects the plants against insect pests and even grazing by cattle. This glycoside is due to a simple dominant allele. A population of *L. corniculatus* consists of 77 plants that possess cyanogenic glycoside and 56 that lack the compound. What is the frequency of the dominant allele that results in the presence of cyanogenic glycoside in this population?

*26. Color blindness in humans is an X-linked recessive trait. Approximately 10% of the men in a particular population are color blind.

(a) If mating is random for the color-blind locus, what is the frequency of the color-blind allele in this population?

(b) What proportion of the women in this population are expected to be color-blind?

(c) What proportion of the women in the population are expected to be heterozygous carriers of the color-blind allele?

*27. The human MN blood type is determined by two codominant alleles, L^M and L^N. The frequency of L^M in Eskimos on a small Arctic island is .80. If the inbreeding coefficient for this population is .05, what are the expected frequencies of the M, MN, and N blood types on the island?

28. Demonstrate mathematically that full sib mating ($F = \frac{1}{4}$) reduces the heterozygosity by $\frac{1}{4}$ with each generation.

29. The forward mutation rate for piebald spotting in guinea pigs is 8×10^{-5}; the reverse mutation rate is 2×10^{-6}. Assuming that no other evolutionary forces are present, what is the expected frequency of the allele for piebald spotting in a population that is in mutational equilibrium?

*30. In German cockroaches, curved wing (cv) is recessive to normal wing (cv^+). Bill, who is raising cockroaches in his dorm room, finds that the frequency of the gene for curved wings in his cockroach population is .6. In the apartment of his friend Joe, the frequency of the gene for curved wings is .2. One day Joe visits Bill in his dorm room, and several cockroaches jump out of Joe's hair and join the population in Bill's room. Bill estimates that 10% of the cockroaches in his dorm room now consists of individual roaches that jumped out of Joe's hair. What will be the new frequency of curved wings among cockroaches in Bill's room?

31. A population of water snakes is found on an island in Lake Erie. Some of the snakes are banded and some are unbanded; banding is caused by an autosomal allele that is recessive to an allele for no bands. The frequency of banded snakes on the island is .4, whereas the frequency of banded snakes on the mainland is .81. One summer, a large number of snakes migrate from the mainland to the island. After this migration, 20% of the island population consists of snakes that came from the mainland.

(a) Assuming that both the mainland population and the island population are in Hardy-Weinberg equilibrium for the alleles that affect banding, what is the frequency of the allele for bands on the island and on the mainland before migration?

(b) After migration has taken place, what will be the frequency of the banded allele on the island?

*32. Calculate the effective size of a population with the following numbers of reproductive adults:

(a) 20 males and 20 females

(b) 30 males and 10 females

(c) 10 males and 30 females

(d) 2 males and 38 females

33. Pikas are small mammals that live at high elevation in the talus slopes of mountains. Populations located on mountain tops in Colorado and Montana in North America are relatively isolated from one another, because the pikas don't occupy the low-elevation habitats that separate the mountain tops and don't venture far from the talus slopes. Thus, there is little gene flow between populations. Furthermore, each population is small in size and was founded by a small number of pikas.

A group of population geneticists propose to study the amount of genetic variation in a series of pika populations and to compare the allelic frequencies in different populations. On the basis of biology and the distribution of pikas, what do you predict the population geneticists will find concerning the within- and between-population genetic variation?

34. In a large, randomly mating population, the frequency of the allele (s) for sickle-cell hemoglobin is .028. The results of studies have shown that people with the following genotypes at the beta-chain locus produce the average numbers of offspring given:

Genotype	Average number of offspring produced
SS	5
Ss	6
ss	0

(a) What will be the frequency of the sickle-cell allele (s) in the next generation?

(b) What will be the frequency of the sickle cell allele at equilibrium?

35. Two chromosomal inversions are commonly found in populations of *Drosophila pseudoobscura*: Standard *(ST)* and Arrowhead *(AR)*. When treated with the insecticide DDT, the genotypes for these inversions exhibit overdominance, with the following fitnesses:

Genotype	Fitness
ST/ST	.47
ST/AR	1
AR/AR	.62

What will be the frequency of *ST* and *AR* after equilibrium has been reached?

*36. In a large, randomly mating population, the frequency of an autosomal recessive lethal allele is .20. What will be the frequency of this allele in the next generation?

37. A certain form of congenital glaucoma results from an autosomal recessive allele. Assume that the mutation rate is 10^{-5} and that persons having this condition produce, on the average, only about 80% of the offspring produced by persons who do not have glaucoma.

(a) At equilibrium between mutation and selection, what will be the frequency of the gene for congenital glaucoma?

(b) What will be the frequency of the disease in a randomly mating population that is at equilibrium?

CHALLENGE QUESTION

38. The Barton Springs salamander is an endangered species found only in a single spring in the city of Austin, Texas. There is growing concern that a chemical spill on a nearby freeway could pollute the spring and wipe out the species. To provide a source of salamanders to repopulate the spring in the event of such a catastrophe, a proposal has been made to establish a captive breeding population of the salamander in a local zoo. You are asked to provide a plan for the establishment of this captive breeding population, with the goal of maintaining as much of the genetic variation of the species as possible in the captive population. What factors might cause loss of genetic variation in the establishment of the captive population? How could loss of such variation be prevented? Assuming that it is feasible to maintain only a limited number of salamanders in captivity, what procedures should be instituted to ensure the long-term maintenance of as much of the variation as possible?

SUGGESTED READINGS

Avise, J. C. 1994. *Molecular Markers, Natural History, and Evolution.* New York: Chapman and Hall.
An excellent review of how molecular techniques are being used to examine evolutionary questions.

Buri, P. 1956. Gene frequency in small populations of mutant *Drosophila. Evolution* 10:367–402.
Buri's famous experiment demonstrating the effects of genetic drift on allelic frequencies.

Hardy, G. H. 1908. Mendelian proportions in a mixed population. *Science* 28:49–50.
Original paper by G. H. Hardy outlining the Hardy-Weinberg law.

Hartl, D. L., and A. G. Clark. 1997. *Principles of Population Genetics,* 3d ed. Sunderland, MA: Sinauer.
An advanced textbook in population genetics.

MacIntyre, R. J., Ed. 1985. *Molecular Evolutionary Genetics.* New York: Plenum.
Contributors treat various aspects of molecular evolution.

Mettler, L. E., T. G. Gregg, and H. S. Schaffer. 1998. *Population Genetics and Evolution,* 2d ed. Englewood Cliffs. NJ: Prentice Hall.
A short, readable textbook on population genetics.

Nei, M., and S. Kumar. 2000. *Molecular Evolution and Phylogenetics.* Oxford: Oxford University Press.
An advanced textbook on the methods used in the study of molecular evolution.

Ou, C., C. A. Ciesielski, G. Myers, C. I. Bandea, et al. 1992. Molecular epidemiology of HIV transmission in a dental practice. *Science* 256:1165–1171.
Study of molecular evolution of HIV in a Florida dental practice.

Provine, W. B. 2002. *The Origins of Theoretical Population Genetics,* 2nd ed. Chicago: Chicago University Press.
A complete history of the origins of population genetics as a field of study.

Saccheri, I., M. Kuussaari, M. Kankare, P. Vikman, W. Fortelius, and I. Hanski. 1998. Inbreeding and extinction in a butterfly metapopulation. *Nature* 392:491–494.
Discusses the role of inbreeding in population extinction of butterflies.

Vial, C., P. Savolainen, J. E. Maldonado, I. R. Amorim, J. E. Rice, R. L. Honeyutt, K. A. Cranall, J. Lundeberg, and R. K. Wayne. 1997. Multiple and ancient origins of the domestic dog. *Science* 276:1687–1689.
Using the molecular clock and mitochondrial DNA sequences, these geneticists estimate that the dog was domesticated more than 100,000 years ago.

Glossary

acceptor arm The arm in tRNA to which an amino acid attaches.

acentric chromatid Lacks a centromere; produced when crossing over takes place within a paracentric inversion. The acentric chromatid does not attach to a spindle fiber and does not segregate in meiosis or mitosis; so it is usually lost after one or more rounds of cell division.

acidic activation domain Commonly found in some transcriptional activator proteins, a domain that contains multiple amino acids with negative charges and stimulates the transcription of certain genes.

acrocentric chromosome Chromosome in which the centromere is near one end, producing a long arm at one end and a knob, or satellite, at the other end.

activator *See* **transcriptional activator protein.**

addition rule Rule of probability stating that the probability of any of two or more mutually exclusive events occurring is calculated by adding the probabilities of the individual events.

additive genetic variance Component of the genetic variance that can be attributed to the additive effect of different genotypes.

adenosine-3′,5′-cyclic monophosphate (cAMP) Modified nucleotide that functions in catabolite repression. Low levels of glucose stimulate high levels of cAMP; cAMP then attaches to CAP, which binds to the promoter of certain operons and stimulates transcription.

adjacent-1 segregation Type of segregation that takes place in a heterozygote for a translocation. If the original, nontranslocated chromosomes are N_1 and N_2 and the chromosomes containing the translocated segments are T_1 and T_2, then adjacent-1 segregation takes place when N_1 and T_2 move toward one pole and T_1 and N_2 move toward the other pole.

adjacent-2 segregation Type of segregation that takes place in a heterozygote for a translocation. If the original, nontranslocated chromosomes are N_1 and N_2 and the chromosomes containing the translocated segments are T_1 and T_2, then adjacent-2 segregation takes place when N_1 and T_1 move toward one pole and T_2 and N_2 move toward the other pole.

A-DNA Right-handed helical structure of DNA that exists when little water is present.

adenine (A) Purine base in DNA and RNA.

allele One of two or more alternate forms of a gene.

allelic frequency Frequency or proportion of a particular allele.

allopolyploidy Condition in which the sets of chromosomes of a polyploid individual possessing more than two haploid sets are derived from two or more species.

allosteric protein Protein that changes its conformation on binding with another molecule.

alternate segregation Type of segregation that takes place in a heterozygote for a translocation. If the original, nontranslocated chromosomes are N_1 and N_2 and the chromosomes containing the translocated segments are T_1 and T_2, then alternate segregation takes place when N_1 and N_2 move toward one pole and T_1 and T_2 move toward the opposite pole.

alternation of generation Complex life cycle in plants alternating between the diploid sporophyte stage and the haploid gametophyte stage.

alternative processing pathway One of several pathways by which a single pre-mRNA may be processed in different ways to produce alternative types of mRNA.

alternative splicing Process by which a single pre-mRNA may be spliced in more than one way to produce different types of mRNA.

Ames test Test in which special strains of bacteria are used to evaluate the potential of chemicals to cause cancer.

amino acid Repeating unit of proteins; consists of an amino group, a carboxyl group, a hydrogen atom, and a variable R group.

aminoacyl (A) site One of three sites in a ribosome occupied by a tRNA in translation. All charged tRNAs (with the exception of the initiator tRNA) first enter the A site in translation.

aminoacyl-tRNA synthetase Enzyme that attaches an amino acid to a tRNA. Each aminoacyl-tRNA synthetase is specific for a particular amino acid.

amniocentesis Procedure used for prenatal genetic testing in which a sample of amniotic fluid is removed from a pregnant woman by insertion of a long sterile needle through the abdominal wall into the amniotic sac.

anaphase Stage of mitosis in which chromatids separate and move toward the spindle poles.

anaphase I Stage of meiosis I in which homologous chromosomes separate and move toward the spindle poles.

anaphase II Stage of meiosis II in which chromatids separate and move toward the spindle poles.

aneuploidy Change from the wild type in the number of chromosomes; most often an increase or decrease of one or two chromosomes.

Antennapedia **complex** Cluster of five homeotic genes in fruit flies that affects development of the adult fly's head and anterior thoracic segments.

antibody Produced by a B cell, a protein that circulates in the blood and other body fluids. An antibody binds to a specific antigen and marks it for destruction by making it easier for a phagocytic cell to ingest the antigen.

anticipation Increasing severity or earlier age of onset of a genetic trait in succeeding generations. For example, symptoms of a genetic disease may become more severe as the trait is passed from generation to generation.

anticodon Sequence of three nucleotides in tRNA that pairs with the corresponding codon in mRNA in translation.

antigen Substance that is recognized by the immune system and elicits an immune response.

antiparallel Refers to a characteristic of the DNA double helix in which the two polynucleotide strands run in opposite directions.

antisense RNA Small RNA molecule that base pairs with a complementary DNA or RNA sequence and affects its functioning.

antiterminator DNA sequence or a protein that inhibits the termination of transcription.

apoptosis Programmed cell death, in which a cell degrades its own DNA, the nucleus and cytoplasm shrink, and the cell undergoes phagocytosis by other cells without any leakage of its contents.

archaea One of the three primary divisions of life. Archaea consist of unicellular organisms with prokaryotic cells.

artificial selection Selection practiced by humans.

attachment site Special site on a bacterial chromosome where a prophage may insert itself.

attenuation Type of gene regulation in some bacterial operons, in which transcription is initiated but terminates prematurely before transcription of the structural genes.

attenuator Secondary structure that forms in the 5′ untranslated region of some operons and causes the premature termination of transcription.

autoimmune disease Characterized by an abnormal immune response to a person's own (self) antigen.

autonomous element Transposable element that is fully functional and able to transpose on its own.

autonomously replicating sequence DNA sequence that confers the ability to replicate; contains an origin of replication.

autopolyploidy Condition in which all the sets of chromsomes of a polyploid individual possessing more than two haploid sets are derived from a single species.

autoradiography Method for visualizing DNA or RNA molecules labeled with radioactive substances. A piece of X-ray film is placed on top of a slide, gel, or other substance that contains DNA labeled with radioactive chemicals. Radiation from the labeled DNA exposes the film, providing a picture of the labeled molecules.

autosome Chromosome that is the same in males and females; nonsex chromosome.

auxotroph Bacterium or fungus that possesses a nutritional mutation that disrupts its ability to synthesize an essential biological molecule; cannot grow on minimal medium but can grow on minimal medium to which has been added the biological molecule that it cannot synthesize.

backcross A cross between an F_1 individual and one of the parental (P) genotypes.

bacterial artificial chromosome (BAC) Cloning vector used in bacteria that is capable of carrying DNA fragments as large as 500 kb.

bacterial colony Clump of genetically identical bacteria derived from a single bacterial cell that undergoes repeated rounds of division.

bacteriophage Virus that infects bacterial cells.

balance hypothesis Proposes that much of the molecular variation seen in natural populations is maintained by balancing selection that favors genetic variation.

Barr body Condensed, darkly staining structure found in most cells of female placental mammals that is an inactivated X chromosome.

basal transcription apparatus Complex of transcription factors, RNA polymerase, and other proteins that assemble on the promoter and are capable of initiating minimal levels of transcription.

base *See* **nitrogenous base.**

base analog Chemical substance that has a structure similar to that of one of the four standard bases of DNA and may be incorporated into newly synthesized DNA molecules in replication.

base-excision repair DNA repair that first excises modified bases and then replaces the entire nucleotide.

base substitution Mutation in which a single pair of bases in the DNA is altered.

B cell Particular type of lymphocyte that produces humoral immunity; matures in the bone marrow and produces antibodies.

B-DNA Right-handed helical structure of DNA that exists when water is abundant; the secondary structure described by Watson and Crick and probably the most common DNA structure in cells.

bidirectional replication Replication at both ends of a replication bubble.

bioinformatics Synthesis of molecular biology and computer science that develops databases and computational tools to store, retrieve, and analyze nucleic acid and protein sequence data.

biotechnology Use of biological processes, particularly molecular genetics and recombinant DNA technology, to produce products of commercial value.

bithorax complex Cluster of three homeotic genes in fruit flies that influences the adult fly's posterior thoracic and abdominal segments.

bivalent Refers to a synapsed pair of homologous chromosomes.

blending inheritance Early concept of heredity proposing that offspring possess a mixture of the traits from both parents.

branch migration Movement of a cross bridge along two DNA molecules.

branch point An adenine nucleotide in nuclear pre-mRNA introns that lies from 18 to 40 nucleotides upstream of the 3′ splice site.

broad-sense heritability Proportion of the phenotypic variance that can be attributed to genetic variance.

caspase Enzyme that cleaves other proteins and regulates apoptosis. Each caspase is synthesized as a large, inactive precursor (a procaspase) that is activated by cleavage, often by another caspase.

catabolite activator protein (CAP) Protein that functions in catabolite repression. When bound with cAMP, CAP binds to the promoter of certain operons and stimulates transcription.

catabolite repression System of gene control in some bacterial operons in which glucose is used preferentially and the metabolism of other sugars is repressed in the presence of glucose.

cDNA library Collection of bacterial colonies or phage colonies containing DNA fragments that have been produced by reverse transcription of cellular mRNA.

cell cycle Stages through which a cell passes from one cell division to the next.

cell line Genetically identical cells that divide indefinitely and can be cultured in the laboratory.

cell theory Theory stating that all life is composed of cells, that cells arise only from other cells, and that the cell is the fundamental unit of structure and function in living organisms.

cellular immunity Type of immunity resulting from T cells, which recognize antigens found on the surface of self cells.

centimorgan Another name for map units.

central dogma Concept that genetic information passes from DNA to RNA to protein in a one-way information pathway.

centriole Cytoplasmic organelle consisting of microtubules; present at each pole of the spindle apparatus in animal cells.

centromere Constricted region on a chromosome that stains less strongly than the rest of the chromosome; region where spindle microtubules attach to a chromosome.

centromeric sequence DNA sequence found in functional centromeres.

centrosome Structure from which the spindle apparatus develops: contains the centriole.

Chargaff's rules Rules developed by Erwin Chargaff and his colleagues concerning the ratios of bases in DNA.

chiasma (pl., chiasmata) Point of attachment between homologous chromosomes at which crossing over took place.

chloroplast DNA (cpDNA) DNA in chloroplasts; has many characteristics in common with eubacterial DNA and typically consists of a circular molecular that lacks histone proteins and encodes some of the rRNAs, tRNAs, and proteins found in chloroplasts.

chorionic villus sampling (CVS) Procedure for prenatal genetic testing in which a small piece of the chorion, the outer layer of the placenta, is removed from a pregnant woman. A catheter is inserted through the vagina and cervix into the uterus. Suction is then applied to remove the sample.

chromatin Material found in the eukaryotic nucleus; consists of DNA and proteins.

chromatin-remodeling complex Complex of proteins that alters chromatin structure without acetylating histone proteins.

chromatin-remodeling protein Binds to a DNA sequence and disrupts chromatin structure, causing the DNA to become more accessible to RNA polymerase and other proteins.

chromatosome A nucleosome and an H1 histone protein.

chromosomal puff Localized swelling of a polytene chromosome; a region of chromatin in which DNA has unwound and is undergoing transcription.

chromosomal scaffold protein Protein that plays a role in the folding and packing of the chromosome, revealed when chromatin is treated with a concentrated salt solution, which removes histones and some other chromosomal proteins.

chromosome deletion Loss of a chromosome segment.

chromosome duplication Duplication of a chromosome segment.

chromosome inversion Chromosome rearrangement in which a chromosome segment has been inverted 180 degrees.

chromosome mutation Difference from the wild type in the number or structure of one or more chromosomes; often affects many genes and has large phenotypic effects.

chromosome rearrangement Change from the wild type in the structure of one or more chromosomes.

chromosome theory of inheritance Theory stating that genes are located on chromosomes.

chromosome walking Method of locating a gene by using partly overlapping genomic clones to move in steps from a previously cloned, linked gene to the gene of interest.

cis configuration Arrangement in which two or more wild-type genes are on one chromosome and their mutant alleles are on the homologous chromosome; also called coupling configuration.

clonal evolution Process by which mutations that enhance the ability of cells to proliferate predominate in a clone of cells, allowing the clone to become increasingly rapid in growth and increasingly aggressive in proliferation properties.

cloning strategy The particular set of methods used to clone a gene or DNA fragment.

cloning vector Stable, replicating DNA molecule to which a foreign DNA fragment can be attached and transferred to a host cell.

cloverleaf structure Secondary structure common to all tRNAs.

coactivator Protein that cooperates with an activator of transcription. In eukaryotic transcriptional control, coactivators often physically interact with transcriptional activators and the basal transcription apparatus.

codominance Type of allelic interaction in which the heterozygote simultaneously expresses traits of both homozygotes.

codon Sequence of three nucleotides that codes for one amino acid in a protein.

coefficient of interference Ratio of observed double crossovers to expected double crossovers.

cohesive end Short, single-stranded overhanging end on a DNA molecule produced when the DNA is cut by certain restriction enzymes. Cohesive ends are complementary and can spontaneously pair to rejoin DNA fragments that have been cut with the same restriction enzyme.

cointegrate structure Produced in replicative transposition, an intermediate structure in which two DNA molecules with two copies of the transposable element are fused.

colinearity Concept that there is a direct correspondence between the nucleotide sequence of a gene and the continuous sequence of amino acids in a protein.

colony *See* **bacterial colony.**

comparative genomics Comparative studies of the genomes of different organisms.

competent cell Capable of taking up DNA from its environment (capable of being transformed).

complementary Refers to the relation between the two nucleotide strands of DNA in which each purine on one strand pairs with a specific pyrimidine on the opposite strand (A pairs with T, and G pairs with C).

complementation Two different mutations in the heterozygous condition are exhibited as the wild-type phenotype; indicates that the mutations are at different loci.

complementation test Test designed to determine whether two different mutations are at the same locus (are allelic) or at different loci (are nonalleleic). Two individuals that are homozygous for two independently derived mutations are crossed, producing F_1 progeny that are heterozygous for the mutations. If the mutations are at the same locus, the F_1 will have a mutant phenotype. If the mutations are at different loci, the F_1 will have a wild-type phenotype.

complete linkage Occurs when genes are located together on the same chromosome and there is no crossing over between them.

complete medium Used to culture bacteria or some other microorganism; contains all the nutrients required for growth and synthesis, including those normally synthesized by the organism. Nutritional mutants can grow on complete medium.

composite transposon Type of transposable element in bacteria that consists of two insertion sequences flanking a segment of DNA.

concept of dominance Principle of heredity discovered by Mendel stating that, when two different alleles are present in a genotype, only one allele may be expressed in the phenotype. The dominant allele is the allele that is expressed, and the recessive allele is the allele that is not expressed.

concordance Percentage of twin pairs in which both twins have a trait.

concordant Refers to a pair of twins both of whom have the trait under consideration.

conditional mutation Expressed only under certain conditions.

conjugation Mechanism by which genetic material may be exchanged between bacterial cells. During conjugation, two bacteria lie close together and a cytoplasmic connection forms between them. A plasmid or sometimes a part of the bacterial chromosome passes through this connection from one cell to the other.

consanguinity Mating between related individuals.

consensus sequence The most common nucleotides found in a series of DNA sequences.

–10 consensus sequence (Pribnow box) Consensus sequence (TATAAT) found in most bacterial promoters about 10 bp upstream of the transcription start site.

–35 consensus sequence Consensus sequence (TTGACA) found in many bacterial promoters approximately 35 bp upstream of the transcription start site.

constitutive mutation Causes the continuous transcription of one or more structural genes.

contig Set of overlapping DNA fragments that have been ordered to form a continuous stretch of DNA sequence.

continuous characteristic Displays a large number of possible phenotypes that are not easily distinguished, such as human height.

continuous replication Replication of the leading strand in the same direction as unwinding, allowing new nucleotides to be added continuously to the 3′ end of the new strand as the template is exposed.

coordinate induction Simultaneous synthesis of several enzymes stimulated by a single environmental factor.

core element Consensus sequence in eukaryotic RNA polymerase I promoters that extends from –45 to +20 and is needed to initiate transcription; rich in guanine and cytosine nucleotides.

core enzyme Part of bacterial RNA polymerase that, during transcription, catalyzes the elongation of the RNA molecule by the addition of RNA nucleotides; consists of four subunits: two copies of alpha (α), a single copy of beta (β), and a single copy of beta prime (β').

corepressor Substance that inhibits transcription in a repressible system of gene regulation; usually a small molecule that binds to a repressor protein and alters it so that the repressor is able to bind to DNA and inhibit transcription.

core promoter DNA sequences immediately upstream of eukaryotic promoter, where the basal transcription apparatus binds.

correlation Degree of association between two or more variables.

correlation coefficient Statistic that measures the degree of association between two or more variables. A correlation coefficient can range from –1 to +1. A positive value indicates a direct relation between the variables; a negative correlation indicates an inverse relation. The absolute value of the correlation coefficient provides information about the strength of association between the variables.

cosmid Cloning vector that combines the properties of plasmids and phage vectors and is used to clone large pieces of DNA in bacteria. Cosmids are small plasmids that carry λ *cos* sites, allowing the plasmid to be packaged into viral coats.

cotransduction Process in which two or more genes are transferred together from one bacterial cell to another. Only genes located close together on a bacterial chromosome will be cotransduced.

cotransformation Process in which two or more genes are transferred together during cell transformation.

coupling *See* **cis configuration.**

CpG island DNA region that contains many copies of a cytosine base followed by a guanine base; often found near transcription start sites in eukaryotic DNA. The cytosine bases in CpG islands are commonly methylated when genes are inactive but are demethylated before initiation of transcription.

cross bridge In a heteroduplex DNA molecule, the point at which each nucleotide strand passes from one DNA molecule to the other.

crossing over Exchange of genetic material between homologous but nonsister chromatids.

cruciform Structure formed if inverted repeats on both strands of double-stranded DNA pair.

C value Haploid amount of DNA found in a cell of an organism.

cyclin A key protein in the control of the cell cycle; combines with a cyclin-dependent kinase (CDK). The levels of cyclin rise and fall in the course of the cell cycle.

cyclin-dependent kinase (CDK) A key protein in the control of the cell cycle; combines with cyclin.

cytokinesis Process by which the cytoplasm of a cell divides.

cytoplasmic inheritance Inheritance of characteristics encoded by genes located in the cytoplasm. Because the cytoplasm is usually contributed entirely by the one parent, cytoplasmically inherited characteristics are usually inherited from a single parent.

cytosine (C) Pyrimidine in DNA and RNA.

deamination Loss of an amino group (NH_2) from a base.

degenerate code Refers to the fact that the genetic code contains more information than is needed to specify all 20 common amino acids.

deletion Mutation in which nucleotides are deleted from a DNA sequence.

deletion mapping Technique for determining the chromsomal location of a gene by studying the association of its phenotype or product with particular chromosome deletions.

delta sequence Long terminal repeat in *Ty* elements of yeast.

denaturation Process of DNA strand separation when DNA is heated.

deoxyribonucleotide Basic building block of DNA, consisting of a deoxyribose sugar, a phosphate, and a nitrogenous base.

deoxyribose sugar Five-carbon sugar in DNA; lacks a hydroxyl group on the 2′-carbon atom.

depurination Break in the covalent bond connecting a purine base to the 1′-carbon atom of the deoxyribose sugar, resulting in the loss of the purine base. The resulting apurinic site cannot provide a template in replication, and a nucleotide with another base may be incorporated into the newly synthesized DNA strand opposite the apurinic site.

determination Process by which a cell becomes committed to developing into a particular cell type.

diakinesis Fifth substage of prophase I in meiosis. In diakinesis, chromosomes contract, the nuclear membrane breaks down, and the spindle forms.

dicentric bridge Structure produced when the two centromeres of a dicentric chromatid are pulled toward opposite poles, stretching the dicentric chromosome across the center of the nucleus. Eventually, the dicentric bridge breaks as the two centromeres are pulled apart.

dicentric chromatid Chromatid that has two centromeres; produced when crossing over takes place within a paracentric inversion. The two centromeres of the dicentric chromatid are frequently pulled toward opposite poles in mitosis or meiosis, breaking the chromosome.

dideoxyribonucleoside triphosphate (ddNTP) Special substrate for DNA synthesis used in the Sanger dideoxy sequencing method; identical with dNTP (the usual substrate for DNA synthesis) except it lacks a 3′-OH group. The incorporation of a ddNTP into DNA terminates DNA synthesis.

dihybrid cross A cross between two individuals that differ in two characteristics—more specifically, a cross between individuals that are homozygous for different alleles at the two loci (*AABB* × *aabb*); also refers to a cross between two individuals that are both heterozygous at two loci (*AaBb* × *AaBb*).

dioecious Refers to the presence of male and female reproductive structures in separate individuals.

diploid Possessing two sets of chromosomes (two genomes).

diplotene Fourth substage of prophase I during meiosis. In diplotene, centromeres of homologous chromosomes move apart, but the homologs remain attached at chiasmata.

directional selection Selection in which one trait or allele is favored over another.

direct repair DNA repair in which modified bases are changed back to their original structures.

discontinuous characteristic Exhibits only a few, easily distinguished phenotypes. An example is seed shape in which seeds are either round or wrinkled.

discontinuous replication Replication of the lagging strand in the direction opposite that of unwinding, which means that DNA must be synthesized in short stretches (Okazaki fragments).

discordant Refers to a pair of twins of whom one twin has the trait under consideration and the other does not.

displaced chromosome duplication Duplication of a chromosome segment in which the duplicated segment is some distance from the original segment.

dizygotic twins Nonidentical twins that arise when two different eggs are fertilized by two different sperm; also called fraternal twins.

D loop Region of mitochondrial DNA that contains an origin of replication and promoters; is displaced during initiation of replication, leading to the name displacement, or D, loop.

DNA fingerprinting Identifying individuals by comparing their DNA sequences.

DNA footprinting Technique used to determine which DNA sequences are bound by a protein.

DNA gyrase *E. coli* topoisomerase enzyme that relieves torsional strain that builds up ahead of the replication fork.

DNA helicase Protein that unwinds double-stranded DNA by breaking hydrogen bonds.

DNA library Collection of bacterial colonies containing all the DNA fragments from one source.

DNA ligase Enzyme that catalyzes the formation of a phosphodiester bond between adjacent 3′-OH and 5′-phosphate groups in a DNA molecule.

DNA methylation Modification of DNA by the addition of methyl groups to certain positions on the bases.

DNA polymerase Enzyme that synthesizes DNA.

DNA polymerase I DNA polymerase in bacteria that removes and replaces RNA primers with DNA nucleotides.

DNA polymerase II DNA polymerase in bacteria that takes part in DNA repair; restarts replication after synthesis has halted because of DNA damage.

DNA polymerase III DNA polymerase in bacteria that synthesizes a new nucleotide strand off the primers.

DNA polymerase IV DNA polymerase in bacteria; probably takes part in DNA repair.

DNA polymerase V DNA polymerase in bacteria; probably takes part in DNA repair.

DNA polymerase α DNA polymerase in eukaryotic cells that initiates replication on the lagging strand.

DNA polymerase β DNA polymerase in eukaryotic cells that participates in DNA repair.

DNA polymerase δ DNA polymerase in eukaryotic cells that replicates the leading strand and continues replication of the lagging strand after initiation by DNA polymerase α.

DNA polymerase ε DNA polymerase in eukaryotic cells that is similar in structure and function to DNA polymerase δ; its precise role in replication is not yet clear.

DNA polymerase γ DNA polymerase in eukaryotic cells that replicates mitochondrial DNA. A γ-like DNA polymerase replicates chloroplast DNA.

DNase I hypersensitive site Chromatin region that becomes sensitive to digestion by the enzyme DNase I.

DNA sequencing Process of determining the sequence of bases along a DNA molecule.

domain Functional part of a protein.

dominance genetic variance Component of the genetic variance that can be attributed to dominance (interaction between genes at the same locus).

dominant Refers to an allele or a phenotype that is expressed in homozygotes (*AA*) and in heterozygotes (*Aa*); only the dominant allele is expressed in a heterozygote phenotype.

dosage compensation Equalization in males and females of the amount of protein produced by X-linked genes. In placental mammals, dosage compensation is accomplished by the random inactivation of one X chromosome in the cells of females.

double fertilization Fertilization in plants; includes the fusion of a sperm cell with an egg cell to form a zygote and the fusion of a second sperm cell with the polar nuclei to form an endosperm.

double-strand-break model Model of homologous recombination in which a DNA molecule undergoes double-strand breaks.

down mutation Decreases the rate of transcription.

downstream core promoter element Consensus sequence [RG(A or T)CGTG] found in some eukaryotic RNA polymerase II core promoters; usually located approximately 30 bp downstream of the transcription start site.

Down syndrome Characterized by variable degrees of mental retardation, characteristic facial features, some retardation of growth and development, and an increased incidence of heart defects, leukemia, and other abnormalities; caused by the duplication of all or part of chromosome 21 (trisomy 21).

Edward syndrome Characterized by severe retardation, low-set ears, a short neck, deformed feet, clenched fingers, heart problems, and other disabilities; results from the presence of three copies of chromosome 18 (trisomy 18).

effective population size Effective number of breeding adults in a population; influenced by the number of individuals contributing genes to the next generation, their sex ratio, variation between individuals in reproductive success, fluctuations in population size, the age structure of the population, and whether mating is random.

egg Female gamete.

egg-polarity genes Genes that determine the major axes of development in an early fruit-fly embryo. One set of egg-polarity genes determines the anterior–posterior axis and another determines the dorsal–ventral axis.

elongation factor G (EF-G) Combines with GTP and is required for movement of the ribosome along the mRNA during translation.

elongation factor Ts (EF-Ts) Protein that regenerates elongation factor Tu in the elongation stage of protein synthesis.

elongation factor Tu (EF-Tu) Protein taking part in the elongation stage of protein synthesis; forms a complex with GTP and a charged amino acid and then delivers the charged tRNA to the ribosome.

end labeling Method for adding a radioactive or chemical label to the ends of DNA molecules.

endosymbiotic theory Proposes that some membrane-bounded organelles, such as mitochondria and chloroplasts, in eukaryotic cells originated as free-living eubacterial cells that entered into an endosymbiotic relation with a eukaryotic host cell and evolved into the present-day organelles; supported by a number of similarities in the structure and sequence between organelle and eubacterial DNAs.

enhancer Sequence that stimulates maximal transcription of distant genes; affects only genes on the same DNA molecule (is cis acting), contains short consensus sequences, is not fixed in relation to the transcription start site, can stimulate most any promoter in its vicinity, and may be upstream or downstream of the gene. The function of an enhancer is independent of sequence orientation.

environmental variance Component of the phenotypic variance that is due to environmental differences among individual members of a population.

episome Plasmid capable of integrating into a bacterial chromosome.

epistasis Type of gene interaction in which a gene at one locus masks or suppresses the effects of a gene at a different locus.

epistatic gene Gene that masks or suppresses the effect of a gene at a different locus.

equilibrium Situation in which no further change takes place; in population genetics, refers to a population in which allelic frequencies do not change.

equilibrium density gradient centrifugation Method used to separate molecules or organelles of different density by centrifugation.

eubacteria One of the three primary divisions of life. Eubacteria consist of unicellular organisms with prokaryotic cells and include most of the common bacteria.

euchromatin Chromatin that undergoes condensation and decondensation in the course of the cell cycle.

eukaryote Organism with a complex cell structure including a nuclear envelope and membrane-bounded organelles. Eukaryotes include unicellular and multicellular forms.

exit (E) site One of three sites in a ribosome occupied by a tRNA. In the elongation stage of translation, the tRNA moves from the peptidyl (P) site to the E site from which it then exits the ribosome.

exon Coding region of a split gene (a gene that is interrupted by introns). After processing, the exons remain in messenger RNA.

expanding trinucleotide repeat Mutation in which the number of copies of a trinucleotide (or some multiple of three nucleotides) increases in succeeding generations.

expected heterozygosity Proportion of individuals that are expected to be heterozygous at a locus when the Hardy-Weinberg assumptions are met.

expressed sequence tag (EST) Unique fragment of DNA from the coding region of a gene, produced by the reverse transcription of cellular RNA. Parts of the fragments are sequenced so that they can be identified.

expression vector Cloning vector containing DNA sequences such as a promoter, a ribosome-binding site, and transcription initiation and termination sites that allow DNA fragments inserted into the vector to be transcribed and translated.

expressivity Degree to which a trait is expressed.

familial Down syndrome Down syndrome caused by a Robertsonian translocation in which the long arm of chromosome 21 is translocated to another chromosome; tends to run in families.

fertilization Fusion of gametes, or sex cells, to form a zygote.

fetal cell sorting Separation of fetal cells from maternal blood. Genetic testing on the fetal cells can provide information about genetic diseases and disorders in the fetus.

F factor Episome of *E. coli* that controls conjugation and gene exchange between *E. coli* cells. The F factor contains an origin of replication and genes that enable the bacterium to undergo conjugation.

F_1 (first filial) generation Offspring of the initial parents (P) in a genetic cross.

first polar body One of the products of meiosis I in oogenesis; contains half the chromosomes but little of the cytoplasm.

fitness Reproductive success of a genotype compared with other genotypes in a population.

5′ cap Modified 5′ end of eukaryotic mRNA, consisting of an extra nucleotide (methylated) and methylation of the 2′ position of the sugar in one or more subsequent nucleotides; plays a role in the binding of the ribosome to mRNA and affects mRNA stability and the removal of introns.

5′ end End of the polynucleotide chain where a phosphate is attached to the 5′-carbon atom of the nucleotide.

5′ splice site The 5′ end of an intron where cleavage takes place in RNA splicing.

5′ untranslated (UTR) region Sequence of nucleotides at the 5′ end of mRNA; does not code for the amino acids of a protein.

fixation Occurs when one allele reaches a frequency of 1, at which point all individuals in the population are homozygous for the same allele.

flanking direct repeat Short, directly repeated sequence produced on either side of a transposable element when it inserts into DNA.

forward mutation Alters a wild-type phenotype.

founder effect Sampling error that arises when a population is established by a small number of individuals; leads to genetic drift.

fragile site Constriction or gap that appears at a particular location on a chromosome when cells are cultured under special conditions. One fragile site on the human X chromosome is associated with mental retardation (fragile-X syndrome) and results from an expanding trinucleotide repeat.

frameshift mutation Alters the reading frame of a gene.

fraternal twins Nonidentical twins that arise when two different eggs are fertilized by two different sperm.

frequency distribution Graphical way of representing values. In genetics, usually the phenotypes found in a group of individuals are displayed as a frequency distribution. Typically, the phenotypes are plotted on the horizontal (x) axis and the numbers (or proportions) of individuals with each phenotype are plotted on the vertical (y) axis.

F_2 (second filial) generation Offspring of the F_1 generation in a genetic cross; the third generation of a genetic cross.

functional genomics Area of genomics that studies the functions of genetic information contained within genomes.

fusion pattern Method of using protein fusion to infer gene function. Two proteins that are separate in one species but exist as a fused protein in another species suggests that the two separate proteins in the first species may be functionally related.

gain-of-function mutation Produces a new trait or causes a trait to appear in inappropriate tissues or at inappropriate times in development.

gametophyte Haploid phase of the life cycle in plants.

gap genes Set of segmentation genes in fruit flies that define large sections of the embryo. Mutations in these genes usually eliminate whole groups of adjacent segments.

gel electrophoresis Technique for separating charged molecules (such as proteins or nucleic acids) on the basis of molecular size or charge or both.

gene Genetic factor that helps determine a trait; often defined at the molecular level as a DNA sequence that is transcribed into an RNA molecule.

gene cloning Inserting DNA fragments into bacteria in such a way that the fragments will be stable and copied by the bacteria.

gene flow Movement of genes from one population to another; also called migration.

gene interaction Interactions between genes at different loci that affect the same characteristic.

gene mutation Affects a single gene or locus.

gene neighbor analysis Analysis of the locations of genes in different species to infer gene function. If two genes are consistently linked in different species, they may be functionally related.

gene pool Total of all genes in a population.

generalized transduction Transduction in which any gene may be transferred from one bacterial cell to another by a virus.

general transcription factor Protein that binds to eukaryotic promoters near the start site and is a part of the basal transcription apparatus that initiates transcription.

gene regulation Mechanisms and processes that control the phenotypic expression of genes.

gene therapy Use of recombinant DNA to treat a disease or disorder by altering the genetic makeup of the patient's cells.

genetic bottleneck Sampling error that arises when a population undergoes a drastic reduction in population size; leads to genetic drift.

genetic correlation Phenotypic correlation due to the same genes affecting two or more characteristics.

genetic counseling Educational process that attempts to help patients and family members deal with all aspects of a genetic condition.

genetic drift Change in allelic frequency due to sampling error.

genetic engineering Common term for recombinant DNA technology.

genetic–environmental interaction variance Component of the phenotypic variance that results from an interaction between genotype and environment. Genotypes are expressed differently in different environments.

genetic map Map of the relative distances between genetic loci, markers, or other chromosome regions determined by rates of recombination; measured in percent recombination or map units.

genetic marker Any gene or DNA sequence used to identify a location on a genetic or physical map.

genetic maternal effect The phenotype of an offspring is determined by the nuclear genotype of the mother. With genetic maternal effect, an offspring inherits genes for the characteristics from both parents, but the offspring's phenotype is determined not by its own genotype but by the genotype of its mother.

genetic variance Component of the phenotypic variance that is due to genetic differences among individual members of a population.

genic balance system Sex-determining system in which sexual phenotype is controlled by a balance between genes on the X chromosome and genes on the autosomes.

genic interaction variance Component of the genetic variance that can be attributed to genic interaction (interaction between genes at different loci).

genic sex determination Sex determination in which the sexual phenotype is specified by genes at one or more loci, but there are no obvious differences in the chromosomes of males and females.

genome Complete set of genetic instructions for an organism.

genomic imprinting The expression of a gene is affected by the sex of the parent that transmitted the gene; if the gene is inherited from the father, its expression is different from that if it is inherited from the mother.

genomic library Collection of bacterial or phage colonies containing DNA fragments that consist of the entire genome of an organism.

genomics Study of the content, organization, and function of genetic information in whole genomes.

genotype The set of genes that an individual possesses.

genotypic frequency Frequency or proportion of a particular genotype.

germ-line mutation Mutation in a germ-line cell (one that gives rise to gametes).

germ-plasm theory Theory stating that cells in the reproductive organs carry a complete set of genetic information.

G_0 (gap 0) Nondividing stage of the cell cycle.

G_1 (gap 1) Stage in interphase of the cell cycle when the cell grows and develops.

G_2 (gap 2) Stage of interphase in the cell cycle that follows DNA replication. In G_2, the cell prepares for division.

G_2/M (gap 2/mitotic) checkpoint Important checkpoint in the cell cycle near the end of G_2. After this checkpoint has been passed, the cell undergoes mitosis.

goodness-of-fit chi-square test Statistical test used to evaluate how well a set of observed values fit the expected values. The probability associated with a calculated chi-square value is the probability that the differences between the observed and the expected values may be due to chance.

group I introns Class of introns in some ribosomal RNA genes that are capable of self-splicing.

group II introns Class of introns in some protein-encoding genes that are capable of self-splicing and are found in mitochondria, chloroplasts, and a few eubacteria.

G_1/S (gap 1/synthesis) checkpoint Important checkpoint in the cell cycle. After the G_1/S checkpoint has been passed, DNA replicates and the cell is committed to dividing.

guanine (G) Purine in DNA and RNA.

guide RNA (gRNA) RNA molecule that serves as a template for an alteration made in mRNA during RNA editing.

gynandromorph Individual that is a mosaic for the sex chromosomes, possessing tissues with different sex-chromosome constitutions.

hairpin Secondary structure formed when nucleotides on the same strand are complementary and pair with each other.

haploid Possessing a single set of chromosomes (one genome).

haploinsufficient gene Must be present in two copies for normal function. If one copy of the gene is missing, a mutant phenotype is produced.

Hardy-Weinberg equilibrium Frequencies of genotypes when the conditions of the Hardy-Weinberg law are met.

Hardy-Weinberg law Important principle of population genetics stating that, in a large, randomly mating population not affected by mutation, migration, or natural selection, allelic frequencies will not change and genotypic frequencies stabilize after one generation in the proportions p^2 (the frequency of AA), $2pq$ (the frequency of Aa), and q^2 (the frequency of aa), where p equals the frequency of allele A and q equals the frequency of allele a.

heat-shock protein Protein produced by many cells in response to extreme heat and other stresses; helps cells prevent damage from such stressing agents.

hemizygous Possessing a single allele at a locus. Males of organisms with XX-XY sex determination are hemizygous for X-linked loci, because their cells possess a single X chromosome.

heritability Proportion of phenotypic variation due to genetic differences. *See* **broad-sense heritability** and **narrow-sense heritability.**

hermaphroditism Condition in which an individual possesses both male and female reproductive structures. True hermaphrodites produce both male and female gametes.

heterochromatin Chromatin that remains in a highly condensed state throughout the cell cycle, found at the centromeres and telomeres of most chromosomes.

heteroduplex DNA DNA consisting of two strands, each of which is from a different chromosome.

heterogametic sex The sex (male or female) that produces two types of gametes with respect to sex chromosomes. For example, in the XX-XY sex-determining system, the male produces both X-bearing and Y-bearing gametes.

heterokaryon Cell possessing two nuclei derived from different cells through cell fusion.

heteroplasmy Presence of two or more distinct variants of DNA within the cytoplasm of a single cell.

heterozygote screening Testing members of a population to identify heterozygous carriers of a disease-causing allele who are healthy but have the potential to produce children with the disease.

heterozygous Refers to an individual that possesses two different alleles at a locus.

highly repetitive DNA DNA that consists of short sequences that are present in hundreds of thousands to millions of copies; clustered in certain regions of chromosomes.

high-mobility-group proteins Small, highly charged proteins that vary in amount and composition in different tissues and different stages of the cell cycle; may play an important role in chromatin structure.

histone Low-molecular-weight protein found in eukaryotes that complexes with DNA to form chromosomes.

Holliday intermediate Structure that forms in homologous recombination; consists of two duplex molecules connected by a cross bridge.

Holliday junction Model of homologous recombination that is initiated by single-strand breaks in a DNA molecule.

holoenzyme Complex of enzyme and other protein factors necessary for complete function.

homeobox Conserved subset of nucleotides in homeotic genes. In *Drosophila,* it consists of 180 nucleotides that encode 60 amino acids of a DNA-binding domain related to the helix-turn-helix motif.

homeotic complex Major cluster of homeotic genes in fruit flies; consists of the *Antennapedia* complex, which affects development of the adult fly's head and anterior segments, and the *bithorax* complex, which affects the adult fly's posterior thoracic and abdominal segments.

homeotic gene Gene that determines the identity of individual segments or parts in an early embryo. Mutations in such genes cause body parts to appear in the wrong places.

homogametic sex The sex (male or female) that produces gametes that are all alike with regard to sex chromosomes. For example, in the XX-XY sex-determining system, the female produces only X-bearing gametes.

homologous genes Evolutionarily related genes, having descended from a gene in a common ancestor.

homologous pair of chromosomes Two chromosomes that are alike in structure and size and that carry genetic information for the same set of hereditary characteristics. One chromosome of a homologous pair is inherited from the male parent and the other is inherited from the female parent.

homologous recombination Exchange of genetic information between homologous DNA molecules.

homoplasmy Presence of only one version of DNA within the cytoplasm of a single cell.

homozygous Refers to an individual that possesses two identical alleles at a locus.

horizontal gene transfer Transfer of genetic information from one species to another in ways other than common descent.

***Hox* gene** Gene that contains a homeobox.

humoral immunity Type of immunity resulting from antibodies produced by B cells.

hybrid dysgenesis Sudden appearance of numerous mutations, chromosome aberrations, and sterility in the offspring of a cross between a male fly that possesses *P* elements and a female fly that lacks them.

hybridization Pairing of two partly or fully complementary single-stranded nucleotide chains.

hypostatic gene Gene that is masked or suppressed by the action of a gene at a different locus.

identical twins Twins that arise when a single egg fertilized by a single sperm splits into two separate embryos.

inbreeding Mating between related individuals that takes place more frequently than expected on the basis of chance.

inbreeding coefficient Measure of inbreeding; the probability (ranging from 0 to 1) that two alleles are identical by descent.

inbreeding depression Decreased fitness arising from inbreeding; often due to the increased expression of lethal and deleterious recessive traits.

incomplete dominance Refers to the phenotype of a heterozygote that is intermediate between the phenotypes of the two homozygotes.

incomplete linkage Linkage between genes with crossing over; intermediate in its effects between independent assortment and complete linkage.

incomplete penetrance The genotype does not always express the expected phenotype; some individuals possess the genotype for a trait but do not express the phenotype.

incorporated error Incorporation of a damaged nucleotide or mismatched base pair into a DNA molecule.

independent assortment Independent separation of chromosome pairs in anaphase I of meiosis; contributes to genetic variation.

induced mutation Results from environmental agents, such as chemicals or radiation.

inducer Substance that stimulates transcription in an inducible system of gene regulation; usually a small molecule that binds to a repressor protein and alters that repressor so that it can no longer bind to DNA and inhibit transcription.

inducible operon Operon or other system of gene regulation in which transcription is normally off. Something must happen for transcription to be induced, or turned on.

induction Stimulation of the synthesis of an enzyme by an environmental factor, often the presence of a particular substrate.

in-frame deletion Deletion of some multiple of three nucleotides, which does not alter the reading frame of the gene.

in-frame insertion Insertion of some multiple of three nucleotides, which does not alter the reading frame of the gene.

inheritance of acquired characteristics Early notion of inheritance proposing that acquired traits are passed to descendants.

initiation codon The codon in mRNA that specifies the first amino acid (fMet in bacterial cells; Met in eukaryotic cells) of a protein; most commonly AUG.

initiation factor 1 (IF-1) Protein required for the initiation of translation in bacterial cells; enhances the dissociation of the large and small subunits of the ribosome.

initiation factor 2 (IF-2) Protein required for the initiation of translation in bacterial cells; forms a complex with GTP and the charged initiator protein and then delivers the charged tRNA to the initiation complex.

initiation factor 3 (IF-3) Protein required for the initiation of translation in bacterial cells; binds to the small subunit of the ribosome and prevents the large subunit from binding during initiation.

initiator protein Protein that binds to an origin of replication and unwinds a short stretch of DNA, allowing helicase and other single-strand-binding proteins to bind and initiate replication.

insertion Mutation in which nucleotides are added to a DNA sequence.

insertion sequence Simple type of transposable element found in bacteria and their plasmids that contains only the information necessary for its own movement.

in situ hybridization Method used to determine the chromosomal location of a gene or other specific DNA fragment or the tissue distribution of an mRNA by using a labeled probe that is complementary to the sequence of interest.

insulator DNA sequence that blocks or insulates the effect of an enhancer; must be located between the enhancer and the promoter to have blocking activity; also may limit the spread of changes in chromatin structure.

integrase Enzyme that inserts prophage, or proviral, DNA into a chromosome.

intercalating agent Chemical substance that is about the same size as a nucleotide and may become sandwiched between adjacent bases in DNA, distorting the three-dimensional structure of the helix and causing single-nucleotide insertions and deletions in replication.

interchromosomal recombination Recombination among genes on different chromosomes.

interference Degree to which one crossover interferes with additional crossovers.

intergenic suppressor mutation Occurs in a gene (locus) that is different from the gene containing the original mutation.

interkinesis Period between meiosis I and meiosis II.

internal promoter Located within the sequences of DNA that are transcribed into RNA.

interphase Period in the cell cycle between the cell divisions. In interphase, the cell grows, develops, and prepares for cell division.

interspersed repeat sequences Repeated sequences at multiple locations throughout the genome.

intrachromosomal recombination Recombination among genes located on the same chromosome.

intragenic mapping Mapping the locations of mutations within a single locus.

intragenic suppressor mutation Occurs in the same gene (locus) as the mutation that it suppresses.

intron Intervening sequence in a split gene; removed from the RNA after transcription.

inverted repeats Sequences on the same strand that are inverted and complementary.

isoaccepting tRNAs Different tRNAs with different anticodons that code for the same amino acid.

isotopes Different forms of an element that have the same number of protons and electrons but differ in the number of neutrons in the nucleus.

junctional diversity Addition or deletion of nucleotides at the junctions of gene segments brought together in the somatic recombination of genes that encode antibodies and T-cell receptors.

karyotype Picture of an individual's complete set of metaphase chromosomes.

kinetochore Set of proteins that assemble on the centromere, providing the point of attachment for spindle microtubules.

Klinefelter syndrome Human condition in which cells contain one or more Y chromosomes along with multiple X chromosomes (most commonly XXY but may also be XXXY, XXXXY, or XXYY). Persons with Klinefelter syndrome are male in appearance but frequently possess small testes, some breast enlargement, and reduced facial and pubic hair; often taller than normal and sterile, most have normal intelligence.

knockout mouse Mouse in which a normal gene has been disabled ("knocked out").

lagging strand DNA strand that is replicated discontinuously.

large ribosomal subunit The larger of the two subunits of a functional ribosome.

lariat Looplike structure created in splicing of nuclear pre-mRNA in which the 5′ end of an intron is attached to a branch point in pre-mRNA.

leading strand Strand that is replicated continuously.

leptotene First substage of prophase I in meiosis. In leptotene, chromosomes contract and become visible.

lethal allele Causes the death of an individual, often early in development, and so the individual does not appear in the progeny of a genetic cross. Recessive lethal alleles kill individuals that are homozygous for the allele; dominant lethals kill both heterozygotes and homozygotes.

lethal mutation Causes premature death.

LINE *See* **long interspersed element.**

linkage group Genes located together on the same chromosome.

linked genes Genes located on the same chromosome.

linker DNA Stretch of DNA separating two nucleosomes.

local variation Variation in secondary structure within a single molecule.

locus Position on a chromosome where a specific gene is located.

lod (logarithm of odds) score The logarithm of the ratio of the probability of obtaining a set of observations, assuming a specified degree of linkage, to the probability of obtaining the same set of observations with independent assortment; used to assess the likelihood of linkage between genes from pedigree data.

long interspersed element (LINE) Long DNA sequence repeated many times and interspersed throughout the genome.

loss-of-function mutation Causes the complete or partial absence of normal function.

Lyon hypothesis Proposed by Mary Lyon in 1961, this hypothesis states that one X chromosome in each female cell becomes inactivated (a Barr body) and suggests that which X becomes inactivated is random and varies from cell to cell.

lysogenic cycle Life cycle of a bacteriophage in which phage genes first integrate into the bacterial chromosome and are not immediately transcribed and translated.

lytic cycle Life cycle of a bacteriophage in which phage genes are transcribed and translated, new phage particles are produced, and the host cell is lysed.

major histocompatibility complex (MHC) antigens Large and diverse group of antigens found on the surfaces of cells that mark those cells as self; encoded by a large cluster of genes known as the major histocompatibility complex. T cells simultaneously bind to foreign and MHC antigens.

malignant tumor Consists of cells that are capable of invading other tissues.

map-based sequencing Method of sequencing a genome in which sequenced fragments are ordered into contigs with the use of genetic or physical maps.

map unit (m.u.) Unit of measure for distances on a genetic map; 1 map unit equals 1% recombination.

maternal blood testing Testing for genetic conditions in a fetus by analyzing the blood of the mother. For example, the level of α-fetoprotein in maternal blood provides information about the probability that a fetus has a neural-tube defect.

mean Statistic that describes the center of a distribution of measurements; calculated by dividing the sum of all measurements by the number of measurements; also called the average.

megaspore One of the four products of meiosis in plants.

megasporocyte Diploid reproductive cell in the ovary of a plant that undergoes meiosis to produce haploid macrospores.

meiosis I First phase of meiosis. In meiosis I, chromosome number is reduced by half.

meiosis II Second phase of meiosis. Events in meiosis II are essentially the same as those in mitosis.

melting *See* **denaturation.**

melting temperature Midpoint of the melting range of DNA.

memory cell Long-lived lymphocyte among the clone of cells generated when a foreign antigen is encountered. If the same antigen is encountered again, the memory cells quickly divide and give rise to another clone of cells specific for that particular antigen.

Mendelian population Group of interbreeding, sexually reproducing individuals.

meristic characteristic Characteristic whose phenotype varies in whole numbers, such as number of vertebrae.

merozygote Bacterial cell that has two copies of some genes— one copy on the bacterial chromosome and a second copy on an introduced F plasmid; also called partial diploid.

messenger RNA (mRNA) RNA molecule that carries genetic information for the amino acid sequence of a protein.

metaphase Stage of mitosis. In metaphase, chromosomes align in the center of the cell.

metaphase I Stage of meiosis I. In metaphase I, homologous pairs of chromosomes align in the center of the cell.

metaphase II Stage of meiosis II. In metaphase II, individual chromosomes align on the metaphase plate.

metaphase plate Plane in a cell between two spindle poles. In metaphase, chromosomes align on the metaphase plate.

metastasis Refers to cells from malignant tumors that separate and travel to other sites, where they establish secondary tumors.

5′-methylcytosine Modified nucleotide, consisting of cytosine to which a methyl group has been added; predominate form of methylation in eukaryotic DNA.

microarray Ordered array of DNA fragments fixed to a solid support, which serve as probes to detect the presence of complementary sequences; often used to assess the expression of genes in various tissues and under different conditions.

microspore Haploid product of meiosis in plants.

microsporocyte Diploid reproductive cell in the stamen of a plant; undergoes meiosis to produce four haploid microspores.

microtubule Long fiber composed of the protein tubulin; plays an important role in the movement of chromosomes in mitosis and meiosis.

migration Movement of genes from one population to another; also called gene flow.

minimal medium Used to culture bacteria or some other microorganism; contains only the nutrients required by prototrophic (wild-type) cells, typically a carbon source, essential elements such as nitrogen and phosphorus, certain vitamins, and other required ions and nutrients.

mismatch repair Process that corrects mismatched nucleotides in DNA after replication has been completed. Enzymes excise incorrectly paired nucleotides from the newly synthesized strand and use the original nucleotide strand as a template when replacing them.

missense mutation Alters a codon in the mRNA, resulting in a different amino acid in the protein.

mitochondrial DNA (mtDNA) DNA in mitochondria; has some characteristics in common with eubacterial DNA and typically consists of a circular molecule that lacks histone proteins and encodes some of the rRNAs, tRNAs, and proteins found in mitochondria.

mitosis Process by which the nucleus of a eukaryotic cell divides.

mitotic spindle Array of microtubules that radiate from two poles; moves chromosomes in mitosis and meiosis.

M (mitotic) phase Period of active cell division; includes mitosis (nuclear division) and cytokinesis (cytoplasmic division).

moderately repetitive DNA DNA consisting of sequences that are from 150 to 300 bp in length and are repeated thousands of times.

modified base Rare base found in some RNA molecules. Such bases are modified forms of the standard bases (adenine, guanine, cytosine, and uracil).

molecular chaperone Molecule that assists in the proper folding of another molecule.

molecular clock Refers to the use of molecular differences to estimate the time of divergence between organisms; assumes a roughly constant rate at which one neutral mutation replaces another.

molecular genetics The study of the chemical nature of genetic information and how it is encoded, replicated, and expressed.

molecular motor Specialized protein that moves cellular components.

monoecious Refers to the presence of both male and female reproductive structures in the same individual.

monohybrid cross A cross between two individuals that differ in a single characteristic—more specifically, a cross between individuals that are homozygous for different alleles at the same locus ($AA \times aa$); also refers to a cross between two individuals that are both heterozygous for two alleles at a single locus ($Aa \times Aa$).

monosomy Absence of one of the chromosomes of a homologous pair.

monozygotic twins Identical twins that arise when a single egg fertilized by a single sperm splits into two separate embryos.

morgan 100 map units.

morphogen Molecule whose concentration gradient affects the developmental fate of surrounding cells.

mosaicism Condition in which regions of tissue within a single individual have different chromosome constitutions.

M-phase promoting factor (MPF) Protein functioning in the control of the cell cycle; consists of a cyclin combined with cyclin-dependent kinase (CDK). Active MPF stimulates mitosis.

multifactorial characteristic Determined by multiple genes and environmental factors.

multiple alleles Presence in a group of individuals of more than two alleles at a locus.

multiple 3′ cleavage sites Refers to the presence of more than one 3′ cleavage site on a single pre-mRNA, which allows cleavage and polyadenylation to take place at different sites, producing mRNAs of different lengths.

multiplication rule Rule of probability stating that the probability of two or more independent events occurring together is calculated by multiplying the probabilities of each of the individual events.

mutagen Any environmental agent that significantly increases the rate of mutation above the spontaneous rate.

mutagenesis screen Method for identifying genes that influence a specific phenotype. Random mutations are induced in a population of organisms, and individual organisms with mutant phenotypes are identified. These individual organisms are crossed to determine the genetic basis of the phenotype and to map the location of mutations that cause the phenotype.

mutation Heritable change in genetic information.

mutation frequency Number of mutations within a group of individual organisms.

mutation rate Frequency with which a gene changes from the wild type to a specific mutant; generally expressed as the number of mutations per biological unit (i.e., mutations per cell division, per gamete, or per round of replication).

narrow-sense heritability Proportion of the phenotypic variance that can be attributed to additive genetic variance.

natural selection Differential reproduction of genotypes.

negative assortative mating Mating between unlike individuals that takes place more frequently than expected on the basis of chance.

negative control Gene regulation in which the binding of a regulatory protein to DNA inhibits transcription (the regulatory protein is a repressor).

negative-strand RNA virus RNA virus whose genomic RNA molecule carries the complement of the information for viral proteins. A negative-strand RNA virus must first make a complementary copy of its RNA genome, which is then translated into viral proteins.

negative supercoiling *See* **supercoiling**.

neutral mutation Changes the amino acid sequence of a protein but does not alter the function of the protein.

neutral-mutation hypothesis Proposes that much of the molecular variation seen in natural populations is adaptively neutral and unaffected by natural selection. Under this hypothesis, individuals with different molecular variants have equal fitnesses.

newborn screening Testing newborn infants for certain genetic disorders; done most commonly for phenylketonuria and other metabolic diseases that can be prevented by early treatment or intervention.

nitrogenous base Nitrogen-containing base that is one of the three parts of a nucleotide.

nonautonomous element Transposable element that cannot transpose on its own but can transpose in the presence of an autonomous element of the same family.

nondisjunction Failure of homologous chromosomes or sister chromatids to separate in meiosis or mitosis.

nonhistone chromosomal proteins Heterogeneous assortment of nonhistone proteins in chromatin.

nonoverlapping genetic code Refers to the fact that generally each nucleotide codes for only one amino acid in a protein.

nonreciprocal translocation Movement of a chromosome segment to a nonhomologous chromosome or region without any (or with unequal) reciprocal exchange of segments.

nonrecombinant gamete Contains only original combinations of genes present in the parents.

nonrecombinant progeny Possesses the original combinations of traits possessed by the parents.

nonreplicative transposition Type of transposition in which a transposable element excises from an old site and moves to a new site, resulting in no net increase in the number of copies of the transposable element.

nonsense codon Codon in mRNA that signals the end of translation; also called stop codon or termination codon. There are three common nonsense codons: UAA, UAG, and UGA.

nonsense mutation Changes a sense codon (one that specifies an amino acid) into a stop codon.

nontemplate strand The DNA strand that is complementary to the template strand; not ordinarily used as a template during transcription.

normal distribution Common type of frequency distribution that exhibits a symmetrical, bell-shaped curve; usually arises when a large number of independent factors contribute to the measurement.

norm of reaction Range of phenotypes produced by a particular genotype in different environmental conditions.

Northern blotting Process by which RNA is transferred from a gel to a solid support such as a nitrocellulose or nylon filter.

nuclear envelope Membrane that surrounds the genetic material in eukaryotic cells to form a nucleus; segregates the DNA from other cellular contents.

nuclear matrix Network of protein fibers in the nucleus; holds the nuclear contents in place.

nuclear pre-mRNA introns Class of introns in protein-encoding genes that reside in the nuclei of eukaryotic cells; removed by spliceosomal-mediated splicing.

nucleoid Bacterial DNA confined to a definite region of the cytoplasm.

nucleoside Ribose or deoxyribose sugar bonded to a base.

nucleosome Basic repeating unit of chromatin, consisting of a core of eight histone proteins (two each of H2A, H2B, H3, and H4) and about 146 bp of DNA that wraps around the core about two times.

nucleotide Repeating unit of DNA, made up of a sugar, a phosphate, and a base.

nucleotide-excision repair DNA repair that removes bulky DNA lesions and other types of DNA damage.

nucleus Space in eukaryotic cells that is enclosed by the nuclear envelope and contains the chromosomes.

nullisomy Absence of both chromosomes of a homologous pair $(2n - 1)$.

NusA factor Protein subunit of bacterial RNA polymerase that facilitates termination of transcription.

Okazaki fragments Short stretches of newly synthesized DNA on the lagging strand that are eventually joined together.

oligonucleotide-directed mutagenesis Method of site-directed mutagenesis that utilizes an oligonucleotide to introduce a mutant sequence into a DNA molecule.

oncogene Dominant-acting gene that stimulates cell division, leading to the formation of tumors and contributing to cancer; arises from mutated copies of a normal cellular gene (proto-oncogene).

one gene, one enzyme hypothesis Idea proposed by Beadle and Tatum that each gene encodes a separate enzyme.

one gene, one polypeptide hypothesis Modification of the one gene, one enzyme hypothesis; proposes that each gene encodes a separate polypeptide chain.

oogenesis Egg production in animals.

oogonium Diploid cell in the ovary; capable of undergoing meiosis to produce an egg cell.

open reading frame (ORF) Continuous sequence of DNA nucleotides that contains a start codon and a stop codon in the same reading frame; is assumed to be a gene that encodes a protein but in many cases the protein has not yet been identified.

operator DNA sequence in the operon of a bacterial cell. A regulator protein binds to the operator and affects the rate of transcription of structural genes.

operon Set of structural genes in a bacterial cell along with a common promoter and other sequences (such as an operator) that control the transcription of the structural genes.

origin of replication Site where DNA synthesis is initiated.

orthologous genes Homologous genes found in different species, because the two species have a common ancestor that also possessed the gene.

outcrossing Mating between unrelated individuals that takes place more frequently than expected on the basis of chance.

overdominance Selection in which the heterozygote has higher fitness than that of either homozygote; also called heterozygote advantage.

ovum Final product of oogenesis.

pachytene Third substage of prophase I in meiosis. The synaptonemal complex forms during pachytene.

pair-rule genes Set of segmentation genes in fruit flies that define regional sections of the embryo and affect alternate segments. Mutations in these genes often cause the deletion of every other segment.

palindrome Sequence of nucleotides that reads the same on complementary strands; inverted repeats.

pangenesis Early concept of heredity proposing that particles carry genetic information from different parts of the body to the reproductive organs.

paracentric inversion Chromosome inversion that does not include the centromere in the inverted region.

paralogous genes Homologous genes in the same species that arose through the duplication of a single ancestral gene.

parental gamete *See* **nonrecombinant gamete.**

parental progeny *See* **nonrecombinant progeny.**

partial diploid Bacterial cell that possesses two copies of genes, including one copy on the bacterial chromosome and the other on an extra piece of DNA (usually a plasmid); also called merozygote.

Patau syndrome Characterized by severe mental retardation, a small head, sloping forehead, small eyes, cleft lip and palate, extra fingers and toes, and other disabilities; results from the presence of three copies of chromosome 13 (trisomy 13).

pedigree Pictorial representation of a family history outlining the inheritance of one or more traits or diseases.

penetrance Percentage of individuals with a particular genotype that express the phenotype expected of that genotype.

pentaploidy Possessing five haploid sets of chromosomes (5n).

peptide bond Chemical bond that connects amino acids in a protein.

peptidyl (P) site One of three sites in a ribosome occupied by a tRNA in translation. In the elongation stage of protein synthesis, tRNAs move from the aminoacyl (A) site into the P site.

peptidyl transferase Activity in the ribosome that creates a peptide bond between two amino acids. Evidence suggests that this activity is carried out by one of the RNA components of the ribosome.

pericentric inversion Chromosome inversion that includes the centromere in the inverted region.

phage *See* **bacteriophage.**

phenocopy Phenotype produced by environmental effects that is the same as the phenotype produced by a genotype.

phenotype Appearance or manifestation of a characteristic.

phenotypic correlation Correlation between two or more phenotypes in the same individual.

phenotypic variance Measures the degree of phenotypic differences among a group of individuals; composed of genetic, environmental, and genetic-environmental interaction variances.

phosphate group A phosphorus atom attached to four oxygen atoms; one of the three components of a nucleotide.

phosphodiester Molecule containing R–O–P–O–R, where R is a carbon-containing group, O is oxygen, and P is phosphorus.

phosphodiester linkage Phosphodiester bond connecting two nucleotides in a polynucleotide strand.

phylogenetic profile The presence-and-absence pattern of genes in different species, which may be used to infer gene function. A presence-and-absence pattern that is the same in different organisms suggests that the genes may be functionally related.

phylogeny Evolutionary relationships among a group of organisms or genes, usually depicted as a family tree or branching diagram.

physical map Map of physical distances between loci, genetic markers, or other chromosome segments; measured in base pairs.

pilus (pl., pili) Extension of the surface of some bacteria that allows conjugation to take place. When a pilus on one cell makes contact with a receptor on another cell, the pilus contracts and pulls the two cells together.

plaque Clear patch of lysed cells on a continuous layer of bacteria on the agar surface of a petri plate. Each plaque represents a single original phage that multiplied and lysed many cells.

plasmid Small, circular DNA molecule found in bacterial cells that is capable of replicating independently from the bacterial chromosome.

pleiotropy A single genotype influences multiple phenotypes.

poly(A)-binding protein (PABP) Binds to the poly(A) tail of eukaryotic mRNA and makes the mRNA more stable. There are several types of PABPs, one of which is PABII.

poly(A) tail String of adenine nucleotides added to the 3′ end of eukaryotic mRNAs after transcription.

polygenic characteristic Encoded by genes at many loci.

polymerase chain reaction (PCR) Method of enzymatically amplifying DNA fragments.

polynucleotide strand Series of nucleotides linked together by phosphodiester bonds.

polypeptide Chain of amino acids linked by peptide bond; also called a protein.

polyploidy Possessing more than two haploid sets of chromosomes.

polyribosome Messenger RNA molecule with several ribosomes attached to it.

polytene chromosome Giant chromosome in the salivary glands of *Drosophila melanogaster;* each polytene chromosome consists of a number of DNA molecules lying side by side.

population The group of interest; often represented by a subset called a sample. Also, a group of individuals of the same species.

population genetics The study of the genetic composition of populations (groups of individuals of the same species) and how a population's collective group of genes changes through time.

positional cloning Method that allows for the isolation and identification of a gene by examining cosegregation of a phenotype with previously mapped genetic markers.

position effect The expression of a gene depends on its location within the genome.

positive assortative mating Mating between like individuals that takes place more frequently than expected on the basis of chance.

positive control Gene regulation in which the binding of a regulatory protein to DNA stimulates transcription (the regulatory protein is an activator).

positive-strand RNA virus RNA virus whose genomic RNA molecule codes directly for viral proteins.

positive supercoiling *See* **supercoiling.**

posttranslational modification Alteration of a protein after translation; may include cleavage from a larger precursor protein, the removal of amino acids, and the attachment of other molecules to the protein.

P (parental) generation First set of parents in a genetic cross.

preformationism Early concept of inheritance proposing that a miniature adult (homunculus) resides in either the egg or the sperm and increases in size during development, with all traits being inherited from the parent that contributes the homunculus.

preimplantation genetic diagnosis Used to select an embryo produced by in vitro fertilization before implantation of the embryo in the uterus.

pre-messenger RNA (pre-mRNA) Eukaryotic RNA molecule that is modified after transcription to become mRNA.

presymptomatic genetic testing Testing people to determine whether they have inherited a disease-causing gene before the symptoms of the disease have appeared.

primary Down syndrome Down syndrome caused by the presence of three copies of chromosome 21.

primary immune response Initial clone of cells specific for a particular antigen and generated when the antigen is first encountered by the immune system.

primary oocyte Oogonium that has entered prophase I.

primary spermatocyte Spermatogonium that has entered prophase I.

primary structure of a protein The amino acid sequence of a protein.

primase Enzyme that synthesizes a short stretch of RNA on a DNA template; functions in replication to provide a 3′-OH group for the attachment of a DNA nucleotide.

primer Short stretch of RNA on a DNA template that provides a 3′-OH group for the attachment of a DNA nucleotide at the initiation of replication.

principle of independent assortment (Mendel's second law) Important principle of heredity discovered by Mendel that states that genes coding for different characteristics (genes at different loci) separate independently; applies only to genes located on different chromosomes or to genes far apart on the same chromosome.

principle of segregation (Mendel's first law) Important principle of heredity discovered by Mendel that states that each diploid individual possesses two alleles at a locus and that these two alleles separate when gametes are formed, one allele going into each gamete.

prion Infectious agent that lacks nucleic acid; believed to replicate and cause infection by altering the shape of proteins produced by cellular genes.

probability Likelihood of a particular event occurring; more formally, the number of times a particular event occurs divided by the number of all possible outcomes. Probability values range from 0 to 1.

proband A person with a trait or disease for whom a pedigree is constructed.

probe Known sequence of DNA or RNA that is complementary to a sequence of interest and will pair with it; used to find specific DNA sequences.

prokaryote Unicellular organism with a simple cell structure. Prokaryotes include eubacteria and archaea.

prometaphase Stage of mitosis. In prometaphase, the nuclear membrane breaks down and the spindle microtubules attach to the chromosomes.

promoter DNA sequence to which the transcription apparatus binds so as to initiate transcription; indicates the direction of transcription, which of the two DNA strands is to be read as the template, and the starting point of transcription.

proofreading Ability of DNA polymerases to remove and replace incorrectly paired nucleotides in the course of replication.

prophage Phage genome that is integrated into a bacterial chromosome.

prophase Stage of mitosis. In prophase, the chromosomes contract and become visible, the cytoskeleton breaks down, and the mitotic spindle begins to form.

prophase I Stage of meiosis I. In prophase I, chromosomes condense and pair, crossing over takes place, the nuclear membrane breaks down, and the spindle forms.

prophase II Stage of meiosis after interkinesis. In prophase II, chromosomes condense, the nuclear membrane breaks down, and the spindle forms. Some cells skip this stage.

proportion of polymorphic loci Percentage of loci in which more than one allele is present in a population.

protein-coding region The part of mRNA consisting of the nucleotides that specify the amino acid sequence of a protein.

protein domain Region of a protein that has a specific shape or function.

protein kinase Enzyme that adds phosphate groups to other proteins.

proteome Set of all proteins encoded by a genome.

proto-oncogene Normal cellular gene that controls cell division. When mutated, it may become an oncogene and contribute to cancer progression.

provirus DNA copy of viral DNA or RNA that is integrated into the host chromosome and replicated along with the host chromosome.

pseudoautosomal region Small region of the X and Y chromosomes that contains homologous gene sequences.

pseudodominance Expression of a normally recessive allele due to a deletion on the homologous chromosome.

Punnett square A shorthand method of determining the outcome of a genetic cross. A grid is drawn, with the gametes of one parent along the upper edge and the gametes of the other parent along the left-hand edge. Within the cells of the grid, the alleles in the gametes are combined to form the genotypes of the offspring.

purine Type of nitrogenous base in DNA and RNA. Adenine and guanine are purines.

pyrimidine Type of nitrogenous base in DNA and RNA. Cytosine, thymine, and uracil are pyrimidines.

pyrimidine dimer Structure in which a bond forms between two adjacent pyrimidine molecules on the same strand of DNA; disrupts normal hydrogen bonding between complementary bases and distorts the normal configuration of the DNA molecule.

quantitative characteristic A continuous characteristic; displays a large number of possible phenotypes, which must be described by a quantitative measurement.

quantitative characteristic locus (QTL) Locus that contributes to the expression of quantitative characteristics.

quantitative genetics Genetic analysis of complex characteristics or characteristics influenced by multiple genetic factors.

quaternary structure of a protein Interaction of two or more polypeptides to form a functional protein.

reading frame Particular way in which a nucleotide sequence is read in groups of three nucleotides (codons) in translation. Each reading frame begins with a start codon and ends with a stop codon.

realized heritability Narrow-sense heritability measured from a response-to-selection experiment.

reannealing In DNA, the process by which two complementary single-stranded DNA molecules pair; also called renaturation.

recessive Refers to an allele or phenotype that is expressed only when homozygous; the recessive allele is not expressed in the heterozygote phenotype.

reciprocal crosses Crosses in which the phenotypes of the male and female parents are reversed. For example, tall male crossed with a short female and a short male crossed with a tall female are reciprocal crosses.

reciprocal translocation Reciprocal exchange of segments between two nonhomologous chromosomes.

recombinant DNA technology Set of molecular techniques for locating, isolating, altering, combining, and studying DNA segments.

recombinant frequency Proportion of recombinant progeny produced in a cross.

recombinant gamete Possesses new combinations of genes.

recombinant progeny Possesses new combinations of traits formed from recombinant gametes.

recombination Sorting of alleles into new combinations.

regression Analysis of how one variable changes in response to another variable.

regression coefficient Statistic that measures how much one variable changes, on average, with a unit change in another variable.

regulator gene Gene associated with an operon in bacterial cells that encodes a protein or RNA molecule that functions in controlling the transcription of one or more structural genes.

regulator protein Produced by a regulator gene, a protein that binds to another DNA sequence and controls the transcription of one or more structural genes.

regulatory element DNA sequence that affects the transcription of other DNA sequences to which it is physically linked (are cis acting).

regulatory gene DNA sequence that encodes a protein or RNA molecule that interacts with DNA sequences and affects their transcription or translation or both.

regulatory promoter DNA sequences located immediately upstream of the core promoter that affect transcription; contains consensus sequences to which transcriptional activator proteins bind.

relaxed state Energy state of a DNA molecule when there is no structural strain on the molecule.

release factor Protein required for the termination of translation; binds to a ribosome when a stop codon is reached

and stimulates the release of the polypeptide chain, the tRNA, and the mRNA from the ribosome.

renaturation *See* **reannealing.**

repetitive DNA Sequences that exist in multiple copies in a genome.

replicated error Replication of an incorporated error in which a change in the DNA sequence has been replicated and all base pairings in the new DNA molecule are correct.

replication Process by which DNA is synthesized from a singled-stranded nucleotide template.

replication bubble Segment of a DNA molecule that is unwinding and undergoing replication.

replication fork Point at which a double-stranded DNA molecule separates into two single strands that serve as templates for replication.

replication licensing factor Protein that ensures that replication takes place only once at each origin; required at the origin before replication can be initiated and removed after the DNA has been replicated.

replication origin Sequence of nucleotides where replication is initiated.

replication terminus Point at which replication stops.

replicative segregation Random segregation of organelles into progeny cells in cell division. If two or more versions of an organelle are present in the original cell, chance determines the proportion of each type that will segregate into each progeny cell.

replicative transposition Type of transposition in which a copy of the transposable element moves to a new site, while the original copy remains at the old site; increases the number of copies of the transposable element.

replicon Unit of replication, consisting of DNA from the origin of replication to the point at which replication on either side of the origin ends.

replusion *See* **trans configuration.**

repressible operon Operon or other system of gene regulation in which transcription is normally on. Something must happen for transcription to be repressed, or turned off.

repressor Regulatory protein that binds to a DNA sequence and inhibits transcription.

resolvase Enzyme required for some types of transposition; brings about resolution—which is crossing over between sites located within the transposable element. Resolvase may be encoded by the transposable element or by a cellular enzyme that normally functions in homologous recombination.

response element Common DNA sequence found upstream of some groups of eukaryotic genes. A regulatory protein binds to a response element and stimulates the transcription of a gene. The presence of the same response element in several promoters or enhancers allows a single factor to simultaneously stimulate the transcription of several genes.

response to selection The amount that a characteristic changes in one generation owing to selection; equals the selection differential times the narrow-sense heritability.

restriction endonuclease Technical term for a restriction enzyme, which recognizes particular base sequences in DNA and makes double-stranded cuts nearby.

restriction enzyme Enzyme that recognizes particular base sequences in DNA and makes double-stranded cuts nearby; also called restriction endonuclease.

restriction fragment length polymorphism (RFLP) Variation in the pattern of fragments produced when DNA molecules are cut with the same restriction enzyme; represents a heritable difference in DNA sequences and can be used in gene mapping.

restriction mapping Determining in a piece of DNA the locations of sites cut by restriction enzymes.

retrotransposon Type of transposable element in eukaryotic cells that possesses some characteristics of retroviruses and transposes through an RNA intermediate.

retrovirus RNA virus capable of integrating its genetic material into the genome of its host. The virus injects its RNA genome into the host cell, where reverse transcription produces a complementary, double-stranded DNA molecule from the RNA template. The DNA copy then integrates into the host chromosome to form a provirus.

reverse chromosome duplication Duplication of a chromosome segment in which the sequence of the duplicated segment is inverted relative to the sequence of the original segment.

reverse mutation (reversion) Mutation that changes a mutant phenotype back into the wild type.

reverse transcriptase Enzyme capable of synthesizing complementary DNA from an RNA template.

reverse transcription Synthesis of DNA from an RNA template.

rho factor Subunit of bacterial RNA polymerase that facilitates transcription termination of some genes.

rho-dependent terminator Sequence in bacterial DNA that requires the presence of the rho subunit of RNA polymerase to terminate transcription.

rho-independent terminator Sequence in bacterial DNA that does not require the presence of the rho subunit of RNA polymerase to terminate transcription.

ribonucleoside triphosphate (rNTP) Substrate of RNA synthesis; consists of a ribose sugar, a nitrogenous base, and three phosphates linked to the $5'$-carbon atom of the sugar. In transcription, two of the phosphates are cleaved, producing an RNA nucleotide.

ribonucleotide Nucleotide containing a ribose sugar; present in RNA.

ribose sugar Five-carbon sugar in RNA.

ribosomal RNA (rRNA) RNA molecule that is a structural component of the ribosome.

ribozyme RNA molecule that can act as a biological catalyst.

RNA-coding region Sequence of DNA nucleotides that encodes an RNA molecule.

RNA editing Process in which the protein-encoding sequence of an mRNA is altered after transcription. The amino acids specified by the altered mRNA are different from those predicted from the nucleotide sequence of the gene encoding the protein.

RNA polymerase Enzyme that synthesizes RNA from a DNA template during transcription.

RNA polymerase I Eukaryotic RNA polymerase that transcribes large ribosomal RNA molecules (18 S rRNA and 28 S rRNA).

RNA polymerase II Eukaryotic RNA polymerase that transcribes pre-messenger RNA and some small nuclear RNAs.

RNA polymerase III Eukaryotic RNA polymerase that transcribes transfer RNA, small ribosomal RNAs (5 S rRNA), and some small nuclear RNAs.

RNA replication Process in some viruses by which RNA is synthesized from an RNA template.

RNA silencing Mechanism by which double-stranded RNA is cleaved and processed to yield small single-stranded interfering RNAs (siRNAs), which bind to complementary sequences in mRNA and bring about the cleavage and degradation of mRNA; also known as RNA interference and posttranscriptional RNA gene silencing. Some siRNAs also bind to complementary sequences in DNA and guide enzymes to methylate the DNA.

RNA splicing Process by which introns are removed and exons are joined together.

Robertsonian translocation Translocation in which the long arms of two acrocentric chromosomes become joined to a common centromere, resulting in a chromosome with two long arms and usually one chromosome with two short arms.

rolling-circle replication Replication of circular DNA that is initiated by a break in one of the nucleotide strands, producing a double-stranded circular DNA molecule and a single-stranded linear DNA molecule, the latter of which may circularize and serve as a template for the synthesis of a complementary strand.

R plasmid (R factor) Plasmid having genes that confer antibiotic resistance to any cell that contains the plasmid.

sample Subset used to describe a population.

sampling error Deviations from expected ratios due to chance occurrences when the number of events is small.

secondary immune response Clone of cells generated when a memory cell encounters an antigen; provides long-lasting immunity.

secondary oocyte One of the products of meiosis I in female animals; receives most of the cytoplasm.

secondary spermatocyte Product of meiosis I in male animals.

secondary structure of a protein Regular folding arrangement of amino acids in a protein. Common secondary structures found in proteins include the alpha helix and the beta pleated sheet.

second polar body One of the products of meiosis II in oogenesis; contains a set of chromosomes but little of the cytoplasm.

segmentation genes Set of about 25 genes in fruit flies that control the differentiation of the embryo into individual segments, affecting the number and organization of the segments. Mutations in these genes usually disrupt whole sets of segments.

segment-polarity genes Set of segmentation genes in fruit flies that affect the organization of segments. Mutations in these genes cause part of each segment to be deleted and replaced by a mirror image of part or all of an adjacent segment.

selection coefficient Measure of the relative intensity of selection against a genotype; equals 1 minus fitness.

selection differential Difference in phenotype between the selected individuals and the average of the entire population.

semiconservative replication Replication in which the two nucleotide strands of DNA separate, each serving as a template for the synthesis of a new strand (all DNA replication is semiconservative).

sense codon Codon that specifies an amino acid in a protein.

sequential hermaphroditism Phenomenon in which the sex of an individual changes in the course of its lifetime; an individual is male at one age or developmental stage and female at a different age or stage.

70S initiation complex Final complex formed in the initiation of translation in bacterial cells; consists of the small and large subunits of the ribosome, mRNA, and initiator tRNA charged with fMet.

sex Gender; male or female.

sex chromosomes Chromosomes that differ morphologically or in number in males and females.

sex determination Specification of sex (male or female). Sex-determining mechanisms include chromosomal, genic, and environmental sex-determining systems.

sex-determining region Y (SRY) gene On the Y chromosome, a gene that triggers male development; also known as the testis-determining factor (TDF) gene.

sex-influenced characteristic Encoded by autosomal genes that are more readily expressed in one sex. For example, an autosomal dominant gene may have higher penetrance in males than in females or an autosomal gene may be dominant in males but recessive in females.

sex-limited characteristic Encoded by autosomal genes and expressed in only one sex. Males and females both carry genes for sex-limited characteristics, but the characteristics appear in only one of the sexes.

sex-linked characteristic Characteristic determined by a gene or genes on sex chromosomes.

Shine-Dalgarno sequence Consensus sequence found in the bacterial 5' untranslated region of mRNA; contains the ribosome-binding site.

short interspersed element (SINE) Short DNA sequence repeated many times and interspersed throughout the genome.

shuttle vector Cloning vector that allows DNA to be transferred to more than one type of host cell.

sigma factor Subunit of bacterial RNA polymerase that allows the RNA polymerase to recognize a promoter and initiate transcription.

signal sequence From 15 to 30 amino acids found at the amino end of some eukaryotic proteins that direct the protein to specific locations in the cell; usually cleaved from the protein.

silencer Sequence that has many of the properties possessed by an enhancer but represses transcription.

silent mutation Alters a codon, but the codon still specifies the same amino acid.

single-nucleotide polymorphism (SNP) Single-base-pair differences in DNA sequence between individual members of a species.

single-strand binding (SSB) protein Binds to single-stranded DNA in replication and prevents it from annealing with a complementary strand and forming secondary structures.

sister chromatids Two copies of a chromosome that are held together at the centromere. Each chromatid consists of a single DNA molecule.

site-directed mutagenesis Produces specific nucleotide changes at selected sites in a DNA molecule.

small cytoplasmic RNA (scRNA) Small RNA molecule found in the cytoplasm of eukaryotic cells.

small interfering RNA (siRNA) Single-stranded RNA molecule (usually from 21 to 25 nucleotides in length) produced by the cleavage and processing of double-stranded RNA; binds to complementary sequences in mRNA and brings about the cleavage and degradation of the mRNA. Some siRNAs bind to complementary sequences in DNA and bring about their methylation.

small nuclear ribonucleoprotein (snRNP) Structure found in the nuclei of eukaryotic cells that consists of snRNA and protein; functions in the processing of pre-mRNA.

small nuclear RNA (snRNA) Small RNA molecule found in the nuclei of eukaryotic cells; functions in the processing of pre-mRNA.

small nucleolar RNA (snoRNA) Small RNA molecule found in the nuclei of eukaryotic cells; functions in the processing of rRNA and in the assembly of ribosomes.

small ribosomal subunit The smaller of the two subunits of a functional ribosome.

somatic-cell hybridization Fusion of different cell types.

somatic hypermutation High rate of somatic mutation such as that in genes encoding antibodies.

somatic mutation Mutation in a cell that does not give rise to a gamete.

somatic recombination Recombination in somatic cells, such as maturing lymphocytes, among segments of genes that encode antibodies and T-cell receptors.

SOS system System of proteins and enzymes that allow a cell to replicate its DNA in the presence of a distortion in DNA structure; makes numerous mistakes in replication and increases the rate of mutation.

Southern blotting Process by which DNA is transferred from a gel to a solid support such as a nitrocellulose or nylon filter.

specialized transduction Transduction in which genes near special sites on the bacterial chromosome are transferred from one bacterium to another; requires lysogenic bacteriophages.

spermatid Immediate product of meiosis II in spermatogenesis; matures to sperm.

spermatogenesis Sperm production animals.

spermatogonium Diploid cell in the testis; capable of undergoing meiosis to produce a sperm.

spindle microtubule Microtubule that moves chromosomes in mitosis and meiosis.

spindle pole Point from which spindle microtubules radiate.

spliced recombinants Possible outcome of homologous recombination, consisting of two heteroduplex DNA molecules, with DNA at each end in combinations different from those originally present.

spliceosome Large complex consisting of several RNAs and many proteins that carry out the splicing of protein-encoding pre-mRNA; contains five small ribonucleoprotein particles (U1, U2, U4, U5, and U6).

spontaneous mutation Arises spontaneously from natural changes in DNA structure or from errors in replication.

sporophyte Diploid phase of the life cycle in plants.

SR proteins Group of serine- and arginine-rich proteins that regulate alternative splicing of pre-mRNA.

S (synthesis) phase Stage of interphase in the cell cycle. In S phase, DNA replicates.

standard deviation Statistic that describes the variability of a group of measurements; the square root of the variance.

stop codon Codon in mRNA that signals the end of translation; also called nonsense codon or termination codon. There are three common stop codons: UAA, UAG, and UGA.

strand slippage Slipping of the template and newly synthesized strands in replication in which one of the strands loops out from the other and nucleotides are inserted or deleted on the newly synthesized strand.

structural gene DNA sequence that encodes a protein that functions in metabolism or biosynthesis or that plays a structural role in the cell.

structural genomics Area of genomics that studies the organization and sequence of information contained within genomes; sometimes used by protein chemists to refer to the determination of the three-dimensional structure of proteins.

submetacentric chromsome Chromosome in which the centromere is displaced toward one end, producing a short arm and a long arm.

supercoiling Coiled tertiary structure that forms when strain is placed on a DNA helix by overwinding or underwinding of the helix. An overwound DNA exhibits positive supercoiling; an underwound DNA exhibits negative supercoiling.

suppressor mutation Hides or suppresses the effect of another mutation at a site that is different from the site of the suppressor.

synapsis Close pairing of homologous chromosomes.

synaptonemal complex Three-part structure that develops between synapsed homologous chromosomes.

synonymous codons Different codons that specify the same amino acid.

syntenic genes Determined to be on the same chromosome by physical-mapping techniques.

tandem chromosome duplication Duplication of a chromosome segment that is adjacent to the original segment.

tandem repeat sequences DNA sequences repeated one after another; tend to be clustered at specific locations on a chromosome.

Taq **polymerase** DNA polymerase commonly used in PCR reactions. Isolated from the bacterium *Thermus aquaticus,* the enzyme is stable at high temperatures, and so it is not denatured during the strand-separation step of the cycle.

TATA-binding protein (TBP) Polypeptide chain found in several different transcription factors that recognizes and binds to sequences in eukaryotic promoters.

TATA box Consensus sequence (TATAAAA) commonly found in eukaryotic RNA polymerase II promoters; usually located from 25 to 30 bp upstream of the transcription start site. The TATA box determines the start point for transcription.

TBP-associated factor (TAF) Protein that combines with the TATA-binding protein to form a transcription factor.

T cell Particular type of lymphocyte that produces cellular immunity; originates in the bone marrow and matures in the thymus.

T-cell receptor Found on the surface of a T cell, a receptor that simultaneously binds a foreign and a self antigen on the surface of a cell.

telocentric chromosome Chromosome in which the centromere is at or very near one end.

telomerase Enzyme made up of both protein and RNA that replicates the ends (telomeres) of eukaryotic chromosomes. The RNA part of the enzyme has a template that is complementary to repeated sequences in the telomere and pairs with them, providing a template for synthesis of additional copies of the repeats.

telomere Stable end of a chromosome.

telomere-associated sequence Sequence found at the end of a chromosome next to the telomeric sequence; consists of relatively long, complex repeated sequences.

telomeric sequence Sequence found at the ends of a chromosome; consists of many copies of short, simple sequences repeated one after the other.

telophase Stage of mitosis. In telophase, the chromosomes arrive at the spindle poles, the nuclear membrane re-forms, and the chromosomes relax and lengthen.

telophase I Stage of meiosis I. In telophase I, chromosomes arrive at the spindle poles.

telophase II Stage of meiosis II. In telophase II, chromosomes arrive at the spindle poles.

temperate phage Bacteriophage that utilizes the lysogenic cycle, in which the phage DNA integrates into the bacterial chromosome and remains in an inactive state.

temperature-sensitive allele Expressed only at certain temperatures.

template strand The strand of DNA that is used as a template during transcription. The RNA synthesized during transcription is complementary and antiparallel to the template strand.

terminal inverted repeats Sequences found at both ends of a transposable element that are inverted complements of one another.

termination codon Codon in mRNA that signals the end of translation; also called nonsense codon or stop codon. There are three common termination codons: UAA, UAG, and UGA.

terminator Sequence of DNA nucleotides that causes the termination of transcription.

tertiary structure of a protein Higher-order folding of amino acids in a protein to form the overall three-dimensional shape of the molecule.

testcross A cross between an individual with an unknown genotype and an individual with the homozygous recessive genotype.

testis-determining factor (*TDF*) gene On the Y chromosome, a gene that triggers male development; also known as the sex-determining region Y (*SRY*) gene.

tetrad The four products of meiosis; all four chromatids of a homologous pair of chromosomes.

tetraploidy Possessing four haploid sets of chromosomes (4n).

tetrasomy Presence of two extra copies of a chromosome (2n + 2).

TFIIB recognition element (BRE) Consensus sequence [(G or C)(G or C) (G or C)CGCC] found in some RNA polymerase II core promoters; usually located from 32 to 38 bp upstream of the transcription start site.

theory of clonal selection Explains the generation of primary and secondary immune responses. Binding of a B cell to an antigen stimulates the cell to divide, giving rise to a clone of genetically identical cells, all of which are specific for the antigen.

theta replication Replication of circular DNA that is initiated by the unwinding of the two nucleotide strands, producing a replication bubble. Unwinding continues at one or both ends of the bubble, making it progressively larger. DNA replication on both of the template strands is simultaneous with unwinding until the two replication forks meet.

30S initiation complex Initial complex formed in the initiation of translation in bacterial cells; consists of the small subunit of the ribosome, mRNA, initiator tRNA charged with fMet, GTP, and initiation factors 1, 2, and 3.

three-point testcross Cross between an individual heterozygous at three loci and an individual homozygous for recessive alleles at those loci.

3′ end End of the polynucleotide chain where an OH group is attached to the 3′-carbon atom of the nucleotide.

3′ splice site The 3′ end of an intron where cleavage takes place in RNA splicing.

3′ untranslated (UTR) region Sequence of nucleotides at the 3′ end of mRNA; does not code for the amino acids of a protein but affects both the stability of the mRNA and its translation.

threshold characteristic Discontinuous characteristic whose expression depends on an underlying susceptibility that varies continuously.

thymine (T) Pyrimidine in DNA but not in RNA.

Ti plasmid Large plasmid from the bacterium *Agrobacterium tumefaciens* that is used to transfer genes to plant cells.

topoisomerase Enzyme that adds or removes rotations in a DNA helix by temporarily breaking nucleotide strands; controls the degree of DNA supercoiling.

totipotent Refers to the potential of a cell to develop into any other cell type.

trans configuration Arrangement in which each chromosome contains one wild-type (dominant) gene and one mutant (recessive) gene.

transcription Process by which RNA is synthesized from a DNA template.

transcriptional activator protein Protein in eukaryotic cells that binds to consensus sequences in regulatory promoters or enhancers and affects transcription initiation by stimulating or inhibiting the assembly of the basal transcription apparatus.

transcriptional antiterminator protein Protein that binds to RNA polymerase and alters its structure so that certain terminators are ignored, allowing transcription to continue past the terminators.

transcription bubble Region of a DNA molecule that has unwound to expose a single-stranded template, which is being transcribed into RNA.

transcription factor Protein that binds to DNA sequences in eukaryotic cells and affects transcription.

transcription start site The first DNA nucleotide that is transcribed into an RNA molecule.

transcription unit Sequence of nucleotides in DNA that codes for a single RNA molecule, along with the sequences necessary for its transcription; normally contains a promoter, an RNA-coding sequence, and a terminator.

transcriptome Set of all RNA molecules transcribed from a genome.

transducing phage Contains a piece of the bacterial chromosome inside the phage coat. *See also* **generalized transduction.**

transductant Bacterial cell that has received genes from another bacterium through transduction.

transduction Type of gene exchange that takes place when a virus carries genes from one bacterium to another. After it is inside the cell, the newly introduced DNA may undergo recombination with the bacterial chromosome.

transesterification Chemical reaction in some RNA splicing reactions.

transfer RNA (tRNA) RNA molecule that carries an amino acid to the ribosome and transfers it to a growing polypeptide chain in translation.

transformant Cell that has received genetic material through transformation.

transformation Mechanism in which DNA found in the medium is taken up by the cell. After transformation, recombination may take place between the introduced genes and the bacterial chromosome.

transforming principle Substance responsible for transformation. DNA is the transforming principle.

transgene Foreign gene or other DNA fragment carried in germ-line DNA.

transgenic mouse Mouse whose genome contains a foreign gene or genes added by employing recombinant DNA methods.

transition Base substitution in which a purine is replaced by a different purine or a pyrimidine is replaced by a different pyrimidine.

translation Process by which a protein is assembled from information contained in messenger RNA.

translocation Movement of a chromosome segment to a nonhomologous chromosome or to a region within the same chromosome; also movement of a ribosome along mRNA during translation.

translocation carrier Individual heterozygous for a translocation.

transmission genetics The field of genetics that encompasses the basic principles of genetics and how traits are inherited.

transposable element DNA sequence capable of moving from one site to another within the genome through a mechanism that differs from that of homologous recombination.

transposase Enzyme encoded by many types of transposable elements that is required for their transposition. The enzyme makes single-strand breaks at each end of the transposable element and on either side of the target sequence where the element inserts.

transposition Movement of a transposable genetic element from one site to another. Replicative transposition increases the number of copies of the transposable element; nonreplicative transposition does not increase the number of copies.

transversion Base substitution in which a purine is replaced by a pyrimidine or a pyrimidine is replaced by a purine.

trihybrid cross A cross between two individuals that differ in a three characteristics (*AABBCC × aabbcc*); also refers to a cross between two individuals that are both heterozygous at three loci (*AaBbCc × AaBbCc*).

triplet code Refers to the fact that three nucleotides encode each amino acid in a protein.

triploidy Possessing three haploid sets of chromosomes (3n).

triplo-X syndrome Human condition in which cells contain three X chromosomes. A person with triplo-X syndrome has a female phenotype without distinctive features other than a tendency to be tall and thin; a few such women are sterile, but many menstruate regularly and are fertile.

trisomy Presence of an additional copy of a chromosome (2n + 1).

trisomy 8 Presence of three copies of chromosome 8; in humans, results in mental retardation, contracted fingers and toes, low-set malformed ears, and a prominent forehead.

trisomy 13 Presence of three copies of chromosome 13; in humans, results in Patau syndrome.

trisomy 18 Presence of three copies of chromosome 18; in humans, results in Edward syndrome.

trisomy 21 Presence of three copies of chromosome 21; in humans, results in Down syndrome.

tRNA charging Chemical reaction in which an aminoacyl-tRNA synthetase attaches an amino acid to its corresponding tRNA.

tRNA introns Class of introns in tRNA genes. Splicing of these genes relies on enzymes.

tRNA-modifying enzyme Creates a modified base in RNA by catalyzing a chemical change in the standard base.

tubulin Protein found in microtubules.

tumor-suppressor gene Gene that normally inhibits cell division. Recessive mutations in such genes often contribute to cancer.

Turner syndrome Human condition in which cells contain a single X chromosome and no Y chromosome (XO). Persons with Turner syndrome are female in appearance but do not undergo puberty and have poorly developed female secondary sex characteristics; most are sterile but have normal intelligence.

two-point testcross Cross between an individual heterozygous at two loci and an individual homozygous for recessive alleles at those loci.

ultrasonography Procedure for visualizing the fetus. High-frequency sound is beamed into the uterus. Sound waves that encounter dense tissue bounce back and are transformed into a picture of the fetus.

unbalanced gametes Gametes that have variable numbers of chromosomes; some chromosomes may be missing and others may be present in more than one copy.

underdominance Selection in which the heterozygote has lower fitness than that of either homozygote.

unequal crossing over Misalignment of the two DNA molecules during crossing over, resulting in one DNA molecule with an insertion and the other with a deletion.

uniparental disomy Inheritance of both chromosomes of a homologous pair from a single parent.

unique-sequence DNA Sequence present only once or a few times in a genome.

universal genetic code Refers to the fact that particular codons specify the same amino acids in almost all organisms.

up mutation Mutation that increasea the rate of transcription.

upstream control element Consensus sequence in eukaryotic RNA polymerase I promoters that extends from 107 to 180 bp upstream of the transcription start site and increases the efficiency of the core element; rich in guanine and cytosine nucleotides.

upstream element Consensus sequence found in some bacterial promoters that contains a number of A−T pairs and is found about 40 to 60 bp upstream of the transcription start site.

uracil (U) Pyrimidine in RNA but not normally in DNA.

variable number of tandem repeats (VNTRs) Short sequences repeated in tandem that vary greatly in number among individuals. Because they are quite variable, VNTRs are commonly used in DNA fingerprinting.

variance Statistic that describes the variability of a group of measurements.

virulent phage Bacteriophage that reproduces only through the lytic cycle and kills its host cell.

virus Noncellular replicating agent consisting of nucleic acid surrounded by a protein coat; can replicate only within its host cell.

Western blotting Process by which protein is transferred from a gel to a solid support such as a nitrocellulose or nylon filter.

whole-genome shotgun sequencing Method of sequencing a genome in which sequenced fragments are assembled into the correct sequence in contigs by using only the overlaps in sequence.

wild type The trait or allele that is most commonly found in natural (wild) populations.

wobble Base pairing between codon and anticodon in which there is nonstandard pairing at the third (3′) position of the codon; allows more than one codon to pair with the same anticodon.

X:A ratio Ratio of the number of X chromosomes to the number of haploid autosomal sets of chromosomes; determines sex in fruit flies.

X-linked characteristic Characteristic determined by a gene or genes on the X chromosome.

X-ray diffraction Method for analyzing the three-dimensional shape and structure of chemical substances. Crystals of a substance are bombarded with X-rays, which hit the crystals, bounce off, and produce a diffraction pattern on a detector. The pattern of the spots produced on the detector provides information about the molecular structure.

yeast artificial chromosome (YAC) Cloning vector consisting of a DNA molecule with a yeast origin of replication, a pair of telomeres, and a centromere. YACs can carry very large pieces of DNA (as large as several hundred thousand base pairs) and replicate and segregate like yeast chromosomes.

Y-linked characteristic Characteristic determined by a gene or genes on the Y chromosome.

Z-DNA Secondary structure of DNA characterized by 12 bases per turn, a left-handed helix, and a sugar–phosphate backbone that zigzags back and forth.

zygotene Second substage of prophase I in meiosis. In zygotene, chromosomes enter into synapsis.

Answers to Selected Questions and Problems

Chapter 1

2. Genetics plays important roles in diagnosing and treating hereditary diseases, in breeding plant and animals for improved production and disease resistance, and in producing pharmaceuticals through genetic engineering.

4. Transmission genetics: inheritance of genes from one generation to the next; gene mapping. Molecular genetics: structure, organization, and function of genes at a molecular level. Population genetics: genes and changes in genes in populations.

6. Pangenesis proposed that information originating from all parts of the body is carried through the reproductive organs to the embryo at conception. Pangenesis allows changes in parts of the body to then be conveyed to the reproductive organs and to the next generation. The germ-plasm theory, in contrast, states that the reproductive cells carry all of the information required to make the complete body; the rest of the body contributes no information to the next generation.

7. The concept of inheritance of acquired characteristics postulated that traits acquired in the course of one's lifetime can be transmitted to offspring. It developed from pangenesis, which postulated that information from all parts of the body are transmitted to the next generation. Thus learning acquired in the brain or larger arm muscles developed through exercise could be transmitted to offspring.

8. Preformationism is the theory that the adult form is already preformed in the sperm or the egg. It meant that all traits would be inherited from only one parent, either the father or the mother, depending on whether the homunculus (the preformed miniature adult) resided in the sperm or the egg.

12. Gregor Mendel.

14. (a) The gene is the fundamental unit of heredity, a unit of information that determines an inherited characteristic.
(b) An allele is a form of a gene.
(c) A chromosome is a structure consisting of DNA and associated proteins that carries a linear array of genes.

(d) DNA (deoxyribonucleic acid) is the molecule that encodes genetic information through the sequence of bases A, C, G, and T.
(e) RNA (ribonucleic acid) encodes genetic information through the sequence of bases A, C, G, and U.
(f) Genetics is the science of heredity.
(g) A genotype is the complement of alleles that determines the phenotype of an individual organism.
(h) A phenotype is the trait expressed by an individual organism.
(i) A mutation is a heritable alteration in an individual organism's genotype and is brought about by permanent alteration in the DNA.
(j) Evolution is genetic change in a population of organisms.

17. Genetics is very old in the sense that hereditary principles have been applied at least since the beginning of agriculture and the domestication of plants and animals. It is very young in the sense that the fundamental principles were not uncovered until Mendel's time, and the advent of molecular biology and recombinant DNA has revolutionized genetics.

Chapter 2

3. Three fundamental events are (1) the cell's genetic information must be copied, (2) the copies of the genetic information must be separated from one another, and (3) the cell must divide.

6.

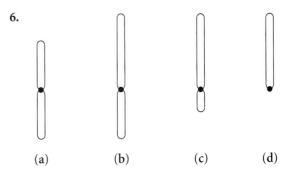

(a)	(b)	(c)	(d)
Metacentric	Submetacentric	Acrocentric	Telocentric

8. Prophase: the chromosomes condense and become visible, the centrosomes move apart, and microtubule fibers are formed from the centrosomes.

Prometaphase: the nucleoli disappear and the nuclear envelope begins to disintegrate, enabling the cytoplasm and nucleoplasm to join; the sister chromatids of each chromosome attach to the microtubles formed from opposite centrosomes.

Metaphase: the spindle microtubules become clearly visible, and the chromosomes arrange themselves on the equatorial plane of the cell.

Anaphase: the sister chromatids separate at the centromeres after the breakdown of cohesion protein, and the newly formed daughter chromosomes move to opposite poles of the cell.

Telophase: the nuclear envelope re-forms around each set of daughter chromosomes; nucleoli reappear; spindle microtubules disintegrate.

9. In the cell cycle, one cell usually produces two cells that contain the same genetic information — that is, the cells are identical with each other and with the parent cell. The S phase gives rise to identical sister chromatids, and mitosis ensures that each cell receives one of these chromatids.

13. Meiosis comprises two cell divisions, thus resulting in the production of four new cells (in many species). The chromosome number of a haploid cell produced by meiosis is half the chromosome number of the original diploid cell. Finally, the cells produced by meiosis are genetically different from the original cell and genetically different from one another.

15.

Mitosis	Meiosis
Single cell division produces two progeny cells that are genetically identical.	Two cell divisions usually produce four progeny cells that are genetically different.
Chromosome number of progeny cells and original cell remain the same.	Daughter cells are haploid (i.e., have half the chromosomal complement of the original diploid cell) as a result of the separation of homologous pairs in anaphase I.
Daughter cells and original cell are genetically identical; no separation of homologous chromosomes and no crossing over.	Crossing over in prophase and separation of homologous pairs in anaphase I produce daughter cells that are genetically different from one another and from the original cell.

Mitosis	Meiosis
Homologous chromosomes do not synapse. In metaphase, individual chromosomes line up on the metaphase plate.	Homologous chromosomes synapse in prophase I. In metaphase I, homologous pairs of chromosomes line up on the metaphase plate; in metaphase II, individual chromosomes line up.
In anaphase, sister chromatids separate.	In anaphase I, homologous chromosomes separate; in anaphase II, sister chromatids separate.

Key differences are that mitosis produces cells that are genetically identical with each other and with the original cell, allowing for the orderly passage of information from one cell to the progeny cells. Meiosis reduces the chromosome number of the original cell by producing daughter cells that do not contain pairs of homologous chromosomes. Meiosis results in genetic variation through crossing over and the random assortment of homologs in anaphase I.

20. (a) 12 chromosomes and 24 DNA molecules
(b) 12 chromosomes and 24 DNA molecules
(c) 12 chromosomes and 24 DNA molecules
(d) 12 chromosomes and 24 DNA molecules
(e) 12 chromosomes and 12 DNA molecules
(f) 6 chromosomes and 12 DNA molecules
(g) 12 chromosomes and 12 DNA molecules
(h) 6 chromosomes and 6 DNA molecules

21. The diploid number of chromosomes is six. (Left) Anaphase of meiosis I; (middle) anaphase of mitosis; (right) anaphase II of meiosis.

24. In a comparison of two organisms, the progeny of the organism whose cells contain the greater number of homologous pairs of chromosomes should be expected to exhibit more variation. The number of different combinations of chromosomes that are possible in gametes is 2^n, where n is equal to the number of homologous pairs of chromosomes. For the fruit fly, with four pairs of chromosomes, the number of combinations is 2^4, or 16, potential variants. For the house fly, with six pairs of chromosomes, the number of combinations is 2^6, or 64, potential variants.

25. (a) Metaphase I

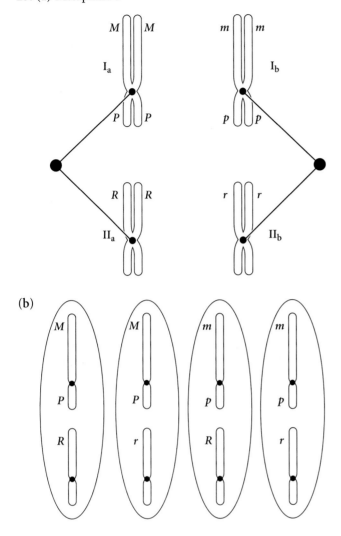

(b)

Chapter 3

1. Mendel was successful for several reasons. He chose a plant, *Pisum sativum*, that was easy to cultivate, grew relatively rapidly, and produced many offspring, which allowed Mendel to detect mathematical ratios. The seven characteristics that he chose to study exhibited only a few distinct phenotypes. Mendel looked at each trait separately and counted the types of progeny. Finally, he adopted an experimental approach and applied the scientific method. From his observations, he proposed hypotheses and was then able to empirically test his hypotheses.

3. The principle of segregation, or Mendel's first law, states that each individual organism possesses two alleles for any one particular trait and that these alleles separate in the formation of gametes.

7. Walter Sutton developed the chromosome theory of inheritance, which states that genes are located on chromosomes. The independent segregation of homologous chromosomes in meiosis provides the biological basis for Mendel's principles of heredity.

12. (a) The parents are *RR* (orange) and *rr* (cream). All the F$_1$ are *Rr* (orange). The F$_2$ are 1 *RR* : 2 *Rr* : 1 *rr* and have an orange-to-cream phenotypic ratio of 3 : 1.
    (b) All the offspring are orange. However, half of them are homozygous for orange fruit (*RR*) and half are heterozygous for orange fruit (*Rr*).
    (c) Half of the offspring are heterozygous for orange fruit (*Rr*) and half are homozygous for cream fruit (*rr*).

13. Because some black female rabbits produce both black progeny and white progeny, these female parents must be heterozygous for the black coat color; they possess a white coat color allele that is recessive to the black coat color allele. The black rabbits that produce only black progeny are most likely homozygous for the black coat color allele.

14. (a) The female parent must have the genotype $i^B i^B$. The male must have the genotype of $I^A i^B$.
    (b) Both parents must be homozygous for the recessive allele, or $i^B i^B$.
    (c) The male parent must be $i^B i^B$. A female who has type A blood could be either $I^A I^A$ or $I^A i^B$. The fact that all offspring are blood type A suggests that the female is $I^A I^A$. It is possible that she is $I^A i^B$ and that chance produced all type A progeny.
    (d) Both parents must be heterozygous for blood type A, or $I^A i^B$.
    (e) Either both parents are homozygous for blood type A ($I^A I^A$) or one parent is homozygous for blood type A ($I^A I^A$) and the other parent is heterozygous for blood type A ($I^A i^B$). The blood types of the offspring will not allow us to determine the precise genotype of either parent.
    (f) The female parent is $i^B i^B$. The male parent is $I^A i^B$.

16. (a) Sally (*Aa*), Sally's mother (*Aa*), Sally's father (*aa*), and Sally's brother (*aa*); (b) $\frac{1}{2}$; (c) $\frac{1}{2}$.

18. Define the hairless allele as *h* and the dominant allele for the presence of hair as *H*. To determine whether the rat terrier with hair is *HH* or *Hh*, a testcross with a hairless (*hh*) rat terrier should be performed. If the terrier is homozygous (*HH*) for the presence of hair, then no hairless offspring will be produced by the testcross. However, if the terrier is heterozygous (*Hh*) for the presence of hair, then half the offspring should be hairless.

21. (a) $\frac{1}{18}$; (b) $\frac{1}{36}$; (c) $\frac{11}{36}$; (d) $\frac{1}{6}$; (e) $\frac{1}{4}$; (f) $\frac{3}{4}$.

22. (a) $(\frac{1}{2})^7 = \frac{1}{128}$
    (b) The children could be all boys [$(\frac{1}{2})^7$] or all girls [$(\frac{1}{2})^7$]; so $(\frac{1}{2})^7 + (\frac{1}{2})^7 = \frac{1}{64}$.

    Parts *c* through *e* will require the use of the binomial expansion:

    $$(a + b)^7 = a^7 + 7a^6b + 21a^5b^2 + 35a^4b^3 + 35a^3b^4 + 21a^2b^5 + 7ab^6 + b^7$$

    Let *a* equal the probability of being a girl and *b* equal the probability of being a boy. The probabilities of both *a* and *b* are $\frac{1}{2}$.

(c) The probability is provided for by the term $7a^6b$. Because the probabilities of both a and b are $^1/_2$, the overall probability is $7(^1/_2)^6(^1/_2) = {}^7/_{128}$.
(d) The probability is provided for by the term $35a^3b^4$. The overall probability is $35(^1/_2)^3(^1/_2)^4 = {}^{35}/_{128}$.
(e) The probability is provided for by the term $35a^4b^3$. The overall probability is $35(^1/_2)^4(^1/_2)^3 = {}^{35}/_{128}$.

24. Parents

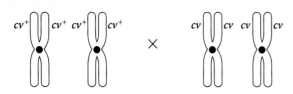

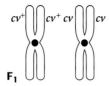

F₁

F₂

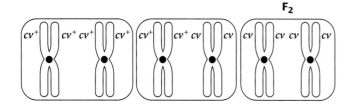

25. (a) In the F_1 black guinea pigs (Bb), only one chromosome possesses the black allele, and so the following numbers of copies will be present: in G_1, one black allele; in G_2, two black alleles; in metaphase of mitosis, two black alleles; in metaphase I of meiosis, two black alleles; after cytokinesis of meiosis, one black allele but in only half of the cells produced by meiosis (the remaining half will not contain the black allele).
(b) In the F_1 brown guinea pigs (bb), both homologs possess the brown allele, and so the following numbers of copies will be present: in G_1, two brown alleles; in G_2, four brown alleles; in metaphase of mitosis, four brown alleles; in metaphase I of meiosis, four brown alleles; in metaphase II of meiosis, two brown alleles; after cytokinesis of meiosis, one brown allele.

28. (a) *AaBbCdDdEe:* $^1/_2$ (*Aa*) × $^1/_2$ (*Bb*) × $^1/_2$ (*Cc*) × $^1/_2$ (*Dd*) × $^1/_2$ (*Ee*) = $^1/_{32}$.
(b) *AabbCcddee:* $^1/_2$ (*Aa*) × $^1/_2$ (*bb*) × $^1/_2$ (*Cc*) × $^1/_2$ (*dd*) × $^1/_4$ (*ee*) = $^1/_{64}$.
(c) *aabbccddee:* $^1/_4$ (*aa*) × $^1/_2$ (*bb*) × $^1/_4$ (*cc*) × $^1/_2$ (*dd*) × $^1/_4$ (*ee*) = $^1/_{256}$.

(d) *AABBCCDDEE:* No offspring with this genotype. The *AaBbCcddEe* parent cannot contribute a *D* allele, and the *AabbCcDdEe* parent cannot contribute a *B* allele. Therefore their offspring cannot be homozygous for the *BB* and *DD* loci.

31. (a) Gametes:

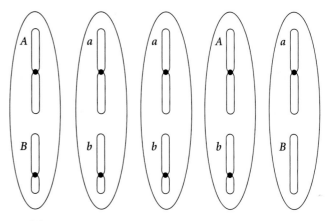

(b) Progeny at G_1:

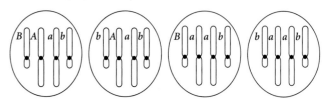

Progeny at G_2:

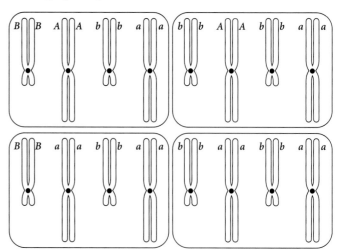

Progeny at metaphase of mitosis:

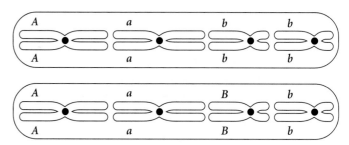

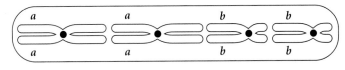

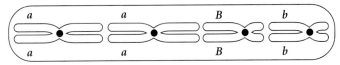

Order of chromosomes on metaphase plate can vary.

Chapter 4

1. Males produce relatively small gametes compared with those produced by females.

5. The pseudoautosomal region is a region of homology between the X and the Y chromosomes and is responsible for pairing of the X and Y chrommomosomes in meiotic prophase I. Genes in this region are present in two copies in both males and females and are thus inherited like autosomal genes, whereas the inheritance of other Y-linked genes is only from father to son.

6. Diploid insects are female; haploid insects are male. An egg that is fertilized by a sperm develops into a female; eggs that are not fertilized develop into males.

10. Males exhibit the phenotypes of all X-linked traits, regardless of whether the X-linked allele is normally recessive or dominant. Males inherit X-linked traits from their mothers and pass on X-linked traits to their daughters (and subsequently to their grandsons) but not to their sons.

14. Y-linked traits appear only in males and are always transmitted from father to son, thus following a strict paternal lineage.

15. (a) Female (X : A = 1.0)
(b) Male (X : A = 0.5)
(c) Male, sterile (X : A = 0.5)
(d) Female (X : A = 1.0)
(e) Male (X : A = 0.5)
(f) Female (X : A = 1.0)
(g) Metafemale (X : A >1.0)
(h) Male (X : A = 0.5)
(i) Intersex (X : A between 0.5 and 1.0)
(j) Female (X : A = 1.0)
(k) Metamale, sterile (X : A < 0.5)
(l) Metamale (X : A < 0.5)
(m) Intersex (X : A between 0.5 and 1.0)

17. (a) Yes; (b) yes; (c) no; (d) no.

18. (a) P X^+X^+ (gray female) crossed with X^yY (yellow male)

F_1 $\frac{1}{2}$ X^+Y (gray males), $\frac{1}{2}$ X^+X^y (gray females)

F_2 $\frac{1}{4}$ X^+Y (gray males), $\frac{1}{4}$ X^yY (yellow males), $\frac{1}{4}$ X^+X^y (gray females), $\frac{1}{4}$ X^+X^+ (gray females)

(b) P X^yX^y (yellow female) crossed with X^+Y (gray male)

F_1 $\frac{1}{2}$ X^yY (yellow males), $\frac{1}{2}$ X^+X^y (gray females)

F_2 $\frac{1}{4}$ X^+Y (gray males), $\frac{1}{4}$ X^yY (yellow males), $\frac{1}{4}$ X^+X^y (gray females), $\frac{1}{4}$ X^yX^y (yellow females)

(c) Backcross: X^+X^y (F_1) crossed with X^+Y (gray male)

F_2 $\frac{1}{4}$ X^+Y (gray males), $\frac{1}{4}$ X^yY (yellow males), $\frac{1}{4}$ X^+X^+ (gray females), $\frac{1}{4}$ X^+X^y (gray females)

(d) Overall genotypic ratios: $\frac{1}{8}$ X^+Y, $\frac{3}{8}$ X^yY, $\frac{1}{16}$ X^+X^+, $\frac{4}{16}$ X^+X^y, $\frac{3}{16}$ X^yX^y

Overall phenotypic ratios: $\frac{1}{8}$ gray males, $\frac{3}{8}$ yellow males, $\frac{5}{16}$ gray females, $\frac{3}{16}$ yellow females

19. If the color blindness is X-linked red-green colorblindness, then John has grounds for suspicion. The color-blind daughter must be homozygous recessive. Normally, their daughter would have inherited John's X chromosome. Because John is not color-blind, he could not have transmitted a color-blind X chromosome to the daughter.

An alternative, remote possibility is that the daughter is XO, having inherited a recessive color-blind allele from her mother and no sex chromosome from her father. In that case, the daughter has Turner syndrome.

Had Cathy had a color-blind son, then John would have no grounds for suspicion. The son would have inherited John's Y chromosome and the color-blind X chromosome from Cathy.

21. Because Bob must have inherited the Y chromosome from his father and his father has normal color vision, a nondisjunction event in the paternal lineage cannot account for Bob's genotype. Bob's mother must be heterozygous X^+X^c, because she has normal color vision and must have inherited a color-blind X chromosome from her color-blind father. For Bob to inherit two color-blind X chromosomes from his mother, the egg must have arisen from a nondisjunction in meiosis II. In meiosis I, the homologous X chromosomes separate; so one cell has the X^+ and the other has X^c. Failure of sister chromatids to separate in meiosis II would then result in an egg with two copies of X^c.

25. (a) Male parent is X^+Y. Because the ratio of long to miniature wings in the male offspring is almost 1 : 1, the female parent must be X^+X^m.
(b) Male parent is X^mY. Because all the male offspring have long wings, the female parent must be X^+X^+.
(c) Male parent is X^mY. Female parent is X^+X^m.
(d) Male parent is X^+Y. Female parent is X^mX^m.
(e) Male parent is X^+Y. Female parent is X^+X^+.

26. P: Z^bZ^b crossed with Z^+W

F_1: $\frac{1}{2}$ Z^bZ^+ (normal males) and $\frac{1}{2}$ Z^bW (bald females)

F$_2$:

	Z^b	W
Z$^+$	¼ Z$^+$Z^b (normal roosters)	¼ Z$^+$W (normal hens)
Z^b	¼ Z^{b}Z^b (bald roosters)	¼ Z^{b}W (bald hens)

28. (a) XX, 1 Barr body
    (b) XY, 0
    (c) XO, 0
    (d) XXY, 1
    (e) XXYY, 1
    (f) XXXY, 2
    (g) XYY, 0
    (h) XXX, 2
    (i) XXXX, 3

30. (a) F$_1$: Males are X^{m}Y s^+s (miniature wings, red eyes)
    Females are X$^+$X^m s^+s

    F$_2$: $\frac{3}{16}$ long wings, red eyes, female
    $\frac{3}{16}$ miniature wings, red eyes, female
    $\frac{3}{16}$ long wings, red eyes, male
    $\frac{3}{16}$ miniature wings, red eyes, male
    $\frac{1}{16}$ long wings, sepia eyes, female
    $\frac{1}{16}$ miniature wings, sepia eyes, female
    $\frac{1}{16}$ long wings, sepia eyes, male
    $\frac{1}{16}$ miniature wings, sepia eyes, male

    (b) F$_1$: Males are X^{m+} s^+s (long wings, red eyes)
    Females are X^{m+}X^m s^+s (long wings, red eyes)

    F$_2$: $\frac{3}{16}$ long wings, red eyes, male
    $\frac{1}{16}$ long wings, sepia eyes, male
    $\frac{3}{16}$ miniature wings, red eyes, male
    $\frac{1}{16}$ miniature wings, sepia eyes, male
    $\frac{3}{8}$ long wings, red eyes, female
    $\frac{1}{8}$ long wings, sepia eyes, female

32. If we assume that color blindness is X linked, then the mother is X^{c}X^c and the father must be X$^+$Y. Normally, all the sons would be color-blind, and all the daughters should have normal vision. The only way that a daughter can be color-blind is if she did not inherit an X$^+$ from her father. The observation that the color-blind daughter is short in stature and has failed to undergo puberty is consistent with Turner's syndrome (XO); so the color-blind daughter is X^{c}O.

Chapter 5

1. In incomplete dominance, the phenotype of the heterozygote is intermediate between the phenotypes of the homozygotes. In codominance, both alleles are expressed and both phenotypes are manifested simultaneously.

2. Dominance is an interaction between two alleles of a single gene or locus, with the dominant allele determining the phenotype. Epistasis results from the interaction of two or more genes at different loci, with the epistatic gene or locus masking the other gene and determining the phenotype.

5. Genomic imprinting refers to a difference in the expression of a gene, depending on whether it was inherited from the male parent or the female parent.

7. Cytoplasmically inherited traits are encoded by genes in the cytoplasm. Because the cytoplasm is usually inherited from a single parent (usually the female parent), reciprocal crosses do not show the same results. Cytoplasmically inherited traits often have great variability, because different egg cells (female gametes) may have differing proportions of cytoplasmic alleles owing to random sorting of mitochondria (or plastids in plants).

10. Continuous characteristics, also called quantitative characteristics, exhibit many phenotypes with a continuous distribution. They result from the interaction of multiple genes (polygenic traits) or the influence of environmental factors on the phenotype or both.

11. (a) The results of the crosses indicate that both cremello and chestnut are pure-breeding traits (homozygous). Palomino is a hybrid trait (heterozygous) that produces a 2 : 1 : 1 ratio when palominos are crossed with palominos. The simplest hypothesis consistent with these results is incomplete dominance, with palomino as the phenotype of the heterozygotes resulting from chestnuts crossed with cremellos.
    (b) Let C^B = chestnut, C^W = cremello, and C^BC^W = palomino.

Cross	Offspring
palomino × palomino	13 palomino, 6 chestnut, 5 cremello
$C^BC^W \times C^BC^W$	C^BC^W C^BC^B C^WC^W
chestnut × chestnut $C^BC^B \times C^BC^B$	16 chestnut C^BC^B
cremello × cremello $C^WC^W \times C^WC^W$	13 cremello C^WC^W
palomino × chestnut $C^BC^W \times C^BC^B$	8 palomino, 9 chestnut C^BC^W C^BC^B
palomino × cremello $C^BC^W \times C^WC^W$	11 palomino, 11 cremello C^BC^W C^WC^W
chestnut × cremello $C^BC^B \times C^WC^W$	23 palomino C^BC^W

12. (a) $\frac{1}{2}$ L^ML^M (type M), $\frac{1}{2}$ L^ML^N (type MN)
    (b) all L^NL^N (type N)
    (c) $\frac{1}{2}$ L^ML^N (type MN), $\frac{1}{4}$ L^ML^M (type M), $\frac{1}{4}$ L^NL^N (type N)
    (d) $\frac{1}{2}$ L^ML^N (type MN), $\frac{1}{2}$ L^NL^N (type N)
    (e) all L^ML^N (type MN)

14. (a) $\frac{1}{4}$ I^AI^B (AB), $\frac{1}{4}$ I^Ai (A), $\frac{1}{4}$ I^Bi (B), $\frac{1}{4}$ ii (O)
    (b) $\frac{1}{4}$ I^AI^A (A), $\frac{1}{4}$ I^Ai (A), $\frac{1}{4}$ I^AI^B (AB), $\frac{1}{4}$ I^Bi (B)

(c) $\frac{1}{4}\ I^A I^A$ (A), $\frac{1}{2}\ I^A I^B$ (AB), $\frac{1}{4}\ I^B I^B$ (B)

(d) $\frac{1}{2}\ I^A i$ (A), $\frac{1}{2}\ ii$ (O)

(e) $\frac{1}{2}\ I^A i$ (A), $\frac{1}{2}\ I^B i$ (B)

19. The child's genotype has an allele for B and an allele for N that could not have come from the mother and must have come from the father. Therefore, the child's father must have alleles for both B and N. George, Claude, and Henry are eliminated as possible fathers because they lack an allele for either B or N.

21. (a) all walnut ($RrPp$)

(b) $\frac{1}{4}$ walnut ($RrPp$), $\frac{1}{4}$ rose ($Rrpp$), $\frac{1}{4}$ pea ($rrPp$), $\frac{1}{4}$ single ($rrpp$)

(c) $\frac{9}{16}$ walnut ($R\_P\_$), $\frac{3}{16}$ rose (R_pp), $\frac{3}{16}$ pea ($rrP\_$), $\frac{1}{16}$ single ($rrpp$)

(d) $\frac{3}{4}$ rose (R_pp), $\frac{1}{4}$ single ($rrpp$)

(e) $\frac{1}{4}$ walnut ($RrPp$), $\frac{1}{4}$ rose ($Rrpp$), $\frac{1}{4}$ pea ($rrPp$), $\frac{1}{4}$ single ($rrpp$)

(f) $\frac{1}{2}$ rose ($Rrpp$), $\frac{1}{2}$ single ($rrpp$)

22. (a) The parents must have been $AABB$ wild type $\times$ $aabb$ yellow.

F_1: all $AaBb \times AaBb$ wild type

F_2: $\frac{9}{16}$ $A\_B\_$ wild type

$\frac{3}{16}$ A_bb amethyst

$\frac{3}{16}$ $aaB\_$ yellow

$\frac{1}{16}$ $aabb$ yellow

(b) Yes, allele a exhibits recessive epistasis, because the aa genotype masks the expression of genes at the B locus. Alleles b and B are hypostatic to a because their expression is masked by the presence of aa.

24. (a) Labrador retrievers vary in two loci: B and E. Black dogs have dominant alleles at both loci ($B\_E\_$), brown dogs have $bbE\_$, and yellow dogs have B_ee or $bbee$. Because all the puppies were black, each of them must have inherited a dominant B allele from the yellow parent and a dominant E allele from the brown parent. The brown female parent must have been $bbEE$, and the yellow male parent must have been $BBee$. All the black puppies were $BbEe$.

(b) Simply mating yellow with yellow will produce all yellow Labrador puppies. Mating two brown Labradors will produce either all brown puppies, if at least one of the parents is homozygous EE, or $\frac{3}{4}$ brown and $\frac{1}{4}$ yellow if both parents are heterozygous Ee.

26. The F_2 are:

$\frac{9}{16}$ $A\_B\_$ disc shaped (like F_1)

$\frac{3}{16}$ A_bb spherical

$\frac{3}{16}$ $aaB\_$ spherical

$\frac{1}{16}$ $aabb$ long

29. (a) The 2 : 1 ratio in the offspring of two spotted hamsters suggests lethality, and the 1 : 1 ratio in the offspring of spotted hamsters mated with hamsters without spots indicates that spotted is a heterozygous phenotype. If we assign S and s to the locus responsible for white spotting, spotted hamsters are Ss; solid-color hamsters are ss; and SS ($\frac{1}{4}$ of the progeny expected from a mating of two spotted hamsters) is embryonic lethal and missing from those progeny, resulting in the 2 : 1 ratio of spotted to solid-color progeny.

(b) Because spotting is a heterozygous phenotype, it is not possible to obtain Chinese hamsters that breed true for spotting.

31. $H^+ H^+ \times H^b H^+ \rightarrow \frac{1}{4}\ H^+ H^+$ males with full hair

$\frac{1}{4}\ H^+ H^+$ females with full hair

$\frac{1}{4}\ H^b H^+$ males with pattern baldness

$\frac{1}{4}\ H^b H^+$ females with full hair

Therefore, $\frac{1}{4}$ of their children will be bald.

35. This is an example of genetic maternal effect; the heterozygous offspring of women with PKU express the phenotype encoded by the mother's genotype. The women with PKU have high levels of phenylalanine in their circulating blood. The phenylalanine reaches the fetus through the placenta. The functional PKU allele of the fetus either is not expressed in embryonic development or is not expressed at high enough levels to metabolize all the excess phenylalanine in the maternal circulatory system, which has a far higher volume than that of the fetal circulatory system. No such problem arises if the father has PKU, because the developing fetus is not exposed to the paternal circulatory system.

Chapter 6

1. The three factors are: (1) mating cannot be controlled, and so setting up controlled mating experiments is impossible; (2) humans have a long generation time, and so tracking the inheritance of traits for more than one generation takes a long time; and (3) the number of progeny per mating is limited, and so phenotypic ratios are uncertain.

2. Autosomal recessive traits: both male and female affected offspring will arise with equal frequency from unaffected parents. The trait often appears to skip generations. Unaffected offspring having an affected parent will be carriers.

Autosomal dominant traits: both male and female affected offspring will arise with equal frequency from a mating in which one parent is affected. The trait does not usually skip generations.

X-linked recessive traits will affect males predominantly and will be passed from an affected father through his unaffected daughter to his grandson. X-linked recessive traits are not passed from father to son.

X-linked dominant traits: will affect both males and females and will be passed from an affected father to all his daughters but not to his sons. An affected mother (usually heterozygous for rare dominant traits) will pass on the trait equally to half of her daughters and half of her sons.

Y-linked traits will show up exclusively in males, and passed from father to son.

3. The two types of twins are monozygotic and dizygotic. Monozygotic twins arise when a single fertilized egg splits into two embryos in early embryonic divisions. They are essentially identical. Dizygotic twins arise from two different eggs fertilized at the same time by two different sperm. They have in common, on average, 50% of the same genes.

6. Genetic counseling provides assistance to a person or to a couple in interpreting the results of genetic testing and diagnosis by providing information about relevant disease symptoms, treatment, and progression; assessing and calculating the various genetic risks that the person or couple face; and helping persons and family members cope with the stress of decision making and facing drastic changes in their lives that may be precipitated by a genetic condition.

8. Amniocentesis samples the amniotic fluid by the insertion of a needle into the amniotic sac, usually performed at about 16 weeks of pregnancy. Chorionic villus sampling, which can be performed several weeks earlier (10th or 11th week of pregnancy), samples a small piece of the chorion by the insertion of a catheter through the vagina. The purpose of these techniques is to obtain fetal cells for prenatal genetic testing.

10. (a)

(b) X-linked recessive. Only males have the trait, and they inherit the trait from mothers who are carriers. The trait is never passed from father to son.
(c) Barring a new mutation or nondisjunction, the probability is zero. Joe cannot pass on his X chromosome (which carries the allele for color blindness) to his son.
(d) The probability is $\frac{1}{4}$. There is $\frac{1}{2}$ probability that their first child will be a boy, and there is an independent $\frac{1}{2}$ probability that the first child will inherit the X chromosome carrying the color-blind allele from the carrier mother. So $\frac{1}{2}(\frac{1}{2}) = \frac{1}{4}$.

(e) Again, the probability is $\frac{1}{4}$. Patty is a carrier—she already has a color-blind son. The reasoning that applies in part d applies here. Each child is an independent event.

12. (a) Autosomal dominant. Both males and females are affected, and both can pass on the trait to both sons and daughters. So the trait must be autosomal, and it must be dominant because affected children are produced in matings between affected and unaffected parents. For rare traits, we can assume that unaffected persons are not carriers. Therefore, all the offspring of a mating between a person affected with recessive traits and an unrelated unaffected person would be expected to be unaffected.
(b) X-linked dominant. Superficially, this pedigree appears similar to the pedigree in part a, in that both males and females are affected, and it seems to be a dominant trait. However, closer inspection reveals that, whereas affected females can pass on the trait to either sons or daughters, affected males pass on the trait only to all daughters.
(c) Y-linked. The trait affects only males and is passed from father to all sons.
(d) Either X-linked recessive or sex-limited autosomal dominant. Because only males have the trait, it could be X-linked recessive, Y linked, or sex limited. We can eliminate Y linkage because affected males do not pass on the trait to their sons. X-linked recessive inheritance is consistent with the pattern of unaffected female carriers producing both affected and unaffected sons and of affected males producing unaffected female carriers but no affected sons. Sex-linked autosomal dominant inheritance is consistent with unaffected heterozygous females producing affected heterozygous sons, unaffected homozygous recessive sons, and unaffected heterozygous or homozygous recessive daughters. The two remaining possibilities of X-linked recessive and sex-limited autosomal dominant could be distinguished if we had enough data to determine whether affected males can have both affected and unaffected sons, as expected from autosomal dominant inheritance, or whether affected males can have only unaffected sons, as expected from X-linked recessive inheritance. Unfortunately, this pedigree shows only two sons from affected males. In both cases, the sons are unaffected, consistent with X-linked recessive inheritance, but two instances are not enough to conclude that affected males cannot produce affected sons.
(e) Autosomal recessive. All the children of the original affected female are carriers. The first cousins in the consanguineous mating in the third generation also are carriers, inheriting the recessive alleles from their carrier parents. The consanguineous mating produced two affected children (one boy and one girl) and four unaffected children.

14. Migraine headaches appear to be influenced by both genetic and environmental factors. Markedly greater concordance in monozygotic twins, who are 100% genetically identical, than in dizygotic twins, who are 50% genetically identical, is indicative of a genetic influence. However, the fact that monozygotic twins show only 60% concordance despite their 100% genetic identity indicates that environmental factors also play a role.

Eye color appears to be purely genetically determined, because the concordance is greater in monozygotic twins than in dizygotic twins. Moreover, the monozygotic twins have 100% concordance for this trait, indicating that environment has no detectable influence.

Measles appear to have no detectable genetic influence, because there is no difference in concordance between monozygotic and dizygotic twins. Some environmental influence can be detected because monozygotic twins show less than 100% concordance.

Clubfoot appears to be influenced by both genetic and environmental factors, by the same reasoning as that for migraine headaches. A strong environmental influence is indicated by the high discordance in monozygotic twins.

Blood pressure is influenced by both genetic and environmental factors, similarly to clubfoot.

Handedness, like measles, appears to have no genetic influence, because the concordance is the same in monozygotic and dizygotic twins. Environmental influence is indicated by the less-than-100% concordance in monozygotic twins.

Tuberculosis lacks an indication of genetic influence, with the same degree of concordance in monozygotic and dizygotic twins. The primacy of environmental influence is indicated by the very low concordance in monozygotic twins.

16. (a) X-linked recessive. Only males have the condition, and unaffected female carriers have affected sons.
(b) The probability is $\frac{1}{4}$. The female III-7 is a carrier; so the probability that the child will inherit her X chromosome with the Nance-Horan allele is $\frac{1}{2}$ and the probability that the child will be a boy is $\frac{1}{2}$.
(c) The probability is $\frac{1}{2}$, because half the boys will inherit the Nance-Horan allele from the III-7 female carrier. All the girls will inherit one Nance-Horan allele from the III-2 affected male, and half of them will get a second Nance-Horan allele from the III-2 female; so half the girls also will have Nance-Horan syndrome.

18. This pedigree shows autosomal recessive inheritance rather than autosomal dominant inheritance. The inheritance can't be dominant, because unaffected persons have affected children. In generation II, two brothers married two sisters; so the members of generation III in the two families are closely related. A single recessive allele in one of the members of generation I was inherited by all four members of generation III. The consanguineous matings in generation III then produced children homozygous for the recessive ectodactyly allele. X linkage is ruled out, because the father of female IV-2 is unaffected.

Chapter 7

1. Recombination means that meiosis generates gametes with allelic combinations that differ from those of the original gametes inherited by an organism. If the organism was created by the fusion of an egg bearing *AB* and a sperm bearing *ab*, recombination generates gametes that are *Ab* and *aB*. Recombination may be caused by loci on different chromosomes that assort independently or by a physical crossing over between two loci on the same chromosome, with breakage and exchange of strands of homologous chromosomes paired in meiotic prophase I.

2. (a) Complete linkage of two genes means that only nonrecombinant gametes will be produced: the recombination frequency is zero.
(b) Independent assortment of two genes will result in 50% of the gametes being recombinant and 50% being nonrecombinant, as would be observed for genes on two different chromosomes. Independent assortment may also be observed for genes on the same chromosome if they are far enough apart that one or more crossovers take place between them in every meiosis.
(c) Incomplete linkage means that more than 50% of the gametes produced are nonrecombinant and fewer than 50% of the gametes are recombinant: the recombination frequency is greater than zero and less than 50%.

5. For genes in coupling configuration, two wild-type alleles are on the same chromosome and the two mutant alleles are on the homologous chromosome. For genes in repulsion, a wild-type allele of one locus and the mutant allele of the other locus are on the same chromosome, and the respective mutant and wild-type alleles are on the homologous chromosome. The two arrangements have opposite effects on the results of a cross. For genes in coupling configuration, most of the progeny either will be wild type for both genes or will be mutant for both genes, with relatively few recombinants that are wild type for one gene and mutant for the other. For genes in repulsion, most of the progeny will be mutant for only one gene and wild type for the other, with relatively few recombinants that are wild type for both or mutant for both.

8. The farther apart two loci are, the more likely the double crossovers between them. Unless there are marker genes between the loci, such double crossovers will be undetected, because the double crossovers and

nonrecombinants give the same phenotypes. The calculated recombination frequency will underestimate the true crossover frequency because the double-crossover progeny are not counted as recombinants.

10. A positive interference value results when the actual number of double crossovers observed is less than the number of double crossovers expected from the single-crossover frequencies. Thus positive interference indicates that a crossover inhibits or interferes with the occurrence of a second crossover nearby. Conversely, a negative interference value, where more double crossovers occur than expected, suggests that a crossover event can stimulate additional crossover events in the same region of the chromosome.

13. (a) With absolute linkage, there will be no recombinant progeny. The F_1 inherited banded and yellow alleles (B^BC^Y) together on one chromosome from the banded yellow parent and unbanded and brown alleles (B^OC^{Bw}) together on the homologous chromosome from the brown unbanded parent. Without recombination, all the F_1 gametes contain only these two allelic combinations, in equal proportions. Therefore, the F_2 testcross progeny will be $\frac{1}{2}$ banded and yellow and $\frac{1}{2}$ unbanded and brown.
(b) $\frac{1}{4}$ banded, yellow; $\frac{1}{4}$ banded, brown; $\frac{1}{4}$ unbanded, yellow; $\frac{1}{4}$ unbanded, brown.
(c) 40% banded, yellow; 40% unbanded, brown; 10% banded, brown; 10% unbanded, yellow.

14. (a) $\frac{1}{4}$ wild-type eyes, wild-type wings; $\frac{1}{4}$ red eyes, wild-type wings; $\frac{1}{4}$ wild-type eyes, white-banded wings; $\frac{1}{4}$ red eyes, white-banded wings.
(b) The genetic distance is equal to the recombination frequency, which equals the number of recombinant progeny divided by the total progeny:
$(19 + 16)/(918 + 19 + 16 + 426) = 4\% = 4$ m.u.

15. To test for independent assortment, we first test for equal segregation at each locus and then test whether the two loci sort independently. Testing for vg, we have
Observed $vg = 224 + 97 = 321$
Observed $vg^+ = 230 + 99 = 329$
Expected vg or $vg^+ = \frac{1}{2} \times 650 = 325$

$$\chi^2 = \Sigma \frac{(\text{observed} - \text{expected})^2}{\text{expected}}$$
$$= (321 - 325)^2/325 + (329 - 325)^2/325$$
$$= 16/325 + 16/325 = 0.098$$

We have $n - 1$ degrees of freedom, where $n = 2$, the number of phenotypic classes; so there is just 1 degree of freedom. From Table 3.4, we see that the P value is between .7 and .8. So these results do not deviate significantly from the expected 1 : 1 segregation.

Similarly, testing for sps, we observe 327 sps^+ and 323 sps, and we expect $\frac{1}{2} \times 650 = 325$ of each:
$$\chi^2 = 4/325 + 4/325 = .025$$

again with 1 degree of freedom. The P value is between .8 and .9; so these results do not deviate significantly from the expected 1 : 1 ratio.

Finally, we test for independent assortment, where we expect 1 : 1 : 1 : 1 phenotypic ratios, or 162.5 of each (o, observed; e, expected).

Observed	Expected	O − E	(O − E)^2	(O − E)^2/E
230	162.5	67.5	4556.25	28.0
224	162.5	61.5	3782.25	23.3
97	162.5	−65.5	4290.25	26.4
99	162.5	−63.5	4032.25	24.8

We have four phenotypic classes, giving us 3 degrees of freedom. The total chi-square value of 102.5 is off the chart; so we reject independent assortment. Instead, the genes are linked, and $RF = (97 + 99)/650 \times 100\% = 30\%$, giving us 30 map units between them.

17. (a) The genotypes of both plants are $DdPp$.
(b) Yes. From the cross of plant A, the map distance is $10/256 = 3.9\%$, or 3.9 m.u. The cross of plant B gives $6/170 = 3.5\%$, or 3.5 m.u. If we pool the data from the two crosses, we get $16/426 = 3.8\%$, or 3.8 m.u.
(c) The two plants have different coupling relations. In plant A, the dominant alleles D and P are coupled; one chromosome is $\underline{D\ P}$ and the other is $\underline{d\ p}$. In plant B, they are in repulsion; one chromosome is $\underline{D\ p}$ and the other is $\underline{d\ P}$.

20. (a) We can calculate the four phenotypic classes and their proportions for e and ro; then each of those classes will be split 1 : 1 for f, because f assorts independently. The recombination frequency between e and ro is 20%, and so each of the recombinants ($e^+ ro$ and $e ro^+$) will be 10%, and each of the nonrecombinants ($e^+ ro^+$ and $e ro$) will be 40%. Each of these two groups will then be split equally between f^+ and f.

$e^+ ro^+ f^+$	normal body color, normal eyes, normal bristles	20%
$e^+ ro^+ f$	normal body color, normal eyes, forked bristles	20%
$e ro f^+$	ebony body color, rough eyes, normal bristles	20%
$e ro f$	ebony body color, rough eyes, forked bristles	20%
$e^+ ro f^+$	normal body color, rough eyes, normal bristles	5%
$e^+ ro f$	normal body color, rough eyes, forked bristles	5%
$e ro^+ f^+$	ebony body color, normal eyes, normal bristles	5%
$e ro^+ f$	ebony body color, normal eyes, forked bristles	5%

(b) The same calculations as those used in part a are used, except that the nonrecombinants are $e^+ ro$ and $e ro^+$ and the recombinants are $e^+ ro^+$ and $e ro$.

$e^+\ ro^+\ f^+$	normal body color, normal eyes, normal bristles	5%
$e^+\ ro^+\ f$	normal body color, normal eyes, forked bristles	5%
$e\ ro\ f^+$	ebony body color, rough eyes, normal bristles	5%
$e\ ro\ f$	ebony body color, rough eyes, forked bristles	5%
$e^+\ ro\ f^+$	normal body color, rough eyes, normal bristles	20%
$e^+\ ro\ f$	normal body color, rough eyes, forked bristles	20%
$e\ ro^+\ f^+$	ebony body color, normal eyes, normal bristles	20%
$e\ ro^+\ f$	ebony body color, normal eyes, forked bristles	20%

21. A recombination frequency of 50% indicates that the genes assort independently. Less than 50% recombination indicates linkage. Starting with the most tightly linked genes, a and g, we look for other genes linked to them and find that only gene d has less than 50% recombination with a and g. So one linkage group consists of a, g, and d. We know that g is between a and d because the a to d distance is 12.

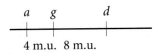

Similarly, we find a second linkage group of b, c, and e, with b in the middle.

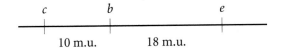

Gene f is unlinked to either of these groups and is on a third linkage group.

22. (a) The nonrecombinants are $Wx\ Sh\ V$ and $wx\ sh\ v$. The double crossovers are $wx\ sh\ V$ and $Wx\ Sh\ v$. Comparing the two, we see that they differ only at the v locus; so v must be the middle locus.
(b) $Wx\!-\!V$ distance: recombinants are $wx\ V$ and $Wx\ v$.

$$RF = (292 + 280 + 87 + 94)/10{,}756$$
$$= 753/10{,}756 = 0.07 = 7\%, \text{ or } 7 \text{ m.u.}$$

$Sh\!-\!V$ distance: recombinants are $sh\ V$ and $Sh\ v$.

$$RF = (1515 + 1531 + 87 + 94)/10{,}756$$
$$= 3227/10{,}756 = 0.30 = 30\%, \text{ or } 30 \text{ m.u.}$$

The $Wx\!-\!Sh$ distance is the sum of these two distances: 7 m.u. + 30 m.u. = 37 m.u.
(c) Expected double crossovers = $RF_1 \times RF_2 \times$ total progeny = $.07(.30)(10{,}756) = 226$.
Coefficient of coincidence = observed double crossovers/expected double crossover = $(87 + 94)/226 = 0.80$.
Interference = 1 − coefficient of coincidence = 0.20.

24.

$b^+\ pr^+\ vg^+$	normal body, normal eyes, normal wings	407
$b\ pr\ vg$	black body, purple eyes, vestigial wings	407
$b^+\ pr^+\ vg$	normal body, normal eyes, vestigial wings	63
$b\ pr\ vg^+$	black body, purple eyes, normal wings	63
$b^+\ pr\ vg$	normal body, purple eyes, vestigial wings	28
$b\ pr^+\ vg^+$	black body, normal eyes, normal wings	28
$b^+\ pr\ vg^+$	normal body, purple eyes, normal wings	2
$b\ pr^+\ vg$	black body, normal eyes, vestigial wings	2

25.

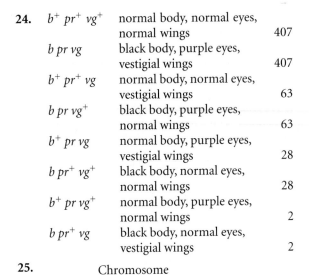

The mutations are mapped to the intervals indicated on the figure above the table. The location of f is ambiguous; it could be in either location shown.

27. Enzyme 1 is located on chromosome 9. Chromosome 9 is the only chromosome that is present in the cell lines that produce enzyme 1 and absent in the cell lines that do not produce enzyme 1. By similar reasoning, enzyme 2 is located on chromosome 4 and enzyme 3 is located on the X chromosome.

Chapter 8

1. Reproduction is rapid, asexual, and produces lots of progeny.

The genome is small and haploid.

Bacteria and viruses are easy to grow in the laboratory.

Techniques are available for isolating and manipulating their genes.

4.

Types of matings	Outcomes
$F^+ \times F^-$	Two F^+ cells
Hfr $\times$ F^-	One F^+ cell and one F^- cell
$F' \times F^-$	Two F' cells

The F factor contains a number of genes that take part in the conjugation process, including genes necessary for the synthesis of the sex pilus. The F factor has an origin of replication that enables the factor to be replicated in the conjugation process.

5. To map genes by conjugation, an interrupted-mating procedure is used. During the conjugation process, an Hfr strain is mixed with an F⁻ strain. The two strains must have different genotypes and must remain in physical contact for the transfer to take place. At regular intervals, the conjugation process is interrupted. The chromosomal transfer from the Hfr strain always begins with a part of the integrated F factor and proceeds in a linear fashion. To transfer the entire chromosome requires approximately 100 minutes. The time required for individual genes to be transferred depends on their positions on the chromosome and the direction of transfer initiated by the F factor. Gene distances are typically mapped in minutes. The genes that are transferred by conjugation to the recipient must be incorporated into the recipient's chromosome by recombination to be expressed.

In transformation, the relative frequency at which pairs of genes are cotransformed indicates the distance between the two genes. Closer genes are cotransformed more frequently. As is the case with conjugation, the donor DNA must recombine into the recipient cell's chromosome. Physical contact of the donor and recipient cells is not needed. The recipient cell takes up the DNA directly from the environment. So, the DNA from the donor strain has to be isolated and broken up before transformation can take place.

A viral vector is needed for the transfer of DNA by transduction. DNA from the donor cell is packaged into a viral protein coat. The viral particle containing the bacterial donor DNA then infects another recipient bacterial cell. The donor bacterial DNA is incorporated into the recipient cell's chromosome by recombination. Only genes that are close together on the bacterial chromosome can be cotransduced. Therefore, the rate of cotransduction, like the rate of cotransformation, gives an indication of the physical distances between genes on the chromosome.

9. In generalized transduction, bacterial genes are transferred from one bacterial cell to another by a virus. In specialized transduction, only genes from a particular locus on a bacterial chromosome are transferred to another bacterium. The process requires lysogenic phages that integrate into specific locations on the host cell's chromosome. When the phage DNA excises from the host chromosome and the excision process is imprecise, the phage DNA will carry a small part of the bacterial DNA. The hybrid DNA must be injected by the phage into another bacterial cell in another round of infection.

Transfer of DNA by transduction requires that the host DNA be broken down into smaller pieces and that a piece of the host DNA be packaged into a phage coat instead of phage DNA. The defective phage cannot produce new phage particles on a subsequent infection, but it can inject the bacterial DNA into another bacterium or recipient. Through a double-crossover event, the donor DNA can become incorporated into the bacterial recipient's chromosome.

10. To determine whether two different mutations occurred at the same gene locus or in different genes, Benzer performed complementation analysis. To conduct the complementation test, Benzer infected cells of *E. coli* K with large numbers of the two mutant phage types. For successful infection, each mutant phage must supply the gene product or protein missing in the other. Complementation will happen only if the mutations are at separate loci. If the two mutations are at the same locus, then complementation of gene products will not take place and no plaques will be produced on the *E. coli* K lawns.

12. A retrovirus is able to integrate its genome into the host cell's DNA genome through the action of the enzyme reverse transcriptase, which can synthesize complementary DNA from either an RNA or a DNA template. Reverse transcriptase uses the retroviral single-stranded RNA as a template to synthesize a double-stranded copy of DNA. The newly synthesized DNA molecule can then integrate into the host chromosome to form a provirus.

14. For 5 years, Smith, by using low doses of antibiotics, selected for bacteria that are resistant to the antibiotics. The doses used killed sensitive bacteria but not resistant bacteria. As time passes, only resistant bacteria will be present in his pigs because any sensitive bacteria have been eliminated by the low doses of antibiotics. In the future, Smith may continue to administer the vitamins, but he should use antibiotics only when a sick pig requires them. In this manner, he will not be selecting for antibiotic-resistant bacteria, and the chances of the antibiotic therapy successfully treating his sick pigs will be greater.

16. The closer genes are to the F factor, the more quickly they will be transferred in a linear fashion. The use of a blender to interrupt the mating process will stop transfer, and the F⁻ strain will have received only those genes carried on the piece of the Hfr strain's chromosome that entered the F⁻ cell before the disruption.

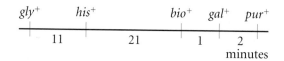

17. In each of the Hfr strains, the F factor has been inserted into a different location in the chromosome.

The orientation of the F factor in the strains varies as well.

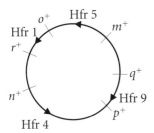

18. The F factor for each Hfr strain has been inserted into a different location on the chromosome, and the orientation of the F factor varies in the different strains. Although most of the selective markers transferred from each Hfr strain to the F⁻ strain are the same, some of the markers for a given Hfr strain are not transferred, because the mating was disrupted before the transfer of that selective marker. The relative positions of the genes to one another in minutes does not vary. So, for the different Hfr strains, the distance in minutes between each gene remains constant. The genes and their relative positions are shown in the adjoining diagram. Times are in minutes.

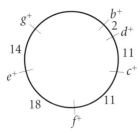

23. We know that the two genes are 8 map units apart. These 8 map units correspond to 8% recombination between the two genes. When the geneticist mixes the two phages ($m^+ c^+ \times m^- c^-$), creating a double infection of the bacterial cell, she should expect 8% of the progeny phages to exhibit recombinant plaque phenotypes. She should expect two types of recombinant plaque phenotypes, $m^+ c^-$ and $m^- c^+$. The combined recombinants should equal a recombination frequency of 8%. The remaining 92% will be a combination of the wild-type phage and the doubly mutant phage.

Plaque phenotype	Expected number
$c^+ m^+$	460
$c^- m^-$	460
$c^+ m^-$	40 (recombinant)
$c^- m^+$	40 (recombinant)
total plaques	1000

25. (a) To determine the recombination frequencies, the recombinant offspring must be identified. The recombination frequency is calculated by dividing the total number of recombinant plaques by the total number of plaques.

Genotype of parent	Progeny genotype	Number of plaques
$h^+ r_{13}^- \times h^- r_{13}^+$	$h^+ r_{13}^+$ (recombinant)	1
	$h^- r_{13}^+$	104
	$h^+ r_{13}^-$	110
	$h^- r_{13}^-$ (recombinant)	2
	total	216
$h^+ r_2^- \times h^- r_2^+$	$h^+ r_2^+$ (recombinant)	6
	$h^- r_2^+$	86
	$h^+ r_2^-$	81
	$h^- r_2^-$ (recombinant)	7
	total	180

The recombination frequency between r_2 and h is $13/180 = 0.072$, or 7.2%. The recombination frequency between r_{13} and h is $3/216 = 0.014$, or 1.4%.

(b)

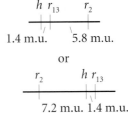

26. (a) By comparing the genotype of the double-recombinant phage progeny with the nonrecombinants, we can predict the gene order.

h^+ t^+ c^+
|————————|————————————————————|

(b) The map distances can be calculated by determining the percent recombination between each gene pair.

$h^+ t^+$: $[(26 + 30 + 5 + 6)/942] \times 100\% = 7.1\%$, or 7.1 map units

$t^+ c^+$: $[(106 + 110 + 5 + 6)/942] \times 100\% = 24.1\%$, or 24.1 map units

h^+ t^+ c^+
|————————|————————————————————|
     7.1 m.u.   24.1 m.u.

(c) Coefficient of coincidence = observed number of double recombinants/expected number of double recombinants.

Coefficient of coincidence = $(6 + 5)/(0.071 \times 0.241 \times 942) = 0.68$

Interference = 1 − coefficient of coincidence = 1 − 0.68 = 0.32

Chapter 9

1. Chromosome rearrangements:

   Deletion: loss of a part of a chromosome.

   Duplication: addition of an extra part of a chromosome.

   Inversion: a part of a chromosome is reversed in orientation.

Translocation: a part of one chromosome becomes incorporated into a different (nonhomologous) chromosome.

Aneupoloidy: loss or gain of one or more chromosomes, causing the chromosome number to deviate from $2n$.

Polyploidy: gain of entire sets of chromosomes, causing the chromosome number to change from $2n$ to $3n$ (triploid), $4n$ (tetraploid), and so forth.

2. The expression of some genes is balanced by the expression of other genes; the ratios of their gene products, usually proteins, must be maintained in a narrow range for proper cell function. Extra copies of a gene cause that gene to be expressed at proportionately higher levels, thereby upsetting the balance of gene products.

5. A paracentric inversion does not include the centromere; a pericentric inversion includes the centromere.

7.

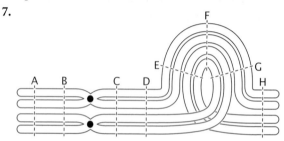

9. Like an inversion, a translocation can produce a phenotypic effect if the translocation breakpoint disrupts a gene or if a gene near the breakpoint is altered in its expression because of relocation to a different chromosomal environment, a position effect.

13. A person with more than one X chromosome maintains only one active X chromosome and inactivates all others in the form of Barr bodies through dosage compensation. Compared with other chromosomes, the Y chromosome is small and contains relatively few genes—none that are essential for human development or viability (after all, half of all people appear to get along perfectly well without a Y chromosome).

14. Primary Down syndrome is caused by spontaneous, random nondisjunction of chromosome 21, leading to trisomy 21. Familial Down syndrome most frequently arises as a result of a Robertsonian translocation of chromosome 21 with another chromosome, usually chromosome 14. If the translocated chromosome segregates with the normal chromosome 21, the gamete will have two copies of chromosome 21 and will result in a child with familial Down syndrome.

15. Uniparental disomy refers to the inheritance of both copies of a chromosome from the same parent. It may arise originally from a trisomy condition, in which the early embryo loses one of the three chromosomes, and the two remaining copies are from the same parent.

17. In autopolyploidy, all sets of chromosomes are from the same species. Autopolyploids arise from mitotic or

meiotic nondisjunction of all the chromosomes. In allopolyploidy, the chromosomes of two different species are contained in one individual. Allopolyploids arise from the hybridization of two related species followed by mitotic nondisjunction.

18. (a) Duplications; (b) polyploidy; (c) deletions; (d) inversions; (e) translocations.

19. (a) Tandem duplication of AB; (b) displaced duplication of AB; (c) paracentric inversion of DEF; (d) deletion of B; (e) deletion of FG; (f) paracentric inversion of CDE; (g) pericentric inversion of ABC; (h) duplication and inversion of DEF; (i) duplication of CDEF, inversion of EF.

22. (a) 15; (b) 24; (c) 32; (d) 17; (e) 14; (f) 14; (g) 40; (h) 18.

23. (a) $1/3$ Notch, white-eyed females, $1/3$ wild-type females, and $1/3$ wild-type males
(b) $1/3$ Notch, red-eyed females, $1/3$ wild-type females, and $1/3$ white-eyed males
(c) $1/3$ Notch, white-eyed females, $1/3$ white-eyed females, and $1/3$ white-eyed males

26. (a)

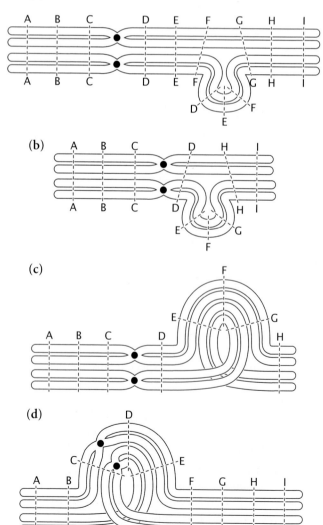

29.

(a)

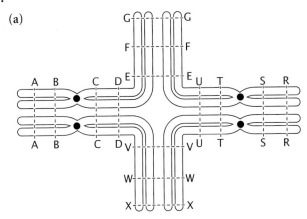

(b) Alternate:

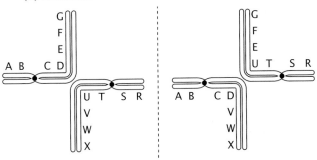

Adjacent-1:

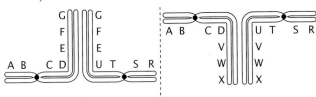

Adjacent-2:

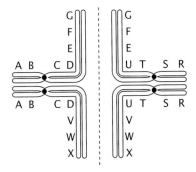

(c) Alternate segregation: gamete contains either both normal or both translocation chromosomes, and either is viable.

$$\underline{A\,B\cdot C\,D\,E\,F\,G} + \underline{R\,S\cdot T\,U\,V\,W\,X}$$

and

$$\underline{A\,B\cdot C\,D\,V\,W\,X} + \underline{R\,S\cdot T\,U\,E\,F\,G}$$

Adjacent-1 segregation: gamete contains one normal and one translocation chromosome, resulting in the duplication of some genes and deficiency of others.

$$\underline{A\,B\cdot C\,D\,E\,F\,G} + \underline{R\,S\cdot T\,U\,E\,F\,G}$$

and

$$\underline{A\,B\cdot C\,D\,V\,W\,X} + \underline{R\,S\cdot T\,U\,V\,W\,X}$$

Adjacent-2 segregation (rare): gamete contains one normal and one translocation chromosome, with the duplication of some genes and deficiency of others.

$$\underline{A\,B\cdot C\,D\,E\,F\,G} + \underline{A\,B\cdot C\,D\,V\,W\,X}$$

and

$$\underline{R\,S\cdot T\,U\,V\,W\,X} + \underline{R\,S\cdot T\,U\,E\,F\,G}$$

32. Familial Down syndrome is caused by a Robertsonian translocation of chromosome 21. Bill and his sister, who are unaffected, are phenotypically normal carriers of the translocation and have 45 chromosomes. Therefore, statement *d* is most likely correct.

33. Mike and Sue's baby could have inherited Tay-Sach's disease by uniparental disomy. A nondisjunction in meiosis II in spermatogenesis could have produced a sperm carrying two copies of the chromosome bearing the Tay-Sach's allele. Fertilization of a normal egg would then produce a trisomic zygote. Loss of the mother's normal chromosome in the first mitotic division would then produce an embryo in which both remaining copies of the chromosome bear the Tay-Sach's allele from Mike.

35. (a) The different types of gametes produced by the man:
   (i)   normal chromosome 13; normal chromosome 22
   (ii) translocated chromosome 13+22
   (iii) translocated chromosome 13+22; normal chromosome 22
   (iv) normal chromosome 13
   (v)   normal chromosome 13; translocated chromosome 13+22
   (vi) normal chromosome 22
   (b) For the gamete types listed in part *a*,
   (i)   13, 13, 22, 22—normal
   (ii) 13, 13+22, 22—translocation carrier
   (iii) 13, 13+22, 22, 22—trisomy 22
   (iv) 13, 13, 22—monosomy 22
   (v)   13, 13, 13+22, 22—trisomy 13
   (vi) 13, 22, 22—monosomy 13
   (c) Half, or 50%.

Chapter 10

1. First, the genetic material must contain complex information. Second, the genetic material must replicate or be replicated faithfully. Third, the genetic material must encode the phenotype.

3. Experiments by Hershey and Chase in the 1950s using the bacteriophage T2 and *E. coli* cells demonstrated that DNA is the genetic material of the bacteriophage. Experiments by Avery, Macleod, and McCarty demonstrated that the transforming material, initially identified by Griffiths, is DNA.

5. Hershey and Chase used the radioactive isotope ^{32}P. The progeny phages released from bacteria infected with ^{32}P-labeled phages emitted radioactivity from ^{32}P. The presence of the ^{32}P in the progeny phages indicated that the infecting phages had passed DNA on to the progeny phages.

7. The three parts of a DNA nucleotide are phosphate, deoxyribose sugar, and a nitrogenous base.

Deoxyguanosine 5'-phosphate (dGMP)

10.

12. DNA is not a static, rigid structure that is invariant. Local variation in DNA structure refers to the variations that exist in a DNA molecule. For instance, B-DNA is described as having an average of 10 bases per turn. However, the actual value may be less than or greater than 10, depending on the environmental conditions.

14. Replication, transcription, and translation—the components of the central dogma of molecular biology.

18. (a) Proteins contain sulfur in the amino acids cysteine and methionine. However, proteins do not typically contain phosphorus (or they contain very limited amounts owing to the phosphorylation of certain proteins by protein kinases). DNA contains much phosphorus, owing to its sugar–phosphate backbone, but no sulfur. Hershey and Chase chose the isotopes ^{32}P and ^{35}S because these radioactive elements would allow them to distinguish between proteins and DNA molecules. Only DNA would contain the isotope ^{32}P, and only proteins would contain the isotope ^{35}S.
(b) No. Both DNA and proteins consist of significant amounts of carbon and oxygen. Hershey and Chase would have been unable to isolate DNA molecules or proteins that contain radioactive isotopes of these elements. The molecules were labeled with the radioactive isotopes in vivo by using phages to infect *E. coli* cells grown in media containing the radioactive isotopes. Because both proteins and DNA contain carbon and oxygen, both molecules in the phage progeny would have received the radioactive isotopes.

20. If each nucleotide pair of a DNA double helix weighs approximately 1×10^{-21} g and the human body contains 0.5 g of DNA, then the number of nucleotide pairs can be estimated as follows:

$$(0.5 \text{ g DNA/human})/(1 \times 10^{-21} \text{ g/nucleotide}) = 5 \times 10^{20} \text{ nucleotide pairs/human}$$

The average distance between each nucleotide pair in B-DNA is 3.4 nm. If a human possesses 5×10^{20} nucleotide pairs, then the reach of that DNA stretched end to end would be

$$(5 \times 10^{20} \text{ nucleotides/human}) \times (3.4 \text{ nm/nucleotide pair}) = 1.7 \times 10^{21} \text{ nm, or } 1.7 \times 10^{9} \text{ km}$$

22. The relations in parts *b, c, e, g,* and *h.*

23. The percentage of thymine (15%) should be approximately equal to the percentage of adenine (15%). The remaining percentage of DNA bases (100% − 15% − 15% = 70%) will consist of cytosine and guanine bases in approximately equal amounts. Therefore the percentages of each of the other bases if the thymine content is 15% are: adenine = 15%; guanine = 35%; and cytosine = 35%.

25. (a) A B-DNA molecule of 1 million nucleotide pairs will have about the following number of complete turns:

(1,000,000 nucleotides)/(10 nucleotides/turn)

= 100,000 complete turns

(b) If the same DNA molecule assumes a Z-DNA configuration, then each turn would consist of about 12 nucleotides. The determination of the number of complete turns in the 1 million nucleotide molecule is:

(1,000,000 nucleotides)/(12 nucleotides/turn) =

83333.3 or 83333 complete turns

27. The central dogma is not consistent with the theory of inheritance of acquired characteristics. The central dogma indicates the flow of information within the cell.

$$DNA \longrightarrow RNA \longrightarrow protein$$

One exception to the central dogma is reverse transcription, whereby RNA codes for DNA. However, biologists know of no process by which information flows from protein back to DNA, which would be required by the theory of inheritance of acquired characteristics.

Chapter 11

1. Supercoiling arises from overwinding (positive) or underwinding (negative) of the DNA double helix when the DNA molecule does not have free ends, as in circular DNA molecules, or when the ends of the DNA molecule are bound to proteins that prevent them from rotating about each other.

3. The nucleosome core particle contains two molecules each of histones H2A, H2B, H3, and H4, forming a protein core, with 145 to 147 bp of DNA wound around the core. A chromatosome contains the nucleosome core and a molecule of histone H1.

6. The centromere is the point of attachment for mitotic and meiotic spindle fibers and is required for movement of the chromosomes in mitosis and meiosis. Centromeres have distinct centromeric DNA sequences to which the kinetochore proteins bind. There is considerable variation in centromere structure. For some species, as in budding yeast, the centromere is compact, consisting of only 125 bp. For other species, including *Drosophila* and mammals, the centromere is larger, consisting of several hundred kilobasepairs of DNA sequence.

7. Telomeres are the stable ends of the linear chromosomes in eukaryotes. They cap and stabilize the ends of the chromosomes to prevent degradation by exonucleases or to prevent the joining of the ends. Telomeres also enable replication of the ends of the chromosome. Telomeric DNA sequences consist of repeats of a simple sequence, usually in the form of $5' - C_n(A/T)_m$.

10. Unique sequences, present in only one or a few copies per haploid genome, constitute most of the protein-coding sequences, plus a great deal of sequence with unknown function.

Moderately repetitive sequences, between a few hundred and a few thousand base pairs long, are present in as many as several thousand copies per haploid genome. Some moderately repetitive DNA consists of functional genes that encode rRNAs and tRNAs, but most is made up of transposable elements and remnants of transposable elements.

Highly repetitive DNA, or satellite DNA, consists of clusters of tandem repeats of short (often fewer than 10 base pairs) sequences present in hundreds of thousands to millions of copies per haploid genome.

11. Most transposable elements have terminal inverted repeats and are flanked by short direct repeats. Replicative transposons use a copy-and-past mechanism by which the transposon is replicated and inserted in a new location, leaving the original transposon in place. Nonreplicative transposons use a cut-and-paste mechanism by which the original transposon is excised and moved to a new location.

12. A retrotransposon is a transposable element that transposes through an RNA intermediate. First, it is transcribed into RNA. A reverse transcriptase encoded by the retrotransposon then reverse transcribes the RNA template into a DNA copy, which then integrates into a new location in the host genome.

13. First, a transposase makes single-stranded nicks on either side of a transposon and on either side of the target sequence. Second, the free ends of the transposon are joined to the free ends of the DNA at the target site by a DNA ligase. Third, the free 3′ ends of DNA on either side of the transposon are used to replicate the transposon sequence, forming the cointegrate. The enzymes normally required for DNA replication, including DNA polymerase, are required for this step. The cointegrate has two copies of the transposon plus the target-site sequence on one side of each copy. Fourth, the cointegrate undergoes resolution, in which resolvase enzymes (such as those used in homologous recombination) catalyze a crossing over within the transposon.

14.

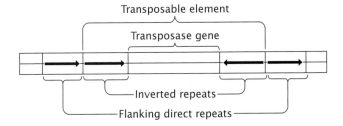

19. The cellular function hypothesis proposes that transposable elements have a function, such as the regulation of gene expression, in the cell or organism.

The genetic variation hypothesis suggests that transposable elements serve to generate genomic variation. A larger pool of genomic variants would accelerate evolution by natural selection.

The selfish DNA hypothesis suggests that transposable elements are simply genomic parasites, serving only to replicate and spread themselves.

20. Prokaryotic chromosomes are usually circular, whereas eukaryotic chromosomes are linear. Prokaryotic chromosomes generally contain the entire genome, whereas each eukaryotic chromosome contains only a part of the genome—that is, the eukaryotic genome is divided among multiple chromosomes. Prokaryotic chromosomes are generally smaller than eukaryotic chromosomes and have only a single origin of DNA replication. Eukaryotic chromosomes are often many times larger than prokaryotic chromosomes and contain many origins of DNA replication. Prokaryotic chromosomes are typically condensed into nucleoids, which have loops of DNA compacted into a dense body. Eukaryotic chromosomes contain DNA and histone proteins packaged into nucleosomes, which are further coiled and packaged into successively higher order structures. The condensation state of eukaryotic chromosomes varies with the cell cycle.

23. (a) Given that each nucleosome contains about 140 bp of DNA tightly associated with the core histone octamer, another 20 bp associated with histone H1, and 40 bp in the linker region, there is one nucleosome for every 200 bp of DNA.

$$\frac{6 \times 10^9 \, \text{bp}}{2 \times 10^2 \, \text{bp/nucleosome}} = \frac{3 \times 10^7 \, \text{nucleosomes}}{(30 \, \text{million})}$$

(b) Each nucleosome contains two each of histones H2A, H2B, H3, and H4. A nucleosome plus one molecule of histone H1 forms the chromatosome. Therefore, there are nine histone protein molecules for every nucleosome:

3×10^7 nucleosomes $\times$ 9 histones $= 2.7 \times 10^8$ molecules of histones complexed to 6 billion bp of DNA

24. More acetylation. Regions of DNase I sensitivity are less condensed than DNA that is not sensitive to DNase I, and the sensitive DNA is less tightly associated with nucleosomes and in a more open state. Such a state is associated with acetylation of lysine residues in the N-terminal histone tails. Acetylation eliminates the positive charge of the lysine residue and reduces the affinity of the histone for the negatively charged phosphates of the DNA backbone.

27. The upper molecule, with a higher percentage of AT base pairs, has a lower melting temperature than that of the lower molecule, which has mostly GC base

pairs. An AT base pair has two hydrogen bonds and thus less stability than does a GC base pair, which has three hydrogen bonds.

30. The flanking direct repeat is in boldface type.
(a) 5′–ATTCGAAC**TGAC** (transposable element) **TGAC**CGATCA–3′
(b) 5′–ATT**CGAA** (transposable element) **CGAA**CTGACCGATCA–3′

31. Such a fly may be homozygous (female) or hemizygous (male) for an allele of the white-eye locus that contains a transposon insertion. The eye cells in these flies cannot make red pigment. In the course of eye development, the transposon may spontaneously transpose out of the white-eye locus, restoring function to this gene and allowing the cell and its mitotic progeny to make red pigment. The number and size of the red spots in the eyes depend on how early in eye development the transposition occurs.

32. Without a functional transposase gene of its own, the transposon would be able to transpose only if another transposon of the same type were in the cell and able to express a functional transposase enzyme. This transposase enzyme would recognize the inverted repeats and transpose both its own element and other nonautonomous copies of the transposon with the same inverted repeats.

33. The length of the flanking direct repeats produced depends on the number of base pairs between the staggered single-strand nicks made at the target site by the transposase.

Chapter 12

2. Meselson and Stahl grew E. coli cells in a medium containing the heavy isotope of nitrogen (^{15}N) for several generations. The ^{15}N was incorporated in the DNA of the E. coli cells. The E. coli cells were then switched to a medium containing the common isotope of nitrogen (^{14}N) and allowed to proceed through a few cellular generations. Samples of the bacteria were removed at each cellular generation. Using equilibrium density gradient centrifugation, Meselson and Stahl were able to distinguish DNAs that contained only ^{15}N from DNAs that contained only ^{14}N or a mixture of ^{15}N and ^{14}N since DNAs containing the ^{15}N isotope are "heavier." The more ^{15}N a DNA molecule contains, the further it will sediment during equilibrium density gradient centrifugation. DNA from cells grown in the ^{15}N medium produced only a single band at the expected position during centrifugation. After one round of replication in the ^{14}N medium, one band was present following centrifugation, but the band was located at a position intermediate to that of a DNA band containing only ^{15}N and a DNA band containing only ^{14}N. After two rounds of replication, two bands of DNA were present. One band was located at a position intermediate to that of a DNA band containing only

^{15}N and a DNA band containing only ^{14}N, while the other band was at a position expected for DNA containing only ^{14}N. These results were consistent with the predictions of semi-conservative replication and incompatible with the predictions of both conservative and dispersive replication.

3.

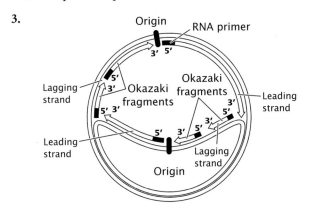

7. The substrates for DNA synthesis are the four types of deoxyribonucleotides: deoxyadenosine triphosphate, deoxyguanosine triphosphate, deoxycytosine triphosphate, and deoxythymidine triphosphate.

10. Primase is a DNA-dependent RNA polymerase. In other words, it is a RNA polymerase. Primase synthesizes the short RNA molecules, or primers, that provide a 3'-OH group to which DNA polymerase can attach deoxyribonucleotides in the initiation of replication. The DNA polymerases require a 3'-OH group to which they add nucleotides, and therefore they cannot initiate replication. Primase does not have this requirement.

13. Both eukaryotic DNA and bacterial DNA are replicated in accord with some basic principles:

Replication is semiconservative.

Replication origins serve as starting points for replication.

A short segment of RNA called a primer provides a 3'-OH group for DNA polymerases to begin synthesis of the new strands.

Synthesis is in the 5'-to-3' direction.

The template strand is read in the 3'-to-5'' direction.

Deoxyribonucleotides are the substrates.

Replication is continuous on the leading strand and discontinuous on the lagging strand.

    Eukaryotic DNA replication differs from bacterial replication in the following ways:

There are multiple origins of replication per chromosome.

Several different DNA polymerases have different functions.

Immediately after DNA replication, nucleosomes are assembled.

16. RecBCD protein unwinds double-stranded DNA and can cleave nucleotide strands.

RecA protein allows a single strand to invade a double-stranded DNA; subsequently, one strand of the DNA helix is displaced.

RuvA and RuvB proteins promote branch migration.

RuvC protein is resolvase, a protein that resolves the Holliday structure by cleavage of the DNA.

DNA ligase repairs nicks or cuts in the DNA generated during recombination.

17. The strands of the extraterrestrial double-stranded nucleic acid must be parallel to one another, unlike antiparallel double-stranded DNA present on Earth. Replication by *E. coli* polymerases can proceed only in the 5'-to-3' direction. If replication is continuous on both strands, the two strands must have the same direction and be parallel.

18. In the initial sample removed immediately after transfer, no ^{32}P should be incorporated into the DNA, because replication in the medium containing ^{32}P has not yet taken place. After one round of replication in the ^{32}P containing medium, one strand of each newly synthesized DNA molecule will contain ^{32}P, whereas the other strand will contain only nonradioactive phosphorus. After two rounds of replication in the ^{32}P containing medium, 50% of the DNA molecules will have ^{32}P in both strands, whereas the remaining 50% will contain ^{32}P in one strand and nonradioactive phosphorus in the other strand.

20. In bidirectional replication, there are two replication forks, each proceeding at a rate of 100,000 nucleotides per minute. So, it will take 5 minutes for the circular DNA molecule to be replicated by bidirectional replication, because each fork can synthesize 500,000 nucleotides (5 minutes $\times$ 100,000 nucleotides per minute) within the time period. Because rolling-circle replication is unidirectional and thus has only one replication fork, 10 minutes will be required to replicate the entire circular molecule.

22.

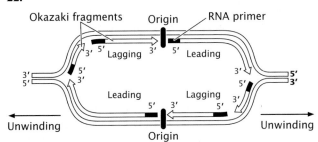

23. (a) The 3' $\rightarrow$ 5' exonuclease activity provides the proofreading function for DNA polymerase I. If this activity were destroyed, more errors would occur in replication.

(b) The 5' $\rightarrow$ 3' exonuclease activity is used to remove the RNA primers that initiate replication. If this activity were destroyed, the primers would not be removed.

(c) The 5' → 3' polymerase activity is used to synthesize DNA to replace the primers. If this activity were destroyed, the primers would not be replaced after their removal.

Chapter 13

1. Both RNA and DNA are polymers of nucleotides that are held together by phosphodiester bonds. An RNA nucleotide contains ribose sugar, whereas a DNA nucleotide contains deoxyribose sugar. The pyrimidine base uracil rather than thymine is found in RNA; DNA, however, contains thymine but not uracil. Finally, an RNA polynucleotide is typically single stranded, even though RNA molecules can pair with other complementary sequences. DNA molecules are almost always double stranded.

Thymine
(DNA only)

Uracil
(RNA only)

Deoxyribonucleotide   Ribonucleotide

3. Promoter, RNA coding region, and terminator.

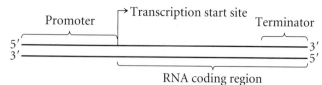

6. RNA polymerase I transcribes rRNA. RNA polymerase II transcribes pre-mRNAs, snoRNAs, and some snRNAs. RNA polymerase III transcribes small RNA molecules such as tRNAs, small rRNAs, and some snRNAs.

8.

Transcription start site

5'–〰〰TTGACA〰〰〰TATAAT〰〰–3'
3'–〰〰AACTGT〰〰〰ATATTA〰〰–5'
        −35   −10
       region   region

An upstream element rich in AT sequences is found in some bacterial promoters and is located upstream of the −35 consensus sequence. However, the typical bacterial promoter consists of the −35 and −10 consensus sequences.

11. Both promoters and enhancers are DNA sequences that affect transcription. Transcriptional activator proteins bind to sequences of regulatory promoters and enhancers and stimulate transcription by interacting with the basal transcriptional apparatus. Promoters are essential for the binding of both general transcriptional factors and RNA. Enhancers increase or stimulate transcription. Promoters are typically adjacent to the transcriptional start site and are highly position dependent, whereas enhancers function at a distance from the gene and function independently of position and direction.

15. The TATA-binding protein (TBP) binds most eukaroytic promoters at the TATA box and positions the active site of RNA polymerase over the transcriptional start site.

16. Characteristics of transcription and replication:

Both utilize a DNA template.

Molecules are synthesized in the 5'-to-3' direction.

Newly synthesized molecules are antiparallel and complementary to the template.

Both use nucleotide triphosphates as substrates.

A complex of proteins and enzymes is necessary for catalysis.

Characteristics of transcription:

Unidirectional synthesis is of only a single strand of nucleic acid.

Initiation does not require a primer.

It is subject to numerous regulatory mechanisms.

Individual genes or small groups of genes are transcribed separately.

Characteristics of replication:

Bidirectional synthesis is of two strands of nucleic acid.

Synthesis initiates from replication origins.

18. Similarities: (1) use of DNA templates, (2) DNA template read in 3'-to-5' direction, (3) 5'-to-3' synthesis of a complementary strand that is antiparallel to the template, (4) both use nucleoside triphosphates as substrate, and (5) actions are enhanced by accessory proteins.

Differences: (1) RNA polymerases use ribonucleotides as substrates, whereas DNA polymerases use deoxyribonucleotides; (2) DNA polymerases require a primer that provides an available 3'-OH group so that synthesis can begin, whereas RNA polymerases do not require primers to begin synthesis; and (3) RNA polymerase synthesizes a copy of only one of the DNA strands, whereas DNA polymerase synthesizes copies of both DNA strands.

21. 5'–AUAGGCGAUGCCA–3'

25. (a) A mutation at the −8 position would probably affect the −10 consensus sequence (TATAAT),which is centered on position −10. This consensus sequence is

necessary for the binding of RNA polymerase. A mutation there would most likely decrease transcription.
(b) A mutation in the −35 region could affect the binding of RNA polymerase to the promoter. Deviations away from the consensus typically down-regulate transcription; so transcription is likely to be reduced or inhibited.
(c) The −20 region is located between the consensus sequences of an *E. coli* promoter. Although the holoenzyme may cover the site, a mutation is not likely to have any effect on transcription.
(d) A mutation in the start site would have little effect on transcription. The position of the start site relative to the promoter is more important than the sequence at the start site.

27. (a)

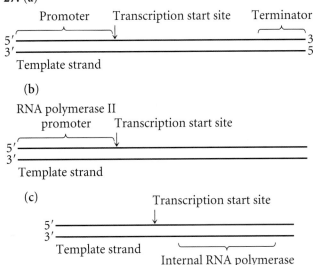

29. (a) Because the rho protein takes part in transcription termination, it should not affect transcription initiation or elongation.
(b) Without the rho protein, transcription would be expected to continue past the normal termination site of rho-dependent terminators, producing some longer molecules than expected.
(c) Only RNA molecules produced from genes using rho-dependent termination should be longer. Genes that are terminated through rho-independent termination should remain unaffected.
(d) Some RNA molecules that are longer than normal would be expected, because the transcription of genes that use rho-dependent termination would not terminate at the normal termination site.
(e) RNA will be copied only from a single strand because rho protein does not affect transcription initiation or elongation.

31. If TBP cannot bind to the TATA box, then genes having these promoters will not be transcribed or will be transcribed at very low levels. Because the TATA box is the most common promoter element for

RNA polymerase II transcription units and is found in some RNA polymerase III promoters, transcription will decline significantly. The lack of proteins encoded by these genes will most likely result in cell death.

34. When TFIIE and TFIIH were individually exchanged, transcriptional levels were altered. However, the paired exchange of TFIIE and TFIIH had no affect on the level of transcription, suggesting that TFIIE and TFIIH must interact to promote transcription or that the absence of their interaction will inhibit transcription. Similar results were obtained by the paired exchange of TFIIB and RNA polymerase, suggesting that interactions between TFIIB and RNA polymerase are needed for transcription. Futhermore, the exchange of TFIIB and RNA polymerase shifted the start point of transcription, indicating that TFIIB plays a role in positioning RNA polymerase at the proper start site for transcription.

Chapter 14

1. According to the concept of colinearity, the number of amino acids in a protein should be proportional to the number of nucleotides in the gene encoding the protein. In bacteria, the concept is nearly fulfilled. However, in eukaryotes, long regions of noncoding DNA sequence split coding regions. So, the concept of colinearity at the DNA level is not fulfilled.

3. The four basic groups are: (1) group I introns, found in rRNA genes and some bacteriophage genes; (2) group II introns, found in protein-encoding genes of mitochondria, chloroplasts, and a few eubacteria; (3) nuclear pre-mRNA introns, found in the protein-encoding genes of the nucleus; and (4) transfer RNA introns, found in tRNA genes.

4. The three principal elements are: (1) the 5′ untranslated region, which contains the Shine-Dalgarno sequence; (2) the protein-encoding region; and (3) the 3′ untranslated region.

6. (a) The 5′ end of eukaryotic mRNA is modified by the addition of the 5′ cap. The cap consists of the extra guanine nucleotide (linked 5′-to-5′ to the mRNA molecule) that is methylated at position 7 of the base and of adjacent nucleotides whose sugars are methylated at the 2′-OH group.
(b) Initially, the terminal phosphate group of the three 5′ phosphates linked to the end of the mRNA molecule is removed. Subsequently, a guanine nucleotide is attached to the 5′ end of the mRNA by 5′-to-5′ phosphate linkage. Next, a methyl group is attached to position 7 of the guanine base. Ribose sugars of adjacent nucleotides also may be methylated at the 2′-OH group.

(c) cap-binding proteins recognize the 5′ cap and stimulate binding of the ribosome to the 5′ cap and to the mRNA molecule. The 5′ cap may also increase mRNA stability in the cytoplasm. Finally, the 5′ cap is needed for efficient splicing of pre-mRNA introns.

8. The spliceosome consists of five small ribonucleoproteins (snRNPs) and additional proteins. Each snRNP is composed of multiple proteins and a single small nuclear RNA (snRNA) molecule. The snRNPs are identified by which snRNA (U1, U2, U3, U4, U5, or U6) each contains. Within the spliceosome, pre-mRNA nuclear introns are spliced.

11. RNA editing alters the sequence of an RNA molecule after transcription either by the insertion, deletion, or modification of nucleotides within the transcript. The guide RNAs (gRNAs) provide templates for the alteration of nucleotides in RNA molecules undergoing editing and are complementary to regions within the preedited RNA molecule. At these complementary regions, the gRNAs base pair to the preedited RNA molecule, and the nucleotides at the paired region undergo alteration.

12. Several modifications to pre-mRNA that produce mature mRNA are: (1) addition of the 5′ cap to the 5′ end of the pre-mRNA; (2) cleavage of the 3′ end of a site downstream of the AAUAAA consensus sequence of the last exon; (3) addition of the poly(A) tail to the 3′ end of the mRNA immediately after cleavage; and (4) removal of the introns, or splicing.

13. Cleavage of the precursor RNA into smaller molecules.

Nucleotides at both the 5′ and the 3′ ends of the tRNAs maybe be removed or trimmed.

Standard bases can be altered by base-modifying enzymes to produce modified bases.

15. Most rRNAs are synthesized as large precursor RNAs that are processed by methylation, cleavage, and trimming to produce the mature mRNA molecules. In *E. coli*, specific bases and the 2′-OH group of the ribose sugars of the 30S rRNA precursor are methylated. The 30S precursor is cleaved and trimmed to produce the 16S rRNA, the 23S rRNA, and the 5S rRNA. In eukaryotes, a similar process takes place. However, small nucleolar RNAs help to cleave and modify the precursor rRNAs.

16. The large size of the dystrophin gene is likely due to the presence of many intervening sequences, or introns, within the coding region of the gene. Excision of the introns through RNA splicing yields the mature mRNA that encodes the dystrophin protein.

18.

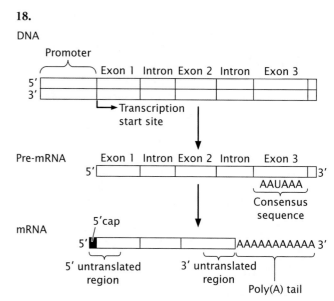

(a) The 5′ untranslated region lies upstream of the translation start site. At the 5′ end of this region is the cap, which serves as the binding site for the ribosome. The 5′ untranslated region may also contain sequences that regulate translation and affect the stability of mRNA.

(b) The promoter is the DNA sequence recognized by the transcription apparatus, which binds to the promoter to initiate transcription.

(c) The AAUAAA consensus sequence lies downstream of the coding region of the gene and determines the location of the 3′ cleavage site in the pre-mRNA molecule.

(d) The transcription start site begins the coding region of the gene and is located from 25 to 30 nucleotides downstream of the TATA box.

(e) The 3′ untranslated region is a sequence of nucleotides at the 3′ end of the mRNA and is not translated into proteins. However, it affects both the translation and the stability of the mRNA molecule.

(f) Introns are noncoding sequences of DNA that intervene within the coding regions of a gene.

(g) Exons are the coding sequences of a gene.

(h) A poly(A) tail is a sequence of adenine nucleotides added to the 3′ end of pre-mRNA. The tail affects mRNA stability.

(i) The 5′ cap functions in the initiation of translation and mRNA stability.

20. (a) Deletion of the AAUAA consensus sequence would prevent binding of the cleavage and polyadenylation factor (CPSF), thus resulting in no polyadenylation of the pre-mRNA, which in turn would affect the stability and translation of the mRNA.

(b) Deletion of the 5′ cap would most likely eliminate splicing of the intron nearest to the 5′ cap. Ultimately, elimination of the cap affects the stability of the pre-mRNA and its ability to be translated.

(c) Polyadenylation increases the stability of mRNA. If the poly(A) tail were eliminated from the pre-mRNA, the mRNA would be quickly degraded by nucleases in the cytoplasm.

23. The stability of mRNA depends on the proteins that bind to the poly(A) tail. If the proteins were unable to bind to the tail, the rate of degradation of the mRNA molecules that contain poly(A) tails would be greatly enhanced.

24. (a) RNAs will be unspliced and intact. The binding of U1 to the 5′ splice site is the first step in the splicing reaction, necessary for all subsequent steps.
(b) RNAs will be unspliced and intact, but with U1 bound to the 5′ splice site. The binding of U2 to the branch site is the second step in the splicing reaction. Without U2, no further reactions can take place.
(c) RNAs will be unspliced and intact, but with U1 and U2 bound. The U6–U4 dimer and U5 form a catalytic complex. A mixture missing any one of these snRNPs will not be catalytically active.
(d) Same as part c.
(e) Same as part c.

Chapter 15

2. Nirenberg and Matthaei used homopolymer and copolymer methods.

Advantages: Using a cell-free protein system, they were able to determine the amino acids encoded by each homopolymer and thus the amino acids specified by UUU, AAA, CCC, and GGG codons. Could predict codons on copolymers on the basis of amino acid sequence.

Disadvantages: The homopolymer method yielded the specifities of only four codons. The copolymer method depended on random incorporation of the nucleotides, which did not always happen. A further problem was that only the bases contained within the codon, not their sequence, could be determined. The redundancy of the code created difficulties because several different codons specified the same amino acid.

Nirenberg and Leder mixed ribosomes bound to short RNAs of known sequences with charged tRNAs. The mixture was passed through a nitrocellulose filter to which the tRNAs paired with ribosome-mRNA complexes adhered. They next determined the amino acids attached to the bound tRNAs.

Advantage: More than 50 codons were identified by this method.

Disadvantage: Not all tRNAs and codons could be identified by this method.

Khorana synthesized RNAs of known repeating sequences and used a cell-free protein-synthesizing system to produce proteins of alternating amino acids.

Advantage: Determined amino acids encoded by repeating sequences.

Disadvantage: Procedure could not specify which codon encodes which amino acid.

4. Synonymous codons encode the same amino acid. A nucleotide at the third position of a codon pairs with a nucleotide in the first position of the anticodon. Unlike the other nucleotide positions in the codon-anticodon pairing, this pairing is often weak, or "wobbles," and nonstandard pairings can result. Because many synonymous codons differ at only the third nucleotide position, in these codons the "wobble" and nonstandard base pairing with the anticodons will likely result in the correct amino acid being inserted in the protein.

5. (a) The reading frame refers to the way in which the groups of three nucleotides, or codons, are read in a sequence of nucleotides. For any sequence of nucleotides, there are potentially three sets of codons—three ways that the sequence can be read in groups of three nucleotides—that can specify the amino acid sequence of a polypeptide.
(b) In an overlapping code, a single nucleotide would be included in more than one codon. The result for a sequence of nucleotides would be that more than one type of polypeptide could be encoded within that sequence.
(c) In a nonoverlapping code, a single nucleotide is part of only one codon. In a sequence of RNA, the result is the production of a single type of polypeptide.
(d) An initiation codon establishes the appropriate reading frame and specifies the first amino acid of the protein chain. Typically, the initiation codon is AUG, however, both GUG and UUG also can serve as initiation codons.
(e) The termination codon signals the termination of translation and the end of the protein molecule. The three types of termination codons—UAA, UAG, and UGA—are also referred to as stop codons or nonsense codons. These codons do not encode amino acids.
(f) A sense codon is a group of three nucleotides that encode an amino acid. Sixty-one sense codons encode the 20 amino acids commonly found in proteins.
(g) A nonsense codon, or termination codon, signals the end of translation. These codons do not encode amino acids.
(h) In a universal code, each codon specifies the same amino acid in all organisms. The genetic code is nearly universal but not completely. Most of the exceptions are in mitochondrial genes.
(i) Most codons are universal (or nearly universal) in that they specify the same amino acids in almost all organisms. However, there are exceptions where a codon has different meanings in different organisms.

7. For each of the 20 different amino acids commonly found in proteins, a corresponding aminoacyl-tRNA synthetase links the amino acid to the tRNA molecule.

10. Three elongation factors have been identified in bacteria: EF-TU, EF-TS, and EF-G. EF-TU joins GTP,

followed by a tRNA charged with an amino acid to form a three-part complex. The charged tRNA is delivered to the ribosome at the A site. The GTP joined to EF-TU is then cleaved to form a EF-TU–GDP complex. EF-TS regenerates EF-TU–GTP. The elongation factor EF-G binds GTP and is necessary for the translocation, or movement, of the ribosome along the mRNA during translation.

14. A number of antibiotics bind to ribosomes and inhibit protein synthesis at different steps in translation. Some antibiotics, such as streptomycin, bind to the small subunit and inhibit translation initiation. Other antibiotics, such as chloramphenicol, bind to the large subunit and block elongation of the peptide by preventing peptide-bond formation. Antibiotics such as tetracycline and neomycin bind to the ribosome near the A site, yet they have different effects. Tetracyclines block the entry of charged tRNAs to the A site, whereas neomycin induces translational errors. Finally, some antibiotics, such as erythomycin, block the translocation of the ribosome along the mRNA.

16. On the basis of the mutant strains' ability to grow on the given substrates, the mutations can be assembled into four groups:

Group 1 mutants (trp-1, trp-10, trp-11, trp-9, trp-6, and trp-7) can grow only on minimal medium supplemented with trpytophan.

Group 2 mutants (trp-3) can grow on minimal medium supplemented with either trpytophan or indole.

Group 3 mutants (trp-2 and trp-4) can grow on minimal medium supplemented with tryptophan, indole, or indole glycerol phosphate.

Group 4 mutants (trp-8) can grow on minimal medium supplemented with the addition of tryptophan, indole, indole glycerol phosphate, or anthranilic acid.

$$\text{precursor} \xrightarrow{\substack{\text{group}\\4}} \text{anthranilic acid} \xrightarrow{\substack{\text{group}\\3}} \text{indole glyerol phosphate} \xrightarrow{\substack{\text{group}\\2}} \text{indole} \xrightarrow{\substack{\text{group}\\1}} \text{tryptophan}$$

18. (a) One, because $4^1 = 4$ codons, which is more than enough to specify two different amino acids; (b) two; (c) three; (d) three; (e) four.

20. (a) Amino–fMet-Phe-Lys-Phe-Lys-Phe–Carboxyl
(b) Amino–fMet-Tyr-Ile-Tyr-Ile–Carboxyl
(c) Amino–fMet-Asp-Glu-Arg-Phe-Leu-Ala–Carboxyl
(d) Amino–fMet-Gly–Carboxyl (The stop codon UAG follows the codon for glycine.)

22. There are two possible sequences:
5′–AUGUGGCAU–3′
DNA template: 3′–TACACCGTA–5′
DNA nontemplate: 5′–ATGTGGCAT–3′

and

5′–AUGUGGCAC–3′
DNA template: 3′–TACACCGTG–5′
DNA nontemplate: 5′–ATGTGGCAC–3′

24. (a) 3′–CCG–5′ or 3′–UCG–5′
(b) 3′–UUC–5′
(c) 3′–AUU–5′ or 3′–UUU–5′ or 3′–CUU–5′
(d) 3′–ACC–5′ or 3′–GCC–5′
(e) 3′–GUC–5′

27. (a) The lack of IF-1 would decrease the amount of protein synthesized. IF-1 promotes the dissociation of the large and small ribosomal subunits. Translation would be initiated but at a lower-than-usual rate because more of the small ribosomal subunits would be bound to the large ribosomal subunits.
(b) No translation would take place. IF-2 is necessary for translation initiation. The lack of IF-2 would prevent fMet-tRNA^{fMet} from being delivered to the small ribosomal subunit, thus blocking translation.
(c) Although translation would be initiated by the delivery of metthionine to the ribosome- mRNA complex, no other amino acids would be delivered to the ribosome. EF-TU, which binds to GTP and the charged tRNA, is necessary for elongation. This three-part complex enters the A site of the ribosome. If EF-TU is not present, the charged tRNA will not enter the A site, thus stopping translation.
(d) EF-G is necessary for the translocation of the ribosome along the mRNA in the 5′-to-3′ direction. When a peptide bond has formed between Met and Pro, the lack of EF-G would prevent the movement of the ribosome along the mRNA; so no new codons would be read. The formation of the dipeptide Met-Pro does not require EF-G.
(e) The release factors recognize the stop codons and promote cleavage of the peptide from the tRNA at the P site. The absence of the release factors would prevent termination of translation at the stop codon.
(f) ATP is required for tRNAs to be charged with amino acids by aminoacyl-tRNA synthetases. Without ATP, the charging would not take place, and no amino acids would be available for protein synthesis.
(g) GTP is required for the initiation, elongation, and termination of translation. If GTP is absent, no protein synthesis will take place.

29. (a) By mutation of the anticodon to 5′–CCA–3′ from 5′–CAU–3′ on tRNA$_i^{Met}$, the initiator tRNA will recognize the codon 5′–UGG–3′, which normally codes only for Trp. If translation in eukaryotes is initiated by the ribosome binding to the 5′ cap of the mRNA followed by scanning, then the first 5′–UGG–3′ codon recognized by the mutated tRNA$_i^{Met}$ will be the start site for translation. If the first 5′–UGG–3′ codon precedes the normal 5′–AUG–3′ codon, then either a protein containing extra amino acids or a shorter protein

containing fewer amino acids could be produced. Finally, truncated proteins could also be produced by the first 5′–UGG–3′ codon being out of the reading frame of the normal coding sequence, in which case a stop codon will likely be encountered before the end of the normal coding sequence and will prematurely terminate translation. The data suggest that translation is initiated by the ribosome scanning the mRNA for the appropriate start sequence.

(b) Very little, if any, protein synthesis would be expected. Translation initiation in bacteria requires the 16S RNA of the small ribosomal subunit to interact with the Shine-Dalgarno sequence. This interaction serves to align the ribosome over the start codon. If the anticodon has been changed such that the start codon cannot be recognized, then protein synthesis is unlikely.

Chapter 16

1. Six levels at which gene expression can be controlled are: (1) alteration or modification of the gene's structure at the DNA level, (2) transcriptional regulation, (3) regulation at the level of mRNA processing, (4) regulation of mRNA stability, (5) regulation of translation, and (6) regulation by posttranslational modification of the synthesized protein.

2.

4. The *lac* operon consists of three structural genes: *lacZ*, *lacY*, and *lacA*. The *lacZ* gene encodes the enzyme β-galactosidase, which cleaves the disaccharide lactose into galactose and glucose. A second function of β-galactosidase converts lactose into allolactose. The *lacY* gene encodes permease, an enzyme necessary for the passage of lactose through the *E. coli* cell membrane. The *lacA* gene encodes the enzyme thiogalactoside transacetylase, whose function in lactose metabolism has not yet been determined. All three genes have in common an overlapping promoter and operator region. Upstream of the lactose operon is the *lacI* gene, which encodes the *lac* operon repressor. The repressor binds at the operator region and inhibits transcription of the *lac* operon by preventing RNA polymerase from successfully initiating transcription. When lactose is present in a cell, the enzyme β-galactosidase converts some of it into allolactose, which then binds to the *lac* repressor, altering its shape and reducing the repressor's affinity for the operator. If the repressor does not bind to the operator, RNA polymerase can initiate the transcription of the *lac* structural genes from the *lac* promoter.

6. Attenuation is the premature termination of transcription before the structural genes of an operon are transcribed. It is due to the formation of a termination hairpin, or attenuator. Two types of secondary structures can be formed by the mRNA 5′ UTR of the *trp* operon. If a hairpin structure is formed by base pairing between region 3 and region 4, then the structural genes will not be transcribed. The hairpin structure formed by the pairing of region 3 with region 4 results in the formation of a terminator that stops transcription. When region 2 pairs with region 3, the resulting hairpin acts as an antiterminator, allowing for transcription to proceed. Region 1 of the 5′ UTR also encodes a small protein and has two adjacent tryptophan codons (UGG).

In bacteria, tryptophan levels affect transcription owing to the coupling of translation with transcription. When tryptophan levels are high, the ribosome quickly moves through region 1 and into region 2, thus preventing region 2 from pairing with region 3. Therefore region 3 is available to form the attenuator hairpin structure with region 4, stopping transcription. When tryptophan levels are low, the ribosome stalls or stutters at the adjacent tryptophan codons in region 1. Region 2 now becomes available to base pair with region 3, forming the antiterminator hairpin. Transcription can now proceed through the structural genes.

7. Antisense RNA molecules are complementary to other DNA or RNA sequences. In bacterial cells, antisense RNA molecules can bind to a complementary region in the 5′ UTR of an mRNA molecule blocking the attachment of the ribosome to the mRNA and thus stopping translation.

9. Changes in chromatin structure can result in either the repression or the stimulation of gene expression. As genes become more transcriptionally active, chromatin shows increased sensitivity to DNase I digestion., suggesting the chromatin structure is more open. The acetylation of histone proteins by acteyltransferase destabilizes the nucleosome structure and increases transcription as well as hypersensitivity to DNase I. The reverse reaction by deacetylases stabilizes nucleosome structure and lessens DNase I sensitivity. Other transcriptional factors and regulatory proteins called chromatin-remodeling complexes bind directly to DNA, altering chromatin structure without acetylating histone proteins. The chromatin-remodeling complex enhances the initiation of transcription by increasing the accessibility of promoters to transcriptional factors.

DNA methylation also decreases transcription. Methylated DNA sequences stimulate histone deacetylases to remove acetyl groups from the histone proteins, thus stabilizing the nucleosome and repressing transcription. Demethylation of DNA sequences is often followed by increased transcription, which may be related to the deacetylation of the histone proteins.

14. The amount of protein synthesized depends on the amount of mRNA available for translation, which in turn depends on the rate of synthesis and degradation of mRNA. Less-stable mRNAs will be degraded and not

be available as templates for translation. The presence of the 5′ cap, 3′ poly(A) tail, 5′ UTR, 3′ UTR, and coding region in mRNA affects stability. Poly(A)-binding proteins bind at the 3′ poly(A) tail, contributing to the stability of the tail and protecting the 5′ cap through direct interaction. After a critical number of adenine nucleotides have been removed from the tail, the protection is lost and the 5′ cap is removed. Its removal enables 5′-to-3′ nucleases to degrade the mRNA.

16. Both bacterial and eukaryotic genes are regulated by repressors and activators. Cascades of gene regulation in which the activation of one set of genes stimulates the activation of another set of genes exists in both eukaryotes and bacteria. The regulation of gene expression at the transcriptional level is common in both types of cells.

   Bacterial genes are often clustered in operons and are coordinately expressed through the synthesis of a single polygenic mRNA. Eukaryotic genes are typically separate, each having its own promoter, and are transcribed into individual mRNAs. The coordinate expression of multiple genes is accomplished through the presence of response elements. Genes sharing the same response element are regulated by the same regulatory factors.

   In eukaryotic cells, chromatin structure plays a role in gene regulation. Chromatin that is condensed inhibits transcription, and so, for a gene to be expressed, the chromatin must be altered to allow for changes in structure. Both the acetylation of histone proteins and DNA methylation are important in these changes.

The process of transcription initiation is more complex in eukaryotic cells than in bacterial cells. In eukaryotes, initiation requires RNA polymerase, general transcription factors, and transcriptional activators. Bacterial RNA polymerase is either blocked or stimulated by the actions of regulatory proteins.

   Finally, in eukaryotes, activator proteins may bind to enhancers at a great distance from the promoter and structural gene. These distant enhancers are much less frequent in bacterial cells.

17. (a) Inactive repressor; (b) active activator; (c) active repressor; (d) inactive activator.

18. (a) The regulator-corepressor complex would normally bind to the operator and inhibit transcription. If a mutation prevented the repressor from binding at the operator, then the operon would never be turned off and transcription would take place all the time.
   (b) In an inducible operon, a mutation at the operator site that blocks binding of the repressor would result in constitutive expression, and transcription would take place all the time.

20. The catabolite activator protein binds to the CAP site of the *lac* operon and stimulates RNA polymerase to bind to the *lac* promoter, thus increasing the levels of transcription from the *lac* operon. If a mutation prevents CAP from binding to the site, then RNA polymerase will bind the *lac* promoter poorly, significantly lowering the levels of transcription of the *lac* structural genes.

23. See the table below.

Genotype of strain	Lactose absent		Lactose present	
	β-Galactosidase	Permease	β-Galactosidase	Permease
lacI⁺ lacP⁺ lacO⁺ lacZ⁺ lacY⁺	−	−	+	+
lacI⁻ lacP⁺ lacO⁺ lacZ⁺ lacY⁺	+	+	+	+
lacI⁺ lacP⁺ lacO^C lacZ⁺ lacY⁺	+	+	+	+
lacI⁻ lacP⁺ lacO⁺ lacZ⁺ lacY⁻	+	−	+	−
lacI⁻ lacP⁻ lacO⁺ lacZ⁺ lacY⁺	−	−	−	−
lacI⁺ lacP⁺ lacO⁺ lacZ⁻ lacY⁺ / lacI⁻ lacP⁺ lacO⁺ lacZ⁺ lacY⁻	−	−	+	+
lacI⁻ lacP⁺ lacO^C lacZ⁺ lacY⁺ / lacI⁺ lacP⁺ lacO⁺ lacZ⁻ lacY⁻	+	+	+	+
lacI⁻ lacP⁺ lacO⁺ lacZ⁺ lacY⁻ / lacI⁺ lacP⁻ lacO⁺ lacZ⁻ lacY⁺	−	−	+	−
lacI⁺ lacP⁻ lacO^C lacZ⁻ lacY⁺ / lacI⁻ lacP⁺ lacO⁺ lacZ⁺ lacY⁻	−	−	+	−
lacI⁺ lacP⁺ lacO⁺ lacZ⁺ lacY⁺ / lacI⁺ lacP⁺ lacO⁺ lacZ⁺ lacY⁺	−	−	+	+
lacI^s lacP⁺ lacO⁺ lacZ⁺ lacY⁻ / lacI⁺ lacP⁺ lacO⁺ lacZ⁻ lacY⁺	−	−	−	−
lacI^s lacP⁻ lacO⁺ lacZ⁻ lacY⁺ / lacI⁺ lacP⁺ lacO⁺ lacZ⁺ lacY⁺	−	−	−	−

25. The *lacI* gene encodes the lac repressor protein, which can diffuse within the cell and attach to any operater. It can therefore affect the expression of genes on the same piece or on different pieces of DNA. The *lacO* gene encodes the operator. It affects the binding of RNA polymerase to DNA and therefore affects the expression of genes only on the same piece of DNA.

26. (a) The data from the strain with no mutation indicates that the *mmm* operon is repressible. The operon is expressed in the absence of *mmm* but is inactive in the presence of *mmm*, which is typical of a repressible operon.

(b) Regulator gene   B

    Promoter   D

    Structural gene for enzyme 1   A

    Structural gene for enzyme 2   C

27. (a) If the ribosome does not bind to the 5′ end of the mRNA, then region 1 of the mRNA 5′ UTR will be free to pair with region 2, thus preventing region 2 from pairing with region 3. Then region 3 will be free to pair with region 4, forming the attenuator or termination hairpin. Transcription of the *trp* structural genes will be terminated, and there will essentially be no gene expression.

(b) In the wild-type *trp* operon, low levels of tryptophan cause the ribosome to pause in region 1 of the mRNA 5′ UTR. The pause permits regions 2 and 3 to form the antiterminator hairpin, allowing transcription of the structural genes to continue. If alanine codons have replaced tryptophan codons, then, under conditions of low levels of tryptophan, the ribosome will not stall. The attenuator will form, stopping transcription. The ribosome will stall when alanine levels are low; so the structural genes will be transcribed only when alanine levels are low.

(c) If region 1 of the mRNA 5′ UTR is free to pair with region 2, then regions 3 and 4 can form the attenuator. An early stop codon will cause the ribosome to "fall off" region 1, which can then form a hairpin structure with region 2. Transcription will not take place, because regions 3 and 4 are now free to form the attenuator.

(d) If region 2 of the mRNA 5′ UTR is deleted, then the antiterminator (2 + 3) cannot be formed. The attenuator (3 + 4) will form and transcription will not take place.

(e) The *trp* operon mRNA 5′ UTR will be unable to form the attenuator (3 + 4) if region 3 contains a deletion, and transcription of the *trp* structural genes will proceed.

(f) Deletions in region 4 will prevent the formation of the attenuator (3 + 4) by the 5′ UTR mRNA, and transcription will proceed.

(g) For the attenuator hairpin to function as a terminator, a string of uracil nucleotides must follow region 4 in the mRNA 5′ UTR. If the string of adenine nucleotides in the DNA is deleted, no string of uracil nucleotides will follow region 4, and transcription will proceed.

31. The *Sxl* gene is necessary for the proper splicing of *tra* pre-mRNA and the production of Tra, a protein needed for the development of female fruit flies. If the *Sxl* gene is absent, no Tra protein will be produced and the embryo develops into a male.

Chapter 17

1. Germ-line mutations are found in the DNA of reproductive cells and may be passed to offspring. Somatic mutations are found in the DNA of an organism's somatic tissue cells and cannot be passed to offspring.

2. Transition mutations result from purine-to-purine and pyrimidine-to-pyrimidine base substitutions. Transversions result from base substitutions of purines for pyramidines and pyrimidines for purines. Transitions are more common because spontaneous mutations that occur in DNA typically result in transition mutations rather than transversions.

3. Expanding trinucleotide repeats result when a DNA insertion mutation increases the number of copies of a trinucleotide repeat sequence. Within a given family, a particular type of trinucleotide repeat may increase in number from generation to subsequent generation, increasing the severity of the mutation in a process called anticipation.

6. Intergenic suppressor mutations restore the wild-type phenotype. However, they do not reverse the original mutation. The suppression is a result of mutation in a gene other than the gene containing the original mutation. Because many proteins interact with other proteins, the original mutation may have disrupted the protein-protein interaction, and the second mutation restores the interaction. Another type of intergenic supresssion is due to a mutation in an anticodon region of a tRNA molecule that allows for pairing with the codon containing the original mutation and the substitution of a functional amino acid in the protein.

7. Mutation frequency is defined as the number of mutations in a population of cells or individuals. The mutation rate is typically expressed as the number of mutations per biological unit, such as per replication or per cell division.

8. Two types of events have been proposed that could lead to DNA replication errors: mispairing due to tautomeric shifts in nucleotides and mispairing through other structures and wobble or flexibility of the DNA molecule. Current evidence suggests that mispairing through other structures and wobble is the most likely cause.

10. Base analogs have structures similar to those of the nucleotides and may be incorporated into DNA in the course of replication. Many of the analogs tend to mispair, which can lead to mutations. DNA replication is required for base-analog-induced mutations to be incorporated into the DNA.

13. The SOS system is an error-prone DNA repair system consisting of at least 25 genes. Induction of the SOS system causes damaged DNA regions to be bypassed, which allows for DNA replication across the damaged regions. However, bypassing damaged DNA results in a less-accurate replication process; thus more mutations will occur.

15. Mismatch repair: repairs replication errors that are the result of base-pair mismatches. Mismatch-repair enzymes recognize distortions in DNA structure due to mispairing and detect the newly synthesized strand by its lack of methylation. The distorted segment is excised, and DNA polymerase and DNA ligase fill in the gap.

Direct repair: repairs DNA damage by directly changing the damaged nucleotide back to its original structure.

Base-excision repair: excises the damaged base and replaces the entire nucleotide.

Nucleotide-excision repair: Relies on repair enzymes to recognize distortions of the DNA double helix. These enzymes excise damaged regions by cleaving phosphodiester bonds on either side of the damaged region. The gap created by the excision is filled by DNA polymerase.

17. The substitution is at the first position of the codon: **GGA** → **UGA**. Because uracil is a pyrimidine and guanine is a purine, the mutation is a transversion.

18. (a) Two codons can encode Phe: UUU and UUC. A single transition could occur at each of the positions of the codon, specifying different amino acids or the same amino acid: Leu, Ser, and Phe.
(b) Cys, Ile, Leu, Tyr, and Val.
(c) Leu, Phe, Pro, and Ser.
(d) Arg, Gln, His, Ile, Leu, Met, Phe, Trp, Tyr, Val, and a stop codon.

20. Amino−Met Thr Gly Asn Gln Leu Tyr Stop−Carboxyl
(a) Amino−Met Thr Gly **Ser** Gln Leu Tyr Stop−Carboxyl
(b) Amino−Met Thr Gly Asn **Stop**−Carboxyl
(c) Amino−Met Thr **Ala Ile Asn Tyr Ile** . . . −Carboxyl
(d) Amino−Met Thr Gly Asn **His** Leu Tyr Stop−Carboxyl
(e) Amino−Met Thr **Thr** Gly Asn Gln Leu Tyr Stop−Carboxyl
(f) Amino−Met Thr **Gly** Asn Gln Leu Tyr Stop−Carboxyl

22. Only two of the six possible Arg codons could be mutated to form Ser codons and then subsequently be mutated at a second position to regenerate an Arg codon.

Original Arg codon	Ser codon	Restored Arg codon
CGU	AGU	AGG or AGA
CGC	AGC	AGG or AGA

24. No, hydroxylamine cannot reverse nonsense mutations. Hydroxylamine modifies cytosine-containing nucleotides and can result only in GC-to-AT transition mutations. In a stop codon, the GC-to-AT transition will result only in a different stop codon.

26. Hydroxylamine adds hydroxyl groups to cytosine, enabling the modified cytosine to occasionally pair with adenine, which could ultimately result in a GC-to-AT transition. So only one base pair in the sequence will be affected. Ultimately, after replication, only one of the strands in double-stranded DNA would have the transition.

Original sequence	Mutated sequence
5′−ATG**T**−3′	5′−AT**A**T−3′
3′−TAC**A**−5′	3′−TA**T**A−5′

29. PFI1 causes both types of transitions: GC to AT and AT to GC.

PFI2 causes transversions or large deletions because mutations caused by PF12 are not reverted by any of the agents.

PFI3 causes GC-to-AT transitions.

PFI4 causes single-base insertions or deletions.

30. The plant breeder should look for plants that have increased levels of mutations either in their germ lines or in their somatic tissues. Tomato plants with defective DNA repair systems may be sensitive to high levels of sunlight and may need to be grown in a reduced-light environment for survival.

Chapter 18

4. Gel electrophoresis acts as a molecular sieve. The gel is an aqueous matrix of agarose or polyacrylamide. DNA molecules are loaded into a slot or well at one end of the gel. When an electric field is applied, the negatively charged DNA molecules migrate toward the positive electrode. Shorter DNA molecules are less hindered by the agarose or polyacrylamide matrix and migrate faster than longer DNA molecules, which must wind their way around obstacles and through the pores in the gel matrix.

5. DNA molecules can be visualized by staining with a fluorescent dye. Ethidium bromide intercalates between the stacked bases of the DNA double helix, and the ethidium bromide-DNA complex fluoresces orange when exposed to an ultraviolet light source. Alternatively, they can be visualized by attaching radioactive or chemical labels to the DNA before it is placed in the gel.

7. In all three techniques, macromolecules are separated by size by gel electrophoresis and are then transferred to the surface of a membrane filter. Southern blotting transfers DNA, Northern blotting transfers RNA, and Western blotting transfers protein.

8. A cloning vector should have (1) an origin of DNA replication so that it can be replicated in a cell; (2) a gene, such as an antibiotic-resistance gene, to select for cells that carry the vector; and (3) a unique restriction site or series of sites at which to cut and ligate a foreign DNA molecule.

11. Many plasmids designed as cloning vectors carry a gene for antibiotic resistance and the *lacZ* gene. The *lacZ* gene in the plasmid has been engineered to contain multiple unique restriction sites. Foreign DNAs are inserted into one of the unique restriction sites in the *lacZ* gene. After transformation, *E. coli* cells carrying the plasmid are plated on medium containing the appropriate antibiotic to select for cells that carry the plasmid. The medium also contains an inducer for the *lac* operon (so that the cells express the *lacZ* gene) and X-gal (a substrate for β-galactosidase that will turn blue when cleaved by β-galactosidase). The colonies that carry plasmids lacking foreign DNA inserts will have intact *lacZ* genes, make functional β-galactosidase, cleave X-gal, and turn blue. Colonies that carry plasmids containing foreign DNA inserts will not make functional β-galactosidase, because the *lacZ* gene is disrupted by the foreign DNA insert, and they will remain white. Thus cells carrying plasmids with inserts will form white colonies.

13. A cosmid is a plasmid vector with a plasmid origin of DNA replication, unique restriction sites for cloning, selectable marker genes, and a lambda *cos* site so that the vector can be packaged into lambda phage particles for efficient delivery into *E. coli* cells. It can accommodate large DNA fragments—as many as 44 kb.

15. A genomic library is created by inserting fragments of chromosomal DNA into a cloning vector. Chromosomal DNA is randomly fragmented by shearing or by partial digestion with a restriction enzyme. A cDNA library is made from mRNA sequences. Cellular mRNAs are isolated, and then reverse transcriptase is used to copy the mRNA sequences to cDNA, which are cloned into plasmid or phage vectors.

19. First, double-stranded template DNA is denatured by high temperature. Then synthetic oligonucleotide primers corresponding to the ends of the DNA sequence to be amplified are annealed to the single-stranded DNA template strands. These primers are extended by a thermostable DNA polymerase so that the target DNA sequence is duplicated. These steps are repeated 30 times or more. Each cycle of denaturation, primer annealing, and extension results in doubling the number of copies of the target sequence between the primers.

PCR amplification is limited in several ways. One limitation is that, to synthesize the PCR primers, the sequence of the gene to be amplified must be known, at least at the ends of the region to be amplified. Another is that the extreme sensitivity of the technique renders it susceptible to contamination. A third limitation is that the most common thermostable DNA polymerase used for PCR, *Taq* DNA polymerase, has a relatively high error rate. A fourth limitation is that PCR amplification is usually limited to DNA fragments of, at the most, a few thousand base pairs; optimized DNA polymerase mixtures and reaction conditions extend the amplifiable length to perhaps 20 kb.

20. In situ hybridization is the hybridization of radiolabeled or fluorescently labeled DNA or RNA probes and DNA or RNA molecules that are still in the cell. This technique can be used to visualize the expression of specific mRNAs in different cells and tissues, as well as the location of genes in metaphase or polytene chromosomes.

23. A knockout mouse has a target gene disrupted or deleted. First, the target gene is cloned. The middle part of the gene is replaced by a selectable marker, typically the *neo* gene that confers resistance to G418. This construct is then introduced back into mouse embryonic stem cells and the cells are selected for G418 resistance. The surviving cells are screened for cells in which the chromosomal copy of the target gene has been replaced by the *neo*-containing construct by homologous recombination of the flanking sequences. These embryonic stem cells are then injected into mouse blastocyst-stage embryos, and these chimeric embryos are transferred to the uterus of a pseudopregnant female mouse. The knockout cells will participate in the formation of many tissues in the mouse fetus, including germ-line cells. The chimeric offspring are interbred to produce offspring that are homozygous for the knockout allele. The phenotypes of the knockout mice provide information about the function of the gene.

25. DNA fingerprinting is the typing of an individual for genetic markers at highly variable loci. The technique is useful in forensic investigations to determine whether a suspect could have contributed to the evidentiary DNA obtained from blood or other bodily fluids found at the scene of a crime. Other applications include paternity testing and identifying bodily remains.

The loci traditionally used for DNA fingerprinting are called variable number of tandem repeat (VNTR) loci; these loci consist of short tandem repeat sequences located in introns or spacer regions between genes. The number of repeat sequences at the locus does not affect the phenotype of the individual in any way; so these loci are highly variable in the population. More recently, tandem repeat loci with smaller repeat sequences of just a few nucleotides, called short tandem repeats (STRs) have been adopted because they can be amplified by PCR.

28. The first three letters are taken from the genus and species name, and the Roman numeral indicates the order in which the enzyme was isolated. Therefore, the enzyme should be named *Ara*I.

30. Here, $4^n = 1,048,500$. So $n = 10$; a 10-bp recognition sequence is most likely.

32. (a) *Bam*H1 GGATCC is $(0.2)(0.2)(0.3)(0.3)(0.2)(0.2) = 0.000144$

$3,000,000,000(0.000144) = 432,000$ times

(b) *Eco*RI GAATTC $= (0.2)(0.3)(0.3)(0.3)(0.3)(0.2) = 0.000324$

$3,000,000,000(0.000324) = 972,000$ times

(c) *Hae*III GGCC $= (0.2)(0.2)(0.2)(0.2) = 0.0016$

$3,000,000,000(0.0016) = 4,800,000$ times

33.

34. (a) Plasmid; (b) phage λ; (c) cosmid.

36. One strategy would be to use the mouse gene for prolactin as a probe to find the homologous pig gene in a pig genomic or cDNA library. A second strategy would be to use the amino acid sequence of mouse prolactin to design oligonucleotides as hybridization probes to screen a pig DNA library. A third strategy would be to use the amino acid sequence of mouse prolactin to design a pair of oligonucleotide PCR primers to PCR amplify the pig prolactin gene.

38. (a) A probe of 18 nucleotides must be based on six amino acids. The sequence of six amino acids with the least degeneracy is Val-Tyr-Lys-Ala-Lys-Trp. This sequence excludes the amino acids Arg and Leu, which have six codons each.

(b) Val and Ala have four codons each, Tyr and Lys have two codons each, and Trp has one codon. Therefore, there are $4 \times 2 \times 2 \times 4 \times 2 \times 1 = 128$ possible sequences.

39. This gene must first be cloned, possibly by using the yeast gene as a probe to screen a mouse genomic DNA library. The cloned gene is then engineered to replace a substantial part of the protein-coding sequence with the *neo* gene. This construct is then introduced into mouse embryonic stem cells, which are transferred to the uterus of a pseudopregnant mouse. The progeny are tested for the presence of the knockout allele, and those with the knockout allele are interbred. If the gene is essential for embryonic development, no homozygous knockout mice will be born. The arrested or spontaneously aborted fetuses can then be examined to determine how development has gone awry in fetuses that are homozygous for the knockout allele.

40. (a) First, we must determine the RFLP genotype of Sally's mother. From the RFLP genotypes of Sally, her siblings, and their father, we deduce that Sally's mother must have had *A2 A3* and *C2 C2*. The linkage relations of these chromosomes are *A1C1* and *A1C3* from the father and *A2C2* and *A3C2* from the mother. The mother passed on an *A2C2* to Sally's brother who has G syndrome; therefore the G-syndrome allele must be linked with *A2C2*. Because Sally inherited the *A2C2* chromosome from her mother, she must have also inherited the G-syndrome allele, assuming that there was no crossover between the *A*, *C*, and *G* loci.

(b) Father: *A1 C1 g*, *A1 C3 g*

Mother: *A2 C2 G*, *A3 C2 g*

Sally's unaffected brother: *A1 C3 g*, *A3 C2 g*

Sally's affected brother: *A1 C3 g*, *A2 C2 G*

Sally: *A1 C1 g*, *A2 C2 G*

Chapter 19

2. A genetic map locates genes or markers on the basis of genetic recombination frequencies. A physical map locates genes or markers on the basis of the physical lengths of DNA sequences. Because recombination frequencies vary from one region of a chromosome to another, genetic maps are approximate. Genetic maps also have lower resolution than physical maps because recombination between loci that are very close to each other is difficult to observe. Physical maps based on DNA sequences or restriction sites have much greater accuracy and resolution, down to a single base pair of DNA sequence.

4. The map-based approach first assembles large clones into contigs on the basis of genetic and physical maps and then selects clones for sequencing. The whole-genome shotgun approach breaks the genome into short sequences, typically from 600 to 700 bp, and then assembles them into contigs on the basis of sequence overlap by using powerful computers.

6. The international collaboration took the ordered, map-based approach, beginning with the construction of detailed genetic and physical maps. Celera took the whole-genome shotgun approach. Celera used the physical map produced by the international collaboration to order sequences in the assembly phase.

10. Homologous sequences are derived from a common ancestor. Orthologs are sequences in different species that are descended from a sequence in a common ancestral species. Paralogs are sequences in the same species that originated by duplication of an ancestral sequence and subsequent divergence.

13. After random mutagenesis with chemicals or transposons, the mutant progeny population is screened for phenotypes of interest. The mutant gene can be identified by cosegregation with molecular markers or by sequencing the position of transposon insertion. To verify that the mutation identified is truly responsible for the phenotype, a mutation can be introduced into a wild-type copy of the gene and the offspring can be searched for the phenotype.

16. Horizontal gene exchange is the transfer of genetic material across species boundaries. In bacteria, horizontal gene exchange may occur through uptake of environmental DNA through transformation, through conjugative plasmids with a broad host range, or through transfection with a bacteriophage with a broad host range.

18. The *Arabidopsis* genome appears to have undergone at least one round of duplication of the whole genome (polyploidy) and numerous localized duplications through unequal crossing over.

21.

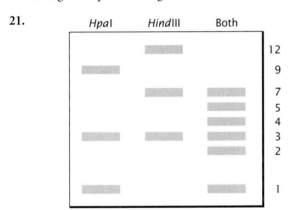

22.

	*Eco*RI	*Sma*I		*Eco*RI	
3 kb		4 kb	2 kb		5 kb

24. Original sequence: 5′–NGCATCAGTA–3′. The base (N) at the 5′ end cannot be determined, because the chain stops in all four lanes.

26.

	Species				
Protein	A	B	C	D	E
P1	+	+	−	−	+
P2	+	+	+	−	−
P3*	+	+	−	+	+
P4	+	−	+	−	+
P5*	+	+	−	+	+

P3 and P5 are either always present or always absent in all species; therefore they are most likely to be functionally related.

30. (a) The minimal genome required might be determined by examining simple, free-living organisms having small genomes to determine which genes they possess in common. Mutations can then be made systematically to determine which genes are essential for these organisms to survive. The apparently nonessential genes (those genes in which mutations do not affect the viability of the organism) can then be deleted one by one until only the essential genes are left; elimination of any additional gene results in loss of viability. Alternatively, essential genes could be assembled through genetic engineering, creating an entirely novel organism.

(b) This synthetic organism would prove that we have acquired the ability to create a new species or form of life. We would then be able to direct evolution as never before. Social and ethical concerns would center on whether human society has the wisdom to temper its power—such novel synthetic organisms might be used to develop pathogens for biological warfare or terrorism. After all, no person or animal would have been previously exposed or have acquired immunity to such a novel synthetic organism. Additionally, the effect of the new organism if it was released or escaped on the ecosystem would be uncertain.

Chapter 20

1. Mitochondrial and chloroplast DNAs have many characteristics in common with eubacterial DNAs. Most mitochondrial and chloroplast chromosomes are small, circular, and lack histone proteins—characteristics that are similar to those of eubacteria but not to those of eukaryotic cells. Chloroplasts and some mitochondria produce polycistronic mRNA, another characteristic in common with eubacteria. Chloroplasts but not most mitochondria typically possess Shine-Dalgarno sequences. Antibiotics that inhibit eubacterial translation also inhibit mitochondrial and chloroplast translation.

   Eukaryotic nuclear genomes are typically composed of linear chromosomes and histone proteins. Eukaryotic nuclear DNA sequences also contain introns. Pre-mRNA introns, which are common in eukaryotes, are not found in chloroplast and mitochondrial genomes.

3. The endosymbiotic theory states that mitochondria and chloroplasts originated from formerly free-living bacteria that became endosymbionts within a larger eukaryotic cell. Both chloroplasts and mitochondria contain genomes that encode proteins, tRNAs, and rRNAs. Chloroplast and mitochondrial ribosomes are similar in size and function to eubacterial ribosomes. Size, circular chromosomes, and other aspects of genome structure in both chloroplasts and mitochondria are similar to those of eubacterial cells. DNA sequences in chloroplast and mitochondrial genomes are most similar to those in eubacteria.

6. A nonuniversal codon is a codon that specifies an amino acid in one organism that is not specified by that codon in most other organisms. Typically,

nonuniversal codons are found in mitochondrial genomes, and the exceptions vary in the mitochondria of different organisms.

8. In mitochondrial translation, the "wobble" at the third codon position is more frequent than in cytoplasmic translation. Most anticodons of the mitochondrial tRNAs can pair with more than one codon. Essentially, the first position of the anticodon can pair with any of the four nucleotides present at the third position of the codon.

10. The chloroplast genome, typically a double-stranded circular DNA whose organization resembles that of eubacterial genomes, contains sequences similar to those of eubacteria. Genes are located on both strands of cpDNA and may contain introns. Typically, the chloroplast genome will encode ribosomal proteins, 5 rRNAs, from 30 to 35 tRNAs, and proteins needed for photosynthesis, as well as proteins not needed for photosynthesis. A large inverted repeat also is found in the genomes of most chloroplasts.

13. The pedigree indicates that the neurological disorder is a cytoplasmically inherited trait. Only females pass the trait to their offspring. The trait does not appear to be sex specific in that both males and females can have the disorder. However, none of the affected males produce offspring that exhibit the neurological disorder, whereas female parents with the disorder pass the trait to all their offspring. These characteristics are consistent with cytoplasmic inheritance.

16.

Chapter 21

1. The ability to clone plants and animals from differentiated cells showed that the nuclei of these differentiated cells still retained all of the genetic information required for the development of a whole organism.

3. Maternally transcribed *bicoid* and *nanos* mRNAs are localized to the anterior and posterior ends of the egg, respectively. After fertilization, these mRNAs are translated, and the proteins diffuse to form opposing gradients: Bicoid protein highest at the anterior and Nanos protein highest at the posterior. Bicoid protein at the anterior activates the transcription of *hunchback*.

Nanos protein at the posterior inhibits the translation of *hunchback* mRNA.

4. Gap genes specify broad regions (multiple adjacent segments) along the anterior–posterior axis of the embryo. Interactions betweem the gap genes regulate transcription of the pair-rule genes.

Pair-rule genes compartmentalize the embryo into segments; each pair-rule gene is expressed in alternating segments. The pair-rule genes regulate the expression of the segment-polarity genes.

Segment-polarity genes specify the organization of segments.

6. Apoptosis is programmed cell death, characterized by nuclear DNA fragmentation, shrinkage of the cytoplasm and nucleus, and phagocytosis of the remnants of the dead cell. Apoptosis is regulated by internal and external signals that regulate the activation of procaspases—cysteine proteases that are activated by proteolytic cleavage. When activated, these caspases activate other caspases in a cascade and degrade key cellular proteins.

7. (a) Somatic recombination: Recombination in somatic cells produces many combinations of variable segments with junction segments and diversity segments.
(b) Junctional diversity: In recombination, the *V-D-J* joining events are imprecise, resulting in small deletions or insertions and frameshifts.
(c) Hypermutation: The *V*-gene segments are subject to somatic hypermutation (accelerated random mutation) that further diversifies antibodies.

9. The multistage theory of cancer states that more than one mutation is required for most cancers to develop. Most retinoblastomas are unilateral, because the likelihood of any cell acquiring two rare mutations is very low, and thus retinoblastomas develop in only one eye. Bilateral retinoblastomas develop in people born with a predisposing mutation, and so only one additional mutational event will result in cancer. Thus the probability of retinoblastoma is higher and likely to develop in both eyes. Because the predisposing mutation is inherited, people with bilateral retinoblastoma have relatives with retinoblastoma.

11. An oncogene stimulates cell division, whereas a tumor-suppressor gene puts the brakes on cell growth. Proto-oncogenes are normal cellular genes that function in cell growth and regulation of the cell cycle: from growth factors such as Sis to receptors such as ErbA and ErbB, protein kinases such as Src, and nuclear transcription factors such as Myc. Tumor suppressors function to inhibit cell-cycle progression: RB and P53 are transcription factors, and NF1 is a GTPase activator.

13. DNA polymerases are unable to replicate the ends of linear DNA molecules. Therefore, the ends of eukaryotic chromosomes shorten with every round of

DNA replication, unless telomerase adds special nontemplated telomeric DNA sequences. Normally, somatic cells do not express telomerase; their telomeres progressively shorten with each cell division until vital genes are lost and the cells undergo apoptosis. Transformed cells (cancerous cells) induce the expression of the telomerase gene to keep proliferating.

15. (a) The products of *bicoid* and *dorsal* affect embryonic polarity by regulating the transcription of target genes.
(b) The product of *nanos* regulates the translation of *hunchback* mRNA.

17. Injecting *bicoid* mRNA in the posterior end of the embryo would cause the transcription of *hunchback* in the posterior regions. Without Nanos protein, the *hunchback* mRNA would be translated, creating high levels of Hunchback protein in the posterior as well as in the anterior. The result would be an embryo with anterior structures on both ends.

19. (a) Females with increased copies of the *bicoid* gene would have higher levels of *bicoid* maternal mRNA in the anterior cytoplasm of their eggs and thus higher levels of Bicoid protein in the embryo after fertilization. The resulting Bicoid protein gradient would extend farther to the posterior, resulting in the enlargement of anterior and thoracic structures.
(b) Conversely, decreased copies of the *bicoid* gene would ultimately result in a reduced Bicoid protein gradient in the eggs. Thus sufficient Bicoid protein concentrations for head structures would be found in a smaller, more anterior part of the embryo, resulting in an embryo with smaller head structures.

22. Light-chain genes undergo recombination to join one *V*-gene segment to one *J*-gene segment in any combination. The number of different possible V and J combinations is given by the product of the number of V segments and the number of J segments for each light chain.

Kappa light chain: $200 \times 4 = 800$

Lambda light chain: $300 \times 6 = 1800$

Total light chains = kappa + lambda = $800 + 1800 = 2600$

24. (a) If the father with unilateral retinoblastoma is heterozygous for an *RB* mutation, then the chance of another child inheriting the mutant *RB* allele is 50%. If the father is homozygous for the *RB* mutation, then the chance of the child having retinoblastoma is nearly 100%.
(b) Because retinoblastoma in this family is most likely an inherited disorder, a child with retinoblastoma is likely to have bilateral retinoblastoma.
(c) The father may have unilateral retinoblastoma because of incomplete penetrance of the mutation in the *RB* gene. Alternatively, it may have been just good fortune (random chance) that one of his eyes was spared the second mutation event that led to retinoblastoma in his other eye.

Chapter 22

1. Discontinuous characteristics have only a few distinct phenotypes. In contrast, a quantitative characteristic shows a continuous variation in phenotype.

3. Many genotypes are possible with multiple genes. Even for the simplest two-allele loci, the number of possible genotypes is equal to 3^n, where n is the number of loci. If each genotype corresponds to a unique phenotype, then we have the same numbers of phenotypes: 27 possible phenotypes for 3 genes and 108 possible phenotypes for 4 genes. Finally, the phenotype for a given genotype may be influenced by environmental factors, leading to an even greater array of phenotypes.

4. A sample is a subset of a population. To be representative of the population, a sample should be randomly selected. A sample must also be sufficiently large to minimize random differences between members of the sample and the population.

7. The magnitude or absolute value of the correlation coefficient reports how strongly the two variables are associated. A value close to $+1$ or -1 indicates a strong association; values close to 0 indicate weak association.

9. V_A: component of variance due to additive genetic variance
V_D: component of variance due to dominance variance
V_I: component of variance due to genic interaction variance
$V_G = V_A + V_D + V_I$: component of variance due to variation in genotype
V_E: component of variance due to environment
V_{GE}: component of variance due to interaction between genes and environment

10. Broad-sense heritability is the part of phenotypic variance that is due to all types of genetic variance, including additive, dominance, and genic interaction variances. Narrow-sense heritability is just that part of the phenotypic variance due to additive genetic variance.

14. The response to selection (R) = narrow-sense heritability (h^2) × selection differential (S). The value of R predicts how much the mean phenotype will change with selection in a single generation.

16. Phenotypic correlations between two traits may be obtained either because the environment influences both traits (environmental correlation) or because genes influence both traits (genetic correlation). Genetic correlation is due to pleiotropy: genes acting on more than one trait.

17. (a) Discontinuous characteristic, because only a few distinct phenotypes are present
(b) Discontinuous characteristic, because there are only two phenotypes (dwarf and normal)
(c) Quantitative characteristic, because susceptibility is a continuous trait that is determined by multiple genes and environmental factors
(d) Quantitative characteristic, because it is determined by many loci
(e) Discontinuous characteristic, because there are only a few phenotypes determined by a single locus

18.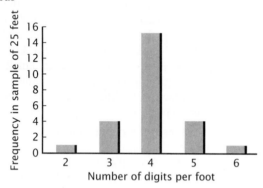

20. In the summer. With a positive correlation, there is a direct association between the variables—as one variable (temperature) increases, the other variable (rainfall) also tends to increase.

22. No, because correlation does not mean causation. A number of other scenarios may be responsible for the observed correlation.

24. (a) $r = 0.86$; (b) 67.8 inches tall.

25. (a) All F_1 progeny will be 10 grams.
(b) $\frac{1}{16}$ with 16 grams
$\frac{4}{16}$ with 13 grams
$\frac{6}{16}$ with 10 grams
$\frac{4}{16}$ with 7 grams
$\frac{1}{16}$ with 4 grams

26.

Height (cm)	Proportion of F_2 progeny
22	$\frac{1}{64}$
20	$\frac{6}{64}$
18	$\frac{15}{64}$
16	$\frac{20}{64}$
14	$\frac{15}{64}$
12	$\frac{6}{64}$
10	$\frac{1}{64}$
	total = $\frac{64}{64}$

27. Six or more loci determine the differences.

29. (a) $h^2 = 0.38$; (b) $H^2 = 0.69$.

31. (a) Group A, because unrelated people have the greatest genetic variance.
(b) No. Siblings raised in the same home should have less environmental variance than does group A (unrelated people).

34. (a) $H^2 = 0.75$.
(b) This estimate may be inaccurate if the environmental variance of the genetically identical population is different from the environmental variance of the genetically diverse population.

35. The regression coefficient = narrow-sense heritability = 0.8.

36. The salesman, because Mr. Jones's determination of heritability was conducted for a population of pigs under one environmental condition: low nutrition. His findings do not apply to any other population or even to the same population under different environmental conditions. High heritability for a trait does not mean that environmental changes will have little effect.

39. $h^2 = 0.75$.

Chapter 23

2. Large population, random mating, and not affected by migration, selection, or mutation.

4. (a) If the frequencies of alleles $A1$, $A2$, and $A3$ are defined as p, q, and r, respectively,
$f(A1A1) = p^2$, $f(A1A2) = 2pq$, $f(A2A2) = q^2$,
$f(A1A3) = 2pr$, $f(A2A3) = 2qr$, and $f(A3A3) = r^2$.
(b) For an X-linked locus with 2 alleles,
$f(X^1X^1) = p^2$ among females;
$p^2/2$ for the whole population
$f(X^1X^2) = 2pq$ among females;
pq for the whole population
$f(X^2X^2) = q^2$ among females;
$q^2/2$ for the whole population
$f(X^1Y) = p$ among males; $p/2$ for the whole population
$f(X^2Y) = q$ among males; $q/2$ for the whole population

7. The proportion of the population due to migrants (m) and the difference in allele frequencies between the migrant population and the original-resident population.

9. The effective population size is the effective number of breeding adults and can be calculated as $N_E = 4 \times$ number of males $\times$ number of females/ (number of males + number of females). The smaller the effective population size, the greater the magnitude of the genetic drift.

13. Mutation increases genetic variation within populations and increases divergence between populations because different mutations arise in each population.

Migration increases genetic variation within a population by introducing new alleles but decreases divergence between populations.

Genetic drift decreases genetic variation within populations because it causes alleles to eventually come to fixation, but it increases divergence between populations because drift occurs differently in each population.

Natural selection may either increase or decrease genetic variation, depending on whether the selection is directional or balanced. It may increase or decrease divergence between populations, depending on whether different populations have similar selection pressures or different selection pressures.

15. The neutral-mutation hypothesis proposes that most molecular variation is adaptively neutral. The balance hypothesis proposes that most genetic variation is maintained by balanced selection, favoring heterozygosity at most loci.

17. The molecular clock is the rate at which nucleotide changes occur in a DNA sequence.

19. $f(EE) = 0.685$; $f(EF) = 0.286$; $f(FF) = 0.029$; $f(E) = 0.828$; $f(F) = 0.172$.

26. (a) $p = 0.1$; (b) 1%; (c) 0.18, or 18%.

27. $f(MM) = 0.648$; $f(MN) = 0.304$; $f(NN) = 0.048$.

30. The new frequency of the gene for curved wings in Bill's room should be 0.56. If random mating is assumed, the frequency of the curved-wing trait should be $q^2 = (0.56)^2 = 0.31$.

32. (a) 40; (b) 30; (c) 30; (d) 7.6.

36. 0.17

Index

Note: Page numbers followed by f indicate figures; those followed by t indicate tables.